Nasal Physiology and Pathophysiology of Nasal Disorders

Özlem Önerci Celebi • T. Metin Önerci
Editors

Nasal Physiology and Pathophysiology of Nasal Disorders

Second Edition

Springer

Editors
Özlem Önerci Celebi
Department of Otoloaryngology Head and Neck Surgery
University of Health Sciences, Istanbul Training and Research Hospital
Istanbul, Turkey

T. Metin Önerci
Faculty of Medicine
Hacettepe University
Ankara, Turkey

ISBN 978-3-031-12388-7 ISBN 978-3-031-12386-3 (eBook)
https://doi.org/10.1007/978-3-031-12386-3

This Springer imprint is published by the registered company Springer Nature Switzerland AG
The registered company address is: Gewerbestrasse 11, 6330 Cham, Switzerland

To our teachers whom we owe all our knowledge and the way to think scientifically, to our colleagues who shared their experience and inspire us to learn, to our students who inspire us to teach.

Preface

It is with great pleasure that we present the second enlarged edition of the book *Nasal Physiology and Pathophysiology of Nasal Disorders*. The first edition was accepted with great enthusiasm and received very good feedback. It was sold out. Since its online publication on 2013, there have been more than 99,000 downloads for the eBook on SpringerLink and it was among the top most downloaded eBooks in its respective eBook Collection in 2019. There is so much new information that we felt the need to make a second edition.

The diseases can be better understood and managed if we know the pathophysiology of the diseases. This is very closely related to the changes in the physiology. The progress only can be made by studying the physiology and pathophysiology of the diseases. The great interest for this book is perhaps due to the wish of our colleagues to learn more and more on this topic.

We want to thank the distinguished authors of this book who are experts in their fields and gave their valuable time to prepare their chapters with the recent information. We tried to cover all aspects of nasal physiology and pathophysiology.

We hope that the second edition with its new, enlarged and updated chapters will fill the gap on this topic.

Istanbul, Turkey
Ankara, Turkey
August, 2021

Özlem Önerci Çelebi
T. Metin Önerci

Contents

1 Mucus, Goblet Cell, Submucosal Gland

Takeshi Shimizu

Abbreviations

AP-1	Activated protein-1
AR	Allergic rhinitis
AZM	Azithromycin
CAM	Clarithromycin
CF	Cystic fibrosis
CREB	cAMP response element-binding protein
CRS	Chronic rhinosinusitis
cysLTs	Cysteinyl leukotrienes
EGFR	Epidermal growth factor receptor
EM	Erythromycin
Foxa2	Forkhead box a2
IL	Interleukin
NF-κB	Nuclear factor κ-B
RA	Retinoic acid
STAT6	Signal transducer and activator of transcription 6
TGF	Transforming growth factor
TLR	Toll-like receptor
TNF	Tumor necrosis factor

Core Messages

Airway mucus is important for the host defense mechanism, acting as a physicochemical barrier to protect the underlying epithelium from pathogens and particles. The major components of mucus are glycoproteins called mucins, which are secreted by epithelial goblet and submucosal gland cells. Mucins are large heterogeneous macromolecules containing hundreds of oligosaccharide chains attached to peptide backbones, which are encoded by several MUC genes. Hypersecretion of mucus is commonly observed in various sinonasal inflammatory conditions such as acute rhinitis, chronic rhinosinusitis, and allergic rhinitis. Mucus hypersecretion, leading to rhinorrhea, contributes significantly to the pathophysiology of these diseases by impairing mucociliary function, resulting in stagnation of pathological mucus that contains various inflammatory mediators and pathogenic microbes.

1.1 Introduction

Airway mucus blankets all mucosal surfaces, providing a physicochemical barrier that protects the underlying epithelium against bacteria, viruses, and inhaled particles and gases. Mucus maintains airway hydration and plays an important role in the innate immune system by trapping foreign and endogenous substances, facilitating clearance by mucociliary activity. Mucus also has antioxidant, antiprotease, and antimicrobial functions. Composed of water, ions, serum protein exudates, epithelial secretions, and glandular

T. Shimizu (✉)
Department of Otorhinolaryngology-Head and Neck Surgery, Shiga University of Medical Science, Otsu, Shiga, Japan
e-mail: shimizu@belle.shiga-med.ac.jp

Ö. Ö. Celebi, T. M. Önerci (eds.), *Nasal Physiology and Pathophysiology of Nasal Disorders*,
https://doi.org/10.1007/978-3-031-12386-3_1

and goblet cell products, mucus contains various defensive components such as glycoproteins (mucins), antibodies, defensin, lysozyme, and lactoferrin.

The mucous blanket is subdivided into two layers: the outer mucous layer and the periciliary fluid layer. The outer mucous layer is a gel layer, produced mainly by secretions of epithelial goblet cells and submucosal glands. The periciliary fluid or sol layer is produced by ion transport through the epithelium (Fig. 1.1). Ciliary beating takes place at 10–14 Hz in this periciliary layer, propelling the outer mucous layer toward the pharynx. Healthy mucociliary transport requires maintenance of a balance among volume and composition of mucus, adequate periciliary fluid, and ciliary beating [1]. Mucus is essential for mucociliary transport. If the mucus is replaced with saline, particles do not move even while cilia beat actively [2].

Hypersecretion of mucus is an important characteristic of sinonasal inflammation that occurs in conditions such as acute rhinitis, chronic rhinosinusitis (CRS), and allergic rhinitis (AR). Hypertrophic, hyperplastic, and metaplastic changes in goblet cells of the surface epithelium and in submucosal gland cells are frequently observed in association with pronounced rhinorrhea. In pathological conditions, mucus hypersecretion and damaged epithelium impair mucociliary clearance, resulting in stagnant pathological mucus containing pathogenic microbes, various inflammatory mediators, and inflammatory cells, a situation that facilitates bacterial colonization and local inflammation, leading to infection and tissue damage.

The major components of mucus are glycoproteins called mucins, which are secreted by epithelial goblet cells and submucosal glands. Mucins are large heterogeneous macromolecules, containing oligosaccharide chains attached to peptide backbones, which are encoded by several MUC genes. Secreted mucins are stored in secretory granules and are released by regulated exocytosis.

This chapter will focus on mucus, goblet cells, and submucosal glands of the nasal mucosa and

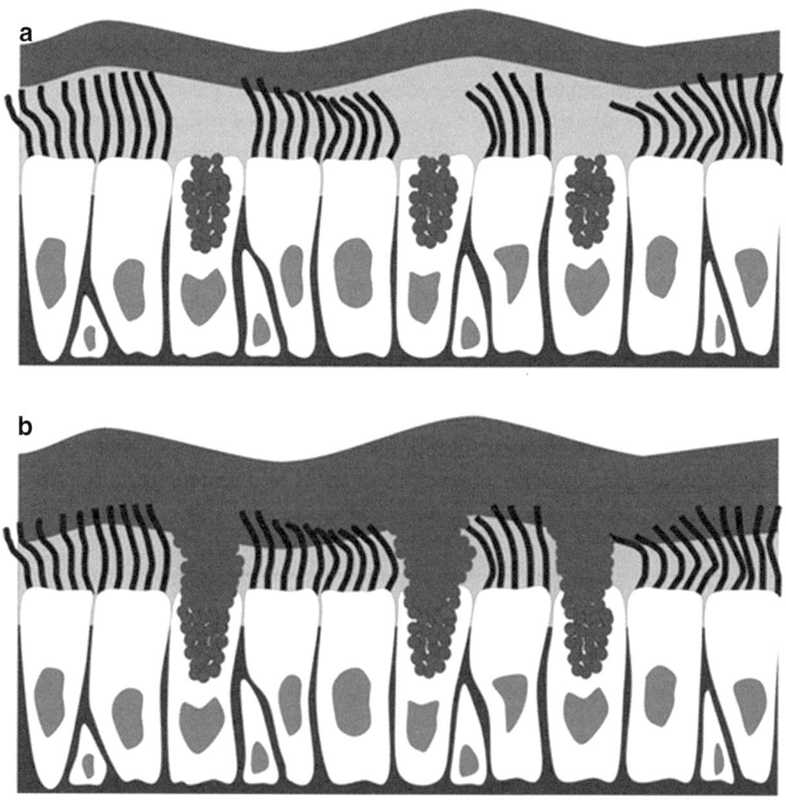

Fig. 1.1 Mucous blanket is subdivided into two layers: the outer mucous layer and the periciliary fluid layer. (**a**) Normal mucous blanket. (**b**) Pathological mucous blanket with hypersecretion of mucus. Mucus hypersecretion increases the viscoelasticity of the mucus, and mucus strands connect the outer mucous layer with the epithelial goblet cells, causing the disruption of the periciliary fluid layer. These changes in mucus impair mucociliary interaction

will summarize the mechanisms regulating mucus secretion and pathological mucus hypersecretion in sinonasal inflammation. Therapeutic strategies to inhibit mucus hypersecretion will be also discussed.

1.2 Mucus Composition

The nasal epithelium is covered by a mucous gel layer. Ciliary beating within the periciliary fluid layer propels the gel layer to the pharynx and is then swallowed together with entrapped particles and pathogens. Mucus is a heterogeneous mixture of water, glucose, various ionic solutes, antimicrobial proteins, cells, and cellular debris. Mucins are major glycoprotein components of mucus that contribute greatly to the viscoelastic and gel-forming properties of the mucous layer [3–5].

Mucins are complex glycoproteins with oligosaccharide chains attached to peptide backbones, which are encoded by MUC genes. Mucins can be divided into two structurally and functionally distinct subfamilies: membrane-bound mucins and secreted mucins. Membrane-bound mucins have transmembrane and cytoplasmic domains that anchor the molecules to the apical cell membrane, where they participate in functions such as structural barrier formation, cellular adhesion, pathogen binding, and signal transduction [6, 7]. Extracellular subunits of membrane-bound mucins can be released from the plasma membrane into the mucus layer by proteolytic cleavage or by shearing forces. Some membrane-bound mucin genes are alternatively spliced to form transcripts that lack a transmembrane domain; these are present in airway secretion [8]. Secreted mucins are stored in secretory granules located in the apical cytoplasm and are released by regulated exocytosis.

Secreted mucins have high molecular weights (more than 1000 kDa) and are heavily glycosylated proteins (composed of 70–90% carbohydrate moieties) with tandemly repeated amino acid sequences that are rich in serine and threonine. Different mucin genes contain various sizes and numbers of tandem repeats, and there are genetic polymorphisms within single mucin genes [9]. Tandem repeats are sites of *O*-glycosylation, and hundreds of oligosaccharide chains are attached to a single core peptide (Fig. 1.2). Mucin glycosylation is determined by tissue-specific glycosyltransferase expression and by host and environmental factors that influence transferase expression. Mucin carbohydrate chains are highly heterogeneous, and this structural diversity may allow mucins to interact with many microorganisms [10]. These carbohydrate moieties are possible sites of attachment for pathogenic bacteria and

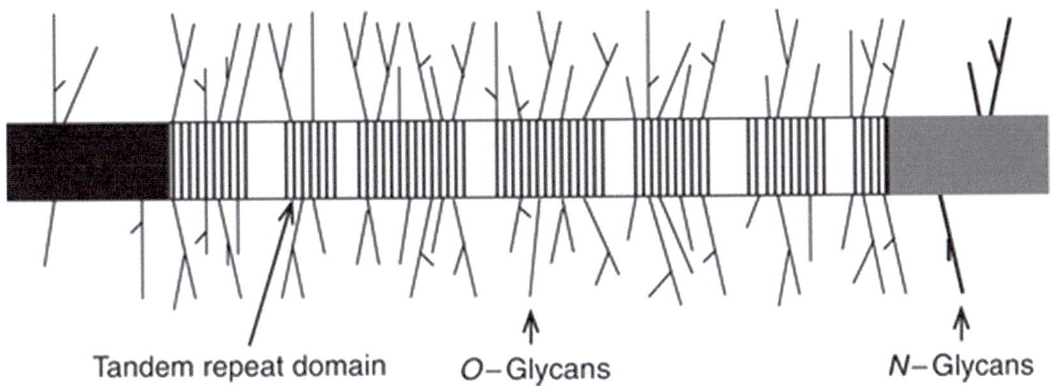

Fig. 1.2 A schematic model of a secreted mucin. Hundreds of *O*-glycans are attached to serine or threonine residues of tandem repeat domains in the MUC protein backbone. Secreted, gel-forming mucins contain cysteine-rich domains that permit oligomerization through the formation of disulfide bonds; they also aggregate through ionic and hydrophobic interactions with proteins and with other mucins

viruses. Several recognition sites for respiratory pathogens have been identified; these promote their entrapment and removal by mucociliary clearance [11–14].

Secreted mucins contain cysteine-rich domains that permit oligomerization through the formation of disulfide bond; they also aggregate through ionic and hydrophobic interactions with proteins and with other mucins. These complex macromolecules provide the viscoelastic and space-occupying properties of the mucus gel layer [15]. Secreted mucins are negatively charged by sulfated and sialylated oligosaccharide chains. Mucins, the most plentiful high-molecular-weight polyanions of the nasal mucosa, interact with and inhibit the effects of cationic inflammatory proteins such as leukocyte elastase and lysozyme [16, 17]. The negatively charged carbohydrates of mucins may protect against proteolysis caused by cationic inflammatory proteins and bacterial enzymes.

1.3 Mucin Genes

The mucin protein backbones are encoded by MUC genes. More than 20 human mucin genes have been identified throughout the respiratory, gastrointestinal, and reproductive tracts [9]. In respiratory epithelium, mainly MUC1, MUC2, MUC4, MUC5AC, MUC5B, MUC7, and MUC8 are expressed, and similar expressions of mucin genes are observed in normal nasal mucosa [18, 19] (Table 1.1). MUC2, MUC5AC, MUC5B, and MUC8 are secreted, gel-forming mucins that contain cysteine-rich domains for oligomerization and are responsible for the viscoelastic property of mucus. They are encoded by a cluster of highly related genes on chromosome 11 and by a similar gene on chromosome 12 [15]. MUC7 is a secreted, non-gel-forming mucin that exists as a monomer and so is not thought to contribute significantly to mucus viscoelasticity.

In the airways, MUC1 and MUC4 are the predominant membrane-bound mucins present on the apical membranes of epithelial cells. Membrane-bound mucins contain a highly glycosylated extracellular domain, a transmembrane domain, and a short cytoplasmic tail. Extracellular units can be released from cells under certain conditions and then potentially contribute to the mucus layer. Some MUC4 mucins secreted into airways are encoded by alternatively spliced transcripts that lack a transmembrane domain [8]. The cytoplasmic tail domain participates in signal transduction and regulates a variety of biological functions [20]. In airway epithelial cells, MUC1 is a receptor for *Pseudomonas aeruginosa* flagellin [21]; MUC1 inhibits flagellin-activated Toll-like receptor (TLR)-5-mediated signaling and interleukin (IL)-8 release [22]. Both MUC1 and MUC4 dimerize with and regulate the epidermal growth factor receptor [23]. Their roles in inflammation and cancer biology have been studied in detail [24–26]. However, their function in nasal mucosa remains to be elucidated.

Recent studies revealed the specific localization of mucins in nasal epithelial cells and submucosal glands. The membrane-bound mucins MUC1 and MUC4 are diffusely expressed at the apical surface of epithelial cells. The secreted, gel-forming mucin MUC5AC is mainly expressed in epithelial goblet cells, while MUC5B is the predominant mucin expressed in mucous cells of the submucosal glands. MUC2 is mainly expressed in epithelial goblet cells, but its expression is much lower than that of MUC5AC. MUC

Table 1.1 Major mucin genes in human nasal mucosa

MUC gene	Mucin subfamily	Chromosome locus	Main tissue localization
MUC1	Membrane-bound	1q21–q24	Epithelial cells
MUC2	Secreted, gel-forming	11p15.5	Goblet cells
MUC4	Membrane-bound	3q29	Epithelial cells
MUC5AC	Secreted, gel-forming	11p15.5	Goblet cells
MUC5B	Secreted, gel-forming	11p15.5	Mucous cells of submucosal glands
MUC7	Secreted, non-gel-forming	4q13.3	Serous cells of submucosal glands
MUC8	Secreted, gel-forming	12q24.3	Goblet cells, mucous cells of submucosal glands

8 is expressed in both epithelial goblet cells and mucous cells of the submucosal glands. The secreted, non-gel-forming mucin MUC7 is expressed in serous cells of the submucosal gland. Similar distributions of MUC genes seem to be found in nasal polyps [27, 28]. MUC2, MUC5AC, and MUC5B expressions are reported to be upregulated during airway inflammation in humans and in animals, and MUC5AC expression is the highest of the three [29, 30]. Therefore, MUC5AC has been the most intensively studied MUC gene with regard to airway mucus secretion.

1.4 Goblet Cells and Submucosal Glands

Mucous cells in the surface epithelium (called goblet cells) and in submucosal glands are the major sources of gel-forming mucins. Epithelial goblet cells contain numerous electron-lucent granules that fill most of the cytoplasm and fuse with the apical cell membrane before secretion (Fig. 1.3). The nucleus, numerous mitochondria, and the rough endoplasmic reticulum are restricted to a small volume in the basal aspect of the cells. Submucosal glands comprise a mixture of mucous cells and serous cells, and the mucous cells, the important source of gel-forming mucin, resemble epithelial goblet cells. These mucous cells can be detected histochemically by Alcian blue and periodic acid Schiff's stains.

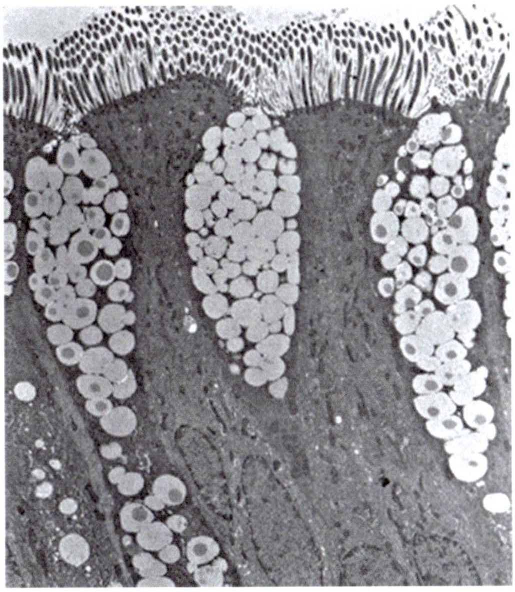

Fig. 1.3 Goblet cells of rat nasal epithelium induced by intranasal instillation of lipopolysaccharides. Numerous electron-lucent granules fill most of the cytoplasm and fuse with the apical cell membrane before secretion

Serous cells of submucosal glands contain discrete electron-dense granules containing secretory products including immunoglobulins, lysozyme, lactoferrin, and other antimicrobial enzymes, all of which are important for normal function of the airway innate immune system [31]. Thus, serous cells are an important source of antimicrobial peptides essential for host defense mechanisms, while mucous cells contribute to the viscoelastic property of airway mucus by producing gel-forming mucins. Human submucosal glands are innervated by parasympathetic and sensory efferent nerves, and the glands express muscarinic receptors. The parasympathetic (cholinergic) nervous system is the neural pathway most active in airway submucosal glands, and stimulation of cholinergic nerves or use of muscarinic receptor agonists induces marked mucus secretion [32, 33].

Mucus hypersecretion is a major characteristic of airway inflammation; it is associated with hypertrophy, hyperplasia, and metaplasia of epithelial goblet and submucosal gland cells. An increased number of hypertrophic goblet cells are commonly observed in the nasal epithelium during experimental inflammation caused by inhalation of irritant gases or allergens or by viral or bacterial infection. In the human nose, the goblet cell density of the inferior turbinate ranges from 5000 to 10,000 cells/mm^2, similar to that of maxillary sinus mucosa [18, 34]. However, changes of goblet cell numbers in patients with allergic rhinitis (AR) and chronic rhinosinusitis (CRS) are controversial.

Submucosal glands contribute to mucus hypersecretion in airway inflammation by secreting mucins, ions, and water; the submucosal gland cell density ranges from 1000 to 2000 cells/mm^2 in human inferior turbinate and maxillary sinus mucosa. In CRS patients, the number of submucosal gland cells increases to 2000–4000 cells/mm^2, and the area occupied by acini of the lamina propria also increases [34].

1.5 Regulation of Mucin Secretion

Mucin secretion is regulated by a multi-step process that includes synthesis and exocytosis. Mucin synthesis includes gene transcription in the nucleus, posttranscriptional modification and transport of mRNA, and translation in the endoplasmic reticulum. In the Golgi apparatus, oligosaccharides chains are then attached to the peptide backbone by glycosylation. The synthesized mucins are stored in secretory granules located in the apical cytoplasm until stimulated for subsequent release by exocytosis.

1.5.1 Mucin Production

Mucin gene expression is induced in response to a wide variety of inflammatory stimuli. MUC5AC is the predominant gel-forming mucin in the human airways and has been extensively studied for evaluation of mucin gene regulation and mucin glycoprotein secretion. A variety of epithelial stimuli (including bacterial products, viral infection, environmental pollutants and chemicals, proteases, inflammatory cytokines, and growth factors) upregulate mucin gene expressions during sinonasal inflammation (Fig. 1.4).

Many bacteria and bacterial products induce mucus hypersecretion in vivo and in vitro. Lipopolysaccharide/TLR-4, peptidoglycan/TLR-2 or TLR-6, and flagellin/TLR-5 signaling induce MUC5AC expression in airway epithelial cells through activation of transcription factors such as nuclear factor-κB (NF-κB), cAMP response element-binding protein (CREB), and

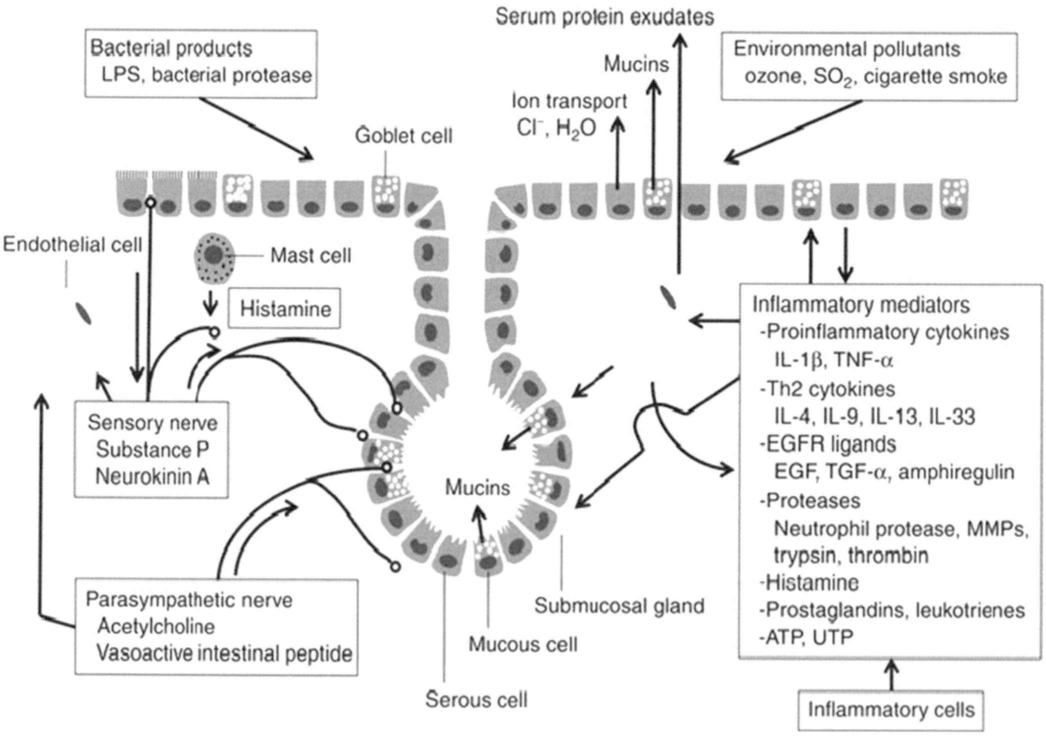

Fig. 1.4 A variety of environmental stimuli, inflammatory mediators, growth factors, and parasympathetic and sensory nerves are involved in mucus production and secretion in nasal mucosa

activated protein-1 (AP-1) [29, 35]. Respiratory viral infections (including influenza virus and respiratory syncytial virus) induce mucin gene expression in mice [36, 37], and rhinovirus stimulates MUC5AC expression in human airway epithelial cells [38]. Environmental pollutants and oxidants (such as cigarette smoke, acrolein, ozone, SO_2, and hydrogen peroxide) stimulate MUC5AC expression and mucin production [39–42]. Airway proteases (including neutrophil elastase, matrix metalloproteases, airway trypsin, and thrombin) stimulate MUC5AC expression and mucin secretion [43–47].

The proinflammatory cytokines IL-1β and tumor necrosis factor (TNF)-α and the Th2 cytokines IL-4, IL-9, and IL-13 stimulate mucin production in vivo [48, 49]. Epithelial cell-derived cytokine IL-33 also induces mucin production in the mouse airway [50]. IL-13 is an important and essential mediator of mucin production in Th2-mediated airway inflammation through direct stimulation of epithelial cells [51]. IL-13 activates signal transducer and activator of transcription 6 (STAT6), and then STAT6-dependent downregulation of the transcription factor Forkhead box a2 (Foxa2) stimulates MUC5AC expression [52, 53]. Foxa2 is a negative regulator of MUC5AC expression, and its deletion induces mucous metaplasia in the mouse lung [54].

Retinoic acid (RA) and its related analogs play an important role in cell growth and cell differentiation. In the airway epithelium, RA is essential for induction and maintenance of mucociliary differentiation [55, 56]. Expression of the gel-forming mucins MUC2 and MUC5AC is RA dependent in cultured airway epithelial cells [57]. Epidermal growth factor receptor (EGFR) signaling is also important for mucin production in a variety of animal models and in human airway epithelial cells. The EGFR ligands EGF, transforming growth factor (TGF)-α, and amphiregulin stimulate MUC5AC expression [58]. Activation of EGFR is critical for in vivo and in vitro induction of mucin production in response to stimulation of airway epithelial cells by allergens, viruses, neutrophil elastase, and cigarette smoke [15, 29, 59–61].

1.5.2 Mucin Exocytosis

Mucin exocytosis from epithelial goblet cells is stimulated by many inflammatory mediators including cholinergic agonists, neuropeptides, prostaglandins, leukotrienes, bacterial products, neutrophil elastase, inhaled pollutants, and nucleosides [62]. The nucleoside ATP is released in response to mechanical, irritant, and inflammatory stimulation of epithelial cells. Extracellular release of ATP activates P2Y2 purinergic receptors on the apical surface of airway epithelial cells, resulting in induction of calcium release and regulated exocytosis [63]. Exocytosis is a complex process controlled by many regulatory molecules, including myristoylated alanine-rich C kinase substrate (MARCKS), which is essential for mucin release in vivo and in vitro [6, 7, 64, 65].

Mucus secretion from submucosal glands is regulated mainly by parasympathetic and sensory nerves. Neurotransmitters and neuropeptides released by these nerves directly stimulate gland secretion [32]. In airway inflammation, parasympathetic activity is stimulated by sensory nerves and histamine. Inflammatory mediators, such as prostaglandins, leukotrienes, and neutrophil elastase, also stimulate gland secretion.

1.5.3 Mucin Glycosylation

In addition to factors that regulate synthesis and release, mucin activity is also regulated by glycosylation [10]. Th2 cytokines and TNF-α alter the glycosylation and sialylation of secreted mucins [66, 67]. The carbohydrate moieties of mucins are potential adhesion sites for bacteria and viruses [68]. Inflammation-associated glycosylation of secreted mucin facilitates the interaction between mucus and microorganisms, leading to entrapment and removal by mucociliary clearance [69, 70].

Increased modification of secreted mucin by sialylation and sulfation is commonly observed in airway inflammation [71–73]. Resulting carbohydrate moieties are negatively charged and have inhibitory effects against cationic inflammatory proteins and bacterial enzymes [16, 17]. Such alterations may be important mechanisms involved in host defense, as they lead to neutralization of these proteolytic enzymes and removal of pathogenic microbes. However, the precise biological function of mucin carbohydrate chains remains to be elucidated.

1.6 Pathophysiological Mucus Hypersecretion

Mucus hypersecretion is a common characteristic of the sinonasal inflammation seen in disorders like chronic rhinosinusitis (CRS) and allergic rhinitis (AR) and is the cause of rhinorrhea. A variety of inflammatory mediators and infiltrating cells are capable of stimulating mucus hypersecretion. Many inflammatory stimuli such as inhaled irritants, neutrophil products, lipopolysaccharide, viral and bacterial infections, and antigen challenge have been used to study mechanisms of mucus hypersecretion in animal models.

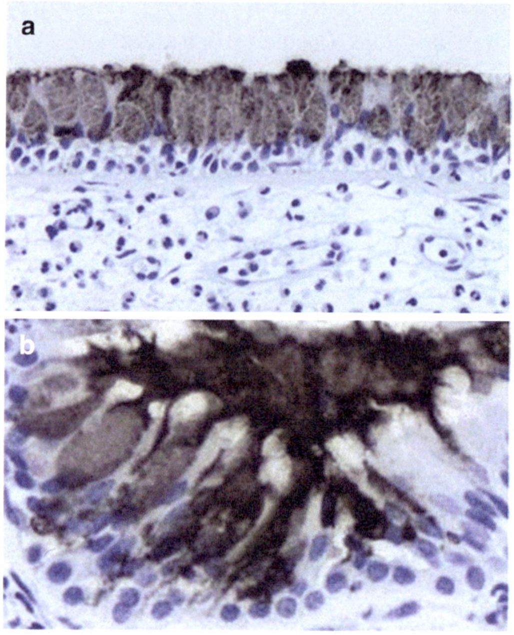

Fig. 1.5 Immunohistochemical staining of MUC5AC in nasal polyp of patient with chronic rhinosinusitis. Mucous granules of epithelial goblet cells (**a**) and the mucus layer of epithelial surface (**b**) are strongly stained. Mucus strands connect the mucous blanket with the epithelial goblet cells (**b**)

1.6.1 Chronic Rhinosinusitis (CRS)

CRS is a common nasal infectious disease with or without nasal polyps, characterized by the following symptoms: anterior and posterior nasal discharge, nasal obstruction, olfactory disturbance, headache, and facial pain. Hypersecretion of mucus in CRS patients may be induced by various inflammatory mediators such as the proinflammatory cytokines IL-1β and TNF-α, bacterial products, and neutrophil products. Mucin gene expression in nasal and sinus mucosa is similar to that in other respiratory epithelia. MUC5AC, MUC5B, and MUC2 are major secreted, gel-forming mucins, and the production of these is upregulated in nasal mucosa and nasal polyps of CRS patients [18, 28] (Fig. 1.5).

Excessive mucin production increases the viscoelasticity of the mucus, and mucus strands connect the mucous blanket with epithelial goblet cells [74] (Figs. 1.1 and 1.5). These changes of the mucus and the damaged epithelium impair mucociliary interaction. Obstruction of the nasal passages caused by inflamed mucosa or nasal polyps and mucociliary dysfunction lead to "mucostasis," an accumulation of stagnant, pathological mucus that contains various inflammatory mediators, inflammatory cells, and pathogenic microbes. Mucostasis triggers mediators, normally host protective, but which in this setting become host invasive, further exacerbating mucus hypersecretion, tissue damage, mucociliary dysfunction, and bacterial infection (Fig. 1.6). Successful treatment of CRS patients involves stopping the self-mediated inflammation caused by stagnant mucus. Nasal blowing, suction and irrigation, antral lavage, and endoscopic sinus surgery are useful treatments to remove stagnant mucus and restore mucociliary clearance.

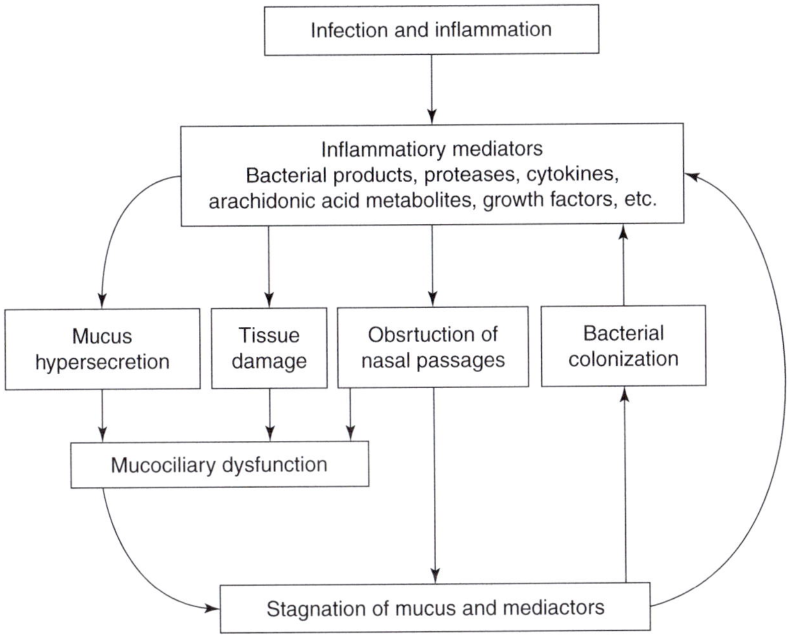

Fig. 1.6 A vicious cycle of self-mediated inflammation caused by the stagnant mucus in the pathogenesis of chronic rhinosinusitis. Mucus hypersecretion and damaged epithelium impair mucociliary function, resulting in "mucostasis," an accumulation of stagnant, pathological mucus that contains various inflammatory mediators and pathogenic microbes. These mediators and microbes exacerbate the local inflammation and further bacterial colonization. For treatment of patients with chronic rhinosinusitis, surgical removal the stagnant mucus is very important to stop the self-mediated inflammation

1.6.2 Allergic Rhinitis (AR)

AR is caused by IgE-mediated Th2 immune responses and is characterized by the following symptoms: sneezing, nasal obstruction, itching, and rhinorrhea. Histamine is a key mediator for allergic rhinitis, and histamine-induced cholinergic nerve stimulation is important for antigen-induced mucus hypersecretion. MUC5AC is a main secreted mucin, and it is found to be upregulated in AR patients [75, 76].

Ovalbumin (OVA)-sensitized animals are commonly used to study the pathophysiologic changes of allergic inflammation. When hypertrophic and metaplastic changes in epithelial goblet cells are induced in a rat model of nasal allergy, intraepithelial mucus production is significantly inhibited by a Th2 cytokine inhibitor and by a cysteinyl leukotrienes (cysLTs) antagonist, indicating that Th2 cytokines and cysLTs (LTs C_4, D_4, and E_4) are important for mucus synthesis in AR [77, 78]. Mucus secretion (goblet cell exocytosis) can be evaluated by measuring the transient decrease of intraepithelial mucus. Histamine stimulates early-phase (1 h after challenge) secretion through H1 receptor on cholinergic nerve terminals, and infiltrating inflammatory cells (eosinophils and/or neutrophils) play a role in late-phase (6 h) secretion. CysLTs are important for both early-phase secretion and late-phase secretion [79].

1.7 Therapeutic Strategies to Inhibit Mucus Hypersecretion

Mucus and mucociliary clearance serve important functions in the host defense system by removing irritants, allergens, pathogens, and dead cells from the airway. However, hypersecretion of mucus impairs mucociliary function and thus becomes part of the pathogenic process, causing symptoms such as nasal obstruction and anterior and posterior nasal discharge. Inhibition of mucus hypersecretion restores mucociliary clearance and improves symptoms. Selection of an appropriate therapeutic strategy is important because mechanisms of mucus synthesis and secretion differ among the diseases, stimuli, and types of inflammation.

1.7.1 Surgical Managements

Hypersecretion of mucus and obstruction of nasal passages impair mucociliary clearance, resulting in stagnation of mucus that contains host-invasive mediators and pathogenic microbes. Stagnation of mucus causes a vicious cycle of self-mediated inflammation in the pathogenesis of CRS (Fig. 1.6). For treatment of CRS patients, surgical removal of the stagnant mucus is very important. Nasal blowing and irrigations are useful for improving the symptoms [80]. Antral puncture and lavage decrease mucus viscoelasticity and improve mucociliary activity [81]. Endoscopic sinus surgery to remove the obstructed mucosa and polyps is very effective for restoring mucociliary clearance [82, 83].

1.7.2 Mucolytic and Mucokinetic Agents

The formation of disulfide bonds to oligomerize gel-forming mucins contributes to the viscoelastic property of mucus. Increased viscoelasticity impairs mucociliary function. Mucolytic agents such as *L*-cysteine and *N*-acetylcysteine have free sulfhydryl groups that dissolve disulfide bonds and therefore decrease the viscoelasticity of nasal discharge and sputum [84, 85]. Mucokinetic agents such as carbocysteine cannot break mucin disulfide bonds, but they improve mucociliary function and suppress goblet cell hyperplasia. These mucolytic and mucokinetic agents are reported to be effective as treatments for CRS patients [86–88].

The release of large amounts of DNA from dead neutrophils contributes to increased mucus viscoelasticity. By decreasing the amount of DNA in sputum, inhaled DNase has become an important treatment for patients with cystic fibrosis (CF) [89]. Inhaled hypertonic saline is also used to improve mucociliary clearance in CF patients [90]; this stimulates water secretion into the airway by creating a temporary osmotic gradient. Hypertonic saline nasal irrigations are reported to be effective as treatment for CRS patients [91].

1.7.3 Macrolide Antibiotics

The 14-member macrolides clarithromycin (CAM), erythromycin (EM), and roxithromycin and the 15-member macrolide azithromycin (AZM) are widely used to treat airway inflammation. Low-dose, long-term macrolide therapy has been reported to be very effective for patients with chronic airway diseases such as diffuse panbronchiolitis, chronic bronchitis, chronic obstructive pulmonary disease, CF, and CRS [92–96]. The effects of these agents depend on anti-inflammatory and immunomodulatory rather than antibacterial actions.

CAM, EM, and AZM inhibit mucus hypersecretion and metaplastic and hypertrophic changes of nasal epithelial goblet cells in vivo and in vitro [97, 98]. They inhibit inflammatory responses of neutrophils, lymphocytes, macrophages, and epithelial cells and suppress gene expression and production of inflammatory cytokines and chemokines [99, 100]. In CRS patients, macrolide therapy reduces anterior and posterior nasal discharge and is effective for treating neutrophilic and lymphocytic inflammation. However, macrolide therapy is not effective for patients with eosinophilic inflammation, characterized by serum and tissue eosinophilia, high serum IgE, multiple polyposis, severe CT findings, and bronchial asthma [96, 100–103].

1.7.4 Anti-inflammatory Agents

Systemic and topical steroids are very effective for reducing mucus hypersecretion in patients with CRS and AR [104, 105]. Glucocorticoids are potent anti-inflammatory agents that inhibit functions of a variety of inflammatory cells. Mucus hypersecretion is suppressed mainly by reductions in the release of secretagogue mediators such as histamine, leukotrienes, neutrophil elastase, and Th2 cytokines.

Anti-inflammatory agents such as antihistamines, cysLTs antagonist, Th2 cytokines inhibitor, and anticholinergic agents are clinically useful for reducing nasal AR symptoms, including anterior and posterior discharge [80, 106]. All have been confirmed in animal model of allergic inflammation as inhibitors of nasal or tracheal mucus hypersecretion.

1.8 Conclusions

Mucin secretion is regulated by a multi-step process that includes synthesis and exocytosis. A wide variety of inflammatory stimuli upregulate mucin gene expression and induce mucus overproduction. During the past decades, there have been significant advances in our understanding of the biological roles of mucus and of mechanisms which regulate mucus hypersecretion, including (1) the structural biology of mucins and mucus, (2) regulation of mucin gene expression, (3) mechanisms of mucin synthesis and exocytosis, (4) epithelial cell differentiation and goblet cell metaplasia, and (5) development of animal models for studying mucus hypersecretion in sinonasal inflammation. Further understanding will provide the best possible therapeutic strategies for the treatment of various causes of mucus hypersecretion in sinonasal inflammation.

References

1. Terran R, Button B, Boucher RC. Regulation of normal and cystic fibrosis airway surface liquid volume by phasic shear stress. Annu Rev Physiol. 2006;68:543–61.
2. Sadé J, Eliezer N, Silberberg A, et al. The role of mucus in transport by cilia. Am Rev Respir Dis. 1970;102:48–52.
3. Perez-Vilar J. Mucin granule intraluminal organization. Am J Respir Cell Mol Biol. 2007;36:183–90.
4. Sheehan JK, Kirkham S, Howard M, et al. Identification of molecular intermediates in the assembly pathway of the MUC5AC mucin. J Biol Chem. 2004;279:15698–705.
5. Thornton DJ, Rousseau K, McGucken MA. Structure and function of the polymeric mucins in airway mucus. Annu Rev Physiol. 2008;70:459–86.
6. Curran DR, Cohn L. Advances in mucous cell metaplasia; a plug for mucus as a therapeutic focus in chronic airway disease. Am J Respir Cell Mol Biol. 2010;42:268–75.
7. Williams OW, Sharafkhaneh A, Kim V, et al. Airway mucus; from production to secretion. Am J Respir Cell Mol Biol. 2006;34:527–36.
8. Moniaux N, Escande F, Batra SK, et al. Alternative splicing generates a family of putative secreted and membrane-associated MUC4 mucins. Eur J Biochem. 2000;267:4536–44.
9. Rose MC, Voynow JA. Respiratory tract mucin genes and mucin glycoproteins in health and disease. Physiol Rev. 2006;86:245–78.
10. Linden SK, Sutton P, Karlsson NG, et al. Mucins in the mucosal barrier to infection. Immunology. 2008;1:183–97.
11. Krivan HC, Ginsburg V, Roberts DD. *Pseudomonas aeruginosa* and *Pseudomonas cepacia* isolated from cystic fibrosis patients bind specifically to gangliotetraosylceramide (asialoGM1) and gangliotriaosylceramide (asialoGM2). Arch Biochem Biophys. 1988;260:493–6.
12. Ramphal R, Carnoy C, Fievre S, et al. *Pseudomonas aeruginosa* recognizes carbohydrate chains containing type 1 (Gal beta 1-3GlcNAc) or type 2 (Gal beta 1-4GlcNAc) disaccharide units. Infect Immun. 1991;59:700–4.
13. Roberts DD, Olson LD, Barile MF, et al. Sialic acid-dependent adhesion of *Mycoplasma pneumonia* to purified glycoproteins. J Biol Chem. 1989;264:9289–93.
14. Suzuki Y, Nagao Y, Kato H, et al. Human influenza A virus hemagglutinin distinguishes sialyloligosaccharides in membrane-associated gangliosides as its receptor which mediates the adsorption and fusion processes of virus infection. J Biol Chem. 1986;261:17057–61.
15. Evans CM, Koo JS. Airway mucus: the good, the bad, the sticky. Pharmacol Ther. 2009;121:332–48.
16. Nadziejko C, Finkelstein I. Inhibition of neutrophil elastase by mucus glycoprotein. Am J Respir Cell Mol Biol. 1994;11:103–7.
17. Van-Seuningen I, Aubert JP, Davril M. Strong ionic interactions between mucins and two basic proteins, mucus proteinase inhibitor and lysozyme, in human bronchial secretions. Int J Biochem. 1992;24:303–11.
18. Ali MS, Pearson JP. Upper airway mucin gene expression: a review. Laryngoscope. 2007;117:932–8.
19. Martinez-Antón A, Debolos C, Garrido M, et al. Mucin genes have different expression patterns in

healthy and diseased upper airway mucosa. Clin Exp Allergy. 2006a;36:448–57.
20. Singh PK, Hollingsworth MA. Cell surface-associated mucins in signal transduction. Trends Cell Biol. 2006;16:467–76.
21. Lillehoj EP, Kim BT, Kim KC. Identification of *Pseudomonas aeruginosa* flagellin as an adhesion for Muc1 mucin. Am J Physiol Lung Cell Mol Physiol. 2002;282:L751–6.
22. Lu W, Hisamatsu A, Koga T, et al. Cutting edge: enhanced pulmonary clearance of *Pseudomonas aeruginosa* by Muc1 knockout mice. J Immunol. 2006;176:3890–4.
23. Song JS, Hyun SW, Lillehoj E. Mucin secretion in the rat tracheal epithelial cells by epidermal growth factor and *Pseudomonas aeruginosa* extract. Korean J Intern Med. 2001;16:167–72.
24. Hattrup CL, Gendler SJ. Structure and function of the cell surface (tethered) mucins. Annu Rev Physiol. 2008;70:431–57.
25. Hollingsworth MA, Swanson BJ. Mucins in cancer: protection and control of the cell surface. Nat Rev Cancer. 2004;4:45–60.
26. Kim KC, Lillehoj EP. MUC1 mucin: a peacemaker in the lung. Am J Respir Cell Mol Biol. 2008;39:644–7.
27. Ali MS, Wilson JA, Bennett M, et al. Mucin gene expression in nasal polyp. Acta Otolaryngol. 2005;125:618–24.
28. Martinez-Antón A, Roca-Ferrer J, Mullol J. Mucin gene expression in rhinitis syndrome. Curr Allergy Asthma Rep. 2006b;6:189–97.
29. Voynow JA, Rubin BK. Mucins, mucus, and sputum. Chest. 2009;135:505–12.
30. Zudhi Alimam M, Piazza FM, Selby DM, et al. Muc-5/5ac mucin messenger RNA and protein expression is a marker of goblet cell metaplasia in murine airways. Am J Respir Cell Mol Biol. 2000;22:253–60.
31. Wine JJ, Joo NS. Submucosal glands and airway defense. Proc Am Thorac Soc. 2004;1:47–53.
32. Fung DCK, Rogers DF. Airway submucosal glands: physiology and pharmacology. In: Rogers DF, Lethem MI, editors. Airway mucus: basic mechanisms and clinical perspectives. Basel: Birkhauser; 1997.
33. Wine JJ. Parasympathetic control of airway submucosal glands: central reflexes and the airway intrinsic nervous system. Auton Neurosci. 2007;133:35–54.
34. Majima Y, Masuda S, Sakakura Y. Quantitative study of nasal secretory cells in normal subjects and patients with chronic sinusitis. Laryngoscope. 1997;107:1515–8.
35. Thai P, Loukoianow A, Wachi S, et al. Regulation of airway mucin gene expression. Annu Rev Physiol. 2008;70:405–29.
36. Barbier D, Garcia-Verdugo I, Pothlichet J, et al. Influenza A induces the major secreted airway mucin MUC5AC in protease-EGFR-ERK-SP1 dependent pathway. Am J Respir Cell Mol Biol. 2012;47:149. https://doi.org/10.1165/rcmb.2011-0405OC.
37. Hashimoto K, Graham BS, Ho SB, et al. Respiratory syncytial virus in allergic lung inflammation increases Muc5ac and gob-5. Am J Respir Crit Care Med. 2004;170:306–12.
38. Hewson CA, Haas JJ, Bartlett NW, et al. Rhinovirus induces MUC5AC in human infection model and *in vitro* via NF-kB and EGFR pathways. Eur Respir J. 2010;36:1425–35.
39. Deshmukh HS, Shaver C, Case LM. Acrolein-activated matrix metalloproteinase 9 contributes to persistent mucin production. Am J Respir Cell Mol Biol. 2008;38:446–54.
40. Kim HJ, Ryu JH, Kim CH, et al. The role of Nox4 in oxidative stress-induced MUC5AC overexpression in human airway epithelial cells. Am J Respir Cell Mol Biol. 2008;39:598–609.
41. Li R, Meng Z. Effects of SO2 derivatives on expressions of MUC5AC and IL-13 in human bronchial epithelial cells. Arch Toxicol. 2007;81:867–74.
42. Shao MX, Nakanaga T, Nadel JA. Cigarette smoke induces MUC5AC mucin overproduction via tumor necrosis factor-alpha-converting enzyme in human airway epithelial (NCI-H292) cells. Am J Physiol Lung Cell Mol Physiol. 2004;287:L420–7.
43. Chokki M, Yamamura S, Eguchi H, et al. Human airway trypsin-like protease increases mucin gene expression in airway epithelial cells. Am J Respir Cell Mol Biol. 2004;30:470–8.
44. Deshmukh HS, Case LM, Wesselkamper SC, et al. Metalloproteinases mediate mucin 5AC expression by epidermal growth factor receptor activation. Am J Respir Crit Care Med. 2005;171:305–11.
45. Shao MX, Nadel JA. Neutrophil elastase induces MUC5AC mucin production in human airway epithelial cells via a cascade involving protein kinase C, reactive oxygen species, and TNF-α-converting enzyme. J Immunol. 2005;175:4009–16.
46. Shimizu S, Shimizu T, Morser J, et al. Role of the coagulation system in allergic inflammation in the upper airways. Clin Immunol. 2008;129:365–71.
47. Voynow JA, Young LR, Wang Y, et al. Neutrophil elastase increases MUC5AC mRNA and protein expression in respiratory epithelial cells. Am J Phys. 1999;276:L835–43.
48. Cohn L, Homer RJ, Marinov A, et al. Th2-induced airway mucus production is dependent on IL-4R-alpha, but not on eosinophils. J Immunol. 1999;162:6178–83.
49. Song K, Lee WJ, Chung KC, et al. Interleukin-1 beta and tumor necrosis factor-alpha induce MUC5AC overexpression through a mechanism involving ERK/p38 mitogen-activated protein kinases-MSK1-CREB activation in human airway epithelial cells. J Biol Chem. 2003;278:23243–50.
50. Kouzaki H, Iijima K, Kobayashi T, et al. The danger signal, extracellular ATP, is a sensor for an airborne allergen and triggers IL-33 release and innate Th2-type responses. J Immunol. 2011;186:4375–87.
51. Whittaker L, Niu N, Temann UA, et al. Interleukin-13 mediates a fundamental pathway for airway epithe-

lial mucus induced by CD4 T cells and interleukin-9. Am J Respir Cell Mol Biol. 2002;27:593–602.
52. Kuperman DA, Huang X, Koth LL, et al. Direct effects of interleukin-13 on epithelial cells cause airway hyperreactivity and mucus overproduction in asthma. Nat Med. 2002;8:885–9.
53. Zhen G, Park SW, Nguyenvu LT, et al. IL-13 and epidermal growth factor receptor have critical but distinct roles in epithelial cell mucin production. Am J Respir Cell Mol Biol. 2007;36:244–53.
54. Wan H, Kaestner KH, Ang SL, et al. Foxa2 regulates alveolarization and goblet cell hyperplasia. Development. 2004;131:953–64.
55. Usui S, Shimizu T, Kishioka C, et al. Secretory cell differentiation and mucus secretion in cultures of human nasal epithelial cells: use of a monoclonal antibody to study human nasal mucin. Ann Otol Rhinol Laryngol. 2000;109:271–7.
56. Yoon JH, Gray T, Guzman K, Koo JS, Nettesheim P. Regulation of the secretory phenotype of human airway epithelium by retinoic acid, triiodothyronine, and extracellular matrix. Am J Respir Cell Mol Biol. 1997;16:724–31.
57. Koo JS, Yoon JH, Gray T, Norford D, Jetten AM, Nettesheim P. Restoration of the mucous phenotype by retinoic acid in retinoid-deficient human bronchial cell cultures: changes in mucin gene expression. Am J Respir Cell Mol Biol. 1999;20:43–52.
58. Takeyama K, Dabbagh K, Lee HM, et al. Epidermal growth factor system regulates mucin production in airways. Proc Natl Acad Sci U S A. 1999;96:3081–6.
59. Burgel PR, Lazarus SC, Tam DC, et al. Human eosinophils induce mucin production in airway epithelial cells via epidermal growth factor activation. J Immunol. 2001;167:5948–54.
60. Shim JJ, Dabbagh K, Ueki IF, et al. IL-13 induces mucin production by stimulating epidermal growth factor receptors and by activating neutrophils. Am J Physiol Lung Cell Mol Physiol. 2001;280:L134–40.
61. Takeyama K, Jung B, Shim JJ, et al. Activation of epidermal growth factor receptors is responsible for mucin synthesis induced by cigarette smoke. Am J Physiol Lung Cell Mol Physiol. 2001;280:L165–72.
62. Davis CW. Goblet cells: physiology and pharmacology. In: Rogers DF, Lethem MI, editors. Airway mucus: basic mechanisms and clinical perspectives. Basel: Birkhauser; 1997.
63. Davis CW, Dickey BF. Regulated airway goblet cell mucin secretion. Annu Rev Physiol. 2008;70:487–512.
64. Li Y, Martin LD, Spizz G, et al. MARCKS protein is a key molecule regulating mucin secretion by human airway epithelial cells *in vitro*. J Biol Chem. 2001;276:40982–90.
65. Singer M, Martin LD, Vargaftig BB, et al. A MARCKS-related peptide blocks mucus hypersecretion in a mouse model of asthma. Nat Med. 2004;10:193–6.
66. Beum PV, Basma H, Bastola DR, et al. Mucin biosynthesis upregulation of core 2 beta 1.6 *N*-acetylglucosaminyltransferase by retinoic acid and Th2 cytokines in a human airway epithelial cell line. Am J Physiol Lung Cell Mol Physiol. 2005;288:L116–24.
67. Delmotte P, Degroote S, Lafitte JJ. Tumor necrosis factor alpha increases the expression of glycosyltransferases and sulfotransferases responsible for the biosynthesis of sialylated and/or sulfated Lewis X epitopes in the human bronchial mucosa. J Biol Chem. 2002;277:424–31.
68. Lamblin G, Lhermitte M, Klein A, et al. The carbohydrate diversity of human respiratory mucins: a protection of the underlying mucosa? Am Rev Respir Dis. 1991;144:S19–24.
69. Ramphal R, Houdret N, Koo L, et al. Differences in adhesion of *Pseudomonas aeruginosa* to mucin glycopeptides from sputa of patients with cystic fibrosis and chronic bronchitis. Infect Immun. 1989;57:3066–71.
70. Shimizu T, Hirano H, Shimizu S, et al. Differential properties of mucous glycoproteins in rat nasal epithelium: a comparison between allergic inflammation and lipopolysaccharides stimulation. Am J Respir Crit Care Med. 2001;164:1077–82.
71. Carnoy C, Ramphal R, Scharfman A, et al. Altered carbohydrate composition of salivary mucins from patients with cystic fibrosis and the adhesion of *Pseudomonas aeruginosa*. Am J Respir Cell Mol Biol. 1991;9:323–34.
72. Schulz BL, Sloane AJ, Robinson LJ, et al. Glycosylation of sputum mucins is altered in cystic fibrosis patients. Glycobiology. 2007;17:698–712.
73. Shimizu T, Takahashi Y, Kawaguchi S, et al. Hypertrophic and metaplastic changes of goblet cells in rat nasal epithelium induced by endotoxin. Am J Respir Crit Care Med. 1996;153:1412–8.
74. Rogers DF. Airway mucus hypersecretion in asthma: an undervalued pathology? Curr Opin Pharmacol. 2004;4:241–50.
75. Ali MS. Nasosinus mucin expression in normal and inflammatory conditions. Curr Opin Allergy Clin Immunol. 2009;9:10–5.
76. Voynow JA, Selby DN, Rose MC. Mucin gene expression (MUC1, MUC2, and MUC5/5AC) in nasal epithelial cells of cystic fibrosis, allergic rhinitis, and normal individuals. Lung. 1998;176:345–54.
77. Shimizu T, Hirano T, Majima Y, et al. A mechanism of antigen-induced mucus production in nasal epithelium of sensitized rats: a comparison with lipopolysaccharides-induced mucus secretion. Am J Respir Crit Care Med. 2000;161:1648–54.
78. Shimizu S, Hattori R, Majima Y, et al. Th2 cytokine inhibitor suplatast tosilate inhibits antigen-induced mucus hypersecretion in the nasal epithelium of sensitized rats. Ann Otol Rhinol Laryngol. 2009;118:67–72.
79. Shimizu T, Shimizu S, Hattori R, et al. A mechanism of antigen-induced goblet cell degranulation in the nasal epithelium of sensitized rats. J Allergy Clin Immunol. 2003b;112:119–25.

80. Suh JD, Kennedy DW. Treatment options for chronic rhinosinusitis. Proc Am Thorac Soc. 2011;8:132–40.
81. Sakakura Y, Majima Y, Saida S, et al. Reversibility of reduced mucociliary clearance in chronic sinusitis. Clin Otolaryngol Allied Sci. 1985;10:79–83.
82. Ikeda K, Oshima T, Furukawa M, et al. Restoration of the mucociliary clearance of the maxillary sinus after endoscopic sinus surgery. J Allergy Clin Immunol. 1997;99:48–52.
83. Min YG, Yun YS, Song BH, et al. Recovery of nasal physiology after functional endoscopic sinus surgery: olfaction and mucociliary transport. ORL J Otorhinolaryngol Relat Spec. 1995;57:264–8.
84. Majima Y, Hirata K, Takeuchi K, et al. Effects of orally administered drugs on dynamic viscoelasticity of human nasal mucus. Am Rev Respir Dis. 1990;141:79–83.
85. Yuta A, Baraniuk JN. Therapeutic approaches to airway mucous hypersecretion. In: Rogers DF, Lethem MI, editors. Airway mucus: basic mechanisms and clinical perspectives. Basel: Birkhauser; 1997.
86. Mainz JG, Schiller I, Ritschel C, et al. Sinonasal inhalation of dornase alfa in CF: a double-blinded placebo-controlled cross-over pilot trial. Auris Nasus Larynx. 2011;38:220–7.
87. Majima Y, Kurono Y, Hirakawa K, et al. Efficacy of combined treatment with S-carboxymethylcysteine (carbocisteine) and clarithromycin in chronic rhinosinusitis patients without nasal polyp or with small nasal polyp. Auris Nasus Larynx. 2012;39:38–47.
88. Wawrose SF, Tami TA, Amoils CP. The role of guaifenesin in the treatment of sinonasal disease in patients infected with the human immunodeficiency virus (HIV). Laryngoscope. 1992;102:1225–8.
89. Fuchs HJ, Borowitz DS, Christiansen DH, et al. Effect of aerosolized recombinant human DNase on exacerbations of respiratory symptoms and on pulmonary function in patients with cystic fibrosis. The pulmozyme study group. N Engl J Med. 1994;331:637–42.
90. Donaldson SH, Bennett WD, Zeman KL, et al. Mucus clearance and lung function in cystic fibrosis with hypertonic saline. N Engl J Med. 2006;19:241–50.
91. Rabago D, Pasic T, Zgierska A, et al. The efficacy of hypertonic saline nasal irrigation for chronic sinonasal symptoms. Otolaryngol Head Neck Surg. 2005;133:3–8.
92. Albert RK, Connett J, Bailey WC, et al. Azithromycin for prevention of exacerbations of COPD. N Engl J Med. 2011;365:689–98.
93. Jaffe A, Francis J, Rosenthal M, et al. Long term-azithromycin may improve lung function in children with cystic fibrosis. Lancet. 1998;351:420.
94. Kudoh S, Azuma A, Yamamoto M, et al. Improvement of survival in patients with diffuse panbronchiolitis treated with low-dose erythromycin. Am J Respir Crit Care Med. 1998;157:1829–32.
95. Shirai T, Sato A, Chiba K. Effect of 14-membered ring macrolide therapy on chronic respiratory tract infections and polymorphonuclear leukocyte activity. Intern Med. 1995;34:469–74.
96. Wallwork B, Coman W, Mackay-Sim A, et al. A double-blind, randomized, placebo-controlled trial of macrolide in the treatment of chronic rhinosinusitis. Laryngoscope. 2006;116:189–93.
97. Shimizu T, Shimizu S. Azithromycin inhibits mucus hypersecretion from airway epithelial cells. Mediat Inflamm. 2012;2012:265714. https://doi.org/10.1155/2012/265714.
98. Shimizu T, Shimizu S, Hattori R, et al. *In vivo* and *in vitro* effects of macrolide antibiotics on mucus secretion in airway epithelial cells. Am J Respir Crit Care Med. 2003a;168:581–7.
99. Cervin A, Wallwork B. Macrolide therapy of chronic rhinosinusitis. Rhinology. 2007;45:259–67.
100. Shimizu T, Suzaki H. Past, present and future of macrolide therapy for chronic rhinosinusitis in Japan. Auris Nasus Larynx. 2016;43:131–6.
101. Haruna S, Shimada C, Ozawa M, et al. A study of poor responders for long-term, low-dose macrolide administration for chronic sinusitis. Rhinology. 2009;47:66–71.
102. Suzuki H, Ikeda K, Honma R, et al. Prognostic factors of chronic rhinosinusitis under long-term low-dose macrolide therapy. ORL J Otorhinolaryngol Relat Spec. 2000;62:121–7.
103. Videler WJ, Badia L, Harvey RJ, et al. Lack of efficacy of long-term, low-dose azithromycin in chronic rhinosinusitis: a randomized controlled trial. Allergy. 2011;66:1457–68.
104. Rudmik L, Schlosser RJ, Smith TL, et al. Impact of topical nasal steroid therapy on symptoms of nasal polyposis: a meta-analysis. Laryngoscope. 2012;122:1431. https://doi.org/10.1002/lary.23259.
105. Weiner JM, Abramson MJ, Puy RM. Intranasal corticosteroids versus oral H1 receptor antagonists in allergic rhinitis: systematic review of randomized controlled trials. BMJ. 1998;317:1624–9.
106. Greiner AN, Meltzer EO. Pharmacologic rationale for treating allergic and nonallergic rhinitis. J Allergy Clin Immunol. 2006;118:985–98.

2 The Coagulation System and Rhinosinusitis

Takeshi Shimizu and Shino Shimizu

Abbreviations

APC	Activated protein C
AR	Allergic rhinitis
BALF	Bronchoalveolar lavage fluid
CRS	Chronic rhinosinusitis
EPCR	Endothelial protein C receptor
FDPs	Fibrin degradation products
GM-CSF	Granulocyte macrophage colony-stimulating factor
IL	Interleukin
LMWH	Low-molecular weight heparin
NP	Nasal polyp
PAF	Platelet-activating factor
PAI-1	Plasminogen activator inhibitor-1
PAR	Protease-activated receptor
PDGF	Platelet-derived growth factor
TAFI	Thrombin activatable fibrinolysis inhibitor
TATc	Thrombin-antithrombin complex
TF	Tissue factor
TFPI	Tissue factor pathway inhibitor
TGF	Transforming growth factor
TM	Thrombomodulin
TNF	Tumor necrosis factor
t-PA	Tissue plasminogen activator
u-PA	Urokinase plasminogen activator
VEGF	Vascular endothelial growth factor

Core Message
Local activation of the coagulation system contributes to the pathophysiology of upper airway inflammation, such as allergic rhinitis (AR) and chronic rhinosinusitis (CRS). Airway inflammation is associated with increased vascular permeability. Leakage of plasma coagulation factors into the tissues induces plasma factor VIIa (FVIIa) to bind tissue factor (TF) expressed on endothelial cells, fibroblasts, epithelial cells, and leukocytes, which ultimately leads to thrombin (FIIa) generation and fibrin deposition. Increased coagulation activity and decreased fibrinolytic activity induce excessive fibrin deposition in human nasal polyp (NP) tissues. Thrombin and coagulation factors play important roles not only in hemostasis and thrombosis but also in inflammation by stimulating the production of cytokines, chemokines, mucin, and extracellular matrix proteins from nasal epithelial cells and from fibroblasts through the protease-activated receptors (PARs). PAR-mediated responses provide a direct link between coagulation and inflammation, and anticoagulant drugs may have a therapeutic potential for the treatment of intractable rhinosinusitis.

T. Shimizu (✉) · S. Shimizu
Department of Otorhinolaryngology-Head and Neck Surgery, Shiga University of Medical Science, Otsu, Shiga, Japan
e-mail: shimizu@belle.shiga-med.ac.jp

Ö. Ö. Celebi, T. M. Önerci (eds.), *Nasal Physiology and Pathophysiology of Nasal Disorders*,
https://doi.org/10.1007/978-3-031-12386-3_2

2.1 Introduction

Airway inflammation is associated with increased vascular permeability and leakage of plasma coagulation factors, leading to the activation of the coagulation system in the extravascular space. Procoagulant activity has been demonstrated in patients with allergic rhinitis (AR) [1], chronic rhinosinusitis (CRS) [2], bronchial asthma [3], and lung fibrosis [4]. The presence of thrombin activity and thrombin-antithrombin complex (TATc) in nasal secretions and in bronchoalveolar lavage fluid (BALF) is clear evidence of local activation of the coagulation system. Airway inflammation enhances the activity of tissue factor (TF), an important initial upstream protein of the extrinsic coagulation cascade, expressed on airway epithelial cells and infiltrating eosinophils [1, 5].

Airway inflammation is characterized by mucus hypersecretion, infiltration of inflammatory cells, and tissue remodeling, such as nasal polyp (NP) formation. Excessive fibrin deposition is detected in the epithelial mucus layer and in the lamina propria of NPs [5, 6], and dense fibrin networks exacerbate airway inflammation by disturbing the mucociliary activity of the epithelial mucus layer and by providing a scaffold for proliferating cells such as fibroblasts and endothelial cells. Aberrant fibrin turnover is induced by an increase in coagulation activity and a decrease in fibrinolysis.

Coagulation factors such as thrombin (FIIa), TF/FVIIa, and FXa play important roles not only in hemostasis and thrombosis but also in inflammation through interactions with protease-activated receptors (PARs; PAR-1, PAR-2, PAR-3, and PAR-4) expressed on epithelial cells, fibroblasts, and vascular endothelial cells [7]. Activated coagulation factor–PAR signaling induces airway inflammation by stimulating the production of cytokines, chemokines, mucin, and extracellular matrix proteins by airway epithelial cells and fibroblasts [1, 8–12].

Figure 2.1 shows the interactions between coagulation and inflammation. The coagulation system is activated in rhinosinusitis by leakage of plasma coagulation factors and by enhanced TF activity expressed on epithelial cells, fibroblasts, endothelial cells, and leukocytes [5]. Fibrin deposition is facilitated by decreased activity of the anticoagulant protein C system [13, 14] and by inhibition of fibrinolysis following the enhanced production of plasminogen activator-1 (PAI-1) [15, 16]. Coagulation modulates airway inflammation by PAR-mediated cytokine/chemokine production and by the anti-inflammatory protein C system. Platelets possess proinflammatory mediators such as thromboxane, histamine, and platelet-activating factor (PAF), and P-selectin expression on their surface induces eosinophil infiltration [17, 18]. This chapter summarizes the current knowledge of the role of coagulation, fibrinolysis, and the anticoagulant system in the pathophysiology of rhinosinusitis. The therapeutic potential of anticoagulant drugs for the treatment of intractable rhinosinusitis is shown.

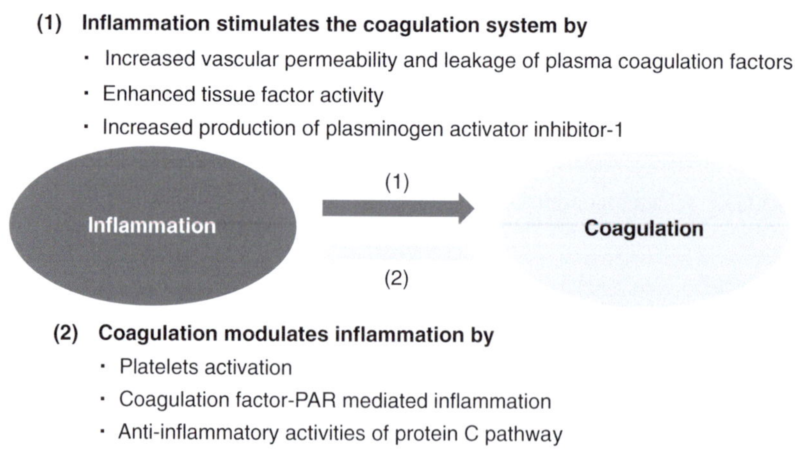

Fig. 2.1 Interaction between the coagulation system and inflammation. Airway inflammation activates the coagulation system, and coagulation modulates inflammation

2.2 The Coagulation System

Hemostasis and thrombosis are controlled by platelet aggregation, coagulation (blood clot forming), fibrinolysis (clot lysing), and the anticoagulant (regulating) system. Platelets immediately form a plug at the site of injury to the blood vessel, and then coagulation factors respond in a cascade to form fibrin strands, which strengthen the platelet plug. Fibrinolysis is the process of fibrin cleavage by plasmin into fibrin degradation products (FDPs), which act to resolve blood clots. Coagulation, fibrinolysis, and the regulating anticoagulant system balance their activities and maintain the homeostasis of the coagulation mechanisms (Fig. 2.2).

2.2.1 The Coagulation Cascade

The coagulation cascade is classically divided into the extrinsic (TF) pathway and the intrinsic (contact activation) pathway. The extrinsic pathway is the most important primary pathway for clot formation in the coagulation cascade. The intrinsic (contact activation) pathway has minor roles in initiating clot formation, and recent research has shown that the contact activation system is more involved in inflammation and innate immunity by activating the complement system and the kallikrein–kinin pathway [19].

Tissue factor (TF) is an important starting upstream protein in the extrinsic coagulation cascade, and it is the most potent stimulator of this cascade [20]. TF is expressed on the cell surface of epithelial cells, fibroblasts, endothelial cells, and leukocytes, including epithelial cells and infiltrating eosinophils in nasal mucosa [1, 5]. Leakage of plasma coagulation factors into tissues induces FVIIa to bind to TF on the cell surface, and this complex binds to factor X (FX), converting it to the activated form, factor Xa (FXa). The coagulation factors are generally ser-

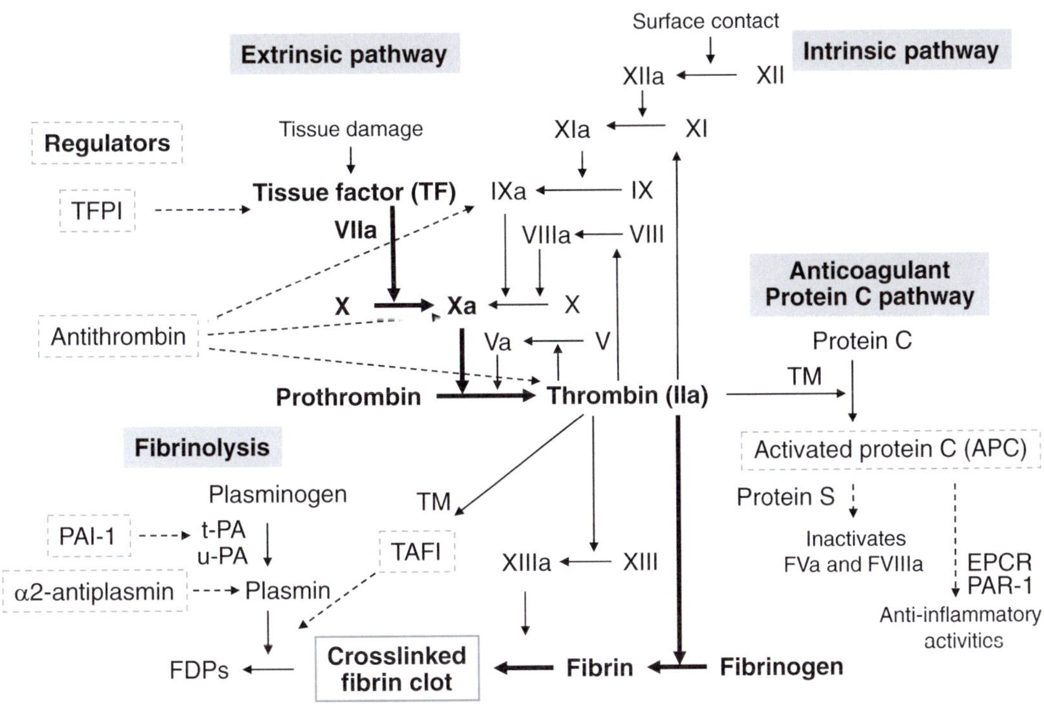

Fig. 2.2 Coagulation cascade. Extrinsic and intrinsic pathways, fibrinolysis, regulators, and the anticoagulant protein C system balance each other's activities and maintain the homeostasis of the coagulation mechanisms. Tissue factor is an important starting protein in the extrinsic coagulation cascade, and thrombin plays a fundamental role by converting fibrinogen to fibrin. Solid arrow: activation, Dotted arrow: Inhibition. EPCR: Endothelial protein C receptor, FDPs: Fibrin degradation products, PAI-1: Plasminogen activator inhibitor-1, t-PA: Tissue plasminogen activator, u-PA: Urokinase plasminogen activator, TAFI: Thrombin activatable fibrinolysis inhibitor, TFPI: Tissue factor pathway inhibitor, TM: thrombomodulin

ine proteases, which act by cleaving downstream proteins. The exceptions are TF, FV, FVIII (glycoproteins), and FXIII (transglutaminase). FXa then leads to eventual thrombin generation and fibrin deposition (Fig. 2.2).

Thrombin (FIIa) plays fundamental roles in the coagulation system. Prothrombin is converted to thrombin by the activation of extrinsic and intrinsic pathways. Thrombin not only converts fibrinogen to fibrin, but it also has feedback activation roles by activating factors V and VIII. The fibrin clot is formed by cross-linking of fibrin monomers and is further stabilized by FXIII, which is activated by thrombin (FXIIIa). Thrombin also activates their inhibitor, protein C, in the anticoagulant protein C pathway in the presence of thrombomodulin (Fig. 2.2).

2.2.2 Regulators

TF pathway inhibitor (TFPI) regulates the initial step of the extrinsic coagulation cascade by targeting the TF-FIIa-FXa complex. TFPI is a serine protease inhibitor that is secreted by endothelial cells, leukocytes, platelets, fibroblasts, smooth muscle cells, and epithelial cells [21]. TFPI is produced locally in response to activation of the coagulation system.

Antithrombin (also called as antithrombin III) is a serine protease inhibitor that inactivates several coagulation factors, such as thrombin (FIIa), FIXa, FXa, FXIa, and FXIIa. Antithrombin is a circulating plasma protein, which is produced in the liver, and thrombin is rapidly bound to antithrombin by forming thrombin–antithrombin complex (TATc). Antithrombin is the major inhibitor, accounting for approximately 80% of the thrombin inhibitory activity in plasma. The anticoagulant heparin accelerates antithrombin activity by the enhanced binding of antithrombin to thrombin and FXa.

The anticoagulant protein C pathway is activated when thrombin binds to thrombomodulin on the cell surface, and the thrombin–thrombomodulin complex converts protein C to activated protein C (APC). APC along with cofactor protein S inactivates FVa and FVIIIa, important procoagulant cofactors in the generation of thrombin. APC also has cytoprotective and anti-inflammatory activities through the endothelial protein C receptor (EPCR) and PAR-1 expressed on airway epithelial cells and endothelial cells [22, 23].

2.2.3 Fibrinolysis

Fibrinolysis is a natural mechanism to prevent excessive fibrin deposition and to resolve clot formation. The major fibrinolytic enzyme, plasmin, cleaves the fibrin mesh into fibrin degradation products (FDPs). Plasmin is formed from plasminogen by tissue plasminogen activator (t-PA) and urokinase plasminogen activator (u-PA). Plasminogen, which is produced in the liver, has affinity for fibrin and is entrapped within the clot when it is formed. Plasminogen activator inhibitor-1 (PAI-1) is a serine protease inhibitor that acts as a principal inhibitor of both t-PA and u-PA. Endothelial cells and many different cells produce t-PA, u-PA, and PAI-1, including airway epithelial cells, fibroblasts, mast cells, and macrophages [24].

Fibrinolysis is also regulated by endogenous antifibrinolytic proteins, α2-antiplasmin, and α2-macroglobulin, which inactivate plasmin. Thrombin activatable fibrinolysis inhibitor (TAFI) is another important regulator of fibrinolysis. TAFI is activated by thrombin and thrombin–thrombomodulin complex and inhibits fibrinolysis by removing the binding and activating sites on fibrin for plasminogen and t-PA [25].

2.3 Activation of the Coagulation System in Rhinosinusitis

Activation of coagulation is initiated by plasma exudation into the tissues in sinonasal inflammation, and coagulation steps start through the interaction of plasma coagulation factors with TF, an initial protein of the coagulation cascade in tissues. Thrombin is generated by the stepwise activation of coagulation factors, and it converts

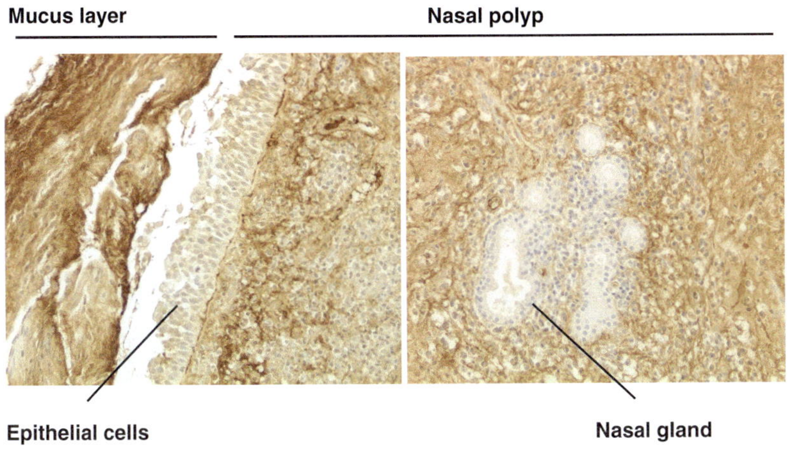

Fig. 2.3 Immunohistochemical staining of fibrin in nasal polyps (NPs) from patients with CRS with NPs. Fibrin is strongly expressed in the epithelial mucus layer and lamina propria of NPs

fibrinogen to fibrin. Enhanced thrombin activity and elevated concentrations of TATc in nasal secretions from patients with AR and CRS indicate the local activation of the coagulation system in rhinosinusitis [1, 2]. Activation of the coagulation system results in excessive deposition of fibrin in the epithelial mucus layer and in the lamina propria of NPs [5] (Fig. 2.3).

2.3.1 Tissue Factor (TF)

TF is an important upstream protein in the extrinsic pathway and plays an essential role in the coagulation cascade. TF and its regulator TFPI are expressed in nasal epithelial cells and subepithelial gland cells, and TF is also expressed in infiltrating inflammatory cells, including eosinophils, in nasal mucosa [5]. TF is a transmembrane glycoprotein, and TF activity is determined as activated FX (FXa) induced by TF/FVIIa complex on cultured cells. Thrombin and tumor necrosis factor (TNF)-α enhance TF activity on cultured airway epithelial cells [2] and on a human eosinophilic leukemia cell line, EoL-1 cells (unpublished data). These results indicate that inflammation activates the local coagulation system through enhanced TF activity on epithelial cells and on infiltrating eosinophils. TF is strongly expressed in the basal area of nasal epithelial cells [5] (Fig. 2.4), and it is reportedly important for the attachment, survival, and proliferation of basal epithelial cells [26, 27].

TFPI, a major regulator of TF-induced coagulation, is released from cultured nasal epithelial cells on stimulation with thrombin and TNF-α. TFPI concentration in nasal secretions is increased in CRS patients with asthma, and it is correlated with both thrombin activity and TATc concentrations in nasal secretions [5]. These results suggest that TPFI is produced locally in response to the activation of the coagulation cascade.

2.3.2 Thrombin

Thrombin activity is determined spectrophotometrically using the synthetic substrate D-Phe-piperonyl-Arg-p-nitroanilide, and it is enhanced in nasal secretions from patients with AR and CRS with asthma [2]. Thrombin is rapidly bound by antithrombin and forms TATc *in vivo*, and the TATc level is a good marker of thrombin generation. The TATc concentration is increased in nasal secretions from patients with AR and CRS with asthma [2]. Increased TATc levels are also reported in nasal lavage fluids and in NP tissues from CRS patients [28], supporting the local activation of thrombin generation and a procoagulant state in rhinosinusitis.

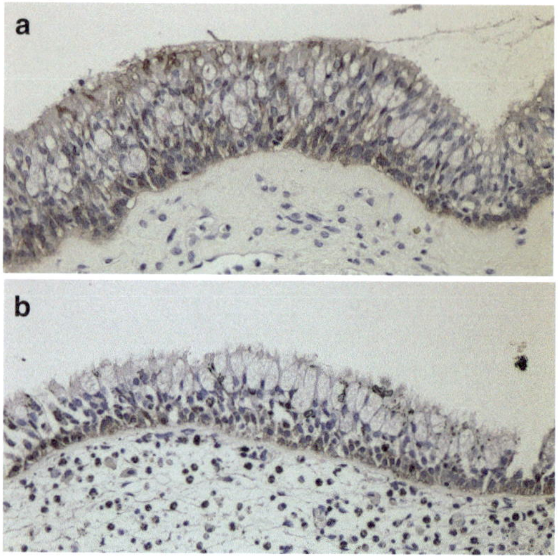

Fig. 2.4 Immunohistochemical staining of tissue factor (TF) in nasal mucosa from patients with CRS with NPs. (**a**) Nasal polyp. (**b**) Inferior turbinate. TF is expressed in epithelial cells and inflammatory cells including eosinophils. Basal areas of epithelial cells are strongly stained

2.4 Coagulation Contributes to the Pathophysiology of Rhinosinusitis

Platelets, the cellular component of blood clots, play important roles in the development of airway inflammation. P-selectin expression of platelets is important for eosinophil infiltration in tissues. Thrombin activates platelets and stimulates the release of proinflammatory mediators, such as thromboxane, histamine, serotonin, and PAF, from platelets [25]. Coagulation factors such as thrombin (FIIa), TF/FVIIa, and FXa contribute to the inflammation through the interaction with PARs expressed on epithelial cells, fibroblasts, and endothelial cells [7].

2.4.1 Coagulation Factors and Protease-Activated Receptors (PARs)

The PAR family consists of four subtypes (PAR-1, PAR-2, PAR-3, and PAR-4); each subtype displays a unique activation site that is recognized by specific proteases [29]. Protease activity of coagulation factors stimulates PARs by proteolytic cleavage and unmasking of an amino-terminal receptor sequence, which acts as a tethered ligand by binding to the body of the receptor to initiate transmembrane signaling (Fig. 2.5). All PARs are expressed in nasal epithelial cells and fibroblasts [1, 9]. PARs agonist peptides, synthetic peptides with the same sequences as the newly formed tethered ligands, can activate the receptor independently. The three PARs, PAR-1, PAR-3, and PAR-4, are thrombin receptors, and FXa may activate PAR-1, PAR-2, and PAR-3[29] (Table 2.1). Thrombin and PAR-1 agonist peptide stimulate the secretion of MUC5AC mucin, PDGF, VEGF, IL-6, IL-8, CCL-2 [1, 2, 7, 8, 30], and granulocyte macrophage colony-stimulating factor (GM-CSF) [31] from cultured airway epithelial cells. Thrombin, FXa, and PAR-1 and PAR-2 agonist peptides stimulate the secretion of transforming growth factor (TGF)-β, fibronectin, eotaxin-1, IL-6, and IL-8 from cultured nasal fibroblasts [9].

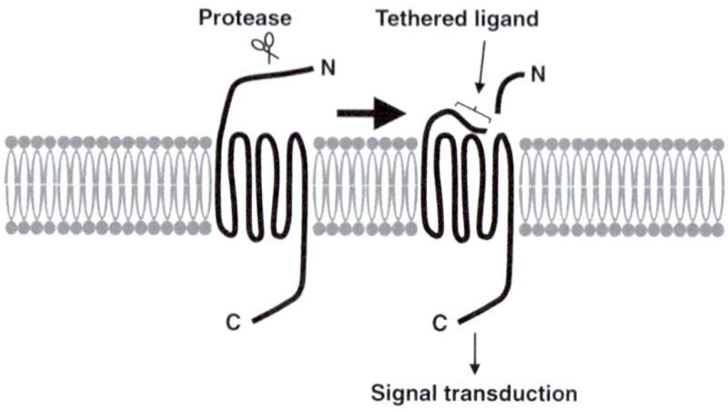

Fig. 2.5 Protease activated receptor-1 (PAR-1). Thrombin stimulates PAR-1 by proteolytic cleavage and unmasking of an amino-terminal receptor sequence, which acts as a tethered ligand by binding to the body of the receptor to initiate transmembrane signaling

2.4.2 Tissue Remodeling

Tissue remodeling is an irreversible histologic change caused by persistent inflammation and aberrant repair mechanisms. Morphological changes of tissue remodeling in rhinosinusitis include epithelial sloughing, thickening of the basement membrane, subepithelial fibrosis, and formation of NPs. Goblet cell metaplasia in asso-

Table 2.1 PARs cleaving proteases

PAR-1	Thrombin, FXa, TF/FVIIa, Activated protein C, Plasmin Trypsin, Chymase, MMP-1-3,8,9,12,13, Cathepsin G, Neutrophil elastase, Proteinase 3, Granzyme A,B,K, Der p 1
PAR-2	FXa, TF/FVIIa, Plasmin Trypsin, tryptase, chymase, cathepsin G,S, Neutrophil elastase, Proteinase 3, Papain, HDM (Der p 1-3,9), Cockroach, *Alternaría,* Japanese cedar pollen
PAR-3	Thrombin, FXa, Trypsin
PAR-4	Thrombin, Trypsin, Cathepsin G Papain, Der p 3

ciation with hypersecretion of mucus is an important characteristic of AR and CRS. MUC5AC mucin is the most predominant gel-forming mucin expressed in airway goblet cells, and it is up-regulated in nasal polyposis. PDGF, VEGF, and TGF-β are profibrotic cytokines that promote tissue remodeling by stimulating the proliferation of vascular endothelial cells, fibroblasts, myocytes, and goblet cells, and by increasing the deposition of extracellular matrix proteins. Overexpressions of PDGF, VEGF, and TGF-β and their receptors are commonly observed in the nasal mucosa of CRS patients [32–35]. IL-6 contributes to tissue remodeling by stimulating mucus production, fibroblast proliferation, and matrix deposition in airways [36]. Fibronectin, an extracellular matrix protein, is cross-linked to the fibrin α chain by FXIIIa [37], and increased expression of fibronectin is reported in NPs [38, 39].

Activated coagulation factors, thrombin and FXa, play important roles in tissue remodeling of rhinosinusitis by leading to fibrin deposition and by stimulating the secretion of MUC5AC mucin, profibrotic cytokines (PDGF, VEGF, TGF-β), IL-6, and extracellular matrix protein (fibronectin) from nasal epithelial cells and from nasal fibroblasts via PAR-1 and PAR-2. Thrombin and FXa are also involved in the activation, infiltration, and survival of inflammatory cells such as neutrophils, eosinophils, and monocytes in rhinosinusitis by stimulating the secretion of IL-8, CCL-2, and GM-CSF from nasal epithelial cells and that of IL-8, and eotaxin-1 from nasal fibroblasts (Fig. 2.6).

2.4.3 Allergic Rhinitis (AR)

The coagulation system is activated by increased plasma exudation into tissues in AR patients. Enhanced thrombin activity and elevated concentrations of TATc are reported in nasal secretions from AR patients [2]. Thrombin activity was significantly increased in nasal secretions from patients with house dust mite (HDM) AR 5 min after allergen provocation with an HDM disc, compared with that before the provocation [1]. Thrombin generation and fibrin deposition are more prominent in the nasal mucosa of ovalbumin-induced AR mice [40]. These results indicate that allergen stimulation induces thrombin generation and resulting fibrin deposition in the nasal mucosa of AR patients.

PAR-2 mediated inflammation is important in allergic inflammation. Activated coagulation factor, FXa, TF/FVIIa, and other proteases can activate PAR-2, both derived from the host (mast cell tryptase and chymase, trypsin, neutrophil elastase, proteinase 3, cathepsin G and S) and from allergens (HDM, *Alternaria*, cockroach, and Japanese cedar pollen) [25]. Allergen-derived protease-induced PAR-2 activation stimulates secretion of epithelial cell-derived cytokines, TSLP and IL-25, from airway epithelial cells, which induce the initiation and development of allergic inflammation. Protease activity of *Alternaria* induces TSLP production from cultured airway epithelial cells, and that of HDM or Japanese cedar stimulates IL-25 production from cultured human nasal epithelial cells via PAR-2 [41–43]. PAR-2 is also expressed on mast cells, eosinophils, and smooth muscle cells in airways. PAR-2 activation induces histamine release from mast cells and degranulation and cytokine release from eosinophils, respectively [44]. FXa and PAR-2 agonists stimulate the secretion of eotaxin-1 and IL-8 from cultured nasal fibroblasts [9], which may induce eosinophil and neutrophil infiltration in nasal mucosa. The

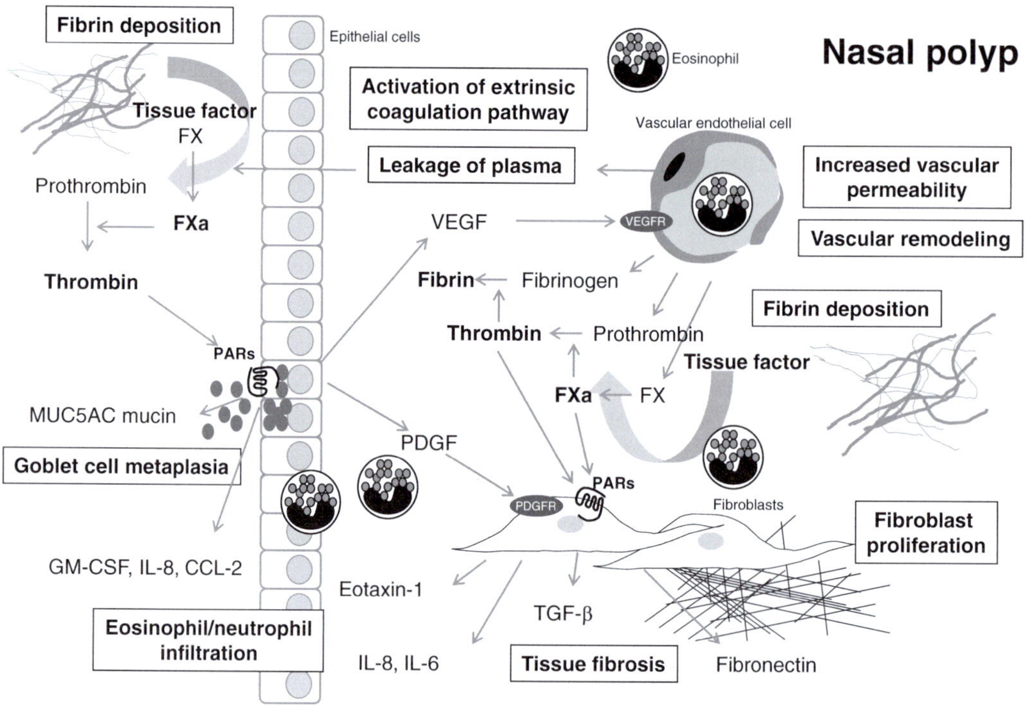

Fig. 2.6 Coagulation factor-PAR-mediated inflammation and tissue remodeling in nasal polyps (NPs). Activated coagulation factors, thrombin and FXa, play important roles in tissue remodeling by leading to fibrin deposition and by stimulating the secretion of MUC5AC mucin, profibrotic cytokines (PDGF, VEGF, TGF-β), IL-6, and extracellular matrix protein (fibronectin) from nasal epithelial cells and from nasal fibroblasts via PAR-1 and PAR-2. Thrombin and FXa are also involved in the activation, infiltration, and survival of inflammatory cells such as neutrophils, eosinophils, and monocytes by stimulating the secretion of IL-8, CCL-2, and GM-CSF from nasal epithelial cells and that of IL-8, and eotaxin-1 from nasal fibroblasts

coagulation system is activated in AR, and FXa and TF/FVIIa can activate PAR-2, but their roles in allergic inflammation have not been fully elucidated (Fig. 2.7).

2.4.4 Excessive Fibrin Deposition

Immunohistochemical analysis has demonstrated excessive fibrin deposition in the epithelial mucus layer and lamina propria of NPs from CRS patients [5] (Fig. 2.3). Excessive fibrin deposition is induced by an increase in coagulation activity and by a decrease in fibrinolysis. Plasmin, a major fibrinolytic enzyme, cleaves the fibrin mesh into FDPs to prevent excessive fibrin deposition. Takabayashi et al. [6] reported reduced fibrinolytic activity in NP. In NP tissues, the protein level of d-dimer (a major FDP) is significantly decreased compared with uncinate tissues from CRS patients or control subjects. Plasmin is generated through cleavage of plasminogen by u-PA and t-PA. The mRNA expression and protein level of t-PA, but not u-PA, are significantly decreased in NP tissues compared with uncinate tissues from CRS patients or control subjects. In addition, t-PA is prominently expressed in epithelial cells and submucosal gland cells in nasal mucosa, and the concentration of t-PA in NP tissues is negatively correlated with that of eosinophil cationic protein (ECP). The Th2 cytokines, IL-4 and IL-13, inhibit the production of t-PA from cultured airway epithelial cells. TAFI is another important regulator of fibrinolysis, activated by thrombin and thrombin-thrombomodulin complex. The TAFI level is significantly increased

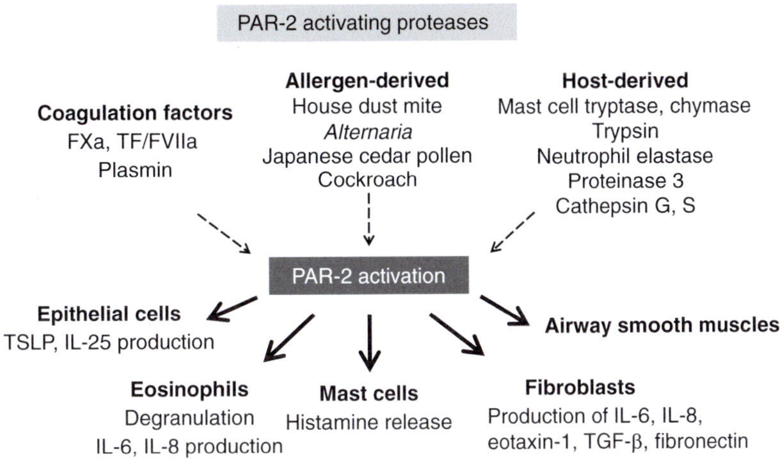

Fig. 2.7 PAR-2-mediated responses are important in allergic inflammation. The coagulation factors, FXa and TF/FVIIa, can activate PAR-2, and FXa and PAR-2 agonists stimulate the secretion of IL-6, IL-8, eotaxin-1, TGF-β, and fibronectin from cultured nasal fibroblasts, but their roles in allergic inflammation have not been fully elucidated

in NP tissues compared with that in uncinate tissues from CRS patients or control subjects, and it is positively correlated with ECP levels in NP tissues [28]. These results indicate that the downregulation of t-PA and upregulation of TAFI may lead to insufficient fibrin degradation in NPs, and that eosinophilic inflammation and Th2 cytokines contribute to the reduced fibrinolysis through the downregulation of t-PA and upregulation of TAFI. Excessive fibrin deposition may accelerate tissue remodeling by providing a scaffold for the proliferating fibroblasts and endothelial cells.

Coagulation factor XIII (FXIII) is activated by thrombin (FXIIIa) and completes fibrin clots by cross-linking the fibrin monomers. Cross-linking of various substrates (α2-antiplasmin and fibronectin) to fibrin affects clot structure, stability, and stiffness, which protect fibrin clots from enzymatic degradation [45]. FXIIIa is a dimer of two active A subunits, and mRNA expression and the protein level of A subunit (FXIII-A) are increased in NP tissues compared with uncinate tissues from CRS patients or control subjects [46]. M2 macrophages are major FXIII-A-producing cells, and the number of M2 macrophages, but not M1macrophages, is increased in NPs [47, 48]. M1 macrophages develop in response to Th1 cytokines or bacterial products such as LPS, and M2 macrophages are induced by exposure to Th2 cytokines, including IL-4 and IL-13. These results indicate that type 2 inflammation plays important roles in excessive fibrin deposition by the upregulation of FXIII-A produced by increased M2 macrophages in NP [46].

2.5 Anticoagulant Treatment of Rhinosinusitis

An activated coagulation system with increased thrombin generation and decreased fibrinolytic activity with excessive fibrin deposition contributes to the pathophysiology of allergic inflammation and tissue remodeling in upper airway inflammation such as AR and CRS [49]. Several studies have shown that thrombin receptor antagonists, TFPI, PAI-1 inhibitors, u-PA, or t-PA attenuate eosinophil recruitment, tissue remodeling, and airway hyperreactivity in the lungs of asthmatic mice by inhibiting the coagulation system or by activating fibrinolysis [21, 50–53]. These results suggest that the coagulation system and fibrinolysis may be potential therapeutic targets in patients with intractable rhinosinusitis. Heparin and APC have been used as anticoagulant drugs in clinical practice, and they also possess anti-inflammatory activities. Recently, potential anti-inflammatory effects of heparin or APC were supported by animal studies and several clinical trials involving patients with inflammatory diseases, including bronchial asthma, lung fibrosis, sepsis, arthritis, burns, ischemia-reperfusion injury, and inflammatory bowel diseases [13, 25, 54–57].

2.5.1 Heparin

Heparin is a glycosaminoglycan and remains one of the most important anticoagulant drugs in clinical practice. Heparin binds to antithrombin and accelerates the inhibitory activity to thrombin, FXa, and other coagulation factors. It is well known that heparin possesses a wide variety of anti-inflammatory activities, including inhibition of inflammatory mediators, suppression of lymphocyte, neutrophil, and eosinophil functions, and inhibition of mast cell degranulation [58–60]. Heparin plays regulatory roles in airway inflammation and subsequent tissue remodeling by binding nonspecifically to many proteins involved in the inflammation process, including cytokines, growth factors, adhesion molecules, tissue destructive enzymes, complement factors, and extracellular matrix proteins [55]. Inhaled heparin attenuates antigen-induced airway hyperreactivity in asthmatic rats [61] and sheep [62], and it inhibits smoke-induced lung injury in sheep [63]. Heparin also attenuates nasal airway pressure and cellular infiltration in nasal lavage fluids in a guinea pig model of AR [64]. Several clinical studies have shown the anti-inflammatory effects of inhaled heparin or nebulized heparin in patients with bronchial asthma and AR, and no adverse effects such as bleeding or thrombocytopenia have been reported [59, 65–69].

The anti-inflammatory effects of heparin on rat nasal epithelium are examined using LPS- or antigen-induced inflammation. Intranasal instillation of heparin attenuates LPS-induced neutrophil infiltration, goblet cell metaplasia, and mucus production in rat nasal epithelium, and the low-molecular weight heparin (LMWH), enoxaparin, shows similar inhibitory effects on LPS-induced changes as those of heparin [70]. LMWH is a fragment of heparin produced by controlled chemical and enzymatic degradation, and its binding to antithrombin accelerates its anticoagulant activity. LMWH has a longer half-life and better bioavailability caused by the reduced binding to plasma proteins and endothelium. LMWH also attenuates antigen-induced neutrophil/eosinophil infiltration, goblet cell metaplasia, and mucus production in rat nasal epithelium [71], and it suppresses TNF-α-induced secretion of IL-8 and MUC5AC mucin from cultured airway epithelial cells [70]. These results indicate that the *in vivo* effects of heparin are caused by a wide range of anti-inflammatory activities including the direct inhibition of the secretion of IL-8 and MUC5AC mucin from airway epithelial cells, together with the inhibition of thrombin generation and fibrin deposition. Topical application of LMWH may provide a new therapeutic strategy for the treatment of intractable rhinosinusitis.

2.5.2 Activated Protein C (APC)

The anticoagulant protein C pathway is activated when the thrombin-thrombomodulin complex converts protein C to APC. APC inhibits coagulation by inactivating FVa and FVIIIa, important coagulation factors in thrombin generation, and it also promotes fibrinolysis by inactivating PAI-1. APC is clinically used for the treatment of patients with congenital or acquired protein C deficiency. APC also has cytoprotective and anti-inflammatory activities through the endothelial protein C receptor (EPCR) and PAR-1 expressed on endothelial cells [56]. Systemic administration of APC improves the clinical outcome of patients with sepsis, and APC inhibits leukocyte adhesion and infiltration into tissues and the secretion of TNF-α and IL-1β from inflammatory cells [72–74].

APC generation is decreased in the lungs of patients with airway inflammation, such as bronchial asthma and lung fibrosis, and reduced APC function is associated with enhanced collagen formation in the lungs of patients with lung fibrosis [75–77]. Intratracheal administration of APC inhibits bleomycin-induced lung fibrosis and production of PDGF, TNF-α, and IL-6 in the mouse lung, and anti-EPCR antibody suppresses these inhibitory activities of APC [57]. Inhaled APC attenuates antigen-induced eosinophil infiltration, Th2 cytokine production, and airway hyperreactivity in mouse lung, and anti-EPCR antibody suppresses these inhibitory activities [14].

Intranasal instillation of APC inhibits thrombin-induced goblet cell metaplasia and mucus production in rat nasal epithelium, and APC inhibits TNF-α or epidermal growth factor (EGF)-induced secretion of MUC5AC mucin from cultured airway epithelial cells [1]. APC inhibits thrombin-induced production of PDGF from cultured airway epithelial cells and from macrophages. EPCR is also expressed in airway epithelial cells and macrophages, and the inhibitory effects of APC on PDGF production are suppressed by anti-EPCR antibody [57]. APC also inhibits the chemotactic activity of eosinophils, neutrophils, and lymphocytes through the EPCR [78–80]. These results show that reduced activities of the protein C system contribute to the pathophysiology of airway inflammation, and APC has a variety of anti-inflammatory activities through its own receptor EPCR on airway epithelial cells, macrophages, and leukocytes. Topical administration of APC may have therapeutic potential for the treatment of intractable rhinosinusitis.

2.6 Conclusions

Coagulation is regulated by multi-step processes, including platelet aggregation, the extrinsic and intrinsic coagulation cascade, fibrinolysis, and regulators such as TFPI, antithrombin, PAI-1, α2-antiplasmin, and the anticoagulant protein C pathway. Thrombin plays central roles in the coagulation cascade by converting fibrinogen to fibrin with positive and negative feedback systems. Airway inflammation activates the coagulation system by facilitating the leakage of plasma coagulation factors into tissue and by enhancing TF activity of epithelial cells and infiltrating eosinophils. Increased coagulation and decreased fibrinolysis result in excessive fibrin deposition leading to tissue remodeling. The coagulation system modulates the inflammation by platelet activation and by coagulation factor-PAR-mediated responses, and the coagulation system contributes to the pathophysiology of rhinosinusitis. Further understanding of the interactions between coagulation and inflammation will provide possible therapeutic strategies, and anticoagulant drugs with anti-inflammatory functions, such as heparin and APC, may have therapeutic potential for the treatment of intractable rhinosinusitis.

References

1. Shimizu S, Shimizu T, Morser J, et al. Role of the coagulation system in allergic inflammation in the upper airways. Clin Immunol. 2008;129:365–71.
2. Shimizu S, Gabazza EC, Ogawa T, et al. Role of thrombin in chronic rhinosinusitis-associated tissue remodeling. Am J Rhinol Allergy. 2011;25:7–11.
3. Gabazza E, Taguchi O, Tamaki S, et al. Thrombin in the airways of asthmatic patients. Lung. 1999;177:253–62.
4. Hernández-Rodríguez N, Cambrey A, Harrison N, et al. Role of thrombin in pulmonary fibrosis. Lancet. 1995;346:1071–3.
5. Shimizu S, Ogawa T, Takezawa K, et al. Tissue factor and tissue factor pathway inhibitor in nasal mucosa and nasal secretions of chronic rhinosinusitis with nasal polyp. Am J Rhinol Allergy. 2015;29:235–42.
6. Takabayashi T, Kato A, Peters AT, et al. Excessive fibrin deposition in nasal polyps caused by fibrinolytic impairment through reduction of tissue plasminogen activator expression. Am J Respir Crit Care Med. 2013;187:49–57.
7. Gabazza E, Taguchi O, Kamada H, et al. Progress in the understanding of protease-activated receptors. Int J Hematol. 2004;79:117–22.
8. Shimizu S, Gabazza E, Hayashi T, et al. Thrombin stimulates the expression of PDGF in lung epithelial cells. Am J Physiol Lung Cell Mol Physiol. 2000;279:L503–10.
9. Shimizu S, Tojima I, Takezawa K, et al. Thrombin and activated coagulation factor X stimulate the release of cytokines and fibronectin from nasal polyp fibroblasts via protease-activated receptors. Am J Rhinol Allergy. 2017;31:13–8.
10. Shinagawa K, Martin J, Ploplis V, et al. Coagulation factor Xa modulates airway remodeling in a murine model of asthma. Am J Respir Crit Care Med. 2007;175:136–43.
11. Howell D, Laurent G, Chambers R. Role of thrombin and its major cellular receptor, protease-activated receptor-1, in pulmonary fibrosis. Biochem Soc Trans. 2002;30:211–6.
12. Asokananthan N, Graham P, Fink J, et al. Activation of protease-activated receptor (PAR)-1, PAR-2, and PAR-4 stimulates IL-6, IL-8, and prostaglandin E2 release from human respiratory epithelial cells. J Immunol. 2002;168:3577–85.

13. Hataji O, Taguchi O, Gabazza E, et al. Activation of protein C pathway in the airways. Lung. 2002;180:47–59.
14. Yuda H, Adachi Y, Taguchi O, et al. Activated protein C inhibits bronchial hyperresponsiveness and Th2 cytokine expression in mice. Blood. 2004;103:2196–204.
15. Cho S, Kang J, Lyttle C, et al. Association of elevated plasminogen activator inhibitor 1 levels with diminished lung function in patients with asthma. Ann Allergy Asthma Immunol. 2011;106:371–7.
16. Oh C, Ariue B, Alban R, et al. PAI-1 promotes extracellular matrix deposition in the airways of a murine asthma model. Biochem Biophys Res Commun. 2002;294:1155–60.
17. Pitchford SC, Momi S, Giannini S, et al. Platelet P-selectin is required for pulmonary eosinophil and lymphocyte recruitment in a murine model of allergic inflammation. Blood. 2005;105:2074–81.
18. Ulfman LH, Joosten DP, van Aalst CW, et al. Platelets promote eosinophil adhesion of patients with asthma to endothelium under flow conditions. Am J Respir Cell Mol Biol. 2003;28:512–9.
19. Long AT, Kenne E, Jung R, et al. Contact system revisited: an interface between inflammation, coagulation, and innate immunity. J Thromb Haemost. 2016;14:427–37.
20. van der Poll T. Tissue factor as an initiator of coagulation and inflammation in the lung. Crit Care. 2008;12(Suppl 6):S3.
21. Adams M. Tissue factor pathway inhibitor: new insights into an old inhibitor. Semin Thromb Hemost. 2012;38:129–34.
22. Suzuki K, Gabazza E, Hayashi T, et al. Protective role of activated protein C in lung and airway remodeling. Crit Care Med. 2004;32:S262–5.
23. Esmon CT. Inflammation and the activated protein C anticoagulant pathway. Semin Thromb Hemost. 2006;32(Suppl 1):49–60.
24. Kim DY, Cho SH, Takabayashi T, et al. Chronic Rhinosinusitis and the coagulation system. Allergy Asthma Immunol Res. 2015;7:421–30.
25. de Boer JD, Majoor CJ, van 't Veer C, et al. Asthma and coagulation. Blood. 2012;119:3236–44.
26. Hajj R, Baranek T, Le Naour R, et al. Basal cells of the human adult airway surface epithelium retain transit-amplifying cell properties. Stem Cells. 2007;25:139–48.
27. Ahmad S, Ahmad A, Rancourt RC, et al. Tissue factor signals airway epithelial basal cell survival via coagulation and protease-activated receptor isoforms 1 and 2. Am J Respir Cell Mol Biol. 2013;48:94–104.
28. Imoto Y, Kato A, Takabayashi T, et al. Increased thrombin-activatable fibrinolysis inhibitor levels in patients with chronic rhinosinusitis with nasal polyps. J Allergy Clin Immunol. 2019;144:1566–1574.e1566.
29. Heuberger DM, Schuepbach RA. Protease-activated receptors (PARs): mechanisms of action and potential therapeutic modulators in PAR-driven inflammatory diseases. Thromb J. 2019;17:4.
30. Mercer P, Johns R, Scotton C, et al. Pulmonary epithelium is a prominent source of proteinase-activated receptor-1-inducible CCL2 in pulmonary fibrosis. Am J Respir Crit Care Med. 2009;179:414–25.
31. Shimizu S, Takezawa-Yasuoka K, Ogawa T, et al. The epidermal growth factor receptor inhibitor AG1478 inhibits eosinophilic inflammation in upper airways. Clin Immunol. 2018;188:1–6.
32. Kouzaki H, Seno S, Fukui J, et al. Role of platelet-derived growth factor in airway remodeling in rhinosinusitis. Am J Rhinol Allergy. 2009;23:273–80.
33. Matsune S, Ohori J, Sun D, et al. Vascular endothelial growth factor produced in nasal glands of perennial allergic rhinitis. Am J Rhinol. 2008;22:365–70.
34. Wittekindt C, Hess A, Bloch W, et al. Immunohistochemical expression of VEGF and VEGF receptors in nasal polyps as compared to normal turbinate mucosa. Eur Arch Otorhinolaryngol. 2002;259:294–8.
35. Zaravinos A, Soufla G, Bizakis J, et al. Expression analysis of VEGFA, FGF2, TGFbeta1, EGF and IGF1 in human nasal polyposis. Oncol Rep. 2008;19:385–91.
36. Rincon M, Irvin CG. Role of IL-6 in asthma and other inflammatory pulmonary diseases. Int J Biol Sci. 2012;8:1281–90.
37. Richardson VR, Cordell P, Standeven KF, et al. Substrates of factor XIII-A: roles in thrombosis and wound healing. Clin Sci (Lond). 2013;124:123–37.
38. Meng J, Zhou P, Liu Y, et al. The development of nasal polyp disease involves early nasal mucosal inflammation and remodelling. PLoS One. 2013;8:e82373.
39. Nakagawa T, Yamane H, Shigeta T, et al. Interaction between fibronectin and eosinophils in the growth of nasal polyps. Laryngoscope. 1999;109:557–61.
40. Hong SN, Zhang YL, Rhee CS, et al. Probable roles of coagulation cascade and fibrinolysis system in the development of allergic rhinitis. Am J Rhinol Allergy. 2019;33:137–44.
41. Kouzaki H, O'Grady S, Lawrence C, et al. Proteases induce production of thymic stromal lymphopoietin by airway epithelial cells through protease-activated receptor-2. J Immunol. 2009;183:1427–34.
42. Kouzaki H, Kikuoka H, Matsumoto K, et al. A mechanism of interleukin-25 production from airway epithelial cells induced by Japanese cedar pollen. Clin Immunol. 2018;193:46–51.
43. Kouzaki H, Tojima I, Kita H, et al. Transcription of interleukin-25 and extracellular release of the protein is regulated by allergen proteases in airway epithelial cells. Am J Respir Cell Mol Biol. 2013;49:741–50.
44. Peters T, Henry PJ. Protease-activated receptors and prostaglandins in inflammatory lung disease. Br J Pharmacol. 2009;158:1017–33.
45. Duval C, Ariëns RAS. Fibrinogen splice variation and cross-linking: effects on fibrin structure/function and role of fibrinogen γ' as thrombomobulin II. Matrix Biol. 2017;60-61:8–15.
46. Takabayashi T, Kato A, Peters AT, et al. Increased expression of factor XIII-A in patients with chronic

rhinosinusitis with nasal polyps. J Allergy Clin Immunol. 2013;132:584–592.e4.
47. Krysko O, Holtappels G, Zhang N, et al. Alternatively activated macrophages and impaired phagocytosis of S. aureus in chronic rhinosinusitis. Allergy. 2011;66:396–403.
48. Peterson S, Poposki JA, Nagarkar DR, et al. Increased expression of CC chemokine ligand 18 in patients with chronic rhinosinusitis with nasal polyps. J Allergy Clin Immunol. 2012;129:119–127.e1-9.
49. Mueller SK, Nocera AL, Dillon ST, et al. Highly multiplexed proteomic analysis reveals significant tissue and exosomal coagulation pathway derangement in chronic rhinosinusitis with nasal polyps. Int Forum Allergy Rhinol. 2018;8:1438–44.
50. Lee SH, Eren M, Vaughan DE, et al. A plasminogen activator inhibitor-1 inhibitor reduces airway remodeling in a murine model of chronic asthma. Am J Respir Cell Mol Biol. 2012;46:842–6.
51. Kuramoto E, Nishiuma T, Kobayashi K, et al. Inhalation of urokinase-type plasminogen activator reduces airway remodeling in a murine asthma model. Am J Physiol Lung Cell Mol Physiol. 2009;296:L337–46.
52. Wagers SS, Norton RJ, Rinaldi LM, et al. Extravascular fibrin, plasminogen activator, plasminogen activator inhibitors, and airway hyperresponsiveness. J Clin Invest. 2004;114:104–11.
53. Howell D, Goldsack N, Marshall R, et al. Direct thrombin inhibition reduces lung collagen, accumulation, and connective tissue growth factor mRNA levels in bleomycin-induced pulmonary fibrosis. Am J Pathol. 2001;159:1383–95.
54. Young E. The anti-inflammatory effects of heparin and related compounds. Thromb Res. 2008;122:743–52.
55. Lever R, Page CP. Novel drug development opportunities for heparin. Nat Rev Drug Discov. 2002;1:140–8.
56. Danese S, Vetrano S, Zhang L, et al. The protein C pathway in tissue inflammation and injury: pathogenic role and therapeutic implications. Blood. 2010;115:1121–30.
57. Shimizu S, Gabazza E, Taguchi O, et al. Activated protein C inhibits the expression of platelet-derived growth factor in the lung. Am J Respir Crit Care Med. 2003;167:1416–26.
58. Niven A, Argyros G. Alternate treatments in asthma. Chest. 2003;123:1254–65.
59. Ahmed T, Garrigo J, Danta I. Preventing bronchoconstriction in exercise-induced asthma with inhaled heparin. N Engl J Med. 1993;329:90–5.
60. Peter K, Schwarz M, Conradt C, et al. Heparin inhibits ligand binding to the leukocyte integrin Mac-1 (CD11b/CD18). Circulation. 1999;100:1533–9.
61. Suchankova J, Mata M, Cortijo J, et al. Effects of bemiparin on airway responses to antigen in sensitized Brown-Norway rats. Eur J Pharmacol. 2005;507:261–71.
62. Ahmed T, Ungo J, Zhou M, et al. Inhibition of allergic late airway responses by inhaled heparin-derived oligosaccharides. J Appl Physiol. 1985;2000(88):1721–9.
63. Murakami K, McGuire R, Cox RA, et al. Heparin nebulization attenuates acute lung injury in sepsis following smoke inhalation in sheep. Shock. 2002;18:236–41.
64. Al Suleimani YM, Dong Y, Walker MJ. Differential responses to various classes of drugs in a model of allergic rhinitis in guinea pigs. Pulm Pharmacol Ther. 2008;21:340–8.
65. Ahmed T, Gonzalez BJ, Danta I. Prevention of exercise-induced bronchoconstriction by inhaled low-molecular-weight heparin. Am J Respir Crit Care Med. 1999;160:576–81.
66. Diamant Z, Timmers MC, van der Veen H, et al. Effect of inhaled heparin on allergen-induced early and late asthmatic responses in patients with atopic asthma. Am J Respir Crit Care Med. 1996;153:1790–5.
67. Bendstrup KE, Jensen JI. Inhaled heparin is effective in exacerbations of asthma. Respir Med. 2000;94:174–5.
68. Bendstrup KE, Gram J, Jensen JI. Effect of inhaled heparin on lung function and coagulation in healthy volunteers. Eur Respir J. 2002;19:606–10.
69. Vancheri C, Mastruzzo C, Armato F, et al. Intranasal heparin reduces eosinophil recruitment after nasal allergen challenge in patients with allergic rhinitis. J Allergy Clin Immunol. 2001;108:703–8.
70. Ogawa T, Shimizu S, Tojima I, et al. Heparin inhibits mucus hypersecretion in airway epithelial cells. Am J Rhinol Allergy. 2011;25:69–74.
71. Ogawa T, Shimizu S, Shimizu T. The effect of heparin on antigen-induced mucus hypersecretion in the nasal epithelium of sensitized rats. Allergol Int. 2013;62:77–83.
72. Esmon CT, Fukudome K, Mather T, et al. Inflammation, sepsis, and coagulation. Haematologica. 1999;84:254–9.
73. Bernard GR, Vincent JL, Laterre PF, et al. Efficacy and safety of recombinant human activated protein C for severe sepsis. N Engl J Med. 2001;344:699–709.
74. Oka S, Gabazza EC, Taguchi Y, et al. Role of activated protein C in Helicobacter pylori-associated gastritis. Infect Immun. 2000;68:2863–9.
75. Schouten M, Van De Pol MA, Levi M, et al. Early activation of coagulation after allergen challenge in patients with allergic asthma. J Thromb Haemost. 2009;7:1592–4.
76. Kobayashi H, Gabazza E, Taguchi O, et al. Protein C anticoagulant system in patients with interstitial lung disease. Am J Respir Crit Care Med. 1998;157:1850–4.
77. Yasui H, Gabazza E, Taguchi O, et al. Decreased protein C activation is associated with abnormal collagen turnover in the intraalveolar space of patients with interstitial lung disease. Clin Appl Thromb Hemost. 2000;6:202–5.
78. Feistritzer C, Sturn DH, Kaneider NC, et al. Endothelial protein C receptor-dependent inhibition of human eosinophil chemotaxis by protein C. J Allergy Clin Immunol. 2003;112:375–81.

79. Feistritzer C, Mosheimer BA, Sturn DH, et al. Endothelial protein C receptor-dependent inhibition of migration of human lymphocytes by protein C involves epidermal growth factor receptor. J Immunol. 2006;176:1019–25.
80. Sturn DH, Kaneider NC, Feistritzer C, et al. Expression and function of the endothelial protein C receptor in human neutrophils. Blood. 2003;102:1499–505.

3 Cilia, Ciliary Movement, and Mucociliary Transport

Mark Jorissen and Martine Jaspers

Core Messages

- Cilia are extensions of the apical membranes and are characterized by a 9 + 2 axonemal structure.
- Ciliary beating depends on the ATPase activity in the dynein arms and is characterized by a specific beating pattern.
- Structural and functional ciliary abnormalities can be the results of external (secondary ciliary dyskinesia) or inherited factors (primary ciliary dyskinesia). An active, coordinated ciliary beating is essential for mucociliary transport. Mucociliary transport is the final result of the functional and ultrastructural organization of the cilia at different levels.

3.1 Cilia

3.1.1 General Description

Cilia are tiny hairlike cell organelles, found on the surface of most cell types in the vertebrate body [1]. There are two types of cilia: motile cilia and non-motile or primary cilia. Primary cilia or non-motile sensory cilia transmit signals to the interior of the cell. The second type of cilia, the motile cilia, is important for the movement of extracellular fluids. Motile cilia are found in the apical surface (ciliated epithelium) of the upper and lower respiratory tract, the oviducts of the female reproductive system, and ependymal cells lining the ventricles of the brain. These epithelial cells contain hundreds of motile cilia that beat together in a concerted manner to propel substances over the epithelial surface. Failure of these cilia to perform their normal function results in respiratory disease, sterility, or hydrocephalus.

Cilia, which line both the upper and lower airways, are covered by a thin layer of mucus and beat in a coordinated fashion propelling particles trapped in the mucus layer to the pharynx. Cilial defects may be either primary or secondary.

3.1.1.1 Ciliated Cells

The respiratory epithelium consists mainly of four cell types: ciliated columnar cells, non-ciliated columnar cells or brush cells with microvilli (m), goblet cells secreting mucin (s), and basal cells (b). The different cell types of the respiratory epithelium are illustrated in Fig. 3.1.

A ciliated cell has a diameter of 5 μm at its apex and carries 100–200 cilia at a density of 6–8/μm^2, interspersed with ±400 short microvilli [2]. The length of a cilium in the nose is 5 μm, in the larger airways 6–7 μm, and 5 μm in the smaller bronchiole [3]. The diameter of the

M. Jorissen (✉) · M. Jaspers
ENT Department, University Hospitals Leuven, Leuven, Belgium
e-mail: mark.jorissen@uzleuven.be; martine.jaspers@med.kuleuven.be

Ö. Ö. Celebi, T. M. Önerci (eds.), *Nasal Physiology and Pathophysiology of Nasal Disorders*, https://doi.org/10.1007/978-3-031-12386-3_3

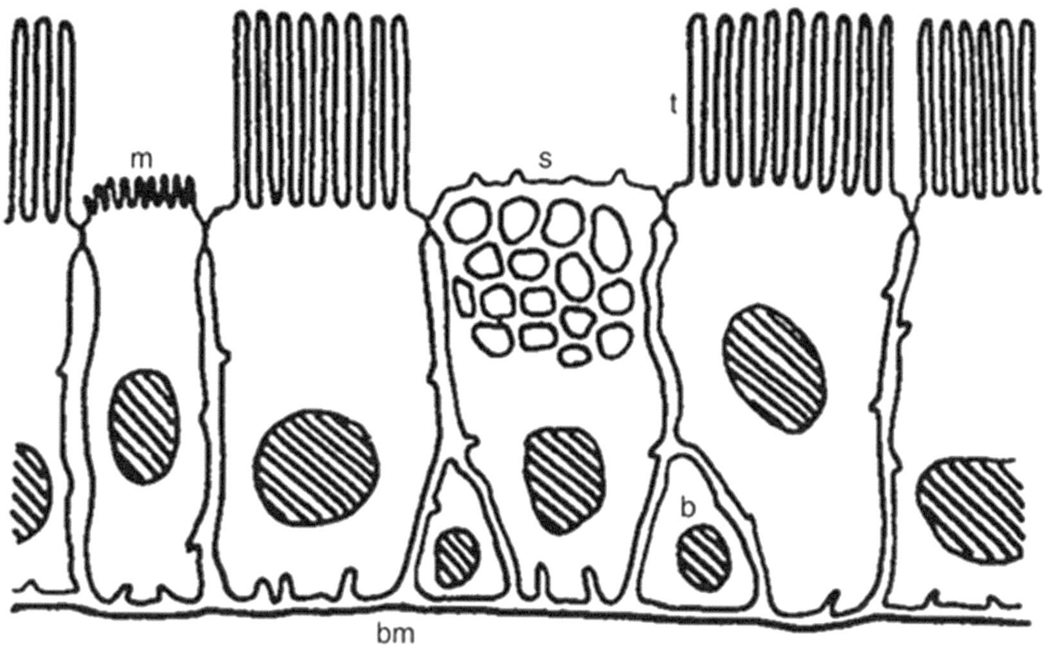

Fig. 3.1 Different cell types of the respiratory epithelium. *m* microvillar cell, *s* secretory cell, *t* ciliated cell, *b* basal cell, *bm* basal membrane

shaft or axoneme measures approximately 0.25 µm at the base and 0.13 µm at the distal segment.

3.1.1.2 Cilia

Cilia are motile hair-like extensions of the epithelial cells, surrounded by the ciliary membrane, a specialized extension of the cell membrane. The length of a cilium varies from 5 to 10 µm and the width between 0.1 and 0.3 µm. Each ciliated epithelial cell has 100–200 motile cilia.

3.1.2 Ciliary Structure

The basic structure of cilia is "9 + 2 microtubuli." The intermicrotubular connections determine the function.

The ultrastructure of a cilium consists of nine outer doublets of microtubules surrounding a central pair of microtubules. This characteristic 9 + 2 organization of microtubules is called an axoneme, as viewed in cross-section with the electron microscope (Fig. 3.2). The two central microtubules are surrounded by a central sheath with spokes directed toward the peripheral microtubular doublets. The outer doublets are connected by nexin links and with the central pair by radial spokes. Each outer doublet microtubule is composed of two subfibers, A and B, of which the A tubule is a complete microtubule with 13 protofilaments, while the B tubule is incomplete and contains only 10 protofilaments. Subfiber A bears inner and outer dynein arms with ATPase activity. The dynein arms are complex structures consisting of several heavy, intermediate, and light chains. The dynein heavy chains contain ATPase domains that act as molecular motors and slide the peripheral microtubular pairs relative to each other. Nexin links limit the relative motion of neighboring doublets and radial spokes control from the central pair. The basal body is a specialized centriole found at the base of the cilium that anchors the cilium in a specific orientation and is thought to be responsible for their formation. Ciliogenesis begins with the generation of basal

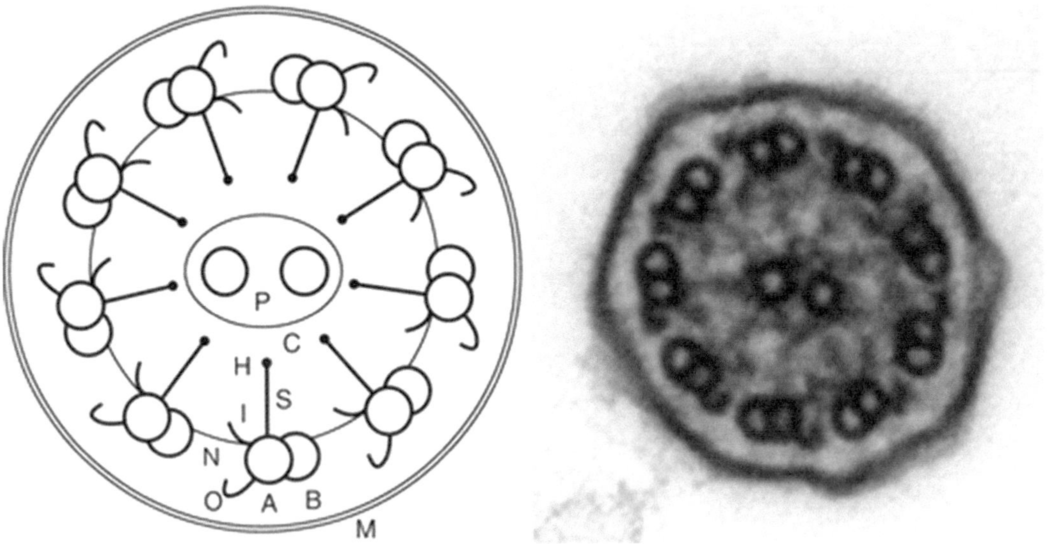

Fig. 3.2 Schematic drawing and photo of normal axonemal structure in transverse section. *A* microtubules A, *B* microtubules B, *C* central sheath, *H* spoke head, *I* inner dynein arm, *M* ciliary membrane, *N* nexin link, *O* outer dynein arm, *P* central pair of microtubules, *S* spoke

bodies in the cytoplasm that then traffic to the apical surface, dock with an anchor to the plasma membrane, and elongate a ciliary axoneme. Basal bodies consist of nine microtubular triplets but do not have central microtubules, as seen by cross-sectional views with electron microscopy. The basal body of a cilium is located just under the apical membrane and it is anchored in the cytoplasm by three types of accessory structures: (1) alar sheets, (2) a laterally directed basal foot, and (3) downward-directed ciliary rootlets at the base (see Fig. 3.3, left; [4]). The basal foot indicates the direction of the effective stroke and is the most reliable reference for measuring ciliary (dis)orientation [5]. Figure 3.3 shows a drawing of a longitudinal section of a ciliary axoneme and cross-sections such as those can be seen at the indicated levels of the cilia. The image of Fig. 3.2 is a cross-section of the main central part of a ciliary axonema.

Ciliary activity is generated by the sliding movements of the microtubules. During a beat, the dynein arms undergo an attachment, retraction, and release with the adjacent doublet, which results in a sliding motion of the microtubule relative to each other. The energy needed for this is delivered by ATP hydrolysis by the ATPase domains of the dynein arms. The basal body anchors the microtubules, and the nexin links, the radial spoke, and probably the cell membrane restricts the degree of sliding between microtubules, thereby converting this sliding into bending [1, 6].

3.1.3 Structural Components: Dynein

Most of our knowledge about outer and inner dynein arm composition originates from studies in Chlamydomonas. Chlamydomonas outer dynein arm is composed of three heavy chains (α, β, γ), two intermediate chains, nine light chains, three docking complex proteins, and at least two associated proteins. The heavy chains are the sites for ATP hydrolysis required for ciliary motility [7]. The dynein heavy chains are composed of a head domain that produces a sliding

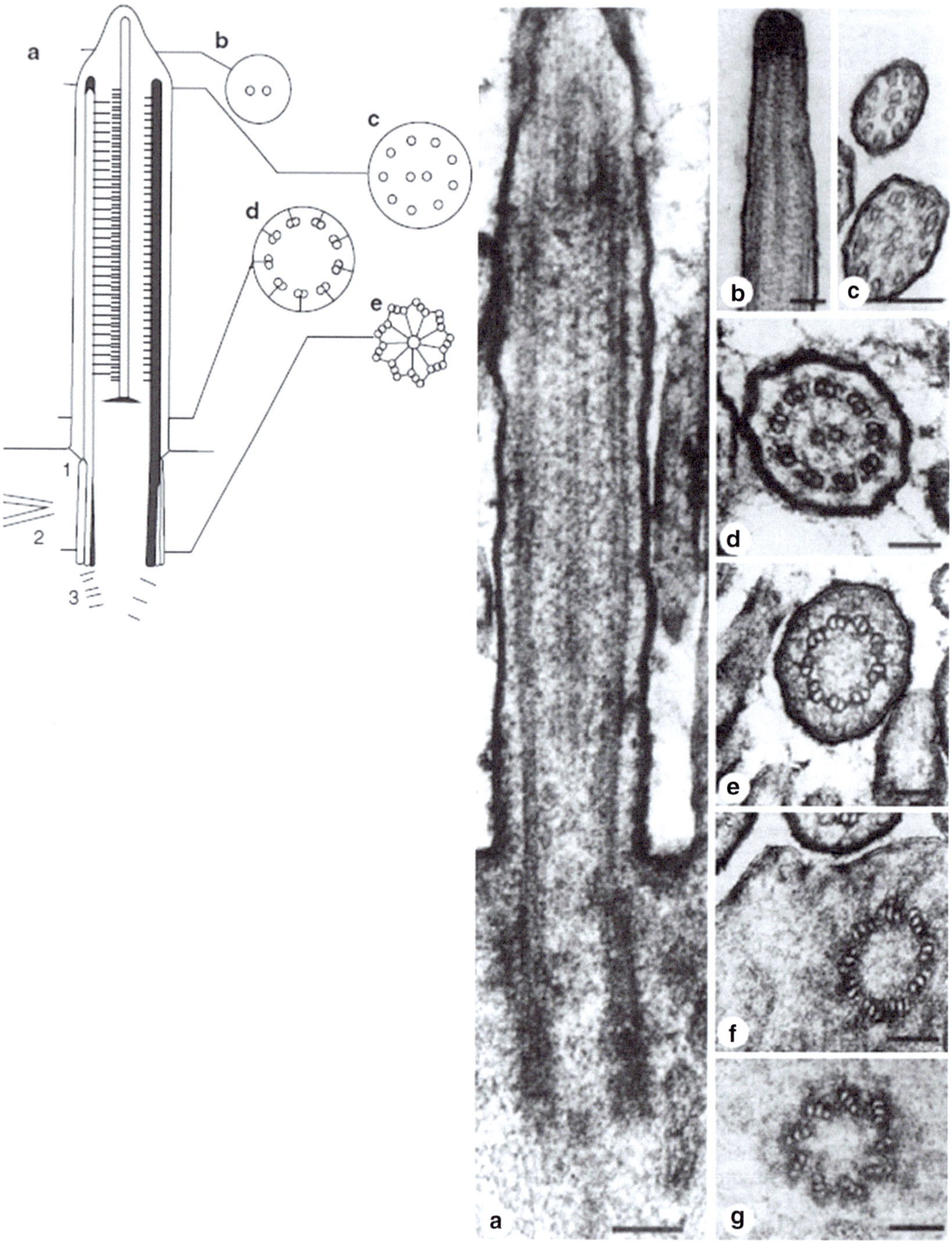

Fig. 3.3 (**a**) Drawing (*left*) and Transmission Electron Microscopy (TEM) photo (*right*) of a longitudinal section of a ciliary axonema; (**b**) longitudinal section of the tip and (**c**) until (**g**) are cross-sections such as those can be seen at the indicated levels namely (**c**) near the tip, (**d**) in the ciliary shaft, (**e**) just above the footplate and (**f**) in the upper part of the basal body; the image of Fig. 3.2 is a cross-section of the main central part of the axonema. **1** fibers, **2** basal foot, **3** rootlets

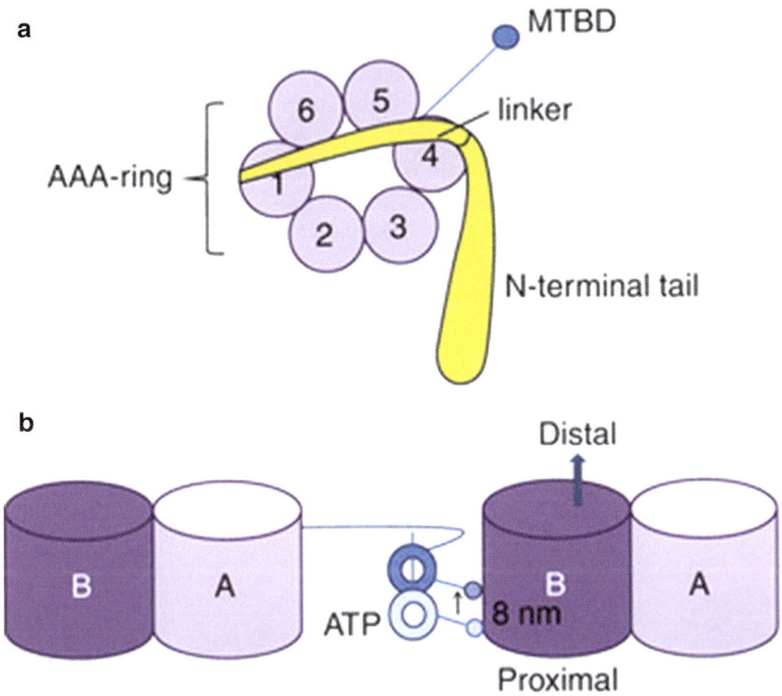

Fig. 3.4 The dynein motor unit or heavy chain consists of six tandemly linked AAA ATPase domains, which form a ring [8], with the linker emanating from AAA1 and the coiled-coil stalk with the microtubule (MT)-binding domain located between AAA4 and AAA5 (**a**). Following ATP hydrolysis, the AAA rings of the dynein motor units were observed to move 8 nm toward the distal end of the axoneme [9]. As the motor is connected temporarily to the adjacent B tubule via the MT-binding domain, located at the tip of the coiled-coil stalk, this would result in the B tubule being dragged distally (**b**)

force through an ATP-driven temporary interaction with an adjacent doublet B tubule and a tail domain that is stably fixed to the outer doublet A tubule (Fig. 3.4).

3.1.4 Structural Abnormalities

Up to 5% abnormalities is normal.

In healthy persons, more than 95% of the cilia are ultrastructurally completely normal. Only in a minority of transverse section of cilia abnormalities are found. The percentage of abnormalities may increase as a result of inflammation, infection, and exposure to toxic agents. This is called secondary ciliary dyskinesia (SCD) to distinguish from the inherited abnormalities: primary ciliary dyskinesia (PCD).

Dynein deficiency remains the most frequent ultrastructural abnormality in PCD. In up to 1/3, no ultrastructural abnormality is found.

Primary (genetic) defects in the structure and function of sensory and motile cilia result in multiple ciliopathies. The most prominent genetic abnormality involving motile cilia (and the respiratory tract) is primary ciliary dyskinesia (PCD). PCD reflects abnormalities in the structure and function of motile cilia. The most common ultrastructural defects related to PCD are the total or partial absence of dynein arms and absence or dislocation of central tubules. Besides, a significant number of PCD patients have cilia with normal ultrastructure but abnormal ciliary mobility (CBF and coordination). Based on the structural abnormalities found in PCD, patients can be classified into different subgroups (see also Figs. 3.5 and 3.6):

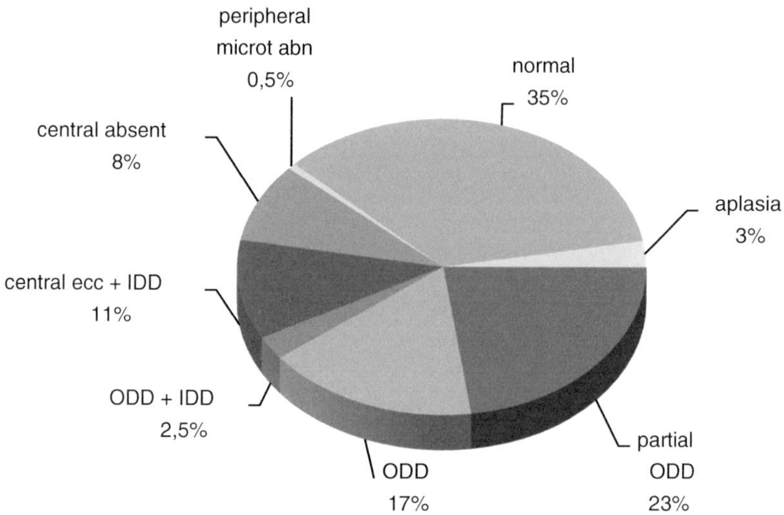

Fig. 3.5 Distribution of different ultrastructural subgroups in PCD based on a series of 312 patients in the UZ Leuven database

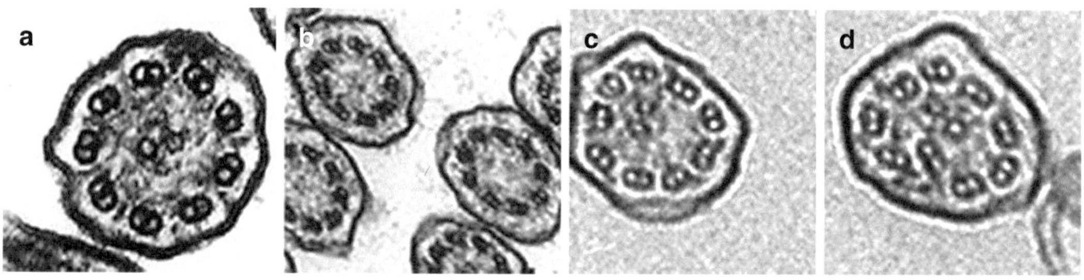

Fig. 3.6 Ultrastructural abnormalities in PCD: (**a**) dynein deficiency, (**b**) absent central pair, (**c**) eccentric central pair, and (**d**) eccentric central pair + transposition

- Outer dynein arms deficiency (ODD).
- Partial outer dynein arms deficiency (part ODD).
- Outer + inner dynein arms deficiency (ODD + IDD).
- Eccentric central pair + inner dynein deficiency.
- Central pair of microtubules absent.
- PCD with normal ultrastructure.
- Ciliary aplasia (no cilia and no basal bodies).

In the majority of patients, these abnormalities can be differentiated from the acquired abnormalities: secondary ciliary dyskinesia (SCD). However, there may be considerable overlap, and in PCD patients, frequently SCD abnormalities are found, because of inflammation and infections. The most frequent ultrastructural abnormalities in SCD are the compound cilia, peripheral microtubular abnormalities, blebs of the axonemal membrane, excess cytoplasm, absence of the axonemal membrane, and ciliary disorientation of the central pair microtubules (see Fig. 3.7).

3.1.5 Genetic Heterogeneity of PCD

The genetic heterogeneity of PCD is predicted by the complexity of the ciliary structure and the different structural component affected. Cilia consist of more than 250 proteins and thus many genes are involved in ciliary structure and function. Currently, mutations in more than 40 different genes coding for axonemal proteins have been described, which explain about 70% of PCD [10, 11]. Mutations in one of these genes known to be associated with PCD (ARMC4,

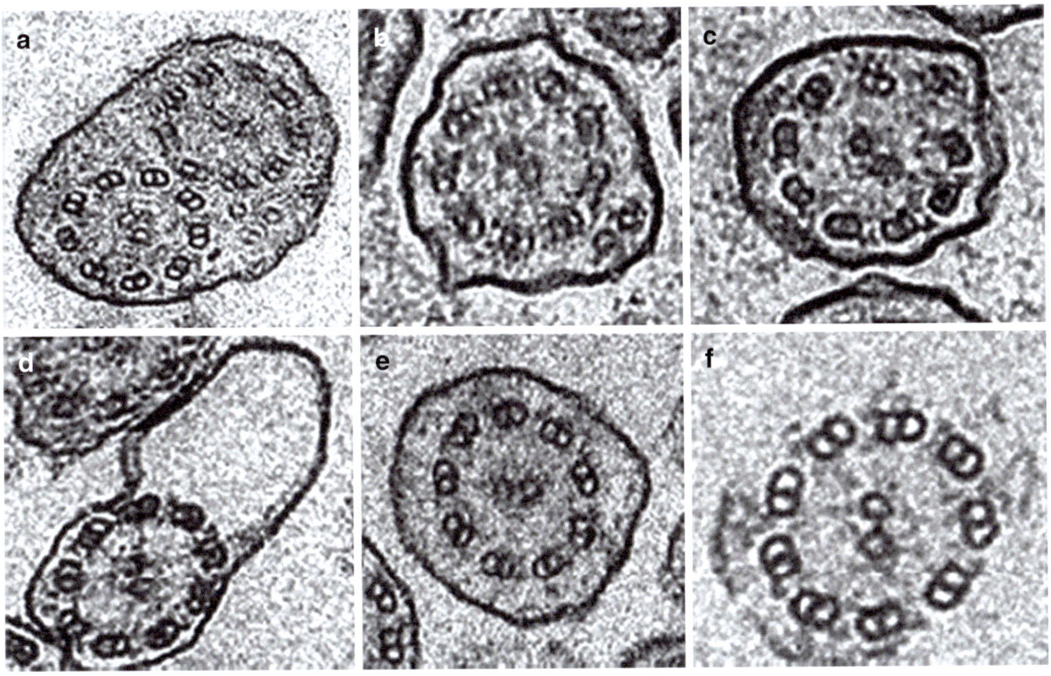

Fig. 3.7 Ultrastructural abnormalities in SCD: (**a**) compound cilia, (**b**) and (**c**) peripheral microtubular abnormalities, (**d**) blebs of the axonemal membrane, (**e**) excess cytoplasm, and (**f**) absence of the axonemal membrane: naked cilium

CCDC103, CCDC114, CCDC151, CCDC39, CCDC40, CCDC65, CCNO, CFAP298, CFAP300, DNAAF1, DNAAF2, DNAAF3, DNAAF4, DNAAF5, DNAH11, DNAH5, DNAH8, DNAH9, DNAI1, DNAI2, DNAJB13, DNAL1, DRC1, GAS8, HYDIN, LRRC56, LRRC6, MCIDAS, NME8, OFD1, PIH1D3, RPGR, RSPH1, RSPH3, RSPH4A, RSPH9, SPAG1, SPEF2, STK36, TEKT1, TTC25, ZMYND10) underlie specific ciliary ultrastructural defects identified by transmission electron microscopy. For instance, DNAH5, DNAI1, and DNAI2 cause outer dynein arm (ODA) defects [12–14], while mutations in radial head spoke proteins (RSPH9, RSPH4A) and the coiled-coil domain-containing proteins (CCDC39, CCDC40) are linked to central pair defects [15, 16]. However, mutations in some genes such as DNAH11 do not result in a detectable ultrastructural defect on transmission electron microscopy [17].

Half of PCD families with ODA defects harbored DNAH5 mutations. DNAH5 encodes a heavy chain of the ODA. The prevalence of DNAI1 mutations is 10–13% in PCD patients with defined ODA defects, but only 2–4% in the whole cohort of PCD patients. All these genes combined explain approximately 70% of PCD cases; therefore, more genes need to be identified [10, 11, 18].

3.2 Ciliary Movement

3.2.1 Ciliary Beat Cycle

The ciliary beat cycle consists of an effective and a recovery stroke.

Respiratory cilia have a rhythmical beating pattern and beat in a synchronous waveform. Every beat cycle consists of two active components: an effective stroke, during which the fully extended cilium moves in a plane perpendicular to the cell surface, and a recovery stroke, during which the bended cilium moves more parallel to the cell surface sideward and backward to its starting position; see Fig. 3.8. The duration of a recovery stroke is two to three times that of an

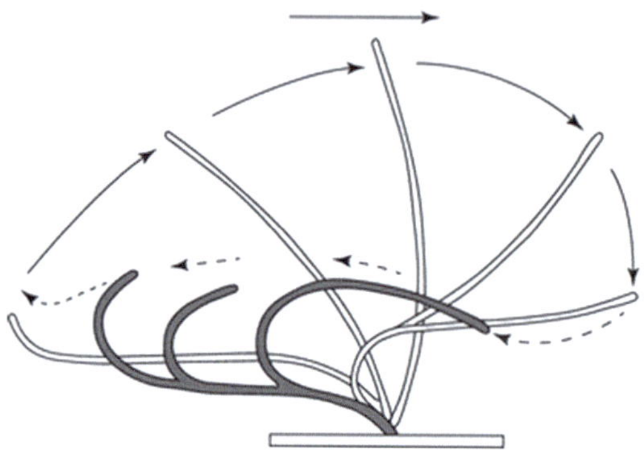

Fig. 3.8 Ciliary beat. The cilium performs an effective stroke (white cilium, *thin arrow*) and stays thereafter for some time in a resting position. The recovery stroke (black cilium, *dotted arrow*) is the start of a new cycle and takes place in a third dimension (courtesy of TESAV)

effective stroke. After the effective stroke, there is a short resting phase before the cilium starts its recovery stroke. The direction of the stroke depends on the orientation of the central microtubules [6].

As mentioned above, the bending of the cilia is produced by sliding the outer microtubule doublets against one another comparable to the actin-myosin system in muscles. The energy needed for the ciliary beat is produced by hydrolysis of ATP by the ATPase domains of the dynein arms. The velocity of the sliding movements and the frequency of the ciliary beat are correlated with the number of dynein arms and the concentration of ATP [19, 20].

A ciliated cell has approximately 200 cilia that beat in a coordinated way, and cilia on adjoining ciliated cells (unit of ciliated cells) are beating simultaneously. Ciliary beating is coordinated by calcium signaling between epithelial cells through gap junctions. Besides the regulatory effect of calcium on the ciliary beat, calcium is also involved in synchronizing the beat among cilia of one single cell as well as between cilia in different cells [21, 22]. Ciliary beat frequency increases from the more peripheral parts of the respiratory tract to the more central parts. In larger airways like the nose, trachea, and main bronchi, the frequency is 13–27 Hz; in smaller airways like the middle ear, small bronchi, and bronchiole, the frequency is 7–12 Hz. The beat frequency increases with temperature and decreases with a reduction in relative humidity of the air.

Regulation of cilia that play a role in mucociliary clearance is complex, and any disturbance can lead to disease.

3.2.2 Factors Influencing Ciliary Activity

Several factors influencing the ciliary beat frequency have been described, including temperature, pH, and osmolarity [23]. A constant medium temperature is essential for the accurate measurements, since CBF is temperature-dependent. Ingels et al. [23] demonstrated a linear relationship between CBF and temperature in the range from 22.5 to 40 °C. Changes in pH and osmolarity do not influence CBF when kept within a certain range. A pH change from 7.5 to 6.5 did not affect CBF, below a pH of 6.5 CBF decreases. Concerning osmolarity, a gradient of 150–225 mM (0.9–1.35%) NaCl did not affect CBF substantially. In hypotonic (0.45%) and hypertonic (1.5%) saline solutions, CBF decreases by 50% compared to the initial frequency [24], while at 3% saline cilia are complete immotile.

Beta-adrenergic influences on the respiratory mucosa are well known to cause enhancement of ciliary activity and mucociliary clearance [25, 26]. However, considering that isoproterenol stimulates secretory function in airways [27], the question remained whether this increase in ciliary activity was due to a direct and specific action on the ciliated cells. Verdugo et al. [28] demonstrated that isoproterenol directly stimulates the activity of ciliated cells of the respiratory epithelium and that this effect was β-adrenergic specific since the observed stimulation could be blocked by propranolol. Toxins derived from bacterial infections reduce ciliary activity of human nasal epithelial cells [29, 30], and reduced ciliary activity can aggravate inflammation. Mallants et al. [31] found that after a toxin-induced decrease, both bacitracin and clindamycin resulted in a complete recovery of CBF, suggesting that topical antibiotic treatment of nasal infections could result in a dual positive effect, namely, treatment of bacterial infection and recovery of the ciliary activity.

3.2.3 Abnormal Beating Patterns in the Context of PCD

Recent studies have confirmed that the ciliary beat pattern is associated with specific ultrastructural defects in PCD [32]. New high-resolution digital high-speed video (DHSV) imaging has allowed the precise beat pattern of cilia to be viewed in three different planes in slow motion or frame by frame. Using this technique, three patterns were identified and correlated with ultrastructural defects.

In the first pattern, the cilia are virtually immotile in large areas. Ciliary movement, when present, is restricted to slow, low-amplitude, stiff flickering motion. This is associated with either an isolated outer dynein arm defect or a combined inner and outer dynein arm defect.

In the second pattern, the cilia have a very abnormal stiff forward power stroke with a markedly reduced amplitude. This pattern is associated with an inner dynein arm defect or a radial spoke defect.

In the third pattern, the cilia beat in a large circular gyrating motion about the base of the cilium. This pattern is associated with transposition defects.

3.3 Mucociliary Transport

3.3.1 Structural and Functional Organization

Ciliary organization comprises four levels.

Mucociliary transport is the final result of the functional and ultrastructural organization of the cilia at different levels: single cilium, interciliary coordination, metachronal waveform, and mucociliary pathways [33].

- First level: Single cilium.

  A single cilium has a specific and well-characterized ultrastructure: a 9 + 2 microtubular organization or axoneme. Morphological investigation at this level is mostly done with transmission electron microscopy (TEM). Ciliary beat frequency (CBF) is the most frequently used parameter of a single ciliary function. Other parameters are the beating pattern, the amplitude, and the beat-to-beat variation (signal consistency [23]; intracellular variability [34]).
- Second level: Ciliary orientation and coordination.

  Cilia have to beat in a coordinated way to produce mucociliary transport, within one ciliated cell and between different cells. Ciliary activity is coordinated when all cilia beat in phase and in the same direction. This intra- and intercellular ultrastructural coordination can be studied using scanning electron microscopy (SEM) and transmission electron microscopy (TEM): ciliary (dis)orientation. Ciliary orientation can be measured in TEM images of transverse sections in which the two central microtubules could be seen. A line is drawn through the central microtubular pair of each transversely sectioned cilium. The angle between this line and a reference line is measured for all cilia seen in one photograph; see Fig. 3.9. The standard deviation of all these measured angles is the ciliary disorientation. A normal value is 15°; >20° may be considered disorientation; and >35° corresponds to random orientation [35–38].
- Third and fourth levels: Metachronal wave form and mucociliary transport pathways.

  Finally, the coordination results in the metachronal wave form which can easily be seen in scanning electron microscopy and which is linked not only to the small phase difference between neighboring cilia within one cell but also intercellularly to a whole surface area. The metachronal waveform and the CBF are regulated by different intraciliary, intracellular, and intercellular mechanisms [39].

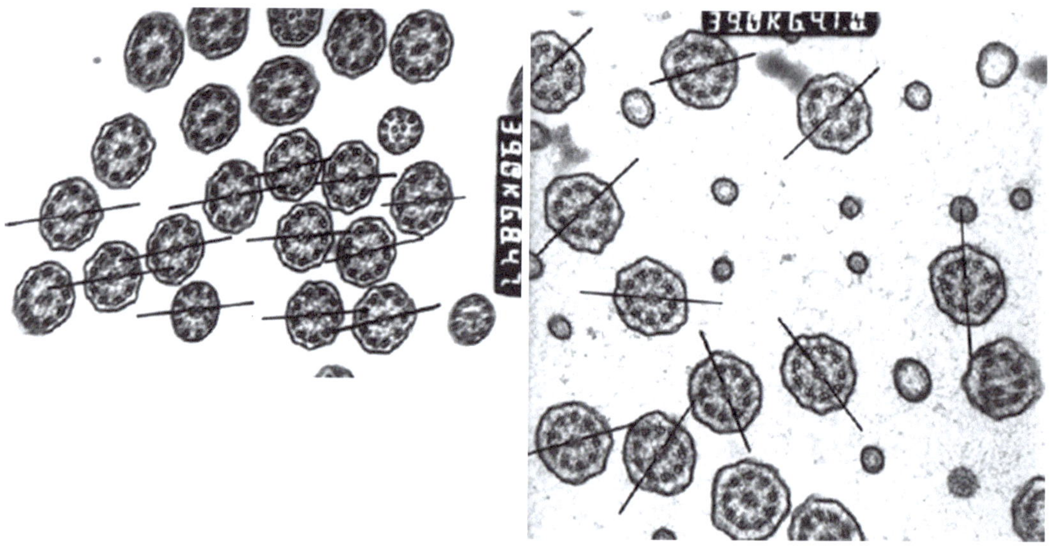

Fig. 3.9 Normal orientation with perfect alignment of all cilia (*left*) and random disorientation of the central pair in the context of PCD (*right*)

At a macroscopical level, this is organized in various streams that can only be measured overall as mucociliary transport. Mucociliary transport is the process by which ciliary activity causes the transport of a thin film of mucus from the upper and lower respiratory tracts toward the digestive tract. Effective and coordinated ciliary beating is of the utmost importance for mucociliary transport.

3.3.2 Mucociliary Transport

Healthy airway surfaces are lined by ciliated epithelial cells and covered with an airway surface liquid, which is composed of two layers: the periciliary layer and the mucus layer. The low viscosity periciliary layer approximates the height of cilia and provides an optimal environment for ciliary beating [40]. The protective mucus layer on top of it is the secretory product of the goblet cells and the submucosal glands. It is a nonhomogeneous, adhesive, viscoelastic gel composed of water, carbohydrates, proteins, and lipids. This mucus layer traps foreign particles like dust, allergens, toxic substances, bacteria, and viruses from the air. Mucus is transported from the respiratory tracts into the pharynx by mucociliary clearance, where it is either swallowed or expelled via coughing. Mucociliary clearance in the airways is driven by the coordinated beating of ciliated cells in the airway epithelium. The permanent clearance of the mucus toward the pharynx is the most important defense mechanism in the upper and lower respiratory tracts. The velocity of mucus clearance is 10–24 mm/min in the trachea, 4.5–7 mm/min in the nose, and 0.5–2 mm/min in the bronchioli. There is great variability between individuals, but for each individual, the clearance rates are fairly constant. Airway diseases may influence mucociliary clearance by changes in the amount and in the viscoelastic properties of the mucus and the periciliary fluid and by changes in the number, the structure, and the activity of the cilia. These changes can be secondary and reversible or primary and nonreversible.

The mucociliary transport pathways are genetically defined and rather specific for each location. The different paranasal sinuses have specific pathways as well as the ostiomeatal complex and the nasal cavity.

3.4 Conclusion

Cilia are extensions of the apical membranes. The cilium itself is characterized by a 9 + 2 axonemal structure. An active, coordinated ciliary beating is essential for mucociliary transport. Ciliary beating depends on the ATPase activity in the dynein arms and is characterized by a specific beating pattern. In healthy persons, 95% of the cilia are ultrastructurally completely normal. Ciliary abnormalities can be the results of external (secondary ciliary dyskinesia) or inherited factors (primary ciliary dyskinesia). Ciliary function and structure are organized at different levels from the individual cilia, over interciliary and intercellular interaction, to the macroscopic level of the ciliated tapestry and mucociliary transport.

References

1. Ferkol TW, Leigh MW. Ciliopathies: the central role of cilia in a spectrum of pediatric disorders. J Pediatr. 2012;160(3):366–71.
2. Rhodin JAG. Ultrastructure and function of human tracheal mucosa. Am Rev Respir Dis. 1966;93:1–15.
3. Jafek BW. Ultrastructure of human nasal mucosa. Laryngoscope. 1983;93:1576–99.
4. Afzelius BA. The immotile-cilia syndrome: a microtubule-associated defect. CRC Crit Rev Biochem. 1985;19:63–87.
5. Satir P, Christensen ST. Overview of structure and function of mammalian cilia. Annu Rev Physiol. 2007;69:377–400.
6. Merkus F, Verhoef J, Schipper N, et al. Nasal mucociliary clearance as a factor in nasal drug delivery. Adv Drug Deliv Rev. 1998;29:13–38.
7. Pazour GJ, Agrin N, Walker BL, et al. Identification of predicted human outer dynein arm genes: candidates for primary ciliary dyskinesia genes. J Med Genet. 2006;43:62–73.
8. Roberts AJ, Numata N, Walker ML, et al. AAA + ring and linker swing mechanism in the dynein motor. Cell. 2009;136:485–95.
9. King SM. Axonemal dyneins winch the cilium. Nat Struct Mol Biol. 2010;17:673–4.
10. Zariwala MA, Omran H, Ferkol TW. The emerging genetics of primary ciliary dyskinesia. Proc Am Thorac Soc. 2011;8:430–3.
11. Leigh MW, Horani A, Kinghorn BA, et al. Primary ciliary dyskinesia (PCD): a genetic disorder of motile cilia. Transl Sci Rare Dis. 2019;4:51–75.
12. Hornef N, Olbrich H, Horvath J, Zariwala MA, et al. DNAH5 mutations are a common cause of primary ciliary dyskinesia with outer dynein arm defects. Am J Respir Crit Care Med. 2006;174:120–6.
13. Zariwala MA, Leigh MW, Ceppa F, et al. Mutations of DNAI1 in primary ciliary dyskinesia: evidence of founder effect in a common mutation. Am J Respir Crit Care Med. 2006;174:858–66.
14. Loges NT, Olbrich H, Fenske L, et al. DNAI2 mutations cause primary ciliary dyskinesia with defects in the outer dynein arm. Am J Hum Genet. 2008;83:547–58.
15. Castleman VH, Romio L, Chodhari R, Hirst RA, et al. Mutations in radial spoke head protein genes RSPH9 and RSPH4A cause primary ciliary dyskinesia with central-microtubular-pair abnormalities. Am J Hum Genet. 2009;84:197–209.
16. Becker-Heck A, Zohn IE, Okabe N, et al. The coiled-coil domain containing protein CCDC40 is essential for motile cilia function and left-right axis formation. Nat Genet. 2011;43:79–84.
17. Knowles MR, Leigh MW, Carson JL, et al. Mutations of DNAH11 in patients with primary ciliary dyskinesia with normal ciliary ultrastructure. Thorax. 2012;67:433–41.
18. Barbato A, Frischer T, Kuehni CE, et al. Primary ciliary dyskinesia: a consensus statement on diagnostic and treatment approaches in children. Eur Respir J. 2009;34:1264–76.
19. Mallik R, Carter BC, Lex SA, et al. Cytoplasmic dynein functions as a gear in response to load. Nature. 2004;427:649–52.
20. Sakakibara H, Kamiya R. Functional recombination of outer dynein arms with outer arm-missing flagellar axonemes of a Chlamydomonas mutant. J Cell Sci. 1989;92:77–83.
21. Sanderson MJ, Chow I, Dirksen ER. Intercellular communication between ciliated cells in culture. Am J Phys. 1988;254:C63–74.
22. Schmid A, Salathe M. Ciliary beat co-ordination by calcium. Biol Cell. 2011;103:159–69.
23. Ingels K, Kortmann M, Nijziel M, et al. Factors influencing ciliary beat measurements. Rhinology. 1991;29:19–26.
24. van de Donk HJ, Zuidema J, Merkus FW. The influence of the pH and osmotic pressure upon tracheal ciliary beat frequency as determined with a new photo-electric registration device. Rhinology. 1980;18:93–104.
25. Wood RE, Wanner A, Hirsch J, Farrell PM. Tracheal mucociliary transport in patients with cystic fibrosis and its stimulation by terbutaline. Am Rev Respir Dis. 1975;111:733–8.
26. Sackner MA, Epstein S, Wanner A. Effect of beta-adrenergic agonists aerosolized by freon propellant on tracheal mucous velocity and cardiac output. Chest. 1976;69:593–8.
27. Melville GN, Horstmann G, Iravani J. Adrenergic compounds and the respiratory tract. A physiologi-

cal and electron-microscopical study. Respiration. 1976;33:261–9.

28. Verdugo P, Johnson NT, Tam PY. Beta-adrenergic stimulation of respiratory ciliary activity. J Appl Physiol. 1980;48:868–71.
29. Yun YS, Min YG, Rhee CS, et al. Effects of alpha-toxin of *Staphylococcus aureus* on the ciliary activity and ultrastructure of human nasal ciliated epithelial cells. Laryngoscope. 1999;109:2021–4.
30. Kim CS, Jeon SY, Min YG, et al. Effects of beta-toxin of *Staphylococcus aureus* on ciliary activity of nasal epithelial cells. Laryngoscope. 2000;110:2085–8.
31. Mallants R, Jorissen M, Augustijns P. Beneficial effect of antibiotics on ciliary beat frequency of human nasal epithelial cells exposed to bacterial toxins. J Pharm Pharmacol. 2008;60:437–43.
32. Chilvers MA, Rutman A, O'Callaghan C. Ciliary beat pattern is associated with specific ultrastructural defects in primary ciliary dyskinesia. J Allergy Clin Immunol. 2003;112:518–24.
33. Jorissen M. Correlations among mucociliary transport, ciliary function and ciliary structure. Am J Rhinol. 1998;12:53–8.
34. Jorissen M, De Brouwer J, Bessems A, Cassiman JJ. Quantitation of ciliary beat frequency by computerized microscope photometry – a preliminary study on suspension cultures of human nasal epithelia showing spontaneous ciliogenesis in vitro. Leitz Sci Tech Info. 1992;10:88–93.
35. Jorissen M, Willems T. The secondary nature of ciliary (dis)orientation in secondary and primary ciliary dyskinesia. Acta Otolaryngol. 2004;124:527–31.
36. Rautiainen M, Collan Y, Nuutinen J. A method for measuring the orientation (beat direction) of respiratory cilia. Arch Otorhinolaryngol. 1986;243:265–8.
37. Rautiainen M, Collan Y, Nuutinen J, et al. Ciliary orientation in the 'immotile cilia' syndrome. Eur Arch Otorhinolaryngol. 1990;247:100–3.
38. Rayner CF, Rutman A, Dewar A, et al. Ciliary disorientation alone as a cause of primary ciliary dyskinesia syndrome. Am J Respir Crit Care Med. 1996;153:1123–9.
39. Wong LB, Miller IF, Yeates DB. Nature of the mammalian ciliary metachronal wave. J Appl Physiol. 1993;75:458–67.
40. Knowles MR, Boucher RC. Mucus clearance as a primary innate defense mechanism for mammalian airways. J Clin Invest. 2002;109:571–7.

Functional Defense Mechanisms of the Nasal Respiratory Epithelium

4

Robert C. Kern and Jennifer R. Decker

Core Messages

- Sinonasal epithelium provides both a physical and immunologic barrier to infection.
- Intracellular junctions, mucus composition, and mucociliary clearance compose the mechanical barrier to pathogen invasion.
- Innate and adaptive immune responses form the immunologic barrier.
- Innate immunity provides the first-line defense to pathogens that circumvent the physical mucosal barrier by recognizing conserved pathogen-associated markers and activating a nonspecific inflammatory response.
- Adaptive immunity confers memory to particular pathogens, providing a faster response to repeated infections.
- Sufficient stimulation of the innate immune system results in activation of and directs the type of subsequent adaptive immune response.
- Fungus and staphylococcal superantigens appear to be disease modifiers in chronic rhinosinusitis rather than the direct cause.
- Dysregulation in the adaptive immune response is the key factor in the pathogenesis of chronic rhinosinusitis.
- Understanding of host-specific sinonasal immune defenses will influence future therapies for CRS.

R. C. Kern (✉)
Department of Otolaryngology – Head and Neck Surgery, Northwestern Memorial Hospital, Chicago, IL, USA
e-mail: r-kern@northwestern.edu

J. R. Decker
Department of Otolaryngology – Head and Neck Surgery, Feinberg School of Medicine, Northwestern University, Chicago, IL, USA

4.1 Overview

The sinonasal epithelium is an important biological point of interface with the external environment, clearing foreign materials without significant collateral tissue inflammation. Multiple components carry out this task, and while they are typically considered separately, they are functionally integrated. The first component is mucus produced by nasal glands and the epithelial goblet cells which trap particulate matter to be swept into the nasopharynx via mucociliary flow. The mucus also contains tonic levels of host defense molecules with antimicrobial properties, limiting microbial survival and proliferation. The next anatomic barrier is the epithelial layer with cells bound together via junctional complexes. Breaching these mechanical barriers brings exogenous agents in contact with receptors that activate the innate response. Secretion of host defense molecules is augmented, and chemokines and cytokines are secreted. The latter initiates inflammation and fosters the accumulation and activation of innate effector cells. If the stimulus is sufficiently strong,

Ö. Ö. Celebi, T. M. Önerci (eds.), *Nasal Physiology and Pathophysiology of Nasal Disorders*,
https://doi.org/10.1007/978-3-031-12386-3_4

dendritic cells are activated to initiate the adaptive response consisting of T and B lymphocytes. The nature of this adaptive response is shaped, in large measure, by input from the epithelial cells. In aggregate, this allows the response to be tailored against the particular pathogen(s) that breach the barrier. Three responses have been broadly defined: Type 1, Type 2, and Type 3 which address viruses, parasites, and bacteria/fungi, respectively. Hence the epithelium plays a pivotal role in maintaining homeostasis as well as initiating and tailoring the immune response across the nasal interface. In CRS, a chronic inflammatory response develops that, in large measure, utilizes these physiological immunodefense mechanisms present in the mucosa. Most commonly, these CRS endotypes are characterized as Type 1, Type 2, and Type 3 immune responses, which trigger the symptoms and remodeling changes associated with these CRS subtypes [1]. The components of this system will be reviewed with a focus on the epithelium, and potential areas of dysfunction will be discussed as these may play a role in the development of CRS.

4.2 Anatomic Barrier

4.2.1 Epithelium Structure

The anterior nasal epithelium consists of stratified squamous epithelium with structural and barrier functions similar to skin. In addition to typical sebaceous and sweat glands, the nasal vestibule contains vibrissa—thick specialized hairs that assist in trapping particles—and many fine hairs for filtering smaller particles [2]. At the internal nasal valves, the sinonasal epithelium transitions to pseudostratified ciliated columnar epithelium; this respiratory-type epithelium is found throughout the remainder of the sinonasal tract [3].

Sinonasal epithelium is respiratory epithelium with the goblet, ciliated, and basal cells bound tightly to each other and the basement membrane.

Sinonasal epithelium consists primarily of ciliated cells, with variable concentrations of mucus-producing goblet cells and glands, and relatively few basal cells. The ciliated cells comprise approximately 70% of the sinonasal epithelium and are responsible for mucous clearance from the sinonasal cavities. Each ciliated cell has hundreds of motile cilia at the apex that beat in unison as well as small, immotile microvilli that increase surface area and prevent drying [4]. Each ciliated cell extends from the basement membrane to the surface of the mucosa with the cilia extending into the overlying mucous layer. Goblet cells comprise approximately 20% of epithelial cells, but their density is irregularly distributed throughout the sinonasal epithelium. Goblet cells have a narrow area of basement membrane attachment and are packed with mucin-containing secretory granules. Their apices extend to the mucosal surface, are also covered with microvilli, and contain a large apical pore for the secretion of mucin. Basal cells comprise less than 5% of the respiratory epithelium. They do not have an epithelial surface component and thus are exposed to fewer inhaled particulates and pathogens. Basal cells contain large numbers of hemidesmosomes that attach them firmly to the basement membrane and function as respiratory cell progenitors [5].

4.2.2 Cilia and Mucus

The sinonasal respiratory epithelium is specialized to effectively trap and remove foreign materials from the nasal cavity before it can cause inflammation or infection. When inspired particulates are removed, a reduction in irritation, allergic response, and pathogen invasion is achieved. The rheologic and immunologic properties of nasal mucous composition combine with ciliary function to produce the phenomenon of mucociliary clearance that maintains a healthy sinus tract.

Respiratory mucus is responsible for trapping and neutralizing inspired environmental particulates. Nasal secretions from the goblet cells and submucous glands are augmented by fluid from the lacrimal glands and transudate from the underlying capillary bed to produce 600–1800 mL of mucous per day. The composition of mucus is designed to trap and bind to particulates and allow for transport by cilia [6]. Nasal mucous exists as a bilayer fluid overlying the epithelium comprised of an outer gel layer and pericellular sol layer. The gel layer is composed primarily of hydrated mucin, a thick mesh of glycoproteins produced by the respiratory goblet cells along with secretions from submucosal mucus glands. As the mucin is secreted, it hydrates and coalesces into mucin rafts overlying the watery pericellular sol layer. The viscous nature of the gel layer allows it to trap inspired pathogens and particulates larger than 0.5 μm [7].

In addition, mucous also contribute to the respiratory host defense system directly. Mucins are known to recognize and bind to microorganism surface adhesins promoting the microorganism–mucin bond and promoting clearance. Mucin is known to bind to Mycoplasma pneumoniae, Streptococcus pneumoniae, Pseudomonas aeruginosa, influenza virus, and Escherichia coli. Mucins may bind and thus concentrate secreted innate immune defense molecules such as lactoferrin and lysozyme, which recognize and kill bacterial pathogens [8].

Mucus contains a variety of innate defense molecules to bind and neutralize pathogens.

Ciliated cells in human respiratory mucosa each contain 50–200 motile cilia extending from the apical surface. In humans, these respiratory cilia are approximately 6 μm in length and are anchored by centriole-like basal bodies [6]. Each cilium consists of an axoneme covered by an extension of the cell membrane. The axoneme is comprised of the "9 + 2" microtubular pattern that is preserved in ciliated cells throughout the body. The nine outer microtubule doublets are each connected radially to the two central microtubules and circumferentially to each other by dynein arms. It is these dynein arms that slide along the microtubules, generating force that is translated into microtubule movement [9].

Ciliary motility generates a forward power stroke with the cilium fully extended, delivering force through the tip into the overlying gel phase of the mucous layer. This is followed by a recovery stroke with the cilium tip bent to sweep through the thin pericellular sol layer as it returns to its starting position [7]. Recovery is followed by a quiescent phase. Coordinated beating of the respiratory cilia creates a metachronous wave that propels the mucin rafts of the gel phase in the direction of the power stroke [10]. The combination of mucus composition and ciliary beating creates the phenomenon of mucociliary transport, which is critical in the maintenance of healthy sinonasal mucosa.

4.2.3 Mucociliary Clearance

The remarkable phenomenon known as mucociliary clearance (MCC) combines the biphasic properties of nasal mucus with coordination of the ciliated cellular beating to effectively clear the mucus blanket with trapped particulates from the nasal cavity. Nasal MCC function responds to environmental factors and is under continual dynamic modulation. Ciliary beat frequency increases as a reaction to stressors including increased breathing, cold air, exercise, or inflammation to accelerate mucus clearance [11]. The mucus blanket is swept out of the sinus cavities in predefined patterns via natural ostia then into the posterior nasopharynx where it is swallowed and neutralized in the stomach.

Messerklinger classically described the defined mucus clearance patterns of the paranasal sinuses in 1966 [12]. In the maxillary sinuses, mucus is swept upwards against gravity along the walls and roof towards the superomedially based maxillary ostium, into the ethmoid infundibulum, and then into the middle meatus. The anterior ethmoid cells each drain through inferiorly based individual ostia into the middle meatus. The posterior ethmoids drain into the superior meatus and then sphenoethmoid recess. The sphenoid sinus sweeps mucus anteriorly toward its natural ostium and also drains into the sphenoethmoid recess. The frontal sinus has the most complex clearance pattern. Mucus is cleared from the medial portion of the sinus in a retrograde fashion superiorly along the posterior wall away from the natural ostium, laterally along the roof to the anterior wall, and then carried inferiorly and medially toward the ostium [13].

Ciliated cells beat in a coordinated metachronous wave that rapidly clears the overlying mucus blanket.

The basis of modern endoscopic sinus surgery is to maintain functional MCC patterns. Surgical goals include relief of obstruction while utilizing mucosal-sparing techniques and preserving natural drainage pathways. Failure to incorporate natural ostia when opening a sinus can lead to recirculation particularly in the maxillary sinuses [14]. Preservation of normal mucosa promotes restoration of natural mucociliary flow patterns. This is particularly true at the narrow nasofrontal duct where disruption of normal mucosa can lead to circumferential scarring with narrowing of the frontal duct, obstruction, and mucostasis.

4.2.3.1 Diseases Affecting Mucociliary Clearance

The sinonasal epithelium utilizes a mechanical barrier, effective mucociliary clearance, and optimal healing to maintain mucosal health. However, disease processes affecting ciliary function, rheologic properties of mucus, or obstruction of natural ostia results in altered MCC effectiveness and resultant infection [15]. These include genetic conditions, allergy, asthma, and acute and chronic rhinosinusitis.

Mucociliary clearance efficiency declines with changes in mucus rheology, ciliary impairment, or narrowed sinus ostia.

Primary ciliary dyskinesia (PCD) is a disorder of the ciliary dynein arms resulting in systemically immotile cilia. These patients may have infertility and are unable to effectively transport mucus out of the sinorespiratory tract. The resultant mucus stasis results in chronic rhinosinusitis, bronchiectasis, and recurrent respiratory infections [16]. Cystic fibrosis (CF) is an autosomal-recessive disorder of the CFTR gene resulting in abnormal electrolyte transport. Clinically, there is an abnormal function of the goblet cells and highly viscous mucus. CF patients have defective MCC clearance due to the inability of the cilia to adequately transport the thick mucus rafts and are subject to bronchiectasis and severe sinopulmonary infections [17]. Nasal polyps are found in approximately 40% of CF patients and show a neutrophilic rather than eosinophilic pattern [18]. However, these patients still display a wide spectrum in the severity of their sinus disease despite having identical CFTR gene mutations [19].

Nasal irritants, allergy, and acute infection result in inflammation and mucosal edema that negatively affect mucociliary clearance. Acute upper respiratory infection has been shown to alter mucus composition, reduce ciliary motility, and create mucosal edema [20]. When MCC function fails to clear sinuses in key areas such as the osteomeatal complex or eustachian tube orifice, associated sinusitis results.

Allergic challenge has been shown to increase transudate levels in nasal mucus. As a result, the depth of the periciliary layer increases, and the cilia tips cannot reach the overlying gel phase to effectively propel the mucin rafts [21]. Swelling of the nasal mucosa in allergic rhinitis is also thought to obstruct sinus ostia leading to poor ventilation and mucostasis [22]. While the role of allergy in the causal development of chronic rhinosinusitis is unclear, failure to address the allergic component clearly lowers the success of surgical intervention in CRS [23].

Failure to address nasal allergies or smoke exposure leads to worse outcomes in the treatment of CRS.

Exposure to cigarette smoke has been shown to significantly alter ciliary beat frequency and secretions resulting in reduced mucociliary clearance [24]. Long-term exposure has been proposed to cause epithelial and submucosal gland hyperplasia, squamous metaplasia, and increase epithelial permeability [25, 26] and to foster the formation of bacterial biofilms [27]. In addition, cigarette smoking and second-hand smoke have been shown to be independently associated with CRS [28, 29]. As such, exposure to cigarette smoke should obviously be avoided in patients with recurrent or chronic sinus disease.

In addition to relatively rare genetic disorders such as PCD and CF, multiple investigations have also demonstrated significant impairment of sinonasal mucociliary clearance in idiopathic CRS. Inflammatory changes seen in chronic rhinosinusitis with and without nasal polyposis can cause secondary ciliary dyskinesia. Possible pathophysiologic explanations include reduced ciliary beating, changing depth or viscosity of respiratory mucus, or alterations in epithelial integrity. CRS patients exhibit blunted ciliary beat frequency responses to adrenergic and cholinergic stimulation in explanted nasal epithelial cultures [30]. Additional studies showed that when removed from the chronic inflammatory environment, cilia resume normal basal and stimulated beat frequency though recovery takes time. This suggests that a chronic inflammatory nasal environment reversibly suppresses normal ciliary function [31, 32].

4.2.4 Epithelial Integrity

Sinonasal cells contain several types of intercellular junctions that serve both communication and attachment functions. Maintenance of epithelial integrity is important for cellular communication and prevention of pathogen invasion or foreign protein exposure to the underlying tissues. Sloughing of epithelial cells leaves a denuded basement membrane and evokes an inflammatory response.

Gap junctions primarily allow cell–cell transfer of ions and signaling molecules; they do not contribute significantly to adhesion [33]. Respiratory cells are held together on their lateral surfaces with adherens junctions and desmosomes, both of which contain cadherin proteins. Hemidesmosomes, which contain integrin proteins, attach cells tightly to the basement membrane and prevent cellular sloughing. The most important junctions, tight junctions, are critical in epithelial barrier integrity against invasion. Tight junctions attach adjacent epithelial cells in a narrow band just beneath their apical surface [34]. The resultant epithelial surface is a watertight barrier to chemicals and particulates including proteins and pathogens at the surface.

Below the basement membrane lies the lamina propria, a network of blood vessels and loose fibrous stroma that contains immune cells responsible for pathogen detection and initial inflammatory response. Here reside the lymphocytes, plasma cells, macrophages, and dendritic cells which monitor the health of the epithelium. Pathogens that overcome the protective barriers of mucus and epithelium activate the innate and adaptive immune systems. Breakdown of the intracellular connections of the epithelium plays an important role in permitting stimulation of an immune response.

4.2.4.1 Diseases Affecting Epithelial Integrity

Invasion of pathogens into the underlying stroma requires disruption of the epithelium and penetration through the basement membrane. Viral pathogens, the initial cause of many upper respiratory tract infections, have developed mechanisms for epithelial disruption that target tight and adherens junctions, allowing for subsequent invasion into the underlying tissues [35, 36]. Respiratory viruses disrupt the upper airway epithelial barrier, resulting in inflammation and predisposing for additional exposure to foreign material and potential tissue invasion [37]. Nasal epithelial explant studies show that viruses loosen epithelial tight junction bonds and penetrate the basement membrane within 24 h [38]. Rhinoviruses, a major cause of the common cold, bind to intercellular adhesion molecule-1 (ICAM-1) on nasal epithelial cells, allowing RNA entry into the cell [39].

Fungi, bacteria, and allergens all carry significant intrinsic protease and virulence activity with the capacity of weakening tight junctions permitting access into the underlying tissues [40, 41]. Intrinsic alterations in tight junction barrier function may be responsible for the increase in epithelial permeability seen in allergic responses [42]. Penetration of antigenic proteins through the epithelium has been proposed as a major factor in asthma and atopy [43].

Respiratory viruses disrupt epithelial tight junctions and increase susceptibility to infection and inflammation.

Chronic inflammatory disorders may weaken the epithelial barrier function and predispose to infection. CRSwNP with a Type 2 cytokine environment exhibits disruption of normal intercellular junctions including decreased levels of the desmosome cadherin proteins DSG2 and DSG3 [44] as well as tight junction proteins claudin and occludin [45]. Disruption of intercellular tight junction connections is seen in nasal polyposis [46]. Decreased production of protease inhibitors such as LEKT1, encoded by the gene SPINK5, has been found in CRSwNP [41]. This molecule acts as a proteinase inhibitor and regulates the processing of epithelial tight junctions critical to maintaining a barrier to infection. In addition, altered intercellular ion transport is seen in CRS and may interfere with the coordination of ciliated cells leading to ineffective mucociliary clearance [47]. Interestingly, it has been recently suggested that epithelial turnover may be altered by a Type 2 cytokine milieu that created a leaky, immature barrier. Furthermore, this leaky barrier may be responsible for the high rate of recurrence seen with T2 CRS. (EPOS 2020) Substantiating this concept are recent single-cell transcriptomic studies, which have suggested that the epithelial changes are epigenetically programmed and will therefore be durable [48]. Periostin is associated with this process and high levels of this protein in the serum may be a sign of the T2 endotype [49].

4.3 Immunologic Barrier Function of the Sinonasal Epithelium

In addition to forming a physical barrier and providing clearance of pathogens from the nasal cavity, sinonasal epithelial cells (SNECs) contribute to the innate and adaptive immune responses. When the various mucosal barrier functions are circumvented by irritant or pathogenic particulates, the mucosal immune system must distinguish between normal commensal organisms and invading pathogens. In addition, the immune system must determine how to mount an appropriate level or intensity of response. The first of these responses, the innate immune system, consists of receptors, defense molecules, and cells that respond to microbial organisms in a relatively nonspecific manner. This innate response is an inborn, germ line-coded defense to generalized pathogen invasion and does not typically exhibit memory. Should the innate immune system be sufficiently challenged, it activates the adaptive arm of the immune system, with a delayed but highly specific T and B cell response capable of memory against specific pathogens. Emerging evidence also indicates that a diminished innate response coupled with a compensatory overactivation of adaptive immunity may play a role in the pathogenesis of the chronic mucosal inflammation seen in CRS.

Pathogen recognition by the innate immune system results in a cascade of cytokines and chemokines that determine the type and strength of the inflammatory response.

4.3.1 Receptor Molecules

Innate immune responses in the sinonasal epithelium are initiated by membrane-bound and cytoplasmic pattern-recognition receptors (PRRs) that recognize highly conserved pathogen-associated molecular patterns (PAMPs) found in various bacteria, mycobacteria, viruses, and parasites as well as necrotic debris from cellular damage [50]. PRRs are also expressed on various types of antigen-presenting cells including dendritic cells, macrophages, and B cells. In addition, SNECs express genes associated with antigen-presenting function [51]. Once the innate immune response is activated, SNECs may also serve as antigen-presenting cells, increasing local tissue inflammatory response. Overall, stimulation of PRRs facilitates both the innate and the adaptive response.

Recognition of PAMPs by PRRs results in the secretion of the endogenous antimicrobial factors that directly aid in pathogen clearance. PRR activation also stimulates SNECs and antigen-presenting cells (APCs) to release inflammatory cytokines and chemokines that attract other innate cellular defenses such as phagocytes. Other receptors detect cellular injury through damage-associated molecular patterns DAMPs [52, 53]. If the combination of cellular damage and PRR activation is sufficiently strong, the resultant innate immune response triggers cytokine patterns that initiate, and determine the nature of, the subsequent adaptive immune response [54] (see Fig. 4.1).

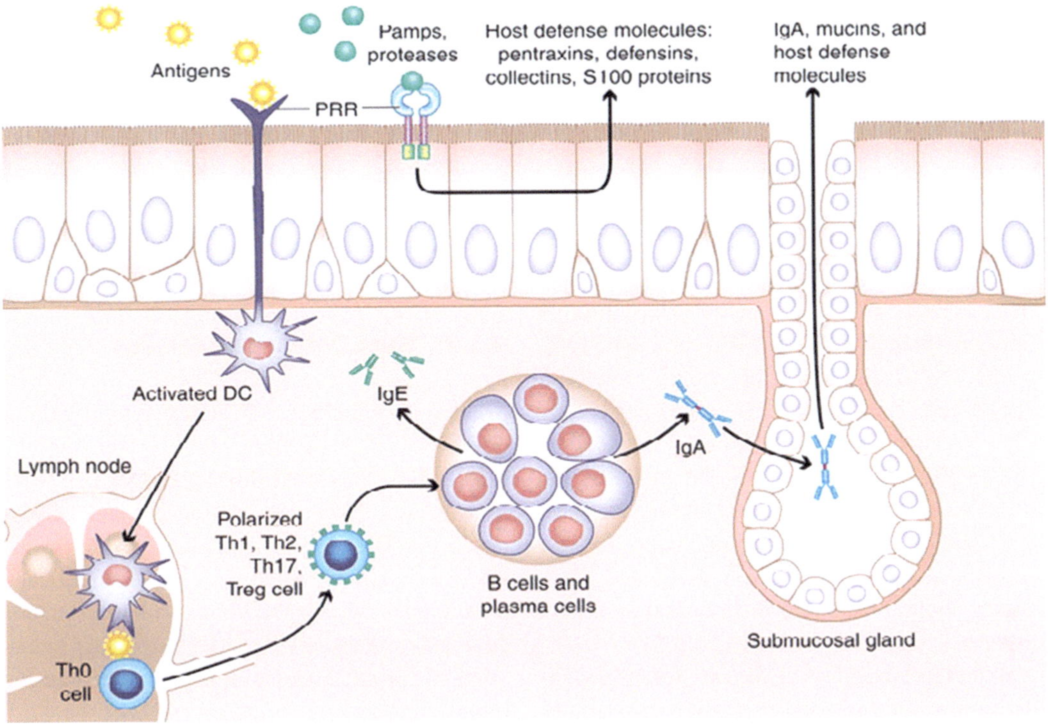

Fig. 4.1 Innate immunity at the epithelial surface involves PAMP activation of epithelial PRRs leading to release of host defense molecules. Exogenous proteases may activate PAR receptors or degrade tight junctional proteins. Secreted IgA, mucins, antiproteases, and host defense molecules are released into the nasal mucus to form a second line of innate defense. Sufficient PRR stimulation activates dendritic cells causing migration to local lymph nodes to present antigen fragments to naïve Th0 lymphocytes. The type and strength of antigen stimulation drive the adaptive immunity through polarized Th1, Th2, Th17, or Treg responses. B cells are stimulated to undergo proliferation, class switch recombination, and differentiation leading to production of IgE, IgA, and other immunoglobulins as well as stimulation of the cells of the adaptive immune system. *See text for details*

PRRs can be separated into three main classes: endocytic, secreted, and signaling. Endocytic PRRs are found on the surface of phagocytes, which recognize pathogenic PAMPs, engulf the pathogens, and present the antigens to lymphocytes of the acquired immune system. One example is the mannose receptor found on macrophages [55].

Secreted PRRs are used as opsonins that trigger the complement cascade or signal phagocytosis. The most abundant of the secreted antimicrobial peptides in the nasal mucosa include lactoferrin, mannose-binding lectin, secretory leukocyte proteinase inhibitor, and lysozyme [56]. These are released into the mucus by nasal epithelial cells in response to activation of PRRs and confer protection by inhibiting epithelial invasion or directly lysing the microorganism [57].

Signaling PRRs trigger the production of antimicrobial peptides and cytokines by epithelial cells in response to PAMPs [58]. Signaling PRRs found in nasal epithelium include the toll-like receptor (TLR) family, the nucleotide binding and oligomerization domain (NOD)-like receptor family (NLRs), and the retinoic acid-inducible-like receptors (RLRs).

Important pathogen recognition receptors (PRRs) include the TLR, NOD, Bitter Taste receptors, and PAR families.

TLRs are transmembrane receptors expressed on various cell types including SNECs that recognize extracellular or intracellular PAMPs such as bacterial lipopolysaccharide (LPS) [59].

TRL2, TRL3, TRL4, TRL9, and possibly others are expressed in sinonasal epithelium and contribute to the host defense. TLR2 responds to gram-positive bacterial as well as fungal PAMPs, TLR 3 recognizes viral replication products, TLR4 recognizes bacterial endotoxin, and unmethylated CpG areas on pathogenic DNA activate TLR9 [33]. TLR activation triggers intercellular signaling through proteins MyD88 and TRIF, which affects gene expression through transcription factors NF-κβ, AP-1, and IRF3 [60].

The NOD-like receptor family includes NOD-1 and -2, which have been shown to recognize bacterial cell walls including staphylococci [61]. NOD levels were increased in CRSwNP and levels decreased after nasal steroid use [62]. RLRs are intracellular receptors important for recognizing RNA derived from RNA and DNA viruses, though their role in sinonasal epithelial response is still being investigated [63].

Bitter taste receptors (T2Rs) are recently identified group of pathogen recognition molecules also present on epithelial cells, function as nonclassical PRRs. They are G protein-coupled receptors (GPCRs), which are widely expressed in host airway epithelial membranes. Airway T2R-mediated immune responses are activated by bacterial quinolones as well as acyl-homoserine lactones [64] secreted by gram-negative bacteria, including Pseudomonas aeruginosa [65]. Linkage studies have demonstrated associations between taste receptor genetics with CRS [66].

The protease-activated receptors (PARs) are a distinct set of receptors found on sinonasal epithelial cells that are activated by endogenous and exogenous proteases including those from bacteria, fungi, and allergens. Once activated, PARs evoke the NF-κβ signaling pathway, the results in chemokine and cytokine production, and phagocytic recruitment and potentially influence the subsequent acquired immune response based on the cytokine milieu [67]. PAR-2 activation by *Staphylococcus aureus* proteases results in increased levels of cytokine IL-8 [68]. Fungal proteases may drive the eosinophilic as well as neutrophilic response via PARs [69]. SNECs secrete antiproteases such as LEKT1 coded by the *Spink 5* gene, which likely acts to protect PARs from both exogenous and endogenous protease stimulation. Reduced levels of LEKT1 have been associated with CRSwNP, suggesting that excessive PAR activation may play a role in polyp pathogenesis [70].

4.3.2 Host Defense Molecules

Sinonasal epithelial cells secrete a multitude of antimicrobial molecules into the surrounding mucus. Enzymes break down pathogen cell walls and include lysozyme, chitinases, and peroxidases. Foreign material is marked for phagocytosis by opsonins such as complement and pentraxin-3. Permeabilizing proteins include defensins and cathelicidins. Defensins are inducible and provide broad antimicrobial activity and inhibit invasion of bacteria and viruses [71]. Cathelicidins are a family of secreted peptides that are active after extracellular cleavage. In humans, cathelicidin LL-37 directly disrupts bacterial membranes [72]. They are chemotactic for effector cells of both the innate and adaptive immune system including neutrophils, monocytes, mast cells, and T cells and may modulate their activity [73]. Collectins include surfactant proteins (SP-A, SP-D) and mannose-binding lectin; these proteins, long studied in lower respiratory mucosa, are also found in sinonasal secretions. They are important in reducing nasal bacterial colonization, inflammation, and infection [74]. Decreased surfactant levels are found in patients with cystic fibrosis, poorly controlled COPD, allergic fungal sinusitis, and CRS [75–77]. Binding proteins include mucin, discussed previously, and lactoferrin, which attaches to foreign material and facilitates its removal by MCC. PLUNC, another secreted antimicrobial, has important anti-biofilm properties. Diminished secretion of many but not all of these host defense molecules has been proposed as a common mechanism broadly underlying the eti-

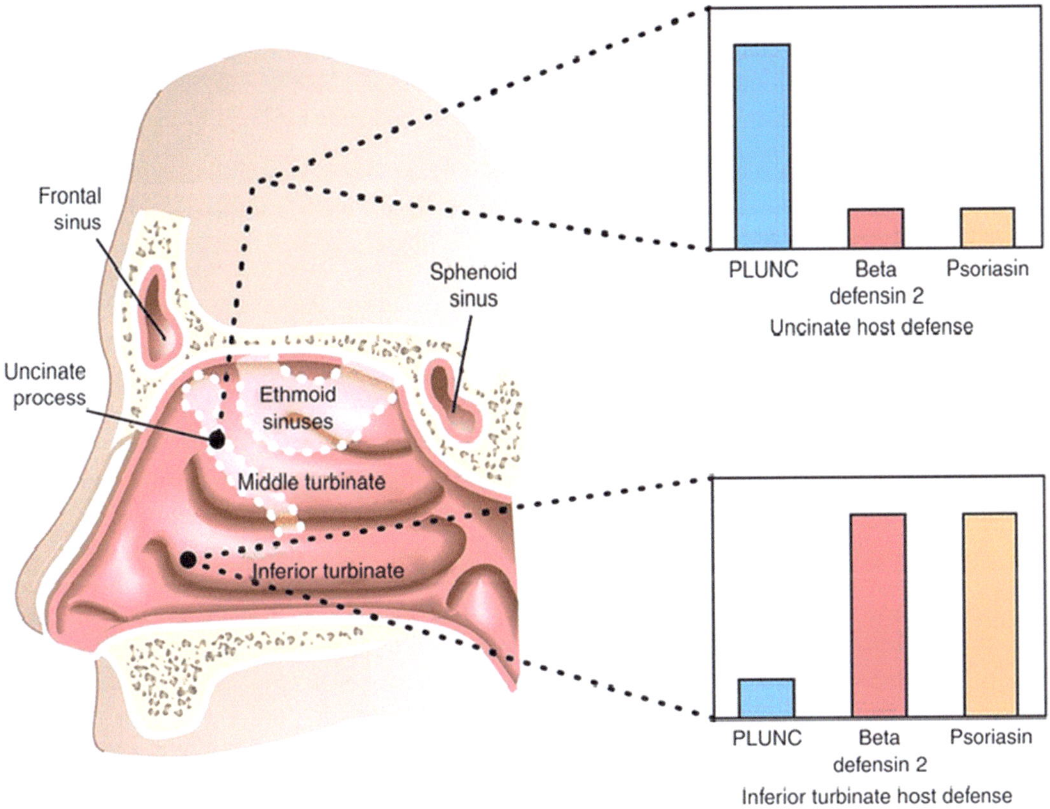

Fig. 4.2 Inferior turbinate and uncinate process tissue from healthy patients were compared for levels of the mucosal innate defense molecules PLUNC, lactoferrin, pentraxin, S100A7, and hBD2. While levels of pentraxin remained the same between the two sites, polarization was noted with the remaining molecules. In nasal uncinate tissue, low levels of PLUNC and lactoferrin but high levels of S100A7 and hBD2 were found. In inferior turbinate tissue, the converse was true, suggesting that host defense molecules show regional specialization in human nasal airways

ology and pathogenesis of CRS [41, 78] (see Fig. 4.2). The cause for this selective reduction is unclear but may be [1] a primary defect in the innate response or [2] a downstream secondary effect of the Type 2 cytokine milieu associated with some forms of CRS [79]. Mechanistic studies to evaluate this have not been completed, but IL-22 and the STAT 3 pathway appear to broadly govern innate nasal mucosal host defense and mechanical barrier integrity [80–82]. Diminished STAT 3 activity in the sinonasal epithelium has been identified in CRS supporting a primary innate defect [83].

A decrease in innate host defense molecules contributes to the pathogenesis of CRS

4.3.3 Epithelial Chemokines and Cytokines

In response to PRR and PAR stimulation, sinonasal epithelial cells produce a variety of cytokines, which are small proteins that regulate inflammation and mediate associated pain, swelling, and vascular dilatation [84]. A partial list include IL-1, TNF, IFN, GM-CSF, eotaxins, RANTES, IP-10, IL-6, IL-8, MDC, SCF, TARC, MCP-4, BAFF, osteopontin, IL-25, IL-33, and TSLP [85–87]. These cytokines have a diverse array of functions including triggering inflammatory responses, activating and recruiting innate effector cells, and facilitating and shaping the adaptive response. To the last point, IL-1, IL-6, and IL-8 have all been identified as particularly important in modulating the transition between the two arms of the immune response [86]. Newly identified epithelial-derived cytokines, including TSLP, IL-25, IL-33, and BAFF, help shape the local adaptive responses more directly. They foster B cell proliferation with immunoglobulin production and shaping the T helper profile via dendritic cell polarization [88, 89]. Defective regulation of these processes has been proposed as playing a role in the pathogenesis of CRSwNP. It has been noted that corticosteroids, a mainstay for treating inflammation and CRS, act in part via the downregulation of epithelial cytokine secretion but spare or even augment the epithelial secretion of host defense molecules [90, 91]. In this fashion, corticosteroids effectively upregulate the innate immune response and downregulate the adaptive response.

Epithelial-derived molecules directly contribute to the innate immune defense and shape-associated B cell and T cell responses.

4.3.4 Epithelial Co-stimulatory Molecules and Inflammatory Enzymes

SNECs express co-stimulatory molecules, particularly homologs of the B7 family of cell-surface ligands [92]. Expression results in downregulation of T cell-acquired response and is induced by TNF-α and IFN-γ [93]. Increased B7 family expression is induced by viral infection and CRS [94]. Reactive oxygen species (ROS) are important in many of the processes discussed: mucin production, epithelial healing, response to toxins, and innate immunity [95]. They can be beneficial, generating hypothiocyanite and peroxide which aid in the microbial killing. In addition, they may be induced by toxins, requiring neutralization by antioxidants in airway epithelial cells. ROS can also interact with reactive nitrogen species (RNS) to create tissue damage in disease [37]. Of particular interest is nitric oxide (NO), an intracellular messenger that mediates inflammation and antimicrobial effects and regulates apoptosis. In particular, NO is produced in high concentrations in the paranasal sinuses and may limit bacterial colonization [96].

4.3.5 Cells of the Innate Immune System

Important cells of the innate immune response that respond to cytokines secreted by the sinonasal epithelium include innate lymphocytes (ILC), dendritic cells, macrophages, neutrophils, eosinophils, natural killer (NK) cells, basophils, and mast cells. The ILCs are divided into three subsets (ILC 1, 2, and 3) and they help tailor the subsequent innate cellular and adaptive response to the address the particular inciting pathogen [1]. Macrophages and neutrophils are primary phagocytes which recognize and bind to pathogens that have been opsonized, or marked as foreign, by antibodies, complement, or collectins [57]. They then engulf the opsonized pathogens and neutralize them by a variety of methods including nitric oxide and radical oxygen species. In addition, macrophages assist in tissue homeostasis, removal of particulates, and tissue repair and influence the adaptive immune response. Two pathways of macrophage differentiation and activation exist: the M1, or classical, pathway and the M2, or alternative, pathway. The M1 pathway is driven by Type 1 cytokines and fosters macrophages directed against intracellular pathogens. The M2 pathway is driven by Type 2 cytokines and fosters macrophages specialized against helminthes, with additional roles in tissue repair and antibody formation [97]. High levels of M1 macrophages are seen in sinus mucosa and polyps of CF patients, while high levels of M2 macrophages are seen in CRSwNP tissues [98, 99]. Neutrophils are among the first cells to respond to infection and are important in early phagocytosis of extracellular microbes. Mucosal recruitment is triggered by PRR stimulation creating chemokines, particularly IL-8, which is also secreted in response to PAR-2 stimulation [100]. Neutrophils kill using free radicals and many of the same antimicrobial peptides that are secreted in the nasal mucus. Their overactivation may contribute to the pathogenesis of CRS, inflammation related to tobacco smoking, and CF-associated polyposis [27, 101].

High numbers of eosinophils are seen in mucosal surfaces in allergy and asthma. Eosinophils are granulocytes that respond to mucosal inflammation by degranulating and releasing stored cytokines. They are prominent in fungal and parasitic responses and release major basic protein, eosinophil-derived neurotoxin, and neutrophil elastase [102]. Eosinophilia is also an important component of CRS where it contributes to mucosal damage and chronic inflammation [103]. Multiple studies have shown that tissue eosinophilia is correlated with severity of CRS and comorbid asthma [104, 105]. The highest levels of tissue eosinophils are seen in CRSwNP, particularly in Western populations [106]. Eosinophil recruitment and activation is in large measure via epithelial cytokines and chemokines, including eotaxins, RANTES, and MCP [36, 107–109]. The regulation of this secretion is in part not only through PRR stimulation but also via Type 2 cytokines IL-4, IL-5, and IL-13 [110, 111]. The primary cellular sources of these Type 2 cytokines include Th 2 and ILC-2 cells. (EPOS 2020) Other factors that may foster eosinophil activation and recruitment include staphylococcal superantigens, IL-25, IL-33, TSLP, SCF, and eicosanoids [112–115].

Upregulation of eosinophils is seen in allergy, asthma, and CRSwNP.

Mast cells are resident cells in the sinonasal mucosa that not only function in innate immunity and tissue repair but also play key roles in the pathogenesis of allergic rhinitis and possibly nasal polyposis. Stem Cell Factor (SCF), secreted by SNECs, is likely important in mast cell recruitment [85]. Mast cell activation results in the secretion of preformed granules including histamine, prostaglandins, serotonin, and serine proteases. In addition, de novo synthesis and secretion of various cytokines, chemokines, and eicosanoids also takes place. They can be induced to phagocytose pathogens as well, though this is not their primary function. Physiological activation occurs typically through stimulation of PRRs. In nasal disease states such as allergic rhinitis, mast cells are activated strongly via surface IgE bound to antigen [116]. In CRSwNP, both IgE-dependent and IgE-independent pathways for mast cell activation likely contribute to pathogenesis [117, 118]. Specifically, mast cells may be able to induce and maintain eosinophilic inflammation leading to polyposis as well as influence the subsequent adaptive immune response.

Dendritic cells (DCs) are key cells in both the innate and adaptive responses via antigen capture, antigen presentation to immature T cells, and the secretion of soluble mediators. DCs phagocytose and thereby sample commensal organisms and pathogens at the sinonasal epithelial surface. Cytokine cross talk between SNECs and DCs helps determine DC polarization [88]. Polarized DCs then migrate to local lymph nodes and present a fragment of the engulfed pathogen, via major histocompatibility complex (MHC) type II on the cell surface, to immature T cells. DC cytokines strongly influence the subsequent T cell profile, secondarily polarizing the helper response [119]. Overall, DCs act as a bridge from the innate to the adaptive immune response. SNECs acting through cytokines that influence DCs to play a significant upstream role in shaping the subsequent adaptive response, which will be discussed below. Defects in this cross talk pathway may foster the development of CRSwNP.

4.3.6 Adaptive Immunity

The adaptive immune response is mediated by T and B cells and includes immunoglobulin and T effector processes that augment the faster, but less specific innate response. Sinonasal epithelial responses connect with ILCs and together play a key upstream role in the adaptive response including the recruitment of cells of the T and B lineage. In addition, as mentioned above, the type, duration, and intensity of PRR activation by PAMP stimulus in SNECs, ILCs, DCs, and other cell types are believed to shape the resultant T lymphocyte profile and to ensue antibody and cell-mediated response at the mucosal surface [120].

B lymphocytes secrete immunoglobulins, or antibodies, to specific antigens, playing an important role in the memory of the adaptive immune response. Immunoglobulins help bind and trap commensal organisms and pathogens, aiding mechanical clearance and facilitating active killing via multiple mechanisms. In the nasal mucosa, B cells respond to antigen presentation by proliferating and undergoing differentiation into mature plasma cells that produce immunoglobulin. In normal mucosal defense, the primary antibody class is secreted IgA from extrafollicular B cells. This response works to limit bacterial colonization with a minimum of inflammation, is T cell independent, and helps maintain mucosal homeostasis. SNECs and other cell types secrete cytokines and chemokines that foster this baseline B cell activity with the capacity for upregulation in response to an immune challenge.

Staphylococcal superantigens and fungal elements act as disease modifiers in CRS.

During frank mucosal infection, secretory IgA is joined by IgG, resulting in the development of a robust inflammatory response. This response exhibits high affinity for the invading pathogens, is T cell dependent, and utilizes immunoglobulins generated by both tissue plasma cells and follicular B cells. Other immunoglobulins play a role in mucosal inflammation including IgM, IgE, and IgD [121]. IgM is an early-response antibody that precedes the development of long-term IgG. IgE is important in allergic response, mast cell activation and survival, and homeostasis as well as defense against pathogens, especially parasitic infections. IgD, though little understood, may influence antigen binding and basophil activation against respiratory bacteria [122].

In chronic inflammatory conditions such as CRS, immunoglobulin profiles are skewed from the normal, apparently in response to bacterial and fungal antigens. CRSwNP appears to show a particularly dysregulated B cell response. Higher levels of IgA, IgE, and IgG are seen in nasal polyp tissues compared to controls and to CRSsNP, and this may have pathophysiological significance. IgE facilitates mast cell degranulation and IgA is a potent activator for eosinophil degranulation [123]. The combined presence of these antibodies with mast cells and eosinophils within nasal polyps may facilitate degranulation and tissue damage. It should be noted that these immunoglobulin levels do not reflect the systemic profile, indicating a localized mucosal response [124]. Not surprisingly, higher levels of immunoglobulin-producing B cells and plasma cells are also found in nasal polyps, and the process of polyp growth may be orchestrated by abnormal local B cell proliferation and recruitment [125]. Evidence suggests that this process may be driven by the epithelial cytokine BAFF, a TNF family member that influences B cell proliferation and class switching [89]. BAFF is found at higher levels in nasal polyps and correlates with the number of B cells within the tissue. In mouse models, excessive BAFF has been associated with the development of autoimmunity. This process has also been documented in recalcitrant CRSwNP with the presence of high levels of local autoantibodies in the polyp tissue [126].

Abnormal B cell proliferation creates inflammation and tissue damage that may lead to polyposis.

Staphylococcal superantigenic toxins (SAGs) have been proposed as disease modifiers of nasal polyposis through the generation of a polyclonal IgE response including IgE directed against the SAGs themselves. The presence of IgE to these toxins within polyp tissue has been correlated with overall increases in polyclonal IgE, eosinophils, asthma, and severity of CRSwNP [126, 127]. It is unclear whether this superantigen-driven process works through BAFF or another, superimposed, pathway.

4.3.7 T Cells and Cytokine Response

Homeostasis across the nasal mucosa is typically maintained via the mechanical barrier, innate immune responses, and tonic IgA secretion. When the mucosal barrier is breached, a protective response is initiated with SNECs, DCs, and other innate immune cells helping to guide the adaptive response and match it to the inciting stimulus. Minor damage is likely handled by activation augmentation of innate responses from SNECs and migrating innate effector cells. A more substantial breach will activate the adaptive response; IL-6 has been proposed as a key cytokine mediating the transition, suppressing innate responses and triggering production of chemokines that promote the adaptive response [89]. At homeostasis, DCs still regularly phagocytize foreign material, but when activated and exposed to sufficient PAMP activation, such as would occur in a mucosal breach, they cease phagocytosis and acquire additional chemokine receptors. Chemokines, stimulated into production during the innate immune response, cause the DCs to migrate to nearby lymph nodes and to secrete the cytokine IL-1 [128]. Antigen from the phagocytosed pathogens is presented to naïve CD4+ T helper (Th) cells in the lymph tissue. These lymphocytes will differentiate into a specific T cell lineage, determining the type of adaptive immune response. This process is further activated by IL-1. The types, duration, and intensity of the PRR activation by PAMP stimulus are believed to influence the resultant cytokine production and shape the resultant T lymphocyte profile [129]. As mentioned above, cytokines from SNECs and other innate cell types play a critical upstream role in this process matching the response to the pathogen.

Signaling cross talk between innate immune cells drives T cell differentiation.

Mature T cells migrate back to the sinonasal mucosa to mediate the adaptive response upon subsequent antigen challenge. T helper lymphocyte responses are divided based on cytokine profiles generated in response to the presented antigen stimulus. Classically, Th1 or Th2 responses were thought to be the primary adaptive sinonasal inflammatory pathways. The Th1 pathway shows high levels of IL-12 and IFN-γ and has a macrophage-rich cellular infiltrate. Th1 responses facilitate defense against intracellular pathogens, particularly viruses and intracellular bacteria including mycobacteria. They appear to be blunted in chronic obstructive pulmonary disease (COPD), psoriasis, Crohn's disease, and CRSsNP [130]. The Th2 pathway results in high levels of cytokines IL-4, IL-5, and IL-13 and has a more eosinophilic cellular response. Th2 responses are important in parasitic infections and are also seen in frequently allergic and asthmatic responses [79]. They are reduced in asthma, atopic dermatitis, ulcerative colitis, and CRSwNP [131]. More recent data indicate that additional Th profiles are important in mucosal immunity. Th17 responses aid in defense against extracellular bacteria and fungi, particularly *Staphylococcus aureus* [132]. This response is fostered primarily by IL-17A as well as cytokines IL-6, TGF-β1, and IL-23 and has a neutrophilic cellular response [133]. Tregs are regulatory lymphocytes that foster immune tolerance with the goal of limiting excessive responses from other Th lineages; Treg differentiation is facilitated by TGF-β [131].

4.3.8 T Cell Response Modulation

Differentiation of CD4+ T cells into a specific lineage is determined in part by innate immune response, co-stimulatory signals, and the cytokine profile [134]. Signaling cross talk between local DCs, SNECs, and resident innate immune cells, including eosinophils, mast cells, NK cells, and macrophages, generates the cytokines that drive the T cell differentiation [135, 136]. In addition, circulating innate lymphoid cells (ILCs) migrate to the local site of immune stimulus and also play a role. These cells are presumably responding to chemokine homing signals emanating from resident cells and are termed "innate" because they recognize foreign substances via PRRs rather than immunoglobulin or T cell receptors. Capable of responding rapidly, ILCs bridge innate and adaptive immunity and may play the pivotal role in orchestration of the adaptive response as Th1, Th2, and Th17 ILC subsets have been described [137]. In terms of pathology, exceptionally high levels of ILC2s have been observed in polyp homogenates from Western CRSwNP patients [138, 139] (see Fig. 4.3).

Innate lymphoid cells play a key role in orchestrating Th1, Th2, and Th17 responses.

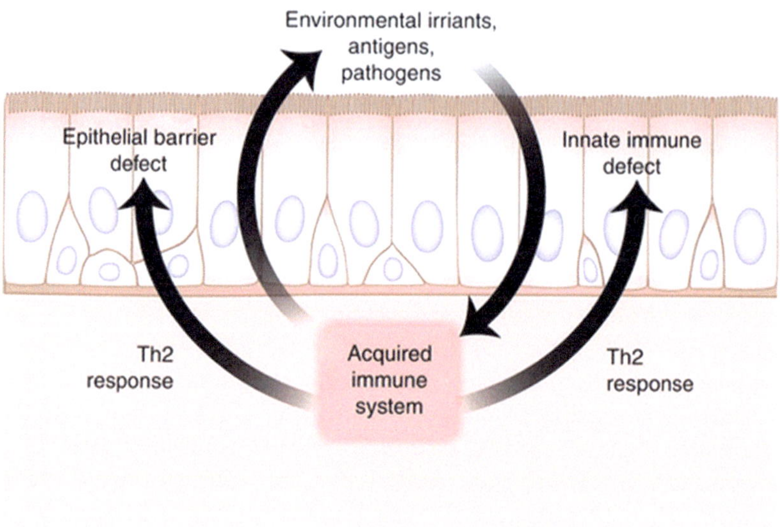

Fig. 4.3 Environmental agents stimulate the immune system inciting an innate response. If strong enough, an adaptive response is recruited as well. Typical protective responses include activation of the T1 and T3 pathways which includes the Th1 and Th17 subsets. If a T2 response (with Th2 or Treg activation) is generated, the innate response may be suppressed

4.3.9 NKT Cells, NK Cells, Cytotoxic T Cells, and Memory T Cells

In addition to the Th subsets discussed above, other T cell subsets play a role in mucosal immunity. Naïve CD8+ T cells differentiate and proliferate following exposure to antigen presented by DCs. Cytotoxic T cells are generated whose primary function is to eliminate intracellular microbes mainly by killing infected cells. These infected cells display microbial antigens on their surface, which the T cells recognize via their T cell receptors (TCR). Although not technically T cells, NK cells have a function similar to cytotoxic T cells but lack TCRs, recognizing foreign proteins by PRRs on their surface. NKT cells have characteristics of both T cells and NK cells with TCRs but with limited variability. Memory T cells are generated along with the effector T subsets and are numerically the predominant subset in nasal polyps [140]. These cells are present in the mucosa and respond to subsequent antigen challenge.

4.4 Conclusion

The sinonasal mucosal defenses are a highly sophisticated interplay involving the local structural cells, resident innate response cells, and circulating innate and adaptive immune cells. In approximately 10% of the Western population, this system fails in that foreign agents, while still cleared, trigger collateral inflammation of the mucosa of varying types and intensities. The associated clinical syndrome is broadly termed "CRS." Recent research in the field of CRS has been geared toward a better understanding of the specific pathway defects in the host. These specific genetic and epigenetic defects in the local immunologic pathways should eventually be associated with the various CRS phenotypes. Ultimately, greater understanding of sinonasal immune defenses will lead to more effective therapies for CRS in the future.

References

1. Fokkens W, Desrosiers M, Harvey R, Hopkins C, Mullol J, Philpott C, et al. EPOS2020: development strategy and goals for the latest European position paper on rhinosinusitis. Rhinology. 2019;57(3):162–8.
2. Yee KK, Pribitkin EA, Cowart BJ, Vainius AA, Klock CT, Rosen D, et al. Neuropathology of the olfactory mucosa in chronic rhinosinusitis. Am J Rhinol Allergy. 2010;24(2):110–20.
3. Wagenmann M, Naclerio R. Anatomic and physiologic considerations in sinusitis. J Allergy Clin Immunol. 1992;90(3):419–23.
4. Busse W, Holgate ST. Asthma and rhinitis. 2nd Ed. Oxford: Blackwell Science Ltd; 2000.

5. Shami SG, Martinez LA, Evans MJ. The role of migrating inflammatory cells in proliferation of lung interstitium and epithelium. Chest. 1986;89(3):170S–3S.
6. Sleigh MA, Blake JR, Liron N. The propulsion of mucus by cilia. Am Rev Respir Dis. 1988;137(3):726–41.
7. Antunes MB, Gudis DA, Cohen NA. Epithelium, cilia, and mucus: their importance in chronic rhinosinusitis. Immunol Allergy Clin N Am. 2009;29(4):631–43.
8. Houdret N, Ramphal R, Scharfman A, Perini J-M, Filliat M, Lamblin G, et al. Evidence for the in vivo degradation of human respiratory mucins during Pseudomonas aeruginosa infection. Biochim Biophys Acta Gen Subj. 1989;992(1):96–105.
9. Satir P, Sleigh MA. The physiology of cilia and mucociliary interactions. Annu Rev Physiol. 1990;52(1):137–55.
10. Gheber L, Priel Z. Synchronization between beating cilia. Biophys J. 1989;55(1):183–91.
11. Braiman A, Priel Z. Efficient mucociliary transport relies on efficient regulation of ciliary beating. Respir Physiol Neurobiol. 2008;163(1–3):202–7.
12. Messerklinger W. On the drainage of the human paranasal sinuses under normal and pathological conditions. 1. Monatsschr Ohrenheilkd Laryngorhinol. 1966;100(1–2):56–68.
13. Stucker FJ. Rhinology and facial plastic surgery. Berlin: Springer; 2009. p. xxix, 946 pp.
14. Gutman M, Houser S. Iatrogenic maxillary sinus recirculation and beyond. Ear Nose Throat J. 2003;82(1):61–3.
15. Cohen NA. Sinonasal mucociliary clearance in health and disease. Ann Otol Rhinol Laryngol Suppl. 2006;115(9_suppl):20–6.
16. Noone PG, Leigh MW, Sannuti A, Minnix SL, Carson JL, Hazucha M, et al. Primary ciliary dyskinesia. Am J Respir Crit Care Med. 2004;169(4):459–67.
17. Wine JJ. The genesis of cystic fibrosis lung disease. J Clin Investig. 1999;103(3):309–12.
18. Hadfield PJ, Rowe-Jones JM, Mackay IS. The prevalence of nasal polyps in adults with cystic fibrosis. Clin Otolaryngol. 2000;25(1):19–22.
19. Cimmino M, Cavaliere M, Nardone M, Plantulli A, Orefice A, Esposito V, et al. Clinical characteristics and genotype analysis of patients with cystic fibrosis and nasal polyposis. Clin Otolaryngol Allied Sci. 2003;28(2):125–32.
20. Houtmeyers E, Gosselink R, Gayan-Ramirez G, Decramer M. Regulation of mucociliary clearance in health and disease. Eur Respir J. 1999;13(5):1177.
21. Baumgarten CR, Togias AG, Naclerio RM, Lichtenstein LM, Norman PS, Proud D. Influx of kininogens into nasal secretions after antigen challenge of allergic individuals. J Clin Investig. 1985;76(1):191–7.
22. Stammberger H, Posawetz W. Functional endoscopic sinus surgery. Eur Arch Otorhinolaryngol. 1990;247(2):63–76.
23. Lane AP, Pine HS, Pillsbury HC. Allergy testing and immunotherapy in an academic otolaryngology practice: a 20-year review. Otolaryngol Head Neck Surg. 2001;124(1):9–15.
24. Cohen NA, Zhang S, Sharp DB, Tamashiro E, Chen B, Sorscher EJ, et al. Cigarette smoke condensate inhibits transepithelial chloride transport and ciliary beat frequency. Laryngoscope. 2009;119(11):2269–74.
25. Dye JA, Adler KB. Effects of cigarette smoke on epithelial cells of the respiratory tract. Thorax. 1994;49(8):825–34.
26. Karaman M, Tek A. Deleterious effect of smoking and nasal septal deviation on mucociliary clearance and improvement after septoplasty. Am J Rhinol Allergy. 2009;23(1):2–7.
27. Goldstein-Daruech N, Cope EK, Zhao K-Q, Vukovic K, Kofonow JM, Doghramji L, et al. Tobacco smoke mediated induction of sinonasal microbial biofilms. PLoS One. 2011;6(1):e15700.
28. García-Rodríguez JF, Corominas M, Fernández-Viladrich P, Monfort JL, Dicenta M. Rhinosinusitis and atopy in patients infected with HIV. Laryngoscope. 1999;109(6):939–44.
29. Tomassen P, Newson RB, Hoffmans R, Lötvall J, Cardell LO, Gunnbjörnsdóttir M, et al. Reliability of EP3OS symptom criteria and nasal endoscopy in the assessment of chronic rhinosinusitis - a GA2LEN study. Allergy. 2010;66(4):556–61.
30. Chen B, Shaari J, Claire SE, Palmer JN, Chiu AG, Kennedy DW, et al. Altered sinonasal ciliary dynamics in chronic rhinosinusitis. Am J Rhinol. 2006;20(3):325–9.
31. Al-Rawi MM, Edelstein DR, Erlandson RA. Changes in nasal epithelium in patients with severe chronic sinusitis: a clinicopathologic and electron microscopic study. Laryngoscope. 1998;108(12):1816–23.
32. Chen B, Antunes MB, Claire SE, Palmer JN, Chiu AG, Kennedy DW, et al. Reversal of chronic rhinosinusitis-associated sinonasal ciliary dysfunction. Am J Rhinol. 2007;21(3):346–53.
33. Yeh T-h, Su M-c, Hsu C-j, Chen Y-h, Lee S-y. Epithelial cells of nasal mucosa express functional gap junctions of Connexin 43. Acta Otolaryngol. 2003;123(2):314–20.
34. Vermeer PD, Einwalter LA, Moninger TO, Rokhlina T, Kern JA, Zabner J, et al. Segregation of receptor and ligand regulates activation of epithelial growth factor receptor. Nature. 2003;422(6929):322–6.
35. Takano K-i, Kojima T, Go M, Murata M, Ichimiya S, Himi T, et al. HLA-DR- and CD11c-positive dendritic cells penetrate beyond well-developed epithelial tight junctions in human nasal mucosa of allergic rhinitis. J Histochem Cytochem. 2005;53(5):611–9.
36. Yeo N-K, Jang YJ. Rhinovirus infection-induced alteration of tight junction and adherens junction components in human nasal epithelial cells. Laryngoscope. 2009;120(2):346–52.
37. Pedersen M, Sakakura Y, Winther B, Brofeldt S, Mygind N. Nasal mucociliary transport, number

of ciliated cells, and beating pattern in naturally acquired common colds. Eur J Respir Dis Suppl. 1983;128(Pt 1):355–65.
38. Glorieux S, Bachert C, Favoreel HW, Vandekerckhove AP, Steukers L, Rekecki A, et al. Herpes simplex virus type 1 penetrates the basement membrane in human nasal respiratory mucosa. PLoS One. 2011;6(7):e22160.
39. Rossmann MG. Viral cell recognition and entry. Protein Sci. 1994;3(10):1712–25.
40. Zulianello L, Canard C, Köhler T, De C, Lacroix J-S, Meda P. Rhamnolipids are virulence factors that promote early infiltration of primary human airway epithelia by Pseudomonas aeruginosa. Infect Immun. 2006;74(6):3134–47.
41. Tieu DD, Kern RC, Schleimer RP. Alterations in epithelial barrier function and host defense responses in chronic rhinosinusitis. J Allergy Clin Immunol. 2009;124(1):37–42.
42. Jeffery PK, Haahtela T. Allergic rhinitis and asthma: inflammation in a one-airway condition. BMC Pulm Med. 2006;6(S1):S5.
43. Xiao C, Puddicombe SM, Field S, Haywood J, Broughton-Head V, Puxeddu I, et al. Defective epithelial barrier function in asthma. J Allergy Clin Immunol. 2011;128(3):549–56.e12.
44. Zuckerman JD, Lee WY, DelGaudio JM, Moore CE, Nava P, Nusrat A, et al. Pathophysiology of nasal polyposis: the role of Desmosomal junctions. Am J Rhinol. 2008;22(6):589–97.
45. Liu S, Yang W, Shen L, Turner JR, Coyne CB, Wang T. Tight junction proteins Claudin-1 and Occludin control hepatitis C virus entry and are downregulated during infection to prevent superinfection. J Virol. 2008;83(4):2011–4.
46. Rogers GA, Beste KD, Parkos CA, Nusrat A, DelGaudio JM, Wise SK. Epithelial tight junction alterations in nasal polyposis. Int Forum Allergy Rhinol. 2011;1(1):50–4.
47. Dejima K, Randell SH, Stutts MJ, Senior BA, Boucher RC. Potential role of abnormal ion transport in the pathogenesis of chronic sinusitis. Arch Otolaryngol Head Neck Surg. 2006;132(12):1352.
48. Ordovas-Montanes J, Dwyer DF, Nyquist SK, Buchheit KM, Vukovic M, Deb C, et al. Allergic inflammatory memory in human respiratory epithelial progenitor cells. Nature. 2018;560(7720):649–54.
49. Lehmann AE, Scangas GA, Bergmark RW, El Rassi E, Stankovic KM, Metson R. Periostin and inflammatory disease: implications for chronic rhinosinusitis. Otolaryngol Head Neck Surg. 2019;160(6):965–73.
50. Janeway CA Jr, Medzhitov R. Innate immune recognition. Annu Rev Immunol. 2002;20:197–216.
51. Lane AP, Saatian B, Yu X-Y, Spannhake EW. mRNA for genes associated with antigen presentation are expressed by human middle meatal epithelial cells in culture. Laryngoscope. 2004;114(10):1827–32.
52. Claeys S, de Belder T, Holtappels G, Gevaert P, Verhasselt B, van Cauwenberge P, et al. Human beta-defensins and toll-like receptors in the upper airway. Allergy. 2003;58(8):748–53.
53. Bianchi ME, Manfredi AA. IMMUNOLOGY: dangers in and out. Science. 2009;323(5922):1683–4.
54. Meylan E, Tschopp J, Karin M. Intracellular pattern recognition receptors in the host response. Nature. 2006;442(7098):39–44.
55. Fraser IP, Koziel H, Ezekowitz RAB. The serum mannose-binding protein and the macrophage mannose receptor are pattern recognition molecules that link innate and adaptive immunity. Semin Immunol. 1998;10(5):363–72.
56. Cole AM, Liao H-I, Stuchlik O, Tilan J, Pohl J, Ganz T. Cationic polypeptides are required for antibacterial activity of human airway fluid. J Immunol. 2002;169(12):6985–91.
57. Ooi EH, Psaltis AJ, Witterick IJ, Wormald P-J. Innate immunity. Otolaryngol Clin N Am. 2010;43(3):473–87.
58. Medzhitov R, Janeway C. Innate immunity. N Engl J Med. 2000;343(5):338–44.
59. Iwasaki A, Medzhitov R. Toll-like receptor control of the adaptive immune responses. Nat Immunol. 2004;5(10):987–95.
60. Vroling AB, Fokkens WJ, van Drunen CM. How epithelial cells detect danger: aiding the immune response. Allergy. 2008;63(9):1110–23.
61. Bnd F, Philpott DJ. Recognition of Staphylococcus aureus by the innate immune system. Clin Microbiol Rev. 2005;18(3):521–40.
62. Månsson A, Bogefors J, Cervin A, Uddman R, Cardell LO. NOD-like receptors in the human upper airways: a potential role in nasal polyposis. Allergy. 2011;66(5):621–8.
63. Schröder M, Bowie AG. An arms race: innate antiviral responses and counteracting viral strategies. Biochem Soc Trans. 2007;35(6):1512–4.
64. Freund JR, Mansfield CJ, Doghramji LJ, Adappa ND, Palmer JN, Kennedy DW, et al. Activation of airway epithelial bitter taste receptors by Pseudomonas aeruginosa quinolones modulates calcium, cyclic-AMP, and nitric oxide signaling. J Biol Chem. 2018;293(25):9824–40.
65. Lee RJ, Cohen NA. Bitter and sweet taste receptors in the respiratory epithelium in health and disease. J Mol Med (Berl). 2014;92(12):1235–44.
66. Cohen NA. The genetics of the bitter taste receptor T2R38 in upper airway innate immunity and implications for chronic rhinosinusitis. Laryngoscope. 2017;127(1):44–51.
67. Hershenson MB. Proteases and protease-activated receptors signalling: at the crossroads of acquired and innate immunity. Clin Exp Allergy. 2007;37(7):963–6.
68. Rudack C, Steinhoff M, Mooren F, Buddenkotte J, Becker K, von Eiff C, et al. PAR-2 activation regulates IL-8 and GRO-? synthesis by NF-?B, but not RANTES, IL-6, eotaxin or TARC expression in nasal epithelium. Clin Exp Allergy. 2007;37(7):1009–22.

69. Shin S-H, Lee Y-H, Jeon C-H. Protease-dependent activation of nasal polyp epithelial cells by airborne fungi leads to migration of eosinophils and neutrophils. Acta Otolaryngol. 2006;126(12):1286–94.
70. Briot A, Deraison C, Lacroix M, Bonnart C, Robin A, Besson C, et al. Kallikrein 5 induces atopic dermatitis–like lesions through PAR2-mediated thymic stromal lymphopoietin expression in Netherton syndrome. J Exp Med. 2009;206(5):1135–47.
71. van Wetering S, Tjabringa GS, Hiemstra PS. Interactions between neutrophil-derived antimicrobial peptides and airway epithelial cells. J Leukoc Biol. 2004;77(4):444–50.
72. Turner J, Cho Y, Dinh N-N, Waring AJ, Lehrer RI. Activities of LL-37, a Cathelin-associated antimicrobial peptide of human neutrophils. Antimicrob Agents Chemother. 1998;42(9):2206–14.
73. Yang D, Chen Q, Schmidt AP, Anderson GM, Wang JM, Wooters J, et al. Ll-37, the neutrophil granule–and epithelial cell–derived cathelicidin, utilizes formyl peptide receptor–like 1 (Fprl1) as a receptor to chemoattract human peripheral blood neutrophils, monocytes, and T cells. J Exp Med. 2000;192(7):1069–74.
74. Crouch E, Hartshorn K, Ofek I. Collectins and pulmonary innate immunity. Immunol Rev. 2000;173(1):52–65.
75. Postle AD, Mander A, Reid KBM, Wang J-Y, Wright SM, Moustaki M, et al. Deficient hydrophilic lung surfactant proteins A and D with normal surfactant phospholipid molecular species in cystic fibrosis. Am J Respir Cell Mol Biol. 1999;20(1):90–8.
76. Ooi EH, Wormald P-J, Carney AS, James CL, Tan LW. Surfactant protein D expression in chronic rhinosinusitis patients and immune responses in vitro to Aspergillus and Alternaria in a nasal explant model. Laryngoscope. 2007;117(1):51–7.
77. Sims MW, Tal-Singer RM, Kierstein S, Musani AI, Beers MF, Panettieri RA, et al. Chronic obstructive pulmonary disease and inhaled steroids alter surfactant protein D (SP-D) levels: a cross-sectional study. Respir Res. 2008;9(1):13.
78. Kern RC, Conley DB, Walsh W, Chandra R, Kato A, Tripathi-Peters A, et al. Perspectives on the etiology of chronic rhinosinusitis: an immune barrier hypothesis. Am J Rhinol. 2008;22(6):549–59.
79. Ramanathan M, Lee W-K, Spannhake EW, Lane AP. Th2 cytokines associated with chronic rhinosinusitis with polyps Down-regulate the antimicrobial immune function of human sinonasal epithelial cells. Am J Rhinol. 2008;22(2):115–21.
80. Wolk K, Witte E, Wallace E, Döcke W-D, Kunz S, Asadullah K, et al. IL-22 regulates the expression of genes responsible for antimicrobial defense, cellular differentiation, and mobility in keratinocytes: a potential role in psoriasis. Eur J Immunol. 2006;36(5):1309–23.
81. Aujla SJ, Chan YR, Zheng M, Fei M, Askew DJ, Pociask DA, et al. IL-22 mediates mucosal host defense against gram-negative bacterial pneumonia. Nat Med. 2008;14(3):275–81.
82. Pickert G, Neufert C, Leppkes M, Zheng Y, Wittkopf N, Warntjen M, et al. STAT3 links IL-22 signaling in intestinal epithelial cells to mucosal wound healing. J Exp Med. 2009;206(7):1465–72.
83. Peters AT, Kato A, Zhang N, Conley DB, Suh L, Tancowny B, et al. Evidence for altered activity of the IL-6 pathway in chronic rhinosinusitis with nasal polyps. J Allergy Clin Immunol. 2010;125(2):397–403.e10.
84. Bachert C, Wagenmann M, Rudack C, Höpken K, Hiltebrandt M, Wang D, et al. The role of cytokines in infectious sinusitis and nasal polyposis. Allergy. 1998;53(1):2–13.
85. Kowalski ML, Lewandowska-Polak A, Wozniak J, Ptasinska A, Jankowski A, Wagrowska-Danilewicz M, et al. Association of stem cell factor expression in nasal polyp epithelial cells with aspirin sensitivity and asthma. Allergy. 2005;60(5):631–7.
86. Lu X, Zhang XH, Wang H, Long XB, You XJ, Gao QX, et al. Expression of osteopontin in chronic rhinosinusitis with and without nasal polyps. Allergy. 2009;64(1):104–11.
87. Nishi Y. Glucocorticoids suppress NF-kB activation induced by LPS and PGN in paranasal sinus epithelial cells. Rhinology. 2009;47(4):413–8.
88. Hammad H, Lambrecht BN. Dendritic cells and epithelial cells: linking innate and adaptive immunity in asthma. Nat Rev Immunol. 2008;8(3):193–204.
89. Kato A, Peters A, Suh L, Carter R, Harris KE, Chandra R, et al. Evidence of a role for B cell–activating factor of the TNF family in the pathogenesis of chronic rhinosinusitis with nasal polyps. J Allergy Clin Immunol. 2008;121(6):1385–92.e2.
90. Schleimer RP. Glucocorticoids suppress inflammation but spare innate immune responses in airway epithelium. Proc Am Thorac Soc. 2004;1(3):222–30.
91. Bobic S, van Drunen CM, Callebaut I, Hox V, Jorissen M, Fokkens WJ, et al. Dexamethasone-induced apoptosis of freshly isolated human nasal epithelial cells concomitant with abrogation of IL-8 production. Rhinology. 2010;48(4):401–7.
92. Kurosawa S, Myers AC, Chen L, Wang S, Ni J, Plitt JR, et al. Expression of the costimulatory molecule B7-H2 (inducible costimulator ligand) by human airway epithelial cells. Am J Respir Cell Mol Biol. 2003;28(5):563–73.
93. Kim J, Myers AC, Chen L, Pardoll DM, Truong-Tran Q-A, Lane AP, et al. Constitutive and inducible expression of B7 family of ligands by human airway epithelial cells. Am J Respir Cell Mol Biol. 2005;33(3):280–9.
94. Heinecke L, Proud D, Sanders S, Schleimer RP, Kim J. Induction of B7-H1 and B7-DC expression on airway epithelial cells by the toll-like receptor 3 agonist double-stranded RNA and human rhinovirus infection: in vivo and in vitro studies. J Allergy Clin Immunol. 2008;121(5):1155–60.

95. Avila PC, Schleimer RP. Airway epithelium. Allergy and allergic diseases. Hoboken: Wiley-Blackwell; 2009. p. 366–97.
96. Lundberg JON, Farkas-Szallasi T, Weitzberg E, Rinder J, Lidholm J, Änggåard A, et al. High nitric oxide production in human paranasal sinuses. Nat Med. 1995;1(4):370–3.
97. Martinez FO, Helming L, Gordon S. Alternative activation of macrophages: an immunologic functional perspective. Annu Rev Immunol. 2009;27(1):451–83.
98. Claeys S, Van Hoecke H, Holtappels G, Gevaert P, De Belder T, Verhasselt B, et al. Nasal polyps in patients with and without cystic fibrosis: a differentiation by innate markers and inflammatory mediators. Clin Exp Allergy. 2005;35(4):467–72.
99. Krysko O, Holtappels G, Zhang N, Kubica M, Deswarte K, Derycke L, et al. Alternatively activated macrophages and impaired phagocytosis of S. aureus in chronic rhinosinusitis. Allergy. 2010;66(3):396–403.
100. Marseglia GL, Pagella F, Klersy C, Barberi S, Licari A, Ciprandi G. The 10-day mark is a good way to diagnose not only acute rhinosinusitis but also adenoiditis, as confirmed by endoscopy. Int J Pediatr Otorhinolaryngol. 2007;71(4):581–3.
101. Polzehl D, Moeller P, Riechelmann H, Perner S. Distinct features of chronic rhinosinusitis with and without nasal polyps. Allergy. 2006;61(11):1275–9.
102. Lee JJ, Jacobsen EA, McGarry MP, Schleimer RP, Lee NA. Eosinophils in health and disease: the LIAR hypothesis. Clin Exp Allergy. 2010;40(4):563–75.
103. Harlin S, Ansel D, Lane S, Myers J, Kephart G, Gleich G. A clinical and pathologic study of chronic sinusitis: the role of the eosinophil. J Allergy Clin Immunol. 1988;81(5):867–75.
104. Bhattacharyya N, Vyas DK, Fechner FP, Gliklich RE, Metson R. Tissue eosinophilia in chronic sinusitis. Arch Otolaryngol Head Neck Surg. 2001;127(9):1102.
105. Szucs E, Ravandi S, Goossens A, Beel M, Clement PAR. Eosinophilia in the ethmoid mucosa and its relationship to the severity of inflammation in chronic rhinosinusitis. Am J Rhinol. 2002;16(3):131–4.
106. Jankowski R, Bene MC, Moneret-Vautrin AD, Haas F, Faure G, Simon C, et al. Immunohistological characteristics of nasal polyps. A comparison with healthy mucosa and chronic sinusitis. Rhinol Suppl. 1989;8:51–8.
107. Jahnsen FL, Haye R, Gran E, Brandtzaeg P, Johansen FE. Glucocorticosteroids inhibit mRNA expression for eotaxin, eotaxin-2, and monocyte-chemotactic protein-4 in human airway inflammation with eosinophilia. J Immunol. 1999;163(3):1545–51.
108. Meyer JE, Bartels J, Görögh T, Sticherling M, Rudack C, Ross DA, et al. The role of RANTES in nasal polyposis. Am J Rhinol. 2005;19(1):15–20.
109. Mullol J, Roca-Ferrer J, Alobid I, Pujols L, Valero A, Xaubet A, et al. Effect of desloratadine on epithelial cell granulocyte-macrophage colony-stimulating factor secretion and eosinophil survival. Clin Exp Allergy. 2006;36(1):52–8.
110. Matsukura S, Stellato C, Plitt JR, Bickel C, Miura K, Georas SN, et al. Activation of eotaxin gene transcription by NF-kappa B and STAT6 in human airway epithelial cells. J Immunol. 1999;163(12):6876–83.
111. Kuperman D, Schleimer R. Interleukin-4, Interleukin-13, signal transducer and activator of transcription factor 6, and allergic asthma. Curr Mol Med. 2008;8(5):384–92.
112. Pérez-Novo CA, Claeys C, Van Zele T, Holtapples G, Van Cauwenberge P, Bachert C. Eicosanoid metabolism and eosinophilic inflammation in nasal polyp patients with immune response to Staphylococcus aureus enterotoxins. Am J Rhinol. 2006;20(4):456–60.
113. Van Zele T, Coppieters F, Gevaert P, Holtappels G, Van Cauwenberge P, Bachert C. Local complement activation in nasal polyposis. Laryngoscope. 2009;119(9):1753–8.
114. Buysschaert ID, Grulois V, Eloy P, Jorissen M, Rombaux P, Bertrand B, et al. Genetic evidence for a role of IL33 in nasal polyposis. Allergy. 2010;65(5):616–22.
115. Foreman A, Holtappels G, Psaltis AJ, Jervis-Bardy J, Field J, Wormald PJ, et al. Adaptive immune responses in Staphylococcus aureus biofilm-associated chronic rhinosinusitis. Allergy. 2011;66(11):1449–56.
116. Stone KD, Prussin C, Metcalfe DD. IgE, mast cells, basophils, and eosinophils. J Allergy Clin Immunol. 2010;125(2):S73–80.
117. Pawankar R, Lee KH, Nonaka M, Takizawa R. Role of mast cells and basophils in chronic rhinosinusitis. Chronic Rhinosinusitis. CRC Press; 2007. p. 109–18.
118. Balzar S, Strand M, Rhodes D, Wenzel SE. IgE expression pattern in lung: relation to systemic IgE and asthma phenotypes. J Allergy Clin Immunol. 2007;119(4):855–62.
119. Pérez Novo CA, Jedrzejczak-Czechowicz M, Lewandowska-Polak A, Claeys C, Holtappels G, Van Cauwenberge P, et al. T cell inflammatory response, Foxp3 and TNFRS18-L regulation of peripheral blood mononuclear cells from patients with nasal polyps-asthma after staphylococcal superantigen stimulation. Clin Exp Allergy. 2010;40(9):1323–32.
120. Hammad H, Lambrecht BN. Dendritic cells and airway epithelial cells at the interface between innate and adaptive immune responses. Allergy. 2011;66(5):579–87.
121. Cerutti A, Chen K, Chorny A. Immunoglobulin responses at the mucosal interface. Annu Rev Immunol. 2011;29(1):273–93.
122. Chen K, Cerutti A. New insights into the enigma of immunoglobulin D. Immunol Rev. 2010;237(1):160–79.
123. Pleass RJ, Lang ML, Kerr MA, Woof JM. IgA is a more potent inducer of NADPH oxidase activation

and degranulation in blood eosinophils than IgE. Mol Immunol. 2007;44(6):1401–8.
124. Pant H, Beroukas D, Kette FE, Smith WB, Wormald PJ, Macardle PJ. Nasal polyp cell populations and fungal-specific peripheral blood lymphocyte proliferation in allergic fungal sinusitis. Am J Rhinol Allergy. 2009;23(5):453–60.
125. Chen K, Xu W, Wilson M, He B, Miller NW, Bengtén E, et al. Immunoglobulin D enhances immune surveillance by activating antimicrobial, proinflammatory and B cell–stimulating programs in basophils. Nat Immunol. 2009;10(8):889–98.
126. Smurthwaite L, Durham SR. Local ige synthesis in allergic rhinitis and asthma. Curr Allergy Asthma Rep. 2002;2(3):231–8.
127. Gevaert P, Holtappels G, Johansson SGO, Cuvelier C, Cauwenberge P, Bachert C. Organization of secondary lymphoid tissue and local IgE formation to Staphylococcus aureus enterotoxins in nasal polyp tissue. Allergy. 2005;60(1):71–9.
128. Bachert C, Claeys SEM, Tomassen P, van Zele T, Zhang N. Rhinosinusitis and asthma: a link for asthma severity. Curr Allergy Asthma Rep. 2010;10(3):194–201.
129. Delves PJ, Roitt IM. The immune system. N Engl J Med. 2000;343(1):37–49.
130. Jones SA. Directing transition from innate to acquired immunity: defining a role for IL-6. J Immunol. 2005;175(6):3463–8.
131. Schleimer RP, Kato A, Peters A, Conley D, Kim J, Liu MC, et al. Epithelium, inflammation, and immunity in the upper airways of humans: studies in chronic rhinosinusitis. Proc Am Thorac Soc. 2009;6(3):288–94.
132. Miller LS, Cho JS. Immunity against Staphylococcus aureus cutaneous infections. Nat Rev Immunol. 2011;11(8):505–18.
133. Tato CM, O'Shea JJ. What does it mean to be just 17? Nature. 2006;441(7090):166–7.
134. Fokkens WJ, Lund VJ, Mullol J, Bachert C, Alobid I, Baroody F, et al. EPOS 2012: European position paper on rhinosinusitis and nasal polyps 2012. A summary for otorhinolaryngologists. Rhinology. 2012;50(1):1–12.
135. Miller SA, Weinmann AS. Common themes emerge in the transcriptional control of T helper and developmental cell fate decisions regulated by the T-box, GATA and ROR families. Immunology. 2009;126(3):306–15.
136. Zhu J, Yamane H, Paul WE. Differentiation of effector CD4 T cell populations. Annu Rev Immunol. 2010;28(1):445–89.
137. Spits H, Di Santo JP. The expanding family of innate lymphoid cells: regulators and effectors of immunity and tissue remodeling. Nat Immunol. 2010;12(1):21–7.
138. Allakhverdi Z, Comeau MR, Smith DE, Toy D, Endam LM, Desrosiers M, et al. CD34+ hemopoietic progenitor cells are potent effectors of allergic inflammation. J Allergy Clin Immunol. 2009;123(2):472–8.e1.
139. Mjösberg JM, Trifari S, Crellin NK, Peters CP, van Drunen CM, Piet B, et al. Human IL-25- and IL-33-responsive type 2 innate lymphoid cells are defined by expression of CRTH2 and CD161. Nat Immunol. 2011;12(11):1055–62.
140. Sanchez-Segura A, Brieva JA, Rodriguez C. T lymphocytes that infiltrate nasal polyps have a specialized phenotype and produce a mixed TH1/TH2 pattern of cytokines. J Allergy Clin Immunol. 1998;102(6 Pt 1):953–60.

Local B-Cell and T-Cell Populations in the Pathophysiology of Chronic Rhinosinusitis with Nasal Polyposis

5

Kent K. Lam and Amber U. Luong

Core Messages

- The local immune system in the airway epithelium of the nasal cavities and paranasal sinuses is well developed with contribution from both innate and adaptive arms to generate different phenotypic and endotypic manifestations of chronic rhinosinusitis.
- T cells consist of a diverse array of subpopulations, including CD8+ cytotoxic T cells and CD4+ helper T cells with Type 1, Type 2, and Type 3 inflammatory responses, each of which participates in the inflammatory cascade through distinct pathomechanisms.
- The plasticity of T cells and resultant inflammatory responses are influenced by a combination of environmental and genetic factors.
- Nasal polyps generally function as tertiary lymphoid organs that demonstrate germinal center-like centers with high activity and differentiation of B cells and the release of immunoglobulins.
- Endotyping of CRSwNP is based on characteristic immunopathologic profiles and has helped improve clinical classifications of CRSwNP, but the process continues to require additional research to fully understand the diverse network of immunologic pathways that lead to sinonasal inflammation and thus nasal polyp development.

5.1 Introduction

Chronic rhinosinusitis (CRS) is a complex and heterogeneous inflammatory disease that involves the mucosal lining of the paranasal sinuses and results in a constellation of symptoms, including nasal congestion, discolored nasal drainage, facial pressure, and smell alterations, for a duration of at least 12 weeks [1, 2]. CRS is typically divided into two phenotypes based on the presence or absence of nasal polyps: CRS with nasal polyposis (CRSwNP) and CRS without nasal polyposis (CRSsNP). Over the past 30 years, research efforts have focused on the inflammatory mechanisms that drive the development and persistence of the two CRS phenotypes. CRSwNP and CRSsNP have traditionally been characterized by distinct immunologic patterns. CRSsNP has classically been characterized by Type 1 inflammation with the presence of interferon (IFN)-γ, T helper 1 (Th1) cells, and neutrophils. CRSwNP, in contrast, has routinely been associated with a skewing

K. K. Lam
Department of Otolaryngology—Head and Neck Surgery, Eastern Virginia Medical School, Norfolk, VA, USA
e-mail: lamkk@evms.edu

A. U. Luong (✉)
Department of Otorhinolaryngology—Head and Neck Surgery, McGovern Medical School, University of Texas Health Science Center at Houston, Houston, TX, USA
e-mail: amber.u.luong@uth.tmc.edu

Ö. Ö. Celebi, T. M. Önerci (eds.), *Nasal Physiology and Pathophysiology of Nasal Disorders*,
https://doi.org/10.1007/978-3-031-12386-3_5

toward Type 2 inflammatory reactions consisting of Type 2 cytokines, T helper 2 (Th2) cells, and eosinophils [3, 4]. The general use of Types 1 and 2 inflammatory profiles to characterize the immunologic expression in CRSsNP and CRSwNP, however, has increasingly become insufficient to fully describe CRS pathogenesis, as a wide diversity of immunopathological mechanisms leading to sinonasal inflammation has been identified.

The ongoing research on CRS pathogenesis highlights the growing emphasis placed upon inflammatory endotypes for particularly CRSwNP, which have become clinically relevant to disease management due to inherent prognostic implications with the development of targeted therapeutics. The immune cells in the adaptive immune system have specifically offered the most consequential breakthroughs in the clinical management of CRSwNP, although both the innate and adaptive arms provide fundamental contributions to the pathophysiology of CRSwNP and exhibit a high degree of cross talk in propagating sinonasal inflammation. This chapter serves as a brief review of the recent insights into the local dysregulation attributed to T cells and B cells, which are central to adaptive immune responses, in the pathogenesis of nasal polyps.

Table 5.1 T-cell subsets and functions in nasal polyps

T-cell subset	Pro-inflammatory roles in nasal polyp formation
CD8+ T cells	• Generate and release cytokines that mediate Type 1, Type 2, and Type 3 inflammatory responses
CD4+ T helper 1 cells	• Generate and release cytokines that mediate Type 1 inflammatory responses in non-eosinophilic and eosinophilic nasal polyps
CD4+ T helper 2 cells	• Generate and release cytokines that mediate Type 2 inflammatory responses • Activate and attract eosinophils to local tissue • Interact with myeloid dendritic cells, innate lymphoid cells, and epithelial-cell derived cytokines to direct Th2 cell polarization • Respond to *Staphylococcus aureus* enterotoxins to direct Th2 cell polarization
CD4+ T helper 17 cells	• Generate and release cytokines that mediate Type 3 inflammatory responses in non-eosinophilic and eosinophilic nasal polyps
CD4+ T follicular helper cells	• Promote extrafollicular B-cell differentiation in nasal polyp tissue
CD4+ regulatory T cells	• Promote inflammatory responses when quantitative or qualitative levels are diminished

5.2 Role of T Cells in CRSwNP Pathophysiology

T cells coordinate the cell-mediated responses in adaptive immunity, maintain the balance between the humoral and cell-mediated pathways, and provide immune regulation through negative feedback. T cells are further subdivided into CD4+ T cells and CD8+ T cells with specific roles in adaptive immunity (Table 5.1). CD8+ T cells, which are commonly referred to as cytotoxic T cells, provide immune defense against intracellular pathogens and tumor-transformed cells. Activated CD8+ T cells regulate cellular toxicity by secreting cytokines, primarily tumor necrosis factor (TNF)-α and IFN-γ, releasing cytotoxic granules with perforin and granzymes, and mediating apoptosis through interactions between their surface Fas ligands and Fas molecules expressed on targeted cells. CD4+ T cells are the classical T helper cells, which consist of several subtypes, including Th1 cells, Th2 cells, T helper 17 cells (Th17), T follicular helper cells (Tfh), and regulatory T cells (Treg). Upon their activation and differentiation, CD4+ T cells modulate the activity of other effector cells involved with the innate, cell-mediated, and humoral immune systems via the secretion of cytokines that contribute to the local inflammatory milieu.

T-cell populations have been well studied for their role in the inflammation of CRSwNP, as T cells are believed to be the predominant lymphocyte population in the mucosa of nasal polyps and thus key drivers of the disease. Overall, the nasal tissues in both CRSwNP and CRSsNP are characterized by significant increases in T-cell counts, but variations in the pathophysiology and thus clinical features between CRSwNP and CRSsNP reflect and likely arise from immunological processes modulated by different populations of T-cell subtypes [5]. In their flow cytometry study, for example, Derycke et al. showed a higher frequency of T cells in the sinonasal mucosa of CRSwNP patients than in the mucosa of both CRSsNP and healthy control patients [6]. Conversely, tissue from CRSsNP patients did not demonstrate increased levels of T cells compared to healthy control tissue. Within the T-cell population, Th1 immune cells, as measured by intracellular IFN-γ immunostaining, composed the predominant subtype in all healthy control, CRSsNP, and even CRSwNP patients, but only CRSwNP tissue was notable for Th2 immune cells. Tissues from control, CRSsNP, and CRSwNP additionally demonstrated other populations of T helper cells, including Th17 and Tfh immune cells. Pant et al. further confirmed that CD4+ T cells are elevated in CRSwNP, but reported that two special CRSwNP subgroups—allergic fungal rhinosinusitis (AFRS) and eosinophilic mucin CRS (EMCRS)—differ from other CRSwNP subtypes with higher populations of CD8+ T cells [7]. In CRSwNP, the high percentage of CD8+ T cells, which demonstrate a terminally differentiated, mature, effector memory phenotype, has also raised the question if CD8+ T cells are involved with the pathogenesis of CRSwNP or appear as a result of inflammation [8].

While the most recognized role of CD8+ T cells involves the mediation of cellular cytotoxicity, CD8+ T cells may participate in the inflammatory pathways of CRSwNP as a source of important cytokines in the local immunologic environment. Like CD4+ T cells, CD8+ T cells have the capacity to differentiate into at least five different effector cell subsets, including Type 1 cytotoxic T cells (Tc1), Type 2 cytotoxic T cells (Tc2), IL-9-producing CD8+ T cells (Tc9), IL-17-secreting CD8+ T cells (Tc17), and CD8+ Treg cells, each with its own distinct expression of cytokines. A study by Ma et al. investigated the cytokine-producing features and also the classic cytotoxic activity of CD8+ T cells in nasal polyp tissue [9]. The results of this study confirmed amplified infiltration of total CD8+ T cells as well increased Tc1, Tc2, and Tc17 subsets in tissue from eosinophilic and non-eosinophilic CRSwNP, when compared to sinonasal mucosa from control patients. A decreased percentage of CD8+ Treg cells in the isolated polyp tissue, however, was also reported. The CD8+ T cells from nasal polyps produced similar or even higher levels of Type 1, Type 2, and Type 3 cytokines compared with their CD4+ T-cell counterparts. Type 3 immunity encompasses innate and adaptive immune responses centered around the Th17 immune cells and the production of IL-17 and IL-22. Additionally, the CD8+ T cells in nasal polyps showed diminished cytotoxic activity with reduced expression of perforin and granzymes compared with their CD8+ T-cell counterparts in the peripheral blood. CD8+ T cells may thus participate in the pathogenesis of CRSwNP by producing pro-inflammatory cytokines, as opposed to a true cytotoxic function, but further research on such a potential inflammatory pathway is warranted.

The usual methods to endotype CRS, however, has not necessarily relied on the presence of CD8+ T-cell populations, but has commonly focused on the various subtypes of CD4+ T cells underlying local inflammation [4]. The balance of T helper cells and the accompanying cytokine patterns have particularly been utilized to characterize these CRS endotypes. As previously mentioned, CRSwNP is traditionally associated with a skewed Type 2 inflammatory profile, in which Th2-associated Type 2 cytokines, including interleukin (IL)-4, IL-5, and IL-13, and immunoglobulin (Ig)-E in nasal polyp tissue, have been implicated as potential factors in disease pathogenesis. CRSsNP, in contrast, has been characterized by a relative skewing away from Type 2 inflammatory responses and toward a mixture of Type 1 and Type 3 patterns [5, 10]. The proposed linkage of Type 2 inflammatory patterns with CRSwNP, nonetheless, has not been completely absolute in certain CRSwNP populations, undermined by the presence of non-Type 2 nasal polyps in cystic fibrosis (CF) as an example [11]. These exceptions to Type 2 inflammatory nasal polyps underscore the challenges and limitations of endotyping CRSwNP subtypes with a universal dichotomous system based on Type 2 and non-Type 2 inflammatory signatures.

Additionally, geographic variations have been found to contribute to differences in CRSwNP endotypes based on the CD4+ T-cell populations, suggesting that genetic and environmental factors also influence the pathophysiology of disease [12–14]. One revealing study supporting these differences in immunologic profiles across different regions in the world was conducted by Wang et al., who evaluated the diversity of CD4+ T-cell-based cytokine profiles in CRS patients from Europe, Asia, and Australia [13]. For this investigation, nasal mucosal concentrations of Type 1, Type 2, and Type 3 cytokines in CRS patients and control subjects at different worldwide recruitment centers were specifically measured with identical methodological tools. The results showed that nasal polyp tissues from CRSwNP patients from Europe, Australia, and Japan demonstrated a Type 2-skewed endotype, whereas those from China mainly demonstrated mixed patterns involving a combination of Type 1, Type 2, and Type 3 inflammatory responses. The eosinophilic endotype, moreover, accounted for over 50% of CRSwNP patients in Europe, Australia, and Japan, whereas less than 30% of CRSwNP patients in China exhibited the eosinophilic endotype. The results from Wang et al. have emphasized the diverse variety in particularly the CD4+ T-cell population involved with CRSwNP pathophysiology. The existence of so many CD4+ T-cell subsets furthermore raises questions about the significance and functional roles of these effector cell subtypes, as summarized in the following.

5.2.1 Th1/Th17 Immune Cells in Nasal Polyps

Th1 and Th17 immune cells are involved in Type 1 and Type 3 inflammation, respectively, and are classically associated with CRSsNP, nasal polyps related to CF, and non-eosinophilic CRSwNP in the Asian population [15, 16]. IL-12 plays a significant role in mediating Th1 cell polarization, resulting in the production of such cytokines as IFN-γ, TNF-β, and IL-2 and the activation of macrophages, neutrophils, CD8+ T cells, B cells, and other downstream effector cells. Th17 cell polarization, on the other hand, is stimulated by transforming growth factor (TGF)-β, IL-23, and IL-6. The activation of Th17 immune cells upregulates the production of primarily not only IL-17 but also IL-17A, IL-22, and IL-26, all of which activate mononuclear phagocytes, recruit neutrophils, and induce epithelial antimicrobial responses [17, 18]. Th1 and Th17 cell polarizations are thus related through their predominantly neutrophilic inflammation in the local nasal polyp microenvironment.

Increasing research, nevertheless, suggests that Th1 and Th17 immune cells are also important drivers of immune signaling in not only non-eosinophilic CRSwNP, but also eosinophilic CRSwNP [16, 19]. Wang et al. have particularly shown that in CRSwNP patients, Th17 immune pathways may contribute to nasal polyp formation by enhancing the expression of Type 2 inflammatory cytokines, including IL-4 and IL-13. The cross talk between the Type 2 and Type 3 immune pathways suggests that Th17 immune cells play an activating role in Th2 cell polarization; though IL-4 and IL-13, nonetheless, have been found to inhibit Th17 immune cells [20]. While the exact pathophysiologic mechanism of Th17 immune cells in CRSwNP remains unclear, Th17 cells are known as important components of the host defense against extracellular pathogens at the mucosal barriers. Microbes and fungi have been linked to the pathogenesis of CRSwNP and AFRS, respectively, and may be influencing the activation of these Th17 responses. In addition, dysfunctional activation of Th17 cells has been linked to the pathogenesis of autoimmune diseases through the disruption of the airway epithelial barrier which may also be contributing to the pathology of CRSwNP [21].

5.2.2 Th2 Immune Cells in Nasal Polyps

Type 2 inflammation in CRSwNP, as directed by Th2 immune cells, relies on its unique cytokine profile, which is characterized by predominantly

IL-4, IL-5, and IL-13, to activate essential downstream effector cells. These effector cells in T-cell immunity include eosinophils, mast cells, and basophils. Th2 immune cells furthermore control a strong humoral immune response by promoting B-cell proliferation, immunoglobulin production, and class switching of immunoglobulins to high levels of IgE. Mucus production, goblet cell metaplasia, and airway hyperresponsiveness, regulated by the Th2 immune mediators, are also hallmarks of Th2 cell polarization. Asthma, atopic dermatitis, and CRSwNP are generally regarded as atopic comorbidities due to a high degree of pathophysiologic similarities in immunologic patterns of Type 2 inflammation occurring at the epithelial cell layers [22].

The functionality of Th2 immune cells is dependent on their initial activation and differentiation from naïve T cells, which in part relies on myeloid dendritic cells (mDCs) to serve as antigen presenting cells. While a second subset of human dendritic cells—plasmacytoid DCs—has also been recognized, plasmacytoid DCs are better known for their role in antiviral immunity and are less effective at antigen presentation. As such, mDCs play an active role in influencing the polarization of T helper cells and are thus found at significantly elevated levels in CRSwNP [23, 24]. Shi et al. have particularly evaluated the specific subsets of mDCs isolated from CRSwNP tissue and their capacity to skew naïve T cells toward either a Th2 phenotype or mixed Th1/Th17 phenotypes based on the predominance of eosinophils in the local polyp tissue [25]. In this study, elevated local Th1 and T17 immune cells were noted in both eosinophilic and non-eosinophilic CRSwNP, but only eosinophilic CRSwNP exhibited increased levels of Th2 immune cells. This Th2 cell polarization in eosinophilic CRSwNP was linked to mDCs that specifically demonstrated an upregulation of two surface markers, OX40 ligand (OX40L) and programmed death ligand-1 (PD-L1). Conversely, mDCs with low expression of OX40L and PD-L1 contributed to the Th1/Th17 cell skewing in non-eosinophilic CRSwNP. Blockade of OX40L and PD-L1 on mDCs from eosinophilic CRSwNP furthermore suppressed Th2 immune cell responses and induced a primary Th1/T17 cell polarization in the local inflammatory milieu. The mechanism by which mDCs can influence Th2 cell polarization in eosinophilic CRSwNP is thus speculated to involve a high expression of OX40L and PD-L1.

Other upstream activators of molecular and cellular mechanisms of Type 2 inflammation now increasingly include cytokines that are produced by innate lymphoid cells (ILCs), which lack antigen-specific receptors and are components of the innate immune system. Three ILC subgroups have been defined according to their profiles of secreted cytokines, which parallel the T helper subgroups in Type 1, Type 2, and Type 3 inflammation. For CRSwNP, ILC2s particularly play an important role in Th2 cell polarization once they are activated by various environmental stress signals at the sinonasal respiratory epithelium. ILC2s have been found in increased numbers in nasal polyp tissue, especially in eosinophilic CRSwNP [26]. Epithelial cell-derived cytokines, such as IL-25, IL-33, and thymic stromal lymphopoietin (TSLP), have been well studied in their potentiation of ILC2s, which induces the expression and release of downstream Type 2 inflammatory cytokines, including IL-4, IL-5, and IL-13 [27–29]. Overall, in the absence of antigen specificity, epithelial cell-derived cytokines and ILC2s effectively influence acquired immune responses that drive Type 2 inflammatory observed in CRSwNP.

Superantigenic enterotoxins produced by the *Staphylococcus aureus* bacteria, furthermore, play a role in the activation of Type 2 inflammation by amplifying local eosinophilic responses and thereby fostering nasal polyp formation [30]. Studies have demonstrated the presence of *S aureus* in a high percentage of CRSwNP patients, but not in the tissue of control or CRSsNP patients [31, 32]. These toxins trigger a massive and uncontrolled immunologic response activating as many as 30% of the T-cell population in affected individuals, compared to the 0.001% activated in a normal antigen-specific immune response [33]. By this mechanism, superantigens bypass the normal steps of antigen recognition and promote polyclonal T-cell proliferation and massive cytokine release, which in the case of typical CRSwNP has a strong Th2 immune component. Many other cell types, including B cells, are affected by this exaggerated immunologic process, resulting in a local polyclonal IgE response in nasal polyp tissue. The pathogenesis of CRSwNP thus likely involves an exogenous antigenic factor that can perpetuate Type 2 inflammatory responses.

5.2.3 T Follicular Helper Cells in Nasal Polyps

Tfh immune cells are a specific subset of CD4+ T cells located within secondary lymphoid organs, where they serve the purpose of supporting B-cell populations to produce high-affinity immunoglobulins as a part of the humoral immune response. In CRSwNP, development of lymphoid cell aggregations has been identified within nasal polyp tissue, which include B-cell and T-cell populations, leading to the formation of ectopic lymphoid structures, or tertiary lymphoid structures [34]. Here, within the germinal center-like structures, Tfh cells further promote B-cell proliferation, local cellular differentiation, and robust immunoglobulin production through somatic hypermutation and class-switch recombination. Tfh immune cells are defined by the master transcription factor B-cell lymphoma 6 (BCL-6), which controls the differentiation of Tfh cells themselves, while IL-21 is the signature Tfh cytokine represented in high levels in CRSwNP [35]. Tfh cells may particularly be stimulated by *S aureus* enterotoxins, as upregulated levels of BCL-6 and IL-21 are demonstrated when nasal polyp tissue is incubated with fragments of *S aureus* enterotoxin B (SEB) in vitro [36]. Additionally, the presence of Tfh immune cells in eosinophilic polyp tissues positively correlate with the local IgE levels, signifying the important role that Tfh immune cells play in the induction of local IgE in CRSwNP [34]. Tfh immune cells thus provide a vital link between T-cell and B-cell activity in the inflammatory responses that contribute to nasal polyp formation and perpetuation.

5.2.4 Regulatory T Cells in Nasal Polyps

Treg cells are important adaptive immune cells with the capacity to modulate immune responses, thereby maintaining tolerance to self-antigens and preventing the over-activation of the immune system. Suppression of the inflammatory responses occurs through the release of soluble cytokines TGF-β and IL-10 and through the upregulation of CTLA-4, which is a T-cell inhibiting protein [37]. Given the protective roles of Treg cells in preventing an overabundance of harmful inflammatory processes, quantitative and qualitative losses of Treg cells have been suggested as important factors that may permit nasal polyp formation in CRSwNP. Studies have demonstrated that cellular counts of Treg cells and expression of regulatory cytokines, TGF-β and IL-10, are both decreased in both eosinophilic and non-eosinophilic CRSwNP patients [9]. Likewise, nasal polyp formation may be associated with depressed levels of mRNA and protein for the gene expression of FoxP3, which is the master transcriptional regulator of Treg cells [14]. Still, other studies, including one by Miljkovic et al., have found that Treg cells are elevated in the nasal tissue of CRSwNP compared to that of CRSsNP; however, these findings may suggest that a dysfunction of Treg cells, as opposed to a drop in cellular number, also affects the inflammatory environment that drives CRSwNP [38]. Additional investigation regarding the exact role of Treg cells in CRSwNP pathogenesis is required, but the current evidence highlights their contributions through the modulation of extreme immune responses in chronic inflammatory diseases like CRSwNP.

5.3 Role of B Cells in CRSwNP Pathophysiology

As T cells have demonstrated an active role in the inflammatory responses in CRSwNP, evidence is increasing that reinforces the importance of B-cell populations in nasal polyp pathogenesis. B cells significantly contribute to the humoral response of the adaptive immune system through the production of immunoglobulins. B cells furthermore serve as antigen-presenting cells, and in response to antigen presentation, undergo differentiation and clonal selection through somatic recombination in germinal centers. Activated B cells can differentiate into immunoglobulin-secreting plasmablasts, plasma cells, or memory B cells. In addition to antibody expression and immune memory, B cells are integral to the activation of T cells through co-stimulation, antigen presentation, and propagation and regulation of immune activity by expressing a diverse array of cytokines.

Increasing evidence has recognized that nasal polyps function as tertiary lymphoid organs, in which Tfh immune cells can locally enhance the differentiation of B cells and thus boost inflammatory responses with the production of immunoglobulins. Early studies in this area demonstrated elevated levels of CD138, which is a specific immune marker for plasma cells, and B-cell activating factor of the TNF family (BAFF) in CRSwNP tissues [5, 39]. BAFF is particularly essential in B-cell immunity for cellular survival and differentiation of B cells to plasma cells. Van Zele et al. further reported that besides elevated CD138 and plasma cell counts, higher levels of naïve B cells with CD19 staining are also characteristic of CRSwNP tissue when compared to non-polyp tissue [40]. B-cell populations in nasal polyp tissue have currently been found to consist of the various stages of differentiation, including naïve B cells, plasmablasts, plasma cells, and memory B cells; in fact, nasal polyp tissue supports a higher frequency of B cells at different stages of maturation than peripheral blood [41, 42]. Even more recently, reports have emerged that the inflammatory environment within nasal polyp tissue upregulates the expression of Epstein-Barr virus-induced protein 2 (EBI2) in plasmablasts, which produce and activate immunoglobulins at high levels [43]. EBI2 is critical for the development of extrafollicular B-cell responses and may be upregulated by ILC2. The relationship between EBI2 and innate lymphoid cells provides a mechanism for B-cell activation by innate immune cells in nasal polyps, highlighting potentially overlapping immunologic pathways that lead to nasal polyp formation [43]. As a whole, the data support the high frequency and infiltration of B-cell subtypes in CRSwNP and suggest a likely role for B cells in the inflammatory disease state.

As B cells are immunoglobulin-producing immune cells, elevated counts of B cells in nasal polyps have also correlated with local elevations of immunoglobulins in CRSwNP, although such high concentrations of immunoglobulins in polyp tissue are not correspondingly reflected in the serum [40]. These immunoglobulins from B cells derived from nasal polyps characteristically include IgG, IgA, and IgE, as the process for B cells to undergo class switch recombination has been found to occur at the local tissue level in CRSwNP [40, 43]. In contrast to allergic rhinitis, which is characteristically defined by an oligoclonal repertoire of IgE to specific antigens, the collection of immunoglobulins in CRSwNP is usually polyclonal. This particular difference between allergic rhinitis and CRSwNP may be partially rooted in the increased expression of recombination activating genes (RAG) 1 and 2 in the sinonasal tissues of CRSwNP. RAG1 and RAG2 critically support local class switch recombination by B cells in polyp tissue and are significantly expressed at elevated rates in CRSwNP [44]. While antigen specificity of these immunoglobulins remains largely unclear, studies have supported the importance of *S aureus* enterotoxins as a robust source of IgE in nasal polyps [40]. Other IgA and IgG in nasal polyp tissue have been found to be specific to autoantigens, including double-stranded DNA (dsDNA) and BP180 [45, 46]. The accumulation of these various immunoglobulins within nasal polyp tissue provides a means for B-cell immunity to promote the harmful inflammatory responses of CRSwNP.

All downstream effector functions that B cells ultimately mediate in CRSwNP pathophysiology are not fully understood, but a variety of immunopathogenic responses from B-cell activation likely occur in the development of the chronic inflammation along the sinonasal mucosa. To start, besides producing large amounts of immunoglobulins in nasal polyp tissue, B cells themselves serve important roles as antigen-presenting cells and produce numerous cytokines that provide essential cross talk with T cell and innate immune cells. Immunoglobulins in nasal polyps are also responsible for the activation of various immune effector cells, such as eosinophils, basophils, and mast cells, which have further been implicated in the pathogenesis of CRSwNP, through interactions with the Fc receptors expressed on the granulocytes. The identification of elevated levels of anti-dsDNA and anti-BP180 autoantibodies in nasal polyp tissue further suggests that unregulated autoimmune responses contribute to chronic inflammation. Recent evidence has shown that activation of the classical, or antibody-mediated, complement pathways at the basement membrane of the nasal polyp epithelium provides an additional mechanism for local immunoglobulin responses to result in local tissue injury [47]. Such an insult to the mucosal barrier integrity can further generate additional inflammatory responses from the innate and adaptive immune cells.

5.4 Summary and Conclusions

Evolving endotypes of CRSwNP are increasingly based upon different inflammatory patterns that underscore the heterogeneous pathophysiologic pathways of disease. Nasal polyps are enriched by high levels of T cells and B cells, which provide important contributions to these inflammatory patterns. T cells demonstrate a significant amount of plasticity. Combined with both environmental and genetic factors, the differentiation of naïve T cells into a variety of specific T-cell subsets, including Th1, Th2, Th17, Tfh, and Treg cells, influence the downstream signaling of effector mechanisms that ultimately shape the clinical manifestations of CRSwNP. Nasal polyps likewise function as tertiary lymphoid organs to generate a robust inflammatory response by B cells and their associated immunoglobulins. There are nonetheless significant gaps in the understanding of the full roles of the diverse array of T cells and B cells involved in the development of CRSwNP. The potential to improve the clinical management of CRSwNP on diagnostic and therapeutic fronts emphasis the need for continued research into the immunologic factors that drive this complex disease.

References

1. Orlandi RR, Kingdom TT, Hwang PH, et al. International consensus statement on allergy and rhinology: rhinosinusitis. Int Forum Allergy Rhinol. 2016;6(Suppl 1):S22–209. https://doi.org/10.1002/alr.21695.
2. Fokkens WJ, Lund VJ, Mullol J, et al. EPOS 2012: European position paper on rhinosinusitis and nasal polyps 2012. A summary for otorhinolaryngologists. Rhinology. 2012;50(1):1–12. https://doi.org/10.4193/Rhino50E2.
3. Kato A. Immunopathology of chronic rhinosinusitis. Allergol Int. 2015;64(2):121–30. https://doi.org/10.1016/j.alit.2014.12.006.
4. Akdis CA, Bachert C, Cingi C, et al. Endotypes and phenotypes of chronic rhinosinusitis: a PRACTALL document of the European Academy of Allergy and Clinical Immunology and the American Academy of Allergy, Asthma & Immunology. J Allergy Clin Immunol. 2013;131(6):1479–90. https://doi.org/10.1016/j.jaci.2013.02.036.
5. Van Zele T, Claeys S, Gevaert P, et al. Differentiation of chronic sinus diseases by measurement of inflammatory mediators. Allergy. 2006;61(11):1280–9. https://doi.org/10.1111/j.1398-9995.2006.01225.x.
6. Derycke L, Eyerich S, Van Crombruggen K, et al. Mixed T helper cell signatures in chronic rhinosinusitis with and without polyps. PLoS One. 2014;9(6):e97581. https://doi.org/10.1371/journal.pone.0097581.
7. Pant H, Beroukas D, Kette FE, Smith WB, Wormald PJ, Macardle PJ. Nasal polyp cell populations and fungal-specific peripheral blood lymphocyte proliferation in allergic fungal sinusitis. Am J Rhinol Allergy. 2009;23(5):453–60. https://doi.org/10.2500/ajra.2009.23.3356.
8. Pant H, Hughes A, Miljkovic D, et al. Accumulation of effector memory CD8+ T cells in nasal polyps. Am J Rhinol Allergy. 2013;27(5):e117–26. https://doi.org/10.2500/ajra.2013.27.3958.
9. Ma J, Shi LL, Deng YK, et al. CD8(+) T cells with distinct cytokine-producing features and low cytotoxic activity in eosinophilic and non-eosinophilic chronic rhinosinusitis with nasal polyps. Clin Exp Allergy. 2016;46(9):1162–75. https://doi.org/10.1111/cea.12758.
10. Van Bruaene N, Perez-Novo CA, Basinski TM, et al. T-cell regulation in chronic paranasal sinus disease. J Allergy Clin Immunol. 2008;121(6):1435–41, 1441 e1-3. https://doi.org/10.1016/j.jaci.2008.02.018.

11. Claeys S, Van Hoecke H, Holtappels G, et al. Nasal polyps in patients with and without cystic fibrosis: a differentiation by innate markers and inflammatory mediators. Clin Exp Allergy. 2005;35(4):467–72. https://doi.org/10.1111/j.1365-2222.2005.02215.x.
12. Cao PP, Li HB, Wang BF, et al. Distinct immunopathologic characteristics of various types of chronic rhinosinusitis in adult Chinese. J Allergy Clin Immunol. 2009;124(3):478–84., 484 e1-2. https://doi.org/10.1016/j.jaci.2009.05.017.
13. Wang X, Zhang N, Bo M, et al. Diversity of TH cytokine profiles in patients with chronic rhinosinusitis: a multicenter study in Europe, Asia, and Oceania. J Allergy Clin Immunol. 2016;138(5):1344–53. https://doi.org/10.1016/j.jaci.2016.05.041.
14. Zhang N, Van Zele T, Perez-Novo C, et al. Different types of T-effector cells orchestrate mucosal inflammation in chronic sinus disease. J Allergy Clin Immunol. 2008;122(5):961–8. https://doi.org/10.1016/j.jaci.2008.07.008.
15. Wu D, Wang J, Zhang M. Altered Th17/Treg ratio in nasal polyps with distinct cytokine profile: association with patterns of inflammation and mucosal remodeling. Medicine (Baltimore). 2016;95(10):e2998. https://doi.org/10.1097/MD.0000000000002998.
16. Ryu G, Bae JS, Kim JH, et al. Role of IL-17A in chronic rhinosinusitis with nasal polyp. Allergy Asthma Immunol Res. 2020;12(3):507–22. https://doi.org/10.4168/aair.2020.12.3.507.
17. Tesmer LA, Lundy SK, Sarkar S, Fox DA. Th17 cells in human disease. Immunol Rev. 2008;223:87–113. https://doi.org/10.1111/j.1600-065X.2008.00628.x.
18. Annunziato F, Romagnani C, Romagnani S. The 3 major types of innate and adaptive cell-mediated effector immunity. J Allergy Clin Immunol. 2015;135(3):626–35. https://doi.org/10.1016/j.jaci.2014.11.001.
19. Jiang XD, Li GY, Li L, Dong Z, Zhu DD. The characterization of IL-17A expression in patients with chronic rhinosinusitis with nasal polyps. Am J Rhinol Allergy. 2011;25(5):e171–5. https://doi.org/10.2500/ajra.2011.25.3645.
20. Wang M, Zhang N, Zheng M, et al. Cross-talk between TH2 and TH17 pathways in patients with chronic rhinosinusitis with nasal polyps. J Allergy Clin Immunol. 2019;144(5):1254–64. https://doi.org/10.1016/j.jaci.2019.06.023.
21. Ramezanpour M, Moraitis S, Smith JL, Wormald PJ, Vreugde S. Th17 cytokines disrupt the airway mucosal barrier in chronic rhinosinusitis. Mediators Inflamm. 2016;2016:9798206. https://doi.org/10.1155/2016/9798206.
22. Gandhi NA, Pirozzi G, Graham NMH. Commonality of the IL-4/IL-13 pathway in atopic diseases. Expert Rev Clin Immunol. 2017;13(5):425–37. https://doi.org/10.1080/1744666X.2017.1298443.
23. Poposki JA, Peterson S, Welch K, et al. Elevated presence of myeloid dendritic cells in nasal polyps of patients with chronic rhinosinusitis. Clin Exp Allergy. 2015;45(2):384–93. https://doi.org/10.1111/cea.12471.
24. Pezato R, Perez-Novo CA, Holtappels G, et al. The expression of dendritic cell subsets in severe chronic rhinosinusitis with nasal polyps is altered. Immunobiology. 2014;219(9):729–36. https://doi.org/10.1016/j.imbio.2014.05.004.
25. Shi LL, Song J, Xiong P, et al. Disease-specific T-helper cell polarizing function of lesional dendritic cells in different types of chronic rhinosinusitis with nasal polyps. Am J Respir Crit Care Med. 2014;190(6):628–38. https://doi.org/10.1164/rccm.201402-0234OC.
26. Tojima I, Kouzaki H, Shimizu S, et al. Group 2 innate lymphoid cells are increased in nasal polyps in patients with eosinophilic chronic rhinosinusitis. Clin Immunol. 2016;170:1–8. https://doi.org/10.1016/j.clim.2016.07.010.
27. Nagarkar DR, Poposki JA, Tan BK, et al. Thymic stromal lymphopoietin activity is increased in nasal polyps of patients with chronic rhinosinusitis. J Allergy Clin Immunol. 2013;132(3):593–600. e12. https://doi.org/10.1016/j.jaci.2013.04.005.
28. Shaw JL, Fakhri S, Citardi MJ, et al. IL-33-responsive innate lymphoid cells are an important source of IL-13 in chronic rhinosinusitis with nasal polyps. Am J Respir Crit Care Med. 2013;188(4):432–9. https://doi.org/10.1164/rccm.201212-2227OC.
29. Artis D, Spits H. The biology of innate lymphoid cells. Nature. 2015;517(7534):293–301. https://doi.org/10.1038/nature14189.
30. Bachert C, Zhang N, Patou J, van Zele T, Gevaert P. Role of staphylococcal superantigens in upper airway disease. Curr Opin Allergy Clin Immunol. 2008;8(1):34–8. https://doi.org/10.1097/ACI.0b013e3282f4178f.
31. Van Zele T, Gevaert P, Watelet JB, et al. Staphylococcus aureus colonization and IgE antibody formation to enterotoxins is increased in nasal polyposis. J Allergy Clin Immunol. 2004;114(4):981–3. https://doi.org/10.1016/j.jaci.2004.07.013.
32. Ramakrishnan VR, Feazel LM, Abrass LJ, Frank DN. Prevalence and abundance of Staphylococcus aureus in the middle meatus of patients with chronic rhinosinusitis, nasal polyps, and asthma. Int Forum Allergy Rhinol. 2013;3(4):267–71. https://doi.org/10.1002/alr.21101.
33. Seiberling KA, Grammer L, Kern RC. Chronic rhinosinusitis and superantigens. Otolaryngol Clin North Am. 2005;38(6):1215–36. ix. https://doi.org/10.1016/j.otc.2005.08.006.
34. Zhang YN, Song J, Wang H, et al. Nasal IL-4(+) CXCR5(+)CD4(+) T follicular helper cell counts correlate with local IgE production in eosinophilic nasal polyps. J Allergy Clin Immunol. 2016;137(2):462–73. https://doi.org/10.1016/j.jaci.2015.07.025.
35. Xiao L, Wei Y, Zhang YN, et al. Increased IL-21 expression in chronic rhinosinusitis with nasal polyps. Clin Exp Allergy. 2015;45(2):404–13. https://doi.org/10.1111/cea.12475.

36. Calus L, Derycke L, Dullaers M, et al. IL-21 is increased in nasal polyposis and after stimulation with Staphylococcus aureus enterotoxin B. Int Arch Allergy Immunol. 2017;174(3–4):161–9. https://doi.org/10.1159/000481435.
37. Palmer C, Mulligan JK, Smith SE, Atkinson C. The role of regulatory T cells in the regulation of upper airway inflammation. Am J Rhinol Allergy. 2017;31(6):345–51. https://doi.org/10.2500/ajra.2017.31.4472.
38. Miljkovic D, Psaltis A, Wormald PJ, Vreugde S. T regulatory and Th17 cells in chronic rhinosinusitis with polyps. Int Forum Allergy Rhinol. 2016;6(8):826–34. https://doi.org/10.1002/alr.21742.
39. Kato A, Peters A, Suh L, et al. Evidence of a role for B cell-activating factor of the TNF family in the pathogenesis of chronic rhinosinusitis with nasal polyps. J Allergy Clin Immunol. 2008;121(6):1385–92., 1392 e1-2. https://doi.org/10.1016/j.jaci.2008.03.002.
40. Van Zele T, Gevaert P, Holtappels G, van Cauwenberge P, Bachert C. Local immunoglobulin production in nasal polyposis is modulated by superantigens. Clin Exp Allergy. 2007;37(12):1840–7. https://doi.org/10.1111/j.1365-2222.2007.02838.x.
41. Miljkovic D, Psaltis A, Wormald PJ, Vreugde S. Naive and effector B-cell subtypes are increased in chronic rhinosinusitis with polyps. Am J Rhinol Allergy. 2018;32(1):3–6. https://doi.org/10.2500/ajra.2018.32.4496.
42. Hulse KE, Norton JE, Suh L, et al. Chronic rhinosinusitis with nasal polyps is characterized by B-cell inflammation and EBV-induced protein 2 expression. J Allergy Clin Immunol. 2013;131(4):1075–83., 1083 e1-7. https://doi.org/10.1016/j.jaci.2013.01.043.
43. Feldman S, Kasjanski R, Poposki J, et al. Chronic airway inflammation provides a unique environment for B cell activation and antibody production. Clin Exp Allergy. 2017;47(4):457–66. https://doi.org/10.1111/cea.12878.
44. Gevaert P, Nouri-Aria KT, Wu H, et al. Local receptor revision and class switching to IgE in chronic rhinosinusitis with nasal polyps. Allergy. 2013;68(1):55–63. https://doi.org/10.1111/all.12054.
45. Tan BK, Li QZ, Suh L, et al. Evidence for intranasal antinuclear autoantibodies in patients with chronic rhinosinusitis with nasal polyps. J Allergy Clin Immunol. 2011;128(6):1198–206. e1. https://doi.org/10.1016/j.jaci.2011.08.037.
46. Jeffe JS, Seshadri S, Hamill KJ, et al. A role for anti-BP180 autoantibodies in chronic rhinosinusitis. Laryngoscope. 2013;123(9):2104–11. https://doi.org/10.1002/lary.24016.
47. Van Roey GA, Vanison CC, Wu J, et al. Classical complement pathway activation in the nasal tissue of patients with chronic rhinosinusitis. J Allergy Clin Immunol. 2017;140(1):89–100. e2. https://doi.org/10.1016/j.jaci.2016.11.015.

6 Mast Cells

Hirohisa Saito

Abbreviations

CPA3	Carboxypeptidase A3
cys-LT	Cysteinyl leukotriene
FcεRI	High-affinity receptor for IgE
GC	Glucocorticoid
GM-CSF	Granulocyte-macrophage colony-stimulating factor
IL	Interleukin
MCs	Mast cells
MC_T	T-type mast cells
MC_{TC}	TC-type mast cells
MIP	Macrophage inflammatory protein
NFAT	Nuclear factor-activated T
NF-κB	Nuclear factor-κB
PAF	Platelet-activating factor
PGD_2	Prostaglandin D_2
SCF	Stem cell factor
TLR	Toll-like receptor
TNF-α	Tumor necrosis factor
TSLP	Thymic stromal lymphopoietin

H. Saito (✉)
Department of Allergy and Clinical Immunology, National Research Institute for Child Health and Development, Tokyo, Japan
e-mail: saito-hr@ncchd.go.jp

Core Message

- Mast cells trigger not only the immediate-type allergic reactions in an IgE-mediated manner but also the late-phase allergic response and chronic allergic inflammation.

6.1 Introduction

Mast cells (MCs) serve as essential effector cells for acute IgE-mediated allergic reactions by releasing histamine and other vasoactive mediators, as seen in allergic rhinitis, for example. MCs are also recognized as important source of a variety of cytokines and chemokines. Thus, MCs trigger not only the immediate-type allergic reactions in an IgE-mediated manner but also the late-phase allergic response and chronic allergic inflammation, thereby regulating the function of other immune cells. MCs are present throughout connective tissues and mucosal surfaces, particularly at the interface with the external environment such as the skin and respiratory tract [1]. The nasal mucosa is the first barrier of the entire respiratory tract that encounters various pathogens or allergens. In this review, we will summarize the roles of MCs in allergic airway diseases by focusing on the role of human MCs in the airways.

Ö. Ö. Celebi, T. M. Önerci (eds.), *Nasal Physiology and Pathophysiology of Nasal Disorders*,
https://doi.org/10.1007/978-3-031-12386-3_6

6.2 Origin and Distribution of MCs

MCs originate from hematopoietic progenitors. Kitamura et al. discovered two different mice strains genetically lacking MCs: *Sl/Sl^d* mice lacking SCF, which turned out to be the mast cell growth factor, and *W/W^v* mice lacking KIT, which is the receptor for SCF. By using these "natural" MC-deficient mice, it was established that immature MC progenitors can migrate from the bone marrow into the tissue through blood circulation, unlike immature granulocytes which are kept in the bone marrow. Then, these cells undergo maturation in the tissues under specific factors like stem cell factor (SCF) present within the microenvironment [2–4].

Phenotypically distinct subsets of MCs are present in rodents, based on their distinct staining characteristics, T-cell dependency, and functions, namely, connective tissue MCs and mucosal MCs [5, 6]. Regarding T-cell dependency, it is well established that mucosal MCs can grow in the presence of interleukin (IL-)3 [7]. However, human MCs do not grow when hematopoietic cells are cultured with IL-3 [8]. Although human IL-3 has a significant sequence homology with murine IL-3, the degree of homology between human and murine IL-3 is almost similar (approximately 26–28% at amino acid sequence) to that between human IL-3 and granulocyte-macrophage colony-stimulating factor (GM-CSF). Also, the receptor structure for IL-3 is distinct between human and mouse. While human has a common β-subunit of the receptors for GM-CSF, IL-3, and IL-5, the mouse has two distinct β-subunits; one is specific for the IL-3 receptor and exists only on mouse MCs, and the other is equivalent to the human common β-subunit [9].

Regarding human MC phenotypes, two types of MCs have been recognized based on the neutral proteases they express. TC-type MCs (MC_{TC}) contain tryptase together with chymase, and other neutral proteases, whereas T-type mast cells (MC_T) contain tryptase but lack the other neutral proteases present in MC_{TC} [10]. Also, MC_{TC} preferentially dwell in the connective tissue such as skin, while MC_T are often found in mucosa such as airway epithelium. In allergic rhinitis and asthma, MCs are known to accumulate within the epithelial compartment of the target organ. In fact, there is a selective increase of MC_T in the epithelial compartment of the nasal mucosa of the patients with allergic rhinitis [11, 12].

Asthma can be divided into two subgroups ("Th2 high" and "Th2 low" asthma) based on epithelial cell gene signatures for the activity of type 2 cytokines such as IL-13 [13]. The patients with Th2 high asthma have more infiltration of MCs into the airway epithelium. These subgroups can be diagnosed based on the level of serum periostin, which production is specifically induced by IL-13, and that the patients with Th2 high asthma subtype are more sensitive to anti-IL-13 therapy [14]. These intraepithelial MCs express both tryptases and carboxypeptidase A3 (CPA3) but not chymase [13]. According to classical definition [10], MC_T were not supposed to express CPA3. However, according to subsequent reports [15, 16], all human MCs, even MC_T may express CPA3. MCs exposed to conditioned media from IL-13-activated epithelial cells showed downregulation of the chymase expression but no change in tryptase or CPA3 expression [13]. This may relate to the reason why MC_T are preferentially found in the mucosa and are deficient in primary immunodeficiency patients [1].

As shown in Table 6.1, MC_{TC} can respond to various non-immunological stimuli such as C5a or substance P, while MC_T do not [1]. Kajiwara et al. reported that MC_T, but not MC_{TC}, express functional receptor for platelet-activating factor (PAF). It was found by searching preferentially expressed genes in lung MCs (MC_T) compared to skin MCs (MC_{TC}). Interestingly, these MC phenotypes, i.e., expression of chymase and receptors for these non-immunological stimuli, are retained over weeks even when these MCs are cultured in the standard MC culture condition (supplemented with SCF and IL-6) [15, 17]. This is contrasting to the results showing that MCs lose chymase by the factor(s) produced in the IL-13-activated epithelial cells [13]. It would be interesting to know whether MCs, which have

Table 6.1 Characteristics of two phenotypes of human mast cells

Phenotype	MC_{TC}	MC_T
Proteases	Tryptase (+++)	Tryptase (++)
	Chymase (+)	Chymase (−)
	Carboxypeptidase A3 (++)	Carboxypeptidase A3 (+)
	Cathepsin G (+)	
Distribution	Skin (++)	Skin (−)
	Intestinal submucosa (+)	Intestinal submucosa (++)
	Intestinal mucosa (−)	Intestinal mucosa (++)
	Alveolar wall (++)	Alveolar wall (−)
	Bronchial subepithelium (+)	Bronchial subepithelium (+)
	Dispersed lung mast cells (−)	Dispersed lung mast cells (++)
	Tonsils (++)	Tonsils (++)
	Nasal mucosa (−)	Nasal mucosa (++)
Relation to pathology	Increased in fibrotic diseases	Increased around the site of T cell aviation
	Unchanged in allergic and parasitic diseases	Increased in allergic and parasitic diseases
	Unchanged in chronic immunodeficiency diseases	Decreased in chronic immunodeficiency diseases
Response to non-immunological stimuli	Substance P (+)	Substance P (−)
	C5a (+)	C5a (−)
	PAF (−)[a]	PAF (+)[a]

Adapted from [1])
[a]Kajiwara et al. [17]

lost chymase by the IL-13-activated epithelial cell-derived factor(s), respond to substance P or PAF.

6.3 Role of MCs in Acute Allergic Reactions

MCs express more than 10^5 high-affinity IgE receptor (FcεRI) per cell. When MCs that have been sensitized with some specific IgE antibody are challenged with the specific allergen, they are activated by cross-linking of FcεRI molecules. Thus, activated MCs evoke immediate-type reaction by releasing their granules in which histamine, neutral proteases, and heparin had been stored. Then, lipid mediators such as cysteinyl leukotriene (cys-LT) or prostaglandin D_2 (PGD_2) are synthesized on their membranes and are released into microenvironment within several minutes.

Released histamine and lipid mediators cause acute allergic symptoms such as nasal discharge, bronchospasms, and urticaria. Histamine plays an essential role in acute skin allergic reactions, whereas cys-LT plays a pivotal role in bronchoconstriction. MCs almost exclusively express PGD_2 synthase compared to all other cell types. Although the role of PGD_2 in immediate-type reaction is unclear, it serves as chemoattractant for eosinophils, basophils, and Th2 cells.

Human MCs also exclusively express tryptase, one of the neutral proteases, among all human cell types. Tryptase constitutes 10% of the MC by protein weight [1]. Proteoglycan (human MCs use "eosinophil" major basic protein instead of proteoglycan molecules) serves as a core protein in the crystalloid structure of the MC granules by binding to heparin and neutral proteases [18]. The MC tryptase acts as trypsin-like enzyme and thereby causes tissue remodeling such as abnormal proliferation of airway smooth muscles [19].

6.4 Role of MCs in Allergic Inflammation

MCs secrete a variety of cytokines and chemokines several hours after allergen-induced degranulation via transcription of these genes. The representative cytokines/chemokines which are produced by activated human MCs are type 2 cytokines such as IL-5, IL-13, and GM-CSF and CC chemokines such as CCL1/I-309, CCL2/monocyte chemoattractant protein-1, CCL3/macrophage inflammatory protein (MIP-)1α, and CCL4/MIP-1β. Activated human MCs also secrete a substantial amount of CXCL8/IL-8 [18, 20]. MCs can store and release some of cytokines such as tumor necrosis factor (TNF)-α during degranulation process. Regarding IL-4 production, the results are reproducible using mouse MCs. However, only a few groups succeeded to immunohistochemically demonstrate the presence of IL-4 on human MCs [21, 22]. In any case, at least in human, basophils are more potent producers of IL-4. Instead, IL-4 potently activates human MC function and maturation. Human MCs can produce a substantial amount of another type 2 cytokine, IL-13, in response to IgE-mediated stimuli, and the IL-13 production is markedly enhanced by preincubation with IL-4 [20]. However, these cytokines and chemokines are not unique to MCs and are produced by other cell types [23]. During antigen stimulation, more type 2 cytokines would be produced by proliferating T cells. Moreover, group 2 innate lymphoid cells (ILC2s) were recently found to produce higher levels of type 2 cytokines [24] and are currently considered to be the culprit for the innate phase of allergic inflammation [25]. In a certain experimental setting in vitro, however, human mast cells seem to produce IL-13, which plays an essential role in the pathogenesis of asthma or other allergic inflammatory diseases, at the level comparable to that produced by human ILC2s. It would be useful if we could dissect the role of mast cells and ILC2s in the late phase asthmatic responses accompanied by eosinophilic inflammation in the same experimental setting (Fig. 6.1).

Although human MCs do not normally produce cytokines in response to other cytokines such as IL-4 without FcεRI cross-linking, it should be noted that IL-33, which are released during necrosis of epithelial-mesenchymal tissue, can stimulate MCs to release a variety of cytokines such as IL-13 [26]. Regarding other innate immune responses, mouse MCs are proven to play an essential role in protection against microbial infection via Toll-like receptors (TLRs) [27–29]. Human MCs can express functional TLR4 after preincubation with IFN-γ. These MCs can produce more TNF-a, CCL5, CXCL10, and CXCL11 compared to IgE dependently activated MCs [30].

Topical use of glucocorticoid (GC) is the first-line therapy for allergic diseases such as asthma and allergic rhinitis. Although GC does not block the degranulation of MCs, these drugs downregulate the gene expression of FcεRI in MCs and thereby downregulate IgE-mediated activation of MCs. More notably, glucocorticoid can inhibit gene expression of a variety of cytokines in MCs. Even in short time incubation, GC blocks the nuclear factor-κB (NF-κB)-dependent gene expression of cytokines, such as IL-13, CXCL8/IL-8, and GM-CSF. On the other hand, GC does not inhibit nuclear factor-activated T (NFAT)-dependent gene expression of cytokines, such as CCL1, CCL3, and CCL4.

Interestingly, an immunosuppressive agent, FK-506 inhibits NFAT-dependent-, but not NF-κB-dependent-, gene expression [31]. If GC and FK-506 are added simultaneously into the reaction buffer for MC activation, the expression of cytokines is almost completely blocked. Among cytokine or growth factor genes, only IgE-mediated amphiregulin gene upregulation is not blocked by preincubation with GC and FK-506. It would be difficult to surpass the effect of GC plus FK-506 even if we could develop a new anti-MC drug.

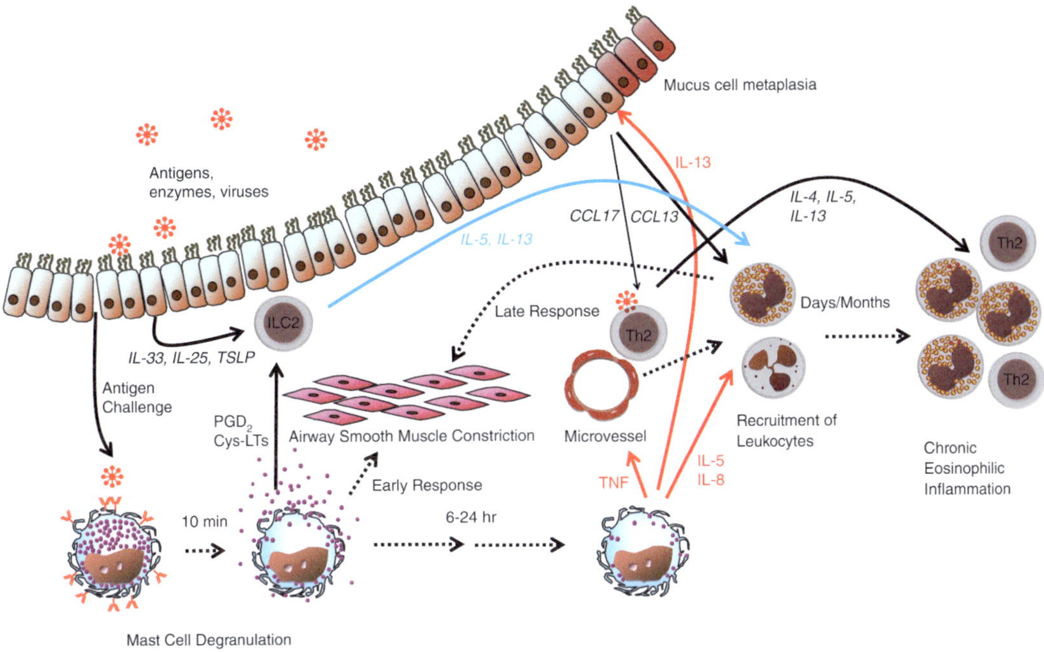

Fig. 6.1 Suggested roles of mast cell-derived and ILC2s-derived cytokines on the late phase allergic reactions. Mast cell-derived cytokines are shown in red, and ILC2s-derived cytokines are shown in blue

6.5 Conclusion

MCs trigger not only the immediate-type allergic reaction in an IgE-mediated manner but also the late-phase allergic response and chronic allergic inflammation, thereby regulating the function of other immune cells. While histamine, tryptase, and PGD_2 released in the immediate-type reaction are unique to MCs (or basophils), most cytokines and chemokines are produced by other cell types as well as MCs. It is necessary to determine the relative role of MCs in the allergic or innate-type inflammation by understanding cytokines/chemokines produced by other immune cell types and epithelial-mesenchymal tissues. The expression of these cytokines is almost completely blocked when GC and FK506 are added simultaneously into the reaction buffer for MC activation. It would be difficult to overwhelm this effect even if we could develop a new anti-MC drug.

References

1. Bradding P, Saito H. Biology of mast cells and their mediators. In: Adkinson F, Bochner BS, Burks AW, Holgate ST, Lemanske RF, O'Hehir R, editors. Middleton's allergy: principles and practice, vol. 1. 8th ed. Philadelphia, PA: Elsevier Saunders; 2013. p. 228–51.
2. Kitamura Y, Go S, Hatanaka K. Decrease of mast cells in W/W^v mice and their increase by bone marrow transplantation. Blood. 1978;52:447–52.
3. Kitamura Y, Go S. Decreased production of mast cells in S1/S1^d anemic mice. Blood. 1979;53:492–7.
4. Kitamura Y, Shimada M, Hatanaka K, Miyano Y. Development of mast cells from grafted bone marrow cells in irradiated mice. Nature. 1977;268:442–3.
5. Befus AD, Pearce FL, Gauldie J, Horsewood P, Bienenstock J. Mucosal mast cells. I. Isolation and functional characteristics of rat intestinal mast cells. J Immunol. 1982;128:2475.
6. Pearce FL, Befus AD, Gauldie J, Bienenstock J. Mucosal mast cells. II. Effects of anti-allergic compounds on histamine secretion by isolated intestinal mast cells. J Immunol. 1982;128:2481.
7. Ihle JN, Keller J, Oroszlan S, Henderson LE, Copeland TD, Fitch F, Prystowsky MB, Goldwasser E, Schrader

JW, Palaszynski E, Dy M, Lebel B. Biologic properties of homogeneous interleukin 3. I. Demonstration of WEHI-3 growth factor activity, mast cell growth factor activity, p cell-stimulating factor activity, colony-stimulating factor activity, and histamine-producing cell-stimulating factor activity. J Immunol. 1983;131:282–7.
8. Saito H, Hatake K, Dvorak AM, Leiferman KM, Donnenberg AD, Arai N, Ishizaka K, Ishizaka T. Selective differentiation and proliferation of hematopoietic cells induced by recombinant human interleukins. Proc Natl Acad Sci U S A. 1988;85:2288–92.
9. Miyajima A. Molecular structure of the IL-3, GM-CSF and IL-5 receptors. Int J Cell Cloning. 1992;10:126–34.
10. Irani AMA, Schecter NM, Craig SS, DeBlois G, Schwartz LB. Two types of human mast cells that have distinct neutral protease compositions. Proc Natl Acad Sci U S A. 1986;83:4464–9.
11. Enerback L, Pipkorn U, Olofsson A. Intraepithelial migration of mucosal mast cells in hay fever. Int Arch Allergy Appl Immunol. 1986;80:44.
12. Pawankar R, Ra C. Heterogeneity of mast cells and T cells in the nasal mucosa. J Allergy Clin Immunol. 1996;98:249.
13. Dougherty RH, Sidhu SS, Raman K, Solon M, Solberg OD, Caughey GH, Woodruff PG, Fahy JV. Accumulation of intraepithelial mast cells with a unique protease phenotype in T_H2-high asthma. J Allergy Clin Immunol. 2010;125:1046–53.
14. Corren J, Lemanske RF, Hanania NA, Korenblat PE, Parsey MV, Arron JR, Harris JM, Scheerens H, Wu LC, Su Z, Mosesova S, Eisner MD, Bohen SP, Matthews JG. Lebrikizumab treatment in adults with asthma. N Engl J Med. 2011;365:1088–98.
15. Kashiwakura J, Yokoi H, Saito H, Okayama Y. T cell proliferation by direct cross-talk between OX40 ligand on human mast cells and OX40 on human T cells: comparison of gene expression profiles between human tonsillar and lung-cultured mast cells. J Immunol. 2004;173:5247–57.
16. Takabayashi T, Kato A, Peters AT, Suh LA, Carter R, Norton J, Grammer LC, Tan BK, Chandra RK, Conley DB, Kern RC, Fujieda S, Schleimer RP. Glandular mast cells with distinct phenotype are highly elevated in chronic rhinosinusitis with nasal polyps. J Allergy Clin Immunol. 2012;130:410–20.e5.
17. Kajiwara N, Sasaki T, Bradding P, Cruse G, Sagara H, Ohmori K, Saito H, Ra C, Okayama Y. Activation of human mast cells through the platelet-activating factor receptor. J Allergy Clin Immunol. 2010;125:1137–45.
18. Nakajima T, Inagaki N, Tanaka H, Tanaka A, Yoshikawa M, Tamari M, Hasegawa K, Matsumoto K, Tachimoto H, Ebisawa M, Tsujimoto G, Matsuda H, Nagai H, Saito H. Marked increase in CC chemokine gene expression in both human and mouse mast cell transcriptomes following Fcε receptor I cross-linking: an interspecies comparison. Blood. 2002;100:3861–8.
19. Brightling CE, Bradding P, Symon FA, Holgate ST, Wardlaw AJ, Pavord ID. Mast-cell infiltration of airway smooth muscle in asthma. N Engl J Med. 2002;346:1699–705.
20. Bischoff SC. Role of mast cells in allergic and non-allergic immune responses: comparison of human and murine data. Nat Rev Immunol. 2007;7:93–104.
21. Bradding P, Feather IH, Howarth PH, Mueller R, Roberts JA, Britten K, Bews JP, Hunt TC, Okayama Y, Heusser CH, Bullock GR, Church MK, Holgate ST. Interleukin 4 is localized to and released by human mast cells. J Exp Med. 1992;176:1381–6.
22. Pawankar R, Okuda M, Yssel H, Okumura K, Ra C. Nasal mast cells in perennial allergic rhinitics exhibit increased expression of the fc epsilonRI, CD40L, IL-4, and IL-13, and can induce IgE synthesis in B cells. J Clin Invest. 1997;99:1492–9.
23. Mukai K, Mindy T, Saito H, Galli S. Mast cells as sources of cytokines, chemokines and growth factors. Immunol Rev. 2018;282:121–50.
24. Moro K, Yamada T, Tanabe M, Takeuchi T, Ikawa T, Kawamoto H, et al. Innate production of T(H)2 cytokines by adipose tissue-associated c-kit(+)Sca-1(+) lymphoid cells. Nature. 2010;463:540–4.
25. Morita H, Moro K, Koyasu S. Innate lymphoid cells in allergic and nonallergic inflammation. J Allergy Clin Immunol. 2016;138:1253–64.
26. Iikura M, Suto H, Kajiwara N, Oboki K, Ohno T, Okayama Y, Saito H, Galli SJ, Nakae S. IL-33 can promote survival, adhesion and cytokine production in human mast cells. Lab Investig. 2007;87:971–8.
27. Krämer S, Sellge G, Lorentz A, Krueger D, Schemann M, Feilhauer K, Gunzer F, Bischoff SC. Selective activation of human intestinal mast cells by Escherichia coli hemolysin. J Immunol. 2008;181:1438–45.
28. Nakajima S, Krishnan B, Ota H, Segura AM, Hattori T, Graham DY, Genta RM. Mast cell involvement in gastritis with or without helicobacter pylori infection. Gastroenterology. 1997;113:746–54.
29. Supajatura V, Ushio H, Nakao A, Akira S, Okumura K, Ra C, Ogawa H. Differential responses of mast cell toll-like receptors 2 and 4 in allergy and innate immunity. J Clin Invest. 2002;109:1351–9.
30. Okumura S, Kashiwakura J, Tomita H, Matsumoto K, Nakajima T, Saito H, Okayama Y. Identification of specific gene expression profile in human mast cells via toll-like receptor 4 and FcεRI. Blood. 2003;102:2547–54.
31. Kato A, Chustz RT, Ogasawara T, Kulka M, Saito H, Schleimer RP, Matsumoto K. Dexamethasone and FK506 inhibit expression of distinct subsets of chemokines in human mast cells. J Immunol. 2009;182:7233–43.

7 Macrophage and Mast Cell

Hideyuki Kawauchi

Core Message

- There are various immunocompetent cells including the so-called macrophages in human nasal mucosa. Those cells are essential for a defense system against various invading pathogens such as bacteria and viruses. Those cells are also key players in the pathogenesis of rhinosinusitis and allergic rhinitis at the epithelial linings of nasal cavity and paranasal sinuses. Among them, macrophages are well known to have immunologically an important role as scavenger cells and antigen-presenting cells (APCs), in order to mount innate and acquired immunity in the upper and lower respiratory tract.

7.1 Part I: General Concept of Macrophage

7.1.1 Origin and Classification of Macrophages

Macrophage lineage cells are produced from pluripotent progenitor cells in the bone marrow [1]. These cells require a combined stimulus from colony-stimulating factor-1 (CSF-1) and factors including interleukin-1 (IL-1), IL-3, GM-CSF, and interferon-gamma so far to differentiate to mature functional macrophages distributed to peripheral tissues through the blood vessel and lymphatic circulation. The definition and nomenclature of tissue-resident macrophage or recruiting inflammatory macrophage are taken into account with evidence that the monocyte subpopulations may possess different propensities to give rise to particular resident populations, particularly in the mucosal surface such as the respiratory and digestive tract. It is clearly demonstrated that blood monocytes are heterogeneous in terms of their expression of key molecules, chemokine receptors, and cell adhesion molecules [2]. But it is yet to remain to categorize the monocyte subsets and how to further divide them in terms of their effector functions with distinct stimuli and locations.

7.1.2 Heterogeneity and Markers

Tissue macrophages have many characteristics, including extensive lysosomes and stellate morphology and location, and they are heterogeneous in terms of function and surface marker expression, although we already know their phagocytic and antigen-presenting cell (APC) function. For example, CD11c in humans is a

H. Kawauchi (✉)
Department of Microbiology, Faculty of Medicine, Shimane University, Izumo, Japan
e-mail: Kawauchi@med.shimane-u.ac.jp

Ö. Ö. Celebi, T. M. Önerci (eds.), *Nasal Physiology and Pathophysiology of Nasal Disorders*,
https://doi.org/10.1007/978-3-031-12386-3_7

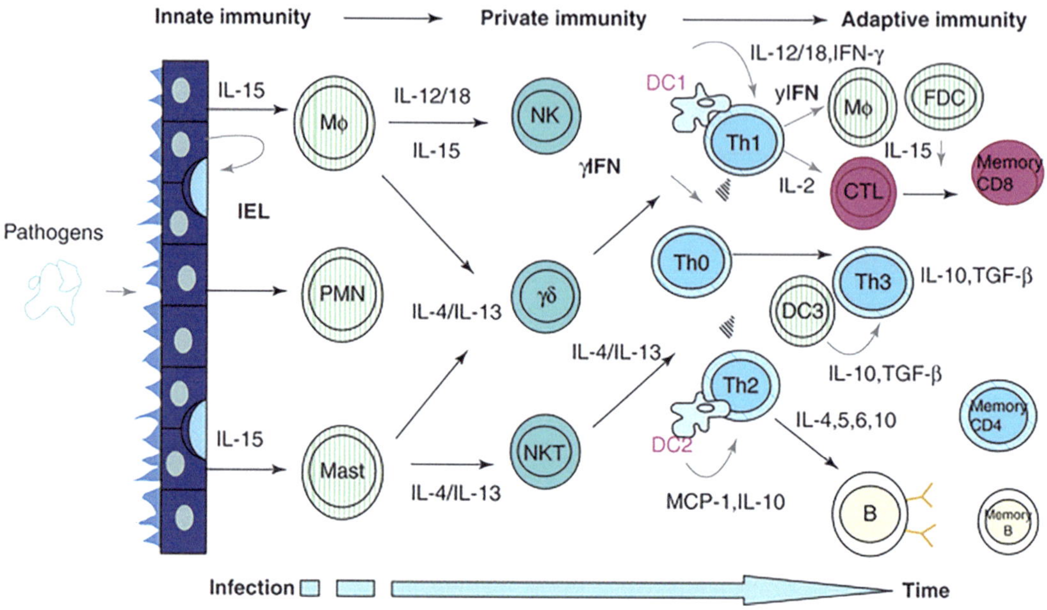

Fig. 7.1 The mechanisms of host defense system from innate to adaptive immunity. *PMN* polymorphonuclear cells, *CTL* cytotoxic lymphocytes, *Mo* macrophages, *DC* dendritic cells

marker for the mononuclear phagocyte system and was later shown to be an active complement receptor 4 (CR4) that is induced during macrophage maturation, although CD11c is clearly not linked to APC function. More importantly, various cell surface molecules in response to Toll-like receptor (TLR) signaling are functionally of particular interest because they determine the ability of macrophage lineage cells to interact with pathogens, and with other cell types, to generate appropriate innate and acquired immune responses (Fig. 7.1). But, there are no markers that are expressed specifically and ubiquitously on all macrophage lineage cells except CSF-1 receptor. Co-stimulatory molecules (CD80, CD86, CD40) are considered to be essential for antigen uptake and antigen presentation from macrophages to T and B cells. And chemokine receptors and integrin family on macrophages may determine the recruitment and locations in tissues. However, for many reasons, surface marker expression cannot be taken as the sole indication of lineage, function, or destiny among macrophages [3].

7.1.3 Recruitment of Macrophages into Peripheral Mucosal Inflammatory Sites

Macrophages are recruited into peripheral mucosal inflammatory sites with a wide range of different stimuli. If microbial infection takes place, neutrophil infiltration precedes and releases toxic agents designed to kill extracellular pathogens, and then macrophages come and evacuate degraded pathogens and apoptotic neutrophils. The tissue-entering process of these cells is called chemotaxis. Chemokines are essential for the recruitment of inflammatory cells into the peripheral mucosal inflammatory sites [4]. Chemokines are subdivided based on the core cysteine motifs that form disulfide bonds to fold the molecule. CC chemokines have two adjacent cysteines, while in CXC chemokines, there is an intervening amino acid. Chemokine receptors are classified in accordance with CCR, CXCR, and CX3CR families. The expression of specific chemokine receptors on different populations of macrophages and dendritic cells provides differ-

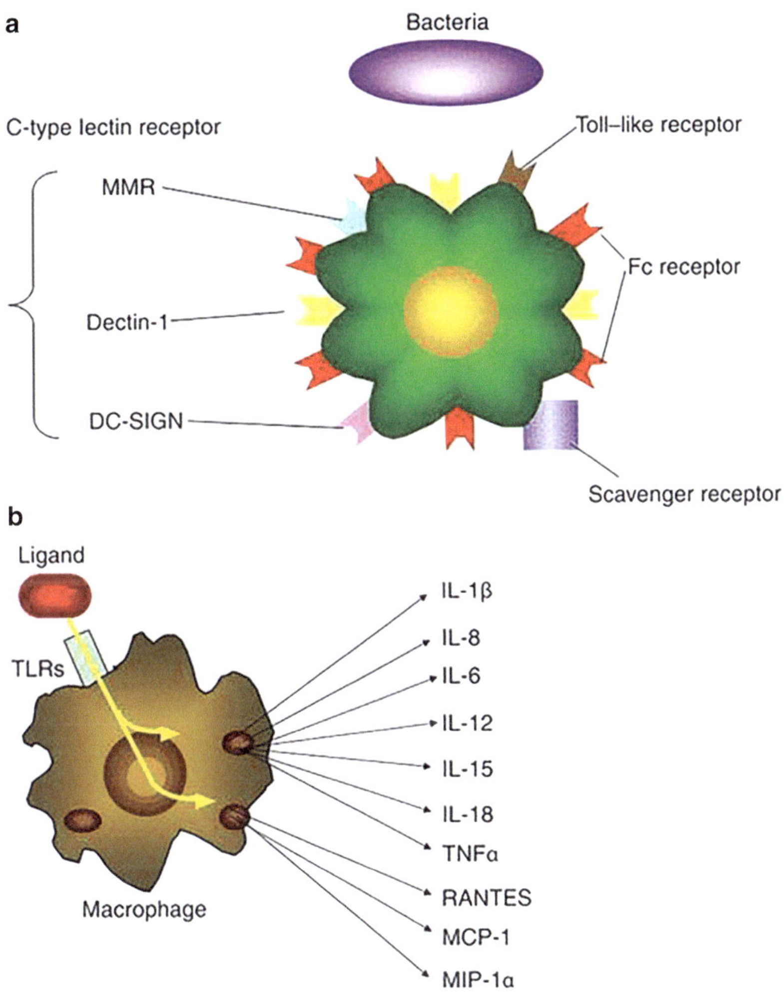

Fig. 7.2 (**a**) Pattern recognition receptors on macrophages. (**b**) Cytokines and chemokines produced by macrophage via Toll-like receptors

ent kinds of effector mechanisms for their differential recruitment in response to different signals and, consequently, might modify inflammatory reactions in mucosal sites such as the respiratory or digestive tract.

7.1.4 Phagocytosis

Phagocytosis is a front-line defense against pathogen attack, so almost by definition, a pathogen is an infectious agent that avoids being killed by phagocytosis. Phagocytosis is a process that requires a mechanism for self–nonself discrimination [5]. Macrophages possess numerous receptors that allow the direct recognition of particles based on novel sugars, lipids, protein sequences, and concentrations of charge that are unique to pathogens (so-called pathogen-associated molecular patterns) (Fig. 7.2a, b). Particles may also be recognized indirectly if they are coated with opsonins such as specific antibodies or complement components.

7.1.5 Antigen Presentation by Macrophages and Dendritic Cells

It is generally accepted that antigens derived from extracellular sources must be taken up, processed by macrophages (phagocytic antigen-presenting cells) and dendritic cells (nonphagocytic or much less-phagocytic antigen-presenting cells), and afterward presented to T

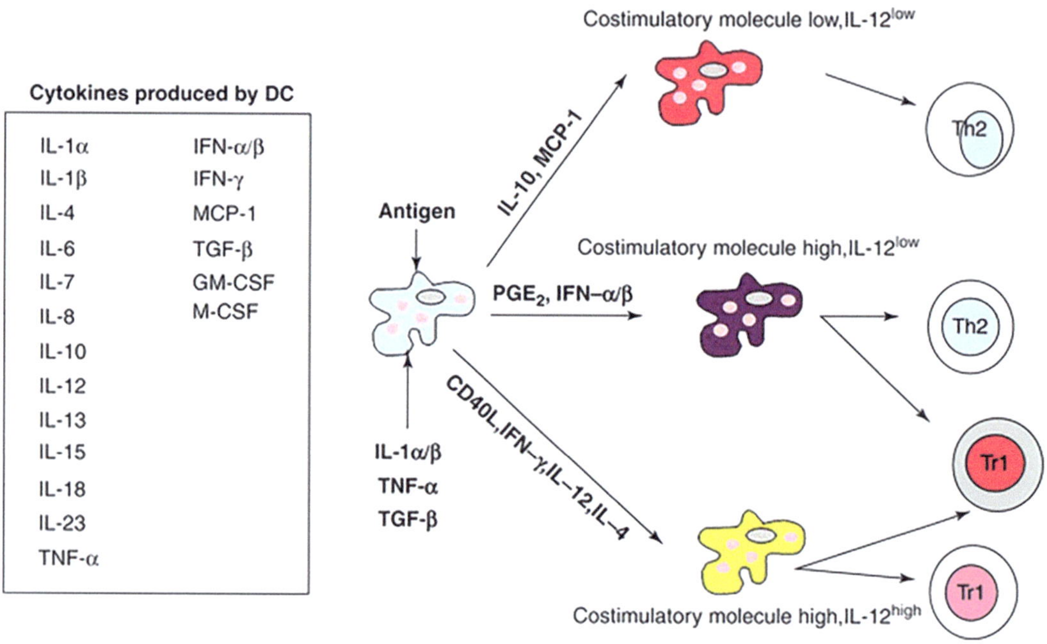

Fig. 7.3 Schematic function of dendritic cells

lymphocytes (Fig. 7.3). The process of uptake, processing, and presentation is now well understood [6, 7]. However, recognition of the antigen major histocompatibility complex (MHC)-II by the T cell receptor is not sufficient to trigger T cell activation. Moreover, T cell activation needs second co-stimulatory signals from the APC in the form of specific cytokines and coreceptors.

7.1.6 Macrophage Activation

Activated macrophages are strongly positive for class II-MHC molecules and adapted to kill microorganisms and tumor cells and present antigen to T lymphocytes. The classical macrophage activating factor, produced by stimulated Th1 lymphocytes and NK cells, is interferon-gamma [8]. Classical macrophage activation, involving a synergistic interaction between interferon-gamma and a pathogen molecule such as lipopolysaccharide (LPS), is just one of the numerous interactions that occur between distinct stimuli. We know that LPS acts on macrophages to initiate a cascade of inflammatory processes that are essential for innate immunity in the upper respiratory tract such as middle and inner ear as well [9, 10]. T cell products are, of course, only part of the story of macrophage activation. Macrophages respond directly to pathogen-associated molecular patterns (PAMPs). They recognize them through the plasma membrane and cytoplasmic receptors such as the Toll-like receptors and intracellular receptors of the NOD-like receptor (NLR) family [11].

7.1.7 Role of Macrophages in Induction of Immune Tolerance

Immunological tolerance is described as no ability of acquired immunity to respond to specific antigens. Central tolerance induction occurs in the thymus for T cells and the bone marrow for B cells. The main mechanism for central tolerance in T cells is the induction of T cell death. Dendritic cells (DCs) are found in abundance in the thymus, where newly produced T cells are educated to become functional $CD4^+$ T or $CD8^+$ T cells and undergo selection to eliminate clones against self. Low-affinity reactive T cells are positively selected and allowed to survive and reach the periphery. The mechanism of peripheral tolerance is a little different from the central one but includes T cell death, anergy, and active suppression by regulatory T cells (Tregs). In this mechanism, DCs could contribute by inducing apoptosis in T cells and by producing IL-10 that induces Tregs [12].

7.2 Part II: Distribution of Macrophages in Murine and Human Nasal Mucosa

There are so many reports as regards the actual distribution of macrophages and dendritic cells in murine and human nasal mucosa, by employing immunohistochemistry with various specific antibodies to those cells. Ichimiya and Kawauchi reported in their article that in nasal mucosa of conventional (CV) mice, Mac-1 positive macrophages, mast cells, and all cell types of lymphocyte subsets were present [13]. But, in the nasal mucosa of specific pathogen-free mice, all cell types were fewer in number than those of CV mice. And they concluded that macrophages and lymphocytes are mobilized to nasal mucosa, responding to continuous antigenic stimuli, and play an important role in the local defense mechanism of the upper respiratory tract. The analysis of macrophages in human nasal mucosa is abundant, and all published articles demonstrated the significant contribution of macrophages to provoke immune responses and control inflammation in nasal mucosa. Albegger investigated to find out macrophages and lymphocytes in the cluster formation of human nasal polyps, employing light and electron microscopy [14]. And his data indicated that cell clusters consisted mainly of macrophages and lymphoid cells. In their study, within the clusters, the cells showed intimate physical contacts being performed by microvilli with varying lengths, suggesting cell-to-cell interaction. These cell clusters may remind us of morphologically those found in vitro and in vivo in the course of immune responses. Jahnsen et al. in their histological study of human nasal mucosa demonstrated the dense network of human leukocyte antigen-DR$^+$ cells with dendritic morphology not only in the epithelium but also in the lamina propria [15]. In addition, they also reported that, in both compartments, these cells could be divided into two main populations based on their phenotypic characteristics: the majority expressed a macrophage-like phenotype (CD11b$^+$CD14$^+$CD64$^+$CD68$^+$RFD7$^+$), whereas the smaller population was predominantly constituted by CD1c$^+$CD11c$^+$ immature DCs. Krysko and Bachert aimed to determine macrophage phenotypes in nasal mucosa of chronic rhinosinusitis with nasal polyp (CRSwNP) and chronic rhinosinusitis without polyp (CRSsNP) and to examine phagocytosis of *Staphylococcus aureus* (*S. aureus*) in these pathologies [16]. They reported that more M2 macrophages were present in CRSwNP than in CRSsNP. This also was positively correlated with increased levels of IL-5, ECP, and locally produced IgE and decreased levels of IL-6, IL-1β, and IFN-γ. In their study, phagocytosis of *S. aureus* by human tissue-derived macrophages was reduced in CRSwNP as compared to macrophages from the control inferior turbinates. Furthermore, they concluded that decreased phagocytosis of *S. aureus* and an M2 activation phenotype in CRSwNP could potentially contribute to the persistence of chronic inflammation in CRSwNP.

7.3 Part III: Modification of Macrophages and Dendritic Cells and Its Clinical Impact on Inflammatory Disorders Such as Allergic Rhinitis

In this part, we would like to introduce a couple of our experimental data in mice as regards how macrophages or dendritic cells are modifying the sinonasal inflammation such as allergic rhinitis and rhinosinusitis. Mature DCs are established as unrivaled APCs in the initiation of immune responses, whereas steady-state DCs are demonstrated to induce peripheral T cell tolerance and consequently attenuate autoimmune-mediated inflammation in animal experiments [17, 18].

In our series of animal experiments, macrophage activation with OK-432 and its effect on allergic rhinitis [19] and regulatory role of lymphoid chemokines CCL19 and CCL21 in the control of allergic rhinitis [20, 21] are introduced as examples, in order, to explain how macrophages or DCs modify the sinonasal inflammation such as allergic rhinitis and rhinosinusitis.

7.3.1 Endogenous IL-12 Induction from Macrophages by OK-432 and Its Effect on the Murine Allergic Rhinitis Model

OK-432, preparation of a low-virulence strain (Su) of *Streptococcus pyogenes* (Group A) killed by a penicillin and lyophilized, is a stiff inducer of Th1 cytokines and brings out anticancer effect in cancer-bearing mice. OK-432 has been reported to consist of many bacterial components, such as peptidoglycan and M-protein. Recently, Toll-like receptor (TLR) family proteins are reported to play a role of recognition of bacterial components and induce interleukin-12 (IL-12) from macrophages. So, we have examined the role of TLR2 for the recognition of OK-432 by macrophages and the effects of OK-432 on allergic rhinitis model. As results, interestingly, IL-12 production by macrophages derived from TLR2 *knockout* mice was markedly reduced in comparison with that of macrophages derived from wild type of mice (Fig. 7.4). Besides, no regulatory effect of OK-432 was observed on allergic rhinitis model in TLR2 *knockout* mice, although nasal symptom of wild type of mice was attenuated upon nasal antigen challenge after systemic sensitization with OK-432 pretreatment (Figs. 7.5 and 7.6). These findings strongly suggest that OK-432 pretreatment provokes macrophage activation to induce IL-12 via TLR2 signaling pathway and conse-

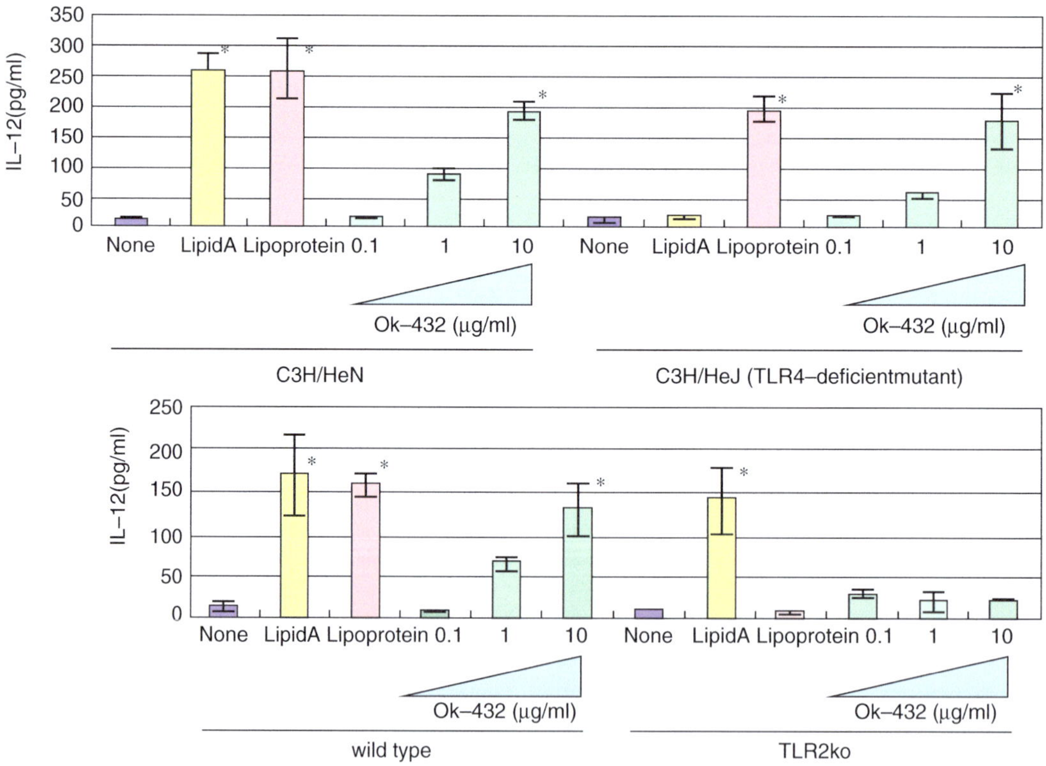

Fig. 7.4 IL-12 production from macrophages with OK-432 stimulation in C3H/HeN, C3H/HeJ, and TLR2 knockout mice

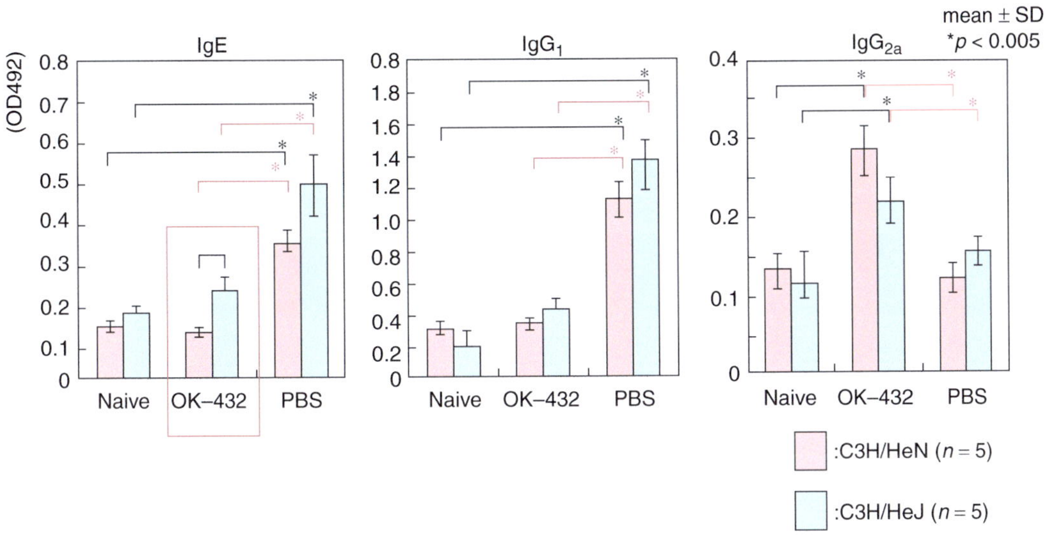

Fig. 7.5 Antigen-specific Th1 and Th2 antibody in serum of C3H/HeN and C3H/HeJ mice after systemic sensitization with OVA and CFA

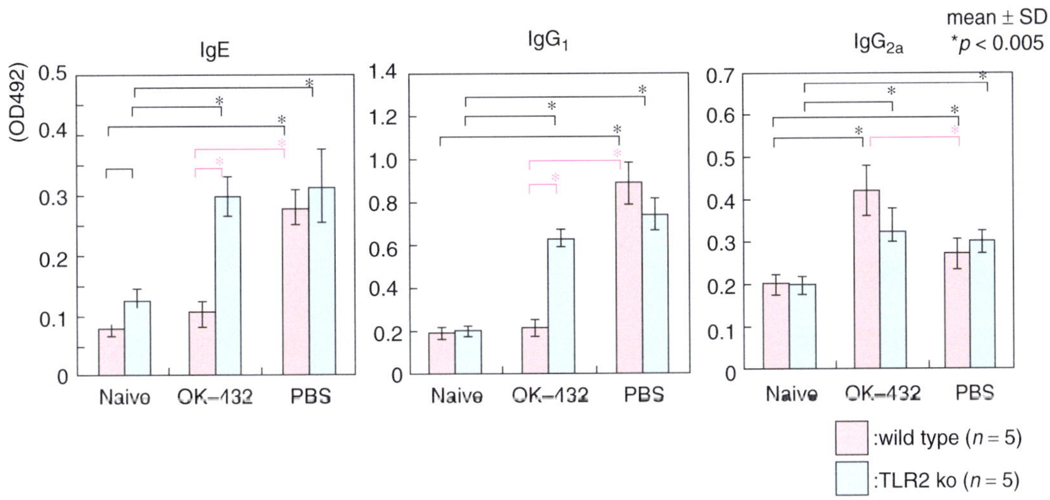

Fig. 7.6 Antigen-specific Th1 and Th2 antibody titers in sera of TLR2 knockout and wild-type mice

quently suppress Th2-mediated allergic inflammation in nasal mucosa (Fig. 7.7).

7.3.2 Regulatory Role of Lymphoid Chemokines CCL19 and CCL21 in the Control of Allergic Rhinitis

The lymphoid chemokines CCL19 and CCL21 are known to be crucial both for lymphoid cell trafficking and for the structural organization of lymphoid tissues such as nasopharynx-associated lymphoid tissue (NALT). However, their role in allergic responses remains unclear, and so our current study aims to shed light on the role of CCL19/CCL21 in the development of allergic rhinitis. After nasal challenge with OVA, OVA-sensitized *plt* (paucity of lymph node T cells) mice, which are deficient in CCL19/CCL21, showed more severe allergic symptoms than did identically treated wild-type mice [22] (Fig. 7.8). OVA-specific IgE production, eosinophil infiltration, and Th2 responses were enhanced in the upper air-

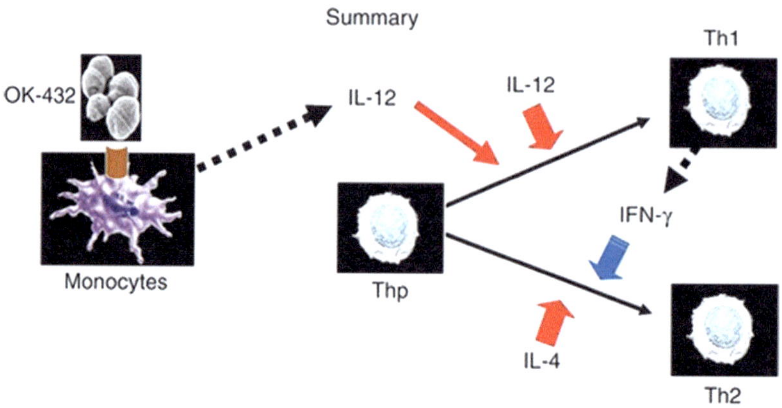

Fig. 7.7 Summary of the effect of OK-432, which is a potent Th1 inducive biological response modifiers, on the murine allergic rhinitis model. (1) OK-432 seems to induce IL-12 production from macrophages via TLR2 and activate Th1 response and consequently downregulate antigen-specific Th2 response. (2) Prophylactic treatment with like as OK-432 (Th1 inducer) may be anticipated to regulate the induction phase of type-I allergic response

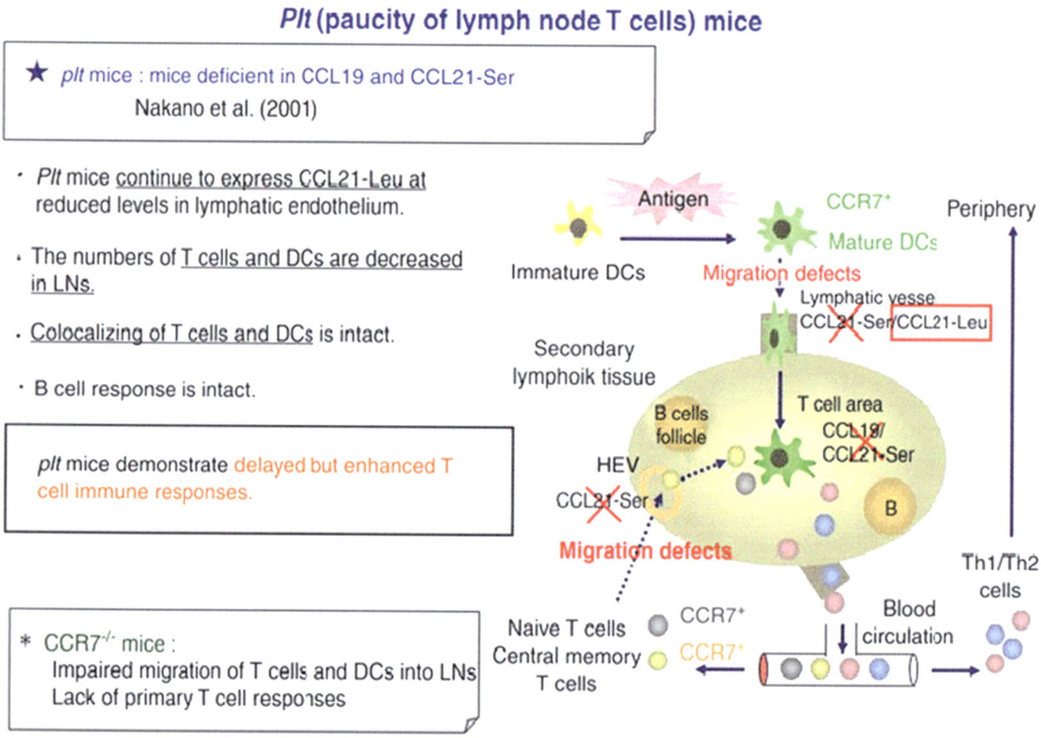

Fig. 7.8 Features of *plt* (paucity of lymph node T cells) mice

way of *plt* mice. Moreover, in *plt* mice, the number of $CD4^+CD25^+$ regulatory T cells declined in the secondary lymphoid tissues, whereas the number of Th2-inducer-type $CD8^+CD11b^+$ myeloid dendritic cells (m-DCs) increased in cervical lymph nodes and NALT. Nasal administration of the plasmid-encoding DNA of CCL19 resulted in the reduction of m-DCs in the secondary lymphoid tissues and the suppression of allergic responses in *plt* mice. These results suggest that CCL19 and

CCL21 act as regulatory chemokines for the control of airway allergic disease and so may offer a new strategy for the control of allergic disease. In a different study to focus on dendritic cells of regional lymph node, we constructed an effective murine model of sublingual immunotherapy (SLIT) in allergic rhinitis, in which mice were sublingually administered with ovalbumin (OVA) followed by an intraperitoneal sensitization and nasal challenge of OVA [20]. Sublingually treated mice showed significantly decreased allergic responses as well as suppressed Th2 immune responses (Fig. 7.9). Sublingual administration of OVA did not alter the frequency of CD4$^+$CD25$^+$ regulatory T cells (Tregs), but led to the upregulation of Foxp3- and IL-10-specific mRNAs in the Tregs of cervical lymph nodes (CLN), which strongly suppressed Th2 cytokine production from CD4$^+$CD25$^-$ effector T cells in vitro. Furthermore, sublingual administration of plasmids encoding the lymphoid chemokines CCL19 and CCL21-Ser DNA together with OVA suppressed allergic responses (Fig. 7.10) [23]. These results suggest that IL-10-expressing CD4$^+$CD25$^+$Foxp3$^+$ Tregs in CLN are involved in the suppression of allergic responses and that CCL19/CCL21 may contribute to it in mice received SLIT (Figs. 7.11a–c).

To summarize our recent data, the important regulatory role of macrophage or dendritic cells and their interaction with T cells in nasal mucosa and its regional lymphoid organ are extensively demonstrated in accordance with human studies on these cells. However, further extensive basic and clinical research is required for pursuing the ideal treatment strategy.

Effect of Lipopolysaccharide (LPS) on Eliciting Phase of Murine Allergic Rhinitis Model in Relation with Toll-Like Receptor

Mast cells which are the key player at the eliciting phase of allergic rhinitis have been reported

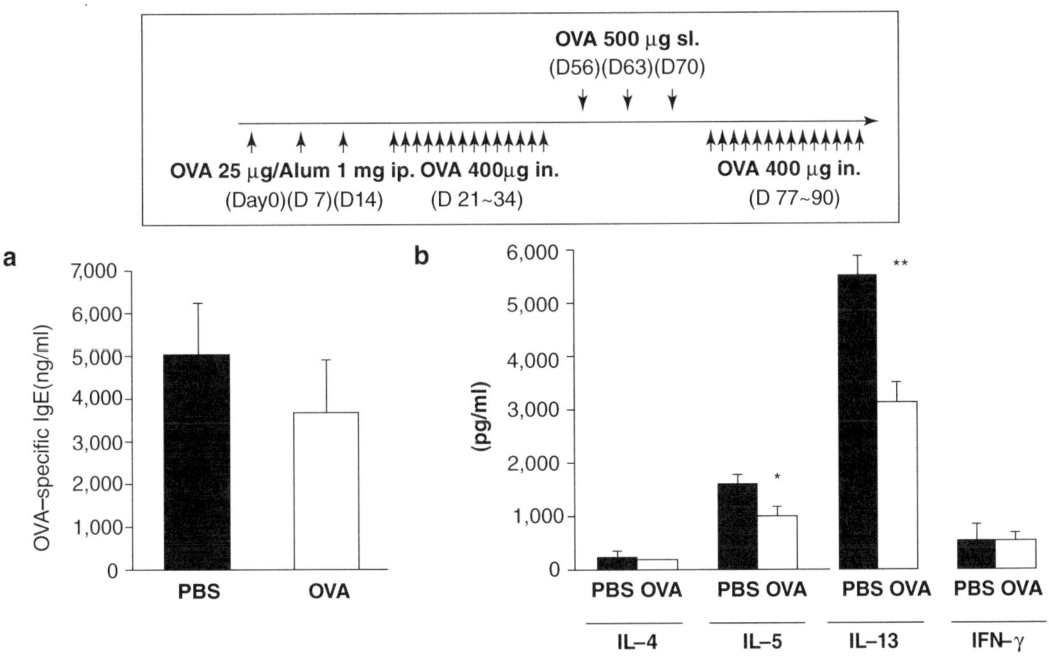

Fig. 7.9 Antigen-specific serum IgE production and Th1/Th2 profile in spleen of mice which received therapeutic sublingual OVA treatment after induction of allergic rhinitis. Mice were sublingually administered with either PBS or OVA after intraperitoneal sensitization and nasal challenges with OVA. Thereafter, the mice received consecutive nasal challenges with OVA again and examined for their allergic responses. (**a**) OVA-specific IgE levels in serum were assayed by sandwich ELISA. (**b**) Culture supernatants of CD4$^+$T cells of spleen obtained from sublingually treated mice with allergic rhinitis were assessed for Th1 and Th2 cytokine production levels by ELISA. These data are representative of two independent experiments containing three to five mice in each group. Significance was evaluated by an unpaired t test. $*p < 0.05$, $**p < 0.01$

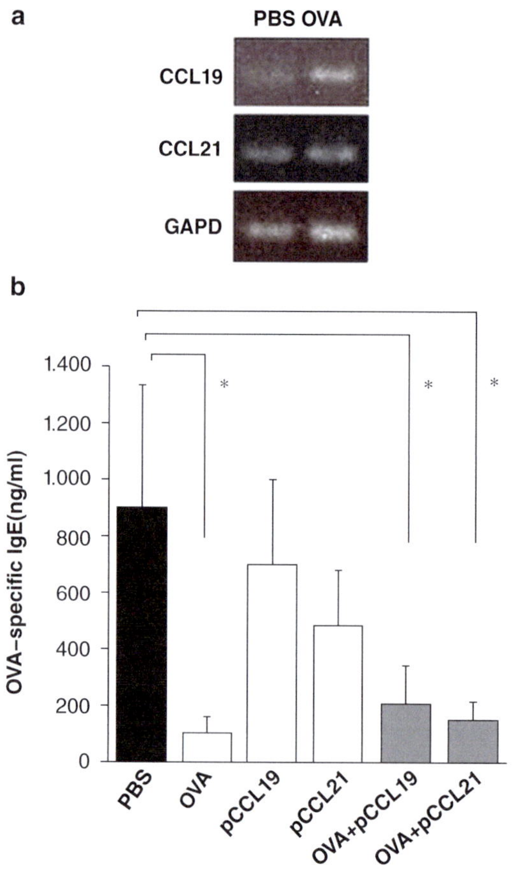

Fig. 7.10 Chemokine expression in CLN of sublingually treated mice and the effect of sublingual administration of pCCL19/pCCL21 with antigen on Th2-mediated allergic responses. (**a**) Semiquantitative RT-PCR was performed to assess mRNA expression pattern of chemokines, CCL19 and CCL21 in whole cells isolated from CLN of mice sublingually treated with either PBS or OVA. (**b**) Mice were sublingually administered with either PBS, OVA alone, 100 mg of pCCL19 together with OVA, and 100 mg of pCCL21 with OVA for total three times before systemic sensitization and nasal challenge. OVA-specific IgE levels in serum were assayed by sandwich ELISA

to produce Th2 cytokines in vitro with lipopolysaccharide (LPS) stimulation via TLR4, but in vivo study remains to be performed [24, 25]. As it is reported in the hygiene hypothesis that neonatal exposure with LPS prevents allergic airway diseases, allergic inflammation is generally reported to be downregulated at the induction phase with the existence of bacterial LPS [26–29]. However, it is still controversial that LPS effect on the eliciting phase of allergic inflammation. Therefore, we investigated the LPS effect on the eliciting phase of murine allergic rhinitis model.

As recently reported by Aoi et al., in our murine allergic rhinitis model, LPS instillation into the nasal cavity with ovalbumin (OVA) resulted in exacerbated nasal symptom and eosinophil infiltration of wild type of mice (LPS-responsive BALB/c strain) [30]. On the other hand, nasal symptom and eosinophil infiltration were evident in C3H/HeN (LPS-responsive) mice, but neither in C3H/HeJ (LPS-non-responsive) mice, nor mast cell-deficient WBB6F1 W/W^v mice. IL-5 production by mast cells in the nasal mucosa of wild type of mice (BALB/c) was enhanced by LPS co-instillation. But it was enhanced neither in those of C3H/HeJ mice nor WBB6F1 W/W^v mice. These data obtained with mast cell-deficient WBB6F1 W/W^v mice may indicate that LPS aggravated nasal symptom, upregulating Th2 cytokine production of mast cells via TLR4.

Therefore, we are summarizing our experimental data as follows. LPS instillation into nasal cavity, at the eliciting phase of murine model of allergic rhinitis, actually exacerbates nasal symptom, which is accompanied by mast cell activation and enhanced Th2 responses. These observations can be extrapolated into the human condition with better understanding the mechanisms of bacterial infection-induced exacerbation of the clinical features of allergic rhinitis. However, LPS concentration should be taken into account how does it affect on the nasal symptom at the eliciting phase as well as the induction of allergic rhinitis. Most our recent experimental data came to the conclusion that a low dose of LPS at eliciting phase of allergic rhinitis can exacerbate allergic nasal symptom but high dose of LPS at eliciting phase of allergic rhinitis conversely downregulate nasal allergic symptom (Unpublished data). So, mast cell can be concluded also the key player as well as macrophage to modify upper respiratory allergic reactions.

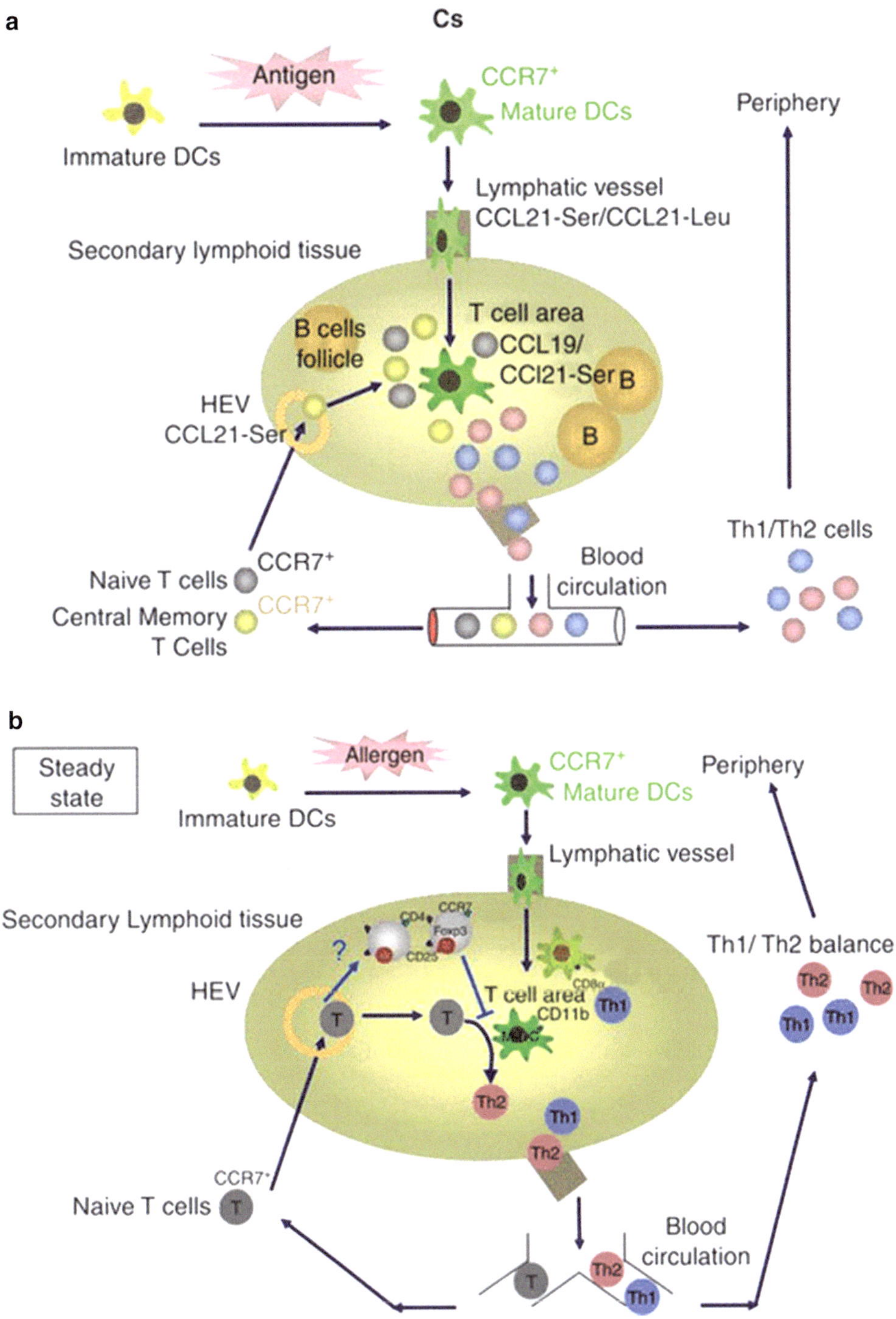

Fig. 7.11 (**a**) Chemokine receptor CCR7 and its ligands CCL19 and CCL21 are involved in the chemotaxis of T cells and DCs. (**b**) In steady state, naturally occurring Tregs have inhibitory effects to the interaction of naïve T cells and m-DCs for the suppression of excessive Th2 differentiation to inhaled allergen. (**c**) The deficiency of CCL19/CCL21 somehow inhibits the accumulation of Tregs, which work as suppressor of Th2 environment induced by m-DCs in the secondary lymphoid tissues, resulted in the establishment of Th2-dominant allergic disease

c

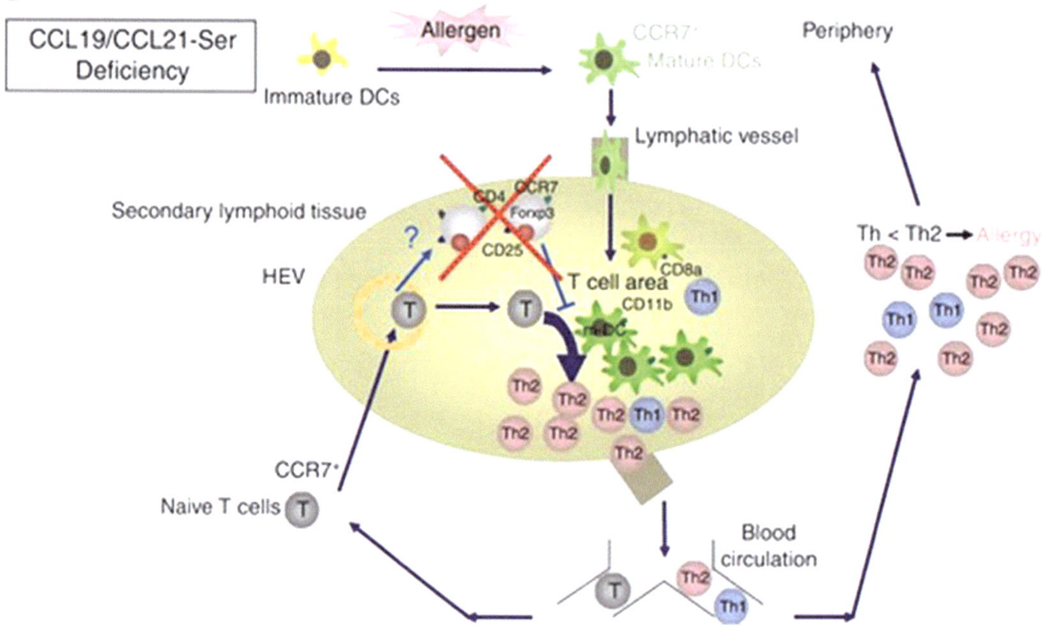

Fig. 7.11 (continued)

References

1. Stanley ER. Lineage commitment: cytokines instruct, at last! Cell Stem Cell. 2009;5(3):234–6.
2. Geissmann F, Manz MG, Jung S, et al. Development of monocytes, macrophages, and dendritic cells. Science. 2010;327(5966):656–61.
3. Taylor PR, Martinez-Pomares L, Stacey M, et al. Macrophage receptors and immune recognition. Annu Rev Immunol. 2005;23:901–44.
4. Sallusto F, Baggiolini M. Chemokines and leukocyte traffic. Nat Immunol. 2008;9(9):949–52.
5. Aderem A, Underhill DM. Mechanisms of phagocytosis in macrophages. Annu Rev Immunol. 1999;17:593–623.
6. Hume DA. Macrophages as APC and the dendritic cell myth. J Immunol. 2008;181(9):5829–35.
7. Jutras I, Desjardins M. Phagocytosis: at the crossroads of innate and adaptive immunity. Annu Rev Cell Dev Biol. 2005;21:511–27.
8. Schroder K, Hertzog PJ, Ravasi T, Schroder K, Hertzog PJ, Ravasi T, et al. Interferon-gamma: an overview of signals, mechanisms and functions. J Leukoc Biol. 2004;75:163–89.
9. Kawai T, Akira S. The role of pattern-recognition receptors in innate immunity: update on toll-like receptors. Nat Immunol. 2010;5:373–84.
10. Kawauchi H, DeMaria TF, Lim DJ. Endotoxin permeability through the round window. Acta Otolaryngol Suppl. 1998;457:100–15.
11. Chen G, Shaw MH, et al. NOD-like receptors: role in innate immunity and inflammatory disease. Annu Rev Pathol. 2009;4:365–98.
12. Lipscomb MF, Masten BJ. Dendritic cells: immune regulators in health and disease. Physiol Rev. 2002;82:97–130.
13. Ichimiya I, Kawauchi H, Fujiyoshi T, et al. Distribution of immunocompetent cells in normal nasal mucosa: comparisons among germ-free, specific pathogen-free, and conventional mice. Ann Otol Rhinol Laryngol. 1991;100(8):638–42.
14. Albegger KW. Cluster formation in human nasal polyps. A light: and electron-microscopic investigation. ORL J Otorhinolaryngol Relat Spec. 1977;39(2):107–12.
15. Jahnsen FL, Gran E, Haye R, et al. Human nasal mucosa contains antigen-presenting cells of strikingly different functional phenotypes. Am J Respir Cell Mol Biol. 2004;30:31–7.
16. Krysko O, Holtappels G, Zhang G, et al. Dendritic cells ameliorate autoimmunity in the CNS by controlling the homeostasis of PD-1 receptor (+) regulatory T cells. Allergy. 2011;66(3):396–403.
17. Mizuno S, Kanai T, Mikami Y, et al. CCR9(+) plasmacytoid dendritic cells in the small intestine suppress development of intestinal inflammation in mice. Immunol Lett. 2012;146(1–2):64–9.
18. Yogev N, Frommer F, Lukas D, et al. Dendritic cells ameliorate autoimmunity in the CNS by controlling the homeostasis of PD-1 receptor (+) regulatory T cells. Immunity. 2012;37(2):264–75.

19. Kawauchi H, Aoi N, Murata A, et al. Clinical application of mucosal immune system for down-regulating nasal allergy. Arerugi. 2009;58(2):103–11.
20. Kawauchi H, Goda K, Tongu M, et al. Short review on sublingual immunotherapy for patients with allergic rhinitis: from bench to bedside. Adv Otorhinolaryngol. 2011;72:103–6.
21. Takamura K, Fukuyama S, Nagatake T, et al. Regulatory role of CCL19 and CCL21 in the control of allergic rhinitis. J Immunol. 2007;179(9):5897–906.
22. Nakano H, Gunn MD. Gene duplications at the chemokine locus on mouse chromosome 4: multiple strain-specific haplotypes and the deletion of secondary lymphoid-organ chemokine and EBI-1 ligand chemokine genes in the plt mutation. J Immunol. 2001;166:361–9.
23. Yamada T, Tongu M, Goda K, et al. Sublingual immunotherapy induces regulatory function of IL-10-expressing $CD4^{+}CD25^{+}Foxp3^{+}$ T cells of cervical lymph nodes in murine allergic rhinitis model. J Allergy. 2012;2012:490905.
24. Masuda A, Yoshikai Y, Aiba K, et al. Th2 cytokine production from mast cells is directly induced by lipopolysaccharide and distinctly regulated by c-Jun N-terminal kinase and p38 pathways. J Immunol. 2002;169(7):3801–10.
25. Rodriguez D, Keller AC, Faquim-Mauro EL, et al. Bacterial lipopolysaccharide signaling through toll-like receptor 4 suppresses asthma-like responses via nitric oxide synthase 2 activity. J Immunol. 2003;171(2):1001–8.
26. Aoi N, Morikura I, Fuchiwaki T, et al. OK432 administration inhibits murine allergic rhinitis at the induction phase, through the macrophage activation with TLR2 signaling pathway. Med Sci. 2018;6(4):107–20.
27. Belderbos M, Levy O, Bont L, Belderbos M, et al. Neonatal innate immunity in allergy development. Curr Opin Pediatr. 2009;21(6):762–9.
28. Bortolatto J, Borducchi E, Rodriguez D, et al. Toll-like receptor 4 agonists adsorbed to aluminium hydroxide adjuvant attenuate ovalbumin-specific allergic airway disease: role of MyD88 adaptor molecule and interleukin-12/interferon-gamma axis. Clin Exp Allergy. 2008;38(10):1668–79.
29. Wang Y, McCusker. Neonatal exposure with LPS and/or allergen prevents experimental allergic airways disease: development of tolerance using environmental antigens. J Allergy Clin Immunol. 2006;118(1):143–51.
30. Aoi N, Fuchiwaki T, Morikura I, Kawauchi H, et al. Nasal administration of lipopolysaccharide exacerbates allergic rhinitis through Th2 cytokine production from mast cells. Allergie. 2021;1(4):216–24.

The Neutrophil and Chronic Rhinosinusitis

8

Martin Y. Desrosiers and Shaun J. Kilty

Core Messages

- Chronic rhinosinusitis (CRS) is a complex heterogeneous inflammatory disease that has traditionally been characterized as primarily "eosinophilic" in nature but in which contributions from other cell types may lead to different sub-phenotypes of disease.
- Neutrophils are key inflammatory cells in the immune response, and while their role is less well understood in CRS, it may nevertheless be significant.
- A better understanding of neutrophilic inflammation in CRS could lead to the development of new therapeutic strategies for CRS.

Descriptions of chronic rhinosinusitis (CRS) are frequently characterized by references to the eosinophilia present in CRS and in allergic diseases. However, the focus in the assessment of inflammation in chronic rhinosinusitis is increasingly shifting away from Th2-dominated mechanisms to a consideration of the contributions by Th1 and Th17 mechanisms as well. This has led to an increased focus of interest in the assessment of lymphocyte subpopulations and of cytokines associated with these inflammatory pathways.

Throughout this, the neutrophil, a somewhat ubiquitous inflammatory cell associated with both Th1 and Th17 patterns of inflammation, has been somewhat ignored. In this chapter, we review the structure and function of the neutrophil and review evidence for its potential implication in chronic rhinosinusitis.

8.1 Histologic Description

Neutrophils are the most abundant leukocyte in humans representing up to 60% of the circulating white blood cells. It is a member of the family of granulocytes, which also includes basophils and eosinophils. Neutrophils are considered an essential component of the innate immune system by virtue of their multiple actions in bacterial killing and sequestration. Neutrophils cause microbial death by three means: phagocytosis, by generating neutrophil extracellular traps (NETs), and through the release of soluble antimicrobials from their primary and specific granules [1].

Neutrophils derived their name from their differential response in staining from other granulocytes with hematoxylin and eosin or Wright's Giemsa staining. Eosinophils intensely capture eosin, giving them their characteristic appearance under the microscope. The abundant azurophilic

M. Y. Desrosiers
Division of Otolaryngology—Head and Neck Surgery, Université de Montréal/Centre Hospitalier de l'Université de Montréal (CHUM) Hôpitale Hôtel-Dieu de Montréal, Montréal, QC, Canada

S. J. Kilty (✉)
Department of Otolaryngology—Head and Neck Surgery, The University of Ottawa/The Ottawa Hospital, Ottawa, ON, Canada

Ö. Ö. Celebi, T. M. Önerci (eds.), *Nasal Physiology and Pathophysiology of Nasal Disorders*,
https://doi.org/10.1007/978-3-031-12386-3_8

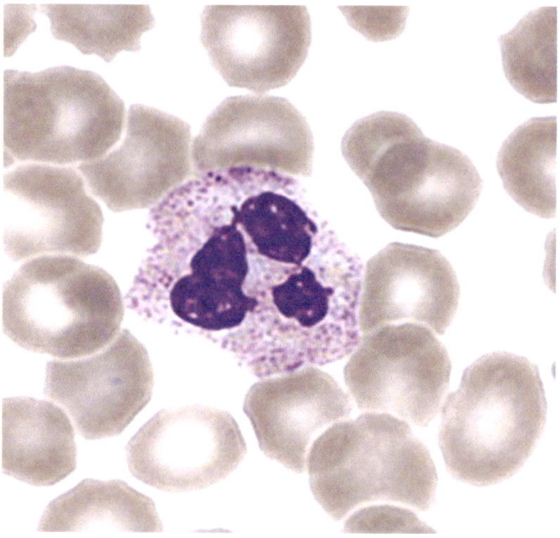

Fig. 8.1 Neutrophil on blood smear, Wright's Giemsa stain. Original magnification 100× oil immersion. Note the characteristic multi-segmented nucleus and mostly lilac-staining specific (secondary) granules and a few azurophilic primary granules. Surrounding cells are mature erythrocytes. (Credit: Ruth F. Padmore MD, FRCPC, PhD, Ottawa, Canada)

granules of the basophils give them a bluish hue. The neutrophil is instead characterized by the neutral aspect of its cytoplasmic staining and its multiply segmented nucleus (Fig. 8.1).

Neutrophils have a diameter of 12–15 μm. In circulating blood, neutrophils are in a quiescent, or resting state, and have a vaguely circular form. However, when activated, the shape changes, with the cell becoming more amoeba-like, with pseudopods extending in search of antigens.

The granules present in neutrophils contain a number of substances with either bactericidal or proteolytic actions. Primary (azurophilic) granules typically contain the bactericidal enzyme defensins and cationic proteins. They also contain proteolytic enzymes, cathepsin G, lysozyme, and myeloperoxidase. The specific (secondary) granules contain lysozyme, lactoferrin, as well as compounds involved in the formation of toxic oxygen species [2].

8.2 Hematologic Progenitors

As granulocytes, neutrophils share their origins with basophils and eosinophils. These are derived from progenitor cells in the bone marrow and then differentiate through myelocytes into promyelocytes and then into the final differentiated cell.

Neutrophils have a relatively short half-life, up to 6 days. This is believed both to prevent the spread of pathogens that may parasitize within neutrophils to facilitate survival and dispersion and also to limit local tissue damage caused by the intense antibacterial activity of the neutrophil in a tissue.

8.3 Physiology and Function

Neutrophils are principally responsible for bacterial killing, which they enact through a variety of mechanisms. Prior to serving this role, neutrophils are primarily present in the circulation in an inactivated form. They then migrate into target tissues to exert their effects following activation by a variety of pro-inflammatory signals.

The transit from the bloodstream to tissue begins with a process called diapedesis. This leukocyte extravasation process occurs when the activated neutrophil approaches the periphery of the blood vessel and then becomes attached to and migrates through the blood vessel wall in a process called diapedesis. This involves an interaction with receptors called intracellular adhesion molecules (ICAMs) and various selectins and integrins.

Arrived at the site of infection, the neutrophil exerts its antibacterial action through the following mechanisms:

1. Phagocytosis and killing of pathogens.
2. Neutrophil extracellular trap (NET).
3. Protease digestion through the release of granule contents.

The activated neutrophil can phagocytose bacteria by engulfing them and killing them within the cell through the generation of toxic substances. This process called the "oxidative burst" creates a high concentration of reactive oxygen species (ROS) through a process involving NADPH oxidase activation and the creation of superoxide dismutase (SOD). Through several steps, this leads to the production of hypochlorous acid (HClO), which may be bactericidal in itself and/or lead to the activation of the necessary proteases [3].

More recently, a role for extracellular trapping and killing of bacteria by neutrophils has been described by the formation of "extracellular NETs." In this mechanism, neutrophils secrete a mesh composed of DNA and various proteins outside the neutrophil, which serves to trap and destroy bacteria (Fig. 8.2). This may also limit the propagation of infection as well. NETs which form intravascularly are responsible for many of the clinical manifestations of sepsis.

The neutrophil also plays a role in limiting potentially negative effects of inflammation through the secretion of various serine proteases. The most well known of these is alpha-1-antitrypsin, which serves to limit the extent of damage caused by neutrophil elastase released from granules in tissue. Individuals with low levels of alpha-1-antitrypsin may have inordinate responses to trauma. In smokers, this may lead to the development of emphysema.

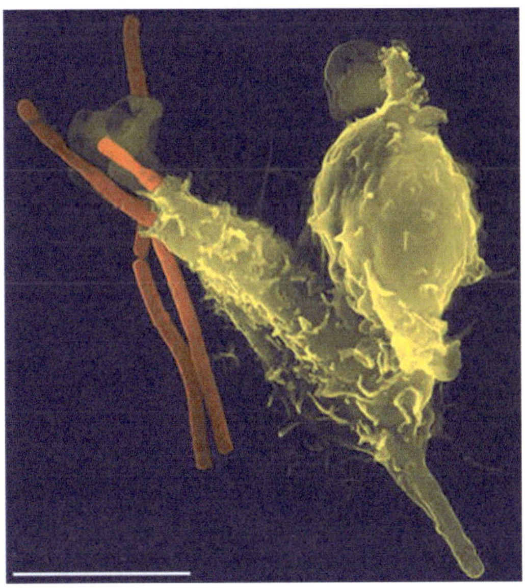

Fig. 8.2 Neutrophil extravasation trap (NET). In this scanning electron microscope image, an Anthrax bacteria (*orange*) is being engulfed by a single neutrophil (*yellow*), by the generation of a NET. (By Volker Brinkmann [CC-BY-2.5 (http://creativecommons.org/licenses/by/2.5)], via Wikimedia Commons)

8.4 Implication in Disease

Low levels of circulating neutrophils, or neutropenia, may occur from genetic causes or from infectious or toxic causes, one of which the most familiar is chemotherapy. Individuals with neutropenia are particularly susceptible to infection by bacterial pathogens.

The neutrophil may also be involved primarily in inflammatory diseases. Familial Mediterranean fever is a relapsing disorder where individuals present with bouts of acute inflammation characterized by hyperthermia, arthralgia, and peritonitis. Due to a dysfunction in the MEFV gene, individuals have a reduced amount or structural malformations of a protein called pyrin which makes up part of the cytoskeleton of leukocytes. Pyrin abnormalities lead to defective inflammation regulation and, subsequently, to inappropriate or prolonged inflammatory activity [4]. It is a disorder not directly linked to neutrophil level, but rather to function, as described for alpha-1-antitrypsin deficiency above.

8.5 Implications in Respiratory Disease

While interest in the role of granulocytes in the development of chronic respiratory disease has principally focused on the role of the eosinophil in the pathogenesis of asthma, evidence is increasing to support the role of the neutrophil in the development of both asthma and chronic obstructive pulmonary disease.

Assessment of a large cohort of individuals with hard-to-treat asthma has identified phenotypes characterized by neutrophilic inflammation in expectorated sputum. This has led to a consideration of neutrophilia in the pathogenesis of steroid resistance in asthma [5, 6].

The neutrophil has also been implicated in chronic obstructive pulmonary disease. Apart from alpha-1-antitrypsin deficiency, as mentioned above, neutrophil elastase has been shown to be increased in individuals with COPD [7]. This is even more pronounced in individuals with COPD and also presents symptoms of bronchitis. Experimental evidence implicating the neutrophil in COPD exists [6]. It has also been demonstrated that the administration of aerosolized neutrophil elastase in mice leads to rapid damage to the surface epithelium with loss of ciliated respiratory, mucosa, and resultant mucosal hyperplasia [8].

8.6 Implication in CRS

As for lower respiratory tract disease, interest in the inflammation in CRS has focused mainly on Th2-mediated inflammation due to the postulated importance of the eosinophil, and therapeutic strategies have focused mainly on the Th2 axis. However, an increasing body of work is emerging to suggest that the neutrophil may play a role in the development of CRS in a subpopulation of patients with this disorder.

The first attention to neutrophils in CRS was derived from reports of predominant neutrophilia in nasal polyps from Asian subjects and in patients with cystic fibrosis [9, 10]. However, later reports have demonstrated pronounced heterogeneity in these groups, with both high-neutrophil and low-neutrophil subgroups being present. Interestingly, a focus on differences between Asian and Caucasian neutrophils in polyps has identified the existence of two groups, one roughly characterized by high IL-5 levels and the second by a predominantly Th1/Th17 activation pattern [11].

Work in our laboratory with cultured sinus epithelial cells harvested from CRS patients and controls without CRS has identified a molecular signature with high spontaneous inflammation present in only a subgroup of CRS cells. Interestingly, simultaneously obtained biopsy specimens from patients where cultures were obtained show similar levels of tissue eosinophilia in both populations. However, neutrophilic infiltrate is present in only the group with the high-inflammation molecular expression signature.

8.7 Therapeutic Implications

Implications for the importance of this finding are characterized by recent studies outlining poor prognosis and lesser response to steroid therapy in individuals with "neutrophilic" CRS. In a study from Brazil, [12]) report that high NFkB activation, a characteristic feature of Th1 axis activity, results in a poorer response to steroids and a rapid recurrence of the disease.

The impact on response to therapy has also potential importance, as a Chinese group recently identified a lesser response to oral steroids in individuals with neutrophilia on biopsy of their nasal polyps. In a similar direction, Al-Mot et al. [13] identified a predominant neutrophilia in the sinus pathology of post-ESS patients who demonstrated a poor response to topical steroid irrigations, whereas patients whose polyps had a low neutrophil level had a favorable response to topical steroid therapy.

8.8 Summary

Taken together, these findings suggest that the neutrophil may play an important role in the subgroup of patients with a "neutrophil" predominant phenotype of CRS. New methods for recognizing the presence of this disease will be important, as this "neutrophil" phenotype is independent of currently used clinical phenotypic markers, the presence or absence of nasal polyposis. This may require novel diagnostic approaches to tailor therapy to individual disease status.

Lastly, recognition of the potential importance of this subgroup may require a different therapeutic approach than has been employed to date for "routine" CRS, possibly requiring the use of alternative, nonsteroid-based anti-inflammatory treatments for the management of disease in these individuals.

References

1. Brinkmann V, Zychlinsky A. Neutrophil extracellular traps: is immunity the second function of chromatin? J Cell Biol. 2012;198:773–83.
2. Hager M, Cowland JB, Borregaard N. Neutrophil granules in health and disease. J Intern Med. 2010;268:25–34.
3. van den Berg JM, Kuijpers TW. Educational paper: defects in number and function of neutrophilic granulocytes causing primary immunodeficiency. Eur J Pediatr. 2011;170:1369–76.
4. Savic S, Dickie LJ, Battellino M, et al. Familial Mediterranean fever and related periodic fever

syndromes/autoinflammatory diseases. Curr Opin Rheumatol. 2012;24:103–12.
5. Gibson PG, Simpson JL, Saltos N. Heterogeneity of airway inflammation in persistent asthma*: evidence of neutrophilic inflammation and increased sputum interleukin-8. Chest. 2001;119(5):1329–36.
6. Moore WC, Meyers DA, Wenzel SE, et al. Identification of asthma phenotypes using cluster analysis in the severe asthma research program. Am J Respir Crit Care Med. 2010;181:315–23.
7. Foreman MG, Campos M, Celedon JC. Genes and chronic obstructive pulmonary disease. Med Clin North Am. 2012;96:699–711.
8. Voynow JA, Fischer BM, Malarkey DE, et al. Neutrophil elastase induces mucus cell metaplasia in mouse lung. Am J Physiol Lung Cell Mol Physiol. 2004;287(6):L1293–302.
9. Rowe-Jones JM, Shembekar M, Trendall-Smith N, et al. Polypoidal rhinosinusitis in cystic fibrosis: a clinical and histopathological study. Clin Otolaryngol Allied Sci. 1997;22:167–71.
10. Wen W, Liu W, Zhang L, et al. Increased neutrophilia in nasal polyps reduces the response to oral corticosteroid therapy. J Allergy Clin Immunol. 2012;129:1522–8.
11. Ba L, Zhang N, Meng J, et al. The association between bacterial colonization and inflammatory pattern in Chinese chronic rhinosinusitis patients with nasal polyps. Allergy. 2011;66(10):1296–303. https://doi.org/10.1111/j.1398-9995.2011.02637.x.
12. Cardoso Pereira Valera F, Queiroz R, Scrideli C, et al. NF-κβ expression predicts clinical outcome for nasal polyposis. Rhinology. 2010;48:408–14.
13. Al-Mot S, Filali-Mouhim A, Rousseau S, et al. Molecular signatures as a new classification scheme for CRS. Otolaryngol Head Neck Surg. 2011;145(2 Suppl):P125.

9 Eosinophils in Rhinologic Diseases

Jens Ponikau, Mary Twarog, David Sherris, and Hirohito Kita

Core Messages

- Eosinophils are the key effector cells in chronic rhinosinusitis (CRS) and also play an important role in allergic rhinitis.
- Eosinophils degranulate in CRS the mucus and not in the tissue.
- Eosinophil granular proteins, especially major basic protein (MBP), mediate the epithelial damage in CRS.
- Secreted protein(s) from the airborne fungus *Alternaria alternata* have been identified as a trigger for the eosinophilic inflammation and degranulation in CRS patients.

9.1 Introduction

The eosinophil granulocyte, although likely first observed by Wharton Jones in 1846 in unstained preparations of peripheral blood, was so named by Paul Ehrlich in 1879 because of the intense staining of its granules with the acidic dye eosin [1]. Since that time the eosinophil has been the subject of extensive investigation. Its occurrence in such disparate conditions as parasitic infections, presumably for the benefit of the human host, and hypersensitivity diseases, perhaps to the detriment of the patient, although paradoxical, has become better understood as a consequence of newer information. Eosinophils are resident and non-pathologic in various organs such as gastrointestinal tract and mammary glands, and they may play roles in the tissue and immune homeostasis of these organs.

However, eosinophils are strikingly absent in the nose and paranasal sinuses in healthy individuals, which is in contrast to their presence in three distinct rhinologic diseases:

1. Chronic rhinosinusitis (CRS).
2. Allergic rhinitis (AR).
3. Upper respiratory viral infection (common cold).

In addition, nonallergic rhinitis with eosinophilia (NARES) has been described as a syndrome. However, it is poorly defined only as a lack of detectable IgE combined with evidence of eosinophils present in the nasal cavity. Coupled with the understanding that CRS is not limited to the sinuses but usually also involves the nasal cavity (and consequently the terminology change from chronic sinusitis to chronic rhinosinusitis), and the similarity of symptoms, no evidence exists to distinguish NARES from a mild or early stage of CRS, and for the purpose of this chapter, is assumed as such.

J. Ponikau (✉) · M. Twarog · D. Sherris
Department of Otorhinolaryngology, Gromo Institute and Sinus Center, Jacobs School of Medicine and Biomedical Sciences, University at Buffalo, The State University of New York, Buffalo, NY, USA
e-mail: jponikau@buffalo.edu

H. Kita
Department of Medicine and Immunology, Mayo Clinic Scottsdale, Scottsdale, AZ, USA

Ö. Ö. Celebi, T. M. Önerci (eds.), *Nasal Physiology and Pathophysiology of Nasal Disorders*, https://doi.org/10.1007/978-3-031-12386-3_9

In the above immune responses, eosinophils are recruited into the sites of inflammation where they produce an array of cytokines and lipid mediators, and release toxic granule proteins. These molecules may regulate immune response, cause tissue damage, and facilitate tissue repair. Eosinophils can also present antigens to naïve and memory T cells and initiate/amplify antigen-specific immune responses. This review summarizes the biological and immunological properties of eosinophils and discusses the roles of eosinophils applied in the field of rhinology, with a focus on chronic rhinosinusitis (CRS).

9.2 Eosinophils at Baseline Condition

9.2.1 Eosinophils Are Resident in Several Tissues at Baseline Condition

The life cycle of the eosinophil is divided into bone marrow, blood, and tissue phases. Eosinophils are produced in bone marrow from pluripotential stem cells. The stem cells differentiate into a progenitor, which is capable of giving rise to mixed colonies of basophils and eosinophils, pure basophil colonies, or pure eosinophil colonies. Among various hematopoietic factors, those important for eosinophil proliferation and differentiation are interleukin (IL)-3, granulocyte-macrophage colony-stimulating factor (GM-CSF), and IL-5. IL-3 and GM-CSF are relatively non-specific and stimulate the proliferation of neutrophils, basophils, and eosinophils. In contrast, IL-5 potently and specifically stimulates eosinophil production [2].

Although the eosinophil is a formed element of the peripheral circulation, it is primarily a tissue-dwelling cell. In healthy individuals, most eosinophils are found in the gut (but not in the esophagus), mammary gland, uterus, thymus, and bone marrow; the gastrointestinal eosinophil is the predominant population of eosinophils [3]. At baseline conditions, eosinophils are present in the gastrointestinal tract independent of adaptive immunity and enteric flora, and the eosinophil levels are regulated by the constitutive expression of eotaxin-1 and eosinophil chemokine receptor, CCR3 [4, 5]. Eosinophils also home into the thymus, mammary gland, and uterus, as controlled by eotaxin-1 [6].

As mentioned above, the eosinophil is absent in the nose and paranasal sinuses and only present in disease stages, suggesting a crucial role as an effector cell in the above diseases.

9.3 Immunoregulatory Roles of Eosinophils

Previously, eosinophils have been considered an end-stage effector cell. However, accumulating evidence suggest that eosinophils can perform various immune regulatory functions likely through presentation of antigens and production and release a range of cytokines and other immunomodulatory molecules.

9.3.1 Eosinophils Present Antigens

Eosinophils possess the ability to internalize, process, and present antigenic peptides within the context of surface-expressed receptors. It also has the capacity to provide costimulatory signals to T cells through surface expression of molecules such as CD80, CD86, and CD40, and ability to physically interact with CD4+ T cells [7]. Similarly, following airway allergen challenge of mice, eosinophils traffic to and accumulate within draining lymph node, where they upregulate MHC II, CD86, and CD54 [8]. Murine eosinophils process and present antigen to T cell clones and hybridomas [9] and to antigen-primed and naïve CD4+ T cells in vitro [10]. In humans, although circulating eosinophils from healthy donors are generally devoid of surface MHC II expression, they are induced to express MHC II [11] and costimulatory molecules [12, 13] with appropriate cytokine stimulation and after transmigration through endothelial cell monolayer [14].

9.3.2 Production of Cytokines and Other Immunomodulatory Molecules by Eosinophils

Eosinophils are a source of a number of regulatory or pro-inflammatory cytokines and chemokines [15, 16]. For example, eosinophils produce cytokines, which are able to act on eosinophils themselves, the so-called "autocrine" cytokines, including IL-3 and GM-CSF [17, 18]. Eosinophils from CRS patients with nasal polyps also express TGF-β1, suggesting that TGF-β1 synthesis by eosinophils may contribute to the structural abnormalities of nasal polyps, such as stromal fibrosis and basement thickening [19]. Indeed, eosinophil-derived TGF-β enhances proliferation and collagen synthesis of lung and dermal fibroblasts [20]. TGF-α produced by cytokine-activated eosinophils increases mucin production by airway epithelial cells [21]. Thus, a number of evidences exist to demonstrate the ability of eosinophils to influence the tissue cells, leading to remodeling of tissues and changes in their physiological properties (e.g., hyperreactivity to exogenous stimuli).

By producing cytokines and chemokines, eosinophils may modulate the functions of other immune cells. Human eosinophils can produce IL-4 [22, 23], and IL-4 protein has been localized to eosinophils in airway [24] and skin [22] tissue specimens from patients with IgE-mediated allergic diseases. Furthermore, when stimulated with eotaxin (CCL11) or RANTES (CCL5), human eosinophils rapidly release stored IL-4 by vesicular transport to the local milieu [25]. Thus, eosinophils can provide a strikingly wide variety of cytokines and chemokines, suggesting that eosinophils are potentially involved in diverse biological responses, from tissue remodeling to activation of resident and infiltrating immune cells.

In addition to producing these cytokines, eosinophils secrete mediators with the potential to promote Th2-type immune responses. Another immunomodulatory factor generated by human eosinophils is one of their granule proteins, namely eosinophil-derived neurotoxin (EDN) (see below for more details). EDN is an RNase A superfamily member and, in addition to its antiviral properties, EDN is a chemoattractant [26] and activator [27] of dendritic cells (DCs). As a consequence, EDN enhances Th2 responses through a TLR2-dependent mechanism [28].

9.3.3 Immunoregulatory Functions of Eosinophils In Vivo

The immunomodulatory functions of eosinophils in vivo are demonstrated in murine models of allergen sensitization and challenge with ovalbumin (OVA) and helminth infection. Eosinophils recruitment into the sites of Th2-type inflammation was considered previously a result of activation of adaptive immune response that produces IL-5 and eotaxin [29]. However, in vivo studies with helminth infection models revealed that an early wave of eosinophil influx into inflammation sites precedes that of lymphocytes [30–32] and that it occurs even in mice deficient in adaptive immunity [33, 34]. Notably, in IL-5/eotaxin double-knockout mice, in which eosinophil numbers in both blood and tissues are severely decreased, IL-13 production of Th2 cells in response to OVA challenge is attenuated. This defect in Th2 cells was restored by eosinophil reconstitution [35], suggesting regulation of adaptive Th2-type immune response by eosinophils.

The roles of eosinophils in asthmatic airway inflammation were subsequently addressed directly by using eosinophil-deficient animals. Both Lee et al. [36] (PHIL mice) and Yu et al. [37] developed mice depleted of eosinophils through different genetic alteration. When sensitized and challenged with OVA, Th2-type airway inflammation and asthma-like pathology (e.g., airway hyperreactivity, airway remodeling, and mucus production) were attenuated in both mouse models, and these responses were restored by reconstitution of eosinophils alone [38] or a combination of eosinophils and antigen-specific T cells [39]. Likewise, airway production of Th2 cytokines and asthma-like pathology were diminished and exposed intranasally to the product of fungus *Aspergillus fumigatus* [40]. This demonstrates that eosinophils are necessary to induce the pathophysiologic changes associated with bronchial asthma.

9.4 Effector Functions of Eosinophils

As summarized in the reviews [1, 15, 29], eosinophils contain numerous highly basic and cytotoxic granule proteins that are released upon activation. They also produce an arsenal of enzymes and lipid mediators, which are implicated in effector functions of eosinophils.

9.4.1 Granule Proteins

Human eosinophil granules contain major basic protein (MBP), eosinophil cationic protein (ECP), eosinophil peroxidase (EPO), and eosinophil derived neurotoxin (EDN). Those proteins are located in the characteristic secondary granules of the eosinophils (Fig. 9.1a, b). MBP is stored in a crystalline stage, forming the characteristic rectangle core of the granule, whereas the other proteins are stored in the surrounding matrix of the granule (Fig. 9.1b). Its name is derived from the fact that MBP is the most basic protein in the humans with a pH of 11.3, and it makes up over 50% of the entire granular protein load of the eosinophil. Human MBP binds to and directly damages and destroys the surfaces of parasites [1], and is also directly toxic to tumor cells and other mammalian cells by disrupting the integrity of lipid bilayers [41].

Human ECP is a basic neurotoxic protein, with antiviral and antiparasitic properties; and human EDN is a powerful neurotoxin that can severely damage myelinated neurons in experimental animals [1]. EDN as well as ECP have antiviral activities and decrease the infectivity in RSV suspensions [42, 43]. When purified EDN or ECP were added to RSV viral suspensions, the viral titer are reduced, dependent on the ribonuclease activities of EDN and ECP [42]. Interestingly, ribonuclease A lacked this antiviral activity, suggesting that ribonuclease activity is necessary but not sufficient for the anti-viral effects of EDN and ECP. Furthermore, in guinea pigs infected with parainfluenza, pretreatment with anti-IL-5 and reduction of eosinophils strikingly increased the viral content in the airways [44], suggesting a potential role for eosinophils in viral immunity and explaining the influx of eosinophils in upper viral infections and the subsequent clinical relevant exacerbation of CRS and asthma during common colds.

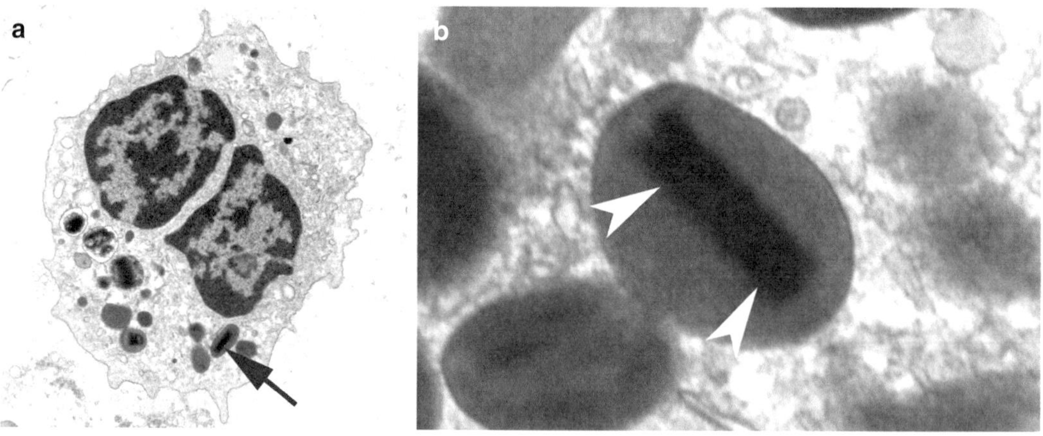

Fig. 9.1 (**a**) Electron microscopy of tissue eosinophil in CRS. Note the black arrow pointing to a characteristic secondary granule (transmission electron microscopy, original magnification ×7000). (**b**) Electron microscopy of an eosinophil secondary granule with its characteristic, rectangle crystal core, which is entirely made up of MBP (transmission electron microscopy, original magnification ×55,000)

EPO is a member of a mammalian peroxidase family. EPO is a central participant in generating reactive oxidants and radical species by activated eosinophils [45]. Eosinophil activation in vivo shows oxidative damage of proteins through bromination of tyrosine residues [46]. Furthermore, eosinophils are a major source of nitric oxide-derived oxidants in specimens from patients with severe asthma [47].

Considerable evidence exists to link these eosinophil granule proteins and human diseases. For example, the concentrations of MBP in the bronchial alveolar lavage (BAL) fluids from patients with asthma and from monkeys are correlated with the severity of bronchial hyperreactivity [48, 49]. MBP has been localized to damaged sites of bronchial epithelium in patients with asthma and chronic rhinosinusitis [50, 51]. Instillation of human MBP and human EPO provokes bronchoconstriction, and MBP increases airway responsiveness to inhaled methacholine [50]. Interestingly, polyglutamic acid antagonizes MBP's ability to increase respiratory resistance and bronchial hyperreactivity in cynomolgus monkeys [52], suggesting that the cationic nature of MBP contributes to the damage and physiologic changes. In vitro, MBP acts as an antagonist for M2 muscarinic receptors. Many eosinophils localized close to nerves with extracellular MBP adhering to the nerves [53]. Finally, neutralization of endogenously secreted MBP, either with a polyanionic peptide or with antibodies to MBP, can prevent antigen-induced bronchial hyperreactivity in guinea pigs [54]. Marked deposition of free EDN is also observed in affected tissues from patients with eosinophilic esophagitis (EoE) [55]. Deposition of EDN is reduced in certain patients with EoE who are treated with anti-IL-5 antibody [56].

Several pro-inflammatory enzymes have been associated with the eosinophil [1]. Arylsulfatase B is located predominantly in the small granules of the eosinophil. β-glucuronidase activity in eosinophils is about twice that in neutrophils, and exposure of eosinophils to opsonized zymosan particles releases up to 24% of the total cellular β-glucuronidase.

9.4.2 Activation of Human Eosinophils Takes Multiple Stages

In early 1980s, increased number of unusual human eosinophils with a specific gravity <1.085 g/mL [57] was reported in peripheral blood of patients with eosinophilic disorders, such as hypereosinophilic syndrome [58] and asthma [59]. These eosinophils, called "hypodense eosinophils," were highly reactive to stimuli and showed increased survival, adhesion, leukotriene synthesis, superoxide production and antibody-dependent cytotoxicity as compared to "normodense eosinophils" [60, 61]. Thus, eosinophils in human blood are not a homogenous population but represent various magnitudes of activation. It was found later eosinophil exposure to activating cytokines, such as IL-3, IL-5, and GM-CSF, leads to development of the hypodense eosinophil.

IL-3, IL-5, and GM-CSF, besides being growth and maturation factors for eosinophils, stimulate several functions of mature human eosinophils. Among human peripheral blood leukocytes, eosinophils are the only cells having detectable levels of IL-5 receptors in agreement with the specific action of IL-5 on human eosinophils [62, 63]. Other Th2 cytokines, such as IL-4 and IL-13, also activate eosinophils. IL-4 upregulates the binding of eosinophils to IgA [64]. IL-4 or IL-13 act synergistically with TNF-α or IL-5 for increased expression of CD69.

9.4.3 Eosinophil Activation in Innate Immunity

Fully activated human eosinophils appear to defend against large, non-phagocytosable organisms, most notably the multicellular helminthic parasites. Some of the mechanisms used by eosinophils in host defense against these organisms may also produce detrimental effects on the host. Several lines of evidence have indicated that bacterial and/or viral infections may exacerbate allergic inflammation. Direct activation of eosinophils by microbe-derived molecules may explain the mechanism.

Importantly, eosinophils are activated by a natural cysteine protease from mite allergens, *Der f 1*, and release granule proteins [65]. Eosinophils also recognize the aspartate protease activity and cysteine protease activity that are produced by fungus *Alternaria alternata* [66] and cockroaches [67], respectively, and they release granule proteins and cytokines. Thus, human eosinophils are equipped with machineries that recognize and respond to proteases, such as those found in microbes and at allergic response sites, resulting in active release of pro-inflammatory mediators.

An association between fungal exposure and asthma has been widely recognized [68]. Moreover, exposure to *Alternaria* is a risk factor for respiratory arrest in patients with asthma [69]. These airborne fungi and their products may contribute to the development and exacerbation of allergic airway diseases. For example, fungal products, e.g., proteases, induce immunologic and inflammatory reactions, resulting in a Th2-like cytokine response and the destruction of mucosal barrier functions [70]. Extracts of *Alternaria alternata* and *Aspergillus* flavus potently induces eosinophil degranulation [71, 72], as a matter of fact they are the only known triggers to induce release of toxic eosinophil Major Basic Protein (MBP) which is detectable in toxic concentration in CRS mucus. *Alternaria* also strongly induces other activation events in eosinophils, including increases in intracellular calcium concentration, cell surface expression of CD63 and CD11b, and production of IL-8 [71]. Interestingly, *Alternaria* does not induce neutrophil activation, suggesting specificity for fungal species and cell type. In addition, when human eosinophils are exposed to live *Alternaria alternata* fungus, eosinophils release their cytotoxic granule proteins into the extracellular milieu and onto the surface of fungal organisms and kill the fungus in a contact-dependent manner [73]. Eosinophils do not express common fungus receptors, such as dectin-1, but use their versatile β2 integrin molecule, CD11b (see below for more details), to recognize and to adhere to a major cell wall component, β-glucan.

The role of IgE in mediating eosinophil activation is controversial. Eosinophils isolated from patients with eosinophilia degranulated in response to anti-IgE antibody or IgE-coated parasites [74, 75]. Eosinophils can potentially express three types of IgE receptors, the low-affinity IgE receptor, lectin-type IgE-binding molecule [76], and high-affinity IgE receptor. It has been claimed that the high-affinity IgE receptor, FcεRI, is present on eosinophils from patients with eosinophilia and that various functions of eosinophils, including degranulation and parasite cytotoxicity, are mediated through this receptor [77]. On the other hand, the number of high-affinity receptors expressed on the surfaces of eosinophils from patients with allergic diseases or airway eosinophilia was minimal or undetectable [78]; and the ligation of IgE FcεRI receptor did not result in detectable eosinophil degranulation [79].

9.5 Differences in the Eosinophilic Inflammation Between Allergic Rhinitis and Chronic Rhinosinusitis

The events and the pathophysiology of allergic rhinitis have been well understood. The inhalation of an allergen where an individual has produced corresponding, circulating IgE antibodies will cause (within 10 min) a crosslink of the IgE FcεRI receptors situated on the mast cells. This will result in an immediate degranulation and histamine release from the mast cells, resulting in histamine-mediated symptoms of sneezing, clear anterior rhinorrhea, and nasal obstruction. About 2–8 h after antigen challenge, eosinophils enter the nasal tissue as part of the so-called late phase allergic reaction (together with further mast cells, B and T lymphocytes), leading to further nasal obstruction.

In contrast to allergic rhinitis, the early phase immediate reaction is missing in CRS, explaining the lack of allergic rhinitis specific, histamine-related symptoms (sneezing, anterior clear rhinorrhea). Another clinical distinction between CRS and AR is that CRS can occur with or without nasal polyps, while AR never leads to nasal polyposis. While some investigators follow the notion that due to different severities in the cytokine pattern, CRS with and without nasal polys should be viewed as two different entities, while others see it as a different spectrum of disease, with inflammatory mucosal thickening on one side of the spectrum, to gross nasal polyps on the other side. The severity of the inflammation can be easily overlooked if patients are given systemic steroids or other anti-inflammatory medication before harvesting the tissue for examination, like pre-operatively. This distinctive eosinophilic inflammation is also very heterogeneous, without eosinophilic infiltration in one area of a nasal mucosal tissue specimen, but with intense eosinophilic infiltration in another area of the same specimen [80]. Thus, reports in which only single biopsies are examined and in which it was unclear whether patients had received steroids before the biopsies were taken need to be interpreted carefully regarding the intensity of the eosinophilic infiltrate.

The eosinophilic inflammation in CRS occurs independently of an IgE-mediated inflammation, as evident by the fact that more than 50% of CRS patients have no detectable IgE-mediated allergies. This in return suggests non-allergic mechanism driving recruitment, migration activation and degranulation. Indeed, very different mechanisms have been identified.

Central to the migration of eosinophils from the vasculature into the tissue is the expression of vascular cell adhesion molecule-1 (VCAM-1), which has been identified in the vascular endothelium in CRS patients [81]. This expression occurred independent of any IgE-mediated allergy and explains the presence of eosinophils in allergic as well as non-allergic patients with CRS. VCAM-1 is known to specifically bind to the VLA-4 (very late-appearing antigen-4) on eosinophils, thus causing selective adhesion and migration of eosinophils from the vasculature to the sinus and nasal tissue. VCAM-1 expression is induced via either IL-4 or IL-13, which share the same receptor on the endothelial cells. IL-4 is present and IL-13 is absent in allergic rhinitis, which is in contrast to non-allergic CRS, where only IL-13 is present in the tissue, but IL-4 is absent [82]. In patients with CRS and allergies, both IL-4 and IL-13 are present. This cytokine pattern indicates that eosinophils are recruited via two distinct cytokines, IL-4 in AR and IL-13 in CRS; however, both diseases can coexist as CRS with allergies (comorbidity), with both IL-4 and IL-13 present.

Significantly elevated levels of IL-5, the key cytokine that mediates eosinophil differentiation, survival and activation, are present in tissue specimens of CRS patients and AR patients, and not in those of healthy controls [2, 83–85]. A majority of the IL-5 staining cells are lymphocytes (68%), followed by eosinophils (18%) and mast cells (14%) [83]. The exact combination of source cells for IL-5 in AR is not known.

The importance of IL-5, IL-13, and eosinophils in the CRS pathophysiology is emphasized though the first approval of a therapeutic monoclonal IL-13 antibody (dupilumab) for the treatment of CRS with nasal polyps. In two large phase 3 trials, dupilumab was efficacious in reducing nasal symptoms including congestion, improved CT scans, and reduced nasal polyp size [86].

In addition, IL-5 antibodies are already FDA approved for eosinophilic asthma, and are currently in clinical trials for CRSwNP.

9.6 What Are the Known Triggers for the Cytokine Response Leading to the Eosinophilic Inflammation?

While many different allergens can lead to the release of IL-5 in AR via the IgE-mediated pathway, any trigger for the non-allergic production of IL-5 and IL-13 in CRS (and other cytokines leading to eosinophilia) was thus far unknown.

This changed when certain molds were found to induced the elevated production of IL-5 in isolated peripheral blood mononuclear cells (PBMCs), which contained lymphocytes and other cells that can serve as antigen-presenting cells from 16 out of 18 CRS patients stimulated with *Alternaria* antigens [87]. More importantly, PBMCs from none of the 15 healthy controls did release IL-5 in response to *Alternaria alternata*. PBMCs from allergic and non-allergic CRS patients produced similar amounts of IL-5, indicating that this reaction is independent from an IgE-mediated allergic reaction. In addition, PBMCs from 33% of CRS patients stimulated with *Cladosporium* and 22% of CRS patients stimulated with *Aspergillus* antigens also show increased production of IL-5; no response is seen with stimulation with *Penicillium* antigen, and none of the healthy control subject responded to any fungal stimulus.

But not only IL-5 was produced by the CRS patients' immune cells in response to *Alternaria*. The mold also induced the release of large amounts of IL-13 in all the CRS patients studied, the cytokine triggering the recruitment of eosinophils from the vasculature into the tissue in CRS. Again, none of the healthy controls were producing any detectable IL-13 [87].

Furthermore, production of interferon-γαμμα (IFN-γ), a Th-1 cytokine which facilitates destruction of parasites by eosinophils, was 5.5 times higher in PBMCs from CRS patients stimulated with *Alternaria* antigen compared with production by healthy control PBMCs [87]. When nasal secretions from nine healthy controls and nine CRS patients were examined, there were no differences in their levels of total *Alternaria* proteins, indicating that both groups had similar levels of *Alternaria* in their nasal mucus. This study was important since it was the first to demonstrate a non-allergic pathway in CRS leading to the production of the crucial cytokines for the eosinophilic inflammation, which was absent in healthy controls. In addition, it also showed a specific trigger for the cytokine production, a common mold being present in nose of every person tested.

Now that a trigger (*Alternaria alternata* (ALT)) had been identified, two more breakthroughs were made: (1) the discovery of innate Lymphocyte Cells Type 2 (ILC-2) and their crucial role in initiating the cytokine cascade leading to Th-2 shifting and eosinophil-mediated immunity, and (2) the discovery of IL-33 as a regulating cytokine which is produced the basal layer of the airway epithelial cells.

When now mice were sensitized to *Alternaria alternata* (ALT), and then challenged with ALT, they produced severe airway eosinophilia, including airway reactivity (asthma). Alternaria challenge caused the release of IL-33 which peaked after 1 h, followed by the release of IL-5 and IL-13 which peaked after 6 h; thus, IL-33 preceded the IL-5/-13 release [88, 89].

The next step was to knock out the IL-33 receptor ($ST2^{-/-}$), which resulted in the elimination of IL-5/-13 release, and showed that IL-5/-13 production was IL-33 dependent [89].

Surprisingly, the control group, which did not get the ALT sensitization, but got only the ALT challenge, produced a similar airway eosinophilia. To further investigate this, mice which had the CD4+ acquired immunity knocked out ($Rag1^{-/-}$) where challenged with ALT, but despite them having no CD4+ acquired immunity produced similar levels of Th2 cytokines IL-5/-13/-33 [89]. Thus, the source for the cytokines had to be innate. Indeed, knocking out the innate lymphocytes (Myd88−/−) completely depleted the ALT-induced IL-5/-13/-33 cytokine production. Thus, the initial source for the cytokines leading to airway eosinophilia was indeed innate, and *Alternaria alternata* was a reliable trigger [89].

Recently discovered innate lymphocyte cells type 2 (ILC-2) were suspected to play a role. Consequently, when ILC-2 knockout mice ($Il7r^{-/-}$) where challenged with ALT the eosinophilic air way inflammation did not occur. However, when ILC-2 cells were isolated from normal mice and transplanted into the ILC-2 knockout mice, the eosinophilia was back to full strength, which demonstrated that ILC-2 cell mediates the initial (innate) Th2 response to ALT [90].

Other allergens by itself, such as *Aspergillus* fumigatus (ASP), or house dust mites (HDM), did not induce any eosinophilic airway inflammation, which was in contrast to ALT. However, when ALT was combined with ASP and HDM, the intensity of eosinophilia more than doubled compared ALT alone. Thus, other allergens were not able to initiate an eosinophilic inflammation like ALT did, but when combined with ALT, they intensified the ALT-induced eosinophilia [91].

The critical involvement of IL-33 in ILC-2 cells in CRS with eosinophilia (ECRS) was then strengthened by showing significant elevation of IL-33 in patients with eosinophilia compared to patients without tissue eosinophils [92, 93], and showing that the ILC-2 cells are responsive to the IL-33 stimulus by releasing IL-13, which controls eosinophil recruitment from the vasculature into the tissue [94].

It was again Alternaria identified as a trigger, which significantly stimulated the release of IL-33 from cultured nasal epithelial cells from ECRS patients, compared with no IL-33 release from non-ECRS patients and other control subjects [95].

A secreted enzyme from Alternaria, which was a serine protease, was identified as a potential molecular culprit to mediate the IL-33-dependent eosinophilic inflammation [96].

Even more importantly, ALT was found to cause a processed version of IL-33 to be released by epithelial cells, which is cleaved and has only a molecular weight of 19 KDa, which is significantly less than the 30 KDa of the natural, full length form [97]. The clinical significance is that the cleaved 19 KDa long processed version is about 30 times as potent than the natural 30 KDa form [97]. Thus, the ALT stimulus causes a significantly more potent version of the crucial IL-33 to be released by epithelial cells.

It was interesting that, while the initial innate response to ALT was mediated by ILC-2 cells, mice exposed to the Alternaria dominated allergen cocktail had developed a CD4+ mediated acquired immune response after 4 weeks of continuous exposure [91].

Thus, their immune system appears to have shifted from a solely innate, ILC-2-mediated initial immunity triggering the eosinophilic inflammation, to a CD4+ Th-2 type immunity which is mediating the long-term IL-5, -13 and -33 cytokine production, which in turn is mediating the eosinophilic inflammation. Importantly, while ALT was the essential trigger, other allergens such as ASP and HDM were able to contribute to ALT-induced inflammation, without being able to incite the eosinophilia itself [91].

9.7 Initiation of IgE Production

If *Alternaria alternata* exposure or stimulation is able to induce Th-2 shifting in mouse models, shifting naïve mammals toward a Th-2 hypersensitivity (allergic) subtype, how would IgE-mediated allergy and aspirin sensitivity fit in?

When naïve mice were exposed intranasally to Ovalbumin (OVA), a very strong allergen, for 8 weeks, they did not respond with an IgE production to OVA. However, when OVA was combined with *Alternaria alternata* (ALT), mice started to produce IgE to OVA. Other combinations, like OVA + Aspergillus (ASP), or OVA + House dust mites (HDM), did not trigger any OVA specific IgE production. However, if ALT, ASP, HDM, and OVA were combined, the OVA-specific IgE production was over 700% higher compared to ALT + OVA alone [91].

In a follow-up study, $JH^{-/-}$ (B cell) knock out mice showed and attenuated response, suggesting that B cells are necessary to mediate this Alternaria-induced pathway to IgE production [98].

This suggests that inhaled allergens work synergistically to trigger IgE production allergens, in this case OVA, with again Alternaria being the key and necessary ingredient for the development of an IgE-mediated allergy to another allergen.

9.8 Aspirin-Exacerbated Respiratory Disease (AERD)

It is clinically well known that certain patients react after Aspirin (ASS) intake with an exacerbation of their asthma symptoms, and the triad of CRS with nasal polyposis, asthma, and aspirin intolerance is named after Samter, who first described it in 1968.

It is now understood that arachidonic acid is metabolized to prostaglandins in a cyclooxygenase (CoX 1+2) dependent pathway. However, if that pathway is blocked by CoX inhibitors, such as Aspirin, the arachidonic acid is metabolized through the 5-Lipooxygenase pathway instead, producing leukotrienes, which significantly magnify existing eosinophilia, but cannot induce the eosinophilia by themselves. Human Th2 cells (in contrast to Th1 cells) selectively express the high-affinity Cysteinyl leukotriene receptor 1 (CysLT1R), which is the receptor for leukotriene D4 (LTD4), and stimulation of Th2 cells with leukotrienes-induced chemotaxis of eosinophils, and IL-13 production that was dependent on CysLT1R [99].

Again, the fungus *Alternaria alternata* was found to induce CysLT1R expression on both innate lymphocyte cells type-2 (ILC-2) as well as Th2 cells, potently induced CysLT1R dependent IL-5, and IL-13, as well as the associated eosinophilia. Additionally, LTD4 then potentiated Alternaria induced eosinophilia and ILC-2 proliferation and accumulation [100].

In a follow-up study, airway challenges with the parent leukotriene C4 (LTC4) potentiated the effect of IL-33 to naive wild-type mice, and led to synergistic increases in airway IL-13 and IL-5 cytokines and resulting eosinophilia, compared to IL-33 alone. This immune response was mediated by innate lymphocyte cells type 2 (ILC-2), independent of the acquired immune system. The synergistic effect of LTC4 with IL-33 was again completely dependent upon CysLT1R. CysLT1R$^{-/-}$ (knockout) mice had reduced lung eosinophils and ILC2 cytokine responses (IL-13/IL-5) after exposure to *Alternaria alternata*, which again induced a robust eosinophilic airway inflammation via the Th2 cytokine pathway. Thus, CysLT1R promotes LTC4 and Alternaria-induced ILC2 activation and eosinophilic airway inflammation [101].

9.9 Eosinophil-Mediated Damage in CRS, But Not in Allergic Rhinitis

CRS patients exhibit severely damaged epithelium and thickened basal membrane, features of airway remodeling seen as also seen in asthma, which is in contrast to the absence of airway remodeling in AR. It has been demonstrated that eosinophilic MBP is capable to produce those changes, and indeed MBP has been localized with the epithelial damages found in CRS [51, 80]. Interestingly, toxic MBP levels have measured in CRS, but free MBP could not be measured in AR mucus, explaining the damage in CRS, and its absence in AR [102]. This suggest that MBP is released in the mucus in CRS but not in AR.

Two prospectively designed histologic studies of tissue and mucus obtained during CRS surgery used extra caution to preserve the mucus. While eosinophils were intact in the tissue and in the epithelium, eosinophilic-rich mucus with clusters of aggregated eosinophils was found in 96% (97/101) and 94% (35/37) of consecutive CRS patients [103, 104]. Another study demonstrated that eosinophils released their toxic MBP in the mucus within these clusters, and not in the tissue [51]. Estimated concentrations of MBP within the clusters, based on digital analysis of the intensity of the MBP staining, were as high as 2 mM and far exceeded those capable of mediating epithelial damage. Overall, the clusters of eosinophils and intense eosinophil degranulation in the mucus suggest that eosinophils merely travel through the CRS tissue to the mucus where they degranulate and release their toxic proteins (Fig. 9.2a, b).

These in vivo observations explain the patterns of damage in CRS, where only the outer layers of tissue are damaged (Fig. 9.3a), suggesting that the damage to the epithelium is inflicted from the outside (luminal side). This epithelial damage may predispose CRS patients to be susceptible for the secondary bacterial infections, leading to acute exacerbations, which are observed clinically, and absent in AR (Fig. 9.3b). Because bacteria always elicit a neutrophilic inflammation in hosts, these acute exacerbations of CRS are presumed to be of bacterial origin. However, bacteria are not known to elicit an eosinophilic inflammation that predominates in CRS, which suggests a nonbacterial etiologic mechanism for CRS.

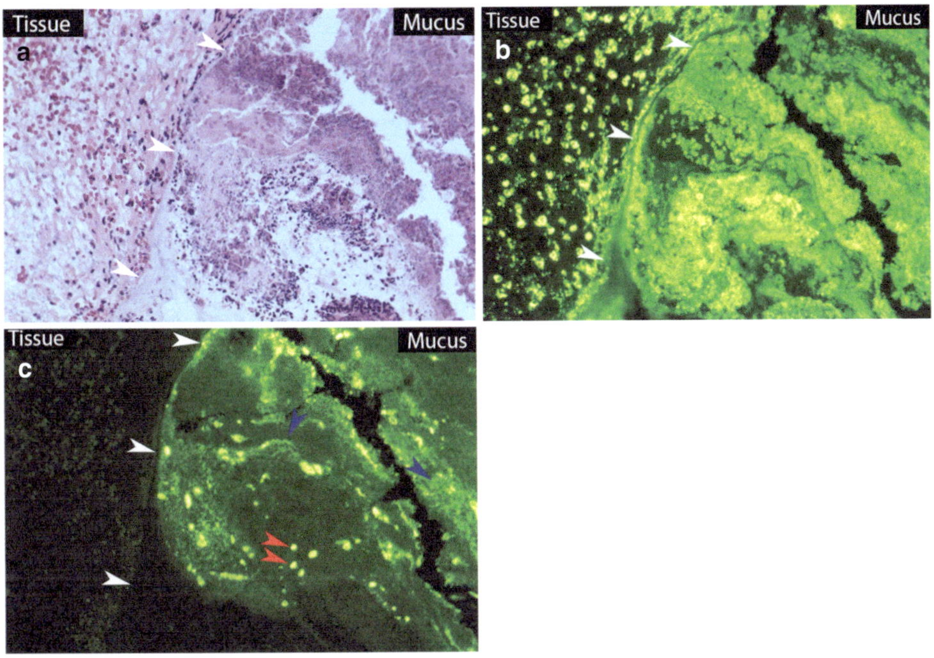

Fig. 9.2 (**a**) CRS tissue and attached eosinophilic mucin show massive eosinophilic migration of eosinophils from the tissue (left side of the image) into the mucus (right side of the image). The white arrows mark the eroded epithelium typical in CRS. The mucus contains large sheets (clusters) of eosinophils and eosinophilic debris (original magnification ×800, HE). (**b**) Serial section of 2a with immunofluorescent staining with an antibody against MBP reveals intact eosinophils in the tissue (left side of the image) and free eosinophil granules. In contrast, once the eosinophils have reached the mucus, MBP is diffusely released in toxic concentrations. Note that MBP staining reaches brightness in the mucus exceeding the one inside the intact tissue eosinophils, indicating continuous deposition of free MBP into the same eosinophilic clusters in the mucus. (original magnification ×800, anti-MBP). (**c**) Serial section of 2a with immunofluorescent staining with an antibody against *Alternaria alternata*. Note the absence of fungal antigens in the mucus. The red arrows mark some examples of fungal hyphae in cross-section; however, disseminated fungal debris is also visible (blue arrows, original magnification ×800, anti-ALT)

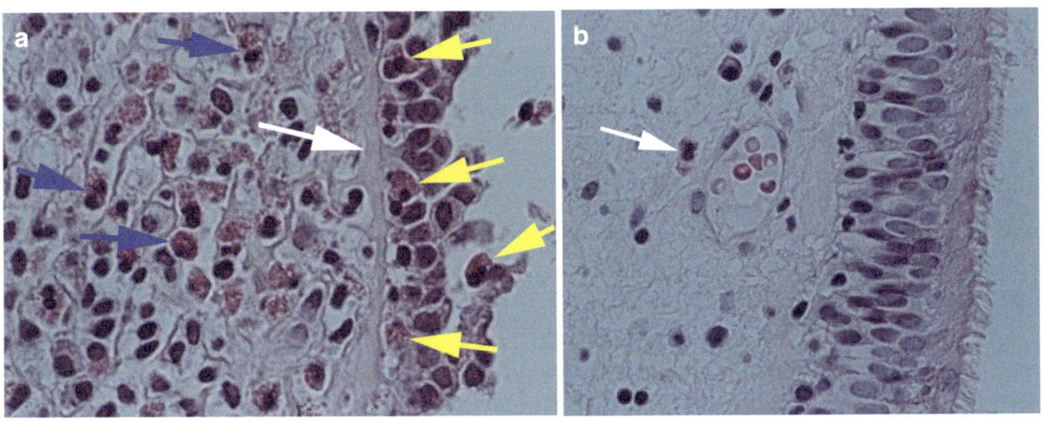

Fig. 9.3 (**a**) Tissue from CRS patient shows intact tissue eosinophils (blue arrows) and eosinophils traveling through the severely damaged epithelium (yellow arrows). Note the missing upper layers of epithelial cells and missing cilia, suggesting that the damage is inflicted from the luminal side. The white arrow highlights the thickened basal membrane. (original magnification ×1000, HE). (**b**) Tissue from AR patient reveals intact epithelium including cilia and scattered eosinophils (white arrow). Note also the absence of basal membrane thickening (original magnification ×400, Hematoxylin Eosin)

9.10 What Causes Eosinophil Degranulation?

In a recent study, eosinophils from healthy people that were incubated with *Alternaria alternata* antigens released significant amounts of eosinophil-derived neurotoxin (EDN) and Major Basic Protein (MBP), the latter known to mediate epithelial damage and basal membrane thickening, being part of airway remodeling in CRS.

When eosinophils from patients with asthma or allergies were used, they even released about 70% more EDN compared to the healthy controls. The fraction from *Alternaria alternata*, which induced the degranulation, had a molecular weight of ≈60 kDa, was highly heat labile, and worked protease-dependent through a G protein-coupled receptor, identified as the beta2- integrin of the CD11b receptor [71]. Other fungal antigens, including *Aspergillus*, *Cladosporium*, and *Candida*, did not induce eosinophil degranulation, nor did neutrophils respond to Alternaria extracts, suggesting the presence of a fungal species and cell type specific novel innate immune response to certain fungi in human.

Another study from India identified also Aspergillus Flavus as a trigger for MBP release, demonstrating that other fungi besides Alternaria alternata (tenuis) can also induce MBP release [72].

Those studies have significance, since no other triggers (except fungal organisms) for eosinophilic MBP release from inhaled allergens or microorganism (bacteria) are known. However, CRS patients have large and toxic amount of MBP in their nasal and paranasal cavity (Fig. 9.2b), especially where fungal antigens can be detected (Fig. 9.2c).

Thus, both innate and acquired immune responses to environmental fungi, such as *Alternaria alternata* may increase production of the cytokines and provide cellular activation signals necessary for the robust eosinophilic inflammation in CRS patients.

9.11 Summary and Future Directions

Eosinophils fulfill distinctive and different function in CRS versus AR, although frequently overlapping clinically. Those differences presumably result in two different pathophysiological mechanisms, mainly distinguishable through the clustering of eosinophils and the subsequent release of the eosinophil specific toxic major basic protein (MBP) into the mucus in CRS. In contrast, in allergic rhinitis, the eosinophils appear to follow more a process of a controlled cell death, without the release of toxic Major Basic Protein (MBP), and without the subsequent epithelial damage. This difference in the degranulation patterns explains the different clinical and pathophysiologic presentation between CRS and AR.

The fungus *Alternaria alternata* (ALT) has emerged as a key trigger for the eosinophilic inflammation. Its antigens induce epithelial cells to release a cleaved version of IL-33, which is about 30 times as potent as the natural occurring IL-33, and which cause the activation of the innate immune system via ILC-2 cells to not only produce crucial IL-13 (eosinophil recruitment) and IL-5 (eosinophil activation and life prolongation), but also shift a naïve immune system toward a Th2-type, allergic subtype, including the initiation of IgE-mediated allergy to other airborne allergens.

ALT also allows other allergens to act synergistic, worsening the eosinophilic inflammation, which they cannot do by themselves alone. This initial innate immunity (ILC-2) appears to be replaced after continuous challenge by the acquired immune system and mediated by CD4+ lymphocytes, resulting in chronicity of airway inflammation. Last by not least, *Alternaria alternata* induces activation and degranulation, including the detrimental MBP release, of human eosinophils.

Thus *Alternaria alternata* is thus far the only trigger known to cause a concerted immune response in epithelial cells and regulatory lymphocytes representing the innate and acquired immune system. In addition, the eosinophils act as effector cells themselves, with ALT causing their degranulation and toxic MBP release. ALT also links together the development of IgE-mediated allergy, Th2-shifting, and aspirin sensitivity, and gives new insides into the mediating receptors.

Understanding the details of those mechanisms and the eosinophil's function has now led to the development and the approval of anti-IL-13 antibody therapy for the treatment of CRS with nasal polyposis, with *Alternaria alternata* being the only known trigger thus far for IL-13 production in CRS patients. Thus, those novel insides into the immunologic reaction produce targets to further improve the care of patients suffering from these chronic, eosinophil-mediated, inflammatory diseases in rhinology.

Take Home Pearls

- Eosinophils show different behavior patterns in chronic rhinosinusitis versus allergic rhinitis, resulting in different pathophysiologies.
- Anti-IL-13 antibody treatment is the first immunologic therapy for CRS with nasal polyps.
- The fungus *Alternaria alternata* causes a concerted immune response, shifting the immune system towards a Th2 subtype.
- Alternaria triggers a concerted immune response, including epithelial cells, innate and acquired lymphocytes, to activate and recruit eosinophils.
- Alternaria triggers activation and degranulation of eosinophils.
- Aspirin sensitivity and the development of IgE-mediated allergy is inducible by Alternaria.

References

1. Gleich GJ, Adolphson CR. The eosinophilic leukocyte: structure and function. Adv Immunol. 1986;39:177–253.
2. Sanderson CJ. Interleukin-5, eosinophils, and disease. Blood. 1992;79:3101–9.
3. Mishra A, Hogan SP, Lee JJ, Foster PS, Rothenberg ME. Fundamental signals that regulate eosinophil homing to the gastrointestinal tract. J Clin Invest. 1999;103:1719–27. PMC:408388.
4. Humbles AA, et al. The murine CCR3 receptor regulates both the role of eosinophils and mast cells in allergen-induced airway inflammation and hyperresponsiveness. Proc Natl Acad Sci U S A. 2002;99:1479–84. PMC:122216.
5. Pope SM, et al. Identification of a cooperative mechanism involving interleukin-13 and eotaxin-2 in experimental allergic lung inflammation. J Biol Chem. 2005;280:13952–61.
6. Gouon-Evans V, Rothenberg ME, Pollard JW. Postnatal mammary gland development requires macrophages and eosinophils. Development. 2000;127:2269–82.
7. Spencer LA, Weller PF. Eosinophils and Th2 immunity: contemporary insights. Immunol Cell Biol. 2010;88:250–6.
8. Duez C, et al. Migration and accumulation of eosinophils toward regional lymph nodes after airway allergen challenge. J Allergy Clin Immunol. 2004;114:820–5.
9. Del Pozo V, et al. Eosinophil as antigen-presenting cell: activation of T cell clones and T cell hybridoma by eosinophils after antigen processing. Eur J Immunol. 1992;22:1919–25.
10. Padigel UM, Lee JJ, Nolan TJ, Schad GA, Abraham D. Eosinophils can function as antigen-presenting cells to induce primary and secondary immune responses to Strongyloides stercoralis. Infect Immun. 2006;74:3232–8. PMC:1479274.
11. Lucey DR, Nicholson-Weller A, Weller PF. Mature human eosinophils have the capacity to express HLA-DR. Proc Natl Acad Sci U S A. 1989;86:1348–51. PMC:286687.
12. Celestin J, et al. IL-3 induces B7.2 (CD86) expression and costimulatory activity in human eosinophils. J Immunol. 2001;167:6097–104.
13. Ohkawara Y, et al. CD40 expression by human peripheral blood eosinophils. J Clin Invest. 1996;97:1761–6. PMC:507241.
14. Yamamoto H, Sedgwick JB, Vrtis RF, Busse WW. The effect of transendothelial migration on eosinophil function. Am J Respir Cell Mol Biol. 2000;23:379–88.
15. Hogan SP, et al. Eosinophils: biological properties and role in health and disease. Clin Exp Allergy. 2008;38:709–50.
16. Moqbel R, Lacy P. New concepts in effector functions of eosinophil cytokines. Clin Exp Allergy. 2000;30:1667–71.
17. Kita H, Ohnishi T, Okubo Y, Weiler D, Abrams JS, Gleich GJ. Granulocyte/macrophage colony-stimulating factor and interleukin 3 release from human peripheral blood eosinophils and neutrophils. J Exp Med. 1991;174:745–8. PMC:2118930.
18. Moqbel R, et al. Expression of mRNA and immunoreactivity for the granulocyte/macrophage colony-stimulating factor in activated human eosinophils. J Exp Med. 1991;174(749):752. PMC:2118946.
19. Ohno I, et al. Eosinophils in chronically inflamed human upper airway tissues express transform-

ing growth factor beta 1 gene (TGF beta 1). J Clin Invest. 1992;89:1662–8. PMC:443044.
20. Levi-Schaffer F, et al. Human eosinophils regulate human lung-and skin-derived fibroblast properties in vitro: a role for transforming growth factor beta (TGF-beta). Proc Natl Acad Sci U S A. 1999;96:9660–5. PMC:22266.
21. Burgel PR, et al. Human eosinophils induce mucin production in airway epithelial cells via epidermal growth factor receptor activation. J Immunol. 2001;167:5948–54.
22. Moqbel R, et al. Identification of messenger RNA for IL-4 in human eosinophils with granule localization and release of the translated product. J Immunol. 1995;155:4939–47.
23. Nakajima H, Gleich GJ, Kita H. Constitutive production of IL-4 and IL-10 and stimulated production of IL-8 by normal peripheral blood eosinophils. J Immunol. 1996;156:4859–66.
24. Nonaka M, et al. Distinct immunohistochemical localization of IL-4 in human inflamed airway tissues. IL-4 is localized to eosinophils in vivo and is released by peripheral blood eosinophils. J Immunol. 1995;155:3234–44.
25. Bandeira-Melo C, Sugiyama K, Woods LJ, Weller PF. Cutting edge: Eotaxin elicits rapid vesicular transport-mediated release of preformed IL-4 from human eosinophils. J Immunol. 2001;166:4813–7.
26. Yang D, Rosenberg HF, Chen Q, Dyer KD, Kurosaka K, Oppenheim JJ. Eosinophil-derived neurotoxin (EDN), an antimicrobial protein with chemotactic activities for dendritic cells. Blood. 2003;102:3396–403.
27. Yang D, et al. Human ribonuclease a superfamily members, eosinophil-derived neurotoxin and pancreatic ribonuclease, induce dendritic cell maturation and activation. J Immunol. 2004;173:6134–42. PMC:2847482.
28. Yang D, et al. Eosinophil-derived neurotoxin acts as an alarmin to activate the TLR2-MyD88 signal pathway in dendritic cells and enhances Th2 immune responses. J Exp Med. 2008;205:79–90. PMC:2234357.
29. Rothenberg ME, Hogan SP. The eosinophil. Annu Rev Immunol. 2006;24:147–74.
30. Sabin EA, Kopf MA, Pearce EJ. Schistosoma mansoni egg-induced early IL-4 production is dependent upon IL-5 and eosinophils. J Exp Med. 1996;184:1871–8. PMC:2192874.
31. Voehringer D, Reese TA, Huang X, Shinkai K, Locksley RM. Type 2 immunity is controlled by IL-4/IL-13 expression in hematopoietic non-eosinophil cells of the innate immune system. J Exp Med. 2006;203:1435–46. PMC:2118302.
32. Voehringer D, Shinkai K, Locksley RM. Type 2 immunity reflects orchestrated recruitment of cells committed to IL-4 production. Immunity. 2004;20:267–77.
33. Sabin EA, Pearce EJ. Early IL-4 production by non-CD4+ cells at the site of antigen deposition predicts the development of a T helper 2 cell response to Schistosoma mansoni eggs. J Immunol. 1995;155:4844–53.
34. Shinkai K, Mohrs M, Locksley RM. Helper T cells regulate type-2 innate immunity in vivo. Nature. 2002;420:825–9.
35. Mattes J, et al. Intrinsic defect in T cell production of interleukin (IL)-13 in the absence of both IL-5 and eotaxin precludes the development of eosinophilia and airways hyperreactivity in experimental asthma. J Exp Med. 2002;195:1433–44. PMC:2193548.
36. Lee JJ, et al. Defining a link with asthma in mice congenitally deficient in eosinophils. Science. 2004;305:1773–6.
37. Yu C, et al. Targeted deletion of a high-affinity GATA-binding site in the GATA-1 promoter leads to selective loss of the eosinophil lineage in vivo. J Exp Med. 2002;195:1387–95. PMC:2193547.
38. Walsh ER, et al. Strain-specific requirement for eosinophils in the recruitment of T cells to the lung during the development of allergic asthma. J Exp Med. 2008;205:1285–92. PMC:2413027.
39. Jacobsen EA, et al. Allergic pulmonary inflammation in mice is dependent on eosinophil-induced recruitment of effector T cells. J Exp Med. 2008;205:699–710. PMC:2275390.
40. Fulkerson PC, Fischetti CA, McBride ML, Hassman LM, Hogan SP, Rothenberg ME. A central regulatory role for eosinophils and the eotaxin/CCR3 axis in chronic experimental allergic airway inflammation. Proc Natl Acad Sci U S A. 2006;103:16418–23. PMC:1637597.
41. Abu-Ghazaleh RI, Gleich GJ, Prendergast FG. Interaction of eosinophil granule major basic protein with synthetic lipid bilayers: a mechanism for toxicity. J Membr Biol. 1992;128:153–64.
42. Domachowske JB, Dyer KD, Adams AG, Leto TL, Rosenberg HF. Eosinophil cationic protein/RNase 3 is another RNase A-family ribonuclease with direct antiviral activity. Nucleic Acids Res. 1998;26:3358–63. PMC:147714.
43. Rosenberg HF, Domachowske JB. Eosinophils, eosinophil ribonucleases, and their role in host defense against respiratory virus pathogens. J Leukoc Biol. 2001;70:691–8.
44. Adamko DJ, Yost BL, Gleich GJ, Fryer AD, Jacoby DB. Ovalbumin sensitization changes the inflammatory response to subsequent parainfluenza infection. Eosinophils mediate airway hyperresponsiveness, m(2) muscarinic receptor dysfunction, and antiviral effects. J Exp Med. 1999;190:1465–78. PMC:2195693.
45. Mitra SN, Slungaard A, Hazen SL. Role of eosinophil peroxidase in the origins of protein oxidation in asthma. Redox Rep. 2000;5:215–24.
46. Wu W, et al. Eosinophils generate brominating oxidants in allergen-induced asthma. J Clin Invest. 2000;105:1455–63. PMC:315470.
47. MacPherson JC, et al. Eosinophils are a major source of nitric oxide-derived oxidants in severe

asthma: characterization of pathways available to eosinophils for generating reactive nitrogen species. J Immunol. 2001;166:5763–72.
48. Gleich GJ. Mechanisms of eosinophil-associated inflammation. J Allergy Clin Immunol. 2000;105:651–63.
49. Gleich GJ, Fryer AD, Jacoby DB. Eosinophil granule proteins and bronchial hyperreactivity. London: Academic; 1993.
50. Gleich GJ, Adolphson CR, Kita H. The eosinophil and asthma. 2nd ed. Oxford: Blackwell Science; 2000.
51. Ponikau JU, Sherris DA, Kephart GM, et al. Striking deposition of toxic eosinophil major basic protein in mucus: implications for chronic rhinosinusitis. J Allergy Clin Immunol. 2005;116:362–9.
52. Gundel RH, Letts LG, Gleich GJ. Human eosinophil major basic protein induces airway constriction and airway hyperresponsiveness in primates. J Clin Invest. 1991;87:1470–3. PMC:295201.
53. Costello RW, Schofield BH, Kephart GM, Gleich GJ, Jacoby DB, Fryer AD. Localization of eosinophils to airway nerves and effect on neuronal M2 muscarinic receptor function. Am J Physiol. 1997;273:L93–103.
54. Costello RW, et al. Antigen-induced hyperreactivity to histamine: role of the vagus nerves and eosinophils. Am J Physiol. 1999;276:L709–14.
55. Kephart GM, et al. Marked deposition of eosinophil-derived neurotoxin in adult patients with eosinophilic esophagitis. Am J Gastroenterol. 2010;105:298–307. PMC:2824254.
56. Straumann A, et al. Anti-interleukin-5 antibody treatment (mepolizumab) in active eosinophilic oesophagitis: a randomised, placebo-controlled, double-blind trial. Gut. 2010;59:21–30.
57. Rothenberg ME, Owen WF Jr, Silberstein DS, Woods J, Soberman RJ, Austen KF, Stevens RL. Human eosinophils have prolonged survival, enhanced functional properties, and become hypodense when exposed to human interleukin 3. J Clin Invest. 1988;81(6):1986–92.
58. Winqvist I, Olofsson T, Olsson I, Persson AM, Hallberg T. Altered density, metabolism and surface receptors of eosinophils in eosinophilia. Immunology. 1982;47(3):531–9.
59. Fukuda T, Dunnette SL, Reed CE, Ackerman SJ, Peters MS, Gleich GJ. Increased numbers of hypodense eosinophils in the blood of patients with bronchial asthma. Am Rev Respir Dis. 1985;132(5):981–5.
60. Lopez AF, Sanderson CJ, Gamble JR, Campbell HD, Young IG, Vadas MA. Recombinant human interleukin 5 is a selective activator of human eosinophil function. J Exp Med. 1988;167(1):219–24.
61. Lopez AF, Williamson DJ, Gamble JR, Begley CG, Harlan JM, Klebanoff SJ, Waltersdorph A, Wong G, Clark SC, Vadas MA. Recombinant human granulocyte-macrophage colony-stimulating factor stimulates in vitro mature human neutrophil and eosinophil function, surface receptor expression, and survival. J Clin Invest. 1986;78(5):1220–8.
62. Ingley E, Young IG. Characterization of a receptor for interleukin-5 on human eosinophils and the myeloid leukemia line HL-60. Blood. 1991;78:339–44.
63. Migita M, et al. Characterization of the human IL-5 receptors on eosinophils. Cell Immunol. 1991;133:484–97.
64. Bracke M, et al. Differential effects of the T helper cell type 2-derived cytokines IL-4 and IL-5 on ligand binding to IgG and IgA receptors expressed by human eosinophils. J Immunol. 1997;159:1459–65.
65. Miike S, Kita H. Human eosinophils are activated by cysteine proteases and release inflammatory mediators. J Allergy Clin Immunol. 2003;111:704–13.
66. Matsuwaki Y, et al. Recognition of fungal protease activities induces cellular activation and eosinophil-derived neurotoxin release in human eosinophils. J Immunol. 2009;183:6708–16. PMC:2843542.
67. Wada K, et al. Inflammatory responses of human eosinophils to cockroach are mediated through protease-dependent pathways. J Allergy Clin Immunol. 2010;126:169–72.e2. PMC:2902556.
68. Bush RK, Prochnau JJ. Alternaria-induced asthma. J Allergy Clin Immunol. 2004;113:227–34.
69. O'Hollaren MT, et al. Exposure to an aeroallergen as a possible precipitating factor in respiratory arrest in young patients with asthma. N Engl J Med. 1991;324:359–63.
70. Kheradmand F, Kiss A, Xu J, Lee SH, Kolattukudy PE, Corry DB. A protease-activated pathway underlying Th cell type 2 activation and allergic lung disease. J Immunol. 2002;169:5904–11.
71. Inoue Y, Matsuwaki Y, Shin SH, Ponikau JU, Kita H. Nonpathogenic, environmental fungi induce activation and degranulation of human eosinophils. J Immunol. 2005;175:5439–47.
72. Kale P, Rudramurthy SM, Panda NK, Das A, Chakrabarti A. The inflammatory response of eosinophil-related fungal rhinosinusitis varies with inciting fungi. Med Mycol. 2015;53(4):387–95. PubMed PMID: 25724204.
73. Yoon J, Ponikau JU, Lawrence CB, Kita H. Innate antifungal immunity of human eosinophils mediated by a beta 2 integrin, CD11b. J Immunol. 2008;181(4):2907–15.
74. Kita H, Adolphson CR, Gleich GJ. Biology of eosinophils. In: Adkinson Jr NF, Bochner BS, Yunginger JW, editors. Middleton's allergy principles and practice. 6th ed. St. Louis: Mosby; 2003.
75. Moqbel R, et al. The effect of platelet-activating factor on IgE binding to, and IgE-dependent biological properties of, human eosinophils. Immunology. 1990;70:251–7. PMC:1384202.
76. Truong MJ, Gruart V, Liu FT, Prin L, Capron A, Capron M. IgE-binding molecules (Mac-2/epsilon BP)expressed by human eosinophils. Implication in IgE-dependent eosinophil cytotoxicity. Eur J Immunol. 1993;23:3230–5.

77. Gounni AS, et al. High-affinity IgE receptor on eosinophils is involved in defence against parasites. Nature. 1994;367:183–6.
78. Seminario MC, Saini SS, MacGlashan DW Jr, Bochner BS. Intracellular expression and release of fc epsilon RI alpha by human eosinophils. J Immunol. 1999;162:6893–900.
79. Kita H, et al. Does IgE bind to and activate eosinophils from patients with allergy? J Immunol. 1999;162:6901–11.
80. Ponikau JU, Sherris DA, Kephart GM, Kern EB, Gaffey TA, Tarara JE, Kita H. Features of airway remodeling and eosinophilic inflammation in chronic rhinosinusitis: is it the histopathology similar to asthma? J Allergy Clin Immunol. 2003;112(6):877–82.
81. Hamilos DL, Leung DY, Wood R, et al. Eosinophil infiltration in nonallergic chronic hyperplastic sinusitis with nasal polyposis (CHS/NP) is associated with endothelial VCAM-1 upregulation and expression of TNF-alpha. Am J Respir Cell Mol Biol. 1996;15:443–50.
82. Hamilos DL, Leung DY, Wood R, et al. Evidence for distinct cytokine expression in allergic versus non allergic chronic sinusitis. J Allergy Clin Immunol. 1995;96:537–44.
83. Hamilos DL, Leung DY, Huston DP, et al. GM-CSF, IL-5 and RANTES immunoreactivity and mRNA expression in chronic hyperplastic sinusitis with nasal polyposis (NP). Clin Exp Allergy. 1998;28:1145–52.
84. Sanchez-Segura A, Brieva JA, Rodriguez C. T lymphocytes that infiltrate nasal polyps have a specialized phenotype and produce a mixed TH1/TH2 pattern of cytokines. J Allergy Clin Immunol. 1998;102:953–60.
85. Sanchez-Segura A, Brieva JA, Rodriguez C. Regulation of immunoglobulin secretion by plasma cells infiltrating nasal polyps. Laryngoscope. 2000;110(7):1183–8.
86. Bachert C, Han JK, Desrosiers M, Hellings PW, Amin N, Lee SE, Mullol J, Greos LS, Bosso JV, Laidlaw TM, Cervin AU, Maspero JF, Hopkins C, Olze H, Canonica GW, Paggiaro P, Cho SH, Fokkens WJ, Fujieda S, Zhang M, Lu X, Fan C, Draikiwicz S, Kamat SA, Khan A, Pirozzi G, Patel N, Graham NMH, Ruddy M, Staudinger H, Weinreich D, Stahl N, Yancopoulos GD, Mannent LP. Efficacy and safety of dupilumab in patients with severe chronic rhinosinusitis with nasal polyps (LIBERTY NP SINUS-24 and LIBERTY NP SINUS-52): results from two multicentre, randomised, double-blind, placebo-controlled, parallel-group phase 3 trials. Lancet. 2019;394(10209):1638–50. PMID: 31543428. https://doi.org/10.1016/S0140-6736(19)31881-1.
87. Shin S-H, Ponikau JU, Sherris DA, et al. Rhinosinusitis:an enhanced immune response to ubiquitous airborne fungi. J Allergy Clin Immunol. 2004;114:1369–75.
88. Kobayashi T, Iijima K, Radhakrishnan S, Mehta V, Vassallo R, Lawrence CB, Cyong JC, Pease LR, Oguchi K, Kita H. Asthma-related environmental fungus, *Alternaria*, activates dendritic cells and produces potent Th2 adjuvant activity. J Immunol. 2009;182:2502–10. PMC2653279.
89. Kouzaki H, Iijima K, Kobayashi T, O'Grady SM, Kita H. The danger signal, extracellular ATP, is a sensor for an airborne allergen and triggers IL-33 release and innate Th2-type responses. J Immunol. 2011;186:4375–87.
90. Bartemes KR, Iijima K, Kobayashi T, Kephart GM, McKenzie AN, Kita H. IL-33-responsive lineage-CD25+CD44hi lymphoid cells mediate innate type 2 immunity and allergic inflammation in the lungs. J Immunol. 2012;188:1503–13.
91. Iijima K, Kobayashi T, Hara K, Kephart GM, Ziegler SF, McKenzie AN, Kita H. IL-33 and thymic stromal lymphopoietin mediate immune pathology in response to chronic airborne allergen exposure. J Immunol. 2014;193(4):1549–59.
92. Ho J, Bailey M, Zaunders J, Mrad N, Sacks R, Sewell W, Harvey RJ. Group 2 innate lymphoid cells (ILC2s) are increased in chronic rhinosinusitis with nasal polyps or eosinophilia. Clin Exp Allergy. 2015;45:394–403.
93. Kouzaki H, Matsumoto K, Kato T, Tojima I, Shimizu S, Shimizu T. Epithelial cell-derived cytokines contribute to the pathophysiology of eosinophilic chronic rhinosinusitis. J Interferon Cytokine Res. 2015;36:169–79.
94. Shaw JL, Fakhri S, Citardi MJ, Porter PC, Corry DB, Kheradmand F, Liu YJ, Luong A. IL-33-responsive innate lymphoid cells are an important source of IL-13 in chronic rhinosinusitis with nasal polyps. Am J Respir Crit Care Med. 2013;188:432–9.
95. Tojima I, Kouzaki H, Shimizu S, Ogawa T, Arikata M, Kita H, Shimizu T. Group 2 innate lymphoid cells are increased in nasal polyps in patients with eosinophilic chronic rhinosinusitis. Clin Immunol. 2016;170:1–8.
96. Snelgrove RJ, Gregory LG, Peiró T, Akthar S, Campbell GA, Walker SA, Lloyd CM. Alternaria-derived serine protease activity drives IL-33-mediated asthma exacerbations. J Allergy Clin Immunol. 2014;134(3):583–592.e6. Epub 2014 Mar 15. PMID: 24636086; PMCID: PMC4152000. https://doi.org/10.1016/j.jaci.2014.02.002.
97. Cohen ES, et al. Oxidation of the alarmin IL-33 regulates ST2-dependent inflammation. Nat Commun. 2015;6:8327. https://doi.org/10.1038/ncomms9327.
98. Drake LY, Iijima K, Hara K, Kobayashi T, Kephart GM, Kita H. B cells play key roles in Th2-type airway immune responses in mice exposed to natural airborne allergens. PLoS One. 2015;10(3):e0121660. https://doi.org/10.1371/journal.pone.0121660.
99. Parmentier CN, Fuerst E, McDonald J, Bowen H, Lee TH, Pease JE, et al. Human T(H)2 cells respond to cysteinyl leukotrienes through selective expres-

sion of cysteinyl leukotriene receptor 1. J Allergy Clin Immunol. 2012;129:1136–42.

100. Doherty TA, Khorram N, Lund S, Mehta AK, Croft M, Broide DH. Lung type 2 innate lymphoid cells express cysteinyl leukotriene receptor 1, which regulates TH2 cytokine production. J Allergy Clin Immunol. 2013;132:205–13.

101. Lund SJ, Portillo A, Cavagnero K, Baum RE, Naji LH, Badrani JH, Mehta A, Croft M, Broide DH, Doherty TA. Leukotriene C4 potentiates IL-33-induced group 2 innate lymphoid cell activation and lung inflammation. J Immunol. 2017;199(3):1096–104. Epub 2017 Jun 30. PMID: 28667163; PMCID: PMC5531601. https://doi.org/10.4049/jimmunol.1601569.

102. Ponikau JU, Winter LA, Kephart GM, Squillace DL, Hershcovitch MD, Moon S, Sherris DA, Kern EB, Gleich GJ, Kita H. An immunologic test for chronic rhinosinusitis based on free intranasal eosinophilic major basic protein. Int Forum Allergy Rhinol. 2015;5(1):28–35.

103. Braun H, Buzina W, Freudenschuss K, et al. "Eosinophilic fungal rhinosinusitis": a common disorder in Europe? Laryngoscope. 2003;113:264–9.

104. Ponikau JU, Sherris DA, Kern EB, Homburger HA, Frigas E, Gaffey TA, Roberts GD. The diagnosis and incidence of allergic fungal sinusitis. Mayo Clin Proc. 1999;74(9):877–84.

10 Biologic Therapies for Chronic Rhinosinusitis

Michael J. Aw and Shaun J. Kilty

Core Points

- Chronic rhinosinusitis (CRS) is a complex heterogeneous inflammatory disease that is characterized by type 1 and type 2 inflammation, with type 2 inflammation predominating in patients with CRS with polyps.
- Despite the efficacy of current medical and surgical therapies, there are many patients for whom disease control remains elusive.
- Biologic therapies are biologically targeted treatments that offer patients with recalcitrant CRS disease, a new treatment option.

Chronic rhinosinusitis (CRS) is a complex multifactorial inflammatory disease of the nose and paranasal sinuses that impacts 5–12% of the worldwide adult population [1–3]. Often diagnosed later in life, the onset of primary CRS occurs generally between 40 and 60 years of age [4]. The diagnosis of CRS is based on both clinical symptoms and mucosal changes observed with either endoscopy or computed tomography (CT) [2]. This disease is associated with significant morbidity and has been associated with a substantially reduced health-related quality of life [4]. In North America, a majority of people with CRS report facial pain (60–92%), nasal congestion (95–100%), loss of olfaction (56–84%) and headache (33–90%) [5]. The costs of managing CRS are not insignificant, with the annual direct cost of CRS treatment ranging from 5560 to 5955 USD per patient, and annual indirect costs estimated at 10,077.07 USD per patient [6, 7]. The annual total direct cost attributed to CRS treatment in the United States has been previously estimated at upwards of from 60.2 billion USD [6].

The commonly used treatments for CRS focus on the control of inflammation by either regulating mucosal inflammation with topical or systemic corticosteroids, reducing planktonic bacterial burden with antibiotics and the improvement of sinus ventilation and access for topical therapies, with endoscopic surgeries. However, despite medical and surgical therapy, a subset of patients, often labelled as *difficult-to-treat*, do not achieve adequate inflammation, and subsequently, symptom control [8, 9]. The use of biologic therapies originally developed for other inflammatory diseases, such as asthma, have demonstrated that they may have an important role to play in CRS care, particularly for *difficult-to-treat* patients. The significant implications on health and healthcare attributed to this prevalent disease warrants the study of disease-modifying agents to better control mucosal inflammation to reduce the burden of disease. Here, we discuss recent advances in CRS treatments.

M. J. Aw
Faculty of Medicine, The University of Ottawa, Ottawa, ON, Canada

S. J. Kilty (✉)
Department of Otolaryngology-Head and Neck Surgery, The University of Ottawa/The Ottawa Hospital, Ottawa, ON, Canada

Ö. Ö. Celebi, T. M. Önerci (eds.), *Nasal Physiology and Pathophysiology of Nasal Disorders*,
https://doi.org/10.1007/978-3-031-12386-3_10

10.1 Pathophysiology

Classically, CRS is dichotomized by the following clinical phenotypes: chronic rhinosinusitis with nasal polyps (CRSwNP) and chronic rhinosinusitis without nasal polyps (CRSsNP) [10]. Specifically, CRSwNP represents 25–30% of CRS cases and is associated with substantial morbidity and disease outcomes [2]. Evidence now suggests this represents an oversimplified understanding of disease and greater emphasis should be placed on CRS endotypes [10, 11]. While the aetiology of CRS remains poorly understood, multiple factors including fungal colonization, bacterial superantigens and host barrier deficiencies have been implicated in the loss of regulation over the normal inflammatory responses, toxic mucosal inflammation and the development and persistence of CRS [12]. More importantly, CRS pathophysiology appears to follow one of the three types of immunological responses, either in isolation or in parallel [13]. Notably, type 2 inflammation, which represents an eosinophilic inflammation, is the most prevalent, strongly associated with CRSwNP [13].

Both fungal and bacterial colonization, namely *Alternaria* and *Staphylococcus aureus*, have been implicated in CRS disease severity. Fungal organisms have been found to inhabit nearly 100% of persons with CRS and these individuals demonstrate exaggerated peripheral blood mononuclear cell responses following *Alternaria* exposure compared to healthy controls [14, 15]. *Alternaria* antigen can elicit eosinophil degranulation and fungal wall components, chiefly chitin, can promote type 2 inflammatory responses [16, 17]. However, antifungal therapies have been largely unsuccessful in treating CRS, which may indicate that fungal infection only represents a major disease modifying factor for a subset of patients [12, 18, 19]. Additionally, the intranasal bacterial microbiome of patients with CRS differs from healthy controls [20]. Specifically, *Staphylococcus aureus* is found in greater abundance. While traditionally associated with eliciting Th17 responses, it is hypothesized that *S. Aureus* enhances type 2 inflammation through superantigenic effects [12]. These superantigenic exotoxins can increase eosinophilic degranulation and bypass traditional antigen-presenting mechanisms to incite massive T cell activation [21, 22]. Furthermore, a significant portion of nasal polyps from CRSwNP patients are colonized with *S. Aureus* [23]. Albeit it remains unclear whether bacterial infection precedes CRS development or merely exacerbates the established inflammation. Currently, it is believed that these microorganisms play an role in disease presentation, but are not directly implicated in CRS development [12].

The prevailing theory for CRS pathophysiology is that deficiencies in host barrier defences permit infectious invasion and the resultant skewing of inflammatory responses [12]. This theory is substantiated by the fact that patients with cystic fibrosis and carriers of the CFTR gene mutation, who have impaired mucociliary function, have high CRSwNP and CRSsNP prevalence [24, 25]. Dysfunction of sinonasal epithelial cells is associated with an impairment of the barrier function, mucociliary clearance, the regulation of eosinophils, dendritic cells, T cells and complement synthesis as well as taste receptor function [26]. Interesting, bitter taste receptors are impaired in a subset of patients with CRS and these receptors have been implicated in responding to bacterial presence [27]. Furthermore, the sinonasal epithelium in CRSwNP is more porous than healthy tissue, which increases its susceptibility to exogenous antigen stimulation [28]. Moreover, the expression of innate antimicrobial proteins such as lactoferrin and S100 are decreased in CRSwNP and CRSsNP and PLUNC are decreased in CRSwNP [29–31]. Ultimately, it is possible that host barrier defects facilitate exogenous antigen exposure, which perpetuates inappropriate inflammatory responses. Notably, the release of epithelium-derived alarmins, IL-33 and thymic stromal lymphopoietin (TSLP) promotes type 2 inflammatory responses [32].

Amongst North American and European populations, eosinophilic type 2 CRS represents the prevailing disease endotype. By contrast, individuals of Asian decent and those with cystic fibrosis often demonstrate dominant type 1 or type 3 inflammatory profiles [10, 33, 34]. It is estimated that 80% of nasal polyps follow a type 2 profile from Western populations, whereas only 20–60% of nasal polyposis is attributed to type 2 inflammation in Eastern Asia [35]. Consequently, CRS endotypes are generally classified as either type 2 or non-type 2 inflammation [10]. In type 2 CRS, exogenous antigen stimulation of nasal epithelium triggers the release of the alarmins IL-25, IL-33 and TSLP [36]. These cytokines stimulate and activate myeloid dendritic cells [37] to migrate to lymphoid tissue and present antigen in the context of IL-4 to favour the differentiation of naïve CD4 T cells into effector Th2 cells [38]. Th2 cells secrete hallmark cytokines associated with type 2 inflammation including IL-4, IL-5, IL-13 and encourage isotype class-switching of B cells towards immunoglobulin E (IgE). Furthermore, IL-25, IL-33 and TSLP encourage IL-4, IL-5 and IL-13 production in type 2 innate lymphoid cells [39], which create a local type 2 inflammatory profile *in situ* [39]. IgE binds to mast cells via FcεRI receptor, which when cross-linked with antigen, then degranulate and release inflammatory markers including histamine, cytokines, pro-angiogenic factors and proteases [40]. Additionally, IL-5 recruits and activates eosinophils which play a pivotal role in nasal tissue remodelling [10, 41]. Tissue remodelling in CRS consists of fibrosis, collagen deposition, osteitis, angiogenesis and mucosal hypertrophy. Clinically, type 2 CRS more frequently impacts the ethmoid sinuses and is associated with headaches, nasal polyposis and anosmia, asthma, allergic rhinitis and or Aspirin Exacerbated Respiratory Disease (AERD) [10].

Non-type 2 CRS is characterized by neutrophil invasion and represents a mix of type 1 (Th1) and type 3 (Th17) inflammation [10]. Following irritant or pathogen exposure, epithelial cells and macrophages secrete IL-6, IL-8, tumour necrosis factor alpha [42] and interferon gamma [13, 43]. The presence of IFN-γ in the context of dendritic cell [37] stimulation of naïve T cells promotes Th1 differentiation, which produces Il-2 and IFN-γ and propagates type 1 inflammation. Alternatively, IL-6 triggers the differentiation of Th17 cells, which produce IL-17 and promote neutrophilic responses [44]. Moreover, IL-6 and IL-8 recruit and activate neutrophils, which release inflammatory mediators and chronically mediate tissue remodelling [10]. Clinically, non-type 2 CRS favours the maxillary sinus and is attributed to pollution and bacterial exposure with purulent nasal discharge [10]. Of note, in the last 20 years, there has been a significant shift in some Asian countries towards greater prevalence of eosinophilic CRS, which further highlights the importance of type 2 CRS endotype [35].

10.2 Current Therapies

Treatment for CRS includes medical and surgical interventions: nasal saline irrigation, corticosteroids, antibiotics and endoscopic sinus surgery [45, 46]. While ineffective as a disease modifying therapy in isolation, nasal saline irrigation represents a cheap and accessible adjuvant therapy for symptom management. Nasal irrigation has been shown to facilitate debris and irritant clearance, enhance intranasal medication delivery and improve mucociliary clearance [47–49]. Topical and oral corticosteroids represent hallmark first-line therapies for CRS [50]. Topical corticosteroid monotherapy may be sufficient for mild cases; however, it may not provide long-term symptom control in more severe manifestations of the disease [50]. It is hypothesized that topical steroids ineffectively penetrate the ostiomeatal complex, which is implicated in sinus ventilation and drainage, due to the mechanism of application and inherent anatomic complexity of the area [51]. Combined topical and oral treatment can reduce polyp size and improve disease symptomatology, including hyposmia and post-nasal drip [51–53]. Yet, a subset of patients fail to meaningfully respond to corticosteroid treatment and while steroids reduce polyp size, they do not effectively ablate nasal polyps. Tuncer et al. treated patients with CRSwNP with dual corticosteroid therapy and discerned that 12% of patients were unresponsive to treatment [52]. Moreover, polyp ablation was unachievable in 88% of patients [52]. While emerging evidence implicating bacteria and the microbiome as negative effectors of the mucosal barrier in CRS, the use of antibiotic treatment remains controversial [54, 55]. Soler et al. conducted a systematic review evaluating eight classes of antimicrobial treatments for CRS using the methodology of the Center for Evidence-Based Medicine [54]. The investigation identified short-term use (<3 week) oral antibiotics and long-term macrolide antibiotics as viable options for routine CRS treatment. Based on insufficient evidence and lack of clinically meaningful effects, topical antibiotics, long-term (>3 weeks) oral/IV antibiotics and antifungals were recommended against [54]. Nonetheless, further understanding of microorganism involvement in CRS is necessary to elucidate the potential role of current or emerging antimicrobial therapies [46]. Surgery is the most invasive approach to disease modification and symptom treatment and is generally considered when CRS is refractory to medical therapy [2]. ESS is generally reserved for more severe presentations since preoperative symptom scores correspond with postoperative success and patients with the most serious preoperative symptom scores generally gain the most benefit from ESS [56–58]. Unfortunately, a significant proportion of patients have refractory polyposis following surgery. Bassiouni and Wormald reviewed 338 operations and found that 19.8% and 22.7% of patients had polyp recurrence postoperatively 6 and 12 months, respectively [59]. Notably, asthma and AERD comorbidity were the most significant risk factors for polyp recurrence. At 40 months postoperatively, nearly 60% of patients with CRSwNP and asthma had polyp recurrence and 90% of patients with the triad of CRSwNP, asthma and NSAID intolerance (AERD) had polyp recurrence. A recent systematic review of 45 studies evaluating postoperative outcomes for CRS, identified asthma (22.6%), AERD (28.7%) and allergic fungal rhinosinusitis (28.7%) as factors associated with polyp relapse in CRSwNP [60]. Moreover, the literature suggests allergic rhinitis may be as prevalent as 50–84% in patients with CRSwNP [61]. Type 2 inflammation and atopy appear to be a common theme amongst difficult to treat patients. Consequently, a significant effort has been made to explore the use of biologics originally designed to treat other inflammatory diseases, particularly asthma, for CRS management.

10.3 Biologic Therapies

10.3.1 Anti-IgE Antibodies

One of the first biologics developed and approved for the treatment of atopic diseases is omalizumab, an anti-IgE humanized monoclonal antibody. IgE, which is produced by IgE class-switched B cells binds to mast cells and basophils via FcεRI, which when cross-linked with antigen degranulate and release inflammatory markers including histamine, cytokines, proangiogenic factors and proteases [40]. Currently, omalizumab is indicated in patients with moderate to severe allergic asthma who are not well controlled with inhaled corticosteroids and inhaled long-acting β2 agonist bronchodilators [62]. There is compelling evidence that omalizumab may represent an effective treatment for eosinophilic CRSwNP. An early randomized double-blinded clinical trial of 24 participants in Belgium found that subcutaneous (SC) omalizumab (max dose 375 mg) every 2 weeks significantly reduced total nasal endoscopic polyp scores (TPS) as early as 8 weeks ($p = 0.001$); this polyp size difference was significantly different than in the placebo group ($p = 0.005$) [63]. Moreover, the Lund-MacKay computed tomography score, a radiologic measure of paranasal sinus inflammation, significantly improved in the study group compared to placebo after 16 weeks [63]. These positive findings have been validated by two recent phase 3 multicentre clinical trials, POLP1 and POLYP2 [64]. Collectively, 265 patients were randomized 1:1 to receive 600 mg SC omalizumab every 2 or 4 weeks and intranasal mometasone or intranasal mometasone alone for 24 weeks. The mean TPS treatment arm differences between the omalizumab and placebo groups were − 1.14 (95% CI = −1.59 to −0.69; $p < 0.0001$) and −0.59 (95% CI = −1.05 to −0.12; $p = 0.0140$) in POLYP1 and POLYP2 respectfully [64]. Sino-Nasal Outcome Test 22 (SNOT-22) scores improved between the two cohorts as early as 4 weeks and were clinically different in both POLYP1 and POLYP2 trials at the 24-week endpoint for the treatment and placebo cohorts (−16.12 [95% CI = −21.86 to −10.38], $p < 0.0001$ and −15.04 [95% CI = −21.26 to −8.82], $p < 0.0001$ respectfully). The University of Pennsylvania Smell Identification Test (UPSIT) [65] scores improved from baseline to 24 weeks in the omalizumab groups and the improvement was significant compared to placebo. Similar findings were reported for loss of smell scores, postnasal drip scores and nasal congestion scores (NCS) [64]. Interestingly, patients with comorbid asthma and AERD experienced comparable improvements in TPS and TPS to those without either comorbidity, when treated with omalizumab. Patients with comorbid asthma in the treatment arm saw significant improvements in asthma quality of life questionnaire scores [66] compared to placebo [64]. Omalizumab appears to be well tolerated, with few adverse events reported, the most common being headache, nasopharyngitis and injection site reaction [63–65]. Overall, these results demonstrate promising support for the use of omalizumab for CRSwNP with comorbid asthma. Currently, omalizumab has been approved in Canada for the treatment of severe CRSwNP refractory to inhaled nasal corticosteroids.

10.3.2 Anti-IL-5 Antibodies

Eosinophilia is a hallmark cellular infiltrate in the majority of CRSwNP cases and is believed to play a significant role in disease propagation and tissue-remodelling associated with polyp formation [10]. Specifically, IL-5 is considered the most important cytokine in the recruitment and activation of eosinophils, which makes it an ideal target for reducing eosinophilia [10, 41]. A number of humanized anti-IL-5 antibodies have been developed including: Reslizumab, Mepolizumab and Benralizumab. A randomized 2-centre study of Reslizumab in CRSwNP determined that a single 1 mg/kg or 3 mg/kg SC dose of Reslizumab was able to significantly reduce blood eosinophilia for up to 8 weeks [67]. However, this effect was transient and there was rebound eosinophilia noted 24 weeks post-injection. Moreover, while there was evidence of improvement in TPS in the treated cohort, there was unconvincing clinical improvement with regards to other symptom improvements [67]. More promising results have been found with the use of mepolizumab. Gevaert et al. demonstrated that 2 IV injections of 750 mg of mepolizumab one month apart were able to reduce nasal burden in 20 patients with CRSwNP refractory to corticosteroids [68]. There was a treatment difference of −1.30 (standard deviation (SD) +/−1.51; $p = 0.028$) in TPS scores between the treatment and placebo cohorts at 8 weeks. Moreover, Mepolizumab use was associated with a significant reduction in blood eosinophils ($p < 0.01$), serum eosinophil cation protein ($p = 0.022$) and serum IL-5Rα ($p < 0.001$) compared to placebo [68]. While there were changes favouring treatment-related improvements in postnasal drip, loss of smell and nasal congestion in the treated group compared to placebo these findings were neither clinically nor statistically significant [68]. A subsequent randomized controlled study of 105 patients with bilateral recurrent CRSwNP requiring surgery found that 750 mg Mepolizumab every 4 weeks (6 doses total) significantly reduced the need for ESS and reduced nasal polyposis severity visual analogue scores (VAS) (treatment difference at week 25 favouring mepolizumab −1.8, 95%CI −2.9 to −0.8; $p = 0.001$) [69]. While the SNOT-22 scores improved over the study period for both the intervention and control groups, there was a significant difference from baseline in the SNOT-22 scores between the cohorts at 25 weeks favouring the mepolizumab group (−13.2, 95%CI −22.2 to −4.2, $p = 0.005$). Additionally, at the study endpoint of 25 weeks, the mepolizumab group demonstrated significant improvements in loss of smell VAS ($p < 0.001$), peak nasal inspiratory flow ($p = 0.027$) and rhinorrhea ($p < 0.001$) compared to placebo [69]. Overall, anti-IL-5 biologics appear well tolerated with no serious adverse events reported from the aforementioned studies. The most common adverse events include upper respiratory tract infections, pyrexia oropharyngeal pain and injection site reactions [67–69]. More recently, Benralizumab has been developed, which may have greater anti-eosinophilic effects by directing targeting IL-5Rα expressing cells independent of ligand as a humanized anti-IL-5Rα monoclonal antibody [70]. Further, the antibody is afucosylated, which enables it to induce cell-mediated cytotoxicity of eosinophils. Benralizumab has demonstrated therapeutic benefits for severe asthmatics; however, no direct CRS clinical trials have been conducted [71]. There is evidence to suggest that patients with comorbid asthma and CRSwNP are more likely to respond to Benralizumab after Mepolizumab treatment compared to those with asthma alone [72]. A small study of 10 patients with eosinophilic CRSwNP and severe asthma received 30 mg SC Benralizumab every 4 weeks. After 24 weeks, all clinical parameters improved compared to baseline [73]. Specifically, the following markers of disease severity decreased including SNOT-22 scores ($p < 0.001$), TPS ($p < 0.001$), Lund-Mackay CT scores ($p < 0.001$) and blood eosinophilia (807.3 ± 271.1 cells/μL to 0 cells/μL, $p < 0.0001$) [73]. Likewise, Matsuno and Minamoto report that Benralizumab treatment exerted more rapid therapeutic action in patients with eosinophilic asthma and comorbid CRSwNP compared to those with asthma alone [66]. Rigorous clinical trials evaluating Benralizumab in CRSwNP are warranted to elucidate whether this biologic is beneficial for CRS alone.

10.3.3 Anti-IL-4/Il-13 Antibodies

Type 2 inflammation in CRS is mediated by effector ILC2 and Th2 cells, with the latter promoting B cell class switching and eosinophil recruitment. Specifically, IL-4 signalling encourages differentiation of naïve T cells to Th2 cells and is necessary for IgE B-cell class switching. Dupilumab is a humanized monoclonal antibody that binds to the alpha subunit of the IL-4 receptor alpha (IL-4Rα). It has been shown to be effective in the treatment of eosinophilic asthma and atopic dermatitis [74, 75]. Targeting IL-4Rα represents a promising target for the treatment of CRSwNP [76]. Moreover, the blockade of IL-4Rα targets two critical type 2 inflammatory cytokines as IL-4 and IL-13 both use this receptor to elicit their effects [74]. IL-13 is implicated in further modulating type 2 inflammation and both cytokines are directly involved in the tissue remodelling seen in CRSwNP [76]. Currently, dupilumab has been studied in CRSwNP in a single Phase 2a study and two phase 3 clinical trials [77, 78]. Two pivotal multicentre, international phase 3 trials, SINUS24 and SINUS52 have elucidated the therapeutic benefit of dupilumab for CRSwNP [78]. In SINUS24, 276 participants were randomized 1:1 to receive 300 mg SC dupilumab every 2 weeks plus nasal steroids or steroids alone for 24 weeks. In SINUS52, 448 participants were randomized 1:1:1 to receive 300 mg SC dupilumab every 2 weeks plus nasal steroids or 300 mg SC dupilumab every 2 weeks for 24 weeks followed by 28 weeks of 300 mg SC dupilumab every 4 weeks or placebo. Both studies demonstrated strong NPS improvement compared to placebo at 24 weeks (the least squares mean change (LMSD) −2·06 [95%CI, −2·43 to −1·69], $P < 0{\cdot}0001$) and (LSMD −1·80 [95%CI −2·10 to −1·51], $P < 0{\cdot}0001$) for SINUS24 and SINUS52 respectfully. At 52 weeks, the collective dupilumab groups continued to demonstrate improved NPS compared to placebo (LSMD −2·40 [95%CI, −2·77 to −2·02], $P < 0{\cdot}0001$) [78]. Similarly, SNOT-22 scores were significantly improved compared to placebo at both 24 weeks for SINUS24 (LSMD −21·12 [95%CI −25·17 to −17·06] $p < 0{\cdot}0001$) and SINUS52 (−17·36 [95%CI −20·87 to −13·85], $p < 0{\cdot}0001$) respectfully. Moreover, both studies demonstrated that at 24 weeks and 52 weeks UPSIT scores, Lund-MacKay CT scores and total symptom scores were significantly improved in the dupilumab cohorts compared to placebo [78]. Interestingly, biomarkers of inflammation including total serum IgE, eotaxin-3, eosinophil cationic protein and nasal IL-5 levels decreased at week 24 in SINUS24 and SINUS52 in the dupilumab groups. However, the improvements in eosinophilia appeared transient as discontinuation of dupilumab were associated with the return of baseline blood eosinophils. Subgroup analyses of persons with comorbid CRSwNP and asthma found that dupilumab treatment improved asthma symptoms and lung function. Moreover, improvements in lung function and asthma control were independent of whether participants had high or low blood eosinophils. Additionally, the data showcased that the effects of dupilumab of CRS symptom control were comparable for those with comorbid AERD and or asthma [78]. With regards to adverse events, dupilumab appears to be well tolerated with the most frequent issues including nasopharyngitis, epistaxis, headache and injection-site reactions being more common in the placebo arms; however, vigilance with new medications is always warranted [77, 78]. Given the high level of evidence, dupilumab represents a viable option for steroid-resistant CRSwNP and has been recently approved for CRSwNP care in Canada.

10.3.4 Future Therapeutic Targets

Relatively recent developments in atopic research have implicated the alarmins, IL-25, IL-33 and TSLP as key regulators of type 2 inflammation. These mediators are expressed by epithelial and innate lymphoid cells and favour activation and recruitment of ILC2s, eosinophils and Th2 cells [79]. Specifically, animal studies have demonstrated that anti-Il-33 treatment reduces mucus thickness, subepithelial collagen deposition and neutrophil infiltration within the nares [42]. Currently, the ECLIPSE phase 2 clinical trial of subcutaneous Etokimab (anti-IL-33 mAb) is being conducted with 106 patients with CRSwNP [80]. Unfortunately, a press report by AnaptysBio, Inc. revealed that Etokimab treatment every 4 or 8 weeks failed to achieve statistically significant improvements in NPS or SNOT-22 scores at the 8-week midpoint [81]. They plan on reassessing their findings at the 16-week study endpoint. While implicated in type 2 inflammation, a murine study reported that anti-IL-33 treatment reduced neutrophilic infiltration but failed to reduce eosinophilic infiltration [42]. As the majority of CRSwNP presentations are eosinophilic, this may explain the failed treatment effect observed with Etokimab. This biologic may have a niche role in the treatment of neutrophilic CRS. TLSP levels are greater in the nares of patients with CRSwNP and have been shown to promote macrophage IL-5 production [37]. A phase 2b clinical trial investigating Tezepelumab (anti-TLSP mAb) in patients with severe asthma with or without nasal polyps found the treatment to be well tolerated. Five hundred and fifty patients were randomized 1:1:1:1 to receive SC Tezepelumab 70 mg every 4 weeks, 210 mg every 4 weeks, 280 mg every 2 weeks or placebo. At the 52-week endpoint, the Tezepelumab-treated cohorts had significant reductions in annualized asthma exacerbation rates, reductions in blood eosinophils, FeNO, IL-5 and IL-13 levels irrespective of nasal polyp status compared to controls [82]. Currently, a multicentred phase 3 trial evaluating the efficacy of Tezepelumab for CRSwNP is underway [83]. As a relatively novel therapeutic avenue, it is important to assess the relative effectiveness of the varying biologics. While many biologics are currently available, no study to date has systematically compared the relative effectiveness of one biologic against another. For example, a case report found that dupilumab clinically improved CRSwNP symptoms in a patient with severe asthma, AERD and CRSwNP, despite Benralizumab treatment failing to provide clinical improvements even though it reduced blood eosinophil levels [84]. Unfortunately, this represents low-level evidence and cannot be generalized to other patients.

10.3.5 Biologic Therapy Guidelines

The cost-effectiveness of biologics dictates the use of guidelines for their use in CRSwNP [85]. In 2014, the estimated annual cost of endoscopic sinus surgery is 3510.31CAD, in contrast, with current annual costs for CRS biologics varying from $31,000 to $40,000 [86, 87]. Scangas et al. (year) used a Markov decision-tree economic model to estimate the long-term cost of endoscopic sinus surgery versus dupilumab as a function of quality-adjusted life years [88]. They estimate dupilumab to be less effective and more costly than surgery even when considering revision surgeries [88]. It is likely that biologic costs will have to reduce significantly before they can be more broadly integrated into a publicly funded CRS treatment model [89]. While the short-term cost of biologic therapy is substantially greater than that of alternative treatments, it is difficult to truly estimate the long-term value of biologic use. Further studies are required to appreciate the long-term implications and impact of biologic management. At this time, biologics should only be considered for patients with CRSwNP with moderate to severe disease who have failed maximal medical and surgical therapy, with sufficient surgery having been previously undertaken. Comorbid type 2 diseases are not required to consider biologic therapy in the setting of CRSwNP. Physicians should evaluate patient responses regularly and annually with a determination made at 16 weeks after treatment onset to determine treatment continuation, using both objective and subjective measures of disease improvement [85].

10.4 Conclusion

CRS represents a prevalent chronic disease negatively impacting a notable proportion of adults and it represents a significant burden from both quality of life and financial perspectives. Despite conventional therapy, difficult-to-treat patients with recurrent CRSwNP represents a population requiring further therapeutic intervention. Biologic therapies, which largely target the signalling mechanisms of type 2 inflammation, represent novel disease modulators which have been shown to be effective at improving objective and subjective disease clinical signs and symptoms and reduce inflammatory markers. There is evidence to suggest that targeting alarmins represents another avenue to regulate CRSwNP. However, current evidence suggests that biologic therapy treatment effects are largely transient, and relapse occurs following termination of biologic treatment. Further evidence is needed to evaluate the long-term effects of biologic therapy and the impact of various combined therapies. Specific endotyping of patients may prove to be beneficial to maximize the benefit seen from biologic therapies.

References

1. Chen Y, Dales R, Lin M. The epidemiology of chronic rhinosinusitis in Canadians. Laryngoscope. 2003;113(7):1199–205. https://doi.org/10.1097/00005537-200307000-00016.
2. Fokkens WJ, Lund VJ, Hopkins C, et al. European Position Paper on Rhinosinusitis and Nasal Polyps 2020. Rhinology. 2020;58(Suppl S29):1–464. https://doi.org/10.4193/Rhin20.600.
3. Husain Q, Sedaghat AR. Understanding and clinical relevance of chronic rhinosinusitis endotypes. Clin Otolaryngol. 2019;44(6):887–97. https://doi.org/10.1111/coa.13455.
4. Stevens WW, Schleimer RP, Kern RC. Chronic rhinosinusitis with nasal polyps. J Allergy Clin Immunol Pract. 2016;4(4):565–72. https://doi.org/10.1016/j.jaip.2016.04.012.
5. Chen S, Zhou A, Emmanuel B, Thomas K, Guiang H. Systematic literature review of the epidemiology and clinical burden of chronic rhinosinusitis with nasal polyposis. Curr Med Res Opin. 2020;36(11):1897–911. https://doi.org/10.1080/03007995.2020.1815682.
6. Caulley L, Thavorn K, Rudmik L, Cameron C, Kilty SJ. Direct costs of adult chronic rhinosinusitis by using 4 methods of estimation: results of the US medical expenditure panel survey. J Allergy Clin Immunol. 2015;136(6):1517–22. https://doi.org/10.1016/j.jaci.2015.08.037.
7. Rudmik L, Smith TL, Schlosser RJ, Hwang PH, Mace JC, Soler ZM. Productivity costs in patients with refractory chronic rhinosinusitis. Laryngoscope. 2014;124(9):2007–12. https://doi.org/10.1002/lary.24630.
8. Fokkens WJ, Lund VJ, Mullol J, et al. EPOS 2012: European position paper on rhinosinusitis and nasal polyps 2012. A summary for otorhinolaryngologists. Rhinology. 2012;50(1):1–12. https://doi.org/10.4193/Rhino50E2.
9. Hong CJ, Tsang AC, Quinn JG, Bonaparte JP, Stevens A, Kilty SJ. Anti-IgE monoclonal antibody therapy for the treatment of chronic rhinosinusitis: a systematic review. Syst Rev. 2015;4:166. https://doi.org/10.1186/s13643-015-0157-5.
10. Lee K, Tai J, Lee SH, Kim TH. Advances in the knowledge of the underlying airway remodeling mechanisms in chronic rhinosinusitis based on the endotypes: a review. Int J Mol Sci. 2021;22(2):910. https://doi.org/10.3390/ijms22020910.
11. Grayson JW, Hopkins C, Mori E, Senior B, Harvey RJ. Contemporary classification of chronic rhinosi-

nusitis beyond polyps vs no polyps: a review. JAMA Otolaryngol Head Neck Surg. 2020;146(9):831–8. https://doi.org/10.1001/jamaoto.2020.1453.
12. Lam K, Schleimer R, Kern RC. The etiology and pathogenesis of chronic rhinosinusitis: a review of current hypotheses. Curr Allergy Asthma Rep. 2015;15(7):41. https://doi.org/10.1007/s11882-015-0540-2.
13. Stevens WW, Peters AT, Tan BK, et al. Associations between inflammatory endotypes and clinical presentations in chronic rhinosinusitis. J Allergy Clin Immunol Pract. 2019;7(8):2812–2820.e3. https://doi.org/10.1016/j.jaip.2019.05.009.
14. Ponikau JU, Sherris DA, Kern EB, et al. The diagnosis and incidence of allergic fungal sinusitis. Mayo Clin Proc. 1999;74(9):877–84. https://doi.org/10.4065/74.9.877.
15. Braun H, Buzina W, Freudenschuss K, Beham A, Stammberger H. 'Eosinophilic fungal rhinosinusitis': a common disorder in Europe? Laryngoscope. 2003;113(2):264–9. https://doi.org/10.1097/00005537-200302000-00013.
16. Inoue Y, Matsuwaki Y, Shin SH, Ponikau JU, Kita H. Nonpathogenic, environmental fungi induce activation and degranulation of human eosinophils. J Immunol. 2005;175(8):5439–47. https://doi.org/10.4049/jimmunol.175.8.5439.
17. Lee CG, Da Silva CA, Dela Cruz CS, et al. Role of chitin and chitinase/chitinase-like proteins in inflammation, tissue remodeling, and injury. Annu Rev Physiol. 2011;73:479–501. https://doi.org/10.1146/annurev-physiol-012110-142250.
18. Rank MA, Adolphson CR, Kita H. Antifungal therapy for chronic rhinosinusitis: the controversy persists. Curr Opin Allergy Clin Immunol. 2009;9(1):67–72. https://doi.org/10.1097/ACI.0b013e328320d279.
19. Ebbens FA, Scadding GK, Badia L, et al. Amphotericin B nasal lavages: not a solution for patients with chronic rhinosinusitis. J Allergy Clin Immunol. 2006;118(5):1149–56. https://doi.org/10.1016/j.jaci.2006.07.058.
20. Feazel LM, Robertson CE, Ramakrishnan VR, Frank DN. Microbiome complexity and Staphylococcus aureus in chronic rhinosinusitis. Laryngoscope. 2012;122(2):467–72. https://doi.org/10.1002/lary.22398.
21. Bachert C, Zhang N, Patou J, van Zele T, Gevaert P. Role of staphylococcal superantigens in upper airway disease. Curr Opin Allergy Clin Immunol. 2008;8(1):34–8. https://doi.org/10.1097/ACI.0b013e3282f4178f.
22. Wang M, Shi P, Chen B, et al. The role of superantigens in chronic rhinosinusitis with nasal polyps. ORL J Otorhinolaryngol Relat Spec. 2008;70(2):97–103. https://doi.org/10.1159/000114532.
23. Ramakrishnan VR, Feazel LM, Abrass LJ, Frank DN. Prevalence and abundance of Staphylococcus aureus in the middle meatus of patients with chronic rhinosinusitis, nasal polyps, and asthma. Int Forum Allergy Rhinol. 2013;3(4):267–71. https://doi.org/10.1002/alr.21101.
24. Wang X, Moylan B, Leopold DA, et al. Mutation in the gene responsible for cystic fibrosis and predisposition to chronic rhinosinusitis in the general population. JAMA. 2000;284(14):1814–9. https://doi.org/10.1001/jama.284.14.1814.
25. Wang X, Kim J, McWilliams R, Cutting GR. Increased prevalence of chronic rhinosinusitis in carriers of a cystic fibrosis mutation. Arch Otolaryngol Head Neck Surg. 2005;131(3):237–40. https://doi.org/10.1001/archotol.131.3.237.
26. Wynne M, Atkinson C, Schlosser RJ, Mulligan JK. Contribution of epithelial cell dysfunction to the pathogenesis of chronic rhinosinusitis with nasal polyps. Am J Rhinol Allergy. 2019;33(6):782–90. https://doi.org/10.1177/1945892419868588.
27. Lee RJ, Kofonow JM, Rosen PL, et al. Bitter and sweet taste receptors regulate human upper respiratory innate immunity. J Clin Invest. 2014;124(3):1393–405. https://doi.org/10.1172/JCI72094.
28. Den Beste KA, Hoddeson EK, Parkos CA, Nusrat A, Wise SK. Epithelial permeability alterations in an in vitro air-liquid interface model of allergic fungal rhinosinusitis. Int Forum Allergy Rhinol. 2013;3(1):19–25. https://doi.org/10.1002/alr.21077.
29. Psaltis AJ, Bruhn MA, Ooi EH, Tan LW, Wormald PJ. Nasal mucosa expression of lactoferrin in patients with chronic rhinosinusitis. Laryngoscope. 2007;117(11):2030–5. https://doi.org/10.1097/MLG.0b013e31812e01ab.
30. Tieu DD, Peters AT, Carter RG, et al. Evidence for diminished levels of epithelial psoriasin and calprotectin in chronic rhinosinusitis. J Allergy Clin Immunol. 2010;125(3):667–75. https://doi.org/10.1016/j.jaci.2009.11.045.
31. Seshadri S, Lin DC, Rosati M, et al. Reduced expression of antimicrobial PLUNC proteins in nasal polyp tissues of patients with chronic rhinosinusitis. Allergy. 2012;67(7):920–8. https://doi.org/10.1111/j.1398-9995.2012.02848.x.
32. Kouzaki H, Matsumoto K, Kato T, Tojima I, Shimizu S, Shimizu T. Epithelial cell-derived cytokines contribute to the pathophysiology of eosinophilic chronic rhinosinusitis. J Interf Cytokine Res. 2016;36(3):169–79. https://doi.org/10.1089/jir.2015.0058.
33. Wang X, Zhang N, Bo M, et al. Diversity of T_H cytokine profiles in patients with chronic rhinosinusitis: a multicenter study in Europe, Asia, and Oceania. J Allergy Clin Immunol. 2016;138(5):1344–53. https://doi.org/10.1016/j.jaci.2016.05.041.
34. Fokkens WJ, Lund VJ, Mullol J, et al. European position paper on rhinosinusitis and nasal polyps 2012. Rhinol Suppl. 2012;23:3 p preceding table of contents, 1–298.
35. Zhang Y, Gevaert E, Lou H, et al. Chronic rhinosinusitis in Asia. J Allergy Clin Immunol. 2017;140(5):1230–9. https://doi.org/10.1016/j.jaci.2017.09.009.
36. Boita M, Bucca C, Riva G, Heffler E, Rolla G. Release of type 2 cytokines by epithelial cells of nasal polyps. J Immunol Res. 2016;2016:2643297. https://doi.org/10.1155/2016/2643297.

37. Nagarkar DR, Poposki JA, Tan BK, et al. Thymic stromal lymphopoietin activity is increased in nasal polyps of patients with chronic rhinosinusitis. J Allergy Clin Immunol. 2013;132(3):593–600.e12. https://doi.org/10.1016/j.jaci.2013.04.005.
38. Omori M, Ziegler S. Induction of IL-4 expression in CD4(+) T cells by thymic stromal lymphopoietin. J Immunol. 2007;178(3):1396–404. https://doi.org/10.4049/jimmunol.178.3.1396.
39. Huang Y, Paul WE. Inflammatory group 2 innate lymphoid cells. Int Immunol. 2016;28(1):23–8. https://doi.org/10.1093/intimm/dxv044.
40. Cao PP, Wang ZC, Schleimer RP, Liu Z. Pathophysiologic mechanisms of chronic rhinosinusitis and their roles in emerging disease endotypes. Ann Allergy Asthma Immunol. 2019;122(1):33–40. https://doi.org/10.1016/j.anai.2018.10.014.
41. Takatsu K, Nakajima H. IL-5 and eosinophilia. Curr Opin Immunol. Jun 2008;20(3):288–94. https://doi.org/10.1016/j.coi.2008.04.001.
42. Kim DK, Jin HR, Eun KM, et al. The role of interleukin-33 in chronic rhinosinusitis. Thorax. 2017;72(7):635–45. https://doi.org/10.1136/thoraxjnl-2016-208772.
43. Hirschberg A, Kiss M, Kadocsa E, et al. Different activations of toll-like receptors and antimicrobial peptides in chronic rhinosinusitis with or without nasal polyposis. Eur Arch Otorhinolaryngol. Jul 2016;273(7):1779–88. https://doi.org/10.1007/s00405-015-3816-1.
44. Tesmer LA, Lundy SK, Sarkar S, Fox DA. Th17 cells in human disease. Immunol Rev. 2008;223:87–113. https://doi.org/10.1111/j.1600-065X.2008.00628.x.
45. Smith KA, Orlandi RR, Rudmik L. Cost of adult chronic rhinosinusitis: a systematic review. Laryngoscope. 2015;125(7):1547–56. https://doi.org/10.1002/lary.25180.
46. Gurrola J, Borish L. Chronic rhinosinusitis: lendotypes, biomarkers, and treatment response. J Allergy Clin Immunol. 2017;140(6):1499–508. https://doi.org/10.1016/j.jaci.2017.10.006.
47. Pynnonen MA, Mukerji SS, Kim HM, Adams ME, Terrell JE. Nasal saline for chronic sinonasal symptoms: a randomized controlled trial. Arch Otolaryngol Head Neck Surg. 2007;133(11):1115–20. https://doi.org/10.1001/archotol.133.11.1115.
48. Fandiño M, Macdonald KI, Lee J, Witterick IJ. The use of postoperative topical corticosteroids in chronic rhinosinusitis with nasal polyps: a systematic review and meta-analysis. Am J Rhinol Allergy. 2013;27(5):e146–57. https://doi.org/10.2500/ajra.2013.27.3950.
49. Hauptman G, Ryan MW. The effect of saline solutions on nasal patency and mucociliary clearance in rhinosinusitis patients. Otolaryngol Head Neck Surg. 2007;137(5):815–21. https://doi.org/10.1016/j.otohns.2007.07.034.
50. Mullol J, Obando A, Pujols L, Alobid I. Corticosteroid treatment in chronic rhinosinusitis: the possibilities and the limits. Immunol Allergy Clin N Am. 2009;29(4):657–68. https://doi.org/10.1016/j.iac.2009.07.001.
51. Vaidyanathan S, Barnes M, Williamson P, Hopkinson P, Donnan PT, Lipworth B. Treatment of chronic rhinosinusitis with nasal polyposis with oral steroids followed by topical steroids: a randomized trial. Ann Intern Med. 2011;154(5):293–302. https://doi.org/10.7326/0003-4819-154-5-201103010-00003.
52. Tuncer U, Soylu L, Aydogan B, Karakus F, Akcali C. The effectiveness of steroid treatment in nasal polyposis. Auris Nasus Larynx. 2003;30(3):263–8. https://doi.org/10.1016/s0385-8146(03)00051-8.
53. Ragab SM, Lund VJ, Scadding G. Evaluation of the medical and surgical treatment of chronic rhinosinusitis: a prospective, randomised, controlled trial. Laryngoscope. 2004;114(5):923–30. https://doi.org/10.1097/00005537-200405000-00027.
54. Soler ZM, Oyer SL, Kern RC, et al. Antimicrobials and chronic rhinosinusitis with or without polyposis in adults: an evidenced-based review with recommendations. Int Forum Allergy Rhinol. Jan 2013;3(1):31–47. https://doi.org/10.1002/alr.21064.
55. Siu J, Mackenzie BW, Klingler L, et al. Sinonasal and gastrointestinal bacterial composition and abundance are stable after 1 week of once-daily oral antibiotic treatment for chronic rhinosinusitis. Int Forum Allergy Rhinol. 2021;11(9):1355–66. https://doi.org/10.1002/alr.22799.
56. Brooks SG, Trope M, Blasetti M, et al. Preoperative Lund-Mackay computed tomography score is associated with preoperative symptom severity and predicts quality-of-life outcome trajectories after sinus surgery. Int Forum Allergy Rhinol. 2018;8(6):668–75. https://doi.org/10.1002/alr.22109.
57. Lal D, Hopkins C, Divekar RD. SNOT-22-based clusters in chronic rhinosinusitis without nasal polyposis exhibit distinct endotypic and prognostic differences. Int Forum Allergy Rhinol. 2018;8(7):797–805. https://doi.org/10.1002/alr.22101.
58. Soler ZM, Hyer JM, Rudmik L, Ramakrishnan V, Smith TL, Schlosser RJ. Cluster analysis and prediction of treatment outcomes for chronic rhinosinusitis. J Allergy Clin Immunol. 2016;137(4):1054–62. https://doi.org/10.1016/j.jaci.2015.11.019.
59. Bassiouni A, Wormald PJ. Role of frontal sinus surgery in nasal polyp recurrence. Laryngoscope. 2013;123(1):36–41. https://doi.org/10.1002/lary.23610.
60. Loftus CA, Soler ZM, Koochakzadeh S, et al. Revision surgery rates in chronic rhinosinusitis with nasal polyps: meta-analysis of risk factors. Int Forum Allergy Rhinol. 2020;10(2):199–207. https://doi.org/10.1002/alr.22487.
61. Rosati MG, Peters AT. Relationships among allergic rhinitis, asthma, and chronic rhinosinusitis. Am J Rhinol Allergy. 2016;30(1):44–7. https://doi.org/10.2500/ajra.2016.30.4252.
62. Agache I, Akdis CA, Akdis M, et al. EAACI Biologicals Guidelines-Recommendations for severe asthma. Allergy. 2021;76(1):14–44. https://doi.org/10.1111/all.14425.

63. Gevaert P, Calus L, Van Zele T, et al. Omalizumab is effective in allergic and nonallergic patients with nasal polyps and asthma. J Allergy Clin Immunol. 2013;131(1):110–6.e1. https://doi.org/10.1016/j.jaci.2012.07.047.
64. Gevaert P, Omachi TA, Corren J, et al. Efficacy and safety of omalizumab in nasal polyposis: 2 randomized phase 3 trials. J Allergy Clin Immunol. 2020;146(3):595–605. https://doi.org/10.1016/j.jaci.2020.05.032.
65. Pinto JM, Mehta N, DiTineo M, Wang J, Baroody FM, Naclerio RM. A randomized, double-blind, placebo-controlled trial of anti-IgE for chronic rhinosinusitis. Rhinology. 2010;48(3):318–24. https://doi.org/10.4193/Rhino09.144.
66. Matsuno O, Minamoto S. Rapid effect of benralizumab for severe asthma with chronic rhinosinusitis with nasal polyps. Pulm Pharmacol Ther. 2020;64:101965. https://doi.org/10.1016/j.pupt.2020.101965.
67. Gevaert P, Lang-Loidolt D, Lackner A, et al. Nasal IL-5 levels determine the response to anti-IL-5 treatment in patients with nasal polyps. J Allergy Clin Immunol. Nov 2006;118(5):1133–41. https://doi.org/10.1016/j.jaci.2006.05.031.
68. Gevaert P, Van Bruaene N, Cattaert T, et al. Mepolizumab, a humanized anti-IL-5 mAb, as a treatment option for severe nasal polyposis. J Allergy Clin Immunol. 2011;128(5):989–95.e1–8. https://doi.org/10.1016/j.jaci.2011.07.056.
69. Bachert C, Sousa AR, Lund VJ, et al. Reduced need for surgery in severe nasal polyposis with mepolizumab: randomized trial. J Allergy Clin Immunol. 2017;140(4):1024–1031.e14. https://doi.org/10.1016/j.jaci.2017.05.044.
70. Ghazi A, Trikha A, Calhoun WJ. Benralizumab--a humanized mAb to IL-5Rα with enhanced antibody-dependent cell-mediated cytotoxicity--a novel approach for the treatment of asthma. Expert Opin Biol Ther. 2012;12(1):113–8. https://doi.org/10.1517/14712598.2012.642359.
71. Nair P, Wenzel S, Rabe KF, et al. Oral glucocorticoid-sparing effect of benralizumab in severe asthma. N Engl J Med. 2017;376(25):2448–58. https://doi.org/10.1056/NEJMoa1703501.
72. Hamada S, Kobayashi Y, Yasuba H. Role of eosinophilic chronic rhinosinusitis in switching to benralizumab treatment in mepolizumab responders. Int J Clin Pharmacol Ther. 2020;58(12):703–8. https://doi.org/10.5414/CP203767.
73. Lombardo N, Pelaia C, Ciriolo M, et al. Real-life effects of benralizumab on allergic chronic rhinosinusitis and nasal polyposis associated with severe asthma. Int J Immunopathol Pharmacol. 2020;34:2058738420950851. https://doi.org/10.1177/2058738420950851.
74. Wenzel S, Ford L, Pearlman D, et al. Dupilumab in persistent asthma with elevated eosinophil levels. N Engl J Med. 2013;368(26):2455–66. https://doi.org/10.1056/NEJMoa1304048.
75. Beck LA, Thaçi D, Hamilton JD, et al. Dupilumab treatment in adults with moderate-to-severe atopic dermatitis. N Engl J Med. 2014;371(2):130–9. https://doi.org/10.1056/NEJMoa1314768.
76. Kariyawasam HH. Chronic rhinosinusitis with nasal polyps: mechanistic insights from targeting IL-4 and IL-13 via IL-4Rα inhibition with dupilumab. Expert Rev Clin Immunol. 2020;16(12):1115–25. https://doi.org/10.1080/1744666X.2021.1847083.
77. Bachert C, Mannent L, Naclerio RM, et al. Effect of subcutaneous dupilumab on nasal polyp burden in patients with chronic sinusitis and nasal polyposis: a randomized clinical trial. JAMA. 2016;315(5):469–79. https://doi.org/10.1001/jama.2015.19330.
78. Bachert C, Han JK, Desrosiers M, et al. Efficacy and safety of dupilumab in patients with severe chronic rhinosinusitis with nasal polyps (LIBERTY NP SINUS-24 and LIBERTY NP SINUS-52): results from two multicentre, randomised, double-blind, placebo-controlled, parallel-group phase 3 trials. Lancet. 2019;394(10209):1638–50. https://doi.org/10.1016/S0140-6736(19)31881-1.
79. Ryu G, Kim DW. Th2 inflammatory responses in the development of nasal polyps and chronic rhinosinusitis. Curr Opin Allergy Clin Immunol. 2020;20(1):1–8. https://doi.org/10.1097/ACI.0000000000000588.
80. AnaptysBio I. Etokimab in adult patients with chronic rhinosinusitis with nasal polyps (CRSwNP). ClinicalTrials.Gov: U.S. Library of Medicine; 2021.
81. AnaptysBio reports top-line data from interim analysis of ECLIPSE phase 2 clinical trial of etokimab in chronic rhinosinusitis with nasal polyps. 2020. https://www.globenewswire.com/en/news-release/2020/08/10/2075615/0/en/AnaptysBio-Reports-Top-Line-Data-from-Interim-Analysis-of-ECLIPSE-Phase-2-Clinical-Trial-of-Etokimab-in-Chronic-Rhinosinusitis-with-Nasal-Polyps.html.
82. Emson C, Corren J, Sałapa K, Hellqvist Å, Parnes JR, Colice G. Efficacy of Tezepelumab in patients with severe, uncontrolled asthma with and without nasal polyposis: a *post hoc* analysis of the phase 2b PATHWAY study. J Asthma Allergy. 2021;14:91–9. https://doi.org/10.2147/JAA.S288260.
83. AstraZeneca. Efficacy and safety of Tezepelumab in participants with severe chronic rhinosinusitis with nasal polyposis (WAYPOINT). Clinicaltrials.gov: U.S. National Library of Medicine; 2021.
84. Suzaki I, Tanaka A, Hirano K, Arai S, Kobayashi H. Successful management of eosinophilic chronic rhinosinusitis complicated by severe asthma using dupilumab, following negative initial results with benralizumab. Allergol Int. 2021;70(1):150–2. https://doi.org/10.1016/j.alit.2020.05.002.
85. Thamboo A, Kilty S, Witterick I, et al. Canadian Rhinology Working Group consensus statement: biologic therapies for chronic rhinosinusitis. J Otolaryngol Head Neck Surg. 2021;50(1):15. https://doi.org/10.1186/s40463-021-00493-2.

86. Au J, Rudmik L. Cost of outpatient endoscopic sinus surgery from the perspective of the Canadian government: a time-driven activity-based costing approach. Int Forum Allergy Rhinol. 2013;3(9):748–54. https://doi.org/10.1002/alr.21181.
87. Anderson WC, Szefler SJ. Cost-effectiveness and comparative effectiveness of biologic therapy for asthma: to biologic or not to biologic? Ann Allergy Asthma Immunol. 2019;122(4):367–72. https://doi.org/10.1016/j.anai.2019.01.018.
88. Scangas GA, Wu AW, Ting JY, et al. Cost utility analysis of dupilumab versus endoscopic sinus surgery for chronic rhinosinusitis with nasal polyps. Laryngoscope. 2021;131(1):E26–33. https://doi.org/10.1002/lary.28648.
89. Patel GB, Peters AT. The role of biologics in chronic rhinosinusitis with nasal polyps. Ear Nose Throat J. 2021;100(1):44–7. https://doi.org/10.1177/0145561320964653.

11 Nasal NO and Its Role in the Physiology of the Nose and Diagnosis

Peter W. Hellings and Glenis K. Scadding

Core Messages

- NO is a gas primarily produced within the paranasal sinus cavities, playing a role in airway homeostasis.
- Nasal NO levels may be normal, increased, or decreased in upper airway diseases.
- Nasal NO represents a screening test for primary ciliary dyskinesia.

11.1 Introduction

Since the last decade, growing attention has been paid to nitric oxide (NO) as a noninvasive marker of inflammation of the airways. Exhaled NO (eNO) measurement has become a routine diagnostic tool in monitoring lower airway inflammation since it had been shown to be a noninvasive parameter for monitoring lower airway inflammation [1]. However, most of the NO is being produced in the sinonasal cavities, without clear insight into its precise role in upper airway homeostasis. Nasal NO (nNO) may be involved in the innate antibacterial effects of the airway mucosa, regulation of ciliary beat frequency, and/or local regulation of blood flow. In addition, nNO is believed to contribute to lower airway homeostasis. The measurement of nNO is nowadays a good screening test for primary ciliary dyskinesia and, therefore, has diagnostic value.

P. W. Hellings (✉)
Department of Otorhinolaryngology, Head and Neck Surgery, University Hospitals Leuven, Leuven, Belgium
e-mail: peter.hellings@uz.kuleuven.ac.be

G. K. Scadding
Department of Allergy and Rhinology, Royal National Throat, Nose and Ear Hospital, London, UK

11.2 Nitric Oxide (NO)

Nitric oxide (NO) is a colorless and odorless gas that is present in the air exhaled through the mouth or nose. NO is produced from arginine and oxygen by nitric oxide synthase (NOS) [2]. Constitutively expressed neuronal and endothelial forms exist as well as an induced form, iNOS, which appears to be upregulated within the respiratory tract in response to pro-inflammatory signals. NO came to prominence for its role in vasodilatation and subsequently as a neurotransmitter and inflammatory mediator. The role of NO in the airways is complex [3], possibly including antibacterial effects, pro-inflammatory effects, and regulation of blood flow and ciliary beat frequency. Exhaled NO (eNO) levels are raised in eosinophilic asthma, and measurement of this has become a standardized, but not yet widespread, tool in the diagnosis and management of asthma. It can potentially provide a rapid, low-cost, objective measure of lower airway inflammation.

Ö. Ö. Celebi, T. M. Önerci (eds.), *Nasal Physiology and Pathophysiology of Nasal Disorders*,
https://doi.org/10.1007/978-3-031-12386-3_11

Far greater levels of NO are produced in the upper than in the lower respiratory tract, with major contributions from the sinuses and to a lesser extent from the nasal mucosa.

11.3 Nasal NO

Measurement of nNO represents a useful tool for research purposes as well as for screening for PCD [4]. Nasal nitric oxide may be normal, raised, or lowered in disease states. However, its measurement may be a useful tool in the diagnosis and management of patients with chronic rhinosinusitis, nasal polyps, and CF, as well as in the diagnosis of PCD [4]. Levels of nNO may follow the clinical changes after medical as well as surgical treatment in patients with CRS with/without nasal polyps [4].

Measuring both bronchial and nasal nitric oxide may assist the combined management of upper and lower airways, but is not routinely performed in clinical practice so far.

High levels of NO are produced constitutively in normal individuals within the paranasal sinuses by calcium-independent nitric oxide synthase, with levels measured at 20–25 ppb. Additionally, nitric oxide is also formed in the nasal mucosa by inducible NOS (iNOS) in response to inflammation. NO and its metabolites are toxic to microorganisms and likely form part of the innate defense mechanism of the respiratory tract. NO may also stimulate cilia beat frequency within the epithelium and regulate nasal vascular tone.

11.4 Technique for Measuring nNO

As for exhaled NO (eNO), nNO can also be measured by chemiluminescence, using noninvasive techniques, providing immediate results. A number of different techniques have been used to ensure sampling from the upper airways only including breath holding and breathing against resistance.

In contrast to measuring eNO, high baseline levels in nNO make background environmental NO levels less of a problem. Conversely, there is a high degree of interindividual variability among healthy controls. Moreover, there is also a significant degree of intraindividual variation over time, meaning that changes of 20–25% or less may be accounted for by normal variation rather than a change in disease status or response to medication [5]. Additionally, the lack of universal standardization of testing procedures means that levels recorded by different study groups vary considerably even among equivalent patient populations. The factors affecting eNO levels such as recent exercise or time of day may similarly affect nNO measurements. Local factors such as nasal volume and patency may also be important.

11.5 Value of nNO Measurement in Clinical Practice

Despite the above limitations, nNO has a number of potentially useful clinical applications. With regard to diagnosis, nNO is useful as a screening tool for patients with possible PCD; levels less than 100 ppb, particularly if these persist following decongestion, should stimulate investigation of mucociliary structure and function. The test is objective and may be easier to perform than a saccharine clearance test in younger children. Similarly, nNO may provide a useful tool in the diagnosis of CF in the context of upper respiratory tract symptoms; levels significantly lower than in controls have been reported in some studies, but not others. nNO has a potential role in the diagnosis and assessment of CRS, especially when associated with NP. Interestingly, despite the increased expression of iNOS in polyp epithelium, low nNO levels have been found in two large studies. Moreover, nNO inversely correlated with endoscopic NP size, CT scores, and clinical severity of the disease. Conversely, in a study involving chronic rhinosinusitis patients with and without polyps, no correlation between nNO and CT scores was found, although patients were again found to have lower baseline nNO than controls.

Low nNO levels in chronic rhinosinusitis are thought to reflect obstruction at the sinus ostium and impairment of gas transfer out from the sinuses. This is supported by the finding of raised nNO following medical and surgical treatment of rhinosinusitis with or without polyps [5].

A number of recent studies have focused on the possible use of humming to improve the sensitivity of nNO measurements. Weitzberg and Lundberg found that humming induced a large increase in nNO [6] and that these increases were not detected in patients with nasal polyps and sinus ostium obstruction. Furthermore, they suggested that the absence of a normal response to humming during nNO measurement could be used to identify allergic rhinitis with sinus ostium obstruction. Whether this adds significant value to standard testing has yet to be fully appreciated.

11.6 Diagnosis of Primary Ciliary Dyskinesia (PCD)

In children with rhinosinusitis presenting with longstanding and persistent anterior rhinorrhea, one may be interested in the evaluation of the function of the mucociliary clearance system for the exclusion of PCD. Normal mucociliary transport is essential for the maintenance of healthy sinuses. In case of infection with secondary ciliary dysfunction and/or congenital dysfunction of the cilia like in PCD, the mucociliary transport is inadequate or has not taken place. In PCD, lack of mucociliary transport may lead to chronic rhinosinusitis and bronchiectasis. In chronic inflammation, mucostasis, hypoxia, microbial products, and toxic inflammatory mediators may induce secondary ciliary changes like in secondary ciliary dyskinesia (SCD), with inadequate mucociliary transport. The mucociliary transport (MCT) mechanism ensures the clearance of entrapped particles in the mucus lining the nasal mucosa toward the hypopharynx. Several nonabsorbable substances have been used for the evaluation of MCT in patients, like saccharin or dyes like methylene blue. As the MCT can only be measured in cooperative patients with patent nasal cavities and in the absence of severe mucosal disease, this test has limited diagnostic value due to its low sensitivity and specificity.

No ideal test is available for the diagnosis of PCD. In case of suspicion of PCD in a patient with rhinosinusitis since birth, familial history of PCD, and associated features of Kartagener syndrome, i.e., situs inversus and infertility, one should consider diagnostic tests of ciliary function by evaluation of CBF, electron microscopic evaluation of the dynein arms of the cilia, and/or evaluation of the cilia after ciliogenesis in vitro. As these techniques are not available in routine ENT practice, one may rely on measuring nasal NO levels as low NO levels have been associated with PCD and therefore represent a good screening tool for PCD.

11.7 Concluding Remarks

Nasal NO measurement represents a useful diagnostic tool in PCD. However, variable baseline levels of nNO and interindividual variability make nNO measurement of little value in the diagnosis and management of uncomplicated rhinitis.

References

1. Alving K, Weitzberg E, Lundberg JM. Increased amount of nitric oxide in exhaled air of asthmatics. Eur Respir J. 1993;6(9):1368–70.
2. Scadding G, Scadding GK. Update on the use of nitric oxide as a non-invasive measure of airways inflammation. Rhinology. 2009;47:115–20.
3. Scadding G. Nitric oxide in the airways. Curr Opin Otolaryngol Head Neck Surg. 2007;15(4):258–63.
4. Scadding G, Hellings P, Alobid I, Bachert C, Fokkens W, van Wijk RG, et al. Diagnostic tools in rhinology EAACI position paper. Clin Transl Allergy. 2011;1(1):2.
5. Ragab SM, Lund VJ, Saleh HA, Scadding G. Nasal nitric oxide in objective evaluation of chronic rhinosinusitis therapy. Allergy. 2006;61(6):717–24.
6. Weitzberg E, Lundberg JO. Humming greatly increases nasal nitric oxide. Am J Respir Crit Care Med. 2002;166(2):144–5.

12 Physiology and Pathophysiology of Sneezing and Itching: Mechanisms of the Symptoms

Murat Songu and T. Metin Onerci

Core Messages

- Despite its being such a commonplace phenomenon suffered the world over, little is known about the sneeze reflex action, which sometimes becomes a sign associated with a series of different medical conditions.
- The sneezing reflex may be divided into two phases. The first is a nasal or sensitive phase, following stimulation of the nasal mucosa by chemical or physical irritants. The efferent or respiratory phase consists of eye closing, deep inspiration, and then a forced expiration with the initial closing of the glottis and increasing intrapulmonary pressure. The sudden dilatation of the glottis gives rise to an explosive exit of air through the mouth and nose, washing out mucosal debris and irritants.
- Clinical studies using positron emission tomography indicate that there is no isolated itching center in the brain but that there are different cortical centers which are involved in the processing of the itch.
- The factors that play role in the etiology of sneeze reflex are rhinitis, photic sneeze reflex, physical stimulants of the trigeminal nerve, central nervous system pathologies, psychogenic sneezing, snatiation reflex, and sexual ideation.

Sneeze is a coordinated protective respiratory reflex which occurs due to stimulation of the upper respiratory tract, particularly the nasal cavity [1]. Apparently, activation of the central and peripheral nervous system plays a major role in the pathophysiology of this process. Sensory nerves of the afferent trigeminal system including myelinated Aδ-fibers and thin, nonmyelinated C-fibers of the nasal mucosa transmit signals generating sensations, including itching and motor reflexes, such as sneezing. These nerves can be stimulated for various reasons. Sneeze reflex frequently accompanies rhinitis of allergic or nonallergic origin. A sneeze can also arise due to bright light or sun (ACHOO syndrome), physical stimulants of the trigeminal nerve, psychogenic or central nervous system pathologies, and even due to a full stomach (snatiation reflex) or sexual ideation. In this chapter, we aimed to review the physiology, pathophysiology, etiology, diagnosis, treatment, and complications of sneezing and itching.

12.1 Historical Perspective

Sneezing has always been a remarkable sign and a noteworthy occurrence throughout the history. In Asia and Europe, the sneezing superstition

M. Songu (✉)
Department of Otorhinolaryngology, Biruni University Faculty of Medicine, İstanbul, Turkey

T. M. Onerci
Department of Otorhinolaryngology, Faculty of Medicine, Hacettepe University, Ankara, Turkey

Ö. Ö. Celebi, T. M. Önerci (eds.), *Nasal Physiology and Pathophysiology of Nasal Disorders*,
https://doi.org/10.1007/978-3-031-12386-3_12

extends through a wide range of race, age, and country. Homer tells in his epic literary work, Odysseia, that Penelope rejoiced greatly when her son Telemachus sneezed when she expressed a wish. In another part, in Odysseia, the Athenian General Xenophon gave a dramatic speech exhorting his fellow soldiers to follow him to liberty or to death against the Persians. He spoke for an hour motivating his army and assuring them a safe return to Athens until a soldier underscored his conclusion with a sneeze. Thinking that this sneeze was a heavenly favorable sign from the gods, the whole army sprang to an attack. Xenophon's death after the destruction of his army by the Persians can be considered the first complication of sneezing in history. Hippocrates has mentioned that sneezing is dangerous only prior to or following a pulmonary disease. Otherwise, it is beneficial, even in patients with fatal illnesses. Celsus of Rome has suggested that sneezing is evidence of convalescence from illness. Aristotle has proposed that sneezing is a holy sign since it arises from lungs which are the principle and most divine parts of the body [2–4]. The Greeks and the Romans took sneezing as a sign of wellness and expressed their good wishes to the person who sneezed using the phrase "live long" or "May Jupiter bless you." In pagan culture, a person who sneezed was believed to get rid of the devil in his body and was congratulated passionately in every medium. It was also believed that sneezing made a person's body open to invasion by Satan and evil spirits or even caused part of one's soul being "thrown out of the body." The remnant of this pagan tradition still exists today in the expression "live a long life" or "God bless you." After the pandemics of the plague in Europe, the view of sneezing changed and it began to be assumed as a sign of a great danger. In the fourteenth century, during the great pandemic of plague, the so-called *black death*, Pope Gregory VII declared the sentence "May God bless you" as a short prayer to be said following every sneeze to protect against the plague [5]. However, the plague eliminated one third of the European population since the Pope, or the powers that be, failed to think of taking measures in order to protect the public against the disease-causing rats or the fleas, which were the real vectors. This short prayer used in the past by the Christian population as a protection against disease is now a common everyday expression which has been passed on to the modern Christian society as an inheritance [6].

In the Talmud, which consists of religious texts including the Jewish law and history, it is considered to be a favorable omen if someone sneezes while praying. It is a sign that just as God looks favorably toward him on earth so too will he look favorably toward him in heaven. According to a common belief that still exists in the Republic of China and Japan, if a person sneezes without a reason, this means that somebody else is talking about him. It is believed that good things are told about him if he sneezes once and bad things if twice. In Naples, the person who sneezes thinks that he is remembered by another person. In the Indian folkloric culture, sneezing before starting work is believed to bring bad luck; therefore, the work is started after a glass of water is drunk to avoid bad luck [6].

12.2 Physiology and Pathophysiology

The largest part of the nasal cavity and the paranasal sinuses are covered by a multilayered ciliated epithelium in the respiratory region. Olfactory cells, sustentacular cells, and submucosal adenoids are located on the roof of the nose including areas of the medial and upper nasal concha and upper septum in the olfactory region [7]. The nasal mucosa is covered by a bilayered secretion film which is produced by submucosal adenoidal and goblet cells. This secretion film includes a low viscous sol film in which cilia move and an apical, highly viscous gel film. This specific assembly serves the physiological cleaning of respiratory air through the mucociliary apparatus. Cholinergic input causes an increased secretion of the submucosal cells. However, goblet cell stimulation has also been shown by substance P (SP) releasing sensory nerves in a rat model [8]. The cavernous tissue of the nasal concha consists of venous sinusoids and regulates the respiratory resistance. Filling of these sinusoids is regulated by sympathetic stimulation as adrenergic fibers are distributed around arteries, arterioles, and veins of the human nasal mucosa [7].

The sneezing reflex may be divided into two phases. The first is a nasal or sensitive phase, following stimulation of the nasal mucosa by chemical or physical irritants. Many distal branches of trigeminal nerve terminate in the facial skin transmitting tactile, pain, and temperature sensations, while some branches distribute in the nasal mucosal epithelium [9]. These branches are myelinated sensory fibers of small diameter which terminate with receptor endings. Some of these receptors are triggered by chemical stimuli, while others are sensitive to tactile and mechanical stimuli [9]. Afferent neural stimuli are transmitted to the trigeminal ganglion via anterior ethmoidal, posterior nasal, infraorbital, and ophthalmic branches of trigeminal nerve [10]. The trigeminal nerve is important for the nociceptive sensory supply of the nasal mucosa in addition to the face, oral mucosa, cornea, and conjunctiva. Itching and sneezing are generated by the activation of trigeminal afferent nerve terminals in the nasal mucosa [11]. These nociceptive nerve fibers consist mainly of two types of fibers: the thin Aδ-fibers that mediate acute perceptions with a quick adaption and activation only during the actual irritation and the nonmyelinated C-fibers which adapt slowly and communicate dull burning, difficult to locate perceptions, which outlast acute pain [12]. In allergic rhinitis, immunologically triggered inflammation results in the recruitment and activation of both types of fibers that result clinically in itching and sneezing [13].

Clinical studies using positron emission tomography indicate that there is no isolated itching center in the brain but that there are different cortical centers which are involved in the processing of the itch [14, 15]. Activation of the anterior gyrus cinguli, the supplementary motor cortex, and the inferior parietal lobe partly explains the connection between itching and the related reflex of scratching [16, 17]. Using functional MRI, the activation of corresponding cortical units following painful trigeminal stimulation has been shown [18]. In this regard, neuropeptides produced in the cell body of C-fiber neurons can also be transported in granule structures within the cytoplasm to nerve terminals in the central nervous system. This leads to "central sensitization," a phenomenon associated with the activation of nociceptive C-fibers [19].

Upon reaching a threshold, the second phase—the efferent or respiratory phase—begins once a critical number of inspiratory and expiratory neurons have been recruited [20]. This consists of eye closing, deep inspiration, and then a forced expiration with initial closing of the glottis and increasing intrapulmonary pressure. The sudden dilatation of the glottis gives rise to an explosive exit of air through the mouth and nose, washing out mucosal debris and irritants. A variety of injuries can occur during a sneeze, especially when a closed-airway sneeze is attempted, and high Valsalva pressure is transmitted to the other systems [21]. Setzen and Platt reported 52 unique sneeze-related injuries in the literature that were categorized into six areas of injury: intrathoracic, laryngeal/pharyngeal, ocular/orbital, intracranial/neurological, otologic, and other.

The number of particles expelled during a forceful sneeze, of which the sizes range from 0.5–5 μm, is estimated to be 40,000. The estimations concerning the speed of a sneeze range between 150 and 1045 km/h (nearly 85% of the sound velocity) [9]. In present days, scenario of pandemic of Covid-19 the protective reflexes namely "Sneeze and Cough" have received great importance but not in terms of protection but in terms of spread of infection [22]. As the flow of air during expiration is turbulent, it causes damage to the Epithelial Lining Fluid present in the respiratory conduit and also gets admixed with the saliva in the oropharynx and oral cavity and mucus in the nose to form droplets of various sizes. The spread of droplet cloud in sneezing may range to 6 m or more as compared to cough hence the concept of 1–2 m of social distancing does not hold good if the patient is sneezing.

12.3 Etiology

The factors that play role in the etiology of sneeze reflex are listed in Table 12.1.

12.3.1 Rhinitis

It is the inflammation of the nasal mucosa causing nasal stuffiness, rhinorrhea, nasal pruritus, and sneezing [23].

12.3.1.1 Allergic Rhinitis

Allergic rhinitis is the inflammation of the mucosa lining the nasal cavity in the form of IgE-dependent type I hypersensitivity reaction [23]. Allergic rhinitis is a common disorder, which represents a considerable burden both on individual patients and on society [24, 25]. Itching and sneezing represent two of the main bothersome symptoms, apart from nasal obstruction and rhinorrhea in allergic rhinitis [26]. In allergy-related nasal inflammation, it can be demonstrated that especially the neurotransmitter SP is released by C-fibers [13, 27]. SP is significantly increased in the nasal lavage of patients with allergic rhinitis, in contrast to healthy subjects, which is interpreted as a sustained stimulation of the sensory system [28]. Exogenically administered SP results in a dose-dependent occurrence of nasal symptoms in asymptomatic patients with allergic rhinitis and controls, without elevation of inflammatory mediators. In addition to SP, other neuropeptides, including calcitonin gene-related peptide (CGRP) and vasoactive intestinal polypeptide (VIP), are increased in nasal lavage fluids after nasal provocation in allergic rhinitis [29]. An important feature of allergic rhinitis is hyperresponsiveness influenced by products of the allergic reaction including eicosanoids; cytokines such as IL-6, Il-1β, and TNF-α; and, most importantly, neurotrophins including nerve growth factor (NGF) and brain-derived neurotrophic factor (BDNF). NGF targets nociceptor fibers, leading to upregulated activity, increased SP content, and dendrite sprouting [30]. The allergen-induced increased BDNF expression in the nasal mucosa significantly correlated with the maximal increase of total nasal symptom score in allergic rhinitis [31] (Fig. 12.1).

Table 12.1 Etiology of the sneeze reflex

(a) Rhinitis
(b) Photic sneeze reflex (ACHOO syndrome)
(c) Physical stimulations of the trigeminal nerve
(d) Central nervous system pathologies
(e) Psychogenic (intractable) sneezing
(f) Snatiation* reflex
(g) Sexual ideation or orgasm

12.3.1.2 Infectious Rhinitis

In more than 50% of cases, rhinovirus is the responsible agent for common cold (nasopharyngitis) which is the most common clinical form of viral infections. Inhalation, close contact with the infected person, contact with the objects contaminated with the virus such as door handles, school desks, household goods or phones, are all ways of contracting the virus. The most common prodromal manifestations include high fever, nasal irritation, and sneezing. Nasal symptoms are also present in influenza (flu) caused by influenza virus; however, they are likely to be overshadowed by malaise, fatigue, myalgia, and high fever. Bacteria may infiltrate the tissue and cause infections during the course of viral rhinitis due to impairment of mucosal integrity and ciliary function. Clinical picture of rhinitis may occur during the course of specific bacterial diseases such as diphtheria, rhinoscleroma, lepra, tuberculosis, syphilis, and glanders. Opportunistic fungal infections which develop in AIDS, chemotherapy, prolonged intensive care unit stay, and disorders of neutrophil number and function such as neutropenia and diabetes may cause rhinitis.

12.3.1.3 Nonallergic Noninfectious Rhinitis

Every sneezing paroxysm does not mean an allergic rhinitis or common cold. There are also many rhinitis clinical pictures of unknown etiology that may cause sneezing and are classified as "nonallergic noninfectious rhinitis." The commonly observed clinical pictures in this group are NARES and vasomotor rhinitis and are usually confused with allergic rhinitis.

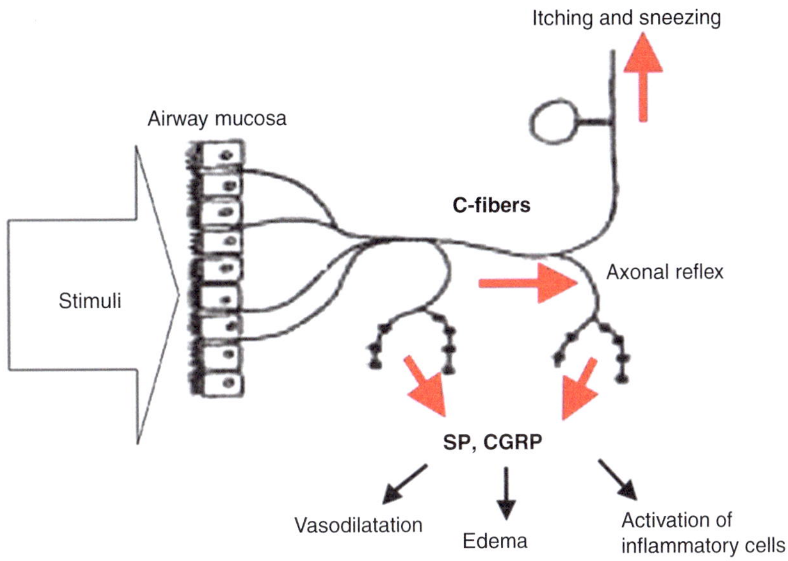

Fig. 12.1 Schematic presentation of the processes involved in "neurogenic inflammation" leading to itching and sneezing. Activation of afferent trigeminal C-fibers by several stimuli including allergen contact may result in an efferent liberation of substance P (*SP*) and calcitonin gene-related peptide (*CGRP*) via axonal reflex. This further leads to vasodilatation, edema, and recruitment/migration of inflammatory cells

NARES (Nonallergic Rhinitis with Eosinophilia Syndrome)

It is a nasal hyperreactivity syndrome, manifested by sneezing paroxysms and watery discharge followed by nasal stuffiness and hyposmia in which the increased eosinophil count exceeds 20% of total leukocyte number in the absence of an IgE-dependent allergy [32]. The suspected etiology is the infiltration of circulating eosinophils into the site of inflammation because of the increased level of substance P as a result of alterations within the nasal mucosa due to irritation, senility, or other factors.

Idiopathic (Vasomotor/Nonallergic and Non-eosinophilic) Rhinitis

It is a short-course hyperreactive nasal mucosal disease of sudden onset manifested by nasal stuffiness, watery nasal discharge or postnasal drip, and sneezing. Eosinophilia is not found in nasal secretions and allergy tests are negative. The most important clinical feature that differentiates vasomotor rhinitis from NARES is the sudden onset of the symptoms which relieve immediately following the disappearance of triggering factors. Environmental factors including cold and dry air, high amount of moisture, dyes, chlorine water, perfume, tangs, cigarette smoke, exhaust gas, and other inhaled irritants may initiate the symptoms of vasomotor rhinitis. A typical example is a sudden onset of nasal stuffiness and short-lasting though forceful sneezing paroxysm in the morning time after being exposed to cold and dry air. Individual factors such as fatigue, stress, and sexual activity may develop the clinical picture as well [33].

12.3.1.4 Others Causes of Rhinitis

Many other clinical pictures classified as nonallergic and noninfectious rhinitis are also common causes of sneezing. *Occupational rhinitis* is the most commonly observed one among these clinical pictures and is triggered by dense cigarette smoke, cold air, air fresheners, formaldehyde, and other chemical irritants in the workplace [32, 34]. *Hormonal rhinitis* may occur in cases of physiologically increased levels of estrogen such as puberty, menstrual cycle, and pregnancy or in case of receiving exogenous estrogen while on oral contraceptives [32, 35]. Inhibiting acetylcholinesterase activity, estrogen leads to edema formation in the nasal mucosa while progesterone causes congestion through vasodilatation in capacitance vessels and sneezing occurs [36]. As a result of sympathetic hypoactivity in hypothyroidism, parasympathetic activity relatively increases leading to vasodilatation in the nasal mucosa and an increase in secretions. Drug-induced rhinitis may be manifested in two clinical patterns. In *medicamentous rhinitis,* many drugs affect nasal mucosa through different mechanisms of action and cause sneezing. It is usually observed in treatments with antihypertensive drugs (reserpine, guanethidine, phentolamine, methyldopa, hydralazine, and prazosin), beta-blockers (propranolol, nadolol), aspirin, and other NSAIDs [37]. *Rebound rhinitis,* on the other hand, develops as a result of prolonged usage of vasoconstrictor drops or sprays [32, 33, 35]. In *geriatric rhinitis*, atrophy in submucosal glands due to senility and severe irritation in sensitive nerve endings result in nasal stuffiness and sticky and thick mucus which may cause sneezing. *Atrophic rhinitis* is a rare pathological condition in etiology of which bacterial infections, deficiency of vitamin A or D, and iron or estrogen deficiencies are suspected. In contrast to clinical conditions concerned, many different factors may develop irritant-induced rhinitis and cause sneezing [33]. Some of these include dust, smoke, perfume, powder, sharp odor, ammonia, inhalation of corrosive gases or chemicals, mechanical obstruction of the nasal cavity, chymic irritation due to cauterization or silver nitrate application, capsaicin, application of airflow into superior nasal meatus by a catheter, and repetitive nasal electrical stimulation [38–40]. Capsaicin, the active ingredient obtained from hot chili peppers, stimulates the nasal small unmyelinated C-fiber afferent nerves to release various tachykinins. These nerves, with their somata in the trigeminal ganglion, transmit the information to the central nervous system through the trigeminal dorsal horn in the medulla and lead to sneezing and a sense of pain [41]. Although various peptides and tachykinins may be involved, it appears that the capsaicin-induced release of substance P is the most potent trigger of the sneezing response [38, 40]. Capsaicin also precipitates sneezing through a local axon reflex [42].

12.3.2 Photic Sneeze Reflex

It seems that *some* people really do sneeze when they look at the sun or actually at any bright light (there is nothing special about the sun). Photic sneeze reflex is also called ACHOO (*a*utosomal dominant *c*ompelling *h*elio-*o*phthalmic *o*utburst) syndrome [43]. This reflex was first described in the medical literature by Sedan in 1954 [44]. It was shown to have an autosomal dominant inheritance pattern and is assumed to affect 17–35% of the world population [45]. Photic sneeze reflex has been reported to be present in 23% of medical students [46]. According to a Swedish study, 24% of blood donors experienced sneezing on visual exposure to strong light [47].

We do not know exactly why this happens, but it might reflect a “crossing” of pathways in the brain, between pupillary light reflex arc and sneezing reflex arc. The reflex can be triggered only after the first exposure to light, never on repetitive stimulation, and many reports cite a refractory period before the reflex can be elicited suggesting that a polysynaptic pathway is involved. The first theory concerning the pathways mentioned belongs to Eckhardt et al. who suggested that stimulation of the optic nerve triggers the trigeminal nerve [48]. The afferent impulses of pupillary light reflex are transmitted via the optic nerve while the efferent impulses are transmitted via the oculomotor nerve. According to this theory, an indirect impulse is transmitted to the ophthalmic division of the trigeminal nerve. This impulse generates the nasal stimulation that causes sneezing by affecting the maxillary division of the trigeminal nerve as well. The second theory of crossing pathways belongs to Watson. Light falling on the retina stimulates afferent fibers to the pretectal nuclei, which then send interneurons to the Edinger-Westphal nuclei. The parasympathetic fibers from the Edinger-Westphal nuclei and the trigeminal afferent fibers from the cornea both pass through the ciliary ganglion, where they may participate in transmission [49]. Parasympathetic generalization may also contribute to photic sneeze. Stimuli which excite primarily one branch of the parasympathetic nervous system tend to activate other branches. Thus, the parasympathetic branches of the oculomotor nerve which are activated to generate pupillary constriction against the bright light cause secretion and congestion in the nasal mucosa by triggering the parasympathetic activation by the pterygopalatine ganglion. This process triggers sneezing [50].

What is the benefit of photic sneeze reflex? Photic sneezing reflex exists in animals for which the smell sensation is vital to survive and can be used to clean the nasal cavity. Animals such as cats and dogs sneeze largely through their nose, while the adults sneeze through their mouth. The reflex arc may also be useful to a limited extent in human beings when it is considered that the nasal respiration is dominant in the neonatal period. Babies have no other way to get rid of the annoying little tickle caused by normal mucus. Young children sometimes have more disgusting ways of dealing with that sensation, but babies just sneeze often with the help of photic sneeze reflex. In conclusion, photic sneeze reflex, which can lead the drivers to have accidents following a sudden exposure to sunlight at the end of a long tunnel, or can cause a plane crash by inactivating the masks of jet pilots, can be considered to be an annoying "holdover" of evolution [51, 52].

Hyden and Arlinger examined whether the tickling inside the nose prior to photic sneeze in cases is associated with a recordable local activity or not, and they stated that no reproducible electrical activity could be recorded [53].

Langer et al. designed a study to study the cortical keystones of photic sneezing, and revealed that photic sneeze might be the result of superior sensitivity to visual stimuli in the visual cortex and of co-activation of somatosensory areas [54]. The 'photic sneeze reflex' is therefore not a classical reflex that occurs only at a brainstem or spinal cord level but, in contrast to many theories, involves also specific cortical areas.

Sevillano et al. assessed the ocular involvement in the pathophysiology of ACHOO syndrome and stated that a dominant autosomal inheritance with mild penetrance was demonstrated, with 67% of the studied subjects showing some degree of prominent corneal nerves [55].

Wand et al. performed a genome-wide association study on photic sneeze reflex in the Chinese population to uncover the underlying genetic markers in a Chinese population of 3417 individuals, and reproducibly identified both a replicative rs10427255 on 2q22.3 and a novel locus of rs1032507 on 3p12.1 in various effect models [56].

There is no recognized management for photic sneeze reflex; however, Bobba et al. offered a practical approach to minimizing the PSR by utilizing the Philtral Pressure Technique [57]. This involved firm digital pressure applied by the patient's index finger transversely to the skin of the sub-philtral region, directed posterosuperiorly onto the maxilla.

12.3.3 Physical Stimulants of the Trigeminal Nerve

Physical or mechanical stimulants in the innervation zone of the trigeminal nerve may trigger sneezing reflex. Some of these stimulants include pulling hair, tearing off eyebrows, or orbital injections administered frequently during ocular surgery under local anesthesia [58, 59].

12.3.4 Central Nervous System Pathologies

The lateral medullary syndrome (LMS), or Wallenberg's syndrome, often results from occlusion or dissection of the vertebral artery. Vertebral artery dissection has been blamed on many different life events, such as sneezing [60]. Paroxysmal sneezing at the onset of LMS is usually interpreted as a cause, since a violent sneeze could potentially result in a vertebral artery dissection causing LMS. Due to inactivation of sneezing center in LMS, sneezing cannot occur although the sensation of sneezing is present [61–63]. Localization of the human sneeze center was described in a patient with right LMS, initially presenting with violent sneezes and followed by brief loss of the sneeze reflex with eventual recovery [64].

Sneezing may commonly accompany temporal lobe and grand mal epilepsy. It may be observed during the aura prior to an epileptic seizure or it may develop as an autonomic reflexive response during the seizure as well [65, 66]. Beverwyck commented upon the analogy of the epileptic seizure with hiccups and sneezing and noted that the physiological and anatomical basis for such a hypothesis remained to be unexplained [67]. In the mid-nineteenth century, Jackson used the term epilepsy "as the name for occasional, sudden, excessive, rapid and local discharges of grey matter" [68]. Jackson further commented upon the healthy and yet random discharge and concluded that "a sneeze is a sort of healthy epilepsy."

12.3.5 Psychogenic (Intractable) Sneezing

Intractable sneezing, first described by Shilkrel in 1949, is a rare pathological condition that has been detected in more than 50 cases in literature up to present [69–71]. Kanner referred to a 13-year-old girl who had incessant sneezing for over 2 months and whose progress was followed by a daily newspaper communique [72]. A diagnosis of hysteria was made and subsequent psychotherapy eliminated the sneezing. Yater referred to similar explosive repetitious episodes and considered them to be a sort of imitation of the true act of sneezing [73].

Psychogenic intractable sneezing occurs mainly in adolescent girls for which a cause may not be found. Organic lesions or causes should always be carefully excluded [71]. Patients are usually refractory to various medications and have an otherwise unremarkable extensive workup [74, 75]. Inspiratory phase is quite short and the amount of nasal mucosal secretion expelled very low. Eyes may remain open during sneezing. It usually develops due to psychogenic factors and is refractory to medical treatment [76]. Approximately 25% of the reported cases resolve without any form of treatment, except counseling of the patient and family [71]. Psychogenic sneezing responds well to psychological measures such as psychotherapy, biofeedback, relaxation exercises, supportive psychotherapy (i.e., explanation of nature of illness, suggestion to overcome symptoms), and behavior therapy (reward when there is symptom reduction, aversion therapy, hypnosis, and relaxation). The role of anxiolytic drugs lies in reducing underlying anxiety and making the patient more amenable to psychotherapy [77].

Medically unexplained physical symptoms usually carry diagnostic difficulties for the physicians [71]. The most important factors that increase these diagnostic difficulties are the possibility of an underlying physical illness and the uncertainty encountered as to how far the investigations for physical causes should go. It was determined that in some somatization patients, organic pathologies were revealed during follow-up. Paradoxically, it is known that repetitive and advanced investigations for any organic etiology in conversion disorder may increase the anxiety and doubts in the family and thus prolong the duration of the illness. In conclusion, one must not assume that every case of paroxysmal sneezing is of psychogenic origin. Due to the nature of such a disorder, these patients should undergo medical evaluations before a psychogenic cause is even considered.

12.3.6 Snatiation* Reflex

An uncontrollable sneezing attack developing as a result of stretching of the stomach following an excessive nutrition is first described by Teebi et al. as a reflex with autosomal dominant inheritance pattern [78]. The mechanism of development is unknown. Snatiation* is a combination of the words "sneeze" and "satiation." Snatiation also stands for "*S*neezing *N*oncontrollably *A*t a *T*une of *I*ndulgence of the *A*ppetite-a *T*rait *I*nherited and *O*rdained to be *N*amed" [79]. This abbreviation was supposed to facilitate the future cases to be evaluated in the same class. Recently, two patients have been reported, who state that several members of their family sneeze on a full stomach [80]. This report doubles the number of families with snatiation reflex in the medical literature.

12.3.7 Sexual Ideation or Orgasm

An association between sexual excitement and sneezing was first described in the nineteenth century [49, 81] followed by a young German otolaryngologist who developed a theory of "nasal reflex neurosis" due to the finding of erectile tissue in both nasal mucous membranes and genital areas [82, 83]. The first report of this phenomenon in the literature describes a 69-year-old man who complains of severe sneezing immediately following orgasm, with no associated psychiatric morbidity [84]. Stromberg in 1975 and Korpas in 1979 described male orgasm as a precipitant for the sneeze reflex [85, 86]. Bhutta described a middle-aged man with uncontrollable fits of sneezing with sexual thought. The patient had no other rhinological symptoms and psychiatric morbidity [87]. Bhutta et al. performed a search of Internet "chat rooms" and found 17 people of both sexes reporting sneezing immediately upon sexual ideation and three people after orgasm. Although Internet reports do not give an accurate incidence, their findings do suggest that it is much more common than recognized. One year later, Bhutta and Maxwell revealed their experience on internet-based media or by spontaneous contact with the authors that sneezing induced by sexual ideation reported in additional 146 cases [80].

12.4 Diagnosis, Differential Diagnosis, and Management of Rhinitis

The evaluation of a patient with sneezing should be individualized according to the duration and severity of the symptom. Laboratory tests are not necessary in the majority of patients, since the diagnosis is usually obvious from the history and physical examination.

It should be remembered that the history of the patient is the most important and determining stage for the diagnosis [88]. The patient should be asked what his/her main complaint is; the duration and frequency of the symptoms like nasal discharge, stuffiness, and pruritus if present; whether the nasal discharge or stuffiness is present on one side or both, perennial or seasonal; whether he/she has allergic complaints, past trauma, past nasal surgery history, known diseases, and drugs used; and also how these symptoms effect the quality of life. In female patients it is also important to ask whether she is pregnant or on oral contraceptives [36]. One of the most common signs of allergic rhinitis in children is a horizontal creasing over the nasal tip. This physical examination sign develops as a result of habitual rubbing, which is also called *allergic salute,* after a duration of at least 2 years, a repeated action in order to relieve pruritus and improve respiration. This habit may turn into facial grimacing in adulthood for social reasons. *Allergic shiner*, on the other hand, is permanent pigmentation on the skin of lower eyelid which present as dark circles at the beginning stage. It develops due to subcutaneous hemosiderin through a capillary leak during periorbital venous stasis as a result of nasal mucosal congestion. *Dennie–Morgan folds* are short semilunar lines or folds found below the inferior eyelid. These lines develop due to venous blood retention cause by continuous spasm of Müller's muscle under the inferior eyelid. *Silky long eyelashes* are another outstanding concomitant sign of allergy. Clinical examination including anterior rhinoscopy and nasal endoscopy provides large information about pathologies related to septum and lateral nasal wall. Allergy is prediagnosed with medical history and physical examination. If the patient has a medical history and complaints that are compatible with allergy, in vivo (prick test, SET, scratch test) and/or in vitro (serum-specific IgE) allergy tests should be performed [89]. The skin prick test is the most common epidermal test. A positive allergen skin test that is compatible with the medical history and findings of physical examination should be assumed to be significant [90]. RAST (radioallergosorbent test) and ELISA (enzyme-linked immunosorbent assay) tests measure the amount of allergen-specific IgE antibodies. Since there is no risk of systemic reactions during the application of these tests, they can safely be performed on pregnant women, on patients with a past history of systemic reaction, during measuring the sensitivity to antigens with a high risk of systemic reaction, on patients with skin diseases, on people who use drugs that may affect the prick skin test results, and in medicolegal cases where objective data are needed and also on children. The changes that develop due to gradually increasing doses of the allergen are followed by nasal provocation test. This test is performed when objective data are needed for occupational rhinitis, and it is rather used for scientific studies. Nasal cytology is not a diagnostic method performed for routine clinical practice and considered as an evaluation that does not provide a sufficient support alone [91]. The use of acoustic rhinometry and rhinomanometry, which are the most common objective nasal airway tests, is confined because they have high costs and is time demanding in terms of application and interpretation [92]. Mucociliary function may be evaluated for differential diagnosis in patients with rhinitis [93]. Radiological examination is not necessary in a patient prediagnosed with rhinitis as long as an additional pathology is not suspected.

The treatment of persistent or recurring sneezing should be directed at the cause whenever possible. The best treatment in patients with allergic rhinitis is to avoid the allergen [94]. Medical treatment or when needed, even immunotherapy when needed, is used in patients who do not benefit from avoidance or environmental control [95–98]. Some patients may benefit from adjunctive surgical treatment. The management of common cold and influenza is symptomatic. Decongestants, antipyretics, bed rest, and increased fluid intake are recommended. Systemic antibiotics are preferred in patients who develop bacterial infections secondary to a viral infection while agent-specific antibiotic treatment is applied in those who develop rhinitis secondary to specific bacteria. It is vital to diagnose the patient and initiate the treatment immediately, particularly in fulminant fungal infections. The initial stage in the treatment of the patients with NARES is avoiding the irritant environmental conditions. Medical treatment is considered in case the initial stage fails to succeed. The success of steroids in early phases decreases in long-term administrations due to decreased steroid receptors on eosinophils. Oral and topical decongestants may be used adjunctive to steroid therapy. Capsaicin, a substance isolated from chili pepper extract, has an initial stimulating effect on C receptors which turns into an inhibiting effect following repetitive applications. Antihistamines are of no use and treatment of vasomotor rhinitis is palliative. Oral and topical decongestants can be applied. Topical corticosteroids are not always beneficial. Ipratropium bromide, which prevents the secretions from serous and seromucous glands inhibiting the cholinergic system, may be effective. Antihistamines are of no use. In the treatment of gestational rhinitis, medication should definitely be avoided for the first 10 weeks, and the treatment should definitely be applied with obstetric advice in the following periods. Isotonic saline sprays may be useful for pregnant women due to their humidifying and mucosal cleaning effects. The first line medical treatment for allergic rhinitis in pregnant women is cromolyn sodium, a mast cell stabilizer. Beclomethasone and triamcinolone are the topical steroids of choice for those who do not benefit from cromolyn sodium. The safest antihistamine during pregnancy is chlorpheniramine and the safest oral decongestant is pseudoephedrine. In the management of rebound rhinitis, the inducing drug should be discontinued and oral or parenteral corticosteroids should be administered in order to relieve the patient's complaints, and the treatment should be supported with topical nasal corticosteroids. Surgery may be necessary if irreversible changes have developed in the inferior conchae. The purpose of the treatment in geriatric rhinitis should be to provide sufficient intracellular moisture. For this purpose, it is appropriate to humidify the nasal mucosa with solutions including a combination of isotonic solution and glycerine and to add guaifenesin to treatment which stimulates the submucosal glands. Isotonic solutions can be combined with glycerine for the initial management of atrophic rhinitis. Other solutions can be antibiotherapy, estrogen support, vitamins A and D administration, iron support, or corticosteroid administration. It has also been recommended to practice surgical closure of one or both nostrils for a period of 1 year, or surgical procedures to narrow the nasal cavity have been recommended. Avoiding irritant substances should be the initial approach for the management of occupational or irritant-induced rhinitis (Table 12.2).

Table 12.2 Differential diagnosis of rhinitis in terms of history and laboratory tests

Feature	Allergic rhinitis	Infectious rhinitis	NARES	Vasomotor rhinitis
Onset of symptoms	Seasonal/perennial	Seasonal	Perennial	Perennial
Symptoms	*Sneezing* Nasal stuffiness Nasal pruritus Nasal discharge Postnasal drip	*Sneezing* Nasal stuffiness Nasal discharge Fever Myalgia	*Sneezing* Nasal stuffiness Nasal discharge	*Sneezing* Nasal stuffiness Nasal discharge Postnasal drip
Triggering allergen	Yes	No	No	No
Triggering irritant	Yes	No	Yes	Yes
Allergy tests	Positive	Negative	Negative	Negative
Nasal cytology	Eosinophilia	Neutrophilia	Eosinophilia	Rare eosinophilia

12.5 Complications of Sneeze Reflex

Since the symptoms of majority of upper respiratory tract infections include cough and sneezing, numerous particles disperse into the air during the course of these diseases. The most important complication of sneezing that affects public health is spread of droplet infections, tuberculosis in particular. The incidence of tuberculosis, which was taken under control through the improvement of efficient treatments in the second half of the twentieth century, began to increase again due to certain factors including the outburst of HIV infection, the decrease in the importance given to disease control, and poverty [99, 100].

Gwaltmey et al. have determined that the intranasal pressure increases up to 176 mmHg during sneezing with the mouth and the nostrils closed [101]. Complications pertaining to this high pressure have been reported in literature. These complications include acute aortic dissection, cerebral venous thrombosis, loss of hearing due to fracture footplate, abortus, orbital emphysema, pneumocephalus, acute wide-angle glaucoma, pneumatocele of the lacrimal sac, intimal tear of the arteriovenous fistula, retinal hemorrhage, and costal fracture reported in a patient with osteoporosis [102–109].

12.6 Conclusion

Sneezing is a phenomenon that is common to all humans and is widespread in the animal kingdom as well. It may play an important role in maintaining health in ways that we do not currently understand. Sneezing, which cannot consciously be controlled, is a protective reflex for the body during which facial, pectoral, and abdominal muscles function concordantly maintaining the respiration. It is rarely a sign of serious illness or impending disaster as feared by previous generations. On the other hand, it can be remarkably annoying. A thorough knowledge of this reflex can be a valuable aid in the diagnosis of other concomitant diseases.

Conflict of Interest The author has no financial relationship with a commercial entity that has an interest in the subject of this manuscript.

References

1. Songu M, Cingi C. Sneeze reflex: facts and fiction. Ther Adv Respir Dis. 2009;3:131–41.
2. Aristotle. Problems, Book XXXIII (trans: Rackham H, Hett WS). Cambridge: Harvard University Press; 1994.
3. Disraeli I. Curiosities of literature. Cambridge: Riverside Press; 1864.
4. Prioreschi P. A history of medicine, Roman medicine, vol. III. Omaha: Horatius Press; 1998.
5. Ovesen L. A regular contribution from Lars Ovesen, which we hope will help the busy scientist in keeping up with the literature fine arts. Eur J Cancer Prev. 2003;12:239–40.
6. Knowlson ST. The origins of popular superstitions and customs. Forgotten Books. London: T. Werner Laruier Clifford's Inn; 1910.
7. Klimek L, Riechelmann H, Saloga J, et al. Anatomie, Physiologie und Immunologie der Nasenschleimhaut, spezifische und unspezifische Abwehrmechanismen. In: Klimek L, Riechelmann H, Saloga J, Mann W, Knop J, editors. Allergologie und Umweltmedizin. Stuttgart-New York: Schattauer; 1997.

8. Kovacs I, Ludany A, Koszegi T, et al. Substance P released from sensory nerve endings influences tear secretion and goblet cell function in the rat. Neuropeptides. 2005;39:395–402.
9. Nishino T. Physiological and pathophysiological implications of upper airway reflexes in humans. Jpn J Physiol. 2000;50:3–14.
10. Wallois F, Macron JM, Jounieaux V, et al. Trigeminal afferences implied in the triggering or inhibition of sneezing in cats. Neurosci Lett. 1991;122:145–7.
11. Widdicombe JG. Nasal pathophysiology. Respir Med. 1990;84(Suppl A):3–9.
12. Torebjork HE, Hallin RG. Perceptual changes accompanying controlled preferential blocking of A and C fibre responses in intact human skin nerves. Exp Brain Res. 1973;16:321–32.
13. Ader R, Cohen N, Felten D. Psychoneuroimmunology: interactions between the nervous system and the immune system. Lancet. 1995;345:99–103.
14. Hsieh JC, Hagermark O, Stahle-Backdahl M, et al. Urge to scratch represented in the human cerebral cortex during itch. J Neurophysiol. 1994;72:3004–8.
15. Pfaar O, Raap U, Holz M, et al. Pathophysiology of itching and sneezing in allergic rhinitis. Swiss Med Wkly. 2009;139(3–4):35–40.
16. Schmelz M, Schmidt R, Bickel A, et al. Specific C-receptors for itch in human skin. J Neurosci. 1997;17:8003–8.
17. Suranyi L. Localization of the sneeze center. Neurology. 2001;57:161.
18. Hummel T, Doty RL, Yousem DM. Functional MRI of intranasal chemosensory trigeminal activation. Chem Senses. 2005;30(Suppl 1):i205–6.
19. Sarin S, Undem B, Sanico A, et al. The role of the nervous system in rhinitis. J Allergy Clin Immunol. 2006;118:999–1016.
20. Batsel HL, Lines AJ. Neural mechanisms of sneeze. Am J Phys. 1975;229:770–6.
21. Setzen S, Platt M. The dangers of sneezing: a review of injuries. Am J Rhinol Allergy. 2019;33:331–7.
22. Singh NK, Kumar N, Singh AK. Physiology of disease transmission of respiratory tract infection: a narrative review. Infect Disord Drug Targets. 2021;21(6):e170721188930.
23. Bousquet J, van Cauwenberge P, Khaltaev N, The ARIA Workshop Group, World Health Organization. Allergic rhinitis and its impact on asthma. J Allergy Clin Immunol. 2001;108(Suppl):147–334.
24. Cingi C, Topuz B, Songu M, et al. Prevalence of allergic rhinitis among the adult population in Turkey. Acta Otolaryngol. 2010b;130:600–6.
25. Ozdoganoglu T, Songu M. The burden of allergic rhinitis and asthma. Ther Adv Respir Dis. 2012;6:11–23.
26. Ozdoganoglu T, Songu M, Inancli HM. Quality of life in allergic rhinitis. Ther Adv Respir Dis. 2012;6:25–39.
27. Kim D, Baraniuk JN. Neural aspects of allergic rhinitis. Curr Opin Otolaryngol Head Neck Surg. 2007;15:268–73.
28. Klimek L, Schäfer D. Wirkung von Substanz-P in der Nasenschleimhaut. Allergologie. 1996;19:524.
29. Mosimann BL, White MV, Hohman RJ, et al. Substance P, calcitonin gene-related peptide, and vasoactive intestinal peptide increase in nasal secretions after allergen challenge in atopic patients. J Allergy Clin Immunol. 1993;92:95–104.
30. Levi-Montalcini R. The nerve growth factor 35 years later. Science. 1987;237:1154–62.
31. Raap U, Fokkens W, Bruder M, et al. Modulation of neurotrophin and neurotrophin receptor expression in nasal mucosa after nasal allergen provocation in allergic rhinitis. Allergy. 2008;63:468–75.
32. Dykewicz MS, Fineman S, Skoner DP, et al. Diagnosis and management of rhinitis: complete guidelines of the joint task force on the parameters in allergy, asthma and immunology. Ann Allergy Asthma Immunol. 1998;81:478–518.
33. Fairbanks DNF, Kaliner M. Nonallergic rhinitis and infection. In: Cummings CW, Fredrickson JM, Harker LA, Krause CJ, et al., editors. Otolaryngology – head and neck surgery. St. Louis: Mosby Year Book; 1998.
34. Stevens HE. Allergic and inflammatory aspects of chronic rhinosinusitis. J Otolaryngol. 1991;20:395–9.
35. King HC, Mabry R. A practical guide to the management of nasal and sinus disorders. 1st ed. New York: Thieme; 1993.
36. Incaudo GA. Diagnosis and treatment of allergic rhinitis and sinusitis during pregnancy and lactation. Clin Rev Allergy Immunol. 2004;27:159–77.
37. Cingi C, Ozdoganoglu T, Songu M. Nasal obstruction as a drug side effect. Ther Adv Respir Dis. 2011;5:175–82.
38. Imamura T, Kambara T. Substance P as a potent stimulator of sneeze responses in experimental allergic rhinitis of guinea pigs. Agents Actions. 1992;37:245–9.
39. Kira S, Nogami Y, Taketa K, et al. Size-selective sampling of oil mist in air and subjective symptoms among machine workers. Ind Health. 1997;35:394–8.
40. Kitajiri M, Kubo N, Ikeda H, et al. Effects of topical capsaicin on autonomic nerves in experimentally-induced nasal hypersensitivity. An immunocytochemical study. Acta Otolaryngol Suppl. 1993;500:88–91.
41. Geppetti P, Fusco BM, Marabini S, et al. Secretion, pain and sneezing induced by the application of capsaicin to the nasal mucosa in man. Br J Pharmacol. 1988;93:509–14.
42. Canning BJ. Neurology of allergic inflammation and rhinitis. Curr Allergy Asthma Rep. 2002;2:210–5.
43. Collie WR, Pagon RA, Hall JG, et al. ACHOO syndrome (autosomal dominant compelling helio-ophthalmic outburst syndrome). Birth Defects Orig Artic Ser. 1978;14:361–3.
44. Sedan J. Photosternutatory reflex. Rev Otoneuroophthalmol. 1954;26:123–6.
45. Morris HH 3rd. ACHOO syndrome: prevalence and inheritance. Cleve Clin J Med. 1987;54:431–3.

46. Everett HC. Sneezing in response to light. Neurology. 1964;14:483–90.
47. Beckman L, Nordenson I. Individual differences with respect to the sneezing reflex: an inherited physiological trait in man? Hum Hered. 1983;33:390–1.
48. Eckhardt LB, McLean JM, Goodell H. The genesis of pain from the eye. Assoc Res Nerv Ment Dis Proc. 1943;23:209–17.
49. Watson WS. Diseases of the nose and its accessory cavities. London: Lewis; 1875.
50. Brubaker AP. The physiology of sneezing. JAMA. 1919;73:585–7.
51. Benbow EW. Practical hazards of photic sneezing. Br J Ophthal. 1991;75:447.
52. Breitenbach RA, Swisher PK, Kim MK, et al. The photic sneeze reflex as a risk factor to combat pilots. Mil Med. 1993;158:806–9.
53. Hydén D, Arlinger S. On light-induced sneezing. Eye (Lond). 2009;23:2112–4.
54. Langer N, Beeli G, Jäncke L. When the sun prickles your nose: an EEG study identifying neural bases of photic sneezing. PLoS One. 2010;5:e9208.
55. Sevillano C, Parafita-Fernández A, Rodriguez-Lopez V, et al. A curious fact: photic sneeze reflex. Autosomical dominant compelling helio-ophthalmic outburst syndrome. Arch Soc Esp Oftalmol. 2016;91:305–9.
56. Wang M, Sun X, Shi Y, Song X, Mi H. A genome-wide association study on photic sneeze reflex in the Chinese population. Sci Rep. 2019;9:4993.
57. Bobba S, Spencer SKR, Fox OJK, et al. Management of the photic sneeze reflex utilising the philtral pressure technique. Eye (Lond). 2019;33:1186–7.
58. Sekizawa SI, Ishikawa T, Sant'Ambrogio G. Asymmetry in reflex responses of nasal muscles in anesthetized guinea pigs. J Appl Physiol. 1998;85:123–8.
59. Wallois F, Bodineau L, Macron JM, et al. Role of respiratory and non-respiratory neurones in the region of the NTS in the elaboration of the sneeze reflex in cat. Brain Res. 1997;768:71–85.
60. Schievink WI. Spontaneous dissection of the carotid and vertebral arteries. N Engl J Med. 2001;344:898–906.
61. Bernat JL, Suranyi L. Loss of ability to sneeze in lateral medullary syndrome. Neurology. 2000;55:604.
62. Hersch M. Loss of ability to sneeze in lateral medullary syndrome. Neurology. 2000;54:520–1.
63. Martin RA, Handel SF, Aldama AE. Inability to sneeze as a manifestation of medullary neoplasm. Neurology. 1991;41:1675–6.
64. Seijo-Martinez M, Varela-Freijanes A, Grandes J, et al. Sneeze related area in the medulla: localization of the human sneeze center? J Neurol Neurosurg Psychiatry. 2006;77:559–61.
65. Penfield W, Jasper H. Epilepsy and the functional anatomy of the human brain. Boston: Little, Brown & Company; 1954.
66. Penfield W, Kristiansen K. Epileptic seizure patterns: a study of the localizing value of initial phenomena in focal cortical seizures. Springfield: Charles C Thomas; 1951.
67. Temkin O. The falling sickness: a history of epilepsy from the Greeks to the beginnings of modern neurology. Baltimore: Johns Hopkins Press; 1945.
68. Jackson JH. On the scientific and empirical investigation of epilepsies. In: Taylor J, editor. Selected writings of John Hughlings Jackson, vol. 1. New York: Basic Books; 1958.
69. Bhatia MS, Khandpal M, Srivastava S, et al. Intractable psychogenic sneezing: two case reports. Indian Pediatr. 2004;41:503–8.
70. Shilkrel HH. Psychogenic sneezing and yawning. Psychosom Med. 1949;11:127–8.
71. Sulemanji MN, Kanbur NO, Derman O, et al. Intractable sneezing: is it always psychogenic? Turk J Pediatr. 2011;53(2):225–8.
72. Kanner L. Child psychiatry. 3rd ed. Springfield: Charles C. Thomas; 1957.
73. Yater WM, Barton WM. Symptom diagnosis: regional and general. 4th ed. New York: D. Appleton-Century Company; 1942.
74. Bergman GE. Psychogenic intractable sneezing in children. J Pediatr. 1984;105:496–8.
75. Lin TJ, Maccia CA, Turnier CG. Psychogenic intractable sneezing: case reports and a review of treatment options. Ann Allergy Asthma Immunol. 2003;91:575–8.
76. Gopalan P, Browning ST. Intractable paroxysmal sneezing. J Laryngol Otol. 2002;116:958–9.
77. Guner SN, Gokcen C, Gokturk B, et al. Haloperidol: a possible medication for the treatment of exacerbation of intractable psychogenic sneezing. Int J Pediatr Otorhinolaryngol. 2010;74(10):1196–8.
78. Teebi AS, Al-Saleh QA. Autosomal dominant sneezing disorder provoked by fullness of stomach (letter). J Med Genet. 1989;26:539–40.
79. Hall JG. The snatiation reflex (letter). J Med Genet. 1990;27:275–8.
80. Bhutta MF, Maxwell H. Further cases of unusual triggers of sneezing. J R Soc Med. 2009;102:49.
81. Mackenzie JN. Irritation of the sexual apparatus as an etiological factor in the production of nasal disease. Am J Med Sci. 1884;87:360–5.
82. Jones E. The life and work of Sigmund Freud. London: Penguin Books; 1974.
83. Young AR. Freud's friend Fliess. J Laryngol Otol. 2002;116:992–5.
84. Anonymous. Paroxysmal sneeze following orgasm. JAMA. 1972;219:1350–1.
85. Korpas J, Tomori Z. Cough and other respiratory reflexes. In: Bollinger CT, editor. Progress in respiratory research, vol. 23. Basel: Karger; 1979.
86. Stromberg BV. Sneezing: its physiology and management. Eye Ear Nose Throat Mon. 1975;54:49–53.
87. Bhutta MF, Maxwell H. Sneezing induced by sexual ideation or orgasm: an under-reported phenomenon. J R Soc Med. 2008;101:587–91.
88. Kramer MF. Diagnosis of persistent rhinitis. MMW Fortschr Med. 2006;148:37–9.

89. Settipane GA, Klein DE. Non-allergic rhinitis: demography of eosinophils in nasal smears, blood eosinophil counts and IgE levels. N Engl Reg Allergy Proc. 1985;6:363–6.
90. Skoner DP. Allergic rhinitis: definition, epidemiology, pathophysiology, detection, and diagnosis. J Allergy Clin Immunol. 2001;108:2–8.
91. Scadding GK. Non-allergic rhinitis: diagnosis and management. Curr Opin Allergy Clin Immunol. 2001;1:15–20.
92. Galen BA. Differential diagnosis: rhinitis. Lippincotts Prim Care Pract. 1997;1(2):129–41.
93. Shaari J, Palmer JN, Chiu AG, et al. Regional analysis of sinonasal ciliary beat frequency. Am J Rhinol. 2006;20:150–4.
94. Cingi C, Unlu HH, Songu M, et al. Seawater gel in allergic rhinitis: entrapment effect and mucociliary clearance compared with saline. Ther Adv Respir Dis. 2010c;4:13–8.
95. Bousquet J, Bachert C, Canonica GW, ACCEPT-1 study group, et al. Efficacy of desloratadine in intermittent allergic rhinitis: a GA(2)LEN study. Allergy. 2009;64:1516–23.
96. Bousquet J, Bachert C, Canonica GW, ACCEPT-2 Study Group, et al. Efficacy of desloratadine in persistent allergic rhinitis – a GA^{2}LEN study. Int Arch Allergy Immunol. 2010;153:395–402.
97. Cingi C, Cakli H, Yaz A, et al. Phototherapy for allergic rhinitis: a prospective, randomized, single-blind, placebo-controlled study. Ther Adv Respir Dis. 2010a;4:209–13.
98. Gunhan K, Unlu H, Yuceturk AV, et al. Intranasal steroids or radiofrequency turbinoplasty in persistent allergic rhinitis: effects on quality of life and objective parameters. Eur Arch Otorhinolaryngol. 2011;268:845–50.
99. Adler JJ, Rose DN. Transmission and pathogenesis of tuberculosis. In: Rom WN, Garay SM, editors. Tuberculosis. New York: Little, Brown and Company; 1996.
100. Centers for Disease Control and prevention. Guidelines for preventing the transmission of Mycobacterium tuberculosis in health-care facilities. MMWR. 1994;43:1–132.
101. Gwaltney JM Jr, Hendley JO, Phillips CD, et al. Nose blowing propels nasal fluid into the paranasal sinuses. Clin Infect Dis. 2000;30:387–91.
102. Baydin A, Nural MS, Guven H, et al. Acute aortic dissection provoked by sneeze: a case report. Emerg Med J. 2005;22:756–7.
103. Birkent H, Durmaz A, Hidir Y, et al. Sneezing-related orbital emphysema and pneumocephalus treated with transnasal endoscopic surgery. J Otolaryngol Head Neck Surg. 2008;37:E167–9.
104. Fonseca AC, Ferreira JJ, Albuquerque L, et al. Sneeze as a precipitating factor of cerebral venous thrombosis. Eur J Neurol. 2007;14:7–8.
105. Gonzalez F. Orbital emphysema after sneezing. Ophthal Plast Reconstr Surg. 2005;21:309–11.
106. Gupta A, Dogra PM, Singh B, et al. Arteriovenous fistula intimal tear: think before you sneeze. Ther Apher Dial. 2010;14:124.
107. Rochels R, Bleier R, Nover A. Compression pneumatocele of the lacrimal sac. Klin Monatsbl Augenheilk. 1989;195:174–6.
108. Sharir M, Huntington AC, Nardin GF, et al. Sneezing as a cause of acute angle-closure glaucoma. Ann Ophthalmol. 1992;24:214–5.
109. Whitehead E. Sudden sensorineural hearing loss with fracture of the stapes footplate following sneezing and parturition. Clin Otolaryngol Allied Sci. 1999;24:462–4.

The Dry Nose

13

Rainer K. Weber, Tanja Hildenbrand,
Detlef Brehmer, and Jochen A. Werner

Core Messages

- An unequivocal definition of the dry nose (DN) is not available. Symptoms range from the purely subjective sensation of a rather dry nose to visible crusting of the (inner) nose (nasal mucosa), and a wide range of combinations are met with. Relevant diseases are termed rhinitis sicca anterior, primary and secondary rhinitis atrophicans, rhinitis atrophicans with foetor (*ozaena*) and empty nose syndrome. The diagnosis is based mainly on the patient's history, inspection of the nose, endoscopy of the nasal cavity, sinuses and nasopharynx, with CT, allergy testing and microbiological swabs being performed where indicated.
- Treatment consists of the elimination of predisposing factors, moistening, removal of crusts, avoidance of injurious factors, care of the mucosa, treatment of infections and, where applicable, correction of overlarge air space.

13.1 Symptoms

One of the chief functions of the nose is to warm and moisten the inspired air, while another is to recover the water in the expired air [1].

The nature of the in- and outflow of the air within the nasal cavity is of decisive importance for this air-conditioning feature. In this context, optimal distribution of the inspired air over the nasal turbinates ensuring intimate contact of the air with the surface-moist mucosal membrane is essential.

The expression dry nose (DN) has not been unambiguously defined. In the main, it is based on relevant anamnestic patient information. ENT specialists often employ the term rhinitis sicca, although here, too, a clear definition is lacking. Symptoms range from the purely subjective sensation of a somewhat dry nose to visible crusting of the nose, and a wide range of combinations are possible:

R. K. Weber (✉) · J. A. Werner
Department of Otorhinolaryngology, Head and Neck Surgery, Rhinology Center Marburg, University Hospital Marburg UKGM, Marburg, Germany
e-mail: weber@uk-gm.de; jochen.werner@uk-gm.de

T. Hildenbrand
Department of Otorhinolaryngology/ENT Department, University Hospital Freiburg, Freiburg, Germany
e-mail: tanja.hildenbrand@uniklinik-freiburg.de

D. Brehmer
Private ENT Clinic Goettingen, Faculty of Health/School of Medicine, University Witten/Herdecke, Goettingen, Germany
e-mail: d.brehmer@hno-praxis-goettingen.de

Ö. Ö. Celebi, T. M. Önerci (eds.), *Nasal Physiology and Pathophysiology of Nasal Disorders*,
https://doi.org/10.1007/978-3-031-12386-3_13

- Sensation of dryness in the nose.
- Itching and mild burning sensation.
- Nasal obstruction.
- Crusting, scabs and 'bogies', possibly associated with an (unpleasant) smell.
- Epistaxis.
- Diminished sense of smell.

13.2 Aetiology

Possible causes of dry nose include a variety of diseases, external and internal factors and environmental conditions:

- Local mechanical irritation.
- Climatic or environmental factors.
  - Dry room or environmental air (relative humidity <50%).
  - Heated room or hot environment.
  - Long-distance flights.
- Workplace conditions.
  - Dry air and clean-room condition [2].
  - Cold and heat.
  - Dusty conditions (e.g. grinding/polishing of plaster, granite, chalk, cement, wood arsenic, nickel carbonyl, tobacco smoke).
- Drugs (cocaine).
- Side effects of medications (see Table 13.1).
- Supportive nasal administration of oxygen [3].
- Symptoms of other diseases (granulomatous, infectious, rheumatic and immunological disorders).
  - Wegener's granulomatosis, sarcoidosis, tuberculosis, syphilis and leprosy (Fig. 13.1).
- Wound healing phase after endonasal sinus surgery and surgery on the nose.
- Anatomical changes to the outer and inner nose, with modification of normal airflow.
- Allergic rhinitis, in particular, house dust mites and moulds.
- Permanent sequelae of surgery on the nose and paranasal sinuses.
- Sequelae of head and neck radiotherapy.
- Patients with obstructive sleep apnoea (OSA) or continuous positive airway pressure (CPAP) treatment in sleep apnoea patients [4]. Moistening led to a reduction in symptoms [5].
- Old age.

Dry nose may be the first symptom of an incipient cold with a runny nose; in such a case, however, it is of only limited duration.

Although an increased susceptibility to infections has frequently been postulated and is patho-

Table 13.1 Medications with the side effect of dry nose

Substance group	Generic name	Indications
Retinoids (1–10%)	Isotretinoin	Severe forms of acne
	Tretinoin	Promyelocyte leukaemia
Doxepin (tricyclic antidepressant)		Depressive conditions, anxiety syndrome, mild withdrawal symptoms in alcoholics and drug-dependent persons, agitation, sleep disorders
Methyldopa (1‰–1%)		Hypertension (of pregnancy)
Sympathomimetics (local)	Dipivefrine (eye)	Glaucoma
	Naphazoline	Diverse forms of rhinitis, only short-term use recommended
	Oxymetazoline	
	Phenylephrine	
	Tetryzoline	
	Tramazoline	
	Xylometazoline	
Antihistaminics, first generation	Clemastine	Urticaria, allergic rhinitis
	Dimenhydrinate	Vertigo, nausea, vomiting
	Dimethindene	Itching, itching dermatoses, allergies
	Diphenhydramine	Difficulty getting to sleep, difficulty staying asleep
	Promethazine	Agitation in underlying psychiatric illnesses, possibly vomiting, nausea and sleep disorders
	Terfenadine	Allergic rhinoconjunctivitis, allergic skin disorders

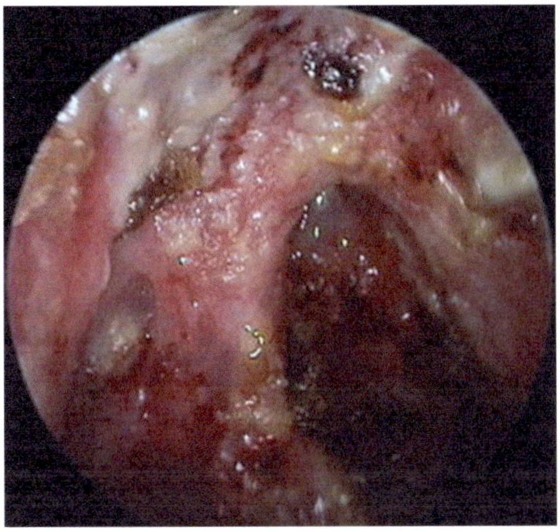

Fig. 13.1 Dry nose in a patient with Wegener's disease

physiologically plausible, it has never been confirmed as the sole cause. The mechanical and functional integrity of the mucous membrane is considered a major natural defence mechanism against infection. Depending on the degree to which it dries out, mucociliary transport and even the epithelial barrier may be impaired.

Epidemiological data on the prevalence of dry nose are not available. However, in particular, when questioned, many people confess to be sufferers and admit to the use of numerous therapeutic measures, mostly ointments and sprays or nasal douches.

In old age, anatomical changes and involution atrophy of the mucosa result in an increase in such complaints as impairment of nasal breathing and dry nose, and the following phenomena may occur:

- Hanging nasal tip and shortening of the columella [6].
- Reduced mucociliary clearance [7–10].
- Reduction in the number of goblet cells and elastic fibres in the nasal mucosa [11].
- Reduced sensitivity of the nasal mucosa [12].
- Enlargement of the nasal cavity resulting from involution atrophy of the nasal mucosa [13, 14].
- Decrease in the body's water content [15].

Altered airflow due to changes in geometry leads to changes in the conditioning situation [14, 16], with the result that in over 60-year-olds, the air-conditioning capacity becomes impaired: both intranasal air temperature and humidity decrease [17].

A familial, i.e. genetic, impairment of nasal air-conditioning has been reported by Sahin-Yilmaz et al. [18], who investigated 47 pairs of twins [11].

13.2.1 Clinical Entities

In common with the symptom itself, a number of individual diseases associated with the symptoms dry nose and crusting are not only unclearly defined but also overlap. In the literature, the following descriptions are to be found:

- Rhinitis sicca anterior (Fig. 13.2).
- Primary rhinitis atrophicans/primary atrophic rhinitis (= PAR) – rhinitis atrophicans with foetor (*ozaena*) (Fig. 13.3).
- Secondary rhinitis atrophicans/secondary or diffuse atrophic rhinitis (= SAR).

13.2.1.1 Rhinitis Sicca Anterior

The term rhinitis sicca anterior defines a chronic inflammation in the region of the anterior part of the nose usually affecting the anterior and caudal septum and/or the corresponding lateral nasal vestibule. Due to irritation (mechanical, finger picking; toxic; persistent secretion; respiration; (air)flow characteristics; dry, hot and dusty environment), drying, superficial erosion and/or ulceration with (thin) crust formation occur.

Patients experience a sensation of dryness, itching and increased crust formation. The crusts are thin and dry and do not extend into the posterior part of the nasal cavity, as is the case with atrophic rhinitis. Usually, there is no foetid smell—only an occasional patient experiences mild foetor caused by bacterial colonisation of the small crusts. Manipulations may give rise to a vicious circle of increased crusting and persistent complaints. In the individual case, continued irritation and manipulation may result in perforation of the nasal septum.

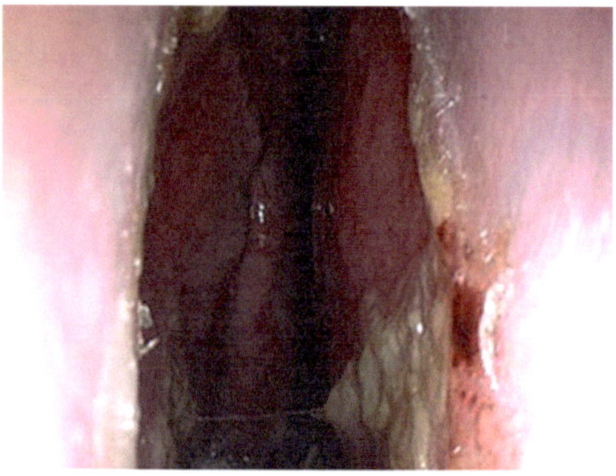

Fig. 13.2 Rhinitis sicca anterior with dry skin in the nasal vestibule and thin yellowish crusts

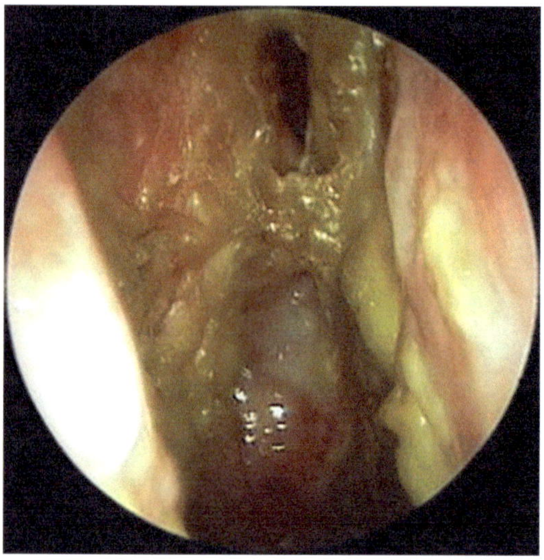

Fig. 13.3 Dry nose with persistent excessive crusting in a female patient with primary atrophic rhinitis

Treatment consists of the application of a protective film to the skin of the nasal vestibule, to allow the chronic inflammation disrupting its integrity to self-heal while avoiding mechanical irritation (Fig. 13.1). This is usually achieved by the application of ointments. Nasal sprays containing dexpanthenol have also been used with success [19].

13.2.1.2 Primary Atrophic Rhinitis (PAR)

Primary atrophic rhinitis is a gradually progressive chronic degenerative condition of the nasal mucosa of unknown aetiology. Progressive atrophy of all the constituents of the mucosa (epithelium, glands and vessels together with osteoclasis) mainly affects the turbinates. The respiratory epithelium undergoes stepwise metaplastic changes. Histologically, atrophy of the serous and mucous glands, loss of the cilia and goblet cells, chronic infiltration of the lamina propria with granulations and scar formation are seen. Some publications also report diminished vascular density together with peri- and endarteritis. The histological changes explain the disturbance to mucociliary transport. Thick secretion that is not transported away promotes the formation of bacterially contaminated crusts that may then be a source of recurrent bacterial Infection. If rhinitis atrophicans is complicated by foetid crust formation, rhinitis atrophicans with foetor (stinking nose, ozaena) results. Microbiological investigation usually detects *Klebsiella ozaenae* (100% of the 45 patients of Moore and Kern [20]) but also *Staphylococcus aureus*, *Proteus mirabilis*, *E. coli* and other bacteria (*Bordetella bronchiseptica* and *Pasteurella multocida*). Endoscopic examination reveals a markedly large and wide nasal cavity and visibly dry mucosa. The turbinates are clearly diminished in size. With ozaena, yellowish-green crusts are found on the mucosa. Apart from crusting, the patient usually also suffers from hypo- or anosmia. Foetor may give rise to social problems.

The cause of PAR is unknown, but both genetic and infectious causes are under discussion. The condition is more commonly seen in association with low socio-economic status, a poor diet and iron deficiency. A much less frequent use of antibiotics in countries with a higher incidence of the condition may have a role to play. The fact that women are more frequently affected suggests a possible endocrinal factor. Over the past years and decades the prevalence of PAR has declined.

13.2.1.3 Secondary Atrophic Rhinitis (SAR)

Secondary atrophic rhinitis develops in the soil of an underlying disease or may result from previous therapeutic measures. Therefore, it is a collective term for several conditions with a different aetiology but a similar clinical feature with the subjective feeling of a dry nose, crusting, gegebenenfalls nasal obstruction, foul smelling and epistaxis as outlined above. In contrast to PAR, there is no osteoclastic activity, so the term diffuse atrophic rhinitis may be more suitable. Pathogenetic factors leading to SAR are as follows:

- Prior radical endonasal surgery. Persistent chronic rhinosinusitis in addition may increase the probability of the development of SAR [21].
- Prior radiotherapy of the head and neck.
- Sjögren's syndrome.
- Extremely rarely: prior trauma.

In the largest collection of cases to date 197 patients with sRA, the following causes were identified [20]:

- Complete removal of the lower and middle nasal turbinates (24%).
- Partial removal of the lower and/or middle nasal turbinates (56%).
- Endonasal sinus surgery without turbinectomy (10%).
- Partial maxillectomy to remove a tumour (6%).
- Nasal trauma requiring surgical reconstruction (1%).
- Granulomatous disease (1%).

Empty nose syndrome (ENS) is defined as a form of iatrogenic SAR, occurring after radical endonasal surgery in particular resection of the nasal turbinates (lower and/or middle), usually in connection with an operation on the nasal sinuses [22–24]. It is characterised by the symptoms nasal and pharyngeal dryness, paradoxical impairment of nasal respiration, dyspnoea and hyposmia, in some cases associated with depression [25].

The disturbed sense of smell is the result of the changed airflow. The dryness of the pharyngeal mucosae is due to the fact that, in contrast to the normal situation, drier intranasal air (no moisture since turbinate mucosa is lacking) results in disrupted airflow in the region of the choanae and also impinges on the posterior wall of the nasopharynx at increased velocity [24]. According to Houser, the pain too is a typical symptom caused by the action of cold air on the mucosa covering the sphenopalatine ganglion [25].

Resection of the lower and middle turbinates reduces the effectiveness of the climatisation function of the nose by 23% [26, 27]. The paradoxical impairment of nasal respiration is explained by the unphysiological airflow, the reduced nasal airway resistance, the lack of areas of functional mucosa together with the simultaneous enlargement of the nasal cavity, and the curtailed contact between air and mucosa [24]. Enlargement of the nasal cross-section reduces the airway resistance and thus the pressure gradient at the air/mucosa surface. In turn, this causes malfunction of the nasopulmonary reflexes, which may lead to a worsening of pulmonary function. In contrast, optimal nasal airway resistance is important for the dilatation of the peripheral bronchioles and for improved alveolar gas exchange.

The risk to develop SAR after resection of the turbinates depends on the extent of resection, individual factors of the patient itself and other external factors which are not clearly defined up to now. Some authors did not find any sign of SAR after total resection of the inferior turbinates [28–31]; others report SAR in 2–22% [27, 32–36].

13.3 Diagnosis

Extensive history taking is always followed by inspection of the outer and inner nose (Table 13.2). This should, for example, identify any anatomical deformations that might cause dry nose by changing the flow of air through the nose. For the purpose of detecting minor lesions in the nasal vestibule in patients with rhinitis sicca anterior, the use of a microscope may prove useful. Endonasal inspection should look for septal deviation or perforation and note the size and shape of the turbinates, the presence and nature of crusts, the humidity of the mucosa, polyps or tumours, postnasal secretion and the nasopharyngeal status.

Table 13.2 Diagnostic workup of dry nose

Medical history
Inspection of the external and inner nose
Endoscopy of the nasal cavity and nasopharynx, where indicated, also of (operated) paranasal sinuses
Where indicated, CT of the paranasal sinuses
Allergy testing
Microbiological swab

A CT of the nasal sinuses is indicated when signs of chronic rhinosinusitis are found or to obtain adjunctive evidence of PAR. Typical signs of PAR in the CT include:

- Thickened mucosa in the paranasal sinuses.
- The osteomeatal complex can no longer be defined due to the destruction of the normal anatomy.
- Hypoplasia of the maxillary sinus.
- Enlargement of the nasal cavity with destruction of the lateral nasal wall.
- Bony destruction of the inferior and middle turbinates.

Testing for allergy is important, for example, in order to diagnose a house mite allergy, which may be associated with the symptom dry nose. When endoscopic examination reveals purulent streaks or crusting, swabs should be taken for a microbiological examination.

The diagnosis is based on the case history, endoscopic findings and, where necessary, adjunctive diagnostic measures.

13.4 Treatment

Treatment of dry nose comprises:

- Elimination or amelioration of triggering or promoting factors.
- Moisturisation (Table 13.3).
- Sufficient daily drinking amount.
- Cleansing (when crusts are present) and care of the mucosa.
- Treatment of obvious infections.
- Where applicable, the elimination of an overlarge endonasal air space.

The individual may have only limited control over environmental factors. The importance of

Table 13.3 Substances for moistening the nose and mucosal care

Nasal douches with saline solution
NaCl solution
Special saline solution
Isotonic–hypertonic, with and without buffering (alkaline)
Nasal ointments
Dexpanthenol
Salt-containing nasal ointments
Diverse other formulations
Nasal oils
Sesame oil
Vitamin A oil
Salt water sprays
Hyaluronic acid nasal spray
Dexpanthenol nasal spray

the latter is obvious when dry nose is no longer experienced during a holiday but reappears when this is over.

By far the most common complaint is a subjective dry nose with no endoscopic findings with the possible exception of a somewhat dry mucosa in the anterior nose. This is the case in almost all those patients exposed to unfavourable climatic or workplace conditions and also the large group of patients with obstructive sleep apnoea syndrome undergoing CPAP treatment. Also affected are patients operated on the nose, before the climatisation function has normalised.

Apart from the strict avoidance of local manipulation, these patients require humidification and care of the dry areas. For this purpose, the market offers a wide range of ointments, oils, sprays and nasal irrigation (Table 13.4).

The nose should be humidified, viscous mucus flushed and liquefied; all inflammation-inducing and inflammation-promoting substances should be cleared out. A protective film should be applied to prevent drying. Transepithelial water loss can be countered by the nasal application of saline solutions or other substances [37].

13.4.1 Nasal Irrigation, Nasal Saline Spray and Inhalation

Nasal irrigations are recommended for a large number of diseases of the nose and nasal sinuses [38]. Precisely, how nasal irrigation works is not

Table 13.4 Basic rules for the treatment of dry nose

Elimination of promoting factors
Environmental and workplace situation
Dietary, iron and vitamin deficiency (?)
Moistening
Local (nasal irrigation, inhalation, nasal spray)
Environment (elevated air humidity)
Systemic: Sufficient liquid intake
Removal of crusts (nasal irrigation, instrumental removal by ENT clinician)
Avoidance of injurious factors
Local (nose picking, cotton carrier, decongestant nose drops, ointments containing potentially injurious substances – Imidazoline derivates, cortisone applied to the skin of the nasal vestibule, …)
Systemic drugs (see Table 13.1)
Care of the mucosa
Oils
Ointments
Occlusion
Treatment of infections
Allergic rhinitis
Ozaena
Chronic rhinosinusitis
Correction of an overlarge air space
Occlusion
Augmentation

clear. It is postulated that the improvement in mucosal function is due to:

- Direct physical cleansing by flushing out thick mucus, crust, debris, allergens, environmental toxins, etc. [39, 40]
- Removal of inflammation mediators.
- Improvement of mucociliary clearance by improving the ciliary beat frequency [41, 42].

In a recent review article published in 2009 [43], nasal irrigation is recommended:

- As adjunctive treatment for chronic rhinosinusitis (Grade A evidence: consistent study results of good quality).
- As adjunctive treatment for allergic rhinitis and viral ARS and follow-up treatment after nasal sinus surgery (Grade B evidence: inconsistent results or limited quality).
- For rhinitis of pregnancy, acute bacterial RS, also sarcoidosis or Wegener's disease (Grade C evidence: consensus recommendations, usual practice, expert opinion, results of case series).

The above shows that on the one hand, nasal irrigation is a common recommendation, while on the other hand, the indication dry nose is not explicitly included in the recommendation, since informative studies that can be integrated into the evidence-based recommendations are very rare.

Nevertheless, nasal irrigations are an important therapeutic option in patients with dry nose. In the case of recurrent crust formation, it is virtually indispensable as an adjunctive aid to instrumental clearing by the ENT physician.

The most commonly employed nasal sprays are salt solutions. In addition to household salt—iodised or non-iodised—pharmaceutical grade salts as well as special nasal spray salts and brines are used [39, 40]. These solutions may be isotonic, hypotonic, hypertonic, unbuffered or buffered. Mildly hypertonic saline solutions (up to approximately 3%), with or without buffering, are all suitable for nasal irrigation. However, it is currently not clear which saline solution is best for what indication.

For isotonic saline solutions (isotonic unbuffered, buffered Emser saline solution), numerous investigations have shown that daily application over the long term produces positive results (prevention and treatment of upper airway infection/rhinosinusitis, aftercare following surgery on nasal sinuses) with no relevant side effects.

In principle, saline sprays serve the same purpose as nasal irrigation. Although no systematic comparison has been reported, the remark by Schmidt that diffuse moistening of the nasal mucosa can be achieved only with irrigation since the spray is merely a punctiform application, is accurate and the potential therapeutic effect must therefore be considered smaller. For the present, the extent to which the admixture of other substances results in a real benefit in the treatment of dry nose remains uncertain.

Inhalation with saline solutions with the aim of moistening the mucosa is also recommended and applied. In view of the resulting diffuse moistening of the mucosa, this can be considered positive in the case of a dry nose. Unfortunately, no meaningful studies are available.

13.4.2 Nasal Ointments

Despite the fact that many patients often use nasal ointments, no meaningful studies on their use in dry nose are available.

A moistening effect is achieved with intranasal use: the application of a nasal ointment reduces nasal water loss—as also does the application of glycerol 10% [37].

Elberg reported on the effect of Emser salt applied in the form of Nisita® Nasal Ointment in 1500 cases including pre- and post-operative applications in patients undergoing operations on the nose and nasal sinuses [44]. Neither pain nor infections were observed with regular application, despite the fact that no antibiotic was given. Follow-up care was reportedly considerably facilitated and abbreviated.

The quality level must, however, as in company-sponsored application studies—which are not considered here—be Grade V evidence.

In comparison with dexpanthenol nasal ointment, dexpanthenol nasal spray proved just as effective, or even tendentially superior, in terms of its effect on mucociliary transport (saccharine test) reported by Verse et al. in a prospective, randomised, open, crossover study [45]. Its advantage vis-à-vis the ointment is presumably the fact that it reaches the upper parts of the nasal cavity.

Topical dexpanthenol is said to reduce transepidermal water loss, to activate in vivo and in vitro fibroblast proliferation and to accelerate the re-epithelialisation process [46].

13.4.3 Nasal Oils

Oils in a not-too-high concentration bring about an improvement in the nasal ciliary beat frequency (CBF). In contrast to Miglyol 840 and thyme oil, sesame oil, soy oil, peanut oil, lavender oil, eucalyptus oil and menthol increased the CBF, the effect being higher at a concentration of the oils of 0.2% than at 2% [47]. According to Riechelmann et al. a mixture of menthol, eucalyptus oil and pine needle oil in concentrations up to 5% had no major negative effect on CBF but did at concentrations of between 7.5 and 10 g/m^3 [48]. With conventional inhalation, concentrations of max. 1% are to be expected.

In a randomised crossover study involving 79 patients with dry nasal mucosa, [49] showed that in comparison with a sodium chloride solution, treatment with sesame oil resulted in a superior moistening effect [49]. Dryness and subjectively impaired nasal respiration were improved significantly better by sesame oil in comparison with saline irrigation. Björk-Eriksson et al. also reported a significant effect of sesame oil (3 × 3 puffs of 25 µl spray daily for 30 days) on the symptoms impaired nasal respiration, dryness (burning sensation, itching, irritation) and crust formation in 20 patients with dry nose and 15 patients post-radiation treatment [50]. A total of five patients reported side effects (one each with unpleasant odour, itching and disturbed nasal respiration and runny nose in two).

13.4.4 Others

Home remedies and self-treatments recommended on the Internet are mostly concerned with achieving moisturising, the application of oils and the prevention of drying, but the efficacy of the respective measures remains unclear. Homoeopathy always recommends an individual constitutional approach to treatment.

In the elderly patient with a dry nose, Slavin recommends moistening the nasal mucosa and looking out for medicament side effects (in particular avoidance of first-generation antihistaminics and decongestive nose drops) [15].

A rough topographical endoscopically orientated classification may be useful for the differential treatment of the dry nose symptom:

- In the case of problems localised in the anterior nose (rhinitis sicca anterior in the widest sense) with a visible lesion and possibly crust-

ing, the first indication is the local application of ointment. Relevant comparative studies are not available. Potentially injurious substances (decongestant medications, cortisone, allergising substances) should be avoided.
- Vague complaints of dry nose in the absence of visible changes to the nasal mucosa would appear the most likely indication for moisturising measures (nasal irrigation, inhalation, moisturising sprays). The question as to whether admixed medicaments can diminish the water loss on expiration needs further investigation.
- Dry nose with visible intranasal crust formation is the domain of nasal irrigation, which is better able to remove crusts than inhalation or sprays.

13.4.5 Treatment of Atrophic Rhinitis

Basic treatment consists of the above outlined measures for dry nose. In the case of atrophic rhinitis moistening measures must be accompanied by removal of any crusts and scabs. For this purpose, not only the commonly employed instrumental removal by the ENT specialist but also nasal irrigation is used. As suitable solutions, the literature mentions not only the classical irrigation solutions (buffered and unbuffered solutions of common salt or special salts) but also solutions of 25% glucose in glycerine and antibiotics [51]. Tap water and other hypotonic solutions are to be rejected.

Bacterial superinfections are treated with specific antibiotics. For ozaena, antibiotic treatment is reported to achieve long-lasting results, e.g. rifampicin 600 mg daily for 12 weeks [51] and ciprofloxacin 2 × 500–750 mg for 8 weeks [52].

Operative measures aim to reduce the size of, or temporarily occlude, the nasal cavity. Although occlusion can resolve the problem of crusting and the considerable social stigma of foetor, it also impairs nasal breathing and the sense of smell. For the diminishment of nasal cavity size using submucosal implantation of tissue, foreign material should not be used despite that some authors describe promising result [53, 54]. [53] implanted Plastipore, a high-density polyethylene sponge with micropores, and reported excellent results in six patients and good results with only minor crusting in two patients and one extrusion after 18 months [53]. Rice used hydroxyapatite for augmentation in one case and reported good results [54]. According to Houser, more suitable materials are the patient's own cartilage (e.g. rib cartilage) or acellular dermis (AlloDerm®) [25]. He treated eight patients with the implantation of AlloDerm®, which resulted in a significant improvement in symptom scores (SNOT 20) after at least 3 months. For treatment planning, the cotton test is suggested: moistened cotton is applied to the area to be augmented for 20–30 min. If the test is positive, the patient can be offered the augmentation. Friedman et al. [55] and Moore and Kern [20] reported some success with acellular dermis, too, in 5 of 10 and 7 of patients, respectively.

13.5 Prophylaxis

Since the uncritical resection of the nasal turbinates represents a significant and frequent factor in the genesis of dry nose, secondary RA and ENS, the following points must be strongly emphasised:

- The main objective of nasal turbinate surgery must be the preservation of functional mucosa while creating an adequately large volume capable of ensuring climatisation and the cleansing of the respired air and also preserving physiological airway resistance [24].
- Without adequate justification the middle turbinate should not be resected. Reduction of the lower nasal turbinate should first be given careful consideration, simultaneous removal of both the lower and middle turbinates should not be done for a non-tumorous condition [56].

13.6 Conclusions

Despite that there is no clear definition of dry nose, many patients complain of this symptom. A carefully taken patient's history and thorough rhinologic examination by an experienced and vigilant clinician are the key elements to optimal outcomes. The turbinates should be handled very cautiously in sinonasal surgery.

Pearls

- There is no clear definition of what a dry nose is.
- Symptoms range from the purely subjective sensation of a rather dry nose to visible crusting of the (inner) nose (nasal mucosa).
- Relevant diseases are rhinitis sicca anterior, primary and secondary rhinitis atrophicans, rhinitis atrophicans with foetor (*ozaen*a) and empty nose syndrome.
- Drugs and environmental factors can induce a dry nose.
- The diagnosis is based mainly on the patient's history, inspection of the nose and endoscopy of the nasal cavity, sinuses and nasopharynx.
- Treatment consists in the elimination of predisposing factors, moistening, removal of crusts, avoidance of injurious factors, care of the mucosa, treatment of infections and, where applicable, correction of an overlarge air space.
- Normal turbinates should be preserved in sinonasal surgery.

References

1. Keck T, Lindemann J. Strömungssimulation und Klimatisierung in der Nase. Laryngorhinootologie. 2010;89 Suppl 1:S1–14.
2. Su SB, Wang BJ, Tai C, Chang HF, Guo HR. Higher prevalence of dry symptoms in skin, eyes, nose and throat among workers in clean rooms with moderate humidity. J Occup Health. 2009;51:364–9.
3. Miyamoto K, Nishimura M. Nasal dryness discomfort in individuals receiving dry oxygen via nasal cannula. Respir Care. 2008;53:503–4.
4. Brander PE, Soirinsuo M, Lohela P. Nasopharyngeal symptoms in patients with obstructive sleep apnea syndrome. Effect of nasal CPAP treatment. Respiration. 1999;66:128–35.
5. Worsnop CJ, Miseski S, Rochford PD. The routine use of humidification with nasal continuous positive airway pressure. Intern Med J. 2009;40:650. https://doi.org/10.1111/j.1445-5994.2009.01969.x.
6. Zankl A, Eberle L, Molinari L, Schinzel A. Growth charts for nose length, nasal protrusion, and philtrum length from birth to 97 years. Am J Med Genet. 2002;111:388–91.
7. Armengot M, Barona R, Garin L, Basterra J. The influence of age, sex and circadian rhythms on the nasal mucosa in the mucociliary clearance. An Otorrinolaringol Ibero Am. 1993;20:581–8.
8. Ho JC, Chan KN, Hu WH. The effect of aging on nasal mucociliary clearance, beat frequency, and ultrastructure of respiratory cilia. Am J Respir Crit Care Med. 2001;163:983–8.
9. Sakakura Y, Ukai K, Majima Y, Murai S, Harada T, Miyoshi Y. Nasal mucociliary clearance under various conditions. Acta Otolaryngol. 1983;96:167–73.
10. Sunwoo Y, Chou C, Takeshita J, Murakami M, Tochihara Y. Physiological and subjective responses to low relative humidity in young and elderly men. J Physiol Anthropol. 2006;25:229–38.
11. Sahin-Yilmaz AA, Corey JP. Rhinitis in the elderly. Curr Allergy Asthma Rep. 2006;6:125–31.
12. Wrobel BB, Bien AG, Holbrook EH, Meyer GE, Bratney NA, Meza J, et al. Decreased nasal mucosal sensitivity in older subjects. Am J Rhinol. 2006;20:364–8.
13. Kalmovich LM, Elad D, Zaretsky U, Adunsky A, Chetrit A, Sadetzki S, et al. Endonasal geometry changes in elderly people: acoustic rhinometry measurements. J Gerontol A Biol Sci Med Sci. 2005;60:396–8.
14. Lindemann J, Brambs HJ, Keck T, Wiesmiller KM, Rettinger G, Pless D. Numerical simulation of intranasal airflow after radical sinus surgery. Am J Otolaryngol. 2005a;26:175–80.
15. Slavin RG. Treating rhinitis in the older population: special considerations. Allergy Asthma Clin Immunol. 2009;5:9.
16. Lindemann J, Keck T, Wiesmiller KM, Rettinger G, Brambs HJ, Pless D. Numerical simulation of intranasal air flow and temperature after resection of the turbinates. Rhinology. 2005b;43:24–8.
17. Lindemann J, Sannwald D, Wiesmiller K. Age-related changes in intranasal air conditioning in the elderly. Laryngoscope. 2008;118:1472–5.
18. Sahin-Yilmaz A, Pinto JM, de Tineo M, Elwany S, Naclerio RM. Familial aggregation of nasal conditioning capacity. J Appl Physiol. 2007;103:1078–81.
19. Kehrl W, Sonnemann U. Dexpanthenol-Nasenspray als wirksames Therapieprinzip zur Behandlung der Rhinitis sicca anterior. Laryngorhinootologie. 1998;77:506–12.
20. Moore EJ, Kern EB. Atrophic rhinitis: a review of 242 cases. Am J Rhinol. 2001;15:355–61.
21. Ly TH, de Shazo RD, Olivier J, Stringer SP, Daley W, Stodard CM. Diagnostic criteria for atrophic rhinosinusitis. Am J Med. 2009;122:747–53.
22. Chhabra N, Houser SM. The diagnosis and management of empty nose syndrome. Otolaryngol Clin N Am. 2009;42:311–30.
23. Payne SC. Empty nose syndrome: what are we really talking about? Otolaryngol Clin N Am. 2009;42:331–7.

24. Scheithauer MO. Nasenmuschelchirurgie und "Empty Nose" Syndrom. Laryngorhinootologie. 2010;89(Suppl 1):S79–102.
25. Houser SM. Surgical treatment for empty nose syndrome. Arch Otolaryngol Head Neck Surg. 2007;133:858–63.
26. Naftali S, Rosenfeld M, Wolf M, Elad D. The air-conditioning capacity of the nose. Ann Biomed Eng. 2005;33:545–53.
27. Passàli D, Lauriello M, Anselmi M, Bellussi L. Treatment of hypertrophy of the inferior turbinate: long-term results in 382 patients randomly assigned to therapy. Ann Otol Rhinol Laryngol. 1999;108:569–75.
28. Cook PR, Begegni A, Bryant WC, Davis WE. Effect of partial middle turbinectomy on nasal airflow and resistance. Otolaryngol Head Neck Surg. 1995;113:413–9.
29. Eliashar R. Total inferior turbinectomy: operative results and technique. Ann Otol Rhinol Laryngol. 2001;110:700.
30. Ophir D. Resection of obstructing inferior turbinates following rhinoplasty. Plast Reconstr Surg. 1990;85:724–7.
31. Talmon Y, Samet A, Gilbey P. Total inferior turbinectomy: operative results and technique. Ann Otol Rhinol Laryngol. 2000;109:1117–9.
32. Courtiss EH, Goldwyn RM. Resection of obstructing inferior nasal turbinates: a 10-year follow-up. Plast Reconstr Surg. 1990;86:152–4.
33. Elwany S, Harrison R. Inferior turbinectomy: comparison of four techniques. J Laryngol Otol. 1990;104:206–9.
34. Martinez SA, Nissen AJ, Stock CR, Tesmer T. Nasal turbinate resection for relief of nasal obstruction. Laryngoscope. 1983;93:871–5.
35. Oburra HO. Complications following bilateral turbinectomy. East Afr Med J. 1995;72:101–2.
36. Odetoyinbo O. Complications following total inferior turbinectomy: facts or myths? Clin Otolaryngol Allied Sci. 1987;12:361–3.
37. Miwa M, Nakajima N, Matsunaga M, Watanabe K. Measurement of water loss in human nasal mucosa. Am J Rhinol. 2006;20:453–5.
38. Brown CL, Graham SC. Nasal irrigation: good or bad? Curr Opin Otolaryngol Head Neck Surg. 2004;12:9–13.
39. Michel O. Nasenspülung bei Rhinosinusitis. Laryngorhinootologie. 2006;85:448–58.
40. Schmidt T. Die tägliche Nasenspülung mit Salzwasser zur Vorbeugung und Behandlung von Atemwegserkrankungen. www.nasespuelen.de. n.d.
41. Boek WM, Keleş N, Graamans K, Huizing EH. Physiologic and hypertonic saline solutions impair ciliary activity in vitro. Laryngoscope. 1999;109:396–9.
42. Talbot AR, Herr TM, Parsons DS. Mucociliary clearance and buffered hypertonic saline solution. Laryngoscope. 1997;107:500–3.
43. Rabago D, Zgierska A. Saline nasal irrigations for upper respiratory conditions. Am Fam Physician. 2009;80:1117–9.
44. Elberg M. Erfahrungen mit Nisita®-Salbe bei Nasenschleimhauteingriffen. MMW Munch Med Wochenschr. 1977;119:445.
45. Verse T, Klöcker N, Riedel F, Pirsig W, Scheithauer MO. Dexpanthenol – Nasenspray vs. Nasensalbe. HNO. 2004;52:611–5.
46. Ebner F, Heller A, Rippke F, Tausch I. Topical use of dexpanthenol in skin disorders. Am J Clin Dermatol. 2002;3:427–33.
47. Neher A, Gstöttner M, Thaurer M, Augustijns P, Reinelt M, Schobersberger W. Influence of essential and fatty oils on ciliary beat frequency of human nasal epithelial cells. Am J Rhinol. 2008;22:130–4.
48. Riechelmann H, Brommer C, Hinni M, Martin C. Response of human ciliated respiratory cells to a mixture of menthol, eucalyptus oil and pine needle oil. Arzneimittelforschung. 1997;47:1034–9.
49. Johnsen J, Bratt BM, Michel-Barron O, Glennow C, Petruson B. Pure sesame oil vs isotonic sodium chloride solution as treatment for dry nasal mucosa. Arch Otolaryngol Head Neck Surg. 2001;127:1353–6.
50. Björk-Eriksson T, Gunnarsson M, Holmström M, Nordqvist A, Petruson B. Fewer problems with dry nasal mucous membranes following local use of sesame oil. Rhinology. 2000;38:200–3.
51. Bahadur S, Take A. Specific chronic infections. In: Gleeson M, editor. Scott-Brown's otolaryngology, head and neck surgery, vol. II. 7th ed. London: Hodder Arnold; 2008.
52. Nielsen BC, Olinder-Nielsen AM, Malmborg AS. Successful treatment of ozena with ciprofloxacin. Rhinology. 1995;33:57–60.
53. Goldenberg D, Danino J, Netzer A, Joachims HZ. Plastipore implants in the surgical treatment of atrophic rhinitis: technique and results. Otolaryngol Head Neck Surg. 2000;122:794–7.
54. Rice DH. Rebuilding the inferior turbinate with hydroxyapatite cement. Ear Nose Throat J. 2000;79:276–7.
55. Friedman M, Ibrahim H, Lee G. A simplified technique for treatment of atrophic and hypotrophic rhinitis. Paper presented at: American Rhinological Society meeting, Boca Raton, May 2002.
56. Rice DH, Kern EB, Marple B, Mabry L, Friedman WH. Determinates in nasal and sinus surgery: the consensus statement. Ear Nose Throat J. 2003;82:82–4.

14 Physiology of the Aging Nose and Geriatric Rhinitis

Yazan Eliyan, Victoria E. Varga-Huettner, and Jayant M. Pinto

Core Messages

- The world's population is rapidly aging due to increasing life expectancy.
- Physiological changes of the nose with age include external structural changes, alterations of the thickness of the respiratory epithelium, decreased ciliary function, blunted vascular responses, decreased intranasal sensitivity and olfaction, reduced immune defense, and decreased ability to humidify the air.
- Such changes contribute to rhinitis in older patients as well as a range of nasal diseases.
- Geriatric rhinitis is poorly understood from a mechanistic standpoint but can be divided by cause into allergic and nonallergic categories.
- Allergic causes can receive standard therapies and symptoms tend to be milder in older patients.
- Nonallergic causes are more difficult to treat and require careful attention to the precise triggers and symptoms; therapies are targeted to the symptom.
- Nonspecific treatments such as humidification, mucolytics, and saline irrigations are generally safe and effective.
- Surgical and medical treatments for geriatric rhinitis are safe and effective, but special considerations of geriatric issues such as polypharmacy, alterations in hepatic and renal function, and side effect profiles must be made.

14.1 Importance of Aging in Rhinology

The world is facing a massive demographic shift in the next 30 years. Both developed and developing nations are experiencing growth in the oldest age groups and declines in children. Indeed, the proportion of older persons was 8% in 1950, 10% in 2000, and is projected to reach 21% in 2050 (http://www.un.org/esa/population/publications/worldageing19502050/). The global median age has increased from 21.5 years in 1970 to over 30 years in 2019. As people have fewer children and those children are more likely to survive, and adults live longer due to advances in hygiene and medicine, the age structure continues to change toward increased numbers and proportions of older adults. For example, 8% are older than 65 in 2019 (https://ourworldindata.org/age-structure). These trends

Y. Eliyan
Pritzker School of Medicine, Chicago, IL, USA
e-mail: Yazan.Eliyan@uchospitals.edu

V. E. Varga-Huettner
Ellis Medical Group, Ear, Nose and Throat Care, Schenectady, NY, USA

J. M. Pinto (✉)
Section of Otolaryngology-Head and Neck Surgery, Department of Surgery, The University of Chicago Medicine and Biological Sciences, Chicago, IL, USA
e-mail: jpinto@surgery.bsd.uchicago.edu

Ö. Ö. Celebi, T. M. Önerci (eds.), *Nasal Physiology and Pathophysiology of Nasal Disorders*,
https://doi.org/10.1007/978-3-031-12386-3_14

are apparent in the United States, Western Europe, and even countries with younger age distributions such as China [1]. Hence, we need to accelerate our understanding of the effects of aging and both normal physiology and disease. Given these demographic shifts, physicians caring for patients with upper airway diseases must be ready to address the special needs of this expanding group of patients. All organ systems are affected by aging in a myriad of ways, making this a population with a significant medical need and a significant challenge for clinicians.

The study of geriatric medicine has its origins in training programs dating back to the 1970s. Programs such as the Department of Veterans Affairs Geriatric Research, Education, and Clinical Centers (GRECC) were organized to increase translational research in geriatric medicine and promote the advancement of clinical care for older people [2]. The creation of the National Institute on Aging (NIA) in 1974 and efforts of private organizations (e.g., the American Federation for Aging Research [AFAR]) have resulted in increased efforts to understand aging and disease and treat or mitigate its effects. Parallel efforts in Otolaryngology-Head and Neck surgery have begun (e.g., the [3]) and more are needed in specific areas such as Rhinology if we are to meet the needs of our older patients. However, our field still needs more research in and clinical focus on otolaryngologic manifestations of aging and the relationship between aging and disease.

One key principle of geriatrics is to prioritize equally prolongation of life and quality of life. Here we address how the anatomy and function of the nose change with age, important diseases of the nose in older adults, and treatment methods allowing us to provide the best rhinologic care for older patients.

14.2 Anatomical Changes of the Aging Nose

14.2.1 Anatomy

Aging affects all portions of the face including the nose, starting with its surface cover, and the skin (Figs. 14.1 and 14.2a, b) [4]. Thinning of the epidermis and decreased collagen production cause the skin to lose its elasticity. This may affect the development of rosacea and rhinophyma, which are more common with age. Anatomical changes in older adults that affect the structure and function of the nose include thinned nasal skin, weakened nasal cartilages, weakened fibrous attachments between upper and lower lateral cartilages, and ptosis of the nasal tip [5]. Edelstein and others have found that the anatomy of the nose undergoes several changes with age,

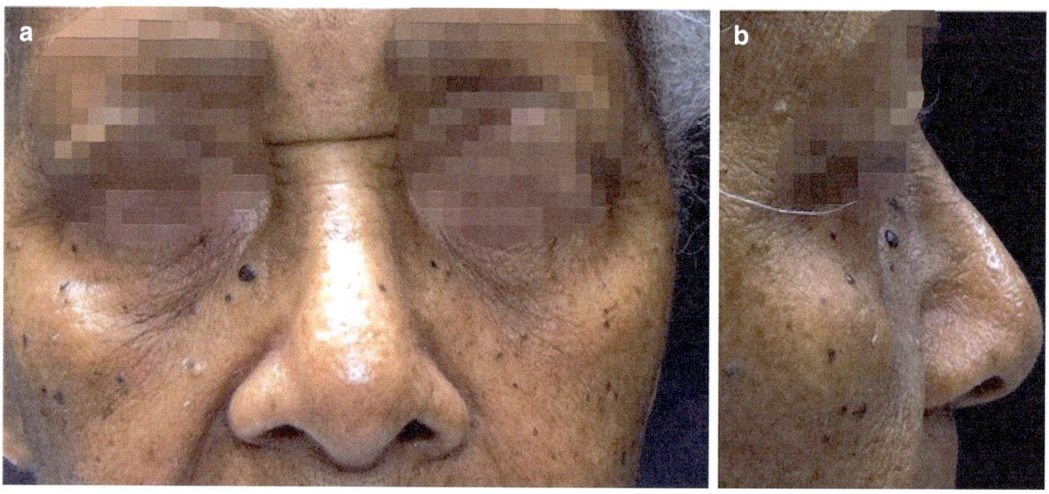

Fig. 14.1 (**a**, **b**) Two examples of the geriatric nose, profile and frontal views. Note the skin changes and external structural effects of aging

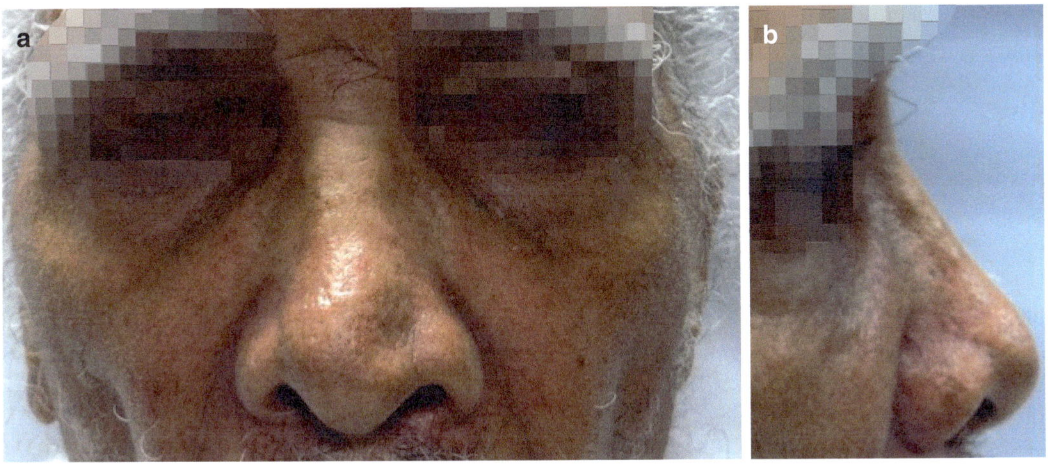

Fig. 14.2 (**a**, **b**) Two examples of the geriatric nose, profile and frontal views. Note the skin changes and external structural effects of aging

including a decrease in the nasolabial angle, a decrease in the height to length ratio of the nose, weakening/separation of the upper and lower nasal cartilages, decreased glycosaminoglycans causing increased porosity of nasal cartilage, retraction of the nasal columella, resorption of bone resulting in maxillary hypoplasia, atrophy of collagen fibers, and attenuated and fragmented fibroelastic attachments [6–9]. These changes make the nose appear longer, deprojected, and underrotated [10]. Some studies have revealed increased cross-sectional area at the internal nasal valve in older adults. For example, although it is difficult to objectively assess patency at the internal nasal valve, Kalmovich et al. attempted to measure endonasal geometry changes in the geriatric population. Using acoustic rhinometry, they found that there was a statistically significant gradual increase of endonasal volumes and minimal cross-sectional areas with age, except in the oldest group of men over 80 years of age [11]. Studies utilizing computed tomography (CT) volumetric analyses have corroborated these findings and demonstrated that intranasal cavity volumes significantly increase with old age [12, 13]. Similarly, Kim et al. demonstrated an increased cross-sectional area at the internal nasal valve associated with age, but no difference in nasal resistance before and after decongestion in any age group greater than 20 years old, supporting their hypothesis that the increased area is related to changes in the non-erectile structures of the nose. In general, however, the sum result is a restriction in nasal airflow, particularly at the nasal valve region, an important site of nasal resistance. In summary, the structural changes seen in the aging nose may result in nasal obstruction and abnormal airflow.

Anatomic changes in the nose are, in part, related to changes in the cartilaginous structures as well as the weakening of soft tissue attachments. The biochemical composition and mechanical properties of cartilage change with age, resulting in a tendency to collapse. For example, there is a decrease in the glycosaminoglycan content of nasal septal cartilage with increasing age, resulting in stiffening with decreased fluid flow through the tissue. Additionally, there appears to be a slight increase in hydroxyproline content with increasing age. These changes are similar to those seen in articular cartilage during the aging process. Although the physical stresses on articular cartilage are much different than that of septal cartilage, the similarity of their biochemical changes may represent systemic effects on cartilage with age [14]. Indeed, histological analysis of nasal cartilage showed that increased age significantly correlated with abnormal reductions in proteoglycan content and decreased chondrocyte activity [15].

Riedler et al. further demonstrated that glycosaminoglycan content, cell size, and cell density found in septal cartilage all significantly decrease with increasing age [16]. Using image analysis of 300 men in Turkey, mean nasal bridge length was the longest in the oldest age group as was the mean nasal tip protrusion [17]. There were notable differences in nasal width and root width between age groups ($p < 0.05$). Similar findings have been identified in other studies, and these make affect surgical decision-making during rhinoplasty or other reconstructive or airway procedures [18]. Differences in facial aging can be analyzed by surface scanning and have been shown to be similar in men and women, but divergent after menopause [19]. Such techniques confirm direct measurements of nasal changes with aging [20] in diverse populations [21].

This information leads to two practical considerations for clinical practice. First, anatomic changes are likely to affect airflow through the nose in older patients, resulting in restrictions that may cause dryness, irritation, and obstruction. These problems may exacerbate existing conditions and contribute in a large way to patient symptoms.

Second, older adults may seek septorhinoplasty to alleviate such functional problems and also for cosmetic rejuvenation. Knowledge of age-related anatomic changes in the nose necessitates specific surgical considerations for performing such operations [22]. Also, careful consideration must be given to these patients as they may have underlying significant medical comorbidities, such as hypertension, coronary artery disease, or diabetes, placing them at higher risk for elective surgery. Cochran et al. noted that many older patients at the time of rhinoplasty had ossified septal cartilage, making it more frequently necessary for auricular or costal cartilage to be harvested [5], potentially lengthening operative time. Medical and anesthesia clearance is prudent prior to proceeding with elective surgery in any older patient and several guidelines exist to make this determination [23, 24]. Nevertheless, nasal surgery in older subjects can be safe and effective when approached carefully [25]. As with all rhinologic patients, psychological motivation for the procedure needs to be assessed. Geriatric patients frequently have to deal with significant life issues, including the death of loved ones, lack of social support, financial challenges, and struggles with quality of life. One must assess whether older patients have taken ample time to consider any cosmetic surgery prior to proceeding. Most importantly, assessment of the functional consequences of rhinoplastic procedures on this patient subset is critical to maintain airflow.

14.3 Physiological Changes with Age

Many subtle changes in the physiology of the nose occur over time and may lead to symptoms and/or disease in older adults. Although we are far from fully understanding comprehensively how the nose changes over time, further study of the known physiological changes will help us to provide better medical and surgical treatment regimens for these patients.

14.3.1 Alterations in Nasal Function

Nasal dryness is a frequent complaint in older patients. This condition presents with crusting, irritation, epistaxis, or obstruction. Several etiologies have been suggested to cause this condition, including mucosal and glandular atrophy, vascular changes which reduce nasal humidification, and medication use (e.g., antihypertensives, which affect vascular regulation in the nose, or first-generation antihistamines, which inhibit cholinergic responses in the nose). Structural changes in the nose, such as increased intranasal cavity volume, may also contribute by causing turbulent airflow which dries the mucosa.

The mucosa itself is altered with age. Schrödter performed biopsies of the middle turbinate in 40 subjects of varying ages and found significant atrophy of the epithelium in the older subjects (Fig. 14.3) [26]. This analysis found thin epithelium and also increased thickness of the basement membrane. Also, the percentage of

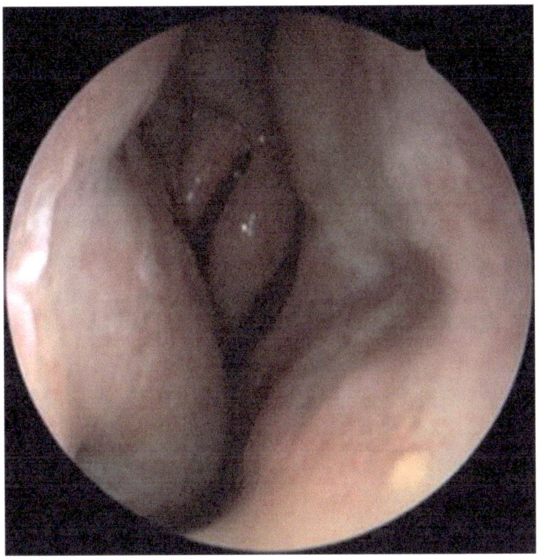

Fig. 14.3 Endoscopic view of a geriatric patient with atrophy of endonasal mucosa

normal ciliated respiratory epithelium declined in the older subjects. Similar studies using electron microscopy are consistent with these findings [27]. Increased expression of caspase 3, an apoptotic marker, indicates that part of the defect may be in the ability of the epithelium to renew itself, and there are other functional aspects of the epithelium (e.g., toxin neutralization) that may be affected [28]. This matches what has observed in nasal endoscopy in clinical practice: a thinner, atrophic epithelium.

Another major alteration affecting this symptom is in one of the main functions of the nose: the ability to warm and humidify air [29]. The two major mechanisms leading to the alteration in nasal conditioning capacity (NCC) are changes in the nasal mucosal temperature (NMT) and the volume of the nasal cavity. During inspiration of air, water evaporates from the nasal mucosa to condition inspired air, leading to heat loss and, consequently, a decrease in NMT [30]. Lindemann found that using in vivo air temperature and humidity measurements at the nasal valve and the region just anterior to the head of the middle turbinate, both temperature and humidity values were significantly lower in older subjects (mean age 70 years) compared to younger subjects (median age 27 years) [31]. The decrease in NMT, which has been suggested to be partly caused by decreased nasal mucosal blood flow [32], could be demonstrating an age-dependent effect due to the fact that blood flow to the nasal mucosa has been shown to decrease with age [33]. Additionally, nasal volumes in the Lindemann et al. study appeared to be larger in the older age group. Increased turbulence of airflow may be one result causing the sensation of nasal obstruction despite larger space (paradoxical nasal blockage). Endonasal geometries are indeed enlarged in older subjects [11–13]. This suggests that NCC in older subjects is compromised. When analyzed via CT imaging, nasal cavity volumes also increased markedly with age, in analyses that adjusted for sex and head size [12]. Thus, a lower ability to warm and humidify air may be present in older subjects, potentially affecting lower airway function. This may also contribute to increased nasal dryness in older adults.

We may take some analogy from changes in the oral mucosa. As with the nose, the lubrication of the oral cavity changes with age. Although there should not be a decrease in saliva production with age, unless altered by medications or systemic conditions, there are changes in the consistency of the saliva, including a decrease in mucin concentration and decreased secretion of protective IgA antibodies [34]. This may lead to increased caries, periodontal disease, and poorer nutritional status.

A second altered physiological function with age is mucociliary clearance. Ciliary beat frequency is decreased in vitro in nasal epithelial cells taken from patients older than 60 years of age [35]. Ho et al. also found decreased nasal mucociliary clearance by the crude saccharin clearance test and the more sensitive measure of ciliary beat frequency by a photometric test [36]. Using the same saccharin clearance test, Paul et al. also found decreased nasal mucociliary clearance among older women compared to younger women [37]. However, there is likely significant interindividual variability in this phenomenon since Sakakura et al. showed that 70% of patients older than age 60 retained saccharin transport times comparable to their younger counterparts [38].

Age-related decline in nasal mucociliary clearance can be attributed to many possible causes. Multiple studies have demonstrated that oxidative stress can decrease ciliary beat frequency [39, 40]. The mechanism by which oxidative stress slows down ciliary beat frequency may be due to the upregulation of PKCε signaling. In addition to oxidative stress, trauma, damage after infection, or exposure to toxins have also been proposed to reduce ciliary function [36]. Diabetes and hypertension, both increased in prevalence in older subjects, are also associated with the decreased ciliary function [41]. In summary, if the cilia are not moving as rapidly, this can lead to symptoms of mucus buildup, rhinitis, inflammation, or infection, from the persistence of organisms and molecules trapped in the mucus layer (Fig. 14.4). Hence, both the structural components and the function of the epithelial lining of the nose in older patients demonstrate significant changes with age that may affect airflow, mucus quality and production, and mucociliary clearance.

A third way nasal function is altered with age is via dysregulation of vascular responses in the nose. Nervous control of vascular tone in the nose is important in the regulation of the critical functions of the nose, to warm and humidify the air, and also may affect diseases involving vasodilation such as allergic and nonallergic rhinitis. Tillmann found that vascular regulatory responses are reduced with age [42]. Fifty-two subjects were acclimated to the laboratory environment for 5 min and baseline measures of perfusion by optical rhinometry were performed. The subjects were then moved to a supine position for 30 min, which should alter blood flow. Older subjects had more rapidly increased perfusion but did not return to baseline in contrast to younger subjects. This study suggests that autonomic control of nasal blood flow is altered in older subjects, potentially affecting humidification (as described above) and the other functions of the nose discussed above. Using liquid crystal thermography exhalation monitoring to measure the nasal cycle, Doty found decreased regulation of the nasal cycle in older adults [43]. Overall, the proportion of subjects exhibiting the alternating rhythmicity associated with the classic nasal cycle decreased with age. No association was present between nasal cycle parameters and scores on the mini-mental state examination (MMSE). The results suggest that the classic nasal cycle may be a marker for age-related central nervous system changes. These phenomena may also be important in nonallergic rhinitis, which is relatively common in older adults.

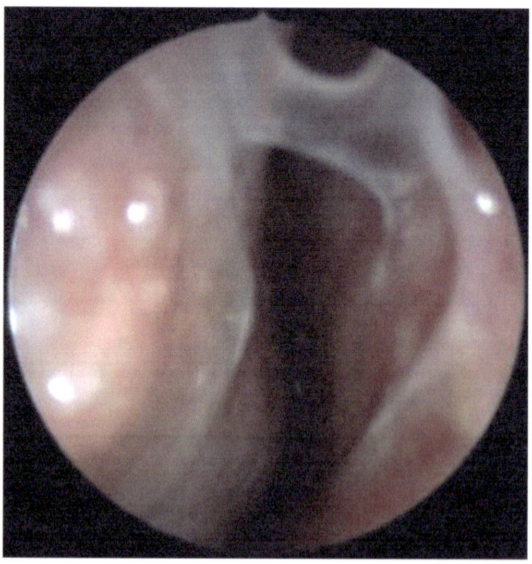

Fig. 14.4 Endoscopic view of a geriatric patient with thickened nasal mucus, a common complaint in this patient population

14.3.2 Olfaction and Nasal Sensitivity

Olfaction, mediated by the olfactory nerve, and sensitivity of the nose, controlled by the somatosensory system and mediated by the trigeminal nerve, both decrease with age. In 1994, the National Health Survey determined that 3.2 million people, or 1.65% of adults in the United States, had self-reported chronic chemosensory problems involving smell and/or taste; 40% of these adults were aged 65 years and older, with an exponential increase seen with increasing age [44]. Using a validated odor identification test, Murphy et al. found that among adults over the age of 53 the prevalence of olfactory impairment was 24.5% [45]. Prevalence rates among adults between the ages of 80 and 97 reached as high as 62.5% [45]. Additionally, adults over 65 years of age have approximately half the intensity of smell and irritant sensation as subjects between the ages of 18 and 25 [46]. In another study, the sensitivity of the nose was assessed by providing tactile stimulation with monofilaments of progressively larger sizes. Adults 50–65 years of age required significantly less stimulation to detect the sensation of the monofilament as compared to adults aged 66 and above [47].

In order to understand how the olfactory system is altered with age, we first need a better understanding of the normal olfactory system. The olfactory mucosa is distributed along the upper portion of the nasal septum, below the cribriform plate, and along the medial aspect of the superior turbinate, as well as along the roof of the middle turbinate [48]. On histologic examination, there is a patchy distribution of olfactory epithelium, which becomes increasingly sparse with aging because olfactory epithelium becomes replaced with respiratory epithelium over time. In support of this finding, several studies have demonstrated increased respiratory metaplasia of the olfactory epithelium with increased age in both mice and humans [49–51]. Other age-related changes in the olfactory epithelium also include increased thinning, as demonstrated in mice models [52, 53]. Robinson et al. demonstrated in a murine model that there is an increase in apoptotic gene expression of olfactory neuron receptors with increasing age, likely resulting in increased cell death [54]. Additionally, Ueha et al. demonstrated that there was a decline in mature and immature olfactory receptor neurons in the aged olfactory mucosa of mice [55]. Over time, even small toxin exposures, including heavy metals, such as manganese, cigarette smoke, or volatile chemicals, may cause harm to the olfactory epithelium [56]. In mice it has also been seen that following chemical damage, there is decreased regeneration of the neuroepithelium of the olfactory bulb with increasing age [49]. Because olfactory stem cells regenerate with time, any alteration of this turnover may also contribute to age-related olfactory loss, either from insult or from inherent aging. Several studies corroborate the suggestion that the basal proliferation of cells in damaged and undamaged epithelium decreases with age [52, 55, 57, 58]. All of these mechanisms may contribute to a decrease in the function of the olfactory system as one ages.

Age-related olfactory loss (presbyosmia) is an important public health problem worldwide [44, 45, 59–63]. In the USA, olfactory complaints from the approximately 14 million individuals over age 55 who are affected lead to over 200,000 physician visits annually [64, 65]. This sensory impairment of aging affects critical functions such as nutrition [66–69], sensation of pleasure [70], detection of environmental hazards [71], mood, cognition, behavior [72–74], sexuality [75–77], quality of life [78], and general well-being [79], and therefore, it poses a profound burden on older adults. Indeed, up to one-third of older subjects report dissatisfaction with their ability to smell [80], and ~50% are unable to detect the standard warning odor in natural gas [81]. Importantly, the decline in olfaction has been linked to several neurodegenerative conditions, such as mild cognitive impairment, dementia (Alzheimer's disease), and neuromotor disorders (Parkinson's disease) [82–91]. If this link involves shared genetic or environmental risk factors, then understanding the pathogenesis of presbyosmia may have broad implications for a wide array of problems, especially regarding other sensory impairments of aging [92]. Thus, the olfactory sensory loss is related to factors that are critical to the physical well-being, social function, and quality of life of older adults. Because olfaction declines over time, the clinical impact will increase as our population ages.

Acquired chemosensory complaints are also important in Otolaryngology. In addition to traumatic injury or post-viral olfactory loss, cancer chemotherapy can negatively impact olfaction [93–95]. Furthermore, t older patients are more susceptible to olfactory decline caused by chemotherapy agents. Because the olfactory epithelium regenerates, one cause of this could be a decrease in stem cells with age in the olfactory mucosa, either inherently or after stimulation of regeneration by injury. Indeed, telomere shortening has been observed in the regenerative response to chemical injury of the olfactory epithelium in a mouse model, providing evidence that telomere shortening impairs the regenerative capacity of this tissue [96]. So, aging may lessen the ability of the olfactory mucosa to survive respiratory insult by toxins. Olfactory decline among chemotherapy patients can have detrimental health consequences. For example, patients undergoing palliative chemotherapy treatment for cancer frequently had complaints of dysgeusia and sensitivity to odors while undergoing treatment. Those patients with severe complaints were found to have lower calorie intake, increased weight loss, and lower quality of life scores, than in patients with less severe complaints [97].

There are many systemic causes of chemosensory dysfunction. Olfactory deficits are seen in neurologic disorders, such as Alzheimer's disease, Huntington's disease, Parkinson's disease, and Korsakoff's psychosis, all of which generally affect the elderly. It is important to counsel patients with hyposmia or anosmia, as they may be unaware of noxious materials in their surroundings. They should obtain natural gas detectors if they have natural gas stoves, ovens, or heat at home as they may not be able to detect a fuel leak. Additionally, they should be aware of dates on food products, label leftovers, and dispose of foods reaching their expiration dates to avoid food spoilage. Lastly, smoke detectors should be checked regularly for function and efforts must be maintained to promote nutrition.

14.3.3 Immunosenescence

Immunosenescence is the term used to describe the decreased function of the immune system with age. Both the adaptive and innate immune systems are impaired [98, 99]. Decreased immune response is believed to contribute to increased infections, autoimmune diseases, and cancers in the elderly population.

There are several effects of immunosenescence in the nose. IgA is an immunoglobulin that is secreted along the respiratory and gastrointestinal tracts, where it helps to neutralize a variety of pathogens. Alford demonstrated that IgA levels can be measured in nasal secretions and comprise approximately 38% of nasal wash proteins in normal patients. He also showed that IgA levels in nasal secretions decrease significantly with age [100]. In a murine model, it has been shown that mucosal immunity wanes prior to systemic immunity in studies where vaccination to cholera toxin was performed by mucosal and subcutaneous methods, respectively, in different ages of mice [101].

IgE levels also decrease with age. Mediaty and Neuber studied 559 individuals with atopic dermatitis, allergic rhinitis or asthma, and insect allergy. In all patients, except those with atopic dermatitis or those with high total serum IgE >300 kU/L, total and specific serum IgE levels were significantly decreased in patients aged greater than 60 compared to their younger counterparts. They hypothesized that there may be more robust mechanisms in atopic dermatitis or conditions resulting in high serum IgE that lead these patients to have a more persistent response. Additionally, the type of atopic disease and the age of disease onset may impact IgE levels [102].

Although not specific to the nose, there are many age-related defects in T cell functioning as well. As age increases, the thymus involutes, thus leading to decreased production of naïve T cells and consequently diminished ability to fight infection [103, 104]. T cell diversity significantly diminished among the older adults in their 70s and 80s, thus reducing their ability to respond to new antigens [105]. Additionally, multiple studies have shown an increased propensity to shift from a Th1 to Th2 subtype with increasing age [106, 107].

Immunosenescence is poorly studied in the upper airway but one can extrapolate findings from the lower airway [108]. In that lung, there are several immune alterations that might facilitate the persistence of asthma, a related airway disease to rhinitis. These include changes in airway neutrophil, eosinophil, and mast cell numbers and function as well as altered antigen presentation, decreased specific antibody responses, and altered cytokine profiles. The sum of these changes might affect susceptibility to upper respiratory tract infections. Indeed, it has been shown that the incidence of nonallergic rhinitis, an inflammatory condition affecting the nasal mucosa, increases with age [109, 110]. In addition, studies have found a high prevalence of nonallergic rhinitis among the elderly population [111, 112]. Although the specific mechanisms underlying the link between immunosenescence and the development of rhinitis in the elderly have not been fully elucidated, it has still been hypothesized that immunosenescence plays some important role in the etiology of the condition [113].

In summary, this is a complex area in which there is little data on the nose specifically, but age-related changes in immune function are likely to impact nasal physiology and disease in older patients.

14.4 Geriatric Rhinitis

14.4.1 Overview

Rhinitis is a pervasive complaint in physicians' offices. It is estimated that 20–40% of subjects living in Western countries are affected by rhinitis. Rhinitis may be allergic or nonallergic in nature, with approximately 50% of rhinitis patients belonging to each group [114]. These forms of rhinitis affect older subjects as well, though because allergy wanes with age, nonallergic rhinitis tends to predominate [115]. There are many potential causes of rhinitis in the geriatric population. Edelstein noted that six nasal complaints were more prominent in older adults: nasal drainage, postnasal drip, sneezing, coughing, olfactory loss, and gustatory rhinitis [6]. Regarding the quality of life in an Italian study, geriatric patients with rhinitis underwent clinical evaluation and responded to the Rhinasthma questionnaire [116]. All patients also underwent skin prick testing, measurement of total IgE level, and nasal cytologic analysis. In the older patients, the epithelial to goblet cell ratio was decreased. The quality of life in older people was more impaired than in young adults. These authors concluded that quality of life is more heavily impaired compared with young adults. In concordance with this conclusion, Song et al. found that among an elderly population, rhinitis was significantly associated with decreased quality of life [112]. However, one study showed that nasal-specific quality of life measures shows no deterioration in older subjects and do not therefore correlate with changes in nasal function [117]. Further investigation of this topic is warranted to elucidate this paradox.

14.4.2 Allergic Rhinitis

Allergy is defined as an immediate type IgE-mediated response to an allergen exposure, where IgE is bound to mast cells and its attachment to an allergen results in degranulation of the mast cell with histamine release. The diagnosis of allergy can be confirmed by skin, in vitro, or provocation testing. The Allergic Rhinitis and its Impact on Asthma (ARIA) consensus categorizes allergic rhinitis as either intermittent allergic rhinitis (IAR) or persistent allergic rhinitis (PAR). IAR denotes that symptoms are present less than 4 days per week or less than 4 consecutive weeks per year. PAR indicates that symptoms are present greater than 4 days per week and for more than 4 consecutive weeks. These can be further categorized as mild or moderate/severe based on the impact of quality of life [118–120].

Although allergy is well studied, the majority of studies performed to date involve either the pediatric population or young adults. In older adults, the incidence and prevalence of allergic rhinitis worldwide actually decrease with age along with atopy as detected by skin prick tests [121–124]. Karabulut et al. noted that only 50% of older adults have positive skin prick tests as compared to 70% of younger adults when matched for allergic symptoms including nasal, eye, pulmonary, and dermatologic symptoms [125]. Subjects from the French cohort aged 65 years and over ($n = 352$) found that respiratory allergy is present in older people and that there is an association between smoking and IgE level independent of allergic reactivity to common allergens in the elderly, highlighting the importance of environmental factors in rhinitis [126]. Population-based studies investigating the effect of smoking on allergic rhinitis have been conflicting, but one meta-analysis determined that smoking was associated with a significantly lower risk of allergic rhinitis among adults [127]. The mechanism behind this may be that those with asthma (many of whom have allergic rhinitis) could be more likely to avoid smoking and irritating their airways [121]. A longitudinal study analyzed allergic sensitivity 15 years after primary testing. Skin prick test (SPT) and evaluation of serum total and specific IgE and nasal eosinophils were conducted on 108 subjects from Palermo, Sicily, in Italy. In general, rhinitis symptoms tended to be milder at follow-up. All parameters examined decreased with time. However, the changes in rhinitis symptoms appear to be related to changes in the nasal eosinophils, independently of SPT and specific IgE [128]. Consistent with these results, a 10-year prospective study of Swedish adults demonstrated that allergen sensitization significantly decreases with age as measured by SPTs and serum IgE levels [129]. In addition, multiple other studies measuring allergen-specific IgE levels showed that the elderly were significantly less likely to be sensitized [124, 130]. Further, Ciprandi et al. showed that among adults with allergic rhinitis, elderly patients had less sensitizations, less severe symptoms, and reduced IgE levels in comparison to adult patients [131]. Thus, in general allergic disease and markers decline with age.

In the Korean National Health and Nutrition Examination Survey (2008–2012), 9.05 of older adults had allergic rhinitis whereas 27.07% had nonallergic rhinitis. Rhinorrhea was significantly increased in the older adults ($p = 0.018$) and sensitization to Dermatophagoides farina was significantly decreased ($p = 0.006$) [124]. In the Korean Longitudinal Study on Health and Aging (KLoSHA, rhinitis was more prevalent than expected, significantly related to impairment in quality of life, and formed close relationships with abdominal obesity, sarcopenia, comorbidities, and urban environments [132].

One practical consideration for these patients is that because elderly patients frequently have atrophic or photodamaged integument, skin testing for allergy may actually be less reliable than in younger adults. There is evidence that there are decreased numbers of mast cells in atrophic skin, thereby making the possibility of a false-negative result higher [133, 134]. In older subjects, an area of skin that is protected from the sun can be considered for skin testing, a histamine control used so that the most dramatic obtainable response can be visualized, and consideration given to in vitro testing if no reliable area of skin is present [135]. Both skin and in vitro testing must always correlate with patient history for accurate diagnosis. Allergic disease in the older patient population, as well as its biological mechanisms, have been reviewed recently [136–138].

14.4.3 Nonallergic Rhinitis

There are many causes of rhinitis which are not allergic in nature but can be equally bothersome to patients and have a marked impact on quality of life. The data regarding this phenomenon are less developed, as compared to those for allergic rhinitis. Studies of patients presenting to allergists have demonstrated that 23–52% of patients with rhinitis have the nonallergic form [139]. A population-based study from Italy demonstrated that the ratio of the prevalence of allergic to nonallergic rhinitis was 1.4 in adults older than 64 years old [121]. Further, this ratio decreased with age, suggesting that with age, a more significant proportion of total rhinitis cases can be attributed to nonallergic rhinitis [121]. No good diagnostic techniques are available to distinguish between the subtypes of nonallergic rhinitis, but a diagnosis is based on history and symptoms and negative allergy testing. Further convoluting the issue is the notion that patients can also be sensitized to allergens at any time, thereby changing the categorization of their rhinitis. Additionally, there is neither a set classification system nor an international consensus on a uniform definition for nonallergic rhinitis. Further complicating this issue is that allergic and nonallergic rhinitis can coexist ("mixed rhinitis"). Interestingly, it has been observed that 70% of patients with nonallergic rhinitis present during adulthood (age >20), whereas 70% of patients with allergic rhinitis initially present during childhood (age <20). There is also a greater propensity for females to be affected, compared to males [140]. Nonallergic rhinitis appears to be mediated in some patients by local IgE production in the nose [141]. The only approved treatment for nonallergic rhinitis currently is intranasal antihistamines which are effective in some cases. Additionally, intranasal steroids can be useful in some circumstances and are recommended by some expert panels [142]. Pollution exposure may explain some degree of chronic rhinitis in older adults [143].

In summary, nonallergic rhinitis is a major disorder in older adults. Due to its complex nature, however, our understanding of its pathophysiology in general and specifically in older adults is limited, resulting in limited ability to treat this problem effectively. A trial and error method of different nasal medications is usually pursued.

14.4.4 Vasomotor Rhinitis

Vasomotor rhinitis is one form of nonallergic rhinitis and is often viewed as an idiopathic diagnosis that is considered when a patient has no evidence of allergy, infection, eosinophilia, hormonal changes, or drug exposure [144]. Symptoms include nasal congestion, nasal drainage or postnasal drip, and perennial symptoms; however, pruritus is rare [145]. There is also typically no cytological evidence of nasal mucosal inflammation [146]. Vasomotor rhinitis is a poorly understood condition but believed that the primary cause is autonomic nervous system dysfunction with either a diminished sympathetic drive or an elevated parasympathetic drive causing increased nasal congestion and resistance [142]. Upregulated expression of transient receptor potential vanilloid 1 (TRPV1) receptor could play a role in the pathogenesis of vasomotor rhinitis [147]. Given the multiple hypotheses regarding the pathogenesis, a number of treatments are used, including intranasal ipratropium, new techniques for vidian neurectomy, and even capsaicin for vasomotor rhinitis [148–151]. The latter is not practical due to the side effects of pain.

14.4.5 Drug-Induced Rhinitis

Many medications affect nasal function and symptoms. Older adults take numerous medications, so they are at high risk of this problem. When evaluating older patients with rhinitis, polypharmacy must be considered as a cause of symptoms. First and foremost, one must consider removing potentially offending medications rather than adding ones. Aspirin, or other nonsteroidal anti-inflammatory drugs (NSAIDs), which are commonly taken by older adults, may cause acute inflammation in the nose in susceptible patients via the inhibition of COX-1. The breakdown of arachidonic acid to the lipoxygenase pathway is favored causing decreased prostaglandin E_2 and an increase in cysteinyl leukotrienes, which include LTC_4 which is thought to be a lead contributor to aspirin-exacerbated asthma [152]. Many other medications can cause rhinitis in older patients. There are also neuromodulatory drugs, such as alpha- or beta-adrenergic antagonists, which work by decreasing sympathetic tone. This causes primarily congestion but also rhinorrhea. Medications such as clonidine and methyldopa fall within this category. Phosphodiesterase-5 inhibitors, which are used for erectile dysfunction, may reportedly impact the erectile tissues of the nose and have been associated with nasal stuffiness and epistaxis [153]. Certain antihypertensives, psychotropic agents, gabapentin, and hormonal treatments, including estrogen therapy, can also cause rhinitis. Their mechanisms of action have not been fully elucidated. Lastly, rhinitis medicamentosa can affect adults who overuse nasal decongestants, causing rebound nasal congestion.

When medication use is thought to be the cause of rhinitis, a pharmacist or geriatrician may be useful in giving recommendations about changing the medication. Polypharmacy is frequently an issue in the geriatric population, and these patients need close monitoring for drug interactions, as well as adverse side effects.

14.4.6 Atrophic Rhinitis

Structural changes in the nasal structure and intranasal function can cause atrophic rhinitis which frequently affects the elderly with a mean age of occurrence of 52–56 years [154]. Symptoms include dryness and crusting of the nasal mucosa, with possible underlying bony resorption [155]. Structural changes in the mucosal glands, vasculature, and connective tissue of the nose as well as cholinergic hyperactivity can all be causes. Histological assessment of patients with atrophic rhinitis shows atrophy of glands, loss of cilia, squamous metaplasia, and inflammatory infiltrate [156]. Atrophic rhinitis has been categorized as either primary or secondary atrophic rhinitis. Although primary atrophic rhinitis has been hypothesized to have multiple possible etiologies, the leading hypothesis seems to be infection and bacterial colonization of the nasal mucosa by *Klebsiella ozaenae* [156, 157]. Secondary atrophic rhinitis is thought to be due to repeated trauma, surgery, irradiation, or granulomatous diseases [156]. Patients with atrophic rhinitis complain of drying of the nose, nasal congestion, and rhinitis and sometimes fetor is noted. Thick nasal secretions increase worsening postnasal drip symptoms and lead to frequent clearing of the throat. Treatment of this condition is difficult as it is impossible to reverse or mitigate the physiological changes. Intranasal steroids, expectorant drugs, and nasal saline sprays are recommended though their efficacy in this group is understudied [158]. Aggressive use of nasal humidification, emollients, and saline irrigation is helpful. The efficacy of the surgical intervention, which aims to reduce the size of the nasal cavity, has shown promise, but still warrants further rigorous investigations [159].

14.4.7 Gustatory Rhinitis

Patients with gustatory rhinitis have profuse watery or mucoid rhinorrhea while eating, with hot and spicy foods often being the trigger. This condition may present in all age groups but has been reported to worsen with age. The pathophysiology is not well understood, and there are many theories as to its cause. One theory is that odors may activate receptors on the mucosa directly, resulting in secretions from goblet cells and submucosal glands. This, however, would not explain why some patients have symptoms while eating foods that are neither hot nor spicy. Another theory is that patients have stimulation of the sensory portion of the trigeminal nerve which causes a reflex parasympathetic response with activation of postganglionic, cholinergic, muscarinic, and parasympathetic fibers. This results in activation of the submucosal glands, causing rhinorrhea [160]. A third theory revolves around a hyperactive nonadrenergic, noncholinergic (NANC) neural system predicated upon the activation of TRPV1 receptors of the upper airway glands by capsaicin [160]. Gustatory rhinitis is a common complaint among rhinologic patients and causes great social discomfort as patients are inhibited from sharing meals with friends. It can be treated easily with intranasal anticholinergics. Research has also investigated surgical interventions and capsaicin treatment as possible alternative therapeutic remedies [160].

14.4.8 Hormonal Rhinitis

Pregnancy induces congestion and rhinorrhea, although the mechanism of rhinitis of pregnancy is not clearly understood [161, 162]. Some suggest that it could be due to hypertrophy of the nasal mucosa caused by placental growth hormone or due to vasodilation of the nasal microvasculature caused by estrogen and progesterone [162]. Through studies of the influence of ovarian hormones on rhinitis, it has been seen that rhinitis may be associated with hormonal peaks, such as those during ovulation or pregnancy, despite the fact that the absolute amount of estrogens present at these times is very different. The number of beta estrogen receptor-positive cells in tissues has been correlated with rhinitic symptoms [163]. One would think that in postmenopausal women, who deal with overall decreased levels of estrogens unless undergoing treatment with exogenous hormones, it would have a large influence on nasal symptoms, but the influence of these hormonal changes has not been elucidated. Some studies demonstrate that when hormones are given exogenously at a steady rate, such as with oral contraceptive pills or hormone replacement therapy, there is no subsequent rhinitis [164]. Consistent with the suggestion that exogenous hormones could have a protective effect on the rhinitis, one study among young women demonstrated the use of hormonal contraceptives was associated with a lower likelihood of developing new-onset rhinitis after puberty [165]. Interestingly, the nose is being studied as a route of administration of hormonal and other drug therapy.

14.4.9 Chronic Rhinosinusitis

Inflammatory causes of rhinitis must be considered in the geriatric population. Chronic rhinosinusitis is the sixth most common chronic condition in people older than 65 [166]. Seventeen percent of patients aged 65–74 years old may suffer from chronic rhinosinusitis [167]. A recent population-based study in Korea demonstrated that the prevalence of chronic rhinosinusitis was significantly higher among adults older than 65 in comparison to adults younger than 65 [168].

Some potential age-related changes that could play a role in the development of chronic rhinosinusitis in the elderly include changes to epithelial barrier function, microbiome characteristics, and patterns of inflammation [169] [170]. In patients over 65 years of age with CRS undergoing sinus surgery, Proteus spp. and Pseudomonas aeruginosa were more commonly cultured compared to younger patients. These data have implications for therapy with antibiotics (where appropriate) and also suggest differences in mucosal immunity with age [171]. Age may indeed affect immune responses in the airway as a CRS endotype with a proinflammatory neutrophilic immune signature was enriched in older patients who had increased mucus levels of IL-1β, IL-6, IL-8, and TNF-α when compared to younger controls [172]. These data show that older patients with CRS have a unique inflammatory signature that corresponds to a neutrophilic proinflammatory response. Neutrophil-driven inflammation may require different therapies and may explain differences in chronic microbial infection or colonization in older patients. Additionally, as some elderly may be relatively immunocompromised because of diabetes mellitus or other underlying illnesses or treatments for medical conditions, some lower-virulence organisms must be considered while treating these patients [167]. Older patients appear to have smaller (and less likely clinically significant) improvements in quality of life after sinus surgery [173]. Standard treatments for this condition are useful in older patients.

14.5 Treatment of the Geriatric Patient with Rhinitis

14.5.1 Overview

Slavin made seminal suggestions for approaching the older patient with rhinitis [174]. First, this condition is neglected and impacts the quality of life. One must consider anatomic changes related to aging before treating. Geriatric issues such as polypharmacy, cognitive dysfunction, changes in body composition, impairment of liver and renal function, and the cost of medications in the face of limited resources are key medical and social issues in this group. One must categorize the etiology of the problem and its related symptoms and treat specific ones. Moisture in general is key to improved nasal symptoms in this age group. Specifics of medications will be reviewed below, but important note should be made of potentially dangerous side effects in this age group. In general, however, several medications are helpful for this problem.

14.5.2 Medical Therapies

In the United States, 58% of adults over age 65 take five to nine medications and 18% take ten or more. In Germany, patients over age 65 take 66% of all prescribed medications, despite representing only 22% of the total population [175]. Polypharmacy has truly become a major issue and careful attention needs to be paid to avoiding drug interactions and adverse side effects from medications. Although rhinitis can be a bothersome condition, its treatment is intended primarily to improve quality of life, and thus, its medical treatment should not interfere with other more pressing medical conditions.

In a cross-sectional study using data from the National Social Life, Health, and Aging Project (NSHAP), a nationally representative sample of community-dwelling, U.S. adults 57–85 years ($n = 2976$) overall prevalence of allergy medication usage was 8.4% (most commonly antihistamines) [176]. Older age was associated with decreased allergy medication use (per decade, OR 0.80; 95% CI, 0.66 to 0.98). Although increased education was associated with increased overall allergy medication use, it was associated with decreased use of allergy medications generally contraindicated in the elderly.

14.5.3 Avoidance

In patients with allergic rhinitis, avoidance measures would make intuitive sense; however, to date, no study has been able to prove that avoidance measures are effective. Despite potentially reducing the number of allergens present, environmental modifications have not been proven to have any impact on rhinitic symptoms [118]. As many of these interventions have no negative sequelae other than cost or logistics, patients may still want to consider them. In patients with known dog or cat allergies, it would make sense to avoid having those animals as pets, but the evidence is weak regarding this impact, as they are likely to encounter cat and dog dander at low levels at home and in public. If patients already have pets, to which they are allergic, they may want to avoid keeping those pets in the bedroom or frequently used living spaces. Patients with dust mite allergies may find it helpful to decrease the amount of dust in the home by having hardwood floors instead of carpeting, washing sheets in hot water between 55 and 60 °C, keeping a HEPA air filter, and vacuuming and cleaning regularly. Cost should be taken into consideration as many elderly patients are on a fixed budget and some of these measures may be expensive, without any evidence of benefit in randomized clinical trials. No single measure of avoidance is likely to be of benefit, but patients may find multiple measures to be helpful [177].

14.5.4 Increasing Nasal Moisture

One of the most important treatments for rhinitis of any form in older adults is to maintain and improve moisture. Nasal saline is a cheap, safe, and effective treatment for patients. Saline increases mucociliary clearance and improves symptoms [178, 179]. Isotonic saline preparations seem to be the most beneficial with the least side effects [180]. They are typically given four to six times daily. Indeed, some studies suggest that intranasal steroids are no better than saline in this age group [181]. Home humidifiers may also provide benefits.

Guaifenesin is a mucolytic agent that has been available for over 50 years. Although its mechanism of action is not entirely understood, it functions as an expectorant, which helps to increase the volume of mucous production, as well as decrease the viscosity of mucous [182]. It can be used in dosages of up to 2400 mg/day [183]. Many clinicians have found this treatment to be helpful in older patients and side effects are minimal.

14.5.5 Emollients

Emollients (petroleum jelly is perhaps the most common) have been found to decrease crusting and dryness of the nose. In a Swedish study, there was a significant decrease in nasal dryness, followed by nasal stuffiness and crusting, during cold weather months in patients receiving sesame oil versus isotonic sodium chloride solution. All emollients such may cause lipoid pneumonitis if aspirated, and so much be used with caution [184].

14.5.6 Oral and Intranasal Antihistamines

Second- and third-generation antihistamines are very effective in treating rhinitis, sneezing, and nasal and ocular pruritus associated with allergic rhinitis. These agents are less likely to cause sedation when compared to first-generation agents, as they are less lipid soluble, and do not readily cross the blood–brain barrier. First-generation antihistamines have been associated with work-related injuries, decreased mentation, and driving accidents [185], problems that might be exacerbated in older subjects. Diphenhydramine is a well-known first-generation antihistamine, which possesses some anticholinergic properties. Older adults are particularly susceptible to these medications and delirium may result even with low dosages [186]; therefore, this drug should be avoided in older patients.

Intranasal antihistamines are well-tolerated treatments for allergic and nonallergic rhinitis and have a low side-effect profile [187]. Patients may also complain of the bitter taste. Intranasal application of azelastine demonstrates six to eight times lower systemic absorption than oral preparations. Sedation is reported to some degree but less when compared to oral second-generation antihistamines. Intranasal antihistamines are safe treatment and provide relief against both the early- and late-phase allergic responses and nonallergic responses [188]. Newer combination intranasal steroid-intranasal antihistamine preparations are now widely available and appear to show improved efficacy [189].

14.5.7 Intranasal Anticholinergics

Intranasal ipratropium bromide is approved for rhinitis [177]. It decreases rhinorrhea and increases the nose's ability to warm and humidify air [190]. Ipratropium bromide functions by antagonizing acetylcholine transport in parasympathetic efferent nerves, thereby decreasing secretion from submucosal glands [191]. In a study by Malmberg et al., ipratropium bromide decreases nasal secretions and was preferred over placebo by older patients [148]. The side-effect profile is favorable: nasal dryness, bleeding, headache, dry mouth/throat, blurred vision, and urinary hesitancy are rarely observed [192].

14.5.8 Corticosteroids

Intranasal corticosteroids are generally first-line treatment for rhinitis (either allergic or nonallergic) in all ages. They function by decreasing the inflammatory response of the nose, and by inhibiting responses of lymphocytes, eosinophils, mast cells, basophils, neutrophils, monocytes, and macrophages [193]. There are numerous formulations available by prescription and now over the counter in the United States and Europe.

There does not appear to be any increased risk of cataract formation or decreased bone mineral density in adults with the current dosages of intranasal corticosteroids due to their low bioavailability [194, 195]. Care should be taken regarding the one common side effect of these medications, epistaxis, which can be more common in older patient.

Oral corticosteroids are used at times in patients with nasal polyposis or acute/chronic rhinosinusitis to decrease congestion; however, these medications come with systemic risks, including effects on the hypothalamic-pituitary axis, cataract formation, glucose abnormalities, and osteoporosis, and increased intraocular pressures [196].

14.5.9 Decongestants

Decongestants are available in topical or oral preparations. Topical preparations, such as oxymetazoline, should only be used for short periods of time as they may lead to rhinitis medicamentosa in some patients. They function by stimulating the endogenous release of norepinephrine which acts on alpha-receptors to cause vasoconstriction in the nose [197]. Oral decongestants, such as pseudoephedrine, also work by stimulating alpha-receptors but have systemic effects such as urinary retention or elevation in blood pressure that can result in serious complications in older adults. Moreover, pseudoephedrine has been known to potentiate arrhythmias, hypertension, myocardial infarction, and stress cardiomyopathy [198]. It should be avoided in patients with insomnia, prostatic hypertrophy, or glaucoma [145]. Thus, it should be avoided in older adults.

14.5.10 Immunotherapy

As with younger patients, allergen-specific immunotherapy for grass pollen or house dust mites is effective and safe in patients aged 60 years or older with allergic rhinitis [199]. Both tablet preparations (grass, dust, ragweed) and subcutaneous therapies are possible.

14.6 Conclusions

As medicine, technology, and societal change allow for longer life expectancies, the importance of caring for older patients will increase worldwide. Understanding the physiology of the aging nose will aid in better diagnosis and treatment of rhinologic conditions that occur in older adults. Careful attention should be paid to polypharmacy, contributing comorbid conditions, and structural and functional changes with aging. All clinicians should strive to improve knowledge and skills required to care for age-related conditions.

Acknowledgments JMP was supported by the National Institute on Aging (R01 AG067497). YE was supported by the Pritzker School of Medicine Summer Research Program.

References

1. Werner CA. The older population: 2010. US Census Bureau [Internet]. 2011. http://www.census.gov/prod/cen2010/briefs/c2010br-09.pdf.
2. Supiano MA, Alessi C, Chernoff R, Goldberg A, Morley JE, Schmader KE, et al. Department of veterans affairs geriatric research, education, and clinical centers: translating aging research into clinical geriatrics. J Am Geriatr Soc. 2012;60(7):1347–56.
3. American Society of Geriatric Otolaryngology. 2007. [updated Aug 2007]. http://geriatricotolaryngology.org/index.html. Accessed Sep 2012.
4. Kahn DM, Shaw RB. Overview of current thoughts on facial volume and aging. Facial Plast Surg. 2010;26(5):350–5.
5. Cochran CS, Ducic Y, DeFatta RJ. Restorative rhinoplasty in the aging patient. Laryngoscope. 2007;117(5):803–7.
6. Edelstein DR. Aging of the normal nose in adults. Laryngoscope. 1996;106(9 Pt 2):1–25.
7. Hirschberg A, Roithmann R, Parikh S, Miljeteig H, Cole P. The airflow resistance profile of healthy nasal cavities. Rhinology. 1995;33(1):10–3.
8. Patterson CN. The aging nose: characteristics and correction. Otolaryngol Clin North Am. 1980;13(2):275–88.
9. Rohrich RJ, Hollier LJ Jr, Janis JE, Kim J. Rhinoplasty with advancing age. Plast Reconstr Surg. 2004;114(7):1936–44.
10. Quatela VC, Pearson JM. Management of the aging nose. Facial Plast Surg. 2009;25(4):215–21.
11. Kalmovich LM, Elad D, Zaretsky U, Adunsky A, Chetrit A, Sadetzki S, et al. Endonasal geometry changes in elderly people: acoustic rhinometry measurements. J Gerontol A Biol Sci Med Sci. 2005;60(3):396–8.
12. Ganjaei KG, Soler ZM, Mappus ED, Worley ML, Rowan NR, Garcia GJM, et al. Radiologic changes in the aging nasal cavity. Rhinology. 2019;57(2):117–24.
13. Loftus PA, Wise SK, Nieto D, Panella N, Aiken A, DelGaudio JM. Intranasal volume increases with age: computed tomography volumetric analysis in adults. Laryngoscope. 2016;126(10):2212–5.
14. Rotter N, Tobias G, Lebl M, Roy AK, Hansen MC, Vacanti CA, et al. Age-related changes in the composition and mechanical properties of human nasal cartilage. Arch Biochem Biophys. 2002;403(1):132–40.
15. Lee JW, McHugh J, Kim JC, Baker SR, Moyer JS. Age-related histologic changes in human nasal cartilage. JAMA Facial Plast Surg. 2013;15(4):256–62. https://www.liebertpub.com/doi/abs/10.1001/jamafacial.2013.825
16. Riedler KL, Shokrani A, Markarian A, Fisher LM, Pepper J-P. Age-related histologic and biochemical changes in auricular and septal cartilage. Laryngoscope. 2017;127(11):E399–407.
17. Özkoçak V, Özdemir F. Age-related changes in the external noses of the Anatolian men. Aesthetic Plast Surg. 2018;42(5):1336–42. Epub 2018 Jun 11. PMID: 29948097. https://doi.org/10.1007/s00266-018-1170-6.
18. Helal HA, Ghanem MAM, Al-Badawy AM, Abdel Haleem MM, Mousa MH. Histological and anthropometric changes in the aging nose. Aesthet Surg J. 2019;39(9):943–52.

19. Windhager S, Mitteroecker P, Rupić I, Lauc T, Polašek O, Schaefer K. Facial aging trajectories: a common shape pattern in male and female faces is disrupted after menopause. Am J Phys Anthropol. 2019;169(4):678–88.
20. Modabber A, Peters F, Galster H, Möhlhenrich SC, Bock A, Heitzer M, Hölzle F, Kniha K. Three-dimensional evaluation of important surgical landmarks of the face during aging. Ann Anat. 2020;228:151435.
21. Lee HY, Cha S, Ban HJ, Kim IY, Park BR, Kim IJ, Hong KW. The age distribution of facial metrics in two large Korean populations. Sci Rep. 2019;9(1):14564.
22. Creighton EW, Boyette JR. Rhinoplasty in the aging patient. Oper Tech Otolaryngol Head Neck Surg. 2020;31(3):238–44. http://www.sciencedirect.com/science/article/pii/S1043181020300440.
23. Palmer RM. Perioperative care of the elderly patient: an update. Cleve Clin J Med. 2009;76(Suppl 4):S16–21.
24. Stefan M, Iglesia Lino L, Fernandez G. Medical consultation and best practices for preoperative evaluation of elderly patients. Hosp Pract. 2011;39(1):41–51.
25. Ban JH, Kwon HJ, Lee KC. Outcomes of endoscopic sinus surgery in an elderly population: comparison with those in an adult population. Clin Otolaryngol. 2010;35(4):300–6.
26. Schrödter S, Biermann E, Halata Z. Histological evaluation of age-related changes in human respiratory mucosa of the middle turbinate. Anat Embryol. 2003;207(1):19–27.
27. Toppozada H. The human nasal mucosa in the menopause: a histochemical and electron microscopic study. J Laryngol Otol. 1988;102:314–8.
28. Getchell ML, Chen Y, Ding X, Sparks DL, Getchell TV. Immunohistochemical localization of a cytochrome P-450 isozyme in human nasal mucosa: age-related trends. Ann Otol Rhinol Laryngol. 1993;102:368–74.
29. Naclerio RM, Pinto J, Assanasen P, Baroody FM. Observations on the ability of the nose to warm and humidify inspired air. Rhinology. 2007;45(2):102–11.
30. Cole P. Respiratory mucosal vascular responses, air conditioning, and thermoregulation. J Laryngol Otol. 1954;68:613–22.
31. Lindemann J, Sannwald D, Wiesmiller K. Age-related changes in intranasal air conditioning in the elderly. Laryngoscope. 2008;118(8):1472–5.
32. Olsson P, Bende M, Ohlin P. The laser Doppler flowmeter for measuring microcirculation in human nasal mucosa. Acta Otolaryngol. 2009;99(1–2):133–9. https://www.tandfonline.com/doi/pdf/10.3109/00016488509119155?needAccess=true
33. Bende M. Blood flow with 133Xe in human nasal mucosa in relation to age, sex and body position. Acta Otolaryngol. 1983;96(1–2):175–9.
34. Ritchie CS. Oral health, taste, and olfaction. Clin Geriatr Med. 2002;18(4):709–17.
35. Kim SW, Mo JH, Kim JW, Kim DY, Rhee CS, Lee CH, et al. Change of nasal function with aging in Korea. Acta Otolaryngol Suppl. 2007;558:90–4.
36. Ho JC, Chan KN, Hu WH, Lam WK, Zheng L, Tipoe GL, et al. The effect of aging on nasal mucociliary clearance, beat frequency, and ultrastructure of respiratory cilia. Am J Respir Crit Care Med. 2001;163(4):983–8.
37. Paul P, Johnson P, Ramaswamy P, Ramadoss S, Geetha B, Subhashini AS. The effect of ageing on nasal mucociliary clearance in women: a pilot study. ISRN Pulmonol. 2013;2013:e598589. https://www.hindawi.com/journals/isrn/2013/598589/.
38. Sakakura Y, Ukai K, Majima Y, Murai S, Harada T, Miyoshi Y. Nasal mucociliary clearance under various conditions. Acta Otolaryngol. 1983;96(1–2):167–73.
39. Burman WJ, Martin WJ. Oxidant-mediated ciliary dysfunction: possible role in airway disease. Chest. 1986;89(3):410–3.
40. Feldman C, Anderson R, Cockeran R, Mitchell T, Cole P, Wilson R. The effects of pneumolysin and hydrogen peroxide, alone and in combination, on human ciliated epithelium in vitro. Respir Med. 2002;96(8):580–5.
41. de Oliveira-Maul JP, de Carvalho HB, Miyuki Goto D, Mendonça Maia R, Fló C, Barnabé V, Reis Franco D, et al. Aging, diabetes and hypertension are associated with decreased nasal mucociliary clearance. Chest. 2013;143(4):1091–7.
42. Tillmann HC, Laske A, Bernasconi C, Stuck BA. Age determines vascular reactivity as measured by optical rhinometry. Eur J Clin Invest. 2009;39(11):1010–6.
43. Mirza N, Kroger H, Doty RL. Influence of age on the 'nasal cycle'. Laryngoscope. 1997;107(1):62–6.
44. Hoffman HJ, Ishii EK, MackTurk RH. Age-related changes in the prevalence of smell/taste problems among the United States adult population. Results of the 1994 disability supplement to the National Health Interview Survey (NHIS). Ann N Y Acad Sci. 1998;855(1):716–22.
45. Murphy C, Schubert CR, Cruickshanks KJ, Klein BEK, Klein R, Nondahl DM. Prevalence of olfactory impairment in older adults. JAMA. 2002;288(18):2307–12.
46. Stevens JC, Plantinga A, Cain WS. Reduction of odor and nasal pungency associated with aging. Neurobiol Aging. 1982;3(2):125–32.
47. Salzano FA, Guastini L, Mora R, Ellepiane M, Salzano G, Santomauro V, et al. Nasal tactile sensitivity in elderly. Acta Otolaryngol. 2010;130(12):1389–93.
48. Paik SI, Lehman MN, Seiden AM, Duncan HJ, Smith DV. Human olfactory biopsy: the influence of age and receptor distribution. Arch Otolaryngol Head Neck Surg. 1992;118(7):731–8.

49. Suzukawa K, Kondo K, Kanaya K, Sakamoto T, Watanabe K, Ushio M, et al. Age-related changes of the regeneration mode in the mouse peripheral olfactory system following olfactotoxic drug methimazole-induced damage. J Comp Neurol. 2011;519(11):2154–74.
50. Holbrook EH, Wu E, Curry WT, Lin DT, Schwob JE. Immunohistochemical characterization of human olfactory tissue. Laryngoscope. 2011;121(8):1687–701.
51. Rosli Y, Breckenridge LJ, Smith RA. An ultrastructural study of age-related changes in mouse olfactory epithelium. J Electron Microsc (Tokyo). 1999;48(1):77–84.
52. Kondo K, Watanabe K, Sakamoto T, Suzukawa K, Nibu K, Kaga K, et al. Distribution and severity of spontaneous lesions in the neuroepithelium and Bowman's glands in mouse olfactory mucosa: age-related progression. Cell Tissue Res. 2009;335(3):489–503.
53. Watanabe K, Kondo K, Takeuchi N, Nibu K, Kaga K. Age-related changes in cell density and the proliferation rate of olfactory ensheathing cells in the lamina propria of postnatal mouse olfactory mucosa. Brain Res. 2006;1116(1):82–92.
54. Robinson AM, Conley DB, Shinners MJ, Kern RC. Apoptosis in the aging olfactory epithelium. Laryngoscope. 2002;112(8 Pt 1):1431–5.
55. Ueha R, Shichino S, Ueha S, Kondo K, Kikuta S, Nishijima H, et al. Reduction of proliferating olfactory cells and low expression of extracellular matrix genes are hallmarks of the aged olfactory mucosa. Front Aging Neurosci. 2018;10:86. https://www.frontiersin.org/articles/10.3389/fnagi.2018.00086/full#B50.
56. Lafreniere D, Mann N. Anosmia: loss of smell in the elderly. Otolaryngol Clin North Am. 2009;42(1):123–31.
57. Ducray A, Bondier J-R, Michel G, Bon K, Millot JL, Propper A, et al. Recovery following peripheral destruction of olfactory neurons in young and adult mice. Eur J Neurosci. 2002;15(12):1907–17.
58. Weiler E, Farbman AI. Proliferation in the rat olfactory epithelium: age-dependent changes. J Neurosci. 1997;17(10):3610–22.
59. Brämerson A, Johansson L, Ek L, Nordin S, Bende M. Prevalence of olfactory dysfunction: the skövde population-based study. Laryngoscope. 2004;114(4):733–7.
60. Landis BN, Konnerth CG, Hummel T. A study on the frequency of olfactory dysfunction. Laryngoscope. 2004;114(10):1764–9.
61. Wysocki CJ, Gilbert AN. National Geographic Smell Survey. Effects of age are heterogenous. Ann N Y Acad Sci. 1989;561:12–28.
62. Rawal S, Hoffman HJ, Bainbridge KE, Huedo-Medina TB, Duffy VB. Prevalence and risk factors of self-reported smell and taste alterations: results from the 2011-2012 US National Health and Nutrition Examination Survey (NHANES). Chem Senses. 2016;41(1):69–76.
63. Schubert CR, Cruickshanks KJ, Fischer ME, Huang G-H, Klein BEK, Klein R, et al. Olfactory impairment in an adult population: the beaver dam offspring study. Chem Senses. 2012;37(4):325–34.
64. NAMCS. Report of the panel of communicative disorders to the National Advisory Neurological and communicative disorders and stroke council. Washington, DC: Public Health Service; 1979.
65. NIDCD. Quick statistics about taste and smell. [Internet]; 2010. [updated Jun 2010]. http://www.nidcd.nih.gov/health/statistics/smelltaste/Pages/stquickstats.aspx.
66. Mattes RD, Cowart BJ. Dietary assessment of patients with chemosensory disorders. J Am Diet Assoc. 1994;94(1):50–6.
67. Mattes RD, Cowart BJ, Schiavo MA, Arnold C, Garrison B, Kare MR, et al. Dietary evaluation of patients with smell and/or taste disorders. Am J Clin Nutr. 1990;51(2):233–40.
68. Miwa T, Furukawa M, Tsukatani T, Costanzo RM, DeNardo LJ, Reiter ER. Impact of olfactory impairment on quality of life and disability. Arch Otolaryngol Head Neck Surg. 2001;127(5):497–503.
69. Duffy VB, Backstrand JR, Ferris AM. Olfactory dysfunction and related nutritional risk in free-living, elderly women. J Am Diet Assoc. 1995;95(8):879–84.
70. Wolfe JM, Kluender KR, Levi DM, Bartoshuk LM, Herz RS, Klatzky RL, et al. Sensation & perception. 2nd ed. Sunderland: Sinauer Associates; 2008.
71. Santos DV, Reiter ER, DiNardo LJ, Costanzo RM. Hazardous events associated with impaired olfactory function. Arch Otolaryngol Head Neck Surg. 2004;130(3):317–9.
72. Herz RS, Schooler JW. A naturalistic study of autobiographical memories evoked by olfactory and visual cues: testing the Proustian hypothesis. Am J Psychol. 2002;115(1):21–32.
73. Schiffman SS, Sattely-Miller EA, Suggs MS, Graham BG. The effect of pleasant odors and hormone status on mood of women at midlife. Brain Res Bull. 1995a;36:19–29.
74. Schiffman SS, Suggs MS, Sattely-Miller EA. Effect of pleasant odors on mood of males at midlife: comparison of African-American and European-American men. Brain Res Bull. 1995b;36:31–7.
75. Bhutta MF. Sex and the nose: human pheromonal responses. J R Soc Med. 2007;100(6):268–74.
76. Jacob S, McClintock MK, Zelano B, Ober C. Paternally inherited HLA alleles are associated with women's choice of male odor. Nat Genet. 2002;30(2):175–9.
77. Siegel JK, Kung SY, Wroblewski KE, Kern DW, McClintock MK, Pinto JM. Olfaction is associated with sexual motivation and satisfaction in older men and women. J Sex Med. 2021;18(2):295–302.
78. Croy I, Nordin S, Hummel T. Olfactory disorders and quality of life--an updated review. Chem Senses. 2014;39(3):185–94.
79. Neuland C, Bitter T, Marschner H, Gudziol H, Guntinas-Lichius O. Health-related and specific

olfaction-related quality of life in patients with chronic functional anosmia or severe hyposmia. Laryngoscope. 2011;121(4):867–72.

80. Wysocki CJ, Pelchat ML. The effects of aging on the human sense of smell and its relationship to food choice. Crit Rev Food Sci Nutr. 1993;33(1):63–82.
81. Cain WS, Stevens JC. Uniformity of olfactory loss in aging. Ann N Y Acad Sci. 1989;561:29–38.
82. Handley OJ, Morrison CM, Miles C, Bayer AJ. ApoE gene and familial risk of Alzheimer's disease as predictors of odour identification in older adults. Neurobiol Aging. 2006;27(10):1425–30.
83. Hawkes C. Olfaction in neurodegenerative disorder. Adv Otorhinolaryngol. 2006;63:133–51.
84. Kovacs T. Mechanisms of olfactory dysfunction in aging and neurodegenerative disorders. Ageing Res Rev. 2004;3(2):215–32.
85. Schubert CR, Carmichael LL, Murphy C, Klein BE, Klein R, Cruickshanks KJ. Olfaction and the 5-year incidence of cognitive impairment in an epidemiological study of older adults. J Am Geriatr Soc. 2008;56(8):1517–21.
86. Wilson RS, Arnold SE, Tang Y, Bennett DA. Odor identification and decline in different cognitive domains in old age. Neuroepidemiology. 2006;26(2):61–7.
87. Wilson RS, Arnold SE, Schneider JA, Tang Y, Bennett DA. The relationship between cerebral Alzheimer's disease pathology and odour identification in old age. J Neurol Neurosurg Psychiatry. 2007a;78(1):30–5.
88. Wilson RS, Schneider JA, Arnold SE, Tang Y, Boyle PA, Bennett DA. Olfactory identification and incidence of mild cognitive impairment in older age. Arch Gen Psychiatry. 2007b;64(7):802–8.
89. Wilson RS, Arnold SE, Schneider JA, Boyle PA, Buchman AS, Bennett DA. Olfactory impairment in presymptomatic Alzheimer's disease. Ann N Y Acad Sci. 2009;1170:730–5.
90. Hüttenbrink K-B, Hummel T, Berg D, Gasser T, Hähner A. Olfactory dysfunction: common in later life and early warning of neurodegenerative disease. Dtsch Arztebl Int. 2013;110(1–2):1–7.
91. Marin C, Vilas D, Langdon C, Alobid I, López-Chacón M, Haehner A, et al. Olfactory dysfunction in neurodegenerative diseases. Curr Allergy Asthma Rep. 2018;18(8):42.
92. Li KZ, Lindenberger U. Relations between aging sensory/sensorimotor and cognitive functions. Neurosci Biobehav Rev. 2002;26(7):777–83.
93. Riga M, Chelis L, Papazi T, Danielides V, Katotomichelakis M, Kakolyris S. Hyposmia: an underestimated and frequent adverse effect of chemotherapy. Support Care Cancer. 2015;23(10):3053–8.
94. Steinbach S, Hummel T, Böhner C, Berktold S, Hundt W, Kriner M, et al. Qualitative and quantitative assessment of taste and smell changes in patients undergoing chemotherapy for breast cancer or gynecologic malignancies. J Clin Oncol. 2009;27(11):1899–905.
95. Steinbach S, Hundt W, Schmalfeldt B, Böhner C, Berktold S, Wolf P, et al. Effect of platinum-containing chemotherapy on olfactory, gustatory, and hearing function in ovarian cancer patients. Arch Gynecol Obstet. 2012;286(2):473–80.
96. Watabe-Rudolph M, Begus-Nahrmann Y, Lechel A, Rolyan H, Scheithauer MO, Rettinger G. Telomere shortening impairs regeneration of the olfactory epithelium in response to injury but not under homeostatic conditions. PLoS One. 2011;6(11):e27801. Epub 2011 Nov 16.
97. Hutton JL, Baracos VE, Wismer WV. Chemosensory dysfunction is a primary factor in the evolution of declining nutritional status and quality of life in patients with advanced cancer. J Pain Symptom Manage. 2007;33(2):156–65.
98. Shaw AC, Panda A, Joshi SR, Qian F, Allore HG, Montgomery RR. Dysregulation of human toll-like receptor function in aging. Ageing Res Rev. 2011;10(3):346–53.
99. Wang L, Xie Y, Zhu LJ, Change TT, Man YG, Li J. An association between immunosenescence and CD4(+) CD25(+) regulatory T cells: a systematic review. Biomed Environ Sci. 2010;23(4):327–32.
100. Alford RH. Effects of chronic bronchopulmonary disease and aging on human nasal secretion IgA concentrations. J Immunol. 1968;101(5):984–8.
101. Koga T, McGhee JR, Kato H, Kato R, Kiyono H, Fujihashi K. Evidence for early aging in the mucosal immune system. J Immunol. 2000;65(9):5352–9.
102. Mediaty A, Neuber K. Total and specific serum IgE decreases with age in patients with allergic rhinitis, asthma and insect allergy but not in patients with atopic dermatitis. Immun Ageing. 2005;2(1):9.
103. Aspinall R, Andrew D. Thymic involution in aging. J Clin Immunol. 2000;20(4):250–6.
104. Fagnoni FF, Vescovini R, Passeri G, Bologna G, Pedrazzoni M, Lavagetto G, et al. Shortage of circulating naive CD8(+) T cells provides new insights on immunodeficiency in aging. Blood. 2000;95(9):2860–8.
105. Naylor K, Li G, Vallejo AN, Lee W-W, Koetz K, Bryl E, et al. The influence of age on T cell generation and TCR diversity. J Immunol. 2005;174(11):7446–52.
106. Jing Y, Gravenstein S, Rao Chaganty N, Chen N, Lyerly KH, Joyce S, et al. Aging is associated with a rapid decline in frequency, alterations in subset composition, and enhanced Th2 response in CD1d-restricted NKT cells from human peripheral blood. Exp Gerontol. 2007;42(8):719–32.
107. Sandmand M, Bruunsgaard H, Kemp K, Andersen-Ranberg K, Pedersen AN, Skinhøj P. Is ageing associated with a shift in the balance between type 1 and type 2 cytokines in humans? Clin Exp Immunol. 2002;127(1):107–14.
108. Busse PJ, Mathur SK. Age-related changes in immune function: effect on airway inflammation. J Allergy Clin Immunol. 2010;126(4):690–9.
109. Brandt D, Bernstein JA. Questionnaire evaluation and risk factor identification for nonallergic

vasomotor rhinitis. Ann Allergy Asthma Immunol. 2006;96(4):526–32.
110. Georgitis JW. Prevalence and differential diagnosis of chronic rhinitis. Curr Allergy Asthma Rep. 2001;1(3):202–6.
111. Jessen M, Janzon L. Prevalence of non-allergic nasal complaints in an urban and a rural population in Sweden. Allergy. 1989;44(8):582–7.
112. Song W-J, Kim M-Y, Jo E-J, Kim M-H, Kim T-H, Kim S-H, et al. Rhinitis in a community elderly population: relationships with age, atopy, and asthma. Ann Allergy Asthma Immunol. 2013;111(5):347–51.
113. Baptist AP, Nyenhuis SM. Rhinitis of the elderly. In: Bernstein JA, editor. Rhinitis and related upper respiratory conditions: a clinical guide [internet]. Cham: Springer; 2018. p. 123–36. https://doi.org/10.1007/978-3-319-75370-6_12.
114. Fokkens WJ. Thoughts on the pathophysiology of nonallergic rhinitis. Curr Allergy Asthma Rep. 2002;2(3):203–9.
115. Sanico A, Togias A. Noninfectious, nonallergic rhinitis (NINAR): considerations on possible mechanisms. Am J Rhinol. 1998;12(1):65–72.
116. Ventura MT, Gelardi M, D'Amato A, Buquicchio R, Tummolo R, Misciagna G, et al. Clinical and cytologic characteristics of allergic rhinitis in elderly patients. Ann Allergy Asthma Immunol. 2012;108(3):141–4.
117. Lindemann J, Tsakiropoulou E, Konstantinidis I, Lindemann K. Normal aging does not deteriorate nose-related quality of life: assessment with "NOSE" and "SNOT-20" questionnaires. Auris Nasus Larynx. 2010;37(3):303–7.
118. Bousquet J, Khaltaev N, Cruz AA, Denbur J, Fokkens WJ, Togias A, et al. Allergy and its impact on asthma (ARIA) 2008 update. Allergy. 2008;63(Suppl 86):8–160.
119. Bousquet J, van Cauwenberge P, Khaltaev N. Allergic rhinitis and its impact on asthma. J Allergy Clin Immunol. 2001;108(5, Supplement):S147–334.
120. Brożek JL, Bousquet J, Agache I, Agarwal A, Bachert C, Bosnic-Anticevich S, et al. Allergic rhinitis and its impact on asthma (ARIA) guidelines—2016 revision. J Allergy Clin Immunol. 2017;140(4):950–8.
121. Cazzoletti L, Ferrari M, Olivieri M, Verlato G, Antonicelli L, Bono R, et al. The gender, age and risk factor distribution differs in self-reported allergic and non-allergic rhinitis: a cross-sectional population-based study. Allergy Asthma Clin Immunol. 2015;11(1):36.
122. Droste JHJ, Kerkhof M, de Monchy JGR, Schouten JP, Rijcken B. Association of skin test reactivity, specific IgE, total IgE, and eosinophils with nasal symptoms in a community-based population study. J Allergy Clin Immunol. 1996;97(4):922–32.
123. Ha J, Lee SW, Yon DK. Ten-year trends and prevalence of asthma, allergic rhinitis, and atopic dermatitis among the Korean population, 2008–2017. Clin Exp Pediatr. 2020;63(7):278–83.
124. Nam JS, Hwang CS, Hong MP, Kim KS. Prevalence and clinical characteristics of allergic rhinitis in the elderly Korean population. Eur Arch Otorhinolaryngol. 2020;277(12):3367–73. https://doi.org/10.1007/s00405-020-06256-5.
125. Karabulut H, Baysal S, Acar B, Babademez MA, Karasen RM. Allergic rhinitis (AR) in geriatric patients. Arch Gerontol Geriatr. 2011;53(3):270–3.
126. Raherison C, Nejjari C, Marty ML, Filleul L, Barberger-Gateau P, Dartigues JF, et al. IgE level and Phadiatop in an elderly population from the PAQUID cohort: relationship to respiratory symptoms and smoking. Allergy. 2004;59(9):940–5.
127. Saulyte J, Regueira C, Montes-Martínez A, Khudyakov P, Takkouche B. Active or passive exposure to tobacco smoking and allergic rhinitis, allergic dermatitis, and food allergy in adults and children: a systematic review and meta-analysis. PLoS Med. 2014;11(3):e1001611.
128. Di Lorenzo G, Leto-Barone MS, La Piana S, Ditta V, Di Fede G, Rini GB. Clinical course of rhinitis and changes in vivo and in vitro of allergic parameters in elderly patients: a long-term follow-up study. Clin Exp Med. 2013;13(1):67–73.
129. Warm K, Backman H, Lindberg A, Lundbäck B, Rönmark E. Low incidence and high remission of allergic sensitization among adults. J Allergy Clin Immunol. 2012;129(1):136–42.
130. Busse PJ, Lurslurchachai L, Sampson HA, Halm EA, Wisnivesky J. Perennial allergen-specific immunoglobulin E levels among inner-city elderly asthmatics. J Asthma. 2010;47(7):781–5.
131. Ciprandi G, Comite P, Ferrero F, Fontana V, Bruzzone M, Mussap M. Serum allergen-specific IgE, allergic rhinitis severity, and age. Rhinology. 2017;54(3):231–8.
132. Song WJ, Chang YS. Respiratory allergies in the elderly: findings from the Korean longitudinal study on health and aging phase I study (2005-2006). Asia Pac Allergy. 2017;7(4):185–92.
133. Bousquet J, Heinzerling L, Bachert C, Papadopoulos NG, Bousquet PJ, Burney PG, et al. Practical guide to skin prick tests in allergy to aeroallergens. Allergy. 2012;67(1):18–24.
134. Scichilone N, Callari A, Augugliaro G, Marchese M, Togias A, Bellia V. The impact of age on prevalence of positive skin prick tests and specific IgE tests. Respir Med. 2011;105(5):651–8.
135. King MJ, Lockey RF. Allergen prick-puncture skin testing in the elderly. Drugs Aging. 2003;20(14):1011–7.
136. Cardona V, Guilarte M, Luengo O, Labrador Horrillo M, Sala-Cunill A, Garriga T. Allergic diseases in the elderly. Clin Transl Allergy. 2011;1(1):11.
137. Di Lorenzo G, Di Bona D, Belluzzo F, Macchia L. Immunological and non-immunological mechanisms of allergic diseases in the elderly: biological and clinical characteristics. Immun Ageing. 2017;14(1):23.

138. Ventura MT, Scichilone N, Paganelli R, Minciullo PL, Patella V, Bonini M, et al. Allergic diseases in the elderly: biological characteristics and main immunological and non-immunological mechanisms. Clin Mol Allergy. 2017;15(1):2.
139. Schroer B, Pien LC. Nonallergic rhinitis: common problem, chronic symptoms. Cleve Clin J Med. 2012;79(4):285–93.
140. Togias A. Age relationships and clinical features of nonallergic rhinitis. J Allergy Clin Immunol. 1990;85:182.
141. Rondón C, Campo P, Togias A, Fokkens WJ, Durham SR, Powe DG, et al. Local allergic rhinitis: concept, pathophysiology, and management. J Allergy Clin Immunol. 2012;129(6):1460–7.
142. Dykewicz MS, Wallace DV, Amrol DJ, Baroody FM, Bernstein JA, Craig TJ, et al. Rhinitis 2020: a practice parameter update. J Allergy Clin Immunol. 2020;146(4):721–67. http://www.sciencedirect.com/science/article/pii/S009167492031023X
143. Burte E, Leynaert B, Marcon A, Bousquet J, Benmerad M, Bono R, Carsin AE, de Hoogh K, Forsberg B, Gormand F, Heinrich J, Just J, Nieuwenhuijsen M, Pin I, Stempfelet M, Sunyer J, Villani S, Künzli N, Siroux V, Jarvis D, Nadif R, Jacquemin B. Long-term air pollution exposure is associated with increased severity of rhinitis in 2 European cohorts. J Allergy Clin Immunol. 2020;145(3):834–42. e6
144. Lal D, Corey JP. Vasomotor rhinitis update. Curr Opin Otolaryngol Head Neck Surg. 2004;12(3):243–7.
145. Shah R, McGrath KG. Nonallergic rhinitis. Allergy Asthma Proc. 2012;33(Suppl 1):S19–21.
146. Settipane RA, Charnock DR. Epidemiology of rhinitis: allergic and nonallergic. Clin Allergy Immunol. 2007;19:23–34.
147. Van Gerven L, Alpizar YA, Wouters MM, Hox V, Hauben E, Jorissen M, et al. Capsaicin treatment reduces nasal hyperreactivity and transient receptor potential cation channel subfamily V, receptor 1 (TRPV1) overexpression in patients with idiopathic rhinitis. J Allergy Clin Immunol. 2014;133(5):1332–9.
148. Malmberg H, Grahne B, Holopainen E, Binder E. Ipratropium (Atrovent) in the treatment of vasomotor rhinitis of elderly patients. Clin Otolaryngol Allied Sci. 1983;8(4):273–6.
149. Fokkens W, Hellings P, Segboer C. Capsaicin for rhinitis. Curr Allergy Asthma Rep. 2016;16(8):60. https://www.ncbi.nlm.nih.gov/pmc/articles/PMC4971034/
150. Marshak T, Yun WK, Hazout C, Sacks R, Harvey RJ. A systematic review of the evidence base for vidian neurectomy in managing rhinitis. J Laryngol Otol. 2016;130(S4):S7–28.
151. Hwang PH, Lin B, Weiss R, Atkins J, Johnson J. Cryosurgical posterior nasal tissue ablation for the treatment of rhinitis. Int Forum Allergy Rhinol. 2017;7:952–6.
152. Varghese M, Glaum MC, Lockey RF. Drug-induced rhinitis. Clin Exp Allergy. 2010;40(3):381–4.
153. Hicklin LA, Ryan C, Wong DK, Hinton AE. Nosebleeds after sildenafil (viagra). J R Soc Med. 2002;95(8):402–3.
154. Settipane RA. Other causes of rhinitis: mixed rhinitis, rhinitis medicamentosa, hormonal rhinitis, rhinitis of the elderly, and gustatory rhinitis. Immonol Allergy Clin N Am. 2011;31(3):457–67.
155. Liston SL, Siegel LG. Nasal and sinus disorders in the elderly: which ones are life-threatening? Geriatrics. 1981;36(2):91–102.
156. Moore EJ, Kern EB. Atrophic rhinitis: a review of 242 cases. Am J Rhinol. 2001;15(6):355–61.
157. Ferguson JL, McCaffrey TV, Kern EB, Martin WJ. Effect of Klebsiella ozaenae on ciliary activity in vitro: implications in the pathogenesis of atrophic rhinitis. Otolaryngol Head Neck Surg. 1990;102(3):207–11.
158. Wallace DV, Dykewicz MS, Bernstein DI, Blessing-Moore J, Cox L, Khan DA, et al. The diagnosis and management of rhinitis: an updated practice parameter. J Allergy Clin Immunol. 2008;122:S1–84.
159. Hildenbrand T, Weber RK, Brehmer D. Rhinitis sicca, dry nose and atrophic rhinitis: a review of the literature. Eur Arch Otorhinolaryngol. 2011;268(1):17–26.
160. Georgalas C, Jovancevic L. Gustatory rhinitis. Curr Opin Otolaryngol Head Neck Surg. 2012;20(1):9–14.
161. Ellegard EK. Pregnancy rhinitis. Immunol Allergy Clin North Am. 2006;26(1):119–35.
162. Caparroz FA, Gregorio LL, Bongiovanni G, Izu SC, Kosugi EM. Rhinitis and pregnancy: literature review. Braz J Otorhinolaryngol. 2016;82(1):105–11.
163. Philpott CM, Wild DC, Wolstensholme CR, Murty GE. The presence of ovarian hormone receptors in the nasal mucosa and their relationship to nasal symptoms. Rhinology. 2008b;46(3):221–5.
164. Philpott CM, Robinson AM, Murty GE. Nasal pathophysiology and its relationship to the female ovarian hormones. J Otolaryngol Head Neck Surg. 2008a;37(4):540–6.
165. Wei J, Gerlich J, Genuneit J, Nowak D, Vogelberg C, von Mutius E, et al. Hormonal factors and incident asthma and allergic rhinitis during puberty in girls. Ann Allergy Asthma Immunol. 2015;115(1):21–7. e2
166. Colclasure JC, Gross CW, Kountakis SE. Endoscopic sinus surgery in patients older than sixty. Otolaryngol Head Neck Surg. 2004;131(6):946–9.
167. Zacharisen MC. Rhinitis in children, adolescents, the elderly, and pregnant women: special considerations. Immunol Allergy Clin North Am. 2000;20(2):425–44.
168. Hwang CS, Lee HS, Kim SN, Kim JH, Park D-J, Kim K-S. Prevalence and risk factors of chronic rhinosinusitis in the elderly population of Korea. Am J Rhinol Allergy. 2019;33(3):240–6.
169. Renteria AE, Mfuna Endam L, Desrosiers M. Do aging factors influence the clinical presentation and management of chronic rhinosinusitis? Otolaryngol Head Neck Surg. 2017;156(4):598–605.

170. Valdés CJ, Tewfik MA. Rhinosinusitis and allergies in elderly patients. Clin Geriatr Med. 2018;34(2):217–31.
171. Leszczyńska J, Stryjewska-Makuch G, Ścierski W, Lisowska G. Bacterial Flora of the nose and paranasal sinuses among patients over 65 years old with chronic rhinosinusitis who underwent endoscopic sinus surgery. Clin Interv Aging. 2020;15:207–15.
172. Morse JC, Li P, Ely KA, Shilts MH, Wannemuehler TJ, Huang LC, Sheng Q, Chowdhury NI, Chandra RK, Das SR, Turner JH. Chronic rhinosinusitis in elderly patients is associated with an exaggerated neutrophilic proinflammatory response to pathogenic bacteria. J Allergy Clin Immunol. 2019;143(3):990–1002. e6
173. Yancey KL, Lowery AS, Chandra RK, Chowdhury NI, Turner JH. Advanced age adversely affects chronic rhinosinusitis surgical outcomes. Int Forum Allergy Rhinol. 2019;9(10):1125–34.
174. Slavin RG. Treating rhinitis in the older population: special considerations. Allergy Asthma Clin Immunol. 2009;5(1):9.
175. Hofer-Dückelmann C. Gender and polypharmacotherapy in the elderly: a clinical challenge. Handb Exp Pharmacol. 2012;214:169–82.
176. Ajmani GS, Wroblewski KE, Baroody FM, Naclerio RM, Pinto JM. Allergy and asthma medication use in home-dwelling U.S. older adults. Allergy Asthma Immunol Res. 2017;7(2):192–8. https://doi.org/10.1002/alr.21856. Epub 2016 Oct 3
177. Tran NP, Vickery J, Blaiss MS. Management of rhinitis: allergic and non-allergic. Allergy Asthma Immunol Res. 2011;3(3):148–56.
178. Chong LY, Head K, Hopkins C, Philpott C, Glew S, Scadding G, Burton MJ, Schilder AG. Saline irrigation for chronic rhinosinusitis. Cochrane Database Syst Rev. 2016;4:CD011995.
179. Head K, Snidvongs K, Glew S, Scadding G, Schilder AG, Philpott C, Hopkins C. Saline irrigation for allergic rhinitis. Cochrane Database Syst Rev. 2018;6(6):CD012597.
180. Thorton K, Alston M, Dye H, Williamson S. Are saline irrigations effective in relieving chronic rhinosinusitis symptoms? A review of the evidence. J Nurse Pract. 2011;7(8):680–6.
181. Carvalho V, Olej B, Rodrigo de Moraes J, Boechat JL. Mometasone furoate is not superior to saline for chronic rhinitis in the elderly. World Allergy Organ J. 2019;12(10):100064.
182. Storms W, Farrar JR. Guaifenesin in rhinitis. Curr Allergy Asthma Rep. 2009;9(2):101–6.
183. Little D. Allergies in the aging. Geriatr Aging. 2005;8(5):52–3.
184. Johnsen J, Bratt BM, Michel-Barron O, Glennow C, Petruson B. Pure sesame oil vs isotonic sodium chloride solution as treatment for dry nasal mucosa. Arch Otolaryngol Head Neck Surg. 2001;127(11):1353–6.
185. Kay GG, Quig ME. Impact of sedating antihistamines on safety and productivity. Allergy Asthma Proc. 2001;22(5):281–3.
186. Ochs KL, Zell-Kanter M, Mycyk MB, Toxikon Consortium. Hot, blind, and mad: avoidable geriatric anticholinergic delirium. Am J Emerg Med. 2012;30(3):514.
187. Chaaban M, Corey JP. Pharmacotherapy of rhinitis and rhinosinusitis. Facial Plast Surg Clin North Am. 2012;20(1):61–71.
188. McNeely W, Wiseman LR. Intranasal azelastine: a review of its efficacy in the management of allergic rhinitis. Drugs. 1998;56(1):91–114.
189. Debbaneh PM, Bareiss AK, Wise SK, McCoul ED. Intranasal Azelastine and fluticasone as combination therapy for allergic rhinitis: systematic review and meta-analysis. Otolaryngol Head Neck Surg. 2019;161(3):412–8.
190. Assanasen P, Baroody FM, Rouadi P, Naureckas E, Solway J, Naclerio RM. Ipratropium bromide increases the ability of the nose to warm and humidify air. Am J Respir Crit Care Med. 2000;162(3 Pt 1):1031–7.
191. Sapci T, Yazici S, Evcimik MF, Bozkurt Z, Karavus A, Ugurlu B, et al. Investigation of the effects of intranasal botulinum toxin type a and ipratropium bromide nasal spray on nasal hypersecretion in idiopathic rhinitis without eosinophilia. Rhinology. 2008;46(1):45–51.
192. Meltzer EO. Intranasal anticholinergic therapy of rhinorrhea. J Allergy Clin Immunol. 1992;90(6 Pt 2):1055–64.
193. Meltzer EO. The role of nasal corticosteroids in the treatment of rhinitis. Immunol Allergy Clin North Am. 2011;31(3):545–60.
194. Allen DB. Systemic effects of intranasal steroids: an endocrinologist's perspective. J Allergy Clin Immunol. 2000;106(4 Suppl):S179–90.
195. Benninger MS, Ahmad N, Marple BF. The safety of intranasal steroids. Otolaryngol Head Neck Surg. 2003;129(6):739–50.
196. Sastre J, Mosges R. Local and systemic effects of intranasal corticosteroids. J Investig Allergol Clin Immunol. 2012;22(1):1–12.
197. Doshi J. Rhinitis medicamentosa: what an otolaryngologist needs to know. Eur Arch Otorhinolaryngol. 2009;266(5):623–5.
198. Zlotnick DM, Helisch A. Recurrent stress cardiomyopathy induced by Sudafed PE. Ann Intern Med. 2012;156(2):171–2.
199. Bozek A. Pharmacological management of allergic rhinitis in the elderly. Drugs Aging. 2017;34(1):21–8.

15 Nutrition and the Upper Respiratory Tract

Jim Bartley

Key Points

- Type II diabetes increases susceptibility to viral and bacterial infections.
- Zinc supplementation reduces the duration and severity of common cold symptoms.
- In some situations, a cow's milk exclusion diet appears to reduce mucus production.
- Probiotic supplementation is beneficial in upper respiratory infections and may have a role in allergic rhinitis management.
- Vitamin D supplementation has a role in the prevention of upper respiratory infections.

15.1 Introduction

Nutritional deficiencies may lead to cellular dysfunction, illness and disease. Type II diabetic subjects are more susceptible to bacterial and viral infections, including pathogenic agents such as COVID-19 [1]. Increasing evidence indicates that probiotics and vitamin D supplementation have a role in the management of upper respiratory infections [2, 3].

J. Bartley (✉)
Department of Otolaryngology—Head and Neck Surgery, Counties Manukau District Health Board, Auckland, New Zealand

15.2 Probiotics

In children, probiotic consumption may decrease the incidence and illness duration of respiratory tract infections and the number of days absent from school or day care in children [4]. Similar conclusions have also been made when adult patients are included in the meta-analysis [2]. *Bifidobacteria* and *Lactobacilli* are found more commonly in the intestinal flora of healthy children when compared to allergic children [5]. Probiotic bacteria in the intestinal microbiota may protect against atopy. The current evidence does not support the routine use of probiotics in allergic disease prevention [6]; however, probiotics may have a role in allergic rhinitis management [7–9]. Increasing evidence indicates that diet influences lung health [10] and allergic rhinitis [11].

The potential underlying mechanisms of probiotics on respiratory tract infections and allergic rhinitis are not well defined. In vitro, certain normal human upper respiratory flora strains, mainly streptococcal species, prevent pathogenic colonization and infection [12]. The possibility exists that the local application of 'healthy bacteria', through the use of nasal sprays, could prevent upper respiratory infections and aid in allergy management [13, 14]. Probiotics influence both innate and adaptive immune responses. They increase interleukin (IL)-10 expression and decrease inflammatory cytokine expression, such

Ö. Ö. Celebi, T. M. Önerci (eds.), *Nasal Physiology and Pathophysiology of Nasal Disorders*,
https://doi.org/10.1007/978-3-031-12386-3_15

as tumour necrosis factor-α, IL-1β and IL-8 [15]. Lactic acid bacteria and Bifidobacteria are the most common types of probiotics [2]. The influence of probiotics on respiratory tract health is an area of ongoing research. Debate continues about the most appropriate probiotic strains and dosage regimens to use in a clinical situation.

15.3 Iron

Bacteria need iron for the purposes of respiration, DNA synthesis and free radical-scavenging mechanisms. The iron-binding proteins, transferrin and lactoferrin help maintain low iron levels in nasal mucus, which helps protect against microbial infection. In the host, iron deficiency significantly impairs cell proliferation and immune function [16]. Iron deficiency is not associated with an increased risk of acute lower respiratory tract infection [17], indicating that it is probably an upper respiratory infection risk factor. The results of iron supplementation in respiratory tract infection management are mixed [16].

15.4 Vitamin A

Vitamin A is required for cellular growth, epithelial integrity, the production of red blood cells and immunity. Vitamin A deficiency increases the susceptibility to a number of illnesses including diarrhoea, measles and lower respiratory infections but not upper respiratory infections. While vitamin A supplementation reduces morbidity and mortality in children, no significant influence on the incidence of respiratory disease or hospitalisations due to diarrhoea or pneumonia has been noted [18, 19].

Cod liver oil, as well as a children's multivitamin/mineral supplement with selenium and other trace metals, reduced paediatric visits for upper respiratory illness during the winter and early spring by 36–58%. Cod liver oil not only contains vitamin A but also contains vitamin D and omega-3 fatty acids, which makes it difficult to totally attribute these results to vitamin A [20].

15.5 Omega-3

Omega-3 and omega-6 oils are essential fatty acids oils, which play a role in the inflammatory response. The results of epidemiological studies investigating maternal fish intake during pregnancy and allergic outcomes in infants/children of those pregnancies are inconsistent. Interventional studies studying oily fish consumption or fish oil supplementation in pregnancy to prevent infant and childhood allergic diseases are also inconsistent [21, 22]. Omega-3 fatty acid supplementation alone does not improve allergic rhinitis symptoms [23]. In Samter's triad (salicylate intolerance, asthma and nasal polyps) patients, high-dose omega-3 supplementation may be useful [24].

15.6 Zinc

Zinc has numerous roles in the immune response. Zinc is crucial for the normal development and function of cells such as neutrophils and natural killer cells that mediate innate immunity. Zinc is also involved in T and B lymphocyte functions as well as Th1 cytokine production. The macrophage, in particular, is adversely affected by zinc deficiency [16]. Zinc taken daily reduces the incidence of the common cold in young children and both the duration and severity of symptoms once one has developed a cold [25]. There has been some discussion of this meta-analysis [26]. In adults, zinc supplementation reduces the duration and severity of common cold symptoms [27]. The role of zinc supplementation in the prevention of lower respiratory infection has not been translated into the possible prevention of upper respiratory bacterial infection [28, 29].

15.7 Milk

Excessive milk consumption has a long association with increased respiratory tract mucus production and asthma [30]. In the human colon, ß-casomorphin-7 (ß-CM-7), an exorphin derived from the breakdown of A1 milk, stimulates mucus production from gut MUC5AC glands [31]. In the presence of inflammation, similar mucus overproduction from respiratory tract MUC5AC glands characterizes many respiratory tract diseases [32, 33]. ß-CM-7 from the bloodstream could stimulate the production and secretion of mucus production from these respiratory glands. A number of studies document that asthma symptoms often improve where milk is excluded from the diet [34–39]. Recent studies indicate that patients with perceived cow's milk hypersensitivity have increased non-type 2 inflammation and airway hyper-responsiveness [40] and that a dairy-free diet reduces the sensation of postnasal mucus [41]. These studies support the clinical observation that in some situations a cow's milk exclusion diet may aid nasal symptom management.

15.8 Vitamin D

Vitamin D (25(OH)D) deficiency is common around the world. Vitamin D is made largely by sun exposure [42]. Vitamin D has important roles in both innate and adaptive immunity [43]. Vitamin D has direct antiviral effects primarily against enveloped viruses; coronavirus is an enveloped virus. Blood vitamin D status can influence the risk of being infected with COVID-19, the seriousness of COVID-19 and mortality from COVID-19 [44]. Vitamin D may have an important role in oral health [45]. Historically, supplementation with cod liver oil (containing vitamin D) reduced upper respiratory tract infection frequency [46, 47]. Vitamin D is a hormone, which may be given in supraphysiological bolus doses. High circulating concentrations after bolus dosing may chronically dysregulate the activity of enzymes responsible for the synthesis and degradation of the active vitamin D metabolite 1,25-dihydroxyvitamin D, resulting in decreased concentrations and limited effectiveness of this metabolite in extra-renal tissues [48]. Vitamin D supplementation is safe and protects against acute respiratory infections. Very deficient individuals and those not receiving bolus doses experience the most benefit [3].

15.9 Type 2 Diabetes

Type II diabetes increases viral and bacterial infection susceptibility [49]. Most Type II diabetes respiratory research has focused on the lower respiratory tract. Obesity can lead to obstructive sleep apnoea and obesity-hypoventilation syndrome. Poor quality sleep is associated with reduced immune function [50]. The adaptive immune response is altered in patients with obesity and type 2 diabetes [51]. Obesity is associated with insulin resistance and chronic low-grade inflammation. Both obesity and Type 2 diabetes are associated with an increased risk of recurrent and secondary infections. Obese individuals have a higher risk of community-acquired pneumonia, cutaneous infections and aspiration pneumonia during hospitalizations [52]. Poor diabetic control is associated with increased infection risk [53]. Obesity is independently associated with high rates of *Staphylococcus aureus* nasal carriage—a proven risk factor for surgical-site infections [49] and for chronic rhinosinusitis [54]. Type 2 diabetic patients are more likely to have nasal polyps, positive *Pseudomonas aeruginosa* and other gram-negative rods isolated from sinus cultures and significantly less clinical improvement 6 months after sinus surgery [55]. Improving Type II diabetic control may be important in managing patients with difficulty sinusitis [53].

15.10 Conclusions

Nutritional deficiencies may lead to cellular dysfunction and disease. Poor diabetic control is associated with increased infection risk. A milk exclusion diet may reduce respiratory tract mucus production. Increasing evidence indicates that probiotic supplementation is beneficial in upper respiratory infections and allergic rhinitis management. Regular vitamin D supplementation has a role in the prevention of upper respiratory infections.

References

1. Daryabor G, Atashzar MR, Kabelitz D, Meri S, Kalantar K. The effects of type 2 diabetes mellitus on organ metabolism and the immune system. Front Immunol. 2020;11:1582.
2. Hao Q, Dong BR, Wu T. Probiotics for preventing acute upper respiratory tract infection. Cochrane Database Syst Rev. 2015;9:CD006895.
3. Martineau AR, Jolliffe DA, Greenberg L, Aloia JF, Bergman P, Dubnov-Raz G, et al. Vitamin D supplementation to prevent acute respiratory infections: individual participant data meta-analysis. Health Technol Assess. 2019;23:1–44.
4. Wang Y, Li X, Ge T, Xiao Y, Liao Y, Cui Y, et al. Probiotics for prevention and treatment of respiratory tract infections in children: a systematic review and meta-analysis of randomized controlled trials. Medicine (Baltimore). 2016;95:e4509.
5. Lambrecht BN, Hammad H. The immunology of the allergy epidemic and the hygiene hypothesis. Nat Immunol. 2017;18:1076–83.
6. Wang HT, Anvari S, Anagnostou K. The role of probiotics in preventing allergic disease. Children (Basel). 2019;6:24.
7. Güvenç IA, Muluk NB, Mutlu FŞ, Eşki E, Altıntoprak N, Oktemer T, et al. Do probiotics have a role in the treatment of allergic rhinitis? A comprehensive systematic review and meta-analysis. Am J Rhinol Allergy. 2016;30:157–75.
8. Zajac AE, Adams AS, Turner JH. A systematic review and meta-analysis of probiotics for the treatment of allergic rhinitis. Int Forum Allergy Rhinol. 2015;5:524–32.
9. Jalali MM, Soleimani R, Foumani AA, Khosravi HG. Add-on probiotics in patients with persistent allergic rhinitis: a randomized crossover clinical trial. Laryngoscope. 2019;129:1744–50.
10. Hanson C, Lyden E, Rennard S, Mannino DM, Rutten EP, Hopkins R, et al. The relationship between dietary fiber intake and lung function in the National Health and nutrition examination surveys. Ann Am Thorac Soc. 2016;13:643–50.
11. Cingi C, Conk-Dalay M, Cakli H, Bal C. The effects of spirulina on allergic rhinitis. Eur Arch Otorhinolaryngol. 2008;265:1219–23.
12. Glück U, Gebbers J-O. Ingested probiotics reduce nasal colonization with pathogenic bacteria (*Staphylococcus aureus, Streptococcus pneumoniae,* and *β-hemolytic streptococci).* Am J Clin Nutr. 2003;77:517–20.
13. Miraglia Del Giudice M, Indolfi C, Capasso M, Maiello N, Decimo F, Ciprandi G. Bifidobacterium mixture (B longum BB536, B infantis M-63, B breve M-16V) treatment in children with seasonal allergic rhinitis and intermittent asthma. Ital J Pediatr. 2017;43:25.
14. Cervin AU. The potential for topical probiotic treatment of chronic rhinosinusitis, a personal perspective. Front Cell Infect Microbiol. 2017;7:530.
15. Maldonado Galdeano C, Cazorla SI, Lemme Dumit JM, Velez E, Perdigon G. Beneficial effects of probiotic consumption on the immune system. Ann Nutr Metab. 2019;74:115–24.
16. Gombart AF, Pierre A, Maggini S. A review of micronutrients and the immune system-working in harmony to reduce the risk of infection. Nutrients. 2020;12:236.
17. Gera T, Sachdev H. Effect of iron supplementation on incidence of infectious illness in children: systematic review. BMJ. 2002;325:1142.
18. Imdad A, Mayo-Wilson E, Herzer K, Bhutta ZA. Vitamin a supplementation for preventing morbidity and mortality in children from six months to five years of age. Cochrane Database Syst Rev. 2017;3:CD008524.
19. Chen H, Zhuo Q, Yuan W, Wang J, Wu T. Vitamin A for preventing acute lower respiratory tract infections in children up to seven years of age. Cochrane Database Syst Rev. 2008;1:CD006090.
20. Linday L. Cod liver oil, young children, and upper respiratory tract infections. J Am Coll Nutr. 2010;29:559–62.
21. Miles EA, Calder PC. Can early omega-3 fatty acid exposure reduce risk of childhood allergic disease? Nutrients. 2017;9:784.
22. Schindler T, Sinn JK, Osborn DA. Polyunsaturated fatty acid supplementation in infancy for the prevention of allergy. Cochrane Database Syst Rev. 2016;10:CD010112.
23. Thien F, Menciahuerta J, Lee T. Dietary fish-oil effects on seasonal hay-fever and asthma in pollen-sensitive subjects. Am Rev Respir Dis. 1993;147:1138–43.
24. Healy E, Newell L, Howarth P, Friedmann PS. Control of salicylate intolerance with fish oils. Br J Dermatol. 2008;159:1368–9.
25. Singh M, Das RR. Zinc for the common cold. Cochrane Database Syst Rev. 2013;2:CD001364.
26. Science M, Johnstone J, Roth DE, Guyatt G, Loeb M. Zinc for the treatment of the common cold: a systematic review and meta-analysis of randomized controlled trials. CMAJ. 2012;184:E551–E61.
27. Wang MX, Win SS, Pang J. Zinc supplementation reduces common cold duration among healthy adults: a systematic review of randomized controlled trials with micronutrients supplementation. Am J Trop Med Hyg. 2020;103:86–99.
28. Rerksuppaphol S, Rerksuppaphol L. A randomized controlled trial of zinc supplementation in the treatment of acute respiratory tract infection in Thai children. Pediatr Rep. 2019;11:7954.
29. Aggarwal R, Sentz J, Miller MA. Role of zinc administration in prevention of childhood diarrhea and respiratory illnesses: a meta-analysis. Pediatrics. 2007;119:1120–30.
30. Bartley J, McGlashan SR. Does milk increase mucus production? Med Hypotheses. 2010;74:732–4.
31. Zoghbi S, Trompette A, Claustre J, El Homsi M, Garzon J, Jourdan G, et al. Beta-Casomorphin-7 regulates the secretion and expression of gastrointestinal mucins through a mu-opioid pathway. Am J Physiol Gastrointest Liver Physiol. 2006;290:G1105–13.

32. Kirkham S, Sheehan J, Knight D, Richardson P, Thornton D. Heterogeneity of airway mucus variations in the amounts and glycoforms of the major oligomeric mucins MUC5AC and MUC5B. Biochem J. 2002;361:537–46.
33. Ding G, Zheng C. The expression of MUC5AC and MUC5B mucin genes in the mucosa of chronic rhinosinusitis and nasal polyposis. Am J Rhinol. 2007;21:359–66.
34. Rowe AH, Rowe A. Bronchial asthma in adults. Calif Med. 1950;72:228–33.
35. Rowe AH, Rowe A. Allergic bronchial asthma. The importance of studies for sensitivity to foods. Calif Med. 1956;85:33–5.
36. Rowe AH, Rowe A, Young J. Bronchial asthma due to food allergy alone in ninety-five patients. JAMA. 1959;169:1158–62.
37. Egger J, Carter CM, Wilson J, Turner MW, Soothill J. Is migraine food allergy? A double-blind controlled trial of oligoantigenic diet treatment. Lancet. 1983;2:865–9.
38. Iacono G, Cavataio F, Montalto G, Florena A, Tumminello M, Soresi M, et al. Intolerance of cow's milk and chronic constipation in children. N Engl J Med. 1998;339:1100–4.
39. Yusoff NAM, Hampton SM, Dickerson JWT, Morgan JB. The effects of exclusion of dietary egg and milk in the management of asthmatic children: a pilot study. J Royal Soc Promot Health. 2004;124:74–80.
40. Tsolakis N, Nordvall L, Janson C, Rydell N, Malinovschi A, Alving K. Characterization of a subgroup of non-type 2 asthma with cow's milk hypersensitivity in young subjects. Clin Transl Allergy. 2019;9:12.
41. Frosh A, Cruz C, Wellsted D, Stephens J. Effect of a dairy diet on nasopharyngeal mucus secretion. Laryngoscope. 2019;129:13–7.
42. Holick M. Vitamin D deficiency. N Engl J Med. 2007;357:266–81.
43. Bartley J. Vitamin D: emerging roles in infection and immunity. Expert Rev Anti-Infect Ther. 2010;8:1359–69.
44. Yisak H, Ewunetei A, Kefale B, Mamuye M, Teshome F, Ambaw B, et al. Effects of vitamin D on COVID-19 infection and prognosis: a systematic review. Risk Manag Healthc Policy. 2021;14:31–8.
45. Uwitonze AM, Murererehe J, Ineza MC, Harelimana EI, Nsabimana U, Uwambaye P, et al. Effects of vitamin D status on oral health. J Steroid Biochem Mol Biol. 2018;175:190–4.
46. Holmes A, Pigott M, Sawyer W, Comstock L. Vitamins aid reduction of lost time in industry. J Indust Eng Chem. 1932;24:1058–60.
47. Holmes A, Pigott M, Sawyer W, Comstock L. Cod liver oil - a five year study of its value for reducing industrial absenteeism caused by colds and respiratory diseases. Indust Med. 1936;5:359–61.
48. Vieth R. How to optimize vitamin D supplementation to prevent cancer, based on cellular adaptation and hydroxylase enzymology. Anticancer Res. 2009;29:3675–84.
49. Falagas ME, Kompoti M. Obesity and infection. Lancet Infect Dis. 2006;6:438–46.
50. Besedovsky L, Lange T, Haack M. The sleep-immune crosstalk in health and disease. Physiol Rev. 2019;99:1325–80.
51. Frydrych LM, Bian G, O'Lone DE, Ward PA, Delano MJ. Obesity and type 2 diabetes mellitus drive immune dysfunction, infection development, and sepsis mortality. J Leukoc Biol. 2018;104:525–34.
52. Andersen CJ, Murphy KE, Fernandez ML. Impact of obesity and metabolic syndrome on immunity. Adv Nutr. 2016;7:66–75.
53. Critchley JA, Carey IM, Harris T, DeWilde S, Hosking FJ, Cook DG. Glycemic control and risk of infections among people with type 1 or type 2 diabetes in a large primary care cohort study. Diabetes Care. 2018;41:2127–35.
54. Vickery TW, Ramakrishnan VR, Suh JD. The role of *Staphylococcus aureus* in patients with chronic sinusitis and nasal polyposis. Curr Allergy Asthma Rep. 2019;19:21.
55. Zhang Z, Adappa ND, Lautenbach E, Chiu AG, Doghramji L, Howland TJ, et al. The effect of diabetes mellitus on chronic rhinosinusitis and sinus surgery outcome. Int Forum Allergy Rhinol. 2014;4:315–20.

16 Physiology of Lacrimal Drainage

Ali Riza Cenk Çelebi and Özlem Önerci Celebi

Core Messages

- There are many factors contributing to lacrimal elimination, but the most important mechanism is the canalicular and sac pump mechanism.
- Canalicular pump is probably more important than the sac pump because, following DCR, tears are still drained through the canaliculi to the nose.
- The pressure gradient between the canaliculi and the sac cannot be produced if the canaliculus is slit open. Therefore, the lacrimal canaliculi should be preserved and should not be damaged.
- Tear elimination is equivalent through the upper and lower canalicular systems. Therefore, attention should be given not to damage both the upper and lower canaliculus.

The lacrimal drainage system works to remove those tears secreted into the palpebral aperture to cover the cornea at a rate of 1.2 μl/min with a total 24-h secretory volume of approximately 10 ml [1]. The tear film travels across the surface of the globe and eyelids, enters the puncta/ampulla, passes through the canaliculi, and enters the lacrimal sac/nasolacrimal duct/nasal passages. With blinking (orbicularis muscle contraction), the closure of palpebral aperture starts from the lateral and proceeds to the medial. This action propels the tears medially toward the lacrimal lake [1].

Factors contributing to lacrimal elimination may include:

- Evaporation of tears from the ocular surface.
- Capillary attraction of the tears.
- Reservoir drainage into the lacrimal sac (so-called Krehbiel flow).
- Siphon effect.
- Microciliation and absorption of tears by the lacrimal sac mucosa.
- Bernoulli's principle and Venturi tube effect.
- Physical forces such as gravity.
- Canalicular and sac pump mechanism.

16.1 Factors in Tear Flow

16.1.1 Evaporation

Much of the tears is lost by direct evaporation from the ocular surface. Low humidity and wind increase this loss. According to Schirmer, almost half of the secreted tears were lost by evaporation [1].

A. R. C. Celebi (✉)
Department of Ophthalmology, School of Medicine, Acibadem University, Istanbul, Turkey

Ö. Ö. Çelebi
Department of Otorhinolaryngology, Faculty of Medicine, University of Health Sciences, Istanbul, Turkey

Ö. Ö. Celebi, T. M. Önerci (eds.), *Nasal Physiology and Pathophysiology of Nasal Disorders*,
https://doi.org/10.1007/978-3-031-12386-3_16

16.1.2 Capillarity

Capillarity or capillary action is the ability of a liquid to flow in a narrow tube without the assistance of gravity. This effect can be seen in the drawing of liquids in a thin tube or in porous materials such as paper. Capillary action can be noticed in the drainage of tears from the eye. The small canaliculi may act like a capillary tube. The canaliculi may draw tears through the punctum and transfer tears through the canaliculi. The fact that trauma to the canaliculi and loss of capillarity does not cause loss of function indicates that capillarity is not the only factor in drawing tears through the punctum [1].

16.1.3 Krehbiel Flow

Krehbiel flow is the flow of the tears from the punctum through the canaliculus due to changes of pressure within the lacrimal sac owing to the effect of the orbicularis tonus on both canaliculi and tear sac when the lids are open [1, 2].

16.1.4 Siphon Effect

The word siphon refers specifically to a tube in an inverted U shape which causes a liquid to flow uphill, above the surface of the reservoir, without pumps, powered by the fall of the liquid as it flows down the tube under the pull of gravity, and is discharged at a level lower than the surface of the reservoir. It is important that while the siphon must touch the liquid in the (upper) reservoir (the surface of the liquid must be above the intake opening), it need not touch the liquid in the lower reservoir and indeed there need not be a lower reservoir—liquid can discharge into midair.

16.1.5 Microciliation and Reabsorption

The internal wall of the lacrimal canaliculi is lined by a stratified epithelium. Epithelial cells are faced by microvilli. The facing of epithelial cells by microvilli gives hints of reabsorption of lacrimal fluid inside the lacrimal ducts [3]. Due to this reabsorption in the nasolacrimal sac and duct, the amount of tears leaving the nasolacrimal duct orifice in the nose is less than the amount of tears entering the puncta.

16.1.6 Bernoulli's Principle and Venturi Tube Effect

The relationship between the velocity and pressure exerted by a moving liquid is described by the Bernoulli's principle: as the velocity of a fluid increases, the pressure exerted by that fluid decreases. The Venturi effect is similar to Bernoulli's principle. The velocity of the fluid increases as the cross-sectional area decreases, with the static pressure correspondingly decreasing. According to the laws governing fluid dynamics, a fluid's velocity must increase as it passes through a constriction to satisfy the principle of continuity, while its pressure must decrease to satisfy the principle of conservation of mechanical energy. An equation for the drop in pressure due to the Venturi effect may be derived from a combination of Bernoulli's principle and the continuity equation.

The canaliculi narrow close to the common canaliculus, and the common canaliculus is a larger structure. Bernoulli principle and Venturi tube effect may play a role in the flow through the canaliculi. According to the Venturi tube effect, narrowing of the canaliculi from lateral to medial increases the speed of flow from lateral to medial and according to Bernoulli's principle movement over a low-pressure area creates a suctional effect. Bernoulli's principle may also play a role in the lacrimal system at the nasal cavity sucking tears from the valve of Hasner area into the nose in addition to the ampulla and just distal to the common internal punctum [4].

16.2 Tear Flow and Elimination

16.2.1 Flow from the Lacrimal Lake Through the Puncta

The tears enter the puncta with three mechanisms [1]:

1. A negative pressure would develop inside the punctum to suck the tears.
2. The small canaliculi may act as capillary tubes and would suction the tears through the small capillary tubes. However, capillarity is not the only factor in drainage because a slit canaliculus where capillarity has been destroyed usually functions well for other reasons.
3. Krehbiel's effect (Reservoir drainage into the lacrimal sac): Krehbiel suggested that even in the resting phase of the blink cycle, tears pass from the punctum through into the canaliculus. This may be due to changes in pressure within the lacrimal sac owing to the effect of the orbicularis tonus on both canaliculi and tear sac when the lids are open. However, if a DCR is performed, this effect may not be seen and therefore it is probably not the intracanalicular suction that is causing this effect, but sac suction [1, 2].

16.2.2 What Canalicular System Is more Important for Tear Elimination: Upper or Lower?

Although it is believed that the upper canalicular system is unimportant, experimental and clinical studies show that tear elimination is equivalent through the upper and lower canalicular systems [5–8]. Surgeons should thus give equal consideration to a patient with lacerations of either the upper or lower canaliculus. Studies by White et al. [8] and Daubert et al. [5] have demonstrated equal tear flow between the upper and lower canalicular systems using radioactive dacryoscintigraphy flow studies. Meyer et al. [7] studied fluorescein dye disappearance in 20 subjects and found that 90% of patients showed minimal or no impairment with monocanalicular (either upper or lower) obstruction.

16.2.3 Flow Through the Canaliculi into the Sac

Although multiple mechanisms may contribute to lacrimal outflow, present evidence suggests that the most important factor is the active palpebral-canalicular pump. It has long been noted that the blinking mechanism readily drains tears even with the head held in an inverted position. When the palpebral blink mechanism is impaired, however, epiphora is common, such as in patients with facial paralysis.

16.2.4 Lacrimal Pump

There are two most popular lacrimal pump theories: one suggested by Jones [9] and the other by Doane [10]. More recently, Becker [11] proposed a tricompartmental model of the lacrimal pump, which in many ways is similar to the Doane model. The lacrimal pump models agree that eyelid closure results in a squeezing of the canaliculi with the nasal movement of tears into the lacrimal sac. The models diverge, however, in the analysis of the changes in the lacrimal sac pressure with eyelid closure and opening.

The first "lacrimal pump" theory is based on classic anatomic studies by Jones [9], describing tendinous and muscular insertions exerting their action on and around the lacrimal sac. The Jones theory for this lacrimal pump involves three components:

1. The deep heads of the pretarsal orbicularis muscle (Horner's muscle).
2. The deep head of the preseptal muscle (Jones' muscle).
3. The lacrimal diaphragm (fascia around the sac).

The tensor tarsi (Horner's) muscle originates on the posterior lacrimal crest and divides to surround the canaliculi. It then becomes continuous with the pretarsal portions of the orbicularis muscle. Since some fibers of this muscle run in a parallel and sometimes spiral manner, the contraction of the muscle can draw the papillae of the puncta in a medial direction. This narrows the ampullae and shortens the canaliculi [12]. The Horner's muscle around the canaliculi pumps tears from the punctum through to the sac [2].

An additional strand of orbicularis muscle from the preseptal area inserting into the lacrimal fascia and posterior lacrimal crest (the deep head of the preseptal orbicularis muscle) was described by Jones and this muscle is named as Jones' muscle. According to Jones, the muscular pull of this preseptal orbicularis muscle (Jones' muscle) on the lacrimal sac draws the lateral wall of the nasolacrimal sac laterally and creates a negative pressure within the sac [13].

With blinking, contraction of the deep preseptal orbicularis fibers (Jones' muscle) draws the lateral wall of the nasolacrimal sac laterally, creating a negative pressure within the sac and allowing the inspiration of tears into the sac. The tears are forced along the canalicular system by contraction of the deep head of the pretarsal muscle (Horner's muscle). When the orbicularis relaxes and the eyelid opens, the sac collapses, forcing tears down the nasolacrimal duct. At the same time, the canaliculi open, siphoning tears into their lumen. Closing the eyelids again pushes and propels the accumulated tears into the lacrimal sac [14].

Doane suggested a different mechanism of tear propulsion through the system. He noted that the puncta came together during the early phases of eyelid closure and occlusion of the puncta occurred as a first step in the tear pump. He postulated that contraction of the pretarsal orbicularis oculi muscle exerts lateral traction on the lacrimal sac wall, compresses the ampulla, and shortens the canaliculi, causing a pressure increase in the canaliculi propelling tear fluid within the canaliculi toward the lacrimal sac (i.e., positive pressure is created during a blink in both the canaliculi and the nasolacrimal sac as a result of muscle contraction occurring in the pretarsal and preseptal orbicularis fibers) [10, 14]. Doane further theorized that as the tension increases on the lacrimal fascia to open the fundus of the sac, the inferior portion closes more tightly, preventing aspiration of air from the nose. As the eyelids open, the puncta initially remain closed by the opposing lid until the end of the opening movement, and partial vacuum forms within the membranous lacrimal conduit. As the eyelid-opening phase of the blink continues, the two lacrimal puncta open and expose the adjacent lacrimal lake to this partial vacuum. Tears rapidly flow into the canaliculi during the 1–3-s interval immediately after the blink. Once again, the canaliculi fill with fluid so that the pumping action of the next blink can continue the lacrimal elimination cycle. With the relaxation of the deep head of the preseptal orbicularis muscle, elastic recoil of the lacrimal fascia collapses the lacrimal sac, expelling any fluid within the sac down into the now patent nasolacrimal duct. Thus, the collapsing lacrimal drainage conduit was believed to push the tears through the system into the nose without the suction phase postulated by Jones [10]. To date, most evidence supports the Doane model [15].

Becker observed that the superolateral wall of the lacrimal sac, which is attached to the deep head of the preseptal orbicularis, moved laterally with lid closure and medially with lid opening. The inferior half of the lateral wall of the lacrimal sac moved medially with lid closure and laterally with lid opening. Becker suggested a tricompartment model of the lacrimal pump that incorporates these findings. With lid closure, the orbicularis muscle contracts, compressing the canaliculi and pulling the superior half of the lateral wall of the lacrimal sac laterally. This creates a lower pressure in the superior sac, allowing tears to be propelled from the canaliculi into the sac. At the same time, the inferior half of the lateral sac wall moves medially, creating a positive pressure in the inferior sac and nasolacrimal duct, thus forcing tears down the duct into the nose. With lid opening, the orbicularis muscle relaxes, allowing the canaliculi to open and the superior half of the lateral sac wall to move medially. The resulting negative intracanalicular pressure allows tears to flow from the lacrimal lake into the canaliculi, and the higher pressure in the superior sac closes the valve of Rosenmüller and forces tears from the superior to inferior sac and proximal nasolacrimal duct. At the same time, the inferior half of the lateral sac wall moves laterally, resulting in negative pressure in the inferior sac and nasolacrimal duct [11]. These observations are in agreement with Doane's model, with the overall lacrimal sac pressure increasing with eyelid closure and reducing with eyelid opening.

These proposed lacrimal sac pumping mechanisms are based on anatomic studies and likely do not have a large role in normal lacrimal elimination, because the system functions quite well with the lacrimal sac completely open, as is the case after dacryocystorhinostomy (DCR). This shows that the canalicular pump is more important than the sac pump [1].

16.2.5 Flow from the Sac to the Nose

The flow of tears from the sac down through the duct has been postulated to be a siphoning effect and a gravitational effect. The fact that tears will flow through the lacrimal system even when one is standing on his head, means that it does not only depend on the gravitational effect but some other active mechanisms as well. The filling of the sac and the increasing pressure in the sac force the tears down through the duct. Each blink expels the fluid through the canaliculi to the sac and the sac expels the fluid to the duct [1].

16.3 Other Factors on Tear Flow

16.3.1 Effect of Respiration

It was postulated that respiration also plays a role in drainage of tears from the duct into the nose and that Bernoulli's principle has some effect on this function. However, since the duct narrows as it approaches to the Hasner valve, Bernoulli's effect is minimal. On the other hand, after DCR operation, common canaliculus opens directly into the nose, and respiration and Bernoulli's principle plays a much more important role in these operated cases [16].

16.3.2 Valves

The function of the valves is to prevent or decrease the retrograde flow of tears and/or air currents. The most important valve is Hasner valve which is located at the lower end of the nasolacrimal duct. Hasner valve prevents air currents from within the nose being drawn up into the lacrimal duct [1].

Rosenmüller valve is located at the common internal punctum and prevents backflow from the sac into the canaliculi. This valve is not a real valve but it functions like a valve because of the anatomic angulation of the canaliculi and common canaliculus. This valve is especially important after DCR operation to prevent backflow of the tears from the nose into the canaliculi [17].

Although other valves at the punctum, the ampulla, and at the junction of the nasolacrimal sac and duct have been described in the literature, these are mainly mucosal folds and do not have much function.

16.3.3 Clinical Principles Derived from the Physiologic Information

The palpebral-canalicular pump mechanism in lacrimal elimination is the major mechanism. The canalicular pump is probably more important than the sac pump because following DCR tears are still drained through the canaliculi to the nose. In facial nerve paralysis, tears will not drain into the nose because orbicularis muscle is not functioning and cannot operate the canalicular pump although there is a patent opening. Therefore, the lacrimal canaliculi should be preserved and should not be damaged. Repeated instrumentation of the lacrimal system or nasolacrimal duct probings may injure the canaliculi and thus permanently impair lacrimal elimination. It is very difficult to restore scarred fibrosed canaliculi. On the other hand, the pressure gradient between the canaliculi and the sac cannot be produced if the canaliculus is slit open. This information should caution clinicians against performing overly aggressive procedures on the lacrimal outflow system [18].

16.3.4 Role of Anatomic Vascular Organization and the Importance of Cavernous Body

The blood vessels residing in nasolacrimal system includes specialized arteries (barrier arteries), venous lacunae (capacitance veins), veins (throttle veins), and arteriovenous anastomoses.

In the physiology of tear outflow control, the cavernous body of the efferent tear ducts is also significant. By expanding and decreasing the size of the cavernous body, the vessels aid in the closing and opening of the lacrimal passage lumen [19].

When the barrier arteries (arteries with an anatomically additional muscular layer) are opened and the throttle veins (veins with a muscle layer of helically arranged smooth muscle cells in the tunica media) are closed, expanding of the cavernous body occurs. However, if the barrier arteries are closed and the throttle veins are opened, blood flow to the convoluted venous lacunae is diminished, allowing blood to flow out of these veins, resulting in cavernous body shrinkage and dilation of the lacrimal passage lumen. When the shunts of the arteriovenous anastomoses are open, direct blood flow between arteries and venous lacunae is possible, avoiding the subepithelially located capillary network and allowing for rapid filling of venous lacunae. The specialized blood vessels allow opening and closing of the lumen of the lacrimal passage, which is effected by the bulging and subsiding of the cavernous body, while also controlling tear outflow. Furthermore, it was proposed that the valves in the lacrimal sac and nasolacrimal duct mentioned by Rosenmüller, Hanske, Aubaret, Béraud, Krause, and Taillefer in the past may be caused by different swelling states of the cavernous body [20].

16.4 Conclusions

There are many factors which are important in lacrimal drainage system. The palpebral-canalicular pump mechanism in lacrimal elimination is the major mechanism. The canalicular pump is probably more important than the sac pump because following DCR tears are still drained through the canaliculi to the nose. Damage to the canaliculi should be avoided since the damage of the canaliculi impairs the lacrimal elimination. After DCR operation, the physiology changes and some physiologic mechanisms may play a more important role. Tear elimination is equivalent through the upper and lower canalicular systems. Therefore, attention should be given not to damage both the upper and lower canaliculus (Figs. 16.1, 16.2, 16.3, 16.4).

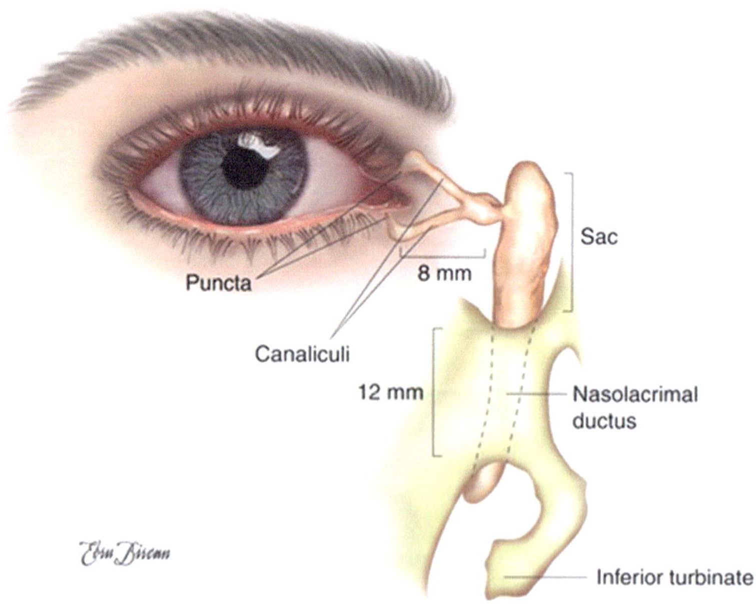

Fig. 16.1 Normal anatomy of the nasolacrimal system (Courtesy of TESAV)

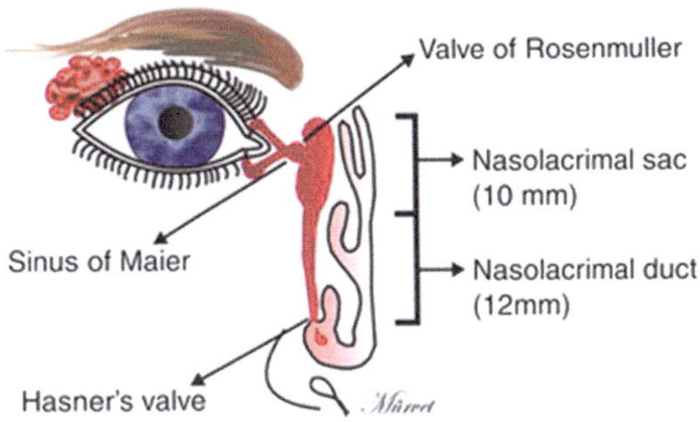

Fig. 16.2 Schematic representation of the nasolacrimal system (Courtesy of TESAV)

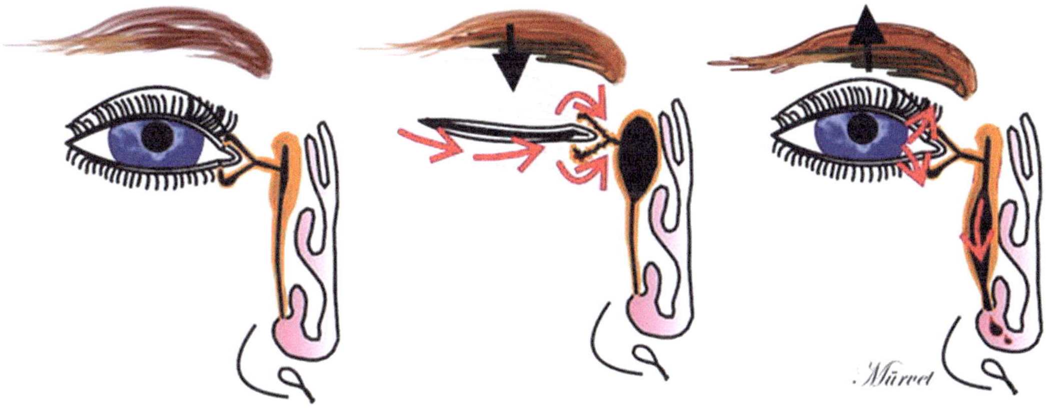

Fig. 16.3 The palpebral-canalicular pump mechanism in lacrimal elimination (Courtesy of TESAV)

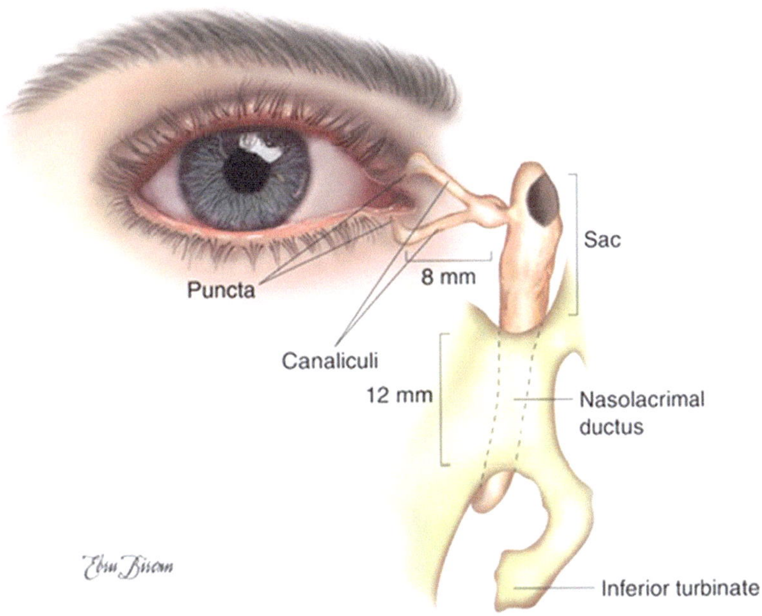

Fig. 16.4 Sump syndrome. If the rhinostomy opening is made too high during DCR operation, the drainage system will not function properly (Courtesy of TESAV)

References

1. Hurwitz JJ. Physiology of the lacrimal drainage system. In: Hurwitz JJ, editor. The lacrimal system. Philadelphia: Lippincott-Raven Publishers; 1996. p. 23–8.
2. Ahl NC, Hill JC. Horner's muscle and the lacrimal system. Arch Ophthalmol. 1982;100:488–93.
3. Paulsen F, Thale A, Kohla G, et al. Functional anatomy of human lacrimal duct epithelium. Anat Embryol. 1998;198:1–12.
4. Sisler HA. One-way flow in the lacrimal drainage system: determinants from the basic sciences. Ann Ophthalmol. 1982;14:76–7.
5. Daubert J, Nik N, Chandeyssoun PA, et al. Tear flow analysis through the upper and lower systems. Ophthal Plast Reconstr Surg. 1990;6:193–6.
6. Linberg JV, Moore CA. Symptoms of canalicular obstruction. Ophthalmology. 1988;95:1077–9.
7. Meyer DR, Antonello A, Linberg JY. Assessment of tear drainage after canalicular obstruction using fluorescein dye disappearance. Ophthalmology. 1990;97:1370–4.
8. White WL, Glover AT, Buckner AB, et al. Relative canalicular tear flow as assessed by dacryoscintigraphy. Ophthalmology. 1989;96:167–9.
9. Jones LT. Anatomy of the tear system. Int Ophthalmol Clin. 1973;13:3–22.
10. Doane MG. Blinking and the mechanics of the lacrimal drainage system. Ophthalmology. 1981;88:844–51.
11. Becker BB. Tricompartment model of the lacrimal pump mechanism. Ophthalmology. 1992;99:1139–45.
12. Jones LT. Epiphora. II. Its relation to the anatomic structures and surgery of the medial canthal region. Am J Ophthalmol. 1957;43:203–12.
13. Jones LT. Epiphora: its relation to the anatomic structures and surgery of the medial canthal region. Trans Pac Coast Oto-Ophthalmol Soc. 1956;37:31–46.
14. Jordan DR, Mawn L, Anderson RL. Surgical anatomy of the ocular Adnexa: a clinical approach. 1st ed. San Francisco: American Academy of Ophthalmology; 2012. p. 30–8.
15. Burkat CN, Hodges RR, Lucarelli MJ, et al. Physiology of the lacrimal system. In: Tasman W, Jaeger EA, editors. Duane's foundations of clinical ophthalmology, vol. 2. Philadelphia: Lippincott Williams&Wilkins; 2006. Chapter 2a.
16. Nik NA, Hurwitz JJ, Ching SH. The mechanism of tear flow after DCR and Jones' tube surgery. Arch Ophthalmol. 1984;102:1643–6.
17. Corin S, Hurwitz JJ, Jaffer N, Botta EP. The true canalicular angle: a mathematical analysis. Ophthal Plast Reconstr Surg. 1990;6:42–5.
18. Myron T, Clinton DM. Lacrimal drainage system. In: Tasman W, Jaeger EA, editors. Duane's clinical ophthalmology, vol. 4. Philadelphia: Lippincott Williams & Wilkins; 2006. Chapter 13.
19. Paulsen F, Garreis F, Schicht M, et al. Anatomie und Physiologie der ableitendenTränenwege [anatomy and physiology of the nasolacrimal ducts]. HNO. 2016;64:354–66.
20. Paulsen F, Thale A, Hallmann U, et al. The cavernous body of the human efferent tear ducts: function in tear outflow mechanism. Invest Ophthalmol Vis Sci. 2000;41:965–70.

Intranasal Trigeminal Perception

17

Philippe Rombaux, Caroline Huart, Basile Landis, and Thomas Hummel

Core Messages

- Intranasal trigeminal system mediates the sensation of temperature, pressure, perception of nasal airflow during breathing, nociception, and participates in the chemosensory perception of odorant stimuli.
- Chemosensory perception is not only mediated by free nerve endings in the nasal mucosa but also by some trigeminal fibers in close contact with solitary chemosensory cells.
- Besides the sensory nerves, the parasympathetic and the orthosympathetic systems play an important role in the normal physiology of the nose.
- Testing the intranasal trigeminal function, both psychophysically and electrophysiologically, is possible and may be used in the assessment of a patient with a chemosensory dysfunction.
- Healthy subjects need to have intact trigeminal and olfactory systems to have a full complete picture of the chemosensory stimulus.
- Olfactory and trigeminal systems interact both at a central and peripheral level.
- In patients with olfactory loss, a compensatory mechanism probably exists between the olfactory and the trigeminal systems.

17.1 Introduction

The intranasal trigeminal system interacts with the olfactory system to provide a complete chemosensory perception of the odorant stimuli. The perception of nasal patency and nasal airflow is also mediated by trigeminal afferents. The trigeminal system is therefore mandatory for both the chemosensory and the somatosensory perception and nasal mucosa with trigeminal nerve endings needs to be preserved in order to assume these two functions.

P. Rombaux (✉) · C. Huart
Department of Otorhinolaryngology, Clinique Universitaires Saint-Luc, Brussels, Belgium

Institute of Neuroscience, Université catholique de Louvain, Brussels, Belgium
e-mail: philippe.rombaux@saintluc.uclouvain.be

B. Landis
Rhinology-Olfactology Unit, Department of Otorhinolaryngology- Head and Neck Surgery, Geneva University Hospitals, Geneva, Switzerland

The Inner Ear and Olfaction Lab, University of Geneva Faculty of Medicine, Geneva, Switzerland

T. Hummel
Department of Otorhinolaryngology, Smell and Taste Clinic, University of Dresden Medical School, Dresden, Germany
e-mail: t.hummel@mail.zih.tu-dresden.de

Ö. Ö. Celebi, T. M. Önerci (eds.), *Nasal Physiology and Pathophysiology of Nasal Disorders*,
https://doi.org/10.1007/978-3-031-12386-3_17

Besides this protective somatosensory function, the intranasal trigeminal system also helps the global chemosensory perception of the olfactory system. Indeed, most of the odorants stimulate the neural olfactory and intranasal trigeminal systems [1].

Finally, the intranasal trigeminal system is also capable of inducing neurogenic inflammation mainly through an axon reflex located in the subepithelial level of the nasal mucosa.

The olfactory (cranial nerve I) and the trigeminal (cranial nerve V) systems interact at different levels and this interaction is essential for the odor sensation [2]. The olfactory system is more dedicated to identification tasks for hedonicity and alimentary behavioral, recognition and memory, behavioral and social comportments than the trigeminal system probably more oriented to protective function and reflexes.

17.2 The Nerves of the Nose

Sensory nerve endings from branches of the trigeminal nerve are located in the epithelia of the nose and sinuses, the eyelids and the cornea, the oral cavity, and the skin. Fibers from the intranasal trigeminal nerve mediate the tactile sensation of temperature, pressure, and perception of nasal airflow during breathing and participate in the chemosensory perception of odorant stimuli. Trigeminal receptors are located throughout the epithelia of the nasal mucosa and contribute to the global perception of odorous stimuli reaching the nasal fossa and the upper airway.

The nasal cavity is innervated by two branches of the trigeminal nerves, i.e., the ophthalmic and the maxillary branches. The ethmoid nerve innervating the anterior nasal mucosa and the external surface of the nasal fossa is part of the ophthalmic division, while the nasopalatine nerve which innervates the posterior part of the nasal cavity is part of the maxillary division. The trigeminal nerve has chemosensory and mechanosensory fibers. Mechanosensory fibers are large fast-conducting Aβ-fibers. Thin and fast-conducting myelinated Aδ-fibers and thin and slow-conducting unmyelinated C-fibers are responsible for thermoreception (cold and warm stimuli) and for nociceptive perception (pain, painful mechanical, noxious chemical stimuli). The sensations mediated by the trigeminal nerve are usually described as burning, stinging, itching, tickling, cooling, and warming feeling. Trigeminal-free nerve endings have receptors which may be activated through several factors such as changes in pressure, temperature, irritants, and humidity. Substance P, calcitonin gene-related peptide (CGRP), and other neuropeptides are found in the trigeminal nerve fibers [3]. Some trigeminal fibers are in close contact with solitary chemosensory cells located in the nasal epithelium and are more responsible for chemosensory perception because they are responsive to both bitter tastants and chemical irritants (Fig. 17.1).

At the receptor level, one of the first described nociceptors was the ion channel receptor family, and characterization of one of these receptors was obtained with the nicotinic acetylcholine receptor. Transient receptor potential (TRP) channels are well expressed on sensory nerves and may influence cell function by mediating the flux of cations across the plasma membrane into the cytoplasm generating action potentials. Ion channels in the TRP family can be opened by many kinds of stimuli, i.e., chemical or physical. The TRP family can be subdivided into six subfamilies and many of them are found at the free nerve ending of the trigeminal nerve such as the vanilloid receptor (TRPV1), the purinergic receptor (P2X), the acid-sensitive ion channels (ASIC/DRASIC), the channel responsive to menthol (TRPM8) (cooling), the channel responsive to changes in heat and eugenol (TRPV3) (warming), and the channel responsive to isothiocyanate (TRPA1), the major compound of mustard oil [4].

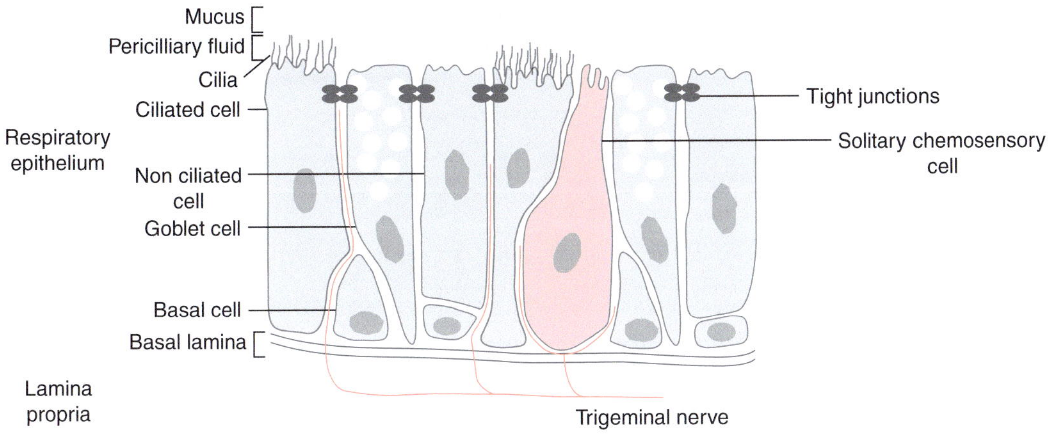

Fig. 17.1 Trigeminal fibers in close contact with solitary chemosensory cells and trigeminal nerve-free endings located in the nasal epithelium and responsible for chemosensory and somatosensory perceptions

Like the skin sensory perception, the unmyelinated C-fibers (slow conduction) are responsible for burning sensations, and the myelinated Aδ-fibers (fast conduction) are responsible for stinging sensations.

The cell bodies of the trigeminal fibers are located in the Gasserian ganglion. Nerve fibers from the cell bodies thereafter participate in the sensory afferent system and project to the trigeminal sensory nucleus that extends from the rostral spinal cord to the midbrain. Interestingly, some individual cells in the ganglion send axons to the olfactory bulb indicating that some interaction exists at this level. Neurons then project to the amygdala and to the ventral posterior medial nuclei of the thalamus. Most of the ascending fibers cross toward the contralateral side with some fibers ascending ipsilaterally (different for the olfactory pathways [5]). The nerve projections terminate in the primary somatosensory cortex (SI) and also in the secondary somatosensory cortex (SII) with a right hemispheric predominance [6–8]. Trigeminal activation also leads to insular cortex activation and to ventral orbitofrontal cortex mainly to the right side explaining at the central level the interactions with other chemosensory systems like taste and olfaction [9].

Besides the sensory nerves, the parasympathetic and the orthosympathetic systems play an important role in the normal physiology of the nose [10]. Parasympathetic nerves have acetylcholine as the major neurotransmitter and act on muscarinic receptors to induce increased glandular secretions and vasodilatation. Vasointestinal peptide (VIP) is another neurotransmitter of the parasympathetic system. The sympathetic system with noradrenaline and neuropeptide Y (NPY) as neurotransmitters acts on adrenergic receptors and induces vasoconstriction and increases nasal airway patency [11, 12].

Pathophysiological mechanisms and nasal symptoms are explained by the interdigitation of these neurologic systems, i.e., the trigeminal sensitive afferent (+ efferent axon reflex), the efferent parasympathetic, and the efferent orthosympathetic systems (autonomic systems) (Fig. 17.2).

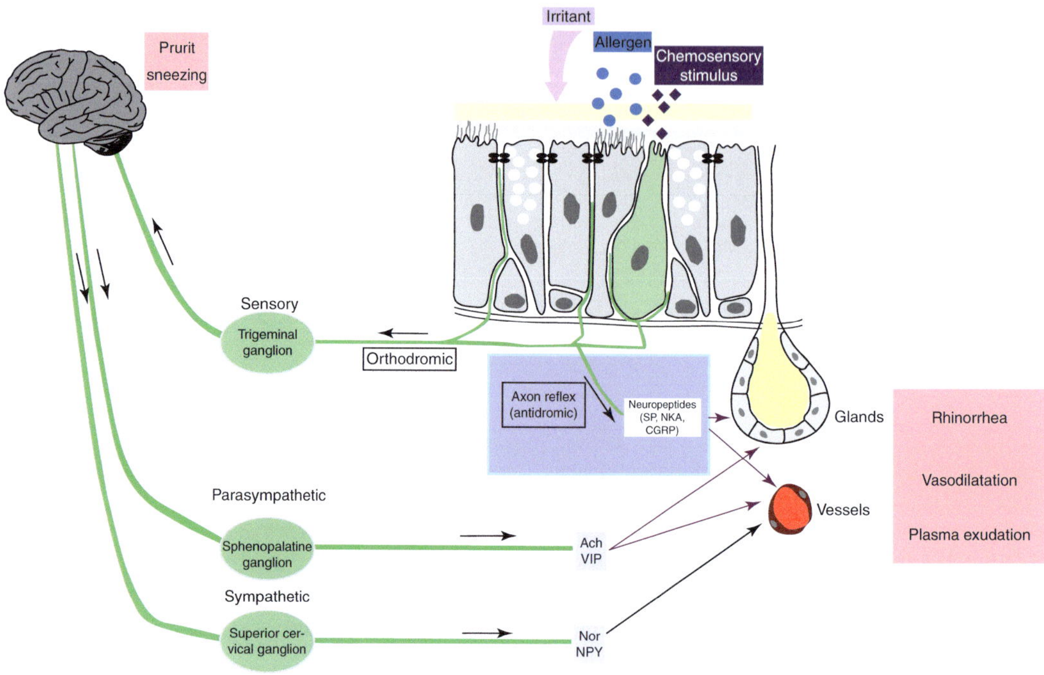

Fig. 17.2 Interdigitation of the neurologic systems found in the nasal mucosa, i.e., the trigeminal sensitive afferent (+ efferent axon reflex), the efferent parasympathetic, and the efferent orthosympathetic systems (autonomic systems)

17.3 Consequences of Activation of Trigeminal Receptor and Nasal Reflexes

The activation of the trigeminal system leads to the perception of potentially noxious stimuli, to a global chemosensory perception of odorant stimuli, and to some nasal reflexes. The nasal fossa may be divided into two parts, the anterior one most dedicated to the chemosensory perception and the posterior one most devoted to mechanosensory functions. This has been demonstrated by Frasnelli et al. where it was clearly stated that anterior nasal mucosa is more sensible to chemosensory stimuli than mechanical stimuli, while the posterior nasal mucosa is equally sensible to both chemosensory and mechanical stimuli [13]. However, thresholds to detect chemosensory stimuli such as CO_2 are lower when the stimulus is given in retronasally compared to orthonasally [14].

Therefore, nasal mucosa should not be seen as a homogenous tissue as it exhibits a varying degree of sensitivity to trigeminal stimuli depending on the stimulus quality and location in the nasal fossa [15].

Activation of trigeminal fibers leads to protective reflexes such as increasing secretions (saliva, tears, nasal mucus), decreasing breathing, sweating initiation, and closure of the nasal passage by augmentation of the turbinate volume.

Trigeminal nerve stimulation also induces many reflexes inducing different responses. The nasal cycle is probably the best known neurologic mechanism leading to a fluctuating congestion–decongestion of the nasal fossa secondary to a changing tone in the vasculature controlled by the autonomic system.

The naso-nasal reflex is supposed to be mediated by the parasympathetic system and explains many exacerbations of rhinorrhea and watery discharge [16].

The naso-ocular reflex is bilateral and mostly contralateral, secondary to chemosensory or tactile or physical stimuli. It induces watery eyes, lacrimation, and redness of the conjunctiva.

The "foot-cooling" reflex is secondary to a cold stimulation at the extremities of the inferior limb inducing in the nose a reduced blood flow and subsequently a nasal decongestion. This is also very similar to the reflexes observed in the nose when cooling of the face induces the same effect. Facial cooling through trigeminal receptors may even induce lower airway symptoms [17].

The naso-cardiovascular reflex is secondary to trigeminal activation in the nose and is responsible for bradycardia and hypotension, may be present during nose surgery, and is of primary importance for the anesthesiologist.

The naso-respiratory reflex or naso-bronchial reflex is present when cold dry air is presented to the subject's nose inducing increased lower airway resistance.

Cold dry air stimulus may also be used to induce both long-lasting painful sensations [18] and secretory response in the nose [19]. This mechanism is thought to be secondary to activation of capsaicin-sensitive fibers; alternatively, the change in the osmotic milieu of the respiratory epithelium may trigger the activation of the nociceptive system. This may play a role in the pathophysiology of nasal hyperactivity and in the non-allergic noninfectious group of rhinitis [20] and would lead to the development of capsaicin-based treatment for the patients suffering from these diseases [21–26]. Capsaicin delivered intranasally has proven its effect in the treatment of the nasal hyperreactivity found in idiopathic rhinitis patients [27].

These responses may be present after single presentation of the stimulus or when repeated application of the stimuli is delivered. C-fibers and Aδ-fibers respond differently to repeated chemical stimulus. If stimuli are repeated, the burning painful sensation driven by C-fibers is increased, and this is the contrary for Aδ-fibers giving the stinging sensation. This is secondary to central nervous summation more than increase in the firing of the nerve fibers at the periphery.

17.4 Neurogenic Inflammation

The activation of sensory nerves and the release of neuropeptides from neuroendocrine cells found in the respiratory mucosa with a subsequent neurogenic inflammation may explain at least partially some diseases of the upper and lower airways [28].

Stimulation of sensory trigeminal fibers may lead to the release of different neuropeptides such as substance P, neurokinin A (NKA), neuropeptide K (NPK), and calcitonin gene-related peptide (CGRP). These neuropeptides are increased in the upper and lower airways of these patients with airway inflammation in a similar way than the inflammatory components usually described as eosinophils or some proinflammatory cytokines [29].

There is a strong evidence that neuroendocrine cells, sensory neurons, and proinflammatory immune cells interact and promote inflammation and airway hyperreactivity. Neurotrophins such as nerve growth factor (NGF) or neurotrophins-3–4 are also linked to the development of a neurogenic inflammation.

In animals, dendrites of intranasal trigeminal nerve endings can be stimulated in an antidromic way. This antidromic stimulation is called the "axon reflex" and leads to the release of inflammatory neuropeptides from the varicosities of the nerve, producing vasodilation, increased vascular permeability, and glandular activation. This phenomenon has been clinically proven in humans where specific activation of the intranasal trigeminal nerve ending produces nasal obstruction, congestion, watery discharge, and sneezing. This axon reflex probably plays a major role in the development of nasal hyperreactivity, non-allergic noninfectious rhinitis known as idiopathic rhinitis, and even allergic rhinitis via the substance P which exacerbates the eosinophilic recruitment after allergen challenge (for review, see [30]).

17.5 Psychophysical Testing of the Intranasal Trigeminal Function

Testing of trigeminal function with psychophysics is based on threshold measurement, rating of suprathreshold stimuli, discrimination tasks, and lateralization tasks [31, 32].

Trigeminal function assessed with psychophysical testing revealed that sensitivity decreases with age [33].

Psychophysical evidence exists for qualitative specificity of the human intranasal trigeminal system. The nasal trigeminal system is less sensitive than the olfactory system for the majority of odorant stimuli. Recognition threshold of trigeminal stimulus such as CO_2 was measured between 32 and 47% v/v for stimuli of 200 ms duration at an airflow of 8 l/min at body temperature. The threshold for detection can be lowered if stimulation duration is increased [14].

Considering *pain ratings*, increase in perceived or painful sensitivity occurs more rapidly for trigeminal stimulus than for olfactory stimuli [34].

The trigeminal and the olfactory systems also have a different contribution on the presentation of *mixed compounds*. In normosmic subjects, trigeminal stimuli are perceived as more intense when they are accompanied by an olfactory stimulus while the olfactory stimulus seems to have no effect when a mixed compound is presented. The trigeminal stimulus may induce an additive or even a hyperadditive effect on the perception after a mixed stimulus presentation [35].

Qualitative discrimination task with trigeminal irritants demonstrates that human are capable to discriminate among different trigeminal stimuli even in the absence of any olfactory stimuli given concomitantly [36], even if this ability seems to decrease with age [37]. In contrast to odor stimulation, trigeminal stimuli can produce increase in pain intensity when repeated stimuli are given with a short interval demonstrating a sensitization effect while on the contrary a desensitization effect exists when repeated stimuli are delivered with long interstimulus interval [38, 39]. Temporal integration of trigeminal information is thus different than olfactory temporal integration. Psychophysical studies with capsaicin have demonstrated a sensitization effect meaning that the subjective pain rating was increased after the second stimulation if the interstimulus interval was less than 1 min. On the contrary with a second stimulation delivered after 4 min, a desensitization effect was observed. This leads to the idea that repetitive delivery to the nasal mucosa was perhaps a treatment for patients with hyperalgia in the nasal fossa or for patients with nonallergic noninfectious rhinitis [38]. However, this mechanism is linked to the type of the stimulus, and sensitization and desensitization in the nasal cavity do not follow the same processes in relation to the molecules studied [40].

Lateralization task revealed that trigeminal stimuli are perceived without error when the subjects blindfolded is asked to determine the side of stimulation and that this ability is lost for olfactory stimuli or when the odor has a mixed property between trigeminal and pure olfactory valence [41]. In others words, pure olfactory stimuli cannot be localized to the nasal cavity while on the contrary pure trigeminal stimuli can be localized. The results are lower in patients with an olfactory dysfunction independent of the cause of the olfactory problem [42].

Subjective ratings of nasal patency are also influenced by the trigeminal system. For example, stimulation of the nasal fossa with menthol is accompanied by an increase of perceived nasal patency [43], while on the contrary, anesthesia of the nasal mucosa leads to a perception of decreased nasal patency even in both cases objective nasal patency did not change.

Many studies have been conducted on anosmic subject, and trigeminal thresholds were found to be higher in anosmic subjects than in control [44]. Age-related decline of intranasal sensitivity was reported with not only psychophysical but also electrophysiological evidence [45].

17.6 Electrophysiology and Functional Imaging

Electrophysiological recordings from the intranasal trigeminal system may be obtained at the peripheral level, i.e., the negative mucosal potential (NMP) and at the central level by recording cortical responses after delivery of an intranasal trigeminal stimulus, i.e., the trigeminal event-related potential (Trigeminal ERP).

The NMP is recorded from the nasal mucosa and is thought to represent the summated receptor potentials of chemical nociceptors in a very similar way to the electro-olfactogram which represents the global activity of olfactory receptor neurons located in the olfactory neuroepithelium [46].

Human NMP may be obtained after CO_2 intranasal stimulation and the amplitude of the NMP is well correlated with the subjective pain rating ([47, 48], b). NMP may also be recorded by stimulating polymodal nociceptors such as TRPA1, TRPV1 and 2. Responses to the NMP are different according to the stimulus used, i.e., CO_2, menthol or ethanol [49] and decrease in response to repetitive stimulation ([47], b).

Trigeminal ERP may be obtained after repetitive stimulation with relatively selective trigeminal stimuli such as CO_2 with an interstimulus interval of 20–40 s and a concentration of 30–60% v/v of CO_2 delivered by an olfactometer [7, 8, 32] or with nicotine [50]. Without producing mechanical sensations (flow embedded in a constant 8 L/Min) and the thermoreceptor (temperature maintained constant at 36–37 °C), the recorded ERP may be viewed as a pure chemosensory component without interfering with mechanoreceptor.

When comparing the electrophysiological responses to subjective rating of the stimulus, the intensity increases more rapidly for trigeminal stimuli than for olfactory stimuli when the concentration of the stimuli increases [51].

Trigeminal ERP can also be meaningful in patients with an olfactory dysfunction. Patients with an olfactory dysfunction usually do not have any olfactory event-related potentials but trigeminal event-related potentials are usually present even if some subtle changes in latency and amplitude may be present [7, 8, 52].

Electrophysiological studies both at the peripheral level (NMP) and at the cortical level (TERP) helped to understand the effects of gender, age, disease, i.e., loss of olfactory function, and drugs [53].

PET-based investigation of cerebral activation following intranasal trigeminal stimulation revealed that olfactory and trigeminal information have common pathways and that CO_2 activated the base of the posterior central gyrus (primary and secondary somatosensory cortex) and the piriform cortex, more in the right hemisphere [54]. This was confirmed by fMRI studies where anterior caudate nucleus, insula, cerebellum, and orbitofrontal cortex were also involved in the processing [55, 56].

17.7 Olfactory and Trigeminal Interaction

Healthy subjects need to have an intact trigeminal and olfactory system to have a full chemosensory perception of the environment [57]. Inhaled chemical compounds have the propensity to stimulate both systems even if relatively selective olfactory and trigeminal stimuli exist. Irritants are thought to stimulate free trigeminal nerve endings in the nasal epithelium but presumably below the level of the tight junction which renders them more sensitive to lipid-soluble stimuli than to water-soluble stimuli. Indeed, lipid-soluble stimuli are more prone to pass across the mucus layer and the tight junction.

Trigeminal nerve fibers are also in close contact with the solitary chemoreceptor cells and respond to chemical stimuli that are water soluble [58]. These cells are found in the respiratory and digestive tracts and have many of the characteristics of the taste cells because they have taste receptors mainly T2R bitter receptor, TRPV1, and TRPM5 channels receptor [59]. The chemosensory stimuli may thus activate the trigeminal nerve fibers surrounding the chemoreceptor cells directly if the stimulus is water soluble and above the level of the tight junction. In contrast, lipid-soluble stimuli may activate the trigeminal-free nerve ending below the level of the tight junction by diffusing across the junctional membrane. There is strong evidence that a cross modal plasticity exists between the two systems and that a mutual interaction exists both in normal and pathological conditions [60].

Much information may be obtained from patients having lost one of the two systems. For example, anosmic patients may be good candidates to deeply study the trigeminal system. Patients with a complete loss of trigeminal function are more difficult to find even if post surgically treated patients (radical surgery for the inferior and middle turbinate's or empty nose syndrome, or patients who had undergone a Gasser ganglion removal) may have some interest to study olfactory abilities without any trigeminal interactions.

Interactions between the two systems exist at the peripheral level (trigeminal nerve with contact at the olfactory neuroepithelium, effect of substance P on the olfactory responses, alteration of receptor activity through modification of nasal permeability or mucus quality), at the olfactory bulb level and more centrally in cortical areas such as the piriform cortex, thalamus, insula, and orbitofrontal cortex. Indeed, some trigeminal collaterals are found in the olfactory bulb explaining the interdigitation of the olfactory and trigeminal systems [61].

When studies are conducted to determine the relative contribution on the perception of trigeminal, olfactory, and mixed stimuli, we can conclude at a relative dominance of the trigeminal system over olfactory sensation and also a dominance of mixed stimuli over either system alone [62].

In healthy subjects, both systems contribute to the complete picture of the chemosensory stimulus. At the periphery, both olfactory and trigeminal sensory information attempt to mutually decrease the other sensory response as there is no need for the peripheral system to catch all the information available from the outside world and entering the nasal fossa. At the central level, both the olfactory and trigeminal information are converging; the resulting percept may a mutual amplification or inhibition of the various sensations depending, among other things, on stimulus quality, intensity, and salience [2] (Table 17.1).

Patients with an olfactory dysfunction have lower trigeminal sensitivity compared to controls [42, 56, 63] and loss of olfactory function leads to a decrease trigeminal sensitivity [47, 64]. Some studies have investigated the olfactory modulation of trigeminally mediated sensations in patients with olfactory loss. These reports have shown that in anosmia, there are mixed sensory adaptation/compensation in the interaction between olfactory and trigeminal systems, where, in acquired anosmia, there is an increased trigeminal activation on mucosal level and a decreased responsiveness at a central level [65]. In congenital anosmia, however, similar responsiveness to trigeminal stimuli was found when compared with healthy subjects [65]. Following these findings, Frasnelli et al. [65] proposed a

Table 17.1 Major differences between olfactory and trigeminal sensory patterns

	Olfactory	Trigeminal
Cranial nerve	I	V
Nerve ending	In the olfactory neuroepithelium Olfactory receptor neuron	In the nasal mucosa Free nerve ending or in contact with solitary chemoreceptor cell
Hemispheric lateralization	Not major Mainly ipsilateral	Right Mainly contralateral
Major stimuli in research	Phenyl ethyl alcohol, H2S, amyl acetate	CO_2, capsaicin, allyl isothiocyanate
Lateralization task	Not possible	Possible
Effect of concentration increase on the subjective rating of the stimulus	Poor	Important
Effect of mixed stimulus on subjective rating	Poor	Important
Threshold	Usually low High sensitive	Usually high Low sensitive

model of mixed sensory adaptation/compensation in the interaction between olfactory and trigeminal system. In this model, the primary trigeminal activation is (1) reduced on a mucosal level due to constant activation of intrabulbar trigeminal collaterals, inducing downregulation in the periphery of the trigeminal system and (2) amplification on a central level in healthy subjects, due to functional integration of olfactory and trigeminal processes. They also hypothesized that in patients suffering from acquired anosmia, missing inhibition via the trigeminal collaterals in the olfactory bulb would lead to a compensatory upregulation in the periphery of the trigeminal nerve in the case of olfactory loss, hence inducing increased peripheral responsiveness. On central levels, however, missing olfactory augmentation of the trigeminal input would not be sufficiently compensated by the increased peripheral trigeminal input, thus leading to decreased amplitude of trigeminal event-related potentials [65]. Finally, recovering would lead to a compensatory mechanism and to an adjustment (Fig. 17.3).

Chemosensory reduction of trigeminal sensitivity in subjects with olfactory dysfunction seems to be specific of the chemosensory pattern as somatosensory information does not seem to be decreased [66].

Finally, patients with absence of trigeminal receptor, those without any remaining nasal mucosa after radical surgery also have a decreased olfactory function and this should be viewed as a plaidoyer against radical surgery if the surgeon wants to maintain intact chemosensory function in the nose [67, 68].

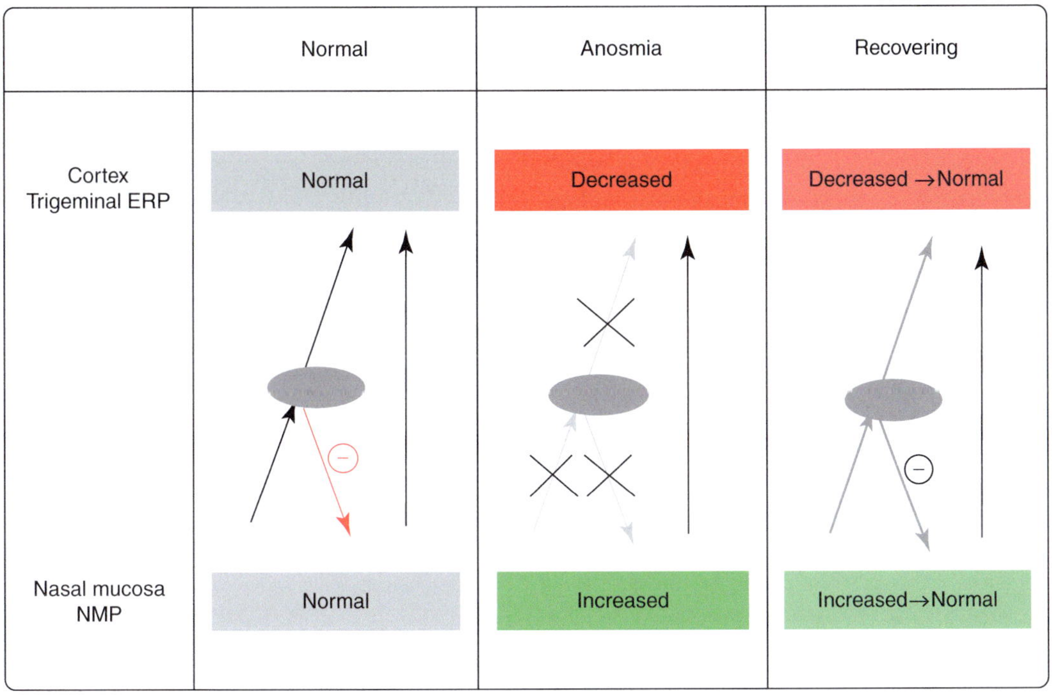

Fig. 17.3 Proposed model for olfactory and trigeminal interaction in normal condition, acquired olfactory dysfunction, and in the recovery phase. *Grey arrows* for olfactory pathways, *black arrows* for trigeminal pathways. In normal condition, constant activation of intrabulbar trigeminal collaterals inducing a downregulation in the periphery of the trigeminal system and amplification on a central level due to functional integration of olfactory and trigeminal processes. In patients suffering from acquired olfactory dysfunction, missing inhibition via the trigeminal collaterals in the olfactory bulb leading to an increased peripheral responsiveness and to a decreased amplitude of central trigeminal event-related potentials. Finally, recovering would lead to a compensatory mechanism and to an adjustment (Adapted from: [66])

17.8 Conclusion

The intranasal trigeminal system is not fully understood and its participation to the global chemosensory perception and to the somatosensory perception, i.e., the nasal patency merits further investigation. The olfactory–trigeminal interaction may be studied through the nature of the stimulus and the subsequent effect, the temporal integration of the stimulus (sensitization vs desensitization), the status of the subject (age, gender, anosmia), or the effects of some drugs (local anesthesia, antagonist receptor activity). Moreover, psychophysical testing of the intranasal trigeminal has not yet been established in clinical routine. It should be pointed out that the intranasal trigeminal system may deeply influence the overall chemosensory perception and that interfering with it would lead to a decreased chemosensory function and to a decrease of olfactory abilities. In the same vein of thoughts, interfering with this system is particularly devastating for the patient in regard to the nasal patency and to the perception of nasal airflow mediated by the somatosensory fibers. Finally, this system plays also an important role in neurogenic inflammation and in the pathogenesis of variants of non-allergic noninfectious rhinitis.

References

1. Doty RL. Intranasal trigeminal chemoreception. In: Doty RL, editor. Handbook of olfaction and gustation. New York: Marcel Dekker; 1995. p. 821–33.
2. Cain WS, Murphy CL. Interaction between chemoreceptive modalities of odour and irritation. Nature. 1980;284(5753):255–7.
3. Finger TE, et al. Ultrastructure of substance P-and CGRP-immunoreactive nerve fibers in the nasal epithelium of rodents. J Comp Neurol. 1990;294:293–305.
4. Bessac BF, Jordt SE. Breathtaking TRP channels: TRPA1 and TRPV1 in airway chemosensation and reflex control. Physiology. 2008;23:360–70.
5. Brand G. Olfactory lateralization in humans: a review. Neurophysiol Clin. 1999;29:495–506.
6. Hari R, Portin K, Kettenmann B, Jousmäki V, Kobal G. Right-hemisphere preponderance of responses to painful CO_2 stimulation of the human nasal mucosa. Pain. 1997;72:145–51.
7. Rombaux P, Mouraux A, Keller T, Hummel T. Trigeminal event related potentials in patients with olfactory dysfunction. Rhinology. 2008a;46(3):170–4.
8. Rombaux P, Guerit JM, Mouraux A. Lateralisation of intranasal trigeminal chemosensory event-related potentials. Neurophysiol Clin. 2008b;38(1):23–30.
9. Anton F, Peppel P. Central projections of trigeminal primary afferents innervating the nasal mucosa: a horseradish peroxidase study in the rat. Neuroscience. 1991;41:617–28.
10. Kaliner M. The physiology and pathophysiology of the parasympathetic nervous system in nasal disease: an overview. J Allergy Clin Immunol. 1992;90:1044–5.
11. Baraniuk JN. Sensory, parasympathetic, and sympathetic neural influences in the nasal mucosa. J Allergy Clin Immunol. 1992;90:1045–50.
12. Baraniuk JN, Lundgren J, Okayama M, Goff J, Mullol J, Merida M, et al. Vasoactive intestinal peptide in human nasl mucosa. Am J Respir Cell Mol Biol. 1991;4:228–36.
13. Frasnelli J, Heilmann S, Hummel T. Responsiveness of human nasal mucosa to trigeminal stimuli depends on the site of stimulation. Neurosci Lett. 2004;362:322–8.
14. Melzner J, Bitter T, Guntinas-Lichius O, Gottschall R, Walther M, Gudziol H. Comparison of the orthonasal and retronasal detection thresholds for carbon dioxide in humans. Chem Senses. 2011;36(5):435–41.
15. Scheibe M, Zahnert T, Hummel T. Topographical differences in the trigeminal sensitivity of the human nasal mucosa. Neuroreport. 2006;17:1417–20.
16. Baraniuk JN, Kim D. Nasonasal reflexes, the nasal cycle and sneeze. Curr Allergy Asthma Rep. 2007;7(2):105–11.
17. Koskela H, Tukiainen H. Facial cooling, but not nasal breathing of cold air, induces bronchoconstriction: a study in asthmatic and healthy subjects. Eur Respir J. 1995;8:2088–93.
18. Lötsch J, Ahne G, Kunder J, Kobal G, Hummel T. Factors affecting pain intensity in a pain model based upon tonic intranasal stimulation in humans. Inflamm Res. 1998;47:446–50.
19. Fontanari P, Burnet H, Zattara-Hartman M, Jammes Y. Changes in airway resistance induced by nasal inhalation of cold dry, or moist air in normal individuals. J Appl Physiol. 1996;81:1739–43.
20. Bernstein J. The role of autonomic nervous system and inflammatory mediators in nasal hyperreactivity: a review. Otolaryngol Head Neck Surg. 1991;105:596–607.
21. Blom H, Severijnen A, Van Rijswijk J, Mulder P, Van Rijk R, Fokkens W. The long-effect of capsaicin aqueous spray on the nasal mucosa. Clin Exp Allergy. 1998;28:1351–8.
22. Lacroix J, Buvelot J, Polla B, Lundberg J. Improvement of symptoms of non-allergic chronic rhinitis by local treatment with capsaicin. Clin Exp Allergy. 1991;21:595–700.
23. Marabini S, Ciabatti P, Polli G, Fusco B, Geppetti P. Beneficial effects of intranasal application of capsaicin in patients with vasomotor rhinitis. Eur Arch Otorhinolaryngol. 1991;248:191–4.
24. Stjarne P, Lundblad L, Anggard A, Lundberg J. Local capsaicin treatment of the nasal mucosa reduces

symptoms in patients with non allergic nasal hyperreactivity. Am J Rhinol. 1991;5:145–51.
25. Taylor-Clark TE, et al. Nasal sensory nerve populations responding to histamine and capsaicin. J Allergy Clin Immunol. 2005a;116:1282–8.
26. Taylor-Clark T, Kollarik M, MacGlashan D, Undem B. Nasal sensory nerve populations responding to histamine and capsaicin. J Allergy Clin Immunol. 2005b;116:1282–8.
27. Van Rijswijk J, Boeke E, Keyzer J, Mulder P, Blom H, Fokkens W. Intranasal capsaicin reduces nasal hyperreactivity in idiopathic rhinitis: a double-blind randomized application regimen study. Allergy. 2003;58:754–61.
28. Lacroix J, Landis B. Neurogenic inflammation of the upper airway mucosa. Rhinology. 2008;46(3):163–5.
29. Shusterman D, Murphy MA, Balmes J. Differences in nasal irritant sensitivity by age, gender, and allergic rhinitis status. Int Arch Occup Environ Health. 2003;76:577–83.
30. Sarin S, Undem B, Sanico A, Togias A. The role of the nervous system in rhinitis. J Allergy Clin Immunol. 2006;118:999–1014.
31. Frasnelli J, Hummel T. Intranasal trigeminal thresholds in healthy subjects. Environ Toxicol Pharmacol. 2005;19(3):575–80.
32. Hummel T. Assessment of intranasal trigeminal function. Int J Psychophysiol. 2000;36:147–55.
33. Wysocki CJ, Cowart BJ. Nasal trigeminal chemosensitivity across the life span. Percept Psychophys. 2003;65:115–22.
34. Cain WS, de Wijk R, Cain WS, Pilla-Caminha G. Human psychophysical and neurophysiological measurements on ethanol. Chem Senses. 1998;23:586–9.
35. Cornetto-Muniz JE, Hernandez SM. Odorous and pungent attributes of mixed and unmixed odorants. Percept Psychophys. 1990;47:391–9.
36. Laska M, Distel H, Hudson R. Trigeminal perception of odorant quality in congenitally anosmic subjects. Chem Senses. 1997;22:456–77.
37. Laska M. Perception of trigeminal chemosensory qualities in the elderly. Chem Senses. 2001;26:681–9.
38. Brand G, Jacquot L. Sensitization and desensitization to allyl isothiocyanate (mustard oil) in the nasal cavity. Chem Senses. 2002;27:593–8.
39. Hummel T, Gruber M, Pauli E, Kobal G. Event-related potentials in response to repetitive painful stimulation. Electroencephalogr Clin Neurophysiol. 1994;92:426–32.
40. Jacquot L, et al. Trigeminal sensitization and desensitization in the nasal cavity: a study of cross interactions. Rhinology. 2005;43:93–8.
41. Kobal G, Hummel T. Brain responses to chemical stimulation of trigeminal nerve in man. Chem Senses. 1990;2:593–8.
42. Hummel T, Futschik T, Frasnelli J, Hüttenbrink KB. Effects of olfactory function, age and gender on trigeminally mediated sensations: a study based on the lateralization of chemosensory stimuli. Toxicol Lett. 2003;140:273–80.
43. Eccles R, Jawad MS, Morris S. Olfactory and trigeminal thresholds and nasal resistance to airflow. Acta Otolaryngol. 1989;108(3–4):268–73.
44. Cornetto-Muniz JE, Cain WS. Trigeminal and olfactory sensitivity: comparison of modalities and methods of measurement. Int Arch Occup Environ Health. 1998;7:105–10.
45. Frasnelli J, Hummel T. Age-related decline of intranasal trigeminal sensitivity: is it a peripheral event? Brain Res. 2003;987:201–6.
46. Thürauf N, et al. The mucosal potential elicited by noxious chemical stimuli with CO_2 in rats: is it a peripheral nociceptive event? Neurosci Lett. 1991;128:297–300.
47. Hummel T, et al. Peripheral electrophysiological responses decrease in response to repetitive painful stimulation of the human nasal mucosa. Neurosci Lett. 1996a;212:37–40.
48. Kobal G. Pain-related electrical potentials of the human nasal mucosa elicited by chemical stimulation. Pain. 1985;22:151–63.
49. Meusel T, Negoias S, Scheibe M, Hummel T. Topographical differences in distribution and responsiveness of trigeminal sensitivity within the human nasal mucosa. Pain. 2010;151(2):516–21.
50. Hummel T, Livermore A, Hummel C, Kobal G. Chemosensory event-related potentials: relation to olfactory and painful sensations elicited by nicotine. Electroencephalogr Clin Neurophysiol. 1992;84:192–5.
51. Lötsch J, March CR, Kobal G. The influence of stimulus duration on the reliability of pain rating after nociceptive stimulation of the nasal mucosa with CO_2. Eur J Pain. 1997;1:207–13.
52. Rombaux P, Mouraux A, Bertrand B, Guerit JM, Hummel T. Assessment of olfactory and trigeminal function using chemosensory event-related potentials. Neurophysiol Clin. 2006;36(2):53–62.
53. Lundström JN, Frasnelli J, Larsson M, Hummel T. Sex differentiated responses to intranasal trigeminal stimuli. Int J Psychophysiol. 2005;57:181–6.
54. Hummel T, Ochme L, Vandenhoff J, Gerber J, Heinke M, Boyle JA, Benthien-Baumann B. PER-based investigation of cerebral activation following intranasal trigeminal stimulation. Hum Brain Mapp. 2009;30(4):1100–4.
55. Boyle JA, Heinke M, Gerber J, Frasnelli J, Hummel T. Cerebral activation to intranasal chemosensory trigeminal stimulation. Chem Senses. 2007;32(4):343–53.
56. Iannilli E, DelGratta C, Gerber JC, Romani GL, Hummel T. Trigeminal activation using chemical, electrical and mechanical stimuli. Pain. 2007b;139(2):376–88.
57. Bouvet JF, Delaleu JC, Holley A. Olfactory receptor cell function is affected by trigeminal nerve activity. Neurosci Lett. 1987;77:181–6.
58. Finger TE, et al. Solitary chemoreceptor cells in the nasal cavity serve as sentinels of respiration. Proc Natl Acad Sci U S A. 2003;100:8981–6.

59. Lin W, et al. TRPM5-expressing solitary chemosensory cells respond to odorous irritants. J Neurophysiol. 2008;99:1451–60.
60. Hummel T, Livermore A. Intranasal chemosensory function of the trigeminal nerve and aspects of its relation to olfaction. Int Arch Occup Environ Health. 2002;75:305–13.
61. Schaefer ML, Böttger B, Silver WN, Finger T. Trigeminal collaterals in the nasal epithelium and olfactory bulb: a potential route for direct modulation of olfactory information by trigeminal stimuli. J Comp Neurol. 2002;444:221–6.
62. Livermore A, Hummel T, Kobal G. Chemosensory evoked potentials in the investigations of interactions between the olfactory and the somatosensory (trigeminal) systems. Electroencephalogr Clin Neurophysiol. 1992;83:201–10.
63. Iannilli E, Gerber J, Frasnelli J, Hummel T. Intranasal trigeminal function in subjects with and without an intact sense of smell. Brain Res. 2007a;1139:235–44.
64. Hummel T, Barz S, Lötsch S, Kettenmann B, Kobal G. Loss olfactory function leads to a decrease of trigeminal sensitivity. Chem Senses. 1996b;21:75–9.
65. Frasnelli J, Schuster B, Hummel T. Interactions between olfaction and trigeminal system: what can be learned from olfactory loss. Cereb Cortex. 2007;17(10):2268–75.
66. Frasnelli J, Schuster B, Zahnert T, Hummel T. Chemosensory specific reduction of trigeminal sensitivity in subjects with olfactory dysfunction. Neuroscience. 2006;142(2):541–6.
67. Huart C, Eloy P, Collet S, Rombaux P. Chemosensory function assessed with psychophysical testing and event-related potentials in patients with atrophic rhinitis. Eur Arch Otorhinolaryngol. 2012;269(1):135–41.
68. Husner A, Frasnelli J, Welge-Lussen A, Reiss G, Zahnert T, Hummel T. Loss of trigeminal sensitivity reduces olfactory function. Laryngoscope. 2006;116:1520–2.

18 Sinus Pain

Jim Bartley

Abbreviations

CNS Central nervous system
CRS Chronic rhinosinusitis.
MOH Medication overuse headache.
TMJ Temporomandibular joint.
TTH Tension- type headache.
TLR Toll-like receptor

Key Points

- Pain perception reflects the functional state of the central nervous system, as well as the site and/or intensity of the peripheral stimuli.
- Central sensitization of the central nervous system is characterized by increased sensitivity to light touch, muscle tenderness, referred pain as well as local reddening and oedema.
- Migraine, tension-type headache and temporomandibular joint pain are manifestations of a sensitized central nervous system.
- Glia, activated by bacterial by-products and non-specific inflammation, produce and release neuroexcitatory agents that can induce central sensitization.
- Acute and chronic sinusitis may induce central sensitization.
- A broad range of differential diagnoses, including nasal and sinus disease, need consideration and exclusion in facial pain patient assessment.

18.1 Introduction

Patients with headache and facial pain that have been attributed to potential nasal or sinus pathology present frequently to otolaryngologists [1]. The classification and diagnostic criteria for different headache and facial pain conditions can be found in the International Association of Pain (IASP) Classification and the International Headache Classification (ICHD-III) [2]. An otolaryngologist, who wishes to understand, and manage patients presenting with 'sinus pain', has to utilize a central sensitization pain model [3]. Neurological, dental, rheumatological and musculoskeletal conditions, as well as nasal and sinus pathologies, need consideration in the differential diagnosis. Many facial pain patients also have associated anxiety and depression issues, which may also need to be addressed [4].

J. Bartley (✉)
Department of Otolaryngology—Head and Neck Surgery, Counties Manukau District Health Board, Auckland, New Zealand

Ö. Ö. Celebi, T. M. Önerci (eds.), *Nasal Physiology and Pathophysiology of Nasal Disorders*,
https://doi.org/10.1007/978-3-031-12386-3_18

18.2 Pain Pathophysiology

Our understanding of pain pathophysiology continues to evolve. In 1983, Clifford Woolf showed that many pain hypersensitivity features that accompanied peripheral tissue injury or inflammation resulted from sensory signalling augmentation in the central nervous system (CNS)—a concept termed 'central sensitization' [5]. Central sensitization introduces the concept that the CNS is continually modifying the degree, duration and spatial extent of pain in a way that reflects the CNS functional state, rather than the site or intensity of peripheral noxious stimuli [3]. Central sensitization is characterized by increased sensitivity to light touch, muscle tenderness, referred pain as well as local reddening and oedema. Migraine, tension-type headache (TTH), temporomandibular joint (TMJ) pain and medication overuse headache (MOH) are different manifestations of central sensitization [3, 6–9].

Most pain research has focused on synaptic plasticity triggered within the CNS by nociceptive inputs [3]. The significance of the glia, gap junctions and membrane excitability has also been recognized. In particular, the glia, their receptors and their secreted signalling factors influence neural function [10–14]. Activated glia produce and release a variety of neuroexcitatory substances. Toll-like receptors (TLRs), particularly TLR2 and TLR4, which respond to endogenous danger signals such as lipopolysaccharides that are associated with bacterial infection, have been implicated in glial cell activation [10, 12, 15]. Functional plastic changes in a large number of cells, including trigeminal neurons, glial cells (satellite cells, microglia, and astrocytes), and immune cells (macrophages and neutrophils), contribute to the sensitization and disinhibition of neurons in the peripheral and CNS, which results in orofacial pain hypersensitivity [16]. The neuropeptide calcitonin gene-related peptide (CGRP), which is implicated in migraine pathology, can initiate and maintain peripheral and central sensitization in the trigeminal nociceptive signalling pathways [17].

18.3 Migraine/Chronic Tension Headache

Current migraine theory centres on sensory processing dysfunction in the brain stem and/ or diencephalic nuclei. The trigeminal nucleus caudalis may be an important migraine generator [18]. Neural events in the brain stem result in the ensuing dilation of blood vessels, which in turn results in pain and further neural activation. The amount of light or sound coming into the body does not change during a migraine attack; the brain's sensory response does. The sensory processing mechanisms in the brain stem and limbic system in migraine patients are often hypersensitive (central sensitization), both before and after an attack [3, 7, 18, 19].

Many patients, who present with symmetrical frontal or temporal headache (or pressure), have a TTH; however, facial pain associated with central sensitization may be lateralized [9, 20, 21]. Mid-segment pain, where patients present with normal sinus CT scans and pain over the maxillary sinuses, has many physical findings consistent with central sensitization. It represents a TTH affecting the midface [20, 21].

A variety of factors can contribute to chronic TTH [18]. Chronic TTH is associated with pain and increased tenderness in the head, neck and shoulder muscles. The severity of the headaches relates directly to muscle tenderness. A common assumption has been that the pain in the head and neck muscles causes headache. However, the neck muscles can also be painful because of central sensitization in the spinal cord or brain stem. People with chronic TTH are often more sensitive not only around the head, neck and shoulder muscles but also elsewhere such as in the low back and calves, indicating an overall central sensitization [3, 21].

18.4 Medication Overuse Headache

In some people, the overuse of medication to treat their migraine/TTH can make their migraine/TTH worse. Medication overuse headache

(MOH) can be a significant clinical challenge [2]. If a person complains of regular migraine attacks or daily headaches requiring regular pain medication (more than twice a week), the headache could be caused by the medications. Almost every medication used to treat TTH and migraine can cause MOH [22]. The general recommendation is patients come off these medications while under medical supervision [6].

18.5 Sinusitis

The International Association of Pain (IASP) Classification and the International Headache Classification (ICHD-III) recognize that acute and chronic rhinosinusitis (CRS) can be associated with facial pain (Tables 18.1 and 18.2) [2]. Many clinical studies show that the majority of patients presenting with 'sinus headache' fulfil the diagnostic criteria for either migraine or TTH [23, 24]. Sinonasal and migrainous disorders may frequently co-exist [25]. The age- and sex-specific prevalence of CRS mirror those of migraine (middle age; women) [26].

A chronic infective process such as sinusitis can induce a central sensitization [10, 12]. TLRs, particularly TLR2 and TLR4, which have been implicated in glial cell activation recognize and respond to endogenous danger signals such as lipopolysaccharides that are released by damaged and dying cells associated with active bacterial infection [10, 15]. Clinical evidence of central sensitization is seen in CRS patients. The tenderness to palpation or percussion over the affected sinuses that is experienced by patients with sinusitis may indicate either excessive peripheral or central sensitization. Muscle tenderness is increased in CRS patients [27].

The quality of clinical evidence supporting the role of sinus surgery in helping patients with facial pain is limited. Facial pain appears an important feature of sphenoid sinusitis [28, 29]. The anterior face of the sphenoid sinus and the superior turbinate are particularly pain-sensitive [30]. A majority of prospective studies looking at the results of endoscopic sinus surgery for facial pain/headache indicate that in a group of CRS patients their facial pain benefits from surgery [31–35]. These studies have limitations, and some of the improvements can be explained using other mechanisms. The natural history of these facial pain/headache conditions is not known, and regression to the mean is also a potential mechanism. Nasal breathing slows the respiratory rate increasing the length of the expi-

Table 18.1 Headache attributed to acute rhinosinusitis

Diagnostic criteria:

A. Any headache fulfilling criterion C
B. Clinical, nasal endoscopic and/or imaging evidence of acute rhinosinusitis
C. Evidence of causation demonstrated by at least two of the following:
   1. Headache has developed in temporal relation to the onset of rhinosinusitis
   2. Either or both of the following:
      (a) Headache has significantly worsened in parallel with worsening of the rhinosinusitis
      (b) Headache has significantly improved or resolved in parallel with the improvement in or resolution of the rhinosinusitis
   3. Headache is exacerbated by pressure applied over the paranasal sinuses
   4. In the case of a unilateral rhinosinusitis, headache is localized and ipsilateral to it
D. Not better accounted for by another ICHD-3 diagnosis

Table 18.2 Headache attributed to chronic or recurring rhinosinusitis

Diagnostic criteria:

A. Any headache fulfilling criterion C
B. Clinical, nasal endoscopic and/or imaging evidence of current or past infection or other inflammatory processes within the paranasal sinuses
C. Evidence of causation demonstrated by at least two of the following:
   1. Headache has developed in temporal relation to the onset of chronic rhinosinusitis
   2. Headache waxes and wanes in parallel with the degree of sinus congestion and other symptoms of the chronic rhinosinusitis
   3. Headache is exacerbated by pressure applied over the paranasal sinuses
   4. In the case of unilateral rhinosinusitis, headache is localized and ipsilateral to it
D. Not better accounted for by another ICHD-3 diagnosis

ratory phase [36]. Slowing breathing and increasing the expiratory phase of the respiratory cycle increase the body's relaxation response [37, 38]. Relaxation techniques are effective treatments for both migraine and TTH [39].

Many patients with purulent secretions visible at nasal endoscopy have no headache or facial pain [20]. Many patients with infective sinusitis and/or nasal polyps do not have facial pain [1, 24]. When so-called sinus pain patients present acutely with symptoms, many have no evidence of infection [20]. There is also no correlation between the severity and location of the pain with the extent or location of mucosal disease [34, 40]. These observations can be explained using a central sensitization model. Sinus infection does not cause pain per se; it simply influences sensory thresholds. In some patients, this is insufficient to cause pain. In other patients, the inflammation will be such that pain processing is influenced. Sinus infection is only one possible factor influencing sensory thresholds; other pathologies can replicate exactly the same symptoms. Patients diagnosed with fibromyalgia and TTH report 'sinus' symptoms [27].

Careful patient counselling and selection are important before operating on 'sinus pain' patients. One should be careful about operating on CRS patients in the absence of significant infective symptoms or signs, and if the sinus CT scan shows no evidence of significant sinus disease. Comorbidities such as anxiety, depression, fibromyalgia, irritable bowel symptoms, neck and low back pain should make the surgeon wary of operating [3].

18.6 Contact Points and Nasal Obstruction

The prevalence of nasal contact points is the same in an asymptomatic population as in a symptomatic population. In symptomatic patients with unilateral pain when a contact point was present it was found on the contralateral side of the pain in 50% of patients [41]. A subgroup of patients may have a contact point headache [42]. Infiltrating the suspicious area with a local anaesthetic should abolish the pain and the pain should disappear for a week [2]. Care needs to be taken in patient counselling and selection.

The relationship between nasal obstruction and headache is controversial. Schonsted-Madsen et al. showed that relief of headache (pain localized to the forehead, glabella or above or around the eyes) was strongly related to relief of nasal obstruction [43]. A potential reason as to why improvement in the nasal airway might lead to improvement in TTH has previously been described (Sect. 18.5). Nasal surgery may also improve sleep quality. Poor sleep quality has been implicated in TTH and TMJ dysfunction [44, 45].

18.7 Temporomandibular Joint Disorders

Chronic TMJ pain is not due to occlusal abnormalities [46, 47]. Psychophysiological forces are important [39, 48]. The TMJ is lubricated by synovial fluid, which also nourishes the avascular cartilage and cartilaginous disc in the middle of the joint. Whenever the joint is compressed, the blood supply to the joint is reduced and the joint has difficulty manufacturing lubricating fluid. The friction within the joint increases and the cartilaginous disc in the middle of the joint begins to stick [49]. In times of stress, people tend to breathe using their upper chest, which tends to lead to a forward head position and increased pressure on the joint or clench and grind their teeth. Increased pressure in the TMJ can lead to clicking and sticking of the cartilage disc [50]. If the friction increases, the ligaments holding the disc in place stretch and the disc moves off the condylar head. In the absence of any major external trauma, before addressing any structural changes within the jaw joint, underlying psychosocial stresses, breathing re-education, as well as musculoskeletal issues, need to be addressed [39, 50–52].

18.8 Causes Removed from the Orofacial Area

A history of neck injury can often be overlooked. The nerve supply to the upper neck and head overlaps in the upper spinal cord. Using first principles, neck pain could be referred to the head. Alternatively, by sensitizing the spinal cord to incoming pain messages, neck pain could lower the pain threshold to other sensory messages being received from the face and head. After a whiplash injury to the neck, areas well away from the original injury site—from the head to the feet—are hypersensitive to normal sensory messages [3]. Migraine and TTH have been associated with food intolerances [53, 54]. The microbiota-gut-brain axis is receiving increasing attention, and the role of gut microbiota in headache and facial pain is an emerging research area [55].

18.9 Psychological Well-Being

Anxiety and depression have significant associations with migraine, TTH, CRS and TMJ disorders [4, 56]. Systems review of the patient often reveals issues with poor short-term memory, palpitations, shortness of breath, irritable bowel symptoms, poor sleep quality, cold hands and feet as well as tingling of the hands in an ulnar nerve distribution. Depressive symptoms are important in deciding on treatment, and in predicting treatment outcomes.

18.10 History Taking

Clinical diagnosis is largely dependent upon an accurate history. The initial differential diagnoses generated depend upon the clinician's experience, the patient's age and gender, and the time course and site of the pain. The differential diagnosis for headaches is extensive. Acute inflammatory causes are usually relatively obvious. Many patients may have symptoms reflecting a sensitized nervous system, such as night sweats, unexplained itch, tinnitus, irritable bowel or bladder symptoms, heavy painful periods and altered sensation on combing their hair, as well as pain problems elsewhere in the body [3]. Poor sleep and sleep quality have been implicated in THH and TMJ dysfunction [44, 45]. The interactions are probably bidirectional [9].

Neuralgias are characterized by sudden, intense, lancinating, burning or stabbing pain lasting only from a few seconds to less than two minutes. This pain is often triggered by sensory or mechanical stimuli. Trigeminal neuralgia is typically seen in older females, unilateral and in the second and/or third divisions of the trigeminal nerve. Rarely, pontine tumours, the base of tongue tumours or multiple sclerosis need to be considered as a secondary cause. If a cough or sneezing makes the headache worse, a posterior fossa lesion may need to be considered.

18.11 Examination

After routine otolaryngological examination, palpation for muscle tenderness and excessive reddening afterwards provides important physical information about the state of the CNS [8]. Many of these tender areas (or trigger points) correspond to traditional Chinese acupuncture points. Acupuncture is effective in the management of TTH [57, 58]. Sometimes there are subtle differences in swelling and redness between the two sides of the face. The pectoralis minor muscle as it inserts into the coracoid process, the mid-point of the upper trapezius muscle (Fig. 18.1), sternocleidomastoid (Fig. 18.2), masseter, temporalis and the suboccipital muscles are usually tender when examined using appropriate palpation techniques [59]. Jaw joint and associated muscle tenderness together with limited and jerky jaw movements and clicking of the jaw joint may be found on clinical examination. The teeth can be examined for excessive wear (bruxism) or percussion tenderness. Alterations in facial sensation are best detected by comparing moving light touch between the two sides of the face in the three divisions of the trigeminal nerve. Posture and cervical range movement should be assessed [9]. An examination of areas away from the head

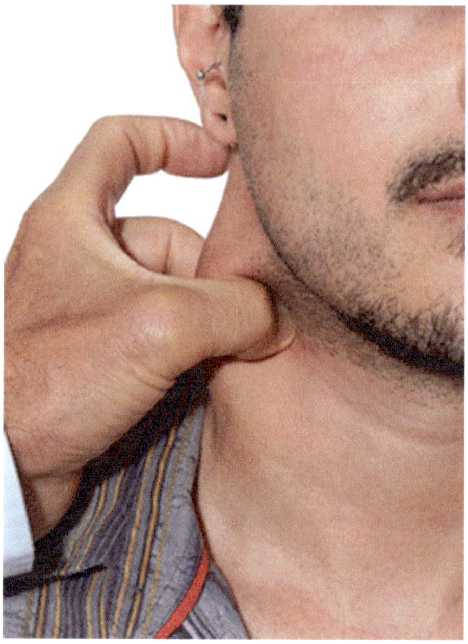

Fig. 18.1 The tender area in the superior midpoint of the shoulder in the trapezius muscle is located if one palpates the superior aspect of the muscle between thumb and forefinger [59]

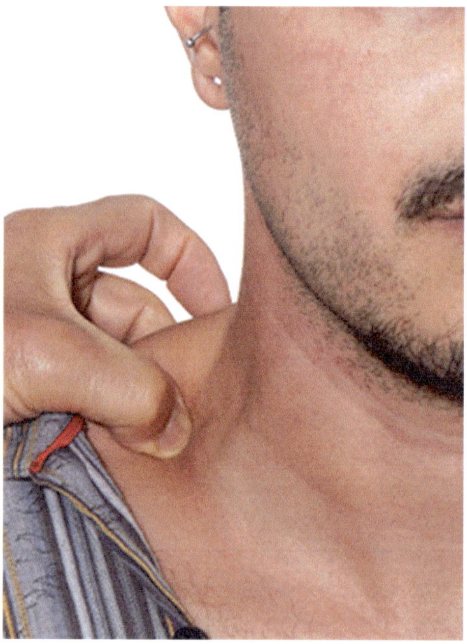

Fig. 18.2 To examine the sternocleidomastoid muscle for tenderness, the muscle needs to be examined up and down its length in a gentle pincer-like grip between thumb and forefinger [59]

and neck often provides useful, additional information. People who are highly stressed tend to take small irregular breaths, largely in their upper chest. The low back, extensor forearm muscles and calves are often tender to palpation in pain patients as well [3]. Clinically, patients with unilateral facial pain are often tender down that side of the body. The muscle tenderness reflects the underlying status of the nervous system.

18.12 Investigations

18.12.1 Radiology

Further urgent investigation and neurological evaluation are warranted for patients presenting with facial pain together with other significant symptoms/signs (Table 18.3). Diagnostic imaging tests (e.g. plain X-rays, dental X-rays, MRI, axial CT scan) may help determine or exclude a cause of pain. The choice and timing of the test vary according to clinical suspicions and the findings on physical examination.

18.12.2 Blood Investigations

Certain blood tests may be useful in evaluating patients presenting with facial pain [60]. Some females will have an iron deficiency. Vegetarians and the elderly can have undiagnosed vitamin B12 deficiencies. People living in countries with temperate climates, who have dark or brown skin,

Table 18.3 Red flags requiring further investigation

- The pain is new or has significantly changed
- There is significant associated nausea and vomiting
- The pain is unusually severe or persistent
- There is accompanying fever
- The pain is made worse by coughing, sneezing or a change in position
- There is a change in strength, coordination or senses
- There is drowsiness, difficulty thinking or concentrating
- The headache is progressively getting worse
- The headache wakes the patient from sleep
- The headache occurs for the first time in childhood or after age 50

often have a significant vitamin D deficiency. Thyroid function needs checking in TTH patients [44]. Granulomatosis with polyangiitis (Wegener's) may need exclusion. In a person aged over 50 years with a rapidly developing headache, an erythrocyte sedimentation rate (ESR) or C-reactive protein (CRP) test is mandatory.

18.13 Management

Depending on the clinical diagnosis, a wide range of treatment options is available.

Many of these interventions are outside the conventional knowledge base of many otolaryngologists, but some knowledge is useful if an otolaryngologist wished to provide comprehensive care and diagnosis. Psychological interventions such as cognitive behavioural therapy and relaxation work can be extremely useful for migraine, TTH and TMJ disorders [8, 44, 51]. Attention to diet, sleep patterns, posture and exercise can be important. Simple analgesics and non-steroidal anti-inflammatory drugs are recommended for the treatment of episodic facial pain. Drugs that are commonly useful are Gabapentin or low-dose Amitriptyline at night. Mirtazapine and Venlafaxine are second-choice drugs.

18.14 Conclusions

The otolaryngologist should be equipped to evaluate, treat or appropriately refer patients presenting with facial pain. Migraine, TTH, TMJ pain and sinus pain share a common underlying pathophysiology. Around the head and neck, other pathologies apart from sinus disease can also be related to facial pain. These factors also need consideration in the diagnostic work up. Because facial pain may involve a range of medical/surgical subspecialties, a multidisciplinary approach is often needed.

References

1. De Corso E, Kar M, Cantone E, Lucidi D, Settimi S, Mele D, et al. Facial pain: sinus or not? Acta Otorhinolaryngol Ital. 2018;38:485–96.
2. Headache classification Committee of the International Headache Society. The international classification of headache disorders, 3rd edition. Cephalalgia. 2018;38:1–211.
3. Woolf CJ. Central sensitization: implications for the diagnosis and treatment of pain. Pain. 2011;152:S2–S15.
4. Jacobson S, Folstein M. Psychiatric perspectives on headache and facial pain. Otolaryngol Clin N Am. 2003;36:1187–200.
5. Woolf C. Evidence for a central component of post-injury pain hypersensitivity. Nature. 1983;306:686–8.
6. Diener HC, Holle D, Solbach K, Gaul C. Medication-overuse headache: risk factors, pathophysiology and management. Nat Rev Neurol. 2016;12:575–83.
7. Burstein R, Noseda R, Borsook D. Migraine: multiple processes, complex pathophysiology. J Neurosci. 2015;35:6619–29.
8. Bendtsen L, Fernandez-de-la-Penas C. The role of muscles in tension-type headache. Curr Pain Headache Rep. 2011;15:451–8.
9. Jay GW, Barkin RL. Primary headache disorders - part 2: tension-type headache and medication overuse headache. Dis Mon. 2017;63:342–67.
10. Milligan ED, Watkins LR. Pathological and protective roles of glia in chronic pain. Nat Rev Neurosci. 2009;10:23–36.
11. Guo W, Wang H, Watanabe M, Shimizu K, Zou S, LaGraize SC, et al. Glial-cytokine-neuronal interactions underlying the mechanisms of persistent pain. J Neurosci. 2007;27:6006–18.
12. Scholz J, Woolf CJ. The neuropathic pain triad: neurons, immune cells and glia. Nat Neurosci. 2007;10:1361–8.
13. Lee S, Zhao YQ, Ribeiro-da-Silva A, Zhang J. Distinctive response of CNS glial cells in oro-facial pain associated with injury, infection and inflammation. Mol Pain. 2010;6:79.
14. Hossain MZ, Unno S, Ando H, Masuda Y, Kitagawa J. Neuron-glia crosstalk and neuropathic pain: involvement in the modulation of motor activity in the orofacial region. Int J Mol Sci. 2017;18:2051.
15. Liu T, Gao Y-J, Ji R. Emerging role of toll-like receptors in the control of pain and itch. Neurosci Bull. 2012;28:131–44.
16. Shinoda M, Hayashi Y, Kubo A, Iwata K. Pathophysiological mechanisms of persistent orofacial pain. J Oral Sci. 2020;62:131–5.
17. Durham PL. Diverse physiological roles of calcitonin gene-related peptide in migraine pathology: modulation of neuronal-glial-immune cells to pro-

mote peripheral and central sensitization. Curr Pain Headache Rep. 2016;20:48.
18. Jay GW, Barkin RL. Primary headache disorders: part I- migraine and the trigeminal autonomic cephalalgias. Dis Mon. 2017;63:308–38.
19. Welch K. Contemporary concepts of migraine pathogenesis. Neurology. 2003;61(8 Suppl 4):S2–8.
20. Jones N. Midfacial pain segment pain: implications for rhinitis and sinusitis. Curr Allergy Asthma Rep. 2004;4:187–92.
21. Fumal A, Schoenen J. Tension-type headache: current research and clinical management. Lancet Neurol. 2008;7:70–83.
22. Cupini LM, Sarchielli P, Calabresi P. Medication overuse headache: neurobiological, behavioural and therapeutic aspects. Pain. 2010;150:222–4.
23. Silberstein S. Headaches due to nasal and paranasal sinus disease. Neurol Clin. 2004;22:1–19.
24. Jones N. Sinus headaches: avoiding over- and misdiagnosis. Expert Rev Neurother. 2009;9:439–44.
25. Eross E, Dodick D, Eross M. The sinus, allergy and migraine study (SAMS). Headache. 2007;47:213–24.
26. Aaseth K, Grande RB, Kvaerner K, Lundqvist C, Russell MB. Chronic rhinosinusitis gives a ninefold increased risk of chronic headache. The Akershus study of chronic headache. Cephalalgia. 2010;30:152–60.
27. Naranch K, Park Y, Repka-Ramirez M, Velarde A, Clauw D, Baraniuk J. A tender sinus does not always mean rhinosinusitis. Otolaryngol Head Neck Surg. 2002;127:387–97.
28. Friedman A, Batra P, Fakhri S, Citardi M, Lanza D. Isolated sphenoid sinus disease: etiology and management. Otolaryngol Head Neck Surg. 2005;133:544–50.
29. Fadda GL, D'Eramo A, Grosso A, Galizia A, Cavallo G. Isolated sphenoid sinus inflammatory disease - a report of 14 cases. Iran J Otorhinolaryngol. 2020;32:101–7.
30. Clerico DM. An experimental study of pain upon stimulation of the nasal and sinus cavities. Am J Otolaryngol. 2014;35:300–4.
31. Moretz W, Kountakis S. Subjective headache before and after sinus surgery. Am J Rhinol. 2006;20:305–7.
32. Phillips J, Vowler S, Salam M. Endoscopic sinus surgery for 'sinus headache'. Rhinology. 2007;45:14–9.
33. Soler Z, Mace J, Smith T. Symptom based presentation of chronic rhinosinusitis and symptom-specific outcomes after endoscopic sinus surgery. Am J Rhinol. 2008;22:287–301.
34. Tarabichi M. Characteristics of sinus related pain. Otolaryngol Head Neck Surg. 2000;122:842–7.
35. Chester A, Antisdel J, Sindwani R. Symptom-specific outcomes of endoscopic sinus surgery: a systematic review. Otolaryngol Head Neck Surg. 2009;140:633–9.
36. Ayoub J, Cohendy R, Dauzat M, Targhetta R, De la Coussaye J, Bourgeois J, et al. Non-invasive quantification of diaphragm kinetics using M-mode sonography. Can J Anaesth. 1997;44:739–44.
37. Cappo B, Holmes D. The utility of prolonged respiratory exhalation for reducing physiological and psychological arousal in non-threatening and threatening situations. J Psychosom Res. 1984;28:265–73.
38. Noble DJ, Hochman S. Hypothesis: pulmonary afferent activity patterns during slow, deep breathing contribute to the neural induction of physiological relaxation. Front Physiol. 2019;10:1176.
39. Carlson C. Psychological considerations for chronic orofacial pain. Oral Maxillofac Surg Clin North Am. 2008;20:185–95.
40. Levine H, Setzen M, Cady R, Dodick D, Curtis P, Schreiber C, et al. An otolaryngology, neurology, allergy and primary care consensus on diagnosis and treatment of sinus headache. Otolaryngol Head Neck Surg. 2006;134:516–23.
41. Abu-Bakra M, Jones N. The prevalence of nasal contact points in a population with facial pain and a control population. J Laryngol Otol. 2001;115:629–32.
42. Peric A, Rasic D, Grgurevic U. Surgical treatment of rhinogenic contact point headache: an experience from a tertiary care hospital. Int Arch Otorhinolaryngol. 2016;20:166–71.
43. Schonsted-Madsen U, Stocksted P, Christensen P, Koch-Hnedriksen N. Chronic headache related to nasal obstruction. J Laryngol Otol. 1986;100:165–70.
44. Sacco S, Ricci S, Carolei A. Tension-type headache and systemic medical disorders. Curr Pain Headache Rep. 2011;15:438–43.
45. Cho SJ, Song TJ, Chu MK. Sleep and tension-type headache. Curr Neurol Neurosci Rep. 2019;19:44.
46. Koh H, Robinson P. Occlusal adjustment for treating and preventing temporomandibular joint disorders. J Oral Rehabil. 2004;31:287–92.
47. Fouda AAH. No evidence on the effectiveness of oral splints for the management of temporomandibular joint dysfunction pain in both short and long-term follow-up systematic reviews and meta-analysis studies. J Korean Assoc Oral Maxillofac Surg. 2020;46:87–98.
48. Suvinen T, Reade P, Kemppainen P, Kononen M, Dworkin S. Review of aetiological concepts of temporomandibular pain disorders: towards a biosocial model for integration of physical disorder factors with psychological and psychosocial illness factors. Eur J Pain. 2005;9:613–33.
49. Nitzan D. The process of lubrication impairment and its involvement in temporomandibular joint disc displacement: a theoretical concept. J Oral Maxillofac Surg. 2001;59:36–45.
50. Bartley J. Breathing and temporomandibular joint disease. J Bodyw Mov Ther. 2011;15:291–7.
51. Aggarwal V, Tickle M, Javidi H, Peters S. Reviewing the evidence: can cognitive behavioral therapy improve outcomes for patients with chronic orofacial pain? J Orofac Pain. 2010;24:163–71.
52. Aggarwal VR, Fu Y, Main CJ, Wu J. The effectiveness of self-management interventions in adults with chronic orofacial pain: a systematic review, meta-analysis and meta-regression. Eur J Pain. 2019;23:849–65.

53. Hadjivassiliou M, Sanders D, Grünewald R, Woodroofe N, Boscolo S, Aeschlimann D. Gluten sensitivity: from gut to brain. Lancet Neurol. 2010;9:318–30.
54. Egger J, Carter CM, Wilson J, Turner MW, Soothill J. Is migraine food allergy? A double-blind controlled trial of oligoantigenic diet treatment. Lancet. 1983;2:865–9.
55. Lin B, Wang Y, Zhang P, Yuan Y, Zhang Y, Chen G. Gut microbiota regulates neuropathic pain: potential mechanisms and therapeutic strategy. J Headache Pain. 2020;21:103.
56. Wasan A, Fernandez E, Jamison R, Bhattacharyya N. Association of anxiety and depression with reported disease severity in patients undergoing evaluation for chronic rhinosinusitis. Ann Otol Rhinol Laryngol. 2007;116:491–7.
57. Gildir S, Tuzun EH, Eroglu G, Eker L. A randomized trial of trigger point dry needling versus sham needling for chronic tension-type headache. Medicine (Baltimore). 2019;98:e14520.
58. Linde K, Allais G, Brinkhaus B, Fei Y, Mehring M, Shin BC, et al. Acupuncture for the prevention of tension-type headache. Cochrane Database Syst Rev. 2016;4:CD007587.
59. Simons DG, Travell JG, Simons LS. Travell & Simons Myofascial pain and dysfunction: the trigger point manual (3rd edition). Philadelphia: Williams and Wilkins; 2018.
60. Gerwin R. A review of myofascial pain and fibromyalgia – factors that promote their persistence. Acupunct Med. 2005;23(3):121–34.

Computational Fluid Dynamics of the Nasal Cavity

19

Ralph Mösges

Core Messages

- Computational fluid dynamics (CFD) is a mature technology used widely in engineering to solve and analyze problems that involve fluid flows.
- Computational fluid dynamics has been used to demonstrate physiologic and pathologic conditions of nasal flow and to support preoperative planning and control of postsurgical outcomes.
- Using high-definition three-dimensional imaging CFD may offer a chance to study the effects of medication on the tissues lining the surface of the nasal cavity.
- CFD may help to design devices for optimal nasal delivery of medications such as nasal spray applicators.
- CFD may become a universal tool not only for research and pharmaceutical development but also for advanced patient care in rhinology.

The nose is not a tube, nor can it be regarded as two pipelines transporting air to the lung.

The nose has multiple functions, enabling the exchange of gas between the circulating blood and the environment, humidification, warming and cleaning of the air, and last but not least it supports the sense of smell as an alarm function but also to find the ideal mate.

The nasal cavity is optimized for all these tasks and only surgical hybris can lead to the assumption that one could easily ameliorate its structure. It is this "Plummer's mentality" that has transformed human beings with slightly obstructed noses into "nasal wrecks" suffering from the empty nose syndrome.

For a long time, otorhinolaryngologists have tried to measure nasal flow under conditions of obstruction, in order to surgically remove what is deemed to be the obstacle to normal nasal breathing. But even in cases where the pre−/post-comparison of nasal flow, measured under defined conditions demonstrated significant improvements, patients were sometimes unhappy with the outcome. Septoplasty and turbinectomy are typical interventions with a high rate of "non-responders" to therapy, at least in the long run. The difference between objective nasal patency and the subjective feeling of obstruction has been very well described in excellent reviews [1]. One of the reasons for this discrepancy may be that we measure nasal flow in our tests at pressures that are by far higher (150 Pa or 200 Pa) than those produced in a normal breathing cycle under resting conditions. However, the test equipment we use these days (and have been using over the last 20 years) does not allow for reliable, reproducible measurements at pressures as low as those occurring in real life. Therefore, especially in

R. Mösges (✉)
Institute of Medical Statistics, Informatics and Epidemiology, University Hospital of Cologne, Köln, Germany
e-mail: ralph@moesges.de

Ö. Ö. Celebi, T. M. Önerci (eds.), *Nasal Physiology and Pathophysiology of Nasal Disorders*,
https://doi.org/10.1007/978-3-031-12386-3_19

cases of obstructed anatomy, other methods may be more adequate to calculate nasal flow and the exchange of warmth and humidity at the borderline, the superficial tissue which is the mucous membrane lining the wall of the nasal cavity.

Physics has developed a tool for this task. It is called computational fluid dynamics (CFD), a mature technology used widely in engineering to solve and analyze problems that involve fluid flows [2]. The mathematical predictions of CFD can also be applied to the nasal airflow. For numerical simulation of the nasal airflow, patients have to undergo computed tomography (CT) or magnetic resonance imaging (MRI) scans of the head [3].

Many of the first flow simulation studies focused on nasal airflow biophysics [4–6]. CFD has already been used for the flow analysis of pathological cases [7–16] and has been proposed as a tool to predict actual surgical outcomes using virtual nasal surgery models [17–21]. A new field of application, for which CFD has been successfully used, is the three-dimensional visualization of the distribution and the consecutive effects of intranasally applied medications such as nasal steroids or decongestive nasal sprays [22–24].

It is the objective of this review, on the one hand, to present a summary of current applications of CFD in rhinology (demonstration of physiologic and pathophysiologic flow distributions in the nose, preoperative planning, and postsurgical assessment of rhinosurgery outcomes); on the other hand, we want to demonstrate that CFD is a powerful tool for the three-dimensional presentation of drug action on the nasal mucous membranes.

19.1 Current Applications of Computational Fluid Dynamics

19.1.1 Demonstration of Physiologic and Pathologic Flow Distributions in the Nose

An adequate example of the three-dimensional visualization of physiologic flow in the nasal cavity is found in the paper published by Ishikawa and colleagues in 2006. They investigated the differences between nasal inspiration and expiration on the basis of computed tomography. They found that during inspiration, nasal flow in the middle meatus was considerably more pronounced and turbulent than during expiration [4]. Tan and coworkers observed a similar flow pattern in their clinical trial. During inspiration, turbulence occurred primarily in the anterior part and on the floor of the nasal cavity. In contrast, no turbulence occurred during expiration. They measured the maximum nasal flow around the plane of the palatine velum during both inspiratory and expiratory phases [5]. Wen and colleagues also investigated the physiologic flow in the nose. They were able to demonstrate that the high velocities in the constrictive nasal valve area region and the high flow appeared close to the septum walls [6].

Moreover, pathophysiologic aberrations in the nasal cavity like septal deviations [7, 10, 14, 15], turbinate hypertrophy [8, 11, 12], nasal bone fracture [9], septal perforation [13], deviation of the external nose [16], and their consequences can be visualized by means of CFD technology.

Sun and colleagues compared, for example, patients with nasal septum deviations with others without anatomical irregularities. They came to the conclusion that flow simulation allows to visualize the differences in nasal flow origination from abnormal anatomy [15]. The Chinese scientist Liu demonstrated the influence of different forms of septum deviation on nasal flow characteristics [14]. His compatriot Guo proved that unilateral infraturbinal hypertrophy also changed the normal anatomy and influenced the aerodynamics of the nasal cavity. According to his work, these changes have a substantial effect on the important functions of the nose-like humidification, warming of the inhaled air, and the sense of smell [11].

Bailie and coworkers validated in a clinical trial in five patients that flow simulations based upon computed tomography can help to analyze the function of the lower and middle turbinate regarding warm up and cool down of inhaled air. Moreover, this flow model may explain that shear stress created by the flow around the turbinates in

Kiesselbach's area may induce nasal bleeding [25]. This clinical trial supported findings published by Pless and colleagues in 2004. They came to the conclusion that both the lower and the middle turbinates are primarily responsible for heat recovery during expiration and that areas of the highest decrease in temperature are characterized by turbulent airflow [26]. Sommer and his working group attributed utmost importance to the middle turbinate for climatization and humidification of inhaled air [27].

Ishikawa and colleagues gave primary attention to olfaction in their publication. In their three-dimensional flow model, they were able to demonstrate that inspiratory airflow is widely distributed in the olfactory region than exhaled air. In contrast, sniffing flow had the widest distribution in the olfactory region, although no increase in nasal flow was noticed in the flow model. They drew the conclusion that recirculation flow strongly promotes olfactory function in the nose [28].

19.1.2 Preoperative Planning and Postsurgical Outcome Assessment in Rhinosurgery

Another stronghold of CFD is the field of preoperative planning and postsurgical outcome assessment in rhinosurgery. As early as 2000, Bockholt and colleagues investigated the potential benefit of nasal flow simulations in the field of rhinosurgery. They came to the conclusion that by setting up a three-dimensional model of the nasal cavity based upon CT slices, planning of a surgical intervention may be optimized and the outcome may be improved for the patient [17]. In their clinical trial, Xiong and colleagues assessed the use of CFD for preoperative planning and postsurgical outcome control. By visualization of nasal airflow before and after a virtual endoscopic intervention, they could simulate patients' outcome [21].

Other research groups confirmed the applicability of CFD to visualize the postsurgical outcome of various rhinosurgical interventions such as rapid maxillary expansion [18], surgery of a hypertrophic turbinate [18], septoplasty, and partial lateral turbinectomy [19].

Another application of CFD technology is the clinical picture of sleep apnea syndrome. Sung and Xu conducted flow simulations in 2006 that resulted in a better understanding of the pathophysiology of obstructive sleep apnea syndrome in adults as well as in children [29, 30]. Bimaxillary surgery with maxillomandibular advancement to widen the post-glossal space is one standardized procedure in the treatment of obstructive sleep apnea syndrome [31]. In a clinical trial in two patients with sleep apnea syndrome conducted by Yu and coworkers, nasal airflow before and after surgical intervention (maxillomandibular advancement) was calculated on the basis of computed tomography slices. CFD demonstrated a postoperative widening of the upper airways with balanced volume flows and pressure patterns. The postsurgical clinical picture of the patient (reduced inspirative energy and better ventilation) confirmed the prognostic value of CFD [32].

19.1.3 Recently Developed Applications of Computational Fluid Dynamics

A new application of CFD is the visualization of drug effects on the nasal mucosa in three-dimensional flow simulations. Garlapati and his working group investigated in their clinical trial the effects of the application of nasal sprays on the mucosa. Based upon magnetic resonance imaging, they could demonstrate that intranasal application of medication is most effective when the patient actively inspires during the application. Surprisingly, the posture of the head had no significant influence on the distribution of the inhaled aerosol [24]. Frank and colleagues complemented this observation with the finding that the posture of the head only influences the distribution of the nasal sprays in the case of absent or minimal inspiratory airflow [23]. Chen and colleagues also investigated the effects of nasal sprays using CFD technology. With the nasal flow model of their patient, they could show that the distribution of nasally applied drugs with a particle diameter of 10 μm was significantly improved after functional endoscopic sinus surgery (FESS), resulting in a moderate nasal flow [22]. Frank and coworkers agree to this observation saying that surgical correction of nasal anatomic deformities (e.g., nasal septum deviation) could improve drug delivery on the nasal mucosa [23].

The application of CFD technology is not only limited to the visualization of drug effects in healthy [24] subjects or patients that underwent surgery [22, 23]. It can also be used to study changes of the mucous membranes in patients treated for symptoms of allergic rhinitis. In two clinical trials, we have studied the anti-obstructive effects of antiallergic medications on the swelling status in a patient suffering from seasonal allergic rhinitis. This example is used to describe the methodology applied in CFD.

19.2 Methodology

19.2.1 Imaging and Grid Generation

19.2.1.1 Using Computed Tomography

The computational grid is generated based on a surface definition by a computer tomographic scan of the human nasal cavity which results in about 300 cuts 1 mm apart or less, a resolution of 512 × 512 pixels or higher per cut, and 2 bytes per pixel for the density resolution.

To allow a better interface detection, blurring must be reduced by sharpening the image, by applying a 3 × 3 × 3 convolution matrix filter, which emphasizes the voxel differences depending on the 3 × 3 × 3 neighborhood around the center voxel. This supports the manual segmentation of the nasal cavity by an experienced ENT specialist, who examines each slice of the three-dimensional image and identifies the region of interest (ROI) with a digital pen tablet. The image is then further preprocessed. A seeded region growing algorithm [33] is used to identify the previously detected ROI by placing seed points inside the fluid volume of the nasal cavity and recursively descending in the neighborhood of them. The identification is based on a lower and an upper threshold depending on the assignment method of the ENT specialist. Based on this segmentation, the Marching Cubes algorithm [34] is used to extract the surface of the nasal cavity yielding a three-dimensional triangle representation. This algorithm is based on the intensity detection along voxel edges and defines vertices along these lines by a bilinear interpolation between the intensities at the corners of the voxels. A set of triangles is defined for such a vertex configuration, which is looked up in a configuration table, containing 256 possible combinations. In a post-processing step, the surface is smoothed using a windowed sinc function [35], removing high-frequency noise in Fourier space by applying a transfer function. In a final step, the surface is split into multiple parts. The nostrils and the throat are separated from the rest of the nasal cavity and are smoothed with a Laplace filter until convergence. This filter performs a relaxation of the mesh and iteratively moves all vertices into one plane. This step allows the proper application of the boundary conditions in the flow simulation. Based on this model, an automatic Cartesian grid generator creates the computational mesh. A minimal bounding cube is initially placed around the surface. This cube is then continuously split into eight smaller cubes until a user-defined level of refinement is reached. During the splitting process, cells outside the fluid domain are removed.

Using a marching cube algorithm for triangulation of the scanning data, an unstructured surface with 200,000 nodes and 420,000 triangles can be obtained. For an exact match of the experimental flow conditions, pipes for in- and outflow are added. Via the grid generation tool, a structured grid of 450,000 nodes in 34 blocks (Fig. 19.2) is generated which has a nested O-topology and additional blocks underneath the turbinates. To ensure a divergence-free solution, the blocks match at their interfaces.

19.2.1.2 Using MRI

T2-weighted MRI is used to visualize detailed internal structures and restricted body functions. It provides a three-dimensional image of the nasal cavity, the sinuses, and the pharynx and allows the assessment of the nasal mucosa membrane swelling. To perform a fluid mechanical analysis of the flow in the human nasal cavity, the surface of the region of interest, i.e., the volume of the nasal cavity, is extracted from MRI data and processed in multiple steps [36]. Since MRI measures the fluid characteristics of different tissues, the distinction between bone and air is generally difficult because they contain no or only a small amount of fluid and give a similar MRI signal, i.e., these areas appear black.

To extract a suitable model of the nasal cavity from MRI data, the first step is segmentation. This is performed manually by an experienced radiologist on a graphic tablet. During the next step, the segmented volume is converted into a triangular mesh surface by a Marching Cubes algorithm. The surface then is smoothed and used for generating a volumetric grid again. This can be done by dividing a large starting cube containing the entire surface into eight smaller ones with the same size each. The small cubes now lying outside of the surface are omitted, while the cubes inside are further divided. The procedure finishes with the so-called Cartesian lattice of approximately 4.1 million cells. For more complex calculations including analysis of humidification, the lattice may have 25 million cells or more.

To simulate nasal airflow, the Navier–Stokes equations, which describe the motion of fluid substances, must be solved for every cell. Therefore, several boundary conditions are required. The applied "no-slip" condition defines the velocity at the walls of the nasal cavity as zero. At the nostrils, a constant inflow is set, which causes a stationary flow field to develop after a sufficient number of simulated time steps. Therefore, the assumption of a quasi-steady flow is needed. The local velocity and pressure are obtained for every cell of the simulated lattice. Since the inflow is set as a fixed input for all four flow simulations, the total flow is identical. The computational fluid dynamics model used implements a Lattice–Boltzmann method. It was validated experimentally by an artificial nose model.

19.2.2 Method of Solution

To simulate the flow field, the three-dimensional Navier–Stokes equations are solved. An explicit five-step Runge–Kutta method of second-order accuracy is used for time integration. The following boundary conditions are imposed. A no-slip isothermal wall condition is assumed and the pressure gradient normal to the wall is set to zero. At the inflow section, a parabolic velocity profile is prescribed with the mean velocity determined from an assumed isentropic expansion from a stagnation state to the local static pressure that is computed from the interior pressure distribution using a vanishing pressure gradient in the streamwise direction. At the outflow plane, the static pressure level is prescribed and a nonreflecting boundary condition of Poinsot and Lele is applied [37].

The simulation was carried out using a Lattice-Boltzmann method and was performed on grids containing about 20×10^6 cells. As for the imposed boundary conditions, a no-slip wall condition proposed by Bouzidi et al. [38] was used. A volume flux of 125 ml/s was prescribed at the inflow boundaries with a von Neumann condition for the velocity. The density was extrapolated in surface normal direction by applying a Dirichlet condition. The outflow boundary condition was based on the formulation by Finck et al. [39] and imposed a constant pressure and extrapolated the velocities. The Reynolds number based on the mean hydraulic diameter of the nostrils and a volume flux of 125 ml/s were calculated for all nasal cavity geometries to guarantee an equal volume flux in all cases (Fig. 19.1).

19.2.3 Virtual Reality-Based Visualization of Flow Simulation Results

Visualization is a fundamental ingredient for gaining insights into the simulation results.

In particular, visualization is indispensable to the understanding of complex, dynamic processes observed in computational fluid dynamics, whose scientific and technical analyses

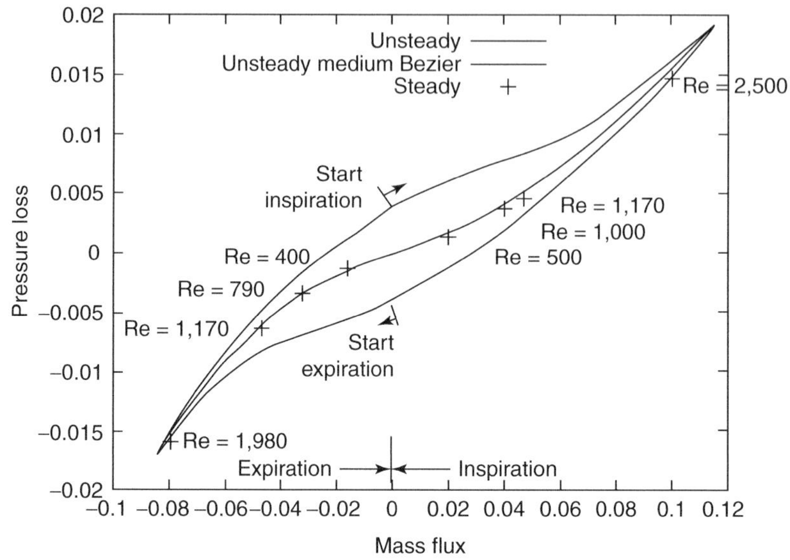

Fig. 19.1 Pressure loss over mass flux; Reynolds numbers indicating turbulent conditions

require a multidimensional representation of flow in space and time. In comparison to 2D or perspective representations, virtual reality-based visualization, i.e., stereoscopic and user-centered projection techniques, allows a much more intuitive comprehension of spatiotemporal correlations. In contrast to animation techniques, virtual reality is inherently interactive and thus allows an interactive exploration of simulation data.

In the context of the project, we are employing different visualization techniques tailored to the needs of the potential users. For the validation of the numerical simulation, we implemented standard visualization techniques like cut planes, isosurfaces, streamlines, and stream ribbons, all primarily based on the visualization toolkit. For the physicians, tailored visualization techniques are still to be determined and to be studied. The added dimensions in the visualization lead to the necessity of a user interface that enables the user to concentrate on the exploration and evaluation of the simulation data via dedicated navigation techniques, speech recognition, and advanced interaction methods for applying the visualization methods. The feedback of users shows that the multimodal user interface supports the visual validation process of the numerical simulation in such a way that the simulation results can be explored in an intuitive manner.

19.3 Application

19.3.1 Nasal Cavity 3D Imaging-Based Modeling: An Assessment Tool for the Anti-Obstructive Potency of Antiallergic Compounds

It was the objective of this study to visualize the anti-obstructive effect of intranasal steroid sprays (INS) in a patient with allergic rhinitis by simulating the nasal airflow with computational fluid dynamics. The patient underwent magnetic resonance imaging (MRI). After nasal allergen challenge, all measurements were repeated 30 min later. During the following 2 weeks, the test subject applied the INS. Then again MRI was performed before and 30 min after allergen challenge.

Figure 19.2 compares the nasal flow before and after allergen challenge at baseline. Before allergen exposure, a pattern of widespread flow distribution over the entire nasal cavity can be noticed (left). After exposure to the allergen, flow to the more cranial parts of the nasal cavity becomes sparse (right).

Figure 19.3 shows the flow patterns in the nasal cavity calculated from the MRI-based model after 14 days of (prophylactic) treatment with an intranasal steroid. Again, the distribution

Fig. 19.2 Nasal airflow at baseline before (*left*) and after (*right*) allergen challenge

Fig. 19.3 Nasal airflow after 2 weeks of treatment with an intranasal steroid before (*left*) and after (*right*) allergen challenge

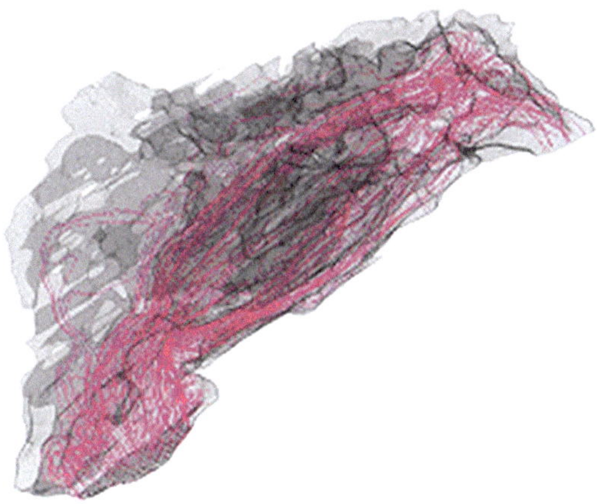

Fig. 19.4 Nasal airflow at baseline after allergen challenge

Fig. 19.5 Nasal airflow after 5 weeks of antihistamine treatment after allergen challenge

of airflow at baseline is widespread over the entire nasal cavity. Velocity of airflow is somewhat reduced after allergen challenge, however, by far not as pronounced as before treatment. We therefore conclude that this demonstrates the anti-obstructive effect of the intranasal steroid.

Similar improvements could be demonstrated for the treatment with an anti-obstructive antihistamine comparing the flow pattern at baseline (Fig. 19.4) with the flow pattern after 5 weeks of treatment (Fig. 19.5).

19.4 Conclusion

Leong and colleagues have pointed out the potential benefits of CFD technology applied to rhinology in their systematic review saying that "this technology has improved understanding of the complex nasal anatomy and the implications of disease and surgery on physiology" [2]. We can only partially support this viewpoint. Nasal flow simulation primarily has served to demonstrate physiologic and pathological flow patterns in the nasal cavity. It has been used in individual cases for preoperative planning and for the assessment of outcomes in rhinosurgery. The origins of this method date back to the 1980s. Research has been ongoing in this field for more than 20 years without establishing the method in regular care. This is due to the fact that standards are still lacking and consensus on the interpretation of the findings and on reliable outcome parameters has not been reached. CFD technology can be a valuable but extremely complex and by consequence costly research tool for sophisticated problems like drug distribution in the nasal cavity or the visualization of drug effects on the nasal mucosa. There is, however, hope that these limitations will be overcome in future with further improvements in imaging and software technology.

References

1. Baraniuk JN. Subjective nasal fullness and objective congestion. Public Health. 2011;8(1):62–9. https://doi.org/10.1513/pats.201006-042RN.
2. Leong SC, et al. A review of the implications of computational fluid dynamic studies on nasal airflow and physiology. Rhinology. 2010;48(2):139–45.
3. Bailie N, et al. An overview of numerical modelling of nasal airflow. Rhinology. 2006;44(1):53–7.
4. Ishikawa S, et al. Visualization of flow resistance in physiological nasal respiration: analysis of velocity and vorticities using numerical simulation. Arch Otolaryngol Head Neck Surg. 2006;132(11):1203–9.
5. Tan J, et al. Numerical simulation of normal nasal cavity airflow in Chinese adult: a computational flow dynamics model. Eur Arch Otorhinolaryngol. 2012;269(3):881–9.
6. Wen J, et al. Numerical simulations for detailed airflow dynamics in a human nasal cavity. Respir Physiol Neurobiol. 2008;161(2):125–35.
7. Chen XB, et al. Assessment of septal deviation effects on nasal air flow: a computational fluid dynamics model. Laryngoscope. 2009;119(9):1730–6.
8. Chen XB, et al. Impact of inferior turbinate hypertrophy on the aerodynamic pattern and physiological functions of the turbulent airflow—a CFD simulation model. Rhinology. 2010;48(2):163–8.
9. Chen XB, et al. Assessments of nasal bone fracture effects on nasal airflow: a computational fluid dynamics study. Am J Rhinol Allergy. 2011;25(1):e39–43.
10. Garcia GJ, et al. Septal deviation and nasal resistance: an investigation using virtual surgery and computational fluid dynamics. Am J Rhinol Allergy. 2010;24(1):e46–53.
11. Guo Y, et al. A computational fluid dynamics study of inner flow through nasal cavity with unilateral hypertrophic inferior turbinate. Lin Chung Er Bi Yan Hou Tou Jing Wai Ke Za Zhi. 2009;23(17):773–7.
12. Lee HP, et al. Changes of airflow pattern in inferior turbinate hypertrophy: a computational fluid dynamics model. Am J Rhinol Allergy. 2009;23(2):153–8.
13. Lee HP, et al. Effects of septal perforation on nasal airflow: computer simulation study. J Laryngol Otol. 2010;124(1):48–54.
14. Liu T, et al. Effects of septal deviation on the airflow characteristics: using computational fluid dynamics models. Acta Otolaryngol. 2012;132(3):290–8.
15. Sun XZ, et al. Analysis of the character of self-adaptation of nasal structure in patients with nasal septum deviation. Zhonghua Er Bi Yan Hou Tou Jing Wai Ke Za Zhi. 2008;43(5):351–4.
16. Zhu JH, et al. Inspirational airflow patterns in deviated noses: a numerical study. Comput Methods Biomech Biomed Engin. 2013;16(12):1298–306. https://doi.org/10.1080/10255842.2012.670850.
17. Bockholt U, et al. Rhinosurgical therapy planning via endonasal airflow simulation. Comput Aided Surg. 2000;5(3):175–9.
18. Iwasaki T, et al. Improvement of nasal airway ventilation after rapid maxillary expansion evaluated with computational fluid dynamics. Am J Orthod Dentofac Orthop. 2012;141(3):269–78.
19. Ozlugedik S, et al. Numerical study of the aerodynamic effects of septoplasty and partial lateral turbinectomy. Laryngoscope. 2008;118(2):330–4.
20. Wexler D, Segal R, Kimbell J. Aerodynamic effects of inferior turbinate reduction: computational fluid dynamics simulation. Arch Otolaryngol Head Neck Surg. 2005;131(12):1102–7.
21. Xiong GX, et al. Computational fluid dynamics simulation of airflow in the normal nasal cavity and paranasal sinuses. Am J Rhinol. 2008;22(5):477–82.
22. Chen XB, et al. Drug delivery in the nasal cavity after functional endoscopic sinus surgery: a computational fluid dynamics study. J Laryngol Otol. 2012;126(5):487–94.
23. Frank DO, et al. Computed intranasal spray penetration: comparisons before and after nasal surgery. Int Forum Allergy Rhinol. 2013;3(1):48–55.
24. Garlapati RR, et al. Indicators for the correct usage of intranasal medications: a computational fluid dynamics study. Laryngoscope. 2009;119(10):1975–82.
25. Bailie N, et al. A model of airflow in the nasal cavities: implications for nasal air conditioning and epistaxis. Am J Rhinol Allergy. 2009;23(3):244–9.
26. Pless D, et al. Numerical simulation of air temperature and airflow patterns in the human nose

during expiration. Clin Otolaryngol Allied Sci. 2004;29(6):642–7.
27. Sommer F, Kroger R, Lindemann J. Numerical simulation of humidification and heating during inspiration within an adult nose. Rhinology. 2012;50(2):157–64.
28. Ishikawa S, et al. Flow mechanisms in the human olfactory groove: numerical simulation of nasal physiological respiration during inspiration, expiration, and sniffing. Arch Otolaryngol Head Neck Surg. 2009;135(2):156–62.
29. Sung SJ, et al. Customized three-dimensional computational fluid dynamics simulation of the upper airway of obstructive sleep apnea. Angle Orthod. 2006;76(5):791–9.
30. Xu C, et al. Computational fluid dynamics modeling of the upper airway of children with obstructive sleep apnea syndrome in steady flow. J Biomech. 2006;39(11):2043–54.
31. Prinsell JR. Maxillomandibular advancement surgery for obstructive sleep apnea syndrome. J Am Dent Assoc. 2002;133(11):1489–97; quiz 1539–40
32. Yu CC, et al. Computational fluid dynamic study on obstructive sleep apnea syndrome treated with maxillomandibular advancement. J Craniofac Surg. 2009;20(2):426–30.
33. Adams R, Bischof L. Seeded region growing. IEEE Trans Pattern Anal Mach Intell. 1994;16(6):641–7.
34. Lorensen WE, Cline HE. Marching cubes: a high resolution 3D surface construction algorithm. ACM Siggraph Comput Graph. 1987;21(4):163–9.
35. Taubin G, Zhang T, Golub G. Optimal surface smoothing as filter design. In: Computer vision—ECCV'96. Berlin: Springer; 1996. p. 283–92.
36. Eitel G, et al. Numerical simulation of nasal cavity flow based on a lattice-Boltzmann method. In: New results in numerical and experimental fluid mechanics VII. Berlin: Springer; 2010. p. 513–20.
37. Wheeler SM, Corey JP. Evaluation of upper airway obstruction—an ENT perspective. Pulm Pharmacol Ther. 2008;21(3):433–41.
38. Bouzidi M, Firdaouss M, Lallemand P. Momentum transfer of a Boltzmann-lattice fluid with boundaries. Phys Fluids. 2001;13:3452.
39. Finck M, Hänel D, Wlokas I. Simulation of nasal flow by lattice Boltzmann methods. Comput Biol Med. 2007;37(6):739–49.

Physiology and Pathophysiology of Nasal Breathing

20

Achim G. Beule, Giorgi Gogniashvili, and Gunter H. Mlynski

Core Messages

- Normal respiratory function in the nose requires low physiological nasal airflow resistance as well as adequate contact between the air and a large mucosal surface in a slit-shaped flow channel.
- Narrowing of the nasal airways and/or severe turbulence can cause a pathological rise in nasal resistance.
- Since resistance to flow rises exponentially with increasing stenosis, even slight narrowing in the area of the isthmus can lead to severe nasal obstruction. This fact is often overlooked.
- In terms of respiratory function, the nasal flow channel can be divided into an inflow area, a functional area, and an outflow area.
- In the inspiratory flow direction, the inflow area consists of the vestibulum, the isthmus, and the anterior cavum. After passing over the head of the inferior turbinate and the septal erectile tissue in the anterior cavum, the airstream is directed to the turbinate region and distributed over its entire cross-sectional surface. The degree of turbulence is regulated.
- In the functional area, the nasal turbinates not only represent a large surface, but through their adaptation in shape and size, they also narrow the nasal cave to an uniform slit-shaped space between the septum and the lateral wall of the cavum, which promotes warming up, humidification, and cleansing of the inspired air.
- The inspiratory outflow area consists of the nasopharyngeal meatus, the choanae, and the nasopharynx. Here the airflow becomes increasingly laminar and is redirected to the deeper air passages.
- A large thermal energy and humidity gradient are required between the mucosa and the airstream for effective warming and humidification of the air in the nose. The nasal cycle assures these conditions.
- The division of the nose into two sides by the septum makes it possible to alternate between working and resting phases to improve gradients of thermal energy and humidity exchange over a long time period.
- The nasal septum is rarely completely straight within the generally asymmetrical human skull. By means of a "physiological deviation" of the septum and adaptation of the con-

A. G. Beule (✉)
Department of Otorhinolaryngology, University Hospital Münster, Münster, Germany

Department of Otolaryngology, Head and Neck Surgery, University Medicine Greifswald, Münster, Germany
e-mail: achimgeorg.beule@ukmuenster.de

G. Gogniashvili
Khujadze-Gogniashvili ENT Clinic, Tbilisi, Georgia

G. H. Mlynski
Stolpe auf Usedom, Germany
e-mail: stolpe@mlynski.com

Ö. Ö. Celebi, T. M. Önerci (eds.), *Nasal Physiology and Pathophysiology of Nasal Disorders*,
https://doi.org/10.1007/978-3-031-12386-3_20

figuration of the turbinates, a symmetrical slit-shaped space results on each side of the nose. Cyclical congestion and decongestion of the turbinates make it possible to have in both nasal sides adequate working and resting phases.

- A more intense involvement of physiological aspects in functional rhinology is essential to improve long-term results of functional rhinosurgery.

20.1 Preliminary Remarks

For an undisturbed gas exchange in the lung, a temperature of 37 °C and a relative humidity of 100% are essential. The respiratory function of the nose is to climatize up to 25.000 L of inspired air every day [1]. For that, an important requirement is a physiologic resistance of nasal breathing, because increased nasal obstruction will result in habitual partially or completely mouth-bypass breathing. From physiology, the appropriate relation of physical activity and required airflow achieving a sufficient oxygen supply during inspiration is known (Fig. 20.1). During physiological activities, like walking or climbing stairs, a maximal inspiration flow up to 500 mL/s is needed. This range of airflow is called "physiological breathing range". During strong physical work airflow will increase up to >1.000 L, within the "stress breathing range" (Fig. 20.1).

To exert the respiratory function of the nose, the necessary breathing volumes should be achieved without any mouth-bypass breathing. Both nasal sides attributed to this respiratory capacity during the nasal cycle using a constantly changing degree of congestion. The degree of congestion is constantly changed also in relation to physical activity (see Sect. 20.5). Within the physiological breathing range, resistance on both sides of the nose (see Sect. 20.2.1), the sections of turbulent flow (see Sect. 20.4), and the inspiratory nasal valve collapse (see Sect. 20.2.2) should behave physiological.

During increasing physical activity, maximal inspiratory velocities up to 1000 mL/s become mandatory. Because endonasal resistance increases exponentially with increasing nasal flow (see Sect. 20.2.1) and physiological nasal valve function impedes high airflow (see Sect. 20.2.2), endonasal air velocities of 1000 mL/s cannot be provided through the nose alone. As a consequence, in the stress breathing range during heavy and very heavy physical activity respiration switched subsequently to mouth breathing (Fig. 20.1).

Fig. 20.1 Schematic representation of the relationship between physical activity and maximal inspiratory airflow velocity (Flow [mL/s]). Classification into a "physiological breathing range" during normal physical activity and "stress breathing range" during strong physical activity, including their relationship to nasal and mouth breathing

Physical activity	Maximum inspir. flow	Breathing range	Ratio mouth-nose breathing
Very heavy Very fast climb stairs	**>1000 ml/s**	**Stress breathing range**	**Mouth breathing**
Heavy Fast climb stairs	**750 ml/s**		**Mouth-bypass-breathing**
Moderate climb stairs	**500 ml/s**	**Physiological breathing range**	**Nose breathing**
Slight walk	**250 ml/s**		
No rest	**125 ml/s**		

20.2 Nasal Airway Resistance

The flow resistance is a consequence of the friction-related energy loss. With laminar flow, the particles flowing forward rub against each other due to their different flow velocities. Friction and thus the resistance decrease as the cross-sectional area of the flow channel increases, because the distance between the flowing particles increases. The cross-sectional area of a channel is therefore indirectly related to its flow resistance:

$$Resistance \sim \frac{1}{Cross - Sectional\ Area}$$

Friction also arises at the wall of the airflow channel. Therefore, endonasal resistance is also determined in relation to the size of the endonasal surface (mucosal coverage of the airflow channel). The larger the endonasal surface, the larger the nasal resistance:

$$Resistance \sim Wall\ Area$$

Turbulent flow sections also have an influence on the resistance. In turbulence, lateral movements cause the flowing air particles to bump against one another and against the wall of the channel. This leads to a high loss of energy and thus to an increase in resistance:

$$Resistance \sim Turbulent\ Flow\ Portions$$

Within the nose, nasal resistance increases exponentially with increasing flow due to the increasing sections of turbulent airflow. Figure 20.2 indicates the development of resistance in relation to nasal airflow during in- and expiration in a patient with pathologic high resistance at nasal obstruction on the right side and physiological low nasal resistance on the left side.

20.2.1 Physiological and Pathological Nasal Airflow Resistance

A **physiologically low resistance** in the nasal flow channel enables an unhindered airflow. For the deeper airways, it has the function of an upstream resistor to protect the mucosa against excessive flow.

In patients with **pathologically low nasal resistance**, this effect as protective upstream resistor is insufficient.

A **physiologically increased nasal resistance** is observed due to congestion during the resting phase of the nasal cycle (see Sect. 20.5). Nasal obstruction is only perceived, if the contralat-

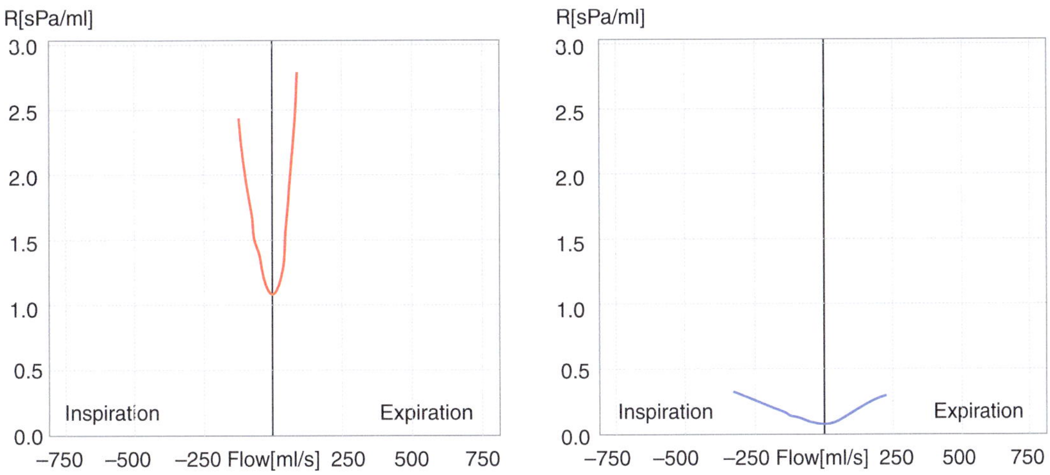

Fig. 20.2 Exponential increase of nasal resistance with increasing flow due to increasing turbulent flow portions during in- and expiration. Red line indicates the right side: pathological increased resistance due to lots of turbulent portions with complaints of nasal obstruction. Blue line indicates the left side: physiological low nasal resistance due to low portions of turbulence without subjective nasal obstruction

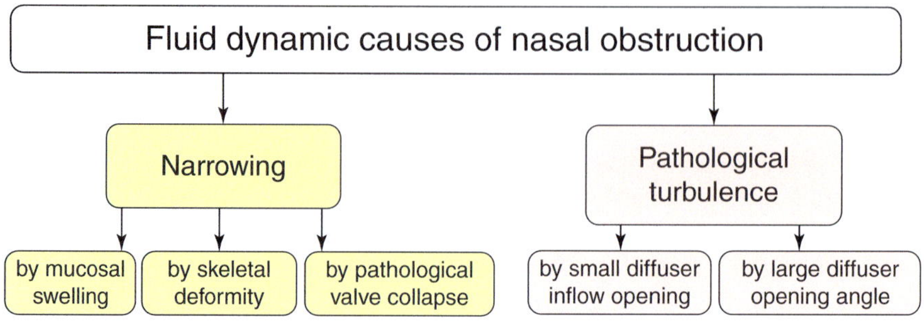

Fig. 20.3 Flow dynamic aetiologies of nasal obstruction

eral side in its working phase does not allow a sufficient airflow for oxygen supply within the physiological breathing range.

Pathologically increased nasal resistance may uni- or bilaterally cause the symptom of nasal obstruction due to several flow dynamic aetiologies (Fig. 20.3):

- Stenosis due to
  - Swelling due to inflammation, trauma, or tumour.
  - Skeletal constricting deformations, such as septal deviation, isthmus stenosis, or other.
  - Pathological nasal inspiratory collapse (see Sect. 20.2.2).
- Pathological turbulence (see Sect. 20.5).

In narrowings, the friction is increased because the flowing particles become more compressed. The relationship of resistance and cross-sectional area is not linear, but exponential (Fig. 20.4). As a result, in a wide flow channel, a large reduction in cross-sectional area only leads to a slight increase of resistance, but in case of an already narrow cross-sectional area, even a small reduction in size causes a large increase of resistance.

20.2.2 Short-Time Regulation of Nasal Airway Resistance

The short-time regulation of the nasal respiratory flow by changing the resistance is the suction (Bernoulli phenomenon) of the nasal valve in the

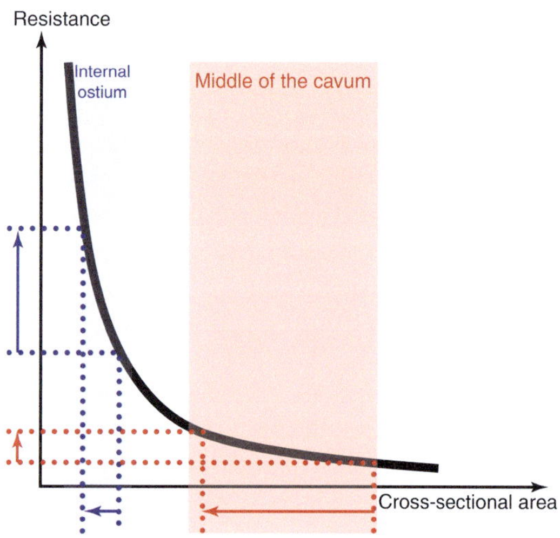

Fig. 20.4 Different effects of narrowing of the cross-sectional area within the nasal airflow channel on nasal resistance in the internal ostium (blue area) and the middle of the nasal cavum (red area)

nasal entrance first described by Mink [2] with the effect of increasing resistance. There are two physiological narrowings with a movable lateral wall: the internal ostium ("Isthmus "or "inner nasal valve") and the external ostium ("outer nasal valve"). The local flow velocity increases in the constrictions of a flow channel. According to the Bernoulli phenomenon, this results in negative pressure on the wall of the flow channel (principle of the water jet pump). When the flow rate is high, the mobile lateral nasal wall is sucked in and moves medially, thus considerably increasing the resistance in a fraction of a second. This mechanism is only possible during inspiration, because during expiration the pressure in the vestibulum is positive.

This **physiological nasal valve collapse** is a protective passive mechanism for the respiratory mucosa of the nose and the subsequent airways, preventing pathologically increased airflow velocities and possible damage due to shear forces.

In contrast, **pathological nasal valve collapse** is observed if medial movement of the lateral nasal wall already occurs at lower nasal airflow velocities. This can be caused not only by an additional pathological narrowing of the already physiologically narrow flow channel but also by a loss of stability of the lateral vestibular wall.

20.2.3 Long-Term Regulation of the Nasal Resistance

For long-term regulation of the nasal resistance, cross-sectional diameter of the nasal airflow channel is modulated via change of endonasal mucosal congestion. Changes take minutes and may persist for hours during in- and expiration. It is controlled by the vegetative nervous system and is the basis for the physiological regulation of resistance in the nasal cycle (see Sect. 20.5).

20.3 Nasal Functional Architecture: The Correlation of the Structure and Respiratory Function

From a functional and fluid dynamic perspective, the nose is a very complicated structure. Researchers have been attempting to investigate the nasal flow channel for more than 100 years [2–20].

For understanding the respiratory function of the nose in connection with the aerodynamics, it is helpful to divide the nose into different areas and to compare anatomical structures of the nose with form elements, whose impact on flow is known from fluid physics (Fig. 20.5 and Table 20.1).

The schema presented in Fig. 20.5 is derived from Bachmann [12, 13] and was then refined after we conducted extensive flow experimental studies [20–22].

The functional area with the turbinates is the principal location for the respiratory function. Here, the air is warmed up, moistened, and cleansed [23, 24]. In inspiratory direction, the inflow area is made up of the nasal vestibulum, the nasal isthmus ("internal valve"), and the ante-

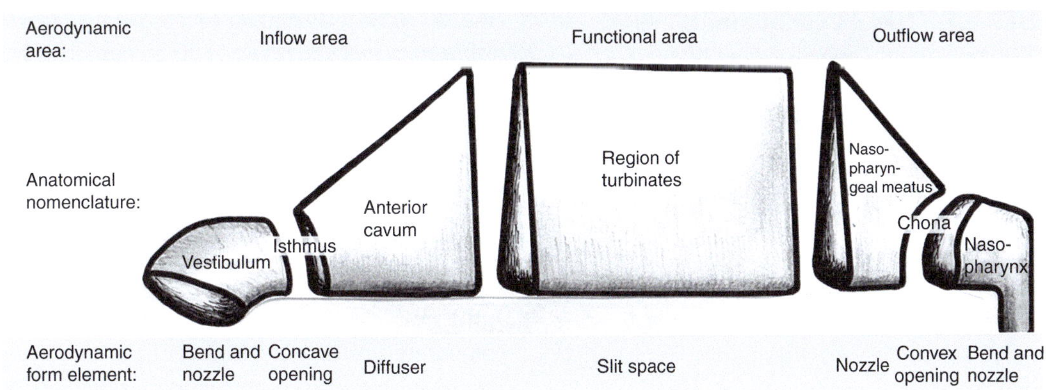

Fig. 20.5 Schematic presentation of dividing the nose into aerodynamic areas and comparative aerodynamic form elements in the inspiratory flow direction

Table 20.1 Fluid dynamic effects of aerodynamic form elements that may be compared to anatomical structures in the nose

Aerodynamic form element	Effect on airflow
Bend	Changing of flow direction
Concave opening	Divergence of flow paths
Convex opening	Convergence of flow paths
Nozzle	Decreasing turbulence, accelerating flow velocity
Diffuser	Increasing turbulence, decelerating flow velocity
Slit-shaped space	Facilitates exchange of heat, humidity, and cleansing between the mucosa and the respiratory airflow

rior cavum with the head of the inferior turbinate and the erectile tissue of the septum. The nasopharyngeal meatus, the choanae, and the epipharynx constitute to the outflow area.

20.3.1 Inflow Area

The function of the inflow area is to configure the airstream in a way to facilitate sufficient contact with the mucosa in the functional area. Figure 20.6 illustrates the course and character of flow in the inspiratory airflow direction in the inflow area of a nasal model [20].

20.3.1.1 Nasal Vestibulum

The nasal vestibulum is shaped like a short, bent tube, and thereby, it acts equivalent to a bend. The curvature is caused by the relative position of the external and internal openings: The ostium externum (inflow opening of the bend) is approximately horizontal, and the ostium internum (outflow opening of the bend) is more vertically located. The fluid dynamic effect of this bend redirects the inspiratory airflow from anterior and inferior towards the area of turbinates (Fig. 20.6).

In this process, the relative position of the vestibulum to the cavum is of significance. In a drooping nose with a small nasolabial angle, the vestibulum is rotated with the alar cartilages pointing downwards. During inspiration, this leads to a very high flow course within the cavum (Fig. 20.7b). As a result, the mucosa of the inferior turbinate cannot contribute to the respiratory function. If the nasolabial angle is too large, the vestibulum and the alar cartilage are rotated upwards. As a consequence, the airstream runs very low through the cavum, and thus, the mucosa of the upper turbinate is not capable of contributing to respiratory and olfaction function (Fig. 20.7c).

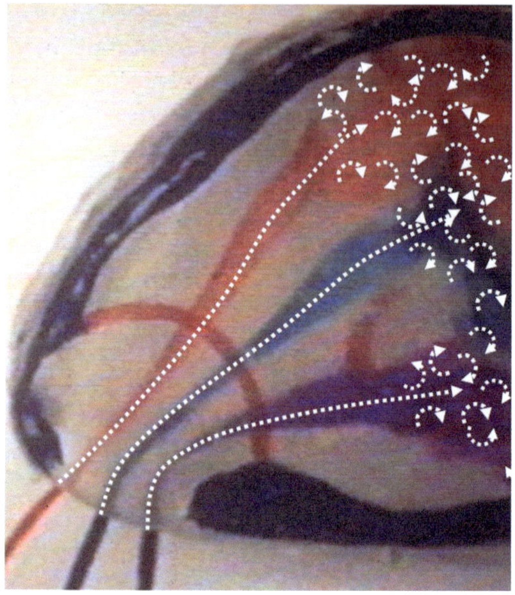

Fig. 20.6 Flow experimental representation of inspiratory flow in the inflow area. White dotted arrows: local flow direction. Laminar flow: only forward movements, turbulent flow: increasing sideways movements. The flow is directed towards the turbinate region (bend effect) and is preserved laminar (nozzle effect). Laminar flow can be recognized from the sharp demarcation between the flowing medium and the coloured particles added for visualization. With increasing turbulence, a diffuse colouration develops as a result of lateral movements of the flowing particles

When the vestibulum is malpositioned, the entire nasal flow channel is not employed for the respiratory function of the nose. This results in a functional airstream constriction, similar to an anatomical constriction. It results in an increased airway resistance as well. Therefore, correction of an inadequate or excessive nasolabial angle is necessary not only for aesthetic reasons but also from a functional point of view.

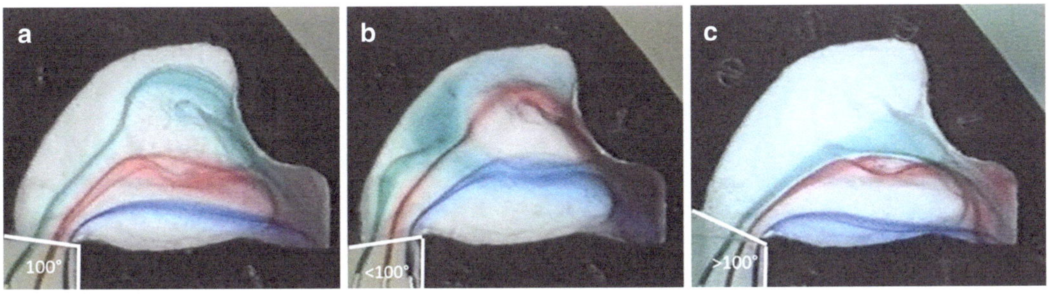

Fig. 20.7 Course of inspiratory flow in nasal models with various positions of the vestibulum to the cavum. (**a**) Normal vestibulum position: air is distributed over the entire region of turbinates. (**b**) Vestibulum rotated downwards (like at a drooping nose) with pathologically diminishes nasolabial angle: airflow runs only through the upper part of the nasal cavum. (**c**) Vestibulum rotated upwards with pathologically enlarged nasolabial angle: airflow runs only through the lower part of the nasal cavum

In cases presenting with an abnormal nasolabial angle, the rhinosurgical challenge is to reconstruct a normal position of the alar cartilages. This means that the cephalic portion of the alar cartilage must be positioned on the caudal edge of the lateral cartilage. Excessive elevations of the tip of the nose for aesthetic considerations should be avoided from a functional point of view.

Since the external ostium is larger than the internal valve area, the vestibulum has also a nozzle effect in inspiratory direction. In a nozzle, progressive restriction of the cross-sectional area causes reduction in turbulent flow. This guarantees that laminar flow can pass through the following narrowest part of the nose, the internal nasal valve (Fig. 20.6). This is of importance, since turbulent flow behaviour in this narrowing would result in a high level of resistance to flow.

The nozzle effect of the vestibulum must be preserved during rhinosurgical procedures. That means that one must not enlarge the internal valve in an effort to reduce resistance too much. The internal ostium must remain proportionately smaller than the external ostium. Excessive enlargement leads to a ballooning phenomenon that creates highly turbulent flow in the cavum as a result of eliminating the nozzle effect in the vestibulum.

The function of the vestibulum as a nozzle results also in an acceleration of local flow velocity, which causes a negative pressure on the mobile lateral vestibular wall. Because of this Bernoulli phenomenon the nasal wing collapses and increases abrupt the resistance. The "nasal valve collapse" is physiologically, if it acts at high breathing flow velocities (>500 mL/s). If a resistance increase caused by valve collapse occurs only in flow velocities above 500 mL/s, the effect is known as "physiological nasal valve collapse". It is an upstream resistor to protect the mucosa in the deeper airways against excessive airflow. By contrast, "pathological nasal valve collapse" arises already within the "Physiological Breathing Range".

Most cases of nasal valve collapse result from constrictions of the vestibulum, particularly within the internal ostium. The negative pressure on the wall of a flow channel depends on the local flow velocity, which is influenced by the width of the channel: the narrower the channel, the higher the local flow velocity, and the higher the negative pressure. The rhinosurgical concept in these cases is the enlargement of the narrowing to a normal width.

20.3.1.2 Internal Nasal Ostium

The inner nasal ostium is the narrowest region along the nasal flow channel and is therefore also known as the "Isthmus nasi". Due to the high friction within the constriction (see Sect. 20.2.1), it has the largest share in the formation of flow resistance in the nose [10, 11, 25, 26].

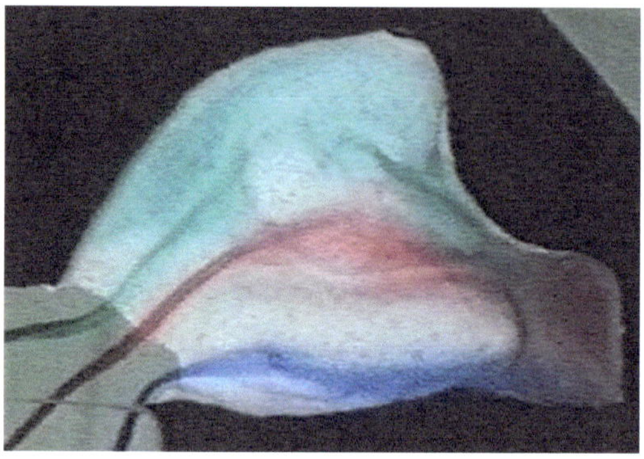

Fig. 20.8 Fluid dynamic experimental representation of the inspiratory airflow in a nose model without a vestibulum. The concavely curved shape of the inner ostium causes a divergence of the flow paths in the cavum

In inspiratory flow direction, the inner ostium serves as an opening for the air flowing from the vestibulum to the cavum. With its concave curvature, it influences the direction of airflow. It has the same influence on the distribution of the airflow as a concave lens does on light rays: it creates a divergence of the subsequent flow and thus contributes to the distribution of the flow over the entire surface in the area of the turbinates (Figs. 20.7 and 20.8) [20, 21, 26].

Given the relationship between form and function, resections of the caudal edge of the lateral cartilage should be performed in such a way that the concave shape of the inner ostium is preserved.

20.3.1.3 Anterior Nasal Cavum

Due to its cross-sectional increase in the direction of the inspiratory flow, the anterior cavum has the fluid dynamic effect of a diffuser. Turbulence is generated and regulated here (see Sect. 20.4).

From a functional point of view, another positive effect of the expansion of the cross-sectional area in the anterior cavum is a decrease in the flow velocity. The lower flow velocity in the area of the turbinates leads to an increase in the contact time for air with the mucous membrane, which is advantageous for the conditioning and cleaning of the inhaled air.

20.3.2 Functional Area

The functional area in the nose is the area of the turbinates. The turbinates not only enlarge the surface of the respiratory mucosa but also narrow the nasal cavum into a slit-shaped space. As a result, the nasal cavum is actually not a cavity at all, but a slit-shaped space [26]. This slit-shaped space is an important prerequisite for the breathing function of the nose (see Sect. 20.1). Analysis of CT images through the middle of the nasal cavum (Fig. 20.9 shows some examples) shows that the width of the slit remains fairly constant over the entire cross section of the cavum even with strong septal deviations (Fig. 20.9). The lateral walls of the cavum are always asymmetrical. The septum divides the nasal cavum into two unequally wide and differently structured cavities. The shape and size of the turbinates adapt to the available space between the lateral walls of the cavum and the septum, thus creating a continuous, uniform slit-shaped space. With its shape and variable thickness, the septum also helps create a slit.

At the end of the nineteenth century, Zuckerkandl (1882; [27]) discovered from his large collection of human skulls that the septum in the always asymmetrical skulls is normally not straight, but deviated. He introduced the term "physiological septal deviation". The joint observation of almost all ENT specialists that not every deviation of the septum leads to an obstructed nose confirms the validity of the concept of a physiological deviation [28, 29]. In the literature, an incidence of septal anomalies in the normal population of up to 90% is indicated [30–34]. It can be assumed that some of them are physiological deviations, since experience shows that the incidence of nasal obstruction is considerably lower.

We define "physiological septum deviation" as a curved septum without significant narrowing of the slit-shaped space and therefore without pathological airway resistance.

Due to the uniform width of the slit, the airflow can be distributed over the entire cross-sectional area (Fig. 20.10a). Since the airflow

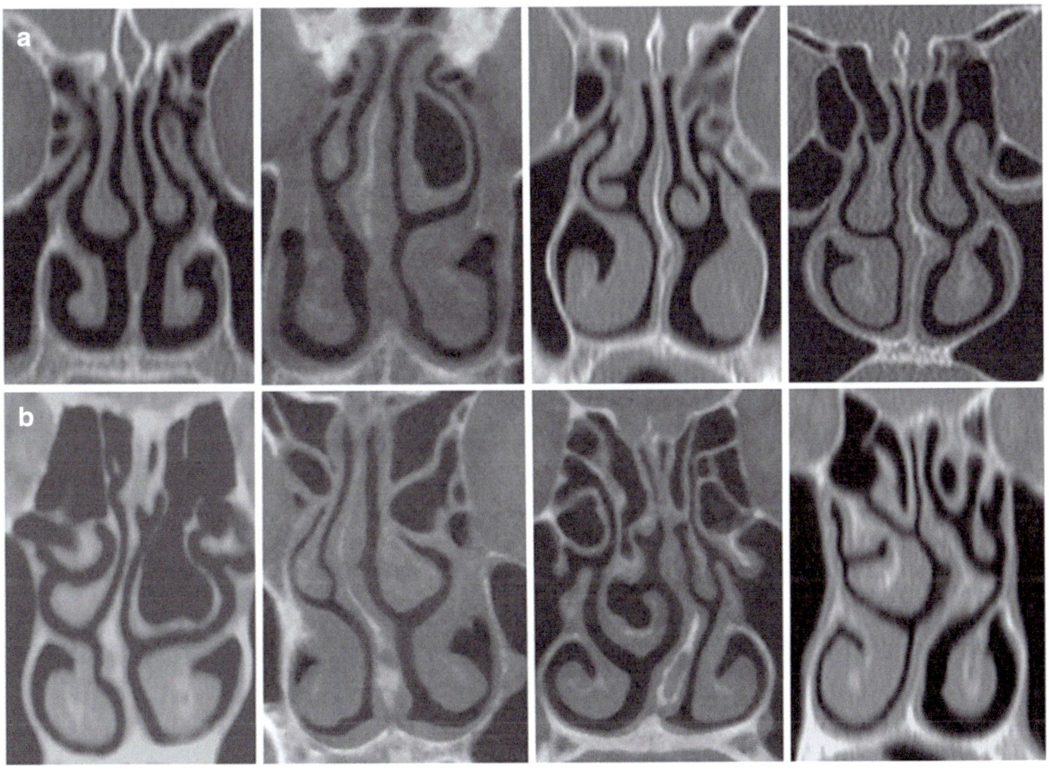

Fig. 20.9 Coronal CT images through the functional area of different noses. The turbinates adapt to the space provided by the lateral wall of the nasal cavum and the septum (**a**) with slight asymmetry of the two sides of the nose and slight septal deviations and (**b**) even with strong cavum asymmetry and severe deviations

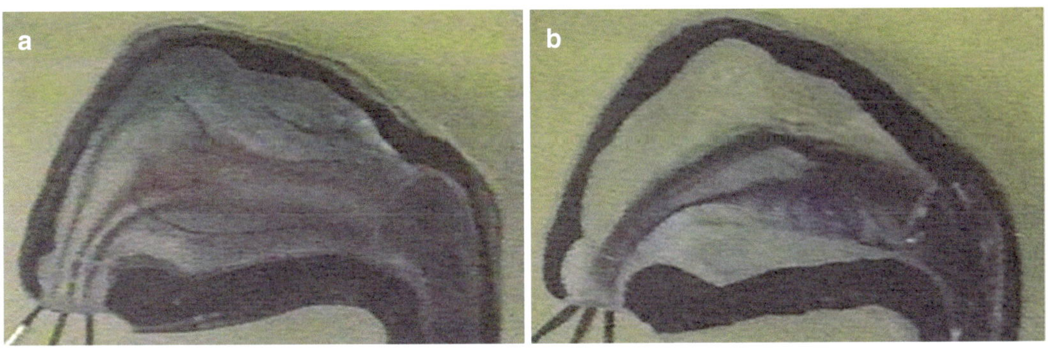

Fig. 20.10 Flow experimental representations of the inspiratory airflow in nose models (**a**) before an (**b**) after significant reduction in the size of the middle turbinate. Flow is distributed in (**a**) across the entire turbinate region, but in (**b**) it is limited almost exclusively to the wide space in the centre of the nasal cavity. As a result, the inferior and superior turbinates are unable to contribute to respiratory function

follows the paths of least resistance, local enlargement, as seen after extensive surgical reduction of the size of the turbinates, results in a significant disruption of the nasal airflow. In these cases, the air flows almost exclusively through the enlarged space (Fig. 20.10b), so that large areas of the mucous membrane of the middle and superior turbinate cannot contribute to the respiratory function, as the air no longer flows in these areas [20].

These results have two important implications for rhinosurgery [26]:

- It is necessary to keep the slit-shaped space as even as possible or, if necessary, to reconstruct the slit-shaped space (e.g. in cases of atrophic rhinitis or empty nose syndrome).
- In the case of a compensatory enlargement of the turbinate in connection with a septal deviation, the surgical intervention on the turbinate should be very gentle and, if possible, not include a resection. This is especially true for post-pubertal septal deviations, since the enlargement of the turbinates in these cases is almost always caused by compensatory swelling rather than compensatory hyperplasia. After the septum correction, the turbinates often adapt to the new space between the septum and the lateral cavum wall with the extent of their swelling [8, 35]. Especially in pre-pubertal septal deviations, in particular, the lower turbinates grow far into the concavity of the deviation. In these cases, lateral positioning of the inferior turbinate is often an appropriate treatment.
- Physiological septal abnormalities need to be identified as such and should not be surgically corrected. Figure 20.11 demonstrates that inadequate surgery in these cases can lead to non-physiological, excessively wide nasal spaces. Unsatisfactory long-term results after septoplasty can be attributed to septoplasties in patients with physiological septal deviations [36–38].
- Pathological septal deviations should not be fully straightened. The mobilized septum should be positioned midway between the turbinates on both lateral sides of the nose. This converts the pathological deviation into a less extensive physiological deviation. As a consequence, an excessive reduction of the turbi-

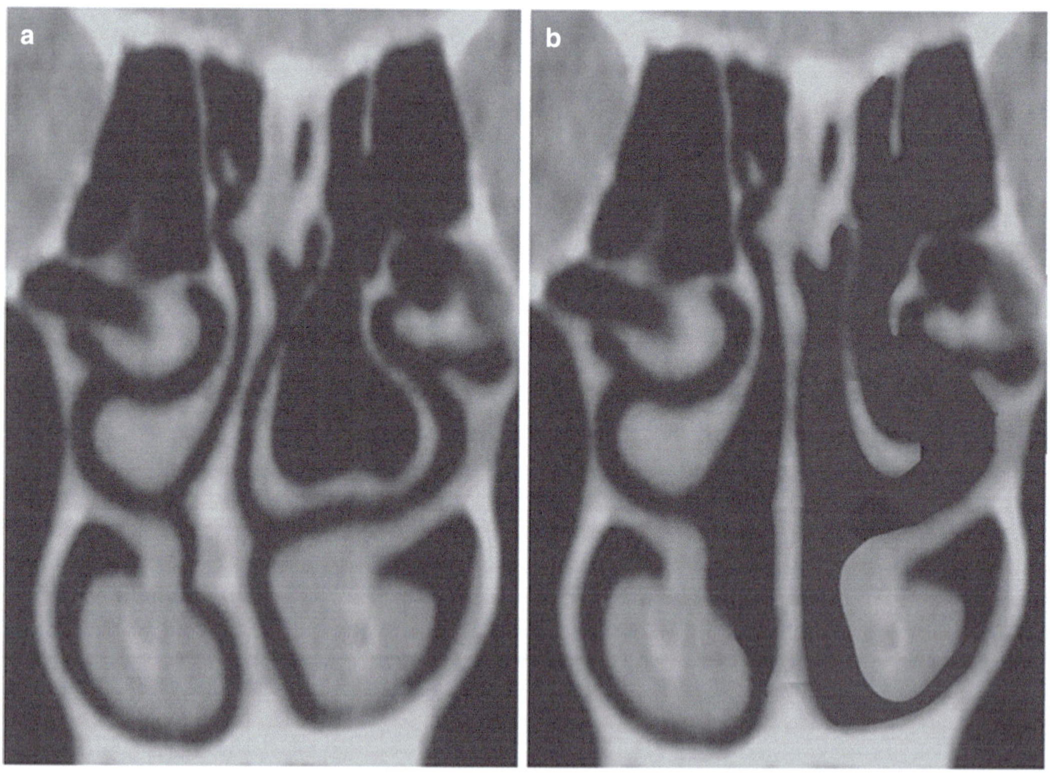

Fig. 20.11 (**a**) Coronal CT scan through the middle of the nose of a patient with a pre-pubertal septal deviation. Since there is no constriction in the slit-shaped space, it is a physiological septal deviation. (**b**) Imitated complete straightening of the septum with resection of the lateral wall of the concha bullosa and reduction in size of the inferior turbinate on the opposite (concave) side of the septal deviation

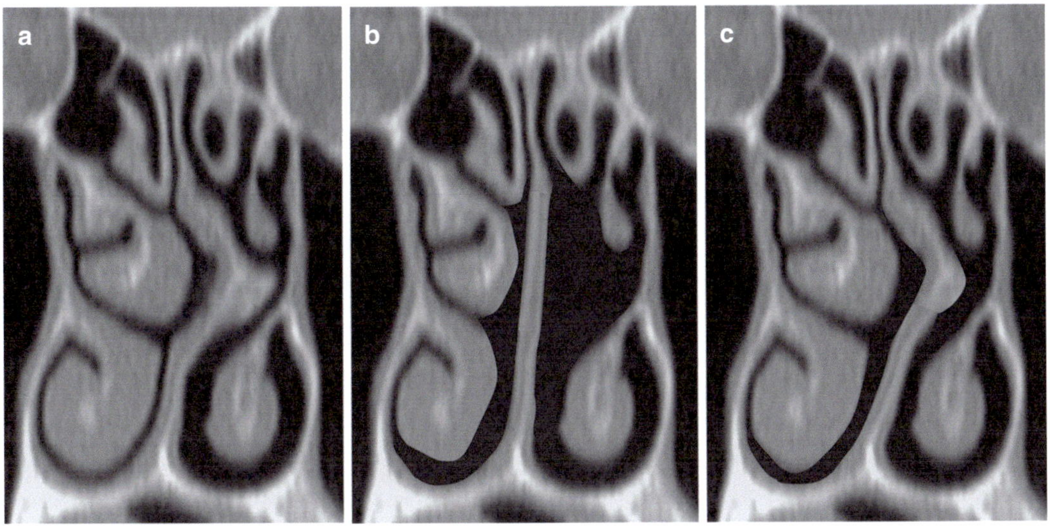

Fig. 20.12 (**a**) Coronal CT scan through the middle of the nose of a patient with a pre-pubertal septal deviation. Because there are narrowings in the slit-shaped space (right side between the inferior turbinate and septum, left side within the middle nasal passage due to a spur), it is a pathological septum deviation (pathologically increased resistance, objectified by objective rhinologic diagnostics, see Chap. 27). (**b**) Imitated complete straightening of the septum with size reduction of the middle and inferior turbinates on the opposite (concave) side of the septal deviation. As a result, a very large space is created, which leads to a disturbance of the flow distribution, as shown in Fig. 20.11b. (**c**) Imitated conversion of the pathological into a physiological deviation, with only removal of the spur, a slight shift of the lower septum to the left and a slight lateral positioning of the right inferior turbinate. A physiological slit-shaped space is created

nate is not necessary. In some cases, lateral positioning of the lower turbinate is efficient to create a symmetrical slit space (Fig. 20.12).

The goal of surgical interventions to relieve nasal congestion caused by septal deviations must not be a completely straight septum and small turbinates to create a wide flow channel in the nose. The frequent occurrence of sicca symptoms after functional rhinosurgery [26, 34, 39–47] should provide the impetus for a more precise consideration of physiological aspects in functional rhinosurgery. Instead, the aim should be the maintenance or reconstruction of the slit-shaped space for optimal respiratory function.

20.3.3 Outflow Area

The outflow area of the nose configures the inspiratory airflow so that it is adapted for passage into the lower airways:

20.3.3.1 Nasopharyngeal Meatus

In the nasopharyngeal meatus, the cross-sectional areas become smaller in the inspiratory direction. From a fluid dynamic point of view, this structure has the effect of a nozzle that reduces the turbulent sections within the airflow. In the subsequent deeper airways, the resistance must be low, and therefore, the flow should be laminar.

20.3.3.2 Choana

In the direction of inspiration, the choana is a convex opening between the cavum and the nasopharynx. This leads to convergence of endonasally distributed airflow paths (see Table 20.1). The airflow becomes narrower and adapts to the dimensions of the lower airways.

20.3.3.3 Nasopharynx

Because it is formed like a bend, the nasopharynx redirects the flowing air towards the lower airways.

20.4 Generation and Regulation of Turbulence in the Nose

With laminar flow, all air particles move forward parallel to the wall of the nasal flow channel, so that only particles flowing along the wall come into contact with the mucous membrane and can thus be warmed up, moistened, and cleaned. Centrally flowing particles are not conditioned. For a sufficient contact between air and mucous membrane in the nose, both forward and lateral movements of air particles due to turbulent airflow portions are important prerequisites.

In the nose, turbulent airflow sections are created in the anterior nasal cavum. In this area the cross-sectional area between the inner nasal ostium and the centre of the nasal cavity increases. A channel with an increasing cross-sectional area is called a diffuser (see Sect. 20.3.1.3). In a diffuser, the extent of creation of turbulence depends on the extent of the increase in cross-sectional area. This parameter and thus the degree of turbulence are characterized by the opening angle of the diffuser ɸ. A large diffuser opening angle leads to strong turbulence. Conversely, a smaller diffuser opening angle leads to less turbulence (Fig. 20.13).

By congestion and decongestion during the nasal cycle, not only the long-term regulation of endonasal flow resistance is achieved (see Sect. 20.2.3) but also the regulation of turbulence during the nasal cycle. Fig. 20.14b shows that on the right side the nasal cavum at the end of the diffuser is wide due to a decongestion of both, the septal erectile tissue and the head of the inferior turbinate. As a result, the increase in cross section in the nasal diffuser (diffuser opening angle ϕ) becomes large and the formation of turbulence is pronounced. This corresponds to a working phase in the nasal cycle, in which sufficient air–mucosal contact is required. On the left side of the nose, the "erectile tissue" is congested, the diffuser opening angle ϕ is small. This means that the airflow within the left side of the nose is largely more laminar. This is important in the resting phase. Less air–mucous membrane contact favours the storage of heat energy and moisture.

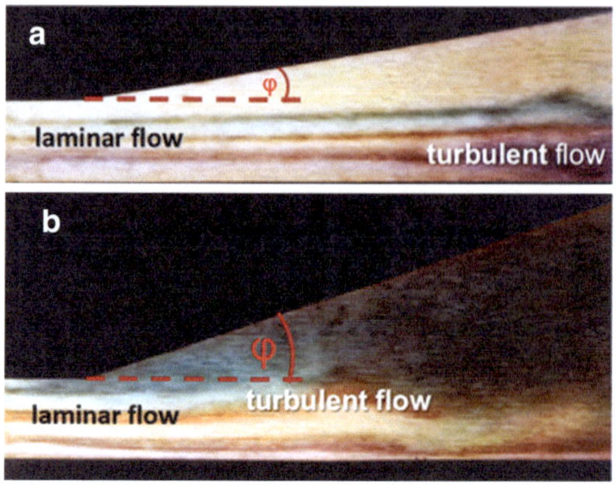

Fig. 20.13 Creation of turbulence in diffusers with different opening angles ϕ. With laminar flow, the coloured flow paths are clearly bordered due to the exclusively parallel, forward-flowing particles. A diffuse colouring as a result of sideways motions indicates turbulent flow. Within a diffuser with a small opening angle (**a**), turbulence only occurs at the end of the diffuser, while with a large opening angle (**b**), turbulent flow portions are already formed at the beginning of the diffuser

The regulation of the turbulence behaviour in the nasal diffuser during the change of nasal swelling in the nasal cycle is shown in Fig. 20.15 by flow experiments.

Inside a pipe, the transition from laminar to turbulent airflow occurs abruptly. In the nose, due to its specific aerodynamic channel structure, the increase in turbulence occurs continuously with increasing nasal airflow. In this large "transition area" (Fig. 20.16) the nasal airflow is not completely laminar and also not completely turbulent. This is an important prerequisite for the respiratory function of the nose: with purely turbulent flow, the heat and fluid supply of the mucous membrane would quickly be exhausted, while with purely laminar the flowing air would not be adequately conditioned (see above). In the nose, we can register a reasonable ratio of laminar and turbulent flow portions. Figure 20.16 shows that the flow is completely turbulent only with a very small volume of flowing air. With increasing flow, turbulent flow portions increase continuously. The more the air volume flows, the more efficiently the air–mucous membrane contact becomes possible by sideways move-

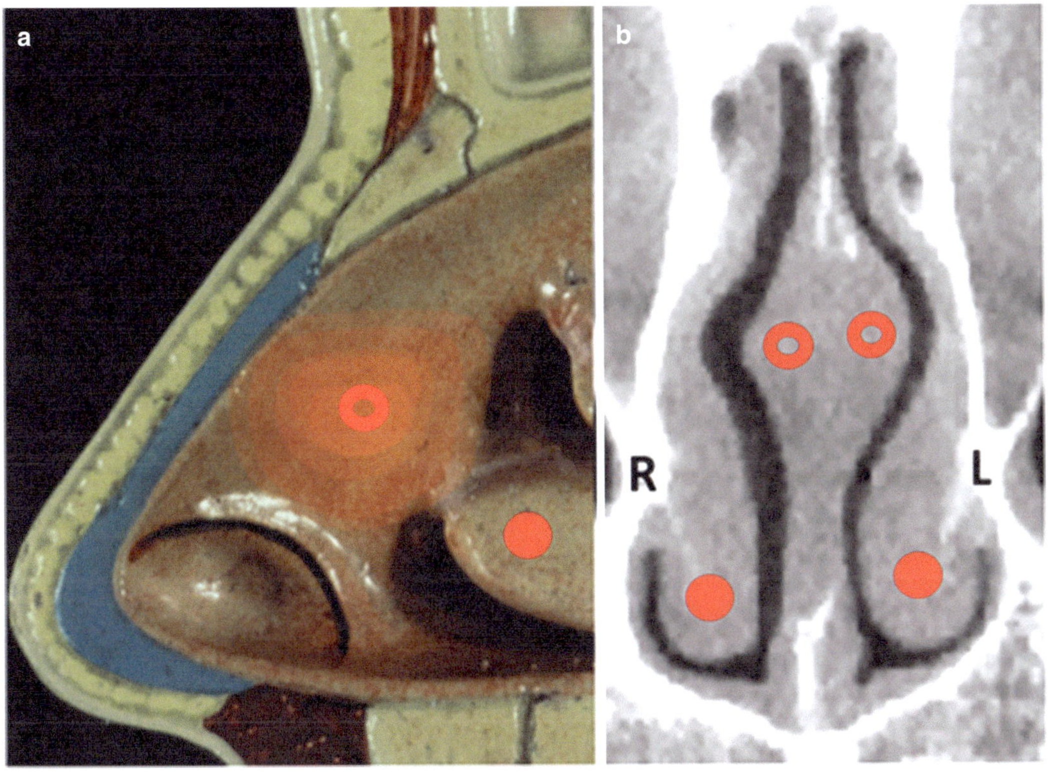

Fig. 20.14 Regulation of the turbulence in the nasal diffuser by congestion and decongestion of the lower turbinates (●) and the septal erectile tissue (○). (**a**) Model of an anterior cavity with its structures for regulating turbulence: the head of the lower turbinate and the septal erectile tissue. (**b**) CT scan of the end of the nasal diffuser. Changes in the cross-sectional area of the nasal airways and thus the degree of turbulence due to decongestion (right side) and obstruction (left side) within the nasal cycle

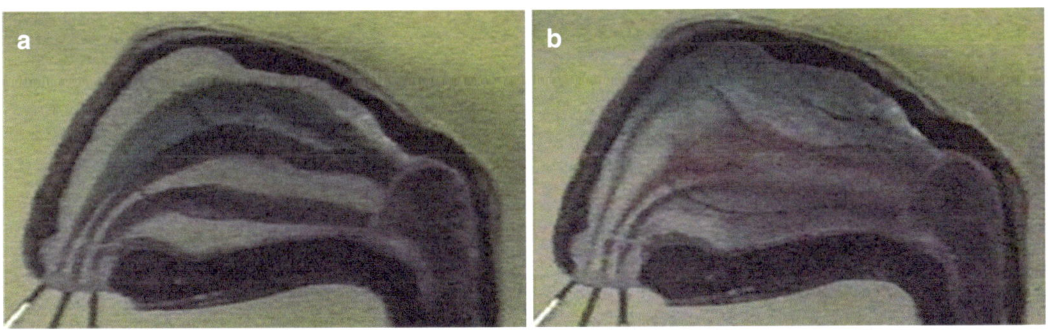

Fig. 20.15 As a consequence of turbulence regulation in the anterior cavum, the inspiratory airstream in the functional area of the nose is predominantly laminar during the resting phase (**a**) and predominantly turbulent during the working phase (**b**)

ments. Only at maximum airflow it is completely turbulent. The figure also shows that the flow is less turbulent before decongestion (resting phase) than after decongestion (working phase).

A steep transition from laminar to complete turbulent airflow, like to see in Fig. 20.17, is assessed as pathological.

Patients with complaints due to endonasal dry mucosal surfaces and incriminating creation of

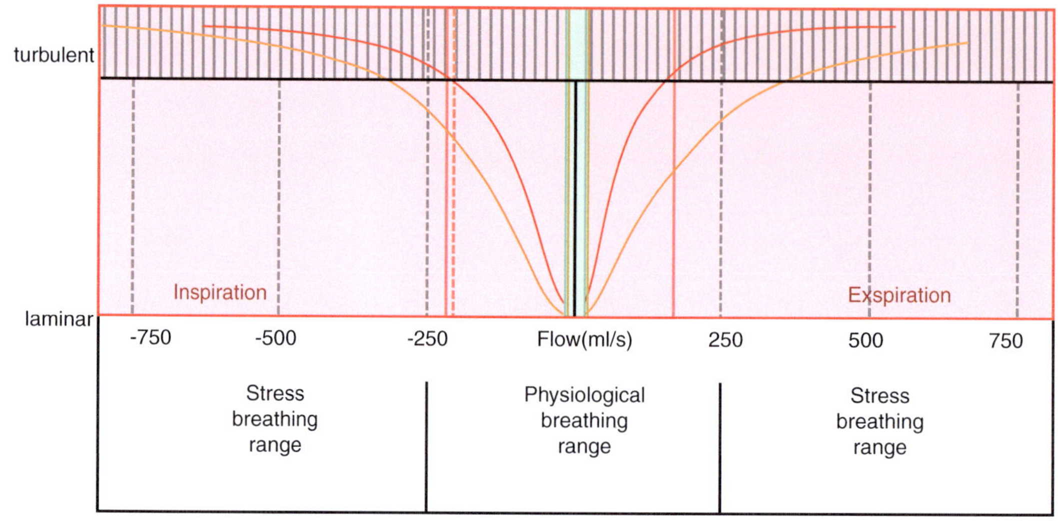

Fig. 20.16 Physiological turbulence behaviour with transition from laminar to turbulent flow behaviour in a nasal cavum in relation to flow. Light red line: before, dark red line: after decongestion. In the physiological breathing area, the flow behaviour is predominantly in the transition range to completely turbulent airflow. Green area: completely laminar flow, yellow area: "transition range" from laminar to turbulent flow, red area: completely turbulent flow

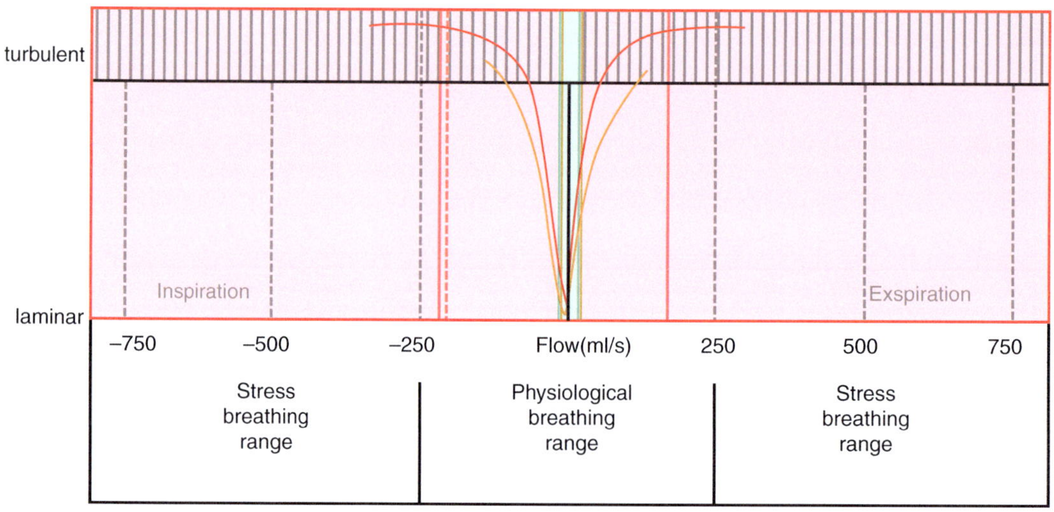

Fig. 20.17 Pathological transition from laminar to turbulent airflow in a nasal side. Within the physiological breathing range predominantly complete turbulent airflow sections can be observed. Green area: completely laminar flow, yellow area: transition from laminar to turbulent flow, red area: completely turbulent flow

endonasal crusts become more and more frequent. Turbulence as cause of these complaints have up to now been only considered in cases with excessive dry mucosa, like atrophic rhinitis and Empty Nose Syndrome. Because the extent of turbulent airflow portions could not be objectively diagnosed, these pathophysiological factors have been neglected in the diagnostic of nasal functional disturbances. Using rhinoresistometry the increase of turbulent streaming portions can be objectified and graphically depicted during inspiration and expiration (see Chap. 27).

Pathological turbulence in the nose has two negative consequences: dryness of the endonasal mucosa and an increase in nasal resistance. Dryness with crusts in the nose are of greater clinical importance due to their negative impact on quality of life. Endonasal dryness and atrophic rhinitis with their sicca syndrome lead more and more frequently to consultations of the affected patients with the rhinologist.

The influence of pathologically increased turbulence on the increase in nasal resistance is much smaller than the effects of local constrictions, since the width of the airflow channel is indirectly proportional to the resistance in the fourth to fifth power. In addition, a nose with a high degree of turbulence is often characterized by a very large width, which is associated with a rather small nasal resistance. The aetiology of the subjectively complained nasal obstruction in these patients is endonasal dryness with crust formation as a result of the pathological turbulence.

20.5 The Nasal Cycle

The nasal cycle as a cyclical change in endonasal resistance due to congestion and decongestion of the nasal erectile tissue was first described by Kayser [48]. It is a basic requirement for the respiratory function of the nose [1, 26, 49–53].

A large amount of energy is required to warm up and humidify the inhaled air. Although energy and moisture are partially recovered during exhalation, the storing capacity is limited for a steady respiratory function during a long time period. The necessary uninterrupted air conditioning function of the nose is possible due to the nasal cycle and the division of the nose into two sides by the septum. The reciprocal congestion and decongestion of the erectile tissue on both sides of the nose, controlled by the central nervous system, ensures that sufficient air can flow during the working phases according to the oxygen requirement and sufficiently warmed up und humidified. During the resting phase, the mucous membrane has sufficient time to regenerate heat, energy, and moisture.

Changes in mucosal swelling not only enable the alternation of working and resting phases but also modify the entire nasal airflow according to the physical activity by changing the flow resistance and thus the breathing flow corresponding to the required oxygen demand (see Sect. 20.1).

In Fig. 20.18 the adaption to oxygen supply during various stages of physical activity is illustrated:

- During physical inactivity a classical type with periodic reciprocal changes of working and resting phases exists. The left nasal side starts in the working phase, thus the right in the complete resting phase with only minimal flow being registered. After about three hours the right nasal side changes into the working phase and the left nasal side converts into a

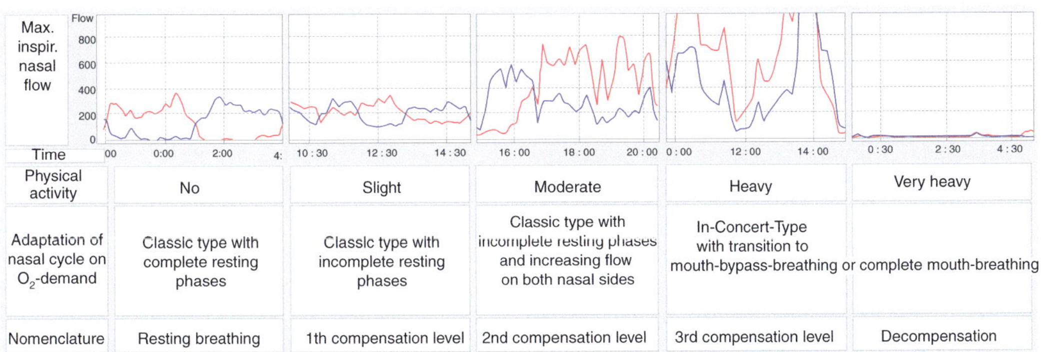

Fig. 20.18 Representation of the temporal change of maximal inspiratory flow on the right (red) and left (blue) nasal side in relation to the extent of physical activity. Adaptation of maximal inspiratory flow by the nasal cycle as necessary for a sufficient oxygen supply during different physical activities and their nomenclature

complete resting phase. This is an example of a classical type of the nasal cycle.

- Also during slight physical activity the classical type persists, but resting phases become incomplete. The nasal side in its resting phase contributes with a relatively small airflow to the oxygen supply. This classical type with incomplete resting phases is called "1. stage of compensation" according to increasing oxygen requirement.
- During moderate physical activity the classical type persists with incomplete resting phases, but with increasing nasal airflow on both nasal sides ("2. stage of compensation" according to increased oxygen requirement).
- During heavy physical activity, the nasal cycle converts into an "in-concert-type". Both nasal sides become synchronously decongested for working phases to achieve a transnasal airflow as high as possible. This is the third stage of compensation according to highly increased oxygen demand.
- Flow resistance increases exponentially with increasing airflow (see Sect. 20.2.1). If the resistance is sufficiently high, nasal airflow is limited. Therefore, with extreme physical activity, mouth-bypass breathing occurs increasingly (see Sect. 20.1). Since the nose is then partially bypassed, it can only perform its respiratory function insufficiently. With total mouth breathing, the respiratory function of the nose is completely eliminated. This is called decompensation of nasal breathing during maximally increased oxygen demand.
- Which level of physical activity leads to which level of compensation varies enormously between individuals. This depends on both endonasal and extranasal factors, including age, gender, body mass index, physical constitution, and extranasal diseases, particularly of the cardiopulmonary system.

Figure 20.19 illustrates in an example of a healthy subjects with non-obstructed nasal breathing how this compensation mechanism of the nasal cycle reacts to the oxygen demand during different physical activities. Illustrated are transnasal airflow of both nasal sides in the upper graph and synchronous measurements for:

- Heart rate as indicator of physical activity.
- Nasal respiratory minute volume as parameter for total transnasal airflow.
- Breathing frequency.

In the lower graph, these continuously measured data are related to the physical activity based on a study protocol.

Figure 20.19a shows the adaptation of the nasal cycle during 24 h with breathing at rest up to heavy physical stress. The curves for heart rate (measure of physical stress) and nasal minute volume run synchronously: with increasing physical activity nasal flow increases. The curves in Fig. 20.19b were recorded with the same subject on the following day for about 6 h with very heavy physical stress. As a result of extreme physical stress (recognized by the very high heart rate), complete mouth breathing can be recognized by the drop in the nasal minute volume to zero. Because nasal breathing is bypassed completely, no breathing rate can be registered.

The division of the nose by the septum into two parts becomes understandable taken the nasal cycle into account. For a steady, sufficient supply of oxygen with simultaneous conditioning of the breathing air, it is necessary for the nose to continuously exercise its respiratory function. Due to the division into two parts, one side of the nose can always condition air, while the other side stores heat, energy, and moisture.

The bisection of the nose and the nasal cycle are important prerequisites for the respiratory function of the nose. However, these conditions only make sense if both sides of the nose

- are sufficiently wide so that they can take over a working phase in the nasal cycle with a physiologically low breathing resistance and
- are only so wide that they can physiologically increase the resistance by swelling the nasal side for a resting phase. Too wide nasal sides, which can no longer be adequately closed for a resting phase due to swelling, are forced to work continuously, and thus, exhaustion (sicca symptoms) is inevitable.

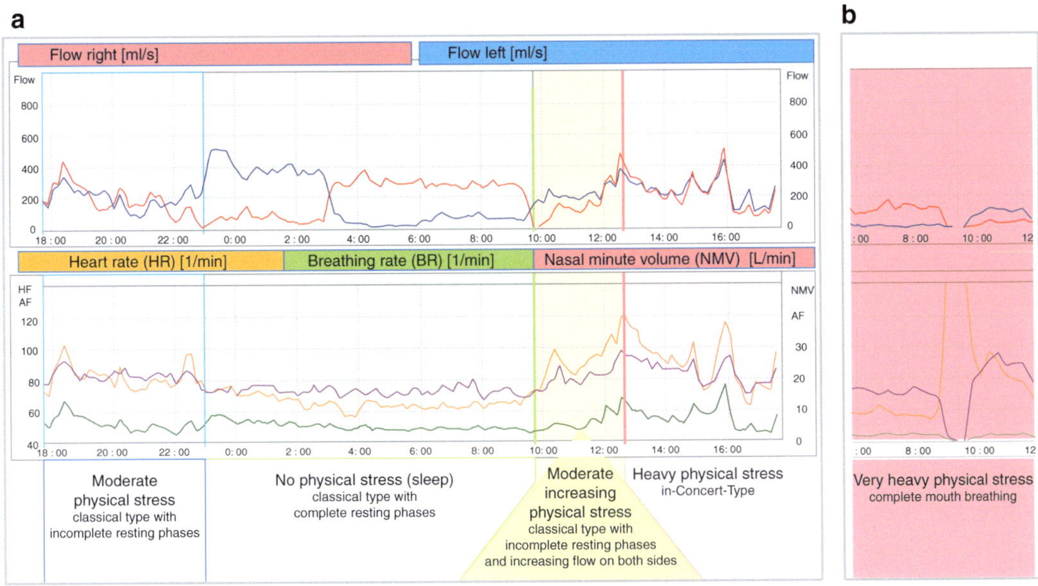

Fig. 20.19 Graphical illustration of the result of a long-term rhinometric examination of a healthy subject with non-obstructed nasal breathing and the simultaneous physical activity from a study protocol. (**a**) During moderate physical stress (blue background), during sleep (green background), during moderately increased physical stress (yellow background), and during heavy physical stress (light red background). (**b**) Measuring results of the same subject during very heavy physical stress (intensive red background). Upper Part of the figure: *X*-axis: time; *Y*-axis: Nasal airflow velocity at maximal inspiration in mL/s for the right (red) and left (blue) nasal side. Lower Part of the figure: *X*-axis: time; *Y*-axis: Heart rate (HR) orange, breathing rate (AR): violet, nasal respiratory minute volume in mL/s (NMV): green

20.6 Conclusions

- The long-term results after functional rhinosurgery are not satisfactory. Frequently reported postoperative complaints are insufficiently improved nasal breathing and dry mucous membranes up to symptoms of an empty nose syndrome. One reason for this is the insufficient consideration of the physiological importance of structures that are essential for the respiratory function in the nose.
- In recent years, our understanding of the relationship between efficient breathing and the very complex structure of the nose has increased significantly. The transfer of this knowledge into rhinosurgical practice is not sufficient. More satisfactory long-term results should be achieved in functional rhinosurgery through a greater inclusion of physiological aspects.
- In functional rhinosurgery, reducing pathological breathing resistance by expanding the nasal flow channel is an essential part. This almost always involves straightening the septum in conjunction with reducing the size of the turbinates.
- When the breathing resistance is sufficiently normalized, oral bypass breathing is eliminated and the respiratory function of the nose is reactivated.
- However, the patient is only symptom-free after surgery if the structures in the nose that are essential for its respiratory function have been preserved or reconstructed.
- If there is a physiological septal deviation, a septoplasty with turbinate reduction is not indicated as the physiological slit-shaped nasal flow channel will be broadened too much. Therefore, before a septoplasty, it should be determined whether the existing deviation is causing the pathological skeletal

resistance or whether it is a physiological deviation. In the latter case, the real reason for the patient's discomfort should be identified (e.g. narrowing of the isthmus with pathological valve collapse, sagging nose, etc.) and corrected.

- Correction of the septum is required in the event of pathological deviations. However, the septum should not be fully straightened. It should be placed in the middle between the asymmetrical lateral nasal walls so that the cavum is approximately the same size on both sides. This is an important requirement for an undisturbed nasal cycle and thereby for the respiratory function.
- The slit-shaped cavum should be sufficiently wide on both sides of the nose so that the inhaled air can flow unhindered for the working phase after decongestion. For the resting phase, the nose should not be too wide so that it can be closed by swelling. Both are prerequisites for an undisturbed nasal cycle.
- With their shape and size, the turbinates contribute significantly to the narrowing of the cavum to a slit-shaped flow channel. With their cavernous bodies and the respiratory mucosa, the turbinates are important structures for the respiratory function of the nose. For these reasons, the turbinates should be preserved as much as possible.
- In cases in which the turbinates are enlarged for compensatory reasons (e.g. to narrow a cavum that is too wide on the concave side of the deviation), septoplasty should not be routinely combined with surgical reduction of the turbinate size. In these cases, it is important to distinguish between turbinate enlargement due to swelling or hyperplasia by also performing an examination after decongestion.
- When treating the compensatory swelling of turbinates, resections are seldom necessary, since the turbinates can adapt with the extent of their swelling in shape and size to the available space.
- In pre-pubertal septal deviations, it is often observed that the turbinate has grown far medially in the area of the concave deviated septum. In these cases, a lateral positioning of the lower turbinate in the region of septal deviation is indicated in order to create a symmetrical slit-shaped channel.
- The reconstruction of the inflow area requires special attention. With drooping noses, the alar cartilage should be rotated and fixed up to a normal nasolabial angle. This restores an even distribution of the airflow over the entire cross section in the turbinate region. From a functional point of view, no overcorrection should be made.
  - If the valve angle is too small or too large (tension and saddle noses), the height of the septum in the cartilaginous nose should be normalized. Therewith, the nozzle effect of the nasal vestibulum also normalizes.
  - During the operation of crooked noses, the diffusers in both nasal sides should be reconstructed as symmetrically as possible, because that is an important prerequisite for regulating turbulence during the nasal cycle.
  - If possible, the head of the lower turbinate should not be reduced because of its function in regulating turbulence. If necessary, a lateral positioning is often sufficient.

References

1. Beule AG. Physiology and pathophysiology of respiratory mucosa of the nose and the paranasal sinuses. GMS Curr Top Otorhinolaryngol Head Neck Surg. 2010;9:Doc07.
2. Mink P. Physiologie der oberen Luftwege. Leipzig: Vogel; 1920.
3. Paulsen E. Experimentelle Untersuchungen über die Strömung der Luft in der Nasenhöhle. Sonderber D Akad d Wiss Wien. 1882;85:352–73.
4. Kayser R. Über den Weg der Atemluft durch die Nase. Z Ohrenheilk. 1889;20:96–109.
5. Rethi L. Experimentelle Untersuchungen über die Luftströmung in der normalen Nase sowie bei pathologischen Veränderungen derselben und des Nasenrachenraumes. Sonderber D Akad d Wiss Wien. 1900;109:17–36.
6. Burchardt. Die Luftströmung in der Nase unter pathologischen Verhältnissen. Arch Laryng. 1905;17:123–46.
7. Hellmann K. Untersuchungen über die Nase als Luftweg. Z Hals-Nas-Ohr-Heilk. 1926;15:354–7.

8. Takahashi K. Vorläufige Mitteilung über die Erforschung des Luftweges in der Nase des Menschen. Z Laryng Rhinol. 1922;11:203.
9. Der TJ. Weg der Atemluft in der menschlichen Nase. Arch Ohr- Nas-u Kehlk-Heilk. 1939;146:41–63.
10. Masing H. Experimentelle Untersuchungen über den Strömungsverlauf im Nasenmodell. Arch Klin Exp Ohren Nasen Kehlkopfheilk. 1967;189:371–81.
11. Fischer R. Die Physik der Atemströmung in der Nase. Berlin: FU Berlin; 1969.
12. Bachmann W. Die Funktionsdiagnostik der behinderten Nasenatmung. Berlin/Heidelberg/New York: Springer; 1982.
13. Bachmann W. Diagnostik der behinderten Nasenatmung: Teil 1: Funktionelle Anatomie und Strömungsphysik. HNO Aktuell. 1982;4:34–8.
14. Naito K, Iwata S, Kondo M, Ohoka E. Human respiratory airflow through an artificial nasal model: pressure/flow relationship. Auris Nasus Larynx. 1989;16(2):89–97.
15. Hess MM, Lamprecht J, Horlitz S. Experimentelle Untersuchung der Strombahnen in der Nasenhaupthöhle des Menschen am Nasen-Modell. Laryngorhinootologie. 1992;71(9):468–71.
16. Simmen D, Scherrer JL, Moe K, Heinz B. A dynamic and direct visualization model for the study of nasal airflow. Arch Otolaryngol Head Neck Surg. 1999;125(9):1015–21.
17. Mlynski G, Low J. Die Rhinoresistometrie--eine Weiterentwicklung der Rhinomanometrie. Laryngorhinootologie. 1993;72(12):608–10.
18. Grutzenmacher S, Robinson DM, Grafe K, Lang C, Mlynski G. First findings concerning airflow in noses with septal deviation and compensatory turbinate hypertrophy--a model study. ORL J Otorhinolaryngol Relat Spec. 2006;68(4):199–205.
19. Grutzenmacher S, Robinson DM, Lang C, Lebe E, Knape U, Mlynski G. Investigations of the influence of external nose deformities on nasal airflow. ORL J Otorhinolaryngol Relat Spec. 2005;67(3):154–9.
20. Mlynski G, Grutzenmacher S, Plontke S, Mlynski B, Lang C. Correlation of nasal morphology and respiratory function. Rhinology. 2001;39(4):197–201.
21. Mlynski G. Aerodynamik der Nase-Physiologie und Pathophysiologie. HNO-Praxis Heute. 2000;20:61–81.
22. Mlynski G. A method for studying nasal airflow by means of fluid dynamic experiments. Z Med Phys. 2000;10:207–14.
23. Bergmark RW, Gray ST. Surgical Management of Turbinate Hypertrophy. Otolaryngol Clin N Am. 2018;51(5):919–28.
24. Sommer F, Lindemann J, Scheithauer MO, Hoffmann TK, Veit JA. Nasal turbinate surgery. HNO. 2017;65(5):443–56.
25. Dishoeck H. Inspiratory nasal resistance. Acta Otolaryngol. 1942;30:431–9.
26. Mlynski G, Pirsig W. Funktionell ästhetische Rhinochirurgie unter Berücksichtigung physiologischer Aspekte. Stuttgart/New York: Georg Thieme Verlag; 2018.
27. Zuckerkandl E. Normale und pathologische Anatomie der Nase und ihrer pneumatischen Anhänge. Wien: Wilhelm Braumüller; 1882.
28. Altissimi G, Gallucci L, Simoncelli C. Positional rhinomanometry in hypertrophic vasomotor chronic rhinitis: considerations before and after functional surgery of turbinates. Acta Otorhinolaryngol Ital. 1992;12(4):363–9.
29. Hanif J, Jawad SS, Eccles R. A study to assess the usefulness of a portable spirometer to quantify the severity of nasal septal deviation. Rhinology. 2003;41(1):11–5.
30. Podoshin L, Gertner R, Fradis M, Berger A. Incidence and treatment of deviation of nasal septum in newborns. Ear Nose Throat J. 1991;70(8):485–7.
31. Sooknundun M, Deka RC, Kacker SK, Verma IC. Nasal septal deviation at birth and its diagnosis. Indian J Pediatr. 1986;53(1):105–8.
32. Uygur K, Yariktas M, Tuz M, Döner F, Ozgan A. The incidence of septal deviation in newborns. Kulak Burun Bogaz Ihtis Derg. 2002;9:117–20.
33. Gola R, Cheynet F, Guyot L, Bellot-Samson V, Richard O. Traumatismes du nez perinataux et de la petite enfance. Etiopathogenie, consequences et principes therapeutiques. Rev Stomatol Chir Maxillofac. 2002;103(1):41–55.
34. Gogniashvilli G, Steinmeier E, Mlynski G, Beule AG. Physiologic and pathologic septal deviations: subjective and objective functional rhinologic findings. Rhinology. 2011;49(1):24–9.
35. Kim DH, Park HY, Kim HS, Kang SO, Park JS, Han NS, et al. Effect of septoplasty on inferior turbinate hypertrophy. Arch Otolaryngol Head Neck Surg. 2008;134(4):419–23.
36. Mlynski G. Restorative procedures in disturbed function of the upper airways—nasal breathing. GMS Curr Top Otorhinolaryngol Head Neck Surg. 2005;4:Doc07.
37. Toyserkani NM, Frisch T. Are too many septal deviations operated on? A retrospective patient's satisfaction questionnaire with 11 years follow-up. Rhinology. 2012;50(2):185–90.
38. Sundh C, Sunnergren O. Long-term symptom relief after septoplasty. Eur Arch Otorhinolaryngol. 2015;272(10):2871–5.
39. Valsamidis K, Titelis K, Rachovitsas D, Konstantinidis I, Markou K, Triaridis S. Long-term evaluation of nasal septoplasty followed by inferior turbinate cauterization for the treatment of nasal obstruction using objective and subjective methods. Int Arch Otorhinolaryngol. 2018;22(3):284–90.
40. Stoksted P. Long term results, following plastic septum surgery. Int Rhinol. 1969;7:53–61.
41. Dommerby H, Rasmussen OR, Rosborg J. Long-term results of septoplastic operations. ORL J Otorhinolaryngol Relat Spec. 1985;47(3):151–7.
42. Haraldsson PO, Nordemar H, Anggard A. Long-term results after septal surgery--submucous resection ver-

sus septoplasty. ORL J Otorhinolaryngol Relat Spec. 1987;49(4):218–22.
43. Grymer L, Rosborg J. The aging nose (long-term results following plastic septal surgery). J Laryngol Otol. 1987;101:363–5.
44. Fjermedal O, Saunte C, Pedersen S. Septoplasty and/ or submucous resection? 5 years nasal septum operations. J Laryngol Otol. 1988;102:796–8.
45. Illum P. Septoplasty and compensatory inferior turbinate hypertrophy: long-term results after randomized turbinoplasty. Eur Arch Otorhinolaryngol. 1997;254(Suppl 1):S89–92.
46. Dinis PB, Haider H. Septoplasty: long-term evaluation of results. Am J Otolaryngol. 2002;23(2):85–90.
47. Mlynski G. Surgery of the nasal septum. Facial Plast Surg. 2006;22(4):223–9.
48. Kayser R. Die exakte Messung der Luftdurchgängigkeit der Nase. Arch Laryng Rhinol (Berl). 1895;8
49. Cole P, Chaban R, Naito K, Oprysk D. The obstructive nasal septum. Effect of simulated deviations on nasal airflow resistance. Arch Otolaryngol Head Neck Surg. 1988;114(4):410–2.
50. Lang C, Grutzenmacher S, Mlynski B, Plontke S, Mlynski G. Investigating the nasal cycle using endoscopy, rhinoresistometry, and acoustic rhinometry. Laryngoscope. 2003;113(2):284–9.
51. Kim JK, Yoon JH, Kim CH, Nam TW, Shim DB, Shin HA. Particle image velocimetry measurements for the study of nasal airflow. Acta Otolaryngol. 2006;126(3):282–7.
52. Braun T, Rich M, Berghaus A, Kramer MF. Effects of oxymetazoline nasal spray on the nasal cycle assessed by long-term rhinoflowmetry. Rhinology. 2012;50(4):370–5.
53. Rohrmeier C, Schittek S, Ettl T, Herzog M, Kuehnel TS. The nasal cycle during wakefulness and sleep and its relation to body position. Laryngoscope. 2014;124(6):1492–7.

Function of the Turbinates: Nasal Cycle

21

Achim G. Beule and Rainer K. Weber

Core Messages

- Spontaneous changes in nasal airway resistance in the two separate nasal passages due to congestion and decongestion of nasal venous sinuses are called the nasal cycle. One respiratory function of the nose is to sufficiently condition the respirated air which goes along with the nasal cycle.
- It is important to consider the nasal cycle and other physiologic changes in the congestion of the nasal mucosa when making a clinical assessment of a patient complaining of nasal obstruction. The indication faor surgery should be based on clinical history, on examination by anterior rhinoscopy and nasal endoscopy, and by considering these physiologic variations in addition to measurements of nasal airway patency. Long-term rhinoflowmetry offers a new possibility for investigating nasal patency for up to 72 h.

A. G. Beule
Department of Otorhinolaryngology, University Hospital Münster, Münster, Germany

Department of Otolaryngology, Head and Neck Surgery, University Medicine Greifswald, Greifswald, Germany
e-mail: achimgeorg.beule@ukmuenster.de

R. K. Weber (✉)
Department of Otolaryngology, Municipal Hospital Karlsruhe, Karlsruhe, Germany

Sinus Academy, Karlsruhe, Germany
e-mail: rainerweber@rainerweber.de

21.1 Definition of Nasal Cycle

The respiratory function of the nose is to sufficiently condition the respirated air, which is maintained by supplying the mucosa with thermal energy and fluid for humidification. This is supplied by blood circulation and in coherence with the nasal cycle [1]. The erectile tissue enables the turbinates to cyclically congest and decongest. We distinguish between three different patterns of congestion and decongestion during the nasal cycle: During the classical nasal cycle, one side of the nose is in its working phase conditioning the air, with an unimpeded air passage and increased turbulence. At the same time, the contralateral side is in its resting phase, saving energy and moisture by high airway resistance and low turbulence [2]. This type of nasal cycle is the physiologic ideal, enabling the resting side to regain the full mucosal potential to clean and humidify the air during the next cycle. In cases of increased physical activity or decreased respiratory function of the nose, a decongestion on both sides can be observed, called the in-concert-type or parallel type [3]. If the erectile tissue is severely malfunctioning, e.g. after excessive resection or due to atypical central nervous impulses to the erectile tissue, the congestion on both sides is irregular. Congestion and decongestion are no longer responsive to the physical or respiratory activity of the subject and the duration of a congestive

Ö. Ö. Celebi, T. M. Önerci (eds.), *Nasal Physiology and Pathophysiology of Nasal Disorders*,
https://doi.org/10.1007/978-3-031-12386-3_21

phase is less than 30 min. to correctly attribute this term. Some authors suggest the use of the term acyclic; however, this fourth type I up to now is rarely used in the literature. To understand the possible influencing factors resulting in one of these three types, the regulation of the nose needs to be remembered:

The airflow through the nose is regulated by the activity of the erectile venous tissue of the nasal mucosa [4]. The nasal epithelium has a very complex vascular nature with a submucosal plexus of venous sinuses lining the nasal mucosa. These venous sinuses from erectile tissue are well developed in the anterior part of the nasal septum and the inferior turbinate [5]. These submucosal cushions of the venous sinuses expand and shrink depending on the degree of congestion, hence altering the calibre of the nasal passages and influencing the nasal airflow.

An enlargement of this tissue leads to a reduction of the nasal lumen and increases the flow resistance. The cyclic congestion and decongestion of the nasal mucosa is called nasal cycle [4] and is observed in about 80% of the people [6, 7], and only about 20–40% seem to show a classical type [3, 6, 8, 9]. As the nasal cycle had already been described by Kayser in 1895 [9], the nasal cycle was investigated using several techniques, including rhinoscopy [9, 10], acoustic rhinometry [2, 11–20], rhinomanometry [21–26], rhinoresistometry [2], radiologic tomographic imaging [27], computed tomography [28], or magnetic resonance imaging [29–32].

Rhinomanometry reveals an opposite swelling behaviour of both nasal sides, while the total resistance of the two nasal sides, the nasal passage, and the respiratory work remain relatively constant [8, 25, 26, 33, 34].

Magnetic resonance imaging could show that even the ethmoid mucosa is involved in the nasal cycle, however, only to a lower degree [30]. Also the tubal function changes with the nasal passage resistance in a homolaterally concordant way [35].

By means of radiologic tomographic procedures, Masing [27] could show that in cases of nasal obstruction or non-existing nasal breathing, the classic nasal cycle can no longer be observed but paradox or irregular turbinate movements occur, which confirms the observations made by others [3, 23, 25, 26]. In the anterior and posterior parts of the nose, the changes were comparable so that the hydraulic diameter of the whole nose seemed to be equally maintained. In cases of significant septal deviation, in-concert or no nasal cycle could be found. In only two of six cases, a unilateral turbinate movement was observed on the obstructed side. The nasal cycle stopped with the beginning of acute rhinitis [27] and may be abolished by use of topical decongestants [36].

Rhinoresistometry and acoustic rhinometry could reveal the changes of the nasal turbulence occurring in the context of the nasal cycle [2]. During the resting phase, a laminar airflow was observed. During the working phase (decongested side), turbulences were also found with low speed. The increasing turbulence was caused by an increased cross-section surface of the anterior nasal area. Hereby, the turbinates and the mucosa of the nasal septum are decongested.

Long-term rhinoflowmetry is a complementary tool in addition to established rhinological diagnostics. The nasal flow can be investigated up to 72 h. Therefore, it appears to be an ideal tool to measure the cyclic alterations of the nasal cycle [1]. Technical details have been published for commercially independent reports on its function [37]. Long-term rhinoflowmetry is especially helpful if nasal obstruction is reported during nighttime or in specific environment to objectify these complaints [38].

Keerl et al. [39] were the first to realize visualization by means of nasal endoscopy and dynamic description with timelapse video. In analogy to the clinical and rhinomanometric observations that the total resistance of both nasal sides is almost constant [8, 20, 33, 40], the congestive procedures on one side occur nearly parallel to the decongestive procedures of the contralateral side (Fig. 21.1a–e). Most of the time, continuous swelling patterns are found with congestion of one side and decongestion of the contralateral side. These swelling changes are relatively rapid with 15 min in our analysis for a cyclic duration of 5 h.

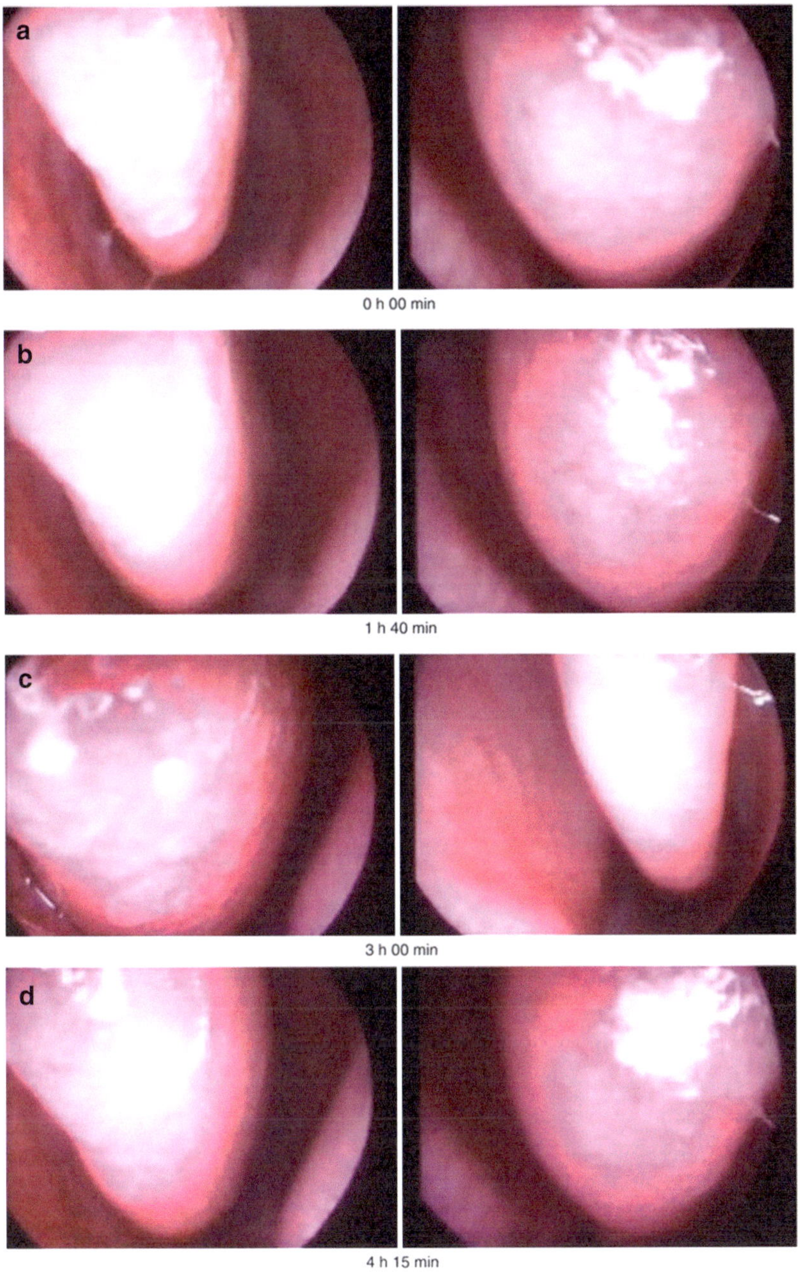

Fig. 21.1 (**a**–**e**) Swelling of inferior turbinate according to the nasal cycle in a healthy young man (parallel endoscopy using a 0° endoscope): (**a**) Decongested mucosa on the right and congested mucosa on the left side at the beginning. (**b**) Change of swelling. (**c**) Maximal congestion on the right and maximal decongested mucosa on the left side. (**d**) Reciprocal change of swelling. (**e**) Same situation at the end of the nasal cycle as at the beginning

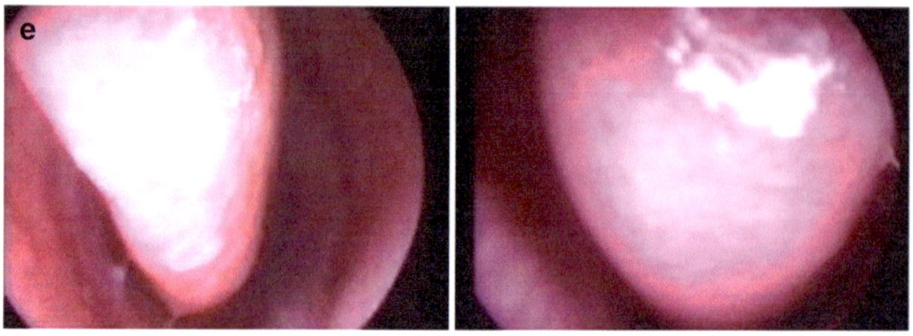

Fig. 21.1 (continued)

With the endoscopic description in timelapse video, the extent of the congesting process could be evaluated precisely. The process of congestion and decongestion occurs relatively rapidly so that the turbinates remain most of the time in a kind of plateau phase of submaximal to maximal congestion or decongestion. Based on our clinical evaluations in healthy subjects, most of the nasal cycle changes occur within 5 min. The size of the turbinates varies between very small and very large with complete obstruction of the nasal cavity in the visible area. During the course of the cycle, complete decongestion of both turbinates never occurred. For a short phase, a similar middle congestion status of both turbinates can be found.

The diagnosis of hyperplastic turbinates is often made in the clinical routine, but with the background of the mentioned results, it must be reconsidered. Hyperplasia means first the enlargement of tissue by cellular multiplication. During the nasal cycle, the healthy turbinate continuously changes from a low to a high degree of congestion. Internal and external factors additionally influence the extent of the swelling situation. As the physiology of a normal turbinate allows each size from minimal to maximal, there is no reference for the diagnosis of hyperplasia or hypertrophy or for definition of a standard normal size at one given time point. Even if one side of the nose is completely obstructed because of the maximally congested inferior turbinate, the other decongested side allows a sufficient nasal air passage based on the physiologic nasal cycle due to the relatively unchanged total nasal resistance of the nose. Only the bilateral massive swelling of the inferior turbinates seems to be unphysiological [39, 41]. It seems to be appropriate demanding to limit the use of the term of hyperplasia only for enlarged turbinates with atypical changes of the surface (e.g. polypous, mulberry-like changes) and to speak about congested turbinates in all other cases.

Besides the congesting and decongesting processes, regular changes of the secretory production could be observed endoscopically [3, 41, 42]. The congesting turbinate seemed to become increasingly humid so that maximally congested situation led to small droplets. Reason is the activation of the parasympathic nervous system. During the decongesting phase, the mucosa dries more and more, partly with resulting small dry layers. In this context, the dehydrating effect of nasal breathing accompanies the sympathomimetic and reduced parasympathic tonus. However, in cases of rhinitis, the interaction of secretory function and nasal cycle has been shown to be affected as well [3].

21.2 Types of Nasal Cycles

Three types of nasal cycles are described [1]: the classical reciprocal type, the in-concert type (simultaneous reduction and increasing of the nasal passage on both sides), and the irregular type. Recently, the term acyclic nose has been introduced [3, 6].

In addition, Kern [43] describes three types of non-cycle noses: there is no fluctuation of the nasal passage, the fluctuation is moderate only on one side, and the fluctuation does not change from one side to the other.

The cyclic duration is very variable with an interval of about 1–6 h [4, 8, 10, 39, 41] but use of rhinoflowmetry has demonstrated that cyclic duration may take up to 11 h [36, 38, 44].

More recent quantitative studies using numerical parameters seem to show that true periodicity and reciprocity of nasal airflow exist only in 21–39% of the population, reciprocity being the truly reciprocal changes in airflow between right and left passages and periodicity being the regularity of changes in airflow occurring with time in each nasal passage. Otherwise, these studies used measurement periods no longer than 8 h. So the timeperiod of examination may be too short to discover the nasal cycle as was outlined by Grützenmacher.

To characterize the nasal cycle apart from its type, duration of the working or resting phase, as well as the magnitude of maximal nasal flow, called amplitude, is currently evaluated [38]. These parameters may also be expressed as ratio of both nasal sides, as calibration and normalization of the measurement used are still challenging. Recently, a more similar distribution of nasal cycle measured by rhinoflowmetry has been reported after septoplasty in comparison to a preoperative measurement [45].

Also in children, a nasal cycle can be revealed [16, 46–48]. For children up to the age of 6, it amounts to a mean of 1 h and 20 min and is not influenced by physical activity. Even infections do not significantly influence the nasal cycle. As studies with a longer observation time are still not available (possibly due to ethical consideration), these findings may be controversially discussed.

Even if the distribution of the nasal cycle types changes, a fluctuation of the nasal passage is found in laryngectomized patients (25% classical type, 40% irregular type, 35% in-concert type) [14].

According to Ingels et al. [49], no correlation is found between the ciliary beating rate (CBR) of the nasal epithelia and the rhinomanometrically measured passage of the nasal cavity. The CBR of different cells in the same biopsies was clearly different.

In contrast to this, Doyle and van Cauwenberge [50] could detect a higher mucociliary clearance rate with increased nasal passage by means of the saccharin test. Also Soane et al. [51] found a significant acceleration of the mucociliary clearance on the free side in comparison to the obstructed nasal side with a factor of 2.5:1 using the gamma scintigraphy.

21.2.1 Pearls

- Spontaneous changes in nasal airway resistance due to congestion and decongestion of nasal venous sinuses are called nasal cycle.
- Three types of nasal cycles are described with a duration of about 1–11 h.
- Consider the nasal cycle and other physiologic changes in the congestion of the nasal mucosa when making a clinical assessment of a patient complaining of nasal obstruction.
- Long-term rhinoflowmetry offers a new possibility for investigating nasal patency for up to 72 h.

21.3 Origin, Function, and Regulation of the Nasal Cycle

Finally, the origin and the function of the nasal cycle are still under scientific discussion: As first suggested by Mlynski [52] and subsequently refined [38, 44, 53], the decongested side is in the working phase of the nose [1]. The respirated air passes through the decongested nasal side; the air is conditioned, i.e. warmed up, moistured, and cleaned. The congested side is in its resting phase.

The regulation of the nasal cycle is probably performed by the central sympathomimetic tonus [4, 5, 54–58]. This tonus is controlled by two centres in the brain stem. They are connected and dominate the tonus via right and left cervical sympathomimetic nerve fibres finally asymmetrically. Consequently, medical conditions affecting this central nervous function of pace making may be detected via the nasal cycle [54, 55].

Changes in the nasal patency that are related to positional changes may be explained by two different mechanisms: an increased central venous pressure when changing from a standing to a lying position and a reflectory change of the nasal vasomotoric tonus when a lateral lying position is taken. The increase of the venous pressure causes an increased filling of the nasal venous sinusoids and an increase in the nasal resistance. The passive hydrostatic effect adds to each asymmetry in the context of the nasal cycle. The side with the highest degree of congestion generally has the highest increase of swelling and is completely obstructed [59]. Lying down leads to a temporary disturbance of the nasal cycle with congestion of the mucosa and increased flow resistance. This effect is especially pronounced in the bottom nasal side and mediated via thoracic receptors [60, 61] or those in the area of the axilla [62]. The amplitudes of the changes in the resistance in the nasal cycle are higher in supine and lateral position [59, 61, 63, 64].

The reciprocal changes of the nasal patency when taking a lateral lying position are induced by the pressure in the area of the shoulder girdle, the lateral thorax, and the hip that are the most sensitive regions. During local anaesthesia in the area of the skin, surface of the axilla cannot suppress this reflectory reaction; this is possible by an intercostal neural blockade [64]. This leads to the effect that the upper side of the nose is mainly open for nasal passage. An explanation may be that this avoids closure of the inferior side of the nose when lying on the ground so that nasal breathing is still possible.

The diagnosis of nasal obstruction and the nasal patency have to bear in mind the physiologic nasal cycle as well as changes of the nasal respiratory resistance in dependence of the surrounding conditions (temperature, probably humidity, irritating substances) [53, 65], the physical activity, body position [61, 66], or pharmacological influences [36, 67, 68].

Physical activity leads to a reduced nasal resistance [28, 38, 44, 66, 69]. According to some studies, the nasal resistance mostly increases with cold temperatures [65, 70]. Schlegel [71] reports about the main decrease of the resistance with cold temperatures and an increase with warmth. The change of the humidity from 20% to more than 90% has no influence on the nasal patency [70].

The nasal cycle may influence the nasal allergy provocation test [72–74]. Gotlib et al. [73, 75] found that the determination of the bilateral reduction of the cross-section surface in the area of the inferior turbinates is more sensitive than the determination of the one of the more reactive side. However, the risk of a spontaneous unilateral total increase of the nasal resistance must be considered.

There is some evidence that the nasal cycle is associated with ultradian rhythms of the cerebral hemispheres which affect cognitive function of the brain [57, 76–78]. This includes atypical findings during psychiatric conditions including hallucinations [77, 79]. Alternating dominance of cerebral hemispheric activity can be demonstrated in humans by EEG, and relative changes in electro-cortical activity seem to have a direct correlation with the nasal cycle [79–81]. These studies found a greater EEG activity on the side contralateral to the decongested side of the nose; a significant improvement in spatial and verbal cognitive function performed by the contralateral (in females) and ipsilateral (in males) cerebral hemisphere occurred in unilateral forced nasal breathing [57, 76]. This may explain why some patients with nasal obstruction find this more than just a simple annoyance and may develop medical indication to normalize the nasal cycle using rhinosurgery. The nasal obstruction may have effects on the ability to work during daytime [5].

Patients with upper respiratory tract infections become far more apparent with a significant increase in the amplitude of the changes than healthy people [82].

As physical activity decreases significantly during sleep, long-term rhinoflowmetry enables to detect a more regular or classical type of nasal cycle during night. Currently, it is discussed whether this is an indicator of increased individual resistance or respiratory malfunctioning of the nose and possible indication for surgical correction.

- The regulation of the nasal cycle is probably performed by the central sympathomimetic tonus.
- An abnormal nasal cycle may give an indication for corrective rhinosurgery in the future.

21.4 Conclusion

The nasal cycle and other spontaneous variations in nasal airflow and the two separate nasal passages must be considered when making a clinical assessment for patients complaining of nasal obstruction. In addition to the clinical history, examination by anterior rhinoscopy and endoscopy of the nose, and classical measurements of the nasal airflow like rhinomanometry or acoustic rhinometry, long-term rhinoflowmetry may be a helpful tool in assessing these patients.

References

1. Grutzenmacher S, Lang C, Mlynski R, Mlynski B, Mlynski G. Long-term rhinoflowmetry: a new method for functional rhinologic diagnostics. Am J Rhinol. 2005;19(1):53–7.
2. Lang C, Grutzenmacher S, Mlynski B, Plontke S, Mlynski G. Investigating the nasal cycle using endoscopy, rhinoresistometry, and acoustic rhinometry. Laryngoscope. 2003;113(2):284–9.
3. Pendolino AL, Scarpa B, Ottaviano G. Relationship between nasal cycle, nasal symptoms and nasal cytology. Am J Rhinol Allergy. 2019;33(6):644–9.
4. Eccles R. Neurological and pharmacological considerations. In: Proctor D, Andersen I, editors. The nose. Amsterdam/New York: Elsevier; 1982.
5. Hanif J, Jawad SS, Eccles R. The nasal cycle in health and disease. Clin Otolaryngol Allied Sci. 2000;25(6):461–7.
6. Pendolino A, Lund V, Nardello E, Ottaviano G. The nasal cycle: a comprehensive review. Rhinology Online. 2018;1(1):67–76.
7. Hasegawa M, Kern EB. The effect of breath holding, hyperventilation, and exercise on nasal resistance. Rhinology. 1978;16(4):243–9.
8. Hasegawa M, Kern EB. The human nasal cycle. Mayo Clin Proc. 1977;52(1):28–34.
9. Kayser R. Die exacte Messung der Luftdurchgängigkeit der Nase. Arch Laryngol. 1895;3:101–210.
10. Heetderks D. Observations on the reaction of normal nasal mucous membrane. Am J Med Sci. 1927;664:231–44.
11. Anselmo-Lima WT, Lund VJ. The effects of endoscopic sinus surgery on the nasal cycle as assessed by acoustic rhinometry. Am J Rhinol. 2001;15(3):165–8.
12. Baczek M, Hassmann E, Alifier M, Iwaszko-Krawczuk W. Acoustic rhinometry assessment of the nasal cycle in neonates. Acta Otolaryngol. 2001;121(2):301–4.
13. Corey JP, Gungor A, Nelson R, Fredberg J, Lai V. A comparison of the nasal cross-sectional areas and volumes obtained with acoustic rhinometry and magnetic resonance imaging. Otolaryngol Head Neck Surg. 1997;117(4):349–54.
14. Fisher EW, Liu M, Lund VJ. The nasal cycle after deprivation of airflow: a study of laryngectomy patients using acoustic rhinometry. Acta Otolaryngol. 1994;114(4):443–6.
15. Fisher EW, Scadding GK, Lund VJ. The role of acoustic rhinometry in studying the nasal cycle. Rhinology. 1993;31(2):57–61.
16. Gallego AJ, Cavallari FE, Valera FC, Demarco RC, Anselmo-Lima WT. Study of nasal cycles in children by acoustic rhinometry. Am J Rhinol. 2006;20(6):560–2.
17. Gogniashvili G, Japaridze S, Khujadze M. Investigation of the nasal cycle function through endoscopy, rhinoresistometry and acoustic rhinomety. Georgian Med News. 2009;174:22–5.
18. Gungor A, Moinuddin R, Nelson RH, Corey JP. Detection of the nasal cycle with acoustic rhinometry: techniques and applications. Otolaryngol Head Neck Surg. 1999;120(2):238–47.
19. Sung YW, Lee MH, Kim IJ, Lim DW, Rha KS, Park CI. Nasal cycle in patients with septal deviation: evaluation by acoustic rhinometry. Am J Rhinol. 2000;14(3):171–4.
20. Tan MFM, Whitcroft KL, Mehta N, Schilder A, Leung TS, Andrews PJ. Investigating the nasal cycle using unilateral peak nasal inspiratory flow and acoustic rhinometry minimal cross-sectional area measurements. Clin Otolaryngol. 2019;44(4):518–24.
21. Grillo C, La Mantia I, Triolo C, Scollo A, La Boria A, Intelisano G, et al. Rhinomanometric and olfactometric variations throughout the menstrual cycle. Ann Otol Rhinol Laryngol. 2001;110(8):785–9.
22. Huang ZL, Ong KL, Goh SY, Liew HL, Yeoh KH, Wang DY. Assessment of nasal cycle by acoustic rhinometry and rhinomanometry. Otolaryngol Head Neck Surg. 2003;128(4):510–6.
23. Pendolino AL, Nardello E, Lund VJ, Maculan P, Scarpa B, Martini A, et al. Comparison between unilateral PNIF and rhinomanometry in the evaluation of nasal cycle. Rhinology. 2018;56(2):122–6.
24. Hasegawa M, Kern EB. Variations in nasal resistance in man: a rhinomanometric study of the nasal cycle in 50 human subjects. Rhinology. 1978;16(1):19–29.
25. Stoksted P. The physiologic cycle of the nose under normal and pathologic conditions. Acta Otolaryngol. 1952;42(1–2):175–9.
26. Stoksted P. Rhinometric measurements for determination of the nasal cycle. Acta Otolaryngol Suppl. 1953;109:159–75.

27. Masing H, Wolf G. Der Nachweis des Nasenmuschelzyklus mit Hilfe des Röntgenschichtbildverfahrens. Z Laryng Rhinol Suppl. 1969;48:684–92.
28. Cole P, Haight J, Cooper P, Kassel E. A computed tomographic study of nasal mucosa: effects of vasoactive substances. J Otolaryngol. 1983;12:58–60.
29. Friling L, Nyman HT, Johnson V. Asymmetric nasal mucosal thickening in healthy dogs consistent with the nasal cycle as demonstrated by MRI and CT. Vet Radiol Ultrasound. 2014;55(2):159–65.
30. Kennedy DW, Zinreich SJ, Kumar AJ, Rosenbaum AE, Johns ME. Physiologic mucosal changes within the nose and ethmoid sinus: imaging of the nasal cycle by MRI. Laryngoscope. 1988;98(9):928–33.
31. Webber RL, Jeffcoat MK, Harman JT, Ruttimann UE. MR demonstration of the nasal cycle in the beagle dog. J Comput Assist Tomogr. 1987;11(5):869–71.
32. Zinreich SJ, Kennedy DW, Kumar AJ, Rosenbaum AE, Arrington JA, Johns ME. MR imaging of normal nasal cycle: comparison with sinus pathology. J Comput Assist Tomogr. 1988;12(6):1014–9.
33. Cole P. Upper respiratory airflow. In: Proctor D, Andersen I, editors. The nose. Amsterdam/New York: Elsevier; 1982.
34. Cole P, Niinimaa V, Mintz S, Silverman F. Work of nasal breathing: measurement of each nostril independently using a split mask. Acta Otolaryngol. 1979;88(1–2):148–54.
35. Koch U, Pau H. Beziehung zwischen Nasenwegswiderstand und Tubenfunktion in Abhängigkeit von der Tageszeit. Arch. Oto-Rhino-Laryngol. 1982;235:583–6.
36. Braun T, Rich M, Berghaus A, Kramer MF. Effects of oxymetazoline nasal spray on the nasal cycle assessed by long-term rhinoflowmetry. Rhinology. 2012;50(4):370–5.
37. Kahana-Zweig R, Geva-Sagiv M, Weissbrod A, Secundo L, Soroker N, Sobel N. Measuring and characterizing the human nasal cycle. PLoS One. 2016;11(10):e0162918.
38. Beule AG, Mlynski G. Präoperatives management. In: Mlynski G, Pirsig W, editors. Funktionell-ästhetische Rhinochirurgie. Stuttgart/New York: Thieme; 2018.
39. Keerl R, Weber R, Huppmann A, Draf W. Darstellung zeitabhangiger Veranderungen der Nasenschleimhaut unter Einsatz modernster Morphsoftware. Laryngorhinootologie. 1995;74(7):413–8.
40. Hasegawa M. Nasal cycle and postural variations in nasal resistance. Ann Otol Rhinol Laryngol. 1982;91(1 Pt 1):112–4.
41. Weber R, Keerl R. Einsatz moderner Bilddatenverarbeitung in der klinisch-rhinologischen Forschung. Eur Arch Otorhinolaryngol Suppl. 1996;1:271–96.
42. Melon J. Contribution a l'etude de l'activite secretoire de la muqqueuse nasale. Caracteres de permeabilite de l'epithelium. Composition chimique du mucus. Le mecaanisme de sa formation dans les conditions normales et au cours des hypersecrtions paroxystiques. Acta Otorhinolaryngol Belg. 1968;22(1):5–244.
43. Kern EB. The noncycle nose. Rhinology. 1981;19(2):59–74.
44. Beule AG. Physiology and pathophysiology of respiratory mucosa of the nose and the paranasal sinuses. GMS Curr Top Otorhinolaryngol Head Neck Surg. 2010;9:Doc07.
45. Letztel J. Subjektive und objektive Beurteilung der Nasenatmung des Riechvermögens vor nach operativer Septumkorrekturen Dresden; 2021.
46. van Cauwenberge P. De schommelingen in de neusweerstand bij jonge kinderen. Acta Otorhinolaryngol Belg. 1980;34(2):145–56.
47. van Cauwenberge PB, de Schynkel K, Kluyskens PM. Clinical use of rhinomanometry in children. Int J Pediatr Otorhinolaryngol. 1984;8(2):163–75.
48. Van Cauwenberge PB, Deleye L. Nasal cycle in children. Arch Otolaryngol. 1984;110(2):108–10.
49. Ingels KJ, Meeuwsen F, van Strien HL, Graamans K, Huizing EH. Ciliary beat frequency and the nasal cycle. Eur Arch Otorhinolaryngol. 1990;248(2):123–6.
50. Doyle WJ, van Cauwenberge PB. Relationship between nasal patency and clearance. Rhinology. 1987;25(3):167–79.
51. Soane RJ, Carney AS, Jones NS, Frier M, Perkins AC, Davis SS, et al. The effect of the nasal cycle on mucociliary clearance. Clin Otolaryngol Allied Sci. 2001;26(1):9–15.
52. Mlynski G, Grützenmacher S, Mlynski B, Lang C. Aerodynamik der Nase – Physiologie und Pathophysiologie. In: Ganz H, Iro H, editors. HNO Praxis heute. Heildeberg: Springer; 2000. p. 61–81.
53. White DE, Bartley J, Nates RJ. Model demonstrates functional purpose of the nasal cycle. Biomed Eng Online. 2015;14:38.
54. Eccles R, Eccles KS. Asymmetry in the autonomic nervous system with reference to the nasal cycle, migraine, anisocoria and Meniere's syndrome. Rhinology. 1981;19(3):121–5.
55. Galioto G, Mevio E, Galioto P, Fornasari G, Cisternino M, Fraietta L. Modifications of the nasal cycle in patients with hypothalamic disorders: Kallmann's syndrome. Ann Otol Rhinol Laryngol. 1991;100(7):559–62.
56. Shannahoff-Khalsa DS, Kennedy B, Yates FE, Ziegler MG. Ultradian rhythms of autonomic, cardiovascular, and neuroendocrine systems are related in humans. Am J Phys. 1996;270(4 Pt 2):R873–87.
57. Shannahoff-Khalsa DS, Yates FE. Ultradian sleep rhythms of lateral EEG, autonomic, and cardiovascular activity are coupled in humans. Int J Neurosci. 2000;101(1–4):21–43.
58. Williams M, Eccles R. A model for the central control of airflow patterns within the human nasal cycle. J Laryngol Otol. 2016;130(1):82–8.
59. Cole P, Haight JS. Posture and the nasal cycle. Ann Otol Rhinol Laryngol. 1986;95(3 Pt 1):233–7.
60. Cole P, Haight JS. Mechanisms of nasal obstruction in sleep. Laryngoscope. 1984;94(12 Pt 1):1557–9.
61. Frye RE, Rosin DF, Morrison AR, Leon-Sarmiento FE, Doty RL. Modulation of the ultradian human

nasal cycle by sleep stage and body position. Arq Neuropsiquiatr. 2017;75(1):9–14.
62. Widdicombe JG. The physiology of the nose. Clin Chest Med. 1986;7(2):159–70.
63. Haight JS, Cole P. Is the nasal cycle an artifact? The role of asymmetrical postures. Laryngoscope. 1989;99(5):538–41.
64. Haight JS, Cole P. Unilateral nasal resistance and asymmetrical body pressure. J Otolaryngol Suppl. 1986;16:1–31.
65. Battle E Einflüsse der Temperatur auf den Nasenwiderstand. Rhinomanometrische Untersuchungen am Gesunden. Dissertation, Medizinische Fakultät Zürich; 1989.
66. Rohrmeier C, Schittek S, Ettl T, Herzog M, Kuehnel TS. The nasal cycle during wakefulness and sleep and its relation to body position. Laryngoscope. 2014;124(6):1492–7.
67. Williams RG, Eccles R. Nasal airflow asymmetry and the effects of a topical nasal decongestant. Rhinology. 1992;30(4):277–82.
68. Flanagan P, Eccles R. Physiological versus pharmacological decongestion of the nose in healthy human subjects. Acta Otolaryngol. 1998;118(1):110–3.
69. Eccles R. Nasal airflow in health and disease. Acta Otolaryngol. 2000;120(5):580–95.
70. Salman SD, Proctor DF, Swift DL, Eveering SA. Nasal resistance: description of a method and effect of temperature and humidity changes. Ann Otol Rhinol Laryngol. 1971;80(5):736–43.
71. Schlegel C, Gammert C. Verhalten des Nasenatemwiderstandes in Kälte und Wärme. Otorhinolaryngol Nova. 1991;1:189–93.
72. Brooks CD, Karl KJ, Francom SF. Unilaterality of obstruction after acute nasal allergen provocation. Relation of allergen dose, nasal reactivity and the nasal cycle. Clin Exp Allergy. 1991;21(5):583–7.
73. Gotlib T, Samolinski B, Grzanka A. Effect of the nasal cycle on congestive response during bilateral nasal allergen provocation. Ann Agric Environ Med. 2014;21(2):290–3.
74. Kim JK, Cho JH, Jang HJ, Shim DB, Shin HA. The effect of allergen provocation on the nasal cycle estimated by acoustic rhinometry. Acta Otolaryngol. 2006;126(4):390–5.
75. Gotlib T, Samolinski B, Grzanka A. Bilateral nasal allergen provocation monitored with acoustic rhinometry. Assessment of both nasal passages and the side reacting with greater congestion: relation to the nasal cycle. Clin Exp Allergy. 2005;35(3):313–8.
76. Shannahoff-Khalsa D. Lateralized rhythms of the central and autonomic nervous systems. Int J Psychophysiol. 1991;11(3):225–51.
77. Shannahoff-Khalsa D, Golshan S. Nasal cycle dominance and hallucinations in an adult schizophrenic female. Psychiatry Res. 2015;226(1):289–94.
78. Block RA, Arnott DP, Quigley B, Lynch WC. Unilateral nostril breathing influences lateralized cognitive performance. Brain Cogn. 1989;9(2):181–90.
79. Werntz DA, Bickford RG, Bloom FE, Shannahoff-Khalsa DS. Alternating cerebral hemispheric activity and the lateralization of autonomic nervous function. Hum Neurobiol. 1983;2(1):39–43.
80. Werntz DA, Bickford RG, Shannahoff-Khalsa D. Selective hemispheric stimulation by unilateral forced nostril breathing. Hum Neurobiol. 1987;6(3):165–71.
81. Zelano C, Jiang H, Zhou G, Arora N, Schuele S, Rosenow J, et al. Nasal respiration entrains human limbic oscillations and modulates cognitive function. J Neurosci. 2016;36(49):12448–67.
82. Eccles R, Reilly M, Eccles KS. Changes in the amplitude of the nasal cycle associated with symptoms of acute upper respiratory tract infection. Acta Otolaryngol. 1996;116(1):77–81.

22 Nasal Physiology and Pathophysiology and Their Relationship with Surgery: The Nasal Valves

Oren Friedman and Kevin Wang

Clinical Pearls

- Understanding normal nasal physiology and pathophysiology is the foundation for nasal healthcare.
- The complex interplay between nasal anatomy and physiology dictates nasal health.
- Proper diagnosis and management of nasal disorders requires an understanding of nasal physiology and pathophysiology.
- Expert surgical care of nasal dysfunction requires us to respect the normal anatomical and physiological constructs, and to improve upon them when necessary.
- Conservative surgical therapies as detailed in this chapter allow us to improve our patients' quality of life.

22.1 Introduction

The nose is a prominent facial feature that plays an important role aesthetically as well as physiologically. During nasal surgery, these physiological functions must be preserved or improved if they have been compromised. Basic nasal functions include respiration, air conditioning, filtration, immune defense, and olfaction. The nose provides a certain level of inspiratory resistance that is crucial to comfortable breathing. The changes in resistance at different portions of the nasal cavity allow for the nose to act as the primary regulator of nasal airflow. The air conditioning function is responsible for warming and humidification of the inspired air, thus making it suitable for the pulmonary airways to optimize gas exchange. The nose also acts as a filter and barrier against airborne particles and pathogens. A number of specific and nonspecific mechanisms contribute to the immune defense of the nasal mucosa and protect humans from external pathogens. The sense of olfaction is best appreciated when it is lacking. Recognition of potentially harmful inhalants can be lifesaving, and the joy of smells and flavors greatly enhances the quality of life.

22.2 The Nasal Valves

22.2.1 Overview

The nasal valves are the narrowest portions of the nasal airway, accounting for over half of the total nasal airway resistance. There are a total of four different "nasal valves," including an internal and an external nasal valve on either side of the nose. It is essential to distinguish between

O. Friedman (✉)
University of Pennsylvania Perelman School of Medicine, Philadelphia, PA, USA

K. Wang
Robert Wood Johnson School of Medicine, Piscataway, NJ, USA

Ö. Ö. Celebi, T. M. Önerci (eds.), *Nasal Physiology and Pathophysiology of Nasal Disorders*,
https://doi.org/10.1007/978-3-031-12386-3_22

the internal and external nasal valves, as these represent distinct anatomical and physiological entities [1–3].

22.2.1.1 The Internal Nasal Valve

The area defined as the "internal nasal valve" is the narrowest portion of the nasal airway and it is therefore the primary regulator of nasal airflow. It is the area bound by the nasal septum medially, the caudal edge of the upper lateral cartilage laterally, the floor of the nose inferiorly, and the head of the inferior turbinate inferolaterally. The normal angle between the upper lateral cartilage and nasal septum is 10–15 degrees, but the area is dynamic so an extensive range exists. The cross-sectional area of this region is approximately 55–64 mm^2 [4, 5]. A smaller cross-sectional area or a narrowed angle between the upper lateral cartilage and nasal septum may contribute to increased airway resistance and the sensation of nasal airway obstruction. There are many approaches to increase the cross-sectional area of the nasal valve or increase the angle in order to facilitate comfortable breathing.

22.2.1.2 The External Nasal Valve

The "external nasal valve" is the gateway to the nose. It is located at the nasal vestibule and is the area bound by the alar rim and columella, including the medial crus, nasal spine, and soft tissues covering the nasal sill and floor. The external nasal valve may be more prone to collapse as a result of being narrowed at rest (i.e., vestibular stenosis) or as a result of a weak lateral component. As with the internal nasal valve, these characteristics cause the nasal valve to narrow and collapse due to the negative pressure associated with nasal inspiration [6, 7].

22.2.1.3 Influences on Nasal Valve Strength

The strength and stability of both the internal and external nasal valves are controlled by the external skin covering, the internal mucosal covering, the intrinsic cartilage, and the subcutaneous muscles [8, 9].

22.3 Nasal Physiology

22.3.1 The Nasal Cycle

Without a sound understanding of the nasal cycle, it is easy to confuse normal nasal physiology with nasal obstruction. First described by Kayser in 1895, the nasal cycle is a function of the nasal submucosa present in 72–80% of the human population. With remarkable regularity, cyclical changes in the cross-sectional lumen of the nasal cavities occur approximately every 3 to 4 h. These changes are generally unnoticed by the patient because the combined nasal airway resistance of both sides remains constant [8]. The nasal cycle can result in near-total unilateral nasal obstruction on inspection or imaging, a finding that must not be confused with pathologic nasal obstruction. The function of the nasal cycle has been hypothesized to allow for regeneration of the nasal mucosa on the obstructed side by increasing the water percentage, maintaining humidification of inspired air [9]. This would suggest that warming, humidification, and other functions challenge the nasal mucosa and submucosa so greatly that these functions cannot be maintained in an uninterrupted manner. Unfortunately, no data are available to quantify the functional residual capacity of the nose. It is therefore unknown how much submucosa or mucosa can be lost before function of the nose is compromised. Further research is needed to better elucidate this important issue as well as the purposes and meaning of the nasal cycle.

An important aspect of nasal anatomy related to nasal physiology is the turbinate bones, also known as conchae. The turbinate bones are crucial components of the nasal mucosa, expanding the total surface area and creating turbulence in the air entering the nasal cavity. This causes air to swirl as it moves through the nasal cavity, increasing contact between the inspired air and the nasal mucosa which may aid the individual in perceiving the airflow.

22.3.2 Olfaction

The sense of olfaction combines neural input from the olfactory, trigeminal, glossopharyngeal, and vagus nerves. Olfaction begins with stimulation of the olfactory system by an odorant, which is transported by endonasal airflow towards the olfactory cleft. Sniffing facilitates this transport by increasing inspired airflow and directing airflow to the olfactory epithelium. The sense of olfaction is also an important contributor to the sense of taste—retrograde airflow brings airborne molecules derived from food in the pharynx towards the olfactory epithelium.

22.3.3 Filtering Function

The nasal mucosa efficiently protects the lower airways from inhalation of particulate matter. Multiple studies have investigated both deposition (uptake, absorption, retention) and regional distribution in the nose. Particles with different sizes, shapes, specific weights, and aerodynamic properties will deposit in different locations in the nasal mucosa. About 60% of particles 1 mm in diameter are deposited in the nasal cavity, and the deposition of larger particles is more complete [10]. Smaller particles in the 10–11 μm size range are primarily deposited in the central nasal regions, such as the anterior turbinates, turbinates, and olfactory regions [11]. Most particles will deposit in two main deposition sites. As a result of the transition from laminar to turbulent flow, particles are deposited on the mucosa posterior to the nasal valve. The direction of the endonasal airflow subsequently directs particles to the second site of predominant deposition, the anterior aspect of the middle turbinate [12]. The endonasal distribution of nasal tumors has been correlated with these particle deposition patterns. This deposition pattern may also explain the preponderance of middle turbinate edema in allergic rhinitis patients, and the common initial formation of nasal polyps most commonly at the middle turbinate region.

22.3.4 Humidifying Capacity

The relatively small surface area of the nasal mucosa (160 cm^2) and the underlying submucosa are faced with the formidable task of humidifying and warming about 14,000 L of inspired air that pass through the nasal cavities of a normal active adult in 1 day. The submucosa allows for energy and metabolites to be transported to the mucosa, which acts as the interface for exchange. Air is heated by conduction, convection, and radiation. The heat exchange is efficient, because the blood flow is in the opposite direction to the incoming airflow, a process known as countercurrent exchange. This process is facilitated by the rich vascular network of the nasal septum and turbinates. In comfortable room conditions, about 280 kJ of energy is required to warm the air to 32 °C in the rhinopharynx. To saturate this volume of inspired air with humidity, 1400 kJ of energy and 600 g of water are expended. On expiration, a considerable proportion of warmth and humidity are extracted from the air to conserve energy [13–15]. Some conditions related to impaired humidifying capacity of the nasal mucosa include nasal crusting, bleeding, rhinorrhea, and asthma, especially in cold weather [16].

22.3.5 Airflow, Resistance, and Regulation

To facilitate proper air conditioning and humidification of inspired air, the nose must provide a certain level of resistance. At the entrance of the nasal cavity, the speed of inspired air is about 2–3 m/s and rises to about 12–18 m/s at the internal nasal valve. Here, the airflow makes an upward angulation of approximately 60 degrees, hence the term "upstream resistors" to describe the area of the nasal valve. Posterior to the nasal valves the speed diminishes to 2 to 3 m/s, and the flow becomes more horizontal and eventually tilts downward toward the choanae. More air passes through the middle meatus than the inferior meatus. As the airflow becomes turbulent, warming and humidifying of the inspired air are enhanced. When entering the nasopharynx, the airflow tilts downward, becomes laminar, and increases in velocity to 3 to 4 m/s. These generalizations of intranasal airflow have been concluded from predominantly static models. It is important to understand that the regulation of nasal resistance is a dynamic process.

A liquid or gas flowing through a tube increases its velocity and diminishes the transmural pressure at an area of constriction. This phenomenon, referred to as the Bernoulli effect, explains the observation that the nasal valve collapses to varying degrees on inspiration. The Bernoulli effect can be quantified by Poiseuille's Law, which states that flow through a tube is proportional to the radius of the tube raised to the fourth power—consequently, changes in cross-sectional area of the nasal valve will have profound effects on flow. When nasal valve collapse becomes premature, pathologic airway obstruction ensues. Three factors determine the collapsibility of the nasal valves: the overall cross-sectional area of the nasal valve, the shape of the nasal valve area, and the resilience of the structural skeleton. Per Poiseuille's Law, a smaller nasal valve area causes a greater velocity of inspired airflow and an increased negative transmural pressure, making the nasal valve more prone to collapse. A slit-like nasal valve area and weaker skeletal structures cause lower stability of the nasal valves, making them more prone to collapse. The theoretical ideal would be a nasal valve with a more rounded, large valve area, and a structurally firm skeleton.

The alae represent the softest part of the nasal tip, followed by the columella, interdomal region, and the anterior septal angle [17]. The midpoint of the ala is the location most prone to collapse. This is due to leverage—this point is most distant from the structural fulcra of the external nasal valve: the anterior septal angle and the alar base.

Further studies have shown that the nasal muscles profoundly influence the structural resilience of the nasal valve. An involuntary resting tone adds stiffness to the ala, and this is further increased by voluntary activation of the nasal musculature (flaring). As evidenced by rhinomanometric measurements, the involuntary (resting) muscle tone decreases nasal airway resistance. Voluntary flaring further opens the nasal airway [18, 19]. This phenomenon explains why people with cranial nerve VII paralysis may suffer from new-onset nasal obstruction.

22.3.6 Immune Defense

The mechanisms that protect the nose against irritants, microorganisms, and allergens can be described as nonspecific and specific systems. The nonspecific system includes the filtering function of the nose with the mucociliary transport system. The mucous blanket is produced by the goblet cells and is driven by ciliary movement toward the lateral pharyngeal walls for ingestion. The mean velocity of the mucus flow and particle transport is about 6 mm/min in normal conditions. Inspired microorganisms, irritants, and allergens are trapped. The specific defense mechanisms include the various immunologic, humoral, and cellular responses.

22.4 Nasal Pathophysiology

22.4.1 Physics of Nasal Valve Pathophysiology

Much of nasal valve pathophysiology can be explained by Bernoulli's principle, which describes how airflow through a tube will have the greatest velocity at the narrowest section, or section with the lowest cross-sectional area. With a greater velocity, airflow through that section will have a decreased pressure as well, leading to nasal valve collapse. This is a crucial concept in nasal valve pathology—it explains why a relatively smaller nasal valve or portion of the nasal valve will be more likely to collapse.

The inverse relationship between flow and cross-sectional area can be quantified by Poiseuille's Law, which states that the relationship between flow and tube radius is fourth order. For example, if the radius of a tube were to double, then flow would increase by 16 times. This also means that a decrease in tube radius would lead to an increase in pressure, to the fourth power.

22.4.2 Nasal Valve Collapse

An estimated 13% of the general population suffer from nasal valve collapse. While nasal valve collapse may occur in the absence of any previous trauma or surgical intervention, it is very commonly a result of a failure to preserve the integrity of the nasal valve during cosmetic rhinoplasty and other traditional nasal surgeries. Nasal valve collapse can occur at both the internal and external nasal valves, and may be dynamic, static, or both [20].

22.4.2.1 Internal Nasal Valve Collapse

Internal nasal valve collapse may be categorized as static or dynamic [21, 22]. Static internal nasal valve collapse is a narrowing of the middle third of the nose at rest—that is, the angle between the upper lateral cartilage and nasal septum is anatomically small, resulting in a reduced valve area. Static collapse is often seen as a result of nasal trauma or previous rhinoplasty in which a weakening of nasal support structures leads to an overly narrowed angle between the upper lateral cartilage and nasal septum (i.e., this may result from simple surgical maneuvers, such as skin elevation, mucosal elevation, or separation of the upper lateral cartilages from the septum). Over-resection of the upper lateral cartilages, excessive narrowing of the dorsum, and displacement of the short nasal bones correlate with internal valve collapse [23]. Static internal valve collapse may also result from scarring of the medial segment of the upper lateral cartilage to the nasal septum following the separation of those structures along with their intervening mucosa. Elevation of the skin soft tissue envelope, damage to the nasal dilator muscles, and weakening of the mucosal support of the middle third cartilages all contribute to a weakening of nasal valve support.

Middle third narrowing at rest, that is, static internal nasal valve collapse, may be seen upon simple external inspection of the nose in which a pinched middle third may be visible. This usually presents as a discontinuity along the brow-tip aesthetic line where the middle third is pinched and narrowed while the upper third, which is supported by bone, remains wide. This inverted V deformity has been linked to the separation of the upper lateral cartilages from the overlying nasal bones during rhinoplasty, but can actually be seen more commonly among patients with internal nasal valve collapse, especially those with thin skin, who have either never undergone previous rhinoplasty or who have had a standard dorsal reduction rhinoplasty without disarticulation of the upper lateral cartilages from the nasal bones. Static internal nasal valve collapse may appear clinically as scar tissue, strictures, or webbing in the valve angle (as might occur after separation of the upper lateral cartilage and mucosa from the septum without sparing the mucosal attachments between the two structures). Static narrowing of the middle third of the nose can also be the result of congenital or traumatic weakness or absence of the upper lateral cartilages, or other deformities, such as thickening or twisting of the upper lateral cartilages. In patients with an over-projected tension type nasal deformity, overgrowth of the nasal septum causes the angle between the upper lateral cartilage and the nasal septum to be excessively narrowed. Such patients may also frequently be found to have thin and weak upper lateral cartilages, which tend to add a dynamic component to the internal nasal valve collapse.

Dynamic internal nasal valve collapse is an active narrowing of the upper lateral cartilage and middle third of the nose which occurs only with nasal inspiration through a valve which, at rest, appears of normal size. Dynamic nasal valve collapse, in which the middle nasal third appears normal at rest but narrowed upon gentle nasal inspiration, often results from an inherent weakness of the nasal sidewalls. Thin, weak, detached, or absent upper lateral cartilages cannot provide the necessary strength along the nasal sidewall to withstand the negative pressures created by inspiratory nasal airflow—as a result, the sidewalls of the nose fall in as the negative pressure created by nasal inspiration draws them inward. Previous rhinoplasty in which the upper lateral cartilages have been weakened or detached from the nasal septum, congenitally or developmentally thin upper lateral cartilages, and absent upper lateral cartilages may all contribute to such structural weaknesses of the nasal sidewall. In such cases, the patient may not have obvious findings suggestive of nasal valve collapse upon inspection at rest, such as an inverted V deformity or a pinched middle third, but when asked to inspire gently through the nose, there is an active narrowing of the middle third which becomes obvious to the examining physician. In order to appreciate dynamic nasal valve collapse and its effects on nasal breathing, it is often helpful for the examining physician to apply gentle lateral traction on the cheek adjacent to the nose (i.e., Cottle maneuver) to assess for improvements in nasal breathing that might occur with stiffening of the lateral nasal wall by digital traction. Additionally, in order to pinpoint the precise location of the collapse, it is helpful to introduce a cotton tip applicator or ear curette into the suspected area of collapse and observe improvements in nasal obstructive symptoms when the precise site of obstruction is stiffened with the examiner's help. Application of external nasal dilation devices further helps identify internal nasal valve collapse.

22.4.2.2 External Nasal Valve Collapse

The external nasal valve is an area primarily supported by the lower lateral cartilages and their overlying skin and soft tissue covering, and is defined anatomically by the region between the columella and the alar rim. The size, shape, and strength of the lower lateral cartilages create the nasal vestibular aperture that defines the external nasal valve. In normal individuals, the rigidity of the lateral walls of the external nasal valve is enough to prevent their collapse during inspiration. However, in patients with external nasal valve dysfunction, static narrowing can be seen with vestibular scarring and stenosis or alar rim collapse. This is commonly a result of trauma, soft tissue triangle injury, reconstruction of nasal skin cancer defects, cleft lip repair, alar base narrowing procedures, significant caudal nasal septal deformities in Cottle area 1, or secondary to a variety of other causes. Functionally unfavorably shaped lower lateral cartilages may narrow the external nasal valve aperture and contribute to nasal obstruction simply due to the shape of the cartilages (concave lower lateral cartilages that impinge on the airway, lateral crural cephalic malposition with resultant concavity along alar rim, etc.). For example, in the case of tension type nasal deformities in which the vestibular aperture at the level of the nasal rim is narrowed and pinched, we see the shape of the lower lateral cartilage affecting the size of the nasal vestibule—a tentpole-like over-projection of the tip results in a narrow nasal vestibule secondary to "slit-like nostrils."

Dynamic external nasal valve collapse occurs when the valve appears normal at rest, but the alar rims collapse upon inspiration through the nose. Primary weakness of the lower lateral cartilage and malposition of the lower lateral cartilage (as with vertically oriented lower lateral cartilages) often lead to dynamic external valve collapse—in both situations, the lower lateral cartilage malposition and inadequate soft tissue support at the rim leads to an inability to support or withstand the negative inspiratory forces generated by nasal breathing.

In examining a patient with external valve collapse, it is best to simply observe the nose during quiet breathing and watch the nasal vestibule for narrowing of the alar rim on gentle nasal inspiration. As with internal valve collapse, application of lateral traction with the examiner's hand, a cotton applicator or wax curette, or with an external nasal dilator will help identify the precise area of weakness and may further help demonstrate to the patient what may be achieved with surgical correction of the external nasal valve weakness.

22.4.2.3 The Aging Nose

A common and increasingly more prevalent clinical scenario in which we find nasal obstruction associated with both internal and external nasal valve collapse is in the aging patient. As the nose ages, it undergoes structural changes that result in various weaknesses which lead to nasal valve collapse and obstruction. A significant loss of nasal support occurs with thinning of the nasal bones and skin, thinning and weakening of the upper lateral cartilages, laxity in the supportive attachments between the upper and lower lateral cartilages at the scroll region, weakening of the lower lateral cartilages, and laxity in the supportive fibrous attachments between the lower lateral cartilages and the nasal septum and maxilla. Additionally, just as in the rest of the body, the nasal muscles likely atrophy with age, which may add to the collapsibility of the nasal sidewall. The structural changes associated with aging contribute to the drooping of the nasal tip (tip ptosis), narrowing and weakening of the internal valve, and narrowing and weakening of the external nasal valve – all of which contribute to both functional breathing problems as well as to the aesthetic changes that are typical of the aging nose. As we face a global aging of the population, nasal surgeons around the world will likely see increasing numbers of patients for surgical correction of nasal breathing [24–26].

22.4.2.4 Paradoxical Obstruction

Paradoxical nasal obstruction may result when a patient has severe unilateral structural obstruction, such as septal deviation. It is a condition where the patient becomes habituated to a complete unilateral obstruction on one side, and only notices an obstruction when the nasal cycle shifts to the unobstructed side. As a result, the patient only complains about the healthy side of their nose, oblivious to the unilateral obstruction. This form of obstruction can typically be corrected surgically by valve repair and/or septoplasty.

Another form of nasal airway impairment is also considered paradoxical obstruction. This form occurs when excessive turbinate resection results in too large a nasal cavity. The excessive reduction of nasal airway resistance causes the perception of impaired nasal breathing. This form of paradoxical obstruction, known commonly as Empty Nose Syndrome, and its associated symptoms are particularly difficult to treat.

22.4.2.5 Chronic Rhinitis

Chronic atrophic rhinitis is a debated entity. This multifactorial disease has historically been attributed to bacterial causes. However, with the advent of modern antibiotics, other causes of chronic atrophic rhinitis have been recognized. The hallmark characteristics of this disease are atrophy and loss of function of the nasal mucosa, resulting in impaired air conditioning and humidification of inspired air. Bacterial colonization results in purulent secretions and odor. Atrophy of the intranasal tissues leaves too much space in the nose, and paradoxical obstruction ensues. Debate about this entity has resulted from a lack of data regarding the manifestation, progression, diagnosis, and etiology of the disease. Causes of chronic atrophic rhinitis include reductive turbinate surgery, radiation, infection, and other destructive measures. Dry mucosa, crusting, and pain have been proposed as diagnostic criteria, but these identify only the terminal stages of the disease. Long delays between turbinate surgery and manifestation of end-stage chronic atrophic rhinitis have been reported in the literature. No good test or established criteria exist to diagnose early changes of chronic atrophic rhinitis or to document its progression.

Vasomotor rhinitis represents a dysregulation of the autonomous neural network of the intranasal vasculature. Clinical manifestations include mucosal hypertrophy with nasal obstruction and clear rhinorrhea. Typically a diagnosis of exclusion, this disease responds well to topical intranasal application of albuterol.

Rhinitis medicamentosa results from overuse of decongesting nasal sprays. Typical agents are phenylephrine and oxymetazoline. Extended use of these sympathomimetic agents causes resistance to their vasoconstrictive properties, rebound vasodilation, and congestion of the nasal mucosa with obstruction. With long-term overuse, patients develop clinically relevant long-term changes to the nasal submucosa and mucosa. These patients seem to present more frequently with recurrent airway obstruction after surgical treatment of the turbinates. The diagnosis of rhinitis medicamentosa is established by history, and weaning of the medication is the first-line therapy.

22.4.2.6 Allergic Rhinitis

Allergic rhinitis is present in 10% to 30% of adults and up to 45% of children. Nearly 50% of patients with allergic rhinitis experience symptoms for >4 months of the year as a result of seasonal changes, which introduce allergens that cause increased mucus secretion and blood vessel dilation within the nasal cavity [27]. Other symptoms include nasal itching, rhinorrhea, sneezing, ocular itching, redness, and tearing.

Symptoms of allergic rhinitis are primarily due to a combination of the early and late phase allergic inflammatory response. Repeat exposure to a particular allergen sensitize the host. When the allergen comes in contact with the nasal mucosa of a sensitized host, immunoglobulin E (IgE) receptors on mast cells cross-link, resulting in the degranulation of these cells and the release of histamine and proteases. A wide array of proinflammatory molecules is also released. It is the release of these proinflammatory molecules that causes swelling and mucus secretion seen in allergic rhinitis [28].

22.4.2.7 Nasal Polyposis

Nasal polyps are semitranslucent, pale gray, benign inflammatory lesions of asymmetric size found in the paranasal sinuses or nasal cavity. It is estimated that they may affect as much as 4% of the population. Nasal polyps can reach 3–4 cm in length and may cause symptoms, such as nasal obstruction, nasal discharge, and impairment of olfaction. Although the cause of nasal polyposis is not clear, it is hypothesized to be a result of chronic inflammation due to chronic infection, aspirin intolerance, alteration in aerodynamics with trapping of pollutants, epithelial disruptions, epithelial cell defects, or inhalant and food allergies [29]. Histologically, they are characterized by the infiltration of inflammatory cells, mainly eosinophils. As a result, total IgE concentration is significantly higher in nasal polyp tissue compared with healthy nasal tissue [30].

22.4.2.8 Epistaxis

Also known as a nosebleed, epistaxis occurs when a vessel within the nasal cavity is ruptured. To facilitate proper humidifying and air conditioning of inspired air, the nasal mucosa contains a rich vascular network that can rupture either spontaneously or due to trauma. In the pediatric population, epistaxis occurs most commonly due to digital trauma. Another common cause of epistaxis is the improper use of topical nose sprays causing trauma to the epithelium of the septal mucosa. Epistaxis occurs more frequently in the winter months due to the decrease in humidity and temperature, which can cause a drying effect on the nasal mucosa, increasing the opportunity for mucosal disruption. A patient can be predisposed to epistaxis if they have anatomical deformities, such as septal deflections, bony spurs, or fractures. Additionally, any nasal obstruction that disrupts airflow can have a drying effect on the nasal mucosa, also causing epistaxis.

Systemic epistaxis is commonly due to cardiovascular or hematological disorders, such as hypertension, aberrations in clotting ability, or inherited bleeding disorders. While there is an undeniable association between hypertension and epistaxis, the exact mechanism still remains unclear. The ability to form blood clots is essential to both the prevention and control of epistaxis [31, 32].

22.5 Treatment

In the last 10–15 years, there has been an increased awareness of the nasal valve as a key contributor of nasal airway obstruction, resulting in a flurry of scientific publications on the matter, innovations in therapeutic options, and increasing applications of a multitude of both surgical and nonsurgical treatments to correct the nasal valve contribution to nasal obstruction. Evaluation of the patient begins with a thorough history, eliciting signs that may hint at nasal obstruction and nasal valve collapse. Does the patient mouth breathe, snore, awaken tired? Has the patient used breathing dilator devices in the past or had prior nasal surgery? The answers to these questions will help guide the patient and surgeon in deciding whether there is a nasal valve component to the nasal obstruction and whether nasal surgery might be of benefit.

In the past, surgical techniques described for nasal valve repair have focused on secondary nasal surgery following nasal valve damage as a result of rhinoplasty [32–36]. With a better understanding of nasal valve physiology in the last couple of decades, coupled with refinements in surgical techniques and thorough preoperative examination, valve disorders can be better identified in previously unoperated individuals complaining of nasal obstruction. These patients can initially be treated by the use of external nasal dilator devices or other breathing devices to help them experience the quality of life improvement associated with corrective nasal valve surgery, helping them make a more informed decision before going through with surgery. Additionally, the application of such breathing devices helps in defining the site of obstruction more precisely in order to optimize surgical outcomes [37]. Primary and secondary functional nasal surgery with correction of the dysfunctional nasal valve has been previously shown to significantly improve quality of life in patients complaining of nasal obstruction who have preoperative findings of nasal valve collapse [38–40].

22.5.1 Surgical Techniques

There have been many surgical techniques identified addressing the dysfunctional nasal valve in the past few decades [21]. Although there is no "gold standard" or "one size fit all" technique that can treat all causes and types of nasal valve collapse, there are many useful techniques that can be integrated into a surgical plan depending on the needs of the patient. Nasal valve collapse is the result of a narrow nasal valve or weak nasal structures—thus, the overall goal of nasal valve surgery is to widen the existing nasal valve area and to strengthen the structural support elements that maintain a patent valve area at rest, and minimize dynamic collapse of the nose that results from the negative inspiratory forces of nasal breathing. This section highlights a number of techniques that have been found useful and are used routinely as part of a comprehensive surgical correction of nasal valve dysfunction. There are many others that may be found in the surgical literature.

Both endonasal and external approaches may be utilized for various nasal valve procedures depending on the needs of the patient and surgeon preferences. Nasal valve surgery can be performed with either general anesthesia or local anesthesia with sedation. In order to preserve nasal anatomy, it is important to use as little anesthetic as necessary. These procedures are typically performed on an outpatient basis. It is also essential to allow adequate time for the anesthetic to take effect in order to minimize the bleeding and maximize visualization. As with all surgeries, the general principle is to minimize the aggressiveness of the surgical intervention in order to minimize the potential risks of the procedure, but at the same time, to maximize the benefit to the patient by selecting the proper group of techniques for the proper situation.

22.5.2 Spreader Grafts

Spreader grafts can be used to treat both static and dynamic internal nasal valve collapse. This technique widens the narrowed valve angle, thereby enlarging the nasal valve area, and is advantageous in its ability to avoid affecting the nasal septum, turbinate, and nasal mucosa. In cases of static collapse, the widening of the middle third of the nose also results in a smoothening of the brow-tip aesthetic line. Ideally, spreader grafts are made of septal cartilage, but in cases of previous surgery in which inadequate septal cartilage remains, conchal cartilage or rib cartilage may be utilized. In patients with a prominent dorsum undergoing dorsal reduction, the upper lateral cartilage excess may be folded on itself to lie between the septum and upper lateral cartilage to serve as an "auto-spreader graft" or a "spreader flap." Standard left hemitransfixion incision is made to access the septal cartilage. Mucoperichondrial flaps are elevated on one or both sides of the septum depending on the septal pathology. Septoplasty is performed in standard fashion, and septal cartilage is harvested for use as a spreader graft. The upper lateral cartilages are separated from the dorsal septum to create space for the spreader grafts. The spreader grafts should be long enough to extend from under the nasal bones to the caudal edge of the upper lateral cartilage. They measure approximately 3–5 mm in height, and are the thickness of the septal cartilage. Occasionally, wider grafts may be required in which case a layering of multiple pieces of septal cartilage may be stacked together to provide adequate thickness. Auto-spreader grafts or spreader flaps involve the turning of upper lateral cartilages to lie between the septum and lateral upper lateral cartilage in order to take advantage of the local tissue excess. Spreader grafts have been shown to be an effective treatment for internal nasal valve collapse [41].

22.5.3 Alar Batten Grafts

Alar batten grafts are versatile grafts that may also be used for both internal and external nasal valve collapse depending on where they are positioned. These grafts are typically limited to use in patients with idiopathic or congenital causes of valve collapse. The first step in an alar batten graft is identification of the region of collapse [5]. If collapse is noted in the external nasal valve, along the alar rim, the graft may be placed along the rim of the nose to provide greater strength and an outward curvature to the nasal rim. A marginal incision is made along the inferior margin of the lower lateral cartilage with a 15 blade scalpel, while a double prong skin hook is utilized for countertraction. A sharp scissor is used to dissect a precise pocket along the nasal rim to the alar-facial groove. A cartilage graft measuring 3–10 mm in width by approximately 7–10 mm in length is harvested from either the septum or the conchal bowl and applied to the pocket. The graft should extend from the alar-facial groove to either the dome or to the area just lateral to the soft tissue triangle in order to avoid a sharp edge being seen through the skin of the soft tissue triangle. It may overlap the lateral crus superiorly. If less support is needed, a smaller graft may be used and has been referred to as an alar rim graft. The marginal incision is closed with simple interrupted 5-0 chromic suture.

If collapse is noted in the middle third of the nose, the alar batten graft is placed in the middle third of the nose. An intercartilaginous incision is made with the 15 blade scalpel and a precise pocket is created superficial to the upper lateral cartilage down to the pyriform aperture. The graft is applied to the pocket, directly on the upper lateral cartilage. As the skin thins, these grafts may become visible over time. An alternative technique involves the placement of the graft deep to the upper lateral cartilage, most often at the scroll region (the junction of the upper and lower lateral cartilage) which is commonly the region of greatest collapse. The grafts are then sutured to the overlying cartilage with 2 or 3 throws of 5-0 chromic or PDS suture to avoid movement of the graft. These underlay grafts (similar to lateral crural strut grafts) often hide better than the alar batten grafts. The intercartilaginous incision is then closed with interrupted 5-0 chromic suture.

22.5.4 Butterfly Graft

The "butterfly graft" is a highly effective procedure to correct nasal valve obstruction. It relies on the elastic nature of conchal cartilage to spring open the internal nasal valve. This graft provides an outward force that widens the nasal airway, leading to an increased internal valve angle [5].

Conchal cartilage is harvested through an anterior helical rim incision or postauricularly. A skin incision is made, followed by blunt and sharp dissection to free the conchal cartilage. A 1-centimeter-wide by 2-cm-long cartilage graft is harvested. Cautery to ensure hemostasis of the ear harvest site is performed. Running 6-0 fast absorbing gut suture was used to close the skin incision and a compressive dressing was placed on the donor site to prevent hematoma formation. The ear dressing is removed on postoperative day one.

Intercartilaginous incision is made on both sides of the nose and connected to a complete transfixion incision. Skin–soft tissue elevation is achieved along the nasal dorsum in a standard sub-SMAS plane to the rhinion and a subperiosteal plane from the rhinion to the nasion in order to allow for proper skin redraping. If a significant or exaggerated supratip depression is present, the graft is simply placed in the supratip depression and its ends are secured to the caudal-most aspect of the upper lateral cartilages with a single throw of 5-0 PDS suture on either side. Once the graft is fixed in position, the skin is redraped and the dorsum is inspected and palpated for irregularities. If irregularities are noted, the dorsum is reduced further to create a smooth contour. Frequently, especially in patients with thin skin, crushed cartilage grafts are placed on the nasal dorsum, cephalic to the upper edge of the butterfly graft, to camouflage the edges of the graft and create a smooth dorsal contour. Mucosal incisions are closed with 5-0 chromic suture.

22.5.5 Nasal Valve Flaring Suture

Skin and soft tissue envelope is elevated off the osseocartilaginous understructure of the nose as previously described. Once the incisions are made and the tissues have been elevated, a retractor is placed under the skin flap to expose the upper lateral cartilages. A horizontal mattress stitch is thrown from one upper lateral cartilage to the other and tied tightly over the nasal dorsum. As the suture is tied down, the upper lateral cartilages elevate outward, thereby widening the nasal valve angle and area. Nasal valve flaring sutures may be used alone or in combination with various other techniques in order to maximize the widening of the valvular airway. Incisions are closed as previously described.

22.5.6 Maxillary Expansion

Maxillary expansion is a technique to consider in individuals with maxillary constriction, a narrow maxilla in the lateral dimension compared to other facial bones, because maxillary constriction increases nasal resistance. Rapid maxillary expansion is an orthodontic treatment that can increase the lateral dimension of the maxilla. This treatment is most often applied to children but can also be performed in adults in conjunction with Lefort I osteotomies. Rapid maxillary expansion has been shown to decrease nasal resistance [42], and there is encouraging evidence that it can reduce apneas in young adults with mild to moderate OSA [43, 44]. With regard to the INV, a recent case series demonstrated an increase in INV angle and area after surgical maxillary expansion in adults which was associated with improved daytime sleepiness and subjective nasal obstruction [44].

22.5.7 Medical Treatment

22.5.7.1 Antihistamines

Although first-generation oral antihistamines (diphenhydramine, chlorpheniramine, and brompheniramine) was able to effectively treat allergic rhinitis, these drugs have been associated with strong sedative effects. Second-generation antihistamines, such as fexofenadine, loratadine, and desloratadine, did not have these sedative effects. Several studies have shown improvements in symptoms associated with allergic rhinitis with these antihistamines relative to the placebo group. Because of their low adverse effect profile, these antihistamines are an effective first-line therapy for mild to moderate allergic rhinitis, especially in patients with intermittent symptoms and children.

These agents were followed by a generation of nonsedating antihistamines, such as fexofenadine, loratadine, and desloratadine. Several studies have shown marked improvement in symptom-based outcome data in patients with allergic rhinitis compared to the placebo group. The low adverse effect profile of these agents supports their use as first-line therapy for mild to moderate allergic rhinitis, especially in patients with intermittent symptoms and in children [45]. For chronic rhinitis, no convincing data are available to show treatment efficacy.

22.5.7.2 Intranasal Corticosteroids

Topical corticosteroid sprays are indicated predominantly for the treatment of allergic rhinitis. Currently, U.S. Food and Drug Administration-approved agents are beclomethasone dipropionate, budesonide, ciclesonide, flunisolide, fluticasone propionate, mometasone furoate, and triamcinolone acetonide. These agents have been shown to treat symptoms of allergic rhinitis (and conjunctivitis) more effectively than do oral antihistamines. Beclomethasone, budesonide, triamcinolone, fluticasone, and mometasone have been compared. The symptom-based outcomes of these studies do not appear to clearly favor one agent over another [46]. The overall adverse effect profile of intranasal corticosteroids is favorable and includes minor symptoms, such as dryness, stinging, or burning. Epistaxis is the most frequent major complication, occurring in about 5% of patients. The risk of septal perforation and epistaxis may be reduced by instructing the patient to direct the spray tip away from the septum toward the lateral nasal wall. Adverse systemic side effects, such as suppression of the hypothalamic-pituitary axis and growth retardation, are no longer a relevant issue with contemporary agents.

22.5.7.3 Systemic Corticosteroids

Systemic corticosteroids are typically administered intramuscularly or orally. Intramuscular agents include betamethasone dipropionate, methylprednisolone acetate, betamethasone phosphate, and triamcinolone acetonide. Systemic corticosteroids are third-line agents when antihistamines and intranasal corticosteroids have failed. These agents suppress endogenous cortisol production to a variable degree for 12 days to 3 weeks but are highly effective. Axelsson and Lindholm showed symptomatic improvement in 16 of 17 patients with allergic rhinitis after administration of a single dose of triamcinolone acetonide, while only 2 of 21 patients improved in the placebo group. This relief can last throughout the allergic season [47, 48]. Few data are available to compare oral with intramuscular corticosteroids. One study showed plasma cortisol levels to be suppressed beyond 3 weeks with oral prednisolone, 7.5 mg daily, but not with intramuscular corticosteroids [49]. With adequate screening of patients for diabetes mellitus, glaucoma, hypertension, and osteoporosis, the use of systemic corticosteroids, such as intramuscular triamcinolone acetonide, has become an important and safe treatment of chronic inflammatory nasal disease.

22.6 Conclusion

An understanding of the nasal valves and how they relate to nasal physiology is crucial in making good clinical decisions. Without proper functioning of the nose and nasal valves, the air conditioning and humidifying capacity of the nose is lost, along with the sense of olfaction—prominent parts of our daily lives. A loss of any of these functions leads to a great decrease in quality of life, so it is important to preserve them. It is common to treat valve disorders with surgery. Because the functional residual capacity of the nose is unknown, it is important to practice great reserve with the reduction of functional turbinate tissue, regardless of technique. Resection of turbinate bone alone with preservation of all mucosal and submucosal tissue is an alternative, though this may also scar the submucosa and mucosa and lead to additional problems. Recognition of the nature and location of nasal valve pathologies allows for adequate correction and superb functional results in the majority of cases. Concurrent rhinoplasty and functional endoscopic sinus surgery can be performed safely. The evolution of atraumatic surgical techniques has resulted in improved patient comfort and speedy recovery.

References

1. Cole P. The four components of the nasal valve. Am J Rhinol. 2003;117(2):107–10.
2. Bridger GP. Physiology of the nasal valve. Arch Otolaryngol. 1970;92:543–53.
3. Constantian M. The incompetent external nasal valve: pathophysiology and treatment in primary and secondary rhinoplasty. Plast Reconstr Surg. 1994;93(5):919–31.

4. Vaiman M, Eviatar E, Segal S. Muscle building therapy in treatment of nasal valve collapse. Rhinology. 2004;42(3):145–52.
5. Sinkler MA, Wehrle CJ, Elphingstone JW, Magidson E, Ritter EF, Brown JJ. Surgical Management of the Internal Nasal Valve: a review of surgical approaches. Aesthet Plast Surg. 2021;45(3):1127–36.
6. Aksoy F, Veyseller B, et al. Role of nasal muscles in nasal valve collapse. Otolaryngol Head Neck Surg. 2010;142(3):365–9.
7. Hamilton GS 3rd. The external nasal valve. Facial Plast Surg Clin North Am. 2017;25(2):179–94.
8. Hasegawa M, Kern EB. The human nasal cycle. Mayo Clin Proc. 1977;52:28–34.
9. Pendolino AL, Lund VJ, Nardello E, Ottaviano G. The nasal cycle: a comprehensive review. Rhinology Online. 2018;1:67–76.
10. Fry FA, Black A. Regional deposition and clearance of particles in the human nose. J Aerosol Sci. 1973;4:113–24.
11. Schroeter JD, Tewksbury EW, Wong BA, Kimbell JS. Experimental measurements and computational predictions of regional particle deposition in a sectional nasal model. J Aerosol Med Pulm Drug Deliv. 2015;28(1):20–9.
12. Itoh H, Smaldone GC, Swift DL, Wagner HN. Mechanisms of aerosol deposition in a nasal model. J Aerosol Sci. 1985;16:529–34.
13. Jones N. The nose and paranasal sinuses physiology and anatomy. Adv Drug Deliv Rev. 2001;51:5–19.
14. Anderson SD, Togias AG. Dry air and hyperosmolar challenge in asthma and rhinitis. In: Busse WW, Holgate ST, editors. Asthma and rhinitis. Boston: Blackwell Scientific Publications; 1995. p. 1178–95.
15. Cole P. Modification of inspired air. In: Proctor DF, Andersen I, editors. The nose, upper airway physiology and the atmospheric environment. Amsterdam: Elsevier Biomedical Press; 1982. p. 351–75.
16. Li H, Martin HL, Marcus JR, Frank-Ito DO. Analysis of nasal air conditioning in subjects with unilateral cleft lip nasal deformity. Respir Physiol Neurobiol. 2021;291:103694.
17. Gassner HG, Remington WJ, Sherris DA. Quantitative study of nasal tip support and the effect of reconstructive rhinoplasty. Arch Facial Plast Surg. 2001;3:178–84.
18. Kienstra MA, Gassner HG, Sherris DA, Kern EB. Effects of the nasal musculature on the nasal airway. Am J Rhinol. 2005;19:375–81.
19. Seyed Resuli A, Oktem F, Ataus S. The role of the depressor nasi Septi muscle in nasal air flow. Aesthet Plast Surg. 2020;44(5):1766–75.
20. Motamedi KK, Stephan SJ, Ries WR. Innovations in nasal valve surgery. Curr Opin Otolaryngol Head Neck Surg. 2016;24(1):31–6.
21. Apaydin F. Nasal valve surgery. Facial Plast Surg. 2011;27(2):179–91.
22. Park SS. The flaring suture to augment the repair of the dysfunctional nasal valve. Plast Reconstr Surg. 1998;101(4):1120–2.
23. Samra S, Steitz JT, Hajnas N, Toriumi DM. Surgical Management of Nasal Valve Collapse. Otolaryngol Clin N Am. 2018;51(5):929–44.
24. Toriumi DM. Surgical correction of the aging nose. Facial Plast Surg. 1996;12(2):205–14.
25. Guyuron B. The aging nose. Dermatol Clin. 1997;15(4):659–64.
26. Rohrich R, Hollier L. Rhinoplasty with advancing age: characteristics and management. Otolaryngol Clin N Am. 1999;2(4):755–73.
27. Wheatley LM, Togias A. Clinical practice. Allergic rhinitis. N Engl J Med. 2015;372(5):456–63.
28. Quraishi SA, Davies MJ, Craig TJ. Inflammatory responses in allergic rhinitis: traditional approaches and novel treatment strategies. J Am Osteopath Assoc. 2004;104(5 Suppl 5):S7–15.
29. Naclerio RM, Bachert C, Baraniuk JN. Pathophysiology of nasal congestion. Int J Gen Med. 2010;8(3):47–57.
30. Bachert C, Maurer M, Palomares O, Busse WW. What is the contribution of IgE to nasal polyposis? J Allergy Clin Immunol. 2021;147(6):1997–2008.
31. Byun H, Chung JH, Lee SH, Ryu J, Kim C, Shin JH. Association of Hypertension With the Risk and Severity of Epistaxis. JAMA Otolaryngol Head Neck Surg. 2020;147(1):1–7. https://doi.org/10.1001/jamaoto.2020.2906. Epub ahead of print. Erratum in: JAMA Otolaryngol Head Neck Surg. 2021 Jan 1;147(1):111
32. Fatakia A, Winters R, Amedee RG. Epistaxis: a common problem. Ochsner J. 2010;10(3):176–8.
33. Sheen JH. Spreader graft: a method of reconstructing the roof of the middle nasal vault following rhinoplasty. Plast Reconstr Surg. 1984;73(2):230–9.
34. Clark M, Cook T. The butterfly graft in functional secondary rhinoplasty. Laryngoscope. 2002;112(11):1917–25.
35. Stucker F, Lian T, Karen M. Management of the keel nose and associated valve collapse. Arch Otolaryngol Head Neck Surg. 2002;128(7):842–6.
36. Stucker FJ, Hoasjoe DK. Nasal reconstruction with Conchal cartilage: correcting valve and lateral nasal collapse. Arch Otolaryngol Head Neck Surg. 1994;120(6):653–8.
37. Gruber RP, Lin AY, Richards T. Nasal strips for evaluating and classifying valvular nasal obstruction. Aesthet Plast Surg. 2011;35(2):211–5.
38. Rhee J, Poetker D, Smith T, Bustillo A, Burzynski M, Davis R. Nasal valve surgery improves disease specific quality of life. Laryngoscope. 2005;115:437–40.
39. Saleh A, Younes A, Friedman O. Cosmetics and function: quality-of-life changes after rhinoplasty surgery. Laryngoscope. 2012;122(2):254–9.
40. Most S. Analysis of outcomes after functional rhinoplasty using disease specific quality of life instrument. Arch Facial Plast Surg. 2006;8(5):306–9.
41. Garg LN, Singh NK, Kappagantu KM, Yadav A. Spreader graft placement: an effective procedure for alleviation of internal nasal valve collapse. J Oral Maxillofac Surg. 2021;S0278-2391(21):00534–6.

42. Hershey HG, Stewart BL, Warren DW. Changes in nasal airway resistance associated with rapid maxillary expansion. Am J Orthod. 1976;69(3):274–84.
43. Cistulli PA, Palmisano RG, Poole MD. Treatment of obstructive sleep apnea syndrome by rapid maxillary expansion. Sleep. 1998;21(8):831–5.
44. Abdelwahab M, Yoon A, Okland T, Poomkonsarn S, Gouveia C, Liu SYC. Otol. Head Neck Surg. 2019;161(2):362–7.
45. Dykewicz MS, Fineman S, Skoner DP, et al. American Academy of allergy, asthma and immunology. Diagnosis and management of rhinitis: complete guidelines of the joint task force on practice parameters in allergy, asthma, and immunology. Ann Allergy Asthma Immunol. 1998;81(Pt 2):478–518.
46. Meltzer EO. The treatment of vasomotor rhinitis with intranasal corticosteroids. World Allergy Organ J. 2009;2(8):166–79.
47. Axelsson A, Lindholm B. The effect of triamcinolone acetonide on allergic and vasomotor rhinitis. Acta Otolaryngol. 1972;73:64–7.
48. Mygind N, Laursen LC, Dahl M. Systemic corticosteriod treatment for seasonal allergic rhinitis: a common but poorly documented therapy. Allergy. 2000;55:11–5.
49. Laursen LC, Faurschou P, Pals H, Svendsen UG, Weeke B. Intramuscular betamethasone dipropionate vs. oral prednisolone in hay fever patients. Allergy. 1987;42:168–72.

Nose and Sleep Breathing Disorders

23

Anne-Lise Poirrier, Philippe Eloy, and Philippe Rombaux

Core Messages

- The nose is the input channel for the airflow. Its rigid and erectile structures determine the outline and the output of the airflow in the upper airway. Nose obstruction, due to reversible or nonreversible factors, produces collapsing forces that are manifested downstream in the collapsible pharynx. Moreover, nose pathologies result in unstable oral breathing, decreased activation of nasal ventilatory reflex and reduced lung nitric oxide. Long-term oral breathing impacts the craniofacial growth. The management of nose pathologies could be medical, mechanical (nose dilators) or surgical. Nasal management should be integrated in a multimodal approach, considering the involvement of a multilevel obstruction, and truly reflecting the complexity of sleep-disordered breathing.

A.-L. Poirrier
Department of Otolaryngology, CHU-Liège, ULG, Sart-Tilman B35, Liège, Belgium
e-mail: annelise@poirrier.be

P. Eloy
HNS & ENT Department, CHU-Mont-Godinne, UCL, Yvoir, Belgium

Department of Otorhinolaryngology, Cliniques Universitaires Saint-Luc, Brussels, Belgium

Institute of Neuroscience, Université Catholique de Louvain, Brussels, Belgium
e-mail: philippe.eloy@uclouvain.be

P. Rombaux (✉)
Department of Otorhinolaryngology, Cliniques Universitaires Saint-Luc, Brussels, Belgium

Institute of Neuroscience, Université Catholique de Louvain, Brussels, Belgium

HNS & ENT Department, Cliniques Universitaires Saint Luc, Brussels, Belgium
e-mail: philippe.Rombaux@uclouvain.be

23.1 Introduction

Sleep-disordered breathing (SDB) is a clinical entity that is more and more recognised by physicians since the 1970s. It consists of a wide spectrum of sleep-related breathing abnormalities. Those related to increased upper airway resistance include snoring, upper airway resistance syndrome (UARS) and obstructive sleep apnoea–hypopnoea syndrome (OSAHS) [1].

Snoring is associated with changes in the calibre of the upper airway which reduce flow and increase airway resistance and is a manifestation of increased turbulence in nasal flow [2, 3]. UARS is caused by sleep-related flow limitation and increase in upper airway resistance that precipitates arousals. UARS results in fragmented sleep and excessive daytime sleepiness. Obstructive sleep apnoea (OSA) syndrome is the complete or partial collapse of breathing despite ongoing respiratory effort. In patients with OSA, recurrent obstruction of the pharynx during sleep results in

Ö. Ö. Celebi, T. M. Önerci (eds.), *Nasal Physiology and Pathophysiology of Nasal Disorders*,
https://doi.org/10.1007/978-3-031-12386-3_23

frequent episodes of airflow cessation, leading to significant hypoxemia, fragmentation of sleep and excessive daytime sleepiness. Obstructive sleep apnoea is a leading cause of neuropsychiatric conditions (e.g. sleepiness, depression, cognitive dysfunction), cerebro- and cardiovascular diseases (e.g. pulmonary and systemic hypertension, congestive heart failure, myocardial infarction, stroke), metabolic disorders, sexual dysfunction, loss in work productivity and increased risk of motor vehicle accidents. OSA represents a major public health problem [2].

In the Wisconsin Sleep Cohort, a stratified random sample of Wisconsin state employees aged 30–60 years, the prevalence of OSA was 9% in women and 24% in men. The incidence increases with age and tobacco and alcohol use and is associated with metabolic and anatomical features (obesity, retrognathia, high anteroposterior cervical diameter, macroglossia, large tonsils, hypertrophic tongue base, large neck size, gastroesophageal reflux and nasal obstruction) [1, 2].

In the past, snoring was considered mainly as a common ordinary disorder that only affected men and was regarded as a social annoyance particularly for the bed partner. Nowadays many clinicians are regarding SDB as a spectrum of diseases in which a patient can move from a snorer without apnoea to a snorer with apnoea. These disorders form actually a continuum. They share a common physiopathology: a multilevel airway obstruction [4].

As the nose plays a major role in the physiology of the respiratory tract, it is important to analyse the role of nasal disorders in the pathogenesis of SDB and the effects of rhinologic treatments on snoring and OSA. This topic has not yet received definitive conclusions because of contradicting reports in the literature. The number of patients with polysomnography documented OSA and treated only by nasal surgery is far less important than the number of cases treated with other therapies within the last two decades. The reason is not quite clear, but one could be that the success rate of nasal management alone for SDB is low and the prediction of individual success is not possible [3].

23.2 Nose Anatomy and Physiology

The nose is the input channel for the airflow and the "touchable" beginning of the airways. About 70% of the resistance met by the inspired airflow during its passage through the upper and lower airways is located into the nose [5]. The nose may be roughly divided into outer and inner anatomy. The outer nose is supported by the nasal bones, the paired upper lateral and lower lateral cartilages and the nasal septum and is covered by the subcutaneous tissue and skin. The inner nose includes the nasal septum on the medial wall of the nasal cavity and the turbinates and the osteomeatal complex on the lateral wall. During inspiration, air is spinning into the nose through the nasal valve. It can be divided into external and internal nasal valves [6, 7].

The external nasal valve comprises the alar cartilages, the nasal wing and the columella and has a shape of an inverted "funnel". Its role consists of orientating the airflow into the nasal cavities without generating any resistance [8]. The internal nasal valve is formed by the junction of the upper lateral cartilages with the nasal septum, the septum, the head of the inferior turbinate and the piriform aperture (Fig. 23.1). The normal angle between the upper lateral cartilages and the septum is about 10–15° and represents the nasal region with the smallest cross-sectional area and the greatest resistance to nasal airflow, crucial to determine nasal resistance (R_N) [6]. The internal

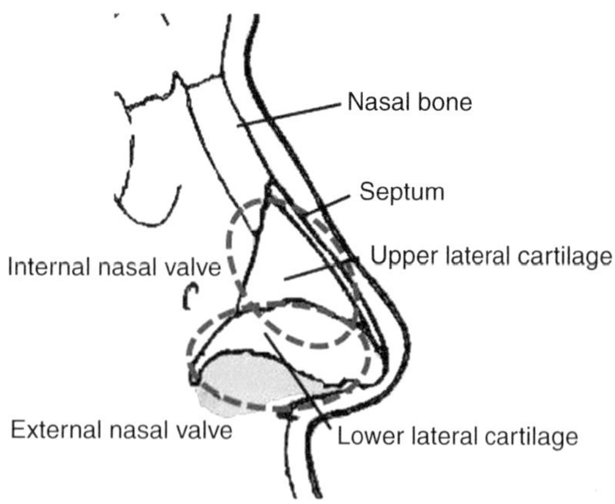

Fig. 23.1 Anatomy of the external nose

nasal valve plays a major role in the physiology of the nose and particularly in air conditioning. Its functioning depends on the shape of the cartilages, the tonus of the dilator muscles and the degree of congestion of the nasal mucosa. The airstream is first directed upward through the internal nasal valve, then bends about 90° posteriorly and flows via the nasopharynx to the lower airways.

The diameter of the valve influences directly the velocity of the airflow. On gentle inspiration, the nasal valve is usually patent. During deep inspiration (exercises or sniffing), the airflow could create a Bernoulli's effect, which accelerates the flow in this narrow cleft and decreases the pressure on each side of the nasal vestibule leading to the collapse of the nasal wing. Patients suffering from a valve collapse may experience nasal obstruction even during normal breathing.

The congestion of the nasal mucosa varies physiologically, spontaneously and alternatively from side to side with time. One side is blocked, while the other side is patent. This alternates every 3–7 h in adults, leading to a spontaneous cycle phenomenon called nasal cycle. Surprisingly, thanks to this alternation of resistance on each side, the total nasal resistance remains constant [9].

The paranasal sinus cavities play also a major role in the physiology of the nose. The sinonasal architecture is organised around the ethmoid bone. The perpendicular plate of the ethmoid articulates medially to the septal cartilage, while the outer wall of the ethmoid, including middle concha, articulates laterally with the vertical plate (ascending process of the frontal bone) of the maxilla. On the lateral nasal wall is the osteomeatal complex (OMC). The OMC comprises the middle turbinate, the uncinate process and the bulla ethmoidalis. In this particular anatomical area drain the secretions from the anterior paranasal cavities, such as the anterior ethmoid cells, the frontal sinus and the maxillary sinus. Anatomical variations of the different structures of the OMC have been described in the literature, such as concha bullosa, paradoxically bent middle turbinate and medially bent uncinate process. In the past ones believed that these anatomical variations were associated to chronic rhinosinusitis. Now most authors do not consider these variations to be responsible of the pathogenesis of chronic sinusitis by themselves.

23.3 Nose Pathologies

All pathologies causing nasal obstruction can cause or worsen SDB [10]. The reasons for nasal obstruction are complex and varied, but the causes can be simplified as nonreversible factors, such as anatomic deformities, and reversible factors, such as mucosal oedema and congestion (Table 23.1).

23.3.1 Nonreversible Factors

Deformity of the nasal septum and/or the nasal pyramid can obviously be associated with uni- or bilateral persistent nasal obstruction. In case of nasal septum deviation, the patient can complain of a uni- or bilateral nasal obstruction depending on the shape, type and location of the deviation

Table 23.1 Causes of nasal obstruction

Nonreversible	Internal/external valve collapse
	Septal deviation, haematoma, perforation
	Other malformation of the nasal framework
	Vestibular synechiae or scars
	Concha hypertrophy
	Nasal polyposis, antrochoanal polyp
	Foreign body, nasal packing
	Benign tumours: angiofibroma, inverted papilloma
	Malignant tumours: squamous cell carcinoma, adenocarcinoma, melanoma
	Meningocele
	Choanal atresia and other craniofacial anomalies
Reversible	Allergic/nonallergic rhinitis: NARES-NANIPER
	Acute or chronic rhinosinusitis with or without polyps
	Drug-induced or occupational rhinitis
	Atrophic rhinitis
	Pregnancy
	Wegener or other granulomatosis

[11]. Anterior nasal septum deviation is more commonly responsible of nasal obstruction than posterior septal deviation [12]. The patient can also complain of a contralateral nasal obstruction, explained by a compensatory hypertrophy of the mucosa of the inferior turbinate. Nasal collapse is another cause of nasal obstruction that is underrated and underestimated by numerous ENT doctors. Nasal obstruction can be revealed during effort, sport or exercises or can be present in a normal and calm breathing. Patients with previous facial nerve palsy or post-traumatic or postsurgical adhesions developed at the level of the nasal vestibule or the columella can present a unilateral nasal collapse. The diagnosis is made by the Cottle manoeuvre or by an anterior and posterior active rhinomanometry and acoustic rhinometry.

23.3.2 Reversible Factors

Nasal obstruction can be caused by a rhinitis. Allergic rhinitis is a very common condition. Bauchau and Durham reported a high heterogeneity of allergic rhinitis incidence among the different European countries and a maximal incidence in Belgium with 29.5% of the population [13]. According to ARIA guidelines, the rhinitis can be intermittent or persistent, mild, moderate or severe [14]. Indoor allergens can cause symptoms during sleep, such as house dust mites, animal dander or fungi.

Nonallergic rhinitis with eosinophils syndrome (NARES) is another type of rhinitis; the eosinophils are present in the nasal smears, and the patient dramatically improves when he uses a nasal topical steroids. There is no sensitisation to any aeroallergens. Loss of smell is a common symptom. This disease can be a precursor of a true nasal polyposis.

Nonallergic noninfectious perennial rhinitis (NANIPER) was called in the past vasomotor rhinitis. The aetiology is unknown, the treatment often disappointing except for nasal obstruction.

Rhinitis medicamentosa is a typical cause of nasal obstruction in a patient who (mis)uses nasal topical decongestant. With time the patient consumes more and more nasal drops. Typically nasal obstruction increases during the night.

Acute and chronic rhinosinusitis with and without polyps are associated with nasal obstruction. Acute rhinosinusitis gives symptoms for a maximum of 6 weeks, whereas chronic rhinosinusitis is symptomatic for more than 12 weeks [15]. Nasal polyposis affects 9% of the general population. It can be restricted to the nose and sinuses or be associated with asthma and aspirin intolerance. Major symptoms in nasal polyposis are nasal obstruction and loss of smell. There are different classifications used to categorise the polyps. In the Caucasian population, nasal polyposis is associated with a chronic inflammatory infiltrate rich in eosinophils. Oedema, epithelial shedding, pseudocyst formation and changes in the extracellular matrix are some histological characteristics of the common nasal polyposis.

23.4 Physiopathology of Nose Obstruction and SDB

23.4.1 Starling Resistor Model

Nasal obstruction produces collapsing forces that are manifested downstream in the collapsible pharynx [16, 17]. In the respiratory model based on a Starling resistor, the nose is a key determinant of upper airway resistance (Fig. 23.2) [18, 19]. Nasal pressure (P_N) is zero (atmosphere reference value) in normal conditions. Nasal resistance (R_N) determines the maximum flow (V_{max}) in the downstream collapsible pharynx. In the pharynx, P_{crit} is the critical value of airway pressure leading to complete collapse and stop of airflow. P_{crit} depends on transmural pressure and external pressure applied by respiratory muscles. The maximum airflow is defined by $V_{max} = (P_N - P_{crit})/R_N$. This equation implies that increase in nasal resistance (R_N) leads to decrease in upper airway flow (V_{max}). Conversely, increase in nasal pressure (P_N) by continuous positive airway pressure (CPAP) device improves upper airway flow (V_{max}) [20].

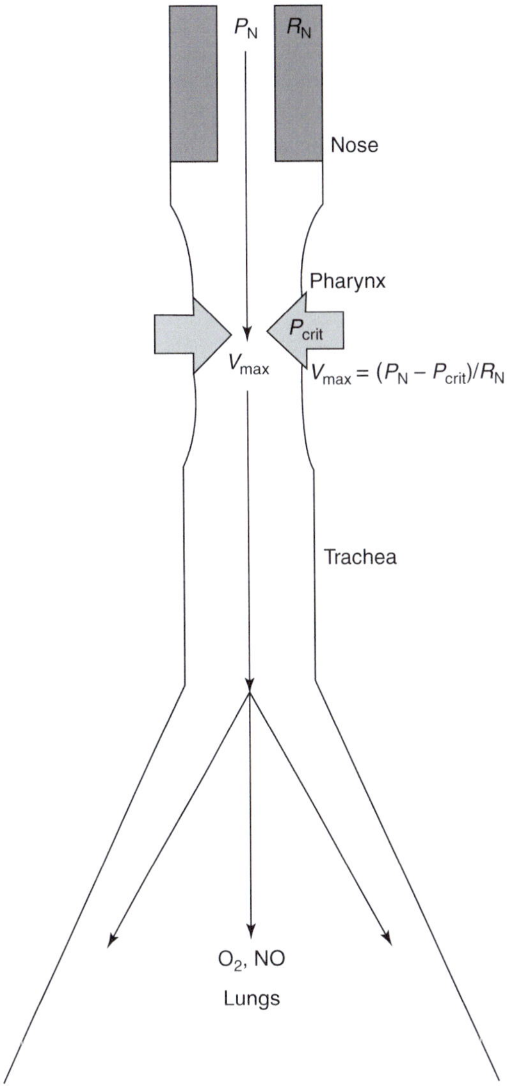

Fig. 23.2 Model of the lungs and upper airway compartments of breathing. The upper airway behaves like a Starling resistor in that obstruction at the inlet produces collapsing forces that are manifested downstream in the collapsible segment, the pharynx. Airflow ceases in the pharynx at a critical value of airway pressure (P_{crit}). Maximum flow (V_{max}) in the pharynx is determined by nasal pressure (P_N) and resistance (R_N) from the equation $V_{max} = (P_N - P_{crit})/R_N$ (Drawing adapted from Ferris et al. [5])

23.4.2 Oral Breathing

Nasal obstruction may lead to mouth breathing and mouth opening, which, in turn, results in inferior movement of the mandible with associated decrease in pharyngeal diameter. The base of the tongue may also fall backwards reducing the posterior pharyngeal space.

Although the precise mechanisms are not fully understood, oral breathing could be an adaptive response once a particular threshold of nasal airflow resistance is exceeded. Combined recording of oral and nasal breathing during sleep indicates that normal subjects partition flow between nasal and oral routes, with the majority of airflow occurring through the nasal route [21]. The route of breathing has profound influence on upper airway resistance during sleep. Oral breathing results in an unstable airway and increases total airway resistance. Oscillation of the soft palate, posterior movement of the jaw angle and posterior retraction of the tongue during mouth opening compromise oral-breathing airflow [16, 22].

23.4.3 Nasal Receptors

A few studies suggest nasal airflow has a stimulant effect on ventilation, probably via nasal mechanoreceptors maintaining respiratory pacing. Application of local anaesthetics to the nasal mucosa increases the episodes of airway occlusion [23, 24] and impairs the arousal response to airway occlusion [25]. The parasympathetic nervous system may play a role in the control of breathing and in the hyperpneic responses associated with airflow obstruction. The parasympathetic nervous system component includes neural receptors in the airways as well as afferent and efferent pathways that travel in the vagus nerves [26].

23.4.4 Nitric Oxide

Another item playing a major role in snoring and OSA is the nitric oxide (NO). Airborne NO is largely produced in the epithelium of the paranasal sinuses and is involved in the regulation of pulmonary function [27, 28]. During inspiration through the nose, high levels of NO follow the airstream to the lower airways and the lungs. Nasally derived NO increases arterial oxygen tension and reduces pulmonary vascular resistance. NO enhances therefore blood flow preferentially in well-ventilated areas of the lung, thus optimising ventilation/perfusion matching [29, 30]. In obstructive sleep breathing disease, nasal NO fails partly to reach the lungs, resulting in ventilation/perfusion mismatch [31]. Lack of NO could also participate in incoordination of pharyngeal and thoracic muscles and in sleep fragmentation. Furthermore, long-term complications of OSA might be due to the repeated temporary dearth of NO in the tissues, secondary to a lack of oxygen [31]. After their passage to the alveoli in the inspired air, both oxygen and NO are removed by haemoglobin and are transmitted to the tissues. Repetitive hypoxia/reoxygenation adversely impacts endothelial function by promoting oxidative stress and inflammation and reducing NO availability. This vicious spiral mediates the cardiovascular manifestations of OSA [32].

23.5 Craniofacial Development

23.5.1 Morphogenic Perspective

Nose function not only has a direct role in upper and lower airway breathing in adults but also has a long-term impact on the development of the anterior skull base and the maxilla. The influence of nasal patency on the development of the anterior skull base and the maxillary bone has been previously demonstrated in mammals [33, 34]. Experimental blockage of rat nostrils resulted after 2–4 months in anatomical changes of the superior maxilla, the skull base and the jaw [33, 34]. Nasal obstruction in monkeys resulted in downward and backward rotation of the mandible and changes in dental occlusion [35]. Likewise, oral breathing may modify craniofacial growth in children [36]. Predominant oral breathing during critical growth periods in children could be inscribed in the bones and lead to breathing disorders [37]. Cephalometric control studies have shown that mouth-breathing children have a higher tendency for clockwise rotation of the growing mandible [38]. Because of mouth breathing, tongue position in the oral cavity is low, and the balance between forces from the cheeks and tongue is different compared with healthy children. This leads to a lower mandibular position and extended head posture. Malocclusion and skeletal discrepancy may be partially corrected after adenotonsillectomy [36]. Similarities in cephalometric studies from OSA adults and mouth-breathing children suggest that the apnoeic pattern develops early in the clinical history of patients with OSA [39]. However, OSA in children differs that in adults. The involvement of nasal resistance is greater in children, with serious consequences for growth and development [40]. In adults, cephalometric measurements of normal subjects and patients have shown a relationship between OSA and transverse dimensions of nasal cavities, limited laterally by the vertical plates of both maxillae. OSA patients have narrower nasal framework and maxillary bone proportions [41]. Craniofacial features in the pathophysiology of OSA could explain ethnic differences in OSA prevalence and severity for a given level of obesity [42, 43].

23.5.2 Phylogenic Perspective

Researchers have speculated that the outer nose may have an evolutionary benefit in human. In addition to an ornamental role for sexual selection, it may play a role in creating a curvilinear airflow pattern [44]. During the course of human evolutionary development, the midface is shortened, and the upper airway is narrowed to form a collapsible and distensible tube. This evolution not only permits the production of spoken language but also results in a predisposition towards upper airway collapse during sleep [45–47]. The development of the human outer nose could be assumed as a compensatory development. The curvilinear airflow pattern provided by the nose adjusts the "angle of attack" of airflow hitting the palate, thus contributing to the pharyngeal opening [44]. From this hypothesis, the external nose could provide an evolutionary benefit in the protection against OSA (Fig. 23.3).

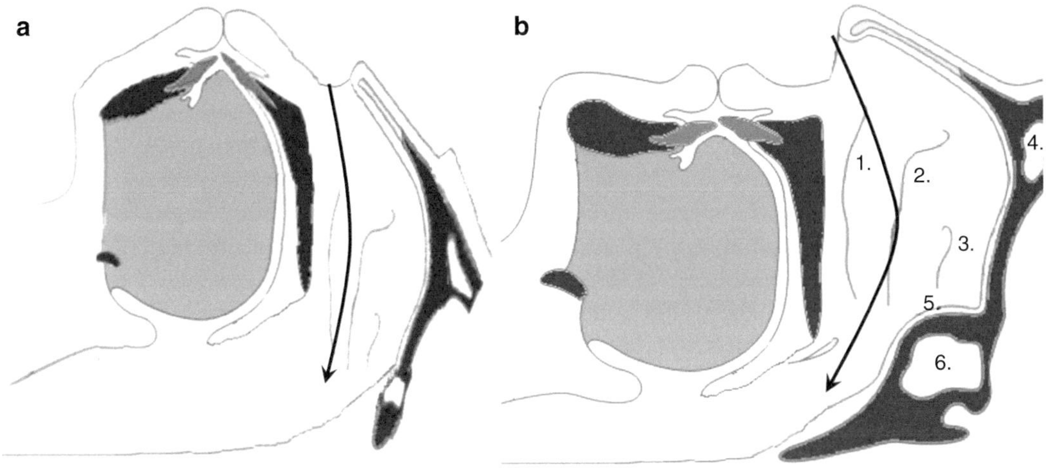

Fig. 23.3 (**a**) In chimpanzee, the upper airway is larger, which lowers the risk of collapse. The nasal airflow is horizontal (*arrow*). (**b**) In normal human, airflow is directed upward through the nasal valve. The outer nose creates a curvilinear airflow pattern (*arrow*). The latter adjusts the "angle of attack" of airflow hitting the palate, thus contributing to the pharyngeal opening. Anatomy of the internal nose: (1) Inferior turbinate, (2) Middle turbinate, (3) Superior turbinate, (4) Frontal sinus, (5) Sphenoethmoidal recess, (6) Sphenoid sinus (Drawing adapted from McNicholas et al. [23] and White et al. [24])

23.6 Patient Evaluation

23.6.1 Clinical Examination

In case of snoring associated or not to obstructive apnoea, a thorough and complete examination of the nose is mandatory. The nasal pyramid must be evaluated, particularly the dorsum, the lateral cartilages and the columella. A nasal valve collapse must be ruled out by the inspection of the external nose and the Cottle manoeuvre.

Then an anterior rhinoscopy evaluates the nasal septum, the shape and colour of the mucosa of the inferior turbinates and the presence of crust, blood, secretions or polyps [48]. Mladina and colleagues defined seven types of septal deviation in a cohort of 2589 adults. They identified three types with vertical crests, one type with a bilateral deformity, two types with horizontal deformities and another type with atypical deformities [11]. Each type may be associated to some degree of nasal obstruction.

Eventually nasal endoscopy must examine the middle meatus, the olfactory cleft and the posterior aspect of the nasal cavity. Nasal polyposis can sometimes be diagnosed with nasal endoscopy only.

23.6.2 Investigations and Functional Testing

Besides the history taking, the patient self-assessment and the anterior rhinoscopy, some investigations must be performed to evaluate the nasal obstruction.

Rhinomanometry and acoustic rhinometry allow for indirect evaluation of nasal anatomy and function [49]. When these are performed in a supine position, these investigations have more value in assessing nasal breathing of patients with sleep disorders [50]. Rhinomanometry uses an intranasal closed loop system to measure nasal airway resistance. Acoustic rhinometry uses acoustic reflections to provide information about cross-sectional area and nasal volumes within a given distance. Acoustic rhinometry gives an anatomic description of a nasal passage, whereas rhinomanometry gives a functional measure of the pressure/flow relationships during the respiratory cycle. Both techniques are proposed to assess the efficacy of different treatments and for assessment of the patient prior to nasal surgery. Rhinomanometry and acoustic rhinometry provide "snap-shot" measurements, which may not be representative of a more chronic condition, since nasal turbinate size and function are dynamic processes that may change considerably over a few hours. It is also important to point out that rhinomanometry and acoustic rhinometry tests do not correlate well with a patient's subjective perception of nasal obstruction. The patient's subjective perception of the degree of nasal obstruction has been shown to be a more sensitive predictor of positive outcome from medical/surgical management than objective anatomic or physiologic measurements alone. Nasal values measured by acoustic rhinometry and rhinomanometry are correlated inversely with polysomnographic values (apnoea–hypopnoea index, oxygen desaturation index) in nonobese patients [50–52]. The association of nasal obstruction measured by posterior rhinomanometry and Mallampati score > 3 is predictive of OSA [53]. Acoustic rhinometry is also important to measure the nasal valve area [54].

Nasal inspiratory peak flow gives a measure of bilateral nasal airflow at maximum effort, but does not reflect a physiological measure of nasal airflow. It is, however, a validated technique to assess the responsiveness of a clinical intervention [55]. It should be associated to lung function evaluation as it is influenced by lower airway as well as upper airway function [56].

The levels of NO in the nose can easily be measured noninvasively. NO is altered in several airway disorders, including allergic rhinitis, ciliary dysfunction and sinusitis. The NO value measured is a sum of NO from the sinus via the ostia and the nasal mucosa. NO measurement is mainly valuable for sinonasal disease [27]. Its significance to sleep disorders is currently experimental [31].

While they provide objective outcome, the measures of nasal function reflect only one aspect of the disease and may thus not encompass all the other aspects. In recent years, there has been a great expansion in the number and use of quality-of-life questionnaires and other patient-based outcomes in health care. Nasal Obstruction Symptom Evaluation (NOSE) score, Sleep Outcomes Survey (SOS), Visual Analogue Scales (VAS), Sinonasal Outcome Test (SNOT) and other surveys have been applied to objectify outcomes in nose and sinus surgery [57–59]. Though subjective, they correlate with objective measurements and integrate general health issues, sleep perception and emotional aspects. They include a cluster of interconnected symptoms associated to the nose function. Septorhinoplasty is remarkably effective in improving sleep-related items of the SNOT-22 questionnaire [60]. Beyond the nasal airflow, questionnaires reflect the patient's perception, suffering and hope. They could help the physician to meet the patient expectations and to provide a reliable follow-up.

Nasal endoscopy and CT scan are two other tools to evaluate the anatomy of the nose and paranasal sinuses. These two examinations are routinely done in all rhinologic work-up.

As obstructive SDB is the consequence of multilevel airway obstruction, nasal evaluation should be integrated with a careful anatomical assessment involving in some cases sleep nasendoscopy, MRI or cephalometry. Lastly, polysomnography remains the gold standard to assess the quality of sleep and to calculate the sleep parameters, including apnoea index, hypopnoea index, apnoea-hypopnoea index, snoring time, amount of REM sleep and sleep latency. Breathing flow can be recorded overnight by means of a thermistor placed at the airway opening (nose and mouth). Inspiratory pressure is indirectly measured by means of chest and abdominal inductance plethysmography belts. Additional devices have been designed to measure nasal pressure [61] or to record mandible movement [62, 63] during sleep.

23.7 Patient Management

23.7.1 Rationale

The rationale to treat nasal obstruction is to improve nasal patency, re-establishing physiological breathing and minimising oral breathing during sleep. The aim of the treatment is also to reduce nasal resistance and improve the negative intraluminal pressure which generates upper airway collapse. Nasal obstruction can be relieved medically or surgically.

23.7.2 Medical Treatment

Only reversible causes of nasal obstruction can be treated with medications. The commonest causes of inflammation of the mucosa of the upper respiratory tract are allergic rhinitis, acute and chronic rhinosinusitis and nasal polyposis.

23.7.2.1 General Treatment of Allergic Rhinitis

As allergic rhinitis (AR) is the best documented disease, we will focus the following paragraph on it. AR is a very common hereditary health problem. It affects 20–40 million US people, approximately 26% of the United Kingdom population, 29.6% of the Belgian population and approximately 10–25% of the population worldwide [64]. It is characterised by inflammation of the upper airway mucous membranes mediated by binding of antigens to specific immunoglobulin E (IgE). The patients suffer from nasal symptoms (itching, sneezing, rhinorrhoea and nasal congestion), ocular symptoms (red itchy eyes) and headache. AR has a negative impact on the patient's quality of life. The patient usually suffers from an impairment of the quality of sleep, daytime fatigue, impaired cognitive function and reduced work productivity and performance [65–67]. AR represents a heavy burden in terms of direct and indirect costs for the patient and the community. There are many drugs on the market to treat it. ARIA proposed some guidelines to use them in a more effective way [68]. H1-antihistamines are certainly the best-known medications to treat AR in adults and children. There are actually two generations of H1-antihistamines: the older ones (the first generation) and the newer ones (the second generation).

First-generation H1-antihistamines are in many countries over-the-counter drugs. A GA(2)LEN position paper recommends to forbid their use over the counter in particular in patients with SDB because they are all sedating and have poor receptor selectivity [69]. They penetrate the blood–brain barrier. Their proclivity to interfere with neurotransmission by histamine at central nervous system H1 receptors potentially leads to drowsiness, sedation, somnolence and fatigue, resulting in impairment of cognitive function, memory and psychomotor performance. In addition, the central H1-antihistaminic effects are primarily responsible for the potentially life-threatening toxicity of first-generation H1-antihistamines overdose. They have been implicated in civil aviation, motor vehicle and boating accidents, deaths from accidental or intentional overdosing in infants and young children and suicide in teenagers and adults. Finally, they exacerbate daytime somnolence because they decrease the quality of sleep and reduce rapid eye movement (REM) sleep. Moreover, they have anticholinergic properties, which can cause dry mouth and make mouth breathing even more uncomfortable in the allergic individual with nasal obstruction [70]. The first generation of H1-antihistamines should therefore be avoided in SDB patients.

The second generation of H1-antihistamines is not associated with fatigue, sedation and dizziness even at high dose. They do not change the structure of the sleep because they have more affinity to the H1 receptor, do not pass the blood–brain barrier and do not have anticholinergic properties [69]. The H1-antihistamines are effective drugs: they improve significantly itching, sneezing and rhinorrhoea, but they are not so effective on nasal congestion. The recommended indications to prescribe an H1-antihistamine are mild to moderate intermittent AR and mild persistent AR. Azelastine, a topical H1-antihistamine, significantly reduces rhinorrhoea and improves subjective sleep, but evidence is lacking on its effects on daytime sleepiness and nasal congestion [71].

Topical intranasal glucocorticoids are considered the gold standard for the treatment of all forms of AR. For the most recent molecules, they have a low systemic bioavailability and a high affinity to the receptors. They have a long-lasting effect with minor adverse events. They are active on sneezing, rhinorrhoea and nasal congestion. A meta-analysis published in 1998 confirmed the place of the intranasal steroids in the treatment of AR [72]. One position paper of the Joint Task Force for the American Academy of Allergy, Asthma and Immunology does not recommend their use over the counter because of the side effects observed in the past with the older generations of topical glucocorticoids [73]. The plasma concentrations of intranasal fluticasone and mometasone are low due to extensive metabolism and clearance by cytochrome P450 enzyme 3A4. Caution is recommended when co-administered with potent CYTP3A4 inhibitors, especially in HIV population. Current antiretroviral regimens often contain the HIV protease inhibitor ritonavir, and co-administration with topical fluticasone results in a dramatic increase in the latter bioavailability. This may result in iatrogenic Cushing's syndrome as alerted in increasing number of case reports [74–77]. Ironically, in patients treated by ritonavir, older generations of topical glucocorticoids appear to be safer options [78]. Apart from these particular cases, second-generation topical glucocorticoids (fluticasone, mometasone) remain the first-line and safest treatment for main patients with allergic rhinitis. Intranasal corticosteroids have broad anti-inflammatory activities. They are the most potent long-term pharmacologic treatment of congestion associated with allergic rhinitis and show some congestion relief in rhinosinusitis and nasal polyposis (Table 23.2).

Topical decongestants reduce congestion associated with allergic rhinitis, but because of the risk of rhinitis medicamentosa, they should not be used for prolonged periods. Oral decongestants reduce nasal congestion but may have adverse effects on sleep, even insomnia, because of their stimulatory effects and their association with systemic side effects.

Oral leukotriene receptor antagonists can be of some help in the management of patients unresponsive to the conventional medications. They effectively reduce rhinorrhoea, congestion and inflammatory mediators [70].

Anticholinergic ipratropium bromide is not considered effective in relieving nasal congestion; however, limited data suggest that sleep and quality of life may be minimally improved with this treatment [88].

23.7.2.2 Management of SDB in Rhinitis Patients

Patients with perennial allergic rhinitis often present with nasal congestion, poor sleep quality, daytime fatigue and loss of productivity. Pharmacologic therapy that reduces nasal congestion should improve these symptoms. In the literature there are a lot of publications related to the management of allergic rhinitis and the

Table 23.2 Effect of medical treatment on sleep-related breathing disorders

Reference	Study	Medication	*n*	Symptoms improvement	PSG improvement
Kerr et al. [79]	Controlled, prospective	Xylometazoline + nasal dilator	10	Yes	No
Craig et al. [80]	Controlled, prospective	Fluticasone	20	Yes	–
Hughes et al. [81]	Controlled, prospective	Budesonide	22	Yes	–
Ratner et al. [82]	Controlled, prospective	Fluticasone vs. montelukast	705	Yes	–
Craig et al. [83]	Controlled, prospective	Fluticasone	32	Yes	No
Kiely et al. [84]	Controlled, prospective	Fluticasone	24	Yes	Yes
Craig et al. [85]	Controlled, prospective (pooled study)	Fluticasone/budesonide/ flunisolide	42	Yes	No
McLean et al. [86]	Controlled, prospective	Xylometazoline + dilator strip	10	No	Yes
Gurevich et al. [87]	Controlled, prospective	Budesonide	26	Yes	–

impact on sleep (Table 23.2). These studies often demonstrate positive effects of the medical treatment on the SDB. However, a majority of these papers are based on subjective assessment (disease-specific quality-of-life measures, quality-of-life questionnaires, general questionnaires, Epworth Sleepiness Scale, Pittsburgh Sleep Quality Index, etc.). Only a few studies have objectively assessed sleep (using polysomnography) in allergic rhinitis. In 20 patients with allergic rhinitis and symptoms of daytime sleepiness, flunisolide significantly improved sleep quality and congestion but not daytime sleepiness [89]. A similar study with fluticasone propionate showed improvement in subjective sleep parameters, but there was no significant change in objective sleep measurements recorded on polysomnography [83]. On the other hand, Kiely et al. have demonstrated a slight decrease in the AHI in snorers with rhinitis treated with fluticasone propionate compared with placebo [84]. Nasal obstruction secondary to allergic inflammation has an impact on sleep quality, and topical corticoid therapy seems to have a positive effect on sleep quality [90]. In one study, 25 patients with seasonal AR and 25 healthy volunteers underwent two consecutive nights of PSG before and during the pollen season [91]. There were statistically significant differences between the two groups in sleep parameters, including increases in the apnoea index, hypopnoea index, apnoea-hypopnoea index, snoring time, amount of REM sleep and sleep latency. Nevertheless, the changes were not considered clinically relevant, as values remained within normal limits. Further research involving objective measures is thus still necessary.

23.7.2.3 Treatment of Nasal Valve Collapse with Nasal Dilators

Nasal valve dysfunction is another underrated and underdiagnosed cause of nasal obstruction. The nasal valve obstruction can be static or dynamic. The diagnosis is made by clinical examination, Cottle manoeuvre, anterior and posterior active rhinomanometry and acoustic rhinometry. An easy way to treat a patient with a nasal valve collapse is to use a nasal dilator. Nasal dilators are an attractive method of decreasing nasal resistance in the valve area with subsequently a probable positive impact on snoring and/or apnoea [92]. Measurements of nasal resistance in awake subjects with a nasal dilator have shown a reduction in resistance, though not uniform, depending on the compliance of the nasal vestibule walls [93]. The dimension of the nasal valve is increased by approximately 30%. Most sleep studies have considered two devices commercially available as nasal dilators: Nozovent®, an internal device, and Breathe Right®, an external device. Other products are now commercially available, like Nasanita®, Airplus®, Respir + ®, Francis alar dilator®, Ognibene dilator® and Side Strip® [94, 95]. There is even a paper on how to bend your own nasal dilator from a plastic-coated paper clip [96]. These devices have been studied in patients with polysomnographic measurements in nine studies (Table 23.3). The conclusions from these studies are that nasal dilators may reduce the subjective sensation of snoring. However, objective measurements of snoring and sleep parameters, such as AHI reveal that nasal dilators are ineffective in the vast majority of the SDB patients. Nasal dilators may be more effective in patients with SDB with concomitant chronic rhinitis [104]. Djupesland et al. found that Breathe Right® was an effective treatment of snoring in a subgroup of patients with morning nasal obstruction and when acoustic rhinometry has revealed a minimal cross-sectional area < 0.6 cm^2 [105]. Based on this information, nasal dilators although ineffective for the vast majority of apnoeic patients may be recommended as a trial for nonapnoeic snorers. Nasal dilators have no side effects and are relatively inexpensive. They may improve CPAP tolerance and reduce the CPAP pressure level [106].

23.7.3 Surgical Management

Surgery concerns nonreversible causes of nasal obstruction: nasal septum deviation, hypertrophy of the mucosa of the inferior turbinates, nasal collapse and nasal polyposis. Two procedures are frequently performed: septoplasty associated or not to turbinates reduction.

Table 23.3 Effect of nasal dilators on sleep-related breathing disorders

Reference	Study	Nasal dilators	*n*	Symptoms improvement	PSG improvement
Hoijer et al. [97]	Noncontrolled, prospective	Nozovent	10	Yes	Yes (36%)
Hoffstein et al. [98]	Noncontrolled, prospective	Nozovent	15	–	No
Liistro et al. [99]	Noncontrolled, prospective	Breathe right	10	–	No
Todorova et al. [100]	Noncontrolled, prospective	Breathe right	30	Yes	Not significant
Gosepath et al. [101]	Noncontrolled, prospective	Breathe right	26	–	Yes (15%)
Bahammam et al. [102]	Controlled, prospective	Breathe right	18	–	No
Schonhofer et al. [103]	Noncontrolled prospective	Nozovent	21	–	No
Pevernagie et al. [104]	Noncontrolled, prospective	Breathe right	12	–	No
Djupesland et al. [105]	Controlled prospective	Breathe right	18	–	Slight if MCA $<0.6\ cm^2$

MCA mean cross-sectional area

Septoplasty involves removing excess septal cartilage and reshaping the cartilage to bring it to the midline. The procedure is usually done under general anaesthesia. Turbinate reduction can be performed with different methods: laser, electrocautery or radiofrequency ablation. The procedure can be done under local or general anaesthesia. Surgery of the nasal valve is not yet extremely popular in SDB. Concerning the nasal polyposis, there is a wide variety of procedures ranging from endoscopic-guided polypectomy [107, 108].

Table 23.4 summarises the effect of surgical procedures on SDB. Most studies were uncontrolled case series [120]. The main surgical procedure was septoplasty, associated or not with turbinoplasty. Only 11 patients (among 420 pooled subjects) underwent septorhinoplasty [109, 111, 119], and the management of the nasal valve was not specifically described. These studies confirmed that current nose surgery improves subjectively the snoring, the daytime sleepiness and the quality of life but failed to improve objective PSG data. The absence of pharyngeal obstruction could predict the success of nose surgery [115]. Conversely, increased nasal resistance could predict the failure of CPAP therapy [121]. Most studies have not demonstrated that reducing nasal obstruction and resistance from various causes and using various techniques (e.g. septoplasty, turbinectomy, polypectomy, turbinoplasty) correlate with a significant reduction in objective OSA indicators, such as the apnoea-hypopnoea index (AHI) or nocturnal oxygen desaturation. Three studies suggested the efficacy of combined nasal and pharyngeal surgery on polysomnography parameters [122, 123] or snoring [124]. Friedman et al. have also suggested that sometimes postoperative polysomnographic data may be worse for mild OSA patients after nasal obstruction relief [125]. They explain this paradoxical effect of nasal surgery by the fact that nasal obstruction relief may allow the patients to sleep in deeper sleep stages. Therefore, apnoea and sleep fragmentation increase because patients sleep more comfortably.

Discrepancies in nose surgery outcome studies may be explained by the variety of surgical procedure, the heterogeneity of patients studied and the variety of outcome measurements (quality-of-life questionnaire, polysomnographic values, subjective snoring) [126]. The pathophysiology of the nose function in sleep-related breathing disorders could explain the relative failure of nose surgery. First, these disorders involve multilevel airway obstruction, including

Table 23.4 Effect of nasal surgery on sleep-related breathing disorders

Reference	Study	Procedure	*n*	Symptoms improvement	PSG improvement
Verse et al. [109]	Controlled, prospective	Septoplasty, septorhinoplasty, FESS	26	Yes	No
Kim et al. [110]	Noncontrolled, retrospective	Septo-turbinoplasty	21	Yes	Yes (19%)
Virkkula et al. [111]	Noncontrolled, prospective	Septo-turbinoplasty, septorhinoplasty	40	No	No
Koutsourelakis et al. [112]	Controlled, prospective	Septoplasty	49	–	No
Li et al. [113]	Noncontrolled, prospective	Septo-turbinoplasty	51	Yes	No
Li et al. [114]	Noncontrolled, prospective	Septo-turbinoplasty	52	Yes	
Morinaga et al. [115]	Noncontrolled, prospective	Septo-turbinoplasty, FESS	35	–	Yes (23%)
Tosun et al. [116]	Noncontrolled, prospective	FESS	27	Yes	No
Li et al. [117]	Controlled, prospective	Septo-turbinoplasty	66	Yes	No
Choi et al. [118]	Noncontrolled, prospective	Septo-turbinoplasty, FESS	22	Yes	No
Sufioglu et al. [119]	Noncontrolled, prospective	Septoplasty, septorhinoplasty, FESS	31	Yes	No

airway length, lateral wall thickness, tongue volume and skeletal structure [127]. One single intervention is therefore unlikely to address the disease. In obese patients, these upper airway anatomic factors may be masked, and obesity is the main etiologic factor for priority handling. Second, usual nose surgery (septoplasty, turbinoplasty) does not attend to correct nose bony framework, which determines the transverse nasal airway dimension, and does not adjust the curvilinear airflow pattern, which is important for the nasopharynx opening. Some researchers speculate that nasal valve surgery combined with a mouth-closing oral appliance may be an ideal therapy for sleep apnoea in nonobese patients [44]. This intervention could address the curvilinear airflow pattern and promote nose breathing. Further studies are, however, necessary to design future surgical algorithms.

Another group of patients that may be considered for nasal surgery are those who have failed CPAP therapy [126]. CPAP therapy remains the first-line therapy of OSA but may cause rhinitis itself and compliance rates ranging from 65 to 80%. Dry nose or mouth in the morning affects 65% of the patients. Sneezing and nasal drip are present in more than 35% of the patients and nasal congestion in 25% [128]. Using a humidifier reduces only poorly the nose side effects [128]. A high nasal resistance is a significant risk factor for nonacceptance of CPAP [129]. Careful evaluation of the nose is mandatory to identify the factors that may be correctable, in order to improve compliance. Septoplasty ± turbinoplasty has been shown to allow for reduced pressure levels of CPAP and easier use of the apparatus [125]. Likewise, radiofrequency turbinate reduction increases CPAP adherence [130]. Early identification and management of OSA patients with high nasal resistance can potentially improve CPAP treatment outcome. However, variable additional factors also impact CPAP compliance, such as individual perception of symptoms and improvement in sleepiness and daily function from initial use of CPAP. For these reasons, larger, well-designed studies are needed to confirm the durability of any beneficial effect on CPAP compliance from nasal surgical procedures for individuals with OSA [131–133].

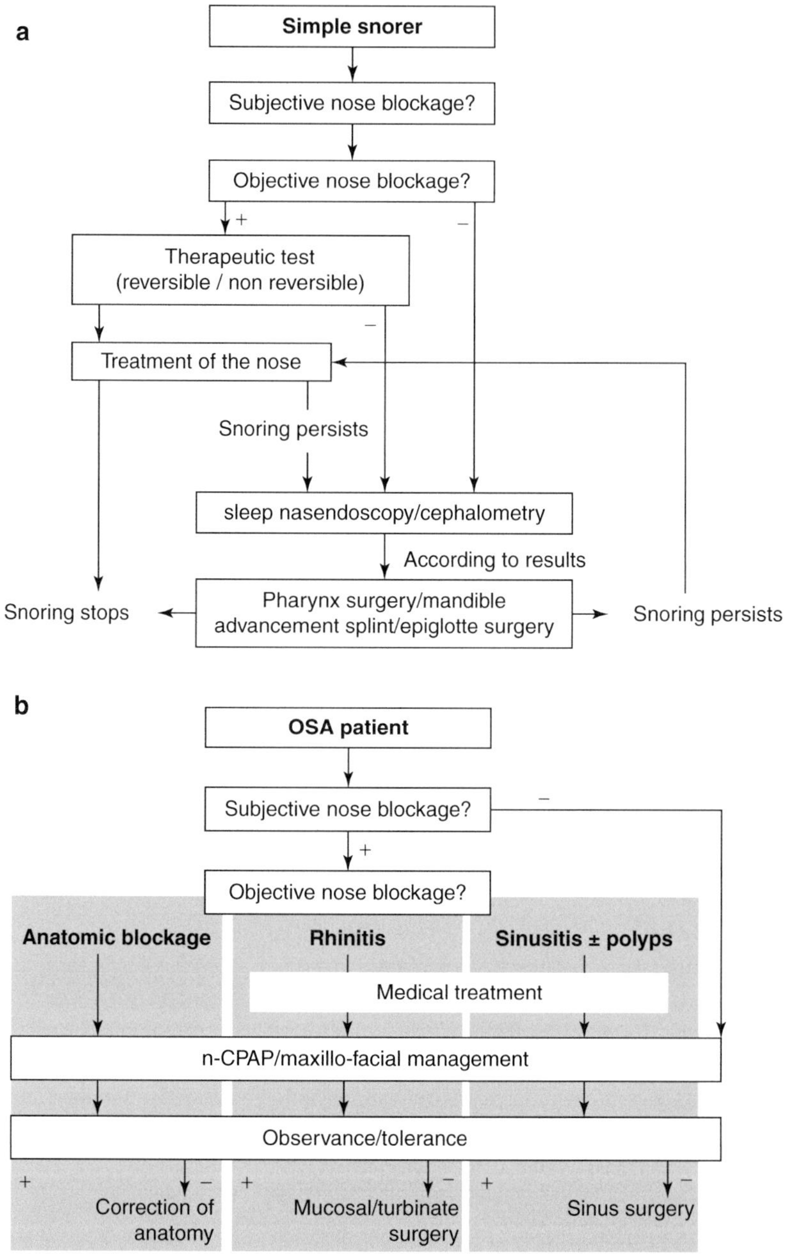

Fig. 23.4 Simplified management scheme for adults with simple snoring (**a**) and OSA (**b**)

To summarise, reducing nasal obstruction and resistance from various causes and using various techniques improve subjectively the snoring, the daytime sleepiness and the quality of life but fails to improve significantly objective data at the polysomnography, such as the apnoea-hypopnoea index (AHI) or nocturnal oxygen desaturation. When some positive effects have been reported, improvement of sleep apnoea occurs only in approximately 15–20% of the patients. Results of nasal surgery in patients with sleep apnoea/hypopnoea are therefore barely predictable. Nevertheless, nasal procedures improve CPAP compliance in individuals with OSA and nasal obstruction requiring high CPAP settings. A simplified management scheme for adults with SDB is proposed in Fig. 23.4.

23.8 Conclusions

Increasing evidence shows that nasal resistance is a contributing risk factor for sleep-related breathing disorders. Nevertheless, nose management alone fails in many cases to address the objective parameters of SDB [109, 134, 135]. Compelling data are lacking concerning the exact role of obstructed nasal breathing in the pathogenesis of obstructive sleep disorders [10, 136]. Under an evidence-based approach, nasal surgery in OSA patients with nasal obstruction effectively ameliorates clinical symptoms of snoring and daytime sleepiness and consequently improves quality of life. However, the efficacy of nasal treatment alone in treating OSA is limited. Nasal management should be integrated in a multimodal approach (diet/smoking cessation/CPAP/mandibular splint/multilevel surgery), truly reflecting the complexity of SDB.

References

1. Young T, Palta M, Dempsey J, Skatrud J, Weber S, Badr S. The occurrence of sleep-disordered breathing among middle-aged adults. N Engl J Med. 1993;328:1230–5.
2. Phillipson EA. Sleep apnea–a major public health problem. N Engl J Med. 1993;328:1271–3.
3. Pirsig W. The nose and sleep-disordered breathing. Sleep Breath. 2003;7:51–2.
4. Primhak R, Kingshott R. Sleep physiology and sleep-disordered breathing: the essentials. Arch Dis Child. 2012;97:54–8.
5. Ferris BG Jr, Mead J, Opie LH. Partitioning of respiratory flow resistance in man. J Appl Physiol. 1964;19:653–8.
6. Rhee JS, Weaver EM, Park SS, et al. Clinical consensus statement: diagnosis and management of nasal valve compromise. Otolaryngol Head Neck Surg. 2010;143:48–59.
7. Spielmann PM, White PS, Hussain SS. Surgical techniques for the treatment of nasal valve collapse: a systematic review. Laryngoscope. 2009;119:1281–90.
8. Courtiss EH, Goldwyn RM. The effects of nasal surgery on airflow. Plast Reconstr Surg. 1983;72:9–21.
9. Kennedy DW, Zinreich SJ, Kumar AJ, Rosenbaum AE, Johns ME. Physiologic mucosal changes within the nose and ethmoid sinus: imaging of the nasal cycle by MRI. Laryngoscope. 1988;98:928–33.
10. Rappai M, Collop N, Kemp S, de Shazo R. The nose and sleep-disordered breathing: what we know and what we do not know. Chest. 2003;124:2309–23.
11. Mladina R, Cujic E, Subaric M, Vukovic K. Nasal septal deformities in ear, nose, and throat patients: an international study. Am J Otolaryngol. 2008;29:75–82.
12. Grymer LF, Hilberg O, Pedersen OF. Prediction of nasal obstruction based on clinical examination and acoustic rhinometry. Rhinology. 1997;35:53–7.
13. Bauchau V, Durham SR. Prevalence and rate of diagnosis of allergic rhinitis in Europe. Eur Respir J. 2004;24:758–64.
14. Brozek JL, Bousquet J, Baena-Cagnani CE, et al. Allergic rhinitis and its impact on asthma (ARIA) guidelines: 2010 revision. J Allergy Clin Immunol. 2010;126:466–76.
15. Fokkens WJ, Lund VJ, Mullol J, et al. EPOS 2012: European position paper on rhinosinusitis and nasal polyps 2012. A summary for otorhinolaryngologists. Rhinology. 2012;50:1–12.
16. Georgalas C. The role of the nose in snoring and obstructive sleep apnoea: an update. Eur Arch Otorhinolaryngol. 2011;268(9):1365–73.
17. McNicholas WT. The nose and OSA: variable nasal obstruction may be more important in pathophysiology than fixed obstruction. Eur Respir J. 2008;32:3–8.
18. Farre R, Montserrat JM, Navajas D. Assessment of upper airway mechanics during sleep. Respir Physiol Neurobiol. 2008;163:74–81.
19. Horner RL. Neural control of the upper airway: integrative physiological mechanisms and relevance for sleep disordered breathing. Compr Physiol. 2012;2:479–535.
20. Gold AR, Schwartz AR. The pharyngeal critical pressure. Chest. 1996;110:1077–88.
21. Fitzpatrick MF, Driver HS, Chatha N, Voduc N, Girard AM. Partitioning of inhaled ventilation between the nasal and oral routes during sleep in normal subjects. J Appl Physiol. 2003a;94:883–90.
22. Fitzpatrick MF, McLean H, Urton AM, Tan A, O'Donnell D, Driver HS. Effect of nasal or oral breathing route on upper airway resistance during sleep. Eur Respir J. 2003b;22:827–32.
23. McNicholas WT, Coffey M, Boyle T. Effects of nasal airflow on breathing during sleep in normal humans. Am Rev Respir Dis. 1993;147:620–3.
24. White DP, Cadieux RJ, Lombard RM, Bixler EO, Kales A, Zwillich CW. The effects of nasal anesthesia on breathing during sleep. Am Rev Respir Dis. 1985;132:972–5.
25. Berry RB, Kouchi KG, Bower JL, Light RW. Effect of upper airway anesthesia on obstructive sleep apnea. Am J Respir Crit Care Med. 1995;151:1857–61.
26. Ko JH, Kuo TB, Lee GS. Effect of postural change on nasal airway and autonomic nervous system established by rhinomanometry and heart rate variability analysis. Am J Rhinol. 2008;22:159–65.
27. Lundberg JO. Nitric oxide and the paranasal sinuses. Anat Rec (Hoboken). 2008;291:1479–84.
28. Lundberg JO, Weitzberg E. Nasal nitric oxide in man. Thorax. 1999;54:947–52.

29. Blitzer ML, Loh E, Roddy MA, Stamler JS, Creager MA. Endothelium-derived nitric oxide regulates systemic and pulmonary vascular resistance during acute hypoxia in humans. J Am Coll Cardiol. 1996;28:591–6.
30. Lundberg JO. Airborne nitric oxide: inflammatory marker and aerocrine messenger in man. Acta Physiol Scand Suppl. 1996;633:1–27.
31. Haight JS, Djupesland PG. Nitric oxide (NO) and obstructive sleep apnea (OSA). Sleep Breath. 2003;7:53–62.
32. Atkeson A, Jelic S. Mechanisms of endothelial dysfunction in obstructive sleep apnea. Vasc Health Risk Manag. 2008;4:1327–35.
33. Paludetti G, Almadori G, Scarano E, Deli R, Laneri de Bernart A, Maurizi M. Nasal obstruction and skull base development: experimental study in the rat. Rhinology. 1995;33:171–3.
34. Scarano E, Ottaviani F, Di Girolamo S, Galli A, Deli R, Paludetti G. Relationship between chronic nasal obstruction and craniofacial growth: an experimental model. Int J Pediatr Otorhinolaryngol. 1998;45:125–31.
35. Yamada T, Tanne K, Miyamoto K, Yamauchi K. Influences of nasal respiratory obstruction on craniofacial growth in young Macaca fuscata monkeys. Am J Orthod Dentofac Orthop. 1997;111:38–43.
36. Peltomaki T. The effect of mode of breathing on craniofacial growth–revisited. Eur J Orthod. 2007;29:426–9.
37. Principato JJ. Upper airway obstruction and craniofacial morphology. Otolaryngol Head Neck Surg. 1991;104:881–90.
38. Harari D, Redlich M, Miri S, Hamud T, Gross M. The effect of mouth breathing versus nasal breathing on dentofacial and craniofacial development in orthodontic patients. Laryngoscope. 2010;120:2089–93.
39. Juliano ML, Machado MA, Carvalho LB, Prado LB, Do Prado GF. Mouth breathing children have cephalometric patterns similar to those of adult patients with obstructive sleep apnea syndrome. Arq Neuropsiquiatr. 2009;67:860–5.
40. Erler T, Paditz E. Obstructive sleep apnea syndrome in children: a state-of-the-art review. Treat Respir Med. 2004;3:107–22.
41. Poirrier AL, Pire S, Raskin S, Limme M, Poirrier R. Contribution of postero-anterior cephalometry in obstructive sleep apnea. Laryngoscope. 2012;122(10):2350–4.
42. Cakirer B, Hans MG, Graham G, Aylor J, Tishler PV, Redline S. The relationship between craniofacial morphology and obstructive sleep apnea in whites and in African-Americans. Am J Respir Crit Care Med. 2001;163:947–50.
43. Ip MS, Lam B, Lauder IJ, et al. A community study of sleep-disordered breathing in middle-aged Chinese men in Hong Kong. Chest. 2001;119:62–9.
44. Stupak HD. The human external nose and its evolutionary role in the prevention of obstructive sleep apnea. Otolaryngol Head Neck Surg. 2010;142:779–82.
45. Davidson TM. The great leap forward: the anatomic basis for the acquisition of speech and obstructive sleep apnea. Sleep Med. 2003;4:185–94.
46. Davidson TM, Sedgh J, Tran D, Stepnowsky CJ Jr. The anatomic basis for the acquisition of speech and obstructive sleep apnea: evidence from cephalometric analysis supports the great leap forward hypothesis. Sleep Med. 2005;6:497–505.
47. Shprintzen RJ. The origin of speech ease: evolution of the human upper airway and its functional implications for obstructive sleep apnea. Editorial commentary: the great leap forward: the anatomic basis for the acquisition of speech and obstructive sleep apnea by Terence M. Davidson. Sleep Med. 2003;4:171–3.
48. Mladina R. The role of maxillar morphology in the development of pathological septal deformities. Rhinology. 1987;25:199–205.
49. Cole P. Acoustic rhinometry and rhinomanometry. Rhinol Suppl. 2000;16:29–34.
50. Virkkula P, Maasilta P, Hytonen M, Salmi T, Malmberg H. Nasal obstruction and sleep-disordered breathing: the effect of supine body position on nasal measurements in snorers. Acta Otolaryngol. 2003b;123:648–54.
51. Virkkula P, Hurmerinta K, Loytonen M, Salmi T, Malmberg H, Maasilta P. Postural cephalometric analysis and nasal resistance in sleep-disordered breathing. Laryngoscope. 2003a;113:1166–74.
52. Yahyavi S, Parsa FM, Fereshtehnejad SM, Najimi N. Objective measurement of nasal airway dimensions and resistance using acoustic rhinometry and rhinomanometry in habitual snorers compared with non-snorers. Eur Arch Otorhinolaryngol. 2008;265:1483–7.
53. Liistro G, Rombaux P, Belge C, Dury M, Aubert G, Rodenstein DO. High Mallampati score and nasal obstruction are associated risk factors for obstructive sleep apnoea. Eur Respir J. 2003;21:248–52.
54. Cakmak O, Coskun M, Celik H, Buyuklu F, Ozluoglu LN. Value of acoustic rhinometry for measuring nasal valve area. Laryngoscope. 2003;113:295–302.
55. Wilson AM, Sims EJ, Robb F, Cockburn W, Lipworth BJ. Peak inspiratory flow rate is more sensitive than acoustic rhinometry or rhinomanometry in detecting corticosteroid response with nasal histamine challenge. Rhinology. 2003;41:16–20.
56. Nathan RA, Eccles R, Howarth PH, Steinsvag SK, Togias A. Objective monitoring of nasal patency and nasal physiology in rhinitis. J Allergy Clin Immunol. 2005;115:S442–59.
57. Hopkins C, Gillett S, Slack R, Lund VJ, Browne JP. Psychometric validity of the 22-item Sinonasal outcome test. Clin Otolaryngol. 2009;34:447–54.
58. Lindsay RW. Disease-specific quality of life outcomes in functional rhinoplasty. Laryngoscope. 2012;122(7):1480–8.

59. Piccirillo JF, Merritt MG Jr, Richards ML. Psychometric and clinimetric validity of the 20-item Sino-nasal outcome test (SNOT-20). Otolaryngol Head Neck Surg. 2002;126:41–7.
60. Poirrier AL, Ahluwalia S, Goodson A, Ellis M, Bentley M, Andrews P. Is the Sino-nasal outcome Test-22 a suitable evaluation for septorhinoplasty? Laryngoscope. 2013;123(1):76–81.
61. Grover SS, Pittman SD. Automated detection of sleep disordered breathing using a nasal pressure monitoring device. Sleep Breath. 2008;12:339–45.
62. Maury G, Cambron L, Jamart J, Marchand E, Senny F, Poirrier R. Added value of a mandible movement automated analysis in the screening of obstructive sleep apnea. J Sleep Res. 2013;22(1):96–103.
63. Senny F, Maury G, Cambron L, Leroux A, Destine J, Poirrier R. The sleep/wake state scoring from mandible movement signal. Sleep Breath. 2012;16:535–42.
64. Storms W. Allergic rhinitis-induced nasal congestion: its impact on sleep quality. Prim Care Respir J. 2008;17:7–18.
65. Kessler RC, Almeida DM, Berglund P, Stang P. Pollen and mold exposure impairs the work performance of employees with allergic rhinitis. Ann Allergy Asthma Immunol. 2001;87:289–95.
66. Marshall PS, O'Hara C, Steinberg P. Effects of seasonal allergic rhinitis on selected cognitive abilities. Ann Allergy Asthma Immunol. 2000;84:403–10.
67. Wilken JA, Berkowitz R, Kane R. Decrements in vigilance and cognitive functioning associated with ragweed-induced allergic rhinitis. Ann Allergy Asthma Immunol. 2002;89:372–80.
68. Bousquet J, Schunemann HJ, Zuberbier T, et al. Development and implementation of guidelines in allergic rhinitis - an ARIA-GA2LEN paper. Allergy. 2010;65:1212–21.
69. Church MK, Maurer M, Simons FE, et al. Risk of first-generation H(1)-antihistamines: a GA(2)LEN position paper. Allergy. 2010;65:459–66.
70. Ferguson BJ. Influences of allergic rhinitis on sleep. Otolaryngol Head Neck Surg. 2004;130:617–29.
71. Golden S, Teets SJ, Lehman EB, et al. Effect of topical nasal azelastine on the symptoms of rhinitis, sleep, and daytime somnolence in perennial allergic rhinitis. Ann Allergy. 2000;85:53–7.
72. Weiner JM, Abramson MJ, Puy RM. Intranasal corticosteroids versus oral H1 receptor antagonists in allergic rhinitis: systematic review of randomised controlled trials. BMJ. 1998;317:1624–9.
73. Passalacqua G, Albano M, Canonica GW, et al. Inhaled and nasal corticosteroids: safety aspects. Allergy. 2000;55:16–33.
74. Kedem E, Shahar E, Hassoun G, Pollack S. Iatrogenic Cushing's syndrome due to coadministration of ritonavir and inhaled budesonide in an asthmatic human immunodeficiency virus infected patient. J Asthma. 2010;47:830–1.
75. Mahlab-Guri K, Asher I, Gradstein S, et al. Inhaled fluticasone causes iatrogenic cushing's syndrome in patients treated with ritonavir. J Asthma. 2011;48:860–3.
76. Samaras K, Pett S, Gowers A, McMurchie M, Cooper DA. Iatrogenic Cushing's syndrome with osteoporosis and secondary adrenal failure in human immunodeficiency virus-infected patients receiving inhaled corticosteroids and ritonavir-boosted protease inhibitors: six cases. J Clin Endocrinol Metab. 2005;90:4394–8.
77. Valin N, De Castro N, Garrait V, Bergeron A, Bouche C, Molina JM. Iatrogenic Cushing's syndrome in HIV-infected patients receiving ritonavir and inhaled fluticasone: description of 4 new cases and review of the literature. J Int Assoc Physicians AIDS Care (Chic). 2009;8:113–21.
78. Foisy MM, Yakiwchuk EM, Chiu I, Singh AE. Adrenal suppression and Cushing's syndrome secondary to an interaction between ritonavir and fluticasone: a review of the literature. HIV Med. 2008;9:389–96.
79. Kerr P, Millar T, Buckle P, Kryger M. The importance of nasal resistance in obstructive sleep apnea syndrome. J Otolaryngol. 1992;21:189–95.
80. Craig TJ, Teets S, Lehman EB, Chinchilli VM, Zwillich C. Nasal congestion secondary to allergic rhinitis as a cause of sleep disturbance and daytime fatigue and the response to topical nasal corticosteroids. J Allergy Clin Immunol. 1998;101:633–7.
81. Hughes K, Glass C, Ripchinski M, et al. Efficacy of the topical nasal steroid budesonide on improving sleep and daytime somnolence in patients with perennial allergic rhinitis. Allergy. 2003;58:380–5.
82. Ratner PH, Howland WC 3rd, Arastu R, et al. Fluticasone propionate aqueous nasal spray provided significantly greater improvement in daytime and nighttime nasal symptoms of seasonal allergic rhinitis compared with montelukast. Ann Allergy Asthma Immunol. 2003;90:536–42.
83. Craig TJ, Mende C, Hughes K, Kakumanu S, Lehman EB, Chinchilli V. The effect of topical nasal fluticasone on objective sleep testing and the symptoms of rhinitis, sleep, and daytime somnolence in perennial allergic rhinitis. Allergy Asthma Proc. 2003;24:53–8.
84. Kiely JL, Nolan P, McNicholas WT. Intranasal corticosteroid therapy for obstructive sleep apnoea in patients with co-existing rhinitis. Thorax. 2004;59:50–5.
85. Craig TJ, Hanks CD, Fisher LH. How do topical nasal corticosteroids improve sleep and daytime somnolence in allergic rhinitis? J Allergy Clin Immunol. 2005;116:1264–6.
86. McLean HA, Urton AM, Driver HS, et al. Effect of treating severe nasal obstruction on the severity of obstructive sleep apnoea. Eur Respir J. 2005;25:521–7.
87. Gurevich F, Glass C, Davies M, et al. The effect of intranasal steroid budesonide on the congestion-related sleep disturbance and daytime somnolence

in patients with perennial allergic rhinitis. Allergy Asthma Proc. 2005;26:268–74.
88. Rabasseda X. A report from the American Academy of Allergy, Asthma and Immunology 2012 annual meeting (March 2–6, 2012 - Orlando, Florida, U.S.A.). Drugs Today (Barc). 2012;48:303–10.
89. Kakumanu S, Glass C, Craig T. Poor sleep and day-time somnolence in allergic rhinitis: significance of nasal congestion. Am J Respir Med. 2002;1:195–200.
90. Rombaux P, Liistro G, Hamoir M, et al. Nasal obstruction and its impact on sleep-related breathing disorders. Rhinology. 2005;43:242–50.
91. Stuck BA, Czajkowski J, Hagner AE, et al. Changes in daytime sleepiness, quality of life, and objective sleep patterns in seasonal allergic rhinitis: a controlled clinical trial. J Allergy Clin Immunol. 2004;113:663–8.
92. Petruson B. Snoring can be reduced when the nasal airflow is increased by the nasal dilator Nozovent. Arch Otolaryngol Head Neck Surg. 1990;116:462–4.
93. Petruson B. Increased nasal breathing decreases snoring and improves oxygen saturation during sleep apnoea. Rhinology. 1994;32:87–9.
94. Ellegard E. Mechanical nasal alar dilators. Rhinology. 2006;44:239–48.
95. Riechelmann H, Karow E, DiDio D, Kral F. External nasal valve collapse - a case–control and interventional study employing a novel internal nasal dilator (Nasanita). Rhinology. 2010;48:183–8.
96. Cheng D, Iriarte GC. The paper clip nasal dilator. Laryngoscope. 1998;108:1247–8.
97. Hoijer U, Ejnell H, Hedner J, Petruson B, Eng LB. The effects of nasal dilation on snoring and obstructive sleep apnea. Arch Otolaryngol Head Neck Surg. 1992;118:281–4.
98. Hoffstein V, Mateika S, Metes A. Effect of nasal dilation on snoring and apneas during different stages of sleep. Sleep. 1993;16:360–5.
99. Liistro G, Rombaux P, Dury M, Pieters T, Aubert G, Rodenstein DO. Effects of breathe right on snoring: a polysomnographic study. Respir Med. 1998;92:1076–8.
100. Todorova A, Schellenberg R, Hofmann HC, Dimpfel W. Effect of the external nasal dilator breathe right on snoring. Eur J Med Res. 1998;3:367–79.
101. Gosepath J, Amedee RG, Romantschuck S, Mann WJ. Breathe right nasal strips and the respiratory disturbance index in sleep related breathing disorders. Am J Rhinol. 1999;13:385–9.
102. Bahammam AS, Tate R, Manfreda J, Kryger MH. Upper airway resistance syndrome: effect of nasal dilation, sleep stage, and sleep position. Sleep. 1999;22:592–8.
103. Schonhofer B, Franklin KA, Brunig H, Wehde H, Kohler D. Effect of nasal-valve dilation on obstructive sleep apnea. Chest. 2000;118:587–90.
104. Pevernagie D, Hamans E, Van Cauwenberge P, Pauwels R. External nasal dilation reduces snoring in chronic rhinitis patients: a randomized controlled trial. Eur Respir J. 2000;15:996–1000.
105. Djupesland PG, Skatvedt O, Borgersen AK. Dichotomous physiological effects of nocturnal external nasal dilation in heavy snorers: the answer to a rhinologic controversy? Am J Rhinol. 2001;15:95–103.
106. Schonhofer B, Kerl J, Suchi S, Kohler D, Franklin KA. Effect of nasal valve dilation on effective CPAP level in obstructive sleep apnea. Respir Med. 2003;97:1001–5.
107. Devars du Mayne M, Pruliere-Escabasse V, Zerah-Lancner F, Coste A, Papon JF. Polypectomy compared with ethmoidectomy in the treatment of nasal polyposis. Arch Otolaryngol Head Neck Surg. 2011;137:111–7.
108. Jankowski R, Pigret D, Decroocq F, Blum A, Gillet P. Comparison of radical (nasalisation) and functional ethmoidectomy in patients with severe sinonasal polyposis. A retrospective study. Rev Laryngol Otol Rhinol (Bord). 2006;127:131–40.
109. Verse T, Maurer JT, Pirsig W. Effect of nasal surgery on sleep-related breathing disorders. Laryngoscope. 2002;112:64–8.
110. Kim ST, Choi JH, Jeon HG, Cha HE, Kim DY, Chung YS. Polysomnographic effects of nasal surgery for snoring and obstructive sleep apnea. Acta Otolaryngol. 2004;124:297–300.
111. Virkkula P, Bachour A, Hytonen M, et al. Snoring is not relieved by nasal surgery despite improvement in nasal resistance. Chest. 2006;129:81–7.
112. Koutsourelakis I, Georgoulopoulos G, Perraki E, Vagiakis E, Roussos C, Zakynthinos SG. Randomised trial of nasal surgery for fixed nasal obstruction in obstructive sleep apnoea. Eur Respir J. 2008;31:110–7.
113. Li HY, Lin Y, Chen NH, Lee LA, Fang TJ, Wang PC. Improvement in quality of life after nasal surgery alone for patients with obstructive sleep apnea and nasal obstruction. Arch Otolaryngol Head Neck Surg. 2008b;134:429–33.
114. Li HY, Lee LA, Wang PC, Chen NH, Lin Y, Fang TJ. Nasal surgery for snoring in patients with obstructive sleep apnea. Laryngoscope. 2008a;118:354–9.
115. Morinaga M, Nakata S, Yasuma F, et al. Pharyngeal morphology: a determinant of successful nasal surgery for sleep apnea. Laryngoscope. 2009;119:1011–6.
116. Tosun F, Kemikli K, Yetkin S, Ozgen F, Durmaz A, Gerek M. Impact of endoscopic sinus surgery on sleep quality in patients with chronic nasal obstruction due to nasal polyposis. J Craniofac Surg. 2009;20:446–9.
117. Li HY, Lee LA, Wang PC, Fang TJ, Chen NH. Can nasal surgery improve obstructive sleep apnea: subjective or objective? Am J Rhinol Allergy. 2009;23:e51–5.
118. Choi JH, Kim EJ, Kim YS, et al. Effectiveness of nasal surgery alone on sleep quality, architecture, position, and sleep-disordered breathing in obstructive sleep apnea syndrome with nasal obstruction. Am J Rhinol Allergy. 2011;25:338–41.

119. Sufioglu M, Ozmen OA, Kasapoglu F, et al. The efficacy of nasal surgery in obstructive sleep apnea syndrome: a prospective clinical study. Eur Arch Otorhinolaryngol. 2012;269:487–94.
120. Li HY, Wang PC, Chen YP, Lee LA, Fang TJ, Lin HC. Critical appraisal and meta-analysis of nasal surgery for obstructive sleep apnea. Am J Rhinol Allergy. 2011;25:45–9.
121. Nakata S, Noda A, Yagi H, et al. Nasal resistance for determinant factor of nasal surgery in CPAP failure patients with obstructive sleep apnea syndrome. Rhinology. 2005;43:296–9.
122. Li HY, Wang PC, Hsu CY, Lee SW, Chen NH, Liu SA. Combined nasal-palatopharyngeal surgery for obstructive sleep apnea: simultaneous or staged? Acta Otolaryngol. 2005;125:298–303.
123. Stow NW, Sale PJ, Lee D, Joffe D, Gallagher RM. Simultaneous tonsillectomy and nasal surgery in adult obstructive sleep apnea: a pilot study. Otolaryngol Head Neck Surg. 2012;147(2):387–91.
124. Carroll W, Wilhoit CS, Intaphan J, Nguyen SA, Gillespie MB. Snoring management with nasal surgery and upper airway radiofrequency ablation. Otolaryngol Head Neck Surg. 2012;146:1023–7.
125. Friedman M, Tanyeri H, Lim JW, Landsberg R, Vaidyanathan K, Caldarelli D. Effect of improved nasal breathing on obstructive sleep apnea. Otolaryngol Head Neck Surg. 2000;122:71–4.
126. Kotecha B. The nose, snoring and obstructive sleep apnoea. Rhinology. 2011;49:259–63.
127. Mohsenin V. Gender differences in the expression of sleep-disordered breathing : role of upper airway dimensions. Chest. 2001;120:1442–7.
128. Pepin JL, Leger P, Veale D, Langevin B, Robert D, Levy P. Side effects of nasal continuous positive airway pressure in sleep apnea syndrome. Study of 193 patients in two French sleep centers. Chest. 1995;107:375–81.
129. Sugiura T, Noda A, Nakata S, et al. Influence of nasal resistance on initial acceptance of continuous positive airway pressure in treatment for obstructive sleep apnea syndrome. Respiration. 2007;74:56–60.
130. Powell NB, Zonato AI, Weaver EM, et al. Radiofrequency treatment of turbinate hypertrophy in subjects using continuous positive airway pressure: a randomized, double-blind, placebo-controlled clinical pilot trial. Laryngoscope. 2001;111:1783–90.
131. Friedman M, Wilson M. Re-redefining success in airway surgery for obstructive sleep apnea. Sleep. 2009;32:17.
132. Weaver TE, Grunstein RR. Adherence to continuous positive airway pressure therapy: the challenge to effective treatment. Proc Am Thorac Soc. 2008;5:173–8.
133. Zonato AI, Bittencourt LR, Martinho FL, Gregorio LC, Tufik S. Upper airway surgery: the effect on nasal continuous positive airway pressure titration on obstructive sleep apnea patients. Eur Arch Otorhinolaryngol. 2006;263:481–6.
134. Kohler M, Bloch KE, Stradling JR. The role of the nose in the pathogenesis of obstructive sleep apnoea and snoring. Eur Respir J. 2007;30:1208–15.
135. Kohler M, Bloch KE, Stradling JR. The role of the nose in the pathogenesis of obstructive sleep apnea. Curr Opin Otolaryngol Head Neck Surg. 2009;17:33–7.
136. Chen W, Kushida CA. Nasal obstruction in sleep-disordered breathing. Otolaryngol Clin N Am. 2003;36:437–60.

Pathophysiology of Obstructive Sleep Apnea

24

Kivanc Gunhan

Core Messages

- Obstructive sleep apnea (OSA) is a common disorder, affecting approximately 12% of the adult population and 10% of children. Patients should be identified and treated accordingly with a fundamental understanding of the pathophysiology and knowledge of common patterns of OSA.
- The disorder is characterized by recurrent episodes of upper airway narrowing and obstruction, and is associated with reductions in ventilation, resulting in recurrent arousals and episodic oxyhemoglobin desaturations during sleep. Significant clinical consequences of the disorder cover a wide spectrum, including daytime hypersomnolence, neurocognitive dysfunction, cardiovascular sequelae (hypertension, stroke, myocardial infarction, and heart failure), metabolic dysfunction, respiratory failure, and cor pulmonale.
- Sleep is essential and an active role player in neural plasticity, such as memory consolidation and synaptic downscaling.
- The major risk factors include obesity, male gender, hormones, age, anatomical factors, genetic factors, posture, and gravity.
- Wakefulness provides compensatory neuronal activation of dilator muscles in an anatomically compromised collapsible pharynx supporting ventilator control. Accordingly, when this activation is lost at sleep onset, the airway narrows and/or collapses. However, the tendency to result (or not to result) in repeated cyclical apneas is the end product of multiple compensatory processes that vary markedly among and within individuals.
- The pathophysiology of OSA is summarized in three steps:
  - The varied structural and functional determinants of an anatomical predisposition for airway closure, an absolutely essential component for OSA.
  - The effects of the sleep itself, which emphasizes on mechanisms underlying both obstructive and central apnea and ventilatory instability.
  - The integration of anatomical deficits with mechanisms underlying central neurochemical control of breathing stability and compensatory neuromuscular control of upper airway caliber, to explain the repetitive nature of OSA.
- The anatomical determinants of upper airway in OSA consists of the unique anatomy of the human airway, the sites of airway collapse, soft tissue and bony abnormalities, obesity and lung volume, airway edema and surface tension, and obesity.
- The mechanical determinants of airway patency are similar to those regulating caliber

K. Gunhan (✉)
Otorhinolaryngology and Head-Neck Surgery, Manisa Celal Bayar University, Manisa, Turkey
e-mail: kivanc.gunhan@cbu.edu.tr

Ö. Ö. Celebi, T. M. Önerci (eds.), *Nasal Physiology and Pathophysiology of Nasal Disorders*,
https://doi.org/10.1007/978-3-031-12386-3_24

of any collapsible tube. There is a critical closing pressure (P_{crit}) of the passive airway, which is defined as the pressure inside the airway at which the airway collapses. With increasing P_{crit}, as the differential between $P_{upstream}$ and P_{crit} decreases, inspiratory airflow limitation will eventually develop and complete airway occlusion occurs.

- Neuromuscular controls of upper airway dynamics during sleep influence the pharyngeal muscle tone and upper airway collapsibility. A variety of defective respiratory control mechanisms are found in OSA, including impaired chemical drive, defective inspiratory load responses, and abnormal upper airway protective reflexes. Tonic and phasic EMG activity of pharyngeal airway dilator muscles (genioglossus and tensor palatine) are progressively reduced from wakefulness to NREM to REM sleep. Changes in proprioceptive and chemoreceptor feedback play a significant role in the dynamics of upper airway caliber. Chemoreceptor influences also have substantial effects on upper airway muscle recruitment, and in the case of CO_2, upper airway motor neurons relative to phrenic motor neurons have a substantially higher threshold for inhibition (via hypocapnia) and activation (via hypercapnia).
- Throughout this review, these mechanisms are discussed with an emphasis on understanding the physiology and pathophysiology of sleep-related breathing disorders and OSA better that are believed to be some of the essential knowledge for the proper management of this multi-facet and complex disorder. The main focus is to review the key pathophysiological factors and their interactions, to highlight recent innovations in our understanding of OSA pathogenesis.

24.1 Pathophysiology of Obstructive Sleep Apnea

Hypnos was the god of Sleep. He resided in Erebos, the land of eternal darkness, beyond the gates of the rising sun living in the dark cave, in the Hades (Underworld), whose entrance was full of poppies and other hypnotic plants; and that through his cave the river of forgetfulness, Lethe, used to flow. From there he rose into the sky each night in the train of his mother Nyx (the goddess of Night).

Hypnos was often paired with his twin brother Thanatos (the god of peaceful Death). He was married to the youngest of the Graces - Pasithea, a deity of hallucination or relaxation and Morpheus and Phobetor (the gods of Dreams) were their sons.—Hesiod, Theogony (Greek epic, C7th B.C.) [1]

In the early ages of humanity sleep was defined as a world of unknown hand in hand with death, whose mystery was full of dreams where everything was forgotten. Compared to other widely studied human behaviors, sleep has been a rather misunderstood phenomenon over the course of human history. Physicians regarded sleep as nothing more than a monolithic loss of consciousness, and believed that it was caused by a lack of blood flow to the brain. Many famous scientists and philosophers, like Hippocrates, Aristotle, Plato, Galen, and Avicenna, had commented over the physiologic and psychological aspects of sleep throughout history, but sleeping and dreaming has been one of the most under-researched areas of human behavior. Despite centuries of speculation and research, we still do not know what sleep really is, or exactly what it is for. Allan Rechtschaffen suggested in 1978 that "If sleep does not serve an absolute vital function, then it is the biggest mistake the evolutionary process ever made." [2]

We have only really begun to study and to understand sleep apnea over the past 40 years. Observations of periodic breathing in sleep were first reported in the mid-1850s, and in the 1870s physicians reported on several cases of obstructed apneas as "fruitless contractions of the inspiratory and expiratory muscles against glottic obstruction with accompanying cyanosis during sleep." [2] During the latter half of the nineteenth century, several cases of obese persons with extreme daytime sleepiness were described [3] and labeled the "Pick Wickian syndrome" after Charles Dickens Fat Boy Joe as described in the Pickwick Papers in 1837. By the mid-1960s, Gastaut et al. recognized obstructive sleep apnea in obese subjects as intermittent airway obstruction with frequent arousals, thereby providing the first comprehensive links between obesity, sleep-induced airway obstruction, sleep fragmentation, and daytime sleepiness [4]. Following these key observations, research proceeded slowly with case reports of obstructive sleep apnea and the occasional use of chronic tracheostomy for treatment in the early 1970s [5].

Not until recently have scientists attempted to understand how and why we sleep. Currently, there are four major theories explaining why we sleep, which are adaptive or evolutionary, energy conservation, restorative, and brain plasticity theories. On the other hand, there have been over 100 diseases defined as sleep disorders.

Disturbance of sleep is one of the most detrimental components to our health. Sleep disorders are extremely common in the general population and can lead to significant morbidity. Sleep disturbances lasting at least several nights per month have been reported by 30% of the population [6]. Sleep disorders may cause or exacerbate preexisting medical and psychiatric conditions and are associated with high rates of depression, anxiety, and impaired daytime functioning [6, 7]. They may also lead to poor occupational performance, motor vehicle accidents, cardiovascular and endocrine disorders, or heightened pain perception [7, 8].

24.1.1 Obstructive Sleep Apnea

The International Classification of Sleep Disorders, third edition (ICSD-3) subdivides sleep disorders into six major categories: insomnia, sleep-related breathing disorders, central disorders of hypersomnolence, circadian rhythm sleep–wake disorders, parasomnias, and sleep-related movement disorders [9]. Sleep-related breathing disorder describes a group of disorders referring to momentary, often cyclical, cessations in breathing rhythm (apneas), or momentary or sustained reductions in the breath amplitude (hypopneas), sufficient to cause significant arterial hypoxemia and hypercapnia. These apneas and hypopneas are specific to the sleeping state and are accompanied by:

1. A compromised, often even completely closed, extrathoracic upper airway "obstructive" event.
2. A marked reduction or cessation of brain stem respiratory motor output "central" event.
3. A combination of central and obstructive events.

These ventilatory inadequacies and their accompanying intermittent hypoxemia often lead to transient arousals from sleep and sleep state fragmentation throughout the night and cause overcompensatory responses of the autonomic nervous system. Obstructive sleep apnea, the most common such disorder, is characterized by the repetitive collapse or partial collapse of the pharyngeal airway during sleep and the need to arouse to resume ventilation. Sleep is thus disrupted, yielding waking somnolence and diminished neurocognitive performance. The recurrent sleep arousal in association with intermittent hypoxia and hypercapnia has been implicated in the occurrence of adverse cardiovascular outcomes and insulin resistance. Despite considerable progress, most patients remain undiagnosed and the principal therapeutic approach, continuous positive airway pressure (CPAP), remains somewhat cumbersome and hence not associated with optimal compliance rates.

Obstructive sleep apnea is a highly prevalent disease, affecting approximately 20% of the adult population [6]. The major risk factors for OSA include obesity, male gender, postmenopausal status, age, anatomical factors, posture and gravity, genetic factors, and hypothyroidism (Table 24.1).

In the presence of an anatomically compromised, collapsible airway, the sleep-induced loss of compensatory tonic input to the upper airway dilator muscle motor neurons leads to collapse of the pharyngeal airway. In turn, the ability of the

Table 24.1 Mechanisms involved in the genesis of OSA

Obesity
High body mass index. Central or visceral obesity is quite important. Predisposing factors: Abdominal circumference >94 cm in men and >80 cm in women; neck circumference >40 cm
Gender
Prevalence is higher in men than in women. Women present greater genioglossus muscle tone, which can be considered a defense mechanism designed to maintain upper airway permeability
Hormones
Estrogen and progesterone promote the maintenance of upper airway permeability (by improving muscle tone), as well as increasing respiratory drive. The androgens induce greater fat deposition and relaxation of the pharyngeal dilator muscles. Menopause also increases the risk of OSA
Age
The activity of the upper airway musculature is decreased with aging
Anatomical factors
Micrognathia and hypoplasia of the mandible are associated with the posterior positioning of the base of the tongue and with upper airway narrowing. Thickening of the lateral pharyngeal walls also causes upper airway narrowing
Genetic factors
Some risk factors, such as craniofacial structure, distribution of body fat, neural control of the upper airways, and central respiratory command, can be inherited
Posture and gravity
The dorsal decubitus position promotes the posterior positioning of the tongue and soft palate, thereby reducing the area of the oropharynx
Other causes
Acromegaly, Down's syndrome, hypothyroidism, genetic syndromes, and deposition diseases (amyloidosis and mucopolysaccharidosis) can promote the narrowing of the upper airways, which are predisposing factors for OSA

sleeping subject to compensate for this airway obstruction will determine the degree of cycling of these events. Several of the classic neurotransmitters and a growing list of neuromodulators have now been identified that contribute to neurochemical regulation of pharyngeal motor neuron activity and airway patency.

Knowledge and understanding of the pathogenic basis, clinical presentation, and diagnosis of OSA are essential for the development of preventive, screening, and therapeutic strategies to reduce the public health burden of the disorder [7, 10].

24.1.2 Influences of Wakefulness on Ventilatory Control

Remarkably, sleep apnea patients experience little or no problems with their breathing or airway patency while awake. In fact, the great majority of people with sleep apnea possess ventilatory control systems that are capable of precise regulation of their alveolar ventilation and arterial blood gases with extremely small variations from the norm throughout the waking hours. In addition, these healthy control systems, while awake, possess sufficiently sensitive feedback and feed forward controls to ensure precise coordination of chest wall and upper airway "respiratory" muscle recruitment so as to provide maximum airway diameter, low airway resistance, and optimum lung volumes and respiratory muscle lengths, regardless of the ventilatory requirement.

To underscore the importance of the "waking stimuli" to breath and to upper airway patency and to ventilatory control, consider the following qualitative influences of sleep on the control of breathing. Electrical activity from medullary inspiratory neurons, EMG activity of diaphragm and abductor muscles of the upper airway in healthy humans and/or in cats, shows reductions in amplitude upon the transition from awake to NREM sleep, usually accompanied by a mild to moderate hypoventilation (2–8 mmHg $PaCO_2$) and two- to fivefold increases in upper airway resistance [11]. Sleep induces consistently greater proportional reductions in the EMG activity in the upper airway versus chest wall pump muscles [12].

$PaCO_2$ can be lowered substantially (using mechanical ventilation) during wakefulness with little or no disruption of breathing pattern; however, in NREM sleep, very small transient reductions in $PaCO_2$ (even only to the waking level) result in significant apnea [13].

Sleep-disordered breathing leading to repeated bouts of ventilatory overshoots and undershoots and accompanying swings in arterial blood gases and intrathoracic pressure takes on many forms. Commonly, sleep-disordered breathing is divided into the so-called central events, denoting an absence or marked reduction in central respiratory motor output to respiratory pump muscles, or "obstructive" events, which are composed of respiratory efforts against a closed upper airway. However, as we discuss below, most cyclical sleep-disordered breathing events are driven by anomalies in both anatomical and neurochemical control of upper airway and/or chest wall respiratory musculature [14].

In short, the process is initiated because the wakeful state provides compensatory neuronal activation of dilator muscles in an anatomically compromised collapsible pharynx; accordingly, when this activation is lost at sleep onset, the airway narrows and/or collapses. However, the tendency to result (or not to result) in repeated cyclical apneas is the end product of multiple compensatory processes that vary markedly among and within individuals. Concepts have continued to evolve as we learn more about the neurophysiological mechanisms governing control of respiratory rhythm and its coupling with upper airway control and states of consciousness and applying these principles to human patients during sleep.

We will discuss the pathophysiology of OSA in three steps. First we detail the varied structural and functional determinants of an anatomical predisposition for airway closure, an absolutely essential component for OSA. The second essential component is sleep. This section emphasizes the effects of the sleeping state on mechanisms underlying both obstructive and central apnea and ventilatory instability. Finally, we attempt to integrate anatomical deficits with mechanisms underlying central neurochemical control of breathing stability and compensatory neuromuscular control of upper airway caliber, to explain the repetitive nature of OSA.

24.1.2.1 Anatomical Determinants of Upper Airway in OSA

The pharyngeal airway is a complex structure that serves several purposes, including speech, swallowing, and respiration. The human pharynx is composed of more than 20 muscles and divided into four sections that include the nasopharynx (from the nasal turbinates to the start of the soft palate), velopharynx (from the start of the soft palate to the tip of the uvula), oropharynx (from the tip of the uvula to the tip of the epiglottis), and hypopharynx (from the tip of the epiglottis to the level of the vocal cords) (Fig. 24.1). The human pharynx can be considered as a collapsible tube that is uniquely susceptible to collapse due to the presence of a floating hyoid bone, a longer airway, and a less direct route for inspired air to travel when compared to other mammals. The presence of soft tissues and bony structures, which increase extraluminal tissue pressures surrounding the upper airway, can predispose the pharynx to collapse. In contrast, the actions of pharyngeal dilator muscles maintain pharyngeal patency due to reflex pathways from the central nervous system and within the pharynx. The presence of these opposing forces suggest that increased pharyngeal collapsibility is due to alterations in anatomically imposed mechanical loads and/or in dynamic neuromuscular responses to upper airway obstruction during sleep [15].

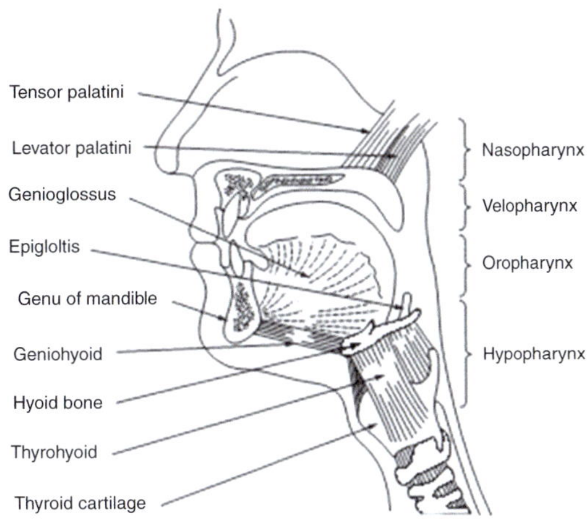

Fig. 24.1 Anatomy of the upper airway and the important muscles controlling the airway patency

Unique Anatomy of the Human Airway

The upper airway is a complex structure required to perform deglutition, vocalization, and respiration. In the human, this structure must also perform tightly controlled and complex motor behaviors required for speech. Upper airway obstruction in sleep is most prevalent in the human in part because the hyoid bone, a key anchoring site for pharyngeal dilator muscles is not rigidly attached to skeletal structures [16]. In other mammals, the hyoid bone is attached to the styloid processes of the skull. Thus, the human pharynx has no rigid support except at its extreme upper and lower ends where it is anchored to bone (upper) and cartilage (larynx); therefore, pharyngeal cross-sectional area will vary with lumen pressure. Humans depend critically on the coordinated actions and interactions of over 20 skeletal muscles that dilate and stent open the oropharynx.

Beyond the hyoid arch, it is also pointed to the anatomical changes in the adult human upper airway during the evolutionary development of speech as a potential major contributor to OSA [17]. Specifically, the gradual decent of the larynx to a position greatly inferior to the oropharynx separated the soft palate from the epiglottis in which the tongue encroaches significantly on the available space.

Sites of Airway Collapse

Studies using nasal pharyngoscopy, computed tomography and magnetic resonance imaging, or pharyngeal pressure monitoring have shown that one or more sites within the oral pharyngeal region are usually where closure occurs in most subjects with OSA, and this region is also smaller in OSA patients versus controls even during wakefulness [18]. Although the retropalatal region of the oropharynx is the most common site of collapse, airway narrowing is a dynamic process, varying markedly among and within subjects and often includes the retroglossal and hypopharyngeal areas [19].

Soft Tissue and Bony Structure Abnormalities

The recent use of quantitative imaging techniques has allowed advances that reveal important differences in both craniofacial and upper airway soft tissue structures in the OSA patient. The reduced size of cranial bony structures in the OSA patient include a reduced mandibular body length, inferior positioned hyoid bone, and retro position of the maxilla, all of which compromise the pharyngeal airspace [20]. Airway length, from the top of the hard palate to the base of the epiglottis, is also increased in OSA patients, perhaps reflecting the increased proportion of collapsible airway exposed to collapsing pressures [21]. As expected, these craniofacial dimensions are primarily inherited, as the relatives of OSA patients demonstrated retroposed and short mandibles and inferiorly placed hyoid bones, longer soft palates, wider uvulas, and higher narrower hard palates than matched controls [22].

Enlargement of soft tissue structures both within and surrounding the airway contributes significantly to pharyngeal airway narrowing in most cases of OSA. An enlarged soft palate and tongue would encroach on airway diameter in the anterior-posterior plane, while the thickened pharyngeal walls would encroach in the lateral plane. Volumetric time overlapped magnetic resonance imaging (MRI) or computer tomography (CT) images strongly implicate the thickness of the lateral pharyngeal walls as a major site of airway compromise, as the airway is narrowed primarily in the lateral dimension in the majority of OSA patients [22]. Furthermore, treatment with CPAP, weight loss, or mandibular advancement all show increases in the lateral pharyngeal dimensions [22]. There are many potential causes of lateral wall thickening in OSA patients. First, as shown in both humans and rodent models, obesity is a major contributor to airway compression through increased area and volume of pharyngeal fat deposits [23]. This excess fat deposition has also been observed under the mandible and within the tongue, soft palate, or uvula. Obesity also gives rise to excess fat-free muscular tissue, thereby increasing the size of many upper airway structures and compressing the lateral airway walls. In children with OSA, tonsillar and adenoid hypertrophy form the major anatomical contributors to airway narrowing [23].

Obesity and Lung Volume

Obesity also contributes indirectly to upper airway narrowing, especially in the hypotonic airway present during sleep, because lung volumes are markedly reduced by a combination of increased abdominal fat mass and the recumbent posture. In turn, the reduced lung volume reduces the "tug" on the trachea induced by the traction exerted via mediastinal structures by negative intrathoracic pressures and by the diaphragm descent, thereby further increasing the thickness of the lateral pharyngeal walls and narrowing the airway [24].

Airway Edema and Surface Tension

Surface tension of the liquid lining the mucosa affects collapsibility of the upper airway in the same way as it has been well documented in the lung's airways [25]. A higher surface tension in the upper airway wall of OSA patients has been reported using a method that quantifies surface tension as the force required to separate two surfaces bridged by a droplet of the liquid under study. Furthermore, in limited studies, surfactant therapy in OSA patients was shown to significantly reduce airway collapsibility and improve apnea–hypopnea index (AHI) by 20–30% [26].

Obesity, Leptin, and Inflammation

Central, or visceral, obesity is associated with the greatest risk for OSA [27]. This suggests that factors other than pure mechanical load may contribute to the pathogenesis of respiratory disturbances during sleep. The concept is now emerging that visceral fat depots, which represent a rich source of humoral mediators and inflammatory cytokines, can impact on neural pathways associated with respiratory control [28]. Perhaps the most well-studied adipocyte-derived factor affecting respiratory control is leptin, which was initially determined to have a primary role of binding to receptors in the hypothalamus to reduce satiety and increase metabolism. Leptin can also act as a respiratory stimulant, and impairment of the leptin signaling pathway, as occurs in leptin-resistant or leptin-deficient states of obesity, causes respiratory depression in mice, and is associated with obesity hypoventilation syndrome in humans [29]. Even though obesity and OSA are associated with elevated circulating levels of leptin, if centers in the brain impacting on respiratory control act in a similar leptin-resistant manner to hypothalamic regions controlling appetite and metabolism, then impaired leptin signaling in the CNS may contribute to respiratory depression as predicted in murine studies [19].

In addition to respiratory control, animal studies show leptin is also critical in lung development and affects the distribution of muscle fiber types in the diaphragm. However, as yet there is no direct evidence that impaired leptin signaling can impact on the control of respiratory muscles of the upper airway, although it may play a role in nocturnal hypoventilation, particularly in REM sleep where respiration is markedly depressed in leptin-deficient mice [19].

24.1.2.2 Mechanical Determinants of Upper Airway Patency

Mechanical determinants of airway caliber of the human pharynx in sleep are similar to those regulating caliber of any collapsible tube [18]. Other well-known biological examples in respiratory physiology include intrathoracic airway collapse upon forced exhalation, collapse of pulmonary capillaries in the lung apex, and collapse of alae nasi at high inspiratory flow rates. A Starling resistor model developed by Schwartz and colleagues consists of a collapsible tube with a sealed box interposed between two rigid segments [28, 30]. The critical closing pressure (P_{crit}) of the passive airway is defined as the pressure inside the airway (P_{in}) at which the airway collapses. The pressure gradient during airflow through the system is defined by $P_{upstream}$, P_{crit}, and remains independent of $P_{downstream}$. Therefore, with increasing P_{crit}, as the differential between $P_{upstream}$ and P_{crit} decreases, inspiratory airflow limitation will eventually develop, and when the P_{us} falls below P_{crit}, complete airway occlusion occurs (Fig. 24.2). Effective therapy for sleep apnea requires that the P_{us} to P_{crit} pressure differential be widened, and this can be accomplished by either:

1. Increasing in P_{us} with appropriate amounts of CPAP applied at the airway opening or
2. Decreasing P_{crit} via either reducing the collapsing pressures on the airway (e.g., weight loss or alteration of cranial-facial anatomy or increasing lung volume) or by augmenting "active" neuromuscular control of airway tone [30].

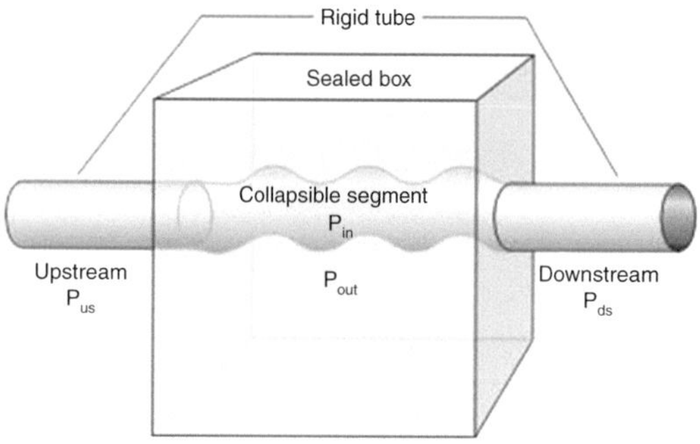

Fig. 24.2 Starling resistor model of obstructive sleep apnea. In the Starling resistor model, the collapsible segment of the tube is bound by an upstream and downstream segment with corresponding upstream pressure (P_{us}), downstream pressure (P_{ds}), and upstream resistance and downstream resistance. Airway occlusion occurs when the surrounding tissue pressure (P_{out}) (composed of pharyngeal muscles and pharyngeal and submucosal fat, mucosal edema, etc.) becomes greater than the intraluminal pressure (P_{in}), resulting in a transmural pressure of zero. In this model of the upper airway, P_{us} is atmospheric at the airway opening, and P_{ds} is the tracheal pressure. The critical closing pressure of the collapsible airway (P_{crit}) is represented by P_{in}. When the P_{crit} is significantly lower than P_{us} and P_{ds}, flow through the tube occurs. When P_{ds} falls during inspiration below P_{crit}, inspiratory airflow limitation occurs and is independent of further decreases in P_{ds}. Under this condition, the pharynx is in a state of partial collapse, and maximal inspiratory airflow varies linearly as a function of the difference between P_{us} and P_{crit}. Finally, when P_{us} falls below P_{crit}, the upper airway is completely occluded

24.1.2.3 Neuromuscular Control of Upper Airway Dynamics in Sleep

Clearly the effects of airway anatomy on airway collapsing pressure in a hypotonic airway are a critical determinant of obstructive apnea. However, several lines of evidence also support neuromuscular factors as significant determinants of airway collapsibility in sleep. First, tonic and phasic EMG activity of pharyngeal airway dilator muscles (genioglossus and tensor palatine) is progressively reduced from wakefulness to NREM to REM sleep and further inhibited coincident with the "phasic" eye movement events in REM [31]. This powerful effect of state has been adequately documented in tracheostomized animal models and recently has been demonstrated in OSA patients in whom the potentially confounding, compensatory responses to sleep-induced changes in upper airway resistance, negative pressure, $PaCO_2$, and respiratory motor output were controlled through the use of either CPAP or positive pressure controlled mechanical ventilation [32]. These state effects on the neuromuscular control of the upper airway likely explain, along with reductions in lung volume why P_{crit} is never positive in the waking state, even in OSA patients.

Second, neuromuscular factors also play a significant role in the dynamic breath-to-breath and intrabreath regulation of upper airway caliber, through changes in proprioceptive and chemoreceptor feedback. During inspiration, the passive pharynx narrows as intraluminal pressure is progressively reduced because of energy lost in overcoming frictional airway resistance and increases in flow velocity secondary to the Bernoulli effect operating in a reduced lumen size [18]. This collapsing effect of a reduced luminal pressure is opposed during inspiration by a reduction in dynamic compliance, i.e., collapsibility, of the airway achieved via reflex activation of pharyngeal dilator muscles. In turn, the reflex activation occurs in response to negative pressure airway mechanoreceptors located principally in the larynx and to a lesser extent in the superficial layers of the pharyngeal wall, with their afferent projections located in the superior laryngeal nerve, and also in glossopharyngeal and trigeminal nerves [11]. Large changes in negative pressure in the isolated upper airway trigger a dual protective reflex, which restores airway patency by both activating airway dilators (to reduce airway compliance) while inhibiting diaphragm EMG activity (which minimizes intraluminal negative pressure) (Fig. 24.3). Vagally mediated feedback influences on laryngeal, tongue, and hyoid muscle via pulmonary stretch receptors also protect against airway collapse as the rate of lung inflation is slowed in the face of increased airway resistance, thereby reflexly activating upper airway motor neurons.

Finally, chemoreceptor influences also have substantial effects on upper airway muscle recruitment, and in the case of CO_2, upper airway motor neurons relative to phrenic motor neurons have been shown to have a substantially higher

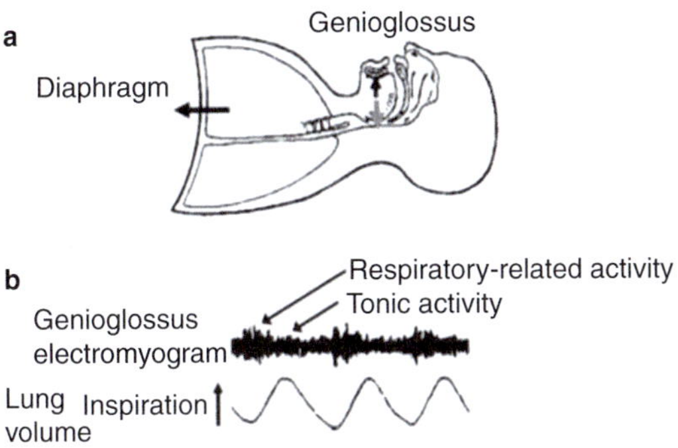

Fig. 24.3 (**a**) Contraction of the diaphragm and chest wall muscles leads to the generation of subatmospheric pressures in the thoracic cavity and airflow into the lungs. For effective lung ventilation, however, this airflow must pass through an open upper airspace. Activation of the pharyngeal muscles, such as the genioglossus muscle of the tongue, helps keep the upper airspace open for effective passage of air. Pharyngeal muscle activation acts to enlarge the airspace. Importantly, reduced pharyngeal muscle tone in sleep can reduce the size of the upper airspace and even promote complete airway obstruction. This tendency for airway collapse is exacerbated by increased weight of the tongue or neck (e.g., caused by obesity), an already anatomically narrow upper airway (e.g., caused by adenotonsillar hypertrophy), or the supine sleeping position. (**b**) The pharyngeal muscles exhibit respiratory-related activity superimposed upon a background of tonic activity. The background tonic muscle tone contributes to baseline airway size and stiffness. The increased pharyngeal muscle activity during inspiration enlarges and further stiffens the airspace to resist the subatmospheric collapsing pressures generated during inspiration

threshold for inhibition (via hypocapnia) and activation (via hypercapnia) [33].

In summary, the evidence to date supports important roles for both anatomical and neural control of dilator muscles to the regulation of upper airway caliber in the sleeping human. The relative contributions of these factors will vary widely among and within individuals with, for example, patterns of fat deposition on the one hand and neurochemical sensitivity for dilator muscle recruitment on the other.

24.1.3 Measurements of Pharyngeal Collapsibility

Quantitative measurements of mechanical and neuromuscular contributions to pharyngeal collapsibility have been difficult to derive during sleep. One approach has been to model the upper airway as a collapsible tube (i.e., a Starling resistor) as mentioned above. The relationship of pressure and flow through collapsible tubes has been well defined in the pulmonary and systemic circulation, the intrathoracic airways, and more recently the upper airway. In the Starling resistor model (Fig. 24.2), the collapsible segment of the tube is bound by an upstream and downstream segment with the corresponding upstream pressure (P_{us}), downstream pressure (P_{ds}), and upstream resistance and downstream resistance. Occlusion occurs when the surrounding pressure (P_{crit}) becomes greater than the intraluminal pressure, resulting in a transmural pressure of zero. In this model of the upper airway, P_{us} is atmospheric pressure at the airway opening and P_{ds} is the tracheal pressure. When the P_{crit} is significantly lower than the P_{us} and P_{ds} ($P_{us} > P_{ds} > P_{crit}$), flow through the tube follows the principles of an Ohmic resistor. When the P_{ds} falls during inspiration below P_{crit} ($P_{us} > P_{crit} > P_{ds}$), inspiratory airflow limitation occurs and is independent of further decreases in P_{ds}. Under this condition, the pharynx is in a state of partial collapse and maximal inspiratory airflow varies linearly as a function of the difference between the P_{us} and P_{crit}. Finally, when the P_{us} falls below P_{crit} ($P_{crit} > P_{us} > P_{ds}$), the upper airway is occluded.

Operationally, P_{crit} in the human upper airway is determined by lowering the nasal pressure until inspiratory airflow ceases. Measurements of P_{crit} have been shown to define a spectrum of upper airway obstruction from normal breathing (P_{crit} < −10 cm cmH_2O), to snoring (P_{crit} range, −10 to −5 cmH_2O), to obstructive hypopneas (P_{crit} range, −5 to 0 cm cmH_2O) and, finally, obstructive apneas (P_{crit} > 0 cmH_2O) during sleep [26]. Patients with the upper airway resistance syndrome (UARS), an entity characterized by flow-limited breathing that results in arousals, has been shown to have P_{crit} levels that are between snoring and hypopneas [34]. (Gold, Marcus) Depending on the methodology, measurements of P_{crit} reflect either the contributions of anatomically imposed mechanical loads on the upper airway or dynamic neuromuscular responses that maintain upper airway patency (Fig. 24.4).

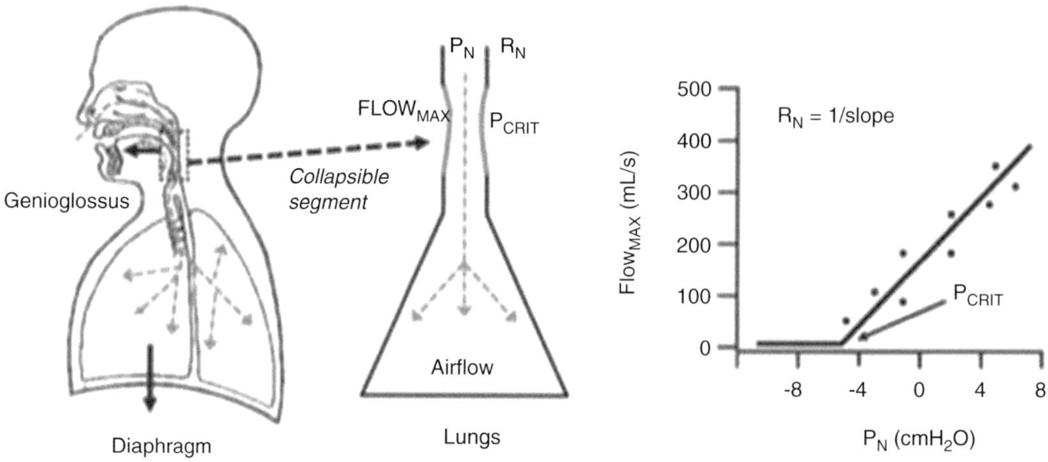

Fig. 24.4 Airflow into the lungs is generated by contraction of the primary respiratory muscles (e.g., diaphragm and intercostal muscles) and modulated by contraction of the secondary respiratory muscles (e.g., genioglossus muscle), which helps keep the upper airway open for effective airflow. The upper airway has been modeled as a collapsible tube where maximum airflow through the collapsible segment of the pharynx ($FLOW_{MAX}$) is determined by upstream nasal pressure (P_N) and resistance (R_N). Airflow ceases in the collapsible pharyngeal segment of the upper airway at a particular value of critical pressure (P_{CRIT}). Maximum airflow through the upper airway is determined by $(P_N - P_{CRIT})/R_N$. This relationship is linear, and experimentally induced decreases in P_N decrease $FLOW_{MAX}$ and ultimately elicit complete airway collapse

24.1.4 Contribution of Anatomic Factors to OSA

OSA is known to be associated with alterations in upper airway anatomy. Structural changes, including tonsillar hypertrophy, retrognathia, and variations in craniofacial structures, have been linked to an increased risk of sleep apnea, presumably by increasing upper airway collapsibility [17, 35, 36]. Ethnic differences in craniofacial features are one potential mechanistic explanation for observed differences in OSA prevalence and severity for a given level of obesity [35]. During wakefulness, CT and MRI studies have demonstrated increased fatty tissue deposition and submucosal edema in the lateral walls of the pharynx, both of which narrow the pharyngeal lumen and may predispose to obstruction during sleep [37].

Based on the presence of upper airway anatomic alterations in OSA patients, it has been proposed that structural or mechanical alterations are a primary determinant of upper airway obstruction during sleep [10]. Recent data suggest that structural alterations in the lateral pharyngeal walls and tongue aggregate on a familial basis, suggesting genetic susceptibility to OSA [38]. In addition, experimental data in the absence of neuromuscular activity demonstrate a reduction in maximal pharyngeal area and elevated P_{crit} in OSA subjects compared with normal subjects [30]. Furthermore, obesity, jaw position, acromegaly, tonsillar hypertrophy, and a smaller bony enclosure surrounding the pharynx have been demonstrated to predispose toward pharyngeal collapsibility [10]. These studies imply that upper airway structural differences distinguish OSA patients from normal subjects, and may predispose to upper airway obstruction when protective neuromuscular mechanisms wane at sleep onset [39].

Obesity, the major risk factor for OSA, has been linked with elevations in neck circumference and increased amounts of peripharyngeal fat, which could narrow and compress the upper airway [40]. Furthermore, increased parapharyngeal fat has been correlated with increased sleep apnea severity [27]. The compressive effects of fatty tissue deposited around the pharynx may increase upper airway collapsibility, and possibly offset the effects of dilator muscles that maintain airway patency. Obesity may also increase pharyngeal collapsibility through reductions in lung volumes, particularly decreases in functional residual capacity, which are accentuated with the onset of sleep. Reductions in function residual capacity may increase pharyngeal collapsibility through reductions in tracheal traction on the pharyngeal segment. Conversely, increases in lung volumes result in increased tracheal traction and stabilize the upper airway during inspiration [41]. In OSA patients, increases in lung volumes have been shown to decrease continuous positive airway pressure (CPAP) requirements and OSA severity, suggesting corresponding improvements in pharyngeal collapsibility [42, 43].

24.1.5 Contribution of Neuromuscular Factors to OSA

It should be noted, however, that anatomically imposed mechanical loads on the upper airway might not be sufficient to produce pharyngeal collapse during sleep. For example, women have been shown to have a smaller pharynx and oropharyngeal junction than men, despite having a lower prevalence of OSA [17]. Furthermore, measurements of P_{crit} under conditions of low neuromuscular activity, which reflect upper airway mechanical loads, demonstrate significant overlap between OSA and normal subjects [44]. Thus, nonstructural (i.e., neuromuscular) factors must also play a role in protecting the upper airway. In fact, changes in upper airway neuromuscular activity during sleep were originally described by Remmers et al., who demonstrated that genioglossal electromyogram (EMGGG) activity was reduced at apnea onset and increased with arousal when airway patency was restored [7, 45]. Subsequently, it was recognized that upper airway obstruction could trigger a variety of neuromuscular responses that restore upper airway patency by recruiting muscles that dilate and elongate the airway. Various pharyngeal muscle groups are important in stabilizing the upper airway throughout the respiratory cycle (tonic activity, e.g., tensor palatini) and in dilating the airway during inspiration (phasic activity, e.g., genioglossus). Pharyngeal motor output is modulated by a number of factors that include wake vs sleep state-dependent mechanisms, local mechanoreceptor responses to negative pressure, and ventilatory control mechanisms.

In OSA patients during wakefulness, elevated genioglossal and tensor palatini muscle activity have been observed and are significantly lowered with the application of positive nasal pressure [46]. In contrast, normal subjects had lower levels of genioglossal and tensor palatini muscle activity that were not further reduced when positive nasal pressure was applied. These observations suggested that increased upper airway dilator muscle activity compensates for a more anatomically narrow upper airway in the OSA patient. Thus, reductions in upper airway muscle activity with sleep onset through serotonergic, cholinergic, noradrenergic, and histaminergic pathways may lead to upper airway obstruction and have been hypothesized to be due to the loss of a "wakefulness stimulus" that may be greater in OSA patients than in healthy control subjects [46].

Pressure-sensing mechanisms play a prominent role in modulating upper airway neuromuscular activity during wakefulness and sleep. A negative pressure reflex within the upper airway serves to stabilize the upper airway during inspiration. At least three lines of evidence suggest that the negative pressure reflex is primarily mediated by mechanoreceptors within the pharynx. First, there is a tight relationship between EMGGG and pharyngeal pressure independent of the central respiratory pattern generator within the brainstem [47]. Second, topical anesthesia to the pharyngeal mucosa attenuates the relationship between genioglossal muscle activity and pharyngeal pressure with an increased number of obstructive apneas and hypopneas during sleep in normal subjects and loud snorers, and/or increased duration of apneic episodes [48]. Third, marked decreases in EMGGG activity with corresponding increases in pharyngeal collapsibility have been observed in patients when breathing through a tracheostomy compared to nasal breathing suggesting that negative pressure within the pharynx during inspiration stabilizes upper airway patency [11].

It is possible that OSA results from trauma to the upper airway due to repetitive collapsing and opening of the upper airway over time, resulting in muscle and neuronal fiber injury. Upper airway sensory pathways may be impaired in OSA patients because temperature, two-point discrimination, and vibratory thresholds are disrupted in OSA patients compared to normal individuals [6]. Sensory receptor dysfunction could attenuate the response of upper airway dilator muscles to the markedly negative airway pressures generated during periods of upper airway obstruction. Further evidence for sensorimotor dysfunction and an upper airway myopathy is provided by graded histopathologic and immunochemical alterations in the palatopharyngeus and muscularis uvulae in OSA patients, relative to asymptomatic snorers and normal subjects [10, 16]. The extent to which such mechanisms are important in the pathogenesis of OSA, however, remains to be established.

24.1.6 Contribution of Neuroventilatory Factors to OSA

Ventilatory control mechanisms may also play a role in modulating pharyngeal collapsibility during sleep. Preactivation of the pharyngeal dilator muscles stabilizes the upper airway prior to the inflow of air and suggests CNS coordination between the upper airway and diaphragm. The CNS is influenced by central and peripheral chemoreceptors with conditions of hypercapnia and hypoxemia increasing central drive to the upper airway and decreasing pharyngeal collapsibility [49]. Increased hypercapnic ventilatory responses, prolonged circulatory times, or low oxygen stores within the body can result in ventilatory instability that leads to the development of periodic breathing [50]. Sleep also unmasks a highly sensitive apneic threshold (the $PaCO_2$ level below which an apnea occurs) that remains within 1–2 mmHg of the normal waking eupneic $PaCO_2$ level. Therefore, a brisk ventilatory response as seen during an arousal in the susceptible individual can result in hypocapnia that is near or at the sleeping apneic threshold and result in a hypopnea or apnea with the reinitiation of sleep [11]. In fact, measurements of loop gain, a measure of ventilatory control instability, have been demonstrated to be high in patients with more severe OSA compared to patients with mild OSA [14, 51]. In contrast, during periods of ventilatory instability, individuals with low levels of mechanical loads on the upper airway appear to be resistant to the development of upper airway obstruction [52].

Nevertheless, whether alteration of loop gain is important in the pathogenesis of upper airway obstruction or is a consequence of OSA has not been established. OSA can develop in normal subjects with the application of negative nasal pressure to the upper airway (lowering the P_{us} near or below the P_{crit} level seen in normal subjects), which reduces the transmural pressure across the upper airway to near or below zero, and results in upper airway obstruction with recurrent obstructive hypopneas and/or apneas [53].

24.1.7 Interaction of Anatomic and Neuromuscular Factors on Pharyngeal Collapsibility

It is likely that a combination of upper airway mechanical loads and disturbances in neuromuscular mechanisms accounts for the pathogenesis of OSA. For example, in a group of OSA subjects, one third of the variability in OSA severity was ascribed to mechanical loads, suggesting that neuromuscular mechanisms accounted for the remaining two thirds [54]. Using techniques to partition the relative contribution of mechanical and neuromuscular factors toward pharyngeal collapsibility, it has been shown that OSA patients during sleep have both an increased mechanical load on the upper airway (passive P_{crit}) and impaired neuromuscular responses to upper airway obstruction (active P_{crit}). As demonstrated in Fig. 24.4, a P_{crit} of approximately -5 cmH_2O represents the disease threshold, above which obstructive hypopneas and apneas occur [55]. In normal subjects, when mechanical loads on the upper airway lowered the P_{crit} below the disease threshold, OSA is not present. In contrast, subgroups of normal subjects elevate mechanical loads that raised the P_{crit} above the disease threshold and placed them at risk of OSA, but are protected through the recruitment of neuromuscular responses that lowered the P_{crit} below the disease threshold. The development of OSA requires a "two-hit" defect, with defects in both upper airway mechanical and neuromuscular responses.

The summary of the above mentioned mechanisms are briefly shown in two schemes below (Fig. 24.5a, b).

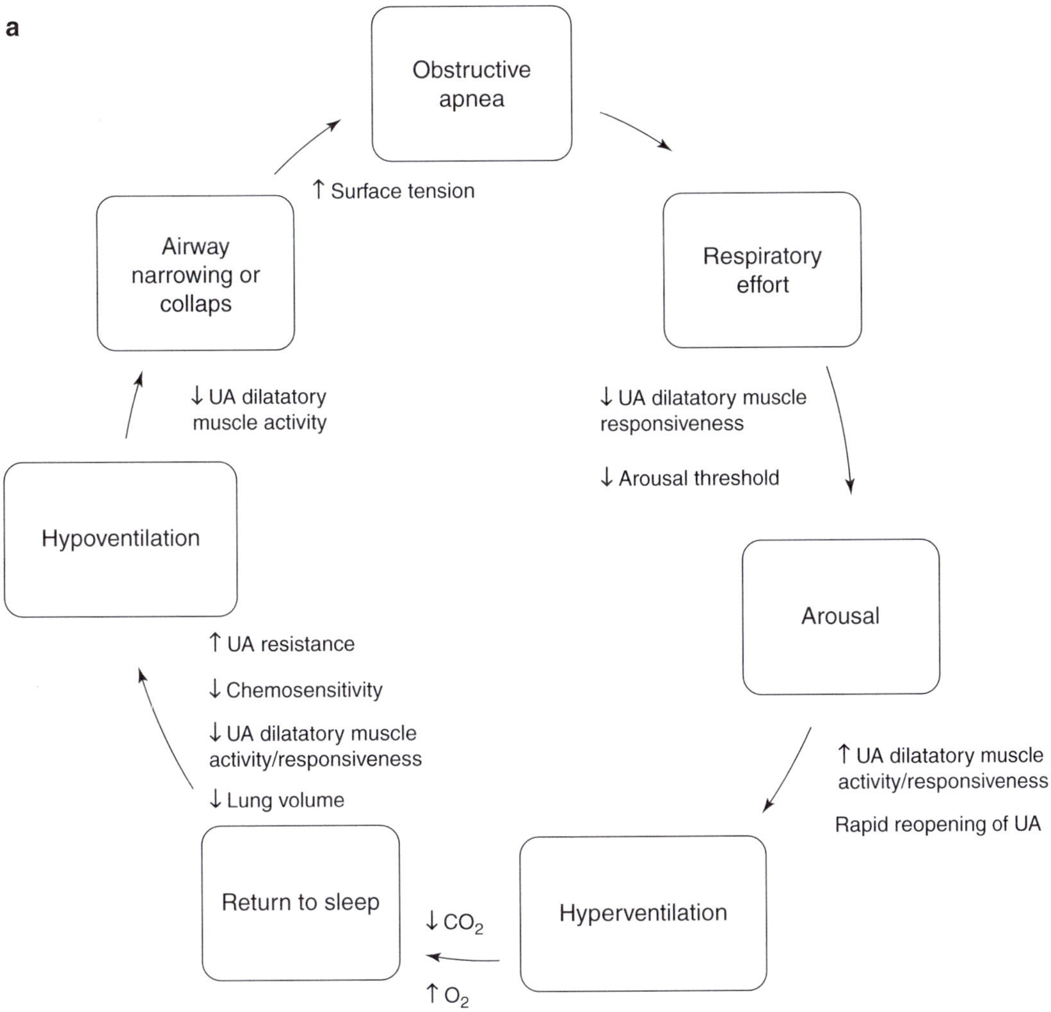

Fig. 24.5 (**a**) A scheme of the protective or beneficence (outside the circle) and unfavorable (inside the circle) physiologic process that are related to the pathophysiological sequences developing in OSA. (**b**) A stepwise scheme of the various pathophysiological events (inside the circle) predisposing or aggravating the clinical findings in the course of OSA. Some of these events are related to each other, such as increasing loop gain, may decrease the upper airway dilatory muscle activity, hence increasing the respiratory response to arousal and prone to cyclic respiration. Dotted lines determine the potential management approach to the current event. *UA* upper airway, *CPAP* continuous positive airway pressure, *ODD* oral dental device

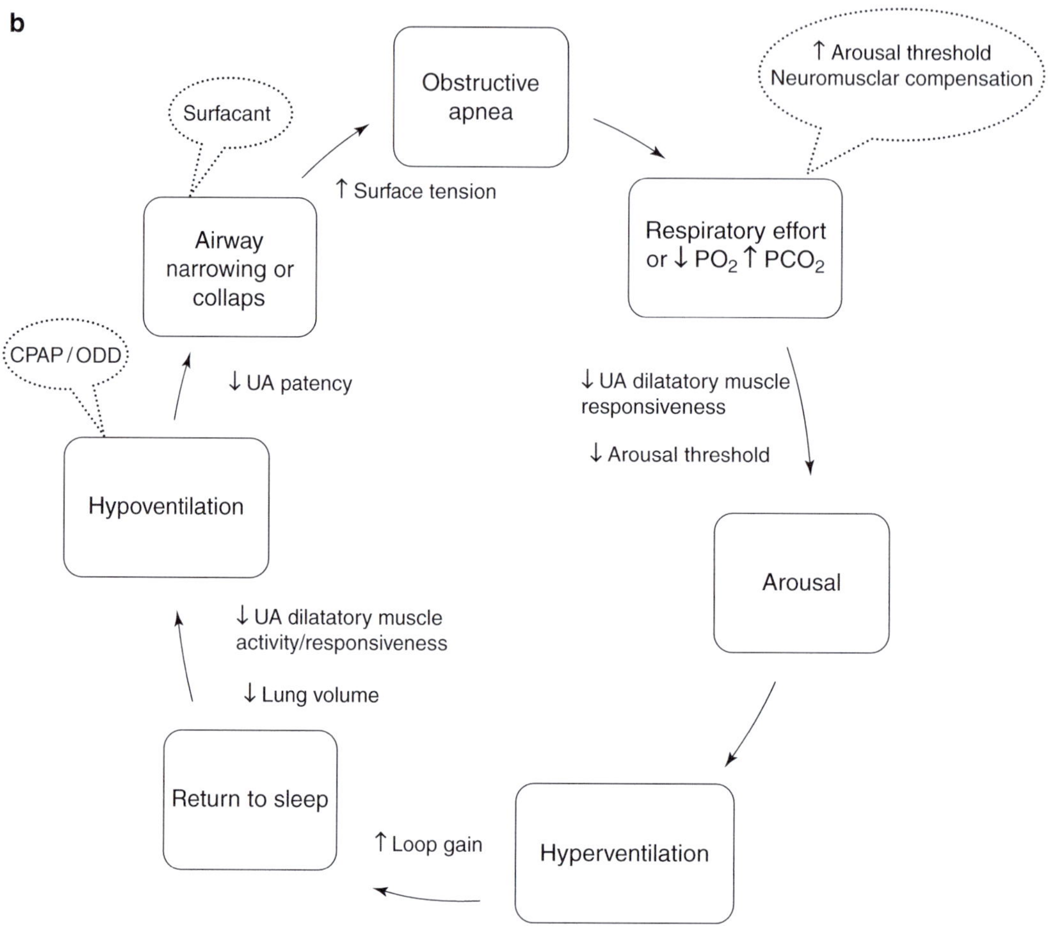

Fig. 24.5 (continued)

24.2 Conclusion

Obstructive sleep apnea is a common sleep disorder that can present in a variety of ways. With a fundamental understanding of the pathophysiology of OSA, the physician can routinely integrate questions into their review of system that will assist in appropriate referral for polysomnography and diagnosis of the disorder. Knowledge of common patterns of OSA may help identify patients and guide therapy.

The pharyngeal muscles are essential for effective lung ventilation because they help maintain an open upper airspace for the unhindered passage of air into the lungs. Sleep, especially rapid eye movement sleep, however, causes fundamental modifications of pharyngeal muscle tone and reflex responses that in normal individuals lead to airway narrowing and hypoventilation. In individuals with already anatomically narrow upper airways, these effects of sleep predispose them to inspiratory flow limitation (hypopneas), airway closure, and OSA. Obstructive sleep apnea is a common disorder, which is associated with serious clinical, social, and economic consequences.

The primary defect is probably an anatomically small or collapsible pharyngeal airway, in combination with a sleep-induced fall in upper airway muscle activity. Current understanding of the pathophysiologic basis of the disorder suggests that a balance of anatomically imposed mechanical loads and compensatory neuromuscular responses are important in maintaining upper airway patency during sleep. OSA develops in the presence of both elevated mechanical loads on the upper airway and defects in compensatory neuromuscular responses.

A sleep history and physical examination is important in identification of patients and appropriate referral for polysomnography. Understanding nuances in the spectrum of presenting complaints and polysomnography correlates are important for diagnostic and therapeutic approaches. Knowledge of common patterns of OSA may help identify patients and guide therapy.

Although obesity is a major risk factor for OSA, roughly 30% of patients with OSA are not obese, emphasizing the need for a high index of suspicion in clinical practice. Further, the prevalence of OSA is 2–3 times greater in men than in women and in older compared to middle-aged individuals. Menopause is a well-established risk factor for OSA in women. OSA can yield major neurocognitive manifestations, including excessive daytime sleepiness/fatigue, impaired cognition, reduced quality of life, and an up to sevenfold increased risk of road traffic accidents. Treatment of OSA leads to improvements in many of these outcome measures. There is evolving evidence to support the role of OSA as an independent risk factor for adverse cardiovascular sequelae. Although some argue that OSA was simply a marker of an unfit patient group, rigorous recent studies have shown that OSA is causally linked to a number of important sequelae. OSA is now a well-established risk factor for hypertension (both incident and prevalent), stroke and probably myocardial infarction, congestive heart failure, and death. OSA has been causally linked to the development of hypertension based on large rigorous cross-sectional and longitudinal epidemiological studies, mechanistic animal studies, and most recently interventional trials [56]. The underlying causes of OSA vary considerably between afflicted individuals. Important components likely include pharyngeal anatomy, pharyngeal dilator muscle responsiveness to respiratory challenges during sleep, the arousal threshold, and the instability of the negative feedback control system regulating ventilation (loop gain).

The pathophysiology of OSA is complex and incompletely understood. A narrowed upper airway is very common among OSA patients, and is usually in adults due to nonspecific factors such as fat deposition in the neck, or abnormal bony morphology of the upper airway. Functional impairment of the upper airway dilating muscles is particularly important in the development of OSA, and patients have a reduction both in tonic and phasic contraction of these muscles during sleep when compared to normals. A variety of defective respiratory control mechanisms are found in OSA, including impaired chemical drive, defective inspiratory load responses, and abnormal upper airway protective reflexes. These defects may play an important role in the abnormal upper airway muscle responses found among patients with OSA. Local upper airway reflexes mediated by surface receptors sensitive to intrapharyngeal pressure changes appear to be important in this respect.

A better understanding of the integrated pathophysiology of OSA should help in the proper management of this important multi-facet disorder and the development of new therapeutic techniques.

References

1. Hesiod, Cadwell RS. Hesiod's Thegony. Newburyport: Focus Classical Library; 1987. p. 52.
2. Rechtschaffen A. The single mindedness and isolation of dreams. Sleep. 1978;1:97–109.
3. Lavie P. Restless nights: understanding snoring and sleep Apnea. New Haven, CT: Yale Univercity Press; 2003.
4. Gastaut H, Tassinari CA, Duron B. Polygraphic study of the episodic diurnal and nocturnal (hypnic and respiratory) manifestations of the Pickwick syndrome. Brain Res. 1967;1:167–86.
5. Guilleminault C, Eldridge FL, Dement WC. Insomnia with sleep apnea: a new syndrome. Science. 1973;181:856–8.
6. Benjafield AV, Ayas NT, Eastwood PR, Heinzer R, Ip MSM, et al. Estimation of the global prevalence and burden of obstructive sleep apnoea: a literature-based analysis. Lancet Respir Med. 2019;7(8):687–98.
7. Pham LV, Schwartz AR. The pathogenesis of obstructive sleep apnea. J Thorac Dis. 2015;7:1348–72.
8. Hou H, Zhao Y, Yu W, Dong H, Xue X, Ding J, et al. Association of obstructive sleep apnea with hypertension: a systematic review and meta-analysis. J Glob Health. 2018;8:010405.

9. Sateia MJ. International classification of sleep disorders-third edition: highlights and modifications. Chest. 2014;146(5):1387–94.
10. Olaithe M, Bucks RS, Hillman DR, Eastwood PR. Cognitive deficits in obstructive sleep apnea: insights form a meta-review and comparison with deficits observed in COPD, İnsomnia, and sleep deprivation. Sleep Med Rev. 2018;38:39–49.
11. Eckert DJ. Phenotypic approaches to obstructive sleep apnoea—new pathways for targeted therapy. Sleep Med Rev. 2018;37:45–59.
12. Orem J, Lovering AT, Dunin-Barkowski W, Vidruk EH. Tonic activity in the respiratory system in wakefulness. Sleep. 2002;25:488–96.
13. Venner A, Todd WD, Fraigne J, Bowrey H, Eban-Rothschild A, et al. Newly identified sleep-wake and circadian circuits as potential therapeutic targets. Sleep. 2019;42:1–14.
14. Eckert DJ, Jordan AS, Merchia P. Central sleep apnea: pathophysiology and treatment. Chest. 2007;131:595–607.
15. Tepedino M, Illuzzi G, Laurenziello M, Perillo L, Taurino AM, et al. Craniofacial morphology in patients with obstructive sleep apnea: cephalometric evaluation. Braz J Otorhinolaryngol. 2020;18:921–9.
16. Bassiouny A, Nasr S, Mashaly M, Ayad E, Qotb M, et al. Electron microscopy study of peripheral nerves in the uvulae of snorers and obstructive sleep apnoea patients. J Laryngol Otol. 2009;123:203–7.
17. Baudouin R, Blumen M, Chaufton C, Chabolle F. Adult sleep apnea and tonsil hypertrophy: should pharyngoplasty be associated with tonsillectomy? Sleep Breath. 2019;23:917–23.
18. Schwab RJ, Kuna ST, Remmers JE. Anatomy and physiology of upper airway obstruction. In: Kryger MH, Roth J, Dement WC, editors. Principles and practice of sleep medicine. Philadelphia, PA: Saunders; 2005.
19. Lu D, Tan L, Wu Y, Cao C, Deng Z. Leptin and leptin receptor gene polymorphisms in obstructive sleep apnea: a HuGE review and meta-analysis. Sleep Breath. 2015;19:1073–8.
20. Flores-Mir C, Korayem M, Heo G, Withmans M, Major MP, et al. Craniofacial morphological characteristics in children with obstructive sleep apnea syndrome: a systematic review and meta-analysis. J Am Dent Assoc. 2013;144:269–77.
21. Ruehland WR, Rochford PD, Pierce RJ, Rhinder J, Jordan AS, et al. Genioglossus muscle responses to resistive loads in severe OSA patients and healthy control subjects. J Appl Physiol. 2019;127:1586–98.
22. Cammaroto G, Stringa LM, Iannella G, Meccariello G, Zhang H, et al. Manipulation of lateral pharyngeal wall muscles in sleep surgery: a review of the literature. Int J Environ Res Public Health. 2020;17:5315.
23. Brennick MJ, Pack AI, Ko K, Kim E, Pickup S, et al. Altered upper airway and soft tissue structures in the New Zealand obese mouse. Am J Respir Crit Care Med. 2009;179:158–69.
24. Winck AD, Heinzmann-Filho JP, Soares RB, da Silva JS, Woszezenki CT, et al. Effects of obesity on lung volume and capacity in children and adolescents: a systematic review. Rev Paul Pediatr. 2016;34:510–7.
25. Kirkness JP, Madronio M, Stavrinou R, Wheatley JR, Amis TC. Surface tension of upper airway mucosal lining liquid in obstructive sleep apnea/hypopnea syndrome. Sleep. 2005;28:457–63.
26. Berry RB, Randall M. Abnormal upper airway sensation in OSA – cause or effect? Sleep. 2005;28:542–4.
27. Turnbull CD, Wang SH, Manuel AR, Keenan BT, McIntyre AG, eta al. Relationships between MRI fat distributions and sleep apnea and obesity hypoventilation syndrome in very obese patients. Sleep Breath. 2018;22:673–81.
28. Schwartz AR, Patil SP, Laffan AM, Polotsky V, Schneider H, Smith PL. Obesity and obstructive sleep apnea: pathogenic mechanisms and therapeutic approaches. Proc Am Thorac Soc. 2008;5:185–92.
29. Caballero-Eraso C, Shin MK, Pho H, Kim LJ, Pichard LE, et al. Leptin acts in the carotid bodies to increase minute ventilation during wakefulness and sleep and augment the hypoxic ventilatory response. J Physiol. 2019;597:151–72.
30. Carberry JC, Jordan AS, White DP, Wellman A, Eckert DJ. Upper airway collapsibility (Pcrit) and pharyngeal dilator muscle activity are sleep stage dependent. Sleep. 2016;39:511–21.
31. Matsuda M, Ogawa T, Sitalaksmi RM, Miyashita M, Ito T, et al. Effect of mandibular position achieved using an oral appliance on genioglossus activity in healthy adults during sleep. Head Face Med. 2019;15:26–33.
32. Chen KS, Zu M, Zhang Z, Chang WC, Gaj T, et al. A hypothalamic switch for REM and non-REM sleep. Neuron. 2018;97:1168–76.
33. Trinder J, Jordan AS, Nicholas CL. Discharge properties of upper airway motor units during wakefulness and sleep. Prog Brain Res. 2014;212:59–75.
34. Gold AR, Marcus CL, Dipalo F, et al. Upper airway collapsibility during sleep in upper airway resistance syndrome. Chest. 2002;121:1531–40.
35. Perri RA, Kairaitis K, Cistulli P, Wheatley JR, Amis TC. Surface cephalometric and anthropometric variables is OSA patients: statistical models for the OSA phenotype. Sleep Breath. 2014;18:39–52.
36. Perri RA, Kairaitis K, Wheatley JR, Amis TC. Anthropometric and craniofacial sexual dimorphism in obstructive sleep apnea patient: is there male-female phenotypical convergence? J Sleep Res. 2015;24:82–91.
37. Walsh JH, Leigh MS, Paduch A, Maddison KJ, Philippe DL, et al. Evaluation of pharyngeal shape and size using anatomical optical coherence tomography in individuals with and without obstructive sleep apnoea. J Sleep Res. 2008;17:230–8.
38. Tanizawa L, Chin K. Genetic factors in sleep-disordered breathing. Respir Investig. 2018;56:111–9.
39. Fogel RB, Trinder J, White DP. The effect of sleep onset on upper airway muscle activity in patients

with sleep apnoea versus controls. J Physiol. 2005;564:549–62.
40. Tom C, Roy B, Vig R, Kang DW, Aysola RS, et al. Correlations between waist and neck circumferences and obstructive sleep apnea characteristics. Sleep Vigil. 2018;2:111–8.
41. Nieman GF, Satalin J, Kollisch-Singule M, Andrews P, Aiash H, et al. Physiology in medicine: understanding dynamic alveolar physiology to minimize ventilator-induced lung injury. J Appl Physiol. 2017;122:1516–22.
42. Heinzer RC, Stanchina ML, Malhotra A. Effect of increased lung volume on sleep disordered breathing in sleep apnoea patients. Thorax. 2006;61:435–9.
43. Heinzer RC, Stanchina ML, Malhotra A. Lung volume and continuous positive airway pressure requirements in obstructive sleep apnea. Am J Respir Crit Care Med. 2005;172:114–7.
44. Patil SP, Schneider H, Marx JJ. Neuromechanical control of upper airway patency during sleep. J Appl Physiol. 2007;102:547–56.
45. Remmers JE, deGroot WJ, Sauerland EK. Pathogenesis of upper airway occlusion during sleep. J Appl Physiol Respir Environ Exerc Physiol. 1978;44:931–8.
46. Zhao D, Li Y, Xian J, Qu Y, Zhang J, et al. Relationship of genioglossus muscle activation and severity of obstructive sleep apnea and hypopnea syndrome among Chinese patients. Acta Otolaryngol. 2016;136:819–25.
47. Kubin L. Neural control of the upper airway: respiratory and state-dependent mechanisms. Compr Physiol. 2016;15:1801–50.
48. Ehsan Z, Mahmoud M, Shott SR, Amin RS, Ishman SL. The effects of anesthesia and opioids on the upper airway: a systematic review. Laryngoscope. 2016;126:270–84.
49. Faizal WM, Ghazali NNN, Khor CY, Badruddin IA, Zainon MZ, et al. Computational fluid dynamics modeling of human upper airway: a review. Comput Methods Programs Biomed. 2020;296:105627.
50. Pal T, Dutta PK, Maka S. Modulation-demodulation hypothesis of periodic breathing in human respiration. Respir Physiol Neurobiol. 2018;252:28–37.
51. Orr JE, Malnotra A, Sands SA. Pathogenesis of central and complex sleep apnoea. Respirology. 2017;22:43–52.
52. Wellman A, Jordan AS, Malhotra A. Ventilatory control and airway anatomy in obstructive sleep apnea. Am J Respir Crit Care Med. 2004;170:1225–32.
53. Ceratl V, Silva H, Carvalho C, Costa-Pereira A, Azevedo I, et al. Model for prediction of pediatric OSA: proposal for a clinical decision rule. Laryngoscope. 2015;12:2823–7.
54. Younes M. Contributions of upper airway mechanics and control mechanisms to severity of obstructive apnea. Am J Respir Crit Care Med. 2003;168:645–58.
55. Lin H, Wang C, Zhang H, Ziong H, Li Z, et al. Threshold of the upper airway cross-section for hypopnea onset during sleep and its identification under waking condition. Respir Res. 2019;20:280–9.
56. Morgan TD, Remmers JE. Phylogeny and animal models: an uninhibited survey. In: Kushida CA, editor. Obstructive Sleep Apnea. New York: Informa Healthcare; 2007.

Rhinomanometry

25

Zeynep Onerci Altunay

Core Messages

- Rhinomanometry allows objective assessment of nasal resistance, the ratio of transnasal pressure over transnasal airflow measured during nasal respiration.
- Many aspects of the study of nasal physiology have been studied with the aid of rhinomanometry.
- Rhinomanometry assesses the overall effect of nasal airway dimension and shape on the passage of air through the nose.

25.1 Introduction

Rhinomanometry is the simultaneous measurement of airflow through the nose and pressure across the nose during breathing. Figure 25.1 shows plots of transnasal pressure and flow during respiration. As the patient inspires, the curves go downward showing a decrease in pressure and the corresponding movement of air in the direction of the lungs. As the patient changes to expiration, the curves move upward, corresponding to pressure increasing and causing the movement of air out of the nose. Dividing the maximum pressure reached during normal inspiration by the highest flow gives a nasal resistance value that correlates with the symptom of nasal obstruction in symptomatic patients. This objective test has been crucial in increasing understanding in many areas of nasal physiology. While the extent of its use varies in different parts of the world, it is still used in research and in clinical assessment of nasal function.

Resistance calculated at the maximum pressure and flow correlates with the symptom of nasal obstruction.

This chapter will provide a framework to put the role of rhinomanometry in context among the other tools used to objectively assess nasal function. The methods of rhinomanometry will be described. The research role that rhinomanometry has played in discoveries in nasal physiology will be covered, including the nasal cycle, changes with growth, posture, and exercise; changes to the downside of the nose when a patient is lying down; and resistance in the normal nose and the nose with disturbed breathing function. The chapter ends with a summary of the clinical applications for which rhinomanometry has been used.

Z. O. Altunay (✉)
Department of Otorhinolaryngology, Head and Neck Surgery, University of Health Sciences Haseki Training and Research Hospital, İstanbul, Turkey

Ö. Ö. Celebi, T. M. Önerci (eds.), *Nasal Physiology and Pathophysiology of Nasal Disorders*,
https://doi.org/10.1007/978-3-031-12386-3_25

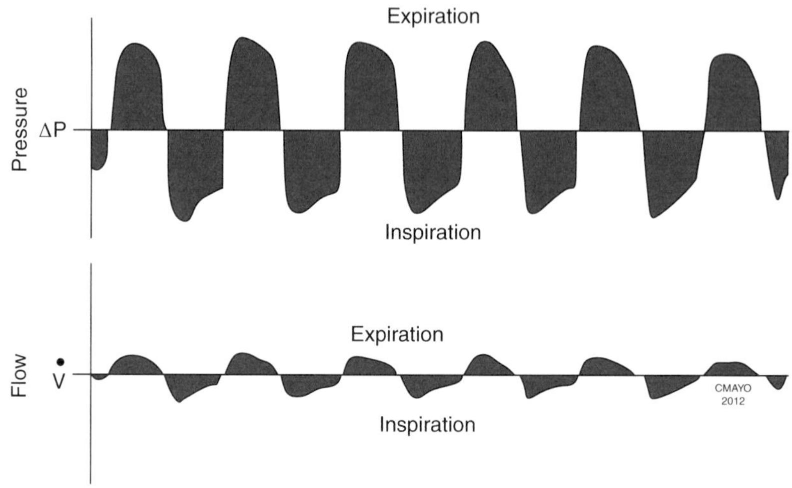

Fig. 25.1 The plot of pressure and flow during respiration. As the patient inspires, the curves go downward showing a decrease in pressure and the corresponding movement of air in the direction of the lungs. As the patient changes to expiration, the curves move upward corresponding to pressure increasing and the movement of air out of the nose

25.2 The Context of Rhinomanometry in Assessing Nasal Respiratory Function

25.2.1 What Are the Functions Associated with Nasal Respiration?

The movement of the diaphragm and lungs results in the movement of air through the nose. The passage of air through the nose is beneficial for the lungs because of the warming, humidification, and protective functions of the nose. The mucous layer in the nose can trap particulates, allowing the cilia to sweep them away. In addition, elements of the immune system are able to have contact with antigens, stimulating protective responses. Furthermore, air is delivered to the olfactory area providing the enhancement of taste and the protective function of detecting potentially harmful substances or organisms. The measurement of some of the important physiologic components of nasal respiration, warming, humidification, mucociliary clearance, and olfaction is covered in other chapters in this book as noted above.

As air passes through the nose, comfortable nasal breathing corresponds with a dimension and shape of the nasal airway that in general optimizes the above functions. Rhinomanometry assesses the overall effect of nasal airway dimension and shape on the passage of air through the nose.

> Comfortable nasal breathing corresponds with a dimension and shape of the nasal airway that in general optimizes the multiple functions of the nose.

25.2.2 What Anatomic Elements are Encountered During Nasal Respiration?

Measurement of the nasal airway assesses the nasal airstream which may be variably affected by the physical dimension of the components of the nasal airway, including the vestibule, valve area, turbinates, and sinus openings. The presentation of these anatomic elements as air is *inspired* through the nose is different than the shape of the presenting surfaces as air is *expired*. This may reflect different functions of the two phases.

The first part of inspiration is the passage of air through the vestibule in a curve first superiorly and then toward the nasal valve area. The airstream then passes through a narrower area referred to as the internal valve. This corresponds to the area under the caudal end of the upper lat-

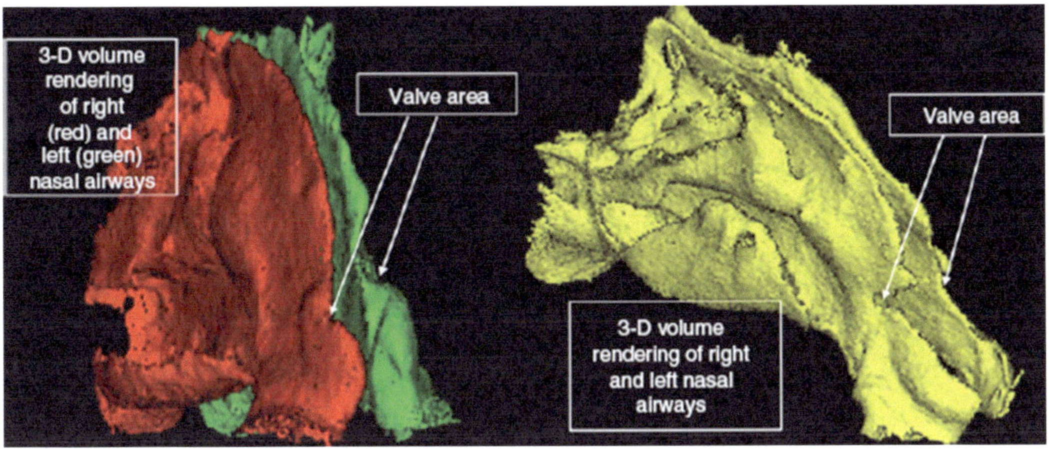

Fig. 25.2 Volume rendering of the nasal airway. Using 3D image analysis software and CT scans, the nasal airway is portrayed. This gives the same image early anatomists accomplished with wax castings. Note the narrow part of the airway at the valve area

eral cartilages and is clearly seen on casts or volume renderings of the nasal airway (Fig. 25.2). As the airstream leaves the valve area, it is dispersed in a wider distribution, thus providing contact with more surface area of the turbinates than if the airstream had passed unimpeded. The turbinates provide the working surface of the nose. By having protuberant curved surfaces, they act as flanges in the airstream providing increased surface areas on which to contact the air for humidification, warming, and filtering. The air then passes on to the choana and turns again, this time in the direction of the larynx and tracheobronchial tree.

As the inspired airstream passes the uncinate, the natural os of the maxillary sinus is protected from exposure to the passing airstream. Openings from the frontal sinus, anterior ethmoid cells, and posterior ethmoid cells are similarly sheltered by presenting baffles though maxillary sinus accessory openings and postsurgical openings can present additional openings to the inspired air. Computational fluid dynamics (CFD) studies have suggested only minimal, if any, effect of single maxillary sinus openings on the respiratory airstream in a typical (unoperated) nose, whereas the presence of an accessory os can result in some airflow into the maxillary sinus [1, 2].

When considering any possible impact of the nasal passage during expiration, retention of heat is usually mentioned. This may be facilitated by having the air pass over the turbinates before encountering the restriction of the valve area. The baffles that sheltered the ostia in the inspiratory direction now may function to catch the expired air. Some associate this phenomenon with nitric oxide from the maxillary sinus serving a regulatory function in respiration. Computational fluid dynamics (CFD; see below) in 3D models showed an increase in maxillary sinus airflow rate with high expiratory flow rates simulating nose blowing [2].

As the airstream leaves the valve area, it is dispersed in a broader distribution, thus providing contact with more surface area of the turbinates than if the airstream had passed unimpeded.

25.2.3 Objective Measurement of Nasal Respiratory Function

Of the various anatomic elements present, measurements of the airstream are primarily influenced by the dimension of the valve and turbinates, by the relative thickness of the mucosal lining, and, at times, by the respiratory effort of the patient.

The objective methods of assessing the nasal respiratory passages include (a) measurement of the dimension of the airway, (b) measurement of the nasal airflow alone, and (c) rhinomanometry, the simultaneous measurement of the transnasal pressure and airflow.

25.2.3.1 Measurement of the Dimension of the Airway

The assessments of the dimensions of the nasal airway are not measurements of the airflow through the nose during respiration. Imaging with CT or MRI and then doing a 3D reconstruction of the airway will demonstrate the airway dimensions in different parts of the nose. This can be helpful especially when used in conjunction with computational fluid dynamics (CFD).

Acoustic rhinometry also measures the airway dimension by calculations done on sound waves reflected back by intranasal structures. While this is not an assessment of the flow of air through the nose, it can be useful for measuring relative airway dimensions as well as changes with time, treatment, or various interventions [3].

Computational fluid dynamics (CFD) uses imaging, typically a CT scan done at one point in time, to generate a 3D model of the airway and then apply fluid dynamic modeling to that airway [4]. By using different transnasal pressures representing a respiratory cycle, the software can calculate the relative flow velocity in a number of anatomic sites in the nasal airway for various points in time during respiration. This capability offers exciting possibilities for the future study of the impact of various anatomic variations or pathology on the airstream. Those studying nasal physiology will be faced with the task of finding the meaning of the plethora of different flow vectors that result from CFD analysis of the nasal airway. To identify the meaningful parameters derived from the large amount of data is a tantalizing possibility for future study that will be facilitated by ever increasing computer processing speed and data handling.

25.2.3.2 Measurement of the Nasal Airflow Alone (Peak Flow Measurement)

Measuring only nasal airflow is popular especially with physicians who already use similar equipment to monitor their asthmatic patients by measuring peak lower airway flows. While this has the limitations of some dependence on patient effort, it has been relatively popular because of its simplicity and the ready availability of the equipment [5]. It has been demonstrated that physicians would like to have a simple tool for objective assessment of results in allergic rhinitis [6]. Since the rate of flow changes throughout the respiratory cycle, taking the measurement at some constant point can help decrease the variability of results and provide a standard for comparison. In this case, the "constant" point is the "peak" airflow reached with maximal effort. Peak nasal inspiratory flow can be measured by modifying the peak flow device for nasal inspiration. The measurements can be affected by valve collapse occurring at higher airflows that may not occur at normal physiologic flows [7]. Nonetheless, this method has been popular, and normative values have been collected [8–11].

Both peak nasal inspiratory flow (PNIF) and peak nasal expiratory flow (PNEF) measurements have been used. There is some debate about the variability of results [12, 13], but the tests have been shown to be useful, particularly for challenge testing in patients with allergic rhinitis [14].

25.2.3.3 Rhinomanometry: The Simultaneous Measurement of the Transnasal Pressure and Airflow

Collecting simultaneous pressure and flow values allows the calculation of nasal resistance or conductance. Calculation of the ratio of pressure to flow could be done at any one of many simultaneous pressure-flow values along the continuously changing curve during respiration (Fig. 25.3). Using a specific airflow value at which to measure the pressure-flow values is an

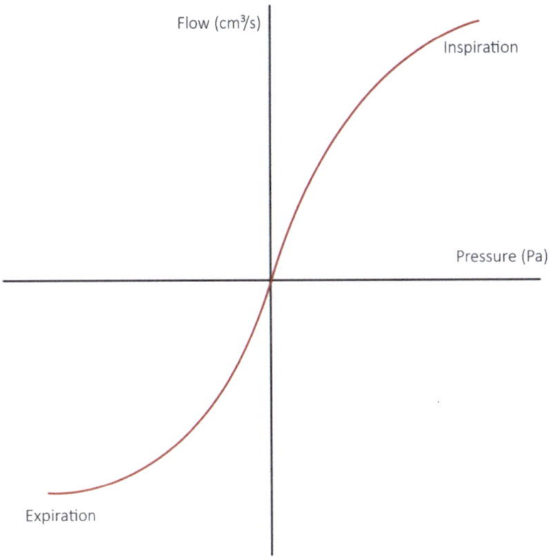

Fig. 25.3 The plot of pressure versus flow. Each point represents the simultaneous measurement of pressure and the corresponding flow value. Pressure values are on the *x*-axis and flow values on the *y*-axis. The sigmoid shape of the curve shows that in general there is a gradual increase in the pressure to flow ratio as one goes further out the curve toward the maximum values reached in normal respiration. Thus, for a given patient, the resistance value reported can vary depending on the point on the curve that is selected for calculating the result

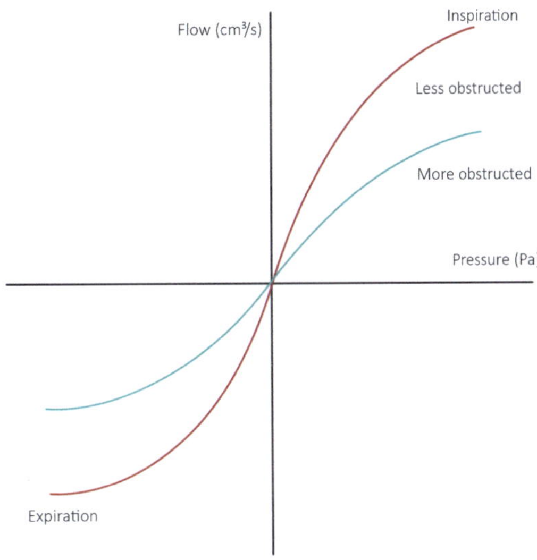

Fig. 25.4 The sigmoid pressure-flow curves for two different patients. The curve that is closer to the *x*-axis (pressure) represents the more obstructed nasal airway with higher resistance values

important element allowing consistent comparisons. Viewing the entire sigmoid pressure-flow curve also allows the observation of the position and amount of curvature that reflects the amount of flow the patient is generating for the range of pressures occurring in the course of their nasal breathing. Rhinomanometry is used (except in rare studies) to assess the pressure and flow across the entire nasal airway, from nasal entrance to nasopharynx.

The use of CFD analyses done from CT images inspires thoughts of using microsensors to unobtrusively detect the pressure and flow changes during respiration for multiple sites in the nasal airway. Just as Lindemann [15] had actual measurements using tiny thermocouples to validate the corresponding CFD calculations they did for temperature at many sites in the nose, multiple localized pressure and flow measurements could verify the results of CFD analyses that yield multiple differing flow vectors at different anatomic sites in the nasal airway. Such validation of CFD, if combined with actual pressure measurements for a given patient, could move it out of the category of assessing airway dimensions to the category of yielding measured information about nasal airflow.

> The plot of pressure and flow during inspiration and expiration yields a sigmoid curve that is closer to the *x*-axis (pressure axis) when nasal obstruction is greater.

25.2.4 Rhinomanometry for Measurement of Nasal Respiratory Function

As noted in the introduction, when rhinomanometry is performed, continuous measurement of transnasal pressure shows a rising and falling curve in the positive and then negative direction throughout each respiratory cycle (Fig. 25.1). As the changing pressure drives an accelerating and then decelerating flow of air, a plot of airflow shows a similar positive and negative excursion. Plotting pressure (*x*-axis) versus flow (*y*-axis) during inspiration and expiration yields a sigmoid curve that is closer to the *x*-axis when obstruction is greater (Figs. 25.3 and 25.4). Vogt

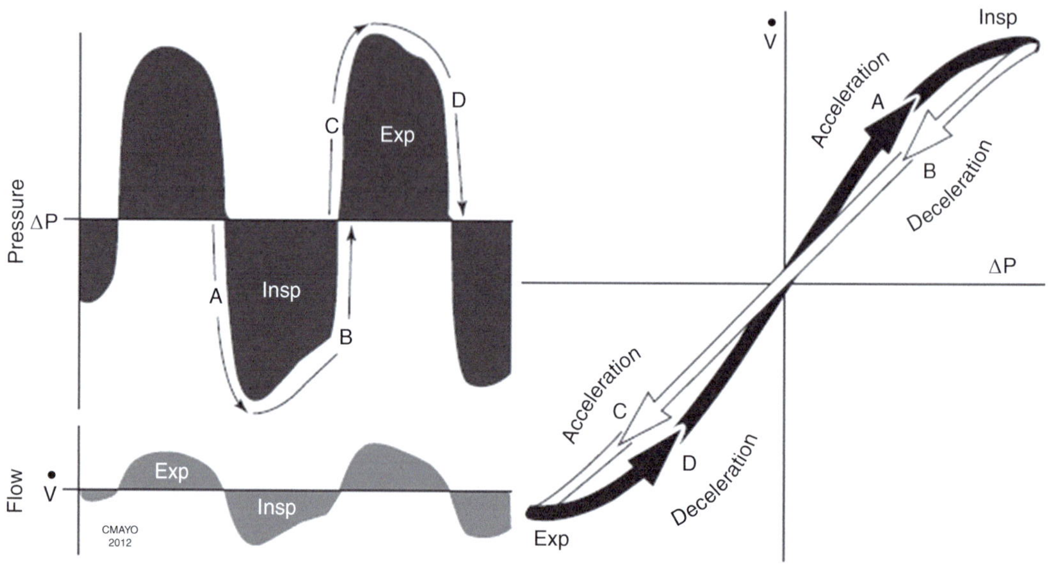

Fig. 25.5 The path of the pressure-flow curve away from the origin during inspiration (the accelerating limb, *A*) does not follow the same curve on the path back to the origin (decelerating phase, *B*). The same is true for the expiratory limb (*C*, *D*)

pointed out that the path of the pressure-flow curve away from the origin during inspiration (the accelerating limb) often does not follow the exact same curve on the path back to the origin (decelerating phase). The same is true for the expiratory limb (Fig. 25.5).

25.3 How Is the Measurement Done with Rhinomanometry?

25.3.1 Different Techniques: Most Common Method

The most commonly employed method of doing rhinomanometry is called anterior masked rhinomanometry. The different methods of rhinomanometry are distinguished by the location of the pressure detection and the apparatus for flow measurement. Table 25.1 lists the different types (methods) of rhinomanometry and the methods of pressure and flow detection that define them.

For measurement of the nasal airway of a patient with a nasal septal perforation, only the total airway is measured because the septal perforation would cause an error in unilateral pressure assessment. Anterior rhinomanometry thus is not used in patients with nasal septal perforations. Since the total airway is not measured when the anterior method is used, it is necessary to derive the total airway values by adding the right and left flow values for each corresponding pressure value along the pressure-flow curve (Fig. 25.6).

25.3.2 Transnasal Flow Measurement: Anterior or Posterior Method

Flow through the nasal airway is most commonly measured by attaching a flowmeter at the outlet of the mask which is sealed tightly on the patient's face. The usual flowmeter consists of pressure detection on either side of a resistive element. Originally nozzles were used to measure the flow through each nostril. When using a mask (or body plethysmograph), unilateral measurements can be done by occluding the opposite nostril with tape.

Since the total airway is not measured when the anterior method is used, it is necessary to derive

Table 25.1 The different types of rhinomanometry

Type of rhinomanometry	Flow detection	Pressure detection	Side(s) that can be directly measured
Anterior masked with full face mask	Device on outlet of full face mask	Catheter with sealed connection to non-measured nostril	Unilateral
Anterior masked with partial face mask	Device on outlet of partial face mask	Catheter with sealed connection to non-measured nostril	Unilateral
Anterior with nozzle	Device connected to nozzle held to nostril opening	Nozzle held to non-measured nostril	Unilateral
Posterior with full face mask	Device on outlet of full face mask	Catheter by nasopharynx—either transoral or transnasal	Total or unilateral
Posterior with partial face mask	Device on outlet of partial face mask	Catheter by nasopharynx—either transoral or transnasal	Total or unilateral
Body plethysmograph	Movement of chest inside body plethysmograph	Posterior catheter by nasopharynx—either transoral or transnasal or anterior catheter to non-measured nostril	Total or unilateral

Anterior masked rhinomanometry is the type most commonly employed

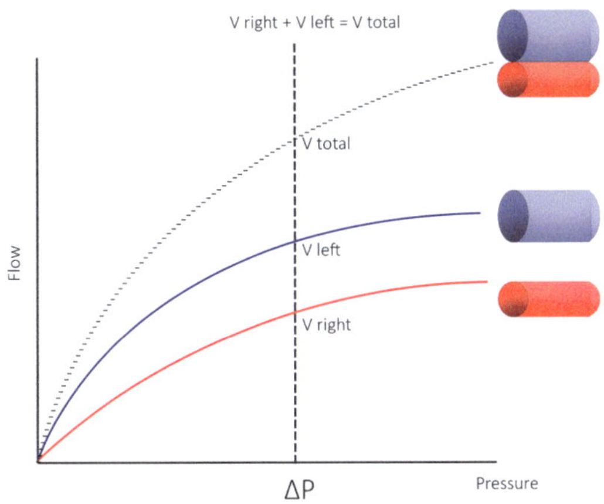

Fig. 25.6 For a given pressure value, the total flow is equal to the right-sided flow plus the left-sided flow, both measured at that pressure value. Flows are only additive if measured at the same pressure. This is analogous to two electrical currents being additive at the same voltages

the total airway values by adding the right and left flow values for each corresponding pressure value along the pressure-flow curve (Fig. 25.6).

25.3.3 Transnasal Pressure Measurement for Posterior Rhinomanometry

Measurement of transnasal pressure requires pressure detection in two sites, outside the nose and in the nasopharynx. Measurement of pressure outside the nose is easily done when the patient is wearing a mask by measuring the pressure inside the mask. Measurement in the nasopharynx can be done in several ways. As shown in Table 25.1, in anterior rhinomanometry, the nasopharyngeal pressure is detected using a tube sealed over the opposite nostril, turning the unmeasured nasal passage into an extension of the tube (Fig. 25.7). In posterior rhinomanometry, the nasopharyngeal pressure is measured by a catheter that is held in the back of the oropharynx with the lips sealed or by a tube passed to the nasopharynx along the floor of the nose (Fig. 25.8). The first of these methods can take extra time to learn for some patients.

It is also possible to measure a segment of transnasal pressure using a double catheter with the two openings on each side of the segment to be measured or by passing a catheter only partially along the floor of the nose. This methodology has only been employed in research but could potentially assess resistance at particular areas of the anatomic dimension of the airway, e.g., at a site suspected to be causing the symptom of nasal obstruction. This calculation uses the approximation of assuming a constant flow along the length of the nasal airway. Haight and Cole passed a catheter progressively further along the nasal airway while measuring the pressure at its tip and found that the greatest change in pressure and resistance occurred at the nasal valve [16].

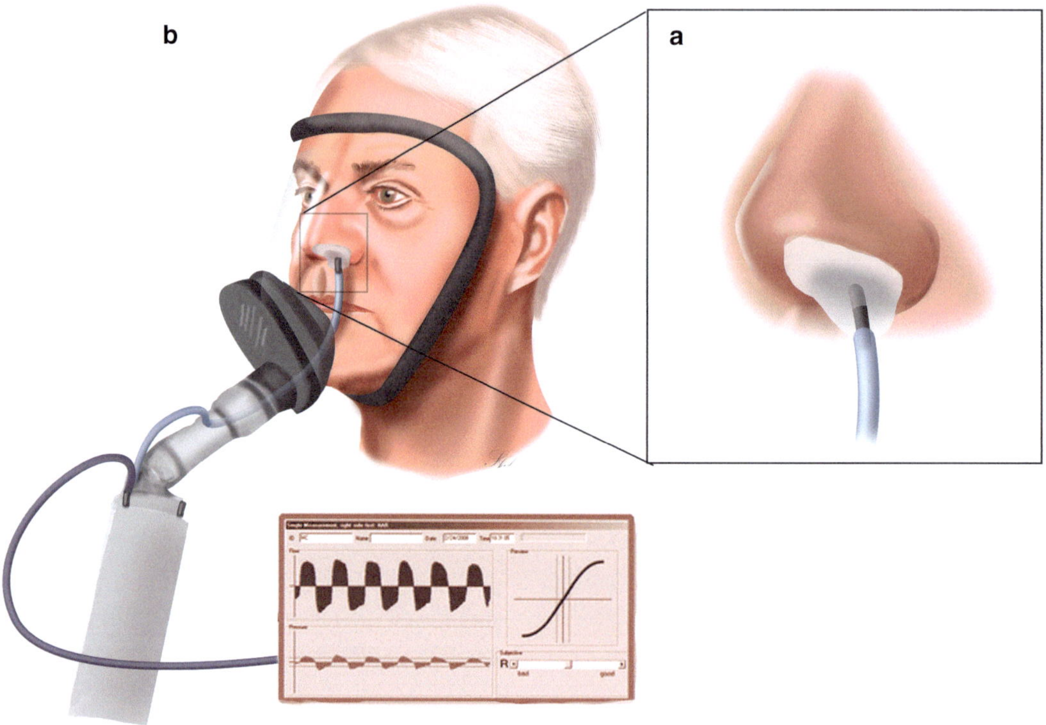

Fig. 25.7 Depiction of anterior masked rhinomanometry. A full face mask (**b**) is being used to avoid any distortion by the nasal alae. The pressure detection for the nasopharyngeal pressure is done with the tube that is sealed to the left nostril (**a**)

25.3.4 Nasal Resistance or Conductance

Resistance at a given point during the cycle of pressure and flow values can be obtained by dividing pressure by the corresponding flow at that point. Conductance is used by some and is the ratio of flow over pressure, the inverse of resistance. Typically resistance (or conductance) values are taken from inspiration, though some devices also report expiratory values.

Since rhinomanometry measures the simultaneous flow and pressure for the entire length of the nasal airway, it is generally thought that it primarily reflects the minimal effective cross-sectional airway. Figure 25.9 shows an example in which the cross-sectional area of an airway is smallest posteriorly rather than in the valve area. In this example, the right valve area has a smaller cross section than the left valve area, but the cross sections further posteriorly are smaller still with the left being the least. In this patient, the left side, which had the smallest overall cross section, is the same side that has the higher measured resistance and the same side where the patient felt the greatest obstruction.

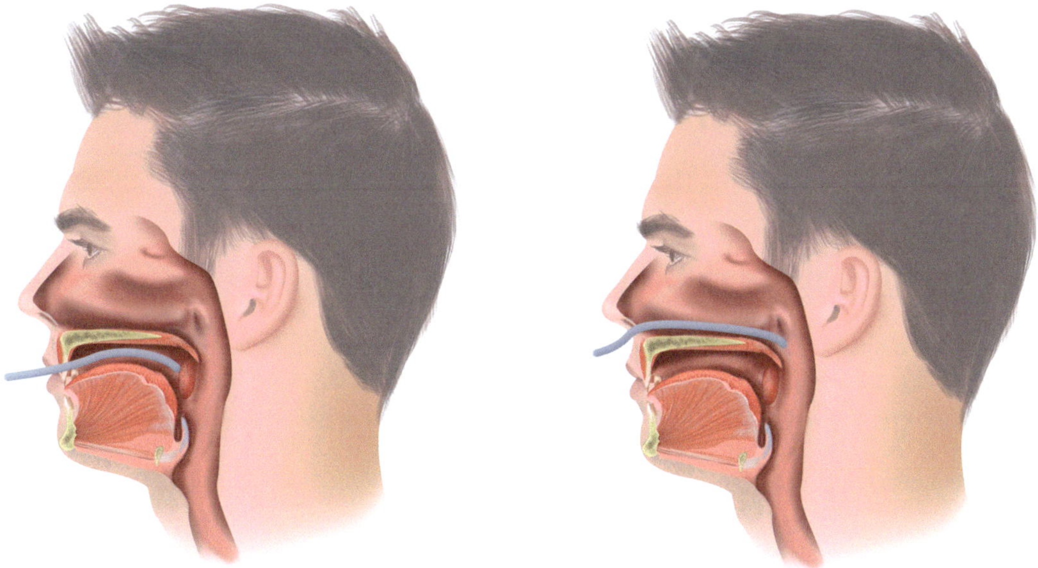

Fig. 25.8 Two methods of measuring the nasopharyngeal pressure in posterior rhinomanometry. The figure on the *left* shows the pressure detection tube being held in the oropharynx with the lips sealed (*A*) and the patient holding the soft palate open (*B*). The figure on the *right* shows the pressure catheter (*C*) passing along the floor of one of the nasal passages back to the nasopharynx. The small dimension of the tube is considered to have negligible effect on the airflow measurement on that side

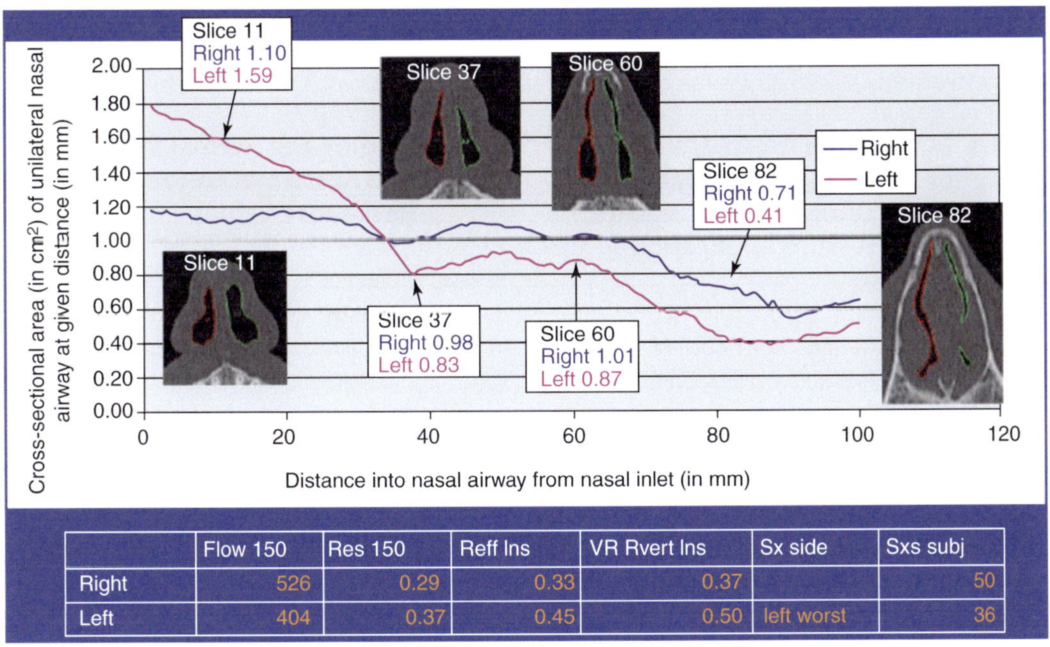

	Flow 150	Res 150	Reff Ins	VR Rvert Ins	Sx side	Sxs subj
Right	526	0.29	0.33	0.37		50
Left	404	0.37	0.45	0.50	left worst	36

Fig. 25.9 The plot of the cross-sectional area of right (blue) and left (pink) nasal airway as one goes further back (along the *x*-axis in mm) in the nasal airway. A 3D reconstruction was done from high-resolution CT scans, and successive cross-sectional areas were calculated perpendicular to the center vector of airflow through the nasal airway. Note that at the valve area (30 mm in) the right-sided cross-sectional area is smaller, but that (at 80–90 mm) the smallest overall cross-sectional area occurs posteriorly on the opposite (left) side of the nose

25.4 Rhinomanometry Has Been Instrumental in Understanding Elements of Nasal Physiology

25.4.1 Measuring Changes That Occur in the Passage of Air Through the Nose with Growth and with Age

Children have smaller nasal passages and thus higher average nasal resistance. Nasal resistance has been shown to decrease as children grow to adulthood. Interestingly, Thulesius found nasal resistance to decrease as adults aged [17].

25.4.2 Measuring the Nasal Cycle

Unilateral nasal resistance measurements have been used to document the periodicity of the nasal cycle. One side of the nose is put at rest as the other is open and doing the work of humidifying, warming, and filtering the air. In some patients, it was found to be fairly regular, and in others, it was shown to be rather irregular [18, 19].

25.4.3 Discovering the Cause of Downside Obstruction When Lying on One's Side (or with Pressure Application in Yoga)

When asked why the downside of the nose becomes more obstructed when lying on one's side, many will say it is due to "gravity." Rhinomanometry was used to demonstrate that this is not the case. Haight [20, 21] mapped the pressure receptors on the side of the body that when activated cause relative congestion of the tissues on that side of the nose. This phenomenon is also known to Yoga practitioners who apply pressure with a hand placed in the axilla to enhance the breathing through the opposite nostril.

In the nasal cycle, one side of the nose is put at rest as the other is open and is doing the work of humidifying, warming, and filtering the air.

25.4.4 Quantitating Airway Change with Recumbency

Hasagawa has used rhinomanometry to demonstrate the significant increase in nasal resistance that can occur with recumbency [22]. Just as our cardiovascular system has to make appreciable adjustments to maintain the same blood flow to our brain and extremities when we change to recumbency, the same regulatory parasympathetic/sympathetic pathways affect the relative congestion of the nasal tissues, particularly in certain individuals, resulting in increased nasal resistance and obstruction in the recumbent position.

25.4.5 Assessing Nasal Airway Change with Exercise and CO_2

Studies using rhinomanometry have shown the opening of the nose with exercise [23]. Measurements of nasal resistance revealed the increase in nasal obstruction occurring as increased amounts of CO_2 is delivered in the inspired air [24].

25.4.6 Finding the Normal Range and Abnormal Range of Nasal Resistance Values

If nasal resistance is measured in a standardized fashion for a large group of people, it is possible to show the distribution of "normal" resistance values for that population. This has been done for the sides of the nose as well as the total nose. By then comparing the nasal resistance of a patient against this distribution of normal values, one can determine if the patient has nasal resistance that is far outside the normal range [25].

25.4.7 Measuring Disturbance in Nasal Respiratory Function

By measuring a large group of patients who complained of the symptom of nasal obstruction, it was possible to describe the range of resistance values that are "abnormal" [26]. The significance of an abnormal unilateral resistance value must be considered in the light of the variation that occurs with the nasal cycle in the non-decongested nasal airway. Measuring the unilateral nasal airway after thorough decongestion can eliminate a major portion of the contribution of the nasal cycle in many individuals, but it will also change the overall range of "normal" and "abnormal" values to lower resistance ranges [25]. The total resistance of the nasal airway is relatively constant [22] through the course of the nasal cycle in the non-decongested nose. Some investigators have therefore suggested the use of total resistance as a value to measure the degree of nasal obstruction.

Pressure receptors on the downside of the body cause the downside nasal airway to have higher resistance.

25.4.7.1 When Is Disturbance in the Nasal Airstream Significant?

If an abnormal value of nasal resistance is measured, is this always of significance? By "significance" in patients, we usually mean that they are experiencing a symptom or condition that warrants treatment. Like an abnormal audiogram, it is the patient's choice as to whether any condition confirmed or found by a test is treated. Like any test, it is possible to have an abnormal result, but for a patient not to feel that they have sufficient symptoms to be treated.

25.4.7.2 Studying the Correlation of Elevated Resistance with the Symptom of Nasal Obstruction

There continues to be active debate about whether objective measurements of the nasal airway correlate with the symptom of nasal obstruction [27–33]. There has also been interesting work about the sensation of nasal obstruction being related to cold receptors that are stimulated by menthol-like compounds [34]. If there is more resistance to airflow, then is it the narrower airway causing less flow and thus less cold receptor stimulation that causes the sensation of obstruction?

Elevated values of nasal resistance have been shown to correlate with the symptom of nasal obstruction [26, 35, 36]. Several studies have looked at which parameter derived from the pressure-flow curve data obtained by rhinomanometry would best correlate with symptoms. Two studies [35, 37] found the maximal resistance during normal respiration to be a parameter that correlated with symptoms better than other parameters. Phillip Cole (personal communication) explained this best, noting that the greatest time during the respiratory cycle (Fig. 25.1) was spent at the extremes of the pressure and flow curves; thus, it would follow that a parameter from this location would have the greatest correlation with patient's symptoms.

In general, recumbency increases nasal resistance.

The variability of the nasal cycle and "subjective" symptoms introduces some noise in demonstrating this correlation. It is most easily shown for larger values of unilateral obstruction and in patients who are experiencing symptoms (as opposed to studies on patients who had no symptoms of nasal obstruction). When studies have been done looking for a correlation with the sensation of obstruction in subjects who are not experiencing obstruction, there is more "noise" (variation) making the correlation less clear [38]. Most subjects with nasal obstruction are able to distinguish the side with the higher resistance and to give a grading of their obstruction that correlates with other patients who are experiencing obstruction of their nose [35]. This ability to perceive the side of the highest resistance has been quantitated and found to be best when there is more than a slight difference in resistance between the sides of the nose at the time of the test [39].

25.4.7.3 Providing Objective Assessment When Crusting and Dysfunction of Nasal Lining Occur Due to Disturbance in the Airstream

When considering the symptoms of nasal obstruction, the question arises as to whether a patient can have a nasal airway that is too open and a corresponding measure of nasal resistance that is too low.

While this is not a common scenario in the measurement of nasal resistance, patients with noses that appear widely patent, dry, and crusty can be shown to have lower resistance. This would suggest that a surgeon's goal of lowering resistance when treating the nasal airway needs to be tempered in this case by maintaining the normal physiologic range of nasal resistance for the unilateral and total nasal airways. This is consistent with the avoidance of disrupting nasal physiology by such procedures as the total removal of turbinate tissues.

Another interesting application of rhinomanometry that can be applied in this context is the measurement of nasal resistance in a patient who complains of symptoms suggesting the type of nasal dysfunction found in patients with the "empty nose syndrome", but in whom the exam looks reasonable. Normal measured nasal resistance in this context would support looking for other explanations for the patients' symptoms.

25.4.8 Studying the Airflow in Conditions of Varying Temperature and Humidity

Rhinomanometry has shown that nasal resistance increases when a patient breathes colder than normal air [23].

By measuring nasal resistance, studies have looked for whether breathing air of different humidities resulted in any change in amount of nasal obstruction. Ivarsson and Malm found no significant difference in breathing air of different percent humidities [40].

Exercise resulting in a higher pulse rate decreases nasal resistance.

25.5 Clinical Applications of Rhinomanometry

25.5.1 When Things Do Not Add Up During Clinical Assessment

We have all been confronted with the cases in which a patient complains bitterly about nasal obstruction, but we are not able to see pathology that would account for the symptoms. Furthermore, some patients who have only minimal symptoms have what appears to be dramatic anatomic obstruction. It is in these cases that objective testing can be particularly helpful in being the "tiebreaker." In the first example, if airway testing demonstrates a significant nasal restriction, it agrees with the patient's complaints and makes us look further for the cause. If the airway testing shows a widely patent airway, it supports our exam observations and cautions that a procedure to increase the dimension of the airway to try to help this patient's feeling of obstruction would be ill advised.

This use of the test results relies on the knowledge that there is a correlation between measured airway restriction and the symptom of nasal obstruction for many patients, giving us an objective basis for comparison to use with the patient who seems to have contradictory findings. Further clinical examples have been described [41].

25.5.2 For Assessment of Surgical Candidate's Chances of Optimal Outcome

Studies have been done showing the value of rhinomanometric results in optimizing the selection of patients who will be helped by nasal airway surgery [42, 43].

25.5.3 To Analyze Changes in Patients Who Do Not Have Symptomatic Improvement with Surgery

We all want to learn from our patients who continue to have symptoms despite our surgical intervention for their airway. Rhinomanometry, applied as noted in Sect. 25.5.1, can suggest whether it is the still unhappy patient's symptoms that are exceptional (patients with an unusually high resistance threshold for comfort) or whether there is still some measurable obstruction in the airway.

25.5.4 Challenge Testing

Some patients may have reactions to airborne antigens yet have negative skin testing. In these cases, a more direct method of identifying allergens and degree of allergic response can be done with challenge testing [44–48]. Rhinomanometry is done first. Then the patient inhales the challenging antigen. Subsequent rhinomanometry can detect significant change in nasal obstruction caused by the antigen in an allergic patient.

25.6 Summary/Conclusion

Anterior rhinoscopy or endoscopic examination of the nasal airway alone do not tell us about the *function* of the nasal airway. An objective measurement method is needed to have better information. Rhinomanometry may be the answer and it is the measurement of airflow through the nose and pressure across the nose during breathing. During inspirations, the curves go downward with a decrease in pressure and the corresponding movement of air in the direction of the lungs. During expiration, the curves move upward corresponding to pressure increasing and causing the movement of air out of the nose. Dividing the maximum pressure reached during normal inspiration by the highest flow gives a nasal resistance value that correlates with the symptom of nasal obstruction in symptomatic patients. It is of paramount importance of understanding nasal physiology. Besides understanding normal physiology, this objective test, rhinomanometry, plays a significant role in understanding the change of physiology in nasal disturbances, and after its correction.

Acknowledgment The author is grateful to Prof John Pallanch for his support for updating this chapter.

References

1. Hood CM, Schroter RC, et al. Computational modeling of flow and gas exchange in models of the human maxillary sinus. J Appl Physiol. 2009;107(4):1195–203.
2. Zhu JH, Lee HP, et al. Effect of accessory ostia on maxillary sinus ventilation: a computational fluid dynamics (CFD) study. Respir Physiol Neurobiol. 2012;183(2):91–9.
3. Lal D, Gorges ML, et al. Physiological change in nasal patency in response to changes in posture, temperature, and humidity measured by acoustic rhinometry. Am J Rhinol. 2006;20(5):456–62.
4. Zhao K, Dalton P. The way the wind blows: implications of modeling nasal airflow. Curr Allergy Asthma Rep. 2007;7(2):117–25.
5. Timperley D, Srubisky A, et al. Minimal clinically important differences in nasal peak inspiratory flow. Rhinology. 2011;49(1):37–40.
6. Serrano E, Klossek JM, et al. Prospective evaluation of the method of measurement of the peak nasal inspiratory flow (PNIF) in allergic rhinitis. Observational study "Pratic in ORL". Rev Laryngol Otol Rhinol. 2007;128(3):173–7.
7. Barnes ML, Lipworth BJ. Removing nasal valve obstruction in peak nasal inspiratory flow measurement. Ann Allergy Asthma Immunol. 2007;99(1):59–60.
8. da Cunha IC, Ribeiro de Andrade C, et al. Reference values for peak nasal inspiratory flow in children and adolescents in Brazil. Rhinology. 2011;49(3):304–8.
9. Ottaviano G, Scadding GK, et al. Peak nasal inspiratory flow; normal range in adult population. Rhinology. 2006;44(1):32–5.
10. Papachristou A, Bourli E, et al. Normal peak nasal inspiratory flow rate values in Greek children and adolescents. Hippokratia. 2008;12(2):94–7.
11. van Spronsen E, Ebbens FA, et al. Normal peak nasal inspiratory flow rate values in healthy children aged 6–11 years in the Netherlands. Rhinology. 2012;50(1):22–5.
12. Blomgren K, Simola M, et al. Peak nasal inspiratory and expiratory flow measurements—practical tools in primary care? Rhinology. 2003;41(4):206–10.
13. Teixeira RU, Zappelini CE, et al. Peak nasal inspiratory flow evaluation as an objective method of measuring nasal airflow. Braz J Otorhinolaryngol. 2011;77(4):473–80.
14. Wilson AM, Sims EJ, et al. Peak inspiratory flow rate is more sensitive than acoustic rhinometry or rhinomanometry in detecting corticosteroid response with nasal histamine challenge. Rhinology. 2003;41(1):16–20.
15. Lindemann J, Keck T, et al. Nasal air temperature and airflow during respiration in numerical simulation based on multislice computed tomography scan. Am J Rhinol. 2006;20(2):219–23.
16. Haight JS, Cole P. The site and function of the nasal valve. Laryngoscope. 1983;93(1):49–55.
17. Thulesius HL, Thulesius HO, et al. What happens to patients with nasal stuffiness and pathological rhinomanometry left without surgery? Rhinology. 2009;47(1):24–7.
18. Hasegawa M, Kern EB. Variations in nasal resistance (nasal cycle): does it influence the indications for surgery. Facial Plast Surg. 1990;7(4):298–306.
19. Hasegawa M, Kern EB, et al. Dynamic changes of nasal resistance. Ann Otol Rhinol Laryngol. 1979;88(1 Pt 1):66–71.

20. Haight JS, Cole P. Unilateral nasal resistance and asymmetrical body pressure. J Otolaryngol. 1986;16:1–31.
21. Haight JS, Cole P. Is the nasal cycle an artifact? The role of asymmetrical postures. Laryngoscope. 1989;99(5):538–41.
22. Hasegawa M. Nasal cycle and postural variations in nasal resistance. Ann Otol Rhinol Laryngol. 1982;91(1 Pt 1):112–4.
23. Cole P, Forsyth R, et al. Effects of cold air and exercise on nasal patency. Ann Otol Rhinol Laryngol. 1983;92(2 Pt 1):196–8.
24. McCaffrey TV, Kern EB. Response of nasal airway resistance to hypercapnia and hypoxia in man. Ann Otol Rhinol Laryngol. 1979b;88(2 Pt 1):247–52.
25. Pallanch JF, McCaffrey TV, et al. Normal nasal resistance. Otolaryngol Head Neck Surg. 1985;93(6):778–85.
26. McCaffrey TV, Kern EB. Clinical evaluation of nasal obstruction: a study of 1000 patients. Arch Otolaryngol. 1979a;105(9):542–5.
27. Andre RF, Vuyk HD, et al. Correlation between subjective and objective evaluation of the nasal airway. A systematic review of the highest level of evidence. Clin Otolaryngol. 2009;34(6):518–25.
28. Barnes ML, White PS, et al. Re: Correlation between subjective and objective evaluation of the nasal airway. Clin Otolaryngol. 2010;35(2):152–3; author reply 153.
29. Eccles R, Doddi NM, et al. Re: Correlation between subjective and objective evaluation of the nasal airway. Clin Otolaryngol. 2010;35(2):149; author reply 150.
30. Hopkins C. Re: Correlation between subjective and objective evaluation of the nasal airway. Clin Otolaryngol. 2010;35(2):147–8; author reply 148.
31. Hopkins C, Earnshaw J, et al. Re: Correlation between subjective and objective evaluation of the nasal airway. A systematic review of the highest level of evidence. Clin Otolaryngol. 2010;35(4):337–8.
32. Nivatvongs W, Earnshaw J, et al. Re: Correlation between subjective and objective evaluation of the nasal airway. A systematic review of the highest level of evidence. Clin Otolaryngol. 2011;36(2):181–2.
33. Williams J, Kulendra K, et al. Re: Correlation between subjective and objective evaluation of the nasal airway. Clin Otolaryngol. 2010;35(2):150–1; author reply 151–2
34. Eccles R, Jawad MS, et al. The effects of oral administration of (—)-menthol on nasal resistance to airflow and nasal sensation of airflow in subjects suffering from nasal congestion associated with the common cold. J Pharm Pharmacol. 1990;42(9):652–4.
35. Pallanch JF. Comparison of the relative strength of correlation of various rhinomanometric parameters with the symptom of nasal obstruction. Omaha: Triologic Society; 1995.
36. Vogt K, Jalowayski AA, et al. 4-Phase-Rhinomanometry (4PR)—basics and practice 2010. Rhinol Suppl. 2010;21:1–50.
37. Vogt K, Zhang L. Airway assessment by four-phase rhinomanometry in septal surgery. Curr Opin Otolaryngol Head Neck Surg. 2012;20(1):33–9.
38. Clarke RW, Cook JA, et al. The effect of nasal mucosal vasoconstriction on nasal airflow sensation. Clin Otolaryngol Allied Sci. 1995;20(1):72–3.
39. Thulesius HL, Cervin A, et al. The importance of side difference in nasal obstruction and rhinomanometry: a retrospective correlation of symptoms and rhinomanometry in 1000 patients. Clin Otolaryngol. 2012;37(1):17–22.
40. Ivarsson A, Malm L. Nasal airway resistance at different climate exposures: description of a climate aggregate and its use. Am J Rhinol. 1990;4:211.
41. McCaffrey TV. Rhinologic diagnosis and treatment. New York: Thieme; 1997.
42. Sipila J. Rhinomanometry before septoplasty: an approach to clinical material with diverse nasal symptoms. Am J Rhinol. 1992;6:17.
43. Suonpaa J. Do rhinomanometric findings predict subjective postoperative satisfaction? Long-term follow-up after septoplasty. Am J Rhinol. 1993;7:71.
44. Bachmann W. Die behinderte masenatmung. Ein diagnostisches vademekum. Munchen: Dustri-Verlag Karl Feistle; 1987.
45. Doyle WJ, Skoner DP, et al. Reproducibility of the effects of intranasal ragweed challenges in allergic subjects. Ann Allergy Asthma Immunol. 1995;74(2):171–6.
46. Fireman P. Nasal provocation testing: an objective assessment for nasal and Eustachian tube obstruction. J Allergy Clin Immunol. 1988;81(5 Pt 2):953–60.
47. Schumacher MJ, Pain MC. Nasal challenge testing in grass pollen hay fever. J Allergy Clin Immunol. 1979;64(3):202–8.
48. Wang D, Clement P. Assessment of early- and late-phase nasal obstruction in atopic patients after nasal allergen challenge. Clin Otolaryngol Allied Sci. 1995;20(4):368–73.

Acoustic Rhinometry

26

Evren Hizal and Ozcan Cakmak

Abbreviations

AR Acoustic rhinometry
CT Computed tomography
MRI Magnetic resonance imaging

Pearls

- Acoustic rhinometry technique is principally based on the computation of cross-sectional area—distance curves from the analysis of the reflected sound waves by the anatomical structures in the nasal cavity.
- Acoustic rhinometry measurements of the healthy adult nasal cavity are reasonably accurate to the level of the paranasal sinus ostia. Beyond this point, acoustic rhinometry overestimates cross-sectional areas.
- The nasal valve is identified by a pronounced minimum (the first minimum after the nostril) on the acoustic rhinometry area–distance curve. However, the second, third and fourth local minima on the acoustic rhinometry area–distance curve do not correspond to any anatomic structure in the nasal passage. These three minima are caused by acoustic resonances in the portion of the nasal cavity beyond the nasal valve.
- Acoustic rhinometry fails to provide quantitative information about paranasal sinus volume, paranasal sinus ostium size, nasal cavity volume between the nostril and choana and the effects of decongestion on the volume of the nasal mucosa. The diagnostic value of this method is limited with the anterior part of the nasal cavity.
- Clinical studies that do not take the limitations of the technique into account may easily lead to misinterpretations.

26.1 Introduction

Acoustic rhinometry (AR) was introduced as an objective tool for the assessment of the nasal cavity geometry in 1989 by Hilberg et al. [1]. AR measurements require minimum patient cooperation and can be performed practically, quickly and easily. Due to its advantages, the technique is widely accepted in a short time. Clinical applications of acoustic rhinometry include determination of the localization and degree of an intranasal anatomic pathology that affects nasal patency,

E. Hizal
Department of Otorhinolaryngology Head and Neck Surgery, SBU Gulhane Training and Research Hospital, Ankara, Turkey

Department of Audiology, SBU Gulhane Faculty of Health Sciences, Ankara, Türkiye

O. Cakmak (✉)
European and International Board Certified on Facial Plastic Surgery, FACEISTANBUL, Istanbul, Turkey

Ö. Ö. Celebi, T. M. Önerci (eds.), *Nasal Physiology and Pathophysiology of Nasal Disorders*,
https://doi.org/10.1007/978-3-031-12386-3_26

evaluation of the results of a nasal surgery, such as septoplasty, turbinate surgery, assessment of the effects of medications on the nose that are used systemically or topically and comparison of different therapeutic methods. Furthermore, AR gives idea about the reversible component of the nasal obstruction as the measurements before and after decongestion of the nose can be compared. In other terms, AR is potentially useful in the assessment of the nasal cavity geometry, nasal patency and results of various medical and surgical therapies. However, complex anatomy of the nasal cavity, operator mistakes and factors inherent to the AR algorithms and physics may influence the measurement of the area–distance function in the nose and lead to systematic errors. In this chapter, we will try to give some essential information on AR and attempt to cover important aspects of the technique, especially from the clinical point of view.

26.2 History

There have been numerous efforts to understand the nature and functions of the nose throughout the centuries. First records in written history describing the nasal cavity can be found in the Papyrus Ebers of ancient Egypt, because of its functional importance in mummification process. Since then, different methods for the examination of nose have been used. Evolution of the scientific method has given rise to attempts to meet the need for more quantitative evaluation methods. In line with this, a simple nasal patency test was introduced by Zwaardemaker and modified by Glatzel, in which the size of the vapour condensation on a cold metal plate or mirror caused by the expired air through one side of the nose was compared to the other [2]. Evaluation of the sound during forced expiration (introduced by Bruck) or humming (introduced by Spiess) was proposed to give a diagnostic idea about the occluded side of the nose [3]. Twentieth century has witnessed brilliant developments which facilitated the use of more quantifiable and objective nasal evaluation tools, such as rhinomanometry and acoustic rhinometry.

Acoustic waves can be used to determine the location of objects in different media, i.e. gases (air), liquids (water) or solids (earth's crust). Indeed, some animals, such as bats, whales or dolphins, are using sound for object detection for millions of years. The use of sound for object detection in water was first documented by Leonardo Da Vinci, who proposed inserting a tube into water and place an ear to the tube in order to detect vessels [4]. Evolution of the scientific method within decades and development of physical and mathematical techniques have led to the development of acoustics as a science and physical properties of the sound have started to be illuminated. Accumulation of scientific data in turn gave rise to innovative thoughts and technological applications of the knowledge on acoustics have started to emerge. One of those applications was SONAR (sound navigation and ranging) systems which have been used for detecting submarines in World War I. Acoustic waves have also been used for object detection in solids, i.e. seismic surveys that aimed to investigate underground structures in the earth's crust. Through the use of electronics and development of modern computer systems, sound measurement and analysis reached new levels of complexity and accuracy. Acoustic reflections have been used to assess the geometry of upper airways, including pharynx, glottis, trachea and lungs after the 1970s. Acoustic rhinometry was then first introduced by Hilberg et al. in 1989 [1].

26.3 Theoretical Background and Criticism

The basic idea behind the acoustic rhinometry method is similar with other methods of acoustic object location and consists of impacting an incident acoustic wave into a medium to generate a reflected acoustic wave. The size and the location of an object through the route of acoustic waves can be determined by calculation of the amplitude of reflected waves and the time difference between the incident and reflected waves, respectively. However, some phenomena related with the inherent nature of acoustic waves and acoustic properties of the medium in which the wave propagates interfere with the measurements and make calculations complicated. Acoustic waves are longitudinal waves that oscillate along the same direction as they move. During their route, they exhibit some characteristic patterns, like reflection and diffraction. Reflection can be defined as the change in direction of the wave at an interface between two different media. Diffraction is, in general, bending of the waves around small obstacles and scattering of waves past small ostia. Similar effects occur when sound waves travel through a medium with varying acoustic impedance. As the waves propagate within a medium (gas, liquid or solid), they are reflected by structures or dissimilar media on their route. But some of the waves penetrate through those structures or media and continue to propagate. The waves that remain on their route will be reflected again by other structures or dissimilar media. Additional reflections, thus, will be added to the acoustic image and analysis and comparison of the waves that are sent into and reflected back from a heterogeneous or irregular medium will become almost impossible. A solution to this problem, i.e. analysis of the acoustic image with multiple backward reflections, was offered by Ware and Aki in 1969 [5]. By the Ware–Aki algorithm it was then possible to analyse the sound waves that were reflected by different layers through the route of the wave. Ware–Aki algorithm, which is used in acoustic rhinometry technique however, has some assumptions regarding the ideal properties of airway. This algorithm assumes that the sound waves are plane waves, and it does not account for losses (airway wall non-rigidity, viscous losses) or non-planar wave propagation effects [6–8]. In order to understand the reasons of some artefacts and errors on acoustic rhinometry area–distance curves, these assumptions will be explained briefly.

The first reconstruction algorithm used in acoustic reflectometry was developed under the ideal conditions of no losses in the propagating wave and that all frequencies were covered by the acoustic pulse. The assumption of planar wave propagation is fundamental to passage area measurements made with AR. Waves are assumed to propagate along the axis of the airway in one dimension. If the frequency of the sound waves is high and therefore the wavelengths are too short, sound waves do not move along a plane and start to be reflected between the walls of the nasal cavity. This, in turn, causes additional delays in the reflected waves, complicates the relation between incident and reflected waves and eventually affects cross-sectional area and distance computations. In other words, planar wave assumption determines and limits the spatial resolution and the frequency bandwidth of the method, and imposes limitations on the transverse sizes of an airway model [6–8].

Spatial resolution is defined as the smallest axial distance that separates two cross-sectional areas that can still be resolved by AR. In rigid-walled airways, the spatial resolution is approximately equal to one-sixth of the shortest wavelength of the incident sound pulse. The frequency bandwidth of the incident sound pulse is important in determining the spatial resolution of the technique, and hence has a major influence on the accuracy of AR measurements. The limited frequency bandwidth of the AR technique may increase the rise distance, and thereby produce a smoother incline in the area–distance curve [6, 9].

The Ware–Aki algorithm is valid under the condition that the acoustic impedance of the one-dimensional acoustic pathway is continuous. If there is a finite sudden jump in the acoustic impedance, the transformations and the potential functions used in the mathematical formulation of this algorithm are not well defined. In other words, the Ware–Aki algorithm is not suitable for calculating the area–distance function at locations where there are abrupt changes in the acoustic impedance [6–9].

It has been argued that any form of energy loss or sound wave attenuation would reduce the amplitude of the reflected wave, which, in turn, would lead to area underestimation. Viscous forces, transmission losses and internal losses that take place as the sound wave is transmitted through the constriction in an airway model have all been attributed to area underestimations that occur with AR. However, the main reason of area underestimations distal to an anterior constriction seems to be a "barrier effect", i.e. "barrier" created by the anterior constriction reflects most of the incident sound power. In models with constrictions (inserts) of small passage area, the high-frequency components of the acoustic pulse generated by AR equipment do not reach the portion of the model beyond the constriction, because these waves are reflected back from the barrier created by the constriction [6–8, 10, 11]. The effect of a constriction on AR measurements will also be covered in nasal valve section below.

26.4 Acoustic Rhinometry Equipment

Acoustic rhinometry is based on the principle of computation of cross-sectional area–distance curves from the analysis of the reflected sound waves by the anatomical structures in the nasal cavity. In other terms, analysis of the sound waves that are sent into and reflected back from the nasal cavity gives the cross-sectional area at a given distance.

A picture and a schematic representation of acoustic rhinometry unit are shown in Figs. 26.1 and 26.2, respectively. Acoustic waves that are produced by a loudspeaker pass through a sound tube and a nose adapter, which is a detachable pipe that establishes a connection between the AR device and the nose. The sound waves that cross the nose adapter then reach the nose. As the sound waves propagate in the nasal cavity, they are reflected by the anatomical structures. A microphone that is placed within the sound tube detects the waves that are reflected back, and transforms these into electrical signals. These signals are then amplified by an amplifier and converted into numeric data by an analogue-digital converter. Numerical data is then analysed by a computer. Frequency bandwidth of the sound waves that are sent into the nasal cavity may contain all audible frequencies between 20 and 20,000 Hz. However, low-pass filters exclude the frequencies over 10,000 Hz since interferences and diffractions increase as the wavelength of the sound waves travelling in nasal cavity decrease [3, 6–8].

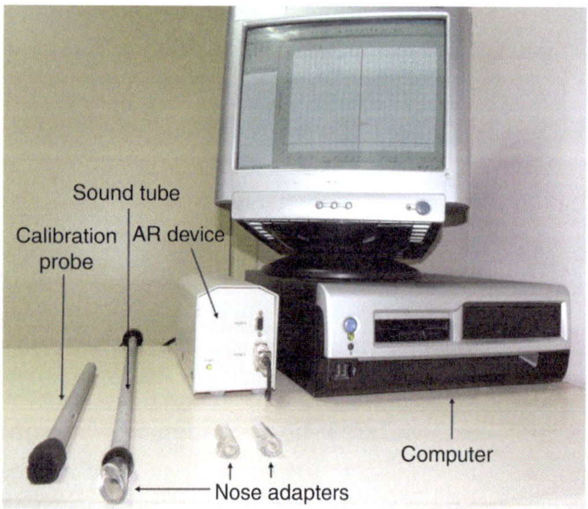

Fig. 26.1 Acoustic rhinometry equipment (Rhinoscan SRE 2000, Interacoustics A/S, Assens, DK)

The changes in cross-sectional area of the airway affect acoustic impedance. Since the nasal cavity is not a straight pipe and has an irregular and complex anatomy, acoustic waves show different reflection patterns at differing cross-sectional areas on their route. Although the mathematical background of the analysis is very complex and will be detailed to some extent later in this chapter, computerized calculations basically give two parameters for a given cross section: Comparison of the amplitudes of the sent and reflected sound waves gives cross-sectional areas, while the time difference between the sent and reflected sound waves gives the distance of a given cross-sectional area to the reference point. These data are then combined and a distance–area curve is obtained. The areas here define the cross-sectional areas that are vertical to the acoustic pathway, i.e. the way that acoustic waves follow in the nasal cavity. Cross-sectional area of a given section is plotted on vertical axis, while distance of that section to the reference point is plotted on horizontal axis of the graph. In order to evaluate the nasal valve region better, vertical axis can be plotted in logarithmic scale.

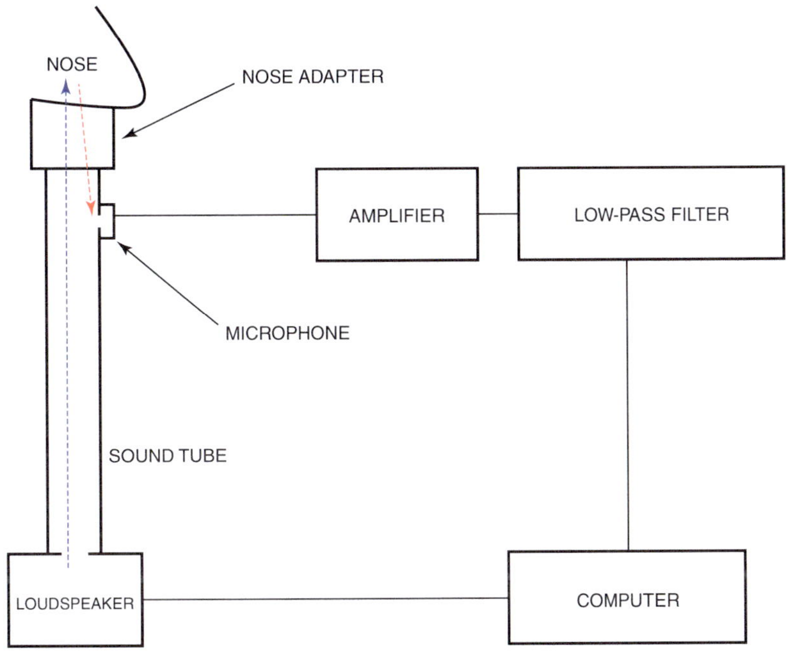

Fig. 26.2 Schematic representation of an acoustic rhinometry circuit

26.5 Test Technique

Classic acoustic rhinometry utilize single impulse and one microphone (transducer) setup. However, there are other methods that use two microphones [12], or devices that use continuous acoustic stimulation [13]. Although the theory is identical to that of the single impulse method and the test technique is similar, a continuous wide-band noise model allows for almost real-time adjustment of the equipment since a visual output is generated more than 20 times per second [14].

By means of the modern computer software the use of acoustic rhinometry is relatively simple. Since external noise, temperature and humidity have potential effects on measurements, testing room should have standard environmental conditions. Test should be applied by experienced staff who are aware of the recommendations for reliable testing [15]. Both sides should be tested separately. Single measurement takes about a few seconds and whole testing process ends in a couple of minutes. After the device is switched on and computer program is opened, calibration is simply done by following the on-screen instructions that are provided by the manufacturer. Calibration should be done every time the device is opened. The patient should have a sitting position and position should not be changed during the tests. Swallowing and breathing should be avoided during the measurements. Any kind of secretions can narrow the airway and affect the measurements. Hence, nasal cavity should be cleaned off from secretions before the test [16]. Several types of nose adapters with different rim shapes are available (Fig. 26.3). A nose adapter that would best fit to the shape of the patient's nostril should be chosen and attached to the probe. To avoid acoustic leak, a medical sealant gel should be applied circumferentially to the edge of the nose adapter, providing an air secure

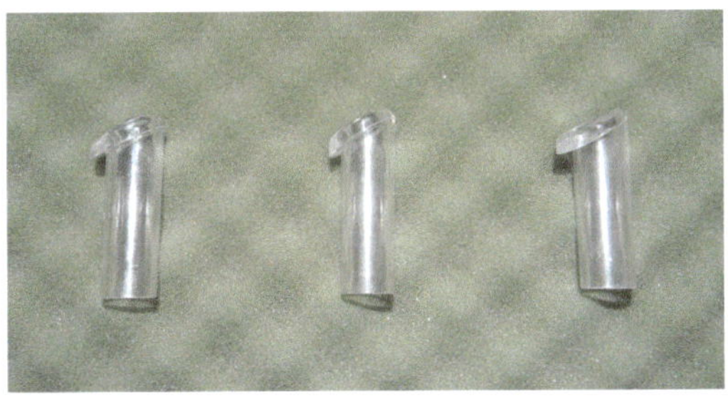

Fig. 26.3 Various types of nose adapters are designed to achieve a better fit to shape of the right or left nostril

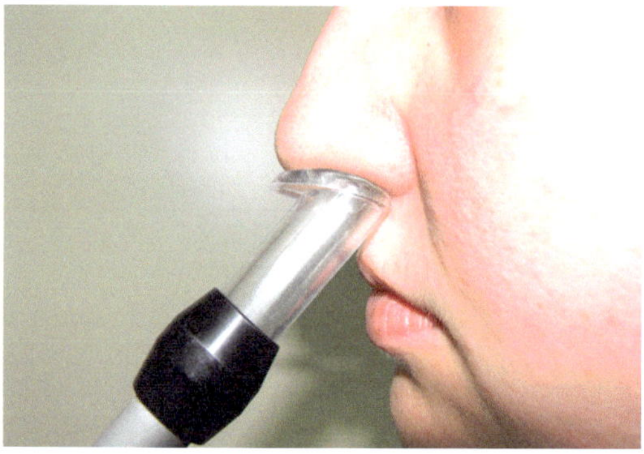

Fig. 26.4 During the measurements, nose probe should be placed accordingly to prevent acoustic leak. Distortion of the nose should be avoided

contact between the nose adapter and the nostril [17]. Nose probe should be placed accordingly to prevent acoustic leak, but distortion of the nose or changes in position of the nose probe should be strictly avoided (Fig. 26.4).

Measurements should be repeated for at least three times to obtain the most correct results. By this way, erroneous measurements due to acoustic leakage or distortion of the vestibule owing to incorrect positioning of the probe can be detected and eliminated.

The operator should be aware of the above mentioned testing principles and instructions. Besides the operator errors, complex anatomy of the nasal cavity, physical limitations of the AR and factors inherent to the AR algorithms may influence the measurements and lead to systematic errors. Nasal cavity has a complex geometry consisting of cartilaginous and bony framework covered with erectile tissue and mucosa, including a narrow segment of nasal valve at the anterior part, and sinus ostia more posteriorly. Narrow segments at anterior parts negatively affect the measurements of more posterior parts and this leads to a potential problem since the narrowest part of the nasal cavity, the nasal valve, is on anterior part of the cavity. Similarly, ostia of the paranasal sinuses affect AR measurements, and cross-sectional areas behind the paranasal sinus ostia are overestimated. In order to understand the results of AR measurements, we will try to take a closer look to the cross-sectional area–distance curve.

26.6 Cross-Sectional Area–Distance Curve

In order to interpret AR results correctly, it is essential to know how some anatomical structures in nasal cavity appear on AR area–distance curves and how these structures' localizations and size affect the area–distance curve.

In a typical AR area–distance curve, there is a minimum on the junction between the nose adapter and the nostril (Fig. 26.5). This minimum is generally not considered as a "minimum" and hence not termed as the "first minimum", since it occurs at "0 cm" which is the start point of the nose. first minimum after the nostril represents the nasal valve region (Fig. 26.5).

There is no clear consensus on interpretation of AR results, even in healthy humans. Inspection of the literature reveals that up to 4 local minima have been commonly observed on the AR area–distance curves and different terms have been used to define these minima. The terms "1st constriction, I-notch, CSA1, MCA, start of isthmus region, start of valve region, or ostium internum" have been used for the first minimum, which was attributed to the nasal valve; "2nd constriction, C-notch, CSA2, inferior concha, or piriform aperture" have been used for the second minimum, which was attributed to the head of the inferior turbinate, and "CSA3" has been used for the third minimum, which was usually attributed to the middle turbinate [3, 10]. However, recent experimental studies with nasal cavity models and clinical studies revealed that the second and third minima on cross-sectional area–distance curves do not represent an anatomical point, and hence, most of these terms might be used inappropriately [6–11, 16, 18, 19].

In healthy humans cross-sectional areas measured by different imaging modalities, such as CT and MRI were compared with AR cross-sectional area measurements and the techniques were found to give comparable results especially on the anterior part of the nasal cavity [9, 10, 20–25]. Regarding the validation of AR curve with imaging modalities, a methodological issue has to be concerned. Areas calculated on AR measurements are the cross-sectional areas that

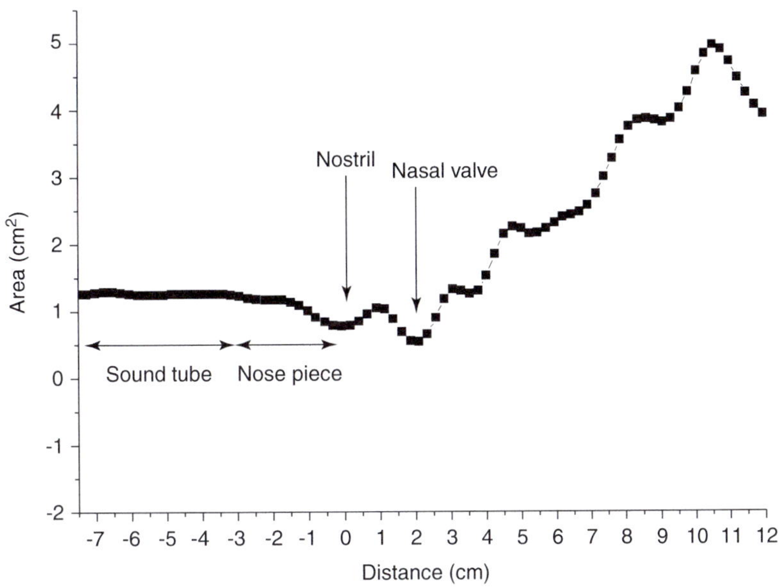

Fig. 26.5 A typical AR cross-sectional area–distance curve. Start point of the nasal cavity (nostril) is accepted as "0 cm". Horizontal axis gives the distance of a given cross-sectional area that is perpendicular to the acoustic axis to the nostril. The cross-sectional area of that section is plotted on vertical axis

are perpendicular to the sound wave propagation axis (acoustic axis). In some of the validation studies, images of the nose were taken perpendicular to the base of the nose, not to the acoustic axis, and the reference point to be used in distance measurements was either chosen as the anterior nasal spine [21, 23], or tip of the nose [25] or even not clearly defined [24]. Studies that do not care to the sloped anatomy of the nasal cavity or use different reference points may lead to significant errors when interpreting the AR curve.

In a study on healthy humans, the actual cross-sectional areas of the nasal cavity together with the actual locations of the nasal valve, the head of the inferior turbinate, the head of the middle turbinate, the openings of the ostia of the maxillary and frontal sinuses and the choanae were calculated from computed tomography sections perpendicular to the curved acoustic axis of the nasal passage [18]. The findings were then compared with the corresponding cross-sectional areas measured by AR. Comparison of the CT- and AR-derived area–distance curves both before and after decongestion revealed that the nasal valve is identified by a pronounced minimum (the first minimum after the nostril) on the CT- and AR-derived area–distance curves. However, neither the head of the inferior turbinate nor the head of the middle turbinate could be distinctly identified on the CT area–distance curves of healthy humans. The same held true for the AR measurements in both cadaver cast models and healthy humans [9, 18]. The second and third minima on the AR area–distance curves did not correspond to the actual locations of the head of the inferior turbinate and the head of the middle turbinate determined from CT, neither before nor after decongestion [9, 18].

In a study that used cast model of the nasal cavity of a cadaver, Cakmak et al. demonstrated that AR was able to detect changes in cross-sectional area larger than approximately 0.19 cm^2 and 0.38 cm^2, at the head of the inferior turbinate and the head of the middle turbinate, respectively [18]. This finding suggests that AR cannot resolve any change in the cross-sectional area of the nasal passage at each of these specific anatomic sites that is smaller than the corresponding limit. In addition, the ability of AR in measuring abrupt changes in cross-sectional area is poor, because of the limited spatial resolution and the long rise distance of the technique [6, 9].

In summary, with the exception of the first minimum after the nostril, which represents the nasal valve, the subsequent minima on the AR area–distance curves for both non-decongested

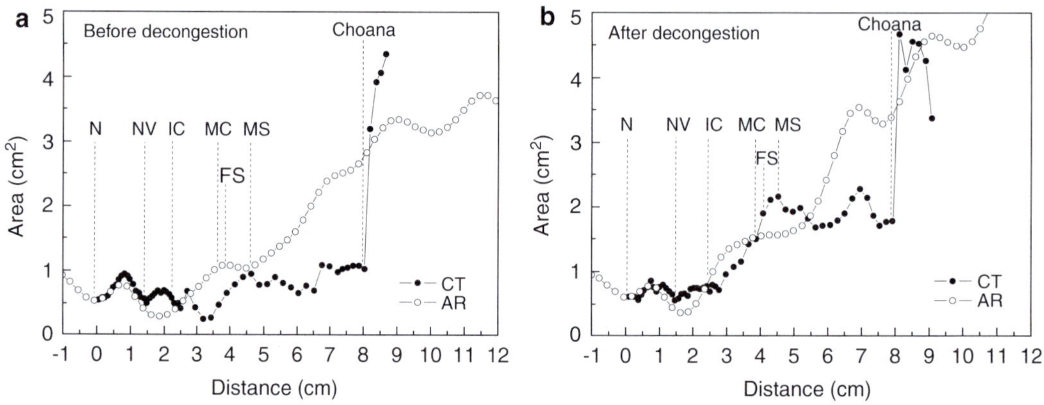

Fig. 26.6 The cross-sectional area–distance curves of a healthy human's nasal cavity as determined by computed tomography and acoustic rhinometry, before (**a**) and after (**b**) decongestion. Actual locations of the anatomical structures are determined on computed tomography sections and depicted on the graphs with vertical dashed lines. The ability of acoustic rhinometry to detect the anatomical structures in nasal cavity can be seen. Acoustic rhinometry fails to quantify the volume change after decongestion (*CT* cross-sectional area–distance curve as determined by computed tomography, *AR* cross-sectional area–distance curve as determined by acoustic rhinometry, *N* nostril, *NV* nasal valve, *IC* head of the inferior concha [turbinate], *MC* head of the middle concha [turbinate], *FS* frontal sinus ostium, *MS* maxillary sinus ostium)

and decongested nasal cavities do not correspond to any anatomic structure in the healthy human nose (Figs. 26.6a, b and 26.7a, b). These minima are formed because of the acoustic resonances in nasal cavity behind the nasal valve region. Effects of some important anatomical landmarks on AR cross-sectional area–distance curves are summarized below.

26.6.1 Nasal Valve

The nasal valve area is widely accepted as the most important part of the nasal passage with respect to its essential role in respiratory physiology. Boundaries of this triangular region are formed by the caudal septum (medial wall), caudal edge of the upper lateral cartilages and head of the inferior turbinate (lateral wall) and floor of the nose (inferior wall). Nasal valve is the narrowest part of the nasal passage and functions as an essential regulator of nasal airflow. The accuracy of AR measurements in the anterior part of the nose, which contains the nasal valve, is substantial in terms of the value of this method in rhinology. Individual anatomical variations of the anterior narrow segment might significantly limit the role of AR as a diagnostic tool for the entire nasal cavity.

As mentioned above, the nasal valve is identified by a pronounced minimum (the first minimum after the nostril) on the AR area–distance curve. Experimental studies on pipe models and nasal cavity models have shown that AR gives an accurate measure of the distance from the nose adapter to the narrow segment that simulates nasal valve, and AR measurements of the anterior nasal passage are reasonably accurate if the nasal valve area is within normal adult ranges [7, 11]. Clinical studies that compare the cross-sectional areas derived by AR and by imaging modalities, such as computed tomography and magnetic resonance imaging, also showed that AR is a valuable method for measuring nasal valve area [10, 18, 20–24]. These studies noted significant correlations between the cross-sectional areas obtained by imaging modalities and AR, with particularly high agreement in the anterior part of the nasal cavity and nasal valve. For the area of the nasal valve, agreement between the AR and imaging techniques was apparent when imaging was obtained perpendicular to the acoustic axis

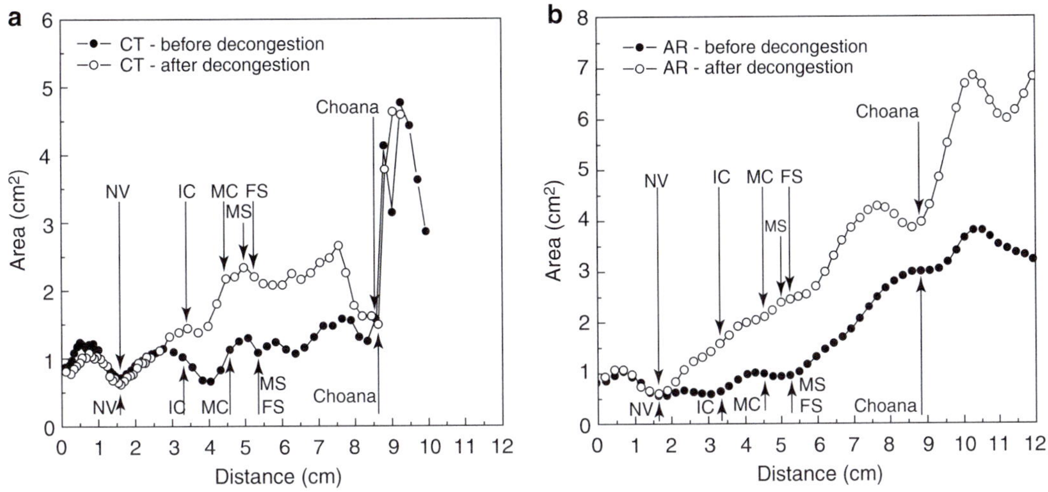

Fig. 26.7 The cross-sectional area–distance curves of a nasal cavity before and after decongestion, as determined by computed tomography (**a**) and acoustic rhinometry (**b**). Actual locations of the anatomical structures before and after decongestion are determined on computed tomography sections and depicted on the graphs with arrows. Acoustic rhinometry fails to detect the localizations of the anatomical structures except for the nostril and nasal valve. (*CT* cross-sectional area–distance curve as determined by computed tomography, *AR* cross-sectional area–distance curve as determined by acoustic rhinometry, *N* nostril, *NV* nasal valve, *IC* head of the inferior concha [turbinate], *MC* head of the middle concha [turbinate], *FS* frontal sinus ostium, *MS* maxillary sinus ostium)

that follows the curve of the nasal passage through the centre of the curved airway [10].

One of the best-recognized problems with acoustic pulse response analysis is its inability to precisely measure the cross-sectional areas beyond narrow apertures. Concomitantly, the accuracy of AR measurements of the nasal cavity depends greatly on nasal passage anatomy, especially that of the narrowest section. In model studies, the cross-sectional area and the length of the narrow segment have been shown to be the factors that most significantly influence the accuracy of AR [6]. When the cross-sectional area and the length of the narrowest part of the passage were relatively small and short, the probability of measurement error was higher. It is well established that the area of a region beyond a severe constriction may not be measured accurately by AR, and a narrowing in the anterior part of the nasal cavity causes errors in AR-derived areas posterior to the site of constriction [1, 7, 16, 26, 27]. The results obtained for living human subjects suggest that, when the nasal valve passage area is within the normal adult range, AR is a valuable method for measuring the cross-sectional areas of the nasal cavity anterior to the paranasal sinus ostia [18, 19]. In other terms, accuracy of AR measurements is closely related with the narrowest section of the nasal passage, the nasal valve, and the passage area of the nasal valve is the most important limiting factor when quantifying the geometry of the anterior nasal cavity with AR. The effects of nasal valve passage area on accuracy of AR measurements were examined by Cankurtaran et al. using simple pipe models with a constriction [7]. These authors demonstrated that the constriction reflects most of the incident sound power. In models with constrictions of small passage area, the high-frequency components of the acoustic pulse generated by AR equipment do not reach the portion of the model beyond the constriction, because these waves are reflected back from the barrier created by the constriction. This finding is of vital importance for AR because the transmitted sound waves probe and hence provide information about, the cross-sectional area posterior to the constriction. Accordingly, an examiner should expect relatively higher degrees of error when measuring the cross-sectional area of a nasal cavity model beyond a constriction of small passage area. Since AR measures the intensity of reflected sound waves and compares this with the intensity of incident waves, AR-measured cross-sectional areas beyond the constriction (a nasal valve that is smaller than normal adult size) are underestimated and the corresponding area–distance curve shows pronounced oscillations [6, 11].

The anatomy of the human nose is complex, and the spectrum of individual differences is broad. For patients with pathologies that narrow the nasal valve, such as septal deviations, polyps, tumours, webs, strictures, or alar cartilage insufficiencies, the value of AR for measuring the entire nasal cavity is limited. All users of this technique must be aware of the effects of nasal valve to prevent misinterpretation of AR findings during clinical assessment.

Together with the nasal valve, the second important anatomical structure that affects the AR measurements is paranasal sinuses.

26.6.2 Paranasal Sinuses

To interpret the AR measurements correctly, it is essential to know how paranasal sinuses affect AR area–distance curves and to what extent the AR measurements give idea on paranasal sinus and their ostia. At this point, the effect of nasal valve on AR measurements should be noticed once more. Model studies revealed that as the area of the nasal valve decrease (<0.283 cm^2), the areas of both the nasal valve and regions posterior to the nasal valve are underestimated and oscillations appear [6, 7, 10, 11]. Clinical studies on healthy humans supported the results of the experimental studies and showed that AR gives reliable results between the nostrils and sinus ostia if the nasal valve area is in normal range [18, 19]. Even if the nasal valve area is within normal range, the areas posterior to the sinus ostia are overestimated and the degree of error increases for the areas that are located more posteriorly.

Clinical and experimental studies revealed that AR measurements behind 5–6 cm, where sinus ostia are located, can include significant mistakes and AR measurements cannot give accurate information about the paranasal sinuses and sinus ostia [6, 8, 11, 19]. The effects of paranasal sinus volume and their ostia on AR measurements have been assessed with model studies, in detail [8, 11]. The pipe models that have been used for that purpose were consisting of a main pipe with a side branch as Helmholtz resonator. The neck diameter (simulating sinus ostium) and the cavity volume (simulating the paranasal sinus) were variable. The results of those studies showed that small ostia had little impact on AR measurements, regardless of sinus volume [8, 11]. However, AR overestimated cross-sectional areas posterior to the simulated sinus ostium when the ostium was large. Overestimation was more pronounced as the diameter of the sinus ostium and volume of the sinus increased. This result suggests that for patients who have a large sinus ostium and large paranasal sinus volume (i.e. after functional endoscopic sinus surgery), the precision of AR measurements beyond the sinus ostium is lower. Paranasal sinus volume can influence the area–distance curve beyond the ostium, but this effect is significant only when the sinus is connected to the nasal cavity by a relatively large opening [11].

Since AR cannot measure the cross-sectional areas on posterior nasal cavity correctly, it also cannot give accurate data on nasal cavity volume. The results of a clinical study revealed that AR overestimates nasal cavity volume by 21% before decongestion and 24% after decongestion, when compared with volume measured by CT [18]. The nasal cavity volume difference with decongestion was 30% more in AR measurements, when compared with CT measurements [18]. In other words, AR overestimates the effect of decongestion on nasal cavity erectile tissue mass.

In order to understand the reasons of the area overestimation behind the sinus ostia, it is essential to review the physical properties of the AR technique once more. The reason for area overestimations is not the acoustic energy loss to the sinuses through the ostia, but it is the interaction between the nasal cavity and paranasal sinuses [19]. The physical principle of AR is based on the reflections of the sound waves that propagate in a pipe [1, 26–28]. As the cross-sectional areas change within the pipe, sound waves are partially reflected and the changes in acoustic impedance at each point constitute a reflection series. Reflection series of the pipe is termed as the "input impulse response of the pipe" and cross-sectional areas are calculated as a function of the distance [1, 29, 30]. Experimental data that include input impulse response is transformed into cross-sectional area–distance curve by the Ware–Aki algorithm [5]. Sound waves that pass through the nasal valve are exposed to multiple reflections at localizations with acoustic impedance changes, such as sinus ostia. Oscillations that are formed by the resonator characteristics of paranasal sinuses are also superimposed with the waves that are reflected from the posterior parts of the nasal cavity. The Ware–Aki algorithm misinterprets superimposed waves and this leads to area overestimations. Previous experimental studies showed that complex acoustic resonances of the paranasal sinuses and nasal cavity and thus the acoustic resonance effects of sinuses and posterior nasal cavity are not accounted in Ware–Aki algorithm, which is still used in AR calculations [8]. The results from healthy humans also showed that the reason of area overestimations behind the paranasal sinus ostia was not sound loss through the sinus ostia to the sinuses, in both non-decongested and decongested cavities [19].

In summary AR does not give reliable data on the dimensions of the paranasal sinus ostia, volumes of the sinuses, nasal cavity volumes between nostril and choana and the effect of the decongestion on nasal mucosa. AR overestimates the cross-sectional areas behind the sinus ostia. The diagnostic value of this method is restricted with the anterior part of the nasal cavity. Thus, the volume measurements in any instance should be done for the area between 0 and 5–6 cm.

26.7 Applications of Acoustic Rhinometry

A simple search on the Medline/PubMed database with the words "acoustic rhinometry" reveals more than 1100 studies between 1989 and 2020. Together with the need for an objective tool to evaluate nasal patency, some attractive factors, such as relative ease of use and low application costs, seem to keep this technique as a popular tool for research. Theoretically, acoustic rhinometry can be used to assess the geometry of nasal airway and the effect of anatomical, physiological or pathological conditions that interfere with nasal patency.

AR has been used to assess the effects of environmental factors (effect of temperature [31, 32], posture [33], nasal cycle [34], inhaled pollutants, gases or particles [3]), pharmacological agents (decongestants [35, 36], antibiotics [37], steroids [38–40], nasal irrigations [41], antiallergic drugs [42–44], systemic drugs [45], nasal challenge testing [46]) and surgical therapies on nasal airway. It also has been used in evaluation of allergic and non-allergic rhinitis, snoring and sleep apnoea [47, 48]. AR can be a useful tool for evaluation of the symptom of nasal obstruction, and for documenting the pre-treatment status and post-treatment outcomes of surgical or medical therapies, for both medical and medicolegal purposes [49–52].

Acoustic rhinometry has been shown to be reproducible in animal studies, both in vivo and post mortem [53, 54]. However, physical and technical improvements for more accurate and applicable results and modification and optimization of the equipment for measurement of small dimensions [55] are necessary to use this technique in animal studies.

Acoustic rhinometry has also been used for measurements in children [56–59]. Due to the uncomplicated and non-invasive nature of the technique, it may prove to be a useful tool in examination of the airways in children. However, the dimensions of the nasal cavity and thus the nasal valve in this population are usually much smaller than those of the adults. Accordingly, limitations of the technique and the validity of measurements should always be kept in mind [60, 61].

26.8 Conclusion

In conclusion, AR is potentially helpful in defining the geometry of nasal cavity, measurement of nasal patency and assessment of the results of nasal medical and surgical interventions. However, AR measurements can include significant mistakes due to the operator's technique and nasal passage anatomy. AR measures the cross-sectional areas in anterior part of the nose with high accuracy and overestimates the cross-sectional areas behind the paranasal sinus ostia. In other terms, diagnostic value of the technique is limited with the anterior part of the nasal cavity. Nasal valve can be identified as a minimum on AR area–distance curves (first minimum after the nostril). The second, third and fourth minima on AR area–distance curves cannot be associated with an anatomical structure in nasal cavity. These minima are formed because of the acoustic resonances in nasal cavity behind the nasal valve region. The cross-sectional areas behind the paranasal sinus ostia are overestimated with AR. This is not because of the sound loss to the sinuses through the sinus ostia, but because of the interactions between the nasal cavity and paranasal sinuses. Acoustic rhinometry cannot give quantitative data about the volumes of the paranasal sinuses and dimensions of the sinus ostia, neither before nor after decongestion. Acoustic rhinometry significantly overestimates the effect of decongestion on nasal mucosa. Clinical studies that do not take the potential errors of AR into account can easily be misinterpreted. Physical limitations should be taken into account to develop better AR equipment and related computer software.

References

1. Hilberg O, Jackson AC, Swift DL, Pedersen OF. Acoustic rhinometry: evaluation of nasal cavity geometry by acoustic reflection. J Appl Physiol. 1989;66(1):295–303.
2. Foxen EH, Preston TD, Lack JA. The assessment of nasal air-flow: a review of past and present methods. J Laryngol Otol. 1971;85(8):811–25.
3. Hilberg O. Objective measurement of nasal airway dimensions using acoustic rhinometry: methodological and clinical aspects. Allergy. 2002;57(Suppl 70):5–39.
4. Leighton TG. Ocean acoustics. In: Fahy F, Walker JG, editors. Fundamentals of noise and vibration. London: Taylor & Francis; 1998. p. 375.
5. Ware JA, Aki K. Continuous and discrete inverse scattering problems in a stratified elastic medium. I. Plane waves at normal incidence. J Acoust Soc Am. 1969;45(4):911–21.
6. Celik H, Cankurtaran M, Cakmak O. Acoustic rhinometry measurements in stepped-tube models of the nasal cavity. Phys Med Biol. 2004;49(3):371–86.
7. Cankurtaran M, Celik H, Cakmak O, Ozluoglu LN. Effects of the nasal valve on acoustic rhinometry measurements: a model study. J Appl Physiol. 2003;94(6):2166–72.
8. Cakmak O, Celik H, Cankurtaran M, Buyuklu F, Ozgirgin N, Ozluoglu LN. Effects of paranasal sinus ostia and volume on acoustic rhinometry measurements: a model study. J Appl Physiol. 2003;94(4):1527–35.
9. Cakmak O, Tarhan E, Coskun M, Cankurtaran M, Celik H. Acoustic rhinometry: accuracy and ability to detect changes in passage area at different locations in the nasal cavity. Ann Otol Rhinol Laryngol. 2005;114(12):949–57.
10. Cakmak O, Coskun M, Celik H, Buyuklu F, Ozluoglu LN. Value of acoustic rhinometry for measuring nasal valve area. Laryngoscope. 2003;113(2):295–302.
11. Cakmak O, Celik H, Cankurtaran M, Ozluoglu LN. Effects of anatomical variations of the nasal cavity on acoustic rhinometry measurements: a model study. Am J Rhinol. 2005;19(3):262–8.
12. Louis B, Glass GM, Fredberg JJ. Pulmonary airway area by the two-microphone acoustic reflection method. J Appl Physiol. 1994;76(5):2234–40.
13. Djupesland PG, Lyholm B. Nasal airway dimensions in term neonates measured by continuous wide-band noise acoustic rhinometry. Acta Otolaryngol. 1997;117(3):424–32.
14. Straszek S. Validation of acoustic rhinometry in laboratory animals. Dan Med Bull. 2008;55:159.
15. Fisher EW, Morris DP, Biemans JM, Palmer CR, Lund VJ. Practical aspects of acoustic rhinometry: problems and solutions. Rhinology. 1995;33(4):219–23.
16. Cakmak O, Celik H, Ergin T, Sennaroglu L. Accuracy of acoustic rhinometry measurements. Laryngoscope. 2001;111(4 Pt 1):587–94.
17. Hamilton JW, McRae RD, Jones AS. The magnitude of random errors in acoustic rhinometry and re-interpretation of the acoustic profile. Clin Otolaryngol Allied Sci. 1997;22(5):408–13.
18. Cankurtaran M, Celik H, Coskun M, Hizal E, Cakmak O. Acoustic rhinometry in healthy humans: accuracy of area estimates and ability to quantify certain anatomic structures in the nasal cavity. Ann Otol Rhinol Laryngol. 2007;116(12):906–16.
19. Tarhan E, Coskun M, Cakmak O, Celik H, Cankurtaran M. Acoustic rhinometry in humans: accuracy of nasal passage area estimates, and ability to quantify paranasal sinus volume and ostium size. J Appl Physiol. 2005;99(2):616–23.
20. Terheyden H, Maune S, Mertens J, Hilberg O. Acoustic rhinometry: validation by three-dimensionally recon-

structed computer tomographic scans. J Appl Physiol. 2000;89(3):1013–21.
21. Hilberg O, Jensen FT, Pedersen OF. Nasal airway geometry: comparison between acoustic reflections and magnetic resonance scanning. J Appl Physiol. 1993;75(6):2811–9.
22. Min YG, Jang YJ. Measurements of cross-sectional area of the nasal cavity by acoustic rhinometry and CT scanning. Laryngoscope. 1995;105(7 Pt 1):757–9.
23. Gilain L, Coste A, Ricolfi F, Dahan E, Marliac D, Peynegre R, et al. Nasal cavity geometry measured by acoustic rhinometry and computed tomography. Arch Otolaryngol Head Neck Surg. 1997;123(4):401–5.
24. Corey JP, Gungor A, Nelson R, Fredberg J, Lai V. A comparison of the nasal cross-sectional areas and volumes obtained with acoustic rhinometry and magnetic resonance imaging. Otolaryngol Head Neck Surg. 1997;117(4):349–54.
25. Dastidar P, Numminen J, Heinonen T, Ryymin P, Rautiainen M, Laasonen E. Nasal airway volumetric measurement using segmented HRCT images and acoustic rhinometry. Am J Rhinol. 1999;13(2):97–103.
26. Hilberg O, Pedersen OF. Acoustic rhinometry: recommendations for technical specifications and standard operating procedures. Rhinol Suppl. 2000;16:3–17.
27. Hilberg O, Lyholm B, Michelsen A, Pedersen OF, Jacobsen O. Acoustic reflections during rhinometry: spatial resolution and sound loss. J Appl Physiol. 1998;84(3):1030–9.
28. Hoffstein V, Fredberg JJ. The acoustic reflection technique for non-invasive assessment of upper airway area. Eur Respir J. 1991;4(5):602–11.
29. Fredberg JJ, Wohl ME, Glass GM, Dorkin HL. Airway area by acoustic reflections measured at the mouth. J Appl Physiol. 1980;48(5):749–58.
30. Jackson AC, Butler JP, Millet EJ, Hoppin FG Jr, Dawson SV. Airway geometry by analysis of acoustic pulse response measurements. J Appl Physiol. 1977;43(3):523–36.
31. Yamagiwa M, Hilberg O, Pedersen OF, Lundqvist GR. Evaluation of the effect of localized skin cooling on nasal airway volume by acoustic rhinometry. Am Rev Respir Dis. 1990;141(4 Pt 1):1050–4.
32. Lundqvist GR, Pedersen OF, Hilberg O, Nielsen B. Nasal reaction to changes in whole body temperature. Acta Otolaryngol. 1993;113(6):783–8.
33. Lal D, Gorges ML, Ungkhara G, Reidy PM, Corey JP. Physiological change in nasal patency in response to changes in posture, temperature, and humidity measured by acoustic rhinometry. Am J Rhinol. 2006;20(5):456–62.
34. Fisher EW, Scadding GK, Lund VJ. The role of acoustic rhinometry in studying the nasal cycle. Rhinology. 1993;31(2):57–61.
35. Fouke JM, Jackson AC. Acoustic rhinometry: effects of decongestants and posture on nasal patency. J Lab Clin Med. 1992;119(4):371–6.
36. Hochban W, Althoff H, Ziegler A. Nasal decongestion with imidazoline derivatives: acoustic rhinometry measurements. Eur J Clin Pharmacol. 1999;55(1):7–12.
37. Samolinski B, Grzanka A, Zawisza E, Arcimowicz M. Acoustic rhinometry in the assessment of the topical treatment of upper respiratory infections with fusafungin. Otolaryngol Pol Pol Otolaryngol. 1998;52(3):327–34.
38. Rimmer J, Greenwood A, Bartlett D, Hellgren J. Nasal steroids improve regulation of nasal patency in asthma and mild rhinitis: a randomised, cross-over trial. Eur Arch Otorhinolaryngol. 2012;269(4):1133–8.
39. Wandalsen GF, Mendes AI, Sole D. Objective improvement in nasal congestion and nasal hyperreactivity with use of nasal steroids in persistent allergic rhinitis. Am J Rhinol Allergy. 2010;24(1):e32–6.
40. Khattiyawittayakun L, Seresirikachorn K, Chitsuthipakorn W, Kanjanawasee D, Snidvongs K. Effects of decongestant addition to intranasal corticosteroid for chronic rhinitis: a systematic review and meta-analysis. Int Forum Allergy Rhinol. 2018;8(12):1445–53.
41. Friedman M, Vidyasagar R, Joseph N. A randomized, prospective, double-blind study on the efficacy of dead sea salt nasal irrigations. Laryngoscope. 2006;116(6):878–82.
42. Yamagiwa M. Acoustic evaluation of the efficacy of medical therapy for allergic nasal obstruction. Eur Arch Otorhinolaryngol. 1997;254(Suppl 1):S82–4.
43. Li AM, Abdullah VJ, Tsen CS, Au CT, Lam HS, So HK, et al. Leukotriene receptor antagonist in the treatment of childhood allergic rhinitis—a randomized placebo-controlled study. Pediatr Pulmonol. 2009;44(11):1085–92.
44. Kim YH, Jang TY. Clinical characteristics and therapeutic outcomes of patients with localized mucosal allergy. Am J Rhinol Allergy. 2010;24(4):e89–92.
45. Aydin E, Hizal E, Onay O, Ozgen B, Turhan B, Zaimoglu M, et al. A double-blind, placebo-controlled, randomized clinical study of the effects of vardenafil on human nasal patency. Am J Rhinol. 2008;22(3):276–9.
46. Mygind N, Dahl R. Challenge tests in nose and bronchi: pharmacological modulation of rhinitis and asthma. Clin Exp Allergy. 1996;26(Suppl 3):39–43.
47. Antila J, Sipila J, Tshushima Y, Polo O, Laurikainen E, Suonpaa J. The effect of laser-uvulopalatopharyngoplasty on the nasal and nasopharyngeal volume measured with acoustic rhinometry. Acta Otolaryngol Suppl. 1997;529:202–5.
48. Invitto S, Calcagni A, Piraino G, Ciccarese V, Balconi M, De Tommaso M, et al. Obstructive sleep apnea syndrome and olfactory perception: an OERP study. Respir Physiol Neurobiol. 2019;259:37–44.
49. Holmstrom M. The use of objective measures in selecting patients for septal surgery. Rhinology. 2010;48(4):387–93.
50. Moore M, Eccles R. Objective evidence for the efficacy of surgical management of the deviated septum

as a treatment for chronic nasal obstruction: a systematic review. Clin Otolaryngol. 2011;36(2):106–13.
51. Andre RF, Vuyk HD, Ahmed A, Graamans K, Nolst Trenite GJ. Correlation between subjective and objective evaluation of the nasal airway. A systematic review of the highest level of evidence. Clin Otolaryngol. 2009;34(6):518–25.
52. Batra PS, Seiden AM, Smith TL. Surgical management of adult inferior turbinate hypertrophy: a systematic review of the evidence. Laryngoscope. 2009;119(9):1819–27.
53. Straszek SP, Taagehoj F, Graff S, Pedersen OF. Acoustic rhinometry in dog and cat compared with a fluid-displacement method and magnetic resonance imaging. J Appl Physiol. 2003;95(2):635–42.
54. Straszek SP, Pedersen OF. Nasal cavity dimensions in guinea pig and rat measured by acoustic rhinometry and fluid-displacement method. J Appl Physiol. 2004;96(6):2109–14.
55. Kaise T, Ukai K, Pedersen OF, Sakakura Y. Accuracy of measurement of acoustic rhinometry applied to small experimental animals. Am J Rhinol. 1999;13(2):125–9.
56. Mostafa BE. Detection of adenoidal hypertrophy using acoustic rhinomanometry. Eur Arch Otorhinolaryngol. 1997;254(Suppl 1):S27–9.
57. Riechelmann H, O'Connell JM, Rheinheimer MC, Wolfensberger M, Mann WJ. The role of acoustic rhinometry in the diagnosis of adenoidal hypertrophy in pre-school children. Eur J Pediatr. 1999;158(1):38–41.
58. Riechelmann H, Rheinheimer MC, Wolfensberger M. Acoustic rhinometry in pre-school children. Clin Otolaryngol Allied Sci. 1993;18(4):272–7.
59. Isaac A, Major M, Witmans M, Alrajhi Y, Flores-Mir C, Major P, et al. Correlations between acoustic rhinometry, subjective symptoms, and endoscopic findings in symptomatic children with nasal obstruction. JAMA Otolaryngol Head Neck Surg. 2015;141(6):550–5.
60. Buenting JE, Dalston RM, Drake AF. Nasal cavity area in term infants determined by acoustic rhinometry. Laryngoscope. 1994;104(12):1439–45.
61. Buenting JE, Dalston RM, Smith TL, Drake AF. Artifacts associated with acoustic rhinometric assessment of infants and young children: a model study. J Appl Physiol. 1994;77(6):2558–63.

27 New Measurement Methods in the Diagnostic of Nasal Obstruction

Gunter H. Mlynski, Giorgi Gogniashvili, and Achim G. Beule

Core Messages

- Accurate preoperative analysis of extent and causes of nasal breathing complaints in combination with a stronger physiological perspective in surgical therapy is needed if we want to make progress beyond unsatisfactory long-term outcomes with inadequate improvement in nasal airway obstruction and frequent postoperative sicca symptoms.
- New measurement methods need to be used to objectify nasal obstruction, since rhinomanometry is only capable of adequately differentiating the degree of obstruction, but not all its causes.
- Rhinoresistometry, a refinement of rhinomanometry, makes it possible to objectively determine not only the degree of obstruction but also to differentiate between swelling, skeletal stenosis, inspiratory collapse of the nasal valves and pathological turbulence as possible fluid dynamics causes of nasal obstruction.
- By combining rhinoresistometry with acoustic rhinometry, it is possible to accurately localize skeletal stenosis and determine the causes of pathological turbulence.
- Rhinoresistometry and acoustic rhinometry only enable the objective measurement of conditions in the nose at the time of measurement. Long-term rhinometry has been developed to overcome this limitation.
- Long-term rhinometry yields information about the nasal cycle in specific conditions and provides an objective measurement of reactive congestion over a 24-h period under the usual conditions of the patient's everyday livings conditions.
- New techniques for the diagnostic evaluation of nasal obstruction make it possible to set better surgical indications prior of functional nasal surgical interventions and to better plan the surgery.
- Rhinoresistometry, acoustic rhinometry and long-term rhinometry now offer the practitioner tools that allow essential postoperative quality control in functional rhinosurgery.

G. H. Mlynski
Stolpe auf Usedom, Germany
e-mail: stolpe@mlynski.com

G. Gogniashvili
Khujadze-Gogniashvili ENT Clinic, Tbilisi, Georgia

A. G. Beule (✉)
Department of Otorhinolaryngology, University Hospital Münster, Münster, Germany

Department of Otolaryngology, Head and Neck Surgery, University Medicine Greifswald, Greifswald, Germany
e-mail: achimgeorg.beule@ukmuenster.de

Ö. Ö. Celebi, T. M. Önerci (eds.), *Nasal Physiology and Pathophysiology of Nasal Disorders*,
https://doi.org/10.1007/978-3-031-12386-3_27

27.1 Preliminary Remarks

Despite several significant advances in the area of procedures for the diagnostic evaluation of nasal obstruction, there has been a striking failure until now to implement routine objective preoperative diagnostic testing prior to functional and aesthetic rhinosurgery, and for postoperative quality control.

To objectify nasal obstruction currently rhinomanometry (RMM) is being used worldwide [1]. This method measures nasal airflow provided pressure and thereby enables the physician to estimate the extent of nasal obstruction. Also, differentiation between a constriction caused by congestion or skeletal stenosis is possible. A critical functional analysis of the important nasal valve area is not possible even though a pathological internal nasal valve is a frequent cause for an obstructed nasal breathing. Moreover, the important endonasal airflow pattern regarding the transition of laminar to turbulent flow is not sufficiently objectified. This is one reason for the fact that pathological turbulence is being seldom analysed as possible pathophysiological factor in the pathophysiology of nasal obstruction and thereby the underlying causes overlooked, resulting in unsatisfactory surgical results.

However, evaluation of nasal airway patency is not sufficient by itself and in some cases may even be misleading. Thus, nasal airway resistance can be decreased by resecting the nasal turbinates, thereby improving airflow. The results of RMM testing might simulate functional improvement [2], even though, in fact, nasal respiratory function has been largely destroyed after resection of the nasal turbinates [3–13].

RMM only provides an evaluation of nasal airflow at one flow velocity. However, nasal resistance varies depending on the flow velocity due to constantly changing endonasal generation of turbulence and also often varies dynamically because of the collapse of the nasal valve from negative pressure as a result of the Bernoulli effect (see Sect. 20.2.2).

For these reasons, we must consider RMM as being of limited value for the diagnostic evaluation of nasal obstruction [2]. Until now, this has the effect of leaving many surgeons sceptical about the overall value of functional testing in the field of functional rhinosurgery. Consequently, RMM has been further refined to rhinoresistometry (RRM) [2, 14–17].

In recent years, acoustic rhinometry (ARM) has been established as an additional analysis tool for rhinologic diagnostic testing (see Chap. 26). It enables measurement of the cross-sectional areas of the nasal flow channel in relation to the distance from the external nasal ostium [1, 18]. Because airway resistance does not solely depend on the magnitude of the cross-sectional area of a flow channel but also on its shape, ARM cannot be used to estimate the magnitude of nasal obstruction. However, ARM permits conclusions about the location of narrowings and possible causes and locations of pathological turbulence in the nose (see Sects. 27.2.1.1 and 27.2.3).

RMM, RRM and ARM only permit an assessment of nasal obstruction at the time of measurement. To gain insight into the rhinologic function of the nose during patient's normal physical activity, long-term rhinometry (LRM) was developed [19, 20]. LRM technique permits side-specific measurement of the nasal cycle over a 24-h period. It yields information on the functionality and adaptive capacity of the nose during periods of increased oxygen demand resulting from physical activity under the typical conditions of a patient's everyday life.

This book devotes a separate Chapter to RMM and to ARM. In this chapter, RRM and LRM will be described in Sects. 27.2.1 and 27.2.2. In addition, we will show how ARM can be used to make deductions about the causes of pathological nasal valve collapse and of pathological nasal turbulences (see Sect. 27.2.1.1). Moreover, we will demonstrate how the combination of RRM an ARM (and in specific cases LRM) makes it possible not only to objectify the extent of nasal obstruction but also to differentiate among its possible causes. This diagnostic algorithm will be illustrated by means of clinical examples.

The subjective sensation of nasal obstruction depends not solely on airway resistance but also on several additional factors, and therefore, it cannot be completely measured even with the most refined methods [21]. Especially because of the known discrepancy between objective measurements of functional impairment and subjective sensations [22–26], the objectification of the different aetiologies of functional impairment is important for effective planning of the appropriate treatment.

A precise preoperative analysis of the extent and the cause of breathing difficulties in the nose in combination with a greater consideration of physiological aspects in surgical therapy is necessary if we want to make progress in long-term results [3–5, 7–9, 13, 27–32]. The latency between surgery and the emerging of sicca symptoms, which often takes several years, reflects the compensatory capacity of the healthy mucosa [33]. This makes it more difficult to recognize the causal relationship with previous surgery, something that will only be possible in the context of prospective cohort studies that look at long-term outcomes [34, 35].

27.2 New Techniques for the Diagnostic Evaluation of Nasal Obstruction

In the following sections, we will describe new techniques for objectifying nasal obstruction and its causes.

27.2.1 Rhinoresistometry (RRM)

Motivated by the inadequacy of information obtained through rhinomanometric testing and the capabilities of modern computer technology, RMM was further refined into RRM [2, 14–17]. Based upon the laws of fluid dynamics, this method uses values measured by rhinomanometry of the pressure difference between the external nasal orifice and the choanal area simultaneously with the airflow velocity to compute diagnostically relevant parameters. The equipment and the measurement procedure used in RRM completely correspond to this used in active anterior RMM. However, for the RRM equipment, the guidelines established by the "International Committee on Objective Assessment of the Upper Airways" [1] need to be followed. In addition, to determine not only the extent of nasal obstruction, RRM permits the differentiation between the possible causes of a nasal obstruction: narrowing caused by swelling, and/or by skeletal stenosis, and/or by inspiratory collapse of the nasal valves, and/or pathologically turbulence behaviour.

The results of RRM are presented by means of graphs and numerical values. The graphs allow the reader to make a "diagnostic at a glance", and the numerical values are used for precise analysis and pre- and post-therapeutic comparisons.

27.2.1.1 Graphical Presentation

Figure 27.1 shows the graphic illustrations used in rhinoresistometry. The findings are presented in red for the right side of the nose and blue for the left side. Measurements taken before mucosal decongestion are shown in a light colour. Measurements taken after decongestion are shown in a dark colour.

Flow-dependent increase in resistance is presented in the upper graph (Fig. 27.1). It is apparent that in both inspiration and expiration, nasal airway resistance rises with increasing flow velocity.

We know from physiology that during moderate physical activity, maximal breathing flow (both sides of the nose combined) of approx., 500 mL/s is required to maintain adequate oxygen supply. During heavy physical activity, the oxygen requirement is met through supplemental bypass breathing through the mouth. As a result, flow velocities rarely exceed 500 mL/s in the nose. For this reason, we describe the range between 0 and 500 mL/s as "physiologic breathing range" (see Sect. 20.1). Experience tells us that during the rhinometric test situations, patients breathe more deeply than necessary. The patient achieves flow velocities of up 800 mL/s. This is reason why we observe very long inspiratory and expiratory curves in rhinomanometric testing. Therefore, in RRM, the inspiratory and expiratory portions of the curves are marked with points to indicate where the flow through the right and left side of the nose together contributes

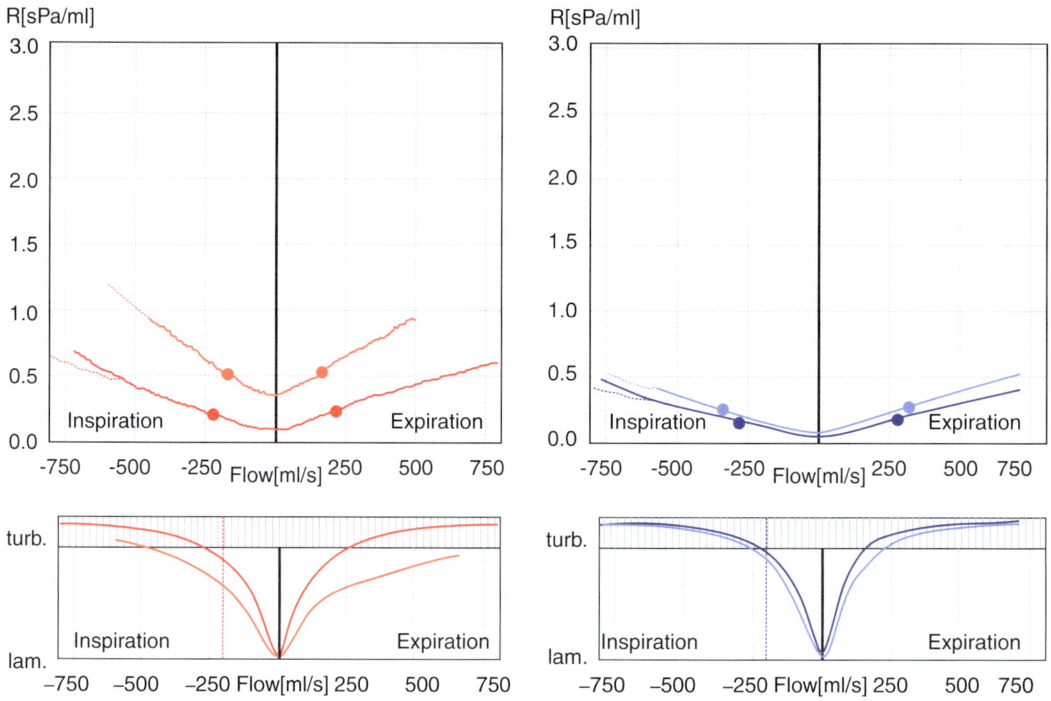

Fig. 27.1 Graphic representation of the findings from RRM. Right side of the nose, red; left, blue. Light-colored curves: before decongestion of the mucosa. Dark-colored curves: after decongestion of the mucosa. Upper graph: nasal airway resistance. Lower graph: turbulence flow behavior of the nasal airstream. *x*-axis: left of the midline: inspiratory flow in mL/s. right of the midline: expiratory flow in mL/s. *y*-axis: upper graph: resistance in sPa/mL. lower graph: turbulent flow behavior: *lam.* laminar, *turb.* turbulent

up to 500 mL/s. This makes it possible to see at a glance how much each side of the nose is contributing to total flow if moderate physical activity is performed and what levels of nasal flow resistance are acting on each side of the nose during "physiologically required airflow".

The graph enables us to evaluate nasal airway resistance for each side of the nose "at a glance": the higher the course of the curve, the more pronounced is the nasal obstruction. In Fig. 27.1 the right side of the nose is moderately obstructed before decongestion. After decongestion the resistance has normalized. On the left side, the resistance is physiological before and after decongestion.

As in RMM, by examining the distance between the curves before and after decongestion, we can differentiate between the portions of nasal airway resistance due to congestion and due to skeletal narrowing. Figure 27.2 shows rhinoresistometric resistance curves of a nose with mucosal swelling on both sides. After decongestion, on the left side the resistance drops to very small values but on the right side an increased resistance remains, indicating an additional skeletal narrowing.

In addition to identifying swelling and skeletal stenosis at a glance, the graph can also be used to identify whether or not there is significant inspiratory nasal valve collapse (NVC). For this purpose, a calculated curve for inspiration is shown as a dotted line, which shows the flow-dependent increase in airway resistance with a stable vestibular sidewall without collapse. The measured and calculated curves are congruent if the nasal valves are not sucked in Fig. 27.2. Deviations of the measured (solid) line from the dotted line indicate abnormalities in the width of the flow channel, such as those caused by NVC (Fig. 27.3).

The greater the extent of NVC, the more pronounced the measured curve will deviate from the calculated curve [36]. A slight deviation at

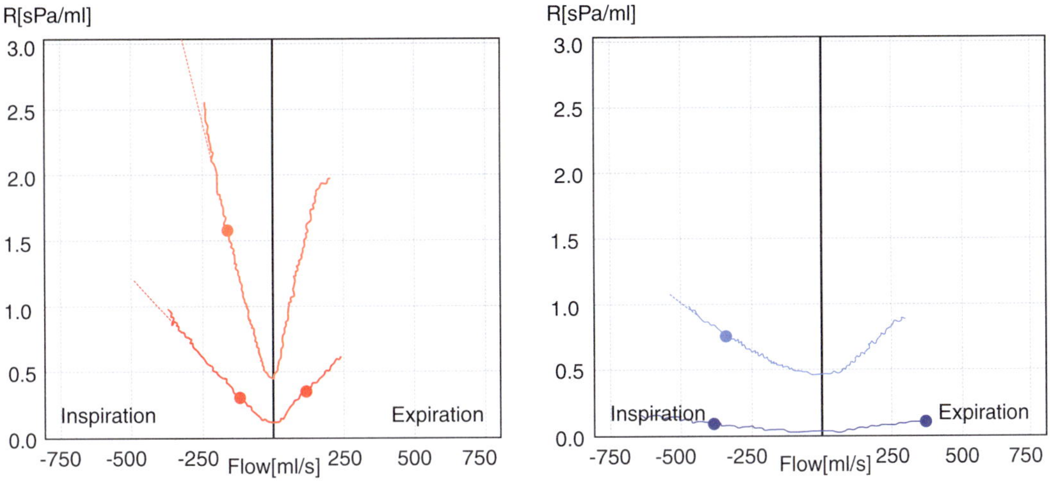

Fig. 27.2 Rhinoresistometry resistance curves of a nose obstructed by mucosal congestion on both sides and additional skeletal constriction on the right

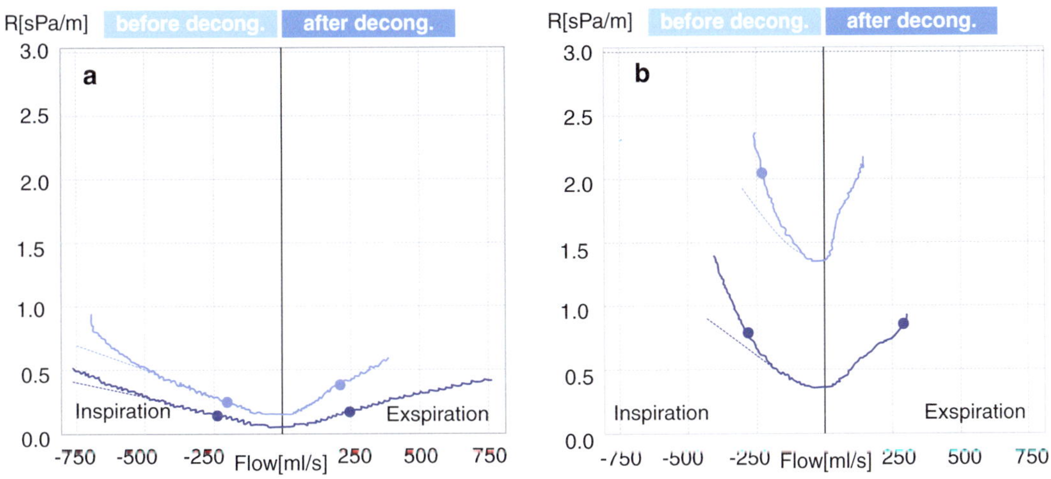

Fig. 27.3 Resistance curves measured by rhinoresistometry (continuous lines) and resistance curves calculated with a stable lateral vestibular wall (dotted lines on the inspiratory side of the curve). (**a**) Before decongestion: no NVC. (**a**) After decongestion and left before decongestion: physiological NVC. Left after decongestion: pathological NVC rhinoresistometric resistance curves of a nose with nasal valve collapse (**a**) physiologic nasal valve collapse; (**b**) pathological nasal valve collapse

high flow velocities (> 500 mL/s) indicates a *physiological* collapse of the nasal valve (Fig. 27.3a). One can identify pathological collapse of the nasal valve based on a significant deviation of the measured from the calculated curves (Fig. 27.3b).

In the RRM, the lower graph (Figs. 27.1 and 27.5) displays the turbulence behaviour of nasal airflow in relation to airflow velocity. The level on the *X*-axis corresponds to pure laminar flow, and the upper blue-grey striped bars correspond to complete turbulence.

At very low flow velocities, the flow is completely laminar in any flow channel and thus in the nose as well [37]. As the velocity of flow increases, both in inspiration and expiration laminar flow portions transition more and more into turbulent flow portions (see 20.4). This "transition range" is important for the respiratory function of the nose. It creates the optimal situation for conditioning the air. It provides adequate mucosal contact by the streaming air without leading to dryness or pathological cooling of the mucosa (see Sect. 20.4). At very high flow velocities, nasal airflow becomes completely turbulent [37–42]. Complete turbulence hardly ever occurs in a normal nose. When high airflow velocities are required to provide adequate oxygen supplies during heavy physical stress, mouth-bypass breathing is unconsciously employed so that the nasal airstream decreases (see Sect. 20.1).

Nasal airflow should become more turbulent when the mucosa is in a decongested state (corresponding to the working phase of the nasal cycle; see Sect. 20.4) as a condition for sufficient mucosal contact than in the congested state (corresponding to a resting phase in the nasal cycle). In the resting phase (decongested mucosa), the flow character should not become purely turbulent up to an airflow velocity of 200 mL/s (Fig. 27.1 bilaterally; Fig. 27.4 right side of the nose; see Sect. 20.4), and thus reduces turbulence in the nose.

At pathological turbulence, there is a rapid transition so that pure turbulence will already develop in the nose at flow velocities <200 mL/s (Fig. 27.4. left side of the nose).

Complete turbulence in the nose can generate elevated airway resistance, and besides it also causes sicca symptoms with a sense of obstruction.

27.2.1.2 Numerical Evaluation

Besides using the graphical representation to get a "diagnosis at a glance" numerical values are calculated for a more precise assessment of extent and causes of obstruction (Fig. 27.5).

Differences between individuals with the same degree of nasal obstruction may have quite varied levels of symptoms depending on age, gender, body mass index, level of the physical fitness and additional medical conditions. Therefore, the reference values presented here

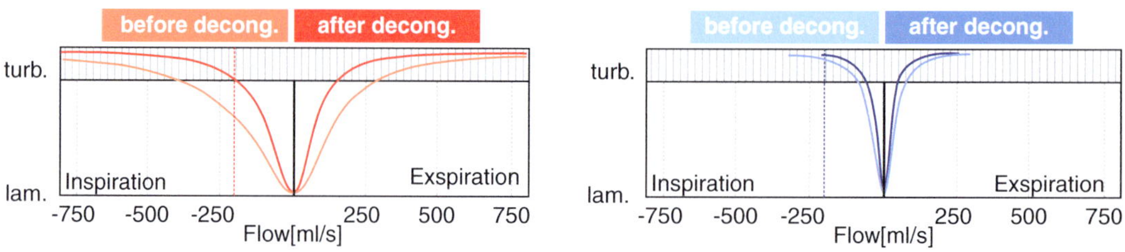

Fig. 27.4 Turbulence curves of rhinoresistometric measurements

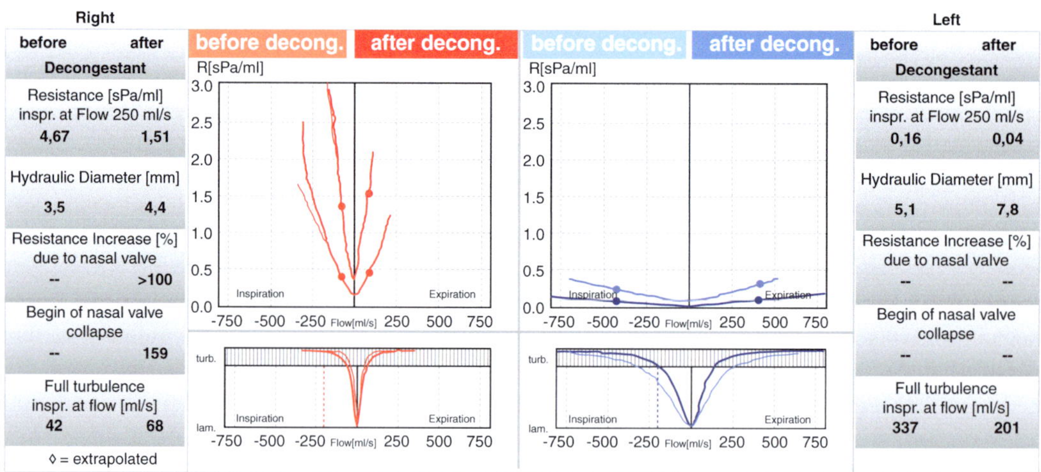

Fig. 27.5 Rhinoresistometric resistance curves and turbulence curves with corresponding numerical values

should not be applied too rigidly. They are simple benchmarks for the purpose of orientation. The values were determined through studies on both rhinologically healthy patients and patients with nasal obstruction [21, 43–45].

Nasal resistance (*R*) is numerically represented at a flow velocity of 250 mL/s before and after decongestion. As reported in Sect. 20.6, an airflow of approx. 500 mL/s is required in the physiological breathing range. If both nasal sides make an equal contribution to this requirement, 250 mL/s would flow per second through each nasal side.

Thus, for the practical evaluation of nasal obstruction, the resistance measured at 250 mL/s is an appropriate value ([42, 46]; see also Sect. 20.1). The extent of nasal obstruction can be estimated according to the reference values in Table 27.1.

The *hydraulic diameter* (d_h) is a measure for the width of the nasal cavum. The nasal cavity is a space with irregular cross sections. Therefore, its width cannot be defined simply by its diameter as it is common with a perfectly round tube. In technological science, the usual practice when dealing with an irregular cross section of a channel is to use the "hydraulic diameter". This is the diameter of a tube of the same length with a round cross section for a channel that has the same resistance to airflow as the irregularly shaped flow channel. The figures shown in Table 27.2 can be used as reference values for estimating the width of the interior of the nose.

The hydraulic diameter of both sides of the nose is calculated before and after the decongestion. Using these values, one can identify a mucosal swelling and a skeletal narrowing as causes for nasal obstruction. The increase in magnitude

Table 27.1 Reference values of nasal obstruction via resistance

Resistance at 250 mL/s	Extent of obstruction on one side of the nose	Interpretation
<0.17 sPa/mL	No obstruction	Physiological resistance
0.17–0.35 sPa/mL	Slight obstruction	
0.36–0.70 sPa/mL	Moderate obstruction	Pathological resistance
>0.70 sPa/mL	Severe obstruction	

Table 27.2 Reference values for the interior of the nose based on the hydraulic diameter

Hydraulic diameter	Width of the nose
<5.0 mm	Too narrow
5.0–6.5 mm	Normal width
>6.5 mm	Too wide

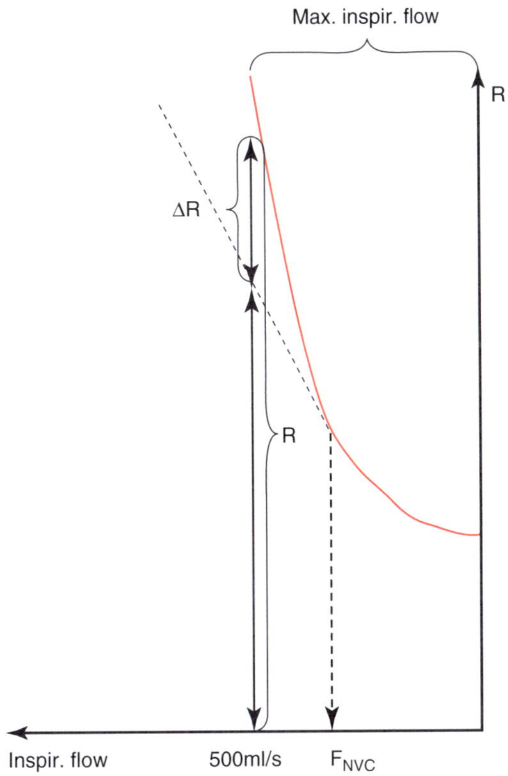

Fig. 27.6 Calculation of numerical value (ΔR) for the increase of nasal resistance by the nasal valve collapse

of the hydraulic diameter by decongestion is an index of the extent of mucosal swelling. If the hydraulic diameter remains low after decongestion, this is evidence of skeletal stenosis. A very large hydraulic diameter indicates a nose that is pathologically too wide.

Numerical values for the NVC are calculated before and after decongestion. Therefore, the calculated course of the flow-dependent increase of resistance without nasal valve collapse (see Sect. 27.2.1.2) is used. The numerical value for the increase of nasal resistance caused by the valve collapse (ΔR) is calculated at a flow of 500 mL/s (Fig. 27.6). Using this parameter, physiological and pathological nasal valve collapse can be differentiated. The reference values are indicated in Table 27.3. In addition, the minimal flow at the starting point of the inspiratory nasal valve collapse (F_{NVC}) is indicated as value. This parameter indicated the flow, at which the calculated curve of nasal resistance starts to differ from the measured one. Physiologically, the nasal valve should not show an inspiratory nasal valve collapse up to 500 mL/s.

The numerical value for the increase in resistance from NVC (ΔR) allows us to estimate the magnitude of the suction phenomena as well as to differentiate between physiological and pathological NVC in accordance with Table 27.3. In addition, the nasal airflow velocity at which collapse begins (F_{NVC}) is quantified (Table 27.4). The nasal valves should not be significantly collapsed in ($\Delta R > 25\%$) at airflow velocities up to the maximum physiologically required nasal airflow of 500 mL/s.

The transition to turbulence in endonasal airflow is numerical classified according to the flow velocity, at which flow is completely turbulent. For the respiratory function of the nose airflow within the physiologic breathing range (see Sect. 20.1) in the transition from laminar to turbulent flow allows a physiological amount of contact to the endonasal mucosa (see Sect. 20.4). In a decongested state, like a working phase of the nasal cycle (see Sect. 20.5), complete turbulence should be achieved at a flow velocity of <200 mL/s. This flow value is indicated as "Full turbulence inspiratory at flow (mL/s)" and corresponds in the graph to a dotted (red or blue) line in the inspiratory section (Figs. 27.4 and 27.5).

Table 27.3 Differentiation between physiologic and pathologic nasal valve collapse (NVC)

ΔR	NVC
≤25%	Physiological
>25%	Pathological

Table 27.4 Classification of turbulence based on the commencement of full turbulent flow at an inspiratory flow

Full turbulence at inspiratory flow (mL/s)	Turbulence behaviour
≥200 mL/s	Physiological
<200 mL/s	Pathological

NVC nasal valve collapse, ΔR refers to the difference between measured and calculated resistance at 500 m/s endonasal flow

27.2.1.3 Objectification of Nasal Obstruction with RRM

The diagnostic possibilities of RRM are illustrated in the following part using a rhinoresistometric result of the patient (cf. Fig. 27.5).

At one glance, the graph of the right nasal side allows detection of a severe obstruction due to the steep course of nasal resistance curve. As aetiology, a skeletal stenosis can be detected because resistance after decongestion persists in an increased matter despite some effect of decongestion. Besides, on this nasal side a pathological NVC (with the steeper course of the measured, continuously illustrated resistance line than the calculated, dotted resistance line) and a pathological behaviour of turbulence generation (fast transition to full turbulence at an inspiratory flow of <200 mL/s) attribute to the increase of nasal resistance.

The corresponding numerical values after decongestion confirm these pathologies. On the right nasal side after decongestion, the flow resistance is pathologically increased with 1.51 sPa/mL, and the nose remains too narrow after decongestion (hydraulic diameter: 4.4 mm). Besides, the contribution of inspiratory nasal valve collapse to nasal resistance at 250 mL/s is larger than 100% and nasal airflow is already completely turbulent at 68 mL/s.

The curve of nasal resistance on the left nasal side is very shallow. With 0.04 sPa/mL, nasal resistance after decongestion is pathologically decreased. This nasal side is pathologically wide (hydraulic diameter: 7.8 mm). Also, after decongestion the behaviour of turbulence is marginal (complete turbulence already at a flow of 201 mL/s).

27.2.2 Acoustic Rhinometry (ARM)

Acoustic rhinometry has been extensively presented in Chap. 26 with its possibilities and limitations.

ARM is an appropriate method to objectify stenosis due to congestion and skeletal deformities [1, 2, 15–17]. It is not suitable to assess the extent of nasal obstruction, because nasal resistance depends not only on the magnitude of the cross-sectional area but also on their configuration. In case of similar cross-sectional areas, the channel with a slit-like configuration has a higher resistance than the more circular one, because the wall area and thereby the friction are larger within the slit-like channel (see Sect. 20.2.1). ARM should therefore only be used in connection with a dynamic measurement method for assessment of resistance, like rhinomanometry or rhinoresistometry. Apart from this limitation, ARM nevertheless accomplishes a valuable contribution to the diagnostic of nasal obstruction. It allows a sufficiently exact assessment of the geometry within the nasal vestibule and the anterior cavum. Shape changes in this for the nasal airflow important inflow area of the nose are common causes for symptoms of the patient (see Sect. 20.3.1).

27.2.2.1 New Parameter in ARM

Using the previously in Sect. 27.2.1 described RRM, an objectification of the effect of a skeletal stenosis or nasal valve collapse on nasal resistance becomes possible. However, information is missing regarding localization of the stenosis or shape changes causing air pathological valve function or pathological turbulence in the nose. However, for effective correction of these shape changes by functional rhinologic surgery all this information is valuable.

To improve the detection of pathological shape changes in the inflow area, the previously calculated parameter was supplemented by two newly introduced ones [2, 15–17]; Fig. 27.8.

- MCA0 for the external ostium (external nasal valve): This inflow opening for the inspiratory air flow is the first physiological constriction, which often causes an increase in the nasal resistance due to an additional permanent pathological constriction. Dynamic stenosis during inspiration (collapse of the outer nasal valve) can also lead to pathological resistance.

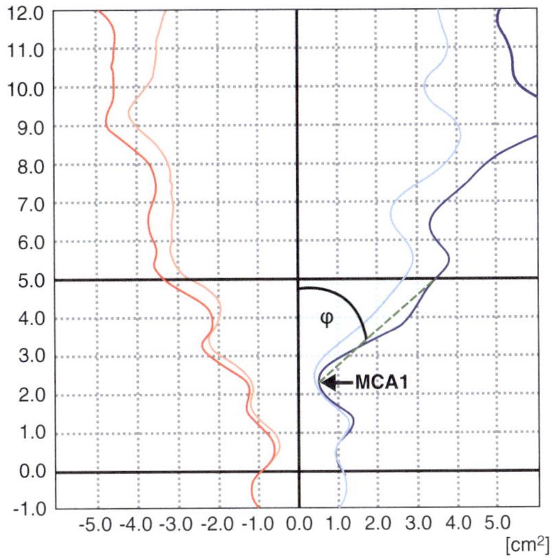

Fig. 27.7 Calculation of the opening angle of the diffuser φ using acoustic rhinometry

Table 27.5 Impact of diffuser opening angle φ on endonasal turbulence

Turbulence in the nose	Diffuser opening angle φ
Slight turbulence formation	<7°
Moderate turbulence formation	7°–9°
Severe turbulence formation	>9°

- Opening angle of the diffuser φ: From the enlargement of the cross-sectional area inside the anterior cavum, an opening angle of the nasal diffuser can be calculated from the areas measured by ARM (Fig. 27.7, Table 27.5). This parameter makes it possible to distinguish two aetiologies of increased pathological turbulence of the nose:
  - A narrow diffuser entrance area (small MCA1).
  - A large increase in the cross-sectional area within the diffuser (see Sect. 20.4).

27.2.2.2 Objectification of Causes for Nasal Obstruction with ARM

The extent of an obstruction cannot be assessed with the ARM (see above). However, the method is suitable to localize pathological constrictions and to objectify the flow dynamic causes of pathological turbulence.

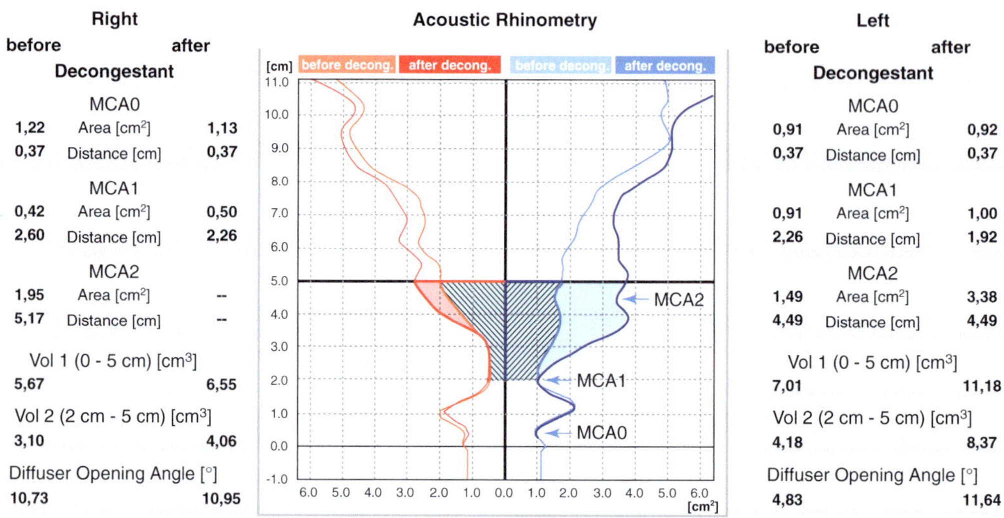

Fig. 27.8 Illustration of acoustic rhinometric assessment including numerical values

In Fig. 27.8 the ARM examination of the same patient as in Fig. 27.5 is illustrated. As a diagnosis at first glance on the right nasal side it is possible.

- To identify the localization of the skeletal stenosis as aetiology for the pathological resistance after decongestion (cf. Fig. 27.8 in the area of MCA1).
- On the right nasal side the narrow entrance of the diffuser (MCA1) before and after decongestion is the cause for the pathologically increased turbulence behaviour (cf. Fig. 27.8).
- The narrow MCA1 causes a high local airflow velocity and thereby a strong Bernoulli effect in the area of the internal nasal valve.
- On the left nasal side a marked increase of the cross-sectional area in the diffuser causes a turbulence, which is on the border to pathological finding (cf. Fig. 27.8).

Information obtained by RRM and ARM are helpful in the preoperative planning of rhinosurgical steps.

In case of the patient with the diagnostic RRM and ARM findings in Figs. 27.5 and 27.7, the surgical aim is to increase the width on the right nasal side by septoplasty. Thereby, the pathologically increased resistance on the right side and the pathologically low resistance on the left side are normalized.

It is not necessary to increase the stiffness of the right nasal wing to correct the pathological valve collapse, caused by a strong Bernoulli phenomenon due to a high local air flow velocity within the narrow MCA1 (see Sect. 20.2). After correcting the constriction, physiological function of the nasal valve is to be expected. The pathological turbulence behaviour is also improved if the entry area of the diffuser (MCA1) becomes larger.

A reduction of the turbinates is contraindicated, because the already large opening angle of the diffuser would result in a further increase of endonasal turbulence. The marginal turbulence behaviour on the left nasal side will be improved by straightening of the nasal septum because the nozzle effect of the vestibule is regained by normalization of the MCA1.

The postoperative result 1 year after surgery is presented in Sect. 27.3.2.2 and Fig. 27.14b.

27.2.3 Long-Term Rhinometry (LRM)

LRM was developed because RMM, RRM and ARM only allow an assessment of nasal obstruction at the time of the measurement [47].

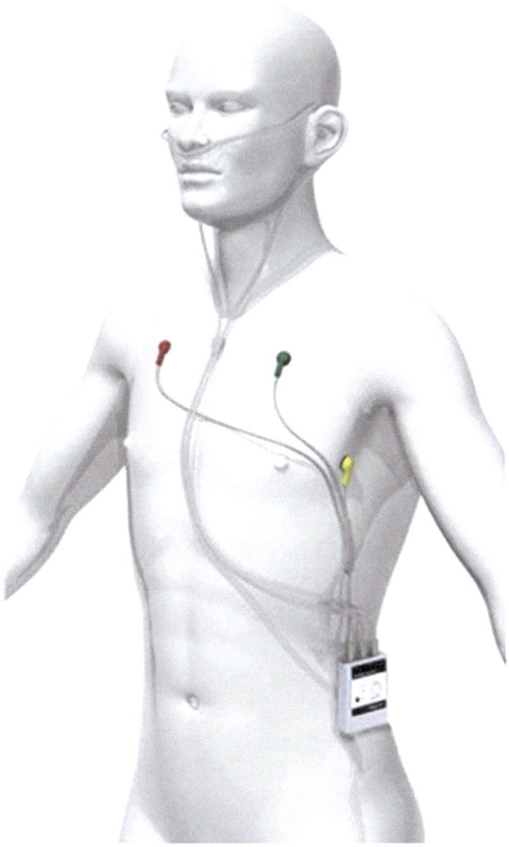

Fig. 27.9 Measurement system for long-term rhinoflowmetry

However, some patients complain about symptoms that occur at other times of the day. LRM makes it possible to measure nasal flow separately for each side of the nose, along with the heart rate to serve as an index for physical activity over a 24-h time period under the patient's everyday life conditions. Nasal flow is measured using standard commercial nasal oxygen cannulas and the heart rate using standard ECG electrodes. Recording and storing is performed by means of a battery-powered portable device (Fig. 27.9). The recording of the nasal cycle enhance the facilities to objectify the nasal respiratory function [48].

Figure 27.10 illustrates the graphical curves resulting from a LRM examination. The upper graph presents the maximal nasal flow values during inspiration separately for each side of the nose in relation to time. The lower graph shows the heart rate for estimating physical activity, respiratory rate, as well as nasal minute volume (NMV) in relation to time.

27.2.3.1 Upper Graph (Figs. 27.10 and 27.11)

The curves for the nasal flow allow a visual assessment of the nasal cycle at a glance. At physical rest or light physical activity, a "classical type" of nasal cycle should occur, with reciprocal alternations between the two sides of the nose in working and resting phases. With increasing physical activity, the simultaneous transition of both sides of the nose into working and resting phases (the "In-concert" type) is a physiological response to increased oxygen demand (see Chap. 20). In this situation, the nasal flow in both sides should reach between 250 and 500 mL/s, as shown in Fig. 27.10. These flow velocities cannot be reached in the presence of pathological nasal air resistance with the deployment of mouth-bypass breathing. An example is shown in Fig. 27.11.

27.2.3.2 Lower Graph (Figs. 27.10 and 27.11)

Heart Rate (Orange Curve)

In LRM, the heart rate is used to determine the level of physical activity. This relationship is known from physiology (Table 27.6). Please note that these values can vary considerably between people, depending on age, gender, level of education and also on medication (e.g. beta blockers). Therefore, these values can only be used relatively.

Since the oxygen requirement depends on physical activity, the heart rate provides an information of the oxygen requirement and thus the necessary respiratory flow. The heart rate is therefore an important measure when assessing the curves for flow, breathing rate and nasal breathing minute volume.

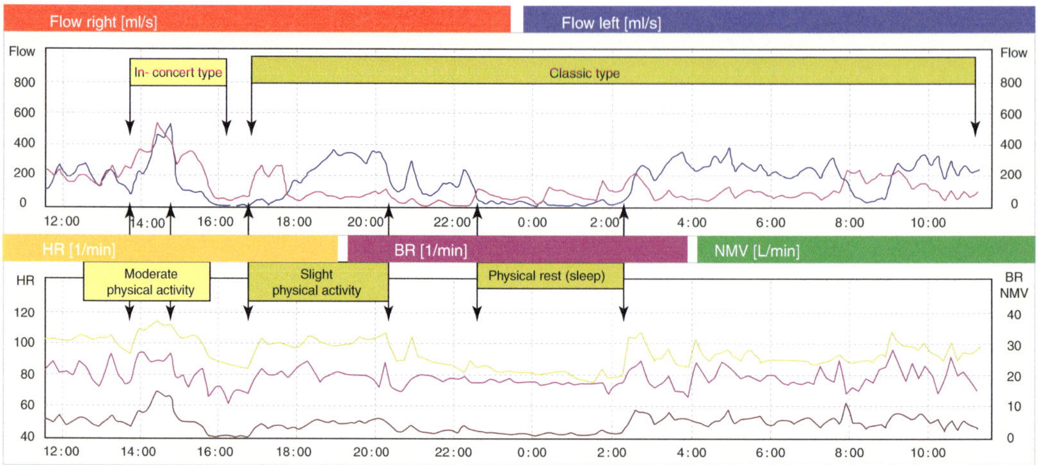

Fig. 27.10 Findings from LRM in a person with normal nasal breathing. To better grasp the relation between physical activity and mucosal congestion in the nose, in this figure the type of cycle and the amount of physical activity were marked. *x*-axis: time of day in hours. *y*-axis: upper graph, nasal respiratory velocity at maximal inspiration in mL/s (red right nose, blue, left nose). Lower graph: orange, heart rate (*HR*); green, nasal respiratory minute volume (*NMV*) in L/min; purple, respiratory rate (*BR*)

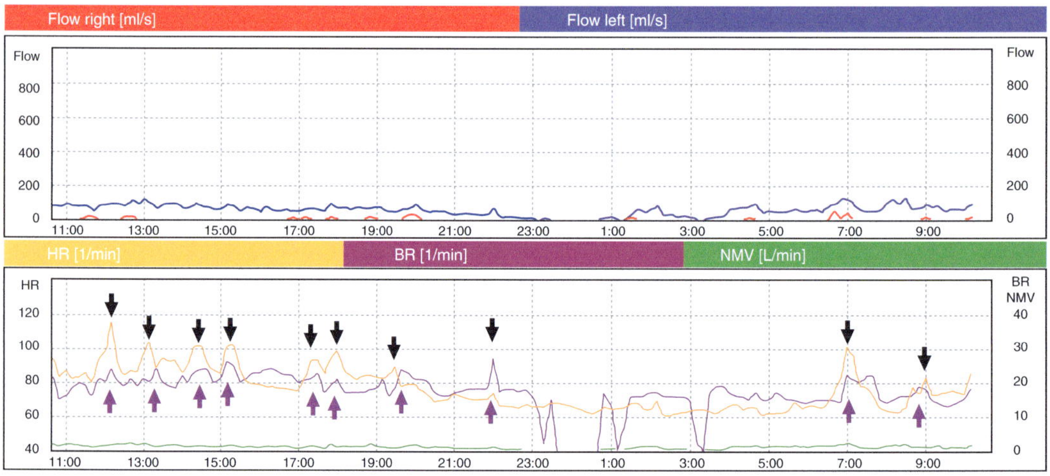

Fig. 27.11 RRM, ARM, and LRM findings in a patient without nasal obstruction

With sufficient nasal breathing, the minute volume and the heart rate should have a similar profile: with increasing activity, the nasal minute volume also increases and vice versa. If, with increasing physical activity, the nasal airflow is no longer sufficient for the current oxygen demand, mouth-bypass breathing occurs. This can be recognized by a falling curve of the nasal respiratory minute volume while the heart rate rises.

Breathing Rate (Purple Curve)

The breathing rate under resting conditions rate is 12–16 breaths per minute.

The breathing rate curve provides an information of the effectiveness of the patient's breathing technique. In conditions of increased oxygen demand under physical stress, there are two breathing techniques available to increase the minute volume:

Table 27.6 Heart rate as indicator of physical activity

Heart rate	Physical activity
<80/min	Physical rest
80–100/min	Slight physical activity
>100–120/min	Moderate physical activity
>120/min	Severe physical activity

Table 27.7 Physiologic reference values of nasal respiratory minute volume during physical activity

Physical activity	Nasal respiratory minute volume
Physical rest	At least 5 L/min
Light physical activity	At least 10 L/min
Moderate physical activity	At least 20 L/min
Severe physical activity	At least 30 L/min

- Increasing the respiratory rate with decreasing tidal volume.
- Decreasing respiratory rate with increased tidal volume. Due to the lower percentage of dead-space ventilation, the latter technique is more effective.

Nasal Minute Volume (NMV) (Green Curve)

Nasal minute volume is the total volume of air that is inhaled through both sides of the nose during 1 min. In combination with the physical activity, one can use the reference values known from physiology in Table 27.7.

If the measured values for the nasal minute volume fall below the indicated reference value (as in Fig. 27.11), one can conclude that mouth-bypass breathing is taking place. This can be a sign of nasal obstruction [49]. Table 27.8 provides an initial evaluation of the obstruction.

27.2.3.3 Indications for Long-Term Rhinometry

LRM is not required for every patient with nasal obstruction. It is indicated when the magnitude of nasal obstruction determined by RMM or RRM does not fully explain the patient's symptoms. Since nasal airway resistance accounts for about 60% of total airway resistance, a deficiency in the respiratory musculature or the cardiovascular system or poor general physical condition may

Table 27.8 Estimation of nasal obstruction based on onset of mouth breathing

Onset of mouth-bypass breathing during	Extent of nasal obstruction
Physical rest	Severe obstruction
Slight physical activity	Moderate obstruction
Moderate physical activity	Slight obstruction
Severe physical activity	No obstruction

lead to symptoms that occur at even normal or slightly elevated levels of nasal airway resistance, which the patient compensates by mouth-bypass breathing. In such cases the LRM shows low levels of baseline nasal respiratory volume per minute that do not increase with increased physical activity.

In diagnosing sicca symptoms and in evaluating the "empty nose syndrome" LRM can be a valuable diagnostic tool. The complete absence of resting phases suggests that either the nose is too wide or that the congestive capacity of the nasal mucosa has been so greatly reduced that closure of the nose permitting a resting phase is no longer possible. In such cases, LRM often shows only minimal ventilation indicated by a very low nasal minute volume, since these patients minimize nasal airflow by unconsciously switching over to mouth-bypass breathing as a way to reduce chronic dryness in their nose by minimizing nasal airflow. By this habit, they artificially create resting phases for both nasal side by means of mouth-bypass breathing. Such LRM findings provide an indication for the treatment option of surgically reducing the nasal cavity width. If resting phases can still be observed in the nasal cycle, this would justify conservative therapy.

27.3 Diagnostic Procedures for Objectifying Nasal Obstruction and Its Causes

27.3.1 Combination of RRM, ARM and LRM

With RRM, ARM and LRM different but additional information about nasal obstructions can be obtained

- RRM provides fluid dynamic information about the extent of an obstruction (airway resistance), about the effects of a narrowing (hydraulic diameter) and about the turbulence behaviour (flow at complete turbulence). Measuring the inspiratory NVC can also estimate the nasal airway resistance due to the Bernoulli phenomenon (ΔR).
- ARM provides insight into the geometry of the inside of the nose up to a depth of 5 cm from the outer nasal ostium [50] and thus provides information about the structure of the inner nose in the anterior region, which is important for air dynamics with regard to stenoses and regulation the airflow in the nasal cycle (see Sect. 20.3).
- LRM provides information about the nasal cycle as an important basis for the respiratory function of the nose. We receive data with which we can recognize the physiological process and the pathological disorders of the nasal cycle under the daily living conditions of the patient. The regulation of the nasal air flow according to the oxygen demand during physical activity can be assessed in terms of efficiency.

The complementarity of these three methods leads to the conclusion that combining them is useful for improving the diagnostic evaluation of nasal obstruction. Every patient complaining of nasal obstruction should undergo a RRM and ARM. A supplemental LRM should be performed:

- If the patient's symptoms occur at times of day or night other than the time of testing.
- If the endonasal findings and results of acoustic and rhinoresistometric testing cannot fully explain the symptoms reported by the patient [48].
- If insight into the nasal cycle is necessary, e.g. in the presence of unexplained fluctuating obstructions, for sicca symptoms and when mouth breathing predominates despite only minor nasal obstruction.

The extent and the causes of nasal obstruction should be objectively determined according to the algorithm presented in Fig. 27.12. The positive impact of standardized decision making in rhinosurgery has recently been demonstrated [51].

27.3.2 Examples

In the following section, we will use clinical examples to demonstrate how the combination of RRM and ARM allows to diagnose the extent and the cause of nasal obstruction. LRM will be additionally included in a few of the examples for didactic purposes, even though it would only be necessary for establishing the diagnosis in examples 6 and 7.

In describing the findings, we will employ the classification of regions in the nose recommended by Cottle (1961) (Table 27.9).

27.3.2.1 Example 1: No Nasal Complaints

Patient: male, 20 years of age

- *History*: No trauma recalled.
- *Complaints*: No rhinologic symptoms.
- *Outer nose*: Normal.
- *Endonasal findings*: Slight septal deviation to the right without any relevant stenosis. Turbinates on the left swollen, after decongestion normally configured. Mucosa normal.
- *Measurement findings*: cf. Fig. 27.13.

Analysis of the Rhinometric Findings

Extent of Obstruction

RRM resistance: *Before decongestion:* very slight obstruction on the right and severe obstruction on the left. *After decongestion:* on both sides no obstruction.

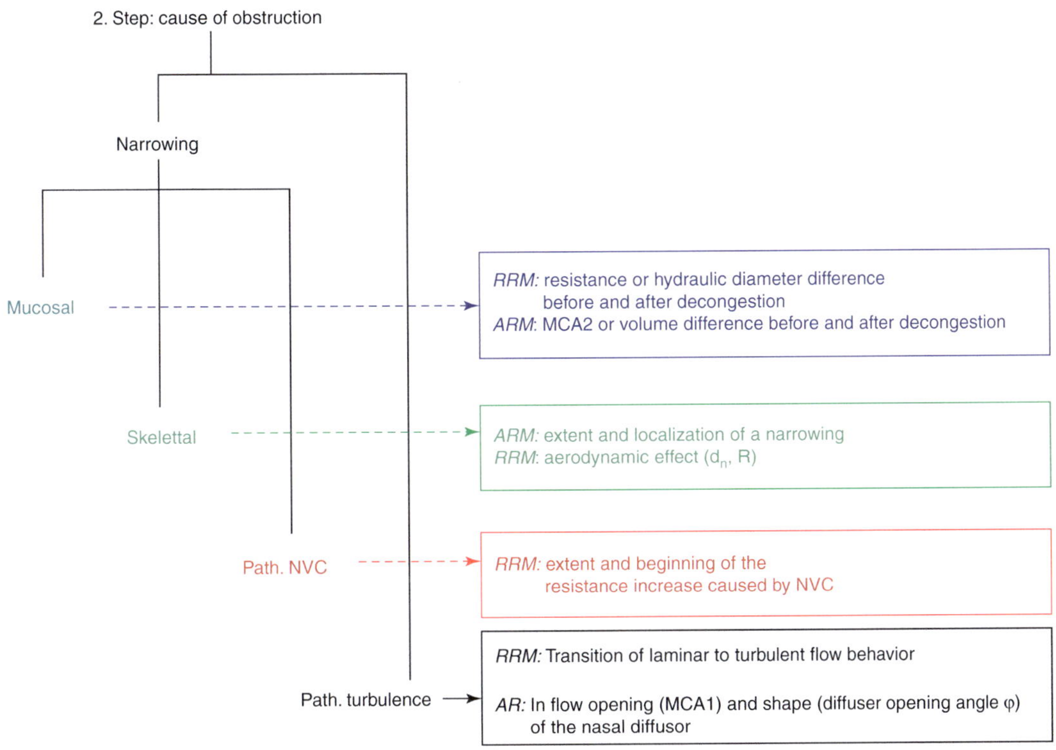

Fig. 27.12 Schematic algorithm: procedure for diagnosing the extent and cause of nasal obstruction using RRM and ARM

Table 27.9 Cottle area and corresponding anatomical endonasal areas

Cottle region	Anatomical area in the nose
Region 1	External nasal ostium (nostrils)
Region 2	Isthmus nasi (nasal isthmus)
Region 3	Region beneath the cartilage and bony pyramid, corresponding in terms of fluid dynamics to the nasal diffuser (see Chap. 20, Sect. 20.3.1.3)

Cause of the Obstruction

RRM: The increase of width due to decongestion on the left from a hydraulic diameter of 3.5 mm to a normal dimension of 5.9 mm indicates a severe congestion without skeletal stenosis. An increased resistance due to inspiratory nasal valve collapse or pathological turbulence is absent.

ARM: Normal curves with regard to both area and distances with sufficient wide physiological stenoses. MCA1 on the left slightly more narrow. Because resistance is not increased, this narrowing and the septal deviation have to be assessed as physiological. The slightly enlarged diffuser opening angle on the right results in a pronounced, but not a pathological turbulence.

LRM: During light and moderate physical activity (heart rate 80–>100/min.) between 11:30 a.m. and 8:30 p.m. in concert type of the nasal cycle. Later on up to 2:00 a.m. with decreasing activity classical type with decreasing flow values up to complete resting phases during sleep between 2:00 a.m. and 8:00 a.m.

Assessment

Normal nasal breathing on both sides in the presence of a physiological deviation of the septum towards the left. Congestion on the left is regarded as resting phase of the nasal cycle.

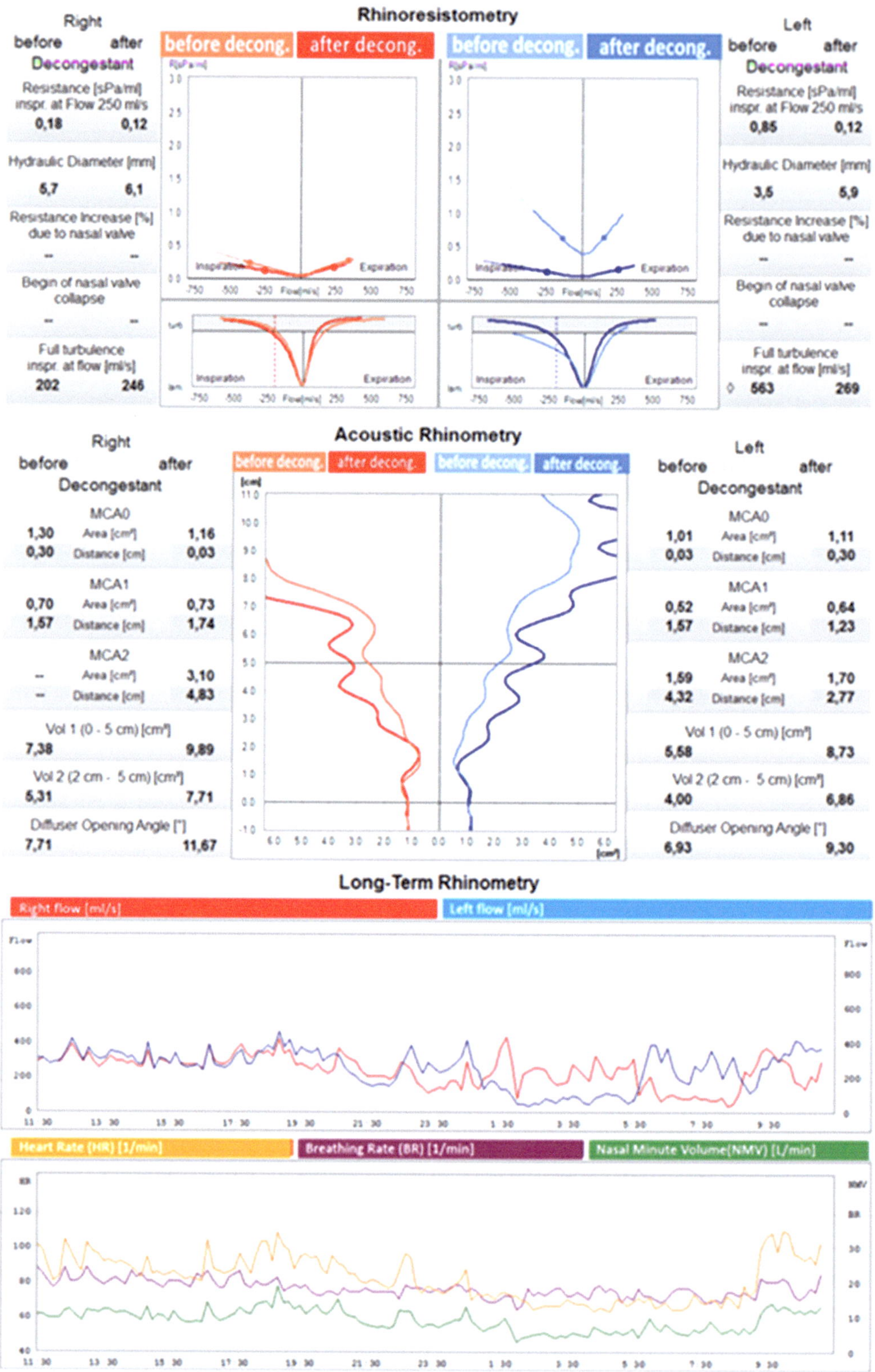

Fig. 27.13 Rhinoresistometric measurements for clinical example 1

27.3.2.2 Example 2: Severe Nasal Obstruction on the Right Due to a Septal Deviation

Patient: male, 24 years of age (same patient like in Figs. 27.5, 27.8 and 27.11)

- *History*: Trauma 2 years ago.
- *Complaints*: Since the trauma, a severe nasal obstruction on the right.
- *Outer Nose*: Normal.
- *Endonasal findings*: Severe septal deviation towards the right in Cottle's regions 2 and 3. Inferior turbinate enlarged on the left, after decongestion normally configured. Mucosa normal.

Rhinometric Findings*:* cf. Fig. 27.14a

- *Analysis of the preoperative rhinometric findings.*

Extent of Obstruction

RRM: resistance: *Before decongestion,* severe obstruction on the right side, no obstruction on the left. *After decongestion,* severe obstruction on the right side and unphysiological low resistance on the left side.

Cause of the Obstruction on the Right

RRM: hydraulic diameter: The increase of the width on the right side from a hydraulic diameter of 3.5 to 4.4 mm by decongestion indicates both, a swelling and a skeletal stenosis. The increase of resistance due to inspiratory NVC at 250 mL/s is >100%. Besides, complete turbulence at 68 mL/s contributes to the severe nasal obstruction. The left nasal side is with the hydraulic diameter of 7.4 mm too wide. This causes a pathologically low resistance as well as a borderline turbulence behaviour.

ARM: The Cottle regions 2 and 3 are after decongestion on the right side very narrow and on the left side too wide. The strong swelling on the left side can be regarded as physiological swelling to achieve a more narrow space for allowing the creation of resting phases in a nasal cavity, which has too wide skeletal dimensions ("compensatory enlargement of the turbinate by swelling").

LRM: This examination was only performed for scientific and illustrative reasons. Clinically, it was not necessary for diagnosis.

During 24 h, no cyclic change of resting and working phases can be observed. The right side does not contribute to oxygen supply and the left nasal side only to a very small amount. Nasal minute volume at 2–3 L/min indicates a permanent mouth-bypass breathing. During rising physical activity no increase of nasal minute volume is observed (*RRM*).

Assessment

The congestion on the right side is regarded as physiologic. After decongestion, on the right side a skeletal stenosis remains as cause for the sever obstruction (septal deviation in Cottle regions 2 and 3). This results in an increased resistance due to a constriction of nasal airflow canal (see Sect. 20.2.1). In addition, this stenosis causes a pathological inspiratory NVC via a Bernoulli phenomenon (see Sect. 20.2.3). Moreover, the constriction in the opening of the diffuser causes pathologically increased turbulence (see Sect. 20.4). Both contribute to the severe nasal obstruction. Stiffening of the nasal wing is not indicated because the correction of the stenosis via a septoplasty will normalize the width of the internal ostium and therefore reduce the underlying pathological high negative pressure because of Bernoulli phenomenon) (see Sect. 20.5). Consequently, pathological NVC will abolish. The narrow ostium internum is also the cause for the pathological turbulence behaviour (see Sect. 20.4). Constant mouth-bypass breathing is required as sufficient nasal minute volume cannot be achieved even with physical rest.

Rhinosurgical Planning

Septoplasty to correct the stenosis on the right side. Thereby, two aims are achieved:

- On the right nasal side in Cottle areas 2 and 3 the stenosis will be enlarged, thus decreasing both resistance and Bernoulli phenomenon. Besides, the opening of the diffuser will be expanded, reducing endonasal turbulence.
- On the left nasal side, cross-sectional area of Cottle areas 2 and 3 will decrease and thereby the pathologically low resistance corrected.

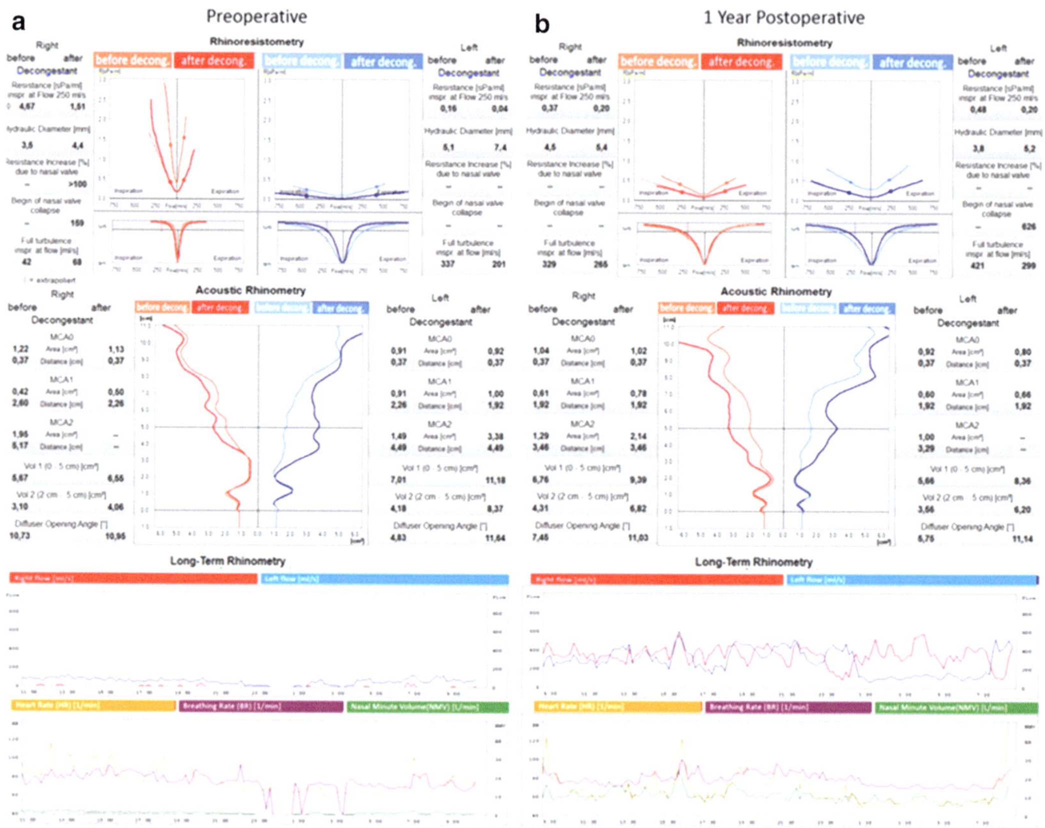

Fig. 27.14 Pre- (**a**) and 1 year postoperative (**b**) rhinoresistometric measurements for clinical example 2

A turbinate reduction of the left inferior turbinate is contraindicated [52, 53]. The too wide cavum will get more narrow in Cottle area 3. Consequently, endonasal congestion will decrease postoperatively within a few days.

Rhinosurgery
Septoplasty without surgery of the turbinate.

One year postoperatively, the patient was re-assessed

- *Complaints*: No nasal obstruction on the right.
- *Outer Nose*: Normal.
- *Endonasal findings*: Septum in the midline in Cottle's regions 2 and 3. Inferior turbinates after decongestion normally configured. Mucosa normal.
- *Measurement findings:* see Fig. 27.14b.

Analysis of Postoperative Rhinometric Findings

RRM: The preoperatively increased resistance on the right and too low resistance on the left has normalized to a very slight obstruction (0.20 sPa/mL) on both sides. Besides, the severe turbulence on the right and the borderline turbulence on the left have normalized to a physiological status. A normalization of NVC can be observed in comparison to preoperatively.

ARM: The septum is within the midline in Cottle areas 2 and 3 and accordingly, both nasal cavities are sufficiently wide.

LRM: A classical nasal cycle is observed with the capability of compensation during rising physical activity, so that the heart rate curve and the curve for the NMV have a largely uniform course (see Sect. 20.5).

27.3.2.3 Example 3: Severe Nasal Obstruction on the Left Due to Septal Deviation

Patient: female, 19 years of age

- *History*: Trauma at the age of 16 years.
- *Complaints*: Severe nasal obstruction on the left. Dry nasal mucosa with nasal crusts (Sicca syndrome).
- *Outer Nose*: Normal.
- *Endonasal findings*: Septal deviation towards the left in Cottle's region. Small valve angle on the left > on the right. Turbinate congestion on both sides (right > left), after decongestion normally configured inferior turbinates. Dry mucosa on the left > on the right.
- *Measurement findings:* cf. Fig. 27.15a.
- *Preoperative analysis of rhinometric findings.*

Extent of Obstruction

RRM: Resistance: *Before decongestion,* severe obstruction on both sides. *After decongestion,* slight nasal obstruction on the right and severe obstruction on the left.

Cause of the Obstruction on the Right

RRM: Based on the increase of width by decongestion from a hydraulic diameter of 3.2 to 5.9 mm, a severe swelling is indicated but no skeletal stenosis. No pathological inspiratory NVC. Complete turbulence at 165 mL/s after decongestion contributes only in a moderate amount to nasal obstruction.

ARM: After decongestion internal ostium right (MCA1 0.61 mm) is only slightly greater compared to the left (MCA1 = 0.59 mm).

Cause of the Obstruction on the Left

RRM: Based on the increase of width by decongestion from a hydraulic diameter of 4.6 to 4.9 mm, both a slight swelling and a severe skeletal stenosis is indicated. Moreover, pathological turbulence (complete turbulence at 72 mL/s) contributes to nasal resistance and thereby to nasal obstruction. No pathological inspiratory NVC.

ARM: In the region of the septal deviation (Cottle area 2) cross-sectional area after decongestion of 0.59 cm^2 is only slightly smaller than the non-obstructed contralateral side (0.61 cm^2).

LRM: not mandatory.

Assessment

The swelling on the right side causes the severe obstruction. This congestion has to be regarded as physiological state due to a resting phase at the time of the measurement. After decongestion, resistance remains slightly increased. On the left, decongestion reveals a persisting skeletal stenosis

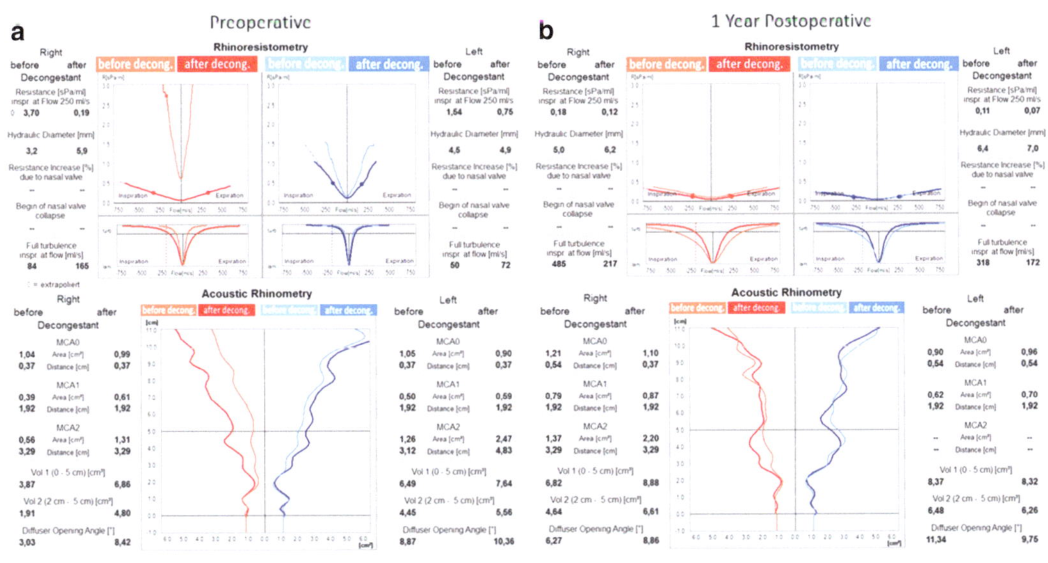

Fig. 27.15 Pre- (**a**) and 1 year postoperative (**b**) rhinoresistometric measurements for clinical example 3

(septal deviation). This cannot be proven by ARM. Taking into account the limitations of ARM (see Sect. 27.2.1.1) and the physiologic laws of fluid dynamics (see Sect. 20.2.1), a slit-like configuration of the internal ostium indicated by the small valve angle with a sufficient cross-sectional area has to be suspected as aetiology of nasal obstruction in this setting (see Sect. 27.2.1.1).

Moreover, the pathologically increased turbulence on the left > right causes mucosal dryness, including sicca syndrome, and contributes to nasal obstruction.

Rhinosurgical Planning

Septoplasty to correct the stenosis will not lead to satisfactory results. While resistance on the left would get better, it would not be normalized. Simultaneously, the slightly increased nasal obstruction on the right would get worse. Consequently, septoplasty needs to be combined with a nasal valve repair on both nasal sides. A turbinate reduction is not necessary, as configuration after decongestion is normal.

Rhinosurgery

Septoplasty and valve repair on both nasal sides without surgery of the turbinates.

One year postoperatively, the patient was re-assessed:

- *Complaints*: No nasal obstruction, no sicca syndrome.
- *Outer Nose*: Normal.
- *Endonasal findings*: Septum in the midline. Sufficiently wide nasal valve angle on both sides. Inferior turbinates after decongestion normally configured. Mucosa normal (no dryness).
- *Postoperative rhinometric findings:* cf. Fig. 27.15b.
- *Analysis of postoperative rhinometric findings*

RRM: The preoperatively increased resistance on the right has normalized. On the left slightly too low. Besides, the pathological turbulence is normalized on the right and clearly improved on the left.

ARM: The septum is within the midline in Cottle areas 2 and 3 and accordingly, both nasal cavities are sufficiently wide. MCA1 indicating the dimensions of the internal ostium is enlarged on both sides in comparison to preoperatively.

27.3.2.4 Example 4: Severe Nasal Obstruction on Both Sides Due to a Tension Nose

Patient: female, 32 years of age

- *History*: No trauma recalled.
- *Complaints*: Severe nasal obstruction on both sides since many years.
- *Outer Nose*: overprojected nasal tip with a high bony and cartilaginous dorsum, an oblique nasal labial angle and the nasal pyramid is narrow and resembles a high, narrow, pointed gothic arch, positive U-phenomenon.
- *Endonasal findings*: Typical tension nose with slit-like internal ostium on both sides and to a lesser extent also external ostium on both sides. Septum within the midline. Turbinate on both sides congested, after decongestion normally configured inferior turbinates. Normal mucosa.
- *Rhinometric findings:* cf. Fig. 27.16a.
- *Analysis of preoperative rhinometric findings:*

Extent of Obstruction

RRM resistance: *Before and after decongestion,* severe obstruction on both sides.

Cause of the Obstruction on Both Sides

RRM: Based on the increase of width by decongestion from a hydraulic diameter of 1.8 to 3.3 mm on the right and from 2.2 to 3.5 mm on the left indicate severe swelling and severe skeletal stenosis on both sides. Complete turbulence at 81 mL/s on the right and at 122 mL/s on the left as well as the increase of resistance by NVC of more than 100% contribute to the severe nasal obstruction.

ARM: The most pronounced stenosis on both sides is localized in the internal ostium (MCA1). But also the external ostia (MCA0) contributes to the increased resistance due to its small dimensions.

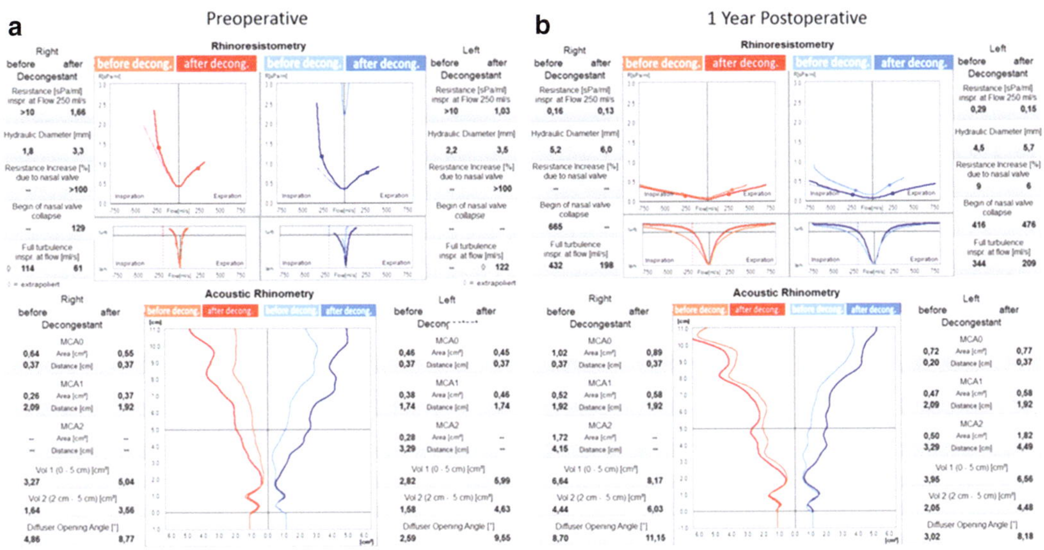

Fig. 27.16 Pre- (**a**) and 1 year postoperative (**b**) rhinoresistometric measurements for clinical example 4

Assessment

The strong swelling on both nasal sides should be clarified by more refined diagnostic testing for allergic and non-allergic rhinitis [54]. The severe obstruction after decongestion is caused on both sides by a stenosis of the internal more than the external ostium. In addition, the narrow internal ostium on both sides leads by the Bernoulli phenomenon to a pathological NVC and by a narrow diffuser entrance to severe pathological turbulence (see Sects. 20.4 and 20.5). Thus, the resistance is further increased.

Rhinosurgical Planning

Correction of the internal and external ostium on both sides is possible by decreasing the height of the septum and reducing overprojection, as well as reducing and reconfiguring the bony pyramid. This will result in extension and rounding of both ostia on both nasal sides. Due to this change, Bernoulli phenomenon will be greatly reduced and the entrance of the nasal diffuser will be enlarged. Accordingly, normalization of the nasal valve and of turbulence behaviour has to be expected. Reduction of turbinates is not required, as configuration after decongestion is normal.

Surgery

Septorhinoplasty to relax and deproject the nose. No stabilization of the lateral nasal wall, no turbinate surgery.

One year postoperatively, the patient was re-assessed:

- *Complaints*: No nasal obstruction.
- *Outer Nose*: Normal. In comparison to preoperatively, *deprojected* nose, physiologic nasal pyramid.
- *Endonasal findings*: Septum in the midline. Sufficiently wide nasal valve angle on both sides, rounded in comparison to preoperatively. Inferior turbinates after decongestion normally configured. Mucosa normal.
- *Rhinometric findings:* cf. Fig. 27.16b.
- *Analysis of postoperative rhinometric findings:*

RRM: The preoperatively increased resistance on both nasal sides is normalized. Besides, the pathological turbulence behaviour is nearly normal on the right and completely physiologic on the left. Pathological NVC has been corrected on both sides; only a physiologic one persists on the left.

ARM: The septum is still within the midline. The entrance areas of both nasal cavities are significantly increased in width, as indicated by MCA0 and MCA1 after decongestion.

27.3.2.5 Example 5: Severe Nasal Obstruction on Both Sides Due to Broad Columella

Patient: male, 40 years of age

- *History*: No trauma recalled. Patient is referred for septoplasty and turbinate surgery.
- *Complaints*: Severe nasal obstruction on both sides since childhood.
- *Outer Nose*: broad columella, small external ostia on both sides, otherwise normal.
- *Endonasal findings*: Slight septal deviation to the right in Cottle areas 2 and 3, no stenosis of endonasal airflow channel within the cavum. Normal mucosa.
- *Measurement findings:* cf. Fig. 27.17a.
- *Preoperative analysis of the findings.*

Extent of Obstruction

RRM resistance: *Before and after decongestion,* severe obstruction on both sides.

Cause of the Obstruction on Both Sides

RRM: Based on the increase of width by decongestion from a hydraulic diameter of 2.8 to 3.1 mm on the right and from 2.5 to 3.5 mm on the left indicate both, severe swelling and severe skeletal stenosis on both nasal sides. On both nasal sides, increase of resistance by NVC of more than 100% contributes to the severe nasal obstruction. Turbulence behaviour is physiologic on both sides.

ARM: The most pronounced stenosis on both sides is localized in the external ostium (MCA0).

Assessment

The aetiology of congestion on both nasal sides should be clarified by more refined diagnostic testing for allergic and non-allergic rhinitis [54]. Septal deviation is regarded as physiologic, because no impact on fluid dynamic parameter of nasal airflow is detectable [54, 55]. The severe obstruction after decongestion is caused on both sides by a stenosis of the external ostia. In addition, the Bernoulli phenomenon at the ostium externum contributes to nasal resistance by NVC.

Rhinosurgical Planning

Correction of the external ostium on both sides by addressing the pathological broad columella will decrease the skeletal, permanent resistance as well as the dynamic increase in nasal resistance caused by pathological NVC. Stiffening of

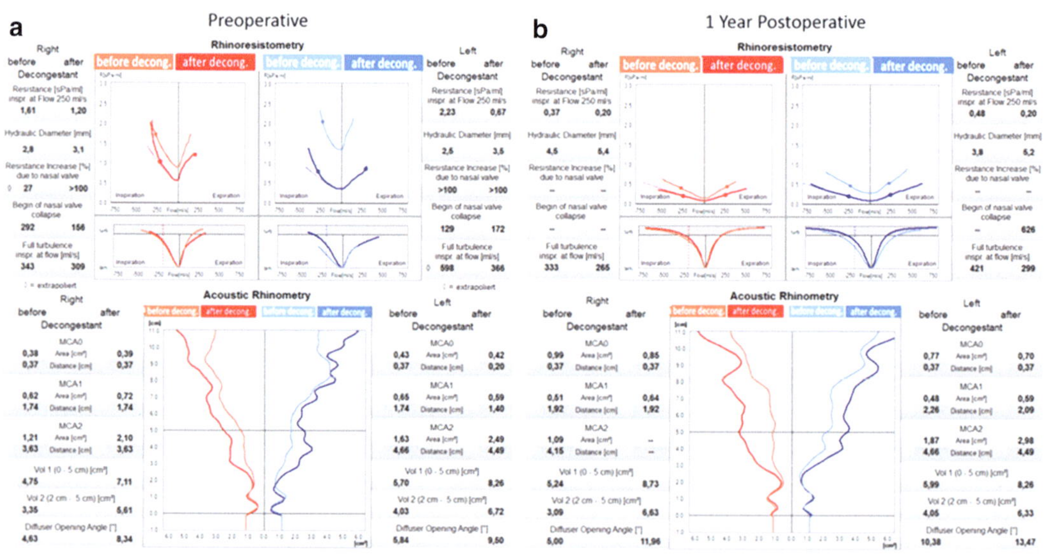

Fig. 27.17 Pre- (**a**) and 1 year postoperative (**b**) rhinoresistometric measurements for clinical example 5

the lateral nasal valve is not indicated, as enlargement of the nasal airflow channel in this segment will reduce the effect of the Bernoulli phenomenon. Reduction of turbinates is not required, as configuration after decongestion is normal.

Surgery

Columella narrowing.

One year postoperatively, the patient was re-assessed:

- *Complaints*: No nasal obstruction.
- *Outer Nose*: Normal. In comparison to preoperatively normal width of the Columella.
- *Endonasal findings*: Septum in the midline. Inferior turbinates after decongestion normally configured. Mucosa normal.
- *Postoperative rhinometric findings:* cf. Fig. 27.17b.
- *Analysis of postoperative rhinometric findings:*

RRM: The preoperatively increased resistance on both nasal sides is nearly normalized. Besides, the pathological NVC has been successfully corrected on both sides. The turbulence behaviour remains physiologic.

ARM: The septum is still within the midline. The ostium externum (MCA0) is clearly enlarged compared to preoperatively.

27.3.2.6 Example 6: Subjective Nasal Obstruction Without Evident Aetiology

Patient: male, 46 years of age

- *History*: No trauma recalled. Multiple prior consultations of different rhinologists because of "insufficient nasal breathing". Concurring assessments after clinical and rhinomanometric examination. RRM (cf. Fig. 27.18): right side: no obstruction, left side: very slight obstruc-

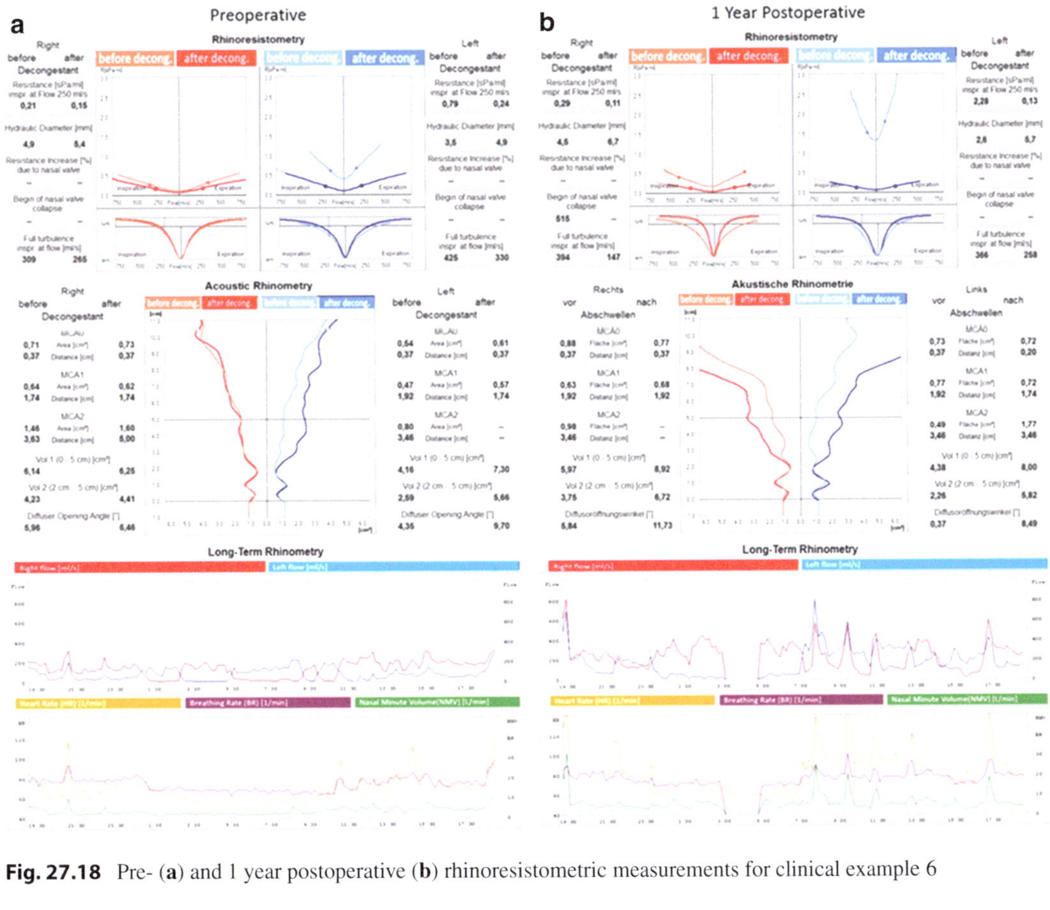

Fig. 27.18 Pre- (**a**) and 1 year postoperative (**b**) rhinoresistometric measurements for clinical example 6

tion. Hence, no pathological endonasal finding as evident aetiology of the nasal obstruction. Therefore, no indication for functional rhinosurgery. However, the patient complains believably on disability by obstruction of his nasal breathing. Consultation of a psychologist: no evidence on aggravation. Decision for a RRM re-examination (Fig. 27.18).

Clinical re-examination

- *Complaints*: Nasal obstruction on both sides, during several years increasing.
- *Outer nose*: Normal.
- *Endonasal findings*: Septum in the midline. Inferior turbinates after decongestion normally configured. Normal mucosa.
- *Rhinometric measurement findings:* cf. Fig. 27.18a.
- *Preoperative analysis of the rhinometric findings:*

Extent of Obstruction

RRM resistance: *Before decongestion* slight obstruction on the right. *After decongestion* on the right no and on the left side slight nasal obstruction.

Cause of the Obstruction

RRM: The increase of width by decongestion on the right from a hydraulic diameter of 4.9 to 5.4 mm objectives a slight congestion, but no skeletal stenosis. The increase of the hydraulic diameter on the left from 3.5 to 4.9 mm by decongestion indicates a severe congestion and a very slight skeletal stenosis. NVC and turbulence are physiologic.

ARM: The physiologic stenoses on both sides seem sufficiently wide.

Assessment

The congestion on the left is interpreted as resting phase of the nasal cycle. The subjective nasal obstruction claimed by the patient cannot be sufficiently explained by clinical examination and rhinometric objective diagnostics.

This discrepancy between subjective complaints and objective assessment is the indication to employ LRM [54, 56].

LRM (cf. Fig. 27.9a, lower graph): During the whole measurement, a classical type of a nasal cycle persists. But nasal airflow, indicated by nasal minute volume, is very low with flow values during the day of about 5 L/min and increasing during physical activity, as indicated by increased heart rate, only up to10 L/min (e.g. 21:00, 11:00, 14:15 and 18:15). This indicates a permanent mouth-bypass breathing.

Assessment Together with Results of the LRM Examination

The constant mouth-bypass breathing objectively confirmed the subjective complaint of nasal obstruction by the patient. The key message is that “this respiratory function is insufficient for this patient”.

Rhinosurgical Planning

The aim is a limited enlargement if the isthmus region on the left > right using functional (wedge-shaped) spreader grafts [17]. Reduction of turbinates is not required, as configuration after decongestion is normal.

Surgery

Septorhinoplasty with functional spreader grafts to enlarge the internal ostium. No turbinate surgery.

One year postoperatively, the patient was re-assessed:

- *Complaints*: No nasal obstruction.
- *External Nose*: Normal.
- *Endonasal findings*: Septum in the midline. Sufficiently wide nasal inflow area on both sides. Inferior turbinates after decongestion normally configured. Mucosa normal.
- *Rhinometric findings:* cf. Fig. 27.18b.

Analysis of Postoperative Rhinometric Findings

RRM: The preoperatively only slightly increased resistance on the left nasal side is decreased and normalized. Also on the right side the resistance is slightly increased, and NVC and turbulence behaviour are normal.

ARM: The nasal inflow area is slightly wider on both nasal sides in comparison to preoperatively.

LRM: Classical type of nasal cycle with significantly increased nasal minute volume in comparison to preoperatively. During physical activity [56], in concert type with sufficient and more pronounced increase of NMV (e.g. at 19:00, 7:45, 10:00, 10:45 12:45, and 16:45).

27.4 Conclusions

- RMM does not allow a sufficient differentiation of nasal obstruction causes. It only provides the rhinosurgeon with insufficient information with regard to the appropriate diagnosis and surgical approach. Despite RMM being world widely currently available, it has not yet been implemented in routine preoperative diagnostics. However, preoperative mismanagement may contribute to an unsatisfying surgical result.
- Therefore, RMM has been refined further to RRM. With this method, not only the extent of nasal obstruction can be objectified but it also allows a differentiation of the four possible causes of nasal obstruction:
  - Mucosal swelling
  - Skeletal stenosis
  - Inspiratory collapse of the nasal valve
  - Pathologically increased degree of turbulence
- With ARM the localization of the decisive narrowings which cause obstruction, the identification of narrowings as cause of a NVC and causes for pathological turbulences in the nose, such as
  - Strong increase of diameter in the nasal diffuser (opening angle).
  - Narrow diffuser entrance may be diagnosed.
- LRM allows objectification of the following:
  - Changes in flow due to physiological swelling, separately in both sides of the nose, depending on physical stress.
  - The nasal cycle as an important precondition for the nasal respiratory function and its disturbances.
  - Pathological swelling under a patient's daily living conditions within the course of 24 h.

A combination of these methods allows a preoperative objectification of the aetiology of a patient's complaints due to nasal obstruction. With this, also the measurement-relevant precondition for a postoperative quality management as a fundament for an evidence-based therapy, similar to developments in other specialities such as otology/neurotology or phonosurgery, is given.

References

1. Clement P, Gordts F. Consensus report on acoustic rhinometry and rhinomanometry. Rhinology. 2005;43:169–79.
2. Mlynski G, Beule A. Diagnostik der respiratorischen Funktion der Nase. HNO. 2008;56:81–99.
3. Dinis PB, Haide H. Septoplasty: long-term evaluation of results. Am J Otolaryngol. 2002;23:85–90.
4. Dommerby H, Rasmussen O, Rosborg J. Long-term results of septoplastic operations. ORL. 1985;47:151–7.
5. Fjermedal O, Saunte C, Pedersen S. Septoplasty and/or submucous resection? 5 years nasal septum operations. J Laryngol Otol. 1988;102:796–8.
6. Gruetzenmacher S, Robinson DM, Grafe K, et al. First findings concerning airflow in noses with septal deviation and compensatory turbinate hypertrophy—a model study. ORL J Otorhinolaryngol Relat Spec. 2006;68:199–205.
7. Grymer L, Rosborg J. The aging nose: long-term results following plastic septal surgery. J Laryngol Otol. 1987;101:363–5.
8. Haraldsson PO, Nordemar H, Anggard A. Long-term results after septal surgery-submucous resection versus septoplasty. ORL J Otorhinolaryngol Relat Spec. 1987;49(4):218–22.
9. Illum P. Septoplasty and compensatory inferior turbinate hypertrophy: long-term results after randomized turbinoplasty. Eur Arch Otorhinolaryngol. 1997;254:89–92.
10. Mlynski G. Surgery of the nasal septum. Facial Plast Surg. 2006;22:223–9.
11. Mlynski G, Grutzenmacher S, Plontke S, et al. Correlation of nasal morphology and respiratory function. Rhinology. 2001;39:197–201.
12. Scheithauer M. Nasenmuschelchirurgie und "Empty nose" Syndrome. Laryngorhinootologie. 2010;89:79–102.
13. Stoksted P. Long term results, following plastic septum surgery. Int Rhinol. 1969;7:53–61.

14. Mlynski G, Löw J. Rhinoresistometrie - eine Weiterentwicklung der Rhinomanometrie. Laryngorhinootologie. 1993;72:608–10.
15. Mlynski G, Gogniashvili G, Beule AG. Objective diagnostic tools and rhinology. In: Recent advances in otolaryngology. New Delhi: Jaypee Brothers Medical; 2014. p. 235–59.
16. Mlynski G, Gogniashvili G, Beule AG. Preoperative and postoperative objective diagnostic approach in functional rhinosurgery. In: Recent advances in otolaryngology: head and neck surgery. 6th ed. New Delhi: Jaypee Brothers Medical; 2017. p. 123–58.
17. Mlynski G, Pirsig W. Funktionell ästhetische Rhinochirurgie. Stuttgart: Thieme; 2018. p. 152.
18. Corey JP. Acoustic rhinometry: should we be using it? Curr Opin Otolaryngol Head Neck Surg. 2006;14:29–34.
19. Gruetzenmacher S, Lang C, Mlynski R, et al. Long-term rhinoflowmetry: a new method for functional rhinologic diagnostic. Am J Rhinol. 2005b;19:53–7.
20. Ohki M, Ogoshi T, Yuasa T, et al. Extended observation of the nasal cycle using a portable rhinoflowmeter. J Otolaryngol. 2005;34:346–9.
21. Gogniashvili G, Steinmeier E, Mlynski G, et al. Physiologic and pathologic septal deviations: subjective and objective functional rhinologic findings. Rhinology. 2011;49:24–9.
22. Andre RF, Vuyk HD, Ahmed A, et al. Correlation between subjective and objective evaluation of the nasal airway. A systematic review of the highest level of evidence. Clin Otolaryngol. 2009;34:518–25.
23. Jones N, Willatt D, Durham L. Nasal airflow: resistance and sensation. J Laryngol Otol. 1989;103:909–11.
24. Kim CS, Moon B, Jung D, et al. Correlation between nasal obstruction symptoms and objective parameters of acoustic rhinometry and rhinomanometry. Auris Nasus Larynx. 1998;25:45–8.
25. McCaffrey T, Kern E, Gordon A, et al. Rhinomanometry for preoperative and postoperative assessment of nasal obstruction. Otolaryngol Head Neck Surg. 1989;101:20–6.
26. Naito K, Cole P, Chaban R, et al. Nasal resistance, sensation of obstruction and rhinoscopic findings compared. Am J Rhinol. 1988;2:65–9.
27. Bohlin L, Dahlqvist A. Nasal airway resistance and complications following functional septoplasty: a 10-year follow-up study. Rhinology. 1994;32:195–7.
28. Gordon AS, et al. Rhinomanometry for preoperative and postoperative assessment of nasal obstruction. Otolaryngol Head Neck Surg. 1989;101:20–6.
29. Jessen M, Ivarsson A, Malm L. Nasal airway resistance and symptoms after functional septoplasty: comparison of findings at 9 months and 9 years. Clin Otolaryngol. 1989;14:231–4.
30. Samad I, Stevens HE, Maloney A. The efficacy of nasal septal surgery. J Otolaryngol. 1992;21:88–91.
31. Shermann A. A study of nasal airway function in the postoperative period of nasal surgery. Laryngoscope. 1977;87:299–303.
32. Tuschen E. Rhinomanometrische und olfaktometrische Untersuchungen nach Resektionen und plastischen Korrekturen der Nasenscheidewand. Bonn: Rheinische Friedrichs-Wilhelm-Universität; 1977.
33. Bhandarkar ND, Smith TL. Outcomes of surgery for inferior turbinate hypertrophy. Curr Opin Otolaryngol Head Neck Surg. 2010;18:49–53.
34. Chen YL, Tan CT, Huang HM. Long-term efficacy of microdebrider-assisted inferior turbinoplasty with lateralization for hypertrophic inferior turbinates in patients with perennial allergic rhinitis. Laryngoscope. 2008;18:1270–4.
35. Liu CM, Tan CD, Lee FP, et al. Microdebrider-assisted versus radiofrequency-assisted inferior turbinoplasty. Laryngoscope. 2009;119:414–8.
36. Gruetzenmacher S, Guenther M, Robinson DM, et al. Investigations for the diagnostic recording of nasal wing collapse. Laryngoscope. 2005a;115:1763–7.
37. Mlynski G, Löw J. Experimentelle Studie zum Strömungsverhalten in der Nase. Teil 2: Untersuchungen im laminaren Bereich bei Heliumatmung. Z Med Phys. 1992;2:224–9.
38. Chen XB, Lee HP, Chong VF, et al. Assessment of septal deviation effects on nasal air flow: a computational fluid dynamics model. Laryngoscope. 2009;119:1730–6.
39. Chen XB, Lee HP, Chong VF, et al. Impact of inferior turbinate hypertrophy on the aerodynamic pattern and physiological functions of the turbulent airflow—a CFD simulation model. Rhinology. 2010;48:163–8.
40. Churchill SE, Shackelford LL, Georgi JN, et al. Morphological variation and airflow dynamics in the human nose. Am J Hum Biol. 2004;16:625–38.
41. Leong SC, Chen XB, Lee HP, et al. A review of the implications of computational fluid dynamic studies on nasal airflow and physiology. Rhinology. 2010;48:139–45.
42. Sawyer K, Brown JS, Hazucha MJ, et al. The effect of exercise on nasal uptake of ozone in healthy human adults. J Appl Physiol. 2007;102:1380–6.
43. Enßen C. Funktionsdiagnostische Untersuchungen gesunder Probanden zur Ermittlung rhinoresistometrischer, rhinomanometrischer und akustisch rhinometrischer Referenzwerte. Dissertation. Greifswald: University Greifswald; 2005.
44. Fiebig K. Untersuchungen zur Bewertung des nasalen Diffusors mittels akustischer Rhinometrie. Dissertation. Greifswald: University Greifswald; 2007.
45. Marschall F. Erstellung von Referenzwerte für die rhinoresistometrisch ermittelten funktionsdiagnostischen Parameter und Vergleich von Untersuchungsbefunden und subjektiven Beschwerden. Dissertation. Greifswald: University Greifswald; 1997.
46. Gehring JM, Garlick SR, Wheatley JR, et al. Nasal resistance and flow resistive work of nasal breathing during exercise: effects of a nasal dilator strip. J Appl Physiol. 2000;89:1114–22.
47. Lang C, Grützenmacher S, Mlynski B. Investigating the nasal cycle using endoscopy, rhinoresistom-

etry, and acoustic rhinometry. Laryngoscope. 2003;113:284–9.

48. Gaberino C, Rhee JS, Garcia GJM. Estimates of nasal airflow at the nasal cycle mid-point improve the correlation between objective and subjective measures of nasal patency. Respir Physiol Neurobiol. 2017;238:23–32.
49. Leiter JC, Baker GL. Partitioning of ventilation between nose and mouth: the role of nasal resistance. Am J Orthod Dentofacial Orthop. 1989;95:432–8.
50. Mlynski R, Grützenmacher S, Mlynski G. Acoustic rhinometry and paranasal sinuses: a systematic study in models, anatomic specimens, and in vivo. Laryngoscope. 2005;115:837–43.
51. Hildebrandt T, Brüning JJ, Lamecker H, Zachow S, Heppt WJ, Schmidt N, Goubergrits L. Digital analysis of nasal airflow facilitating decision support in rhinosurgery. Facial Plast Surg. 2019;35(1):3–8.
52. Sommer F, Scheithauer MO, Hoffmann TK, Grossi AS, Hauck K, Lindemann J. Value of turbinoplasty in rhinosurgery—a controlled randomized study. Rhinology. 2019;57(5):352–7.
53. Van Egmont MMHT, Rovers MM, Hannink G, Hendriks CTM, van Heerbeek N. Septoplasty with or without concurrent turbinate surgery versus non-surgical management for nasal obstruction in adults with a deviated septum: a pragmatic, randomised controlled trial. Lancet. 2019;394(10195):314–21.
54. Beule AG. Funktionen und Funktionsstörungen der respiratorischen Schleimhaut der Nase und der Nasennebenhöhlen. Laryngorhinootologie. 2010;89:15–34.
55. Keck T, Lindemann J. Strömungssimulation und Klimatisierung in der Nase. Laryngorhinootologie. 2010;89:1–14.
56. Olson LG, Strohl KP. The response of the nasal airway to exercise. Am Rev Respir Dis. 1987;135:356–9.

28 Testing of Transport and Measurement of Ciliary Activity

Mark Jorissen and Martine Jaspers

Core Messages

- The mucociliary transport (MCT) can be studied by recording MCT as well as by measuring ciliary activity, but none of these tests are reliable for the diagnosis of inherited abnormalities.
- Most people with PCD have unusually low levels of nasal NO, but a low nNO measurement is nondiagnostic for PCD. Sequential monolayer–suspension cell culture with dedifferentiation and redifferentiation of the ciliated epithelium is the most reliable screening test for the diagnosis of PCD.

28.1 Introduction

Ciliary activity causes the transport of mucus in the airways, which is an essential defense mechanism of the respiratory tract. Inhaled particles, bacteria, and viruses are trapped in the mucus layer that covers the airways and are transported by the beat of the cilia to the nasopharynx, where they are either swallowed or coughed up. Inborn disorders of the mucociliary transport are due to ciliary dysfunction as in primary ciliary dyskinesia (PCD) or to increased viscosity of the respiratory secretions as in cystic fibrosis (CF). Also frequently mucociliary transport is impaired because of inflammation, infection, and exposure to ciliotoxic agents.

Mucociliary transport (MCT) can be studied by recording MCT in vivo, as well as by measuring ciliary activity in vitro. Methods based on nasal ciliary motility for the diagnosis of primary ciliary dyskinesia (PCD) are often hampered by the presence of acquired abnormalities [secondary ciliary dyskinesia (SCD)].

> Mucociliary clearance can be evaluated by using the saccharine and/or the ^{99m}Tc-albumin colloid test as well as by measuring ciliary activity and is important in the diagnosis of primary ciliary dyskinesia.

28.2 Testing of Transport

28.2.1 Testing of Transport In Vivo

The mucociliary transport rate can be measured in vivo either by using the saccharine test [1, 2] or by using the radioisotope technique [3]. If with one of these methods active mucociliary transport can be demonstrated, it is accepted that the diagnosis of PCD is excluded. An abnormal result can certainly not be considered as proof for the disease. It implicates only that further investigation is needed.

M. Jorissen (✉) · M. Jaspers
ENT Department, University Hospitals Leuven, Leuven, Belgium
e-mail: mark.jorissen@uzleuven.be; martine.jaspers@med.kuleuven.be

Ö. Ö. Celebi, T. M. Önerci (eds.), *Nasal Physiology and Pathophysiology of Nasal Disorders*,
https://doi.org/10.1007/978-3-031-12386-3_28

28.2.1.1 Saccharine Test

With the saccharine test, a particle of saccharine (most frequently dipped in a blue color, such as indigo blue powder) is placed on the inferior turbinate and the time elapsed before the patient taste the saccharine is measured. Patients are instructed to swallow at least once per minute. The appearance in the pharynx can be verified by the blue color in the pharynx. With this technique the mean normal mucociliary transport time (MTT) is about 10 min. MTT up to 30 min is still considered normal. If this takes more than 30 min, the test is considered abnormal. For this test cooperation of the individual is needed, since he/she has to report the sweet taste. Also sniffing, sneezing, and blowing the nose are prohibited since these may affect the position of the particles. This limits the use of the test in children. The saccharine has to be placed on the respiratory (ciliated) epithelium. Otherwise no transport will be found.

Most frequently a color (methylene blue, indigo blue, charcoal, etc.) is added to the test, as a visual control. Repeated examination will allow one to verify the transport of the particle and to compare the appearance of the color in the pharynx with the perception of the sweet taste.

When combined with nasal endoscopy, the colored particles can be followed to evaluate the transport pattern. That technique can also be used to follow and study the pathways within the (maxillary) sinuses.

28.2.1.2 Nuclear Testing

Up to now measuring the transport of radiolabeled albumin remains the most reliable method for measuring mucociliary transport.

When a minute amount of radiolabeled ^{99m}Tc-albumin colloid particles [3] is placed on the inferior turbinate or on the nasal septum, the migration can be followed with a gamma camera. Normally within 30 min the majority of the radioactivity must have disappeared from the nasal cavity. The percentage of radioactivity remaining in the nasal cavity can be calculated and in sagittal views the migration of the spot can be measured. It has been shown that the dose of radioactivity is low enough that immotility does not create problems. In contrast to the saccharine test, this test is not influenced by sniffing. A normal test result is considered an exclusion criterion of PCD. If the particles moved insufficiently, further investigation is needed, because dysmotility could be due to upper airway infections or PCD. Moreover, in up to 25% of individuals with SCD and also in controls, no migration of the tracer is found.

Marthin et al. [4] studied an alternative method, the pulmonary radioaerosol mucociliary clearance technique, which has a higher specificity for PCD as secondary dysmotility is much less prevalent in the lower airways. The radioactive-labeled ^{99m}Tc-albumin transportation test is more reliable than the saccharine test but requires expensive equipment and can only be done in specialized centers.

28.2.2 Testing of Transport In Vitro

Bioptic or brushed material can be checked for the presence of cilia under phase contrast microscopy. Real movements such as displacements and rotations of cell clusters or cell sheets within the fluid and movement of particles within the fluid lining the cilia are criteria for the presence of coordinated ciliary activity. Absence of these elements in the presence of ciliary activity is recorded as "uncoordinated ciliary activity." It should always be checked whether cilia are present.

28.3 Testing of Ciliary Activity

28.3.1 In Vivo

Laser light-scattering spectroscopy provides an improved, precise, and simple method to study ciliary activity. The light from a laser beam is directed at a ciliated surface, and due to the Doppler effect, the scattered light returning from the moving cilia has an altered frequency and phase induced by the movement of the reflecting surfaces of the cilia. The scattered light can be detected with a photomultiplier tube, and the spectral structure of its intensity fluctuations can be analyzed to provide quantitative information regarding the frequency and synchrony of the ciliary beat [5].

Other techniques are under development for measuring ciliary activity in vivo.

28.3.2 In Vitro

CBF is 8 Hz at room temperature and 12 Hz at 37 °C.

In vitro brushings from the nasal cavity or biopsy samples taken from the inferior border of the middle turbinate or from the inferior turbinate can be evaluated for coordinated ciliary beating and ciliary beat frequency (CBF). Using microscope photometry [6], CBF can be deduced from fast Fourier transform analysis of the light scattering. Normal CBF values depend on the temperature with normal values around 8 Hz at room temperature and 12 Hz at 37 °C.

The introduction of high-speed cameras (up to 500 Hz) created new possibilities [7, 8]: not only ciliary beat pattern analysis but also amplitude, degree, and speed of ciliary beat cycle. Also field analysis with measurement of ciliary coordination on a whole area of ciliated cells has become available [9].

For clinical use only CBF is used, and one should realize that there is no good correlation between CBF and mucociliary transport velocity [10].

28.4 Additional Testing

28.4.1 Nasal Nitric Oxide

Nasal nitric oxide (nNO) measurement is a good screening tool for PCD.

Nasal nitric oxide (nNO) was found to be tenfold lower in PCD patients than in control patients, and it can be used as an easy screening test for the diagnosis of PCD. It is a noninvasive technique, but as it requires cooperation of the patient (breath holding for stable plateau measurements), it is almost impossible to use it below 5 years of age. nNO measurement cannot be used neither for exclusion of PCD nor for proof of PCD since normal values can be observed in PCD and low values can be caused by other factors, e.g., obstruction [11–15].

28.4.2 Genetic Analysis

Making a diagnosis of primary ciliary dyskinesia (PCD) remains challenging. Molecular diagnosis involves time-consuming tissue culturing, cilia beating measurements, and/or electron microscopy of microtubule structures.

Also at the genetic level, a diagnosis of PCD has been challenging. PCD is an autosomal recessive disease. Mutations in many different genes can result in PCD, and from a genetic point the disease is very heterogeneous. Next-generation sequencing, including the use of whole exome sequencing or whole genome DNA sequencing, has further identified genes causing PCD. So far, mutations in 50 genes (ARMC4, CCDC103, CCDC114, CCDC151, CCDC39, CCDC40, CCDC65, CCNO, CFAP298, CFAP300, CFAP53, DNAAF1, DNAAF2, DNAAF3, DNAAF4, DNAAF5, DNAH11, DNAH5, DNAH6, DNAH8, DNAH9, DNAI1, DNAI2, DNAJB13, DNAL1, DRC1, ENKUR, GAS2L2, GAS8, HYDIN, LRRC56, LRRC6, MCIDAS, MNS1, NEK10, NME8, OFD1, PIH1D3, RPGR, RSPH1, RSPH3, RSPH4A, RSPH9, SPAG1, SPEF2, STK36, TEKT1, TTC12, TTC25, ZMYND10) have been reported to cause PCD. Mutations in these genes account for approximately 70% of PCD cases; therefore, further gene discovery is still expected [16, 17]. The ciliary axoneme is composed of over 250 proteins [18], and a mutation in each of their genes possibly can result in PCD. It is expected that several of them will be found to be mutated in PCD patients.

Next-generation sequencing, especially exome sequencing, has been become the method of genetic analysis of PCD patients [19–22]. Only in PCD families in which the PCD mutations have been identified, Sanger sequencing is still mainly used for genetic analysis of other

Table 28.1 PCD causing genes classified by associated ultrastructural defect

Ultrastructural defect in TEM	Mutated genes
ODA deficiency	ARMC4 CCDC103 CCDC114 CCDC151 DNAH5 DNAH9 DNAI1 DNAI2 DNAL1 LRRC56 NME8 TTC25
ODA + IDA deficiency	CFAP298 CFAP300 DNAAF1 DNAAF2 DNAAF3 DNAAF4 DNAAF5 LRRC6 PIH1D3 SPAG1 ZMYND10
Normal ciliary ultrastructure	CCDC65 DNAH11 DRC1 GAS8 HYDIN SPEF2 STK36 TEKT1
Central pair abnormalities and radial spokes	CCDC39 CCDC40 DNAJB13 RSPH1 RSPH3 RSPH4A RSPH9
Ciliary aplasia	CCNO MCIDAS
Unclear TEM	DNAH6 DNAH8 NEK10 ODF1 RPGR GAS2L2 TTC12

members in that family, such as determining whether a relative is a carrier of that mutation, or not.

Mutant genes causing PCD can be classified by location of their protein products in the ciliary axoneme (e.g., outer dynein arm, radial spoke, central complex, nexin–dynein regulatory complex; [22]) or by associated ultrastructural defect (e.g., absent outer dynein arm, absent outer and inner dynein arms, central pair abnormalities; Table 28.1). However, mutations in some genes (such as DNAH11) do not result in a detectable ultrastructural defect on transmission electron microscopy (TEM).

Worldwide, further studies are being performed to identify candidate genes and to detect disease causing mutations in these genes. Earlier diagnosis could help prevent evolution into irreversible lung damage with bronchiectasis.

28.4.3 Cell Culture

> Evaluation of cilia after ciliogenesis in culture remains the most reliable diagnostic technique for PCD.

Epithelial cells from biopsies can be cultured in vitro. During the growth phase, epithelial cells will dedifferentiate and cilia will get lost completely. A sequential monolayer–suspension culture system can then be used to let epithelial cells redifferentiate [23] into ciliated cells (see Fig. 28.1).

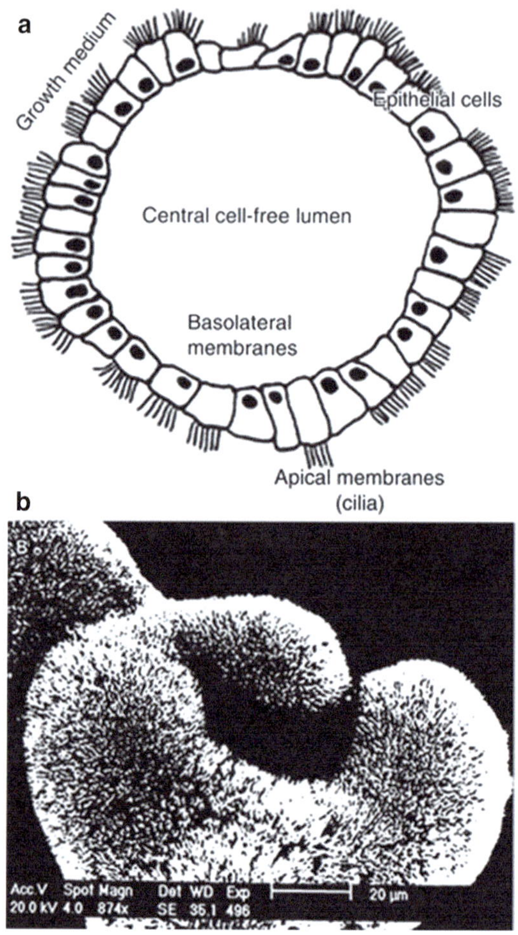

Fig. 28.1 (**a**) Schematic representation of a spheroid in suspension culture. (**b**) SEM picture of spheroid after 6 weeks in suspension

These newly formed cilia do not express the acquired abnormalities, but the inherited abnormalities (PCD) are expressed in the culture system [24]. As the functional abnormalities are also clearly present, this culture can be used for the diagnosis of PCD [25]. PCD with normal ultrastructure would easily be missed with classical transmission electron microscopy on bioptic material [26]. Ciliogenesis in vitro has also been achieved by placing the cells in an air–liquid interface culture system [27].

28.5 Conclusion

The mucociliary transport rate can be measured in vivo using either the saccharine and/or color test or radioisotope transport testing as well as by measuring ciliary activity in vitro. Inborn disorders of the mucociliary transport as in primary ciliary dyskinesia (PCD) result in the absence of mucociliary transport. Nasal nitric oxide (nNO) was found to be tenfold lower in PCD patients. None of these tests are absolutely reliable for the diagnosis of inherited abnormalities. Sequential monolayer–suspension cell culture with dedifferentiation and redifferentiation of the ciliated epithelium improves the reliability of PCD diagnosis.

References

1. Andersen I, Camner P, Jensen PL, et al. A comparison of nasal and tracheobronchial clearance. Arch Environ Health. 1974;29:290–3.
2. Andersen I, Proctor DF. Measurement of nasal mucociliary clearance. Eur J Respir Dis Suppl. 1983;64:37–40.
3. De Boeck K, Proesmans M, Mortelmans L, et al. Mucociliary transport using 99mTc-albumin colloid: a reliable screening test for primary ciliary dyskinesia. Thorax. 2005;60:414–7.
4. Marthin JK, Mortensen J, Pressler T, Nielsen KG. Pulmonary radioaerosol mucociliary clearance in diagnosis of primary ciliary dyskinesia. Chest. 2007;132:966–76.
5. Svartengren K, Wiman L-G, Thyberg P, et al. Laser light scattering spectroscopy: a new method to measure tracheobronchial mucociliary activity. Thorax. 1989;44:539–47.
6. Jorissen M, De Brouwer J, Bessems A, et al. Quantification of ciliary beat frequency by computerized microscope photometry—a preliminary study on suspension cultures of human nasal epithelia showing spontaneous ciliogenesis in vitro. Leitz Sci Tech Inf. 1992;10:88–93.
7. Sanderson MJ, Dirksen ER. A versatile and quantitative computer-assisted photoelectronic technique used for the analysis of ciliary beat cycles. Cell Motil. 1985;5:267–92.
8. Chilvers MA, O'Callaghan C. Analysis of ciliary beat pattern and beat frequency using digital high speed imaging: comparison with the photomultiplier and photodiode methods. Thorax. 2000;55:314–7.
9. Dimova S, Maes F, Brewster ME, et al. High-speed digital imaging method for ciliary beat frequency measurement. J Pharm Pharmacol. 2005;57:521–6.
10. Jorissen M. Correlations among mucociliary transport, ciliary function and ciliary structure. Am J Rhinol. 1998;12:53–8.
11. Lundberg JO, Weitzberg E, Nordvall SL, et al. Primarily nasal origin of exhaled nitric oxide and absence in Kartagener's syndrome. Eur Respir J. 1994;7:1501–4.
12. Bush A, Cole P, Hariri M, et al. Primary ciliary dyskinesia: diagnosis and standards of care. Eur Respir J. 1998;12:982–8.
13. Karadag B, James AJ, Gultekin E, et al. Nasal and lower airway level of nitric oxide in children with primary ciliary dyskinesia. Eur Respir J. 1999;13:1402–5.
14. Lefevere L, Willems T, Lindberg S, Jorissen M. Nasal nitric oxide. Acta Otorhinolaryngol Belg. 2000;54:271–80.
15. Barbato A, Frischer T, Kuehni CE, et al. Primary ciliary dyskinesia: a consensus statement on diagnostic and treatment approaches in children. Eur Respir J. 2009;34:1264–76.
16. Zariwala MA, Omran H, Ferkol TW. The emerging genetics of primary ciliary dyskinesia. Proc Am Thorac Soc. 2011;8:430–3.
17. Leigh MW, Horani A, Kinghorn BA, et al. Primary ciliary dyskinesia (PCD): a genetic disorder of motile cilia. Transl Sci Rare Dis. 2019;4:51–75.
18. Ostrowski LE, Blackburn K, Radde KM, et al. A proteomic analysis of human cilia: identification of novel components. Mol Cell Proteomics. 2002;1:451–65.
19. Berg JS, Evans JP, Leigh MW, et al. Next generation massively parallel sequencing of targeted exomes to identify genetic mutations in primary ciliary dyskinesia: implications for application to clinical testing. Genet Med. 2011;13:218–29.
20. Liu L, Luo H. Whole-exome sequencing identified a novel compound heterozygous mutation of LRRC6 in a Chinese primary ciliary dyskinesia patient. Biomed Res Int. 2018;2018:1854269.
21. Zhang W, Li D, Wei S, et al. Whole-exome sequencing identifies a novel CCDC151 mutation, c.325GT (p.E109X), in a patient with primary ciliary dyskinesia and situs inversus. J Hum Genet. 2019;64:249–52.
22. Lucas JS, Davis SD, Omran H, et al. Primary ciliary dyskinesia in the genomics age. Lancet Respir Med. 2020;8:202–16.
23. Jorissen M, Van der Schueren B, Tyberghein J, et al. Ciliogenesis and coordinated ciliary beating in human nasal epithelial cells cultured in vitro. Acta Otorhinolaryngol Belg. 1989;43(1):67–73.
24. Jorissen M, Willems T, Van der Schueren B, et al. Secondary ciliary dyskinesia is absent after ciliogenesis in culture. Acta Otorhinolaryngol Belg. 2000a;54:333–42.
25. Jorissen M, Willems T, Van der Schueren B, et al. Ultrastructural expression of primary ciliary dyskinesia after ciliogenesis in culture. Acta Otorhinolaryngol Belg. 2000b;54:343–56.
26. Jorissen M, Willems T, Van der Schueren B. Ciliary function analysis for the diagnosis of primary ciliary dyskinesia: advantages of ciliogenesis in culture. Acta Otolaryngol. 2000c;120:291–5.
27. Hirst RA, Rutman A, Williams G, et al. Ciliated air–liquid cultures as an aid to diagnostic testing of primary ciliary dyskinesia. Chest. 2010;138:1441–7.

29 Nasal Defensive Proteins: Distribution and a Biological Function

Hideyuki Kawauchi

Core Messages

- There are various defensive mechanisms in upper respiratory tract mucosal linings, such as mechanical and functional barriers.
- The mechanical barrier of sinonasal mucosa consists of mucus, motile cilia, and respiratory epithelial cells linked by adhesion complexes that include tight junctions.
- The functional barrier in sinonasal mucosa is dynamic and more complex, being equipped with innate and acquired immune response among resident cells on the epithelia and immunocompetent cells in sinonasal submucosa.
- Various types of nasal defensive proteins in human nasal mucosa are responsible for exerting those defensive mechanisms. Those proteins are essential for a defense system against various invading pathogens, such as bacteria and viruses, and modulate allergic or infective chronic inflammations as well.
- Those proteins derived from epithelial cells or recruited inflammatory cells are also one of the important key players in the pathogenesis of rhinosinusitis and allergic rhinitis at the epithelial linings of nasal cavity and paranasal sinuses.
- Those defensive proteins could be classified from constitutional and functional aspects, such as surfactants, mucins, antimicrobial peptides, and inflammatory cell-derived enzymes.
- Chemokines, cytokines, and various antibodies can be dealt with as defensive proteins, to provoke cellular interactions against microbial infections or pathogenesis of sinonasal persistent inflammations.

H. Kawauchi (✉)
Department of Microbiology, Faculty of Medicine, Shimane University, Izumo City, Japan
e-mail: kawauchi@med.shimane-u.ac.jp

29.1 General Concept of Nasal Defensive Proteins and Its Mechanism of Actions

29.1.1 Surfactants

Surfactant proteins are considered to play an important role in surfactant metabolism and host defense mechanisms in mucosal linings of respiratory tract. Surfactant proteins (SPs) are sialoglycoproteins and members of the collectin family. They are now distinguished as SP-A, SP-B, SP-C, and SP-D [1]. They are hydrophilic proteins responsible for innate immunity [1, 2], as SP-A and SP-D bind various pathogens, such as bacteria, viruses, and fungi, at the initial phase of defense line of mucosal linings. On the other hand, SP-B and SP-C are hydrophobic proteins,

Ö. Ö. Celebi, T. M. Önerci (eds.), *Nasal Physiology and Pathophysiology of Nasal Disorders*,
https://doi.org/10.1007/978-3-031-12386-3_29

contributing to surfactant function and packing and organizing of phospholipids [3]. SPs are now elucidated to be expressed in many mucosal sites, such as gastric and intestinal mucosa [4], joints, peritoneum, nasal mucosa [5], maxillary sinus mucosa [6], and middle ear mucosa as well [6], employing various experimental methods with immunohistochemistry, Western blotting, and reverse transcription-polymerase chain reaction (RT-PCR).

29.1.2 Mucins

29.1.2.1 Classification

Mucins are high molecular weight glycoproteins, constituting the major component of mucus secretions in various mucosal surfaces, such as sinonasal cavity, middle ear, and Eustachian tube as well. There are secretory and membrane-associated forms of mucin, and they protect the epithelial surface and trap pathogenic bacteria and viruses for mucociliary clearance. Secretory mucins contribute to the viscid mucus of the respiratory, gastrointestinal, and reproductive tracts and typically form extremely large oligomers through linkage of their protein monomers by disulfide bonds. These proteins are secreted from the cell to form the mucous gel, which becomes an integral part of the mucociliary escalator. In contrast, the membrane-associated mucins have a hydrophobic membrane-spanning domain and have not been observed to form oligomer complexes. The histochemical profile of cellular glycoprotein in normal condition is quite different in regard with airway level, stage of maturation, and species. A majority of surface secretory cells contain a glycoprotein consisting of a protein backbone with sugar side chains, having terminal sialic acid, galactose residues, and a variable content of sulfate esters.

29.1.2.2 Mucin Genes

With the development of molecular biological techniques, the complementary deoxyribonucleic acid (cDNA) sequences of mucin genes can now be obtained and the amino acid sequences of the mucin peptide core deduced. Until now, 21 mucin genes (*MUC1–MUC21*) have been reported, but later on some of them were found as the same gene as already reported [7, 8]. Among them, there are at least eight human mucin genes (*MUC1–MUC4*, *MUC5AC*, *MUC5B*, *MUC6*, and *MUC7*) and one mouse mucin gene (*MUC1*) confirmed to be present in respiratory tract mucosae.

29.1.2.3 Mucin Gene Expression and Upregulation in Animal Model

The specific mechanisms by which pathogens or mucosal irritants induce the mucin gene upregulation are unknown yet. It could be easily considered that they probably act either by increasing the transcription rate or decreasing the rate of degradation of mucin messenger ribonucleic acid (mRNA). Jany observed human *MUC2* gene expression in SO_2-exposed rats [9]. They used rats exposed to 400 ppm SO_2 for 3 h day-1, 5 days a week, for 1–3 weeks, and found increased numbers of goblet cells and visible mucinous secretions in the airway lumen. In addition, they employed the northern blot analysis using the total ribonucleic acid (RNA) extracted from the rat lung and hybridized it with human *MUC2* cDNA (SMUC41) and demonstrated an upregulation of the *MUC2* gene in the SO_2-exposed rats.

29.1.2.4 Localization of Mucin Genes in Human Airways

A number of experimental approaches have been performed with regard to localization of mucin genes in human airways by in situ hybridization. As regard lung or bronchial mucosal linings, the localization of *MUC2* gene expression was reported by Audie et al. [10]. In their study, bronchial mucosae were obtained from four patients, whose lungs had been surgically resected for cancer. A radiolabeled antisense oligonucleotide probe corresponding to the tandem repeat domain of *MUC2* was applied on the tissue specimens, and consequently, it was localized to occasional goblet cells of the surface epithelium and to submucosal gland ducts, but not to their secretory acini. However, on the other hand, oligonucleotide probes to *MUC4*, *MUC5B*, and *MUC5AC* strongly labeled the gland acinar cells. These data indicate that several distinct mucin genes are expressed in the mucosa either by the same or distinct cells of the epithelium and glands. Li et al. [11] demonstrated in nasal mucosa that *MUC2* mRNA transcripts are present in ciliated and basal cells of the surface epithelia, serous and mucous acini of submucosal glands, and occasionally mononuclear inflammatory cells. Voynow et al. precisely compared expression levels of three mucin genes, *MUC*1, *MUC*2, and *MUC*5/5 AC, in the respiratory tract of patients with cystic fibrosis, patients with allergic rhinitis, and normal individuals [12]. Mucin transcript levels in nasal epithelial cells free from inflammation were quantitated by an *MUC* mRNA slot-blot method. Their elegant study revealed that *MUC*5/5 AC mRNA was expressed at five- to tenfold greater levels than *MUC*2 or *MUC*1 for all subjects and *MUC*2 mRNA levels were similar among all subject groups. To be generally considered, in situ hybridization demonstrated that MUC5AC-positive mucous cells are populated in upper respiratory epithelium, whereas MUC5B-positive mucous cells are populated in mucous glands of upper respiratory epithelia. In contrast, middle ear epithelial cells are negative for MUC5AC mRNA transcripts, but spotty MUC5B mRNA transcripts are identified [13, 14]. These data remind us the transitional process of mucin member from the lower airway to the middle ear cavity. Preciado et al. found that MUC5B mucin is predominant in patients with chronic otitis media [15, 16] and confirmed that transitional process. However, other mucins, such as MUC2, may be involved in middle ear mucus of animal models [17], but their amount is limited or undetectable in human [13, 14]. Conclusively, the quantitative study of mucins and its comparison looks notoriously difficult and should be evaluated at the semiquantitative level, because of higher level of glycosylation representing a posttranslational modification.

29.1.3 Antimicrobial Peptides

The body fluids and organized tissues naturally contain a variety of antimicrobial substances that kill or inhibit the growth of microorganism. The sources and activities of a variety of host antimicrobial substances are summarized in Table 29.1. Among them, a low-molecular weight antimicrobial peptide, defensin, is introduced herein in relation to the various receptors contributing to innate immunity.

29.1.3.1 Defensin

Defensins are a family of evolutionarily related vertebrate antimicrobial peptides with a characteristic beta-sheet-rich fold and a framework of six disulfide-linked cysteines. Two main defensing subfamilies, α- and β-defensins, differ in the length of peptide segments between the six cyste-

Table 29.1 Antimicrobial substance of host origin present in body fluids and organized tissues

Substance	Common sources	Chemical composition	Activity
Lysozyme	Serum, saliva, tears	Protein	Bacterial cell lysis
Complement	Serum	Protein–carbohydrate lipoprotein complex	Cell death or lysis of bacteria; participates in inflammation
Basic proteins	Serum	Proteins or basic peptides	Disruption of bacterial plasma membrane
Lactoferrin and transferrin	Body secretions, serum epithelial cells	Glycoprotein	Inhibit microbial growth by binding (withholding) iron
Defensin	Epithelial cells, neutrophils	Oligopeptides	Cell death or lysis of bacteria
Peroxidase	Saliva, tissues, neutrophils	Protein	Act with peroxide to cause lethal oxidation of cells
Fibronectin	Serum, mucosal surfaces	Glycoprotein	Clearance of bacteria
Interferons	Virus-infected cells	Protein	Resistance to virus infections
Interleukins	Macrophages, lymphocytes	Protein	Cause fever; promote activation of immune system

ines and the pairing of the cysteines that are connected by disulfide bonds. Defensins are abundant in cells and tissues that are involved in host defense mechanism of mucosal linings, such as respiratory and intestinal mucosa. In many species, the highest concentrations of α-defensins are found in granules, the storage organelles of leukocytes [18, 19]. β-Defensins are mainly produced by a variety of epithelial cells, such as lung and middle ear [20, 21], and 30 genes from DEFB1 to DEFB136 are currently identified as human β-defensin family. Human β-defensin 1 (DEFB1) is expressed constitutively, whereas β-defensin 2 (DEFB4A) is induced by bacterial molecules [22, 23] as well as cytokines [24]. Paneth cells are another site of high α-defensin concentration and contain defensin-rich secretory granules. Claeys et al. reported that there was no baseline detection of human β-defensin 2 in any sinus mucosal samples, and no upregulation was measured for human β-defensin 2 in paranasal sinus mucosa in patients with chronic sinusitis or nasal polyposis compared with control turbinate mucosa [25]. On the other hand, α- and β-defensins were detected in human nasal mucosa by Lee et al. [26]. They examined the expression of defensins in inferior turbinate mucosa of normal subjects and inferior turbinate mucosa and nasal polyps of patients with chronic sinusitis, employing reverse transcription-polymerase chain reaction (RT-PCR) and immunohistochemistry. According to their results, β-defensin 1 mRNA was expressed in all tissue samples. β-Defensin 2 mRNA was detected in the turbinate mucosa and nasal polyps of patients with chronic sinusitis, but not in normal mucosa. Its expression level was significantly higher in nasal polyps than in turbinate mucosa. α-Defensin 5 and 6 mRNAs were not expressed in any tissues, but α-defensins 1, 2, and 3 were detected in all tissue samples obtained from patients with chronic sinusitis. These results suggest that β-defensin 1 may play a constitutive role in nasal defenses, whereas α-defensins 1, 2, and 3 and β-defensin 2 may be induced in response to local infection or inflammation.

The average concentration of defensins in these epithelial cells reaches the 10–100 μg/mL range [27], but the local concentrations might be higher because of uneven distribution.

Most defensins show antimicrobial activity against bacteria and fungi, especially when tested under low ionic strength conditions [28, 29] and with low concentrations of divalent cations, plasma proteins, and other interfering substances. It should be taken into consideration that under these optimal conditions, antimicrobial activity is observed at concentrations as low as 1–10 μg/mL (low micromole). Permeabilization of target membrane is the crucial step in defensin-mediated antimicrobial activity and cytotoxicity. It is also reported that defensins act as an immunomodulatory molecules, inducing IL-8 in epithelial cells and modulating complement activation [30, 31] and neutrophil apoptosis [32, 33].

29.1.3.2 Defensin Synthesis and Its Regulation

Defensin synthesis and release are regulated by various signals, such as microbial signals, developmental signals, and cytokines. Human β-defensin 1 (DEFB1) is expressed constitutively with low level, whereas β-defensin 2 (DEFB4A) is highly induced by bacterial molecules [23, 24] as well as cytokines [25]. Various signaling pathways are identified to orchestrate β-defensin 2 expression, such as toll/IL-1 receptor (TIR)-dependent NF-κB activation, TIR-dependent MAPK signaling, and NOD-2-dependent NF-κB activation. Wang et al. actually demonstrated that airway epithelia regulate expression of human β-defensin 2 through toll-like receptor 2 [34]. Vora et al. also proved that human β-defensin 2 expression is regulated by toll-like receptor signaling in intestinal epithelial cells and that LPS and peptidoglycan stimulated β-defensin 2 promoter activation in a TLR4- and TLR2-dependent manner, respectively [35].

29.2 Brief Introduction of Toll-Like Receptors in Nasal Epithelial Cells and Signaling Pathway

29.2.1 Distribution of TLRs in Nasopharyngeal Mucosae and Involvement of IL-15 in Inflammation

We employed northern blot assay and RT-PCR to see the TLR distribution (Fig. 29.1) in upper respiratory mucosa and showed that human nasal epithelial cells constitutively expressed mRNA for TLR2, 3, and 6, but not for TLR4 and TLR9 (Fig. 29.2). Lipoprotein used as a pathogen-associated molecular pattern (PAMP) also induced IL-15 production of respiratory epithelial cells, which strictly depend on TLR2 (Fig. 29.3). In Fig. 29.3, in northern blot analysis, IL-15 mRNA was strongly expressed after lipoprotein stimulation. But in contrast, it was not found after lipid A stimulation as a ligand of TLR4. IL-15 concentration in the supernatants of CCL185 was also upregulated after lipoprotein stimulation in a dose-dependent manner. Figure 29.4 shows NF-κB activation of respiratory epithelial cells in response to lipoprotein. In Western blot analysis, phosphorylation of IkB-alpha is detected in epithelial cells 15 and 30 min after lipoprotein stimulation. Luciferase assay and DNA-binding assay reveal that NF-κB activity actually correlated with the lipoprotein concentration.

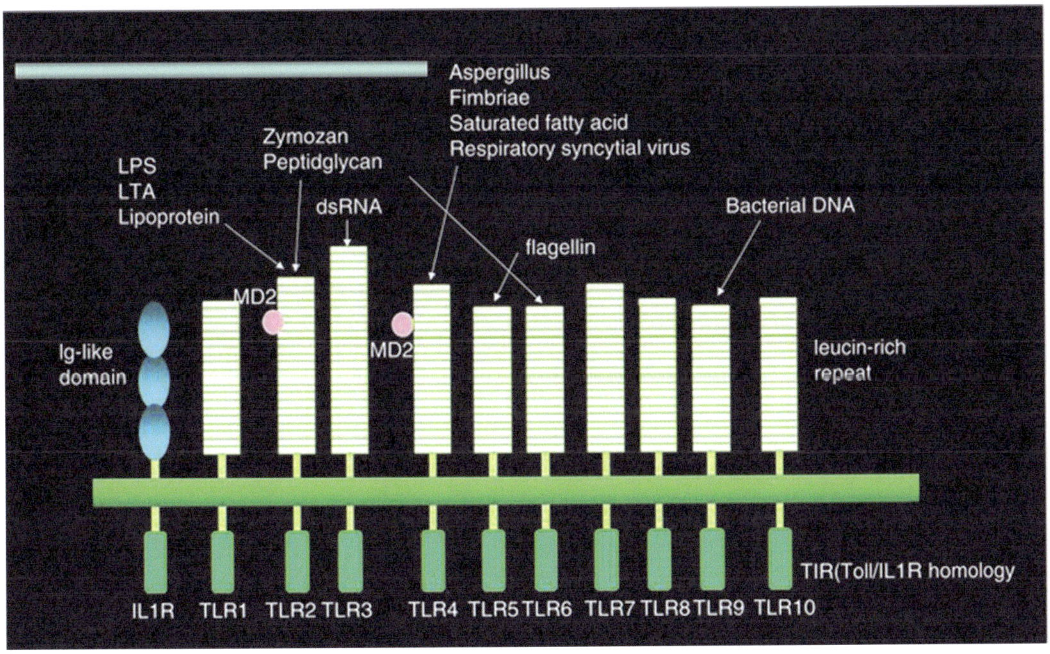

Fig. 29.1 Toll family proteins

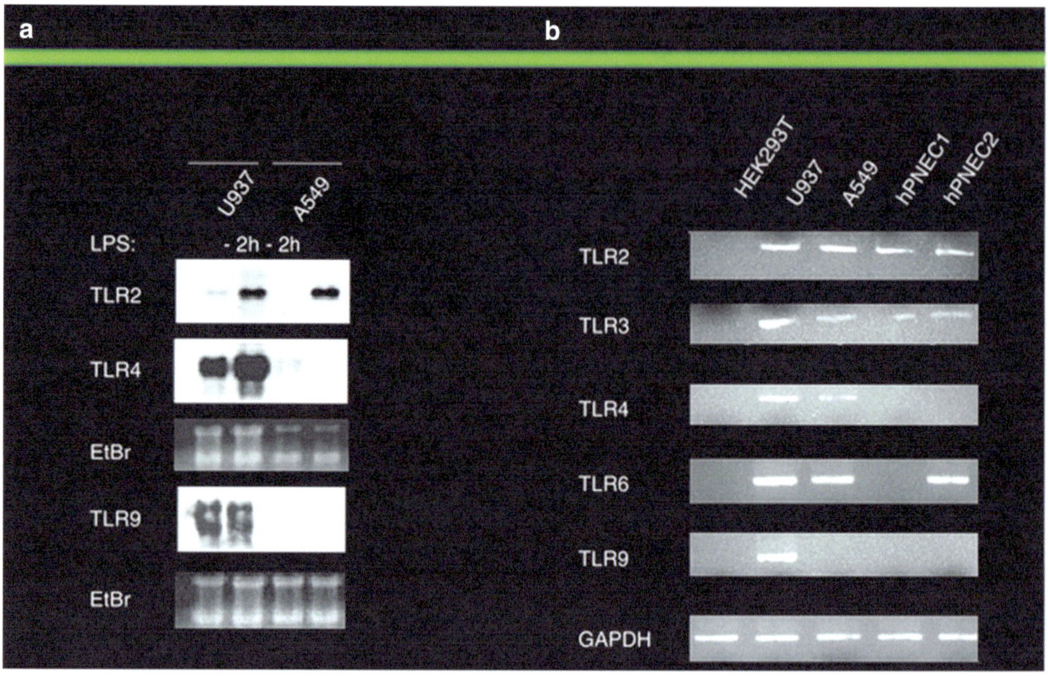

Fig. 29.2 Expression of TLRs on macrophages and nasal epithelial cells. (**a**) Northern blot assay. (**b**) RT-PCR

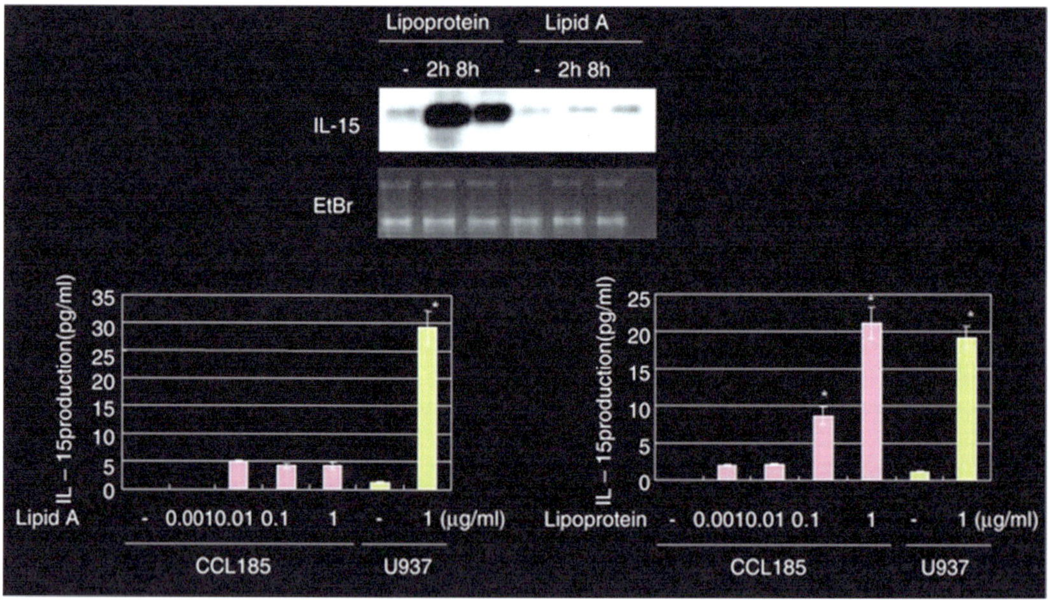

Fig. 29.3 Lipoprotein inducing IL-15 from respiratory epithelial cells

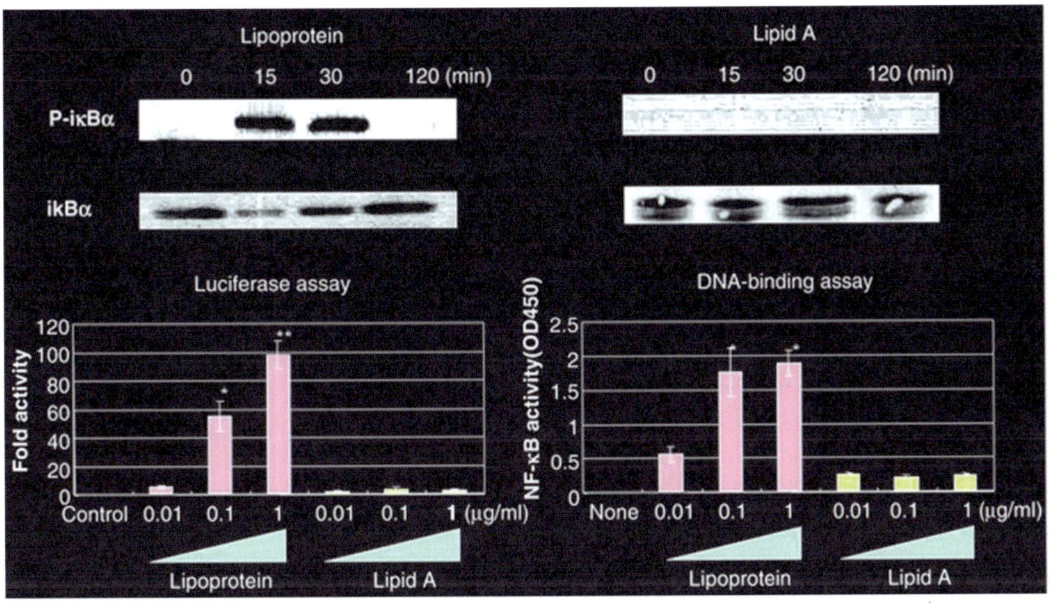

Fig. 29.4 NF-kB activation of respiratory epithelial cells in response to lipoprotein and lipid A

29.3 Conclusion

In this chapter, I have extensively introduced the significance of the so-called defensive proteins on the upper respiratory tract mucosal linings from various aspects. As you can see, there are so many complex factors involved in the defense line of our sinonasal mucosa in accordance with microbial invasion and pathogenesis of sinonasal disorders through intracellular signaling pathways. Further investigation of those mechanisms should be absolutely needed for elucidating pathologies of sinonasal disorders and seeking for promising therapeutic strategies. To further end, I have intensively described antiviral immunity with interferons as an appendix, for your understanding of SARS-CoV-2 virus infection.

Appendix: Antiviral Proteins and Immunity for the Understanding of SARS-CoV-2 Virus Infection

Introduction

We all know viruses are the intracellular parasites, taking place infection at various mucosal linings or nerve tissue, including central nervous system, by invading to resident cells of the mammalian body. Viral life cycle consists of a short period of extracellular phase, and a longer intracellular period as long as viruses keep replication. Interferon, defensin, complement, and antiviral antibodies are, respectively, important as antiviral proteins. And as antiviral cellular responses, natural killer (NK) cells, cytotoxic T lymphocytes (CTLs) are elucidated to, respectively, exert non-specific and specific immune responses [36]. Those immune system composes of three different stages: innate responses (complement, interferon, NK cells) appear with a few days after viral infection and suppress viral replication. Antigen-specific CTL responses develop a few days later, having peak response between days 7 and 10 and resolve ongoing intracellular infection. Finally, antiviral antibodies in serum reach maximum titers and continue for months to years, and specific those antibodies provide a protection against reinfection. The concept of "innate immunity" includes all sorts of defense mechanism that expel infections with little specificity and less clonal expansion. It works without much adaptation and no generation of a long-lasting memory. So, the mammalian innate immune defenses comprise defensins, the complement system, non-specific phagocytic and cytolytic leukocytes, and cytokines, such as the antiviral active interferons.

Defensins

Defensins are already introduced in this chapter, but the details why defensins inhibit viral infection. They are produced by immune cells and skin and mucosal epithelial cells and are consequently present on epithelia and in body fluids. Defensin genes can be induced by viral infection [37]. Their most common antimicrobial function is the formation of destructive pores in membranes of pathogens, including enveloped viruses. Defensins can, however, also block infection by enveloped and non-enveloped viruses alike by aggregating the particles, blocking receptor binding, inhibiting virus entry, particle uncoating or intracellular trafficking, interfering with essential cell signaling, or viral gene expression. Moreover, besides these direct antiviral activities, defensins were shown to attract immune cells and modulate adaptive immune responses.

Complement System

Complement activation is mediated by specific receptors recognizing pathogens or immunocomplexes. Three different pathways are distinguished: the classical pathway (triggered by antigen–antibody complexes), the mannan-binding lectin pathway (triggered by lectin binding of pathogen surfaces), and the alternative pathway (triggered by complement factor C3b-coated pathogen surfaces), respectively. They all activate a cascade of reactions involving more than 20 soluble and cell-bound proteins, resulting in a rapid and massive response. The complement system is able to tag infected cells for destruction by phagocytic cells (opsonization), prime humoral immune responses, and perforate membranes of infected cells by the membrane attack complex [38]. In response to those defense mechanisms, viruses have evolved effective counter weapons, such as incorporation of cellular complement regulatory proteins into particles or expressing specific inhibitors in infected cells.

Interferons

Interferons and innate and adapted antiviral immune response interferons (IFNs) are a group of signaling proteins made and released by host cells in response to the infection or presence of several viruses. In a typical scenario, a virus-infected cell will release interferons, causing nearby cells to upregulate their antiviral activity. Interferons are originally named for their ability to "interfere" with viral replication by protecting cells from virus infections. And, IFNs have various functions. They activate immune cells, such as natural killer cells and macrophages. And they increase host defenses by upregulating antigen presentation by virtue of increasing the expression of major histocompatibility complex (MHC) antigens. More than 20 distinct IFN genes and proteins have been identified in animals and human beings. They are typically divided among three groups, namely, Type I IFN (IFN-alpha and -beta), Type II IFN (IFN-gamma), and Type III IFN. In general, type I and II interferons are responsible for regulating and activating the immune response.

Type I interferons are produced by a variety of immunocompetent cells and exert inhibition of viral replication and cell proliferation. Those also enhance natural killer cell activity to lyse virus-infected host cells. Natural killer cells represent a different lymphocyte lineage that recognize and lyse virally infected cells. They are mainly effective during an early stage of viral infection, since there is no lag phase of clonal expansion for NK as occurs with T and B lymphocytes. Specific immune antiviral mechanisms are both humoral and cellular [39]. Specific antibodies protect against viral infections and play an important role in antiviral immunity, mainly during the early stage of the infection. The most effective antiviral antibodies are neutralizing antibodies which bind to the viral envelope or capsid proteins and block the virus from entering host cell. The main effectors involved in specific antiviral immunity are CD8+ cytotoxic T lymphocytes (CTL). These cells recognize viral antigens presented at the cell surface associated with class I MHC molecules. CTL response is not always beneficial, since the tissue destruction caused by CTL is sometimes greater than the damage done by the virus.

In response to virus infection, plasmacytoid dendritic cells (pDCs) are particularly equipped to synthesize and secrete IFN-a/b, but in principle all nucleated cells can do it. In autocrine and paracrine manner, IFNs trigger a signaling chain leading to the expression of genes for potent antiviral proteins which limit further viral spread. In addition, IFNs initiate, modulate, and enhance the adaptive immune response. The signaling events which culminate in the direct IFN-dependent restriction of virus growth can be divided into three steps, namely, (1) transcriptional induction of IFN synthesis, (2) IFN signaling, and (3) antiviral mechanisms.

Interferon Induction

Nucleic acids are main pathogen-associated molecular patterns (PAMPs) of viruses, being recognized by a number of pattern recognition receptors (PRRs) to initiate induction of IFN genes. Virus-triggered PRRs classes can be divided into the endosomal Toll-like receptors (TLRs) and various intracellular (mostly cytoplasmic) receptors. It is thought TLRs can be activated by viral nucleic acids that had been released from virus particles. Major PAMPs are double-stranded RNA (dsRNA), single-stranded RNA (ssRNA), 5′-tripho-sphorylated RNAs, and double-stranded DNA (dsDNA). dsRNA is an almost ubiquitous by-product of virus infection that is recognized by a number of PRRs. In the endosome, it is recognized by the TLR3, and in the cytoplasm by the RNA helicases RIG-I and MDA-5 during genome transcription and replication. Viral ssRNAs can be recognized in the endosome by TLR7 and TLR8.

In addition to nucleic acids, some viral proteins can provoke a TLR response, such as envelope protein of respiratory syncytial virus and measles virus by activating TLR4 and TLR2, respectively.

All PRRs are triggering signaling chains which culminate in activation the IFN regulatory factor (IRF)-3, the general immune regulatory transcription factor NF-kB, and the stress activated transcription factor AP-1. In a cooperation, they upregulate IFN gene expression. This leads to a "first wave" of IFN production (IFN-beta and IFN-arupha4 in mice) which triggers expression of the transcription factor IRF-7. IRF-7 is a master regulator of IFN gene expression cooperating with IRF-3 for full activity. IRF-7 can be activated in the same way as IRF-3 and is responsible for a positive feedback loop that initiates the synthesis of several IFN-alpha subtypes as the "second wave" IFNs.

While cells with a nucleus are thought to be equipped with the set of intracellular PRRs, expression of TLRs is more restricted to epithelial and immune cells. Myeloid dendritic cells (mDCs), for example, can recognize dsRNA by the classical intracellular RLR pathway and, in addition, by TLR3. pDCs recognize the presence of viral ssRNA or dsDNA by TLR7, TLR8, and TLR9 to transcriptionally activate multiple IFN-alpha genes. IRF-7 is further upregulated in response to IFN and generates a positive feedback loop for high IFN-alpha and IFN-beta production. Furthermore, TLR7 and TLR9 are retained in the endosomes of pDCs to allow prolonged IFN induction signaling.

Type I IFN Signaling

IFN-beta and multiple IFN-alpha subspecies activate a common type I IFN receptor signaling to the nucleus through the so-called JAK–STAT pathway [40, 41]. The signal transducer and activator of transcription (STAT) proteins are latent cytoplasmic transcription factors, which become phosphorylated by the Janus Kinases JAK1 and TYK2. Phosphorylated STAT1 and STAT2 recruit a third factor, IRF9, to form a complex known as IFN-stimulated gene factor 3 (ISGF3), which translocates to the nucleus and binds to the IFN-stimulated response element (ISRE) in the promoter region of ISGF.

Direct and Indirect Antiviral Effects of Type I IFNs

Type I IFNs activate the expression of several hundred STAT-dependent ISGs of which only a fraction has been studied in great detail. IFN-alpha and IFN-beta bind to the type I IFN receptor (IFNAR) and activate the expression of numerous ISGs via the JAK/STAT pathway [42]. Several ISGs contribute in a more indirect manner to the enhancement of both innate and adaptive immune responses. Type I IFNs can directly enhance clonal expansion and memory formation of CD8+ T cells. IFNs promote NK cell-mediated cytotoxicity and trigger the synthesis of other cytokines, such as IFN-gamma or IL-15. These cytokines modulate the adaptive immune response, enhance NK cell proliferation, and support CD8+ T cell memory. Moreover, by upregulating TLRs, MHCs, and costimulatory molecules, IFNs enable APCs (most prominently DCs) to become competent in presenting viral antigens and stimulating the adaptive immune response.

Type II IFN

IFN-gamma categorized type II interferon is a cytokine that is essential and critical for innate and adaptive immunity against viral, some bacterial, and protozoan infections [43]. IFN-gamma is an important activator of macrophages and inducer of major histocompatibility complex class II molecule expression. Aberrant IFN-gamma expression is associated with a number of autoinflammatory and autoimmune diseases. The importance of IFN-gamma in the immune system stems in part from its ability to inhibit viral replication directly, and most importantly from its immunostimulatory and immunomodulatory effects. IFN-gamma is produced predominantly by natural killer (NK) cells and natural killer T (NKT) cells as a part of the innate immune response, and by CD4 Th1 and CD8 cytotoxic T lymphocyte (CTL) effector T cells once antigen-specific immunity develops as part of the adaptive immune response. IFN-gamma is also produced by non-cytotoxic innate lymphoid cells (ILCs), a family of immune cells first discovered in the early 2010s.

References

1. Mason RJ, Greene K, Voelker AR. Surfactant protein A and surfactant protein D in health and disease. Am J Physiol Lung Cell Mol Physiol. 1998;275:1–13.
2. Wright JR. Pulmonary surfactant: a front line of lung host defense. J Clin Invest. 2003;111:1453–5.
3. Weaver TE, Conkright JJ. Function of surfactant proteins B and C. Annu Rev Physiol. 2001;63:555–78.
4. Bourbon JR, Chailley-Heu B. Surfactant protein in the digestive tract, mesentery, and other organs: evolutionary significance. Comp Biochem Physiol. 2001;129:151–61.
5. Kim JK, Kim SS, Rha KW, Kim CH, Cho JH, Lee CH, et al. Expression and localization of surfactant proteins in human nasal mucosa. Am J Physiol Lung Cell Mol Physiol. 2007;292:879–84.
6. Dutton JM, Goss K, Khubchandani KR, Shah CD, Smith RJH, Snyder JM. Surfactant protein A in rabbit sinus and middle ear mucosa. Ann Otol Rhinol Laryngol. 1999;108:915–24.
7. Kerschner JE, Khampang P, Erbe CB, et al. Mucin gene 19 (MUC19) expression and response to inflammatory cytokines in middle ear epithelium. Glycoconj J. 2009;26:1275–84.
8. Li G, Zhang H, Lv J, et al. Tandem repeats polymorphism of MUC20 is an independent factor for the progression of immunoglobulin A nephropathy. Am J Nephrol. 2006;26:43–9.
9. Jany B, Gallup M, Tsuda T, Basbaum C. Mucin gene expression in rat airways following infection and irritation. Biochem Biophys Res Commun. 1991;181:1–8.
10. Audie JP, Janin A, Porchet N, et al. Expression of human mucin genes in respiratory, digestive, and reproductive tracts ascertained by in situ hybridization. J Histochem Cytochem. 1993;41:1479–85.
11. Li D, Wang D, Majumdar S, et al. Localization and up-regulation of mucin (MUC2) gene expression in human nasal biopsies of patients with cystic fibrosis. J Pathol. 1997;181:305–10.
12. Voynow JA, Selby DM, Rose MC. Mucin gene expression (*MUC*1, *MUC*2, and *MUC*5/5AC) in nasal epithelial cells of cystic fibrosis, allergic rhinitis, and normal individuals. Lung. 1998;176:345–54.
13. Lin J, Tsuprun V, Kawano H, et al. Characterization of mucins in human middle ear and Eustachian tube. Am J Physiol. 2001;280(6):1157–67.
14. Lin J, Tsuboi Y, Rimell F, et al. Expression of mucins in mucoid otitis media. J Assoc Res Otolaryngol. 2003;4(3):384–93.

15. Preciado D, Goyal S, Rahimi M, et al. MUC5B is the predominant mucin glycoprotein in chronic otitis media fluid. Pediatr Res. 2010a;68(3):231–6.
16. Preciado D, Kuo E, Ashktorab S, et al. Cigarette smoke activates NF-kappa-B mediated TNF-alpha release from mouse middle ear cells. Laryngoscope. 2010b;120(12):2508–15.
17. Lin J, Ho S, Paparella M, et al. Mucin gene expression in the rat middle ear; an improved method for RNA harvest. Ann Otol Rhinol Laryngol. 1999;108(8):762–8.
18. Ganz T. Extracellular release of antimicrobial defensins by human polymorphonuclear leukocytes. Infect Immun. 1987;55:568–71.
19. Ganz T, et al. Defensins. Natural peptide antibiotics of human neutrophils. J Clin Invest. 1985;76:1427–35.
20. Bals R, Wang X, Wu Z, et al. Human beta-defensin 2 is salt-sensitive peptide antibiotic expressed in human lung. J Clin Invest. 1998;102:874–80.
21. Moon SK, Lee HY, Li JD, et al. Activation of a Src-dependent RAL-MEK1/2-ERK signaling pathway is required for IL-1 alpha-induced upregulation of beta-defensin 2 in human middle ear epithelial cells. Biochim Biophys Acta. 2002;1590:41–51.
22. Moon SK, Lee HY, Pan H, et al. Synergistic effect of interleukin 1 alpha on nontypable *Haemophilus influenzae*-induced up-regulation of human beta-defensin 2 in middle ear epithelial cells. BMC Infect Dis. 2006;6:12–20.
23. Wehkamp J, Harder J, Wehkamp K, et al. NF-kappaB- and AP-1-mediated induction of human beta defensin-2 in intestinal epithelial cells by *Escherichia coli* Nissle 1917: a novel effect of a probiotic bacterium. Infect Immun. 2004;72:5750–8.
24. Kao CY, Chen Y, Thai P, et al. IL-17 markedly upregulates beta-defensin-2 expression in human airway epithelium via JAK and NF-kappaB signaling pathways. J Immunol. 2004;173:3482–91.
25. Claeys S, De Belder T, Holtappels G, et al. Human β-defensins and toll-like receptors in the upper airway. Allergy. 2003;58(8):748–53.
26. Lee SH, Kim JE, Lim HH, Lee HM, Choi JO. Antimicrobial defensin peptides of the human nasal mucosa. Ann Otol Rhinol Laryngol. 2002;111(2):135–41.
27. Harder J, Bartels J, Christophers E, et al. A peptide antibiotic from human skin. Nature. 1997;387:861–2.
28. Lehrer RI, Ganz T, Szklarek D, et al. Modulation of the in vitro candidacidal activity of human neutrophil defensins by target cell metabolism and divalent cations. J Clin Invest. 1988;81:1829–35.
29. Selsted ME, Szklarek D, Ganz T, et al. Activity of rabbit leukocyte peptides against *Candida albicans*. Infect Immun. 1985;49:202–6.
30. Panyutich AV, Szold O, Poon PH, et al. Identification of defensin binding to C1 complement. FEBS Lett. 1994;356:169–73.
31. Van Wetering S, Mannesse-Lazeroms S, Van Sterkenburg MA, et al. Effect of defensins on interleukin-8 synthesis in airway epithelial cells. Am J Physiol. 1997;272:888–96.
32. Nagaoka I, Niyonsaba F, Tsutsumi-Ishii Y, et al. Evaluation of effect of human beta-defensins on neutrophil apoptosis. Int Immunol. 2008;20:543–53.
33. Yang D, Biragyn A, Kwak LW, et al. Mammalian defensins in immunity: more than just microbicidal. Trends Immunol. 2002;23:292–6.
34. Wang X, Zhang Z, Louboutin JP, et al. Airway epithelia regulate expression of human β-defensin 2 through toll-like receptor 2. FASEB J. 2003;17:1727–9.
35. Vora P, Youdim A, Thomas LS, et al. β-Defensin 2 expression is regulated by TLR signaling in intestinal epithelial cells. J Immunol. 2004;173:5398–405.
36. Moskophidis D, Lechner F, Pircher H, Zinkernagel RM. Virus persistence in acutely infected immunocompetent mice by exhaustion of antiviral cytotoxic effector T cells. Nature. 1993;362:758–61.
37. Wilson SS, Wiens ME, Holly MK, Smith JG. Defensins at the mucosal surface: latest insights into defensin-virus interactions. J Virol. 2016;90:5216–521.
38. Holers VM. Complement and its receptors: new insights into human disease. Annu Rev Immunol. 2014;32:433–59.
39. Parkin J, Cohen B. An overview of the immune system. Lancet. 2001;357:1777–89.
40. Barrat FJ, Crow MK, Ivashkiv LB. Interferon target-gene expression and epigenomic signatures in health and disease. Nat Immunol. 2019;20:1574–83.
41. Navarro J, Perez-Ruiz M. Antiviral immunity. Curr Immunol Rev. 2011;7:19–243.
42. Schmid S, Mordstein M, Kochs G, Garcia-Satre A, Tenoever BR. Transcription factor redundancy ensures induction of the antiviral state. J Biol Chem. 2010;285:42013–22.
43. Schoenborn JR, Wilson CB. Regulation of interferon-γ during innate and adaptive immune responses. Adv Immunol. 2007;96:41–101.

30 Olfaction

Huart Caroline, Philippe Eloy, and Philippe Rombaux

Core Message

- This chapter reviews the physiology and the pathology of the olfactory system. The aim is to provide adequate information to clinicians in order to improve their understanding about olfaction and its troubles and to promote adequate management of patients with olfactory disorders.

30.1 Introduction

Olfaction is one of the most ancient senses. Nevertheless, the field of olfaction has received far less attention as compared to other sensory modalities. This is notably due to the technical challenge of working with odorous stimuli and the difficulties of measuring brain activity induced by a chemosensory stimulus.

Although a majority of people consider it as one of the less important senses, this sense plays a major role in our interaction with the environment. Not only olfactory system acts for the detection of potential danger in the environment, such as smoke or gas, but also it influences our nutrition, social behavior, well-being, and memory processes.

This chapter proposes a global view of human olfaction. First we will extend on physiology of olfaction, paying a particular interest to olfactory pathways. Then, we will study pathological situations associated with olfactory dysfunction. More particularly, we will see into detail post-infectious olfactory loss, post-traumatic olfactory loss, and sinonasal-related olfactory disorder.

30.2 Physiology

30.2.1 Embryology

The olfactory placode is induced at the end of the fourth week of pregnancy when the local ectoderm makes direct contact with the prosencephalic vesicle. Some cells of the olfactory placode will differentiate into primary neurosensory cells, further constituting the olfactory neuroepithelium. At the end of the fifth week, these cells will develop axons, reaching the neurons from the anterior wall of the prosencephalon, which becomes the telencephalon. This will induce the

H. Caroline (✉) · P. Rombaux
Department of Otorhinolaryngology, Cliniques universitaires Saint-Luc, Brussels, Belgium

Institute of Neuroscience, Université catholique de Louvain, Brussels, Belgium
e-mail: caroline.huart@saintluc.uclouvain.be

P. Eloy
Department of Otorhinolaryngology, Cliniques universitaires Saint-Luc, Brussels, Belgium

Department of Otorhinolaryngology, CHU Dinant-Godinne (Site Godinne), Yvoir, Belgium
e-mail: philippe.rombaux@uclouvain.be

Ö. Ö. Celebi, T. M. Önerci (eds.), *Nasal Physiology and Pathophysiology of Nasal Disorders*,
https://doi.org/10.1007/978-3-031-12386-3_30

development of the olfactory bulb which begins to differentiate from the telencephalon. At the seventh week, the olfactory bulb individualizes at the tip of each hemisphere. It will then lengthen and come to lie on the cribriform plate of the ethmoid bone at the 12th week of pregnancy. Secondary neurosensory cells will differentiate inside the olfactory bulb and their dendrites synapse with axons of the primary neurosensory cells. Axons of secondary neurosensory cells will group to form the olfactory tract and synapse with cortical olfactory areas of the entorhinal paleocortex and archicortex [1, 2].

30.2.2 Olfactory Pathways

The olfactory system detects odorant molecules dissolved in air and trapped in the airflow passing through the nasal cavity. Nasal turbinates will guide the airflow to the olfactory cleft, allowing the odorant molecules to reach the olfactory neuroepithelium.

30.2.2.1 The Olfactory Neuroepithelium

The olfactory neuroepithelium is located in the upper part of the nasal vaults. It covers the cribriform plate of the ethmoid bone, medially to the middle turbinate and can extend to the superior turbinate, the superior part of the septum, and the middle turbinate [3]. In adult humans, its surface area is 2.5 cm^2 per nasal fossa. The location of the olfactory epithelium is dependent on individual factors and is thought to change with age, resulting from a conversion of olfactory neuroepithelium to respiratory epithelium or due to loss of olfactory neurons with age or from damages (smoke, toxics, chemicals, chronic infection).

The olfactory neuroepithelium is a pseudostratified columnar epithelium covering a lamina propria. It is composed of (1) olfactory receptor neurons (ORNs), (2) supporting cells, (3) basal cells, some of which serve as ORN stem cells for the regeneration of new olfactory sensory neurons throughout life, and (4) the duct of the Bowman's glands (which are located in the lamina propria) (Fig. 30.1).

The ORNs are bipolar cells, with their dendritic extensions directed toward the olfactory cleft and carrying on its surface several cilia that project into the mucus. Odorants are carried through the mucus layer by olfactory binding proteins, and bind to olfactory receptors located on the ORNs. In 1991, Axel and Buck [4] discovered a family of approximately 1000 genes that encode for an equivalent number of olfactory receptors, corresponding to the largest fam-

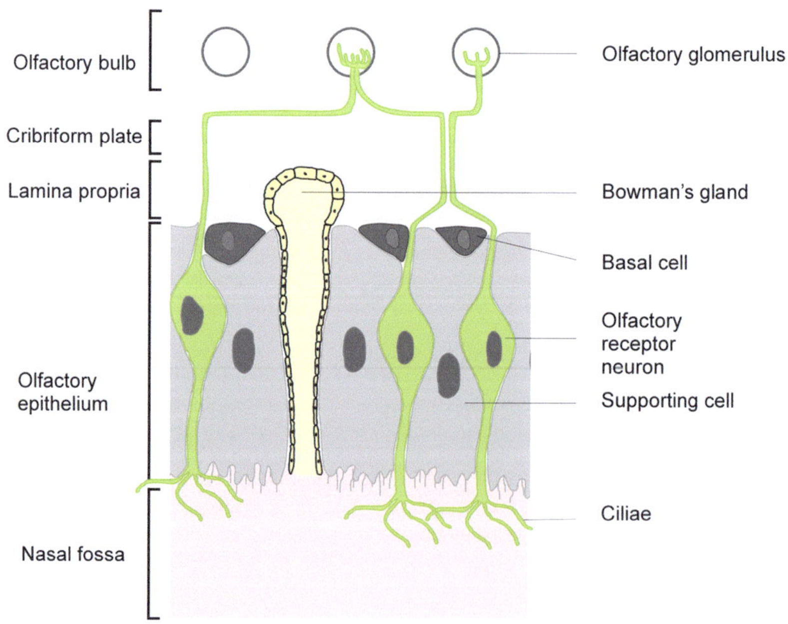

Fig. 30.1 Schematic representation of olfactory neuroepithelium. The olfactory neuroepithelium is composed by olfactory receptor neurons, supporting cells, and basal cells. The dendritic extension of olfactory receptor neurons carries on its surface several cilia, where are located the olfactory receptors. The axons run through the cribriform plate of the ethmoid bone and reach the olfactory bulb where they synapse with mitral cells in spherical structures named glomerulus

ily of genes in the mammalian genome [5], highlighting their important role in physiology. In the majority of mammals most of these genes are functional, but in primate the number of functional genes decreases and is to about 350 in humans [6]. Axel and Buck found that each ORN possesses only one type of odor receptor and each receptor is specialized for a small number of odors. Hence, a given odorant will bind a typical pattern of olfactory receptors. The binding results in the activation of G proteins. The activation of G proteins stimulates the formation of cyclic AMP. Increased levels of cAMP open cyclic nucleotide-gated channels. This causes the opening of the channels and Ca^{2+} influx. This influx activates chloride channels, opening them up, causing Cl^{-}leaves, and finally depolarizing the ORN and generating the action potential.

ORNs axons converge into the olfactory nerves, passing through the cribriform plate of the ethmoid bone and projecting directly to the ipsilateral olfactory bulb where they synapse into spherical structures known as the glomerulus.

30.2.2.2 The First Olfactory Structure: The Olfactory Bulb

The olfactory bulb is ovoid in shape and located in the anterior cranial fossa, above the cribriform plate of the ethmoid bone, and under the frontal lobe. It contains a major structure that is considered as the first olfactory structure: the glomerulus. The glomerulus is the only relay between the periphery and the cortex. Each glomerulus collects ORN axons from the same type of odorant receptor (Fig. 30.2). ORNs axons and dendrites of mitral cells synapse in the glomerulus.

The olfactory bulb has a multilayered cellular architecture. It encompasses 6 different layers: (1) the external layer is composed of ORNs axons; (2) the glomerular layer is composed by glomeruli wherein axons of ORNs synapse with dendrites of mitral cells; (3) the external plexiform layer consists of dendrites of mitral and

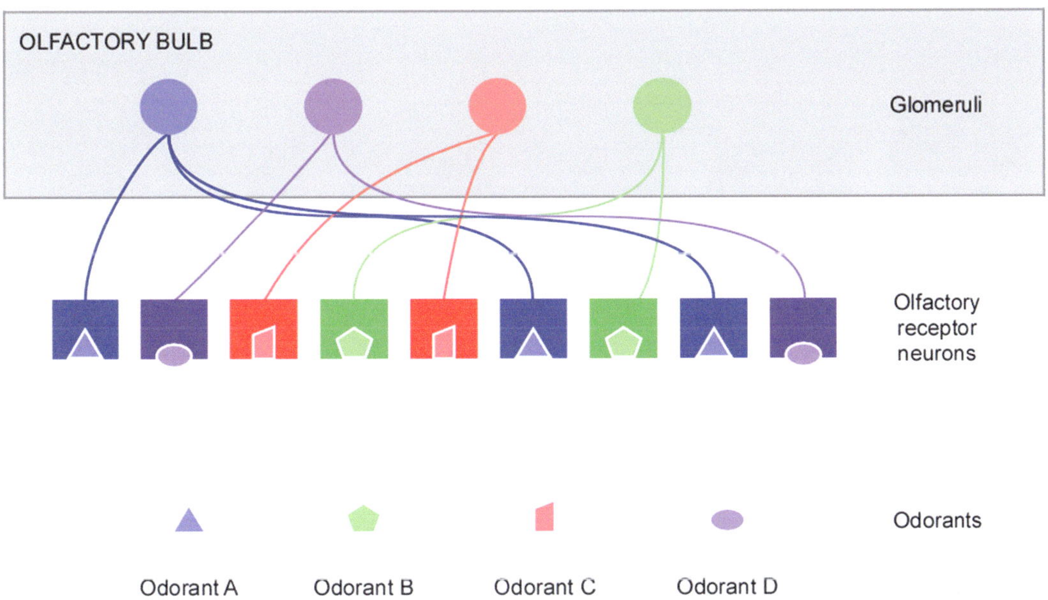

Fig. 30.2 Basic schematic representation of odor coding at the level of neuroepithelium and glomeruli. Odorant molecules bind with specific olfactory receptor neurons. Each olfactory receptor neuron possesses only one type of odorant receptor. Olfactory receptor neurons carrying the same type of receptor send their axon to the same glomerulus at the level of the olfactory bulb

tufted cells; (4) the mitral and tufted cells layer contains cell bodies of mitral and tufted cells (second-order olfactory neuron); (5) the internal plexiform layer; and (6) the granule cell layer contains rows of mitral and tufted axons and granule cells which are interneurons.

Axons of the mitral cells and tufted cells coalesce to form the olfactory tract, located at the base of the forebrain.

Centripetal information is secondary to neuronal activation, with glutamate as the principal neurotransmitter.

30.2.2.3 The Second Olfactory Structure: The Primary Olfactory Cortex

As compared to all other senses, olfaction is particular in that second-order olfactory neurons send information directly to primary olfactory cortex. In humans, the olfactory bulb is connected to the primary olfactory cortex by the fibers of the lateral olfactory tract (LOT). The LOT conveys olfactory information to a wide number of brain areas within the frontal lobe and the dorsomedial surface of the temporal lobe, often referred to primary olfactory cortex.

The primary olfactory cortex comprises the piriform cortex, which covers the uncus, the entorhinal cortex, the anterior olfactory nucleus, the periamygdaloid cortex, the olfactory tubercle, and nucleus. These projections are mainly ipsilateral, but there are also contralateral connections via the anterior commissure [7–9]. Some of the structures of the primary olfactory cortex then project to tertiary highest cognitive centers of the brain. The major projection of the piriform cortex is the thalamus, but it will also project to the insular cortex, the orbitofrontal cortex (neocortex), and the hypothalamus. The entorhinal cortex supplies afferent input to the hippocampus, while the olfactory tubercle connects to the thalamus. The amygdala is the major source of afferents to the hypothalamus (Fig. 30.3). Interestingly, there are many interactions between the secondary olfactory structures: between the anterior olfactory nucleus and the piriform cortex, the piriform cortex and the olfactory tubercle, the piriform cortex and the entorhinal cortex.

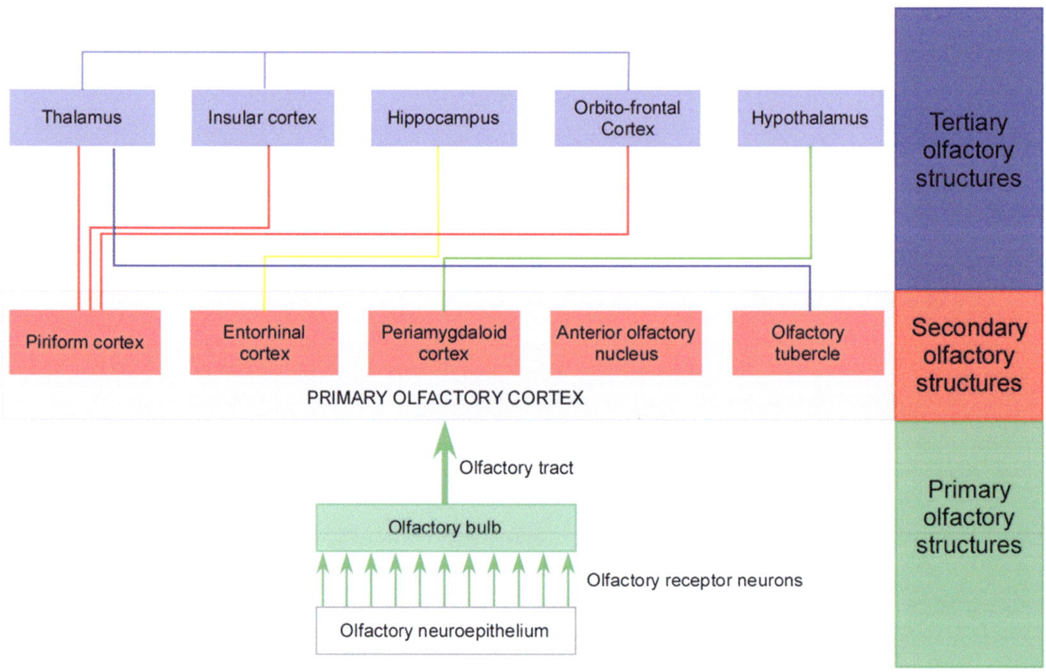

Fig. 30.3 Schematic diagram of major olfactory pathways

30.2.2.4 The Tertiary Olfactory Structures

The tertiary olfactory structures are the thalamus, the hypothalamus, the amygdala, the hippocampus, the orbitofrontal cortex, and the insular cortex.

The thalamus receives information from the piriform cortex and the olfactory tubercle. The hypothalamus, the orbitofrontal cortex, and the insular cortex also receive afferent input from the piriform cortex, while hippocampus is connected to entorhinal cortex. We should also note that there are also some interactions between these tertiary olfactory structures. In this way, the thalamus connects to the orbitofrontal cortex and the insular cortex. Therefore, the orbitofrontal cortex and the insular cortex receive direct input from the piriform cortex and indirect input via the thalamus.

30.2.2.5 Centrifugal Information

Most secondary and tertiary structures have numerous centrifugal fibers leading to the olfactory bulb, with GABA and acetylcholine as principal neurotransmitter. The supposed aim of this centrifugal information is to allow the brain to control the incoming flow of olfactory signals.

30.2.2.6 Properties of Olfactory Pathways

The olfactory pathways are distributed to different brain structures that are involved in the determination of our personal and social behavior. For example, the connections with:

1. Hippocampus and limbic system are thought to influence our memory system.
2. Amygdala system could act on emotional, motivational and craving circuits.
3. Hypothalamus, that mediates feeding regulation, could influence our feeding behavior.
4. Orbitofrontal cortex mediates our conscious perception of odors and could influence our preferences [10].

Hence, odor perception may affect our behavior and plays a major role in our interaction with the environment.

The olfactory system present unique properties as compared to other sensory systems. They are (1) the predominance of ipsilaterality of the olfactory projections, (2) the conduction of odor-evoked signals without an obligatory thalamic relay, and (3) the intimate overlap with limbic regions of the brain [11].

1. Odor processing remains principally ipsilateral [7–9] all the way from the nasal periphery to the primary olfactory cortex. This feature is different for other sensory modalities, such as the visual or auditory systems which, early in the processing pathways, supply sensory information in both hemispheres. This may help the cortex better discriminate and make bilateral odor comparisons and perhaps provide differential access to odor memories.
2. The absence of an obligatory thalamic relay is also in contrast with other sensory modalities in which an incoming signal undergoes thalamic modulation prior to being delivered to the sensory-specific cortex [11]. The absence of thalamic sensory integration in the olfactory pathways would seem to have an evolutionary explanation [11].
3. The connections between the olfactory system and the limbic system appear to be involved in the emotional and memory background to odorant stimuli, our social behavior, and the formation of novel stimulus-reinforced associations [11].

30.2.3 Orthonasal and Retronasal Olfaction

Paul Rozin noted that smell is unique in having a "dual nature"—meaning that it can sense signals originating outside (orthonasal) or inside (retronasal) the body [12].

Orthonasal olfaction refers to odorants originating outside and sniffed in through the nares to reach the olfactory neuroepithelium. This route is used to smell odors from the environment, such as perfumes, food aromas, smoke, predator smell, social odors, or pheromones. Orthonasal olfac-

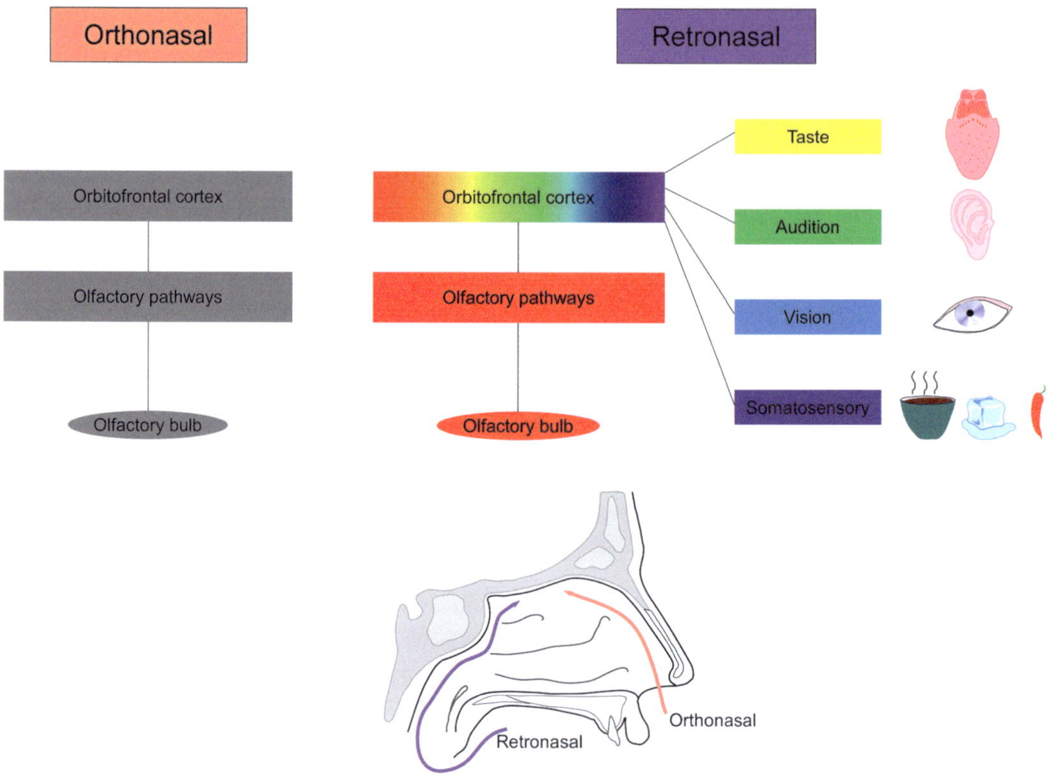

Fig. 30.4 Schematic representation of the central processing of orthonasal and retronasal olfaction. Orthonasal olfaction is processed by the olfactory pathways. On contrast, retronasal olfaction is not only processed by olfactory pathways but is also influenced by other sensory modalities, which are taste, sound, vision and proprioception. These multisensory information are integrated in the orbitofrontal cortex

tion is processed by olfactory pathways and is influenced by the visual pathway.

Retronasal olfaction refers to odorants originating from the back of the mouth and reaching the olfactory neuroepithelium via the nasopharynx. This retronasal stimulation occurs during food ingestion. It is activated only when breathing out through the nose, between mastication and swallowing [10, 13]. The retronasal olfaction, also termed as "flavor," account for an important part of food identification. This explains why a majority of patients suffering from smell disorder also complain of "taste" disorder, although their sense of taste is intact. On contrast to orthonasal olfaction, flavor perception is not only processed by olfactory pathways but is also influenced by almost all sensory modalities, which are taste, touch, sound, and proprioception (for a review see [10]). Indeed, the orbitofrontal cortex receives connections from other sensory neocortical areas (taste, hearing, touch, and vision) [14] (Fig. 30.4). Since it is receiving multisensory input and integrating these different sensory information, the orbitofrontal cortex is an important area to influence our food preferences and choices.

30.2.4 Olfactory and Trigeminal Interactions

The nasal fossa has double innervations from olfactory and trigeminal afferents. Although odorants are defined as volatile compounds having the ability to activate the olfactory system, the vast majority of odorants will actually activate both the olfactory and trigeminal system. Sensations resulting from the activation of the olfactory system are those of odors, while sensations induced by the stimulation of the trigeminal nerve are somatosensory (tactile, thermic, pain, humidity).

Olfactory and trigeminal systems closely interact with each other, and the stimulation of these two systems leads to important overlap in their activation pattern in the brain [15–19]. The interaction between both systems is complex and takes place both at a peripheral or central level (for a review see [20, 21]). This interaction is difficult to predict, but it has a powerful influence on odor perception both at different concentrations of a single stimulus and between mixtures of chemosensory stimuli. According to the literature, the pattern of interaction seems to depend on stimulus quality, intensity, and relative intensity of olfactory and trigeminal components of the mixture (for a review see [21]). Some reports have investigated the olfactory modulation of trigeminal-mediated sensations in patients with olfactory loss demonstrating that a close interaction and many compensatory mechanisms exist [22].

30.2.5 Variability in Normal Olfactory Function

Among other senses the olfactory function decreases over time and it has been described in numerous previous studies that there is a strong decrease in olfactory function above the age of 55 years [23, 24]. Several mechanisms have been proposed to explain this age-related olfactory dysfunction. At a peripheral level, changes in mucociliary movement, mucus composition, submucosal blood flow, and epithelia thickness might disturb the transport of the odorant to the receptor [25]. At the level of the neuroepithelium it is assumed that the regeneration of olfactory receptor neurons decreases over age [26, 27]. At a central level, brain damages due to chronic ischemia or systemic disturbance might also be proposed as a potential cause of age-related olfactory disorder.

Hummel et al. reported that there is a differential change of olfactory functions with aging. Indeed, olfactory thresholds decrease more strongly with age as compared to odor discrimination and odor identification [23, 28]. Since threshold measurements best reflect the function of the peripheral olfactory system than other olfactory tests [29–31], this finding might indicate that age-related change of olfactory function is at least in part due to damage of the olfactory epithelium [23]. Nevertheless, we should also keep in mind that age-related decrease of olfactory function might also be a consequence of side effects of drugs, onset of neurodegenerative diseases, etc.

A sex-related difference in olfactory function has also been widely reported [23, 32–35], with women outperforming men. Several causes have been proposed to explain this phenomenon, such as hormonal effects and congenital factors. However, the origin of this sex-related difference is still unclear.

Finally, some healthy people might present a specific anosmia. That is a physiological condition where a person of otherwise normal olfactory acuity is unable to detect a specific odorant. Specific anosmias have been described for series of odors [36]. It is admitted that specific anosmia has a genetic basis and the occurrence of specific anosmia indicates that specific receptors are necessary for perceiving specific odors [37–41]. One of the most frequent and well-known specific anosmia is androstenone anosmia. The prevalence of this specific anosmia is still a matter of debate. It is usually admitted that about 30% of the population is unable to detect the odor of the androstenone. But there is a high variability in the prevalence reported in the literature, ranging from 1.8 to 75% [42] (for a review see [42]). This might be at least in part explained by the various stimulation methods, criterion for non-detection, and concentrations that were used in the different studies [42].

30.3 Pathology

Although olfaction is often described as one of the less important sense, smell disorders have severe consequences, including impaired quality of life, daily life problems (cooking, detection of potentially dangerous odors) [43], altered food choices and consumption patterns than can negatively impact health (decreased body weight, overuse of salt inducing blood hypertension, overuse of sugar inducing diabetes mellitus, impaired immunity, etc.), and even depression [44].

The incidence of olfactory dysfunction among the population is still a matter of debate. Authors report an incidence of 1–3% of dysfunction among population [24, 45]. Nevertheless a study by Landis et al. reported higher values of olfactory dysfunction among population without sinonasal complaints, with a rate of 4.7% of anosmia and 16% of hyposmia. The frequency of parosmia and phantosmia was reported with a rate of 2.1% and 0.8%, respectively [46]. Recently, the SARS-CoV-2 pandemic has been found to be frequently associated with olfactory dysfunction, causing an anosmia pandemic across the world.

The evaluation of patients suffering from olfactory disorders requires a precise clinical work-up procedure in order to (1) determine the etiology of the olfactory dysfunction, (2) assess olfactory function and, hence, (3) provide an optional treatment, a prognosis, and appropriate counseling to patients. Assessment of olfactory function is reviewed in Chap. 33.

30.3.1 Classification of Olfactory Disorders

30.3.1.1 Quantitative Olfactory Disorders

Quantitative olfactory disorders are hyposmia, hyperosmia, and anosmia (Table 30.1). Hyposmia refers to a decreased ability to smell. This is a common condition. Indeed, Landis et al. reported that up to 16% of the general population is hyposmic [46].

Table 30.1 Classification of smell disorders

Quantitative smell disorders	Qualitative smell disorders
• Hyposmia • Anosmia – Functional anosmia – Specific anosmia • Hyperosmia	• Parosmia • Phantosmia • Olfactory agnosia

Hyperosmia is a rare condition and refers to enhanced ability to smell. It can happen after exposure to toxic vapors [47] or during migraine [48].

Anosmia refers to the lack of ability to smell. It is assumed that about 5% of the general population exhibit functional anosmia [46]. Functional anosmia refers to a significantly reduced ability to smell although some smell sensations can be present.

30.3.1.2 Qualitative Olfactory Disorders

Qualitative olfactory disorders are parosmia, phantosmia, and olfactory agnosia. Parosmia is sensation that a given odor is different than the typical odor for this substance. Parosmia is typically associated with reduced olfactory sensitivity and is particularly frequent in patients suffering from post-infectious olfactory loss (up to 50%) [49]. It is also associated with post-traumatic olfactory loss or sinonasal-related olfactory disorder. Studies found a prevalence of parosmia in 19% [50], 20% [51], and 28% [49] of patients presenting to « Smell and Taste » clinics, while the prevalence of parosmia in the general population is reported to be 2.1 [46] to 4% [52]. It is typically unpleasant. Euosmia is a rare form of parosmia with a pleasant parosmia to selected odorants [53]. The pathophysiology of parosmia is not clear. There are two hypotheses: the central and the peripheral hypotheses. In periphery, loss of olfactory receptor neurons changes the integrity of the olfactory image, resulting in an incomplete and meaningless picture of the odorant. Centrally, it has been proposed that the integration and interpretation of odors are altered [54].

Phantosmia is the perception of an odor when none is present. It may be reed to a wide range of pathologies (post-infectious olfactory loss, post-traumatic olfactory loss, rhinosinusitis, neurologic, etc.).

Finally, olfactory agnosia is defined as the inability to recognize odor sensation.

30.3.2 Etiology of Olfactory Disorders

There are several causes of olfactory dysfunction. The most frequent are chronic rhinosinusitis, post-infectious olfactory loss, and post-traumatic olfactory loss. These three etiologies account for up to two-thirds of the patients with olfactory disorder [55, 56]; therefore, we will largely extend on these 3 pathologies. However, several pathologies might also affect olfactory function, such as neurological disease, metabolic diseases, toxics, and tumoral disease of the sinonasal cavities or brain (Table 30.2). It is therefore essential to investigate about the etiology of olfactory dysfunction. An algorithm for the management of olfactory dysfunction is proposed in Fig. 30.5.

30.3.2.1 Chronic Rhinosinusitis

Chronic rhinosinusitis (CRS) with or without polyps is the most common cause of olfactory dysfunction, accounting for 14–30% of cases [57–60]. Inversely, olfactory impairment is acknowledged as a key symptom for the diagnosis of CRS [61]. Nevertheless up to one quarter of patients with CRS are unaware of their decreased olfactory abilities, probably because the olfactory dysfunction in CRS develops slowly, and in consequence, only a few patients note this disorder [62].

Olfactory dysfunction in CRS is explained by a combination of conductive olfactory loss (i.e., polyps, edema, nasal discharge, etc.) and neurosensory disturbance due to mucus and neuroepithelial alterations resulting from chronic inflammation [63].

In the context of CRS, olfactory dysfunction is mainly quantitative, appears gradually and fluctuates over time. Patients may also report qualitative dysfunction such as parosmia and

Table 30.2 The different etiologies of olfactory disorders

Rhinologic disease
- Chronic rhinosinusitis (with or without nasal polyps)
- Allergic rhinitis [164–166]
- Atrophic rhinitis [167]
- Post-surgical [167, 168]
- Olfactory cleft syndrome

Post-infectious olfactory loss
COVID-19-related olfactory loss
Post-traumatic olfactory loss
Congenital anosmia
Neurologic disorder
- Alzheimer's disease
- Idiopathic Parkinson's disease, …

Tumor
- Intranasal
  - Esthesioneuroblastoma
  - Adenocarcinoma
- Intracranial
  - Gliomas
  - Olfactory meningiomas

Toxic [169–171]
- Metals (cadmium, manganese, mercury, aluminum)
- Gases (formaldehyde, methyl bromide, styrene, chlorine)
- Solvents (toluene, butyl acetate, benzene)
- REF
- Hairdressing chemicals
- Intranasal zinc

Drug induced (for review, see [172, 173])
- Chemotherapy drugs
- Analgesic (antipyrine)
- Local anesthetics (cocaine HCL, procaine HCL, tetracaine HCL, lidocaine)
- General anesthetics
- Antimicrobial (amoxicillin, aminoglycosides, macrolides, doxycycline, pyrazinamide)
- Antirheumatics (mercury/gold salts, D-penicillamine)
- Antithyroids (propylthiouracil, thiouracil)
- Cardiovascular, hypertensives (angiotensin conversion enzyme inhibitors, nifedipine, amlodipine)
- Gastric medication (cimetidine)
- Intranasal saline solutions (with acetylcholine, menthol, zinc sulfate)
- Opiates
- Sympathomimetics

Metabolic/endocrine (for a review, see [174])
- Adrenocortical insufficiency
- Cushing's syndrome
- Hypothyroidism
- Pseudohypoparathyroidism
- Hepatic [175] or renal failure [176]

Psychiatric [177]
Idiopathic

Pathologies that are underlined are deeply commented in the text

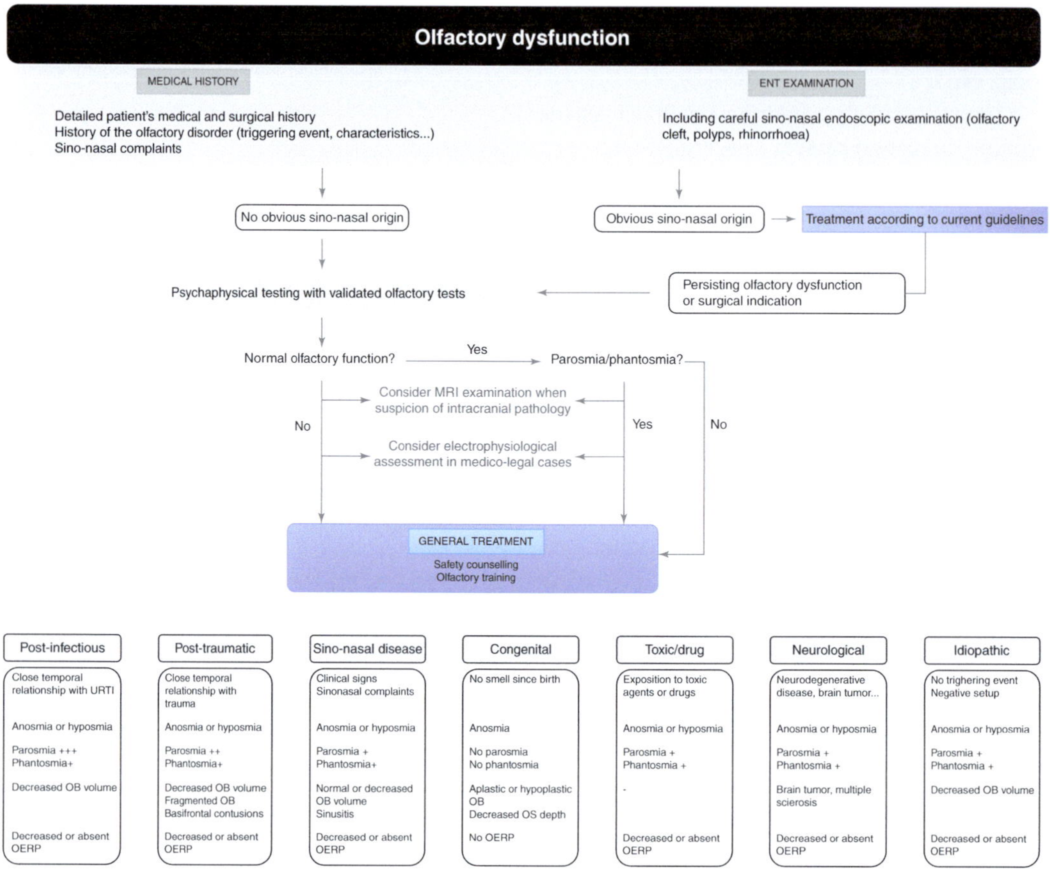

Fig. 30.5 Algorithm for the management of olfactory disorders and summary of the main features of the most frequent olfactory disorders. *OB* olfactory bulb, *OS* olfactory sulcus, *OERP* olfactory event-related potentials

phantosmia. However, these symptoms seem less frequent in sinonasal disease compared to other etiologies (i.e., post-infectious, post-traumatic). Reden et al. [49] reported incidence of parosmia and phantosmia in patients with CRS of 28% and 7%, respectively.

Studies have described that the severity of quantitative olfactory dysfunction is related to the importance of the sinonasal disease, based on endoscopy and CT score [64]. Patients with nasal polyps show a higher incidence of olfactory disturbances and a higher incidence of anosmia than patients with CRS without polyps. This more severe symptomatology may be explained not only by the conductive olfactory loss induced by polyps but also by degenerative changes associated with recurrent infections, scaring, chronic nasal medication, exotoxins, and enhanced secretion of cytokines from *Staphylococcus aureus* infection and neurotoxic cytokines released by a huge eosinophilic population [65–69].

Treatment of CRS should follow current guidelines [61]. Medical treatment notably relies on corticosteroids, either topical or systemic [61]. Corticosteroids are able to improve CRS-related olfactory dysfunction, with usually a higher efficacy of systemic steroids (for a review see [63]). In the last years biological treatments have been developed. Dupilumab is now FDA approved for the treatment of CRS and has favorable olfactory outcome [61, 70]. Surgery can also be proposed to a subset of patients suffering from CRS. Several studies have investigated the effect of surgery on olfactory function and generally showed an improvement of olfactory function [71]. However, there is a large heterogeneity regarding the methodology and only a few studies used validated psychophysical testing to assess olfaction [71].

30.3.2.2 Post-Infectious Olfactory Loss

Post-infectious olfactory loss is defined as a sudden loss of olfactory function following an upper respiratory tract infection (URTI) and was described for the first time more than 20 years ago [72]. The upper respiratory infection subsides over time and leaves the patient with an olfactory dysfunction that persists over a long period. There is a close connection in time between the URTI and the onset of the olfactory disorder [73]. The exact pathogenic agent is rarely determined but is assumed to be viral, and so this disease is known as "post-viral" or "post-infectious" olfactory loss. The exact incidence of olfactory dysfunction following URTI is not known as many patients with URTI do not report their symptoms, so the exact incidence of common cold in the population is unknown. However, post-infectious olfactory loss is diagnosed in approximately one quarter of the patients in groups presenting to specialized centers, such as smell and taste clinics [44, 74–76]. Recently, a new pathogen, the SARS-CoV-2, is emerged. Besides classical respiratory symptoms initially described, it revealed to create smell disorders in a high rate of patients. COVID-19-related olfactory dysfunction seems to differ from other post-infectious olfactory loss. Therefore, we will give a specific focus to it.

Patients with post-infectious olfactory loss are usually women and the disease typically occurs between the fourth and the sixth decades of life [73, 75, 77]. Onset of the URTI is often sudden and awareness of the olfactory dysfunction is present when major symptoms secondary to the infection subside. Many patients also have endoscopic or radiological evidence of rhinosinusitis. It is therefore mandatory to treat this condition and observe the impact of this treatment on the sensorineural disorder. Patients usually complain of moderate to severe olfactory loss but the degree of olfactory loss is usually less severe than in patients with head trauma [78]. Parosmia and phantosmia are also present and range to 10–50% [49, 72, 79]. Seasonal variation in the incidence of post-infectious olfactory loss has been demonstrated with the highest incidences being in March and May [80]. This is probably due to the seasonal variation of viral particles, such as parainfluenza virus type 3 [75, 81, 82].

Diagnosis should be based on (1) history of an olfactory disorder following an URTI and a close temporal relationship between the two, (2) patency of the olfactory cleft at the endoscopic examination, and (3) absence of any other causes, such as toxic exposure (medication taken to treat the URTI and possibly causing an olfactory disorder themselves), an inflammatory process in the nasal fossa (diagnosed with an endoscopic evaluation), or neurological problems, such as neurodegenerative diseases.

The exact mechanism leading to post-infectious olfactory loss is not yet fully understood. Viral particles may damage the olfactory receptor neuron and provoke immune response that also leads to damages in the olfactory neuroepithelium and damage the central olfactory pathways. Viruses are also capable of penetrating the brain via the fovea ethmoidalis. Many viruses may cause olfactory impairment, examples being the influenza virus, parainfluenza virus, respiratory syncytial virus, coxsackievirus, adenovirus, poliovirus, enterovirus, and herpes virus. The exact determination of the viral agent is not useful in the clinic and viral serology is not mandatory. Experimental intranasal infection with influenza virus A leads to increased apoptosis and increased fibrosis in the olfactory neuroepithelium [83, 84]. This mechanism is thought of as a protective one that limits the access of viral particles to the brain. Histopathological findings relating to the olfactory neuroepithelium of patients with post-infectious olfactory loss have revealed that severely affected patients have reduced numbers of ciliated olfactory receptor cells [85]. Moreover, dendrites of the olfactory receptor neurons usually fail to reach the epithelial surface and therefore have no contact with odorant particles. Attempting to correlate the importance of the olfactory neuroepithelial damage with the extent of the olfactory dysfunction as well with the chances of recovery generated conflicting results [85, 86]. Overall, post-infectious olfactory loss is probably secondary to a viral attack both at a peripheral level (olfactory neuroepithelium) and at a central level (olfactory bulb) and these two sites interact both in the pathological condition and in the recovery phase.

The spontaneous recovery of olfactory performance is found, due to the plasticity of our olfactory system, in about one-third of the post-infectious olfactory loss patients [87]. Olfactory function may decline (rare), not change, show some improvement, a major improvement, and improve into the absolute normal range or into the range adjusted for age [88]. Several prognosis factors have been described in the literature. Prognosis seems to be more favorable when the psychophysical tests reveal incomplete olfactory loss (i.e., hyposmia vs anosmia) [88–90]. Age and sex also seem to be important in the assessment of the prognosis since women and younger patients tend to recover more frequently than men and older patients [89]. Another prognostic factor is the duration of the disease [88, 90]. Electrophysiological measures are also predictive of recovery since it was demonstrated that the presence of olfactory event-related potentials at the time of diagnosis is linked to a better outcome in patients with post-infectious olfactory loss [91]. With regard to the meaning of qualitative olfactory disorders, reports have been mixed in relation to the likelihood of recovery [49, 90].

At present there is no medical therapy that has been proven effective. Many drugs have been tried in non-randomized and uncontrolled trials: topical or systemic corticosteroids [92], zinc sulfate [93, 94], quinoxaline derivatives [95], alpha lipoic acid [96], and pentoxifylline [97]. Although early promising results with some molecules have been demonstrated, these medications helped patients achieve partial or full recovery in unpredictable ways [87]. Recent systematic reviews about the usefulness of corticosteroids concluded that there is a paucity of high-quality studies demonstrating the efficacy of oral or topical steroids. Notably, topical steroids do not improve olfactory function, while weak evidence supports oral steroids, that could be considered as an option, in some selected patients after careful consideration of the potential side effects [98, 99]. On contrast, olfactory training is acknowledged as the gold-standard treatment of post-infectious olfactory loss [99]. This method consists in smelling four odorants, for 10 min, twice a day, during at least 12 weeks [100]. Finally, adequate counseling of the patient is of importance to help him cope with his deficit in his everyday life (for nutrition, hazard detection, hygiene, etc.).

30.3.2.3 COVID-19-Related Olfactory Dysfunction

Olfactory loss is a common symptom of COVID-19, a disease caused by the severe acute respiratory syndrome coronavirus 2 (SARS-CoV-2). The SARS-CoV-2 pandemic was therefore associated to a pandemic of olfactory dysfunction. This raised media and public attention toward the, until there usually neglected, sense of smell.

The reported frequencies of olfactory loss vary a lot across the different studies, probably because of the large heterogeneity in methodological approaches. Meta-analyses have determined that the pool frequency olfactory dysfunction (either based on questionnaire or smell tests) was 56 [101]–61% [102]. Importantly, olfactory dysfunction can be reported as the first symptom of COVID-19 in 20% of cases [103]. Olfactory dysfunction can range from anosmia to hyposmia [104]. Patients usually recover within the first month [104]; however, it is estimated that 5% patients will remain with persistent smell loss at 6 months after the onset [105]. Importantly, olfactory dysfunction does not seem to be associated with other nasal symptoms, such as nasal congestion or rhinorrhea [106]. Parosmia frequently occurs a few months after the acute event and has been reported by 27.5% of patients at 4-month follow-up. It usually resolves over time, but 12.5% of patients still report anosmia at 1 year [107].

Several mechanisms have been proposed to explain COVID-19-related olfactory dysfunction [108]. First, it could result from the release of proinflammatory mediators at the level of the olfactory cleft. Indeed, high levels of TNF-α and IL-6 have been reported in COVID-19 patients with olfactory dysfunction. These mediators not only could induce edema of the olfactory mucosa and conductive smell loss but could also impair sensory transduction and induce apoptosis [108–110]. Second, ACE2 and TMPRSS2, the entry proteins of SARS-COV-2, have been found in the olfactory supporting cells. Infection of olfactory supporting cells could lead to a disorganization and hence a dysfunction of the olfactory epithelium [111, 112]. Third, although olfactory sensory neurons do not express ACE2 and TMPRSS2, it is possible that these cells are infected through tight junctions with sustentacular cells [108]. Finally, it has been found that IgA produced against SARS-CoV-2 may block odorant receptors carried at the surface of olfactory sensory neurons [113].

Until now, there is no specific treatment for COVID-19-related olfactory dysfunction and olfactory training is recommended considering its effectiveness in other causes of olfactory dysfunction and safety. Moreover, considering the high recovery rate after COVID-19 olfactory dysfunction, it is also important to reassure the patient regarding its probable outcome.

30.3.2.4 Post-traumatic Olfactory Loss

Head trauma is, according to Nordin's literature review published in 2008, the third most common cause of olfactory disorder [114]. The incidence of olfactory disorder following head injury is difficult to estimate because (1) patients admitted to emergency often fail to receive an assessment of their sense of smell due to the potentially life-threatening nature of head trauma and the frequent occurrence of other injuries requiring immediate medical attention, (2) the time and the resources for such an examination are lacking in emergencies, (3) patients are not able to recognize their loss of olfactory function, more especially when there is an associated neurological deficit, (4) there is no medical follow-up if the olfactory disorder has no subjective impact for the patient, and (5) there is a lack of reports about spontaneously resolving olfactory disorder. For these reasons, the reported incidence is probably underestimated and variable depending on recruitment: in patients seen at a Head Injury Clinic, incidence is estimated between 2 and 12% [115, 116] and between 8 and 20% in Smell and Taste Centers [44, 49, 50, 60, 117, 118].

The patients at most risk of post-traumatic olfactory loss are young male adults. This is thought to be related to increased severity of trauma in this population. In both sexes generally, patients aged over 70 are most at risk [115, 119]. The most common type of trauma is a fall in 61% of patients, followed by car accidents in 20% and assault in 13% [115]. Several risk factors for the development of post-traumatic olfactory loss have been described: (1) the severity of the injury with more severe trauma being at higher risk of olfactory dysfunction [115, 120, 121], (2) the impact location and direction with a higher prevalence of post-traumatic olfactory loss when the front of the back of the head is stuck rather than the side [115, 122], and (3) the age of the patient, with a higher risk in the elderly patients [119]. In addition, it is important to note that in a traumatic context, olfaction might also be affected by the treatment of the injury, possibly complicating the etiological diagnosis (i.e., neurosurgical procedures, facial fracture reductions, usage of drugs, such as opioids and antimicrobial agents).

Typically, patients experience a sudden onset of olfactory symptoms [44], occurring some days after the trauma. Head trauma produces, on average a greater degree of olfactory decrement as compared with other etiologies of olfactory dysfunction [44, 119]. Post-traumatic olfactory disorder patients have anosmia ranging from 48 to 78% and hyposmia ranging from 5 to 27.4% [54, 115, 116, 122]. Qualitative disorders are commonly found in patients with head trauma. Parosmia is reported in 14–35% of cases [49, 60, 117] and phantosmia in 10–41% of the patients [49, 50, 117]. The prevalence of parosmia tends to decrease over time.

Three possibly co-existent lesions are likely to cause post-traumatic olfactory loss [116]. First, injuries to the sinonasal tract with obstruction of the passage to the olfactory cleft can lead to an obstructive post-traumatic olfactory loss. Second, shearing of the olfactory nerves at the cribriform plate might induce an olfactory loss. Notably, olfactory nerves can be injured after: (1) a translational shift of the encephala secondary to postero-anterior coup and contrecoup forces in the case of occipital impact; or (2) fractures is the naso-orbito-ethmoid region involving the cribriform plate. Third, contusions and brain hemorrhage involving olfactory bulbs and/or the olfactory cortex might also produce an olfactory disorder.

The prognosis seems to be reserved: some improvement may be expected in about one-third of the patients, although complete recovery is only achieved in 10–15% [44, 49, 88, 89]. Recovery is most likely to occur within the first 6 months to 1 year after the initial insult [88]. However, late recovery has been described, occurring until 9 years after the trauma [123–125].

No medical treatment has yet been proven to be effective. Olfactory training is currently recommended as gold-standard treatment [100, 126] and appropriate counseling is also mandatory.

30.3.2.5 Congenital Anosmia

Congenital anosmia is defined as the absence of olfactory sensation since birth or early childhood. This condition can be divided in (1) syndromic anosmias (e.g., Kallmann's syndrome [127] and Klinefelter's syndrome [128, 129], congenital insensitivity to pain [130, 131], and ciliary dysfunction [132, 133]) or (2) anosmias without evidence of other defects (isolated anosmia since birth or early childhood), which seems to be more frequent than syndromic anosmia [134–137]. Although 5% of the general population is anosmic [46], congenital anosmia remains a rare cause of olfactory disorder and isolated congenital anosmia account for about 1% of anosmias.

Diagnosis of congenital anosmia is based on anamnesis (patients have no recollection of ever being able to smell), psychophysical and electrophysiological assessment of olfactory function, and imagery. Magnetic resonance imaging is the imaging modality of choice for the assessment of olfactory apparatus in cases of suspected congenital anosmia. Indeed, we know from the literature that in isolated anosmia and in Kallmann's syndrome, the olfactory bulb and olfactory tract can be aplastic or hypoplastic [138–140]. The depth of the olfactory sulcus is also a useful indicator of congenital anosmia since we know that the depth of the olfactory sulcus at the level of the "plane of the posterior tangent through the eyeballs" reflects the presence of olfactory bulbs and tracts [140] and clearly indicates isolated anosmia if it is smaller than 8 mm [141].

When diagnosing a congenital anosmia, it should be discussed to realize a genetic and endocrinological evaluation. Also, since this disease cannot be treated, it is mandatory to give the patient or his/her parents the best information about this disease and to give counseling about everyday life (i.e., gas and smoke alarms, particular attention when cooking, hygiene, etc.).

30.3.2.6 Neurological Disorders

It is well known that some neurodegenerative diseases, such as idiopathic Parkinson's disease and Alzheimer's disease are associated with early olfactory dysfunction [142]. Because of the high prevalence of these neurodegenerative diseases in elderly subjects, the early diagnosis of these diseases constitutes a major public health issue for our aging societies. At present, the early differential diagnosis between idiopathic Parkinson's disease and Parkinsonism associated, for example, to multiple system atrophy, Lewy body disease, or corticobasal degeneration remains difficult, such as the differential diagnosis of mild cognitive impairment that may be the early expression of Alzheimer's disease, but also other forms of neurodegenerative diseases, and late-life depression. Many studies have suggested that the evaluation of the olfactory function can contribute significantly to the early diagnosis of these pathologies.

In idiopathic Parkinson's disease (IPD), olfactory dysfunction was first described by Ansari and Johnson in 1975 [143]. Olfactory disorders are often considered as an early and reliable sign of idiopathic Parkinson's disease (IPD), since they are present in more than 90% of all IPD patients [144, 145]. In accordance with the staging of Braak [146] it has been hypothesized that the early development of olfactory dysfunction is due to the early involvement of olfactory regions in the course of the disease. Several recent studies have shown that people suffering from idiopathic hyposmia or anosmia have an increased risk of developing IPD [147, 148], and that chemosensory event-related potentials are delayed or absent in IPD patients [149]. While IPD is associated with marked olfactory dysfunction leading to anosmia, other causes of Parkinsonism are not associated with strong olfactory dysfunction. For example, olfactory disorders would be moderate in multiple system atrophy, and absent in Parkinson's disease and in vascular Parkinsonism [150].

Olfactory disorders can also constitute one of the first signs of Alzheimer's disease (AD) [151]. The time course of histopathological changes in AD also indicates that olfactory dysfunction should precede cognitive dysfunction. Indeed, it has been shown that the formation of neurofibrillary tangles occurs first in the entorhinal cortex, while cognitive symptoms appear only once neuropathological changes have spread to the hippocampus and temporal neocortex [151]. Olfactory discrimination of AD patients is significantly lower than olfactory discrimination of patients suffering from mild cognitive impairment, which itself is lower than that of age-matched control subjects [152]. In addition, a recent clinical study has shown that the olfactory bulb and olfactory tract volume is decreased in AD patients, and that this atrophy is already present at an early stage of the disease [153]. Since early diagnosis of AD remains problematic, assessment of olfactory function should be useful for the early differential diagnosis of AD. This should be further investigated.

30.3.2.7 Olfactory Cleft Disease

Olfactory cleft disease is defined as an olfactory dysfunction due to a pathologic process limited to the olfactory cleft that can be visualized on clinical or radiological examination. Only few authors have reported this entity [154–156]. Biacabe et al. showed that olfactory disability was the major symptom of olfactory cleft disease and they identified three possible pathologic processes inducing olfactory cleft disease – malformative, inflammatory, and inflammatory associated with anatomical deformities of olfactory cleft boundaries – and hence suggest that computed tomography scanning is useful for the diagnosis of this disease. Finally, they showed that medical therapy was effective in lowering olfactory thresholds in 25% of the cases. Nevertheless, until now, indications of functional endoscopic surgery remain to be defined after failure of medical therapy.

30.3.2.8 Miscellaneous

Several other pathologies that might affect the olfactory function, such as tumors, toxics, drugs, and endocrine disorders, have been described. They are reported in Table 30.1.

30.3.2.9 Idiopathic Olfactory Loss

For a large number of patients, no obvious etiology to the olfactory disorder can be found. These people are thus considered as suffering from idiopathic olfactory loss. The reported prevalence of idiopathic olfactory loss in the literature range from about 20% [44, 49] to one-third [117] of patients suffering from olfactory disorder. These patients not only complain of quantitative olfactory disorder but may also complain of qualitative olfactory dysfunction [49, 157].

Previous work has indicated that idiopathic olfactory loss may be related to sinonasal disease. In fact in a study of 55 patients, almost 1/3 of patients with idiopathic olfactory loss responded to systemic treatment with corticosteroids [92], possibly indicating the presence of inflammation-related dysfunction. Hence, systemic steroid trial could be considered in patients suffering from idiopathic olfactory loss, after careful consideration of possible side effects. Olfactory training seems to be effective and should thus be proposed to patients suffering from idiopathic olfactory loss [126].

Finally, it is important to keep in mind that some patients with idiopathic olfactory loss may develop idiopathic Parkinson's disease or Alzheimer's disease [158]. A recent retrospective study from Haehener et al. on a cohort of 474 patients found that 9.9% of patients with idiopathic smell loss developed Parkinson's disease after an 8-year mean follow-up. This rate increased to 28.6% for patients with combined smell and taste loss [159].

30.3.2.10 Olfaction and Quality of Life

We must note that olfactory dysfunction severely impairs the quality of life of patients, including detection of hazardous events, eating habits and cooking, nutritional intake, and interpersonal relations [43]. Furthermore, it has been demonstrated that patients suffering from olfactory disorders have a higher prevalence of mild to severe depression as compared to the general population [44]. Importantly, the impact on the quality of life is more severe when patients have an associated qualitative olfactory dysfunction [160, 161]. Indeed, parosmia leads to higher rate of mild depression than quantitative olfactory disorders. Finally, since patients reporting an improvement of their olfactory abilities have a better quality of life than patients reporting no improvement [162]; it is essential to investigate about the etiology of olfactory dysfunction in instance to provide an optimal treatment to the patients.

Several studies have also highlighted the link between age-related olfactory loss and nutritional disorders in older people. As reported above, there is a physiological decrease of olfactory function with advancing age. This age-related olfactory disorder negatively impacts the food intake of older people. Indeed, not only older people might have a reduced interest in food and hence reduced food intake, but also they tend to have less varied diet and consequently might develop deficiencies. This is problematic since inadequate diet and malnutrition are associated with a decline in functional status, impaired muscle function, decreased bone mass, immune dysfunction, anemia, reduced cognitive function, poor wound healing, and delay in recovering [163]. This may constitute a major public health issue in our aging population.

30.3.2.11 Counseling of the Patient

Consequences for daily life and coping strategy should be integrated in clinical management of patients, focusing on instructional information about fire alarms, domestic gas, hygiene, etc.

Nutritional recommendations should also be proposed to the patients in order to avoid altered food choices and consumption patterns than can negatively impact health (decreased body weight, overuse of salt inducing blood hypertension, overuse of sugar inducing diabetes mellitus, impaired immunity, etc.).

Finally, as reported above, olfactory training seems to be effective and should thus be recommended to patients [126].

30.4 Conclusion

Describing the olfactory pathways, we have shown that olfactory system has connections with brain areas associated with memory processes, feeding circuits, emotional, motivational, and craving circuits. Hence, it is easy to understand that, although often neglected, the olfactory system plays a preponderant role in our everyday's life and strongly influence consciously or nonconsciously on emotions, social behavior, nutrition, memory, etc. Therefore, we can easily understand that olfactory disorders severely impact our quality of life and that patients suffering from olfactory disorders need a particular support. Indeed, physicians taking care of patients suffering from olfactory disorders must pay a particular attention to the quality of life of patients and to the potential negative impact of olfactory dysfunction on the patient's health (nutrition, detection of danger, depression, etc.).

Olfactory dysfunction due to sinonasal disease can be treated either medically or surgically, according to available guidelines. Unfortunately, medical treatments are still missing today for non-sinonasal causes and olfactory training remains the mainstay of the treatment. Besides this, considering the impact of olfactory dysfunction on everyday life, it is mandatory to provide patients complete information about the nature of their olfactory disorder and their prognosis, as well as advices about occupations, safety at home, and how to make food more palatable and safe to eat.

Nowadays, about 20% of the population is hyposmic. But this number could increase in the future years due to our aging population and COVID-19 pandemic. This might constitute a major public health issue in the future years considering the close relationship between olfactory dysfunction and nutritional disorders in elderly people. Further researches are thus mandatory in order to propose new treatments to recover or to compensate for the olfactory loss.

Take-Home Messages

- Olfaction plays an important role in our daily life and olfactory impairment negatively affects quality of life and well-being.
- Besides physiological age-related decline of olfaction, olfaction can be affected by a wide range of pathologies. Identifying these possible causes is mandatory for the appropriate management and counseling of the patient.
- Sinonasal-related olfactory dysfunction should be treated according to current guidelines available.
- Unfortunately, no medical treatment as yet been proven to significantly impact the outcomes of non-sinonasal olfactory dysfunction.
- Olfactory training remains the gold-standard treatment for olfactory disorders.
- Counseling, psychological, and nutritional support are also mainstay of the treatment.

References

1. Larsen W. Embryologie humaine. 2nd ed. Boom: De Boeck; 2003. p. 568.
2. Drews U. Color atlas of embryology. Stuttgart: Thieme; 1995. p. 383.
3. Leopold DA, Hummel T, Schwob JE, Hong SC, Knecht M, Kobal G. Anterior distribution of human olfactory epithelium. Laryngoscope. 2000;110(3 Pt 1):417–21.
4. Buck L, Axel R. A novel multigene family may encode odorant receptors: a molecular basis for odor recognition. Cell. 1991;65(1):175–87.
5. Zhang X, Firestein S. The olfactory receptor gene superfamily of the mouse. Nat Neurosci. 2002;5(2):124–33.
6. Crasto C, Marenco L, Miller P, Shepherd G. Olfactory Receptor Database: a metadata-driven automated population from sources of gene and protein sequences. Nucleic Acids Res. 2002;30(1):354–60.
7. Royet JP, Plailly J. Lateralization of olfactory processes. Chem Senses. 2004;29(8):731–45.
8. Lascano AM, Hummel T, Lacroix JS, Landis BN, Michel CM. Spatio-temporal dynamics of olfactory processing in the human brain: an event-related source imaging study. Neuroscience. 2010;167(3):700–8.
9. Cleland TA, Linster C. Central olfactory structures. In: Doty RL, editor. Handbook of olfaction and gustation. New York: Marcel Dekker Inc.; 1995. p. 165–80.
10. Shepherd GM. Smell images and the flavour system in the human brain. Nature. 2006;444(7117):316–21.
11. Gottfried JA. Smell: central nervous processing. Adv Otorhinolaryngol. 2006;63:44–69.
12. Rozin P. "Taste-smell confusions" and the duality of the olfactory sense. Percept Psychophys. 1982;31(4):397–401.
13. Taylor AJ, Linforth RST, Harvey BA, Blake A. Atmospheric pressure chemical ionisation for monitoring of flavour release in vivo. Food Chem. 2000;71:327–38.
14. Ongur D, Ferry AT, Price JL. Architectonic subdivision of the human orbital and medial prefrontal cortex. J Comp Neurol. 2003;460(3):425–49.
15. Hummel T, Doty RL, Yousem DM. Functional MRI of intranasal chemosensory trigeminal activation. Chem Senses. 2005;30(Suppl 1):i205–6.
16. Boyle JA, Heinke M, Gerber J, Frasnelli J, Hummel T. Cerebral activation to intranasal chemosensory trigeminal stimulation. Chem Senses. 2007;32(4):343–53.
17. Iannilli E, Gerber J, Frasnelli J, Hummel T. Intranasal trigeminal function in subjects with and without an intact sense of smell. Brain Res. 2007;1139:235–44.
18. Bensafi M, Iannilli E, Gerber J, Hummel T. Neural coding of stimulus concentration in the human olfactory and intranasal trigeminal systems. Neuroscience. 2008;154(2):832–8.
19. Zald DH, Pardo JV. Functional neuroimaging of the olfactory system in humans. Int J Psychophysiol. 2000;36(2):165–81.
20. Brand G. Olfactory/trigeminal interactions in nasal chemoreception. Neurosci Biobehav Rev. 2006;30(7):908–17.
21. Hummel T, Livermore A. Intranasal chemosensory function of the trigeminal nerve and aspects of its relation to olfaction. Int Arch Occup Environ Health. 2002;75(5):305–13.
22. Frasnelli J, Schuster B, Hummel T. Interactions between olfaction and the trigeminal system: what can be learned from olfactory loss. Cereb Cortex.

2007;17(10):2268–75. https://doi.org/10.1093/cercor/bhl135. Epub 2006 Dec 5.PMID: 17150985

23. Hummel T, Kobal G, Gudziol H, Mackay-Sim A. Normative data for the "Sniffin' Sticks" including tests of odor identification, odor discrimination, and olfactory thresholds: an upgrade based on a group of more than 3000 subjects. Eur Arch Otorhinolaryngol. 2007;264(3):237–43.
24. Murphy C, Schubert CR, Cruickshanks KJ, Klein BE, Klein R, Nondahl DM. Prevalence of olfactory impairment in older adults. JAMA. 2002;288(18):2307–12.
25. Rawson NE. Olfactory loss in aging. Sci Aging Knowledge Environ. 2006;2006(5):pe6.
26. Naessen R. An enquiry on the morphological characteristics and possible changes with age in the olfactory region of man. Acta Otolaryngol. 1971;71(1):49–62.
27. Conley DB, Robinson AM, Shinners MJ, Kern RC. Age-related olfactory dysfunction: cellular and molecular characterization in the rat. Am J Rhinol. 2003;17(3):169–75.
28. Hummel T, Heilmann S, Murphy C. Age-related changes of chemosensory functions. In: Rouby C, Schaal B, Dubois D, Gervais R, Holley A, editors. Olfaction, taste, and cognition. New York: Cambridge University Press; 2002. p. 441–56.
29. Hornung DE, Kurtz DB, Bradshaw CB, Seipel DM, Kent PF, Blair DC, et al. The olfactory loss that accompanies an HIV infection. Physiol Behav. 1998;64(4):549–56.
30. Jones-Gotman M, Zatorre RJ. Olfactory identification deficits in patients with focal cerebral excision. Neuropsychologia. 1988;26(3):387–400.
31. Moberg PJ, Agrin R, Gur RE, Gur RC, Turetsky BI, Doty RL. Olfactory dysfunction in schizophrenia: a qualitative and quantitative review. Neuropsychopharmacology. 1999;21(3):325–40.
32. Brand G, Millot JL. Sex differences in human olfaction: between evidence and enigma. Q J Exp Psychol B. 2001;54(3):259–70.
33. Doty RL, Applebaum S, Zusho H, Settle RG. Sex differences in odor identification ability: a cross-cultural analysis. Neuropsychologia. 1985;23(5):667–72.
34. Lundstrom JN, Frasnelli J, Larsson M, Hummel T. Sex differentiated responses to intranasal trigeminal stimuli. Int J Psychophysiol. 2005;57(3):181–6.
35. Lundstrom JN, Hummel T. Sex-specific hemispheric differences in cortical activation to a bimodal odor. Behav Brain Res. 2006;166(2):197–203.
36. Amoore J. Specific anosmias. In: Getchell TV, editor. Smell and taste in health and diseases. New York: Raven Press; 1991. p. 655–64.
37. Gross-Isseroff R, Ophir D, Bartana A, Voet H, Lancet D. Evidence for genetic determination in human twins of olfactory thresholds for a standard odorant. Neurosci Lett. 1992;141(1):115–8.
38. Lancet D, Ben-Arie N, Cohen S, Gat U, Gross-Isseroff R, Horn-Saban S, et al. Olfactory receptors: transduction, diversity, human psychophysics and genome analysis. Ciba Found Symp. 1993;179:131–41.
39. Wysocki CJ, Beauchamp GK. Ability to smell androstenone is genetically determined. Proc Natl Acad Sci U S A. 1984;81(15):4899–902.
40. Amoore JE, Venstrom D, Davis AR. Measurement of specific anosmia. Percept Mot Skills. 1968;26(1):143–64.
41. Menashe I, Man O, Lancet D, Gilad Y. Different noses for different people. Nat Genet. 2003;34(2):143–4.
42. Bremner EA, Mainland JD, Khan RM, Sobel N. The prevalence of androstenone anosmia. Chem Senses. 2003;28(5):423–32.
43. Temmel AF, Quint C, Schickinger-Fischer B, Klimek L, Stoller E, Hummel T. Characteristics of olfactory disorders in relation to major causes of olfactory loss. Arch Otolaryngol Head Neck Surg. 2002;128(6):635–41.
44. Deems DA, Doty RL, Settle RG, Moore-Gillon V, Shaman P, Mester AF, et al. Smell and taste disorders, a study of 750 patients from the University of Pennsylvania Smell and Taste Center. Arch Otolaryngol Head Neck Surg. 1991;117(5):519–28.
45. Hoffman HJ, Ishii EK, Macturk RH. Age-related changes in the prevalence of smell/taste problems among the United States adult population. Results of the 1994 disability supplement to the National Health Interview Survey (NHIS). Ann N Y Acad Sci. 1998;855:716–22.
46. Landis BN, Konnerth CG, Hummel T. A study on the frequency of olfactory dysfunction. Laryngoscope. 2004;114(10):1764–9.
47. Henkin RI. Hyperosmia and depression following exposure to toxic vapors. JAMA. 1990;264(21):2803.
48. Blau JN, Solomon F. Smell and other sensory disturbances in migraine. J Neurol. 1985;232(5):275–6.
49. Reden J, Maroldt H, Fritz A, Zahnert T, Hummel T. A study on the prognostic significance of qualitative olfactory dysfunction. Eur Arch Otorhinolaryngol. 2007;264(2):139–44.
50. Nordin S, Murphy C, Davidson TM, Quinonez C, Jalowayski AA, Ellison DW. Prevalence and assessment of qualitative olfactory dysfunction in different age groups. Laryngoscope. 1996;106(6):739–44.
51. Landis BN, Frasnelli J, Croy I, Hummel T. Evaluating the clinical usefulness of structured questions in parosmia assessment. Laryngoscope. 2010;120(8):1707–13.
52. Nordin S, Bramerson A, Millqvist E, Bende M. Prevalence of parosmia: the Skovde population-based studies. Rhinology. 2007;45(1):50–3.
53. Landis BN, Frasnelli J, Hummel T. Euosmia: a rare form of parosmia. Acta Otolaryngol. 2006;126(1):101–3.
54. Leopold D. Distortion of olfactory perception: diagnosis and treatment. Chem Senses. 2002;27(7):611–5.
55. Murphy C, Doty RL, Duncan HJ. Clinical disorders of olfaction. In: Doty RL, editor. Handbook of olfaction and gustation. New York: Marcel Dekker; 2003.

56. Rombaux P, Mouraux A, Collet S, Eloy P, Bertrand B. Usefulness and feasibility of psychophysical and electrophysiological olfactory testing in the rhinology clinic. Rhinology. 2009;47(1):28–35.
57. Holbrook EH, Leopold DA. An updated review of clinical olfaction. Curr Opin Otolaryngol Head Neck Surg. 2006;14(1):23–8.
58. Mott AE, Leopold DA. Disorders in taste and smell. Med Clin North Am. 1991;75(6):1321–53.
59. Raviv JR, Kern RC. Chronic sinusitis and olfactory dysfunction. Otolaryngol Clin North Am. 2004;37(6):1143–57.
60. Seiden AM, Duncan HJ. The diagnosis of a conductive olfactory loss. Laryngoscope. 2001;111(1):9–14.
61. Fokkens WJ, Lund VJ, Hopkins C, Hellings PW, Kern R, Reitsma S, et al. European position paper on rhinosinusitis and nasal polyps 2020. Rhinology. 2020;58(Suppl S29):1–464.
62. Nordin S, Monsch AU, Murphy C. Unawareness of smell loss in normal aging and Alzheimer's disease: discrepancy between self-reported and diagnosed smell sensitivity. J Gerontol B Psychol Sci Soc Sci. 1995;50(4):P187–92.
63. Rombaux P, Huart C, Levie P, Cingi C, Hummel T. Olfaction in chronic rhinosinusitis. Curr Allergy Asthma Rep. 2016;16(5):41.
64. Litvack JR, Mace JC, Smith TL. Olfactory function and disease severity in chronic rhinosinusitis. Am J Rhinol Allergy. 2009;23(2):139–44.
65. Bernstein JM, Allen C, Rich G, Dryja D, Bina P, Reiser R, et al. Further observations on the role of *Staphylococcus aureus* exotoxins and IgE in the pathogenesis of nasal polyposis. Laryngoscope. 2011;121(3):647–55.
66. Joshi H, Getchell ML, Zielinski B, Getchell TV. Spectrophotometric determination of cation concentrations in olfactory mucus. Neurosci Lett. 1987;82(3):321–6.
67. Litvack JR, Fong K, Mace J, James KE, Smith TL. Predictors of olfactory dysfunction in patients with chronic rhinosinusitis. Laryngoscope. 2008;118(12):2225–30.
68. Vento SI, Simola M, Ertama LO, Malmberg CH. Sense of smell in long-standing nasal polyposis. Am J Rhinol. 2001;15(3):159–63.
69. Wang JH, Kwon HJ, Jang YJ. *Staphylococcus aureus* increases cytokine and matrix metalloproteinase expression in nasal mucosae of patients with chronic rhinosinusitis and nasal polyps. Am J Rhinol Allergy. 2010;24(6):422–7.
70. Bachert C, Zinreich SJ, Hellings PW, Mullol J, Hamilos DL, Gevaert P, et al. Dupilumab reduces opacification across all sinuses and related symptoms in patients with CRSwNP. Rhinology. 2020;58(1):10–7.
71. Hummel T, Whitcroft KL, Andrews P, Altundag A, Cinghi C, Costanzo RM, et al. Position paper on olfactory dysfunction. Rhinol Suppl. 2017;54(26):1–30.
72. Henkin RI, Larson AL, Powell RD. Hypogeusia, dysgeusia, hyposmia, and dysosmia following influenza-like infection. Ann Otol Rhinol Laryngol. 1975;84(5 Pt 1):672–82.
73. Seiden AM. Postviral olfactory loss. Otolaryngol Clin North Am. 2004;37(6):1159–66.
74. Bonfils P, Corre FL, Biacabe B. Semiology and etiology of anosmia: apropos of 306 patients. Ann Otolaryngol Chir Cervicofac. 1999;116(4):198–206.
75. Sugiura M, Aiba T, Mori J, Nakai Y. An epidemiological study of postviral olfactory disorder. Acta Otolaryngol Suppl. 1998;538:191–6.
76. Cain WS, Gent JF, Goodspeed RB, Leonard G. Evaluation of olfactory dysfunction in the Connecticut Chemosensory Clinical Research Center. Laryngoscope. 1988;98(1):83–8.
77. Rombaux P, Martinage S, Huart C, Collet S. Post-infectious olfactory loss: a cohort study and update. B-ENT. 2009;5(Suppl 13):89–95.
78. Duncan HJ, Seiden AM. Long-term follow-up of olfactory loss secondary to head trauma and upper respiratory tract infection. Arch Otolaryngol Head Neck Surg. 1995;121(10):1183–7.
79. Leopold DA, Hornung DE, Youngentob SL. Olfactory loss after upper respiratory tract infection. In: Getchell M, Doty RL, Bartoshuk LM, Snow J, editors. Smell and taste in health and disease. New York: Raven Press; 1991. p. 731–4.
80. Konstantinidis I, Haehner A, Frasnelli J, Reden J, Quante G, Damm M, et al. Post-infectious olfactory dysfunction exhibits a seasonal pattern. Rhinology. 2006;44(2):135–9.
81. Suzuki M, Saito K, Min WP, Vladau C, Toida K, Itoh H, et al. Identification of viruses in patients with postviral olfactory dysfunction. Laryngoscope. 2007;117(2):272–7.
82. Wang JH, Kwon HJ, Jang YJ. Detection of parainfluenza virus 3 in turbinate epithelial cells of postviral olfactory dysfunction patients. Laryngoscope. 2007;117(8):1445–9.
83. Mori I, Goshima F, Imai Y, Kohsaka S, Sugiyama T, Yoshida T, et al. Olfactory receptor neurons prevent dissemination of neurovirulent influenza A virus into the brain by undergoing virus-induced apoptosis. J Gen Virol. 2002;83(Pt 9):2109–16.
84. Mori I, Nishiyama Y, Yokochi T, Kimura Y. Virus-induced neuronal apoptosis as pathological and protective responses of the host. Rev Med Virol. 2004;14(4):209–16.
85. Doty RL. The olfactory vector hypothesis of neurodegenerative disease: is it viable? Ann Neurol. 2008;63(1):7–15.
86. Yamagishi M, Fujiwara M, Nakamura H. Olfactory mucosal findings and clinical course in patients with olfactory disorders following upper respiratory viral infection. Rhinology. 1994;32(3):113–8.
87. Hummel T. Perspectives in olfactory loss following viral infections of the upper respiratory tract. Arch Otolaryngol Head Neck Surg. 2000;126(6):802–3.
88. London B, Nabet B, Fisher AR, White B, Sammel MD, Doty RL. Predictors of prognosis in

patients with olfactory disturbance. Ann Neurol. 2008;63(2):159–66.
89. Reden J, Mueller A, Mueller C, Konstantinidis I, Frasnelli J, Landis BN, et al. Recovery of olfactory function following closed head injury or infections of the upper respiratory tract. Arch Otolaryngol Head Neck Surg. 2006;132(3):265–9.
90. Hummel T, Lotsch J. Prognostic factors of olfactory dysfunction. Arch Otolaryngol Head Neck Surg. 2010;136(4):347–51.
91. Rombaux P, Huart C, Collet S, Eloy P, Negoias S, Hummel T. Presence of olfactory event-related potentials predicts recovery in patients with olfactory loss following upper respiratory tract infection. Laryngoscope. 2010;120(10):2115–8.
92. Heilmann S, Huettenbrink KB, Hummel T. Local and systemic administration of corticosteroids in the treatment of olfactory loss. Am J Rhinol. 2004;18(1):29–33.
93. Aiba T, Sugiura M, Mori J, Matsumoto K, Tomiyama K, Okuda F, et al. Effect of zinc sulfate on sensorineural olfactory disorder. Acta Otolaryngol Suppl. 1998;538:202–4.
94. Henkin RI, Schecter PJ, Friedewald WT, Demets DL, Raff M. A double blind study of the effects of zinc sulfate on taste and smell dysfunction. Am J Med Sci. 1976;272(3):285–99.
95. Quint C, Temmel AF, Hummel T, Ehrenberger K. The quinoxaline derivative caroverine in the treatment of sensorineural smell disorders: a proof-of-concept study. Acta Otolaryngol. 2002;122(8):877–81.
96. Damm M, Vent J, Schmidt M, Theissen P, Eckel HE, Lotsch J, et al. Intranasal volume and olfactory function. Chem Senses. 2002;27(9):831–9.
97. Gudziol V, Hummel T. Effects of pentoxifylline on olfactory sensitivity: a postmarketing surveillance study. Arch Otolaryngol Head Neck Surg. 2009;135(3):291–5.
98. Yan CH, Overdevest JB, Patel ZM. Therapeutic use of steroids in non-chronic rhinosinusitis olfactory dysfunction: a systematic evidence-based review with recommendations. Int Forum Allergy Rhinol. 2019;9(2):165–76.
99. Hura N, Xie DX, Choby GW, Schlosser RJ, Orlov CP, Seal SM, et al. Treatment of post-viral olfactory dysfunction: an evidence-based review with recommendations. Int Forum Allergy Rhinol. 2020;10(9):1065–86.
100. Sorokowska A, Drechsler E, Karwowski M, Hummel T. Effects of olfactory training: a meta-analysis. Rhinology. 2017;55(1):17–26.
101. Pang KW, Chee J, Subramaniam S, Ng CL. Frequency and clinical utility of olfactory dysfunction in COVID-19: a systematic review and meta-analysis. Curr Allergy Asthma Rep. 2020;20(12):76.
102. Hajikhani B, Calcagno T, Nasiri MJ, Jamshidi P, Dadashi M, Goudarzi M, et al. Olfactory and gustatory dysfunction in COVID-19 patients: a meta-analysis study. Physiol Rep. 2020;8(18):e14578.
103. Borsetto D, Hopkins C, Philips V, Obholzer R, Tirelli G, Polesel J, et al. Self-reported alteration of sense of smell or taste in patients with COVID-19: a systematic review and meta-analysis on 3563 patients. Rhinology. 2020;58(5):430–6.
104. Niklassen AS, Draf J, Huart C, Hintschich C, Bocksberger S, Trecca EMC, et al. COVID-19: recovery from chemosensory dysfunction. A multicentre study on smell and taste. Laryngoscope. 2021;131:1095–100.
105. Lechien JR, Chiesa-Estomba CM, Beckers E, Mustin V, Ducarme M, Journe F, et al. Prevalence and 6-month recovery of olfactory dysfunction: a multicentre study of 1363 COVID-19 patients. J Intern Med. 2021;290:451–61.
106. Haehner A, Draf J, Drager S, de With K, Hummel T. Predictive value of sudden olfactory loss in the diagnosis of COVID-19. ORL J Otorhinolaryngol Relat Spec. 2020;82(4):175–80.
107. Renaud M, Thibault C, Le Normand F, McDonald EG, Gallix B, Debry C, et al. Clinical outcomes for patients with anosmia 1 year after COVID-19 diagnosis. JAMA Netw Open. 2021;4(6):e2115352.
108. Zugaj M, van Ditzhuijzen NS, Golebski K, Fokkens WJ. The effect of coronaviruses on olfaction: systematic review. Rhinology. 2021;59(3):226–35.
109. Cazzolla AP, Lovero R, Lo Muzio L, Testa NF, Schirinzi A, Palmieri G, et al. Taste and smell disorders in COVID-19 patients: role of interleukin-6. ACS Chem Nerosci. 2020;11(17):2774–81.
110. Torabi A, Mohammadbagheri E, Akbari Dilmaghani N, Bayat AH, Fathi M, Vakili K, et al. Proinflammatory cytokines in the olfactory mucosa result in COVID-19 induced anosmia. ACS Chem Nerosci. 2020;11(13):1909–13.
111. Brann D, Tsukahara T, Weinreb C, Logan DW, Datta SR. Non-neural expression of SARS-Cov-2 entry genes in the olfactory epithelium suggests mechanisms underlying anosmia in COVID-19 patients. bioRxiv. 2020;10:25.
112. Fodoulian L, Tuberosa J, Rossier D, Landis BN, Carleton A, Rodriguez I. SARS-CoV-2 receptor and entry genes are expressed by sustentacular cells in the human olfactory neuroepithelium. BioRxiv. 2020;2020:013268.
113. Root-Bernstein R. Anosmia-hyposmia and dysgeusia in COVID-19 may be due to SARS-CoV-2 protein mimicry of olfactory receptors. Rhinol Online. 2020;3(3):148–51.
114. Nordin S, Bramerson A. Complaints of olfactory disorders: epidemiology, assessment and clinical implications. Curr Opin Allergy Clin Immunol. 2008;8(1):10–5.
115. Swann IJ, Bauza-Rodriguez B, Currans R, Riley J, Shukla V. The significance of post-traumatic amnesia as a risk factor in the development of olfactory dysfunction following head injury. Emerg Med J. 2006;23(8):618–21.

116. Reiter ER, DiNardo LJ, Costanzo RM. Effects of head injury on olfaction and taste. Otolaryngol Clin North Am. 2004;37(6):1167–84.
117. Bramerson A, Nordin S, Bende M. Clinical experience with patients with olfactory complaints, and their quality of life. Acta Otolaryngol. 2007;127(2):167–74.
118. Mori J, Aiba T, Sugiura M, Matsumoto K, Tomiyama K, Okuda F, et al. Clinical study of olfactory disturbance. Acta Otolaryngol Suppl. 1998;538:197–201.
119. Harris R, Davidson TM, Murphy C, Gilbert PE, Chen M. Clinical evaluation and symptoms of chemosensory impairment: one thousand consecutive cases from the nasal dysfunction clinic in San Diego. Am J Rhinol. 2006;20(1):101–8.
120. Eftekhari M, Assadi M, Kazemi M, Saghari M, Mojtahedi A, Fard-Esfahani A, et al. Brain perfusion single photon emission computed tomography findings in patients with posttraumatic anosmia and comparison with radiological imaging. Am J Rhinol. 2006;20(6):577–81.
121. Renzi G, Carboni A, Gasparini G, Perugini M, Becelli R. Taste and olfactory disturbances after upper and middle third facial fractures: a preliminary study. Ann Plast Surg. 2002;48(4):355–8.
122. Cross DJ, Flexman JA, An/zai Y, Morrow TJ, Maravilla KR, Minoshima S. In vivo imaging of functional disruption, recovery and alteration in rat olfactory circuitry after lesion. Neuroimage. 2006;32(3):1265–72.
123. Mueller CA, Hummel T. Recovery of olfactory function after nine years of post-traumatic anosmia: a case report. J Med Case Reports. 2009;3:9283.
124. Sumner D. Post-traumatic anosmia. Brain. 1964;87:107–20.
125. Zusho H. Posttraumatic anosmia. Arch Otolaryngol. 1982;108(2):90–2.
126. Hummel T, Rissom K, Reden J, Hahner A, Weidenbecher M, Huttenbrink KB. Effects of olfactory training in patients with olfactory loss. Laryngoscope. 2009;119(3):496–9.
127. Kallmann FJ, Schoenfeld WA, Barrera SE. The genetic aspects of primary eunochoidism. Am J Ment Defic. 1944;48:203–36.
128. Pawlowitzki IH, Diekstall P, Schadel A, Miny P. Estimating frequency of Kallmann syndrome among hypogonadic and among anosmic patients. Am J Med Genet. 1987;26(2):473–9.
129. Hazard J, Rozenberg I, Perlemuter L, Kestenbaum S, Vendrely E, Raoul O, et al. Gonadotropin responses to low dose pulsatile administration of GnRH in a case of anosmia with hypogonadotropic hypogonadism associated with gonadal dysgenesis 47 XXY. Acta Endocrinol. 1986;113(4):593–7.
130. Goldberg YP, MacFarlane J, MacDonald ML, Thompson J, Dube MP, Mattice M, et al. Loss-of-function mutations in the Nav1.7 gene underlie congenital indifference to pain in multiple human populations. Clin Genet. 2007;71(4):311–9.
131. Weiss J, Pyrski M, Jacobi E, Bufe B, Willnecker V, Schick B, et al. Loss-of-function mutations in sodium channel Nav1.7 cause anosmia. Nature. 2011;472(7342):186–90.
132. McEwen DP, Koenekoop RK, Khanna H, Jenkins PM, Lopez I, Swaroop A, et al. Hypomorphic CEP290/NPHP6 mutations result in anosmia caused by the selective loss of G proteins in cilia of olfactory sensory neurons. Proc Natl Acad Sci U S A. 2007;104(40):15917–22.
133. Kulaga HM, Leitch CC, Eichers ER, Badano JL, Lesemann A, Hoskins BE, et al. Loss of BBS proteins causes anosmia in humans and defects in olfactory cilia structure and function in the mouse. Nat Genet. 2004;36(9):994–8.
134. Assouline S, Shevell MI, Zatorre RJ, Jones-Gotman M, Schloss MD, Oudjhane K. Children who can't smell the coffee: isolated congenital anosmia. J Child Neurol. 1998;13(4):168–72.
135. Jafek BW, Gordon AS, Moran DT, Eller PM. Congenital anosmia. Ear Nose Throat J. 1990;69(5):331–7.
136. Leopold DA, Hornung DE, Schwob JE. Congenital lack of olfactory ability. Ann Otol Rhinol Laryngol. 1992;101(3):229–36.
137. Yousem DM, Geckle RJ, Bilker W, McKeown DA, Doty RL. MR evaluation of patients with congenital hyposmia or anosmia. AJR Am J Roentgenol. 1996;166(2):439–43.
138. Klingmuller D, Dewes W, Krahe T, Brecht G, Schweikert HU. Magnetic resonance imaging of the brain in patients with anosmia and hypothalamic hypogonadism (Kallmann's syndrome). J Clin Endocrinol Metab. 1987;65(3):581–4.
139. Truwit CL, Barkovich AJ, Grumbach MM, Martini JJ. MR imaging of Kallmann syndrome, a genetic disorder of neuronal migration affecting the olfactory and genital systems. AJNR Am J Neuroradiol. 1993;14(4):827–38.
140. Abolmaali ND, Hietschold V, Vogl TJ, Huttenbrink KB, Hummel T. MR evaluation in patients with isolated anosmia since birth or early childhood. AJNR Am J Neuroradiol. 2002;23(1):157–64.
141. Huart C, Meusel T, Gerber J, Duprez T, Rombaux P, Hummel T. The depth of the olfactory sulcus is an indicator of congenital anosmia. AJNR Am J Neuroradiol. 2011;32:1911–4.
142. Mesholam RI, Moberg PJ, Mahr RN, Doty RL. Olfaction in neurodegenerative disease: a meta-analysis of olfactory functioning in Alzheimer's and Parkinson's diseases. Arch Neurol. 1998;55(1):84–90.
143. Ansari KA, Johnson A. Olfactory function in patients with Parkinson's disease. J Chronic Dis. 1975;28(9):493–7.
144. Hawkes CH, Shephard BC, Daniel SE. Is Parkinson's disease a primary olfactory disorder? QJM. 1999;92(8):473–80.
145. Herting B, Schulze S, Reichmann H, Haehner A, Hummel T. A longitudinal study of olfactory func-

tion in patients with idiopathic Parkinson's disease. J Neurol. 2008;255(3):367–70.
146. Braak H, Rub U, Gai WP, Del Tredici K. Idiopathic Parkinson's disease: possible routes by which vulnerable neuronal types may be subject to neuroinvasion by an unknown pathogen. J Neural Transm. 2003;110(5):517–36.
147. Ponsen MM, Stoffers D, Twisk JW, Wolters E, Berendse HW. Hyposmia and executive dysfunction as predictors of future Parkinson's disease: a prospective study. Mov Disord. 2009;24(7):1060–5.
148. Haehner A, Hummel T, Hummel C, Sommer U, Junghanns S, Reichmann H. Olfactory loss may be a first sign of idiopathic Parkinson's disease. Mov Disord. 2007;22(6):839–42.
149. Welge-Lussen A, Wattendorf E, Schwerdtfeger U, Fuhr P, Bilecen D, Hummel T, et al. Olfactory-induced brain activity in Parkinson's disease relates to the expression of event-related potentials: a functional magnetic resonance imaging study. Neuroscience. 2009;162(2):537–43.
150. Katzenschlager R, Lees AJ. Olfaction and Parkinson's syndromes: its role in differential diagnosis. Curr Opin Neurol. 2004;17(4):417–23.
151. Djordjevic J, Jones-Gotman M, De Sousa K, Chertkow H. Olfaction in patients with mild cognitive impairment and Alzheimer's disease. Neurobiol Aging. 2008;29(5):693–706.
152. Westervelt HJ, Bruce JM, Coon WG, Tremont G. Odor identification in mild cognitive impairment subtypes. J Clin Exp Neuropsychol. 2008;30(2):151–6.
153. Thomann PA, Dos Santos V, Toro P, Schonknecht P, Essig M, Schroder J. Reduced olfactory bulb and tract volume in early Alzheimer's disease—a MRI study. Neurobiol Aging. 2009;30(5):838–41.
154. Biacabe B, Faulcon P, Amanou L, Bonfils P. Olfactory cleft disease: an analysis of 13 cases. Otolaryngol Head Neck Surg. 2004;130(2):202–8.
155. Liu JF, Ni DF. Three cases report of olfactory cleft disease. Zhonghua Er Bi Yan Hou Tou Jing Wai Ke Za Zhi. 2006;41(4):274–5.
156. Trotier D, Bensimon JL, Herman P, Tran Ba Huy P, Doving KB, Eloit C. Inflammatory obstruction of the olfactory clefts and olfactory loss in humans: a new syndrome? Chem Senses. 2007;32(3):285–92.
157. Rombaux P, Potier H, Markessis E, Duprez T, Hummel T. Olfactory bulb volume and depth of olfactory sulcus in patients with idiopathic olfactory loss. Eur Arch Otorhinolaryngol. 2010;267(10):1551–6.
158. Koss E, Weiffenbach JM, Haxby JV, Friedland RP. Olfactory detection and identification performance are dissociated in early Alzheimer's disease. Neurology. 1988;38(8):1228–32.
159. Haehner A, Masala C, Walter S, Reichmann H, Hummel T. Incidence of Parkinson's disease in a large patient cohort with idiopathic smell and taste loss. J Neurol. 2019;266(2):339–45.
160. Frasnelli J, Hummel T. Olfactory dysfunction and daily life. Eur Arch Otorhinolaryngol. 2005;262(3):231–5.
161. Neuland C, Bitter T, Marschner H, Gudziol H, Guntinas-Lichius O. Health-related and specific olfaction-related quality of life in patients with chronic functional anosmia or severe hyposmia. Laryngoscope. 2011;121(4):867–72.
162. Miwa T, Furukawa M, Tsukatani T, Costanzo RM, DiNardo LJ, Reiter ER. Impact of olfactory impairment on quality of life and disability. Arch Otolaryngol Head Neck Surg. 2001;127(5):497–503.
163. Ahmed T, Haboubi N. Assessment and management of nutrition in older people and its importance to health. Clin Interv Aging. 2010;5:207–16.
164. Apter AJ, Mott AE, Cain WS, Spiro JD, Barwick MC. Olfactory loss and allergic rhinitis. J Allergy Clin Immunol. 1992;90(4 Pt 1):670–80.
165. Cowart BJ, Flynn-Rodden K, McGeady SJ, Lowry LD. Hyposmia in allergic rhinitis. J Allergy Clin Immunol. 1993;91(3):747–51.
166. Guilemany JM, Garcia-Pinero A, Alobid I, Cardelus S, Centellas S, Bartra J, et al. Persistent allergic rhinitis has a moderate impact on the sense of smell, depending on both nasal congestion and inflammation. Laryngoscope. 2009;119(2):233–8.
167. Huart C, Eloy P, Collet S, Rombaux P. Chemosensory function assessed with psychophysical testing and event-related potentials in patients with atrophic rhinitis. Eur Arch Otorhinolaryngol. 2012;269(1):135–41.
168. Landis BN, Hummel T, Lacroix JS. Basic and clinical aspects of olfaction. Adv Tech Stand Neurosurg. 2005;30:69–105.
169. Upadhyay UD, Holbrook EH. Olfactory loss as a result of toxic exposure. Otolaryngol Clin North Am. 2004;37(6):1185–207.
170. Amoore J. Effects of chemical exposure on olfaction in humans. Whashington, DC: Hemisphere Publishing; 1986.
171. Schwartz BS, Doty RL, Monroe C, Frye R, Barker S. Olfactory function in chemical workers exposed to acrylate and methacrylate vapors. Am J Public Health. 1989;79(5):613–8.
172. Schiffmann SS. DRugs influencing taste and smell perception. In: Getchell TV, Doty RL, Bartoshuk LM, Snow JB, editors. Smell and taste in health and disease. New York: Raven Press; 1991. p. 845–50.
173. Nores JM, Biacabe B, Bonfils P. Olfactory disorders due to medications: analysis and review of the literature. Rev Med Interne. 2000;21(11):972–7.
174. Schiffman SS. Taste and smell losses in normal aging and disease. JAMA. 1997;278(16):1357–62.
175. Temmel AF, Pabinger S, Quint C, Munda P, Ferenci P, Hummel T. Dysfunction of the liver affects the sense of smell. Wien Klin Wochenschr. 2005;117(1–2):26–30.
176. Frasnelli JA, Temmel AF, Quint C, Oberbauer R, Hummel T. Olfactory function in chronic renal failure. Am J Rhinol. 2002;16(5):275–9.
177. Turetsky BI, Hahn CG, Borgmann-Winter K, Moberg PJ. Scents and nonsense: olfactory dysfunction in schizophrenia. Schizophr Bull. 2009;35(6):1117–31.

Olfactory Impairement in Disease and Aging

31

Ayşe Elif Özdener-Poyraz
and Mehmet Hakan Özdener

Core Messages

- Olfactory dysfunction can present as reduced, altered, or complete loss of the sense of smell.
- Etiologies of olfactory dysfunction include diseases, such as certain viral infections or neurodegenerative disorders, normal aging processes, and adverse effects of medication. In some cases, olfactory dysfunction may be permanent.
- Olfactory dysfunction significantly reduces a person's quality of life and may cause serious complications, such as inability to smell spoiled foods. Thus, health care providers should regularly assess olfactory function, especially in patients at high risk for olfactory dysfunction due to underlying conditions.
- It is important to investigate whether a person's olfactory dysfunction is caused by medications, such as enalapril and amlodipine. If so, the medication should be discontinued, if clinically possible, or switched to an alternative medication that does not have the same risk of adverse effects.
- Function of the chemosensory system specifically olfactory function should be evaluated in routine exams by ENT specialists.

31.1 Introduction

The sense of smell is critical for most mammals in terms of identification and evaluation of food, mate, and territories, and in general for the maintenance of a good quality of life. The olfactory sensory system is able to detect and discriminate an enormous variety of volatile molecules with great sensitivity and specificity. Tens of thousands of chemicals can be detected, many at concentrations as low as a few parts per trillion [1, 2]. This feat is accomplished through anatomical, cellular, and molecular features that are designed to amplify, encode, and integrate a vast array of incoming olfactory information.

The neurological systems responsible for olfactory function represent perhaps the most diverse, complex, and adaptable components of the nervous system. Losses in olfaction result from changes at both the anatomical and molecular level. This loss can result from aging, toxins, infectious agents, environmental factors, and a variety of diseases. Chronic inflammation of the nasal lining can cause damage to nasal tissues and cause anosmia, which can be permanent. One study shows that chronic allergic rhinitis may lead to hyposmia and anosmia in 5–22% of

A. E. Özdener-Poyraz
School of Pharmacy and Health Sciences, Fairleigh Dickinson University, Florham Park, NJ, USA

M. H. Özdener (✉)
Monell Chemical Senses Center, Philadelphia, PA, USA
e-mail: hozdener@monell.org

Ö. Ö. Celebi, T. M. Önerci (eds.), *Nasal Physiology and Pathophysiology of Nasal Disorders*,
https://doi.org/10.1007/978-3-031-12386-3_31

people [3]. Other disorders associated with smell impairment include Alzheimer's disease, cleft palate, diabetes, head trauma, influenza, malnutrition and zinc deficiency, Parkinson's disease, schizophrenia, HIV, and Sjögren's syndrome [4].

Interest in olfactory dysfunction has increased tremendously in the past decade, in part because chemosensory impairment is an early symptom of many neurodegenerative diseases. Several features of the olfactory system make it particularly relevant to understanding neurodegeneration/regeneration. First, while being vulnerable to environmental and infectious exposure, the olfactory system has the unique property of ongoing replacement of the olfactory sensory neurons under physiological conditions and following injury. Second, the receptor neurons residing in the periphery are developmentally related to the central nervous system (CNS) yet are accessible relatively noninvasively via biopsy from living subjects. Third, olfactory performance can easily be tested in a variety of ways that provide insight into the function of brain areas involved in olfactory detection, identification, and memory.

A better understanding of olfactory neurobiology is needed to develop ways to treat olfactory dysfunction, which affects both quality of life and personal safety. At least 3,000,000 Americans suffer from chemosensory disorders, and that number is likely to grow as the aging segment of the population increases. Olfactory impairment affects patients' acceptance of food and drink, causes difficulties in daily life, and has an impact on safety by impairing detection of spoiled foods, natural gas, smoke, and other volatile compounds in the environment.

Although smell and taste pathways are different, the ability to taste is strongly affected by smell, and those suffering from anosmia or other olfaction problems also have difficulties perceiving the flavor of food and drink [5]. In fact, a person smells 75% of a food's flavor [4]. A study of 750 patients showed that 68% of them reported decreased quality of life due to olfactory dysfunction [6]. Another study of 225 participants showed that patients reported incidences of burning food or ingestion of spoiled food (62.8%), mood changes or depression (35.9%), changes in communication habits (58.1%), and negative impact on sexual behavior (56.8%) because of olfactory dysfunction [7]. Over time, patients can suffer from nutritional deficits and lose weight. Those suffering from olfactory disturbances may also resort to adding excess sugar and salt in their diet to improve the taste of foods. This can lead to new onset or worsening of preexisting hypertension, diabetes, heart failure, and renal dysfunction.

Understanding the neurobiology of this sensory system may help us develop new diagnostic measures and treatments for neurodegenerative disease, as well as improving the quality of life for millions of people who are handicapped by the inability to detect the odors and flavors around them. This review provides a brief overview of the olfactory system and then covers recent progress in understanding olfactory function and dysfunction caused by disease, aging, and drugs.

31.2 Overview of the Olfactory System

31.2.1 Anatomy and Cellular Features

The olfactory epithelium is localized to the interior surface of the nasal cavity and consists primarily of three basic cell types: olfactory receptor neurons (ORNs), supporting cells, and basal cells [8]. Postmortem biopsies, anatomical studies, and explant cultures of ORNs from different parts of the nasal cavity show that sensory epithelium extends from the olfactory cleft down onto the superior aspect of the medial turbinate [9–11]. The turbinate structures are cartilaginous ridges covered with respiratory epithelium, a nonsensory ciliated columnar epithelial tissue also populated with mucus-secreting goblet cells. This structure increases the surface area available for both warming and humidifying incoming air, as well as funneling volatile chemicals up into the sensory epithelium.

Human ORNs have a morphology generally similar to those of other vertebrates, although there is variation across species [12]. The ORN comprises a cell body, an axon, and an apical dendrite terminating in a knob containing multiple nonmotile cilia. The cilia project into the mucus overlying the nasal epithelium, where their receptors have direct contact with volatile chemicals in the air. The axon projects through the cribriform plate to synapse with the dendrites of mitral cells in the olfactory bulb. The mitral cells project via the olfactory nerve (cranial nerve I) to the entorhinal cortex, as well as regions involved in emotion and memory, such as the amygdala and hippocampus. Several types of interneurons modulate mitral cell activity, including periglomerular cells, tufted cells, and granule cells (dopaminergic/GABAergic interneurons that are involved in signal processing and modulation) [13, 14].

31.2.2 Regeneration of Olfactory Neurons

The exposure of ORNs to the external environment makes these primary sensory neurons vulnerable to injury from environmental insults, such as toxins, infectious agents, and trauma. However, unlike CNS neurons and other sensory neurons, ORNs are able to be replaced after injury. The average life cycle of an ORN is approximately 30–120 days [15]. This replacement process begins with a population of multipotent basal neuroepithelial precursor cells, which undergo successive stages of differentiation to a fully mature ORN. It is also thought that these cells may differentiate along a non-neuronal pathway, although there may be separate populations of precursor cells as well [16–19].

Following olfactory nerve injury or toxic exposure, reconstitution of neuroepithelial cells and establishment of connections within the olfactory bulb can be enhanced by growth factors, including retinoic acid, IGF-1, TGF-α, TGF-β, and FGF2 [20–24]. Apoptotic cell death has been observed in cells representing all stages of neurogenesis (e.g., in proliferating neuronal precursors, immature ORNs, and mature ORNs), implying that apoptotic regulation of neuronal numbers may occur at multiple stages of the neuronal lineage [25, 26].

Remarkably, central components of the olfactory system may also be replaced. A population of proliferating stem cells arising in the subventricular zone migrate into the olfactory bulb, where they differentiate into granule cells [27, 28]. This process appears to continue throughout life, although the impact on olfaction and the mechanisms controlling proliferation and differentiation of these cells are unknown.

31.2.3 Human Olfactory Epithelium

Human olfactory mucosa is a specialized neuroepithelium in the superior turbinate of the nasal cavity. The epithelium is a self-renewing tissue that can generate new neurons from stem and progenitor cells within the basal layers [9, 10]. Mesenchymal-like stem cells are present in the underlying lamina propria and are also likely to participate in regeneration [11]. The loss of this tissue leads to decreased olfactory acuity, hyposmia, and irreversible anosmia. In adults the olfactory mucosa is often patchy, with interspersed respiratory epithelium that suggests intermittent damage and incomplete regeneration throughout life [29]. Causes of damage are diverse and include viral infection, toxic effects, senescence with apoptosis, and degeneration of dopaminergic systems in Alzheimer's and Parkinson's diseases [8].

31.2.4 Transduction Mechanisms

The past decade has witnessed tremendous advances in our understanding of the initial events in olfactory transduction. A cornerstone in this progress is the discovery of a large family of genes encoding the 7-transmembrane domain in G-protein-coupled receptors that apparently represent odorant receptors. Each olfactory receptor is responsive to a range of stimuli. Odorant binding leads to a depolarizing current within the cilia of ORNs, which ultimately triggers action potentials transmitted via the axon to collectively provide the neural code deciphered by higher brain centers [30]. Most of the olfactory receptor proteins are linked to the stimulatory guanine nucleotide-binding protein (G_{olf}). When stimulated, G_{olf} activates the enzyme adenylate cyclase to produce the second messenger cyclic adenosine monophosphate (cAMP) [11]. cAMP diffuses through the cell and activates cellular depolarization via the opening of cyclic nucleotide-gated ionic channels [12].

Some odorants also activate cyclic guanosine monophosphate (cGMP), which is believed to play a role in modulating the sensitivity of ORNs, such as during adaptation [12]. In addition, recent evidence indicates that some odorants may activate phospholipase C to produce the second messenger inositol trisphosphate (IP_3), which may also modulate the activity of the cAMP pathway via the activity of phosphoinositol-3 kinase [13].

Medications, diseases, or disorders that interfere with or alter the ability of these transduction pathways to operate can influence olfactory performance. Also, as this tissue is available via biopsy from live subjects, functional studies may suggest specific genetic or pathologic alterations related to early symptoms or causes of neurodegenerative disease [31].

31.2.5 Diagnosis and Treatment of Olfactory Disorders

Although diagnosis of taste and smell disorders has improved considerably over the last two decades, treatment of these disorders is still limited to conditions with discernible and reversible causes [32, 33]. Olfactory function is commonly measured based on either detection sensitivity (threshold), identification, or discrimination of odors. These tests can be performed easily and may be useful tools to improve diagnosis of conditions in which olfactory loss is an early symptom, such as Alzheimer's dementia, Parkinson's disease, and other neurodegenerative disorders. Unlike other sensory neurons and neurons in the CNS, ORNs are unique in their ability to be replaced after injury. These primary receptor neurons are developmentally related to the CNS yet are accessible for study from living patients. Studies of the cellular processes underlying the regenerative capacity of the peripheral and central components of the olfactory system are leading to new therapeutic approaches in the treatment of spinal cord injury [34, 35] and stroke [30, 36] and may also contribute to the treatment or prevention of neurodegenerative diseases.

31.3 Common Causes of Olfactory Dysfunction

Several features of the olfactory system are particularly susceptible to age- and disease-associated changes that may lead to functional deficits. Changes at both the anatomical and molecular level may contribute to this loss. Local injury from physical or chemical causes; damage to neural projections; disturbance of the cycle of neurogenesis resulting from general malnutrition, infectious diseases, metabolic disturbances, drugs, or radiation; and alteration of the composition of the mucus due to medication are possible major causes to consider as the pathogenesis of olfactory dysfunction [37].

Aging, chronic sinusitis infections, neurodegenerative diseases, and viral infections, as well as head injuries and nasal obstructions, are the most common causes of olfactory disorders [37]. Olfactory impairment is a common occurrence in aging and may be an early signal of neurodegenerative diseases. At least one-third of older people report dissatisfaction with their sense of taste or smell [29, 38].

31.3.1 Age-Associated Olfactory Loss

Detailed information about the prevalence of chemosensory disorders has been limited. A report by the National Institutes of Health (NIH) covering the late 1970s indicated that more than two million US adults had a smell or taste disorder. Another survey conducted by the National Geographic Society in 1987 showed that 1% of their 1.2 million respondents could not smell 3 or more of the

6 odorants in a "scratch and sniff" test. This study indicated that age was an important factor: decline began in the second decade of life [39]. A study by the NIH in collaboration with the National Center for Health Statistics, conducted in 1993 to acquire information on the prevalence of smell/taste problems, showed an overall prevalence of 2.7 million (1.4%) US adults with olfactory problems. Prevalence rates increased exponentially with age; almost 40% of 1.5 million respondents reporting a chemosensory problem were 65 or more years of age [32, 40].

The ability to detect, discriminate, and identify odors is most sensitive to age- and disease-related dysfunction [39, 41]. And the elderly population is increasing due to improved health care throughout the world. Despite the widespread age-related prevalence of olfactory loss, remarkably little is known about the specific mechanisms responsible, and no treatments are currently available.

The impact of sensory loss on elders is not only physiological but also emotional. Taste and smell are fundamental sensory systems responsible for the perception of aroma and flavor. Olfactory dysfunction can also lead to changes of dietary habits that may in turn exacerbate disease states or contribute to nutritional deficiencies [33, 37].

The susceptibility of elderly people to aging and diseases, particularly neurodegenerative diseases, varies. Therefore, complaints about sensory functions should be seriously considered as possible indicators of neurodegenerative disease or another underlying condition (e.g., medication side effects).

The composition of mucus is critical to proper ORN function and may change with hydration, which is often reduced in the elderly, as well as from age-related diseases or associated medications. Changes at the level of the ORN may also reduce and/or alter olfactory function. For instance, age-related loss of selectivity was observed in a study of odorant response characteristics of ORNs dissociated from biopsies [31], and age-related changes in ion channel distribution [42, 43] or other components of the intracellular signaling cascades [44] could result in receptor cell dysfunction.

Although the extent of olfactory epithelium is reduced with aging [31, 45], whether this accounts for age-related olfactory loss remains unclear: studies indicate that olfactory function is not affected even with substantial reduction of olfactory epithelial area [31, 45]. Other anatomical changes, such as altered vascular and mucosal composition and peptidergic innervation, could lead to reduced sensitivity through indirect mechanisms or changes in the transport and clearance of odorants [46]. Sensory dysfunction may be a consequence of chronic disease, such as diabetes, cancer, radiation, surgery, or dentures. However, in most cases the cause of olfactory loss is unknown, and the development of treatments will require a better understanding of the mechanisms underlying this sensory impairment.

31.3.2 Chronic Rhinosinusitis (CRS)

Many olfactory dysfunctions are related to CRS, nasal polyps, and allergies [47]. Olfactory dysfunction is a frequent complaint in CRS patients. CRS is an inflammatory disease that occurs in the nasal cavity and sinus mucosa [29]. In most patients, CRS is accompanied by olfactory or taste dysfunctions that can compromise their quality of life. Many of patients with CRS are known to have olfactory impairment [48, 49].

31.3.3 Neurodegenerative Diseases

Olfactory dysfunction can be the direct or indirect result of pathological processes at any point in the olfactory pathway. Accumulating data indicate that neurodegenerative disorders of CNS areas involved in olfactory processing may contribute to olfactory dysfunction [50–52]. Nevertheless, the precise mechanisms that connect these diseases with olfactory loss are still unclear [53]. While these reports suggest olfactory involvement and potential utility in diagnostic approaches for these diseases, no studies have been done to directly investigate the neuropathology or cell/molecular basis for olfactory impairment.

Since the first observation of olfactory function impairment in Parkinson disease [54] and senile dementia [55], olfactory function testing has revealed compromised olfactory function in a number of neurodegenerative diseases, such as Alzheimer's disease [56–59], Parkinson disease [60, 61], Huntington's disease [62], HIV-associated dementia [63, 64], and amyotrophic lateral sclerosis [65]. Using and olfactory brushing procedure to detect the pathologic prion proteins α-synuclein, β-amyloid, tau, and TDP-43 in patients with neurodegenerative disorders demonstrated the presence of proteins associated to neurodegeneration in the cells of the olfactory mucosa [66].

31.3.4 Alzheimer's Disease (AD)

Olfactory impairment and neuroanatomical changes in the central portions of the olfactory system occur early in the development of AD, and olfactory testing has been explored as a diagnostic aide [45, 67, 68]. Patients with AD perform more poorly on tests of odor identification [59, 69] and exhibit altered olfactory evoked response potentials compared to age-matched controls [70]. Olfactory tests alone, however, are insufficient to discriminate AD from Parkinson's disease, and careful consideration of cognitive function is required to ensure reliable results [71]. It is generally accepted that the classic AD neuropathology occurs in the entorhinal cortex very early in the development of the disease [72], and this observation led some to speculate that a causative agent might enter the brain via the nasal epithelium.

Some have investigated whether histological studies of biopsies of the olfactory neuroepithelium might be useful as an early diagnostic tool for AD. However, more comprehensive studies indicate that the density of plaques and tangles in the olfactory bulb is less severe, and studies of the peripheral olfactory epithelium have been inconsistent. In general, phosphorylated tau and neurofilament proteins are not observed in the perikarya of the olfactory neurons but are evident within the axons and dendrites of these cells [73, 74]. Several studies have reported AD-specific neuropathology within the olfactory epithelium [68, 75, 76], but others using different markers have noted similar features in non-AD and healthy olfactory tissue from elderly controls. In addition, studies that have included tissue from patients with other types of dementia or neurological disease have failed to identify any marker present in the olfactory epithelium that would serve to reliably distinguish AD from other conditions, such as Parkinson's disease or vascular dementia [77, 78]. Consistent with these findings, while we have observed some functional differences in preliminary studies of ORNs obtained via biopsy from patients with early-stage AD, these neurons appeared normal morphologically and were able to respond to odorants [31]. Further studies of olfactory neuronal cell function may prove more useful than histology alone in understanding altered cellular metabolism or signaling that heralds the onset of AD.

31.3.5 Down's Syndrome

Patients with Down's syndrome display neuropathologic features similar in some respects to those seen in AD. Likewise, individuals with Down's syndrome had significant deficits in olfactory functioning compared to the control groups [79]. The Alcohol Sniff Test, a rapid screen for olfactory function, revealed olfactory deficits in children with Down's syndrome [80]. Another study also indicated that olfactory deficits may provide a sensitive and early indicator of the deterioration and progression of the brain in older patients with Down's syndrome [81].

31.3.6 Parkinson's Disease (PD)

Impaired olfactory function is a well-documented abnormality in patients with PD [82]. The cellular and molecular mechanisms for this deficit are unknown but likely relate to impairment at several levels of the olfactory system. Olfactory impairment in PD was not related to degree of motor dysfunction or disease duration but was related to disease severity [61, 83–87]. Interestingly, Hawkes et al. proposed that idiopathic PD may start in the olfactory system prior to damage in the basal ganglia [88]. Another study indicated that olfactory testing may be useful for differential diagnosis between PD and progressive supranuclear palsy (PSP) [89]. Both PD and PSP have similar motor symptoms, and PSP is commonly misdiagnosed as PD, although they are distinct neuropathologic entities [31, 89].

In one study, olfactory dysfunction was observed in patients with an abnormal reduction in striatal dopamine transporter binding who subsequently developed clinical Parkinsonism. None of 23 normosmic relatives of these patients developed signs or symptoms of Parkinsonism [90]. These observations indicate that olfactory deficits may precede clinical motor signs in PD and support a practical clinical application in the early diagnosis/prognosis of the disease.

PD-related olfactory dysfunction may relate to the function of dopamine receptors in both central [13, 91–93] and peripheral components of the system [13, 94–96]. Centrally, dopamine modulates synaptic activity in the olfactory bulb and entorhinal cortex, influences the activity of several ion channels and enzymes involved in olfactory transduction, and has been reported to induce apoptosis and modulate differentiation of olfactory neurons in vitro [90, 97, 98]. These effects are mediated via D2 receptors in the periphery [98] and D1 and D2 receptors on mitral/tufted and juxtaglomerular cells in the olfactory bulb [99]. The dopaminergic granule cells in the olfactory bulb derive from stem cells that migrate from the subventricular zone throughout life. These stem cells are being studied as a potential source for dopaminergic replacement cells via transplant [100].

31.3.7 Huntington's Disease

Patients with Huntington's disease exhibit significant deficits in odor identification, but odor recognition memory was not affected [67, 101]. In an animal model of the disease, the olfactory system exhibited early and significant accumulation of huntingtin-containing aggregates, which may account for the early olfactory impairment [102].

31.3.8 Multiple Sclerosis

Olfactory dysfunction may also be an early indicator of disease progression in multiple sclerosis [103]. The Cross-Cultural Smell Identification test utilized in patients with multiple sclerosis indicated that these patients scored significantly worse than control groups. They also found significant correlations among smell alteration, symptoms of anxiety and depression, and severity of neurological impairments [104]. In several studies, neuropathology based on plaque numbers was directly related to olfactory function [104, 105]: as plaque numbers declined or increased in the inferior frontal and temporal lobes, olfactory function declined or improved [106–108].

31.3.9 Creutzfeldt–Jakob Disease (CJD)

This rapidly progressive, fatal neurodegenerative disorder is believed to be caused by pathologic prion protein (PrPSc), which may be found in the neuroepithelium of the olfactory mucosa in patients with CJD [109]. Taste and smell loss were reported as an early sign of CJD [110]. This result indicates that olfactory biopsy may provide diagnostic information, and further studies are warranted.

31.3.10 Viral Infections

Animal models have shown that neurotropic viruses, such as murine coronavirus [111], Borna disease virus [112], pseudorabies virus [113], herpes simplex virus type 1 and 2 [114, 115], Zika virus [116], human coronavirus OC43 [117], and adenovirus [118], reach the CNS after having contact with the nasal mucosa. Vesicular stomatitis virus infects the olfactory epithelium and then spreads along the olfactory nerves into the glomeruli of the olfactory bulb, progressively spreading to the brain [119]. The presence of various viruses has been confirmed in the nasal discharge of patients with post-viral infection olfactory dysfunction, such as rhinovirus, parainfluenza virus, Epstein-Barr virus, and coronavirus [120]. Significant recovery was not observed after 24 weeks in almost all the patients.

31.3.11 Human Immunodeficiency Virus (HIV) Infection and AIDS

HIV-associated dementia is a leading cause of neurodegenerative disorders among individuals under 30 years [121]. HIV infects the CNS; 7–25% of patients with CNS infection develop dementia, and at least 50% of these patients develop mild neurocognitive impairment. Several studies have shown that patients with neurocognitive impairment caused by HIV infection had diminished odor sensitivity [63, 64, 122]. Impaired olfactory function may serve as early marker of HIV-associated neurological impairment [123] and could be helpful to evaluate the impact of therapeutic agents. HIV-positive patients had significantly impaired menthol detection compared to controls. It is likely that chemosensory losses found in patients with HIV reflect both central and peripheral deficits [124].

Of 207 HIV-infected patients, 70% of them ($n = 144$) reported that chemosensory complaints were associated with a poor quality of life [125]. Wasting with reduced caloric intake is an increasingly common clinical manifestation of AIDS. The perceptions of taste and smell play an important role in stimulating caloric intake, and flavor enhancement of food can have a significant positive impact on nutritional status in hospitalized patients [124]. Significant taste and smell losses in HIV-infected patients may be of clinical significance in the development or progression of HIV-associated wasting and are thus worthy of clinical consideration and treatment [126].

31.3.12 Coronavirus-19 (SARS-CoV-19, COVID-19) Infection

Accumulating reports from around the world confirm the high prevalence of different degrees of smell and taste loss in COVID-19-positive patients. Currently, loss of taste or smell is one of the six most reliable symptoms of COVID-19 [127]. While it is well known that coronaviruses can cause chemosensory dysfunction, the underlying pathophysiological mechanism is unknown and may be distinct for each chemosensory system. Since the start of the pandemic most studies have focused specifically on cellular and molecular mechanisms of coronavirus-induced smell loss.

The novel coronavirus (SARS-CoV-2) is a highly pathogenic, Coronaviridae single-stranded, positive-sense RNA virus responsible for the present outbreak of COVID-19. The RNA virus genome is covered by an envelope comprising spike proteins and a lipid membrane. The infection cycle is initiated by attachment of the spike proteins to angiotensin-converting enzyme 2 (ACE-2), a host cell membrane protein that serves as a receptor for viral entry by endocytosis. In addition, TMPRSS2, a 70-kDa member of the serine protease family, becomes activated upon proteolytic cleavage and serves as a co-receptor for the SARS-CoV-2 spike glycoprotein, a step that facilitates virus recognition of the host cells.

Whether and how the SARS-CoV-2 virus may infect sensory system cells and cause olfactory dysfunction still is a question of major interest and concern. Sudden loss of smell and taste could be the only features in asymptomatic newly infected individuals and could also serve as early symptoms of the disease. In fact, the American Academy of Otolaryngology–Head and Neck Surgery recognized that smell and/or taste loss could be the only and/or early symptom of newly infected individuals and recommended self-isolation for anyone with these symptoms [128]. The WHO has also recognized them as key symptoms of COVID-19 [129, 130].

31.3.13 Influenza Virus Infection

Upper respiratory tract viral infections, such as influenza, parainfluenza, and rhinovirus, are among the most common causes of olfactory dysfunction [131]. Post-viral olfactory disorder develops after infection with the common cold and is a relatively severe and prolonged disorder without rhinosinusitis [132]. Changes in olfactory perception in patients with the influenza or parainfluenza type 3 viruses have been reported. Seasonal changes in the incidence of olfactory loss have been reported with respect to influenza and parainfluenza type 3 infections, occurring most frequently in winter and spring, respectively [120].

31.3.14 Medications and Olfactory Dysfunction

When assessing patients for reduced ability, complete inability, or distorted ability to smell (hyposmia, anosmia, dysosmia/parosmia, respectively), it is important to consider their medications. Because olfaction is a strong indicator for quality of life, drug-induced olfactory changes can also lead to medication nonadherence, which compromises the management of acute and chronic conditions.

Table 31.1 lists some common medications associated with olfactory change, including antibiotics, cardiovascular drugs, decongestant nose sprays, and intranasal zinc products. In 2009, the US Food and Drug Administration (FDA) issued a warning to consumers to discontinue the zinc-containing nasal spray Zicam, which was marketed to shorten the duration of the cold. The FDA received over 100 reports from doctors and consumers citing a causative relationship between the zinc nasal spray and anosmia. In some cases, this adverse drug effect was permanent [133, 134].

Most of the information available about specific drugs causing olfactory changes is from case reports rather than from large, randomized, controlled trials. One study identified 71 drugs that

Table 31.1 Drugs associated with changes in smell [135, 136, 139, 150–156]

Drug class	Drug names	Effect on olfactory system
Cardiovascular		
ACE inhibitors	Enalapril	Reduced
Calcium-channel blockers	Diltiazem, amlodipine, felodipine, nifedipine	Reduced
HMGCo-A reductase inhibitors	Atorvastatin, lovastatin, pravastatin	Altered
$Alpha_1$ blocker	Doxazosin	Altered
Antiarrhythmic	Tocainide, amiodarone	Altered
Antibiotics		
Macrolides	Azithromycin, clarithromycin	Reduced
Penicillin	Amoxicillin	Reduced
Fluoroquinolones	Ciprofloxacin, levofloxacin	Reduced
Tetracycline	Doxycycline	Altered
Miscellaneous		
Antihistamine and anti-allergy	Intranasal zinc, fluticasone, prednisone	Reduced
Thyroid	Levothyroxine	Altered
Opioid analgesic	Morphine	Reduced
Blood viscosity reducer	Pentoxifylline	Enhanced
Phosphodiesterase-5 inhibitor	Sildenafil	Reduced
Immunomodulatory	Alpha interferon, methotrexate	Interferon reduced, methotrexate enhanced
Antifungal	Terbinafine	Reduced
Topical	Silver nitrate	Reduced
Antidepressant	Duloxetine	Altered

probably alter olfaction in humans [135]. Another study showed that long-term use of opioid and nonopioid analgesics caused significant reduction in olfactory function compared to age-matched controls not taking analgesics [136].

Patients may develop drug-induced olfactory disturbances for several reasons. Medications may damage olfactory mucosa regeneration, deplete trace metals, such as zinc, or cause neurotoxicity that impairs smell pathways. Topical silver nitrate and silver sulfadiazine are used as an antibiotic in thermal injury. Prolonged use of these drugs can lead to silver deposition in the nasal mucosa, leading to anosmia. Drugs can also cause impairment of neuronal and mucosal tissue potassium and calcium ion efflux, causing a reduction in impulse conduction [4]. Calcium channel blockers, such as nifedipine and diltiazem, inhibit calcium-mediated impulse transmission to the olfactory bulb, which can lead to anosmia in some patients [137, 138]. Additional pathways that can impact olfaction include reduced access to receptors due to mucosa dryness; altered chemical or ionic environment in the region of the receptor due to changes in the components of saliva or mucus or in agonizing or antagonizing receptor sites; and changes in neurotransmitter function [139].

As part of the natural aging process that causes pharmacokinetic and pharmacodynamic changes, people gradually lose their ability to smell. Older adults are also at an increased risk of developing drug-induced olfactory changes and other adverse drug events, because they take more prescription and nonprescription medications compared to younger adults [140, 141]. Between 1988 and 2010, the proportion of adults 65 and older taking at least five medications tripled, from 12.8 to 39.0%, primarily due to increased use of cardiovascular and antidepressant medications [142]. One survey study found that dysosmia was higher in those taking more than four medications per day [7].

Because numerous pathologies can lead to olfactory dysfunction, a causal relationship must be established between the medication and the patient's symptoms before the event is deemed an adverse drug reaction. The Naranjo Algorithm, also known as the Adverse Drug Reaction Probability Scale, can be used to assess whether there is a causal relationship. The algorithm is a series of 10 questions with assigned probability scores that help determine the likelihood of an adverse drug event [143]. If it is decided that the person's smell dysfunction is indeed caused by a medication, this medication should be discontinued or switched to an equally efficacious and safer alternative.

31.4 Summary and Future Directions

The olfactory system is a fundamental sensory system responsible for the perception of flavor and fragrance. Olfaction is critical for most mammals for the maintenance of a good quality of life. Scientists and physicians have paid increasing attention to the olfactory system and its function during the last decade, largely because of (a) advances in our understanding of the histocompatibility basis for the receptor mechanisms, (b) evidence that ORNs undergo neurogenesis and both programmed and induced cell death, (c) important technical and practical developments in psychophysical testing, and (d) the accessibility of these cells relatively noninvasively from living subjects [144–148]. These developments have led to standardized olfactory testing to assess detection sensitivity (threshold) and ability to identify odors. These tests can be easy to perform in the clinic and may be useful in improving diagnosis [149].

Although diagnosis of smell disorders caused by aging and neurodegenerative diseases has improved considerably over the last two decades, treatment of these disorders is still limited to conditions with discernible and reversible causes. Sensory complaints are often overlooked by the medical community. Understanding the biological bases for olfactory system disorders can help us develop new approaches to improve the quality of flavor experience for those with impaired ability. In addition, studies of olfaction and ORN function may lend new insights into the etiology of neurodegenerative disease. Future research is needed for a better understanding of chemosensory mechanisms, establishing improved diagnostic procedures, and disseminating knowledge about chemosensory disorders among practitioners and the public [71, 72].

References

1. Walker JC, Hall SB, Walker DB, Kendal-Reed MS, Hood AF, Niu XF. Human odor detectability: new methodology used to determine threshold and variation. Chem Senses. 2003;28:817–26.
2. Angioy AM, Desogus A, Barbarossa IT, Anderson P, Hansson BS. Extreme sensitivity in an olfactory system. Chem Senses. 2003;28:279–83.
3. Li C, Yousem D, Doty R, Kennedy D. Neuroimaging in patients with olfactory dysfunction. AJR Am J Roentgenol. 1994;162:411–8.
4. Ackerman BH, Kasbekar N. Disturbances of taste and smell induced by drugs. Pharmacotherapy. 1997;17:482–96.
5. Mattes R, Cowart B, Schiavo M, Arnold C, Garrison B, Kare M, et al. Dietary evaluation of patients with smell and/or taste disorders. Am J Clin Nutr. 1990;51:233–40.
6. Deems DA, Doty RL, Settle RG, Moore-Gillon V, Shaman P, Mester AF, et al. Smell and taste disorders, a study of 750 patients from the University of Pennsylvania Smell and Taste Center. Arch Otolaryngol Head Neck Surg. 1991;117:519–28.
7. Merkonidis C, Grosse F, Ninh T, Hummel C, Haehner A, Hummel T. Characteristics of chemosensory disorders—results from a survey. Eur Arch Otorhinolaryngol. 2014;272:1403–16.
8. Graziadei PPC. Cell dynamics in the olfactory mucosa. Tissue Cell. 1973;5:113–31.
9. Rawson NE, Gomez G, Cowart B, Brand JG, Lowry LD, Pribitkin EA, et al. Selectivity and response characteristics of human olfactory neurons. J Neurophysiol. 1997;77:1606–13.
10. Nibu KI, Li G, Zhang X, Rawson NE, Restrepo D, Kaga K, et al. Olfactory neuron-specific expression of NeuroD in mouse and human nasal mucosa. Cell Tissue Res. 1999;298:405–14.
11. Gomez G, Rawson NE, Hahn CG, Michaels R, Restrepo D. Characteristics of odorant elicited calcium changes in cultured human olfactory neurons. J Neurosci Res. 2000;62:737–49.
12. Schild D, Restrepo D. Transduction mechanisms in vertebrate olfactory receptor cells. Physiol Rev. 1998;78:429–66.
13. Trombley PQ, Shepherd GM. Synaptic transmission and modulation in the olfactory bulb. Curr Opin Neurobiol. 1993;3:540–7.
14. Brünig I, Sommer M, Hatt H, Bormann J. Dopamine receptor subtypes modulate olfactory bulb γ-aminobutyric acid type A receptors. Proc Natl Acad Sci USA. 1999;96:2456–60.
15. Astic L, Saucier D. Neuronal plasticity and regeneration in the olfactory system of mammals: morphological and functional recovery following olfactory bulb deafferentation. Cell Mol Life Sci. 2001;58:538–45.
16. DeHamer MK, Guevara JL, Hannon K, Olwin BB, Calof AL. Genesis of olfactory receptor neurons in vitro: regulation of progenitor cell divisions by fibroblast growth factors. Neuron. 1994;13:1083–97.
17. Calof AL, Bonnin A, Crocker C, Kawauchi S, Murray RC, Shou J, et al. Progenitor cells of the olfactory receptor neuron lineage. Microsc Res Tech. 2002;58:176–88.
18. Calof AL, Mumm JS, Rim PC, Shou J. The neuronal stem cell of the olfactory epithelium. J Neurobiol. 1998;36:190–205.
19. Huard J, Youngentob S, Goldstein B, Luskin M, Schwob J. Adult olfactory epithelium contains multipotent progenitors that give rise to neurons and non-neural cells. J Comp Neurol. 1998;400:469–86.
20. Costanzo R. Regeneration of olfactory receptor cells. Ciba Found Symp. 1991;160:233–42.
21. Schwob JE. Neural regeneration and the peripheral olfactory system. Anat Rec. 2002;269:33–49.
22. Federico G, Maremmani C, Cinquanta L, Baroncelli GI, Fattori B, Saggese G. Mucus of the human olfactory epithelium contains the insulin-like growth factor-I system which is altered in some neurodegenerative diseases. Brain Res. 1999;835:1095–8.
23. Amoore JE. Specific anosmia: a clue to the olfactory code. Nature. 1967;214:1095–8.
24. Yee KK, Rawson NE. Retinoic acid enhances the rate of olfactory recovery after olfactory nerve transection. Dev Brain Res. 2000;124:129–32.
25. Doty R. Olfaction. Annu Rev Psychol. 2001;52:423–52.
26. Holcomb JD, Mumm JS, Calof AL. Apoptosis in the neuronal lineage of the mouse olfactory epithelium: regulationin vivoandin vitro. Dev Biol. 1995;172:307–23.
27. Englund U, Björklund A, Wictorin K. Migration patterns and phenotypic differentiation of long-term expanded human neural progenitor cells after transplantation into the adult rat brain. Dev Brain Res. 2002;134:123–41.
28. Fricker RA, Carpenter MK, Winkler C, Greco C, Gates MA, Björklund A. Site-specific migration and neuronal differentiation of human neural progenitor cells after transplantation in the adult rat brain. J Neurosci. 1999;19:5990–6005.
29. Pelchat ML. You can teach an old dog new tricks: olfaction and responses to novel foods by the elderly. Appetite. 2000;35:153–60.
30. Arlotta P, Magavi SS, Macklis JD. Induction of adult neurogenesis: molecular manipulation of neural precursors in situ. Ann N Y Acad Sci. 2003;991:229–36.
31. Rawson NE, Gomez G, Cowart B, Restrepo D. The use of olfactory receptor neurons (ORNs) from biopsies to study changes in aging and neurodegenerative diseases. Ann N Y Acad Sci. 1998;855:701–7.

32. Spielman A. Chemosensory function and dysfunction. Crit Rev Oral Biol Med. 1998;9:267–91.
33. Bromley SM. Smell and taste disorders: a primary care approach. Am Fam Physician. 2000;61:427–36.
34. Huang H, Chen L, Wang H, Xiu B, Li B, Wang R, et al. Influence of patients' age on functional recovery after transplantation of olfactory ensheathing cells into injured spinal cord injury. Chin Med J. 2003;116:1488–91.
35. Resnick D, Cechvala C, Yan Y, Witwer B, Sun D, Zhang S. Adult olfactory ensheathing cell transplantation for acute spinal cord injury. J Neurotrauma. 2003;20:279–85.
36. Parent JM. Injury-induced neurogenesis in the adult mammalian brain. Neuroscientist. 2003;9:261–72.
37. Schiffman S. Taste and smell in disease (second of two parts). N Engl J Med. 1983;308:1337–43.
38. Winkler S, Garg AK, Mekayarajjananonth T, Bakaeen LG, Khan E. Depressed taste and smell in geriatric patients. J Am Dent Assoc. 1999;130:1759–65.
39. Wysocki C, Gilbert A. National Geographic Smell Survey: effects of age are heterogeneous. Ann N Y Acad Sci. 1989;561:12–28.
40. Hoffman H, Ishii E, Macturk R. Age-related changes in the prevalence of smell/taste problems among the United States adult population: results of the Disability Supplement to the National Health Interview Survey (NHIS). Ann N Y Acad Sci. 1994;1998(855):716–22.
41. Deems DA, Doty RL. Age-related changes in the phenyl ethyl alcohol odor detection threshold. Trans Pa Acad Ophthalmol Otolaryngol. 1987;39:646–50.
42. Thibault O, Hadley R, Landfield PW. Elevated postsynaptic Ca^{2+} i and L-type calcium channel activity in aged hippocampal neurons: relationship to impaired synaptic plasticity. J Neurosci. 2001;21:9744–56.
43. Tanaka Y, Ando S. Age-related changes in the subtypes of voltage-dependent calcium channels in rat brain cortical synapses. Neurosci Res. 2001;39:213–20.
44. Palego L, Giromella A, Mazzoni MR, Marazziti D, Naccarato AG, Giannaccini G, et al. Gender and age-related variation in adenylyl cyclase activity in the human prefrontal cortex, hippocampus and dorsal raphe nuclei. Neurosci Lett. 2000;279:53–6.
45. Loo AT, Youngentob SL, Kent PF, Schwob JE. The aging olfactory epithelium: neurogenesis, response to damage, and odorant-induced activity. Int J Dev Neurosci. 1996;14:881–900.
46. Ohta Y, Ichimura K. Changes in epidermal growth factor receptors in olfactory epithelium associated with aging. Ann Otol Rhinol Laryngol. 2000;109:95–8.
47. Glezer I, Bruni-Cardoso A, Schechtman D, Malnic B. Viral infection and smell loss: the case of COVID-19. J Neurochem. 2020;157(4):930–43.
48. Kohli P, Naik AN, Harruff EE, Nguyen SA, Schlosser RJ, Soler ZM. The prevalence of olfactory dysfunction in chronic rhinosinusitis. Laryngoscope. 2017;127:309–20.
49. Litvack JR, Mace JC, Smith TL. Olfactory function and disease severity in chronic rhinosinusitis. Am J Rhinol Allergy. 2009;23:139–44.
50. Doty RL. Olfaction in Parkinson's disease. Parkinsonism Relat Disord. 2007;13:S225–8.
51. Averback P. Two new lesions in Alzheimer's disease. Lancet. 1983;2:1203.
52. Hyman BT, Arriagada PV, Van Hoesen GW. Pathologic changes in the olfactory system in aging and Alzheimer's disease. Ann N Y Acad Sci. 1991;640:14–9.
53. Dibattista M, Pifferi S, Menini A, Reisert J. Alzheimer's disease: what can we learn from the peripheral olfactory system? Front Neurosci. 2020;14:440.
54. Ansari KA, Johnson A. Olfactory function in patients with Parkinson's disease. J Chronic Dis. 1975;28:493–7.
55. Waldton S. Clinical observations of impaired cranial nerve function in senile dementia. Acta Psychiatr Scand. 1974;50:539–47.
56. Koss E, Weiffenbach JM, Haxby JV, Friedland RP. Olfactory detection and identification performance are dissociated in early Alzheimer's disease. Neurology. 1988;38:1228–32.
57. Morgan CD, Nordin S, Murphy C. Odor identification as an early marker for Alzheimer's disease: impact of lexical functioning and detection sensitivity. J Clin Exp Neuropsychol. 1995;17:793–803.
58. Murphy C, Gilmore MM, Seery CS, Salmon DP, Lasker BR. Olfactory thresholds are associated with degree of dementia in Alzheimer's disease. Neurobiol Aging. 1990;11:465–9.
59. Rezek DL. Olfactory deficits as a neurologic sign in dementia of the Alzheimer type. Arch Neurol. 1987;44:1030–2.
60. Korten JJ, Meulstee J. Olfactory disturbances in Parkinsonism. Clin Neurol Neurosurg. 1980;82:113–8.
61. Doty RL, Deems DA, Stellar S. Olfactory dysfunction in Parkinsonism: a general deficit unrelated to neurologic signs, disease stage, or disease duration. Neurology. 1988;38:1237–44.
62. Moberg PJ, Pearlson GD, Speedie LJ, Lipsey JR, Strauss ME, Folstein SE. Olfactory recognition: differential impairments in early and late Huntington's and Alzheimer's diseases. J Clin Exp Neuropsychol. 1987;9:650–64.
63. Razani J, Murphy C, Davidson TM, Grant I, McCutchan A. Odor sensitivity is impaired in HIV-positive cognitively impaired patients. Physiol Behav. 1996;59:877–81.
64. Hornung DE, Kurtz DB, Bradshaw CB, Seipel DM, Kent PF, Blair DC, et al. The olfactory loss that accompanies an HIV infection. Physiol Behav. 1998;64:549–56.
65. Ahlskog JE, Waring SC, Petersen RC, Esteban-Santillan C, Craig UK, O'Brien PC, et al. Olfactory dysfunction in Guamanian ALS, Parkinsonism, and dementia. Neurology. 1998;51:1672–7.

66. Brozzetti L, Sacchetto L, Cecchini MP, Avesani A, Perra D, Bongianni M, et al. Neurodegeneration-associated proteins in human olfactory neurons collected by nasal brushing. Front Neurosci. 2020;14:145.
67. Moore AB, Paulsen JS, Murphy C. A test of odor fluency in patients with Alzheimer's and Huntington's disease. J Clin Exp Neuropsychol. 1999;21:341–51.
68. Arnold SE, Smutzer GS, Trojanowski JQ, Moberg PJ. Cellular and molecular neuropathology of the olfactory epithelium and central olfactory pathways in Alzheimer's disease and schizophrenia. Ann N Y Acad Sci. 1998;855:762–75.
69. McCaffrey RJ, Duff K, Solomon GS. Olfactory dysfunction discriminates probable Alzheimer's dementia from major depression: a cross-validation and extension. J Neuropsychiatry Clin Neurosci. 2000;12:29–33.
70. Warner M, Peabody C, Flattery J, Tinklenberg J. Olfactory deficits and Alzheimer's disease. Biol Psychiatry. 1986;21:116–8.
71. Thesen T, Murphy C. Age-related changes in olfactory processing detected with olfactory event-related brain potentials using velopharyngeal closure and natural breathing. Int J Psychophysiol. 2001;40:119–27.
72. Mesholam RI, Moberg PJ, Mahr RN, Doty RL. Olfaction in neurodegenerative disease: a meta-analysis of olfactory functioning in Alzheimer's and Parkinson's diseases. Arch Neurol. 1998;55:84–90.
73. Talamo BR, Rudel RA, Kosik KS, Lee VMY, Neff S, Adelman L, et al. Pathological changes in olfactory neurons in patients with Alzheimer's disease. Nature. 1989;337:736–9.
74. Reyes PF, Deems DA, Suarez MG. Olfactory-related changes in Alzheimer's disease: a quantitative neuropathologic study. Brain Res Bull. 1993;32:1–5.
75. Yamagishi M, Ishizuka Y, Seki K. Pathology of olfactory mucosa in patients with Alzheimer's disease. Ann Otol Rhinol Laryngol. 1994;103:421–7.
76. Lee JH, Goedert M, Hill WD, Lee VMY, Trojanowski JQ. Tau proteins are abnormally expressed in olfactory epithelium of Alzheimer patients and developmentally regulated in human fetal spinal cord. Exp Neurol. 1993;121:93–105.
77. Crino PB, Martin JA, Hill WD, Greenberg B, Lee VMY, Trojanowski JQ. Beta-amyloid peptide and amyloid precursor proteins in olfactory mucosa of patients with Alzheimer's disease, Parkinson's disease, and down syndrome. Ann Otol Rhinol Laryngol. 1995;104:655–61.
78. Yamagishi M, Ishizuka Y, Seki K. Definitive diagnosis of Alzheimer's disease using olfactory mucosal biopsy. Nippon Jibiinkoka Gakkai Kaiho. 1994;97:51–60.
79. Nijjar RK, Murphy C. Olfactory impairment increases as a function of age in persons with down syndrome. Neurobiol Aging. 2002;23:65–73.
80. Davidson TM, Freed C, Healy MP, Murphy C. Rapid clinical evaluation of anosmia in children: the alcohol sniff test. Ann N Y Acad Sci. 1998;855:787–92.
81. Murphy C, Jinich S. Olfactory dysfunction in Down's syndrome. Neurobiol Aging. 1996;17:631–7.
82. Sobel N, Thomason ME, Stappen I, Tanner CM, Tetrud JW, Bower JM, et al. An impairment in sniffing contributes to the olfactory impairment in Parkinson's disease. Proc Natl Acad Sci USA. 2001;98:4154–9.
83. Adler CH, Gwinn KA, Newman S. Olfactory function in restless legs syndrome. Mov Disord. 1998;13:563–5.
84. Doty RL, Bromley SM, Stern MB. Olfactory testing as an aid in the diagnosis of Parkinson's disease: development of optimal discrimination criteria. Neurodegeneration. 1995;4:93–7.
85. Wszolek Z, Markopoulou K. Olfactory dysfunction in Parkinson's disease. Clin Neurosci. 1998;5:94–101.
86. Tissingh G, Berendse HW, Bergmans P, DeWaard R, Drukarch B, Stoof JC, et al. Loss of olfaction in de novo and treated Parkinson's disease: possible implications for early diagnosis. Mov Disord. 2001;16:41–6.
87. Fullard ME, Morley JF, Duda JE. Olfactory dysfunction as an early biomarker in Parkinson's disease. Neurosci Bull. 2017;33:515–25.
88. Hawkes CH, Shephard BC, Daniel SE. Is Parkinson's disease a primary olfactory disorder? QJM. 1999;92:473–80.
89. Doty RL, Golbe LI, Mckeown DA, Stern MB, Lehrach CM, Crawford D. Olfactory testing differentiates between progressive supranuclear palsy and idiopathic Parkinson's disease. Neurology. 1993;43:962–5.
90. Berendse HW, Booij J, Francot CMJE, Bergmans PLM, Hijman R, Stoof JC, et al. Subclinical dopaminergic dysfunction in asymptomatic Parkinson's disease patients' relatives with a decreased sense of smell. Ann Neurol. 2001;50:34–41.
91. Church AC, Bunney BS, Krieger NR. Neuronal localization of dopamine-sensitive adenylate cyclase within the rat olfactory tubercle. Brain Res. 1982;234:369–76.
92. Baker H, Farbman AI. Olfactory afferent regulation of the dopamine phenotype in the fetal rat olfactory system. Neuroscience. 1993;52:115–34.
93. Mijnster M, Isovich E, Flügge G, Fuchs E. Localization of dopamine receptors in the tree shrew brain using [3H]-SCH23390 and [125I]-epidepride. Brain Res. 1999;841:101–13.
94. Herve D, Levi-Strauss M, Marey-Semper I, Verney C, Tassin JP, Glowinski J, et al. G(olf) and G(s) in rat basal ganglia: possible involvement of G(olf) in the coupling of dopamine D1 receptor with adenylyl cyclase. J Neurosci. 1993;13:2237–48.
95. Berkowicz DA, Trombley PQ. Dopaminergic modulation at the olfactory nerve synapse. Brain Res. 2000;855:90–9.

96. Vargas G, Lucero MT. Dopamine modulates inwardly rectifying hyperpolarization-activated current (I(h)) in cultured rat olfactory receptor neurons. J Neurophysiol. 1999;81:149–58.
97. Coronas V, Féron F, Hen R, Sicard G, Jourdan F, Moyse E. In vitro induction of apoptosis or differentiation by dopamine in an immortalized olfactory neuronal cell line. J Neurochem. 1997;69:1870–81.
98. Koster N, Norman A, Richtand N, Nickell W, Puche A, Pixley S, et al. Olfactory receptor neurons express D2 dopamine receptors. J Comp Neurol. 1999;411:666–73.
99. Wilson D, Sullivan R. The D2 antagonist spiperone mimics the effects of olfactory deprivation on mitral/tufted cell odor response patterns. J Neurosci. 1995;15:5574–81.
100. Parati E, Bez A, Ponti D, Sala S, Pozzi S, Pagano S. Neural stem cells. Biological features and therapeutic potential in Parkinson's disease. J Neurosurg Sci. 2003;47:8–17.
101. Nordin S, Paulsen JS, Murphy C. Sensory- and memory-mediated olfactory dysfunction in Huntington's disease. J Int Neuropsychol Soc. 1995;1:281–90.
102. Menalled LB, Sison JD, Dragatsis I, Zeitlin S, Chesselet MF. Time course of early motor and neuropathological anomalies in a knock-in mouse model of Huntington's disease with 140 CAG repeats. J Comp Neurol. 2003;465:11–26.
103. Lucassen EB, Turel A, Knehans A, Huang X, Eslinger P. Olfactory dysfunction in multiple sclerosis: a scoping review of the literature. Mult Scler Relat Disord. 2016;6:1–9.
104. Zivadinov R, Zorzon M, Monti Bragadin L, Pagliaro G, Cazzato G. Olfactory loss in multiple sclerosis. J Neurol Sci. 1999;168:127–30.
105. Zorzon M, Ukmar M, Bragadin LM, Zanier F, Antonello RM, Cazzato G, et al. Olfactory dysfunction and extent of white matter abnormalities in multiple sclerosis: a clinical and MR study. Mult Scler. 2000;6:386–90.
106. Doty R, Li C, Mannon L, Yousem D. Olfactory dysfunction in multiple sclerosis. N Engl J Med. 1997;336:1918–9.
107. Doty RL, Li C, Mannon LJ, Yousem DM. Olfactory dysfunction in multiple sclerosis: relation to longitudinal changes in plaque numbers in central olfactory structures. Neurology. 1999;53:880–2.
108. Doty RL, Li C, Mannon LJ, Yousem DM. Olfactory dysfunction in multiple sclerosis: relation to plaque load in inferior frontal and temporal lobes. Ann N Y Acad Sci. 1998;855:781–6.
109. Zanusso G, Ferrari S, Cardone F, Zampieri P, Gelati M, Fiorini M, et al. Detection of pathologic prion protein in the olfactory epithelium in sporadic Creutzfeldt–Jakob disease. N Engl J Med. 2003;348:711–9.
110. Reuber M, Al-Din ASN, Baborie A, Chakrabarty A. New variant Creutzfeldt–Jakob disease presenting with loss of taste and smell. J Neurol Neurosurg Psychiatry. 2001;71:412–3.
111. Perlman S, Jacobsen G, Afifi A. Spread of a neurotropic murine coronavirus into the CNS via the trigeminal and olfactory nerves. Virology. 1989;170:556–60.
112. Shankar V, Kao M, Hamir AN, Sheng H, Koprowski H, Dietzschold B. Kinetics of virus spread and changes in levels of several cytokine mRNAs in the brain after intranasal infection of rats with Borna disease virus. J Virol. 1992;66:992–8.
113. Babic N, Mettenleiter TC, Ugolini G, Flamand A, Coulon P. Propagation of Pseudorabies virus in the nervous system of the mouse after intranasal inoculation. Virology. 1994;204:616–25.
114. Allavena RE, Desai B, Goodwin D, Khodai T, Bright H. Pathologic and virologic characterization of neuroinvasion by HSV-2 in a mouse encephalitis model. J Neuropathol Exp Neurol. 2011;70:724–34.
115. Barnett EM, Cassell MD, Perlman S. Two neurotropic viruses, herpes simplex virus type 1 and mouse hepatitis virus, spread along different neural pathways from the main olfactory bulb. Neuroscience. 1993;57:1007–25.
116. Ozdener M, Donadoni M, Cicalese S, Speilman A, Garcia-Blanco A, Gordon J, et al. Zika virus infection in chemosensory cells. J Neurovirol. 2020;26:371–81.
117. St-Jean JR, Jacomy H, Desforges M, Vabret A, Freymuth F, Talbot PJ. Human respiratory coronavirus OC43: genetic stability and neuroinvasion. J Virol. 2004;78:8824–34.
118. Lemiale F, Kong W, Akyürek LM, Ling X, Huang Y, Chakrabarti BK, et al. Enhanced mucosal immunoglobulin A response of intranasal adenoviral vector human immunodeficiency virus vaccine and localization in the central nervous system. J Virol. 2003;77:10078–87.
119. Lundh B, Kristensson K, Norrby E. Selective infections of olfactory and respiratory epithelium by vesicular stomatitis and Sendai viruses. Neuropathol Appl Neurobiol. 1987;13:111–22.
120. Suzuki M, Saito K, Min WP, Vladau C, Toida K, Itoh H, et al. Identification of viruses in patients with postviral olfactory dysfunction. Laryngoscope. 2007;117:272–7.
121. Wilkie FL, Goodkin K, Khamis I, Van Zuilen MH, Lee D, Lecusay R, et al. Cognitive functioning in younger and older HIV-1-infected adults. J Acquir Immune Defic Syndr. 2003;33:S93–105.
122. Brody D, Adler L, Kim T, Angrist B, Rotrosen J. Effects of buspirone in seven schizophrenic subjects. J Clin Psychopharmacol. 1990;10:68–9.
123. Serby M, Larson P, Kalkstein D. Olfaction and neuropsychiatry. In: Serby M, Chobor K, editors. Science of olfaction. New York: Springer; 1992. p. 559–84.
124. Heald AE, Schiffman SS. Taste and smell. Neglected senses that contribute to the malnutrition of AIDS. N C Med J. 1997;58:100–4.
125. Heald A, Pieper C, Schiffman S. Taste and smell complaints in HIV-infected patients. AIDS. 1998;12:1667–74.

126. Graham CS, Graham BG, Bartlett JA, Heald AE, Schiffman SS. Taste and smell losses in HIV infected patients. Physiol Behav. 1995;58:287–93.
127. Passarelli PC, Lopez MA, Bonaviri GNM, Garcia-Godoy F, DâAddona A. Taste and smell as chemosensory dysfunctions in COVID-19 infection. Am J Dent. 2020;33:135–7.
128. Wong D, Gendeh H, Thong H, Lum S, Gendeh B, Saim A, et al. A review of smell and taste dysfunction in COVID-19 patients. Med J Malays. 2020;75:574–81.
129. Ibekwe T, Fasunla A, Orimadegun A. Systematic review and meta-analysis of smell and taste disorders in COVID-19. OTO Open. 2020;4:2473974X20957975.
130. World Health Organization. Coronavirus. https://www.who.int/health-topics/coronavirus#tab=tab_3. Accessed 1 Dec 2020.
131. Doty RL. Treatments for smell and taste disorders: a critical review. Handb Clin Neurol. 2019;164:455–79.
132. Sugiura M, Aiba T, Mori J, Nakai Y. An epidemiological study of postviral olfactory disorder. Acta Otolaryngol Suppl. 1998;538:191–6.
133. Alexander TH, Davidson TM. Intranasal zinc and anosmia: the zinc-induced anosmia syndrome. Laryngoscope. 2006;116:217–20.
134. Jafek BW, Linschoten MR, Murrow BW. Anosmia after intranasal zinc gluconate use. Am J Rhinol. 2004;18:137–41.
135. Lötsch J, Knothe C, Lippmann C, Ultsch A, Hummel T, Walter C. Olfactory drug effects approached from human-derived data. Drug Discov Today. 2015;20:1398–406.
136. Mizera L, Gossrau G, Hummel T, Haehner A. Effects of analgesics on olfactory function and the perception of intranasal trigeminal stimuli. Eur J Pain. 2017;21:92–100.
137. Berman JL. Dysosmia, dysgeusia, and diltiazem. Ann Intern Med. 1985;102:717.
138. Levenson J, Kennedy K. Dysosmia, dysgeusia, and nifedipine. Ann Intern Med. 1985;102:135–6.
139. Doty RL, Bromley SM. Effects of drugs on olfaction and taste. Otolaryngol Clin N Am. 2004;37:1229–54.
140. Atkin PA, Veitch PC, Veitch EM, Ogle SJ. The epidemiology of serious adverse drug reactions among the elderly. Drugs Aging. 1999;14:141–52.
141. Atkin P, Shenfield G. Medication-related adverse reactions and the elderly: a literature review. Advers Drug React Toxicol Rev. 1995;14:175–91.
142. Charlesworth CJ, Smit E, Lee DSH, Alramadhan F, Odden MC. Polypharmacy among adults aged 65 years and older in the United States: 1988–2010. J Gerontol A Biol Sci Med Sci. 2015;70:989–95.
143. National Institute of Diabetes and Digestive and Kidney Diseases. Adverse drug reaction probability scale (Naranjo) in drug induced liver injury. Bethesda: National Institute of Diabetes and Digestive and Kidney Diseases; 2019.
144. Rawson NE, Gomez G. Cell and molecular biology of human olfaction. Microsc Res Tech. 2002;58:142–51.
145. Amoore JE. Evidence for the chemical olfactory code in man. Ann N Y Acad Sci. 1974;237:137–43.
146. Menashe I, Man O, Lancet D, Gilad Y. Different noses for different people. Nat Genet. 2003;34:143–4.
147. Wysocki CJ, Dorries KM, Beachamp GK. Ability to perceive androstenone can be acquired by ostensibly anosmic people. Proc Natl Acad Sci USA. 1989;86:7976–8.
148. Beauchamp GK, Yamazaki K. Chemical signalling in mice. Biochem Soc Trans. 2003;31:147–51.
149. Murphy C. Loss of olfactory function in dementing disease. Physiol Behav. 1999;66:177–82.
150. Tuccori M, Lapi F, Testi A, Ruggiero E, Moretti U, Vannacci A, et al. Drug-induced taste and smell alterations: case/non-case evaluation of an Italian database of spontaneous adverse drug reaction reporting. Drug Saf. 2011;34:849–59.
151. Wang J-J, Chen J, Doty RL. Impact of antibiotics on smell dysfunction. World J Otorhinolaryngol Head Neck Surg. 2018;4:33–8.
152. Mayet AY. Loss of smell (anosmia) and taste (ageusia) in a patient treated with pegylated interferon alfa and ribavirin. Curr Ther Res. 2007;68:271–7.
153. Zargari O. Methotrexate, hyperosmia, and migraine. Dermatol Online J. 2006;12:28.
154. Baskoy K, Ay SA, Altundag A, Kurt O, Salihoglu M, Deniz F, et al. Is there any effect on smell and taste functions with levothyroxine treatment in subclinical hypothyroidism? PLoS One. 2016;11:e0149979.
155. Yoshida K, Fukuchi T, Sugawara H. Dysosmia and dysgeusia associated with duloxetine. BMJ Case Rep. 2017;2017:bcr2017222470.
156. Doty RL, Philip S, Reddy K, Kerr KL. Influences of antihypertensive and antihyperlipidemic drugs on the senses of taste and smell: a review. J Hypertens. 2003;21:1805–13.

Electron Microscopy and the Nose

32

Mürvet Hayran

Core Messages

- Electron microscopy is a very important tool for evaluating the ultrastructural features of the nose and helping diagnose diseases related to nasal structures.
- Biopsies should be small (less than 2 mm) and fixed in seconds to prevent autolysis and to obtain optimum diffusion of the fixatives.
- Cilia are hairlike extensions of the apical plasma membrane containing microtubules. The basal body is a microtubule-organizing center located in the apical region of the ciliated cell. The existence of the characteristic "9 + 2" organization of the axonemes of the cilia and the presence of the basal bodies are important for the normal function of the cilia.
- The goblet cell is common in the airway epithelium. The parasympathetic nervous system does not control release from goblet cells. Rather, these cells respond to physical and chemical irritants; however, mediators have not yet been clearly identified.
- Basal cells are stem cells from which other cell types arise. They lie on the basement membrane and do not reach the lumen.
- The olfactory segment is the region at the roof of the nasal cavity. In humans, the olfactory region is a small area formed by a modified pseudostratified epithelium. The olfactory epithelium is composed of olfactory receptor cells, supporting or sustentacular cells, basal cells, and brush cells. In contrast with the other regions of the nasal cavity, there are no goblet cells in this area.

32.1 Electron Microscopy and the Nose

32.1.1 The Electron Microscope

The electron microscope is a type of microscope that uses an *electron beam* accelerated under high vacuum instead of light source to create an image of the specimen. In studying the microscopic anatomy of the nose, it is essential to understand the requirements and capabilities of the electron microscope and which tissues and cells should be clearly observed by electron microscopy.

The electron microscope uses an accelerated electron beam, emitted by a cathode and controlled by a series of electrostatic and electromagnetic lenses [1, 2]. Components of the electron microscope are (Fig. 32.1):

1. *Electron optical column.*
2. *Electron gun* that consists of an electron source to produce electrons, such as a tungsten filament.
3. *Magnetic lenses* to demagnify the beam.

M. Hayran (✉)
Department of Anatomy, Ekonomi University Faculty of Medicine, İzmir, Turkey
e-mail: mtuncel@hacettepe.edu.tr

Ö. Ö. Celebi, T. M. Önerci (eds.), *Nasal Physiology and Pathophysiology of Nasal Disorders*,
https://doi.org/10.1007/978-3-031-12386-3_32

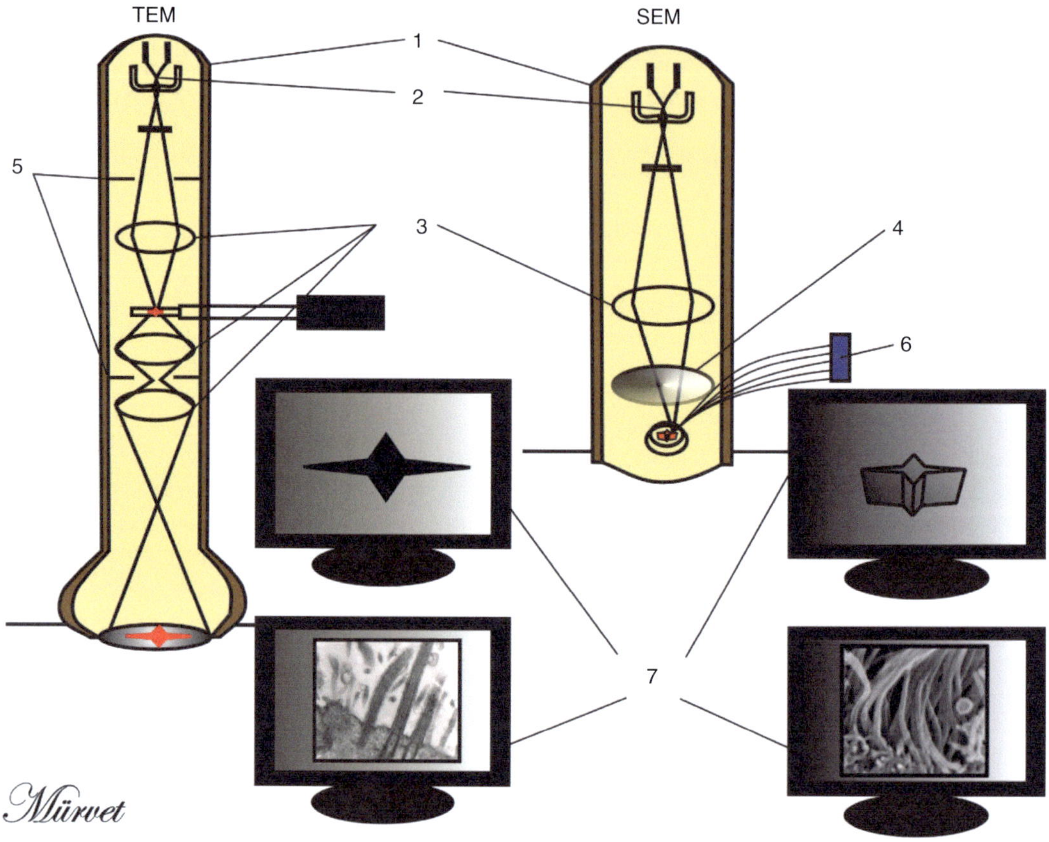

Fig. 32.1 Diagram of standard transmission electron microscopy (*TEM*) and scanning electron microscopy (*SEM*), showing major components: *1* electron optical column, *2* electron gun, *3* magnetic lenses, *4* magnetic coil, *5* apertures, *6* detector, and *7* digital imaging systems

4. *Magnetic coils* to control and modify the beam.
5. *Apertures* to define the beam and prevent electron spray.
6. *Detectors* to collect, detect, and display the signal.
7. *Digital imaging systems* that produce an image from the signal.

There are also vacuum systems consisting of vacuum pumps and a vacuum chamber.

There are two basic types of electron microscopy, *transmission electron microscopy* (TEM) and *scanning electron microscopy* (SEM).

TEM uses an electron beam that transmits through ultrathin (60–90 nm) sections that are glutaraldehyde-fixed and usually double-stained. TEM produces two-dimensional images on a *fluorescent screen*, photographic film, or CCD (charge-coupled device) camera. TEM can detect structures by the transmission of the electron beam and discriminate details of 0.2 nm. However, the quality of the obtained image mainly depends on the preparation of the biological sample [1].

SEM obtains topographic, three-dimensional images with a resolution of about 2 nm. The lens system of SEM produces a small focused spot of electrons that are then scanned over the specimen surface by a *deflection coil*. SEM is able to produce an image by detecting secondary electrons and backscattered electrons generated from the specimen. A *secondary electron detector* in the SEM builds the image by mapping the signals of a finely focused electron beam that is scanned on the sample surface [1, 3, 4].

Both TEM and SEM can provide only black-and-white images. Although the original image is monochrome, micrographs can be colored digitally to emphasize details. In recent years, digital imaging systems provide incredible opportunities to obtain high-quality electron micrographs.

Biological materials usually require processing before being viewed by electron microscopy [1, 3]. The tissue preparation technique varies depending on the type of microscope and the type of specimen. Low-vacuum SEMs and the environmental scanning electron microscope (ESEM) overcome both these limitations [5].

The stages for tissue preparation for TEM are as follows

(a) *Fixation*: can be achieved by perfusion and microinjection or immersion using various fixatives including aldehydes.
(b) *Post fixation*: performed in OsO_4.
(c) *Dehydration*: done with a graded series of alcohol.
(d) *Epoxy resin block preparation*: treat with propylene and embed in epoxy resin.
(e) *Semi-thin sectioning*: one- to two-micrometer-thick semi-thin sections obtained from the epoxy resin blocks should be stained with methylene blue or azure for light microscopy.
(f) *Ultrathin sectioning*: ultramicrotome, an instrument for cutting extremely thin sections, is used for ultrathin sectioning of tissue. Ultrathin sections are obtained from selected areas and then double-stained with uranyl acetate/lead citrate.

During SEM sample preparation, after the fixation step, the specimens must be dried. Electron microscopists prefer to use *Critical Point Drying*. By removing carbon dioxide after the transition from the liquid to the gas phase at the critical point, the specimen can be dried without structural damage. Specimens must be mounted onto a holder that can be inserted into the scanning electron microscope. The last step prior to sample imaging is coating the samples. The objective of this coating is to increase its conductivity in the scanning electron microscope and to prevent the buildup of high-voltage charges on the specimen. Typically, specimens are coated with a thin layer of gold, gold-palladium, or platinum [3].

In addition to visual inspection, a grading system can be used to quantitatively evaluate the samples to compare different experimental conditions. The data can then be analyzed statistically to make further evaluations. This grading system was established based on similar principles of methods used for evaluating different tissue samples [6–10].

It is important to pay close attention during the sample preparation to get small-sized (less than 2 mm) biopsy materials and to fix them in seconds to prevent autolysis and obtain optimum diffusion of the fixatives [3].

32.1.2 Microscopic Anatomy of the Nose

The nose humidifies, filters, and warms the air we breathe as well as provides the sense of olfaction. The nose is considered to have two parts: *the external nose* and *the nasal cavity*.

The *external nose* consists of the skin and a framework of compact bone and hyaline cartilage that forms a projection covered by skin. Electron microscopic observation of this part does not show any regional specifications. The skin consists of two main layers. The outer layer is the *epidermis*, which is composed of a keratinized and stratified squamous epithelium (Figs. 32.2 and 32.3) (see Sect. 32.1.2.1). The inner layer is the *dermis*, which is dense connective tissue including epithelial derivatives of the skin such as hair follicles and sweat and sebaceous glands [11]. The supporting framework is composed of nasal bones, the frontal process of the maxillae, and the nasal part of the frontal bone and septum, as well as major and minor alar cartilages. Bone is also a connective tissue characterized by a mineralized extracellular matrix containing mainly type I collagen along with other non-collagenous matrix proteins [1]. Bones of the external nose consist of layers of relatively thick compact bone with a layer of spongy bone covered by periosteum, which is a sheath of dense fibrous connective tissue containing *osteoprogenitor cells*.

The type of cartilage that contributes to the framework of the nose is *hyaline cartilage*

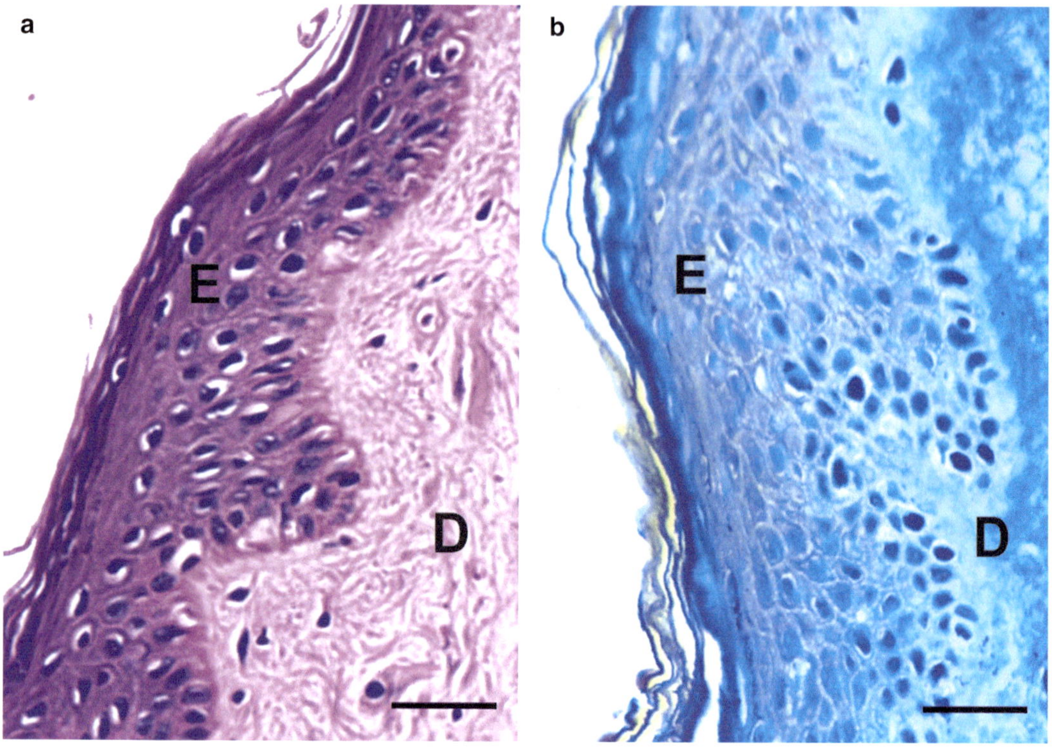

Fig. 32.2 The epidermis, which is composed of a keratinized stratified squamous epithelium (E), and the dermis (D) (scale bar: 25 μm). (**a**) Light micrograph of the epidermis of the external nose (paraffin block, stain: H&E). (**b**) Light micrograph of the epidermis of the external nose (araldite block, stain: methylene blue). The specimens were obtained from a fresh frozen cadaver from a microscopic anatomy lab at Hacettepe University, Faculty of Medicine, Department of Anatomy. *H&E* hematoxylin and eosin

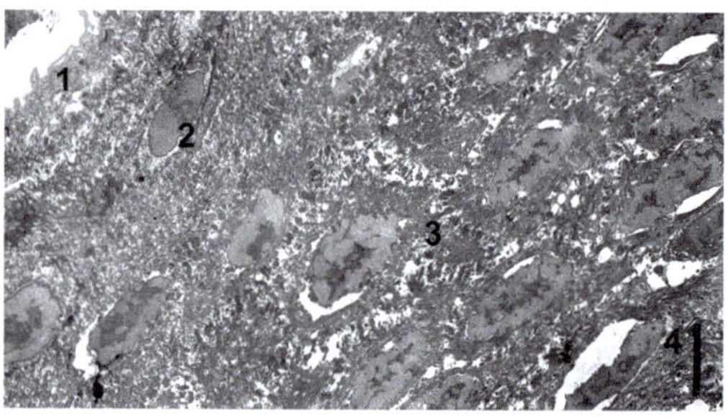

Fig. 32.3 Electron micrograph (TEM) keratinized stratified squamous epithelium of the epidermis of the external nose (scale bar: 5 μm) (Araldite block, stain: uranyl acetate/lead citrate). The specimen was obtained from a fresh frozen cadaver from a microscopic anatomy lab at Hacettepe University, Faculty of Medicine, Department of Anatomy. TEM transmission electron microscopy, *1* stratum corneum, *2* stratum spinosum, *3* stratum granulosum, and *4* stratum basale

(Fig. 32.4). The matrix of the hyaline cartilage consists of collagen, predominantly type II fibrils and other cartilage-specific collagen molecules [1]. The chondrocytes are either rounded or ellipsoidal (Figs. 32.4 and 32.5). The plasma membrane is folded into a moderate number of microvilli. Numerous cytoplasmic filaments and coarse granules of glycogen are prominently present in the cytoplasm. The *Golgi complex* is also prominent and frequently contains dilated vesicles enclosing small dense particles. A small amount of rough endoplasmic reticulum is present, while unattached ribosomes are not numerous. There are a few lipid droplets in the cytoplasm. The nuclei are ovoid and usually contain a single large nucleolus (Fig. 32.5). Mitochondria are small and not very numerous.

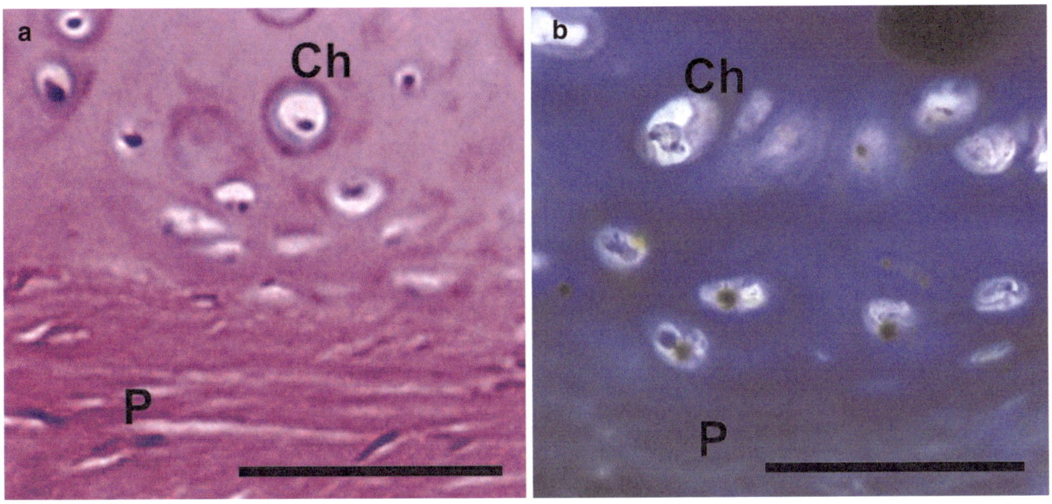

Fig. 32.4 The hyaline cartilage, rounded or ellipsoidal chondrocytes (Ch), and fibroblast-like cells of the perichondrium (P). (**a**) Light micrograph (paraffin block, stain: H&E) (scale bar: 100 μm). (**b**) Light micrograph (Araldite block, stain: methylene blue) (scale bar: 100 μm). *H&E* hematoxylin and eosin

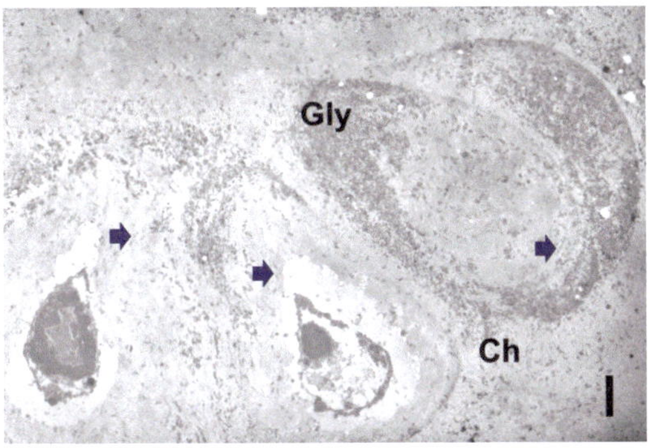

Fig. 32.5 Electron micrograph (TEM) of the hyaline cartilage, rounded or ellipsoidal chondrocytes (*Ch*) (scale bar: 2 μm) (Araldite block, stain: uranyl acetate/lead citrate). The specimen was obtained from a fresh frozen cadaver at a microscopic anatomy lab at Hacettepe University, Faculty of Medicine, Department of Anatomy. *TEM* transmission electron microscopy, *Arrows* cytoplasmic filaments, *Gly* coarse granules of glycogen

The matrix is composed mostly of collagen fibrils and matrix granules. Frequently, granules appear to be linked together by extremely fine intergranular fibrils, usually less than 50 Å thick, which connect the projections of adjacent granules. Infrequently, clusters of membrane-bounded matrix vesicles are observed between collagen fibrils of the matrix [12].

The *perichondrium*, a firmly attached dense connective tissue composed of fibroblast-like cells, surrounds the hyaline cartilage (Fig. 32.4).

The *nasal cavity* is divided into paired chambers separated by a bony and cartilaginous septum. Each chamber is divided into three regions:

(a) Vestibule of the nasal cavity.
(b) Respiratory region.
(c) Olfactory region.

32.1.2.1 Vestibule of the Nasal Cavity

The nasal cavity extends from the *nares* anteriorly to the choanae posteriorly. Just behind the nares, the nasal cavity widens and forms the vestibule [13]. It is lined with keratinized stratified squamous epithelium and the dermis (Fig. 32.6) that contains connective tissue elements, many hair follicles (hairs in this region are called *vibrissae*) (Figs. 32.7 and 32.8), and sebaceous glands and sweat glands (Figs. 32.9 and 32.10) (see Sect. 32.1.2).

The stratified squamous epithelium consists of several layers of cells. The flattened cells form its outer layer and the deepest cells are columnar. The epidermis is composed of four distinct layers. These are the stratum corneum, stratum spinosum, stratum granulosum, and stratum basale

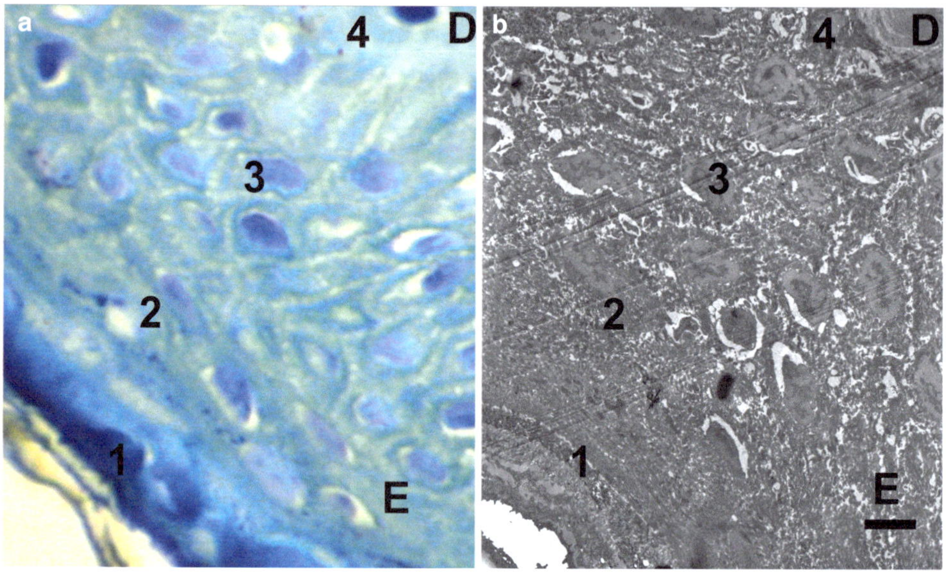

Fig. 32.6 (**a**) Light micrograph of the epidermis of the vestibule (scale bar: 5 μm) (Araldite block, stain: methylene blue). (**b**) Electron micrograph (TEM) of the epidermis of the vestibule (scale bar: 5 μm) (Araldite block, stain: uranyl acetate/lead citrate). Both specimens were obtained from fresh frozen cadavers at a microscopic anatomy lab at Hacettepe University, Faculty of Medicine, Department of Anatomy. *TEM* transmission electron microscopy, *E* epidermis, *D* dermis, *1* stratum corneum, *2* stratum spinosum, *3* stratum granulosum and *4* stratum basale

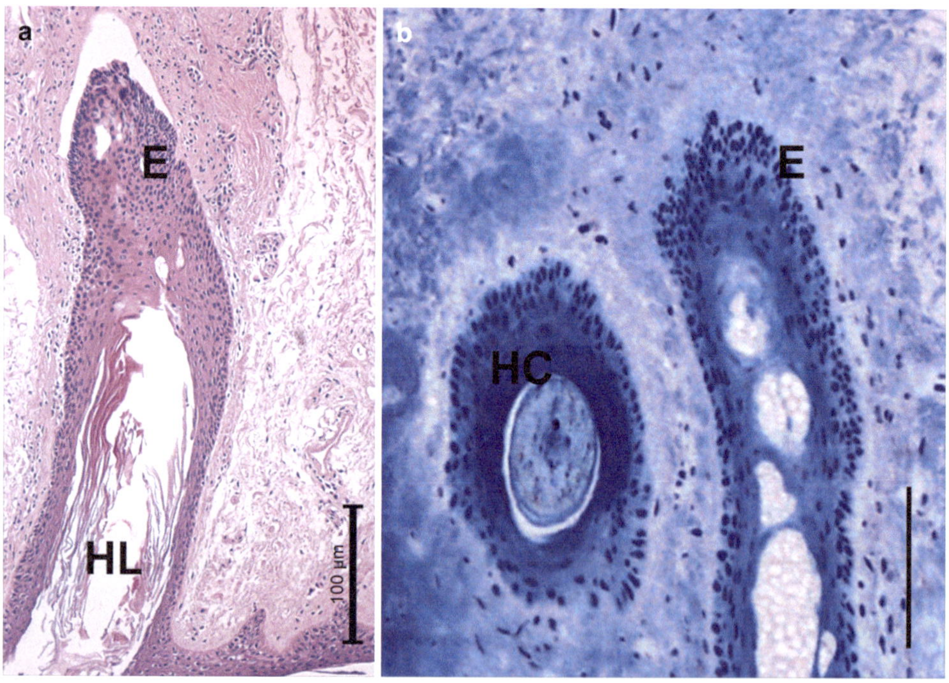

Fig. 32.7 Light micrograph from the dermis of the vestibule (scale bar: 100 μm). (**a**) Hair follicle and vibrissae, longitudinal section (paraffin block, stain: H&E). (**b**) Hair follicle and vibrissae, cross (*HC*) and longitudinal section (*HL*) (Araldite block, stain: methylene blue) (scale bar: 100 μm). The specimens were obtained from fresh frozen cadavers at a microscopic anatomy lab at Hacettepe University, Faculty of Medicine, Department of Anatomy. *H&E* hematoxylin and eosin, *E* epithelial cells

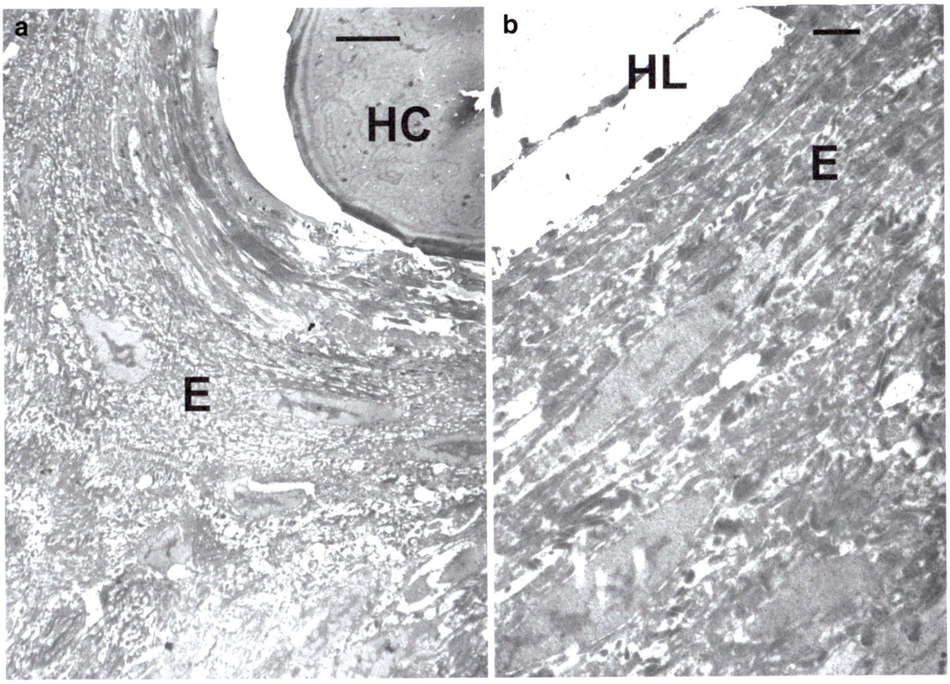

Fig. 32.8 Electron micrographs (TEM) from the dermis of the vestibule. (**a**) Cross and (**b**) longitudinal sections of the vibrissae (scale bar: 5 μm) (Araldite block, stain: uranyl acetate/lead citrate). The specimen was obtained from a fresh frozen cadaver at a microscopic anatomy lab at Hacettepe University, Faculty of Medicine, Department of Anatomy. *TEM* transmission electron microscopy, *E* epithelial cells

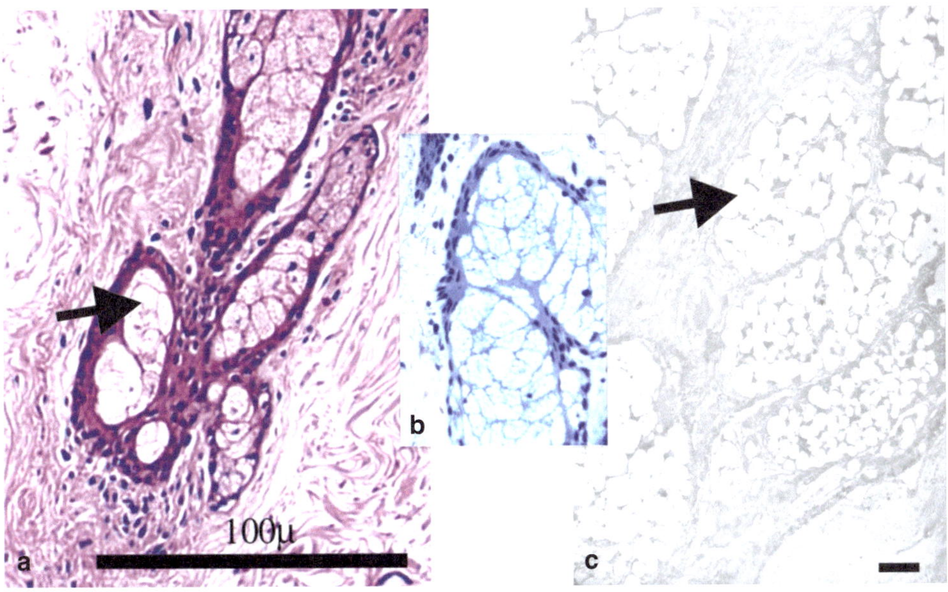

Fig. 32.9 The sebaceous glands (arrows) from the dermis of the vestibule (scale bar: 5 μm). (**a**) Light micrograph (paraffin block, stain: H&E). (**b**) Light micrograph (Araldite block, stain: methylene blue). (**c**) Electron micrographs (TEM) (Araldite block, stain: uranyl acetate/lead citrate). *H&E* hematoxylin and eosin, *TEM* transmission electron microscopy

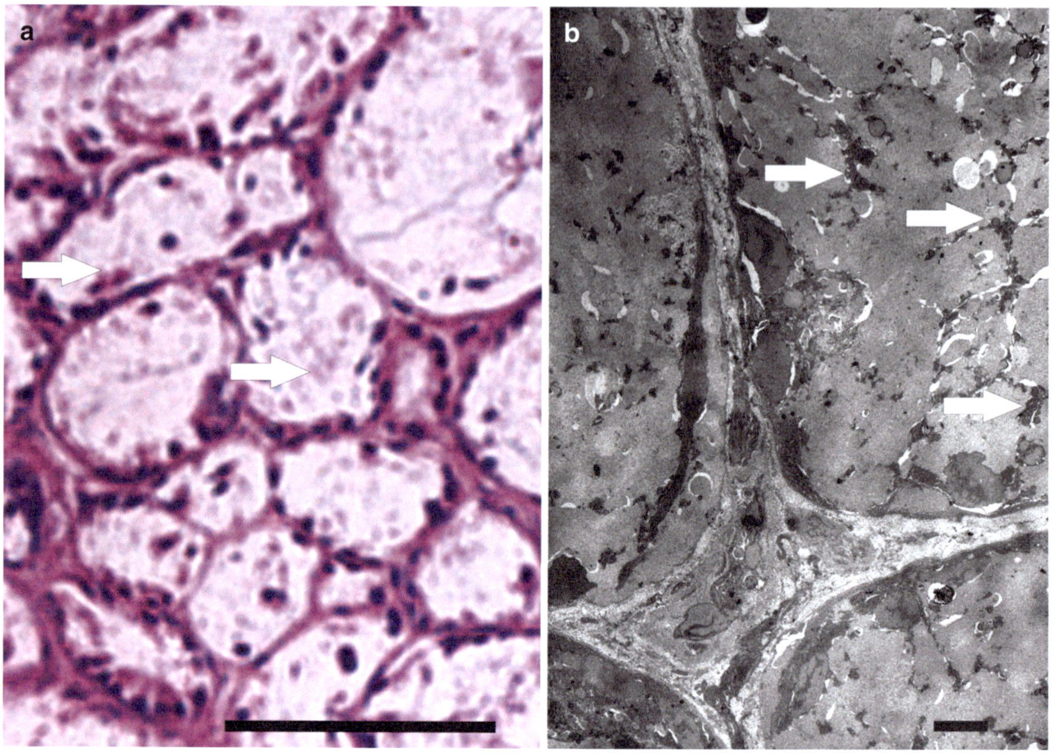

Fig. 32.10 The sweat glands and their myoepithelial cells (arrows) from the dermis of the vestibule. (**a**) Light micrograph (paraffin block, stain: H&E) (scale bar: 5 μm). (**b**) Electron micrographs (TEM) (Araldite block, stain: uranyl acetate/lead citrate) (scale bar: 100 μm). *H&E* hematoxylin and eosin, *TEM* transmission electron microscopy

(stratum germinativum) (Fig. 32.6). The cells of the stratum corneum are anucleate corneal cells (squamous), called corneocytes or cornified cells. The corneocytes are flattened cells that lack nuclei and cytoplasmic organelles. The cells contain aggregated keratin filaments. The upper spinous layer and granular cell layer also contain smaller lamellate granules called lamellar, membrane-coating granules (MCGs or Odland bodies). These are numerous within the upper spinous layer. They play an important role in providing the barrier and intercellular cohesion functions of the stratum corneum. They release their lipid components into the intercellular space. The basal and spinous cells together are called *the Malpighian layer*, which includes cells such as melanocytes, *Langerhans cells,* and *Merkel cells*. When outer cells become damaged, cell division occurs within the basal layer. The cells move outwards to the stratum corneum, passing through the stratum spinosum. The characteristics of these cells then transdifferentiate to become the cells of the stratum corneum. There are biochemical and signaling interactions between the epithelial cells, including desmosomes, adherens junctions, gap junctions, and tight junctions [14].

Posteriorly, where the vestibule ends, the stratified squamous epithelium becomes thinner and undergoes a transition to the pseudostratified epithelium that characterizes the respiratory region. At this site the sebaceous glands end [15]. At the level of the limen nasi, the lining of the nasal cavity gradually changes from squamous epithelium to non-ciliated cuboidal or columnar epithelium. At the level of the *inferior turbinate*, the epithelium continues as *pseudostratified ciliated columnar epithelium* [16].

32.1.2.2 Respiratory Region of the Nasal Cavity

The mucosa of the respiratory region warms, moistens, and filters inspired air. The lamina propria of the respiratory region has a rich, vascular network that includes a complex set of capillary loops. The nasal mucosa microvasculature is composed of arterioles, venules, capillaries, and cavernous sinuses. Both arterioles and venules run parallel to the long axis of the nasal concha. The capillaries and cavernous sinuses are particularly abundant and interconnect with numerous short anastomoses to form a rich dense network [17]. The arrangement of the vessels allows the inhaled air to be warmed by blood flow through the part of the capillaries closest to the surface. These capillaries provide a mechanical heat exchange system. Submucosal capillaries and venules have fenestrated endothelial linings and relatively porous basement membranes, facilitating the transit of fluid and white blood cells to the mucosal surface. The lamina propria then becomes distended with fluid, resulting in a marked swelling of the mucous membrane and consequent obstruction of the air passage. This makes breathing difficult. The mucosa contains large venous-like spaces known as *swell bodies*, which may become congested during allergic reactions or infections. The morphological view of fenestrated endothelia might change in response to alterations in the physiological conditions. It is essential to define the different shapes and courses of the muscle cells responsible for constriction and dilatation of nasal swell bodies for proper clinical diagnoses [18].

The nasal vasculature is controlled by dense innervations. Myelinated nerve bundles and small axons are found in the arterial wall located in the adventitia. Veins also have nerve structures, but they are fewer and are found in the muscle layer. Therefore, no axons are present in capillaries. The differences in the density of axons indicate that these vessels are controlled by neural structures and play an important role in the swelling of the nasal mucosa [19].

Seromucous glands are one of the main components of the human nasal mucosa. Their secretion contributes to the moistening function of the goblet cells in the respiratory epithelium. The terminal segments of the glands are surrounded by contractile myoepithelial cells in a basketlike fashion. These cells, in particular, show a high number of mitochondria [20]. This innervation pattern is important in understanding the control of different physiological glandular functions. Unmyelinated nerve fibers have typical neuronal components such as neurofilaments, neurotubules, and mitochondria in their cytoplasm [21].

Inhaled agents contact the nasal mucosa and cause a local immune response. Because of the nature of these local immune responses, nasal mucosal antibody production is best achieved via direct stimulation of IgA-committed, nasal-associated lymphoid tissue-derived B cells [22].

The respiratory region constitutes most of the volume of the nasal cavities. The medial wall of the respiratory region, the *nasal septum*, is smooth, but the lateral walls contain three shelf-like, bony projections called turbinates or conchae. The turbinates increase surface area to more efficiently warm inspired air. This air is also filtered by the mucus-covered walls of the nasal cavity. Particles trapped in this layer of mucus are transported to the pharynx by means of coordinated sweeping movements of cilia and are subsequently swallowed. Therefore, these motile cilia play a critical role in mucociliary clearance. This segment is lined by a ciliated, pseudostratified columnar epithelium. The pseudostratified respiratory epithelium actually consists of one layer of cells, but their nuclei frequently lie at different levels, and some cells do not reach the epithelial surface. Hence, the epithelium looks stratified even if all the cells rest on a basement membrane located between the epithelial cells and the loose lamina propria. Basal cells, situated close to the basement membrane, replace the ciliated cells or the goblet cells when needed [1, 23]. The lamina propria is attached to the *periosteum* of the adjacent bone. The submucosa contains blood vessels, venous plexus, glandular elements, sensory nerves, and immune system cells.

The ciliated, pseudostratified columnar epithelium of the respiratory region is composed of five cell types: ciliated columnar cells, non-ciliated columnar cells, goblet cells, basal cells,

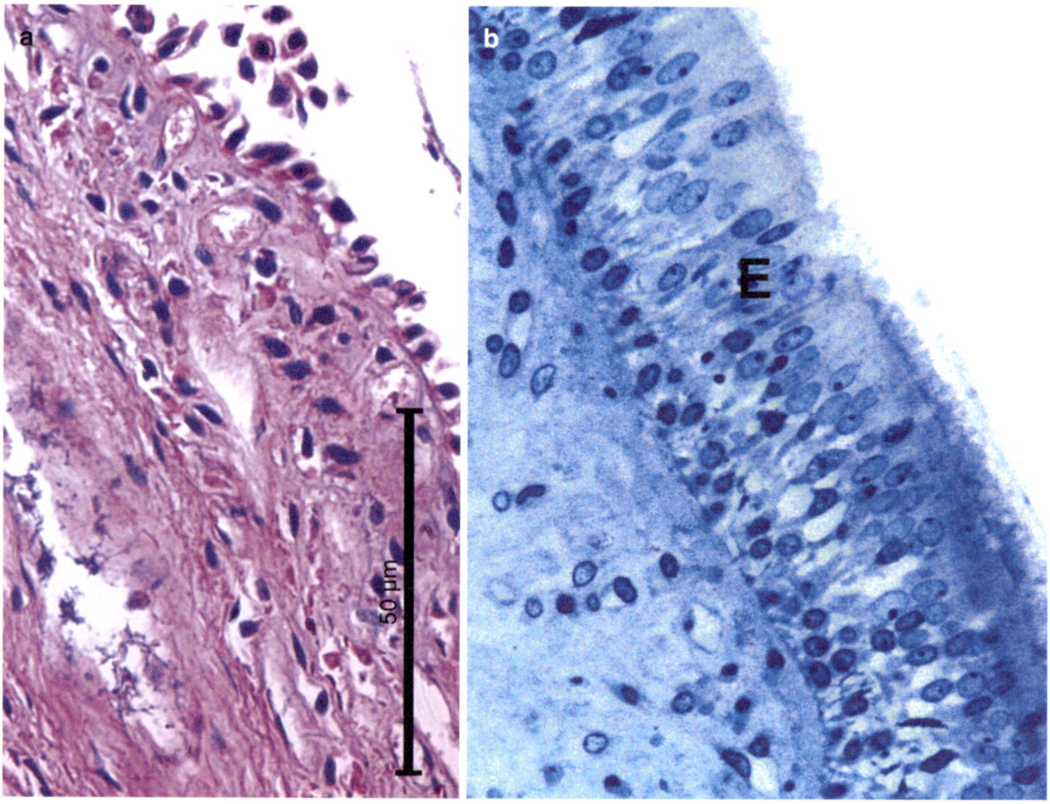

Fig. 32.11 The pseudostratified columnar epithelium (*E*) of the respiratory region. (**a**) Light micrograph of the epidermis of the external nose (paraffin block, stain: H&E) and (**b**) light micrograph of the epidermis of the external nose (Araldite block, stain: methylene blue) (scale bar: 50 μm). The specimens were obtained from fresh frozen cadavers at a microscopic anatomy lab located at Hacettepe University, Faculty of Medicine, Department of Anatomy. *H&E* hematoxylin and eosin

and brush cells (Fig. 32.11). All of these cells can be investigated under TEM (Fig. 32.12). However, because some of these cells do not reach the surface of the epithelium, only three types (ciliated, non-ciliated, and goblet cells) can be seen during SEM studies (Fig. 32.13).

The anterior one-third of the nasal cavity is non-ciliated. Cilia begin just behind the front edge of the inferior turbinate. The posterior part of the nasal cavity, as well as the paranasal sinuses, is densely covered by cilia [24]. However, the epithelium on the tips of the nasal turbinates is cuboidal and sparsely ciliated. Additionally, mild squamous metaplasia may be seen around these areas [16].

Ciliated columnar cells, the predominant cell type at the surface, rest on the basement membrane and project both *cilia* and *microvilli* from

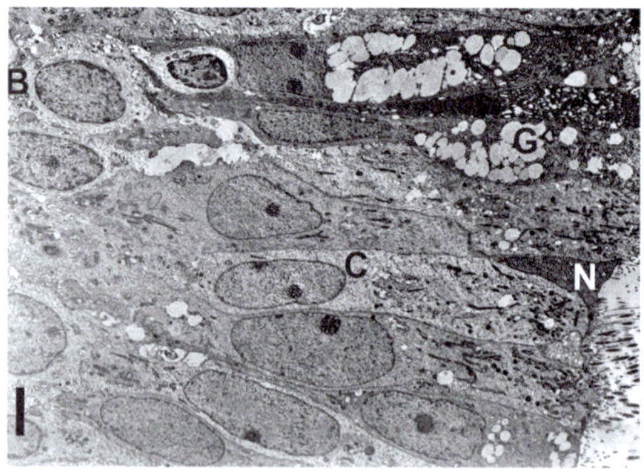

Fig. 32.12 Electron micrograph respiratory epithelium (TEM) (scale bar: 5 μm) (Araldite block, stain: uranyl acetate/lead citrate). The specimen was obtained from a fresh frozen cadaver at a microscopic anatomy lab at Hacettepe University, Faculty of Medicine, Department of Anatomy. *TEM* transmission electron microscopy. *B* basal cell, *C* ciliated cell, *G* goblet cell, *N* non-ciliated cell

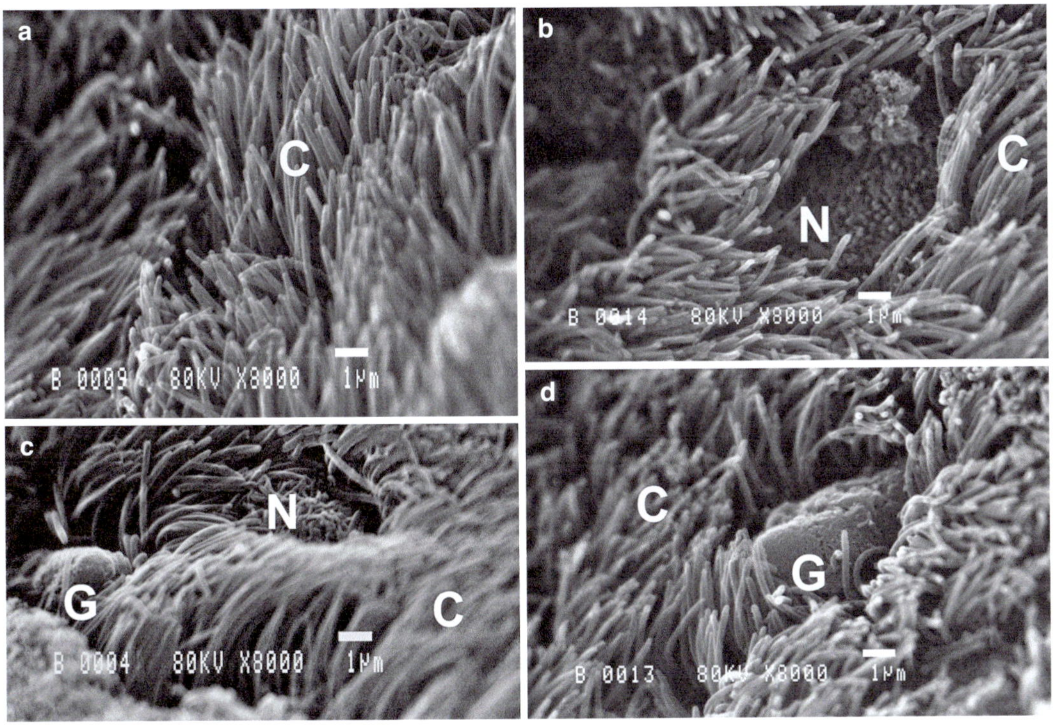

Fig. 32.13 (**a**–**d**) Different areas on the nasal mucosal surface. Electron micrograph of the respiratory epithelium (SEM) (scale bar: 1 μm). Surface invaginations of three types of cells: ciliated (*C*), non-ciliated (*N*), and goblet (*G*) cells. *SEM* scanning electron microscopy

their apical surface into the nasal lumen [23] (Fig. 32.14). Ciliated and *non-ciliated columnar cells* interact with neighboring cells by tight junctions and by interdigitations of the cell membrane. The cytoplasm contains many mitochondria located apically (Fig. 32.15), indicating a highly active metabolism [24].

Each ciliated cell contains approximately 1000 motile cilia [23]. Cilia are hairlike extensions of the apical plasma membrane containing an *axoneme*, a microtubule-based internal structure. Ultrastructurally, cilia are anchored to basal bodies below the cell surface. The *basal body* is a centriole-derived microtubule-organizing center located in the apical region of the ciliated cell. The basal body consists of nine short *microtubule triplets* arranged in a ring. The basal bodies are associated with several accessory structures (alar sheets, basal feet, and striated rootlets) that

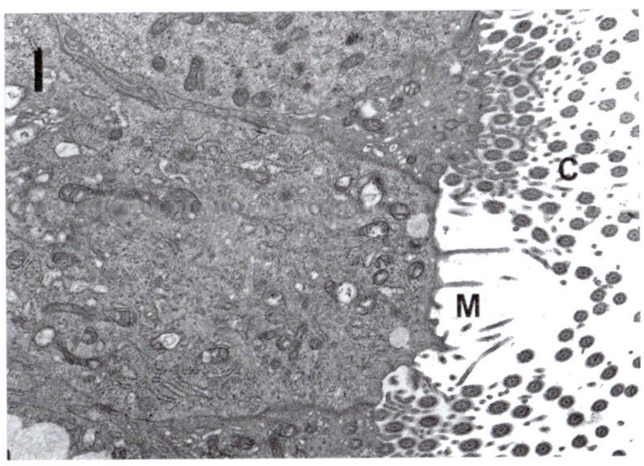

Fig. 32.14 Electron micrograph (TEM) of ciliated columnar cells, the predominant cell type on the surface, resting on the basement membrane and projecting both cilia (*C*) and microvilli (*M*) from their apical surface (scale bar: 1 μm) (Araldite block, stain: uranyl acetate/lead citrate). The specimen was obtained from a fresh frozen cadaver at a microscopic anatomy lab at Hacettepe University, Faculty of Medicine, Department of Anatomy. *TEM* transmission electron microscopy

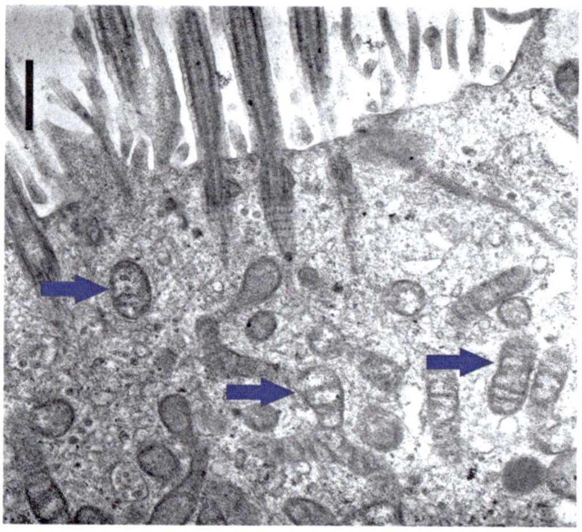

Fig. 32.15 Electron micrograph (TEM) of the apical cytoplasm of columnar cells with numerous mitochondria (arrows) (scale bar: 500 nm) (Araldite block, stain: uranyl acetate/lead citrate). The specimen was obtained from a fresh frozen cadaver at the microscopic anatomy lab at Hacettepe University, Faculty of Medicine, Department of Anatomy. *TEM* transmission electron microscopy

anchor them in the cytoplasm. Axonemes are formed by microtubules arranged in a characteristic "9 + 2" pattern. Nine outer pairs of microtubules make a cartwheel pattern at the periphery of the axoneme, surrounding two single microtubules in the center (the inner microtubules) (Fig. 32.16). Each of the paired microtubules of the ciliary axoneme is continuous with two of the triplet microtubules of the basal body. When examining a cross-section at high resolution, each outer microtubule pair (*doublet*) can be seen to be regularly arranged *dynein arms* (*ciliary dynein*). Dynein is a microtubule-associated motor protein. Each microtubule pair is linked to adjacent pairs by an elastic substance called nexin. Although the two central microtubules are separate, they are partially enclosed by a central sheath projection. *Radial spokes* extend from each of the nine doublets toward the two central microtubules [1, 23] (Fig. 32.17).

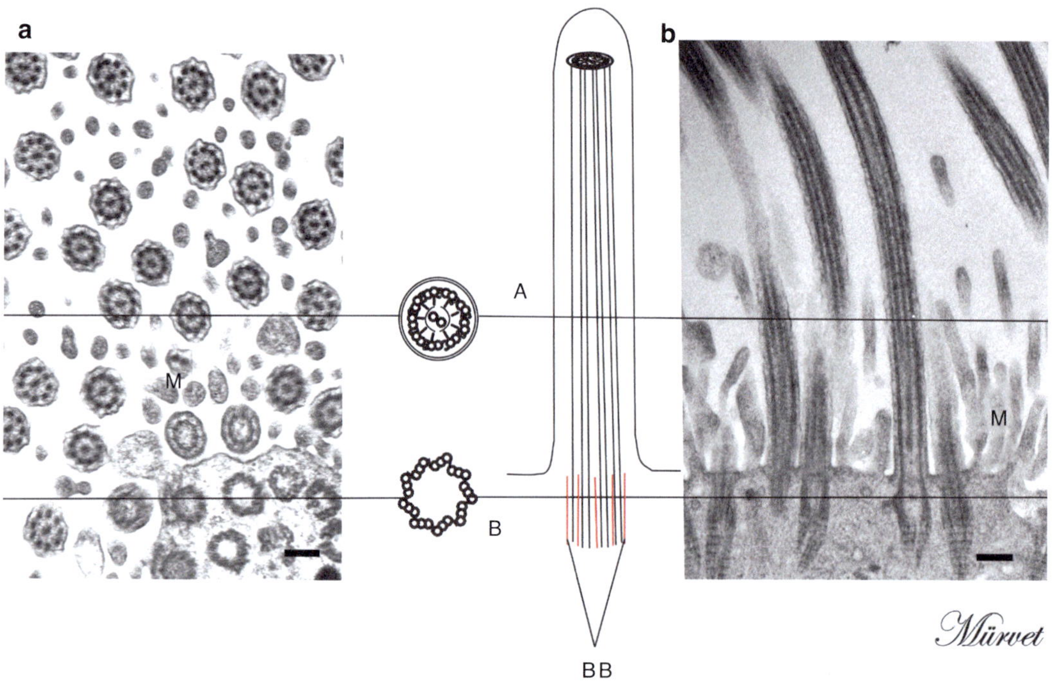

Fig. 32.16 Diagram showing the basic structure of cilia and electron micrographs of the cross (**a**) and longitudinal (**b**) sections of the cilia (scale bar: 200 nm). Line A and line B indicate cross-section levels of the cilia, indicating the cilia and the basal body, respectively. *M* microvilli

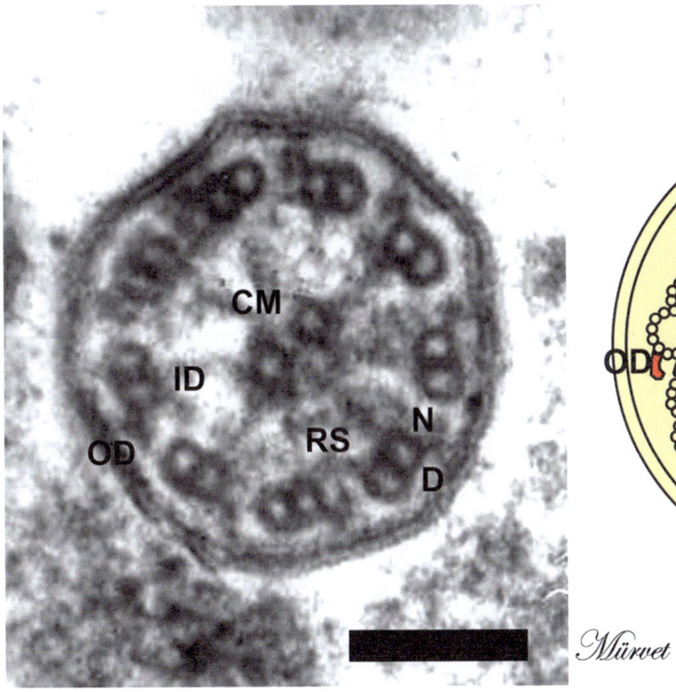

Fig. 32.17 Cross-section of the cilia showing the organization of the axoneme (scale bar: 100 nm), microtubule pairs (doublet) (*D*), inner doublet (*ID*), outer dynein (*OD*) arms (comprising ciliary dynein), and an elastic substance called nexin (*N*). The two central microtubules (*CM*) are separate; however, they are partially enclosed by a central sheath projection. Radial spokes (*RS*) extend from each of the nine doublets toward the two central microtubules

All columnar cells, ciliated and non-ciliated, are covered by *microvilli*, short and slender fingerlike cytoplasmic processes containing a core of actin filaments (Figs. 32.13 and 32.18). They are uniformly distributed over the entire apical surface and increase the surface area of the epithelial cells, thus promoting exchange processes across the epithelium [1]. The microvilli also prevent drying of the surface by retaining moisture that is essential for ciliary function [24].

Another characteristic cell type of the airway epithelium is the goblet cell (Fig. 32.19). The moistening and protecting of the nasal mucosa by secretion are mainly provided by mucous and seromucous glands. However, goblet cells also contribute to nasal secretion. The release mechanisms from goblet cells are not controlled by the parasympathetic nervous system. It is considered to be in response to physical and chemical irritants, but the mediators have not yet been clearly identified. The surface epithelial cells are joined by tight junctions, but ultrastructural studies have shown discontinuity of tight junctions around filled goblet cells [24].

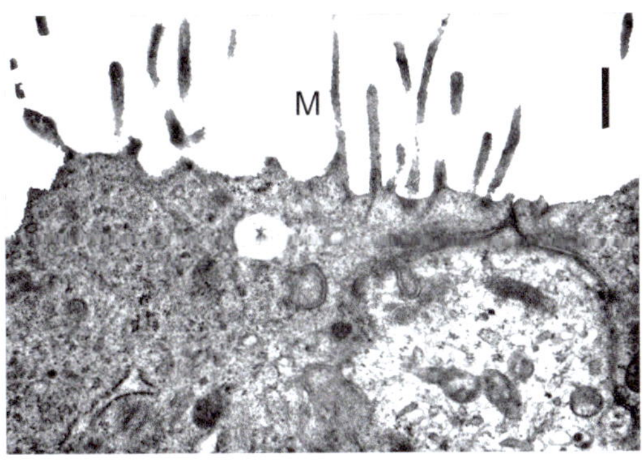

Fig. 32.18 Electron micrograph of a non-ciliated cell of the respiratory epithelium (TEM) (scale bar: 500 nm). *TEM* transmission electron microscopy, *M* microvillus

Basal cells are stem cells from which the other cell types arise. They lie on the basement membrane and do not reach the lumen (Fig. 32.12). Small granule cells and cells that resemble basal cells contain secretory granules.

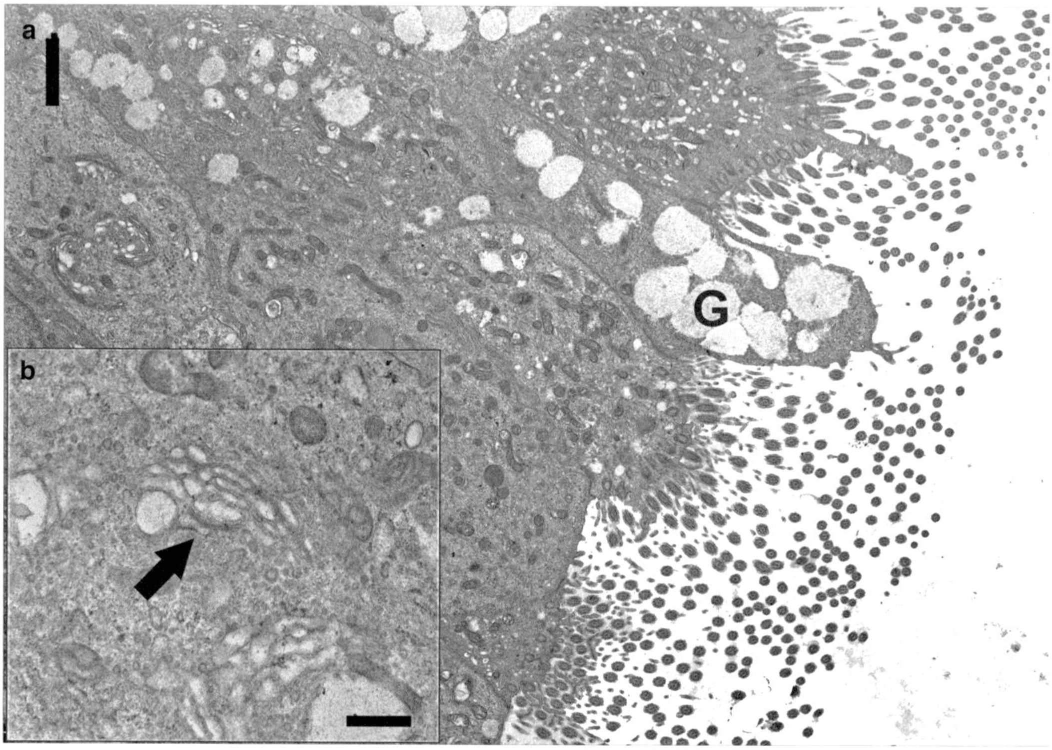

Fig. 32.19 (**a**) Electron micrograph of the respiratory epithelium (TEM) (scale bar: 2 μm). Characteristic secretory cell called a goblet cell (*G*) and (**b**) Golgi complex (arrow) (scale bar: 500 nm) (Araldite block, stain: uranyl acetate/lead citrate). The specimen was obtained from a fresh frozen cadaver at a microscopic anatomy lab at Hacettepe University, Faculty of Medicine, Department of Anatomy. *TEM* transmission electron microscopy

Brush cells, a general name for those cells in the respiratory tract, bear short, blunt microvilli.

The *vomeronasal organ* of Jacobson (VNO) is the paired embryonic remnant that is situated under the lower anterior side of the nasal septum. It forms a tubular sac with a diameter of approximately 0.2–0.6 cm. Columnar epithelium with microvilli on their apical surface lines this tubular structure. In many vertebrates, the VNO is highly developed to establish intense olfactory sensibility [13]. Differences in the frequency of morphological patterns of the VNO between the sexes may be one of the factors leading to variations in pheromone perception between men and women [25, 26]. Even with a large number of literature on the human VNO, there is little consensus on its persistence and functionality in humans. While their precise function is unknown, it is believed to be associated with pheromone recognition and food flavor perception [16, 26–28].

32.1.2.3 Olfactory Region of the Nasal Cavity

The olfactory region is located on the roof of the nasal cavity. Human olfaction is not as highly developed as it is in other animals. In humans, the olfactory region is formed by a modified pseudostratified epithelium occupying a small area. The olfactory epithelium is composed of *olfactory receptor cells*, *supporting* or *sustentacular cells*, basal cells, and brush cells. In contrast with the other regions of the nasal cavity, there are no goblet cells in this segment.

Olfactory receptor cells are bipolar neurons (Fig. 32.20) that are spindle-shaped neurosensory cells. At one end they have sensitive hairlike protrusions and nerve fibers at the other end. They have numerous microvilli and long slender cilia on their apical surface. Olfactory receptor cells have a single dendritic process-forming olfactory vesicle. The cilia have typical basal bodies and rise from the olfactory vesicle to the epithelial

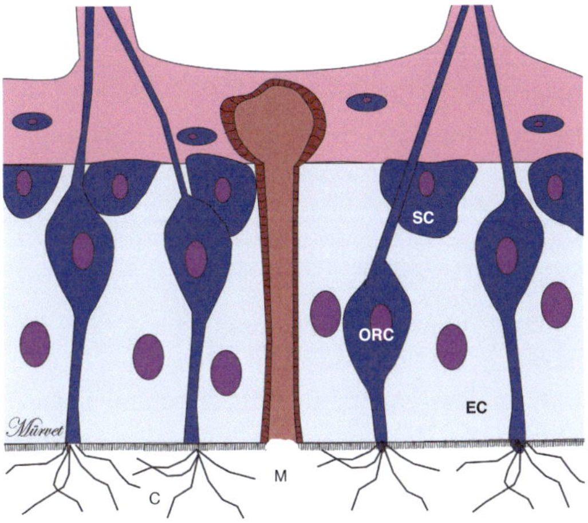

Fig. 32.20 Diagram of the olfactory epithelium. The olfactory receptor cells (*ORC*) have specialized cilia (*C*), supporting cells (sustentacular cell) (*SC*), and epithelial columnar cells (brush cells) (*EC*) with large, blunt microvilli (*M*) on their apical surfaces

surface. The cilia are nonmotile or have limited motility. The basal pole of the olfactory receptor cell gives rise to unmyelinated axonal processes that leave the epithelial compartment. They are grouped into bundles and pass through the cribriform plate of the ethmoid bone to form the olfactory bulb of the brain. Mitochondria and smooth-surfaced endoplasmic reticula are abundant in their cytoplasm. They also possess lipofuscin granules, which cause the yellowish color of human mucosa. Adherens junctions are present between these cells and the olfactory cells, but the gap and tight junctions are absent [1].

Airborne chemicals diffuse across the mucous membrane and reach the cilia, leading to stimulation of the olfactory receptor neuron. Each olfactory receptor makes a specific kind of olfactory receptor protein. The receptors, when stimulated, activate signaling cascades that eventually generate action potentials.

The other cell type in this mucosa is the *supporting cell* (sustentacular cell). Their function is to provide both metabolic and physical support to the olfactory cells, similar to that of glial cells [1, 24].

The olfactory epithelium also contains a limited number of brush cells. As previously noted, these cells are detected in the epithelium of the other regions responsible for air passage conduction. They are columnar cells that exhibit large, blunt microvilli at their apical surface, a feature from which their name is derived. The basal surface of the brush cell makes synaptic contact with nerve fibers that penetrate the basal lamina. The nerve fibers are the terminal branches of the trigeminal nerve (cranial nerve V) that functions in general sensation rather than olfaction. Brush cells appear to be involved in the transduction of general sensory stimulation of the mucosa [24].

Basal cells are the progenitors of the other mature cell types. These are small rounded cells located close to the basal lamina. Their nuclei are frequently invaginated and lie at a level below those of the olfactory cell nuclei. The cytoplasm contains few organelles, a feature consistent with their role as a reserve or stem cell. They proliferate and differentiate into supporting cells.

The lamina propria of the olfactory mucosa is directly contiguous with the periosteum of the underlying bone. This connective tissue contains numerous blood and lymphatic vessels, unmyelinated olfactory nerves, myelinated nerves, and olfactory glands (*Bowman's glands*). The olfactory glands, a characteristic feature of the mucosa, are branched tubuloalveolar serous glands that secrete via ducts to the olfactory surface. Lipofuscin granules are prevalent in gland cells, and in combination with the lipofuscin granules in the supporting cells of the *olfactory epithelium*, they give the mucosa its natural yellow-brown color. In the lamina propria, short ducts composed of cuboidal cells lead away from the glands. As the ducts pass through the basal lamina into the olfactory epithelium, the ductal cells become squamous and are then difficult to discern under light microscopy [24]. The serous secretion of the olfactory glands serves as a trap and solvent for odoriferous substances. The constant flow from the glands rids the mucosa of remnants of detected odoriferous substances so that new scents can be continuously detected as they arise [24].

32.1.3 Clinical Orientation of the Microscopic Anatomy of the Nose

Examination of samples with electron microscopy is particularly aimed to evaluate alterations and damages to the ultrastructural features of the nose. Many diseases related to nasal structures are the subject of electron microscopic studies, including primary ciliary dyskinesia, allergic rhinopathy, and chronic inflammatory hyperplasia. Electron microscopy is also useful for the diagnosis in these cases. Ciliary impairment is the most common cause of obstructive nasal diseases. "Secretion" and "obstruction" are predominant clinical symptoms in rhinology affecting patients with disorders of the nose [18]. The surface characteristics of the cells should be investigated under SEM in cases of ciliary dysfunction. TEM studies are useful for many other diagnoses. For many years, the cytological examination of nasal secretions has been included among laboratory diagnostic tests and performed with various methodologies (blowing, washing, or scraping). However, the results obtained differ in terms of their reliability and information. Using electron microscopy, the study of the nasal mucosa can be extended to the epithelial cytostructure with a detailed depiction of even the thinnest ultrastructural components (cytoplasm and organelles, cilia, and intercellular junctions) [29].

32.2 Conclusion

Electron microscopic investigation of the ultrastructural features of the nasal structures, especially nasal mucosal cells, provides important information for both clinical diagnosis and research. Nasal epithelial cells can be used as indicators when the detailed investigation of the epithelium is needed, not only for nasal structures but also for the whole air pathway. Epithelial changes, inflammation, or other pathological conditions including mediator release and receptor expression can be observed. In addition to the biopsy materials, specimens can be established from minimally invasive nasal brushings [30]. Diagnosis for some undifferentiated neoplasms of the nose and nasal sinuses should also be performed by electron microscopy. Ultrastructural histopathology of human olfactory dysfunction can also be used for the classification and diagnosis of patients with olfactory disorders.

Both clinicians and researchers can take advantage of being aware of the benefits of electron microscopy and the ultrastructural features of the nose. This awareness offers new horizons for future research and more advanced diagnoses of patients.

References

1. Ross MH, Pawlina W. Histology, a text and atlas with correlated cell and molecular biology. 6th ed. Philadelphia: Lippincott Williams & Wilkins; 2011.
2. Voutou B, Stefanaki EC. Electron microscopy: the basics. In: Physics of advanced materials. 2008. http://www.mansic.eu/documents/PAM1/Giannakopoulos1.pdf. Accessed 15 Jun 2012.
3. Bozzola JJ, Russell LD. Electron microscopy. 2nd ed. Sudbury: Jones and Bartlett Publishers; 1999.
4. Potter UJ, Love G. Scanning electron microscopy. In: Robinson RK, et al., editors. Encyclopedia of food microbiology. San Diego: Academic Press; 1999.
5. Slowko W. Secondary electron detector with a microporous plate for environmental SEM. Vacuum. 2001;63(4):457–61.
6. Görgülü A, Palaoğlu S, İsmailoğlu Ö, Tuncel M, Sürücü MT, Erbil M, et al. Effect of melatonin on cerebral edema in rats. Neurosurgery. 2001;49(6):1434–41.
7. Hazer DB, Berker M, Narin F, Ileri-Gurel E, Basak AT, Seringec N, et al. Effects of pravastatin on cellular ultrastructure and hemorheology in rats after traumatic head injury. Clin Hemorheol Microcirc. 2010;46(1):1–11.
8. Kaptanoğlu E, Tuncel M, Palaoğlu S, Konan A, Demirpençe E, Kılınç K. Comparison of the effect of the melatonin with methylprednisolone in experimental spinal cord injury. J Neurosurg. 2000;93(1 Suppl):77–84.
9. Kırkalı Z, Esen AA, Hayran M, Gençbay A, Gidener S, Güven H, et al. The effect of extracorporeal electromagnetic shock waves on the morphology and contractility of rabbit ureter. J Urol. 1995;154:1939–43.
10. Taşdemir O, Katırcıoğlu F, Küçükaksu S, Göl K, Hayran M, Keçeligil T, et al. Warm blood cardioplegia; ultrastructural and hemodynamic study. Ann Thorac Surg. 1993;56:305–11.

11. Jones N. The nose and paranasal sinuses physiology and anatomy. Adv Drug Deliv Rev. 2001;51:5–19.
12. Anderson HC, Sajdera SW. The fine structure of bovine nasal cartilage. Extraction as a technique to study proteoglycans and collagen in cartilage matrix. J Cell Biol. 1971;49(3):650–63.
13. Stenberg SS. Histology for pathologists. 2nd ed. Philadelphia: Lippincott Williams & Wilkins; 1997.
14. McGrath JA, Eady RAJ, Pope FM. Anatomy of the organization of human skin. In: Rook A, Burns T, et al., editors. Rook's textbook of dermatology. Chichester: Wiley Blackwell; 2010. http://www.blackwellpublishing.com/content/BPL_Images/Content_store/Sample_chapter/0632064293/DermChap3.pdf.
15. Ross MH, Romrell LJ, Kaye GI. Histology, a text and atlas. 3rd ed. Philadelphia: Williams & Wilkins; 1995.
16. Uraih LC, Maronpot RR. Normal histology of the nasal cavity and application of special techniques. Environ Health Perspect. 1990;85:187–208.
17. Chen W. Scanning electron microscopic observation of the nasal mucosa microvasculature. Zhonghua Er Bi Yan Hou Ke Za Zhi. 1989;24(6):322–84.
18. Grevers G, Kastenbauer E. Functional morphology of nasal blood vessels in humans. Acta Otolaryngol. 1996;116(2):312–5.
19. Riederer A, Grevers G, Welsch U, Herzmann S. Electron microscopy studies of vascular innervation of nasal mucosa in the human. Laryngorhinootologie. 1997;76(7):405–10.
20. Mir-Salim PA, Merker HJ, Jahnke V, Berghaus A. Glands of the human nasal mucosa—electron microscopy and immunohistochemical studies. Laryngorhinootologie. 1998;77(6):322–7.
21. Knipping S, Holzhausen HJ, Mir-Salim PA, Riederer A, Berghaus A. Electron microscopy studies of innervation of nasal mucosa glands in humans. Laryngorhinootologie. 2000;79(3):146–50.
22. Heritage PL, Underdown BJ, Arsenault AL, Snider DP, McDermott MR. Comparison of murine nasal-associated lymphoid tissue and Peyer's patches. Am J Respir Crit Care Med. 1997;156:1256–62.
23. Lane AP. Nasal anatomy and physiology. Facial Plast Surg Clin North Am. 2004;12:387–95.
24. Mygind N, Dahl R. Anatomy, physiology and function of the nasal cavities in health and disease. Adv Drug Deliv Rev. 1998;29:3–12.
25. Baum JB. Contribution of pheromones processed by the main olfactory system to mate recognition in female mammals. Front Neuroanat. 2012;6:20.
26. Meredith M. Human vomeronasal organ function: a critical review of best and worst cases. Chem Senses. 2001;26:433–45.
27. Bhatnagar K, Smith TD. The human vomeronasal organ. III. Postnatal development from infancy to the ninth decade. J Anat. 2001;199:289–302.
28. Carvalho MFP, Alves AL, Barros MD. Study on the morphology and frequency of the vomeronasal organ in humans. Int J Morphol. 2008;26(2):283–8.
29. Bozzo C, Fenu G, Stomeo F, Meloni F, Cau M, Montella A. Cytomorphologic and ultrastructural study of nasal mucosa by means of brushing: a comparison between asymptomatic and rhinitic subjects. Rhinology. 2005;43:261–5.
30. McDougal CM, Blaylock MG, Douglas JG, Brooker RJ, Helms PJ, Walsh GM. Epithelial cells in nasal epithelial cells as surrogates for bronchial epithelial cells in airway inflammation studies. Am J Respir Cell Mol Biol. 2008;39:560–8.

33 Genetic Background of the Rhinologic Diseases

Mehmet Gunduz, Eyyup Uctepe, and Esra Gunduz

Core Messages

- It is expected that the prevalence of allergic rhinitis in the Western world will reach 50% within the next 15 years.
- Exposure early in childhood to microbes and infections leads to modification of the immune system, a reduced risk of IgE sensitization, and a decreased risk of AR throughout life.
- People who have persistent nonallergic rhinitis based on the allergy skin test and serum-specific IgE test result might have a localized form of allergic rhinitis.
- Information on the genetics of allergic diseases is valuable not only for analyzing the molecular basis of allergic diseases but also for investigating new drugs.
- Environmental effects on various genetic variants as well as epigenetics determine the fate of chronic nasal diseases.
- CFTR genotype affects the progression of airway obstruction in CF.
- The studies demonstrated the role of various modifier genes such as ADIPOR2, EDNRA, IFRD1, IL-8, MBL2, TCF7L2, MSRA, SERPINA1, and TGF-b1 in CF for pathologies of pulmonary function, liver disease, intestinal obstruction, diabetes, and infection.
- Variants in the promoter region (−509) and first exon (codon 10) of TGF-b1 are correlated with poor lung function.
- ΔF508 homozygosity was associated with clinical severity of paranasal sinus diseases and with the presence of polyps on endoscopy.

M. Gunduz (✉) · E. Gunduz
Department of Otolaryngology Head and Neck Surgery, Faculty of Medicine, Wakayama Medical University, Wakayama, Japan

E. Uctepe
Acıbadem Labmed Ankara Tissue Typing Laboratory, Ankara, Turkey

33.1 Introduction

Rhinologic diseases are very commonly seen pathologies all over the world. They affect millions of people and cause huge social and economic burdens. In this chapter, the genetic origin of the main rhinologic diseases is discussed. These are allergic rhinitis, chronic sinusitis, vasomotor rhinitis, cystic fibrosis, and nasal polyps.

33.2 Architecture and Function of the Sinuses

The paranasal sinuses and turbinates are formed from the primordial ridge during fetal development. Each sinonasal structure develops from these ethmoturbinals separate from the inferior turbinate. Although there is a fairly consistent pattern to the formation of these structures, resulting in a series of oblique structures that attach to the ethmoid bulla, the extent and com-

Ö. Ö. Celebi, T. M. Önerci (eds.), *Nasal Physiology and Pathophysiology of Nasal Disorders*,
https://doi.org/10.1007/978-3-031-12386-3_33

plexity of pneumatization can be variable. There is only the ethmoid sinus at birth, while the maxillary, frontal, and sphenoid sinuses are not fully formed and expand out from these primary structures into their relevant cranial bones during childhood and adolescence. Although the structure of the sinuses is well known, there are different opinions to explain the purpose and functions of the paranasal sinuses. Among these, lightening the weight burden of the cranium, adorning vocal resonance, or constituting thermal insulation are the most common ideas. Also, a recent theory proposes that they form a "crumple zone" for the crucial structures of the head such as the brain and eyes to diminish damaging forces resulting from sudden trauma [1].

It is shown that cranial sinuses produce nitric oxide. This molecule has many functions in the immune system such as killing microorganisms (bacteria, viruses, and fungi) and tumor cells. So the sinuses assist the innate immune defense of the airway. Proper functioning of the sinuses depends on sufficient drainage of their produced mucus and normal ventilation. In this regulation, mucociliary clearance is a highly coordinated effort that consists of millions of beating cilia to direct any mucus or particles toward the ostia of the sinus. This process is quite prone to obstruction during swelling or inflammation of the nasal mucosa. Therefore, it causes a decrease in ciliary beat frequency and results in poor drainage from the sinus ostia [2]. This situation leads to stagnation of sinus secretions and a decrease in oxygen level in the sinus, subsequently reducing mucociliary clearance and nitric oxide production.

33.3 Histopathologic Features of the Nasal and Sinus Mucosa

The nasal mucosa is composed of a ciliated pseudostratified columnar epithelium which includes ciliated and nonciliated columnar epithelial cells, goblet cells, and basal cells [3, 4]. These cell types lie on a basement membrane which includes types I, III, and IV collagen fibrils; but, generally, some of these cells cannot reach the luminal surface and lead to the illusion of a stratified layer. Just beneath the basement membrane is a cell-free zone, which includes fibronectin and collagen types III and V and a submucosal layer consisting of glands, inflammatory and interstitial cells, extracellular matrix, nerves, and blood vessels. There are three types of glands within this layer: mucous, seromucous, and serous glands. These glands, throughout epithelial goblet cells, synthesize the mucus that overlies the epithelium and serve an antimicrobial function. Furthermore, they transport particulate matter, antigens, and bacteria by mucociliary clearance. Serous glands synthesize secretory IgA, which is essential in mucosal defense [5]. The submucosal gland area constitutes almost 25% of the lamina propria, while this ratio is only 15% in individuals without nasal allergies. Normally, lymphocytes, macrophages, and mast cells are the basic cells in the nasal mucosa, and nasal mast cells are usually found immediately beneath the basement membrane.

33.4 Allergic Rhinitis and Its Genetic Background

33.4.1 Introduction

Allergic rhinitis (AR) is a growing health and social problem worldwide and the most common type of chronic rhinitis. Most of the children with AR become symptomatic and are diagnosed before 6 years of age. Recent data indicate that 10–30% of adults and up to 40% of children in developed countries suffer from this disease. The prevalence of AR has increased considerably after 1950 in Western populations. Moreover, it is thought that within the next 15 years, the prevalence of allergic rhinitis in the Western world will reach 50%. Almost 80% of all patients with AR have symptoms before 20 years of age. In some patients, symptoms of AR can be detected before 2 years of age [6]. Current data suggest that 44–87% of patients with rhinitis might have mixed rhinitis, both allergic and nonallergic types [7].

Environmental and lifestyle factors are important in this prevalence increase. Because it is hard to consider changes in the genetic composition of a population in a short time, one of the most powerful hypotheses to elucidate this prevalence increase is the hygiene hypothesis. It suggests that exposure early in childhood to microbes and infections leads to an altered immune system, a reduced risk of IgE sensitization, and a decreased risk of AR throughout life. AR is a debilitating disease on individual and social levels, and it has a major negative impact on morbidity, lost working days, high cost for medical treatment, and reduced school performance. Allergic rhinitis occurs when inhaled allergens interact with IgE antibodies on cells in the airway. AR can sometimes be a restricted disease without a systemic response of IgE sensitization, like an increase in serum-specific IgE levels or a positive allergen skin test result. Patients with AR may initially present symptoms like nonallergic rhinitis. These patients may have nasal symptoms for longer than 2 years but negative skin test results and absence of serum-specific IgE response to allergic disease's nasal markers. On the other hand, some of the patients with persistent nonallergic rhinitis may have increased eosinophil number, increased specific and total IgE levels, as well as increased T cell counts in their nasal lavage sample compared with healthy control subjects. These findings must alert us that people who have persistent nonallergic rhinitis based on the allergy skin test and serum-specific IgE test result might have a localized form of allergic rhinitis [8]. In these cases, examination of the nasal mucosa and nasal challenge tests may help to identify these patients.

33.4.2 Allergic Rhinitis and Asthma

It must be emphasized that AR should be evaluated as a global systemic disease. Allergic inflammation does not necessarily limit itself to the nasal airway, but the attending physician must also be aware of the possible comorbidities of AR, including asthma, conjunctivitis, sinusitis, and otitis media [9]. Allergic asthma and rhinitis are comorbid conditions that are associated pathophysiologically and epidemiologically [10, 11]. Epidemiological studies have shown that up to 80% of patients with asthma demonstrate symptoms of rhinitis, while approximately 20–40% of patients with AR have clinical asthma. Both are airway diseases in which IgE antibody sensitization to aeroallergens is a prominent feature. The link between asthma and rhinitis in particular has gained much interest. There is evidence that systemic trafficking of inflammatory cells from local inflammation in one portion of the respiratory tract can induce inflammatory changes in the other; segmental bronchial allergen challenge in patients with AR has been shown to result in both bronchial and nasal inflammatory responses [10]. There is growing evidence that rhinitis is a risk factor for the development of asthma, independent of atopy. Additionally, the airway mucosa of the nose and bronchi has many similarities, and the clinical and pathophysiological changes in asthma and AR are often very comparable. Although there are still some differences that should be highlighted, the strong relationship between rhinitis and asthma has introduced the concept of "the united airway disease."

33.4.3 Allergic Rhinitis and Genetics

On the basis of genetic studies, multiple groups have tried to identify a susceptibility gene for allergy using the candidate gene approach and/or genome-wide screening. Each of these two approaches suggested genetic heterogeneity of allergic diseases. Candidate genes identified so far are associated with various diseases in different ethnic groups and their function is being investigated. Based on the information accumulated thus far and the information on the human genome sequence, future advances in research on genetic factors for allergic diseases will likely lead to the establishment of more effective prophylaxis and therapy for these diseases.

There is now overwhelming support for a genetic component to allergic diseases. Information on the genetics of allergic diseases is valuable not only for analyzing the molecular basis of allergic diseases but also for investigating new drugs [12].

Recent rapid improvements in the field of genetics have disclosed several pathways that are crucial for AR pathogenesis and perhaps more importantly have shown that AR behaves like typical complex diseases. That is conditions in which various genetic variants that are individually mild may be capable of major phenotypic effects when acting in concert within a certain environmental context. Researches in the genetics of AR have made much progress over the last decade and are expected to advance even further in the near future, as increasingly powerful analytical tools are being developed to unlock the complexities of genetic diseases. However, some formidable challenges still remain for AR geneticists: i.e., the identification of all the genes involved in the disease, the mechanisms underlying the phenotypic heterogeneity of AR, the difficulties in replicating associations between genotype and phenotype across populations, and last, but not least, understanding how environmental and developmental factors interact with genetic determinants to affect disease susceptibility [13].

Shirkani et al. (2019) conducted a study in the northeast of Iran in which the IL-13 polymorphism (rs20541, Exo 4, G > A, Arg130Gln) and IL-4 polymorphism (rs2243250 = C-590 T, promoter, T > C) are co-associated with AR and sensitivity to aeroallergens [14]. In another study, a significant association of rs20541 (Arg130Gln) with serum total IgE levels in Chinese adult patients with allergic rhinitis was found [15]. They determined that the patients with a gln/gln genotype showed much higher serum total IgE levels as compared with an arg/arg genotype.

In a Korean population composed of 295 patients with allergic rhinitis and 418 controls, Li et al. determined a significant association between allergic rhinitis and 3 SNPs of the FOXJ1 gene:-460C-T (rs880213), 1805G-T (rs1868823), and 3375G-C (rs3192453) [16]. Haplotype analysis disclosed that the main haplotype, CGG of the 3 SNPs, respectively, was significantly associated with allergic rhinitis ($P = 0.000018$). There was no association between the SNPs and serum IgE levels. Based on these data, it was suggested that dysregulation of FOXJ1 may influence T cell activity.

Amarin et al. (2017) found that an intronic single-nucleotide polymorphism (rs13217795) in FOXO3 gene is associated with asthma and allergic rhinitis. Under the general model, the odds of the asthma phenotype were 1.46 (0.64 to 3.34) and 3.42 (1.37 to 8.57) times higher in heterozygotes and derived allele homozygotes, respectively, compared to ancestral allele homozygotes [17].

Guo et al. (2018) reviewed 29 case–control studies in a meta-analysis. They found that E237G (B vs. A: OR = 1.28, 95% CI = 1.06–1.53, $P < 0.001$, I2 = 63.1%) and −109 C/T (BB vs. AA + AB: OR = 1.58, 95%CI = 1.26–1.98, $P < 0.001$, I2 = 66.4%) were risk factors for allergic diseases [18].

Haagerup et al. (2001) performed a genome-wide scan to identify susceptibility genes for allergic rhinitis in affected sib-pair families [19]. From 100 Danish nuclear families selected for allergy, families having at least two full sibs with clinical allergic rhinitis and enhanced specific IgE against at least 1 of the 11 tested allergens were chosen. A total of 33 sib-pair families were qualified for the genome-wide scan that was undertaken using 446 microsatellite markers. The collected data were analyzed by nonparametric multipoint linkage analysis using the maximum likelihood score (MLS) approach. One major candidate region on chromosome 4q24-q27 (lod score of 2.83) was revealed. There is evidence of linkage between allergic rhinitis and the region spanning from the marker D4S1651 to D4S2394, which is approximately 13 cM and which potentially harbors a specific susceptibility gene. Another eight regions including 2q12-q33, 3q13, 4p15-q12, 5q13-q15, 6p24-p23, 12p13, 22q13, and Xp21 showed evidence of linkage detected by lod scores in the range of 1.04–1.63. Among these eight regions, two regions were in line with previous findings associating chromosomes 2q33 and 6p23 with asthma and asthma-associated phenotypes.

Fine mapping in three independent Danish populations was performed by Brasch-Andersen et al. (2006), consisting of a total of 236 sib-pair families with clinical allergy, including the 33 sib-pair families with allergic rhinitis originally studied by Haagerup et al. (2001) [20]. Analysis of 28 microsatellite markers on chromosome 3q displayed significant linkage to 3q13.31 for rhinitis (MLS = 5.55, identity by descent, or IBD = 63.9%) and for atopy (MLS = 3.71, IBD = 61.7%). An MLS of 5.1 (IBD = 67.3%) was obtained when sib pairs with both rhinitis and atopy were analyzed.

33.4.4 Molecular Targets for AR Therapy

The worldwide efforts to discover susceptibility genes in allergic diseases are motivated by the conviction that the identification of disease genes may facilitate the design of new classes of anti-inflammatory compounds. Molecules concerned with the allergic reaction, such as cytokines, chemokines, their receptors, major histocompatibility complex molecules, and transcription factors, could provide a candidate drug target. Current medications for AR such as antihistamines and leukotriene receptor antagonists are aimed at symptom relief and block the production of proinflammatory cytokines to suppress allergic inflammation. However, SYK (*s*pleen t*y*rosine *k*inase) inhibitors can block the allergen-induced release of all mast cell mediators like eicosanoids and cytokines. Therefore, SYK kinase is a desirable therapeutic target for acute and chronic allergic inflammation. SYK kinase inhibitors are now entering clinical trials. A series of 2,4-diaminopyrimidine compounds were developed as SYK kinase inhibitors using cell-based structure–activity relationships with primary human mast cells. One of these compounds, referred to as R112, was suitable for intranasal delivery and was tested for safety and efficacy in allergic rhinitis patients. In a park environment, R112 showed remarkable relief of acute allergic rhinitis symptoms with rapid onset of action. These results indicate the clinical significance of inhibiting SYK kinase in allergic upper airway disorders [21]. Figure 33.1 shows the SYK kinase pathway and its relation with FcεR.

Kato et al. (2018) showed that JTE-852, a novel spleen tyrosine kinase inhibitor, significantly blocked effects on antigen-induced allergic reactions in rat models, indicating that JTE-852 in oral dosage form would improve a wide range of allergic signs. However, the current anti-allergic drugs, on the other hand, failed to display significant suppression in several models [22].

Ramasamy et al. reported an association of three loci for either AR or grass sensitization through the evaluation of genome-wide meta-analysis of genetic variants. The HLA variant rs7775228, which cis-regulates HLA-DRB4, was strongly associated with grass sensitization and weakly with AR. Variants in a locus near chromosome 11 open reading frame 30 (C11orf30) and leucine-rich repeat containing 32 (LRRC32), which was previously associated with atopic dermatitis and eczema, were also strongly associated with both phenotypes (rs2155219). The third genome-wide significant variant was rs17513503, which is located near transmembrane protein 232 (TMEM232) and solute carrier family 25, member 46 (SLC25A46). They also observed strong associations of both AR and IgE sensitization to grass with a common (minor allele frequency, 47%) polymorphism in the 11q13.5 locus [23].

Lu et al. (2011) found that the CT/CC genotypes in IL-4 C-590 T were associated with a significantly decreased risk of mite-sensitized PER (adjusted odds ratio (OR) = 0.64, 95% confidence interval (CI) 0.45–0.92), compared to the TT genotype [24]. Moreover, persistent allergic rhinitis (PER) patients with CT/CC genotypes had significantly lower serum levels of total IgE than those with TT genotype ($P = 0.001$). However, there was no significant association of the IL-13 and IL-4RA polymorphisms with mite-sensitized PER ($P = 0.05$).

Kruse et al. (2012) reported genome-wide significant linkage to a novel AR locus at 1p13 and

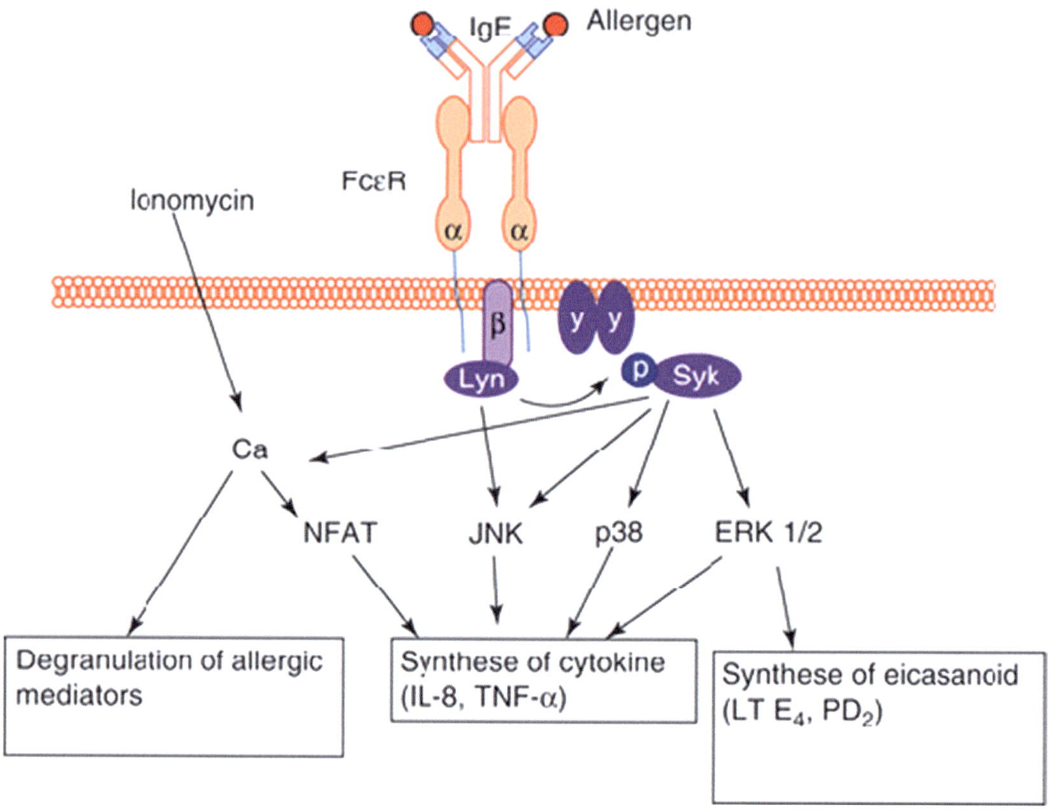

Fig. 33.1 Allergens bind FcεRI and trigger the Syk-mediated signaling pathway resulting in mast cell degranulation, eicosanoid mediator synthesis, and cytokine production. Ionomycin, a calcium mobilization agent, can also induce degranulation of mast cells

suggestive linkage to two novel regions at 1q31-q32 and 20p12 [25]. The locus has previously been demonstrated to have a suggestive linkage with asthma [26]. Possible candidate genes are the glutathione S-transferase M1 (GSTM1) [27] and acidic mammalian chitinase (CHIA) genes located at 1p12-p13, earlier shown to be associated with asthma [28]. Chromosome 20p12 is also a novel finding in regard to AR. However, it has previously been found to have a linkage with atopy as well as asthma [26].

Several studies have found an association of filaggrin (FLG), a filament-associated protein that binds to keratin fibers in epithelial cells with various allergic conditions and diseases including atopy as well as atopic dermatitis (AD), asthma, and AR [29].

At 2q13-q14, the interleukin 1 gene cluster (IL-1A, IL-1B, and IL-1RN) was previously associated with AR [30].

At 20p13, a disintegrin and metalloproteinase domain 33 (ADAM33) gene residing almost 3 Mb from the maximum linkage signal has previously been associated with AR. Toll-like receptors 7 and 8 (TLR7 and TLR8) located at Xp22 confer susceptibility to several allergic diseases and among these are AR [25].

Hussein et al. showed that the genotype and allele frequencies of the TLR2 Arg753Gln and TLR4 Asp299Gly polymorphisms are not significantly different between asthmatic children or allergic rhinitis as compared to controls ($P > 0.05$ for each) or even when compared further with IgE level [31]. However, it is shown that the mutant allele of TLR2 or TLR4 polymorphisms was significantly associated with the moderate to severe groups compared to the mild group in both atopic asthmatics and allergic rhinitis groups ($P > 0.001$ for each). In conclusion, their study demonstrated a lack of association of TLR2 and TLR4 polymorphisms with asthma and allergic rhinitis but suggests a significant association between these genetic variants and the disease severity.

To date, two allergy vaccine-containing TLR agonists (Pollinex Quattro and AIC) have been investigated in clinical trials. Preseasonal subcutaneous injection of Pollinex Quattro contains monophosphoryl lipid, a TLR4 agonist and AIC contains, CpG motifs activating the TLR9 cascade have been found safe and efficacious in control of nasal symptoms of patients with allergic rhinitis. Other molecules like CRX-675 (a TLR4 agonist), AZD8848 (a TLR7 agonist), VTX-1463 (a TLR8 agonist), and 1018 ISS and QbG10 (TLR9 agonists) are currently being investigated for allergic rhinitis and asthma [32].

33.4.5 Allergic Rhinitis and Epigenetics

The immune system is heavily affected by environmental changes. One of the good and current examples of this situation is the remarkable increase of all immune diseases with urbanization. Likewise, the rising prevalence of immune diseases in infancy indicates that there may be an essential early period of sensitivity. During fetal life, essential arrangements occur such as the structure, function, and response patterns of many systems. So, elucidating early events may offer important insights into the pathogenesis of the disease, as well as the pathways of environmental influence.

A variety of environmental exposures in pregnancy such as maternal diet [33], pollutants like cigarette smoke [34], and microbial exposure [35] have been shown to alter immune function. Conspicuously, the same determinants were identified as potential immune modifiers in epidemiological studies of allergic disease.

The effects of diet are potentially complex. More studies are needed to examine the effects of related dietary nutrients such as vitamins B2, B6, and B12, methionine, and choline, which may be implicated in epigenetic effects through their effects on folate-mediated one-carbon metabolism.

Belinsky et al. also showed that exposure to the ultrafine particulate matter found in pollution may also change DNA methylation in maternal and fetal DNA and may be associated with altered inflammatory response pathways.

There is current exciting evidence that some pathogens can affect the epigenetic profile of the host cell, influencing or mimicking mechanisms that participate in DNA methylation and histone modification [36]. The studies also showed that allergy protection by in utero microbial exposure in rural farming environments is correlated with enhanced neonatal Treg (regulatory–suppressor T cells; these cells downregulate the immune system, thus keeping tolerance to self-antigens) function, FOXP3 expression, and associated epigenetic effects (hypomethylation) of the FOXP3 gene [37].

Several specific nutritional changes have been correlated with the rising allergic propensity, including a decrease in omega-3 polyunsaturated fatty acids (n-3PUFA), soluble fiber, antioxidants, and other vitamins [38], based on epidemiological associations and immunological effects. Similarly, antioxidants have been shown to have effects on T cell regulation and induction of IL-12 production by antigen-presenting cells [39]. So, this could induce the development of Th1 and repress Th2 responses. Devereux et al. showed an immunomodulatory role for maternal dietary antioxidants during pregnancy; this, combined with evidence that oxidative stressors can trigger epigenetic changes and alter disease risk (see below), indirectly suggests a role for these pathways [40].

A recent study in mice also showed that exposure to diesel exhaust particles augments the production of IgE after allergen sensitization through the hypermethylation of IFNγ and hypomethylation of the IL-4 locus [41]. In placental tissue, nicotine has also been shown to alter cytokine production via NFκB [42].

The current and remarkable rise in allergic diseases and the very early onset of disease indicate that in utero events have a more essential effect on immune development and allergic susceptibility than genomic inheritance. Above all, as epigenetic modifications are commonly reversible, this may provide for the development of novel therapeutic compounds that may be efficacious in arresting or even reversing the allergy epidemic [33].

A recent large gene expression microarray study of unstimulated CD4+ T cells found no differences between allergic and nonallergic individuals [43].

The existence of monozygotic (MZ) twins who are discordant for intermittent allergic rhinitis (IAR) suggests disease mechanisms that are independent of genetics [44]. Sjogren et al. for the first time performed mRNA and microRNA expression microarray analyses of CD4+ T cells from MZ twins discordant for IAR. They analyzed the CD4+ T cells outside the pollen season, after in vitro allergen challenge and during the pollen season without in vitro allergen challenge. The allergen challenge in vitro resulted in significant differences in mRNA and protein levels between the allergic and healthy twins. The cytokines IL-4, IL-5, and IL-13 were increased in the supernatants from allergic twins. In their study, they performed microRNA expression arrays of all the twin pairs but found no significant differences. Next, blood samples were taken during pollen season from four MZ twin pairs discordant for IAR [45]. Consequently, they identified disease-relevant mRNAs and proteins that differed between the discordant MZ twins. No differences in microRNA expression were determined. Because the MZ twins are almost genetically identical, the observed altered expression of essential disease-associated genes and proteins may have epigenetic causes [45].

The elevation in the levels of HDAC1, 3, and 11 in the nasal epithelia have been shown in different studies related to allergic rhinitis and HDAC inhibitors have been found efficient in decreasing the symptoms of allergic rhinitis. Their effectiveness is attributed to restoration of the expression of TWIK-related potassium channel-1, regulation of cytokine profile along with normalization of Th1/Th2 imbalance. Also, it is shown that some miRNAs including miR-135a, miR-146a, miR-181a, miR-155 downregulated and miRNA19a upregulated in allergic rhinitis patients [46].

In conclusion, the role of genetic as well as epigenetic factors in the pathophysiology of nasal diseases, including allergic rhinitis, is becoming more evident. Environmental effects on genetic variants and epigenetics determine the fate of chronic nasal diseases. Clarification of such relationships and mechanisms will lead to novel molecular diagnostic and therapeutic approaches for these kinds of diseases.

33.5 Role of Genetics in Chronic Rhinosinusitis

33.5.1 Introduction

The first contact of the respiratory system with the external environment occurs in the nose, which is responsible for air filtering, humidification, and temperature regulation. Because of this close contact of the respiratory mucosa with a great variety of allergens and pathogens, upper respiratory illnesses are one of the most frequent diseases in humans. Chronic rhinosinusitis (CRS) is characterized by a chronic inflammatory condition of the sinonasal mucosa, but it is mostly defined by its clinical manifestation rather than the inflammation pattern. CRS presents with chronic symptoms such as nasal congestion, anterior or posterior nasal drainage, hyposmia, and facial pain [47]. It is highly prevalent. According to an analysis of the 2008 National Health Interview Survey data, almost 1 in 7 adults suffering from rhinosinusitis [48]. Additionally, it has a huge effect on the quality of life and healthcare expense in terms of antibiotic prescriptions filled, lost work days, and lost school days [49]. The disease is characterized by chronic inflammation of the sinonasal mucosa, and because inflammation of the nasal and sinus mucosa often coincides, it is named “rhinosinusitis” in current literature. It is an attempt to stress the concept that patients present with symptoms attributable not only to the sinuses but also to nasal inflammation that might often, but not always, be present.

The term chronic refers to symptoms persisting for more than 12 weeks with no definitive resolution. Also, the most recent consensus definitions subclassify CRS into CRS without nasal polyposis (CRSsNP), CRS with nasal polyposis (CRSwNP), and allergic fungal rhinosinusitis (AFRS) [50]. A study showed that the inflammatory mediator profile in the nasal mucosa of patients with CRS without nasal polyps (CRSsNP) and CRS with nasal polyps (CRSwNP) is similar. It further strengthens the idea that rhinitis and sinusitis can indeed be classified as one disorder entity and supports the use of the term rhinosinusitis. Although primary allergic or upper respiratory tract infectious inflammation in the nose can cause sinus ostia obstruction and subsequent sinusitis, it remains unproved from prospective studies whether and how often this occurs.

Historically, chronic sinusitis has been considered a single unimodal clinical disease. But in recent years chronic sinusitis has been recognized as comprising several diseases with different causes, with each one characterized by a definite histological pattern and gene and protein expression. Identification of specific disease subgroups and their etiologies is important for correct differential diagnosis and to apply appropriate therapeutic intervention.

33.5.2 Chronic Sinusitis (CS)

For diagnosis of CS, it is important to see symptoms of nasal irritation, anterior and posterior rhinorrhea, and nasal congestion with the accompanying presence of pressure or pain in a "sinus" distribution area that lasts for more than 12 weeks [51]. But validation with nasal endoscopy or computed tomography (CT) imaging is important because there can be significant overlapping of these symptoms that are very likely to be of migraine or midfacial pain syndrome origin [52]. Chronic sinusitis has been divided into two subgroups depending on the presence or absence of NPs [51]. Some researchers determined a significant correlation between nasal polyposis and the presence of tissue eosinophilia. However, both the presence and degree of eosinophilia in NPs can be quite variable, and researchers did not observe eosinophilia in a large patient group with idiopathic nasal polyposis [53]. These results support the opinion that certain forms of CS such as allergic fungal sinusitis (AFS) [54] and aspirin-exacerbated respiratory disease (AERD) [55] may be more likely to produce NPs and present rich eosinophilic infiltrate into the sinus cavity. Polyposis can present as a complication in any form of CS. So it should not be used as the basis for diagnosis or decisions regarding treatment. Evaluation of the presence or absence of NPs can be especially important in clinical practice for identifying patients who are more or less likely to have eosinophilic disease and thereby also identifying patients who are more or less likely to respond to eosinophil-targeted medicines. However, it should be emphasized that, in practice, CS presents as a spectrum of disorders in which the level of eosinophilia and predilection for polyposis exist on a continuum. Although we will stress the different features of the pathology of eosinophilic and noneosinophilic forms of these diseases, in daily life patients can generally present with variable overlapping symptoms.

It is demonstrated in a recent study that omalizumab and mepolizumab treatments improved endoscopic nasal polyp score (EPS) and symptoms score in patients with CRSwNP when compared with placebo. Reductions in the size of nasal polyps have been observed after reslizumab treatment in patients with raised intranasal interleukin-5 levels. Also, a 70% reduction in EPS compared with 20% in the placebo group has been observed after dupilumab treatment ($p < 0.001$) [56].

33.5.3 Remodeling Theories

When we look at the histological investigation of CRS, two different remodeling types are seen. One of them is CRSsNP, which is associated with fibrosis, basement membrane thickening, subepithelial edema, goblet cell hyperplasia, and mononuclear cell infiltration. The other is CRSwNP, which is associated with an edematous stroma with albumin deposition, pseudocyst formation, and subepithelial and perivascular inflammatory cell infiltration.

Remodeling is a dynamic balance between production and degradation of extracellular matrix (ECM) and is regulated by various factors among which TGF-b has a central role. The Treg cells are one of the most important factors in the remodeling process. TGF-b is also an essential factor in the remodeling process of the airways. It is responsible for the attraction and induction of proliferation of fibroblasts, and it also causes upregulation of ECM synthesis. A study showed that TGF-b1 and 2 protein concentrations, TGF-b receptor I and III mRNA expression, and the numbers of activated pSmad 2-positive cells were significantly lower in patients with CRSwNP than in control subjects. However, in patients with CRSsNP, TGF-b1 protein concentration, TGF-b receptor II and III mRNA expression, and the number of activated pSmad 2-positive cells were significantly higher than in control subjects [57]. Indeed, the upregulation of the TGF-b signaling pathway in patients with CRSsNP causes excessive collagen deposition associated with fibrosis, while its downregulation at the protein level in patients with CRSwNP causes edema formation and a lack of collagen production (Table 33.1).

Past investigations have focused on inflammatory differences, but recent information from studies comparing patients with CRSsNP and patients with CRSwNP showed that TGF-b proteins and their signaling might be suitable markers to distinguish between the different CRS subtypes.

Table 33.1 ARIA classification of allergic rhinitis

"Intermittent" rhinitis	*"Persistent" rhinitis*
Symptoms are present:	Symptoms are present:
≤4 days a week	≥4 days a week
Or ≤ 4 consecutive weeks	And ≥ 4 consecutive weeks
"Mild rhinitis"	*"Moderate/severe" rhinitis*
None of the following items are present:	≥1 of the following items are present:
Sleep disturbance impairment of daily activities, leisure, and/or sport impairment of work or school work, troublesome symptoms	Sleep disturbance impairment of daily activities, leisure, and/or sport impairment of work or school work, troublesome symptoms

Adapted from Bousquet et al. (2001)

33.5.4 Chronic Infectious Sinusitis

There is a significant loss of barrier and innate immune functions in all types of CS, making these patients highly prone to frequent and longstanding episodes of acute sinusitis. As a consequence, all forms of CS generally copresent with anaerobic bacteria, gram-negative organisms, *Staphylococcus aureus*, and other bacterial colonization in the sinuses. However, chronic infection (i.e., an episode of acute sinusitis persisting beyond 12 weeks despite antibacterial therapy) is less often the cause of CS, and, when present, the clinician should consider whether there is an underlying immune deficiency, HIV infection, immotile cilia syndrome, or cystic fibrosis. At the present time, the investigation into the presence of allergic atopy or anatomic abnormalities is recommended for patients with CS because these factors might have a causative or modifying role on the disease. The exact significance of these factors in relation to sinus inflammation is unknown, but their impairment might cause a decreased level of patient improvement (Table 33.2) [58].

33.5.5 Noneosinophilic Sinusitis (NES)

Idiopathic noneosinophilic sinusitis (NES) is thought to result from chronic or recurrent blockage of the sinus ostia caused by anatomic predisposition like septal deviation viral rhinitis, allergic rhinitis, or other causes. As a consequence, these processes cause recurrent and prolonged bacterial infections, possibly in asso-

Table 33.2 Clinical manifestation of chronic sinusitis

1. Infectious	Idiopathic
	Associated with immune deficiency
2. Eosinophilic	AFS
	CHES
	AERD
3. Noneosinophilic	Idiopathic
	Induced by an anatomic tendency
	Induced by chronic rhinitis
4. Cystic fibrosis	

ciation with barotrauma of the sinus cavities and harm to the respiratory epithelium, ciliary destruction, prominent mucous gland, and goblet cell hyperplasia similar to the bronchial epithelium in asthma disease, bacterial colonization, and biofilm formation [59]. A mononuclear cell infiltrate with few neutrophils is observed in the inflammatory component of this form of sinusitis. If neutrophils are presented, it indicates recent infection, persistent infection, or CF. Prominent remodeling with dense deposition of collagen and other matrix proteins is characteristic of NES.

33.5.6 Molecular Basis of NES

Knowledge about the development of this disease is very limited, and as such, there are very few studies that have investigated whether there is a genetic component to NES. One study identified the plasminogen activator inhibitor 1 (PAI-1) gene as a possible candidate [60]. Subsequently, recent observations related to the expression of PAI-1 and the thrombotic/fibrinolytic pathways further suggest a role for PAI-1 in CS [61]. The 4G allele of PAI-1 is involved in the arrangement of fibrosis in asthmatic patients and especially in airway remodeling that leads to irreversible obstruction [62]. In this study, overrepresentation of the 4G allele was observed in the NES group compared with the control group of subjects without sinus disease (0.53 vs. 0.45) [60]. In a recent study on subjects (excluding asthma, atopy, or aspirin intolerance and thus were more likely to have NES), it was found that patients with CS had an increased prevalence of a GG genotype at position 2174 of the IL-6 promoter compared with a control group without sinus disease (odds ratio, 2.65) [63]. There is an effect of IL-6 on the differentiation of naive CD4 T1 cells to TH17 lineage, so this result conforms with the finding of an increased TH17 signature in NE-NPs [64]. IL-6 is also important for plasma cell differentiation, and this function also is consistent with the previously discussed role for humoral immunity in patients with NES.

Different studies showed that eosinophilic CRSwNP (ECRSwNP), and non-eosinophilic CRSwNP (NECRSwNP) types had distinct gene expression motifs [8, 9]. Interferons and acute inflammatory cytokines (IL1 and IL6) were predicted as upstream regulators in NECRSwNP in this study [65].

33.5.7 Chronic Hyperplastic Eosinophilic Sinusitis (CHES)

CHES is an inflammatory disease. It is characterized by intensive eosinophils accumulation in sinuses and also rarely accompanied by NP tissue. It participates a lot of histological and immunologic features with asthma and frequently associate with asthma, suggesting that CHES and asthma might include similar idiopathic immune process, each of them influences the upper and lower airways, respectively [66].

33.5.8 Molecular Basis of CHES

The close relation between CHES and NP formation has been shown in more than 30 studies addressing a possible genetic linkage. Several genes from the inflammatory pathway were found to relate/correlate with CHES, thus indicating a role for dysregulation of cytokine production in the pathogenesis of CHES. A few studies have replicated the relation between IL-1a and CS, similar to the relation between IL-1a and NPs. They found a greater risk related to the G allele at position 14,858 in exon 5 of the gene [67]. Several studies have also found that the 2308 G-to-A polymorphism in the TNF-a gene is associated with CS and NPs [68] however, other studies were unable to replicate these findings [67].

A study in patients with asthma reported that a C-to-T polymorphism at position 2590 of the IL-4 promoter may be connected to an increased risk of asthma [69]. Two cohort studies have also shown this polymorphism to be associated with the development of NPs [69]. IL-1ra, IL-1 receptor-like 1, IL-33, and matrix metalloproteinase 9 have been determined as other relevant inflammatory genes. The MHC genetic region is very significant for the development of CHES. Various genes within this region have been especially associated with defects in antigen presentation as a cause of disease. A 2- to five-fold increase in disease risk is associated with the HLA-DQA1* 0201 allele [70]. Also, more than 12 other HLA alleles have been described in disease pathogenesis, but many of them have not been replicated in later studies. Tokunaga et al. (2017) compared the gene expression profile of nasal polyps from patients who had ECRS and nasal polyps from patients who had non-ECRS. They demonstrated that the expression of transient receptor potential vanilloid 3 (TRPV3) was statistically different between these two groups [71].

33.5.9 Aspirin-Exacerbated Respiratory Disease (AERD)

AERD, also known as the Samter triad, is a condition characterized by NPs, asthma, and aspirin sensitivity [72]. Aspirin intolerance presents in 20% of adult asthmatic patients. Incidence increases to 30% if asthma is also associated with CS or nasal polyposis [73]. The key features of this disorder are its association with severe and extensive pansinusitis and its propensity to develop de novo in adulthood. In patients who use aspirin regularly, if pansinusitis is present on CT scan examination, it may indicate aspirin sensitivity [55]. Nasal polyposis in AERD is aggressive with multiple polyps. Polypoid changes are characterized by rapid growth and frequent recurrence after surgery [74]. Despite the diffuse involvement of the sinuses with inflammatory tissue, the patient is asymptomatic and compared with acute sinusitis or NES, patients seldom suffer from headaches or "sinus pressure." However, anosmia is one of the consistent complaints and perhaps causes the greatest morbidity.

33.5.10 Molecular Basis of AERD

Increased leukotriene synthesis is typical for AERD, so genes related to leukotriene synthesis or response have been the center of attraction for many researchers. LTC4S is the most important rate-limiting enzyme in leukotriene synthesis. Some studies have found an A-to-C base substitution at position 2444 of the LTC4S promoter, which increases LTC4S expression [75]. Some researchers subsequently identified a relationship between this base change and AERD [76], while others have not [77]. Thus, the significance of this substitution is still unclear. 5-lipoxygenase is the first enzyme in the leukotriene synthesis pathway. There are multiple variants of the tandem repeat GGGCGG within the promoter of the 5-lipoxygenase gene and Sp1 [78]. One study found an increased risk for the development of AERD related to this gene (odds ratio, 5.0) [79]. Additionally, another study showed that this polymorphism gives rise to the severity of airway hyperresponsiveness in patients with AERD [80]. There are also other proteins in the leukotriene pathway associated with AERD such as CysLT1 and CysLT2 receptors. A recent study summarized that the arachidonic acid metabolism signaling pathway genes LTC4S, ALOX5, CYSLTR1, and CYSLTR2 have the strongest evidence for a role in AERD pathogenesis [81]. Additionally, novel genetic polymorphisms, such as P2RY12 and dipeptidyl peptidase 10 gene, and epigenetic predispositions of AERD were determined [82].

In a GWAS, it is demonstrated that HLA-DPB1, rs3128965, DPP10 rs17048175 in a Korean population and TSLP rs1837253 in a Japanese population are associated with the phenotypes of NERD [83].

33.5.11 Allergic Fungal Sinusitis (AFS)

Sometimes, mold living commensally in the sinuses can cause activation of innate immune pathways and synergistically evoke robust TH2 lymphocyte and eosinophilic inflammatory responses. Initially, AFS was only attributed to Aspergillus species, but it is now known that many fungi species can be associated with AFS, including Cladosporium, Alternaria, Penicillium, Curvularia, and Bipolaris [84]. The key feature of AFS is specific IgE sensitization, which is demonstrated by skin prick tests or serum immunoassays and measurement of increased total serum IgE concentrations. AFS generally develops in young, immunocompetent, and atopic subjects [54]. AFS has some features that distinguish it from other forms of eosinophilic sinusitis, such as its often being unilateral and limited to one or a few sinuses. Dense material fills and expands the sinuses and can typically be detected with CT scan [85]. The mucous and inflammatory responses frequently occupy a space in the nasal cavity and lead to expansion into proximate tissue. This blocks the sinus ostia and subsequently causes bone absorption with resultant expansion into the orbits and cranium [86].

33.5.12 Genetics of AFS

There is only one study that shows a genetic linkage with AFS. In a study of 74 subjects including 44 enrolled with AFS, a weakly significant association of disease was determined with the MHC class II allele HLA-DQB1*03 [87], however, many subjects from the control group had at least one fungal species in the skin prick test.

33.6 Genetics of Cystic Fibrosis and Pathophysiology in Airways

33.6.1 Introduction

Cystic fibrosis (CF) is the most common lethal autosomal recessive genetic disorder, with a rate of approximately 1 in 2500 live births among Caucasians. There are approximately 80,000 children and young adults with CF in the world. Genetic and nongenetic factors contributing to the disease and its variants have been widely investigated. Though the major gene responsible for the pathophysiology of CF is the cystic fibrosis transmembrane conductance regulator (CFTR) gene, recent research suggests that variations in other so-called modifier genes have an important influence on phenotypic differences in this disease. In recent years, multiple candidate modifier genes have been investigated, in particular, genes that are involved in the control of infection, immunity, and inflammation [88].

33.6.2 Rationale for Cystic Fibrosis

Subjects with CF typically present with the disease in the lungs, sweat gland, pancreas, intestine (which is especially important during the newborn period), liver, and male reproductive tract [89]. CFTR controls chloride across the apical membranes of polarized epithelia [90]. Disruption in CFTR function inhibits the transport of sodium, chloride and water across epithelial tissues and so it causes insufficient hydration of mucous secretions in CF patients. Certain organs are eventually damaged from/by a blockage in the luminal space and follow recurrent cycles of inflammation and fibrosis [89, 91]. Many CF patients suffer from intestinal malabsorption and an abnormal nutritional status due to obstruction of the exocrine pancreas. The major cause of death in CF patients is complications arising from obstructive lung disease, a condition that occurs in approximately 90% of patients [92].

33.6.3 Genetics and CF

Lung function measurements are notably different among CF patients with identical CFTR genotypes (e.g., F508del homozygotes) [93]. In fact, an analysis of almost 88,000 patients in the CFTR2 database showed a low correlation between CFTR mutations and FEV1. There are only a few mutations that cause a milder pancreatic phenotype (e.g., p. Arg455Glu) [94, 95]. In aggregate, these studies show that factors other than the CFTR genotype affect the progression of airway obstruction in CF.

Recurrence of complications in affected siblings at rates higher than in unrelated patients indicates a genetic effect, but care must be taken to account for the effect of a similar environment for siblings. A more powerful approach is to compare monozygous (MZ) and dizygous (DZ) twin pairs for concordance for qualitative traits and correlation for quantitative traits. When MZ pairs show a stronger correlation than DZ pairs for a clinical feature, it shows that genetic factors may be responsible [96].

A higher correlation between composite measures of lung function and body mass index (BMI) was observed in 29 MZ versus 12 DZ twin pairs. This was the first twin-based assessment of the contribution of gene modifiers to CF disease severity and suggested genetic control of this trait [97]. Analysis of lung function and weight for height as independent measures did not show significant differences between the MZ and DZ twin pairs. Another comparison of 38 MZ pairs with six same-sex DZ pairs and 61 same-sex sibling pairs under 3 years of age demonstrated the heritability of lung function based on FEV1 measurements ranging from 0.54 to 1.0 [98]. Variance analysis of 231 pairs of affected siblings showed an insignificantly higher estimate of heritability for the FEV1 measures (0.68–1.0) [98]. In aggregate, these studies show that genetic modifiers have an essential role in determining FEV1, a key measure of lung function, which is correlated with survival.

Collaco et al. recruited 134 MZ twins and 272 DZ twins and siblings when living together and after moving apart to estimate the relative effect of genetic and environmental factors on FEV1 among CF patients. Differences in lung function between MZ twin pairs while living together in the same house supplied an estimate of the effect of unique environmental and stochastic contributions. Changing the home environment to independent living was used to assess the effect of a shared environment. The effect of genetic factors was estimated by comparing the similarities in lung function measures in MZ and DZ twin pairs when living together and subsequently when living apart. These methods showed that genetic and nongenetic factors had approximately equal effects on lung function. Analysis of 58 MZ twins and 568 DZ twins and siblings showed similar estimates for the genetic and nongenetic contributions to lung function variance [99].

33.6.4 Modifier Genes in CF

Two independent studies with more than 500 patients combined showed that more than nine genes can be involved in modifying some features of the CF phenotype. Several recent studies provide detailed lists of all the CF-related/modifier genes that have been studied thus far [99, 100]. These studies demonstrated the role of various modifier genes such as *MBL2*, *EDNRA*, and *TGF-β1* in lung function; *MBL2* in age at first *P. aeruginosa* infection; *MSRA* in meconium ileus; *TCF7L2* in CF-related diabetes; *SERPINA1* in CF-related liver disease.

Three studies in CF patients showed an earlier age of infection with Pseudomonas aeruginosa (Pa) to be related to mannose-binding lectin (MBL) deficiency genotypes. Lung disease severity, which is measured by FEV1 and infection status, is correlated with and two of them are changed by the age of the patient and by CFTR genotype. In aggregate, MBL2 genotype was found to be related to infection status more than the other variables [101]. Hence, deficits in MBL causes/can cause a predisposition to early infection with *Pa*, which leads to more severe lung disease than that observed in patients of the same age and CFTR genotype but who do not have MBL deficiency.

The Genetic Modifier Study (GMS), one of the largest CF genetic modifier studies to date, analyzed 808 F508del homozygotes drawn from the extremes of lung function (highest 30 percentile and lowest 30 percentile) and reported that alleles in the promoter (−509) and first exon (codon 10) of TGF-b1 are correlated with worse lung function [102]. This finding was studied in 498 patients with different CFTR genotypes and was separately confirmed when a haplotype composed of the opposite alleles at −509 and codon

10 was correlated with improved lung function [103]. Six studies including over 2500 CF patients determined a relationship between TGFb1 and CF lung function (see Table 33.1) while one study including 118 patients did not [104] and another involving 171 patients [105] found a relation between worse lung function and the opposite alleles than those reported by Drumm et al. and Bremer et al. [102, 103].

Three SNPs in the highest ranking gene, the interferon-related developmental regulator 1 gene (IFRD1), were identified in the whole GMS sample and showed a relationship using transmission-based methods in the family-based CF Twin and Sibling Study (TSS) [98]. IFRD1 acts via transcriptional mechanisms to alter neutrophil function in response to bacterial infection, as demonstrated by cell- and mouse-based studies.

It was demonstrated that variants in the interleukin-8 (IL-8) gene correlated with lung function. This result supported the idea that modification of CF lung disease may be caused by an altered neutrophil response to infection [106].

There are other mechanisms, which seemingly contribute to CF lung pathology as demonstrated by evidence that variants in the endothelin receptor type A (EDNRA) gene correlate with lung disease severity. Correlation between a variant in the 3′ untranslated region of EDNRA was identified in 709 F508del homozygous patients in the GMS study and replicated in three independent samples of CF patients. Also, alleles of the EDNRA variant are associated with differences in RNA transcript level, which indicates a possible functional role. Given that variation of EDNRA has been implicated in vasoconstrictive diseases as a result of effects on smooth muscle function, it was hypothesized that this gene may modulate CF lung disease by changing smooth muscle tone in the airways and vascular system [107].

33.6.5 Cystic Fibrosis and Nasal Findings

Clinical manifestations in the upper airways (UAW) occur in almost 100% of CF patients, appearing as recurrent sinusitis, rhinitis, and/or nasal polyposis [108, 109]. The frontal sinuses seldom develop in these patients, perhaps because of the early occurring/earlier occurring disorder of sinusitis which hinders pneumatization [110]. Sinusitis onset and nasal polyposis commonly occur between 5 and 14 years of age, with adult onset being unusual.

Most patients with CF (over 90%) [111, 112] develop chronic and recurring rhinosinusitis with or without nasal polyps. Modified mucus composition and viscoelasticity cause decreased mucociliary clearance and blockage in paranasal sinus drainage ostia, thereby promoting local inflammation, hypoxia and increased carbon dioxide partial pressure. Mucosal edema generally develops after impaired ciliary function and bacterial colonization, usually by Staphylococcus aureus and Pseudomonas aeruginosa [113, 114].

Franco et al. reported a relation between nasosinusal symptoms and cystic fibrosis. They found (the?) most common symptoms like cough (45%), oral breathing (44%), sleep disorders (42%), and nasal obstruction (37%) in CF patients. Twenty-eight patients (28%) had purulent nasal discharge and 41% had medial bulging of the nasal lateral Wall [115]. It is reported that nasosinusal involvement may worsen pulmonary disorder [116]. Hence, otorhinolaryngologists should investigate these patients in more detail for signs of pulmonary diseases. A recent study in Brazil [117] demonstrated more attention to the nasosinusal findings of CF patients, because CF is genetically very heterogeneous, with many types of mutations and a wide diversity in clinical presentations [118].

ΔF508 homozygosity was found more frequently in the patients undergoing sinus surgery (58%) compared with a control population (48%) [119]. Lastly, a study reported that ΔF508 homozygosity was associated with clinical severity of paranasal sinus diseases and with the presence of polyps on endoscopy in 113 patients [120].

33.6.6 CF and Nasal Polyposis

Nasal polyposis in CF patients was first described almost 50 years ago [121] but there is a little known about its pathophysiology [122]. The prevalence of nasal polyposis varies by population [117]. The incidence of nasal polyps has been observed in 6–48% of cases [123] depending on cystic fibrosis was diagnosed. Nearly 4% of patients already have symptomatic nasal polyposis when their diagnosis of CF is established and it is expected that nearly 14% of patients will undergo surgical intervention for their nasal polyp disease [113].

Weber et al. showed that nasal polyps were estimated in 39.1% of CF patients and, interestingly, all of them were older than 6 years of age, presenting with recurrent pneumonia in 82.6%, pancreatic insufficiency in 87%, and malnutrition in 74%. No correlation was seen between nasal polyps and sweat chlorine concentration, genotype, clinical signs of severity, and nasal symptoms. Nasal polyps regressed in seven patients treated with topical steroids, while six patients showed complete resolution [124].

Some researchers reported that patients with nasal polyposis had better pulmonary function, however a higher rate of Pseudomonas aeruginosa colonization, more hospitalizations, and more prevalence of allergy to Aspergillus fumigatus than the comparison group. They found no statistically different genotype distribution between the group with polyposis and the control group. But they also emphasized that the prevalence of the compound heterozygous genotype is higher within the nasal polyposis group than within controls [113].

33.7 Role of Genetics in Nasal Polyposis

33.7.1 Introduction

The nasal polyp is one of the final manifestations of chronic inflammation. Nasal polyposis is a chronic inflammatory disorder of the upper respiratory tract that 1–4% of the human population suffers from [125]. The lamina propria of nasal polyps usually presents great numbers of eosinophils and lymphocytes. In chronic inflammation, inflammatory cells produce neuropeptides, cytokines, and growth factors. These molecules lead to an extensive network of cellular interactions. In addition, resident structural cells can synthesize many of these molecules. Fibroblasts, epithelial cells, and endothelial cells help to organize the inflammatory process in nasal polyps [126].

Recently, it has been shown that there are pro-inflammatory cytokines such as tumor necrosis factor-α (TNF-α) and interleukin-1b (IL-1b) in the epithelial and endothelial cells of nasal polyps. Also, cell adhesion molecules such as very late antigen-4 (VLA-4) have been found on the surface of eosinophils, while integrins such as vascular cell adhesion molecule-1 (VCAM-1) have been shown on the surface of the small venules of the nasal polyp. Lastly, the presence of chemokines such as regulated upon activation of normal T cell expressed and secreted (RANTES), eotaxin, and IL-8 in the epithelium of the nasal polyps has been determined.

The nasal polyp tissue and the nasal mucosa have a sufficient collection of inflammatory molecules to combat efficiently against different agents such as allergens, bacteria, fungi, chemical particles, and viruses that come into the nose from the external environment. One of the most significant cells to offer an immune response may be the lymphocyte subpopulations. The percentages of TH_1 lymphocytes (which produce IL-2 and interferon-α [INF- α]) and TH_2 lymphocytes (which produce IL-4 and IL-5 cytokines) in the nasal pharyngeal tonsillar lymphocytes and peripheral blood lymphocytes have been determined in patients with nasal polyposis [127]. These same researchers have described the lymphocyte subpopulations and cytokines in nasal polyps [128].

33.7.2 Mucosal Irritation and the Role of Staphylococcal Exotoxin

As the nasal polyp symbolizes a final point in chronic inflammation, it is difficult to describe the initial events that trigger the inflammatory process in the lateral wall of the nose. Some substances, such as allergens, bacteria, viruses, air pollutants, and fungal elements, enter the submucosa of the lateral wall of the nose and damage the airway epithelium. These irritants lead to changes in some of the possible modifications of the respiratory epithelium that may take place after the entrance of these particles. These changes include the following: first, the synthesis of inflammatory eicosanoids, which are potent cell activators and chemoattractants; second, proinflammatory cytokines such as TNF-α and IL-1, which have major effects on growth, differentiation, migration, and activation of inflammatory cells; and, third, specific cell adhesion molecules, which have an essential role in managing the inflammatory cell. Lastly, major histocompatibility class II antigens have a crucial role in antigen presentation to T cells [129] and are also responsible for consequent activation of T cells. Figure 33.2 shows the possible changes in respiratory epithelium after the entrance of bacteria, viruses, allergens, and fungal elements.

Various cytokine subtypes are produced by the stimulation of epithelial cells by these elements. Shortly after exposure, activation of specific inflammatory cells occurs. Hence, the early growth of nasal polyposis may be the effect of stimulation of the epithelium by allowing irritants to change or damage the surface epithelium metabolically or physically. A cascade of inflammatory alterations takes place after this surface epithelium is damaged (Fig. 33.2).

A superantigen concept for massive nasal polyposis has been postulated. S. aureus is the most common bacterial species found in the nasal mucus. It has been shown in different studies that these bacteria synthesize exotoxins and that the corresponding variable-β region of the T cell receptor is also upregulated in polyp lymphocytes [130]. Based on these results, it is postulated that toxin-producing Staphylococci cause preliminary damage to the lateral wall of the nose. These exotoxins can act as superantigens, which lead to the proliferation of lymphocytes, which in turn synthesize cytokines that are associated with the massive proliferation of inflammatory cells that are observed in massive nasal polyposis.

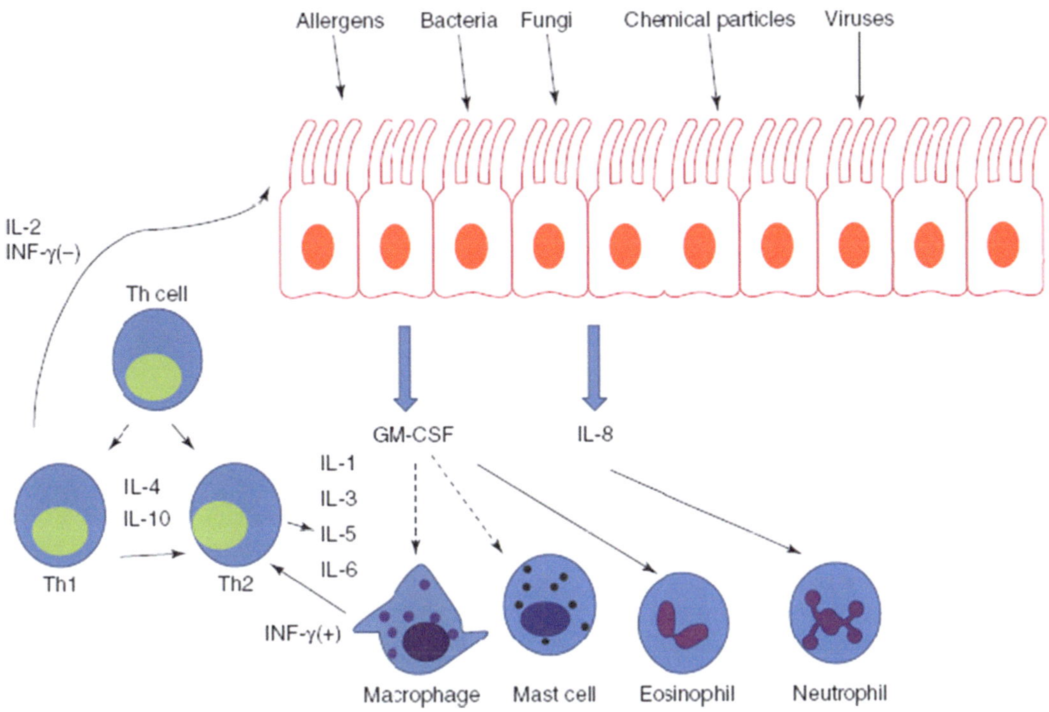

Fig. 33.2 Schematic presentation of epithelial damage in nasal polyp

33.7.3 Proinflammatory Cytokines Produced in Nasal Polyps

TNF-α and IL-1β cytokines play a role in the second process involved in the development of nasal polyposis after the initial mucosal irritation. The basic function of these two cytokines in the upregulation of endothelial adhesion molecules implicated in inflammatory reactions. TNF-α and IL-1β increase the production of endothelial adhesion molecules such as intracellular cell adhesion molecule-1 (ICAM-1) and VCAM-1. It has lately been shown by in vitro studies and animal experiments that certain adhesion molecules are important for adherence of eosinophils to endothelium and their subsequent extravasations.

Eotaxin and RANTES, which are cysteine/cysteine chemokines, attract and stimulate eosinophils in vitro and direct eosinophils into inflammatory lesions. There is strong evidence supporting the hypothesis that cytokines released from activated CD4 T cells mostly account for the restricted accumulation and activation of eosinophils in allergy-related disorders. It has been reported that these T cells produce some cytokines such as IL-4 and IL-13, which are also known as TH_2 cytokines. These cytokines play a role in favored extravasations of eosinophils through selective stimulation of VCAM-1 and IL-5. Also, granulocyte-macrophage colony-stimulating factor (GM-CSF) and IL-3 are essential for eosinophil activation and survival [131]. However, many studies propose that TH_1 cells are the main cells in nasal polyps and their release of INF-γ and IL-2 are highly present in nasal polyps. Thus, the recruitment of eosinophils may be linked to both TH_1 and TH_2 cytokines [132].

The specific localization of the eosinophil onto the vascular endothelial surface of the nasal polyp occurs due to the interaction of VLA-4 on eosinophils and VCAM-1 on venule endothelial cells. Eosinophil migration occurs within the nasal polyp venules after blood flow slowing down, and the subsequent transepithelial migration of these eosinophilic cells into the lamina propria of the nasal polyp occurs via the influence of chemokines such as RANTES and eotaxin. One study suggests that the eosinophil is the main cell in the nasal polyp, where eosinophils constitute up to 80% of the inflammatory cells [133].

Lymphocytes are enormously widespread cells accompanied by eosinophils found in the lamina propria of the nasal polyp. It is believed that protracted survival of these cells occurs via the autocrine upregulation of cytokines within the lamina propria of the nasal polyp. For instance, at least three cytokines (IL-3, GM-CSF, IL-5) are shown to decrease the apoptosis of eosinophils [134]. They have an impact on the long-term survival of eosinophils and their activation. Among these, IL-5 appears to have the most powerful effect in increasing the survival of eosinophils in the nasal polyp. Also, it was determined that the eosinophil itself can react by producing similar cytokines in an autocrine upregulation pattern. This vicious cycle of autocrine upregulation increases the recruitment of more eosinophils into the nasal polyp so that the chronic inflammatory state of eosinophils is extended.

33.7.4 Eosinophils and Electrophysiology of Respiratory Surface Epithelium

Airway mucus secretion is stimulated by eosinophilic cationic protein and inhibited by eosinophilic major basic protein (MBP) [135]. Just over a decade ago, Jacoby and colleagues (1988) showed that MBP increased net chloride secretion [136]. Also, MBP significantly facilitates sodium flux into the epithelial cell. Although there was a large change of chloride in and out of the cell, the net flux of chloride was not clearly determined. Finally, the short-circuit current seemed to be increased significantly with MBP compared with the control group.

One of the potential new strategies for the management of nasal polyps is based on the effect of amiloride and other sodium channel blocking agents (such as furosemide) on water movement into and out of nasal mucosa. Amiloride notably reduced sodium absorption and the short-circuit current. Hence, amiloride or furosemide may be useful as topical agents that could reduce sodium absorption into the cell and thereby reduce cellular and subcellular edema. It is approved by the findings from the bioelectric studies suggesting that nasal polyp epithelial cells have a normal luminal chloride channel which was controlled by increased chloride permeability after isoproterenol administration. Amiloride caused a larger decrement in sodium absorption in nasal polyp cells than in cells from the inferior turbinate mucosa. Amiloride is a specific blocker of the apical sodium channel and reduces the basal voltage and basal short-circuit current. These results show that sodium absorption may be increased in nasal polyps.

The mediators such as MBP produced by inflammatory cells of nasal polyps may increase sodium absorption, which could cause water retention in the epithelium of the lamina propria of polyps. The efficacy of corticosteroid treatment for nasal polyps depends on the inhibition of the synthesis of multiple cytokines. Decreased expression of the CFTR protein in remodeled human nasal epithelium from non-CF patients was demonstrated [137]. In normal adult pseudostratified human nasal surface epithelium, the CFTR is localized to the apical domain of the ciliated cells, whereas in CF, the mutated DF 508 CFTR gene causes an abnormal cytoplasmic location of the CFTR protein. Airway epithelial damage, in CF or non-CF patients, may induce a remodeling of the surface epithelium characterized by a change in the morphologic structure from normal columnar pseudostratified to basal hyperplasia, mucus cell hyperplasia, or squamous metaplasia. These histological findings are found in human polyp epithelium in the non-CF patient. Thus, abnormally low expression of the CFTR protein not only may be caused by the CFTR gene mutation in CF but also may be associated with airway surface epithelial differentiation and remodeling as occurring in nasal polyps from non-CF patients.

33.7.5 Medical Treatment of Chronic Rhinosinusitis with Massive Nasal Polyposis Based on the Molecular Biology of Inflammation

Patients with CRS and massive nasal polyposis typically have eosinophilic and lymphocytic infiltration in the lateral wall of the nose. In the phases of inflammation, there is a complex interaction between cytokine molecules. It gives rise to increased numbers and survival of eosinophils and lymphocytes in the nose. Hence, a rational approach to the medical treatment of this chronic inflammatory disorder can be properly achieved only after a complete understanding of the cytokine network (Table 33.3).

The major pathologic aspect of CRS with and without nasal polyposis is chronic inflammation. Hence, the application of specific anti-inflammatory drugs such as corticosteroids, which are the most commonly utilized drugs in the treatment of CRS, particularly with nasal polyposis, is useful. Antileukotriene therapy has also been found useful in the management of nasal polyposis. This drug may be especially effective in the aspirin-sensitive patient who has CRS with nasal polyposis.

Erythromycin and clarithromycin have prominent effects against neutrophils and some inflammatory cytokines, and interest in the potential anti-inflammatory effects of macrolide antibiotics has increased in the last 50 years. However, there are many reports of increasing bacterial resistance to macrolides for many significant species that specifically cause upper respiratory tract infection.

Microorganisms can stick on various surfaces and shape a three-dimensional constitution known as biofilm. After a biofilm has been formed on the mucosal surface, the bacteria har-

Table 33.3 Inflammation changes in different CRS subtype

CRS with NP		CRS without NP	
TGF-β1 ↑	Edema	TGF-β1↑↑↑↑	Fibrosis
T_{reg} ↑	T_H2 +	T_{reg} ↑↑↑↑	T_H1
	T_H2 –		

bored in the biofilm are less exposed to the immune response and less vulnerable to antibiotics. One study mentions that the use of furosemide and amiloride was found to be valuable in the postoperative treatment of CRS with nasal polyposis [138].

Anti-IgE therapy is a compelling new therapeutic molecule for the neutralization of IgE and the inhibition of IgE synthesis [139]. Monoclonal anti-IgE therapy may be a logical approach in the treatment of chronic hyperplastic sinusitis when allergy is a major factor in a patient with IgE-mediated hypersensitivity.

33.8 Vasomotor Rhinitis and Its Genetic Background

33.8.1 Introduction

Rhinitis is an inflammation of the nasal area and generally characterized by rhinorrhea, nasal congestion, sneezing, and/or nasal itching [140]. It is classified into subtypes of allergic, nonallergic, occupational, hormonal (pregnancy and hypothyroidism), drug induced, and food ingestion induced [141]. Vasomotor rhinitis (VMR) is the most common type of chronic nonallergic rhinitis (NAR). Millions of people suffer from vasomotor rhinitis and it causes uncomfortable symptomatology. VMR is an idiopathic condition diagnosed after exclusion of infection, allergy, eosinophilia, hormonal changes (such as pregnancy), and exposure to drugs. Hence, sometimes scientists have described it as a "wastebasket diagnosis" [142, 143].

Certain odors, alcohol, spicy foods, emotions, and environmental factors such as temperature, barometric pressure changes, and bright lights exacerbate these symptoms [144] (Fig. 33.3). Allergic and nonallergic rhinitis have notably overlapping symptoms, but the causes appear to be entirely different [144].

Skin testing or in vitro tests for allergen-specific IgE are usually used for allergic rhinitis (AR) or VMR diagnosis. Also, a patient may have both allergic and nonallergic components and this is named "mixed rhinitis." These patients must be recognized properly because positive testing for a specific allergen may cause the clinician to ignore the role and management of non-allergic factors.

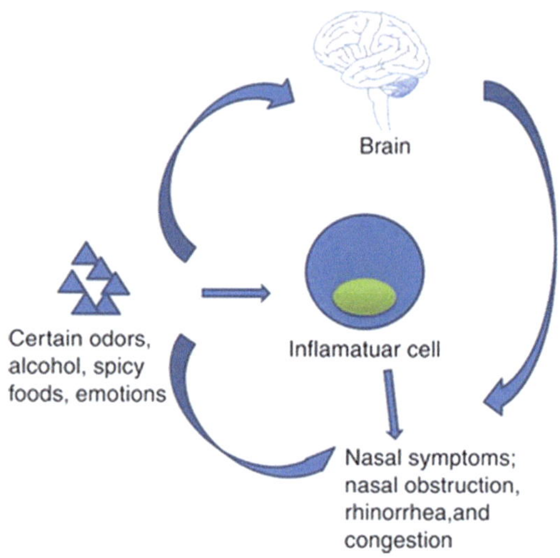

Fig. 33.3 Possible mechanism of VMR

33.8.2 Epidemiology

Approximately 19 million people suffer from NAR in the United States and a further 26 million experience mixed rhinitis.

33.8.3 Pathophysiology of VMR

Several hypotheses have been suggested for the pathophysiology of VMR.

33.8.3.1 Trauma

Surgical and nonsurgical trauma has been accounted to cause VMR as a long-term complication.

33.8.3.2 Autonomic Dysfunction

Patients with vasomotor rhinitis are clinically separated into two subgroups: "runners," who present "wet" rhinorrhea, and "dry" patients, who show nasal obstruction and airflow resistance with minimal rhinorrhea. Many researchers have endeavored to explain the pathogenic mechanisms for these subgroups. An imbalance in the autonomic input to the nasal mucosa has been held responsible for VMR since the 1950s. Recent theories postulate that increased cholinergic glandular secretory activity is responsible for runners, while nociceptive neurons with increased sensitivity to generally innocent stimuli are responsible for dry patients. Recent studies have suggested that VMR is due to a hypoactive sympathetic nervous system rather than a hyperactive parasympathetic system [145]. There are some factors that trigger symptoms of VMR such as changes in temperature or humidity, smoke, alcohol, odors, perfumes, sexual arousal, and emotional factors [143]. A study determined nasal hyperreactivity to cold air using anterior rhinomanometry [146]. Numata et al. determined nasal hyperreactivity to histamine using acoustic rhinometry [147].

33.8.3.3 Cytokines and VMR

Chen et al. showed that there were no significant differences in levels of IL-10, IL-13, or IL-16 between vasomotor rhinitis and normal controls. But the level of IL-12 in vasomotor rhinitis was lower than that of normal controls. Further research is needed on the role of IL-12 in vasomotor rhinitis [148].

33.8.3.4 Light and Electron Microscopic Findings

Giannessi et al. recently studied microscopic and ultrastructural alterations in the nasal mucosa of VMR patients [149]. VMR patients who underwent inferior turbinate reduction showed abnormal epithelium in 80–90% of the nasal surface with light microscopy. They observed decreases in epithelial thickness and loss of ciliated and goblet cells on the nasal surface. Also, ultrastructural studies supported light microscopic findings. They detected ciliary loss, lack of tight junctions, loss of vibratile cilia, loss of goblet cells and ciliated cells, and a marked expansion of the intercellular spaces.

33.8.3.5 Neuropeptides

Some scientists have investigated the neurogenic and molecular mechanisms of VMR. Tai and Baraniuk [150] suggest that sensory nerve endings and autonomic dysfunction have a role in VMR. Stimulation of nasal sensory nerves caused sensations of pain and stuffiness. Type C nociceptive nerves synthesize some neuropeptides such as substance P (SP) and calcitonin gene-related peptides, and it increased plasma extravasation and glandular secretion. Groneberg et al. [151] studied the neuropeptide content of mucosal parasympathetic, sympathetic, and sensory nerves of patients with toxic rhinitis caused by chronic cigarette smoke exposure. They measured concentrations of calcitonin gene-related peptides, SP, vasoactive intestinal peptide, and neuropeptide tyrosine (NPY) and determined significantly increased concentrations of vasoactive intestinal peptide and NPY in the nasal mucosa of toxic rhinitis compared with normal subjects. Also, the level of SP expression was increased. SP is generally distributed in nerve fibers near submucosal glands and blood vessels, whereas NPY is found near submucosal blood vessels. SP has many functions in the body such as increasing plasma extravasation, glandular secretion vasodilatation, and mucociliary clearance. Vasoactive intestinal peptide (VIP) is a neurotransmitter that has a role in the inhibitory noncholinergic airway nervous system and it always dilates bronchus and vasculature. Increased levels of VIP may lead to hypersecretion. Groneberg et al. (2003) postulated that a separate subclass of nerves might be responsible for the pathophysiology of toxic rhinitis and that major changes in the content of mucosal nerves occur in toxic rhinitis [151].

Schierhorn et al. (2002) investigated ozone-induced releases of SP and neurokinin A, and ozone stimulation was found to increase SP and neurokinin A levels [152]. Also, it was reported that the ozone-induced increase in neuropeptides in allergic patients was higher as compared with nonallergic patients. Ozone might increase sensory nerve activity in the upper airways and as a result increase neuropeptide release in the upper airways.

33.8.3.6 Nitric Oxide

The function of nitric oxide (NO) in the pathogenesis of VMR was studied by Giannessi et al. (2003) and by Ruffoli et al. (2000) [149, 153]. Three isoforms of NOS have been shown in the human nasal mucosa. Nicotinamide adenine dinucleotide phosphate (NADPH) is used by all isoforms of NOS as a cofactor. Hence, NADPH-diaphorase histochemistry is used to investigate NOS in tissues. The damaged epithelium containing cells with marked reactivity to NADPH-diaphorase was found in the nasal respiratory epithelium in VMR by Giannessi et al. [149]. Basal cells in VMR presented strong NADPH-diaphorase activity, while NOS activity was negative in basal cells of normal subjects.

NO is known to have cytostatic and cytotoxic effects against microbes and cancer cells. Inducible NOS stimulation could synthesize a high degree of NO, and it could cause decreased viability of normal tissue and necrosis. High-level NOS expression in the nasal epithelium could lead to a constant high level of NO and result in constant epithelial damage. This has been postulated as one of the possible pathogeneses of VMR. Repression of mucociliary clearance decreased the number of tight junctions, and disruption in the basement membrane continuities might permit environmental agents to interact directly with the subepithelial structures. Consequently, symptomatic VMR is caused by increased responsiveness to the afferent trigeminal fibers, and recruitment of secretory and vascular reflexes could occur in the nasal respiratory mucosa.

Cervin et al. (1999) studied the functional effects of NPY receptors on blood flow and NO concentrations in the human nose by using dose-dependent effects of the intranasal application of NPY [154]. They showed that application of NPY leads to vasoconstriction and a decrease in NO levels. Braat et al. (2002) showed that pollution and meteorologic factors are linked with the severity of symptoms in VMR patients [155]. They determined that minimum daily temperature and the levels of ozone and NO had the highest association with the severity of symptoms.

33.8.3.7 Nasal Secretory Proteins

The protein analysis of nasal washes to differentiate VMR from other forms of rhinitis was studied by Iguchi et al. (2002) and Tosun et al. (2002) [156, 157]. Iguchi et al. (2002) investigated control, VMR, and perennial AR subjects [156]. The total protein and albumin level in AR was higher than the total protein and albumin level in NAR ($P < 0.01$ for both). It is shown that the difference in total protein and albumin concentration between normal control subjects and NAR was also statistically significant ($P < 0.05$ for both). The control subjects had the lowest total protein and albumin levels in their nasal lavage. They also found a protein with a molecular weight of 26 kDa. The identity of this protein has not been determined yet, but it is believed to derive from the nasal glands since its secretion can be provoked in normal volunteers with pilocarpine nasal spray. The average level of this protein was significantly higher in AR subjects compared with control subjects ($P < 0.01$) and NAR subjects ($P < 0.05$). The level of the 26-kDa protein in NAR was higher than in control subjects but it was not statistically significant. It is suggested that increased vascular permeability led to increased albumin concentration in nasal discharge. The level of the 26-kDa protein is therefore enhanced due to increased gland secretion. In the NAR group, vascular permeability may have been increased over control subjects, but gland secretion was minimal, and the 26-kDa protein level remained low. Hence, the presence of the 26-kDa protein can be used to distinguish AR from NAR.

The gel electrophoretic assessments of proteins in nasal washings of patients with AR and VMR were studied [157]. The average total level of proteins, 66-kDa proteins, and 26-kDa proteins was determined to be higher in AR compared with those from VMR. The lowest rate of these proteins was seen in the control group. The differences in the mean concentration of proteins in AR, VMR, and control groups were statistically significant ($P < 0.05$).

Aust et al. (1997) studied the gene expression of eight types of mucin in control and vasomotor inferior turbinates, and the only difference observed between normal and VMR turbinates was a minor decrease in mucin 1, transmembrane (MUC 1) gene expression in the vasomotor group [158]. They suggested whether this decrease could begin abnormal neurogenic signals that lead to an increase in nasal secretions in VMR.

33.8.3.8 Acid Reflux

It is suggested that there is a relation between laryngopharyngeal reflux and VMR based on autonomic dysfunction [145]. Patients with VMR and those with extraesophageal manifestations of gastroesophageal reflux show findings of autonomic dysfunction. Patients with both VMR and esophageal reflux show a considerably greater degree of autonomic dysfunction compared with patients with only VMR. Shaker et al. investigated the intrapharyngeal distribution of gastric acid refluxate [159] in reflux laryngitis, VMR, and control subjects using dual pharyngeal and esophageal probes. The study determined that the number and extent of reflux events in the esophagus and lower pharynx are indistinguishable between patients and control subjects.

33.9 Conclusions

Rhinologic diseases are very common worldwide. It is known that these diseases have a significant genetic background. A lot of development in the field of genetics has been achieved over the last decade, and it is expected to advance even further in the next, as increasingly powerful analytical tools are being developed to solve the complexities of genetic diseases. These progress in information about the genetics of allergic diseases and new tools help not only for illuminating the molecular basis of these diseases but also for developing new therapies. In the future, many rhinologic diseases with chronic progress will have a chance to be treated through new developments in the field of genetics.

References

1. Kellman RM, Schmidt C. The paranasal sinuses as a protective crumple zone for the orbit. Laryngoscope. 2009;119(9):1682–90.
2. Chen B, Shaari J, Claire SE, Palmer JN, Chiu AG, Kennedy DW, et al. Altered sinonasal ciliary dynamics in chronic rhinosinusitis. Am J Rhinol. 2006;20(3):325–9.
3. National Institute of Allergy and Infectious Diseases Task Force. Asthma and other allergic diseases. Washington, DC: US Department of Health, Education and Welfare; 1979.
4. Mygind NJH. In: AB Kay, editor. Structure and function of the upper airways, 1st ed. London: Blackwell Science; 1997.
5. White MV, Kaliner MA. Mediators of allergic rhinitis. J Allergy Clin Immunol. 1992;90(4 Pt 2):699–704.
6. Wright AL, Holberg CJ, Martinez FD, Halonen M, Morgan W, Taussig LM. Epidemiology of physician-diagnosed allergic rhinitis in childhood. Pediatrics. 1994;94(6 Pt 1):895–901.
7. Settipane RA, Charnock DR. Epidemiology of rhinitis: allergic and nonallergic. Clin Allergy Immunol. 2007;19:23–34.
8. Saltoun C, Avila PC. Advances in upper airway diseases and allergen immunotherapy in 2007. J Allergy Clin Immunol. 2008;122(3):481–7.
9. van Cauwenberge PWJ-B, Van Zele T, Van HH. Allergic rhinitis allergy. Ghent: Elsevier Ltd; 2006.
10. Wallace DV, Dykewicz MS, Bernstein DI, Blessing-Moore J, Cox L, Khan DA, et al. The diagnosis and management of rhinitis: an updated practice parameter. J Allergy Clin Immunol. 2008;122(2 Suppl):S1–84.
11. Bousquet J, Khaltaev N, Cruz AA, Denburg J, Fokkens WJ, Togias A, et al. Allergic rhinitis and its impact on asthma (ARIA) 2008 update (in collaboration with the World Health Organization, GA(2)LEN and AllerGen). Allergy. 2008;63(Suppl 86):8–160.
12. Toda M, Ono SJ. Genomics and proteomics of allergic disease. Immunology. 2002;106(1):1–10.
13. Vercelli D. Discovering susceptibility genes for asthma and allergy. Nat Rev Immunol. 2008;8(3):169–82.
14. Shirkani A, Mansouri A, Farid Hosseini R, Jabbari Azad F, Alsadat Mahmoudian R, Montazer M, et al. The role of interleukin-4 and 13 gene polymorphisms in allergic rhinitis: a case control study. Rep Biochem Mol Biol. 2019;8(2):111–8.
15. Wang M, Xing ZM, Lu C, Ma YX, Yu DL, Yan Z, et al. A common IL-13 Arg130Gln single nucleotide polymorphism among Chinese atopy patients with allergic rhinitis. Hum Genet. 2003;113(5):387–90.
16. Li CS, Chae SC, Lee JH, Zhang Q, Chung HT. Identification of single nucleotide polymorphisms in FOXJ1 and their association with allergic rhinitis. J Hum Genet. 2006;51(4):292–7.
17. Amarin JZ, Naffa RG, Suradi HH, Alsaket YM, Obeidat NM, Mahafza TM, et al. An intronic single-nucleotide polymorphism (rs13217795) in FOXO3 is associated with asthma and allergic rhinitis: a case-case-control study. BMC Med Genet. 2017;18(1):132.
18. Guo H, Peng T, Luo P, Li H, Huang S, Li S, et al. Association of FcepsilonRIbeta polymorphisms with risk of asthma and allergic rhinitis: evidence based on 29 case-control studies. Biosci Rep. 2018;38(4):BSR20180177.

19. Haagerup A, Bjerke T, Schoitz PO, Binderup HG, Dahl R, Kruse TA. Allergic rhinitis--a total genome-scan for susceptibility genes suggests a locus on chromosome 4q24-q27. Eur J Hum Genet. 2001;9(12):945–52.
20. Brasch-Andersen C, Haagerup A, Borglum AD, Vestbo J, Kruse TA. Highly significant linkage to chromosome 3q13.31 for rhinitis and related allergic diseases. J Med Genet. 2006;43(3):e10.
21. Masuda ES, Schmitz J. Syk inhibitors as treatment for allergic rhinitis. Pulm Pharmacol Ther. 2008;21(3):461–7.
22. Kato T, Ohta T, Iwasaki H, Kobayashi H, Matsuo A, Hata T, et al. JTE-852, a novel spleen tyrosine kinase inhibitor, blocks antigen-induced allergic reactions in rats. J Vet Med Sci. 2018;80(3):465–72.
23. Ramasamy A, Curjuric I, Coin LJ, Kumar A, McArdle WL, Imboden M, et al. A genome-wide meta-analysis of genetic variants associated with allergic rhinitis and grass sensitization and their interaction with birth order. J Allergy Clin Immunol. 2011;128(5):996–1005.
24. Lu MP, Chen RX, Wang ML, Zhu XJ, Zhu LP, Yin M, et al. Association study on IL4, IL13 and IL4RA polymorphisms in mite-sensitized persistent allergic rhinitis in a Chinese population. PLoS One. 2011;6(11):e27363.
25. Kruse LV, Nyegaard M, Christensen U, Moller-Larsen S, Haagerup A, Deleuran M, et al. A genome-wide search for linkage to allergic rhinitis in Danish sib-pair families. Eur J Hum Genet. 2012;20(9):965–72.
26. Denham S, Koppelman GH, Blakey J, Wjst M, Ferreira MA, Hall IP, et al. Meta-analysis of genome-wide linkage studies of asthma and related traits. Respir Res. 2008;9:38.
27. Ober C, Hoffjan S. Asthma genetics 2006: the long and winding road to gene discovery. Genes Immun. 2006;7(2):95–100.
28. Bierbaum S, Nickel R, Koch A, Lau S, Deichmann KA, Wahn U, et al. Polymorphisms and haplotypes of acid mammalian chitinase are associated with bronchial asthma. Am J Respir Crit Care Med. 2005;172(12):1505–9.
29. Rodriguez E, Illig T, Weidinger S. Filaggrin loss-of-function mutations and association with allergic diseases. Pharmacogenomics. 2008;9(4):399–413.
30. Joki-Erkkila VP, Karjalainen J, Hulkkonen J, Pessi T, Nieminen MM, Aromaa A, et al. Allergic rhinitis and polymorphisms of the interleukin 1 gene complex. Ann Allergy Asthma Immunol. 2003;91(3):275–9.
31. Hussein YM, Awad HA, Shalaby SM, Ali AS, Alzahrani SS. Toll-like receptor 2 and toll-like receptor 4 polymorphisms and susceptibility to asthma and allergic rhinitis: a case-control analysis. Cell Immunol. 2012;274(1–2):34–8.
32. Aryan Z, Rezaei N. Toll-like receptors as targets for allergen immunotherapy. Curr Opin Allergy Clin Immunol. 2015;15(6):568–74.
33. Prescott S, Saffery R. The role of epigenetic dysregulation in the epidemic of allergic disease. Clin Epigenetics. 2011;2(2):223–32.
34. Noakes PS, Holt PG, Prescott SL. Maternal smoking in pregnancy alters neonatal cytokine responses. Allergy. 2003;58(10):1053–8.
35. Prescott SL, Wickens K, Westcott L, Jung W, Currie H, Black PN, et al. Supplementation with Lactobacillus rhamnosus or Bifidobacterium lactis probiotics in pregnancy increases cord blood interferon-gamma and breast milk transforming growth factor-beta and immunoglobin A detection. Clin Exp Allergy. 2008;38(10):1606–14.
36. Arbibe L, Kim DW, Batsche E, Pedron T, Mateescu B, Muchardt C, et al. An injected bacterial effector targets chromatin access for transcription factor NF-kappaB to alter transcription of host genes involved in immune responses. Nat Immunol. 2007;8(1):47–56.
37. Schaub B, Liu J, Hoppler S, Schleich I, Huehn J, Olek S, et al. Maternal farm exposure modulates neonatal immune mechanisms through regulatory T cells. J Allergy Clin Immunol. 2009;123(4):774–82 e5.
38. Devereux G, Litonjua AA, Turner SW, Craig LC, McNeill G, Martindale S, et al. Maternal vitamin D intake during pregnancy and early childhood wheezing. Am J Clin Nutr. 2007;85(3):853–9.
39. Utsugi M, Dobashi K, Ishizuka T, Endou K, Hamuro J, Murata Y, et al. c-Jun N-terminal kinase negatively regulates lipopolysaccharide-induced IL-12 production in human macrophages: role of mitogen-activated protein kinase in glutathione redox regulation of IL-12 production. J Immunol. 2003;171(2):628–35.
40. Breton CV, Byun HM, Wenten M, Pan F, Yang A, Gilliland FD. Prenatal tobacco smoke exposure affects global and gene-specific DNA methylation. Am J Respir Crit Care Med. 2009;180(5):462–7.
41. Liu J, Ballaney M, Al-alem U, Quan C, Jin X, Perera F, et al. Combined inhaled diesel exhaust particles and allergen exposure alter methylation of T helper genes and IgE production in vivo. Toxicol Sci. 2008;102(1):76–81.
42. Dowling O, Rochelson B, Way K, Al-Abed Y, Metz CN. Nicotine inhibits cytokine production by placenta cells via NFkappaB: potential role in pregnancy-induced hypertension. Mol Med. 2007;13(11–12):576–83.
43. Hansel NN, Cheadle C, Diette GB, Wright J, Thompson KM, Barnes KC, et al. Analysis of CD4+ T-cell gene expression in allergic subjects using two different microarray platforms. Allergy. 2008;63(3):366–9.
44. Bell JT, Spector TD. A twin approach to unraveling epigenetics. Trends Genet. 2011;27(3):116–25.
45. Sjogren AK, Barrenas F, Muraro A, Gustafsson M, Saetrom P, Wang H, et al. Monozygotic twins discordant for intermittent allergic rhinitis differ in mRNA and protein levels. Allergy. 2012;67(6):831–3.

46. Yang J, Zhong W, Xue K, Wang Z. Epigenetic changes: an emerging potential pharmacological target in allergic rhinitis. Int Immunopharmacol. 2019;71:76–83.
47. Tomassen P, Van Zele T, Zhang N, Perez-Novo C, Van Bruaene N, Gevaert P, et al. Pathophysiology of chronic rhinosinusitis. Proc Am Thorac Soc. 2011;8(1):115–20.
48. Pleis JR, Lucas JW, Ward BW. Summary health statistics for U.S. adults: National Health Interview Survey, 2008. Vital Health Stat 10. 2009;(242):1–157.
49. Bhattacharyya N. Contemporary assessment of the disease burden of sinusitis. Am J Rhinol Allergy. 2009;23(4):392–5.
50. Hamilos DL. Chronic rhinosinusitis: epidemiology and medical management. J Allergy Clin Immunol. 2011;128(4):693–707. quiz 8-9
51. Slavin RG, Spector SL, Bernstein IL, Kaliner MA, Kennedy DW, Virant FS, et al. The diagnosis and management of sinusitis: a practice parameter update. J Allergy Clin Immunol. 2005;116(6 Suppl):S13–47.
52. Perry BF, Login IS, Kountakis SE. Nonrhinologic headache in a tertiary rhinology practice. Otolaryngol Head Neck Surg. 2004;130(4):449–52.
53. Bachert C, Gevaert P, Holtappels G, Johansson SGO, van Cauwenberge P. Total and specific IgE in nasal polyps is related to local eosinophilic inflammation. J Allergy Clin Immun. 2001;107(4):607–14.
54. Schubert MS, Goetz DW. Evaluation and treatment of allergic fungal sinusitis. I. Demographics and diagnosis. J Allergy Clin Immunol. 1998;102(3):387–94.
55. Mascia K, Borish L, Patrie J, Hunt J, Phillips CD, Steinke JW. Chronic hyperplastic eosinophilic sinusitis as a predictor of aspirin-exacerbated respiratory disease. Ann Allergy Asthma Immunol. 2005;94(6):652–7.
56. Iqbal IZ, Kao SS, Ooi EH. The role of biologics in chronic rhinosinusitis: a systematic review. Int Forum Allergy Rhinol. 2020;10(2):165–74.
57. Van Bruaene N, Derycke L, Perez-Novo CA, Gevaert P, Holtappels G, De Ruyck N, et al. TGF-beta signaling and collagen deposition in chronic rhinosinusitis. J Allergy Clin Immunol. 2009;124(2):253–9.
58. Rosenfeld RM, Andes D, Bhattacharyya N, Cheung D, Eisenberg S, Ganiats TG, et al. Clinical practice guideline: adult sinusitis. Otolaryngol Head Neck Surg. 2007;137(3 Suppl):S1–31.
59. Payne SC, Early SB, Huyett P, Han JK, Borish L, Steinke JW. Evidence for distinct histologic profile of nasal polyps with and without eosinophilia. Laryngoscope. 2011;121(10):2262–7.
60. de Alarcon A, Steinke JW, Caughey R, Barekzi E, Hise K, Gross CW, et al. Expression of leukotriene C4 synthase and plasminogen activator inhibitor 1 gene promoter polymorphisms in sinusitis. Am J Rhinol. 2006;20(5):545–9.
61. Shimizu S, Gabazza EC, Ogawa T, Tojima I, Hoshi E, Kouzaki H, et al. Role of thrombin in chronic rhinosinusitis-associated tissue remodeling. Am J Rhinol Allergy. 2011;25(1):7–11.
62. Cho SH, Hall IP, Wheatley A, Dewar J, Abraha D, Del Mundo J, et al. Possible role of the 4G/5G polymorphism of the plasminogen activator inhibitor 1 gene in the development of asthma. J Allergy Clin Immunol. 2001;108(2):212–4.
63. Kosugi EM, de Camargo-Kosugi CM, Weckx LL, Guerreiro-da-Silva ID, Gregorio LC. Interleukin-6 -174 G/C promoter polymorphism and nasal polyposis. Rhinology. 2009;47(4):400–4.
64. Zhang N, Van Zele T, Perez-Novo C, Van Bruaene N, Holtappels G, DeRuyck N, et al. Different types of T-effector cells orchestrate mucosal inflammation in chronic sinus disease. J Allergy Clin Immunol. 2008;122(5):961–8.
65. Okada N, Nakayama T, Asaka D, Inoue N, Tsurumoto T, Takaishi S, et al. Distinct gene expression profiles and regulation networks of nasal polyps in eosinophilic and non-eosinophilic chronic rhinosinusitis. Int Forum Allergy Rhinol. 2018;8(5):592–604.
66. Braunstahl GJ, Overbeek SE, Kleinjan A, Prins JB, Hoogsteden HC, Fokkens WJ. Nasal allergen provocation induces adhesion molecule expression and tissue eosinophilia in upper and lower airways. J Allergy Clin Immunol. 2001;107(3):469–76.
67. Endam LM, Cormier C, Bosse Y, Filali-Mouhim A, Desrosiers M. Association of IL1A, IL1B, and TNF gene polymorphisms with chronic rhinosinusitis with and without nasal polyposis a replication study. Arch Otolaryngol. 2010;136(2):187–92.
68. Batikhan H, Gokcan MK, Beder E, Akar N, Ozturk A, Gerceker M. Association of the tumor necrosis factor-alpha-308 G/A polymorphism with nasal polyposis. Eur Arch Otorhinolaryngol. 2010;267(6):903–8.
69. Park SK, Heo KW, Jung H, Yea SS, Yang YI. Expression of cyclooxygenase-2 and 5-lipoxygenase in nasal polyps associated with interleukin-4 promoter polymorphism -590. Otolaryngol Head Neck Surg. 2006;135(6):928–32.
70. Molnar-Gabor E, Endreffy E, Rozsasi A. HLA-DRB1, -DQA1, and -DQB1 genotypes in patients with nasal polyposis. Laryngoscope. 2000;110(3 Pt 1):422–5.
71. Tokunaga T, Ninomiya T, Kato Y, Imoto Y, Sakashita M, Takabayashi T, et al. The significant expression of TRPV3 in nasal polyps of eosinophilic chronic rhinosinusitis. Allergol Int. 2017;66(4):610–6.
72. Samter M, Beers RF Jr. Intolerance to aspirin. Clinical studies and consideration of its pathogenesis. Ann Intern Med. 1968;68(5):975–83.
73. Vally H, Taylor ML, Thompson PJ. The prevalence of aspirin intolerant asthma (AIA) in Australian asthmatic patients. Thorax. 2002;57(7):569–74.
74. Szczeklik A, Sanak M. Genetic mechanisms in aspirin-induced asthma. Am J Respir Crit Care Med. 2000;161(2):S142–S6.
75. Sanak M, Pierzchalska M, Bazan-Socha S, Szczeklik A. Enhanced expression of the leukotriene C-4 syn-

thase due to overactive transcription of an allelic variant associated with aspirin-intolerant asthma. Am J Respir Cell Mol Biol. 2000;23(3):290–6.
76. Kawagishi Y, Mita H, Taniguchi M, Maruyama M, Oosaki R, Higashi N, et al. Leukotriene C4 synthase promoter polymorphism in Japanese patients with aspirin-induced asthma. J Allergy Clin Immunol. 2002;109(6):936–42.
77. Van Sambeek R, Stevenson DD, Baldasaro M, Lam BK, Zhao J, Yoshida S, et al. 5′ flanking region polymorphism of the gene encoding leukotriene C4 synthase does not correlate with the aspirin-intolerant asthma phenotype in the United States. J Allergy Clin Immunol. 2000;106(1 Pt 1):72–6.
78. In KH, Asano K, Beier D, Grobholz J, Finn PW, Silverman EK, et al. Naturally occurring mutations in the human 5-lipoxygenase gene promoter that modify transcription factor binding and reporter gene transcription. J Clin Investig. 1997;99(5):1130–7.
79. Choi JH, Park HS, Oh HB, Lee JH, Suh YJ, Park CS, et al. Leukotriene-related gene polymorphisms in ASA-intolerant asthma: an association with a haplotype of 5-lipoxygenase. Hum Genet. 2004;114(4):337–44.
80. Kim SH, Bae JS, Suh CH, Nahm DH, Holloway JW, Park HS. Polymorphism of tandem repeat in promoter of 5-lipoxygenase in ASA-intolerant asthma: a positive association with airway hyperresponsiveness. Allergy. 2005;60(6):760–5.
81. Dahlin A, Weiss ST. Genetic and epigenetic components of aspirin-exacerbated respiratory disease. Immunol Allergy Clin N Am. 2016;36(4):765–89.
82. Le Pham D, Lee JH, Park HS. Aspirin-exacerbated respiratory disease: an update. Curr Opin Pulm Med. 2017;23(1):89–96.
83. Park BL, Kim TH, Kim JH, Bae JS, Pasaje CF, Cheong HS, et al. Genome-wide association study of aspirin-exacerbated respiratory disease in a Korean population. Hum Genet. 2013;132(3):313–21.
84. Robson JMB, Hogan PG, Benn RAV, Gatenby PA. Allergic fungal sinusitis presenting as a paranasal sinus tumor. Aust NZ J Med. 1989;19(4):351–3.
85. Mukherji SK, Figueroa RE, Ginsberg LE, Zeifer BA, Marple BF, Alley JG, et al. Allergic fungal sinusitis: CT findings. Radiology. 1998;207(2):417–22.
86. Lydiatt WM, Sobba-Higley A, Huerter JV Jr, Leibrock LG. Allergic fungal sinusitis with intracranial extension and frontal lobe symptoms: a case report. Ear Nose Throat J. 1994;73(6):402–4.
87. Schubert MS, Hutcheson PS, Graff RJ, Santiago L, Slavin RG. HLA-DQB1 *03 in allergic fungal sinusitis and other chronic hypertrophic rhinosinusitis disorders. J Allergy Clin Immunol. 2004;114(6):1376–83.
88. Cutting GR. Modifier genetics: cystic fibrosis. Annu Rev Genomics Hum Genet. 2005;6:237–60.
89. Welsh MR, Ramsey BW, Accurso FJ, Cutting GR. Cystic fibrosis. In: Scriver CB, Beaudet AL, Valle DS, Sly WS, editors. The metabolic and molecular bases of inherited disease. New York: McGraw Hill; 2001. p. 5121–88.
90. Anderson MP, Gregory RJ, Thompson S, Souza DW, Paul S, Mulligan RC, et al. Demonstration that Cftr is a chloride channel by alteration of its anion selectivity. Science. 1991;253(5016):202–5.
91. Cutting G. Cystic fibrosis. In: Rimoin DL, Connor JM, Pyeritz RE, Korf BR, editors. Emery and Rimoin's principles and practice of medical genetics. Philadelphia: Churchill Livingstone; 2007. p. 1354–94.
92. Cystic Fibrosis Foundation. Cystic Fibrosis Foundation Patient Registry annual data report 2005. Cystic Fibrosis Foundation; 2005.
93. Kerem E, Corey M, Kerem BS, Rommens J, Markiewicz D, Levison H, et al. The relation between genotype and phenotype in cystic-fibrosis - analysis of the Most common mutation (Delta-F508). N Engl J Med. 1990;323(22):1517–22.
94. Gan KH, Veeze HJ, van den Ouweland AM, Halley DJ, Scheffer H, van der Hout A, et al. A cystic fibrosis mutation associated with mild lung disease. N Engl J Med. 1995;333(2):95–9.
95. De Braekeleer M, Allard C, Leblanc JP, Simard F, Aubin G. Genotype-phenotype correlation in cystic fibrosis patients compound heterozygous for the A455E mutation. Hum Genet. 1997;101(2):208–11.
96. Falconer DS, Genet AH. Inheritance of liability to certain diseases estimated from incidence among relatives. Ann Hum Genet. 1965;29:51–71.
97. Mekus F, Ballmann M, Bronsveld I, Bijman J, Veeze H, Tummler B. Categories of deltaF508 homozygous cystic fibrosis twin and sibling pairs with distinct phenotypic characteristics. Twin Res. 2000;3(4):277–93.
98. Vanscoy LL, Blackman SM, Collaco JM, Bowers A, Lai T, Naughton K, et al. Heritability of lung disease severity in cystic fibrosis. Am J Respir Crit Care Med. 2007;175(10):1036–43.
99. Stanke F, Becker T, Kumar V, Hedtfeld S, Becker C, Cuppens H, et al. Genes that determine immunology and inflammation modify the basic defect of impaired ion conductance in cystic fibrosis epithelia. J Med Genet. 2011;48(1):24–31.
100. Knowles MR. Gene modifiers of lung disease. Curr Opin Pulm Med. 2006;12(6):416–21.
101. McDougal KE, Green DM, Vanscoy LL, Fallin MD, Grow M, Cheng S, et al. Use of a modeling framework to evaluate the effect of a modifier gene (MBL2) on variation in cystic fibrosis. Eur J Hum Genet. 2010;18(6):680–4.
102. Drumm ML, Konstan MW, Schluchter MD, Handler A, Pace R, Zou F, et al. Genetic modifiers of lung disease in cystic fibrosis. N Engl J Med. 2005;353(14):1443–53.
103. Bremer LA, Blackman SM, Vanscoy LL, McDougal KE, Bowers A, Naughton KM, et al. Interaction between a novel TGFB1 haplotype and CFTR genotype is associated with improved lung function in cystic fibrosis. Hum Mol Genet. 2008;17(14):2228–37.

104. Brazova J, Sismova K, Vavrova V, Bartosova J, Macek M Jr, Lauschman H, et al. Polymorphisms of TGF-beta1 in cystic fibrosis patients. Clin Immunol. 2006;121(3):350–7.
105. Arkwright PD, Laurie S, Super M, Pravica V, Schwarz MJ, Webb AK, et al. TGF-beta(1) genotype and accelerated decline in lung function of patients with cystic fibrosis. Thorax. 2000;55(6):459–62.
106. Hillian AD, Londono D, Dunn JM, Goddard KA, Pace RG, Knowles MR, et al. Modulation of cystic fibrosis lung disease by variants in interleukin-8. Genes Immun. 2008;9(6):501–8.
107. Darrah R, McKone E, O'Connor C, Rodgers C, Genatossio A, McNamara S, et al. EDNRA variants associate with smooth muscle mRNA levels, cell proliferation rates, and cystic fibrosis pulmonary disease severity. Physiol Genomics. 2010;41(1):71–7.
108. Cepero R, Smith RJH, Catlin FI, Bressler KL, Furuta GT, Shandera KC. Cystic-fibrosis - an otolaryngologic perspective. Otolaryngol Head Neck Surg. 1987;97(4):356–60.
109. Ramsey B, Richardson MA. Impact of sinusitis in cystic-fibrosis. J Allergy Clin Immunol. 1992;90(3):547–52.
110. Ledesma-Medina J, Osman MZ, Girdany BR. Abnormal paranasal sinuses in patients with cystic fibrosis of the pancreas. Radiological findings. Pediatr Radiol. 1980;9(2):61–4.
111. Umetsu DT, Moss RB, King VV, Lewiston NJ. Sinus disease in patients with severe cystic-fibrosis - relation to pulmonary exacerbation. Lancet. 1990;335(8697):1077–8.
112. Weber A, Kiefer J, Peters S, Schneider M, Bargon J, May A. Eosinophilic cationic protein as a marker of nasal inflammation in patients with cystic fibrosis. Laryngoscope. 1999;109(10):1696–702.
113. Cimmino M, Cavaliere M, Nardone M, Plantulli A, Orefice A, Esposito V, et al. Clinical characteristics and genotype analysis of patients with cystic fibrosis and nasal polyposis. Clin Otolaryngol Allied Sci. 2003;28(2):125–32.
114. Batsakis JG, El-Naggar AK. Cystic fibrosis and the sinonasal tract. Ann Otol Rhinol Laryngol. 1996;105(4):329–30.
115. Franco LP, Camargos PA, Becker HM, Guimaraes RE. Nasal endoscopic evaluation of children and adolescents with cystic fibrosis. Braz J Otorhinolaryngol. 2009;75(6):806–13.
116. Daniel S. Infection and inflammation CF: management of the basics upper airway diseases. Paediatr Respir Rev. 2006;7(Suppl 1):S154–5.
117. Sakano E, Ribeiro AF, Barth L, Condino Neto A, Ribeiro JD. Nasal and paranasal sinus endoscopy, computed tomography and microbiology of upper airways and the correlations with genotype and severity of cystic fibrosis. Int J Pediatr Otorhinolaryngol. 2007;71(1):41–50.
118. Gentile VG, Isaacson G. Patterns of sinusitis in cystic fibrosis. Laryngoscope. 1996;106(8):1005–9.
119. Moss RB, King VV. Management of sinusitis in cystic fibrosis by endoscopic surgery and serial antimicrobial lavage. Reduction in recurrence requiring surgery. Arch Otolaryngol Head Neck Surg. 1995;121(5):566–72.
120. Jorissen MB, De Boeck K, Cuppens H. Genotype-phenotype correlations for the paranasal sinuses in cystic fibrosis. Am J Respir Crit Care Med. 1999;159(5 Pt 1):1412–6.
121. Lurie MH. Cystic fibrosis of the pancreas and nasal mucosa. Ann Otol Rhinol Laryngol. 1959;68(2):478–86.
122. Hulka GF. Head and neck manifestations of cystic fibrosis and ciliary dyskinesia. Otolaryngol Clin N Am. 2000;33(6):1333–41, vii-viii
123. Shwachman H, Kulczycki LL, Mueller HL, Flake CG. Nasal polyposis in patients with cystic fibrosis. Pediatrics. 1962;30:389–401.
124. Weber SA, Ferrari GF. Incidence and evolution of nasal polyps in children and adolescents with cystic fibrosis. Braz J Otorhinolaryngol. 2008;74(1):16–20.
125. Pawankar R. Nasal polyposis: an update: editorial review. Curr Opin Allergy Clin Immunol. 2003;3(1):1–6.
126. Liu CM, Hong CY, Shun CT, Hsiao TY, Wang CC, Wang JS, et al. Inducible cyclooxygenase and interleukin 6 gene expressions in nasal polyp fibroblasts - possible implication in the pathogenesis of nasal polyposis. Arch Otolaryngol. 2002;128(8):945–51.
127. Bernstein JM, Ballow M, Rich G. Detection of intracytoplasmic cytokines by flow cytometry in adenoids and peripheral blood lymphocytes of children. Ann Oto Rhinol Laryngol. 2001;110(5):442–6.
128. Bernstein JM, Ballow M, Rich G, Allen C, Swanson M, Dmochowski J. Lymphocyte subpopulations and cytokines in nasal polyps: is there a local immune system in the nasal polyp? Otolaryngol Head Neck Surg. 2004;130(5):526–35.
129. Salik E, Tyorkin M, Mohan S, George I, Becker K, Oei E, et al. Antigen trafficking and accessory cell function in respiratory epithelial cells. Am J Respir Cell Mol Biol. 1999;21(3):365–79.
130. Bernstein JM, Ballow M, Schlievert PM, Rich G, Allen C, Dryja D. A superantigen hypothesis for the pathogenesis of chronic hyperplastic sinusitis with massive nasal polyposis. Am J Rhinol. 2003;17(6):321–6.
131. Sun Q, Jones K, McClure B, Cambareri B, Zacharakis B, Iversen PO, et al. Simultaneous antagonism of interleukin-5, granulocyte-macrophage colony-stimulating factor, and interleukin-3 stimulation of human eosinophils by targetting the common cytokine binding site of their receptors. Blood. 1999;94(6):1943–51.
132. Hamilos DL, Leung DYM, Wood R, Cunningham L, Bean DK, Yasruel Z, et al. Evidence for distinct cytokine expression in allergic versus nonallergic chronic sinusitis. J Allergy Clin Immunol. 1995;96(4):537–44.

133. Bernstein JM, Gorfien J, Noble B, Yankaskas JR. Nasal polyposis: immunohistochemistry and bioelectrical findings (a hypothesis for the development of nasal polyps). J Allergy Clin Immunol. 1997;99(2):165–75.
134. Bates ME, Liu LY, Esnault S, Stout BA, Fonkem E, Kung V, et al. Expression of interleukin-5- and granulocyte macrophage-colony-stimulating factor-responsive genes in blood and airway eosinophils. Am J Respir Cell Mol Biol. 2004;30(5):736–43.
135. Lundgren JD, Davey RT Jr, Lundgren B, Mullol J, Marom Z, Logun C, et al. Eosinophil cationic protein stimulates and major basic protein inhibits airway mucus secretion. J Allergy Clin Immunol. 1991;87(3):689–98.
136. Jacoby DB, Ueki IF, Widdicombe JH, Loegering DA, Gleich GJ, Nadel JA. Effect of human eosinophil major basic protein on ion transport in dog tracheal epithelium. Am Rev Respir Dis. 1988;137(1):13–6.
137. Dupuit F, Kalin N, Brezillon S, Hinnrasky J, Tummler B, Puchelle E. CFTR and differentiation markers expression in non-CF and delta F 508 homozygous CF nasal epithelium. J Clin Invest. 1995;96(3):1601–11.
138. Passali D, Bernstein JM, Passali FM, Damiani V, Passali GC, Bellussi L. Treatment of recurrent chronic hyperplastic sinusitis with nasal polyposis. Arch Otolaryngol Head Neck Surg. 2003;129(6):656–9.
139. Bez C, Schubert R, Kopp M, Ersfeld Y, Rosewich M, Kuehr J, et al. Effect of anti-immunoglobulin E on nasal inflammation in patients with seasonal allergic rhinoconjunctivitis. Clin Exp Allergy. 2004;34(8):1079–85.
140. International Consensus Report on the diagnosis and management of rhinitis. International Rhinitis Management Working Group. Allergy. 1994;49(19 Suppl):1–34.
141. Dykewicz MS, Fineman S, Skoner DP, Nicklas R, Lee R, Blessing-Moore J, et al. Diagnosis and management of rhinitis: complete guidelines of the Joint Task Force on Practice Parameters in Allergy, Asthma and Immunology. American Academy of Allergy, Asthma, and Immunology. Ann Allergy Asthma Immunol. 1998;81(5 Pt 2):478–518.
142. Smith TL. Vasomotor rhinitis is not a wastebasket diagnosis. Arch Otolaryngol Head Neck Surg. 2003;129(5):584–7.
143. Corey JP. Vasomotor rhinitis should not be a wastebasket diagnosis. Arch Otolaryngol Head Neck Surg. 2003;129(5):588–9.
144. Druce HM, Middleton E Jr, Ellis EF, Yunginger JW, Reed CE, Adkinson NF, Busse WW. Allergic and nonallergic rhinitis. St. Louis: Mosby; 1998.
145. Loehrl TA, Smith TL, Darling RJ, Torrico L, Prieto TE, Shaker R, et al. Autonomic dysfunction, vasomotor rhinitis, and extraesophageal manifestations of gastroesophageal reflux. Otolaryngol Head Neck Surg. 2002;126(4):382–7.
146. Braat JPM, Mulder PG, Fokkens WJ, van Wijk RG, Rijntjes E. Intranasal cold dry air is superior to histamine challenge in determining the presence and degree of nasal hyperreactivity in nonallergic noninfectious perennial rhinitis. Am J Respir Crit Care Med. 1998;157(6):1748–55.
147. Numata T, Konno A, Hasegawa S, Hanazawa T, Nagata H, Motosugi H, et al. Pathophysiological features of the nasal mucosa in patients with idiopathic rhinitis compared to allergic rhinitis. Int Arch Allergy Immunol. 1999;119(4):304–13.
148. Chen J, Kong W, Zhou Y, Xiang J, Shu H, Shi Q, et al. The expression of serum IL-10,12,13,16 in patients with allergic rhinitis and vasomotor rhinitis. Lin Chung Er Bi Yan Hou Tou Jing Wai Ke Za Zhi. 2007;21(20):913–5.
149. Giannessi F, Fattori B, Ursino F, Giambelluca MA, Soldani P, Scavuzzo MC, et al. Ultrastructural and ultracytochemical study of the human nasal respiratory epithelium in vasomotor rhinitis. Acta Otolaryngol. 2003;123(8):943–9.
150. Settipane RA, Lieberman P. Update on nonallergic rhinitis. Ann Allergy Asthma Immunol. 2001;86(5):494–507; quiz –8
151. Groneberg DA, Heppt W, Cryer A, Wussow A, Peiser C, Zweng M, et al. Toxic rhinitis-induced changes of human nasal mucosa innervation. Toxicol Pathol. 2003;31(3):326–31.
152. Schierhorn K, Hanf G, Fischer A, Umland B, Olze H, Kunkel G. Ozone-induced release of neuropeptides from human nasal mucosa cells. Int Arch Allergy Immunol. 2002;129(2):145–51.
153. Ruffoli R, Fattori B, Giambelluca MA, Soldani P, Giannessi F. Ultracytochemical localization of the NADPH-d activity in the human nasal respiratory mucosa in vasomotor rhinitis. Laryngoscope. 2000;110(8):1361–5.
154. Cervin A, Onnerfalt J, Edvinsson L, Grundemar L. Functional effects of neuropeptide Y receptors on blood flow and nitric oxide levels in the human nose. Am J Respir Crit Care Med. 1999;160(5):1724–8.
155. Braat JPM, Mulder PG, Duivenvoorden HJ, Van Wijk RG, Rijntjes E, Fokkens WJ. Pollutional and meteorological factors are closely related to complaints of non-allergic, non-infectious perennial rhinitis patients: a time series model. Clin Exp Allergy. 2002;32(5):690–7.
156. Iguchi Y, Yao K, Okamoto M. A characteristic protein in nasal discharge differentiating nonallergic chronic rhinosinusitis from allergic rhinitis. Rhinology. 2002;40(1):13–7.
157. Tosun F, Sezen I, Gerek M, Ozkaptan Y, Yapar M, Caliskaner Z, et al. Electrophoretic evaluation of nasal discharge in patients with allergic rhinitis and vasomotor rhinitis. Am J Rhinol. 2002;16(3):141–4.
158. Aust MR, Madsen CS, Jennings A, Kasperbauer JL, Gendler SJ. Mucin mRNA expression in normal and vasomotor inferior turbinates. Am J Rhinol. 1997;11(4):293–302.
159. Shaker R, Bardan E, Gu CM, Kern M, Torrico L, Toohill R. Intrapharyngeal distribution of gastric acid refluxate. Laryngoscope. 2003;113(7):1182–91.

34 Vomeronasal Organ

Cemal Cingi, Aytuğ Altundağ, and İsmail Koçak

Core Messages

- There are still debates about the nature and role of the vomeronasal organ in the human nose.
- The vomeronasal organ's remnant appears to be a blind-ending tube lined by pseudostratified epithelium and associated with submucosal glands.
- Vomeropherin pregna-4,20-diene-3,6-dione's local and systemic effects help the human VNO's functioning and its implications for autonomic and psychophysiological functions. Some researchers, on the other hand, assume that the human VNO has epithelia that may function as a chemical sensory organ, but that there are no links between the VNO and the central nervous system.

The vomeronasal organ (VNO) is an auxiliary olfactory system peripheral sensory organ. In most amphibians, reptiles, and mammals, it is a paired organ found at the base of the nasal septum or on the roof of the mouth [1].

The VNO is an accessory olfactory organ that receives pheromones, which are chemical stimuli that induce behavioral, reproductive, or neuroendocrine responses in individuals of the same species [2]. In the year 1703, Frederik Ruysch discovered the vomeronasal cavities in humans. On each side of the anterior part of the nasal septum of a young cadaver, he identified a "canalibus nasalibus." Kölliker examined the role of the vomeronasal cavities in the nasal septum of dead fetuses, infants, and adults in great detail. The cavity's opening is evident as a pit on the septum's surface [3].

In several mammalian species, Ludwig Lewin Jacobson described the vomeronasal organ in great detail. He did, however, point out that the vomeronasal structure does not evolve in humans [4].

According to a recent Bulgarian survey, VNO is present in around 27% of the adult population (men, 53%; women, 47%) [5].

34.1 VNO Anatomy

The researchers describe a blind-ending tube with submucosal glands and a pseudostratified epithelium on all sides. It appears that this struc-

C. Cingi (✉)
Department of Otolaryngology-Head Neck Surgery, Osmangazi University, Eskisehir, Turkey

Medical Faculty, ENT Department, Eskisehir Osmangazi University, Meselik Kampüsü, Eskisehir, Turkey
e-mail: ccingi@ogu.edu.tr

A. Altundağ
Department of Otolaryngology, Istanbul Surgery Hospital, Istanbul, Turkey

İ. Koçak
ENT Department, Liv Hospital, Istanbul, Turkey
e-mail: info@drismailkocak.com

Ö. Ö. Celebi, T. M. Önerci (eds.), *Nasal Physiology and Pathophysiology of Nasal Disorders*,
https://doi.org/10.1007/978-3-031-12386-3_34

ture is the adult human vestige of the vomeronasal organ [6]. The vomeronasal cavities were found at the base of the nasal septum's most anterior portion, which can be seen on computed tomography. According to histological research, the vomeronasal cavities consisted of a pit connected to a duct running posteriorly under the nasal mucosa. In the duct, there were several glands that contained mucus [3]. There are some differences in how VNO, VNO trap, and VNO cavity are defined and identified. Researchers have identified the variations of the nasopalatine fossa (NPF) and the nasopalatine recess in some studies (NPR). The endoscopic view of the vomeronasal pit poses several doubts. The NPF and the NPR are two distinct but variable structures found near the VNO field. The NPF isn't a trap for vomit. The vomeronasal duct opening could be hidden by a septal mucosal pit. The VNO is a submucosal structure that lies 2–8 mm above the NPR and cannot be detected macroscopically or endoscopically [7]. However, in another study, endoscopy showed vomeronasal cavities in some adults, but they lacked sensory neurons and nerve fibers [8]. The following are the main findings of a report on the human VNO: (1) A VNO is detectable in about two-thirds of the population, and bilateral VNOs are present in about 40% of investigated subjects, (2) its location on the left and right nasal septum is nearly symmetrical, and (3) the VNO's detectability is unrelated to age or gender [9].

VNO has also not been reliably observed during fetal development in recent studies, and its existence is not dependent on the gender or age of the fetus. By weeks 6–7, the VNO is present in human fetuses, and its ducts have opened into the nasal cavity by the 28th week. The VNO was discovered to be made up of olfactory epithelium, lamina propria, and a dense vascularization network when it was examined histologically. The VNO epithelium comprises a population of cells with neural properties that have the ability to differentiate into a chemosensory epithelium at first, but the neural population declines significantly as the fetal age increases. The findings of these studies back up the hypothesis that the neuroepithelium present in the early stages of fetal development is regressing [10, 11].

34.2 VNO Histology

The human VNO was discovered to have several forms. Tall cells with discontinuous cilia on their free surface made up the epithelium, which was variable in thickness both medially and laterally [12]. The anteroposterior and superoinferior locations of human VNOs in relation to the anterior nasal spine and the nasal cavity floor were also distinct [13].

The epithelium that lines the human VNO varies from that which lines VNOs in other animals and from that which lines the olfactory or respiratory epithelium in humans. Many elongated cells present a microvillar surface to the organ's lumen, but the majority are not like microvillar vomeronasal sensory organs (VSNs) in other mammals. They don't have axons that leave the epithelium or make synaptic interaction with axons in the epithelium, according to research. If these cells are chemosensitive, there is no clear way for them to communicate with the brain [14].

Several immunomarkers have been discovered to dye these special elongated bipolar microvillar cells [15]. The vomeronasal epithelium's (VNE) histology tended to be highly varied. Slender bipolar cells made up portions of stratified, respiratory, and standard pseudostratified vomeronasal epithelia. The human VNE does not behave like the mature olfactory epithelium, according to mainly negative immunohistochemical findings for OMP [16]. These cells have physiological properties that are similar to those found in other mammalian species' chemosensory receptor cells. A small subset of VNE cells had neuron-like activity, as shown by the existence of certain bipolar cells positive for both protein gene product (PGP) 9.5 and soybean lectin. In the VNE of adult humans, immunohistochemistry was used to look for three molecular markers: neuron-specific enolase (NSE) and PGP 9.5 for neurons and neuroendocrine cells, and olfactory marker protein for olfactory receptor neurons. Immunoreactive cells for NSE and PGP 9.5 have been found in the VNE [17].

Human and chimp VNOs resembled nonhomologous ciliated gland ducts found in other primates in terms of structure. The human/chimp VNO, on the other hand, differs from the VNOs of other primates or nonhomologous epithelial systems in the following ways: (1) bilateral epithelial tubes; (2) superiorly displaced location in the same plane as the paraseptal cartilages; (3) homogeneous, pseudostratified columnar morphology with ciliated regions; and (4) mucous-producing structures inside the epithelium [18].

However, the authors of a recent study discovered serial or semiserial portions of 30 fetuses aged 7–18 weeks. Calretinin and S100 protein staining showed the terminal nerve along the anterior edge of the ethmoid's perpendicular lamina, as well as the VNO along the lamina's posterior edge. The terminal nerve was made up of 1–2 nerve bundles that passed through the cribriform plate's anterior end, while the VNO was made up of 2–3 bundles behind the olfactory nerves. The terminal nerve ran along the backside of the nasal branch of the anterior ethmoidal nerve, crossing it. On the lateral surfaces of the ethmoid crista galli, multiple clusters of small ganglion cells were discovered, which could be the sources of both the terminal nerve and the VNO [11].

34.3 Genes Related to VNO

TRPC2, a mouse gene that is required for VNO function, is a human pseudogene [19]. TRPC2 is only expressed in the VNO, so its loss of selective pressure can be used as a molecular marker for when the VNO became vestigial [20]. Human olfactory mucosa contains transcripts of the V1RL1 vomeronasal receptor, which may indicate the fact that the accessory olfactory system has been integrated into the primary olfactory system in humans (and some other mammals) [21]. A second mouse receptor subclass was discovered in the olfactory epithelium in 2006. Some of the trace amine-associated receptors (TAAR) in mouse urine, including one putative mouse pheromone, are activated by volatile amines present in the urine [22].

Humans have orthologous receptors, suggesting that there is a mechanism for detecting human pheromones [23]. According to research, humans have the genes for at least six of the same pheromone receptors as mice [24, 25].

34.4 VNO Responses

There appears to be a process in or near the VNO pit that selectively generates an electrical reaction to small amounts of certain chemicals. The word "vomeropherin" has been proposed as a generic term for substances that stimulate the VNO in any species and as a name for chemicals that evoke this response [26]. The "electrovomeronasogram" (EVG) reported from the VNO pit region in awake human subjects is the first form of response. It gets its name from the electro-olfactogram (EOG), which can be reported from the olfactory epithelium's surface in response to odor stimulation [23]. It has also documented preliminary evidence that bipolar cells aspirated from the human VNO pit exhibit electrical responses to certain "vomeropherins" as a second form of response. These are the EVG-inducing steroids that are linked to skin chemicals that this group claims are human pheromones [15].

34.5 VNO Function

Almost all social animals are known to interact using pheromones [27]. Furthermore, studies have shown that people can deduce a person's sex by smelling them [28]. In addition to the conventional olfactory system, also known as the primary olfactory system, an accessory olfactory system has evolved in the vast majority of terrestrial animals to detect pheromones rather than other environmental odorants [29].

Human smegma and vaginal secretions, as well as human apocrine glands, have been found to contain pheromones or vomeropherins [30]. Many experiments have been conducted in order to weigh the evidence for and against human VNO function, to distinguish this problem from the question of pheromone signaling, and to provide a working definition of "pheromone." Humans have VNO, which is considered to be nonfunctional since the vomeronasal receptor and signal transduction genes in humans are pseudogenes [31]. In addition to a functional vomeronasal-pituitary pathway and an effect on gonadotropin pulsatility in adult humans, the vomeropherin also has autonomic reflex effects after VNO stimulation [32].

Cranial nerve 0 and VNO fibers are embryologically independent of the olfactory nerve, according to studies in adult brains and fetuses [10, 11]. The ubiquitous hypothalamic-pituitary-gonadal axis is thought to play a role in the unconscious perception of particular odors that influence the autonomic and reproductive hormonal systems [33, 34]. Furthermore, research indicates that it may play a role in detecting pheromones for mate selection and neuromodulation of reproductive functions. VNO has been blamed for intra-specific chemical communication by pheromones, which are chemical messengers secreted externally by one animal and detected by another [35, 36].

In the male human VNO, vomeropherin pregna-4,20-diene-3,6-dione (PDD) has a dose-dependent local effect. Then comes a mild parasympathomimetic effect, which involves a 10% rise in vagal tone and a decrease in the frequency of electrodermal activity events. Furthermore, local delivery of PDD to the male human VNO reduces serum LH and testosterone levels. This research backs up the human VNO's mechanism and the effects it has on autonomic and psychophysiological processes, as well as neuroendocrine secretions [37]. Occlusion or absence of the VNO did not affect the perceptual measurements or functional processing of the putative human pheromone androstadienone, according to another study. [38].

As a result, the human VNO has epithelia that could potentially act as a chemical sensory organ; however, the VNO receptor genes in humans are nonfunctional pseudogenes. Furthermore, although the human VNO includes sensory neurons, there seem to be no interactions between the VNO and the central nervous system [39].

References

1. Meredith M. Human vomeronasal organ function: a critical review of best and worst cases. Chem Senses. 2001;26(4):433–45.
2. Witt M, Hummel T. Vomeronasal versus olfactory epithelium: is there a cellular basis for human vomeronasal perception? Int Rev Cytol. 2006;248:209–59.
3. Trotier D, Eloit C, Wassef M, Talmain G, Bensimon JL, Døving KB, et al. The vomeronasal cavity in adult humans. Chem Senses. 2000;25(4):369–80.
4. Trotier D, Døving KB. Anatomical description of a new organ in the nose of domesticated animals by Ludvig Jacobson (1813). Chem Senses. 1998;23:743–54.
5. Stoyanov G, Moneva K, Sapundzhiev N, Tonchev AB. The vomeronasal organ - incidence in a Bulgarian population. J Laryngol Otol. 2016;130(4):344–7.
6. Johnson EW, Eller PM, Jafek BW. Calbindin-like immunoreactivity in epithelial cells of the newborn and adult human vomeronasal organ. Brain Res. 1994;638(1–2):329–33.
7. Bhatnagar KP, Smith TD, Winstead W. The human vomeronasal organ: part IV. Incidence, topography, endoscopy, and ultrastructure of the nasopalatine recess, nasopalatine fossa, and vomeronasal organ. Am J Rhinol. 2002;16(6):343–50.
8. Trotier D. Vomeronasal organ and human pheromones. Eur Ann Otorhinolaryngol Head Neck Dis. 2011;128(4):184–90.
9. Knecht M, Kühnau D, Hüttenbrink KB, Witt M, Hummel T. Frequency and localization of the putative vomeronasal organ in humans in relation to age and gender. Laryngoscope. 2001;111(3):448–52.
10. Dénes L, Pap Z, Szántó A, Gergely I, Pop TS. Human vomeronasal epithelium development: an immunohistochemical overview. Acta Microbiol Immunol Hung. 2015;62(2):167–81.
11. Jin ZW, Cho KH, Shibata S, Yamamoto M, Murakami G, Rodríguez-Vázquez JF. Nervus terminalis and nerves to the vomeronasal organ: a study using human fetal specimens. Anat Cell Biol. 2019 Sep;52(3):278–85. https://doi.org/10.5115/acb.19.020.
12. Kunwar P, Bhatnagar KP, Tımothy D, Smith TD. The human vomeronasal organ III: postnatal development from infancy to the ninth decade. J Anat. 2001;199:289–302.
13. Smith TD, Buttery TA, Bhatnagar KP, Burrows AM, Mooney MP, Siegel MI. Anatomical position of the vomeronasal organ in postnatal humans. Ann Anat. 2001;183(5):475–9.
14. Stensaas LJ, Lavker RM, Monti-Bloch L, Grosser BI, Berliner DL. Ultrastructure of the human vomeronasal organ. J Steroid Biochem Mol Biol. 1991;39(4B):553–60.
15. Monti-Bloch L, Jennings-White C, Berliner DL. The human vomeronasal system: a review. Ann N Y Acad Sci. 1998b;855:373–89.
16. Witt M, Georgiewa B, Knecht M, Hummel T. On the chemosensory nature of the vomeronasal epithelium in adult humans. Histochem Cell Biol. 2002;117(6):493–509.
17. Takami S, Getchell ML, Chen Y, Monti-Bloch L, Berliner DL, Stensaas LJ, et al. Vomeronasal epithelial cells of the adult human express neuron-specific molecules. Neuroreport. 1993;4(4):375–8.
18. Smith TD, Bhatnagar KP, Shimp KL, Kinzinger JH, Bonar CJ, Burrows AM, et al. Histological definition of the vomeronasal organ in humans and chimpan-

zees, with a comparison to other primates. Anat Rec. 2002;267(2):166–76.
19. Rodriguez I, Mombaerts P. Novel human vomeronasal receptor-like genes reveal species-specific families. Curr Biol. 2002;12(12):R409–11.
20. Liman ER, Innan H. Relaxed selective pressure on an essential component of pheromone transduction in primate evolution. Proc Natl Acad Sci U S A. 2003;100(6):3328–32.
21. Rodriguez I, Greer CA, Mok MY, Mombaerts P. A putative pheromone receptor gene expressed in human olfactory mucosa. Nat Genet. 2000;26(1):18–9.
22. Liberles SD, Buck LB. A second class of chemosensory receptors in the olfactory epithelium. Nature. 2006;442(7103):645–50.
23. Pearson H. Mouse data hint at human pheromones. Nature. 2006;442(7102):495.
24. Rodewald A, Mills D, Gebhart VM, Jirikowski GF. Steroidal pheromones and their potential target sites in the vomeronasal organ. Steroids. 2019;142:14–20.
25. Vasuki AK, Fenn TK, Devi MN, Hebzibah TD, Jamuna M, Sundaram KK. Fate and development of human Vomeronasal organ - a microscopic fetal study. J Clin Diagn Res. 2016;10(3):AC08–11.
26. Monti-Bloch L, Grosser BI. Effect of putative pheromones on the electrical activity of the human vomeronasal organ and olfactory epithelium. J Steroid Biochem Mol Biol. 1991;39(4B):573–82.
27. Comfort A. Likelihood of human pheromones. Nature. 1971;230(5294):432–3 passim.
28. Russell MJ. Human olfactory communication. Nature. 1976;260(5551):520–2.
29. Meisami E, Bhatnagar KP. Structure and diversity in mammalian accessory olfactory bulb. Microsc Res Tech. 1998;43(6):476–99.
30. Gower DB, Ruparelia BA. Olfaction in humans with special reference to odorous 16-androstenes: their occurrence, perception and possible social, psychological and sexual impact. J Endocrinol. 1993;137(2):167–87.
31. Doty RL. Olfaction. Annu Rev Psychol. 2001;52:423–52.
32. Berliner DL, Monti-Bloch L, Jennings-White C, Diaz-Sanchez V. The functionality of the human vomeronasal organ (VNO): evidence for steroid receptors. J Steroid Biochem Mol Biol. 1996;58(3):259–65.
33. Biehl MJ, Raetzman LT. Developmental origins of hypothalamic cells controlling reproduction. Semin Reprod Med. 2017;35(2):121–12.
34. Salazar I, Barrios AW, SáNchez-Quinteiro P. Revisiting the Vomeronasal system from an integrated perspective. Anat Rec (Hoboken). 2016;299(11):1488–91.
35. Baum MJ, Bakker J. Roles of sex and gonadal steroids in mammalian pheromonal communication. Front Neuroendocrinol. 2013;34(4):268–84.
36. Morozova SV, Savvateeva DM, Svistushkin VM, Toporkova LA. The role of the vomeronasal system in the formation of the human sexual behaviour. Vestn Otorinolaringol. 2017;82(2):90–4.
37. Monti-Bloch L, Diaz-Sanchez V, Jennings-White C, Berliner DL. Modulation of serum testosterone and autonomic function through stimulation of the male human vomeronasal organ (VNO) with pregna-4,20-diene-3,6-dione. J Steroid Biochem Mol Biol. 1998a;65(1–6):237–42.
38. Frasnelli J, Lundström JN, Boyle JA, Katsarkas A, Jones-Gotman M. The vomeronasal organ is not involved in the perception of endogenous odors. Hum Brain Mapp. 2011;32(3):450–60. https://doi.org/10.1002/hbm.21035.
39. Grammer K, Fink B, Neave N. Human pheromones and sexual attraction. Eur J Obstet Gynecol Reprod Biol. 2005;118(2):135–42.

35

Physiology of the Nasal Cartilages and Their Importance to Rhinosurgery

Wolfgang Pirsig

Core Messages

- The complex framework of the human nasal cartilages, unique among mammals, is the mobile portal to the respiratory system providing conditions for the passage of airstreams and the generation of nasal resistance and turbulence.
- The septodorsal (septolateral) cartilage is the dominating structure for nasal and midfacial growth and is decisive for the appearance and several functions of the nose. Together with the erectile lining of the nasal cavities, the cartilaginous framework enables air-conditioning and the acting of the nasal cycle.
- The anterior nose with the nasal valve as its crucial functional region is the narrowest part of the upper airways and provides two-thirds of the total airaway resistance.
- Due to its protruding exposition, it is especially vulnerable to external injuries and may react as a protecting crumpled zone. This exposition also is the reason why most of the obstructing structures of the nasal airways are diagnosed in the anterior nose.
- Long-term observations have shown that the most effective outcome after functional and aesthetic rhinosurgery is achieved by focusing on restoring the physiological functions of the anterior nose, the site of the most resistive nasal segments.

The human nose, with its prominent bridge, its elongated tip, and its downturned nostrils, is unique. Besides sniffing strange odours, it acts as a vital air-conditioning unit, warming, cleaning, and moistening the air we breathe in before it reaches the delicate lungs. Assisting this—and also adding resonance to the voice—are the nasal sinuses, but the price we pay for possessing these valuable cavities is an all too common susceptibility to local infections.

This precise summing up of the appearance and functions of the nose including its Achilles' heel tendon, the diseases, was published in 1985 by the zoologist Desmond Morris in his outstanding book *Bodywatching* [1].

In the following chapter, I'll focus on some functional aspects of the cartilaginous framework of the human nose with an emphasis on its vulnerability and long-term results following surgical treatment. Why do we have a nose of such a protruding shape, which is divided by the septum into two parallel halves? Why is this mobile organ built up by a complex framework of hyaline cartilages and covered by muscles? Why do we need this nose at all, although we can survive by breathing through the mouth? I asked anatomists, physiologists, rhinosurgeons, biologists, and also engineers for stream technology. The answers varied, several ended with a question

W. Pirsig (✉)
Department of Otorhinolaryngology, University Hospital Ulm, Ulm, Germany
e-mail: wolfgang.pirsig@uni-ulm.de

Ö. Ö. Celebi, T. M. Önerci (eds.), *Nasal Physiology and Pathophysiology of Nasal Disorders*,
https://doi.org/10.1007/978-3-031-12386-3_35

mark, but in all of them the terms 'turbulence', 'resistance', and 'air-conditioning' were somehow mentioned.

35.1 Part I: Anatomical Considerations

35.1.1 Nomenclature

The lower two-thirds of the nose are built up by a framework of the following hyaline cartilages: septal cartilage (quadrangular cartilage in adulthood), a partition separating the two nasal cavities; soft tissue part of the septum is the septal turbinate (septal body or intumescentia septi or septal tuberculum); two triangular (upper lateral) cartilages as expansions of the septal cartilage forming together the T-bar-shaped framework also known as septodorsal or septolateral cartilage; two paraseptal (vomeronasal) cartilages, lying along the inferior margin of the caudal septal cartilage, attached to the vomer posteriorly and to the maxillary crest anteriorly; two alar (lower lateral or lobular) cartilages composed of a medial and lateral crus melted together at the dome on the tip; and some sesamoid (accessory) cartilages in the soft tissue area between triangular cartilage, lateral crus and piriform aperture, in the so-called hinge area.

In this chapter, the terms septum, septal, septodorsal, paraseptal, triangular, alar and sesamoid cartilages will be used.

35.1.2 Intrauterine Development

A two-tube system and a still unruffled entrance are recognisable in the fourth month of foetal development of the cartilaginous nose (Fig. 35.1). The cartilaginous framework consists of a T-bar-shaped bilateral vault fused in the midline to the septal cartilage. The complete sidewall of the cartilaginous nasal capsule is superiorly connected with the spheno-ethmoidal cartilage and dorsally with the septal cartilage which posteriorly merges into the cartilaginous anterior skull base with the crista galli. Both vaults are separated by the supraseptal groove. Caudally, the margin of the sidewall bends medially to join with the inferior turbinate. Thus, the palatine bone, the vomer and the paraseptal cartilage are visible.

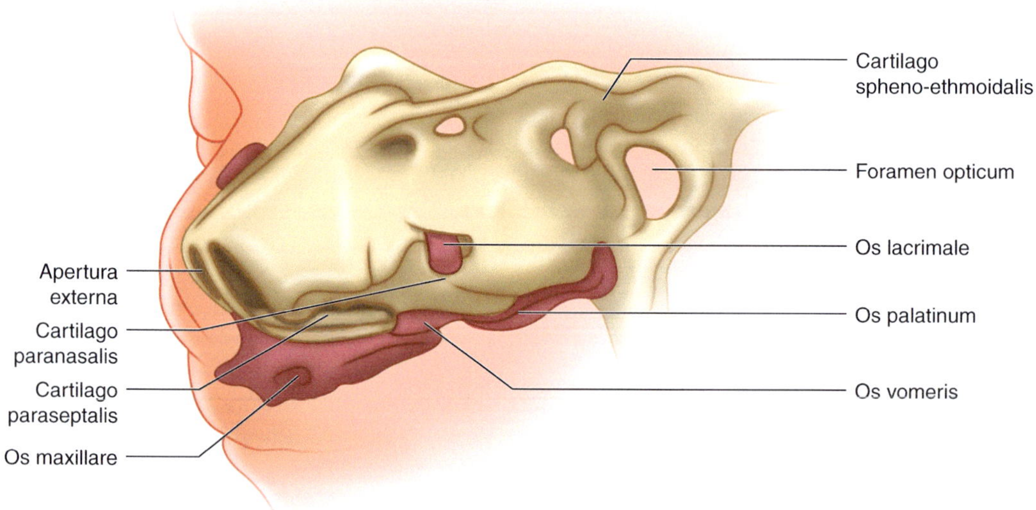

Fig. 35.1 Nasal skeleton of a foetus (8 cm) in oblique side view [2]

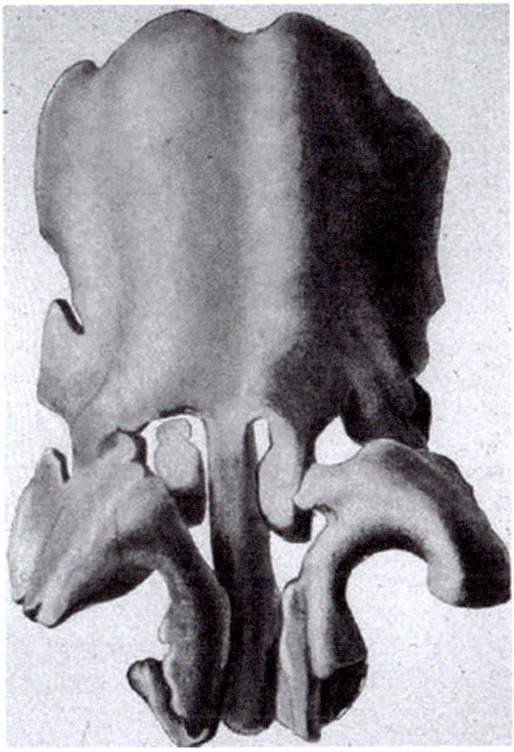

Fig. 35.2 Model of the cartilaginous nasal capsule of a foetus (27.5 cm, 30–32 weeks) [2]

In the sixth foetal month, ingrowth of connective tissue divides the cartilaginous nasal capsule into the individual septal cartilage, lateral cartilages and alar cartilages. Thus, the complex cartilaginous nasal portal is developed which later forms the cartilaginous framework of the mobile nose.

In the model of the foetal cartilaginous capsule aged 30–32 weeks (Fig. 35.2), we look at the two vaults of the lateral cartilages fused together with the septal cartilage in the supraseptal groove. The median portion bifida septi remains as the rest of the fusion of the vaults. Two semicircular-shaped cartilaginous bars have been almost completely separated from the caudal septal cartilage. They encompass the nasal vestibule medially, dorsally and laterally to form the definite alar cartilages. In the medial upper angle of the alar cartilage, an isolated piece of cartilage is visible on the right-hand side, while on the left-hand side, it is still a process in connection with the left sidewall. Later, these pieces become the sesamoid cartilages. The same happens to small processes of the lateral sidewall which are concealed by the semicircular cartilage in Fig. 35.2. The paraseptal cartilages are concealed by the caudal part of the semicircular cartilages. The triangular cartilages are not yet differentiated. In the newborn, they are still continuous with the cartilaginous anterior skull base. Their caudal margins are firmly connected to the alar cartilages and to the piriform aperture. In the newborn, the alar cartilages are developed as separate structures as in the adult and overlap with their cranial margins and the caudal margins of the triangular cartilages.

35.1.3 Postnatal Development

The septal cartilage of the newborn is progressively transformed from posteriorly, cranially and caudally into a unit of posterior bone and remaining anterior quadrangular cartilaginous plate as clearly depicted by Virchow in 1857 [3]. This old knowledge has recently been supported by Kim et al. [4] who evaluated the anatomical correlations among components of the nasal septum using computed tomography. They found again that the area of the cartilaginous septum decreases with age, while the area of the perpendicular plate increases with age at the expense of the area of the septal cartilage. However, the area of the total nasal septum remains constant.

In the first decade of life, the triangular cartilages show regression from the cephalic to the caudal end under the nasal bones until being transformed into the approximately triangular shape of the adult nose. Only a small cartilaginous remnant of the triangular cartilage remains overlapped by the caudal margin of the nasal bone. Caudally, the triangular cartilages are overlapped by the cranial margins of the alar cartilages.

Very rarely the regression process of the triangular cartilage is retarded or even impeded. This may happen in the case of a median nasal fistula or of a nasal dermoid cyst which can cranially end at the crista galli. Nasal bifidity is another congenital malformation with impeded regression of the triangular cartilages.

Potter et al. [5] studied 35 adults, white cadaveric specimens and specially focused on the caudal attachment of the triangular cartilage to the septal cartilage, where usually a small cleft facilitates mobility between the caudal edge of the triangular cartilage and the septum, the so-called weak triangle of Converse. The attachment ranged from no cartilaginous connection (68%) to complete fusion (32%), i.e. a coincident location of the anterior septal angle and the caudal edge of the triangular cartilage.

In the adult, the relationship between the triangular cartilage and the lateral crus of the alar cartilage shows four variations [6]. Most often Dion et al. found an overlap of the caudal margin of the triangular cartilage by the cranial margin of the lateral crus. Less frequently are the relationships 'end to end', 'scroll' and 'opposed scroll'.

Each cartilage of the cartilaginous vault is encased in its own fibrous capsule, whose fibres decussate, form a fibrous band or aponeurosis and join the capsule of adjacent cartilage. This aponeurosis, acting as a flexible membrane, allows freedom of movement between the neighbouring cartilages. According to Hinderer [7], the most distinctive fibres are:

1. Those between the terminal ends of the triangular cartilage and septum that supply the mobility necessary for valve action between these two structures.
2. Those between the caudal end of the septum and the medial crura of the columella which form the membranous septum.
3. Those between the caudal margin of the triangular cartilages and the cranial border of the alar cartilages.

A permanent continuity between the encasing fibrous capsule of the triangular cartilage and the periosteum of the nasal bone was found by Bruintjes et al. [8], supporting the clinical observation of the firm connection between the nasal bones and the triangular cartilages.

Additional mobility of the interacting nasal cartilages is provided by the thin layer of seven muscles covering the external nasal pyramid. Besides their mimic function, some muscles act as dilators of the valve region or openers of the nostrils or provide stability for the lateral nasal wall [9]. Especially, the dilator naris accompanies each nasal inspiration, thus directly varying with ventilation, nasal resistance, hypoxia and hypercapnia. It stabilises the anterior nasal airway and precedes diaphragmatic contractions and ceases to act with mouth or tracheostomal breathing [10].

The framework of nasal cartilages with their fibrous connections and covering muscle layer acts as a shield and portal to the erectile lining of the nasal cavities. Eugene Kern from Rochester, USA, termed the nasal mucosa 'the organ of the nose' to sum up all its many functional tasks which are described in other chapters of this book. While the nasal vestibule is covered with squamous epithelium continuing some millimetres around the edge of the caudal margin of the triangular cartilages, the nasal mucosa with ciliated cylindrical epithelium starts in the posterior valve region [11, 12]. Figure 35.3 shows the histological section through the posterior valve

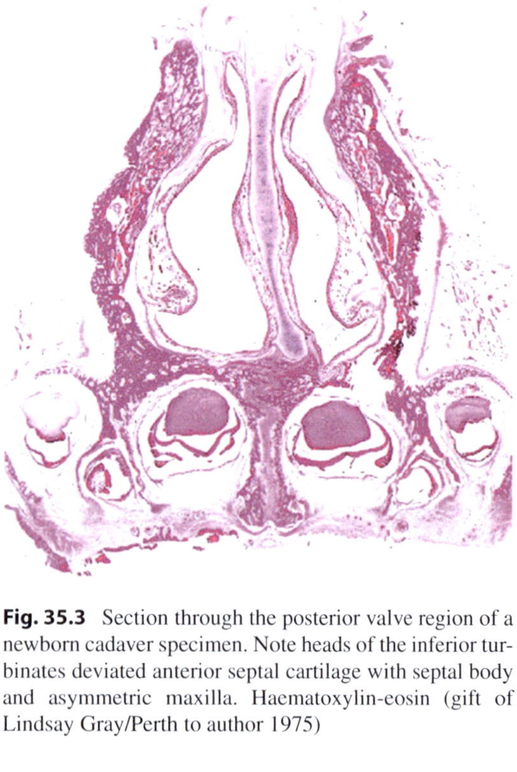

Fig. 35.3 Section through the posterior valve region of a newborn cadaver specimen. Note heads of the inferior turbinates deviated anterior septal cartilage with septal body and asymmetric maxilla. Haematoxylin-eosin (gift of Lindsay Gray/Perth to author 1975)

region of a newborn. The heads of the inferior turbinates catch our eyes as elevations from the mucosal lining of the lateral walls. The slightly deviated cartilaginous septum with its mucosal lining presents its thickening in the middle covered by the thick pad of erectile mucosa, the septal body or intumescentia septi. The different thickness of the vertically cut septal cartilage is clearly visible: the thick cranial part merges caudally with a thin segment to end in the broad deviated cartilaginous foot embedded in the premaxillary bone. This pattern of different cartilaginous thicknesses remains persistent throughout its lifetime as investigated by van Loosen et al. [13].

35.2 Part II: Functional Aspects

In this book, nasal functions like breathing, resistance, turbulence and nasal cycle are treated in special chapters in detail. Thus, I can confine to a few comments on functions where nasal cartilages are essentially involved like in the nasal valve region or the vulnerable anterior septum.

The anterior septodorsal cartilage, the paraseptal and the alar cartilages with the sesamoid cartilages form the framework of the nasal lobule and thus the portal to the upper airways. Soft tissue connections between the cartilages of varying thickness and several small muscles acting on their outside enable mobility and/or stability of the lateral nasal walls. Thus, the entrance to the nose can be dilated and narrowed to modify the inspiratory and expiratory airstreams. The entrance of the airflow is comparable with two flat, oval, hollowed structures, the nasal vestibules, which terminate in the aperture between the septum and the caudal end of the triangular cartilages, both parts of the nasal valve region. The vestibule is at an oblique angle to the cavum [11]. According to Cole's studies [10, 14] the nasal valve region consists of four anatomical airflow-resistive components: the aperture between the septum and triangular cartilage, the bony entrance to the nasal cavum which is occupied by erectile tissues of both lateral (head of inferior turbinate) and septal nasal walls (including the septal body) that modulate the cross-sectional area of the airway and airflow resistance. Two-thirds of the total nasal resistance during inspiration is provided by the nasal valve region. This has clearly been proved by Haight and Cole [15] using pressure-sensing cannulas. As two-thirds of the total airway resistance to breathing is created in the nose—one-third is provided by the open mouth [16]—the nasal valve region is the main resistor of the total airway. The narrow nasal valve region with the smallest cross-sectional area of the nasal cavum accelerates the inspired air, creating turbulence. Valve constrictions disrupt the laminar characteristics of the inspired air as it enters the body of the cavum and thereby enhance exchanges with the nasal mucosa of heat, water and noxious materials. In the widened nasal cavum, airflow decreases its speed [14].

In clinical praxis, the physician has to face several patients complaining of breathing problems which are caused by a disturbed anatomy of the nasal valve region, especially patients with a tension nose or a saddle nose. A patient with a tense nose presents with a prominent, curved, small dorsum due to a too large and too high septum. Other characteristics are the increased projection of the tip, slitlike nostrils with an elongated columella and elongated thin nasal alae. The feet of the medial crura are cranially shifted, thus broadening the base of the columella. The nasolabial angle is enlarged and often the upper incisivi are showing. The valve angle is less than 10° which results in a bilateral alar collapse during inspiration.

In the cartilaginous saddle nose, the septum is too short mostly due to fractures or perforations. The caudal septodorsal cartilage is depressed which causes an enlargement of the valve angle to more than 20°. The nostrils with ballooning alae are elliptically distorted; the columella is shortened and often retracted. Although the nasal cavities look very wide, the main complaint of the patients is insufficient breathing due to too much turbulence of the airstreams.

The fact that the nasal septum divides the nose into two parallel halves is the condition for cyclic activities of the erectile nasal mucosa. Kayser in

1895 [17] first reported on the spontaneous, cyclical congestion and decongestion in the nasal cavities. The anatomic conditions for this mucosal behaviour are mainly provided by its capacitance vessels. The 'working phase' of one nasal cavity characterised by decongestion of the cavernous tissues alternates with the 'resting phase' characterised by congestion of the mucosa [14, 18, 19]. Changes in flow and resistance during the nasal cycle have been studied using rhinomanometry [18, 20]. By means of acoustic rhinometry, Fisher et al. [21] measured the changes in the cross-sectional areas in the nasal cavities due to congestion and decongestion.

Using the combination of endoscopic imaging, rhinomanometry and acoustic rhinometry, Lang et al. [19] observed a periodic change in turbulence behaviour in addition to the known cyclical changes in flow resistance and nasal width. In the resting phase, the mainly laminar flow was found. During the working phase, the onset of turbulence occurred already at low velocities. The increase of turbulence during the working phase was created by the increase in cross-sectional area in the anterior cavum due to decongestion of the mucosa of the head of the inferior turbinate and the septal body. A special insight into the behaviour of the nasal cycle is measurable using the new method of long-term rhinoflowmetry which yields information about the nasal cycle over a 24-h period [22]. This method is also of practical value in Sleep Medicine to investigate sleep-disordered breathing.

The nasal cartilages are insofar involved in the process of the nasal cycle as they provide the anatomical trigger zones for the transition from laminar to turbulent airflow, namely, the anterior nasal cavum between the nasal valve region and the head of the middle turbinate.

Bruintjes et al. [8] studied the kinematics of the lateral nasal wall which is made up of three parts: (a) the osseous-cartilaginous chain of the nasal bone, triangular cartilage and the lateral crus of the alar cartilage; (b) the hinge area with the sesamoid cartilages between the lateral piriform aperture and the lateral margin of the lateral crus; and (c) the ala, the part between piriform aperture and alar cartilage, not supported by cartilage. While part (a) is relatively stable and part (b) much more compliant, they found part (c) to be the most compliant part of the lateral nasal wall. They also studied the muscles influencing the lateral nasal wall and its compliance. With this knowledge of the mechanical properties of the lateral nasal wall, they were able to analyse pathological clinical conditions, which may occur at the level of the nasal valve and at the level of the vestibule or nostril. Thus, they could explain, for instance, the physiological alterations caused by facial nerve palsy where muscle denervation may lead to alar collapse [23].

Loss of alar stability with nasal obstruction during inspiration is often caused by inadequate surgical procedures for the nasal lobule. Mostly the continuity of the osseous-cartilaginous chain, 'nasal bone-triangular cartilage-lateral crus of alar cartilage', is destroyed [24]. Detailed knowledge of these anatomical connections could help to reduce such adverse surgical sequelae.

The alar collapse associated with the drooping tip of elderly people may result from nasal muscle atrophy, a change in cartilage resilience and stretching of the intercartilaginous fibrous tissue with loss of cartilage overlap in the intercartilaginous junction [25]. The surgical procedure of the so-called rhinolift could reduce the breathing problems of elderly people. Via intercartilaginous incisions in the limen nasi, the cephalic margins of the lateral alar crura are bilaterally partly resected. After elevating the dorsal skin of the nasal pyramid and excision of an oval piece of skin in the nasal root, both mobilised alar cartilages can be lifted and fixed, thus widening the angles of the nasal valves.

The elastic nasal cartilages are useful elements for protective functions. The 9-year-old boy in Fig. 35.12 is a good example to explain the function of the septal cartilages as a 'crumpled zone' in case of severe anteroposterior load. Between 1970 and 1972, I performed septorhinoplasty according to the techniques of Cottle, Goldman and Masing in 92 children with a mean age of 10.5 years [27]. The indication was obstructed nasal breathing due to established post-traumatic nasal deformities. Intraoperatively, fractures and defects were visible in the anterior septodorsal cartilage with involvement of the nasal valves and sometimes of alar cartilages in 44/92 noses. These noses were damaged by loads from frontal and/or below. About 23/92 noses showed fractures in the septodorsal cartilage like the previous group and in addition fractures of the nasal bones and perpendicular plate were caused by the mainly anteroposterior load. This means that 67/99 or 72% of the children suffered nasal obstruction due to damaged nasal cartilages in the anterior nose.

The build-up of the septal cartilage in regions of different thicknesses [13] throughout the whole life is one component to react more elastically to front load. The second component of more compliance is the vaultlike construction of the septodorsal and alar cartilages with their joint-like fibrous interconnections [8]. Note the distortion of the caudal septodorsal cartilage of the boy in Fig. 35.12 (3): a 90° angle of the caudal septum with fractures in the caudal edge and depressed edges of the triangular cartilages. Van Velzen et al. [28] published the prepared septal cartilage of a 4-year-old boy who died in a frontal accident. The fracture lines in the cartilaginous septum followed the thin regions of the cartilage as sites of minor resistance in reaction to the load.

In idealised and patient-specific models, Lee et al. [29] recently published the reactions of the human septal cartilage exposed to anteroposterior load. They found the maximum stress areas in the nasal septum in the vicinity of the bony-cartilaginous junction and the anterior nasal spine, which are consistent with clinical experience. The findings of their study also suggest that the septum does function as a 'crumpled zone', absorbing a significant amount of stress before it is transmitted to the skull.

The extreme variant of a damaged crumpled zone nose is the classic boxer's nose, the so-called rubber nose, mainly formed by the alar cartilages. The septodorsal cartilage is shrunken to a minimum; nasal bones and anterior nasal spine are pressed down to the level of the piriform aperture or resorbed.

35.3 Part III: Alterations of Nasal Cartilages

The functions of the nasal cartilages can best be recognised in children and adults with a disturbed cartilaginous framework. The growing nose is influenced by genetic and epigenetic influences such as oxygen supply, nutrition, hormones, medication, infections and injuries including nasal surgery as a controlled trauma, to name a few. In this part, examples are presented describing some long-term influences on the nasal cartilages citing literature and own case reports.

35.3.1 Lacking Septodorsal Cartilage

Already in utero, the septodorsal cartilage, being composed of the septal and both triangular cartilages, develops as the dominating structure for nasal and midfacial growth. This was shown in 1791 by Soemmerring [30] who published a newborn's skull with a lacking septodorsal cartilage and the skull of a healthy newborn for comparison (Fig. 35.4). In the deformed newborn, the nasal bones and the incisive parts of the maxilla are not developed. The size of the piriform aperture is reduced to one-third in width compared with the healthy newborn. The height and width of the maxilla are reduced, while the contours of the orbital cavities are distorted compared with the healthy newborn.

35.3.2 Lacking Alar Cartilages

The following case shows that congenitally lacking both alar cartilages did not impede the growth of the nose and midface apart from the complete lacking nasal lobule (Fig. 35.5). The 15-year-old boy came from a family with no genetic nasal disorders and an uneventful pregnancy of his mother. There was a complete absence of both alar cartilages. The pseudo-columella was formed by skin covering the caudal septal cartilage. The

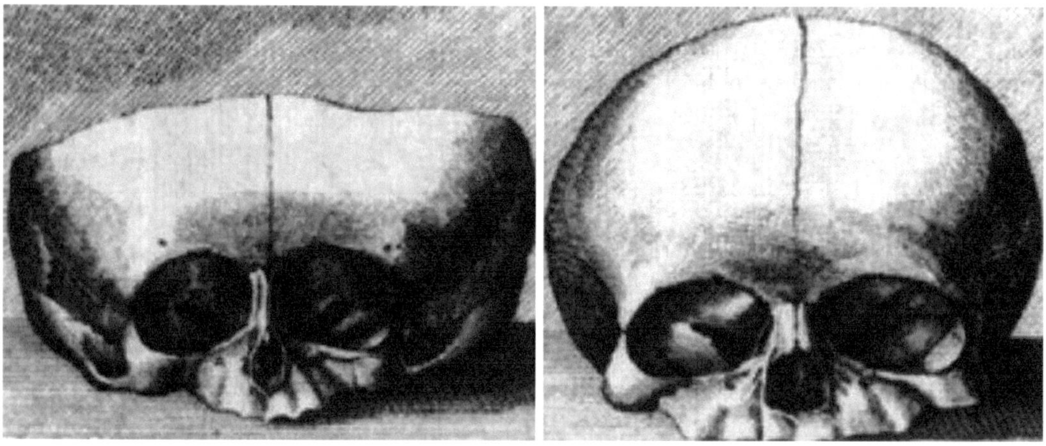

Fig. 35.4 Skulls of newborns lacking septodorsal cartilage (*left*), with normal midface (*right*) [30]

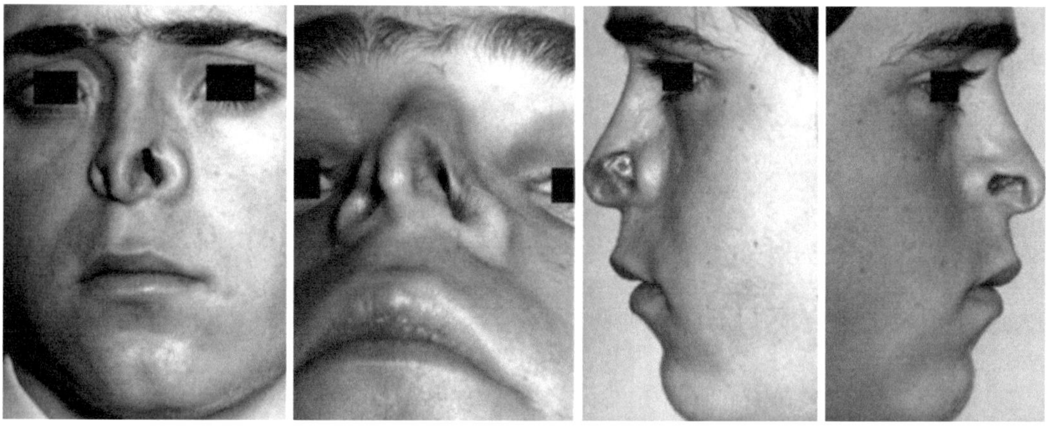

Fig. 35.5 Fifteen-year-old boy with congenital lack of both alar cartilages [31]

caudal margins of the triangular cartilages were covered by thick skin on the right and thin skin on the left side. All the other nasal structures were inconspicuous. The boy's breathing was normal [31].

35.3.3 Physiological Septal Deviation

The anatomical term 'physiological septal deviation' was introduced by Zuckerkandl [32] who defined this type of septal deformation as a bent septum within the asymmetrical human skull. He published several examples on beautiful lithographs (Fig. 35.6). Comparable histologic sections of his own cases were published by Gray [33] who kindly left me a few of them like in Fig. 35.3. In more recent publications [34, 35] on this topic, the incidence of the physiological septal deviation is reported between 30% and 75% for children and between 13% and 96% in adults, with strikingly less patients reporting to suffer from subjective symptoms [36]. From the physiological point of view, it is practical to characterise the physiological septal deviation by normal endonasal resistance [34]. Many of these physiological septal deviations develop during foetal life in connection with asymmetrical maxillary growth. Often the non-pathological role of this deviation is not recognised and a septal operation is indicated, although it is useless or even wors-

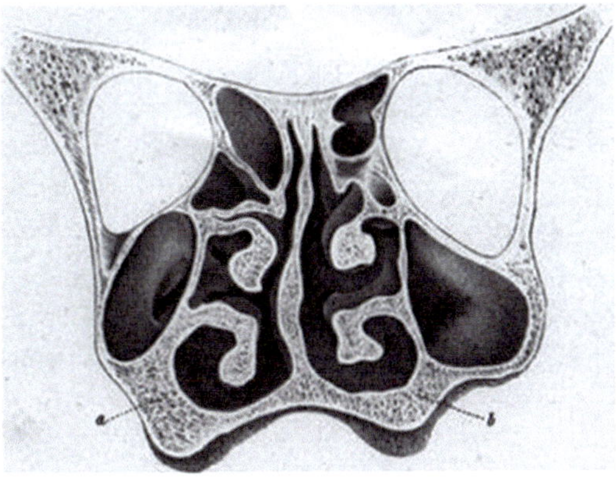

Fig. 35.6 Physiological septal deviation in an asymmetrical skull; The thicker spongiosa (**a**) compared to the thinner one (**b**) narrows the right maxillary sinus [32]

ens the nasal symptoms. A septal body may be mistaken as pathological septal deviation or the real cause of nasal obstruction such as a problem of the nasal valve or an inspiratory ala collapse is failed to be recognised.

Until now, the diagnosis of the 'physiological septal deviation' has mainly been made by nasal endoscopy and the subjective history of the patient to be free of nasal breathing problems. In a prospective cohort study, Gogniashvili et al. [34] investigated 158 patients between 18 and 40 years using rhinoresistometry, acoustic rhinometry, endoscopy and visual analogue scales for subjective complaints. They defined the normal one-sided nasal resistance as less or equal to 0.35 sPa/cm^3 at a flow velocity of 250 cm^3/s. Applying this benchmark, the unselected group of non-rhinological patients was differentiated into 144 subjects with physiological nasal resistance and 44 with pathological septal deviation. This means an incidence of 72.2% physiological septal deviation in an unselected cohort which fits well with the data mentioned above.

35.3.4 Congenital Nasal Deviations

The histological depiction of a slight septal deviation in a newborn without signs of an acute injury is documented in Lindsay Gray's Fig. 35.3. In neonates, two types of nasal deviation are observed: a septal dislocation that can easily be replaced in the midline and a nasal deviation that cannot be replaced by manipulation. The first type is considered as nasal trauma during delivery. The second type had been published by several authors since the end of the nineteenth century [37] and more detailed by Cottle [38] who generally concluded: 'For these, expectant waiting is recommended unless there is complete inability to breathe and eat. One sees remarkable improvement in the appearance and development of these noses without surgical intervention. During the first decade, however, some will require surgical aid'. To find out the incidence of these two types of nasal deviation, we investigated 3425 children, born in 2 years between 1980 and 1981, in the Obstetric Department of the University of Ulm, Germany. A total of 110 (3.23%) of these neonates, belonging to the Caucasian population, showed deviated nasal structures. 81 (2.37%) of these dislocated septa could easily be replaced by a closed reposition. In 29 (0.86% of 3425) newborns, repositioning the deviated nasal structure was impossible, thus leaving the baby with a deviated septum and bony pyramids like the neonate in Fig. 35.7 (left) who presents nasal deviation to the left, oblique columella and asymmetrical nostrils. Typically is also the oblique bony pyramid with asymmetrical slope and length on both sides.

Over a period of 11–12 years, 14 children out of 29 newborns with non-replaceable nasal deviation were prospectively followed by the author and a second otorhinolaryngologist [37]. No child had a history of nasal trauma in the meantime. The results show that the newborns' noses, which deviated intrauterinely due to some unknown reasons, did not spontaneously restore in each case. Nine children had a completely straight bony pyramid, symmetric nostrils and reported subjectively normal breathing. Four children had proven nasal allergy and presented bilateral hypertrophy of the inferior turbinates. Eight of the nine septa showed slight

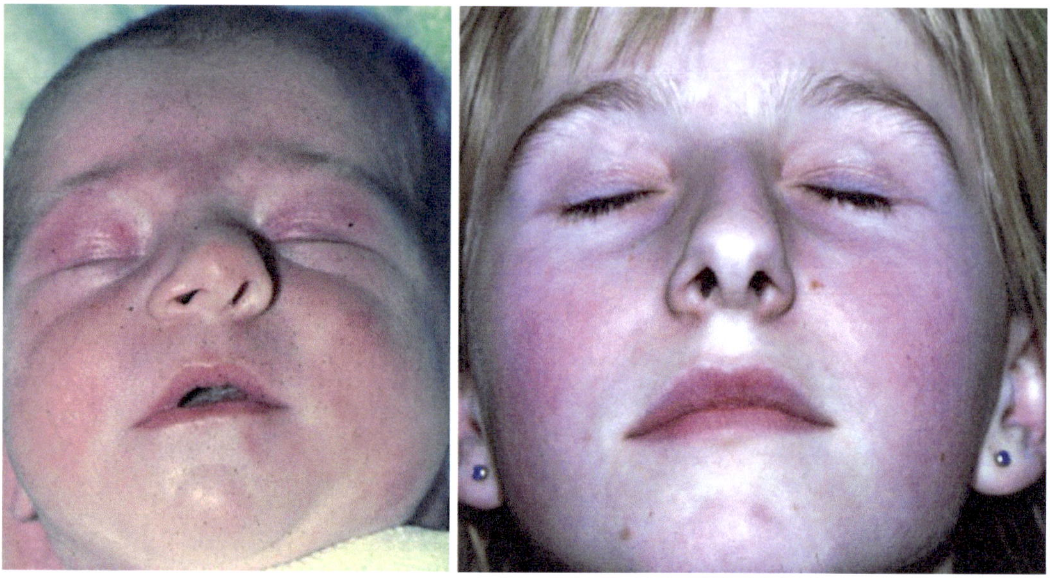

Fig. 35.7 Female newborn with a non-replaceable nasal deviation (*left*). 12 years later, the girl presents with an inverted C-shaped nasal deviation and nasal obstruction (*right*). Note the longer right and the short steeper left nasal wall ([39]; unpublished)

deviations, spurs or crests. Eight of these nine children had malocclusion, two of them having been treated orthodontically. We considered these eight children as having a 'physiological septal deviation' in an asymmetrical skull, as described by Zuckerkandl in 1882 [32]. However, five girls of the 14 children showed a deviation of the nasal pyramid to the same side as found at birth, in one girl markedly (Fig. 35.7—right), in a second girl only minor and in three cases slightly. In four of these five girls, a longitudinal deviation, corresponding to the deviation of the bony pyramid, was evaluated by nasal endoscopy and measured by acoustic rhinometry. One of the five girls complained of moderate obstructive breathing, which was caused by allergic rhinitis. All five girls showed malocclusion, two of them being under orthodontic therapy.

35.3.5 Acromegaly

Acromegaly is an endocrine disease due to growth hormone excess originating from somatotrophic adenoma of the pituitary gland. Patients complain about the coarsening of the facial contours caused by a bony proliferation of the skull and mandible and by excessive nasal growth. Our groups in Ulm and Zurich [40] investigated the growth mechanism of the septal cartilage in six acromegalic patients. Small strips of septal cartilage were obtained during septoplasty from healthy adults or during transnasal hypophysectomy from acromegalic patients. Growth activity in five different areas of the septal cartilage was measured by in vitro incorporation of 35 S-labelled sulphates and 3 H-labelled thymidines. The growth activities in the posterior area, which is situated anterior to the septoethmoidal junction, were significantly enhanced compared to a control group of hormonally normal adults. Growth activities in the anterior caudal end and in the suprapremaxillary area were not different in both groups. This indicates that growth hormone excess in acromegaly enhances human septal growth by stimulating the growth activities in the posterior area.

In another study [41], five different enzymatic pathways were analysed in these septal areas. Cathepsin D, an acid proteinase, was not influenced by the augmented growth hormone level in acromegaly, whereas cathepsin B, a neutral proteinase, showed its highest activity in the caudal prolongation and the posterior area and was significantly increased in all areas in acromegaly. Beta-hexosaminidase activity was highest in the central and posterior area and caudal prolongation of the septum. In acromegaly, a significant increase in its activity was found in the suprapremaxillary and posterior areas. Acid phosphatase activity was highest in the caudal prolongation of the septum, but its activity was significantly increased in all tested areas in acromegaly. Alkaline phosphatase activity could only be found in the posterior area and the caudal prolongation in healthy adults. However, in acromegaly, this enzyme could be detected in the central area and the posterior end of the suprapremaxillary area, suggesting an altered process of mineralisation. Thus, a distinct local pattern of enzymes related to intercellular substance metabolism and mineralisation can be demonstrated in the septal cartilage of healthy adults and acromegalic patients.

35.3.6 Damaged or Lacking Triangular Cartilage

What happens when parts of the septodorsal cartilage are damaged or removed? Verwoerd and Verwoerd-Verhoef [42] found that the behaviour of hyaline cartilage of the human nose appeared to be comparable to that of other mammals, especially rabbits. Their results can be supported by the following own observation. Because of histologically suspected sarcoma, the left nasal bone and triangular cartilage in a 6-year-old boy had to be resected in 1973 [39]. Fortunately, a curable circumscribed osteomyelitis was diagnosed. Following this boy 7 years later, we found a shortened left nostril with the bony pyramid deviating to the left and the nasal tip deviating to the non-operated side. The left piriform aperture was positioned higher and the left nasal process of the maxilla was reduced. The left inferior turbinate was smaller than the right-sided one, and the caudal end of the septal cartilage slightly deviated to the right (Fig. 35.8). Thus, the resection of one

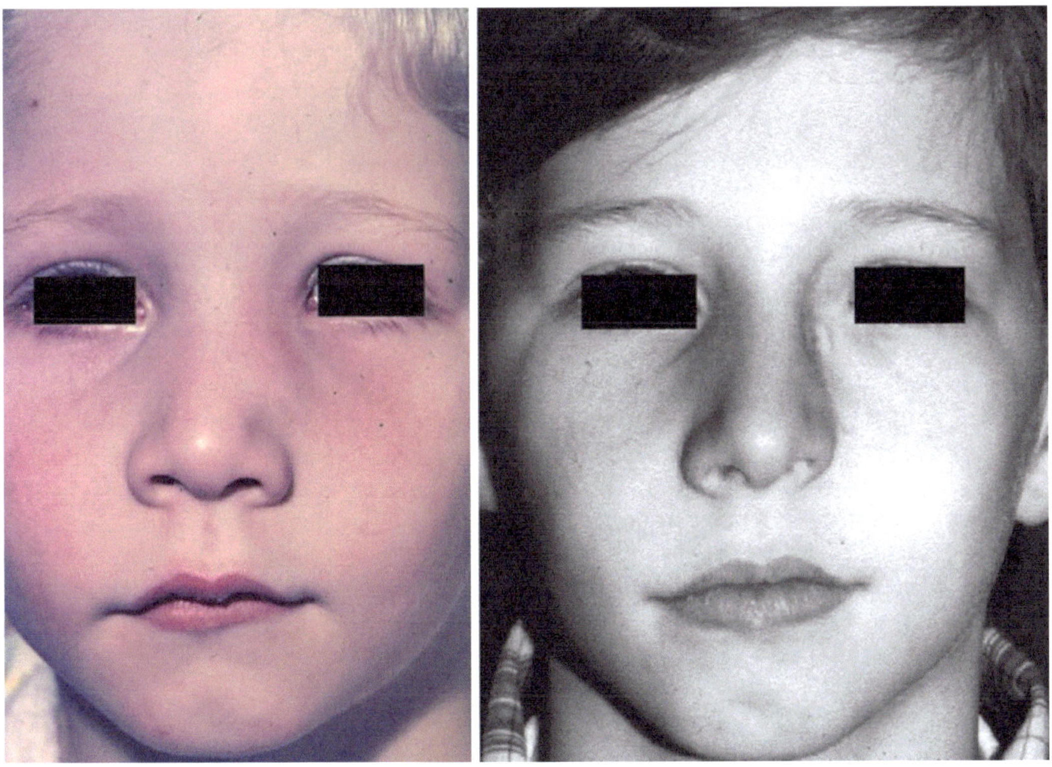

Fig. 35.8 6-year-old boy before resection of left triangular cartilage (*left*); midfacial growth inhibition at age of 13 years (*right*) [39]

triangular cartilage and the nasal bone in connection with an infection affected the growth of nasal and midfacial structures. These findings are similar to Poublon's results from the Rotterdam group after unilateral resection of the triangular cartilage in growing rabbits [43].

35.3.7 Some Histological Aspects of Traumatised Nasal Cartilage

As there is a detailed description of this topic by Verwoerd-Verhoef in this book about the sequelae of different kinds of injury to the nasal cartilage in man and especially in growing animals, I confine to three topics associated with the pathophysiology of the human nasal cartilages.

The first aspect is the effects of incomplete fractures which are the main reasons for the more or less obstructive bending of the septal cartilage. During nasal surgery these bends are often difficult to transform into a straight septal plate, a problem which Hunter Fry associated with the disturbed interlocked stresses within the hyaline nasal cartilage [44]. Ten Koppel et al. [45] added new convincing data to this problem, more discussed below. Figure 35.9 shows the biopsy from a vertical strip of septal cartilage with an incomplete fracture. The 8-year-old boy had a nasal trauma some years ago and underwent a septoplasty alio loco because his anterior septum obstructed both anterior nasal cavities [46]. In Fig. 35.9, a scar of fibrous tissue is filling the cartilaginous defect, measuring approximately 30% of the thickness of the intact cartilage. According to the finds of Fry [44] and Ten Koppel et al. [45], one should expect a bending of the cartilage to the other side. One possible explanation for the bending to the opposite direction may be that the power of the granulation tissue which first fills the incomplete fracture during healing is strong enough to achieve the bending to the unexpected direction. In the 1960s and 1970s when septoplasty was often performed only by unilateral elevation (tunnelling) of the mucoperichondrium

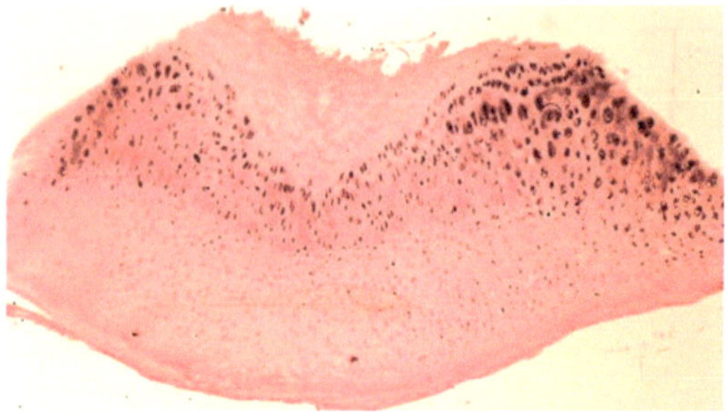

Fig. 35.9 Histological section of a biopsy of septal cartilage with a vertical incomplete fracture filled with scar tissue in an 8-year-old boy. Haematoxylin-eosin [46]

from the septum, recurrences of septal bending were observed, although the septum looked straight at the end of the operation [47]. One reason was that small incomplete cartilaginous fractures on the side with the mucoperichondrium left attached could cause bending because the elevated mucoperichondrium had changed the balance of the interlocked stresses within the cartilage in an unpredictable manner. That's why many rhinosurgeons bilaterally elevate the mucoperichondrium to better recognise pathologies and incomplete scars of the septal cartilage.

The second aspect is the partial regeneration of pieces of septal cartilage within its traumatised inner perichondrium in children. The nasal perichondrium is built up of a thick outer layer and a thin inner layer. The inner layer usually remains connected with the hyaline cartilage when properly elevating the mucoperichondrium from the septal cartilage during surgery, because it contains fibres which end inside the cartilage. In case of damage by surgical or nonsurgical trauma, the inner layer has the potential to create new cartilage [26, 46, 48]. Figure 35.10 shows the histological section from a piece of destroyed anterior septal cartilage removed during septoplasty in a 12-year-old boy. In the right upper corner, two pieces of new cartilage are visibly grown within the torn perichondrium of the damaged septal cartilage after the untreated nasal injury at the age of 3 years [46].

In Fig. 35.11, the histological section shows an accumulation of regenerated hyaline cartilag-

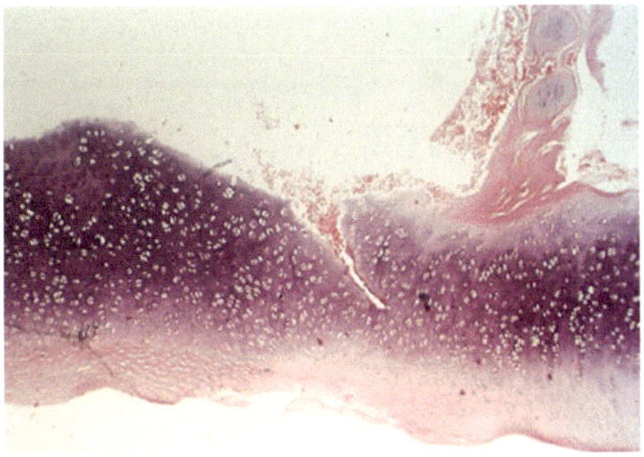

Fig. 35.10 Histological section of a biopsy of a traumatised septal cartilage at the age of 3 years, taken from the 12-year-old boy. In the upper right corner, two pieces of regenerated cartilage grow within the damaged perichondrium. Haematoxylin-eosin [46]

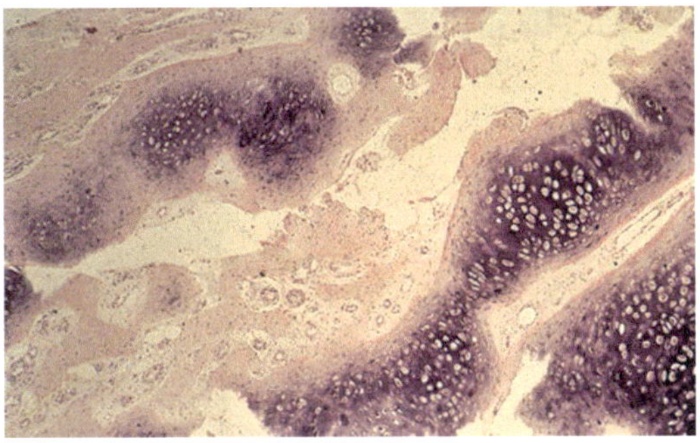

Fig. 35.11 Histological section of a biopsy from a cartilage-like layer attached to the fractured anterior septum of a 9-year-old boy. Islands of regenerated cartilage and vessels in the scared perichondrium. Haematoxylin-eosin [26]

inous islands embedded in connective tissue with vessels. The biopsy was taken from the cartilage-like layer firmly attached to the remnant of the anterior septal cartilage of a 9-year-old boy during septoplasty. These islands had developed in the multiply torn perichondrium of the damaged anterior septum within 8 years of impeded nasal growth. The boy suffered a severe frontal nasal injury at the age of 1 year which caused a hypoplastic nose with obstruction of the anterior nose (Fig. 35.12). Unfortunately, such a layer of 'floppy cartilage' can only be used as 'filling tissue', but not for anterior septal reconstruction [26].

The third histological example demonstrates how the lack of nasal cartilage leads to the atrophy of nasal erectile lining. It is well known that submucous septal resection may induce atrophy of the septal mucoperichondrium and septal perforation while interposing pieces of cartilage between the mucoperichondrial flaps can markedly reduce this tendency to atrophy [49]. Nasal septa from cadavers with previous submucous septal resection were histologically investigated and compared with age-matched nonoperated septa [49, 50]. In areas where the cartilaginous septum had been removed, most of the secretory epithelium was replaced by squamous cells. The submucous layer was markedly reduced in thickness (Fig. 35.13). The submucous vessels were reduced in number and size, and the submucosal glands were partly atrophic. The site of the former septal cartilage was filled with dense connective tissue containing only a few vessels. These finds demonstrate the importance of interposing autogenic cartilage plates between the elevated perichondrium flaps at the end of septoplasty to reduce the propensity for mucosal atrophy.

35.3.8 Transposition Technique

In cases of severe destruction of the anterior septal cartilage due to a frontal trauma, an established technique to reconstruct the caudal septum and the valve area is to remove the remnants of the anterior septum and replace them with a boomerang-shaped cartilaginous or bony part from the posterior septum. In the late 1960s, I learnt this procedure from Helmut Masing in Erlangen, Germany, who termed it 'transposition technique'. After satisfying results in adults, we used this technique in an 11-year-old boy suffering from bilateral nasal obstruction after a frontal nasal injury some years ago. Via the hemitransfixion incision, an 'empty columella' was found with a few small cartilaginous remnants isolated from the scars between the mucoperichondrial flaps. From the posterior septum which was not yet ossified, boomerang-shaped cartilage was harvested. This transplant was anteriorly fixed

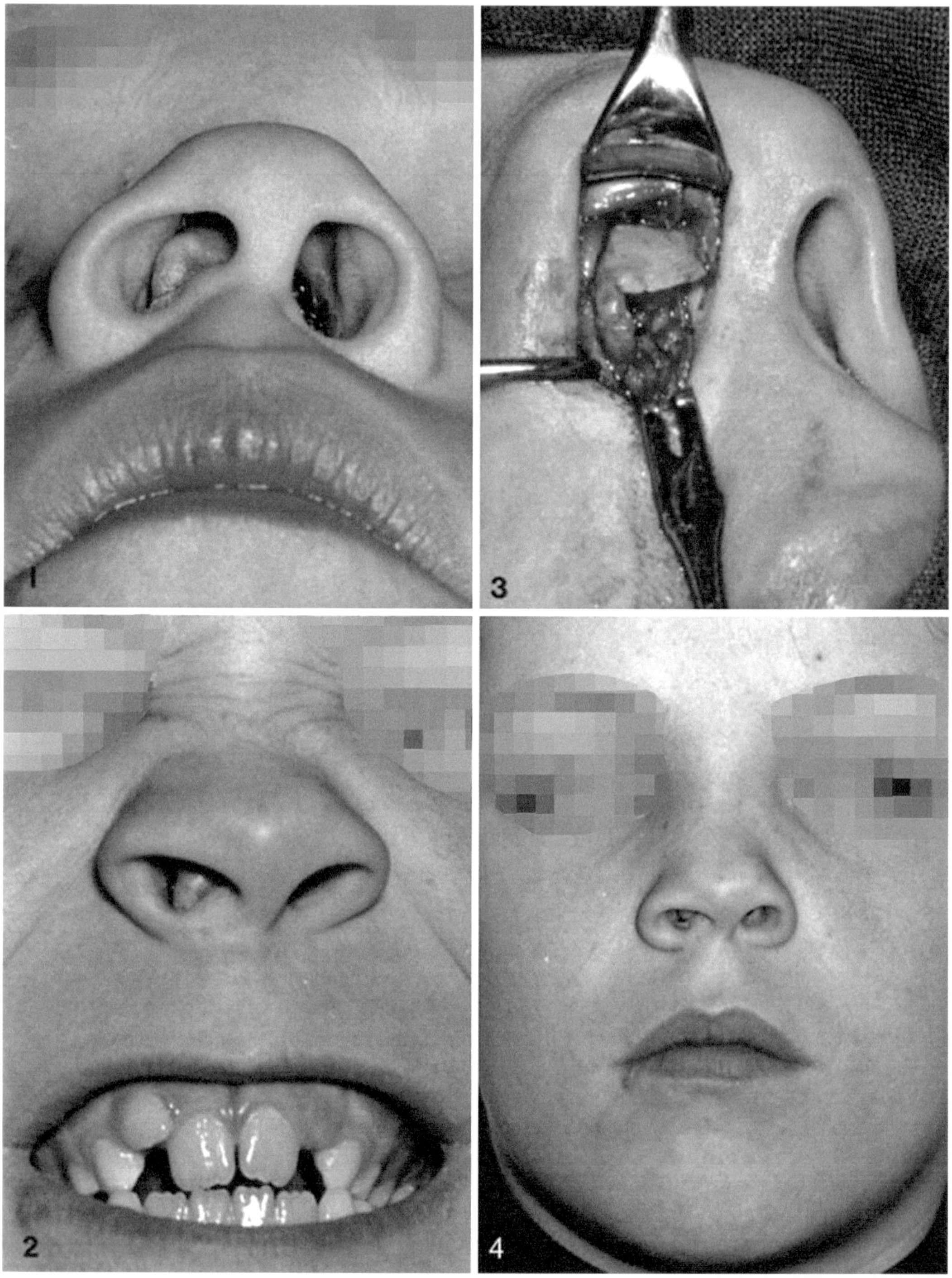

Fig. 35.12 Three pre- and one intraoperative photographs of the 9-year-old boy in Fig. 35.11. Figures of base (*1*), with damaged bite (*2*), intraoperative (*3*), and frontal view (*4*). (Pirsig, unpublished)

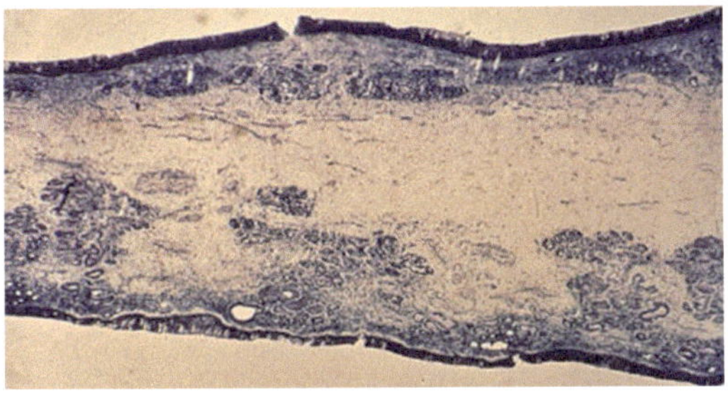

Fig. 35.13 Histological section through septal remnant after submucous septal resection 10 years ago in a 51-year-old man; Giemsa staining (Pirsig, unpublished)

between the hypoplastic anterior nasal spine and the supra-tip region under the nasal dorsum to restore the valve angles. The cartilaginous remnants from the anterior septal region were inserted into the posterior septal region. A follow-up after 11 years postoperatively including an X-ray film of the nose (Fig. 35.14—right) revealed the ossification of the posterior autogenic transplant. This was one reason why the transplant did not grow within the 11 years postoperatively. As a consequence, I prefer to use autogenic cartilage from the ear to reconstruct the destroyed anterior septum in children instead of harvesting cartilage from the posterior septum (see back-to-back technique). The nasal appearance after 16 postoperative years is shown in Fig. 35.14. The young man reported normal breathing all the years. The nose with a slightly depressed lobule showed a retracted columella and a slight maxillary retrusion as signs of growth inhibition.

35.3.9 Frontal Nasal Trauma

As mentioned above, the septodorsal cartilage is the dominating structure for nasal and midfacial growth. This also means if the septodorsal cartilage is markedly damaged by mechanical loads or diseases, this is mirrored in the whole nose and often in adjacent midfacial tissues. Thus, as to the long-term outcome of a septorhinoplasty, the crucial effect of the nasal reconstruction depended on the repair of the damaged septodorsal cartilage and not of the distorted nasal bony structures. A mostly satisfying repair is much easier to perform in the nose injured by a lateral load with lateral distortions of the nasal tissues than in the nose after an anteroposterior load with infraction and dislocation of the bony and cartilaginous nose. In Fig. 35.15, we see a girl with an untreated frontal nasal trauma at the age of 6 years (left). The nose shows a slight bony deviation to the right side, a minor C-shaped bending of the dorsum, and a small saddle. The harmony of the midfacial proportions is not yet disturbed. Her photograph at age 17 (Fig. 35.15—middle) shows an underdeveloped nose, still a 'child's nose', with a marked bony deviation, the saddling more pronounced and the lobule hypoplastic. The maxilla is retruded. The midfacial harmony is severely disturbed. On the sketch (Fig. 35.15—right), drawn during the open approach at age 17, the pathologies of the bony and cartilaginous nasal structures are clearly visible: distorted, asymmetric, infractured nasal bones, asymmetric piriform aperture, scars in the fractures of the bent triangular cartilages and the deformed and fractured caudal septal cartilage. A closed reposition at the time of the nasal trauma would probably have prevented the development of the crooked bony pyramid, but not the formation of the numerous scars due to the incomplete and complete fractures of 'the T- bar-shaped septodorsal cartilage', to use a term of Carel Verwoerd from Rotterdam, the Netherlands. This is an example of an acutely injured nose which cannot properly be treated by immediate surgery because the whole septodorsal framework is irreversibly disturbed by the anteroposterior load!

35.3.10 Lateral Nasal Trauma

In Fig. 35.16, we see the preoperative photographs of a girl of 11 years who had an untreated mainly lateral nasal trauma at the age of 6 years and complained of permanent mouth breathing and severely reduced olfaction. At school, she was teased as 'butter witch'. A functional and aesthetic septorhinoplasty was performed including paramedian, lateral and transverse osteotomies and

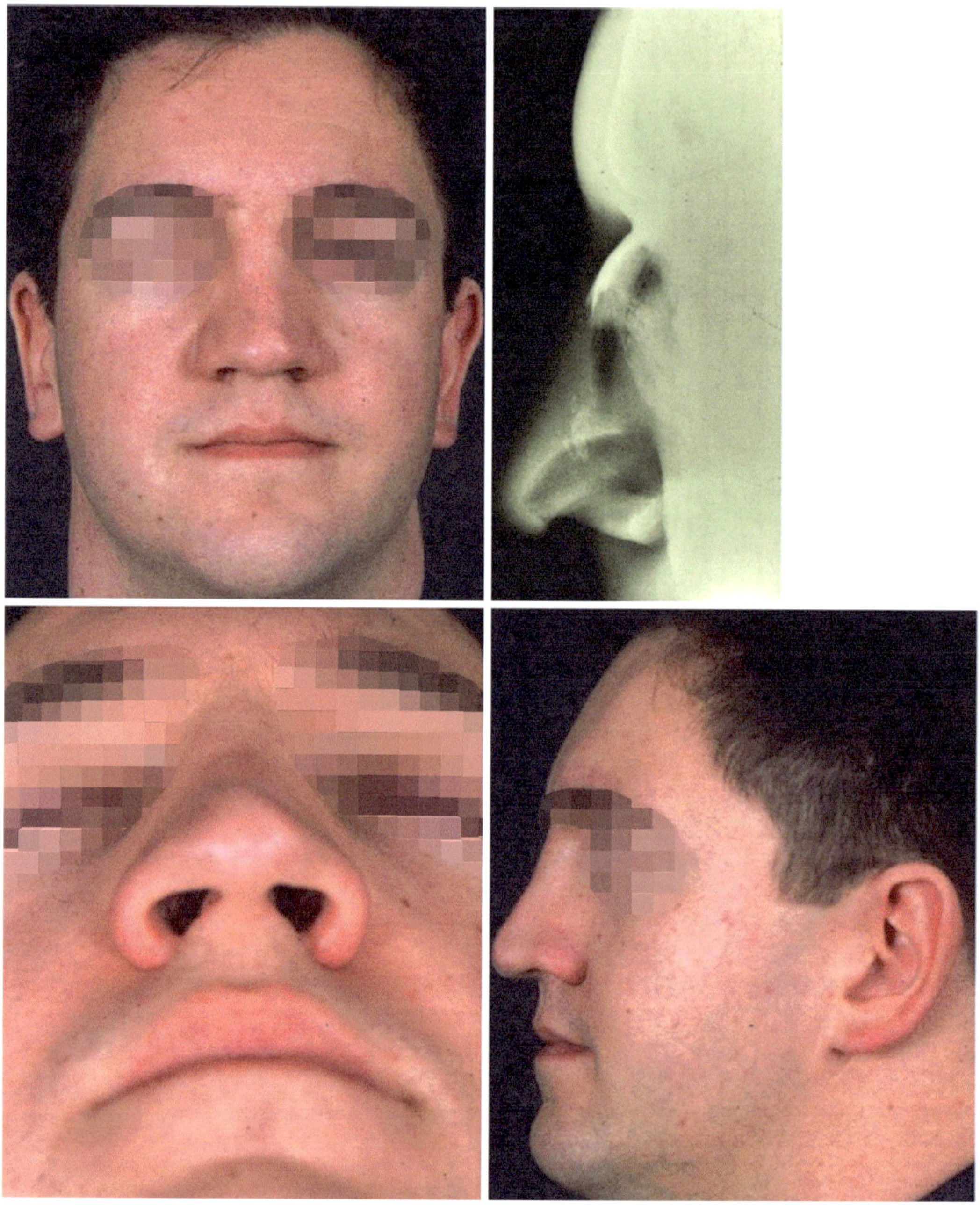

Fig. 35.14 Young man, 16 years after a septoplasty at the age of 11 using the transposition technique. Ossification of the transplant visible on X-ray film 11 years postoperatively (Pirsig, unpublished)

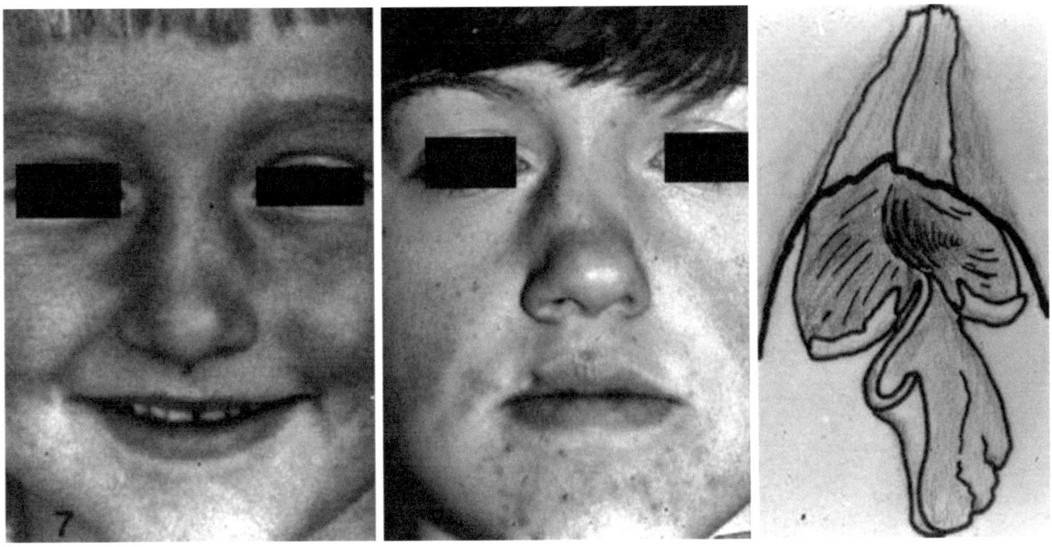

Fig. 35.15 Untreated nasal trauma in a 6-year-old girl (*left*), at 17 years (*middle*), and drawing with pathologic finds of the nasal bones and septodorsal cartilage (*right*) [39]

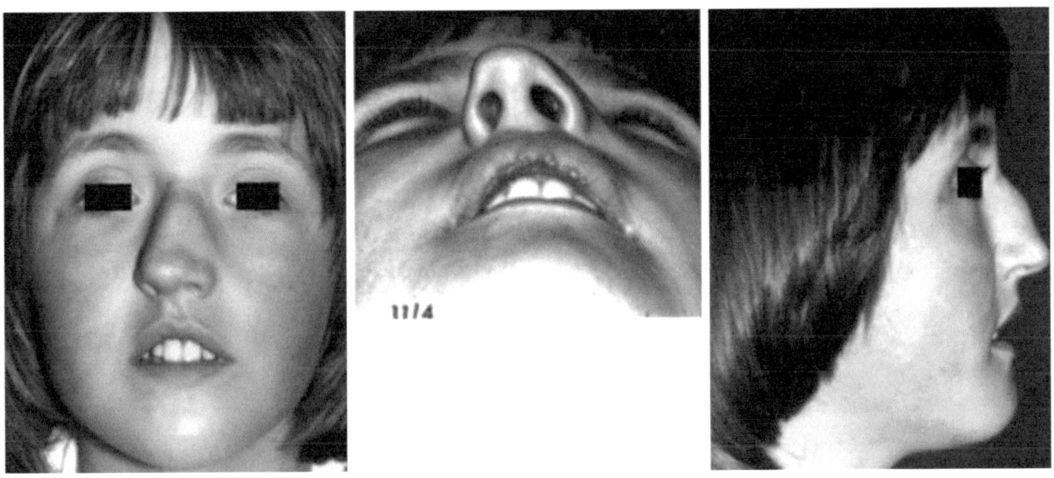

Fig. 35.16 Girl of 11/4 years preoperatively [51]

removal of a cartilaginous-bony hump [51]. In Fig. 35.17 we look into the face of a self-confident young woman with an inconspicuous nasal appearance, 8 years after nasal surgery. Her sense of olfaction had markedly improved. If this girl at the time of her nasal injury had been treated with a closed nasal reposition, most of her emotional and physical suffering due to her nose problem would probably have been avoidable. This example demonstrates how the osteotomised nasal bones follow the position of the reconstructed septodorsal cartilaginous framework. It also underlines the experience that nasal osteotomies do not impede nasal growth in children. Further, it shows that the outcome of repaired noses damaged by a lateral load is much better than the long-term results after reconstructing noses damaged by a frontal load as seen in Fig. 35.14.

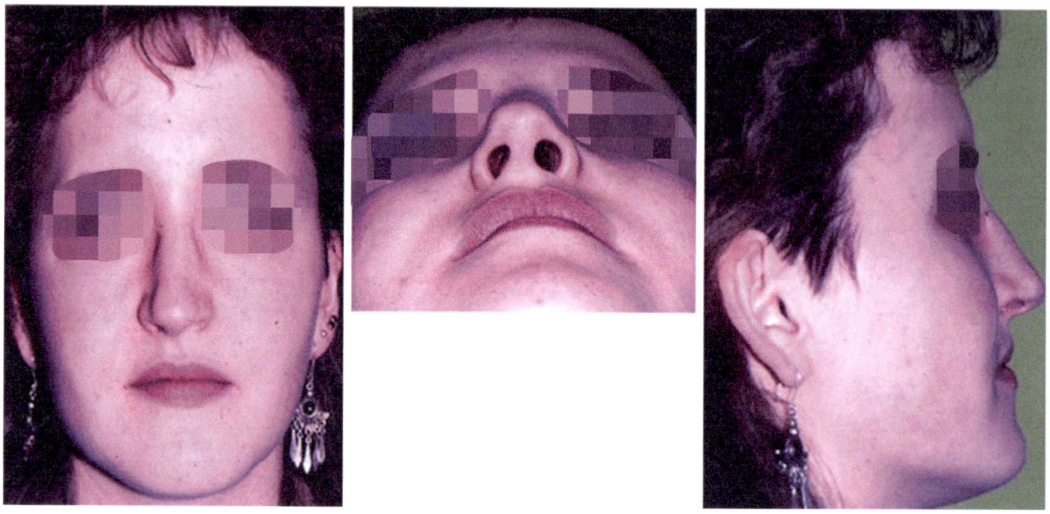

Fig. 35.17 Young woman as in Fig. 35.16, 8 years after functional and aesthetic septorhinoplasty (Pirsig, unpublished)

35.3.11 Scoring the Nasal Cartilages

Cartilage can be shaped by scoring, gridding or cross hedging to straighten a convexity or sculpt it. In an ex vivo experiment on nasal septal cartilage of adult rabbits, Ten Koppel et al. [45] demonstrated that there is a clear linear relationship between the depth of the incision and the resulting degree of cartilage bending when incisions are made up to half of the cartilage thickness. If the incision surpasses half of the cartilage, the resulting bending becomes unpredictable. In addition, their in vivo experiments showed 10 weeks after surgery that the scored cartilage of the healed septum maintained the imposed shape and its degree of bending in all animals towards the non-scored side. The authors conclude that with the results of this model, the effect of cartilage scoring can be better predicted during rhinosurgery. This may hold true for adult nasal cartilages.

On the other side, we know from van Loosen et al. [13] who investigated septa from birth to 62 years that the thickness of the septal cartilage is considerably variable in both the anteroposterior and cranial-caudal direction. This pattern of cartilaginous thickness remains persistent throughout the lifetime, but cannot exactly be evaluated during surgery with the mucoperichondrium attached or elevated from the septal cartilage. The surgeon can only recognise the cartilaginous thickness at defined sites by cutting through the whole cartilage. If only one side of the cartilaginous surface is incised, one cannot find out where half of the septal cartilage is reached (see Fig. 35.9). Therefore, a vertical scoring of exactly the same depth over a length of 3 cm, for instance, will result with slight differences in bending due to the diverse thickness of the septal cartilage over this length in the valve region. This means the predictability of the amount of bending becomes questionable.

Another reason for the unpredictability of the scoring effect is connected with the incomplete wound healing of the septal cartilage [46, 52, 53]. After healing of an incompletely incised cartilaginous surface, a scar of connective tissue fills the cartilaginous gap, which influences the amount and direction of bending. Figure 35.18 shows the part of distorted septal cartilage (3.7 cm long) removed during revision septoplasty at the age of 14 years [39]. The boy had been operated 6 years before alio loco because of traumatic septal deviation. The previous surgeon had unilaterally scored the septal surface by oblique, nearly parallel incomplete incisions which resulted in the markedly distorted piece of

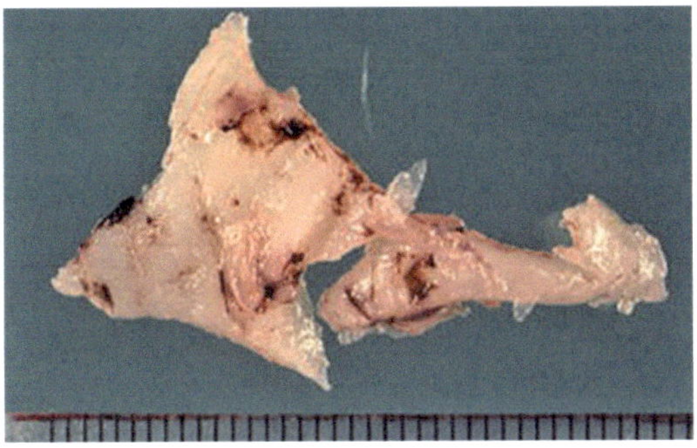

Fig. 35.18 Remnants of the anterior septal cartilage of a 14-year-old boy. The cartilage had been scored by 10 incisions 6 years ago, which resulted in partial resorption and unpredictable bending of the surface [39]

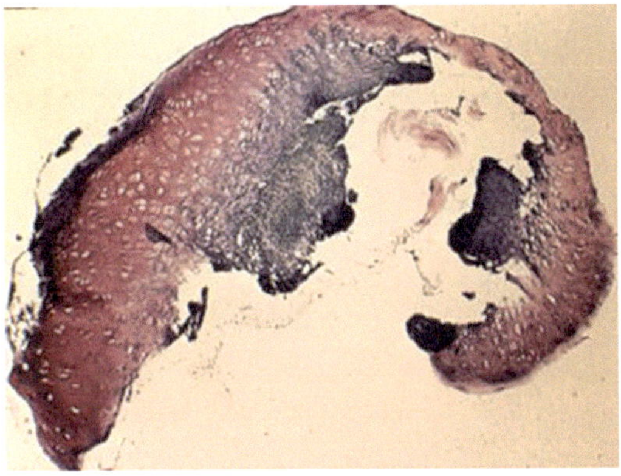

Fig. 35.19 Septal abscess 10 days ago. Histological section through a biopsy of the partially necrotising septal cartilage; toluidine blue [46]

septal cartilage removed at revision septoplasty. One can recognise the remnants of the parallel scorings and their diverse mode of healing ranging from resorption with small cartilaginous defects to hardly visible scars on the cartilaginous surface. This irregular surface pattern is due to the different thicknesses of the septal cartilage and its incomplete wound healing following scoring. That's why I don't recommend scoring septal cartilage in children.

35.3.12 Nasal Septal Abscess

Among the acquired nasal injuries during childhood, the septal abscess can not only destroy the septal cartilage but also affect midfacial tissues. The growth inhibition is more pronounced the earlier the abscess happened, especially in the first decade. The majority of septal abscess is caused by acute nasal injury. Blood vessels are disrupted in the mucoperichondrium, which is not torn because it is thicker and more elastic in children than in adults. The resulting hematoma is highly susceptible to infection. White blood cells invade the cartilage, create an acid pH and destroy it within some hours. This happens due to cathepsin D, a necrolytic and autolytic collagen-degrading enzyme with the optimum acid pH. This enzyme is normally distributed all over the healthy septal cartilage. This finding helps to explain the rapidity of the cartilage destruction (Fig. 35.19) in many cases of the septal abscess [46].

Immediate action is required. After puncture for bacterial culture, the abscess is drained through a hemitransfixion incision. Pus and necrotic tissues are removed. The defect is immediately reconstructed by transplantation of autogenic ear cartilage to avoid dorsal saddling and columellar retraction. The incision is left partially open for drainage. Loose internal dressings are applied; antibiotics are administered systemically. It has been shown that each septal abscess will result in some nasal growth inhibition. However, the immediate transplantation of autogenic cartilage can mostly prevent the typical saddle nose formation [54, 55].

Figure 35.20 shows that the amount of growth inhibition caused by a septal abscess depends on the age of the affected child: the earlier the nasal injury, the more pronounced is the damage to the nose and midface [55]. The three adolescent girls came for nasal surgery, all at the age of 16 years. All had a history of a drained septal abscess. During surgery in all of them, a subtotal loss of the septal cartilage was revealed. The nasal injury occurred at 3 years (left), 5 years (middle) and 7.5 years (right). The differences in nasal length, height, tip projection and maxillary retrusion are striking.

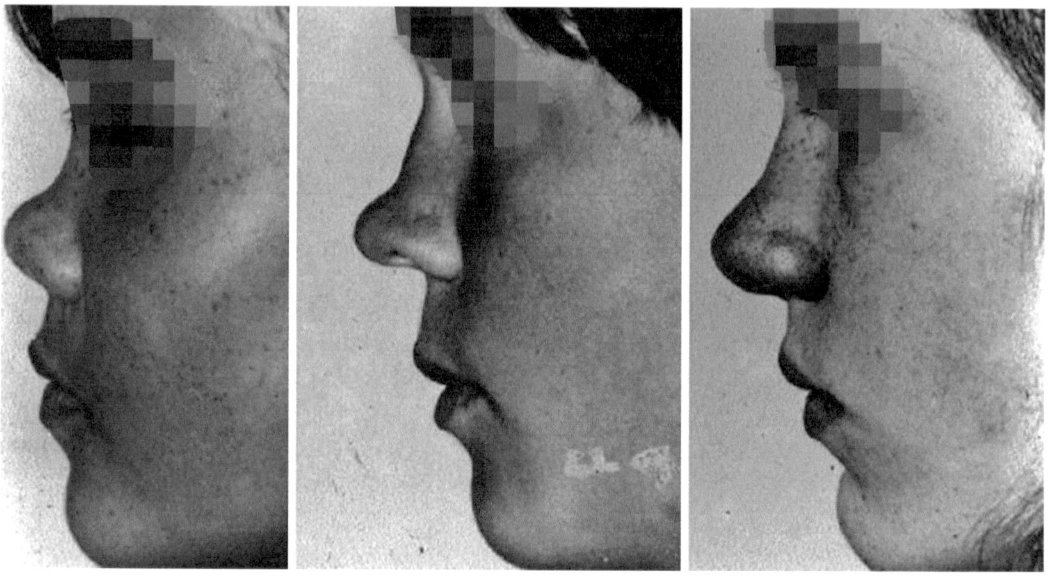

Fig. 35.20 Three female adolescents aged 16 years with a drained septal abscess at age 3, 5 and 7.5 years, respectively [55]

35.4 Part IV: Remarks on Nasal Reconstruction

35.4.1 Anterior Nose and Nasal Cavities

The nasal cartilages form a complex triangular framework for a mobile nasal lobule which acts as the portal to the upper airways. The eye-catching shape of this cartilaginous pyramid may provoke emotions ranging from delight at first sight of a beautiful person to frightful reactions setting eyes on destroyed nasal remains. The increase of the nose in size also mirrors the development of *Homo sapiens* into different populations during mankind's settling on the whole earth from Africa. Or, as Desmond Morris summed it up: 'the human nose grew taller and longer as mankind spread out and away from its hot moist Garden of Eden, keeping its air-conditioning function up to scratch' [1].

The composition of this cartilaginous framework is so unique to each individual and may concern rhinosurgeons because they cannot predict the outcome of their surgical procedures. Essential functions for the whole airways are triggered and controlled in the anterior nose. On the other hand, the protruding position in the midface makes the nose more vulnerable to external damage. No wonder the anterior nose is also the site of most nasal obstructions caused by cartilaginous and bony distortions as mentioned above. Although prospective studies on the incidence of rhinosurgical mistakes and complications are lacking, the adverse results are most often associated with the surgery of the nasal cartilages. Cartilages heal following their intrinsic laws and do not behave the way the surgeons want. In particular complications of septoplasty are due to wrong indications as a consequence of an incorrect or incomplete analysis and interpretation of the anatomical structures and the nasal functional tests [49].

Therefore, clinical diagnostics should especially focus on the finds of the anterior nose, supported by endoscopy, rhinomanometry, acoustic rhinometry, rhinoresistometry and long-term study of the nasal cycle [19, 35]. Cole and coworkers [10, 56], who contributed many basic data on the functions of the anterior nose, concluded from their studies as to nasal treatments that it is seldom necessary to extend septal and/or turbinate surgery far beyond the piriform aperture in the treatment of nasal obstruction [14]. For many patients, I can fully support this statement.

To restore the disturbed structures of the anterior nose, our surgical options are septorhinoplasty to form a straight anterior septum and correction of the nasal valve regions and enlarged erectile tissues, for instance, by turbinoplasty of the inferior turbinate. The septal turbinates should better be preserved. The reconstruction of the valve region is sometimes more effective for breathing than a septoplasty alone. If transplants or implants are used, they should be cartilage-like as to elasticity, thus avoiding the creation of an immobile and vulnerable anterior nose. This also means to prefer autogenic tissues. If an autogenic bone is used, it should be a boomerang-shaped piece instead of a rigid L-shaped bone.

The aim to reconstruct the nose posteriorly to the valve regions is to create physiological slit-like nasal cavities providing a proper nasal resistance, turbulence and nasal cycle for breathing, air-conditioning and olfaction. This cannot be achieved by performing one schematic surgical procedure, but only by applying several technical options tailored for the individual pathological nasal finds. This means for the septum that need not be reconstructed as a straight plate in the middle and posterior nose, but it should be placed approximately in the middle between the always asymmetrical lateral nasal walls. It is of utmost importance to create an adequate distance of the septum to the erectile tissues of the lateral walls which enables the achievement of the above-mentioned functions [35]. This may also mean to leave a physiological septal deviation as it is grown or to transform a crooked septum into a physiological septal deviation. Figure 35.21 shows an example of this 'philosophy'. The crooked and airway obstructing septum to the left impacted by the medially deformed right-sided middle turbinate was surgically corrected and is still slightly deviated to the left. In addition, the right-sided inferior turbinate was submucously reduced. After 3 months the slitlike nasal cavities enabled normal breathing with a bilateral nasal cycle.

35.4.2 One Option to Treat a Nasal Valve Stenosis

Several procedures have been published to treat a nasal valve problems [57, 58]. The following technique has successfully been used since 1975 by the author. The indication is valve stenosis caused by a mostly congenitally too long caudal end of the triangular cartilage, often without a returning of the lower margin and a valve angle less than 10°.

A rhomboid piece of skin (marked red in Fig. 35.22) is excised from the cul-de-sac. After elevation of the nasal mucosa from the posterior aspect of the triangular cartilage, the cranial surface of the caudal end of the triangular cartilage (here depicted with a tiny returning) is freed from connective tissue and excised cranially from the remaining triangular cartilage (marked blue in Fig. 35.22). Closure of the incision using 5-0 sutures creates a slightly curved new part of the nasal valve region with an angle larger than 20° that acts as a bend which transforms inspiratory laminar airstreams into more turbulent ones. The efficiency of correcting a disturbed nasal valve region can be increased by adding the anterior turbinoplasty in case of an enlargement of the anterior inferior turbinate as shown by acoustic rhinometry [59]. Especially in the case of the physiological septal deviation, both methods may be sufficient to solve the functional breathing problem without touching the septum.

35.4.3 Back-to-Back Technique to Reconstruct the Anterior Septum

The severe destruction of the anterior nasal septum from trauma, including septal abscess and perforation, frequently produces saddling of the cartilaginous nasal dorsum with enlarged angles of the nasal valve. Functional and aesthetically acceptable long-term results of anterior septal and nasal valve reconstruction could be achieved in 26 patients after a mean follow-up of 36 months

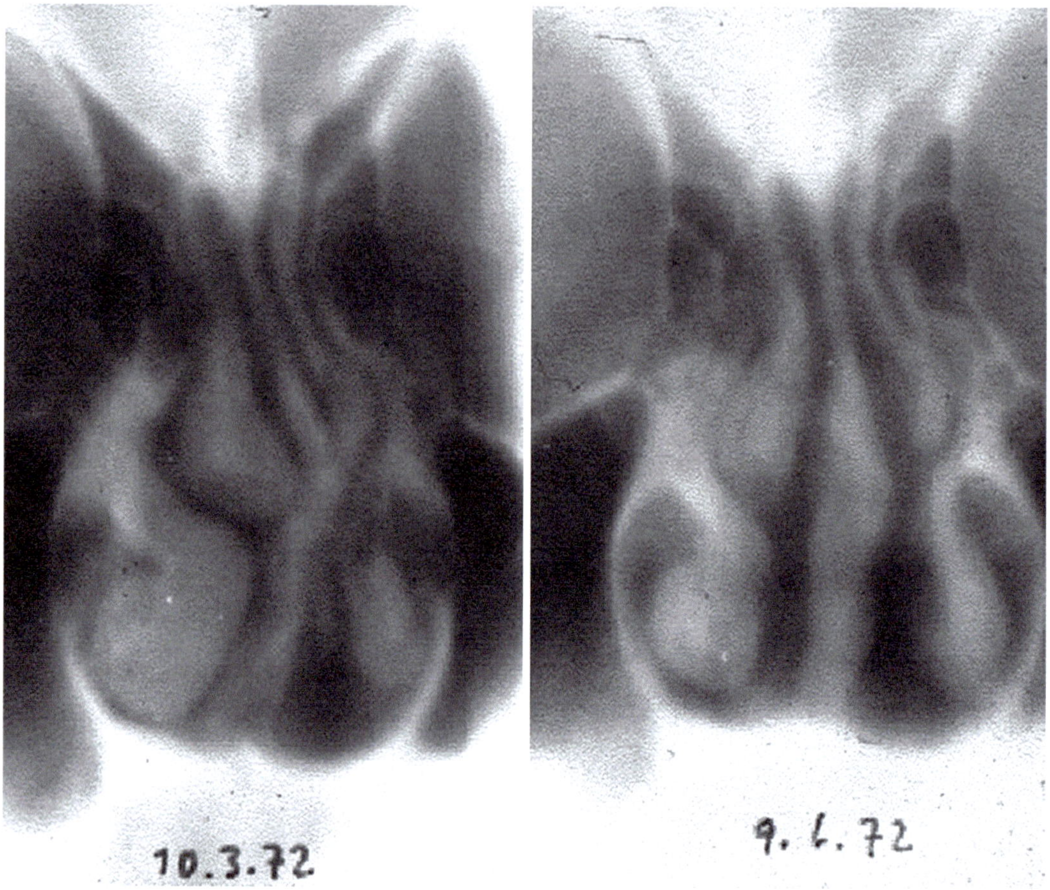

Fig. 35.21 X-ray images of a patient with septal deviation. *Left*: preoperatively. *Right*: 3 months after septoplasty and correction of the right-sided middle and inferior turbinates resulting in a physiological septal deviation and bilateral slitlike cavities (Pirsig 1972, unpublished)

using a straight and balanced back-to-back autogenic ear cartilage introduced by the author in 1986 [60].

Ear cartilage grafts from the cymba-cavum concha complex were harvested through an anterolateral approach (Fig. 35.23). A special incision was used to divide the concave ear cartilage into two halves while preserving the posterior perichondrium. The graft was folded and fixed with guide sutures in its final position between the hypoplastic anterior nasal spine and the caudal end of the cranial septodorsal cartilage remnant. Thus, a viable, stable, balanced back-to-back graft of 2.5–3 cm length was created, long enough to reconstruct the anterior septum and the nasal valve and to correct part of the saddle nose deformity. The rest of the ear cartilage was used to fill the remaining cartilaginous saddle. At follow-up, the back-to-back grafts showed no macroscopic signs of resorption. Graft position and shape had remained intact, and all noses were adequately projected and mobile. All patients but one felt satisfied with the functional and aesthetic result. The saddle completely disappeared in two-thirds of the patients. Nasal breathing considerably improved in 21 patients, remained the same in 4 patients, and worsened in 1 patient.

Our long-term study also showed that even 42% of the patients with a large septal perforation—which was not closed—reported improved nasal breathing and marked reduction of previous nasal symptoms because the valve region had been reconstructed by the back-to-back transplant.

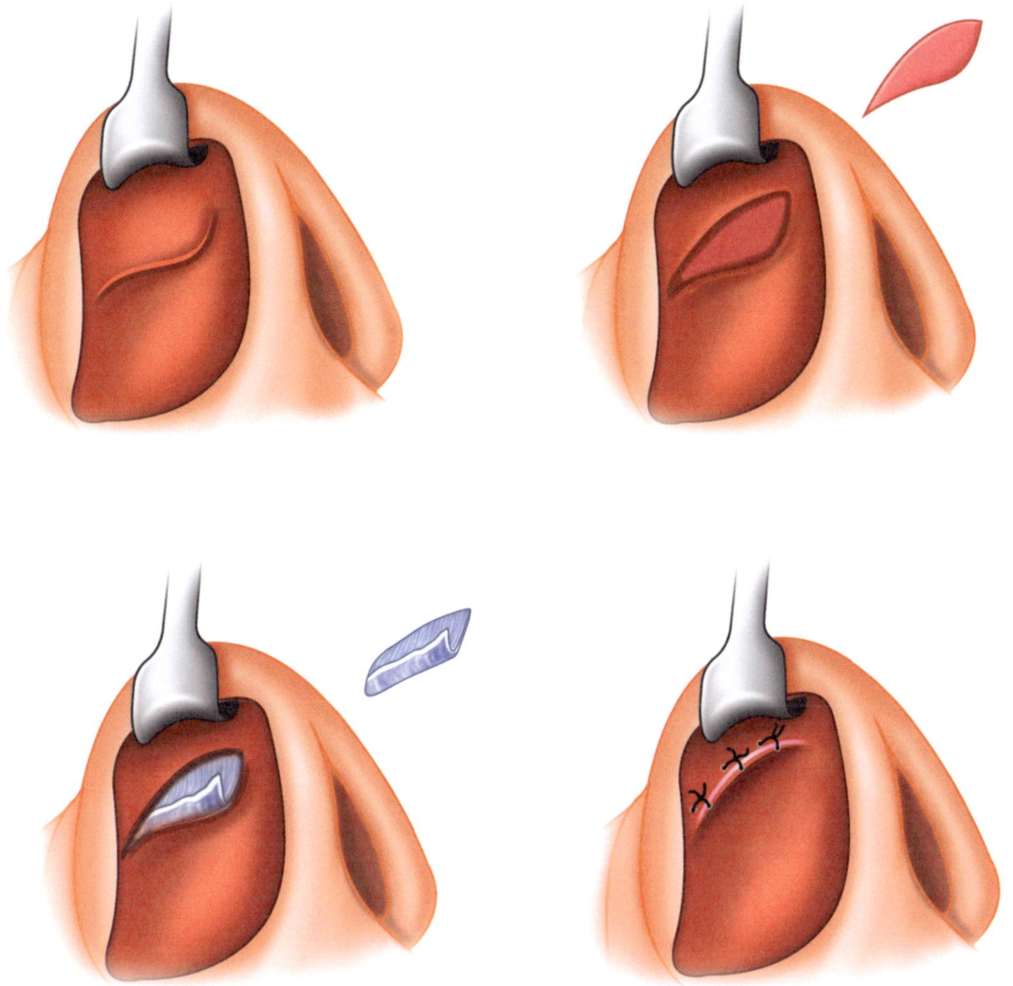

Fig. 35.22 Steps to enlarge the too small valve angle by shortening part of the caudal end of the triangular cartilage (Pirsig 1975, unpublished)

35.4.4 Closure of Septal Perforation in a Child

The worst sequela of the septal abscess is the septal perforation, especially during growth. We had to face this relatively rare sequela due to nose picking in early childhood in a Caucasian 7-year-old boy. He suffered nearly daily epistaxis, crusting and permanent mouth breathing. After insufficient conservative treatments, we decided to perform a pilot study on a 9-year-old boy to close the septal perforation of 1 cm in diameter in the areas II and III according to Cottle (Fig. 35.24). Four bipedicled mucosal advancement flaps introduced by Fairbanks [61] and Schultz-Coulon [62] were used to reconstruct the mucosal lining. To fill the cartilaginous defect, a piece of autogenic mainly hyaline cartilage was taken that had been grown from a composite graft of demineralised bovine bone matrix (DBBM), enrolled in a pedicled perichondrial flap of the boy's right pinna. The Rotterdam group of Verwoerd and Verwoerd-Verhoef had shown the feasibility of this new type of composite graft

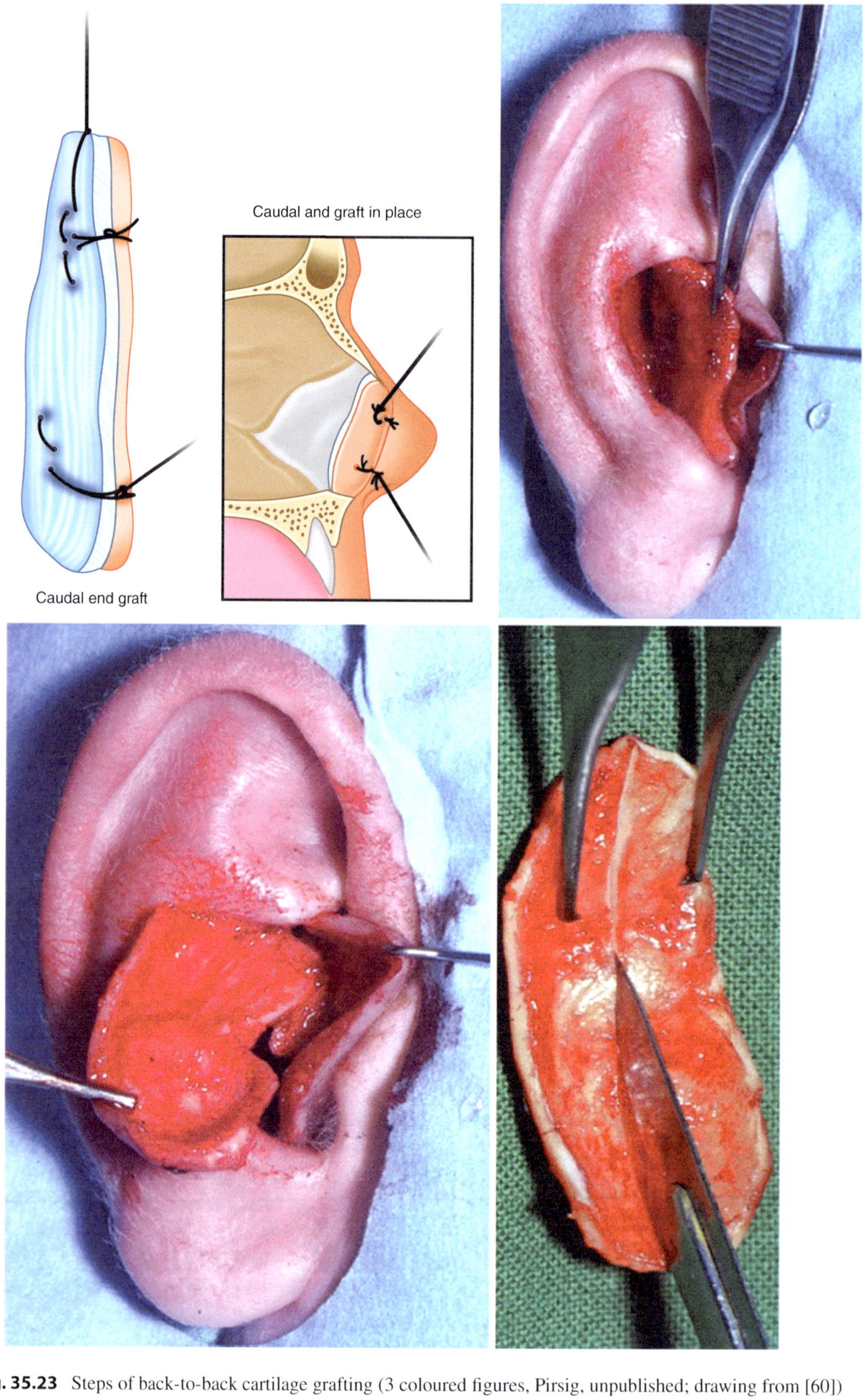

Fig. 35.23 Steps of back-to-back cartilage grafting (3 coloured figures, Pirsig, unpublished; drawing from [60])

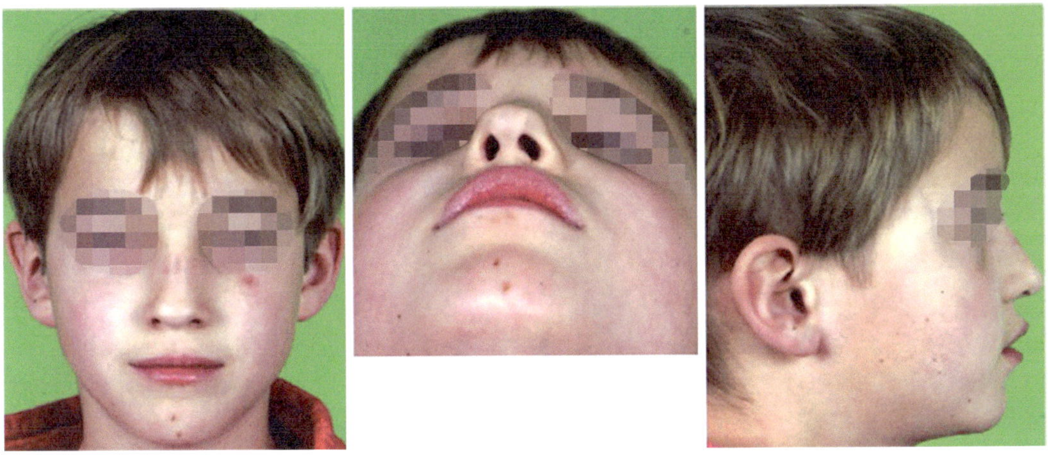

Fig. 35.24 Nine-year-old boy with septal perforation, preoperatively (Pirsig, unpublished)

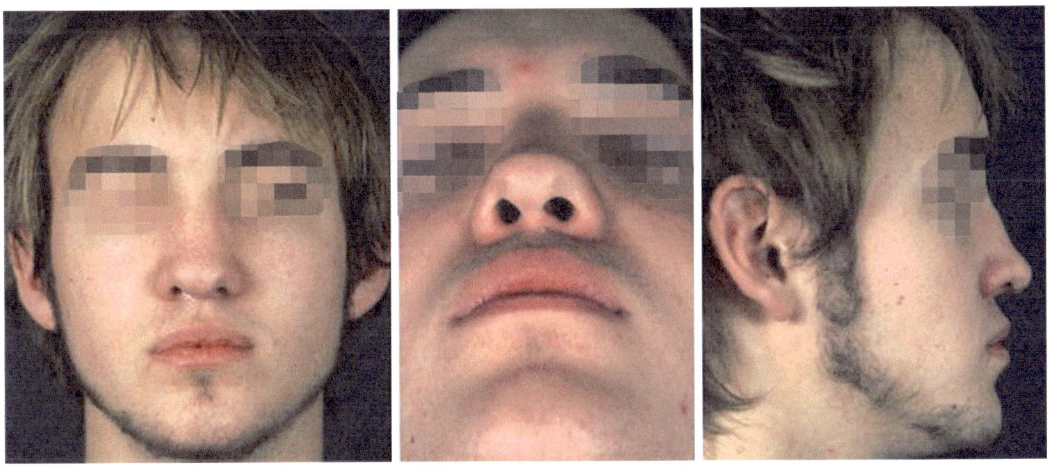

Fig. 35.25 Seventeen-year-old adolescent, 8 years after the closure of septal perforation (Pirsig, unpublished)

applied in a two-step procedure for the reconstruction of defects in the cricoid ring [63] and anterior laryngeal wall of growing rabbits [64]. Furthermore, they could show that this neocartilage provided a valuable substitute for the lost parts of the cartilage and appeared capable of growth. The operation to close the septal perforation by implanting the transformed xenogenic DBBM in the septal defect of the boy of 9 years was successful. Details of the surgical procedure and histological findings were published 2 years later [65].

I could follow the adolescent over 8 postoperative years (Fig. 35.25). He had no breathing problems and epistaxis over all the years, but a dry nose which he treated with saline douches. The nasal length was adequate, but the lobule showed growth inhibition, minimal cartilaginous sagging which was not visible preoperatively and a retracted columella. The septum was straight with ciliary activity on the sites of the former perforation. The maxillary retrusion was marked but already visible at the age of 9 years when signs of septal growth inhibition due to the perforation were already obvious. In a final step under local anaesthesia, I tried to improve the nasal appearance using pieces of ear cartilage. During this surgery, I elevated the right mucoperichondrium from the septal cartilage to get a look at the implant. There was a complete connection of the transformed cartilage with the original septal remnant. The surface of the implant was slightly tuberous and solid. Unfortunately, I could not evaluate whether the implant had grown.

35.5 Conclusions

Nasal cartilages function together in a complex anatomical framework connected by a web of connective tissues and partly covered by a layer of fine muscles. Their protruding position in the centre of the face may act as an eye-catcher, and their mobility may serve as a crumpled zone protecting the head against external frontal stress. The anterior nose with its two parallel tubes is the ideal portal to the respiratory system. In the valve regions, we find the narrowest cross-sectional areas of the nose, where two-thirds of the total respiratory resistance are generated. In these regions, the laminar airstreams are transformed into turbulent ones. Thus, together with the erectile lining of the nasal cavities, the anterior nose provides the tools for the air-conditioning of the respiratory system and for the acting of the nasal cycle.

No wonder such a complex and exposed construction like the anterior nose is very vulnerable to all types of damaging influences and especially to mechanical injuries. The septodorsal cartilage is the dominating structure for nasal shape and midfacial growth. Particularly, its anterior part is most often involved in nasal injuries and its reconstruction mainly influences the surgical long-term outcome. Thus, rhinosurgeons should predominantly focus on the rehabilitation of this anterior nose with its complex cartilaginous framework and erectile lining. Diagnostic methods like endoscopy, acoustic rhinometry, rhinoresistometry and imaging allow a better insight into the structural and functional characteristics of the damaged soft- and hardware of the nose. Too mechanistic thinking for nasal reconstruction should be replaced by a concept of applying several technical options tailored for the individual pathological nasal finds and the requirements of nasal physiology. In this chapter, I presented some of my personal experiences and how I got an insight into the complexity of the nasal cartilages through long-term follow-up of the patients. One example is the acceptance of a physiological septal deviation which acts in harmony with the lateral walls instead of creating a straight antero-posterior septal plate just for optical beauty.

References

1. Morris D. Bodywatching. A field guide to the human species. London: Jonathan Cape; 1985.
2. Peter K. Atlas der Entwicklung der Nase und des Gaumens beim Menschen mit Einschluss der Entwicklungsstoerungen. Jena: Gustav Fischer; 1913.
3. Virchow R. Untersuchungen ueber die Entwicklung des Schaedelgrundes im gesunden und krankhaften Zustande und über den Einfluss derselben auf Schaedelform, Gesichtsbildung und Gehirnbau. Berlin: Georg Reimer; 1857.
4. Kim J, Cho JH, Kim SW, Kim BG, Lee DC, Kim SW. Anatomical variation of the nasal septum: correlation among septal components. Clin Anat. 2010;23:945–9.
5. Potter JK, Rogers T, Finn R. J Oral Maxillofac Surg. 2000;58:867–76.
6. Dion MC, Jafek BW, Tobin CE. The anatomy of the nose. Arch Otolaryngol. 1978;104:145–50.
7. Hinderer KH. Fundamentals of anatomy and surgery of the nose. Birmingham: Aesculapius Publishing Company; 1971.
8. Bruintjes TD, van Olphen AF, Hillen B, Huizing EH. A functional anatomic study of the relationship of the nasal cartilages and muscles to the nasal valve area. Laryngoscope. 1998;108:1025–32.
9. Bruintjes TD, van Olphen AF, Hillen B. Review of the functional anatomy of the cartilages and muscles of the nose. Rhinology. 1996;34:66–74.
10. Cole P. The respiratory role of the upper airways. A selective clinical and pathophysiological review. Mosby Year Book: St. Louis; 1993.
11. Bachmann W, Legler U. Studies on the structure and function of the anterior section of the nose by means of luminal impressions. Acta Otolaryngol. 1972;73:433–42.
12. Wustrow F. Schwellkörper am Septum nasi. Z Anat Entwickl Gsch. 1951;116:139–42.
13. Van Loosen J, Verwoerd-Verhoef HL, Verwoerd CDA, van Velzen D. The significance of regional variations in thickness of the human nasal septum. In: van Loosen J. Thesis Erasmus University Rotterdam. chapter 5; 2000. p. 55–66.
14. Cole P. The four components of the nasal valve. Am J Rhinol. 2003;17:107–10.
15. Haight JSJ, Cole P. The site and function of the nasal valve. Laryngoscope. 1983;83:49–55.

16. Swift AC, Campell IT, McKown TM. Oronasal obstruction, lung volumes and arterial oxygenation. Lancet. 1988;1:73–5.
17. Kayser R. Die exacte Messung der Luftdurchgaengigkeit der Nase. Arch Laryngol. 1895;3:101–20.
18. Hasegawa M, Kern EB. Variations in nasal resistance in man: a rhinomanometric study of the nasal cycle in 50 human subjects. Rhinology. 1978;16:19–29.
19. Lang C, Grützenmacher S, Mlynski B, Plontke S, Mlynski G. Investigating the nasal cycle using endoscopy, rhinoresistometry, and acoustic rhinometry. Laryngoscope. 2003;113:284–9.
20. Stoksted P. Rhinometric measurements for determination of the nasal cycle. Acta Otolaryngol Suppl. 1953;109:159–75.
21. Fisher EW, Scadding GK, Lund VJ. The role of acoustic rhinometry in studying the nasal cycle. Rhinology. 1993;31:57–61.
22. Mlynski G, Pirsig W. Funktionell-ästhetische Rhinochirurgie unter besonderer Berücksichtigung physiologischer Aspekte. Stuttgart: Thieme; 2018.
23. May M, West JJ, Hinderer KH. Nasal obstruction from facial palsy. Arch Otolaryngol. 1977;103:389–91.
24. Kasperbauer JL, Kern EB. Nasal valve physiology: implications in nasal surgery. Otolaryngol Clin N Am. 1987;20:699–719.
25. Krmpotic-Nemanic J, Kostovic I, Rudan P. Morphological and histological changes responsible for the droop of the nasal tip in advanced age. Acta Otolaryngol. 1971;71:278–81.
26. Pirsig W, Lehmann I. The influence of trauma on the growing septal cartilage. Rhinology. 1975;13:39–46.
27. Pirsig W, Knahl R. Rhinoplasty in children: a follow-up study in 92 cases (in German). Laryngol Rhinol. 1974;53:250–65.
28. Van Velzen D, van Loosen J, Verwoerd CDA, Verwoerd-Verhoef HL. Persistent pattern of variations in thickness of the nasal septum: implications for stress and trauma as illustrated by a complex fracture in a 4-year.old boy. In: otolaryngology in ASEAN countries. Adv Otorhinolaryngol. 1997;51:46–50.
29. Lee SJ, Liong K, Lee HP. Deformation of nasal septum during nasal trauma. Laryngoscope. 2010;120:1931–9.
30. Soemmerring ST. Abbildungen und Beschreibungen einiger Missgeburten. Mainz: Universitaetsbuchhandlung; 1791.
31. Pirsig W. Die Aplasie und Hypoplasie des Nasenflügels in der Moche-Periode und heute (in German). Auris Nasus Larynx (Tokyo). 1989;16(Suppl I):47–52.
32. Zuckerkandl E. Normale und pathologische Anatomie der Nase und ihrer pneumatischen Anhänge. Wilhelm Braumüller: Wien; 1882.
33. Gray L. Deviated nasal septum; incidence and etiology. Ann Otol Rhinol Laryngol Suppl. 1978;5:7–17.
34. Gogniashvili G, Steinmeier E, Mlynski G, Beule AG. Physiologic and pathologic septal deviations: subjective and objective functional rhinologic findings. Rhinology. 2011;49:24–9.
35. Mlynski G. Restorative procedures in disturbed function of the upper airways - nasal breathing. GMS Curr Top Otorhinolaryngol Head Neck Surg. 2005;4:Doc07.
36. Hanif J, Jawad SS, Eccles R. A study to assess the usefulness of a portable spirometer to quantify the severity of nasal septal deviation. Rhinology. 2003;41:11–5.
37. Pentz S, Pirsig W, Lenders H. Long term results of neonates with nasal deviation: a prospective study over 12 years. Int J Pediatr Otorhinolaryngol. 1994;28:183–91.
38. Cottle MH. Nasal surgery in children. Eye Ear Nose Throat Mon. 1951;30:32–8.
39. Pirsig W. Growth of the deviated septum and its influence on midfacial development. Facial Plast Surg. 1992;8:224–32.
40. Vetter U, Gammert C, Pirsig W, Landolt A, Heinze E. Growth activities of the nasal septal cartilage in acromegaly. Rhinology. 1984;22:125–31.
41. Vetter U, Helbing G, Pirsig W, Heinze E, Gammert C, Landolt A. Human nasal septal cartilage: local distribution of different enzyme activities in healthy adults and acromegalic patients. Laryngoscope. 1985;95:469–73.
42. Verwoerd CDA, Verwoerd-Verhoef HL. Rhinosurgery in children: developmental and surgical aspects of the growing nose. GMS Curr Top Otorhinolaryngol Head Neck Surg. 2010;9:5. ISSN 1865–1011.
43. Poublon RML. The cartilaginous nasal dorsum and the postnatal growth of the nose. Thesis Erasmus University Rotterdam, chapter 10. Delft: Eburon Publishing; 1987. p. 55–60.
44. Fry H. Nasal skeletal trauma and the interlocked stresses of the nasal septal cartilage. Br J Plast Surg. 1967;20:146–58.
45. Ten Koppel PG, van der Veen JM, Hein D, van Keulen F, van Osch GJ, Verwoerd-Verhoef HL, et al. Controlling incision-induced distortion of nasal septal cartilage: a model to predict the effect of scoring of rabbit septa. Plast Reconstr Surg. 2003;111:1948–57; discussion 1958-1959
46. Pirsig W. Morphologic aspects of the injured nasal septum in children. Rhinology. 1979;17:65–76.
47. Eitschberger E, Merklein C, Masing H, Pesch HJ. Deviations of septum cartilage after unilateral separation of mucoperichondrium in rabbits. Arch Otorhinolaryngol. 1980;228:135–48.
48. Duynstee MLG, Verwoerd-Verhoef HL, Verwoerd CDA, van Osch GJVM. The dual role of perichondrium in cartilage wound healing. Plast Reconstr Surg. 2002;110:1073–9.
49. Schwab JA, Pirsig W. Complications of septal surgery. Facial Plast Surg. 1997;13:3–14.
50. Bewarder F, Pirsig W. Long-term results of submucous septal resection (in German). Laryngol Rhinol Otol. 1978;57:922–31.
51. Pirsig W. Rhinoplasty and the airway in children. Facial Plast Surg. 1986;3:225–34.
52. Verwoerd CDA, Verwoerd-Verhoef HL, Meeuwis CA. Stress and wound healing in the cartilaginous septum. Acta Otolaryngol (Stockh). 1989;107:441–5.
53. Verwoerd-Verhoef HL, Koppel PGJ, van Osch GJVM, Meeuwis CA, Verwoerd CDA. Wound healing of car-

tilage structures in the head and neck region. Int J Pediatr Otorhinolaryngol. 1998;43:241–51.
54. Huizing EH. Long term results of reconstruction of the septum in the acute phase of a septal abscess in children. Rhinology. 1984;22:55–63.
55. Pirsig W. Historical notes and actual observations on the nasal septal abscess especially in children. Int J Pediatr Otorhinolaryngol. 1984;8:43–54.
56. Cole P, Chaban R, Naito K, Oprysk D. The obstructive nasal septum: effect of simulated deviations on the nasal airflow resistance. Arch Otolaryngol Head Neck Surg. 1988;114:410–2.
57. Bloching MB. Disorders of the nasal valve area. GMS Curr Top Otorhinolaryngol Head Neck Surg. 2007;6:Doc07.
58. Kern EB. Surgical approaches to abnormalities of the nasal valve. Rhinology. 1978;16:165–89.
59. Lenders H, Pirsig W. Diagnostic value of acoustic rhinometry: patients with allergic and vasomotor rhinitis compared. Rhinology. 1990;28:5–16.
60. Pirsig W, Kern B, Verse T. Reconstruction of anterior nasal septum: back-to-back autogenous ear cartilage graft. Laryngoscope. 2004;114:627–38.
61. Fairbanks DNF. Closure of nasal septal perforation. Arch Otolaryngol. 1980;106:509–13.
62. Schultz-Coulon HJ. Experiences with the bridge-flap technique for the repair of large nasal septal perforations. Rhinology. 1994;32:25–33.
63. Bean JK, Verwoerd-Verhoef HL, Meeuwis J, Verwoerd CDA. Reconstruction of the growing cricoid with composite graft of demineralized bovine bone and autogenous perichondrium, a comparative study in rabbits. Int J Pediatr Otorhinolaryngol. 1993;25:163–72.
64. Bean JK, Verwoerd-Verhoef HL, Verwoerd CDA. Reconstruction of the anterior laryngeal wall with a composite graft of demineralized bovine bone and autogenous perichondrium. ORL J Otorhinolaryngol Relat Spec. 1994;56:224–9.
65. Pirsig W, Bean JK, Lenders H, Verwoerd CDA, Verwoerd-Verhoef HL. Cartilage transformation in a composite graft of demineralized bovine bone matrix and ear perichondrium used in a child for the reconstruction of the nasal septum. Int J Pediatr Otorhinolaryngol. 1995;32:171–81.

Physiology and Pathophysiology of the Growing Nasal Skeleton

36

H. L. Verwoerd-Verhoef, G. J. V. M. van Osch, and C. D. A. Verwoerd

Core Messages

- The outcome of surgical interventions in the midfacial region is largely dependent on the quality of wound healing of the tissues. However, in children, the second aspect of paramount importance is the impact on further growth.
- The knowledge of the anatomy of the nasal skeleton from birth to adolescence and current data concerning the 'normal' development of the midfacial profile is mandatory for physicians working in this field.
- Surgery of the nasal skeleton in children of different ages should reinstate form and function, optimise further growth and minimise the risks for abnormal development. As to restoring normal growth, clinical observations have still insufficiently produced convincing evidence.
- Results of animal experiments have largely contributed to understanding developmental mechanisms of the midface, the way they are influenced by various surgical interventions like partial resections and fractures of the cartilaginous and bony skeleton, and finally the feasibility to restore growth by surgery.
- Key issues are (1) the dominant role of specific growth zones in the cartilaginous nasal septum, the shift of its support from the sphenoid to the anterior rim of the perpendicular plate, the connection with the premaxilla (via the anterior nasal spine) and the connection of the upper lateral cartilages with the nasal bones; (2) the poor wound healing capacity of growing and maturing nasal cartilage and its deformation due to the release of interlocked stresses in the tissue.
- The clinical long-term results as far as nasal growth is concerned after nasal surgery (like septoplasty) should be studied more extensively at different ages up till after puberty.
- The process of cartilage tissue engineering is making progress when autologous cells from the septum or auricle are seeded together with stem cells in a biodegradable scaffold.
- The emerging technology of 3D and 4D printing offers the opportunity to design personalised implants for nasal reconstruction.

36.1 Physiology of the Growing Nasal Skeleton

36.1.1 Introduction

Rhinosurgical procedures are common in the adult patient group, and techniques have been developed and improved based on the experience with large numbers of patients. Most common is the septoplasty which is performed to correct a symptomatic

H. L. Verwoerd-Verhoef (✉) · G. J. V. M. van Osch
C. D. A. Verwoerd
Erasmus MC Rotterdam, Rotterdam, The Netherlands

Ö. Ö. Celebi, T. M. Önerci (eds.), *Nasal Physiology and Pathophysiology of Nasal Disorders*,
https://doi.org/10.1007/978-3-031-12386-3_36

deviated septum of congenital, developmental, or traumatic aetiology. Performing such procedures in the paediatric age group is controversial because of concerns about retarding or otherwise altering nasal and midfacial growth patterns [1]. Children as "category" include patients from 0 years of age to adolescence. In this period the nose is characterized by an increase in dimensions and an ongoing development of the supporting midfacial skeleton. Consequently, no "standard" anatomy exists for children as for adults. The age-specific anatomy, the vulnerability of the immature skeleton and the morphogenetic processes providing a "normal" adult nose, should be respected when nasal surgery is considered for a child. The aim should be to restore the anatomy in the short term, whereas a "normal" facial profile at the adult age has to be the—equally important—objective in the long run. However, which developmental processes should be restored and how? Anecdotal clinical evidence in small series or single case reports suggests that nasal trauma, septal infection or even childhood septoplasty can cause severe morphological and functional disturbances of the nose later in life. The restriction of clinical observations is the temporariness without, in general, the possibility to follow up the developmental processes for a longer period, or to analyse the effects of various well-defined injuries and treatment modalities in children of different age groups. Here comes the value of research in experimental animals.

Experimental studies demonstrated the morphogenetic mechanisms which might be held responsible for a normal development of the nose and upper jaw (the inferior wall of the nose!) and the developmental effects of lesions and surgical interventions of various parts of the nasal skeleton. Clinical evidence, facial morphogenesis in children, and the interaction between wound healing, growth and surgical procedures, as studied in animal experiments, all contribute to the current practice of nasal surgery in children.

> The age-specific anatomy of the midface makes rhinosurgery in children different from the procedure in adults.

Knowledge of the age-specific anatomy and developmental processes is essential for proper diagnosis and treatment of midfacial lesions in children. It became clear that the dimensional aspects of nasal growth, the evolution of the growth rate during childhood and the age at which the growth spurt ends, are related to gender [2]. The growth rate of the facial skeleton (nose, upper and lower jaw) is higher compared to that of the brain skull. In the first years of life the growth rate is also faster, and only gradually decreases in adulthood. The craniofacial ratio at birth is said to be 8:1 compared to 4:1 at 5 years of age and 2:1 in adulthood [3]. The human nasal septum is the dominating structure to determine the size and shape of the visible nose and, thus, influences the appearance of the face. The anatomy and dimensions of the nose are changing with increasing age. The baby's face acquires an adult profile. The anatomical development of the nose with tissue maturation of the cartilaginous and bony parts, including their age-specific characteristics, are described below as they have to be considered very important and studied extensively before treatment or surgery is performed. Up to now, these morphogenetic processes in the septodorsal cartilage and in the sutures of the nasal pyramid and maxilla have been observed as more or less independent growth centres. The effects of frequently occurring injuries of the midfacial skeleton on the further development of the face, however, point to a strong correlation between the various parts. In an anatomical study on cleft, lip and palate it was phrased by Hall and Precious as follows: "Growth of the nasal septal cartilage outstrips the growth of other skeletal and soft tissues in the midface to such an extent that it is the *pacemaker* for growth of the face and anterior portion of the skull" [4].

It appeared that the release of interlocked stresses within the cartilage and the poor wound healing after fractures or loss of septum cartilage are key factors in maldevelopment of the midface, and also form a serious risk factor in surgical reconstruction at a young age [5, 6]. These biological aspects of surgical intervention have to be taken into account when the ultimate aim is to restore the normal form and function of the nose and midface.

Up to three decades ago a review of the clinical literature reporting results of nasal trauma and surgery in children demonstrated that definite conclusions were still hampered by various elements such as a lack of differentiation for the age of the patients at the moment of injury or surgery, a too short follow-up period after surgical treatment and incomplete documentation of the surgical procedures applied [6].

Since then several larger clinical studies have been published presenting the results of a series of young patients who presented a nasal disorder and mostly underwent septoplasty without or with other procedures in an endonasal or external approach [7–15]. In most of these studies, the documentation had been improved and the follow-up period was extended to sometimes several years, however, with a very large dispersion (scatter) and mean value. Nevertheless, the outcome of paediatric septoplasty in these studies is varying from "not interfering with the normal growing nasal process" [12] to "a solid appreciation of long-term outcomes and effects on growth remain elusive" [16], which sounds as a conservative warning. An impeding factor in the interpretation of all these results described for children may be that the moments of age at the surgery have been added up from generally 4 years to mid-10s in one group with a mean value. The results of the surgery have never been studied or measured (by anthropometry or cephalometry) per age group for every year. Mostly the number of patients in each category would then probably be too small for statistics. Moreover, this could explain that rhinoplasty in children is complicated by high revision and aesthetic dissatisfaction [13]. "Good results following surgery is *no* guarantee for long-term satisfaction" [13]. A variation in the analysis of these clinical reports seems to have added to the confusion of tongues in various review articles [17–19].

There is still work to do!

Studies in young experimental animals like rabbits [20–23], guinea pigs [25, 24], rats [26], ferrets [27], cats [28], dogs [29] and primates [30, 31] have contributed substantially to understanding the effects of trauma and surgery on the growing nasal and midfacial structures; observations in larger series of pure-bred mammals could be defined by geometrical measurements and data statistically processed. Despite the differences between human and animal anatomy, the constituting elements of the facial skeleton show sufficient significant similarities to allow comparisons relevant to the notion of the pathophysiology of the human face and provide suggestions for treatment. In growing rabbits as well as in children the nose and upper jaw grow faster and over a longer period than the brain skull.

36.1.2 The Facial Profile and Nasal Skeleton of the Newborn

In the newborn, the dimensions of the splanchnocranium (maxilla, nasal pyramid, and mandible) are small in proportion to the size of the neurocranium (brain skull)! The large dimensions of the latter are related to the rapid development of the brain during pregnancy which continues into the first years of life. The facial profile of the neonate (Fig. 36.1) shows smaller vertical dimensions of the midface, less frontal projection of the nasal dorsum and a larger nasolabial angle compared to the facial proportions in fully grown individuals [32].

In the newborn, the cartilaginous and bony nasal skeleton demonstrate a few specific aspects. Septal cartilage and upper lateral cartilages on both sides form the three-dimensional septodorsal cartilage, a supporting structure for the nasal dorsum resembling a T-bar configuration [33]. The cartilaginous septum is based on the sphenoid (Fig. 36.2). The upper lateral cartilages extend under the nasal bones to merge with the cartilaginous anlage of the anterior cranial base. The nasal bones are fibrously connected in the sutures to the frontal and maxillary bones. At their ventral rim, the periosteum of the nasal bones is firmly connected to the perichondrium of the underlying upper lateral cartilages. The nasal bones are the product of extra-cartilaginous ossification of cephalic mesenchyme. The first anlage of the vomer is also represented by islands of mesenchymal bone formation reaching from palatal bone to cartilaginous septum

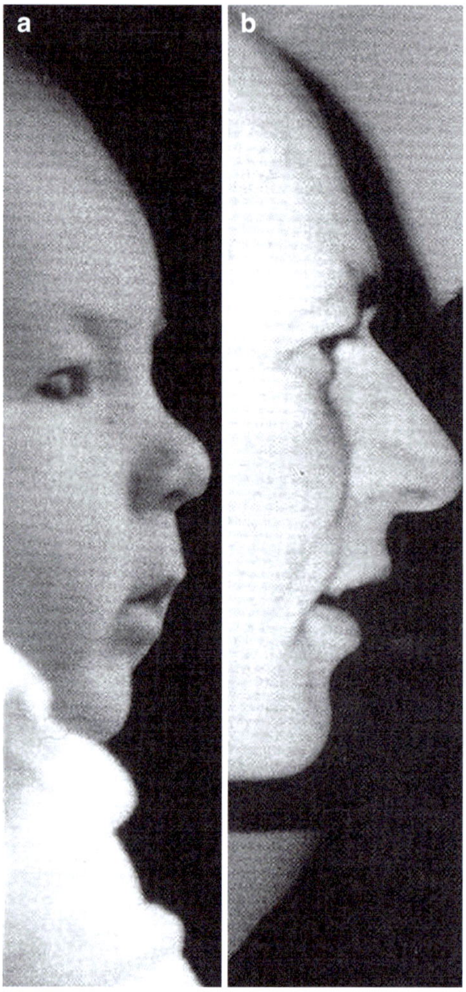

Fig. 36.1 The facial profile of a 3-month-old boy (**a**) and his father; (**b**) Note the proportional differences between the facial and brain skull of father and son. The baby face shows smaller vertical dimensions, a less frontal projection of the nose and a larger nasolabial angle

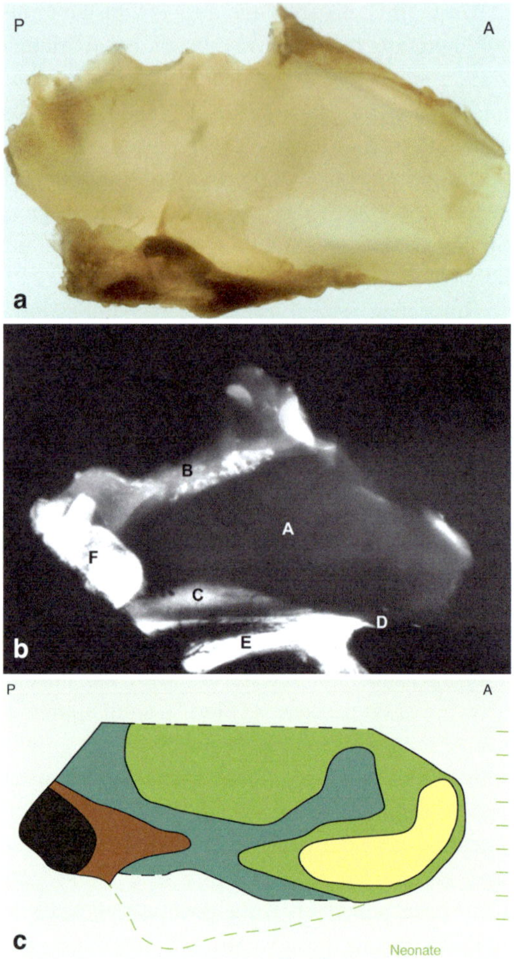

Fig. 36.2 A specimen of neonatal cartilaginous nasal septum; (**a**) indentations at the border with the anterior cranial base concur with ossified parts lost during preparation; (*A*) anterior and (P) posterior side; (**b**) Lateral radiograph of a human neonate including the nasal skeleton, adjacent skull base and upper jaw; (*A*) cartilaginous nasal septum, (*B*) lamina cribrosa, (*C*) vomeral wing, (*D*) anterior nasal spine, (*E*) palate, (*F*) sphenoid; (**c**) Schematic representation of the thinner and thicker zones in the neonatal human septum (superior part of the septum, adjacent to the anterior cranial base, is not included). The thinnest frontal part (*yellow, 400–500 mu*) is bounded by the columellar rim (*light green, 500–1500 mu*); sphenodorsal and sphenospinal zones of thicker cartilage (*dark green-brown, 1500–3000 mu*); sphenoid (*black*)

with extensions of ossifying mesenchyme along both sides of the septum cartilage. Enchondral ossification of the septum cartilage may be observed as early as the first months after birth near the anterior cranial base [34]. The cartilaginous septum shows a specific pattern of thinner and thicker zones; the thickness of the septum was found to vary between 0.4 and 3.5 mm. Two areas of thick cartilage are extending from the sphenoid in antero-superior direction to the nasal dorsum (sphenodorsal zone) and the anterior nasal spine (sphenospinal zone), respectively (Fig. 36.2c). The thinnest part is found anteriorly between these two zones and the slightly thickened caudal rim.

In young children the cartilaginous nasal septum is based on the sphenoid, whereas the upper lateral cartilages extend under the nasal dorsum to merge with the cartilaginous cranial base. Dimensions and anatomical features change with increasing age.

In young rabbits, a similar pattern of thicker and thinner areas has been demonstrated in the elongated septum [35, 36]. The thicker areas play a specific and important role in the postnatal development of the nose and upper jaw as will be discussed in the paragraph on acquired malformations of the septum. Also in young rabbits, the septal and upper lateral cartilages form a T-bar, based on the anterior skull base.

36.1.3 Midfacial Development from Neonate to Adolescent

In children, the development of the cartilaginous nasal skeleton is a complex process including the proliferation of chondroblasts, an increase of intercellular matrix, tissue maturation, partial regression, and enchondral ossification of the septodorsal cartilage. The dimensional growth of the septum cartilage shows its highest rate in the newborn and is slowing down gradually after the age of 2 years [37]. From that time onwards formation of new cartilage continues but is balanced by simultaneous loss of cartilage through enchondral ossification [38]. Mitotic activity of chondroblasts and expansion of the intercellular matrix might compensate for the loss of cartilage by ossification till the ratio between bony and cartilaginous parts has been changed to its definite state (Fig. 36.3). Consequently, the sagittal dimensions of the bony perpendicular plate are increasing relative to the cartilaginous part of the septum [39]. The growing perpendicular plate will intervene between septal cartilage and sphenoid. The septum cartilage is later firmly connected to the thickened caudal rim of the perpendicular plate. The junction of the cartilaginous septum and perpendicular plate—an important surgical landmark—will change from an intracranial position in young children to the extracranial location at the anterior margin of the nasal bones in adults. The basis of the septum cartilage shifts from the sphenoid to the anterior rim of the perpendicular plate (Fig. 36.4a).

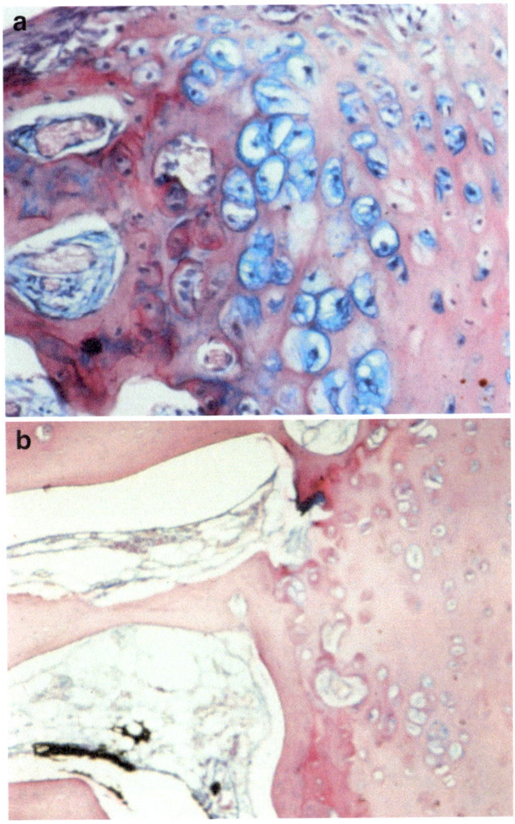

Fig. 36.3 The interface between septum cartilage (*right*) and bony perpendicular plate (*left*); (**a**) active enchondral ossification in a young child; (**b**) no signs of ossification in the adult stage

The vomeral bone which first anlage was already present in the neonate will develop into a definite part of the osseous nasal skeleton [40]. The vomer is enclosing the basal rim of the septum cartilage by two bony layers (vomeral wings) which converge inferiorly in the medial, unpaired bony plate separating the inferior part of the nasal cavities and extending in a posterior direction to the choanal edge of the nasal septum. The moment at which the ossifying front of the perpendicular plate should reach the vomeral wings is not clearly defined and may even surpass the

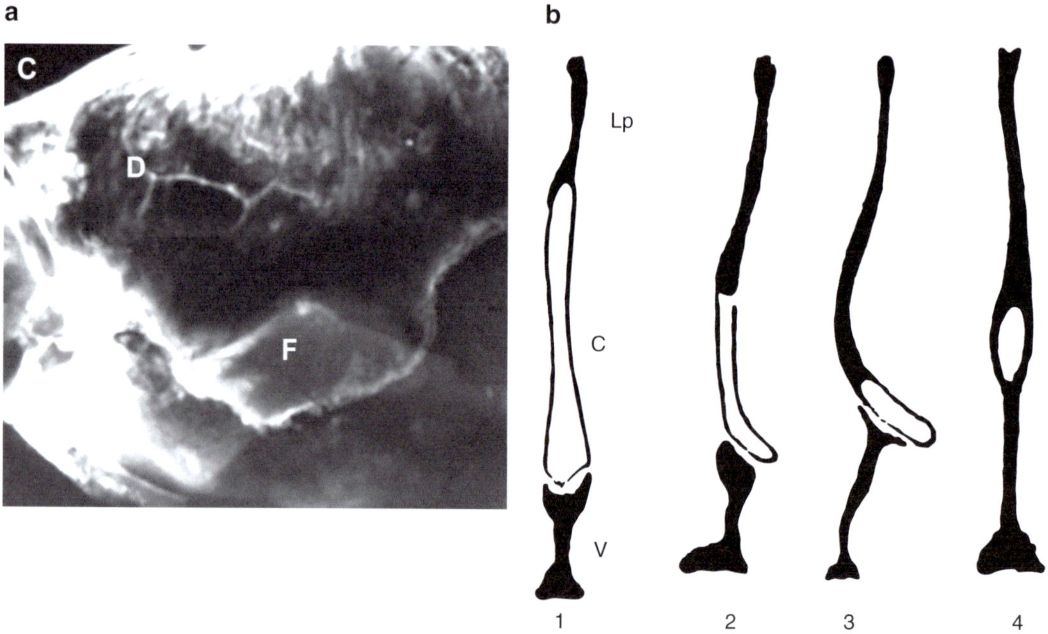

Fig. 36.4 Detail of a human nasal septum (at the age of 17 years); (**a**) Lateral radiograph with the overlap of vomeral wing and perpendicular plate; (C) anterior skull base; (D) lamina perpendicularis; (F) vomeral wing; (**b**) schematic representation (in frontal sections) of various modalities of the septo-ethmoido-vomeral junction (after Takahashi 1987); (1) normal situation, Lp = lamina perpendicularis, c = cartilage v = vomer; (2) asymmetrical development of vomer and cartilage; (3) asymmetrical development of vomer with formation of vomeral spine; (4) sphenoid tail

age of 10 years, while overlap between wings and plate has been found in adolescence. Cartilage may remain present in the bony canal formed by vomeral wings and perpendicular plate (vomeral tunnel) for a longer period and extend to the sphenoid, as the sphenoid tail, but ultimately ossify in most individuals. Asymmetry of the vomeral wings (the ala on one side is larger than on the other side) may be observed when the sphenoid tail bulges out into one nasal cavity, sometimes in combination with a vomeral spine (Fig. 36.4b). Variations in the septovomeral junction are very common and a symmetrical development is an exception rather than the rule. In a study on human foetuses of 5 months old, it was observed by Takahashi that around 25% demonstrated an abnormality of the septovomeral junction which could increase to almost 40% at birth [41]. These deformities were ascribed to "an imbalance of the '*overdeveloping*' septum (cartilage) and the pressure of the surrounding structures", the last-mentioned being the developing bony facial skeleton.

Progressive ossification of the septum cartilage results in an expanding perpendicular plate starting from the area of the anterior cranial base into the ventrocaudal direction. The ventral rim of the perpendicular plate shifts gradually inferiorly and, therefore, is in children not a reliable point of orientation in relation to the anterior skull base. The junction between cartilaginous septum, perpendicular plate and vomer may demonstrate considerable variation but is thought to have been established between 10 and 14 years of age.

The length of the upper lateral part of the dorsoseptal cartilage will be reduced on both sides to finally around 5–8 mm. The progress of this reduction shows individual variation. While the dimensions of the nasal skeleton increase in size, the relation between cartilaginous and bony parts is going to be altered. This remodelling will result in a more anterior position of the cartilaginous part of the septum. It is an important issue which should be understood by the doctor at the moment of diagnosis and before treatment of the child might be started.

Anatomical data do not give information pertinent to the developmental mechanisms. Animal experiments are necessary to analyse these morphogenetic mechanisms and the way they are affected by injury or surgery.

36.1.4 Postnatal Development of the Midfacial Skeleton in Mammals

The skulls of mammals demonstrate essentially similar components (Fig. 36.5). Various components however may show very different dimensions in different species. An example is a proportion between the skeletal components of the upper jaw. In the human skull the premaxilla is small compared with the maxilla whereas in rodents, like the rabbit, the anteroposterior dimensions of the premaxilla exceed those of the maxilla. In children, the osteogenic activity of the premaxillo-maxillary suture is restricted to the first years of life contrary to rabbits in which sutural growth contributes to the lengthening of the upper jaw and continues till sexual maturity.

Also in young rabbits, the septodorsal cartilage extends under the nasal bones to the anterior cranial base, later followed by a reduction of the posterior part of the cartilage leaving only the most anterior part in situ. The cartilaginous septum remains in direct contact with the sphenoid. Only a small part will demonstrate ossification to form the perpendicular plate. Obvious differences compared to the human anatomy are demonstrated by the vomer. The vomer is a product of mesenchymal ossification along both sides of the cartilaginous septum and between the palatal bone and the inferior margin of the septum The cartilaginous septum, anteriorly connected to the anterior nasal spine, shows thinner and thicker parts similar to those described for the human nasal septum (Fig. 36.6). An centro-anterior area of thin cartilage is surrounded by sphenodorsal and sphenospinal zones of thick cartilage and anteriorly bordered by a slightly thickened anterior rim. In growing rabbits it was demonstrated that the "extra" growth of the nose and maxilla depends primarily on the growth of the septodorsal carti-

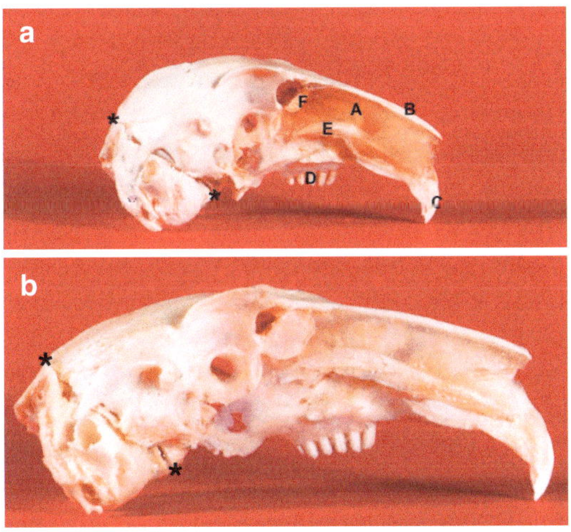

Fig. 36.5 Lateral aspect of the rabbit septum and skull (after removal of the right part of the nose and upper jaw); (**a**) at the age of 4 and 24 weeks; (**b**) It demonstrates the "extra" growth of the nasal skeleton and upper jaw up to the adult stage, compared to the dimensional development of the brain skull. *-* Line between lambdoid suture and spheno-occipital suture. (*A*) cartilaginous nasal septum, (*B*) nasal bone, (*C*) incisors, (*D*) molar complex, (*E*) vomeral wing, (*F*) perpendicular plate

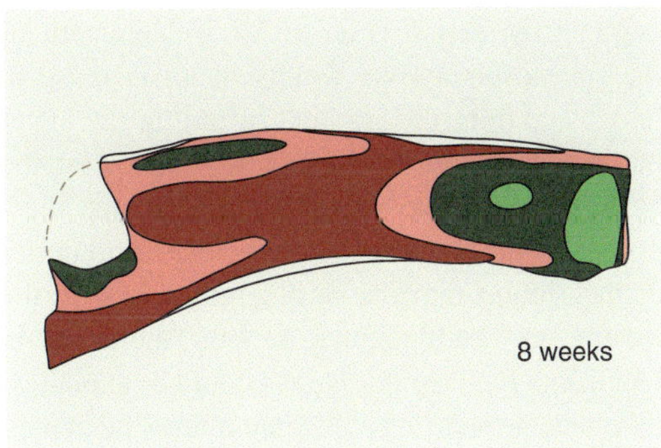

Fig. 36.6 Schematic presentation of regional differences in thickness of the cartilaginous nasal septum in an 8-week-old rabbit. The sphenospinal zone of thick cartilage extends from the sphenoid to the anterior nasal spine, whereas the sphenodorsal zone is extending under the nasal dorsum. In rabbits the lower lateral cartilages have a common medial crus, which is connected by a thin cartilaginous membrane to the slightly thickened anterior rim of the "real" cartilaginous sptum; this thickened rim and the sphenodorsal and sphenospinal zones of thick cartilage enclose an area of very thin cartilage; light green: *50–250 mu, dark green:250–450 mu, light red 450–650 mu, dark red 650–850 mu*

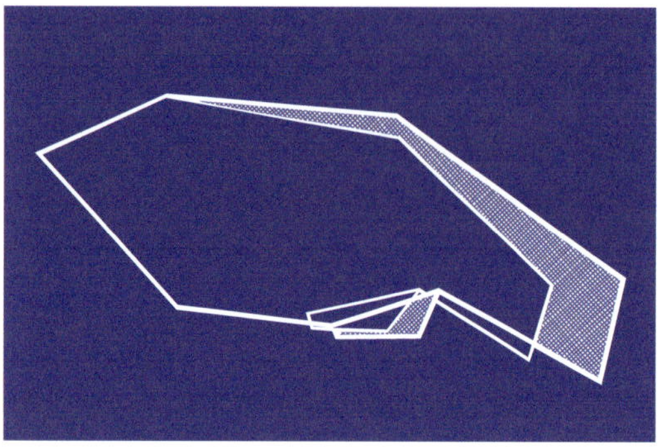

Fig. 36.7 Graphic representation of the 'extra' growth of the midfacial skeleton (with grid code) of the rabbit skulls between 4 and 24 weeks after birth; the line connecting lambdoid suture and the spheno-occipital suture (viz. Fig. 36.5a) has been made equal for both series (n = 20); elongation of the upper jaw and the nose and forward shift of the molar complex result in the adult proportions between facial and brain skull

lage (Fig. 36.7). The sphenodorsal zone of thicker cartilage is responsible for lengthening the upper part of the T-bar, the upper lateral cartilages, and indirectly of the overlying nasal bones. These upper laterals were found to stabilise the growing septum cartilage in a median position. Also in the rabbit, the extension of the lateral cartilages under the nasal bones will be gradually reduced by ongoing regression (Fig. 36.8). Enlargement of the sphenospinal zone results in an increase in length and thus, the gradual shifting of the lower part of the septum and upper jaw. In the rabbit, the perpendicular plate will be limited to the most posterior part of the septum, whereas the majority of the septum remains cartilaginous. Also in the mouse, the cartilaginous septum increases in length much more rapidly than could be explained by caudal growth, implying that interstitial expansion is the more important contributor to septum development [42]. Equally important as the septodorsal cartilage is for the growing midface, are the sutures and their bone formation for the facial skeleton, the nose included.

36.1.5 Dimensional Growth of the Nose and Maturation

Post-mortem anatomical studies suggested a phase of rapid growth directly after birth with a gradual deceleration after 5 years with the greatest velocity in the first 2 years [37, 43]. Conclusions drawn from a study in the Aegean region of Turkey revealed that nasal height and nasal bridge length reached full maturation in females already at 12 years of age and in boys at around 15 years [44]. Nasal growth has further been studied by measuring cohorts of children and calculating "standards" for various age groups, differentiating between boys and girls. Next to these horizontal studies, a few vertical studies have been published based on measurements of the same child at increasing ages. Such a vertical study demonstrated in boys a period of accelerated growth, most frequently observed around the age of 13 years [2]. In young girls periods of accelerated growth were found to occur between the age of 6 and 8 years. Growth spurts have not been demonstrated in horizontal studies including large cohorts of children. In the last two decennia most data on postnatal growth of the maxilla and the nose have been derived from anthropometric or lateral cephalographs in children of 7 years and older [2, 45–47]. In a more recent review, the steepest descending slope of midfacial growth velocity is reported to take place at the average age of 13.4 years for adolescent girls and 14.7 years for boys [48]. It was

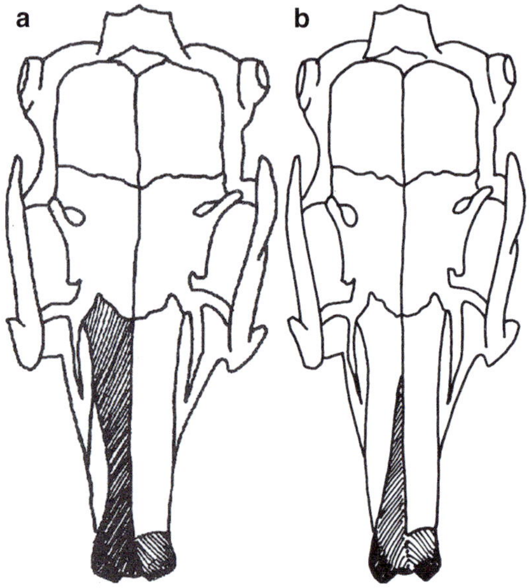

Fig. 36.8 Schematic representation of the regression of the upper lateral cartilage from (**a**) 4 to (**b**) 24 weeks; the left nasal bone is removed (courtesy of Dr R.M.L. Poublon)

suggested that 98% of white, adolescent girls are "nasally mature" at the age of 15.8 years and 98% of the boys at the age of 16.9 years.

36.2 Pathophysiology of the Growing Nasal Skeleton

36.2.1 Congenital Anomalies

36.2.1.1 Midfacial Clefts and the Nasal Septum

Craniofacial growth is a dynamic process balancing form and function. This equilibrium is easily disturbed at various levels by passive or active change of one of the mechanisms involved, prenatally or postnatally. The impact of the interaction with the adhering muscles on the developing midfacial skeleton has been recognized in recent years [50, 4, 49]. A spectrum of midfacial malformations may be observed as part of a syndrome or as solitary deformity which may point to an interaction between the growth of the nasal septum and the premaxilla-maxilla. One of the most well-known congenital facial deformities is the cleft lip, alveolus and palate. In adult human skulls with facial clefts, which were known to be untreated, a specific pattern of growth disturbances could be observed [52, 51]. Unilateral clefts showed a deviation of the premaxilla to the non-cleft side, whereas the maxillary part on the cleft side had collapsed medially and fell behind compared to the non-cleft side (retrognathism). In addition, these skulls showed a specific malformation of the nasal septum with (a) a deviation of the perpendicular plate to the cleft side and (b) a disjunction between the perpendicular plate and vomer (Fig. 36.9). The vomer, only connected to the palatal margin of the cleft, tends to a more horizontal position to meet the deviated perpendicular plate, suggesting a broadening of the nasal floor on the non-cleft side. From this study, it was concluded that these growth anomalies of the maxilla and nasal framework were part of a cleft syndrome, specific for each type of cleft [40, 53]. The collapse of the upper alveolar arch and maxillary retrognathism were also observed in adult members of a North-Borneo tribe who were born with cleft lip, alveolus and palate, and did not undergo surgical treatment [54]. In a study on lateral cephalograms of untreated unilateral cleft patients with superimposed images of cleft and non-cleft sides, the abnormal position of the maxilla could not be confirmed, most probably due to the over projection of the non-affected side [55].

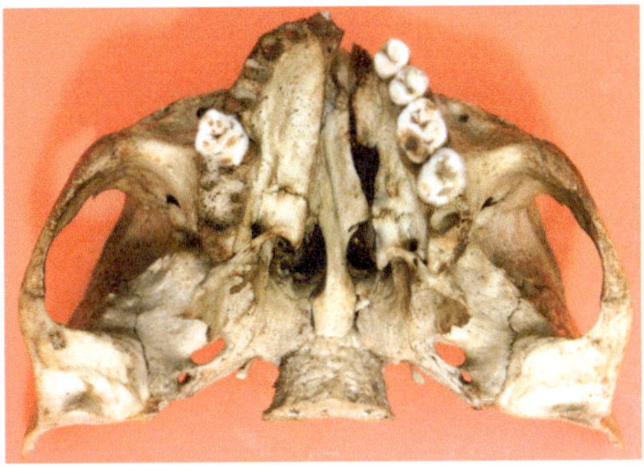

Fig. 36.9 Caudal view of a human skull with unilateral cleft of alveolus and palate (Dept. of Anatomy and Embryology, Amsterdam UMC). The maxillary segment on the cleft side demonstrates a medial collapse and a retroposition (asymmetrical position of the left and right tuber maxillae); deviation of vomer and premaxilla to the non-cleft side

Whether a combination of midfacial anomalies associated with a cleft alveolus and palate might represent a syndrome of craniofacial anomalies developing as a reaction to a cleft, was investigated in growing rabbits [56]. Unilateral clefts of the alveolus and palate, produced surgically in young growing rabbits, appeared to affect the further development of the nose and (pre) maxilla into adulthood (Fig. 36.10a). On the cleft side the molar complex, lacking a connection with the growing septum, does not move forward and will show—in the adult stage—a retroposition (retrognathism) compared to the non-cleft side. Here, the premaxilla, maxilla and nasal bones will grow to normal length, and gradually rotate and deviate to the non-cleft side as previously reported for human skulls [52]. The results of these animal experiments indicate that the cleft syndrome as identified in 'untreated' human

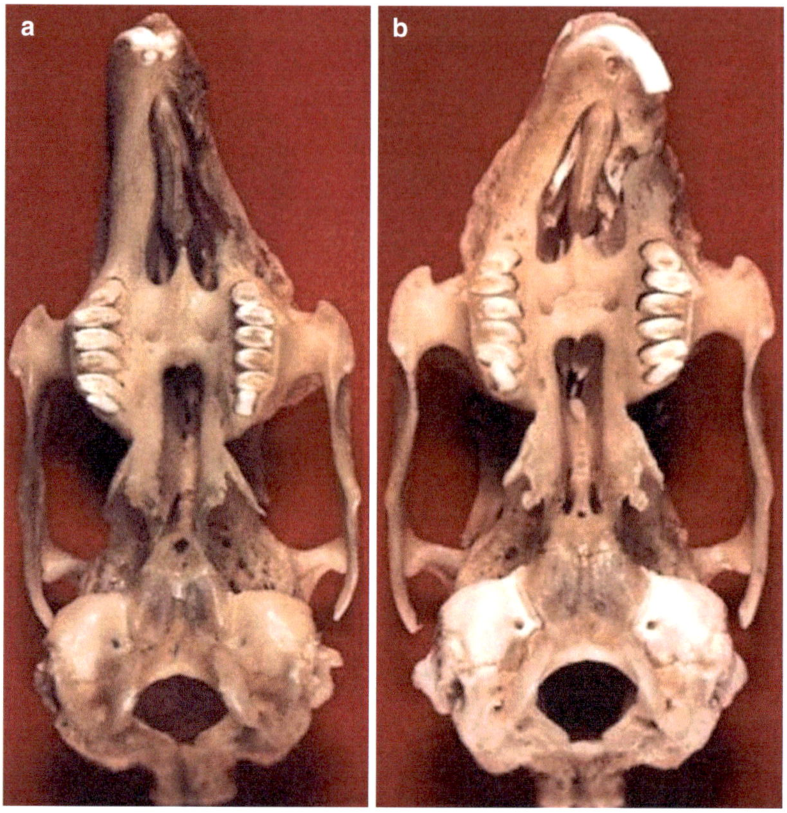

Fig. 36.10 Caudal aspect of the skulls of adult rabbits (24 weeks of age) with maldevelopment following surgical procedures at the age of 4 weeks; (**a**) following a surgical execution of an unilateral cleft lip and alveolus (resection of the premaxilla-maxillary suture): deviation of the premaxilla from the midline, retroposition of the maxilla, zygoma and pterygoid process on the cleft side; (**b**) after a similar cleft has been made and closed by primary osteoplasty: marked deviation to the operated side with shortening of the upper jaw, anteroposition of the maxilla including molar complex, zygoma and pterygoid process

skulls with facial clefts should be considered an adaptation to the altered developmental mechanics in the developing midfacial skeleton.

The interaction between the growing cartilaginous and bony skeleton is modified when midfacial sutures are missing as in the presence of facial clefts; similar cleft syndromes were found in animals and untreated patients.

Another interesting observation has been done when in young rabbits the premaxillary-maxillary suture was resected and replaced by non-sutural bone (Fig. 36.10b). Following this intervention, normal lengthening of the ipsilateral upper jaw failed. Secondary effects were a progressive deviation of the premaxilla to the non-growing side and an excessive forward shift of the molar complex on the operated side. These observations confirm disappointing midfacial development as observed after osteoplasty of an alveolar cleft in young children. The outcome of the abovementioned experiments refers to the role of the cartilaginous nasal septum in the postnatal development of the upper jaw. The growing and lengthening of the cartilaginous nasal septum, connected on both sides to the premaxilla-maxilla, seems to be responsible for the lengthening of the upper jaw and secondly, for a shift in the anterior direction of the upper jaw relative to the cranial base. In the presence of a unilateral cleft, the "mechanical" balance between both sides is disturbed.

Although the results of orthodontic and surgical treatment of facial clefts are improving, the secondary cleft nose in these patients can demonstrate some of the following features: asymmetry of the tip, columella, nostril, ala and nostril floor; deflection of the caudal part of the septum cartilage to the non-cleft side; stenosis of the vestibule on the cleft side; hypoplasia of the maxilla on the cleft side; the variable collapse of the maxilla with asymmetry of the piriform aperture; underdevelopment of the maxilla with retroposition of the anterior nasal spine; deviation of the cartilaginous and bony nasal dorsum; and deviation of the posterior part of the nasal septum to the cleft side [57].

36.2.1.2 Congenital Malformation of the Nose

Nasal congenital anomalies are extremely rare and range from bifidity of the nasal tip or dorsum to nasal aplasia with or without proboscis [50]. Reports on nasal dysplasia, which is mostly indicating a unilateral malformation, are scattered through the literature. The nasal cavity is missing and pneumatization of the maxillary, ethmoidal and frontal sinuses has failed. Exploration reveals no cartilage, but just solid bone. The affected half of the maxilla is hypoplastic and regularly associated with other malformations such as cleft lip, palate or coloboma. The last variety to be mentioned is a duplication of the nasal dorsum, which can occur in different forms and as part of a syndrome (Fig. 36.11).

36.2.2 Acquired Anomalies of the Nose

36.2.2.1 The Nose of the Neonate

Various studies refer to the importance of the nasal septum for midfacial development. In humans, intrauterine exposure to warfarin appeared to cause early calcification of the septal cartilage; subsequent nasal and midfacial hypoplasia was demonstrated in a cephalometric study, suggesting that even in utero midfacial growth is retarded by interference with the nasal septum [58]. In another anatomical study, the morphological interaction between the nasal septum and nasofacial skeleton was found to be correlated and maintained throughout ontogeny [49]. The nasal septal cartilage was demonstrated to be the key growth factor.

The neonate's nose may show a slight or more pronounced deviation, with luxation of the lower septal border into the nasal cavity, acquired during the passage through the birth canal (Fig. 36.12). In most cases, this anomaly will restore spontaneously but infrequently the deformity has to be corrected by manipulation of the neonatal nasal septum which is still mostly cartilaginous from sphenoid to the columella. Already

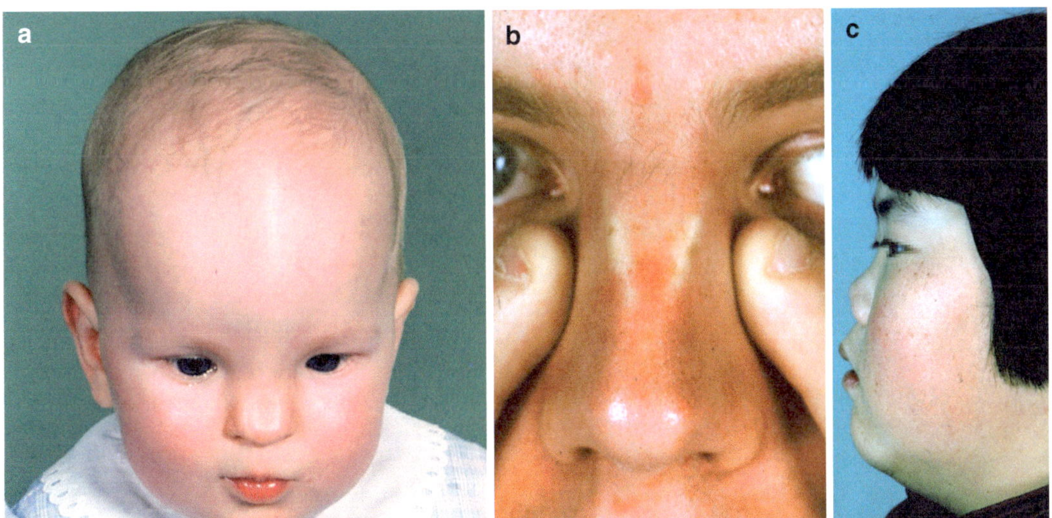

Fig. 36.11 Congenital anomalies of the nose; (**a**) Incomplete fusion of the left and right anlage of the cartilaginous and bony nasal skeleton in a 3-year-old girl and (**b**) an adult man; (**c**) Under development of the cartilaginous and bony nasal pyramid

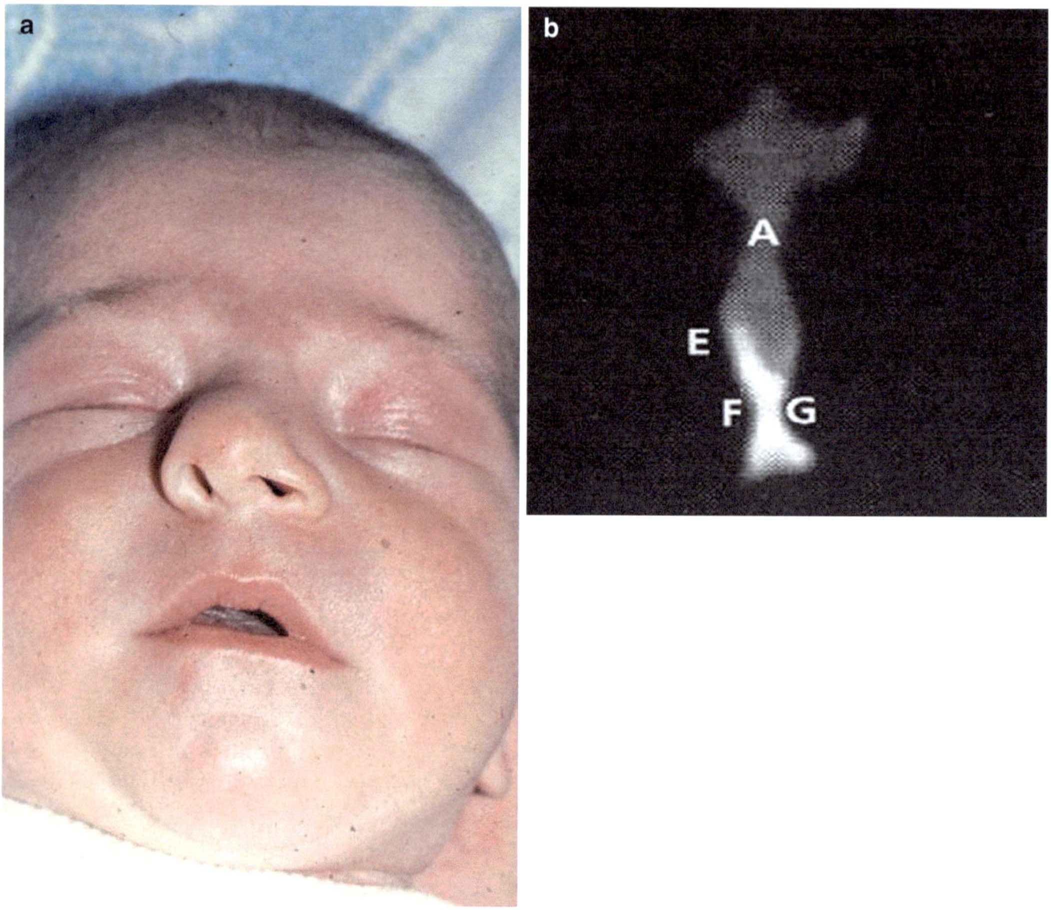

Fig. 36.12 A newborn baby with a deviation of the cartilaginous nasal dorsum and columella (**a**) due to a bending of the cartilaginous septum at the weak zone superior to the thick basal rim or a dislocation of the basal rim from the vomeral gutter, acquired during the passage through the birth canal (courtesy of Prof. W. Pirsig); (**b**) Asymmetrical anlage of the vomer in a CT scan (frontal plane) of a neonatal septum; (A) cartilaginous nasal septum, (E) ala vomeris, (F, G) inferior parts of the vomer

in 1978, Gray found some type of septum anomaly in 57% of 2380 neonates [59]. The septal deformity was explained by Takahashi to be an inevitable condition resulting from the autonomic growth force of the septum cartilage [60].

36.2.2.2 Trauma at a Young Age: Bone and Cartilage Lesions

Any kind of mechanical or surgical trauma at a young age, destroying the septodorsal cartilage, initiates an irreversible disturbance of the cartilage from its genetically based developmental orientation. A lesion may vary from fracture and dislocation, with later deviation, to haematoma and abscess formation and loss of septum cartilage, with various associated sequelae. The regional differences in the thickness of the septum cartilage, described above, seem to explain that fractures preferably occur at sites with the least mechanical strength, and thus the thinnest parts of the septum. Septal deviations could, therefore, be related to the 'weak' areas [61]. Dislocation of the caudal end of the septum, the most frequent disorder in young children, is caused by a fracture running through the thin area lying cephalic of the columellar rim. The thin area above the thick basal rim is another vulnerable fracture-prone region. This preference for specific fracture lines was described by several authors in children as well as adult patients; it was demonstrated in the so-called C-fracture running horizontally and superiorly along the thick basal rim of the septum, then more horizontally through the thin portion of the perpendicular plate, while ending in the thinner zone just under the nasal dorsum [62, 63]. Another weak point is the area of thin cartilage, semi-parallel to the caudal rim and just anterior to the connection between the septum and upper lateral cartilages, which may easily fracture when the caudal end of the septum gets dislocated (viz. Fig. 36.2c).

36.2.2.3 Fractures: Direct Effects and Follow-Up

Although injuries of the nose appear to occur frequently in young children—and the incidence still seems to be growing—only a relatively small number are seen by the general practitioner, and even less are diagnosed by an ORL-specialist. Consequently, the long-term follow-up of these cases is rare and certainly not routine. In the years following such a "minor" trauma, however, a progressive deformity of the midface may be observed with increasing deviation or distortion of the nose, eventually in combination with underdevelopment of the maxilla, while the final effects can only be defined after the adolescent growth spurt [8].

After a traumatic accident, fractured parts of the bony midface and the septum will lead to, often disproportionate, disfigurement of the facial profile. In particular, the less obvious lesions of the cartilaginous septum are seldom diagnosed and during further growth the fractured parts will bend and dislocate increasingly, thus leading to spine formations, local overlapping edges and finally severe deviations [64]. A midfrontal trauma may cause multiple fractures with larger and smaller pieces of cartilage resulting in a highly distorted septum and finally an underdeveloped and distorted nose (Fig. 36.13). The direction of deviations is mostly unpredict-

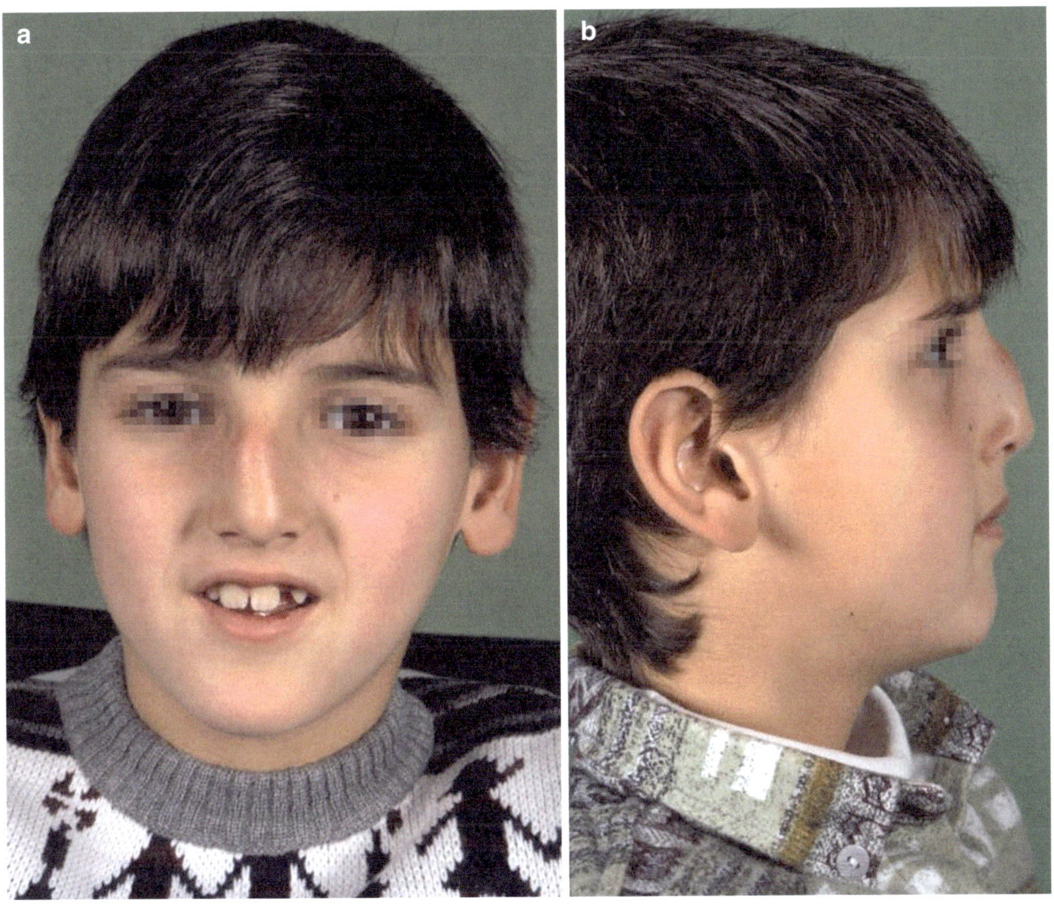

Fig. 36.13 Underdevelopment and deviation of the cartilaginous nasal pyramid, multiple and irregular deviations of the cartilaginous septum after a previous injury; (**a**) boy at the age of 8 years and (**b**) 13 years. Could a progressive anomaly have been prevented?

able because it depends on the direction of the deforming force and on the reaction of the traumatized tissues involved. Captivating evidence for the significance of the nasal septum in midfacial growth is supplied by a rather large study of 21 monozygotic twins ranging in age from 18 to 22 years [65]. In only one of each twin pair, a deformity was found in the anterior part of the septum. The nose of the affected siblings was demonstrated by cephalometry to be shorter than in their 'normal' other half.

Animal Experiments

Poor wound healing of cartilage is the main reason for the occurrence of disorders of the nasal skeleton, with or without nasal airway obstruction. Experimental studies in young rodents and other mammals have consistently demonstrated that removal of full-thickness (cartilage + mucosal lining) portions of the growing nasal septum may result in growth retardation of nose, maxilla, premaxilla and palate [22, 66–68]. Later studies with larger series of experiments in young rabbits using submucosal techniques could fill in some open ends of these earlier investigations as discussed in Sect. 36.2.2.4 [69]. The growth of the rabbit skull was studied from 0 to 24 weeks after birth. The facial skeleton appeared to grow faster and over a longer period than the brain skull like in humans (Fig. 36.7). The growth rate increased rapidly in the first 8 weeks with a gradual decline thereafter till around the 22nd week of age. Twenty-four weeks was the final date of the experimental period. To achieve a complete image of the significance and developmental role of the bony and cartilaginous parts, the first experiments were performed on the nasal bones, maxilla and lateral nasal wall. Transverse osteotomy of the nasal bone anterior to the frontonasal suture on one or two sides did not disturb nasal growth in length nor did extirpation of the frontonasal suture alter the growth of the snout. The snout showed no asymmetry, and a small bony defect remained only present in a small minority of the animals. Even subtotal resection of the nasal bone did not result in growth disorders as long as the underlying upper lateral cartilage was undamaged [68]. The same restriction applied to partial resection of the lateral nasal wall, which in rabbits can only be performed through an opening of the nasal cavity from above after mobilisation of the nasal bone without disturbing the upper lateral cartilage on that side [70]. Although the FESS procedure in children is not completely analogous it was demonstrated that enlargement of the entrance to the maxillary sinus via this method did not disturb the normal growth of the rabbit's nose.

Experiments including resection of the premaxillo-maxillary suture(s), also in combination with resection of the midpalatal suture—comparable to the cleft upper jaw and palate—gave similar results as the cleft syndrome described for the human patients and skulls (see Sect. 36.2.1.1). The retroposition of the maxillary portion on the cleft side is explained by the failing connection to the growing septum at the anterior nasal spine. Consequently, the septum will only pull forward the premaxillary complex on the unaffected side, and gradually deviate to that side. Malocclusion is the result. Closure of an alveolar defect with non-sutural bone, like it was performed in children with cleft alveolus and palate in the seventies and eighties, demonstrated clearly that the development of the mid-face became severely disturbed [56].

36.2.2.4 Septum Haematoma and Abscess: Loss of Septum Cartilage

From many human case studies described in the literature, it is evident that the sequelae of a nasal injury are frequently a septum haematoma and, due to contamination and infection, a subsequent septum abscess. A haematoma may mostly, when not treated adequately, cause damage to the cartilage while an abscess invariably causes destruction of the nasal cartilage leading to septum perforation or a submucosal defect of the septum cartilage. Saddling deformity with loss of tip support and growth retardation with a short nasal dorsum are then regular findings (Fig. 36.14a,b). Grymer and Bosch reported about one twin sibling, who endured septum demolition due to a septum abscess at the age of 7 years, and was followed up with its twin for more than 10 years

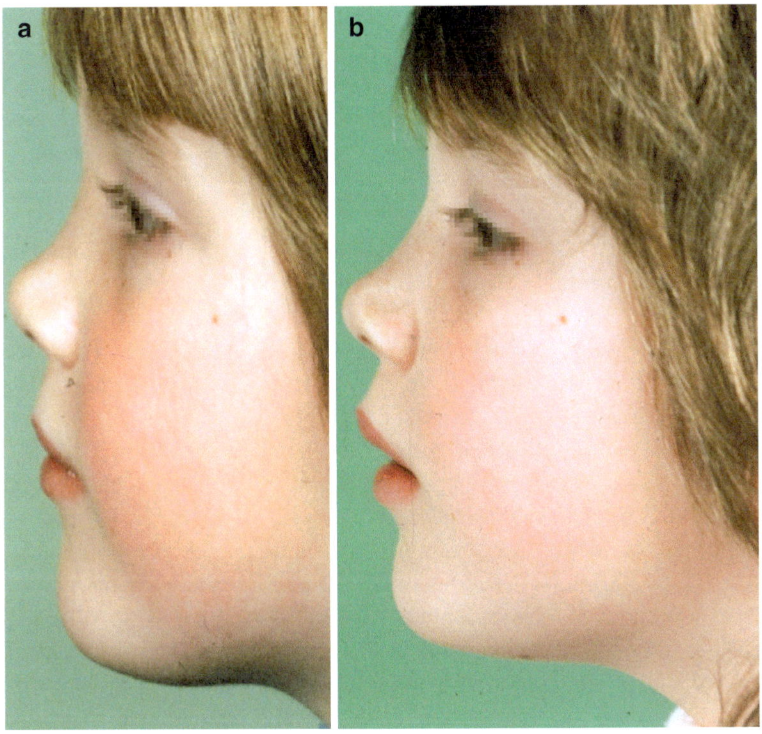

Fig. 36.14 Underdevelopment of the cartilaginous and bony nasal pyramid after a previous septal abscess; (**a**) girl at the age of 5 years and (**b**) 10 years. Increased nasolabial angle, with a prominent tip (supported by the alar cartilages)

[71]. Despite the treatment with homologous cartilage from the tissue bank, the affected boy developed a severe saddle nose with an upward displacement of the anterior part of the maxilla, decreased nasal projection and a retrognathical position of the maxilla compared to the unaffected sibling. The authors' observations suggested that the nasal septum has a great influence on midfacial development determining vertical as well as anteroposterior maxillary growth. And secondly, that bank cartilage grafting is not adequate in preventing malformations of the midface during growth. Patients with a history of facial trauma develop progressive nasal deformation and/or nasal obstruction but the younger the age of the child the more impact such a lesion may have in the long run [3]. Also the extent of the injury fracture, haematoma, abscess and perforation, surgical treatment—will play a role in the final result after the adolescent growth spurt. And these results are often disappointing, thus many of these patients have to undergo second or even more surgical procedures later in life [72]. From a series of 241 children with a history of nasal injury, 40% were diagnosed with pathological septal findings, with the longest follow-up period after the trauma of 10 years [73]. Septum deviation associated with snoring and abnormal rhinomanometric values appeared in 55% of the patients. In a group of 16 paediatric patients, ranging from 2 to 14 years, major sequelae (62.5%) like dorsum, tip or pyramid deformation, septum deviation with nasal obstruction, functional vault deformity and septum perforation, were observed, whereas the other patients (37.5%) all showed minor sequelae without any airway compromise [74]. Another consecutive series of 20 children (2 months to 15 years) were admitted to the hospital for treatment of nasal trauma followed by a nasal abscess (12 patients) universally associated with the destruction of the septum cartilage [75]. Nasal obstruction, however, was the most common symptom (19 patients). Recently a structured review of PubMed, EMBASE and the Cochrane Databases comprising 81 citations, dating from 1920 up to now, regarding nasal septal abscess was published [76]. It was concluded that it is a serious

condition that necessitates urgent surgical management in order to prevent severe complications, and even the growing child should be treated with early reconstruction essential for normal development of the midface. In this whole sequence, the lack of knowledge of the developmental mechanics is part of the problem.

Animal Experiments

Role of Cartilaginous Septum in Midfacial Growth: Experimental Evidence

To improve the definition of the role of the septodorsal cartilage for midfacial growth a large series of animal experiments in which various parts of the cartilage were resected, resected and reimplanted, crushed and reimplanted, and replaced by artificial materials or tissue-engineered cartilage, was executed [77–80]. In this study, the most elementary and basic experiments of the former series were repeated in each new series. Surgical manipulation of the nasal septum with the removal of cartilage resulted in midfacial malformations; the severity was depending on the extension of the cartilage defect.

> The dorsoseptal cartilage is the dominant growth center of the midface. Specific growth zones are stimulating the development of the nasal skeleton and upper jaw.

Upper Laterals

In neonatal rabbits, the upper lateral cartilages are extending from the nasal tip to the anterior cranial base, grossly comparable to the morphology in human newborns (viz Fig. 36.8). Upper lateral cartilages and septum form an unpaired T-bar construct like in humans. This construction is supposed to possess much more mechanical strength and exert more "pressure" in the period of growth than the septum alone. The growing septodorsal cartilage could therefore mechanically stimulate the lengthening of the upper jaw and nasal bones which are firmly connected. In the adult rabbit the upper lateral cartilages, having lost the connection with the anterior cranial base, have adopted a more triangular shape and extend only halfway under the nasal bones. Partial to subtotal resection of upper lateral cartilage in young animals, on the other hand, would lead to specific anomalies: a decrease in length and curvature of the nasal bone, a deviation of the nasal tip to the non-operated side and a considerable diminishing of the size of the turbinate on the operated side [80]. This type of deformity could also be found in young patients with injury to one upper lateral cartilage at a young age [3].

Mucosal Elevation

In the sequence of submucosal septal interventions the first step was the elevation of the mucosa on one or both sides, the so-called tunnelling; being executed carefully it did not affect septal and nasal growth in rabbits [81]. At a microscopic level, however, a short period of swelling of the lamina propria of the perichondrium with exudate formation could be observed, followed by chondroblast proliferation, sometimes resulting in some extra cartilage production within the perichondrium [82].

Release of Interlocked Stresses

Interruption of the septum cartilage over the anteroposterior length by a single and complete incision, comparable to a single total fracture, demonstrated an immediate overlap of the cut edges [53]. It is explained by the release of interlocked stresses present in the hyaline cartilage, which was earlier described by Gibson and Davis for human rib cartilage, and later by Fry for nasal septum cartilage in vitro [83, 5]. Forces were supposed to be locked within the matrix of the cartilage; these should be in a state of balance between tension (in the three-dimensional collagen network) and pressure (evoked by water-binding hydrophilic proteins). Hyaline cartilage of the nasal septum contains a collagen network of fibres aligned perpendicular to the surface, and ending in alignment with the surface forming an inner and transitional layer of the perichondrium: the most proliferative cambium layer (Fig. 36.15a). The cartilage cells (chondroblasts-chondrocytes) with their surrounding proteoglycans-containing

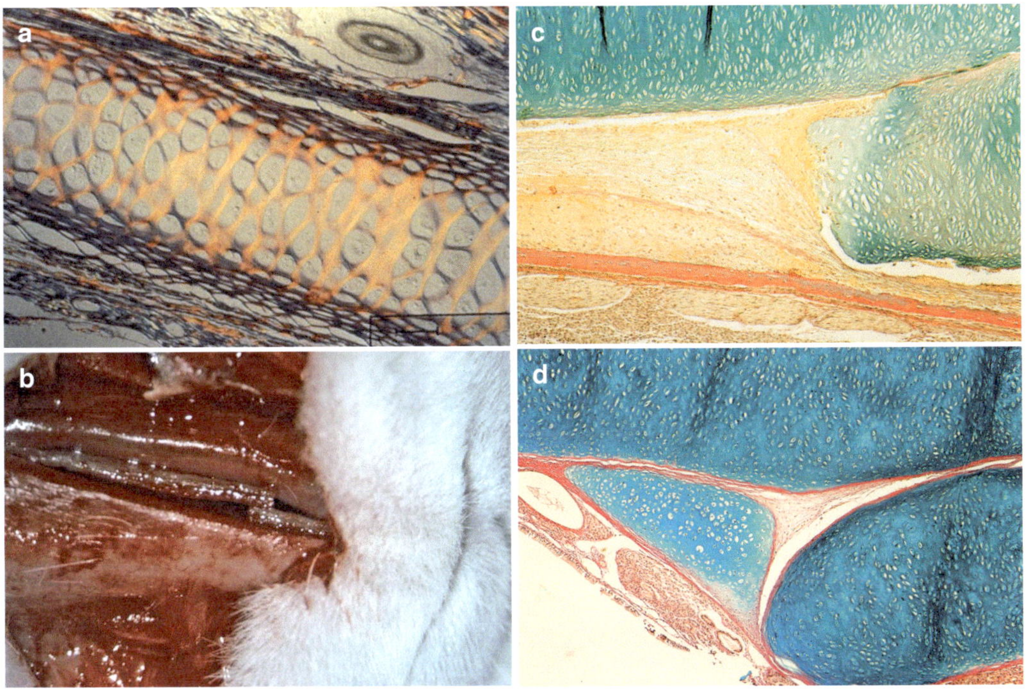

Fig. 36.15 Histological section of septum cartilage; (**a**) Sirius Red staining specifically demonstrating the dense layer of collagen fibres in the perichondrium and the network of collagen fibres in which the chondrocytes are enclosed; binding of water by intercellular proteins generates interlocked stresses within the collagen mantle and network; (**b**) After transection of the septal cartilage, release of interlocked stresses is responsible for an immediate overlap of the cut ends as observed in vivo; (**c**) Reaction at the wound surface with exsudate, necrosis and swelling of the elevated perichondrium, at the level of the vomer wing; (**d**) 20 weeks after a surgical intervention; the separated parts are reconnected side to side by their perichondrium (*dehiscence is artefact*); the 'free space under the elevated perichondrium is filled with newly-formed cartilage (PAS Alcian blue staining)

matrix are detained within the fibrous network. By fracture, incision or scoring of cartilage the above-mentioned stresses might be released, thus altering the form of the cartilage. Also in the nasal cartilage of rabbits this interlocked stress release resulting in a local duplicature of the cut ends could be found (Fig. 36.15b–d). This phenomenon could not be undone because within a short time the overlapping ends were attached side-to-side by fibrous tissue originating from the perichondrium and the interlocked stresses would then recur [32].

The poor wound healing capacity of cartilage and the release of interlocked stresses after loss or fracture of the septum have definite effects on later development of the midface.

A relation between the depth of scoring and the immediate warping of cartilage in an isolated situation was studied separately [84]. The resected septa were immerged in a bath of saline. Scoring for less than 15% of the total thickness was not followed by bending of the cartilage, but incisions up to 50% of the thickness appeared to be directly related to the degree of warping; differences between young and mature cartilage could not be observed. Effects on the mechanical strength of resected cartilage or on midfacial development could, of course, not be measured.

Submucosal Resection of Parts of the Septum Cartilage: Effects on the Developing Nasal Skeleton

Submucosal resection of the middle one-third of the septum cartilage of 4-week-old rabbits (per series n = 10), interrupting the anteroposterior axis, had marked consequences for midfacial growth: a decreased length of the premaxilla and maxilla and malocclusion of the molar complexes and incisors resulting in underdevelopment of the midfacial skeleton (Fig. 36.16). Even when this surgical procedure was performed at a later stage, in rabbits at the age of 9 weeks with a follow-up of 16 weeks until the animals had reached adulthood, digital measurements (Screen Calipers version 3.3, Iconico.com Software) confirmed the above-mentioned observations [1] The septal defects in these experiments will interrupt both zones of thicker cartilage, earlier described as the sphenodorsal and sphenospinal zones (Fig. 36.17). Growth of the first is considered to contribute to the lengthening of the bony nasal dorsum, whereas the second should "push forward" the connected premaxilla-maxilla. Discontinuity of one or both zones can evidently explain different types of midfacial maldevelopment.

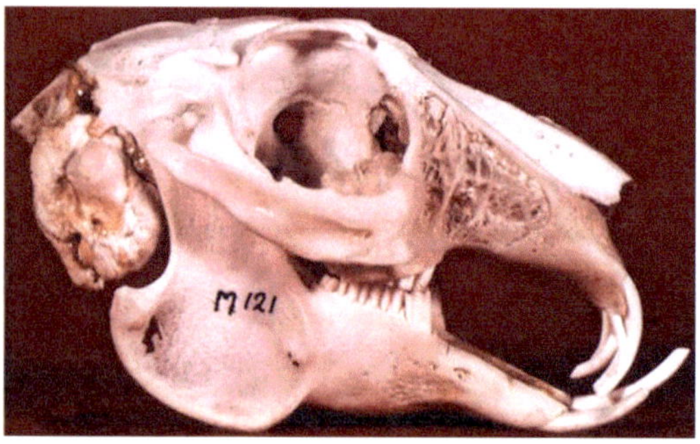

Fig. 36.16 Lateral view of an adult rabbit skull 20 weeks after resection of the middle one-third of the septum cartilage; shortened nasal dorsum with slight 'saddle'; malocclusion with crossbite of incisors

Histology

Histological examination of the septum after submucosal resection of the middle-one third of the septal cartilage demonstrates specific reactions of the elevated outer perichondrium, the transected

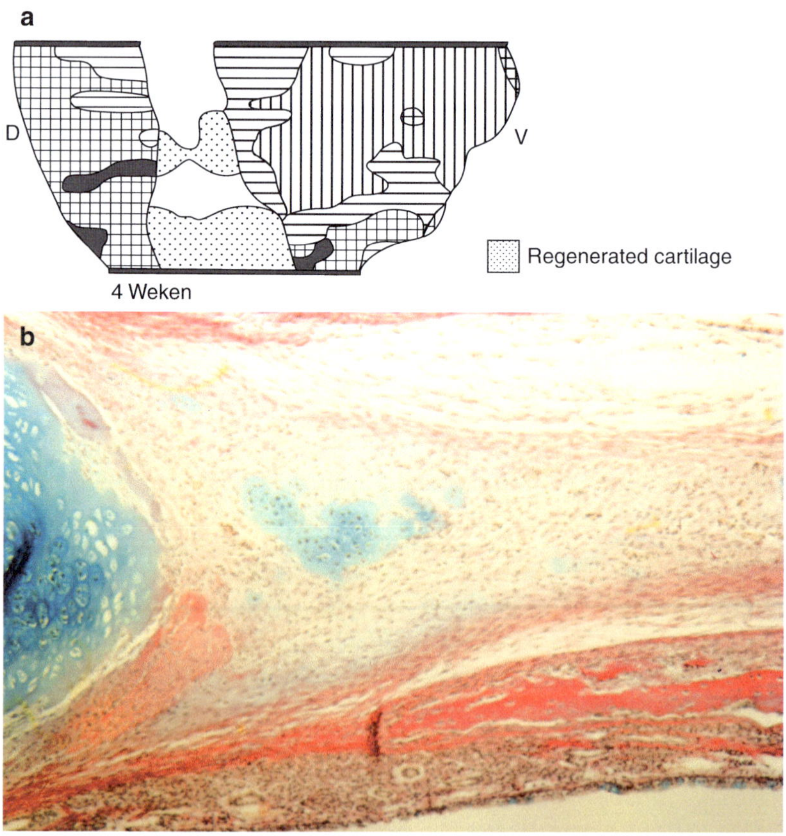

Fig. 36.17 (**a**) Schematic reconstruction of a septum 4 weeks after submucosal resection of 1cm in the middle part of the septum; regenerated cartilage is observed which locally might bridge the defect; the majority of newly formed cartilage is found between the two vomer wings which keep the defect open (courtesy of Dr C.A. Meeuwis); (**b**) Histologic section at the level of the vomer one week after surgery; precursor cells from the inner perichondrium (cambium) invade the open space and the first chondroblasts are formed; the original septum cartilage shows signs of necrosis: the chondrocytes have died, the intercellular matrix remains, the adjacent area shows highly activated cells

cartilage and adhering cambium. The wound reaction of the cut ends of the cartilage implies a ± 3 mm deep zone of regressive changes and cell death bordering the cut surface, with an adjacent zone of high cellular activity with enlarged dedifferentiating chondrocytes and a few mitoses, as well as first signs of redifferentiation finally resulting in the formation of new cartilage. After the first 2–4 days the necrotic area will be marked with viable cartilage by a layer of fibres. The necrotic cartilage is invaded and resorbed by polymorphonuclear cells and macrophages originating from the outer perichondrium [85]. In the meantime cells migrating from the cambium layer grow over the cut cartilage, showing marked mitotic activity and redifferentiation into cartilage (Fig. 36.17). This new cartilage may even be recognized 20 weeks later in the adult stage by its large number of relatively small chondrocytes. Cells migrating from the outer perichondrium will soon cover the newly formed cartilage which shows hardly any further growth. The role of the outer and inner perichondrium in wound healing of cartilage has been further analysed by in vivo and in vitro experiments and will be discussed later [86].

Where the septal cartilage has been resected, cells from the elevated (outer) perichondrium seem to migrate into the underlying "free space" initially filled with exudates. In the following weeks the "free space" narrows and the elevated perichondrium from both sides will reconnect, locally enclosing islands of thin cartilage probably formed by migrating from parts of the cambium which might have been connected to the elevated perichondrium (Fig. 36.18).

Even in adult rabbits, in which submucosal resection of a central basal segment of the nasal septum (1.0 × 2.5 cm) was removed, islands of isogenic chondrocytes could be identified, 7 months later [87]. This regrown cartilage morphologically resembled but was not identical to native septum cartilage. In contrast, healing of split cricoid cartilage did not occur easily in older rabbits: no mitotic activity or new cartilage formation was then observed [88]. It is suggested that the potential of the septal perichondrium is much stronger than that of the thinner perichondrium covering the cricoid.

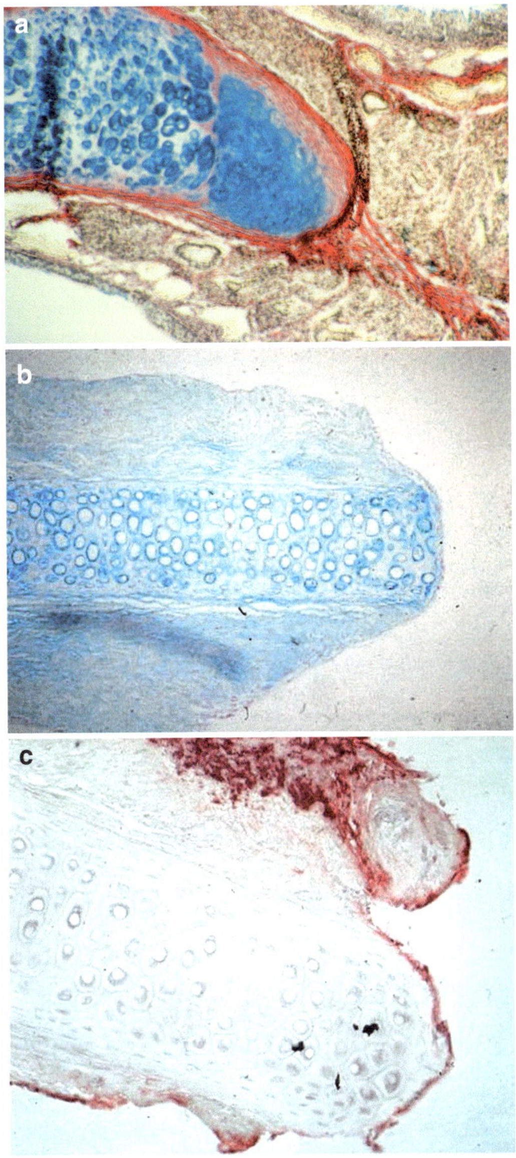

Fig. 36.18 Histologic section of a wound 6 weeks after resection of the middle part of the septum cartilage; (**a**) the cut end has been encapsulated by a fibrotic layer originating from the outer perichondrium; in the defect, the perichondrium layer on both sides have fused; (**b**) isolated piece of ear cartilage with intact perichondrium, cultured for 3 days, demonstrates fibrous cells originating from the outer perichondrium growing over the cut surface of the cartilage (Alcian blue staining; original magnification x75); (**c**) immunohistochemical staining for TGF ß1, present in the outer layer of the perichondrium and in the overgrowth, at day 7 of culturing

Submucosal resection of the dorsal half of the middle part of the septum in young animals results in a shortening of the nasal bones in the

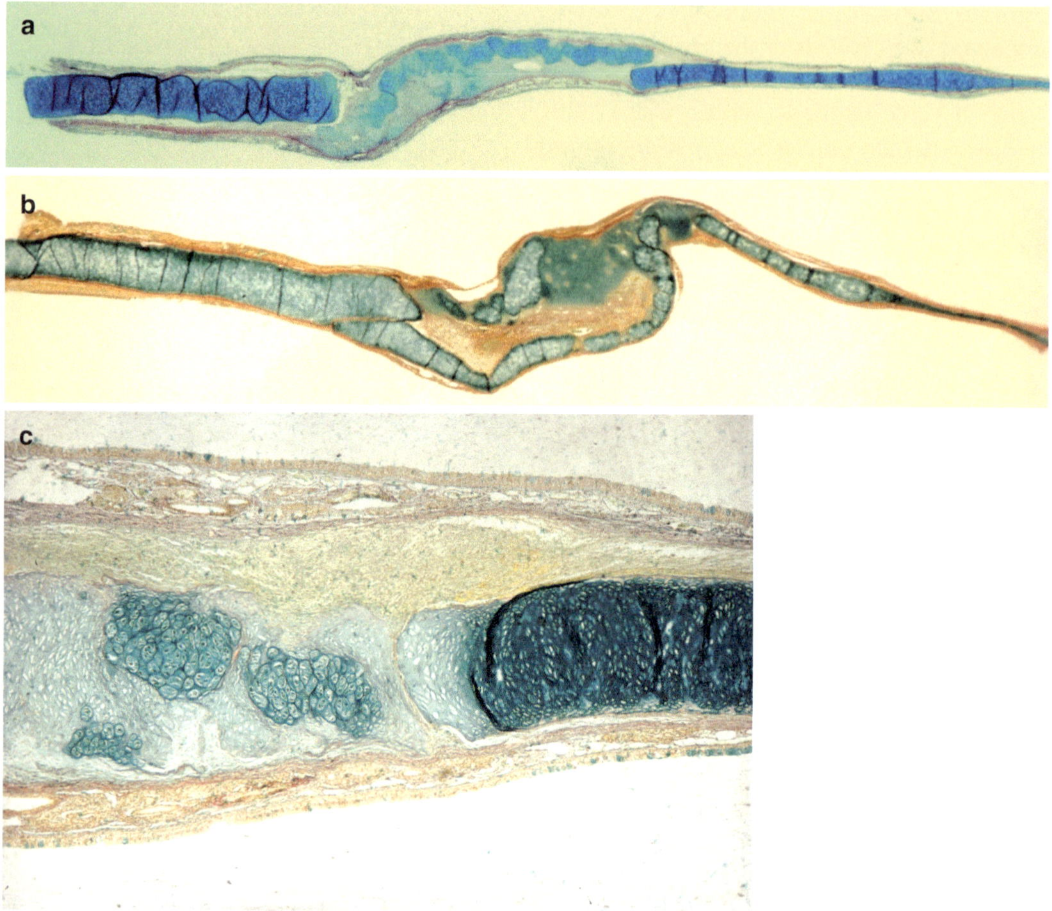

Fig. 36.19 Histologic sections of nasal septum with a crushed implant 2 days after surgery (**a**) and 20 weeks later (**b**) with deformation of the septum due to disorganised and insufficient repair of the cartilage; (**c**) detail at 7 days after crushing and grafting: the necrotic part of the stump is demarcated, invading cells from the perichondrium differentiate into chondroblasts, and proliferating viable cells form larger islands of cartilage through mitotic activity

adult stage due to the interruption of the sphenodorsal zone of thick cartilage [89]. Growth of the uninterrupted sphenospinal zone is responsible for the normal lengthening of the premaxilla and maxilla. If resection of the middle part of the septum is actually restricted to the sphenospinal zone, no growth disturbance of the nose and upper jaw could be demonstrated. The resected part appeared to be enclosed between the vomeral wings on both sides [77]. The outer perichondrium is here attached to the vomeral periosteum which may prevent their collapse after resection of cartilage leaving free space for cartilage's new formation. Then, overgrowth of the amputated ends of the septum cartilage by the outer layer of the perichondrium may be retarded or prevented (Fig. 36.19). Geometrical measurements of this experimental series demonstrated no statistically significant differences with the control group [90]. It may be concluded that the effects of septum resection in rabbits are depending on the dimensions and location of the created defects. The larger the deficit of tissue the greater the interference with normal growth, in particular in the most important growth areas. The experimental results make clear that inadequate wound healing of cartilage is at the moment limiting the possibilities of surgery to restore normal facial development.

36.2.3 Repair and Reconstruction of the Nasal Septum

36.2.3.1 Wound Healing of Hyaline Cartilage and Perichondrium

In Vitro Experiments

The wound healing capacity of hyaline cartilage and its perichondrium in the head and neck region, humans as well as animals, has been investigated since the midst of the last century [92, 91]. The role of perichondrium was found to be twofold since it appeared to be composed of two layers: an inner layer (cambium) which is involved in the production of new cartilage cells and an outer, more fibrous, layer responsible for nutrition and protection [86]. This was observed in explants of rabbit ear cartilage, cultured under various conditions for up to 21 days, with perichondrium intact or after removal of the outer layer or both layers. After 3 days the cartilage explants with an intact perichondrium demonstrated fibroblasts, originating from the outer perichondrium layer, growing over the bare cut ends of the cartilage (Fig. 36.18b,c). When the outer layer was resected leaving the cambium layer in situ fibrous overgrowth could not be observed and the new cartilage was formed at the cut ends. Pieces of cartilage denuded from inner and outer perichondrium did not show any reaction. It was concluded that the efficacy of neochondrogenesis lies in the inner perichondrium layer, the cambium. It seems evident that wound healing of cartilage would be improved when the overgrowth of fibres from the outer layer of perichondrium could be delayed or prevented favouring cartilage new formation by the cambium.

Animal Experiments

In experimental animals such as rats, cats, guinea pigs, ferrets, beagles, chimpanzees, and particularly rabbits reconstruction of the septum after cartilage resection with re-implantation of resected tissue or homologous cartilage, or grafting with other materials was tested. As in most research models and series, not all outcome is uniform. For example, in the cat with a very short nose, cephalometric measurements performed in the lateral sagittal plane after septoplasty did not show any growth disturbance but septum deviations can only be made visible in the transversal plane. It might lead to incorrect conclusions and advice regarding septoplasty at a young age [28]. Experiments with 6-week-old rabbits and 8-week-old beagles revealed that following surgical creation of a cleft lip, alveolus and palate using identical experimental protocols, overall craniofacial aberrations were similar for both species [93]. From septoplasty procedures in chimpanzee monkeys of unknown age, it was concluded that by delaying the time of surgery the impact of growth deceleration could be decreased [31].

Histology

In-vivo cartilage wound healing is hindered by chondrocyte death at the lesion site [94–96]. In injured cartilage, dead cells and intercellular matrix at the lesion site are not easily eliminated. A band of avital tissue can hinder the integrative repair of the lesion, even though the cells adjacent to the layer of necrosis increase the expression of growth factors and synthesis of matrix components as well as mitotic activity. The hampered healing of cartilage wounds is the cause of deformation and growth disturbance. Positive results have been suggested for the use of a polydioxinone plate as a 'splint' between the perichondrial layers on both sides to prevent deformation [97]. Using a very sharp scalpel might help to revive the wound edges since cell death is less extensive at sharp cut edges [95]. In experimental laboratory studies, the application of enzymes to degrade the avital tissue in the traumatized area has been demonstrated to improve integrative bonding of two pieces of cartilage, which thus far has not been tested in a clinical setting [98–100]. Furthermore, the application of metabolically active cells at the lesion site that can degrade the old matrix at the wound edge and lay down a newly formed matrix has been proposed [101].

It should be understood that this newly generated cartilage is not always of the same quality as the original tissue. The matrix is often more fibrous and does not have the same fibre architecture and mechanical properties. Therefore, the repair tissue might be inadequate to bring forward normal nasal growth.

36.2.3.2 Implantation of Grafts

In general, cartilage grafts from the nasal septum, auricle or cricoid will survive when embedded in a physiological condition (environment), the more so when the cartilage graft is covered by perichondrium. If the embedding is inadequate cartilage necrosis and resorption will take place. The majority of experimental investigations of the midface were performed on rabbits in different designs.

Animal Experiments

As discussed previously, submucoperichondrial resection of a middle one-third of the nasal septum in young pure-bred rabbits resulted in all studies to retarded snout growth rather in length than in height [78]. Reimplantation of the resected, and thus autologous, cartilage could not prevent growth anomalies after a follow-up of 20 weeks when the animals are adult: reduced length, sometimes saddling of the nose and septal deformities were noted. Inadequate healing of the graft with a good binding to the original cartilage had resulted in septal spines and deviations. Crushing of resected pieces of septum cartilage followed by reimplantation showed similar sequelae: the crushed tissue may partially survive with dedifferentiation of new chondroblasts but is also the victim of necrosis and resorption (Fig. 36.19). Such a new construct appeared not to be suitable for reconstruction purposes in the growing animal [102].

When insufficient material for grafting is available in the patient, alternatives should be considered. Artificial materials like proplast and Gore-Tex were tested and found absolutely not suitable for reconstruction because of their disastrous effect on the growing septum. Infection and rejection of the graft resulting in severe midfacial malformations were observed in the growing animals [79].

Tissue Engineering

In addition to the use of artificial materials only, the combination of cells and carrier materials as used in the field of tissue engineering may offer advantages. New tissue can be formed by implanting materials that stimulate endogenous repair, by combining materials and cells in situ or by creating constructs with cells cultured on materials in the laboratory before implantation in the patient. Each year new materials and technologies become available but the use of artificial materials, either with or without seeded cells, for cartilage reconstruction in the head and neck region remains challenging because the risk of infection and rejection is particularly high. Therefore, natural materials seem a better choice for the reconstruction of the nasal septum. One of the first natural materials investigated extensively for cartilage formation is demineralised bovine bone [103–105]. This material is full of natural growth factors that can attract cells and induce cartilage formation. A promising method for cartilage implantation in the head and neck region was found when a composite graft consisting of a piece of the demineralised bovine bone matrix was enveloped in a flap of auricular perichondrium (Fig. 36.20a). In young rabbits, this compound graft was transformed into a piece of hyaline cartilage (Fig. 36.20b). This newly formed cartilage was then removed from the ear and transplanted into a septum or cricoid defect which apparently survived and could integrate well with the surrounding margins of the original cartilage [103]. One case has been published in which a boy who lost most of his septum cartilage at a young age (around 3 years of age) was treated at the age of 9 with this method and showed a very good, developed nose at the age of 19 years [106]. Although positive results could be demonstrated, the irregular tendency of early ossification and above all the growing incidence of mad cow disease gave reason to interrupt this research at that moment.

Based on the fact that autologous cartilage grafts are still the best option and that the pio-

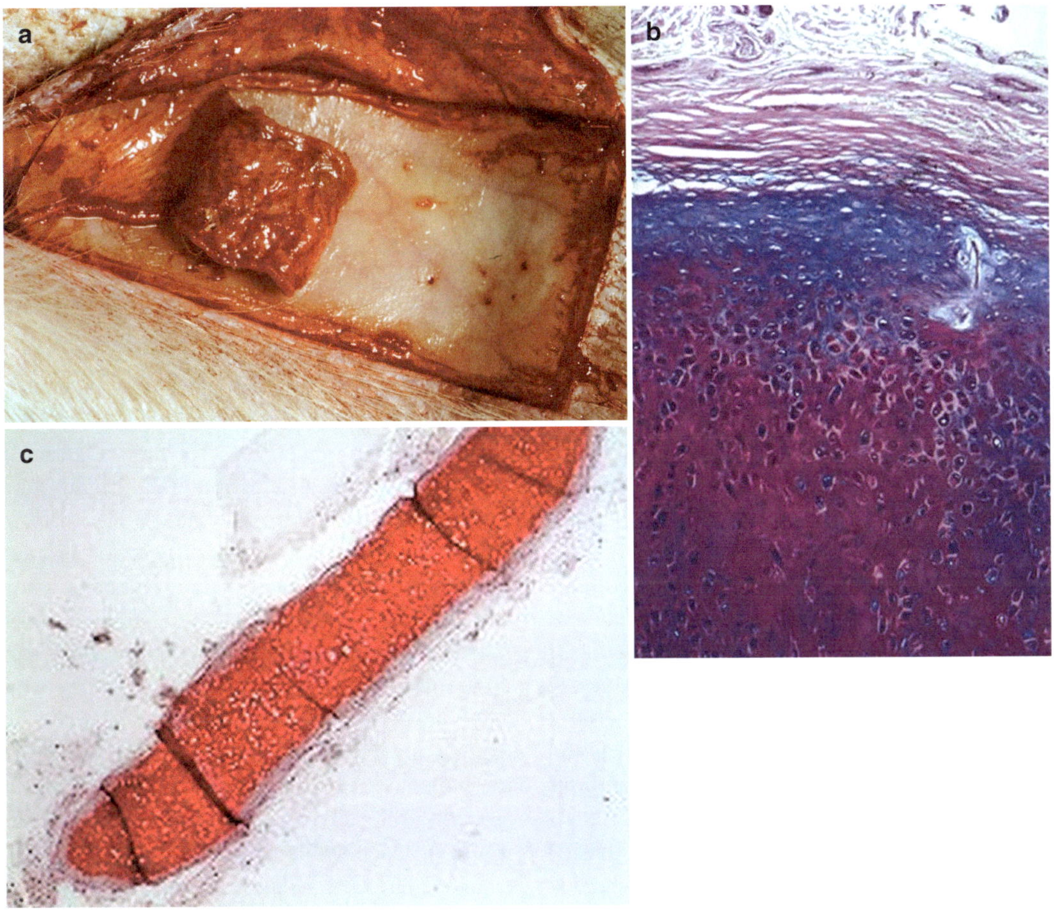

Fig. 36.20 Tissue engineering in vivo; (**a**) a piece of demineralised bovine bone matrix is wrapped in elevated rabbit ear perichondrium and left for 6–8 weeks under the skin; (**b**) histology of human engineered cartilage in vivo (courtesy of Prof. W. Pirsig). (**c**) Tissue engineering in vitro: construct generated from culture-expanded auricular chondrocytes of an 18-year-old donor; scaffold-free flat constructs were made by seeding the cells on a membrane and by culturing them for 14 days in the lab followed by implantation subcutaneously in athymic mice for 6 weeks; a new piece of cartilage-like tissue, positive for collagen type II on this immunohistochemical staining, was produced

neering studies with demineralised bone matrix indicated a promise of engineering solid cartilage in vivo, the interest in decellularised cartilage grafts was renewed. The use of decellularised cartilage grafts has several advantages including good mechanical properties and low foreign body reaction. A very recent study with decellularised porcine nasal septal cartilage grafts in a rabbit septal replacement model confirmed the good biocompatibility and stability of the graft over a period of up to 6 months [107]. However, since the repopulation of decellularized hyaline cartilage is limited because of the dense cartilage matrix, the stability of such a graft in the long term remains questionable. A solution to allow cells to migrate in and repopulate the decellularized carrier is offered by a method to generate small channels in the cartilage matrix based on selective enzymatic digestion of elastin in decellularized auricular cartilage [108, 109]. Whether these decellularised allografts or xenografts of cartilage that possess the functional properties of native tissue and can be repopulated by autologous cells, will perform well in clinical practice—even for paediatric patients—needs to be investigated in future studies. The development of tailor-made scaffolds from natural materials such as collagen, collagen with glycosaminoglycans or hyaluronic acid-based materials might overcome some of the problems inherent to the use of allogenic or xenogenic tissue grafts. These materials can be combined with growth factors [110]. The disadvantage of these materials, often used as hydrogels, is their weak mechanical properties. This shortcoming can be addressed with new approaches such as the incorporation of second networks to form so-called interpenetrating polymer network (IPN) hydrogels [111]. An alternative approach to improve mechanical properties would be to use composites with firm but flexible materials such as for example described by Zhou and colleagues. They reported the use of permanent support in the form of a coiled wire embedded into a porous collagen scaffold in sheep to maintain the construct's size and ear-specific shape [112]. For all materials, it is required to tune the rate of formation of new tissue with degradation of the carrier material in order to prevent too much foreign body reaction. Another option to improve the mechanical properties is to culture the materials with cells for several weeks in order to let these cells produce a matrix that will provide more mechanical stability.

The cells required to generate cartilage matrix can be harvested from several sources. The nasal septum would be the first choice. Nasal septal chondrocytes have been demonstrated to grow well in culture and to be able to form a new cartilage matrix [114, 113]. They have recently been used in clinical trials for the repair of articular cartilage defects, without any safety issues [115]. In case the quantity of available cartilage from the septum itself is insufficient, the auricle may be a suitable source. A small biopsy will suffice to harvest cells and expand the number of cells in culture to obtain the number required to make a graft [116–118]. Chondrocytes from the auricle have proven to be very good cartilage matrix producers (Fig. 36.20c). The amount of cartilage matrix generated by these cells is higher than from cells derived from a nasal septum. Previously, it was shown, however, that the molecular profiles of cells from auricle and septum are different, even after the expansion of cells in the laboratory for several weeks [119]. It is to be expected that the quality of the matrix and thus the functional behaviour of the graft produced by cells from the nasal septum and auricle will differ as well, although this has never been examined thoroughly. Moreover, it should be investigated whether this has clinical consequences. For the moment auricular cartilage grafts are being used frequently for nose reconstructions, with variable and temporary success. In a growing nose, however, the long-term consequences have not yet been explored.

The harvest of cartilage for the isolation of cells will always cause donor-site morbidity; the defect will not heal properly due to the intrinsic limited repair capacity of cartilage. An alternative cell source may be found in stem cells. They can be isolated with relatively low donor-site morbidity from the perichondrium of the ear, bone marrow or from adipose tissue. Moreover, in contrast to cartilage cells, stem cells can be extensively expanded in number while retaining the potency to differentiate into cartilage. In spite of all the efforts, it appeared very challenging to produce stable cartilage using stem cells, as the generated cartilage is transient and will be remodelled into bone [121, 120]. Since stem cells are considered multipotential and capable of forming many different tissues, ongoing research efforts aim to unravel ways to stimulate stem cells to produce stable hyaline cartilage [122, 123]. High costs and regulations concerning advanced-therapy medicinal products [Regulation EC, no. 1,394/2007] make long cell culture procedures less applicable for use in patients. One option to combat this challenge is to combine stem cells with freshly isolated nasal septal chondrocytes. In a set-up with 20% nasal septum chondrocytes and 80% stem cells, it was shown feasible to engineer cartilage constructs with properties similar to those of constructs containing chondrocytes only [124]. Because this requires a lower amount of chondrocytes, expansion in culture is not needed. The cartilage matrix will be produced by the chondrocytes, whereas the stem cells secrete factors that stimulate the chondrocytes to produce matrix reducing the earlier-mentioned risk of bone formation by stem cells [124]. Moreover, the use of allogeneic stem cells in this set-up was demonstrated to be safe and successful for articular cartilage repair, allowing the production of an off-the-shelf product.

Next to a mismatch in mechanical properties between the implant and the native nasal septum, a mismatch in the shape of the implant or unprecise surgical implantation may be a cause of poor results in nasal reconstruction. Computational technology becomes available for cartilage application where it can play a role in surgical planning and reconstruction as well as 3D-printing of personalized implants as was reviewed by Shi and Huang [125]. 3D-printing has now entered the field of Otorhinolaryngology [126]. 3D-printing allows the fabrication of implants with precise internal and external 3D-architecture. This enables the generation of patient-matched shapes and sizes. 3D-printing was used to improve the fitting of silicone nasal septum prosthesis in patients based on CT images of the septal defect [127]. 3D printing is also employed for the reconstruction of septal defects with resorbable implants. 3D-printed resorbable grafts of polycaprolactone have been implanted in rabbits where retention of the initial shape and position of the implant, with a minimal inflammatory reaction, was demonstrated after 3 months [128]. In the following multi-centred clinical trial including 20 patients with caudal septal deviations that underwent septoplasty, an improvement in the Nasal Obstruction Symptom Evaluation Score as well as in nasal cavity cross-sectional areas and septal deviations on the CT scans was reported [129]. No complications were observed, patient satisfaction was good and surgeons indicated that the grafts were easy to handle and manipulate in the surgical procedure. The short follow-up of only 3 months is a limitation of this study and long-term results as well as comparison with conventional technology should demonstrate the value of these 3D-printed implants. 3D-printing was also applied to print a mould of a patient-specific implant that was then injected with a hydrogel derived from matrix of decellularized cartilage and human adipose tissue-derived mesenchymal stem cells, and subsequently implanted subcutaneously in the mouse to form tissue [130]. The possibility to print multiple materials can be used to better control mechanical and degradation behaviour. Moreover, 3D-bioprinting of cells in hydrogels offers the potential to precisely place cells in desired positions and patterns. The hydrogels used as bioinks have weak mechanical properties and are, therefore, often combined with other materials to provide sufficient mechanical properties. Scaffolds in the shape of a nose were bioprinted with a gelatin methacrylate (gelMA) bioink combined with cellulose [131], a gellan gum bioink and poly-(ethylene glycol) diacrylate that was UV-crosslinked [132] or a gellan gum/alginate bioink with BioCartilage®, a commercial cartilaginous ECM [133]. The development of biocompatible bioinks is one of the current challenges. Furthermore, the newest developments in 4D-printing technology are entering the field of medical applications and offer opportunities to design constructs that change their form or function in time in response to an external stimulus [134]. This emerging technology, which depends on biomaterials that are sensitive to environmental stimuli such as temperature, pH, light, electric or magnetic current or mechanical stress to change the form, might offer opportunities for the reconstruction of growing structures such as the paediatric nasal septum, in the future.

> A key issue in current research—in vivo and in vitro—is the promotion of repair of defects and fractures by tissue engineered cartilage.

36.3 Conclusions and Clinical Epilogue

In the previous paragraphs current data on the normal growth of the nose in children have been presented:

- The development of the facial profile from baby to adult.
- The evolution of the growth rate of the external nose in relation to age and sex.
- The development of the cartilaginous and bony nasal skeleton.
- The specific organisation of the cartilaginous nasal septum in thinner and thicker areas.
- Growth and maturation of involved tissues, enchondral and mesenchymal ossification.

In the latest decennia animal experiments have elucidated:

- The role of the septodorsal cartilage as the dominant growth centre in the midface with specific growth zones within the cartilaginous septum, and upper lateral cartilages, responsible for the growth of the nasal bones and premaxilla-maxilla.
- The nasal septum as the *pacemaker* for midfacial growth [4].
- The effects of various lesions of the septodorsal cartilage, nasal bones and facial clefts on the developing midfacial skeleton, as well as the feasibility of surgical interventions to prevent these late effects.
- The potential of the cambium (inner perichondrium) layer to form new cartilage, which is restricted by overgrowing outer perichondrium.
- Prevention of untidy overgrowth by the outer perichondrium during repair of defects or fractures in cartilage is needed for restoration of the conditions for normal development.

In the clinical domain, nasal surgery in children has been practised for several years.

Septal haematoma and septal abscess are recognized as acute indications for surgical treatment. Early drainage of a haematoma or abscess could save the septal cartilage from necrosis. In addition to drainage of the abscess and antibiotic treatment the implantation of preserved bone has been advocated to prevent saddling of the nasal dorsum. There is no evidence that preserved bone will restore normal growth. Occasionally a central submucosal defect is found leaving intact the thicker areas—growth and support zones—in the septal cartilage. The sequelae of nasal trauma in children are currently not sufficiently recognized and the late effects of fractures on further growth are often not taken into full consideration. It seems that most children are not referred to an ORL specialist in case of acute trauma.

When fractures and dislocation are diagnosed in children, repositioning and realignment of dislocated fractured parts of the bony and cartilaginous nasal pyramid seems indicated at short notice. Uncertainties about further nasal growth should be discussed with the patient and/or parents. This equally applies to elective nasal surgery in case of progressive and severe functional or aesthetic problems.

Various studies of paediatric patients have reported successful results after rhinosurgical procedures at a young age [9, 11, 135–137]. The number of patients, however, was usually limited and the follow-up was too short. To date, there is no consensus among clinicians with regard to the optimal age for surgical management of pathology of the nasal skeleton. Reconstruction with autologous cartilage (septal, costal or auricular) grafts in an open approach has been advocated as preferential for older children, although growth disturbances remain a risk.

Significant progress has been made in the preclinical domain—from human anatomy to experimental surgery in animal models—in recognizing critical factors of cartilage wound healing. New developments might reveal methods that combine a patient's own (stem) cells with scaffolds, that can be 3D printed to provide a personalized shape, to reconstruct the growing nose in the future. For clinicians the next step should be to acquire more clinical evidence, differentiating for the age of the child and including an adequate follow-up, to evaluate the later nasal development in comparison with the outcome of the animal model studies [76, 6].

References

1. Wong KK, Filatov S, Kibblewhite DJ. Septoplasty retards midfacial growth in a rabbit model. Laryngoscope. 2010;120:450–3.

2. Meng HP, Goorhuis J, Kapila S, Nanda RS. Growth changes in the nasal profile from 7 to 18 years of age. Am J Orthod Dentofac Orthop. 1988;94:317–26.
3. Pirsig W. The influence of trauma on the growing nose. In: Mladina R, Passali D, editors. Current concepts in pediatric rhinology. Senese: Siena; 2000.
4. Hall BK, Precious DS. Cleft lip, nose and palate: the nasal septum as the pacemaker for midfacial growth. Oral Surg Oral Med Oral Pathol Oral Radiol. 2013;115:442–7.
5. Fry HJ. Nasal skeletal trauma and the interlocked stresses of the nasal septal cartilage. Br J Plast Surg. 1967;20:146–58.
6. Verwoerd CDA, Kubba H, Verwoerd-Verhoef HL. Pediatric rhinology: developmental aspects and surgery. In: Georgalas C, Fokkens W, editors. Rhinology and Skull Base surgery. Stuttgart/New York: Thieme Verlag; 2013.
7. Triglia JM, Cannoni M, Pech A. Septorhinoplasty in children: benefits of the external approach. J Otolaryngol. 1990;19:274–8.
8. Pentz S, Pirsig W, Lenders H. Long-term results of neonates with nasal deviation: a prospective study over 12 years. Int J Pediatr Otorhinolaryngol. 1994;28:183–91.
9. Béjar I, Farkas LG, Messner AH, Crysdale WS. Nasal growth after external septoplasty in children. Arch Otolaryngol Head Neck Surg. 1996;122:816–21.
10. El Hakim H, Crysdale WS, Abdollel M, Farkas LG. A study of anthropometric measures before and after external septoplasty in children. Arch Otolaryngol Head Neck Surg. 2001;127:1362–6.
11. Dispenza C, Saraniti C, Dispenza F, Caramanna C, Salzano FA. Management of nasal septal abscess in childhood: our experience. Int J Pediatr Otorhinolaryngol. 2004;68:1417–21.
12. Adil E, Goyal N, Fedok FG. Corrective nasal surgery in the younger patient. JAMA Facial Plast Surg. 2014;16:176–82.
13. Bae JS, Kim E-S, Jang YJ. Treatment outcome of pediatric rhinoplasty. Int J Pediatr Otorhinolaryngol. 2013;77:1701–10.
14. Costa DB, Anselmo-Lima WT, Tamashiro E, Enoki C, Valera FC. The impact of Metzenbaum septoplasty on nasal and facial growth in children. Braz J Otorhinolaryngol. 2013;79:454–9.
15. Tasca I, Compadretti GC. Nasal growth after pediatric septoplasty at long-term. Am J Rhinol Allergy. 2011;25:e7–12.
16. Johnson MD. Management of pediatric nasal surgery. Facial Plast Surg Clin N Am. 2017;25:211–21.
17. Calvo-Henriquez C, Carlos-Neves J, Arancibia-Tagle D, Chiesa-Estomba C, Lechlen JR, Mayo-Yánez M, Martinez-Capoccioni G, Martin-Martin C. Does pediatric septoplasty compromise midfacial growth? A systematic review. Eur Arch Otorhinolaryngol. 2020;277:1565–74.
18. Cingi C, Muluk NB, Ulusoy S, Lopatin A, Sahin E, Passali D, Bellussi L, Atilla H, Hanci D, Altintoprak N, Rusetski Y, Sarafoleanu C, Mladina R, Kalogjera L, Manea C. Septoplasty in children. Am J Rhinol Allergy. 2016;30:e42–7.
19. Lawrence R. Pediatric septoplasty: a review of the literature. Int Pediatr Otorhinolaryngol. 2012;76:1078–82.
20. Fry HJ. The aetiology of so-called "septal deviations" and their experimental production in the growing rabbit. 1968;21:419–22.
21. Sarnat BG. Postnatal growth of the nose and face after resection of the septal cartilage in the rabbit. Oral Surg 1968;26:712–27.
22. Sarnat BG. Normal and abnormal growth; some experimental and clinical considerations. Angle Orthod. 1983;53:263–89.
23. Sarnat BG. Some factors related to experimental snout growth. J Craniofac Surg. 2008;19:1308–14.
24. Hellquist R. Facial skeleton growth after periosteal resection. An osteometric, roentgenographic and histologic study in the rabbit and guinea pig. Scand J Plast Reconstr Surg. 1972;10(suppl 10):28–33.
25. Stenström SJ, Thilander BL. Effects of nasal septal cartilage resection on young guinea pigs. Plast Reconstr Surg. 1970;45:160–70.
26. Copray JC. Growth of the nasal septal cartilage of the rat in vitro. J Anat. 1986;144:99–111.
27. Cupero TM, Middleton CE, Silva AB. Effects of functional septoplasty the facial growth of ferrets. Arch Otolaryngol Head Neck Surg. 2001;127:1367–9.
28. Haye R, Freng A. Experimental septoplasty in the growing cat. Acta Otolaryngol. 1986;102:113–7.
29. Bernstein L. Early submucous resection of nasal septal cartilage. Arch Otolaryngol. 1973;97:273–8.
30. Chierici G, Harvold EP, Dawson WJ. Primate experiments on facial symmetry. J Dent Res. 1970;49:847–51.
31. Siegel MI. A longitudinal study of facial growth in Papio cynocephalus after resection of the cartilaginous nasal septum. J Med Primatol. 1979;8:122–7.
32. Verwoerd CDA, Verwoerd-Verhoef HL. Rhinochirurgie bei Kindern: Entwicklungsphysiologische und chirurgische Aspekte der wachsenden Nase (Rhinosurgery in children; developmental and surgical aspects of the growing nose). Laryngorhinootologie. 2010;89(suppl):46–71.
33. Poublon RML, Verwoerd CDA, Verwoerd-Verhoef HL. Anatomy of the upper lateral cartilages in the human newborn. Rhinology. 1990;28:41–5.
34. Van Loosen J, Verwoerd-Verhoef HL, Verwoerd CDA. The nasal septal cartilage in the newborn. Rhinology. 1988;26:161–5.
35. Tonneyck-Müller I, van der Werf F. Development of the septum nasi in rabbits. Acta Morphol Neerl Scand. 1982;20:379–91.
36. Tonneyck-Müller I, van der Werf F. Die Entwicklung des Septum nasi beim Kaninchen; III Quantitative Beobachtungen an Zellen und Interzellularsubstanz. Acta Morphol Neerl Scand. 1984;22:133–41.
37. Van Loosen J, van Zanten GA, Howard CV, Verwoerd-Verhoef HL, van Velzen D, Verwoerd

CDA. Growth characteristics of the human nasal septum. Rhinology. 1996;34:78–82.
38. Verwoerd CDA, Verwoerd-Verhoef HL. Rhinosurgery in children: basic concepts. Facial Plast Surg. 2007;23:219–30.
39. Schultz-Coulon HJ, Eckermeier L. Postnatal growth of the nasal septum. Acta Otolaryngol. 1976;82:131–42.
40. Verwoerd CDA, van Loosen J, Schütte HE, Verwoerd-Verhoef HL, van Velzen D. Surgical aspects of the anatomy of the vomer in children and adults. Rhinol Suppl. 1989a;9:87–96.
41. Takahashi R. The formation of the nasal septum and the etiology of the septal deformity. Acta Otolaryngol Suppl. 1987;443:1–160.
42. Wealthall RJ, Herring SW. Endochondral ossification of the mouse nasal septum. Anat Rec A Discov Mol Cell Evol Biol. 2006;288(11):1163–72.
43. Van Loosen J, Baatenburg de Jong RJ, van Zanten GA, Engel T, Lanjewar DN, van Velzen D. A cephalometric analysis of nasal septal growth. Clin Otolaryngol. 1997;22:453–8.
44. Akgüner M, Baratçu A, Karaca C. Adolescent growth patterns of the bony cartilaginous framework of the nose: a cephalometric study. Ann Plast Surg. 1998;41:66–9.
45. Farkas LG, Posnick JC, Hreczko TM. Growth patterns of the face. Cleft Palate Craniofac J. 1992;29:308–15.
46. Ochoa BK, Nanda RS. Comparison of maxillary and mandibular growth. Am J Orthod Dentofac Orthop. 2004;125:148–59.
47. Zankl A, Eberle L, Molinari L, Schinzel A. Growth charts for nose length, nasal protrusion and philtrum length from birth to 97 years. Am J Med Genet. 2002;111:388–91.
48. Van der Heijden P, Korsten-Meijer AG, van der Laan BF, Wit HP, Goorhuis-Brouwer SM. Nasal growth and maturation age in adolescents. Arch Otolaryngol Head Neck Surg. 2008;134:1288–93.
49. Goergen MJ, Holton NE, Grünheid T. Morphological interaction between the nasal septum and nasofacial skeleton during human ontogeny. J Anat. 2017;230:689–700.
50. Stricker M, Raphael B, van der Meulen J, Mazzola R. Craniofacial development and growth. In: Stricker M, editor. Craniofacial malformations. Edinburgh: Churchill Livingstone; 1990.
51. Atherton JD. Morphology of the facial bones in skulls with unoperated unilateral cleft palate. Cleft Palate J. 1967;4:18–30.
52. Van Limborgh J. Some aspects of the cleft-affected face. In: Hotz R, editor. Early treatment of cleft lip and palate; international symposium 1964, University of Zurich Dental Institute. Berne: Huber; 1964.
53. Verwoerd CDA, Verwoerd-Verhoef HL, Meeuwis CA. Stress and wound healing of the cartilaginous nasal septum. Acta Otolaryngol. 1989b;107:441–5.
54. Innis CO. Some preliminary observations on unrepaired hare-lips and cleft palates in adult members of the Dusan tribes of North Borneo. Br J Plast Surg. 1962;15:173–81.
55. Capelozza LC, Taniguchi SM, daSilva OG. Craniofacial morphology of adult unoperated complete unilateral cleft lip and palate patients. Cleft Palate Craniofac J. 1993;30:376–81.
56. Verwoerd CDA, Urbanus NA, Verwoerd-Verhoef HL. Growth mechanisms in skulls with facial clefts. Acta Otolaryngol. 1979b;87:335–9.
57. Verwoerd CDA, Mladina R, Nolst Trenité GJ, Pigott RW. The nose in children with unilateral cleft lip and palate. Int J Pediatr Otorhinolaryngol. 1995;32(suppl):45–52.
58. Howe AM, Hawkins JK, Webster WS. The growth of the nasal septum in the 6-9 week period of foetal development - warfarin embryopathy offers a new insight into prenatal facial development. Aust Dent J. 2004;49:171–6.
59. Gray L. Deviated nasal septum; incidence and etiology. Ann Otol Rhinol Laryngol Suppl. 1978;87(3 Pt 3 Suppl 50):3–20.
60. Takahashi R. The evolution of the nasal septum and the formation of septal deformity. Rhinol Suppl. 1988;6:1–23.
61. Mladina R. Classification of septal deformities. In: Mladina R, Passali D, editors. Current concepts in pediatric rhinology. Senese: Siena; 2000.
62. Harrison D. Nasal injuries: their pathogenesis and treatment. Br J Plast Surg. 1979;32:57–64.
63. Van Velzen D, van Loosen J, Verwoerd CD, Verwoerd-Verhoef HL. Persistent pattern of variations of the nasal septum: implications for stress and trauma as illustrated by a complex fracture in a 4-year-old boy. Adv Otorhinolaryngol. 1997;51:46–50.
64. Manning SC. A 3-year-old child with a severely deviated septum and airway obstruction. Arch Otolaryngol Head Neck Surg. 1999;125:699–700.
65. Grymer LF, Pallisgaard C, Melsen B. The nasal septum in relation to the development of the nasomaxillary complex: a study in identical twins. Laryngoscope. 1991;101:863–8.
66. Kremenak CR, Searls JC. Experimental manipulation of midfacial growth: a synthesis of five years of research at the Iowa maxillofacial growth laboratory. J Dent Res. 1971;50:1499–501.
67. Pinkston DR, Schubkegel AJ, Zimmelman MB, Smith RJ. The effects of surgery on mid facial growth in the rabbit. Am J Rhinol. 1995;9:115–4.
68. Verwoerd CDA, Urbanus NA, Mastenbroek GJ. The influence of partial resections of the nasal septal cartilage on the growth of upper jaw and nose: and experimental study in rabbits. Clin Otolaryngol. 1980;5:291–302.
69. Verwoerd CDA, Verwoerd-Verhoef HL. Rhinosurgery in children: developmental and surgical aspects of the growing nose. In: Rettinger G, editor. Current topics in otorhinolaryngology head and neck surgery Vol. IX, rhinologic functions – functional rhi-

nosurgery. Mönchengladbach: Rheinware Verlag; 2011.
70. Verwoerd-Verhoef HL, Verwoerd CDA. Surgery of the lateral nasal wall and ethmoid: effects on sinonasal growth. An experimental study in rabbits. Int J Pediatr Otorhinolaryngol. 2003;67:263–9.
71. Grymer LF, Bosch C. The nasal septum and the development of the midface. A longitudinal study of a pair of monozygotic twins. Rhinology. 1997;35:6–10.
72. Derkay CS. A conservative role for septoplasty in young children. Arch Otolaryngol Head Neck Surg. 1999;125:702–3.
73. Blahova O. Late results of nasal septum injury in children. Int J Pediatr Otorhinolaryngol. 1985;10:137–41.
74. Alvarez H, Osorio J, De Diego JI, Prim MP, De la Torre C, Gavilan J. Sequelae after nasal septum injuries in children. Auris Nasus Larynx. 2000;27:339–42.
75. Canty PA, Berkowitz RG. Hematoma and abscess of the nasal septum in children. Arch Otolaryngol Head Neck Surg. 1996;122:1373–6.
76. Alshaikh N, Lo S. Nasal septal abscess in children: from diagnosis to management and prevention. Int J Pediatr Otorhinolaryngol. 2011;75:737–44.
77. Meeuwis J, Verwoerd-Verhoef HL, Verwoerd CDA. Normal and abnormal growth after partial resection of the cartilaginous septum. Acta Otolaryngol. 1993;113:379–82.
78. Nolst Trenité GJ, Verwoerd CDA, Verwoerd-Verhoef HL. Reimplantation of autologous septal cartilage in the growing nasal septum part I. The influence of resection and reimplantation of septal cartilage upon nasal growth: an experimental study in rabbits. Rhinology. 1987;25:225–36.
79. Nolst Trenité GJ, Verwoerd CDA, Verwoerd-Verhoef HL. Reimplantation of autologous septal cartilage in the growing nasal septum part II. The influence of reimplantation of rotated or crushed autologous septal cartilage on nasal growth: an experimental study in rabbits. Rhinology. 1988;26:25–32.
80. Verwoerd CDA, Verwoerd-Verhoef HL. Rhinosurgery in children. In: Nolst Trenité GJ, editor. Rhinoplasty. The Hague: Kugler Publications; 2005.
81. Verwoerd CDA, Urbanus NA, Nijdam DC. The effects of septal surgery on the growth of nose and maxilla. Rhinology. 1979a;17:53–63.
82. Verwoerd CDA, Verwoerd-Verhoef HL, Meeuwis CA, van der Heul RO. Wound healing of the nasal septal perichondrium in young rabbits. ORL J Otorhinolaryngol Relat Spec. 1990;52:180–6.
83. Gibson T, Davis WB. The distortion of autogenous cartilage graft; its cause and prevention. Br J Plast Surg. 1958;10:257–74.
84. Ten Koppel PG, van der Veen JM, Hein D, van Keulen F, van Osch GJ, Verwoerd-Verhoef HL, Verwoerd CDA. Controlling incision-induced distortion of nasal septal cartilage: a model to predict the effect of scoring of rabbit septa. Plast Reconstr Surg. 2003;111:1948–57.
85. Verwoerd CDA, Verwoerd-Verhoef HL, Meeuwis CA, van der Heul RO. Wound healing of autologous implants in the nasal septal cartilage. ORL J Otorhinolaryngol Relat Spec. 1991;53:310–4.
86. Duynstee ML, Verwoerd-Verhoef HL, Verwoerd CDA, van Osch GJ. The dual role of perichondrium in cartilage wound healing. Plast Reconstr Surg. 2002;110:1073–9.
87. Kaiser ML, Karam AM, Sepehr A, Wong H, Liaw LL, Vokes DE, Wong BJ. Cartilage regeneration in the rabbit nasal septum. Laryngoscope. 2006;116:1730–4.
88. Verwoerd-Verhoef HL, ten Koppel PG, van Osch GJ, Meeuwis CA, Verwoerd CDA. Wound healing of cartilage structures in the head and neck region. Int J Pediatr Otorhinolaryngol. 1998b;43:241–51.
89. Rhys Evans PH, Brain DJ. The influence of nasal osteotomies and septum surgery on the growth of the rabbit snout. J Laryngol Otol. 1981;95:1109–19.
90. Verwoerd CDA, Verwoerd-Verhoef HL. Nasal malformations; developmental and surgical aspects. In: Cauwenberge V, et al., editors. The nose. The Hague: Kugler Publications; 1998.
91. Engkvist O, Skoog T, Pastacaldi P, et al. The cartilaginous potential of the perichondrium in rabbit ear and rib: a comparative study in vivo and vitro. Scand J Plast Reconstr Surg. 1979;13:275–7.
92. Peer LA. Fate of autogenous septal cartilage after transplantation in human tissues. Arch Otolaryngol. 1941;34:696–709.
93. Bardach J, Kelly KM. Role of animal models in experimental studies of craniofacial growth following cleft lip and palate repair. Cleft Palate J. 1988;25:103–13.
94. Bos PK, van Osch GJ, Frenz DA, Verhaar JA, Verwoerd-Verhoef HL. Growth factor expression in cartilage wound healing: temporal and spatial immunolocalization in a rabbit auricular cartilage wound model. Osteoarthr Cartil. 2001;9:382–9.
95. Tew SR, Kwan AP, Hann A, Thomson BM, Archer CW. The reactions of articular cartilage to experimental wounding: role of apoptosis. Arthritis Rheum. 2000;43:215–25.
96. Verwoerd-Verhoef HL, Bean JK, van Osch GJ, ten Koppel PG, Meeuwis JA, Verwoerd CDA. Induction in vivo of cartilage grafts for craniofacial reconstruction. Am J Rhinol. 1998a;12:27–31.
97. Boenisch M, Nolst Trenité GJ. Reconstruction of the nasal septum using polydioxanone plate. Facial Plast Surg. 2010;12:4–10.
98. Bos PK, DeGroot J, Budde M, Verhaar JA, van Osch GJ. Specific enzymatic treatment of bovine and human articular cartilage: implications for integrative cartilage repair. Arthritis Rheum. 2002;46:976–85.
99. Janssen LM, In der Maur CD, Bos PK, Hardillo JA, van Osch GJ. Short-duration enzymatic treatment promotes integration of a cartilage graft in a defect. Ann Otol Rhinol Laryngol. 2006;115:461–8.

100. Van de Breevaart Bravenboer J, In der Maur CD, Bos PK, Feenstra L, Verhaar JA, Weinans H, van Osch GJ. Improved cartilage integration and interfacial strength after enzymatic treatment in a cartilage transplantation model. Arthritis Res Ther. 2004;6:469–76.
101. Pabbruwe MB, Kafienah W, Tarlton JF, Mistry S, Fox DJ, Hollander AP. Repair of meniscal cartilage white zone tears using a stem cell/collagen-scaffold implant. Biomaterials. 2010;31:2583–91.
102. Verwoerd-Verhoef HL, Meeuwis CA, van der Heul RO, Verwoerd CDA. Histologic evaluation of crushed cartilage grafts in the growing nasal septum of young rabbits. ORL J Otorhinolaryngol Relat Spec. 1991;53:305–9.
103. Bean JK, Verwoerd-Verhoef HL, Meeuwis JA, Verwoerd CDA. Reconstruction of the growing cricoid with a composite graft of demineralised bovine bone matrix and autogenous perichondrium. Int J Pediatr Otorhinolaryngol. 1993;25:163–72.
104. Reddi AH, Gay R, Gay S, Miller EJ. Transitions in collagen types during matrix-induced cartilage, bone, and bone marrow formation. Proc Natl Acad Sci U S A. 1977;74:5589–92.
105. Ten Koppel PGJ, van Osch GJ, Verwoerd CDA, Verwoerd-Verhoef HL. Efficacy of perichondrium and the trabecular demineralized bone matrix for generating cartilage. Plast Reconstr Surg. 1998;102:2012–20.
106. Pirsig W, Bean JK, Lenders H, Verwoerd CDA, Verwoerd-Verhoef HL. Cartilage transformation in a composite graft of demineralized bovine bone matrix and ear perichondrium used in a child for the reconstruction of the nasal septum. Int J Pediatr Otorhinolaryngol. 1995;32:171–81.
107. von Bomhard A, Elsaesser A, Riepl R, Pippich K, Faust J, Schwarz S, Koerber L, Breiter R, Rotter N. Cartilage regeneration using decellularized cartilage matrix: long-term comparison of subcutaneous and intranasal placement in a rabbit model. J Craniomaxillofac Surg. 2019;47:682–94.
108. Lehmann J, Nürnberger S, Narcisi R, Stok KS, van der Eerden BCJ, Koevoet WJLM, Kops N, Ten Berge D, van Osch GJ. Recellularization of auricular cartilage via elastase-generated channels. Biofabrication. 2019;11(3):035012.
109. Nürnberger S, Schneider C, van Osch GVM, Keibl C, Rieder B, Monforte X, Teuschl AH, Mühleder S, Holnthoner W, Schädl B, Gahleitner C, Redl H, Wolbank S. Repopulation of an auricular cartilage scaffold, AuriScaff, perforated with an enzyme combination. Acta Biomater. 2019;86:207–22.
110. Rosso F, Marino G, Giordano A, Barbarisi M, Parmeggiani D, Barbarisi A. Smart materials as scaffolds for tissue engineering. J Cell Physiol. 2005;203:465–70.
111. Dhand AP, Galarraga JH, Burdick JA. Enhancing biopolymer hydrogel functionality through interpenetrating networks. Trends Biotechnol. 2020;39(5):519–38.
112. Zhou L, Pomerantseva I, Bassett EK, Bowley CM, Zhao X, Bichara DA, et al. Engineering ear constructs with a composite scaffold to maintain dimensions. Tissue Eng Part A. 2011;17:1573–81.
113. Farhadi J, Fulco I, Miot S, Wirz D, Haug M, Dickinson SC, Hollander AP, Daniels AU, Pierer G, Heberer M, Martin I. Precultivation of engineered human nasal cartilage enhances the mechanical properties relevant for use in facial reconstructive surgery. Ann Surg. 2006;244:978–85.
114. Van Osch GJ, Marijnissen WJ, van der Veen SW, Verwoerd-Verhoef HL. The potency of culture-expanded nasal septum chondrocytes for tissue engineering of cartilage. Am J Rhinol. 2001a;15:187–92.
115. Mumme M, Barbero A, Miot S, Wixmerten A, Feliciano S, Wolf F, Asnaghi AM, Baumhoer D, Bieri O, Kretzschmar M, Pagenstert G, Haug M, Schaefer DJ, Martin I, Jakob M. Nasal chondrocyte-based engineered autologous cartilage tissue for repair of articular cartilage defects: an observational first-in-human trial. Lancet. 2016;388(10055):1985–94.
116. Mandl EW, van der Veen SW, Verhaar JA, van Osch GJ. Multiplication of human chondrocytes with low seeding densities accelerates cell yield without losing redifferentiation capacity. Tissue Eng. 2004;10:109–18.
117. Van Osch GJ, van der Veen SW, Verwoerd-Verhoef HL. In vitro redifferentiation of culture-expanded rabbit and human auricular chondrocytes for cartilage reconstruction. Plast Reconstr Surg. 2001b;107:433–40.
118. Van Osch GJ, Mandl EW, Jahr H, Koevoet W, Nolst-Trenite G, Verhaar JA. Considerations on the use of ear chondrocytes as donor chondrocytes for cartilage tissue engineering. Biorheology. 2004;41:411–21.
119. Hellingman CA, Koevoet W, van Osch GJVM. Can one generate stable hyaline cartilage from adult mesenchymal stem cells? - a developmental approach. J Tissue Eng Regen Med. 2012;6:e1–11. https://doi.org/10.1002/term.502.
120. Farrell E, Both S, Odorfer KI, Koevoet W, Kops N, O'Brien FJ, et al. In-vivo generation of bone via endochondral ossification by in-vitro chondrogenic priming of adult human and rat mesenchymal stem cells. BMC Musculoskelet Disord. 2011;12:31.
121. Scotti C, Tonnarelli B, Papadimitropoulos A, Scherberich A, Schaeren S, Schauerte A, et al. Recapitulation of endochondral bone formation using human adult mesenchymal stem cells as a paradigm for developmental engineering. Proc Natl Acad Sci U S A. 2010;107:7251–6.
122. Hellingman CA, Verwiel ETP, Slagt I, Koevoet W, Poublon RML, Nolst-Trenité GJ, van Osch GJ, et al. Differences in cartilage forming capacity of expanded human chondrocytes from ear and nose and their gene expression profiles. Cell Transplant. 2011;20:925–40.
123. Narcisi R, Cleary MA, Brama PA, Hoogduijn MJ, Tüysüz N, ten Berge D, van Osch GJ. Long-term

expansion, enhanced chondrogenic potential, and suppression of endochondral ossification of adult human MSCs via WNT signaling modulation. Stem Cell Reports. 2015;4(3):459.

124. Pleumeekers MM, Nimeskern L, Koevoet WL, Karperien M, Stok KS, van Osch GJ. Cartilage regeneration in the head and neck area: combination of ear or nasal chondrocytes and mesenchymal stem cells improves cartilage production. Plast Reconstr Surg. 2015;136(6):762e–74e.
125. Shi B, Huang H. Computational technology for nasal cartilage-related clinical research and application. Int J Oral Sci. 2020;12(1):21.
126. Di Gesù R, Acharya AP, Jacobs I, Gottardi R. 3D printing for tissue engineering in otolaryngology. Connect Tissue Res. 2020;61(2):117–36.
127. Onerci Altunay Z, Bly JA, Edwards PK, Holmes DR, Hamilton GS, O'Brien EK, Carr AB, Camp JJ, Stokken JK, Pallanch JF. Three-dimensional printing of large nasal septal perforations for optimal prosthetic closure. Am J Rhinol Allergy. 2016;30(4):287–93.
128. Park SH, Yun BG, Won JY, Yun WS, Shim JH, Lim MH, Kim DH, Baek SA, Alahmari YD, Jeun JH, Hwang SH, Kim SW. New application of three-dimensional printing biomaterial in nasal reconstruction. Laryngoscope. 2017;127(5):1036–43.
129. Kim DH, Yun WS, Shim JH, Park KH, Choi D, Park MI, Hwang SH, Kim SW. Clinical application of 3-dimensional printing technology for patients with nasal septal deformities: a multicenter study. JAMA Otolaryngol Head Neck Surg. 2018;1144:1145–52.
130. Yi H-G, Choi Y-J, Jung JW, Jang J, Song T-SH, Chae S, Ahn M, Choi TH, Rhie J-W, Cho D-W. Three-dimensional printing of a patient-specific engineered nasal cartilage for augmentative rhinoplasty. J Tissue Eng. 2019;10:2041731418824797.
131. Shin S, Park S, Park M, Jeong E, Na K, Youn HJ, Hyun J. Cellulose nanofibers for the enhancement of printability of low viscosity gelatin derivatives. Bioresources. 2017;12(2):2941–54.
132. Wu D, Yu Y, Tan J, Huang L, Luo B, Lu L, Zhou C. 3D-bioprinting of gellan gum and poly (ethylene glycol) diacrylate based hydrogels to produce human-scale constructs with high-fidelity. Mater Des. 2018;160:486–95.
133. Kesti M, Eberhardt C, Pagliccia G, Kenkel D, Grande D, Boss A, Zenobi-Wong M. Bioprinting complex cartilaginous structures with clinically compliant biomaterials. Adv Funct Mater. 2015;25(48):7406–17.
134. Lui YS, Sow WT, Tan LP, Wu Y, Lai Y, Li H. 4D-printing and stimuli-responsive materials in biomedical aspects. Acta Biomater. 2019;92:19–36.
135. Christophel JJ, Gross CW. Pediatric septoplasty. Otolaryngol Clin N Am. 2009;42:287–94.
136. DeRosa J, Smit JR. Septal abscess in a 14-month-old child: diagnosis, management and discussion of reconstructive options. Int J Ped Otorhinolaryngol Extra. 2009;73:496.
137. Menger DJ, Tabink IC, Nolst Trenité GJ. Nasal septal abscess in children. Reconstruction with autologous cartilage grafts on polydioxanone plate. Arch Otolaryngol Head Neck Surg. 2008;134:842–7.

37 Physiologic Concerns During Rhinoplasty

E. B. Kern

Core Messages

- The prime physiologic concern during cosmetic rhinoplasty centers on the breathing function, correcting a present breathing dysfunction or preventing a postoperative breathing disturbance while improving the patient's appearance.
- The nasal valve is the most important part of the nose (internal and external) causing breathing disturbances.
- Managing the entire nasal valve area is the main issue in protecting the patient's breathing function during cosmetic rhinoplasty.

37.1 Introduction

Truth is as you see it, and as I see it, after evaluating thousands of rhinologic patients, the most common symptom that I observed after failed rhinoplasty was difficulty breathing. This statement is based on my almost 50 years of working in otorhinolaryngology (primarily rhinology) and the findings of almost 9000 breathing test patients (anterior mask rhinomanometry) and numerous papers presented to the profession over many years from data collected on patients at the Mayo Clinic in Rochester, Minnesota, USA, between the years 1972 and 2003 [1–12]. After prolonged study and writing regarding nasal physiology, I choose to summarize (simplify) and emphasize four primary functions of the nose [13–21]. These primary functions are listed as follows:

1. Olfaction
2. Defense (sneeze, mucociliary transport, defensive proteins, and the immune system)
3. Respiration (breathing, providing optimal nasal resistance and charging the inspired air with warmth and moisture, proper humidity, so oxygen and carbon dioxide exchange occurs optimally at the alveolar level)
4. Cosmesis (appearance) Because a patient's symptoms can be so disturbing, my prime physiologic concern during cosmetic rhinoplasty centers on breathing function. Of course, during rhinoplasty, cosmesis (appearance) is the usual first concern of the patient, yet for the surgeon, the two essential concerns must be improving the appearance and correcting a present breathing dysfunction or preventing a postoperative breathing disturbance. The dictum of "do not sacrifice breathing function for appearance" is the mantra quietly singing in the back of the mind in my role as a "functional" nasal surgeon.

E. B. Kern (✉)
Department of Otorhinolaryngology, State University of New York, Buffalo, NY, USA

Rhinology and Facial Plastic Surgery, Mayo Clinic Medical School, Rochester, MN, USA
e-mail: ekern@mayo.edu, kern.eugene@mayo.edu

Ö. Ö. Celebi, T. M. Önerci (eds.), *Nasal Physiology and Pathophysiology of Nasal Disorders*,
https://doi.org/10.1007/978-3-031-12386-3_37

How many of us have experienced that sad and disquieting moment in our practice when the patient, even years after primary rhinoplastic surgery, complains bitterly about disturbed sleep [22, 23], mouth breathing, shortness of breath, or other symptoms of post-rhinoplasty nasal airway obstruction [24, 25]? In other words, it is the physiologic respiratory (breathing) concerns during 10% (approximate numbers) in experienced hands, after viewing websites and contacting plastic, facial plastic, and cosmetic societies by phone, I estimate that the number of rhinoplasty operations performed in the USA is about 500,000 operations per year.

Rhinoplasty that I will address in this chapter with emphasis on those anatomic component parts of the nose that require meticulous attention during rhinoplasty to avoid producing a postoperative breathing disturbance. It is essential to preserve or improve nasal breathing and not impair any function while performing the rhinoplasty.

By definition, in my view, a rhinologist is a physician who is interested in patients who have disturbances in the nose and paranasal sinuses. Surgery is but one treatment option for managing the varied symptoms seen in our rhinologic patients. I will review some of the important concepts concerning the nasal valve (internal and external nasal valve) since it is the nasal valve that is the most critical area for nasal breathing and by extension, it is the nasal valve area that is the most important site of my physiologic concern while performing a primary cosmetic rhinoplasty.

37.2 Anatomy of the Nasal Valve Area

Since my primary physiologic concern during cosmetic rhinoplasty centers on the breathing function and the nasal valve is the most critical area for breathing, a review of both the anatomy and physiology of the nasal valve area is imperative. Regarding the anatomy, an arbitrary subdivision of the nasal valve for practical clinical thinking has been forwarded in the literature, suggesting that we think of the nasal valve area as having two component parts. First, the internal nasal valve (most of our discussion will center around the internal nasal valve as it is composed of a number of diverse anatomic parts) and an external nasal valve, which is primarily composed of the lower lateral cartilages and surrounding soft tissues, which are covered both dorsally and ventrally by the skin. Both valves (internal and external) are still subjected to the laws of fluid (air is the fluid) dynamics [13–15, 26–38].

37.2.1 Internal Nasal Valve

First, let's look at the anatomy of the external nose. The component parts of the external nose include a nasal skeletal structure (bone and cartilage) conveniently divided into three distinct and separate anatomic parts including:

1. Upper bony portion (paired nasal bones).
2. Middle third is composed of the single upper lateral extension of the septal cartilage (the upper lateral cartilage is one cartilage and should not be thought of as two separate cartilages); this upper lateral cartilage is also termed the roof or triangular cartilage (Fig. 37.1). At its distal end (caudal end), the

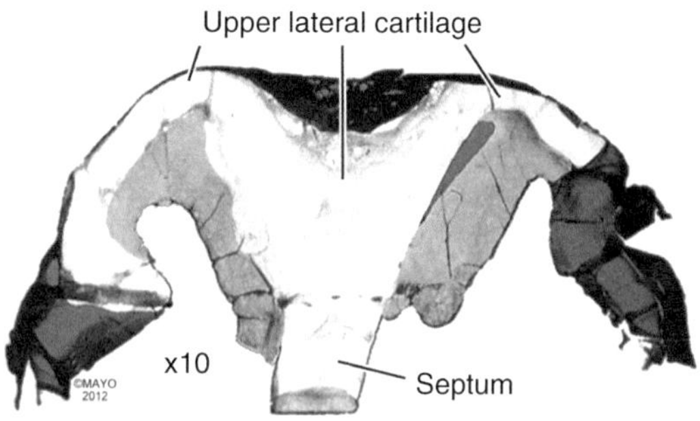

Fig. 37.1 Photograph showing that the upper lateral cartilage (roof cartilage) is an upper lateral extension of the septal cartilage in humans. The photograph was taken after a rhinectomy that the author performed on a patient with invasive nasal cancer (By permission of Mayo Foundation for Medical Education and Research. All Rights Reserved)

upper lateral cartilage is usually separated by a narrow cleft from the nasal septum and projects beneath the paired lower lateral cartilages.
3. Lower third of the external nose is composed of paired lower lateral cartilages (also called lobular, rim, alar, or great alar cartilages).

For emphasis, remember that this single upper lateral cartilage is attached proximally beneath the paired nasal bones and is attached distally beneath the lower lateral cartilages [26, 32, 33]. At its distal (caudal) end, the upper lateral cartilage frequently curls back upon itself forming a portion that is called the "scroll" also termed "returning" which is in contact with the proximal (cephalic) end of the lower lateral cartilages [26, 32, 33, 39]. The upper lateral cartilage also extends laterally out to the bony margin of the piriform aperture formed, in part, by the frontal (ascending) process of the maxilla. This external nasal skeletal structure (bone and cartilage) is actually covered *externally* by the soft tissues of the skin, muscles, perichondrium, periosteum, and neurovascular bundles. *Internally* (intranasal) these structures have mucocutaneous (skin and mucosa) coverings.

The entire internal nasal valve is more than the relationship between the distal end of the upper lateral cartilage and its angle formed with the nasal septum. Because of the complexity of the anatomy and the various papers on the subject, I think, for practical and functional reasons, the nasal valve should be thought of as an area. I think of the *nasal valve area* as a three-dimensional inverted cone-shaped structure that extends from the region of the distal (caudal) end of the upper lateral cartilage in its relationship to the nasal septum medially, including the premaxillary wing region, the floor of the nose, and laterally to the bony piriform aperture including the soft fibrofatty areolar tissue of this region of the internal nasal valve and is bounded posteriorly by the head of the inferior turbinate (Figs. 37.2 and 37.3).

Various workers have investigated the internal nasal valve and determined the approximate site using various methods, but the overwhelming

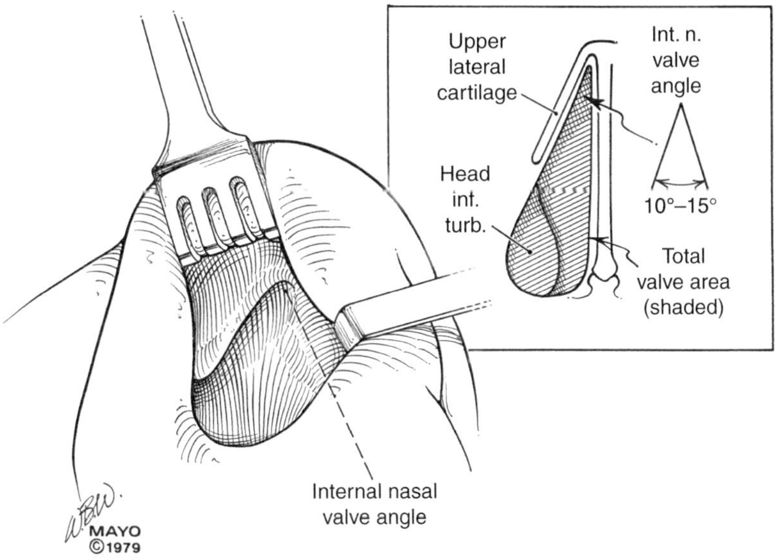

Fig. 37.2 This illustration demonstrates the clinical internal nasal valve angle and in the upper right corner of the illustration a representative view of the total internal nasal valve area which includes the upper lateral cartilage, the septal cartilage (the septal turbinate "swell body" is not shown here), the floor of the nose, the piriform aperture including the frontal (ascending) process of the maxilla, and the head of the inferior turbinate are all shown. The internal nasal valve angle is represented as ranging from 10 to 15 ° although some authors have different findings [40] (By permission of Mayo Foundation for Medical Education and Research. All Rights Reserved)

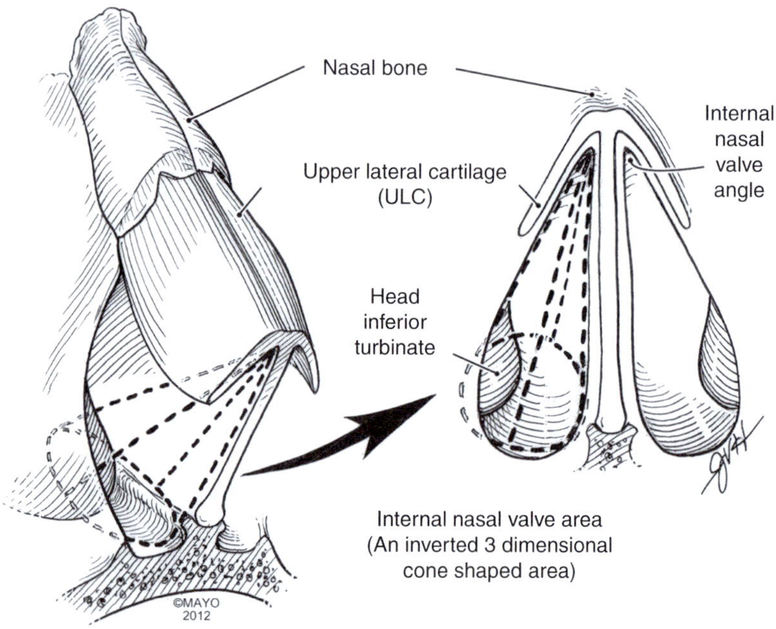

Fig. 37.3 Illustration demonstrates that the internal nasal valve area can be thought of as a three-dimensional structure (*an inverted cone*) with a triangular portion (*bottom*) of the cone fitting into the apex of the internal nasal valve angle (By permission of Mayo Foundation for Medical Education and Research. All Rights Reserved)

consensus is that most of the upper airway resistance is provided by the internal nasal valve and about one third of that resistance is due to the cartilaginous vestibule and approximately two thirds of that nasal resistance is due, in large measure, to the congestive capabilities of the anterior end (head) of the inferior turbinate. This turbinal "swell" body has a counterpart on the septum ("septal turbinate, swell body, or septal body") located on the nasal septum approximately adjacent to the anterior end (head) of the inferior turbinate [38, 40]. The finding of the septal turbinate (swell body) was first described in the 1600s, and these findings have been reported and supported by investigations in humans [26, 38, 40]. This entire *internal nasal valve area* ranges somewhere between 55 and 64 square millimeters.

The narrowest portion of the nasal airway is the apex of the internal nasal valve area at the nasal valve angle which is the specific triangular slit-like portion between the caudal (distal) end of the upper lateral cartilage and the nasal septum (Figs. 37.2 and 37.3). This nasal valve angle was first described over 100 years ago by Mink and is continuously re-reported to be approximately 10–15° in the leptorrhine (Caucasian) nose, as pointed out by Miman and coworkers [40]. In their prospective comprehensive study of 248 nasal cavities using direct endoscopic examination along with acoustic rhinometry and rhinomanometry, they discovered different types of nasal valve angles, ranging "between 22.5 and 52°" [40]. Miman et al. studied normal asymptomatic control subjects and proposed a classification related to the upper lateral cartilage's caudal border stated as convex, concave, or twisted and the valve angle as blunt, sharp, or occupied by the septal body. The septal body has been noted and reported before with a prevalence ranging from 52% to 66% of individuals [38, 41]. Miman et al. also discussed another study whose authors used computed tomography to study the nasal valve angle and that was reported to be 11.4° ± 2.6°. After a closer study of the other data obtained from computed tomography (while Miman et al. used direct endoscopic examination along with acoustic rhinometry and rhinomanometry), Miman et al. realized that positioning and placement of the lines for measurement are subjective and even a slight alteration of line placement could significantly alter the results by at least 15–20°. In light of their findings, Miman et al. concluded that the literature's emphasis on a normal nasal valve angle of 10–15° needs challenge and must be open for further study and discussion.

It is the entire *internal nasal valve area* that is the primary airflow inflow regulator; accounting for the majority of the inspiratory resistance to inspiratory airflow and the head of the inferior turbinate is the posterior portion of the *internal nasal valve area*. The head of the inferior turbinate has a significant and at times a dominant role (think acute viral or acute allergic rhinitis) as an inflow regulator in the Caucasian (leptorrhine), Black (platyrrhine) nose, and the Asian (mesorrhine) nose. Remember that the functional airway unit is the entire *internal nasal valve area* and any component anatomic part including the upper lateral cartilage; the anterior nasal septum, the premaxillary wing region, the floor of the nose, the piriform aperture, lateral fibrofatty areolar tissue, frontal (ascending) process of the maxilla and the head of the inferior turbinate, or the skin or mucosa (mucocutaneous coverings); or any combination of those structures when anatomically or pathologically altered from "normal" may contribute to dysfunction of the entire *internal nasal valve area. In other words, any anatomic part of the internal nasal valve area* may be surgically altered during primary rhinoplasty and therefore vulnerable to surgical alteration that could have negative consequences for breathing function postoperatively [6, 42–61].

37.2.2 External Nasal Valve

The external nasal valve is that portion of the nasal airway (ala of the nose) that is sprung open by the tensile strength of the paired lower lateral cartilages and the covering soft tissues and skin of the ala. The external nasal valve becomes a problem for the patient if the lower lateral cartilage is resorbed secondary to nonsurgical trauma, surgical trauma with over-resection of the lower lateral cartilages or changes in the tensile strength of the cartilage due to the atrophy of aging or loss of muscle tone secondary to facial palsy or other neurologic disorders [33, 62–65].

37.3 Physiology of the Nasal Valve Area

Today, 40 years after Williams merely speculated that the nasal valve functions as a type of inflow device controlling the rate and depth of respiration, the complete function of the nasal valve is still unknown [66]. Hinderer suggested that the nasal valve controls inspiratory air currents changing them from a column to a sheet of air, thereby giving shape, velocity, direction, and resistance to the inspired air [32, 52]. It has been previously suggested that the head of the inferior turbinate [39] and the septal turbinate [26, 38, 67] work in concert and both are important components of the internal nasal valve area because they provide the necessary resistance to breathing during inspiration. The nasal valve has been considered [13, 14] to function as a Starling resistor influenced by Bernoulli forces so that when the airflow increases (accelerates), the interior nasal pressure decreases allowing the lateral nasal walls to collapse inward (towards the nasal septum). Both the nose and a Starling resistor consist of a semirigid tube with a short collapsible segment (the internal and external nasal valve). The portion of the internal nasal valve at the head of the inferior turbinate can be considered the upstream segment, whereas that portion posterior to the head of the inferior turbinate and the septal body is the downstream segment (Fig. 37.4). In other words, the semirigid tube conducts pressure changes to the collapsible segment, which is influenced by several factors including the conducted pressure, the extramural pressure, Bernoulli forces, and the elasticity of the collapsible cartilaginous segment. In the nose, negative inspiratory pressure is transmitted from the nasopharynx to the internal nasal valve area, which then narrows. The degree of narrowing depends on three variables:

1. The pressure difference between the internal nasal pressure during breathing and the atmospheric pressure (transmural pressure).
2. The flexibility of the valve area's collapsible cartilaginous segment.
3. The size of the valve area.

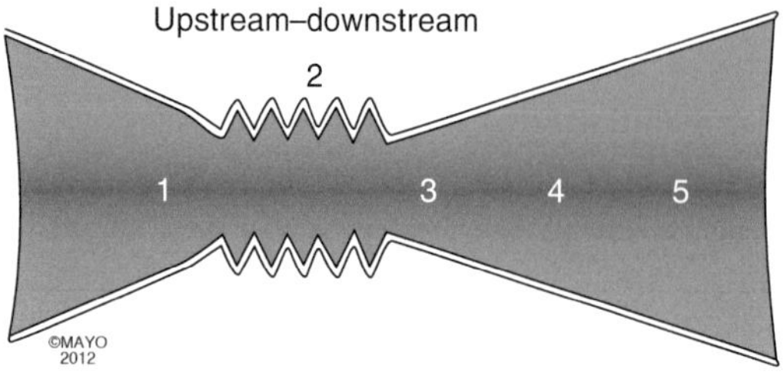

Fig. 37.4 Illustration is a graphic representation forwarding the idea that the nasal valve (internal and external) can potentially function as a collapsible segment dividing the nose into upstream—cartilaginous portion (*Cottle areas 1 and 2*)—and downstream "bony portion" (*Cottle areas 3, 4, and 5*). According to Cottle, the nasal septal areas are as follows: *area 1* refers to the nasal septum at the caudal end of the nasal septum. *Area 2* refers to the nasal septum at the valve area (now comprising the internal valve (including the anterior head of the inferior turbinate) and external valve (including the lower lateral cartilages)). *Area 3* refers to the nasal septum at the attic (the region of the septum beneath the nasal bones). *Area 4* refers to the nasal septum and its relationship to the turbinates (except for the most anterior head of the inferior turbinate, which is certainly part of the internal nasal valve and at the same time is the most posterior part of the internal nasal valve). *Area 5* refers to the nasal septum and its relationship to the choana [26, 32, 33, 67] (By permission of Mayo Foundation for Medical Education and Research. All Rights Reserved

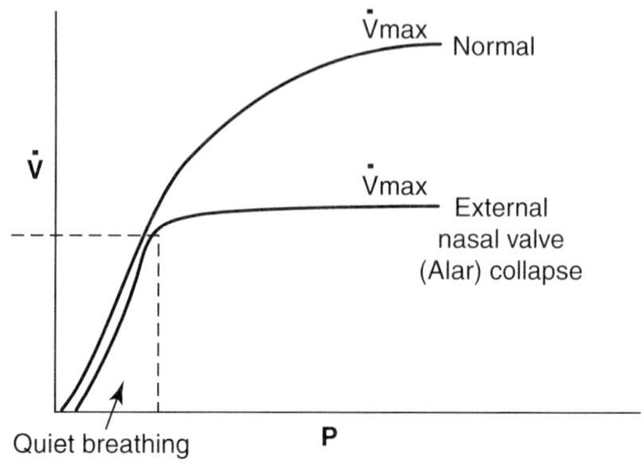

Fig. 37.5 Nasal pressure (*x*-axis) and nasal flow (*y*-axis) are seen on an idealized rhinomanometric curve in two situations. A *normal* functioning nose. From a patient with *external nasal valve* (*alar*) *collapse* demonstrating limitation of flow at high inspiratory (negative to atmospheric) pressure. At maximal inspiratory flow (Vmax), the nose collapses and no further increase in pressure (negative inspiratory pressure) can cause a further increase in airflow through the nose. In other words, when the valve (internal and/or external) collapses, there is plateauing of the curve since no matter how hard a subject tries to increase the inspiratory effort (negative pressure during inspiration), there is a cessation of airflow secondary to (internal and/or external) nasal valve collapse

In the normal nose, the nose does not collapse during quiet breathing because the tensile strength of the cartilage (upper lateral cartilage, lower lateral cartilage, and septal cartilage) support of the nose offsets the closing forces of the transmural pressure and the Bernoulli effect. The resistance to airflow is dependent on the skeletal (cartilage and bone) and mucocutaneous nasal structures that are anterior (upstream) and posterior (downstream) to the head of the inferior turbinate. At maximal inspiratory flow rates, a greater inspiratory effort (increased negative pressure) fails to increase airflow because the external nasal valve (ala) is collapsed (Fig. 37.5). The external nasal valve does not remain collapsed as the process is reversed during expiration.

The external nasal valve (ala) is really the airway opening supported by the paired lower lateral (great alar or alar) cartilages. The critical sites for collapse are either at the external nasal valve (lower lateral cartilage) or at the internal nasal valve (nasal valve area composed of the upper lateral cartilage, nasal septum, including the premaxillary wing region, the floor of the nose, piriform aperture, and the head of the inferior turbinate). Some observers believe that the septal turbinate (septal "swell" body) is part of the nasal valve area and I tend to agree, but these findings tend to support the notion that total understanding of the valve area and its function is still far from fully understood [30, 33, 38, 40]. Nonetheless, it is necessary to have some practical understanding of the nasal valve area and its critical sites in order to perform surgery without disturbing nasal breathing function.

These critical sites are seen in Fig. 37.6. A clinical un-instrumented view of many of the structures of the internal and external nasal valve can be seen in Fig. 37.7. The alar muscles are also involved in valvular function and aid in pre-

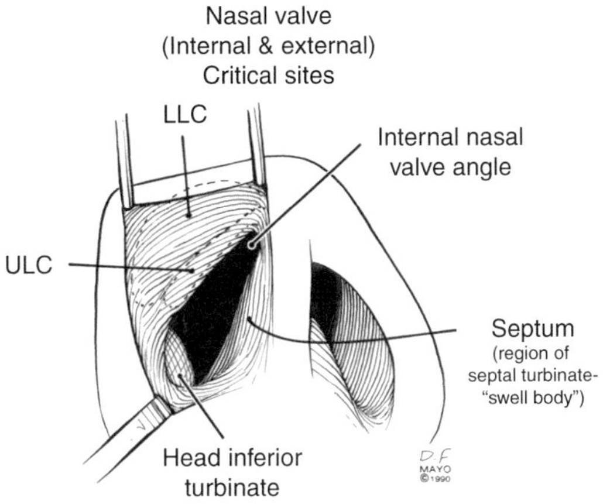

Fig. 37.6 Illustration depicts the most "critical sites" of "Physiologic concern during rhinoplasty" that the surgeon must understand and avoid sacrificing structural support in these "critical sites"; otherwise breathing dysfunction will most probably occur postoperatively. These "critical sites" include the following: the lower lateral cartilage (*LLC*), the major component of the *external nasal valve*; the *internal nasal valve* including the upper lateral cartilage (*ULC*), the septum, and the valve angle between the upper lateral cartilage (*ULC*) and the septum; and the head of the inferior turbinate. If any of these "critical sites" are sacrificed for esthetic reasons, then the reconstruction must follow at the time of surgery; otherwise, a postoperative breathing dysfunction will often follow. If the internal nasal valve angle is blunted or scarred, then postoperative breathing dysfunction will almost invariably follow (By permission of Mayo Foundation for Medical Education and Research. All Rights Reserved)

venting the collapse of the valve at high inspiratory flow rates. Rigidity (scar) or flaccidity (over-resection) of cartilage structure including both the upper and the lower lateral cartilages and soft tissues would affect valvular function and breathing. Even minor changes in the diameter of the airway can significantly increase nasal airway resistance in concert with Poiseuille's laws where the airflow is proportional to the pressure changes times the radius to the fourth power divided by the length. In other words, small changes in the radius to the fourth power are extremely significant.

Using acoustic rhinometry and active anterior rhinomanometry, Zambetti and associates [68]

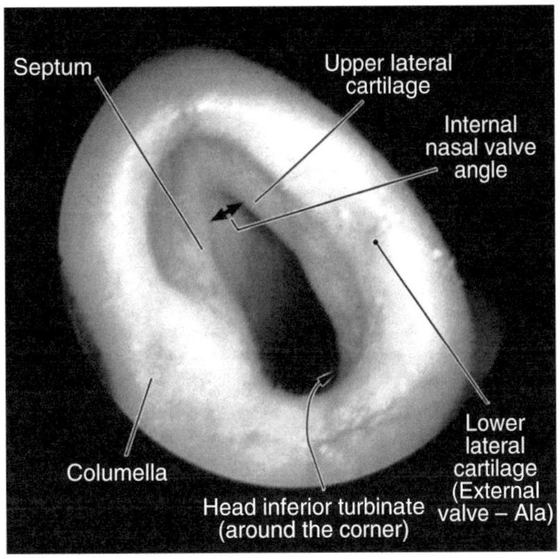

Fig. 37.7 Photograph is an un-instrumented clinical view of the left side of a person with a normal nose. Many structures of the internal and external nasal valve are seen and labeled (By permission of Mayo Foundation for Medical Education and Research. All Rights Reserved)

proved and supported the long-held clinical observation that small deformities in the anterior portion of the nose (upstream) like a scar in the valve angle can more profoundly and adversely affect nasal breathing than larger deformities in the interior of the nose (downstream) when they said, "Modest nasal cross-sectional area reductions posterior to the nasal valve do not cause substantial variations in nasal resistance…" [68].

The nasal valve area must function in concert with the structures upstream and downstream to provide nasal pulmonary respiratory balance over the wide range of physiologic demands. Since the head of the inferior turbinate is part of the internal nasal valve, it is affected by the vascular changes of the capacitance vessels in the stroma, which is termed the nasal cycle [41, 69, 70]. Actually, the nasal cycle is a normal physiologic phenomenon that includes the alternating congestion and decongestion of the turbinates with changes in uninasal resistance while the total nasal resistance remains relatively constant (Fig. 37.8).

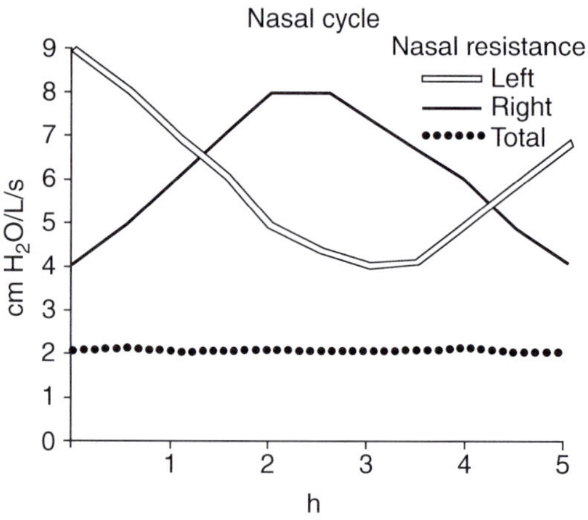

Fig. 37.8 The nasal cycle occurs in approximately 80% of the adult population. It can be described as the normal physiologic alternating congestion and decongestion of the nasal turbinates producing changes in one-sided nasal (uninasal) resistance. Nasal resistance is calculated from the measurements of the trans-nasal pressure changes during breathing (inspiration and expiration) divided by the airflow during breathing (inspiration and expiration). In other words, resistance = pressure divided by flow. Note that each side (uninasal resistance right and left side) usually alternates (resistance values) with the other side, after the passage of time, while the total nasal resistance remains relatively constant. The patient experiences one side of the nose as obstructed (plugged) and after some period of time the other side becomes obstructed (plugged) while breathing through both sides (both noses) feels unobstructed

37.4 Pathophysiology of the *Nasal Valve Area*

It is critical that nasal surgeons particularly rhinoplasty surgeons understand the pathophysiology of internal and external nasal valves since any disturbance in any portion of the critical sites in the nasal valve area (external valve and/or internal valve) can produce disturbed nasal breathing during exercise and/or at rest resulting in a very unhappy patient. The abnormalities responsible for nasal airway dysfunction include:

1. The *external nasal valve*: dysfunction may be due to a flaccid external valve that collapses at low inspiratory (negative pressures) airflow rates which can occur when there is a loss of cartilage (lower lateral cartilage) support due to surgical or non-surgical trauma, the effects of aging with atrophy of the lower lateral cartilages. A flaccid external nasal valve can also occur secondary to facial palsy. Avoid "over-resection" of the lower lateral cartilages since over-resection may lead to loss of structural support with early collapse during inspiration and postoperative breathing difficulties (Table 37.1).
2. The *internal nasal valve*: dysfunction may be due to valve area narrowing due to structural changes in the valve angle (upper lateral cartilage and/or septal cartilage) or any one of the structural components of the valve area (Table 37.2).
3. Combinations of structural skeletal (cartilage) changes and/or the absence of skeletal support (especially secondary to surgery) including avulsion of the upper lateral cartilage [71] which may be associated with early collapse of the valve during nasal airway breathing; however, often the airway obstruction is so severe that the patient is unable to breathe through the nose and becomes a mouth breather.

After blunt trauma [72], the second most likely cause of nasal valve abnormalities (with breathing problems) is nasal surgery [26, 33, 45, 71, 73–75]. Narrowing of the internal nasal valve angle by a medially displaced upper lateral cartilage after "hump" removal and infracture, additionally, over-resection of the "hump" leaving an

Table 37.1 Structures of the *external* nasal valve

1	Lower lateral cartilage (LLC)
2	Fibroareolar soft tissues of ala

Table 37.2 Structures of the *internal* nasal valve area

1	Nasal septum (including premaxillary wings)
2	Upper lateral cartilage (ULC)
3	Piriform aperture
4	Fibroareolar lateral soft tissues
5	Frontal (ascending) process of the maxilla
6	Head of the inferior turbinate
7	Mucosa and skin coverings of these structures

incomplete infracture, results in the "open roof" deformity [26, 32, 76] and early collapse during inspiration in addition to posttraumatic neurogenic pain syndromes. Uncorrected septal pathology or over-resection of septal cartilage resulting in a "flaccid" septum with "flutter" is another cause of a collapsed internal nasal valve. Over-resection of the lower lateral cartilages may also result in "flaccid" collapse during inspiration. This "flaccid" collapse is also seen with loss of muscular action (as in facial paralysis) or with the atrophy of aging secondary to a decreased tensile strength of the cartilages and soft tissues producing a "droopy tip" [77].

37.5 A Brief Word About the Preoperative Evaluation and Discussion

When rhinoplasty is considered, I take a considerable amount of time to develop a personal relationship (rapport) with the patient while obtaining a general medical history, supplemented with a rhinologic questionnaire [78] which succinctly covers the essentials including a history of smoking and alcohol use, attention to allergic conditions (including seasonal and perennial allergic rhinitis, allergies to medications, soaps, surgical solutions, latex, tape), bleeding tendencies, use of steroids (topical and/or systemic), diabetes, hypertension, current and recent medications (including aspirin, aspirin-like drugs—nonsteroidal anti-inflammatory drugs), peripheral vascular disease, and previous operations of any kind.

In addition to physical nasal examination (looking for a Cottle sign [79]) also use the endoscope (0°, 4 mm) for an intranasal examination before and after topical decongestion to study the mucosa and the interior of the nose back to the nasopharynx. I usually obtain the following tests and consultations (when medically indicated):

1. Minnesota Multiphasic Personality Inventory (MMPI) is obtained on all patients prior to surgery looking for evidence of personality disorders, schizophrenia, and psychosis [80].
2. Olfaction testing is performed on *all* patients.
3. Rhinomanometry (anterior mask technique) is also performed on all patients prior to surgery based on techniques previously described [1–3, 8, 10, 81]. Other surgeons suggest the use of acoustic rhinometry (first introduced by Hilberg and associates in 1989) as an objective "evaluation of the nasal cavity geometry by acoustic reflections" [82–84].
4. CT scanning (direct coronal) is ordered when indicated.
5. Preoperative photographs are ordered on ALL patients even prior to sinus surgery as the patients can occasionally "forget" how their face and nose appeared prior to surgery.
6. Blood studies and other medical consultations (including allergy evaluation) are ordered when indicated.

I always ask for the patient's parent(s) or significant other to be present before surgery, and I always welcome questions from the patient and family before ending the interview. The preoperative discussion is critical to establishing the necessary rapport with the patient and the family to better manage complications should they occur postoperatively. Succinctly, for cosmetic rhinoplasty, I ALWAYS discuss the following in detail:

1. Goal(s) of the operation.
2. Risk(s) (to life) of the operation and the anesthesia.
3. Possible complications of the operation.
4. Possible need for nasal septal surgery [63, 85, 86] to treat a current breathing disorder or the possible need for obtaining graft material from the patient such as cartilage and/or bone from the septum, ear, or rib for reconstruction and grafting [87–90].

It is also wise to prepare the patient for possible revision surgery (second operation) before you perform the first operation as the patient may require a secondary or "touch up" operation at some time in the future. Happily, most patients are more satisfied with the results of rhinoplasty perhaps even more than their surgeon [91, 92].

The important details of the preoperative discussion have been presented previously in the literature [93]:

> Trust your feelings. If the results of the MMPI are abnormal or you are concerned that the patient has significant emotional issues that preclude a "normal" doctor patient relationship, seek psychiatric consultation (because your inner voice tells you there is something "wrong" and you have an uncomfortable feeling about the patient, listen to your inner voice and trust your feelings). Obtain psychiatric consultation for your patient with a specific psychiatrist (who you have pre-selected, cultivated and is interested in cosmetic patients, to rule out body dysmorphic disorder (BDD) or other significant psychiatric illness. ([80]; [76])

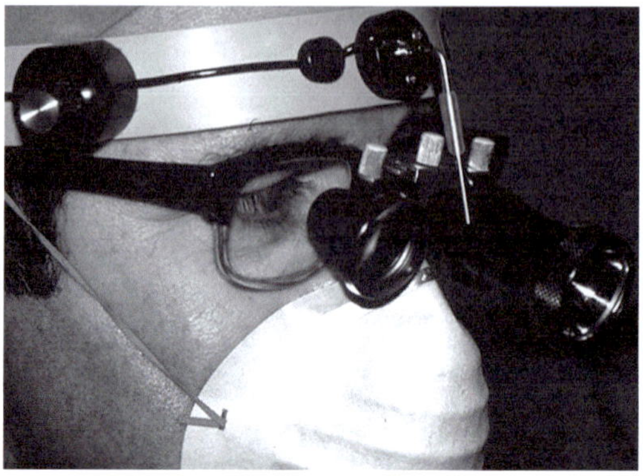

Fig. 37.9 Performing rhinoplasty using magnification with 2–2.5 power loops is extremely useful for the meticulous tissue handling required for surgery of the nasal valve (internal and external) area

37.6 Prevention of Nasal Valve Area Breathing Complications During Rhinoplasty

In addition to esthetic improvements, the surgeon's challenge is preventing the nasal valve (internal and/or external valve) collapse during the performance of the cosmetic rhinoplasty. That includes proper placement of incisions [94], which should always be placed in the skin (not in the mucosa). The surgeon must correct existing pathology when present, including all component parts of the nasal valve (internal and/or external), and the surgeon must prevent surgical maneuvers that could compromise the nasal airway postoperatively.

In my view, the goals of cosmetic rhinoplasty are to:

1. Improve appearance.
2. Improve nasal breathing function (if needed).
3. Avoid postoperative iatrogenic breathing dysfunction.

Small deformities in the internal valve region (upstream) can produce significant nasal airway symptoms. I strongly believe that nasal surgery must be exacting and precise, especially when operating in the nasal valve area and magnification with 2–2.5 power loops are extremely useful for the meticulous tissue handling required for delicate surgery of the internal valve (Fig. 37.9).

It is well known, based on the post-rhinoplasty complications seen in the literature and from my own experience, that both the internal nasal valve and external nasal valve are at risk of being compromised during the performance of a "routine" rhinoplasty. A successful outcome of cosmetic rhinoplasty includes both an esthetic improvement and a normal breathing nose, which includes correction of an abnormality or prevention of breathing dysfunction. The pathophysiologic features of post-rhinoplasty breathing disturbances almost always involve the internal or external nasal valve.

I will cover specific surgical maneuvers that risk causing nasal valve area (internal and/or external nasal valve) breathing problems postoperatively.

37.6.1 Incisions

To gain adequate exposure and access to the skeletal structures (cartilage and bone), various incisions have been utilized including the transfixion (hemitransfixion and full or complete transfixion), *inter*cartilaginous (IC), *intra*cartilaginous, rim, marginal, "slot dome," trans-columellar (external approach), vestibular, or various external incisions for osteotomies [26, 32, 33]. Of these incisions, only the endonasal intercartilaginous incision, which is a skin incision and not a mucosal incision, impinges directly on the internal nasal valve. This incision is created by eversion and exposure of the upper lateral cartilage and the lower lateral cartilage at the cartilaginous junction by incising the intervening vestibular skin. This incision is frequently joined to a transfixion incision to allow mobilization of the columella and access to the bony and cartilaginous dorsum. When performed with meticulous atraumatic technique and closed under accurate suture approximation, the ensuing healing scar does not adversely affect internal nasal valve function. However, should the tissues be handled roughly or closure performed imprecisely, the risk of subsequent scar contracture with stenosis of the valve angle is markedly increased [94]. If indicated, detachment of the upper lateral cartilage from the nasal septum through parallel intranasal mucosal and skin incisions increases the risk of scarring if not performed carefully or closed meticulously. This is especially true of skin incisions at the apex of the internal nasal valve angle. Certainly, if it is necessary to separate the upper lateral cartilage from its septal attachment, detachment should be performed submucosally (intraseptally) beneath the underlying skin (sometimes called "junction tunnels") and beneath the more posteriorly placed mucosa to minimize postoperative scar contracture. All intranasal skin incisions must be suture approximated to promote accurate healing by primary intention.

37.6.2 Approaches

Exposure of the nasal tip (lower lateral cartilages) and the nasal dorsum may be achieved through a variety of approaches including retrograde, cartilage splitting, and delivery [32, 33, 95], or the external (trans-columellar) approach. The retrograde and delivery approaches exposing the lower lateral cartilages necessitate an intercartilaginous incision. In addition, these two approaches (retrograde and delivery) require severing the attachments between the upper lateral and the lower lateral cartilages. This maneuver has the potential to decrease the structural rigidity of the internal nasal valve. Dissection with meticulous atraumatic technique and precise suture closure of incisions ensure minimal tissue damage. The cartilage-splitting incision (probably less common today) and the incision for the external (trans-columellar) rhinoplasty approach [96] (probably more common today) are performed away from the valvular region and do not have an adverse potential impact on the internal or external valve area with secondary scarring.

37.6.3 Excisions

In the entire scope of nasal surgery, there are only a few surgical options:

1. Removal of tissue (bone and/or cartilage).
2. Adding tissue (biodegradable grafts of bone and/or cartilage are preferred over inorganic implants) [9, 26, 32, 33, 87, 88].
3. Repositioning of tissue (releasing tension).
4. Combinations of the above.

With the goals of tip rotation, tip projection, and tip definition, the majority of cosmetic nasal operations require some degree of modification of the lower lateral cartilages [33, 95]. This change of the lower lateral cartilage usually involves excision of the cephalic border of these lower lateral cartilages. Extreme or total resection of the lateral crus of the lower lateral cartilages often leads to loss of structural support of the nasal ala with the subsequent inspiratory collapse of the external valve. In addition, "excessive" cephalic trimming in this region can lead to the characteristic pinched appearance causing the tip to appear overly bulbous, thus distorting the symmetry of the nasal base. In addition, interruption of the intact strip of the lower lateral cartilage necessarily results in diminished structural support, increasing the risk of inspiratory collapse of the nasal ala in rhinoplasty. Therefore, unless absolutely necessary, cephalic removal of the lower lateral cartilages should be performed conservatively, meaning depending on the gender, age, skin quality, ethnicity, and thickness and resiliency of cartilage; "conservative" means leaving 6 to 8–10 mm of cartilage behind and/or combined with modeling of the lower lateral cartilages by suture techniques which minimize the risk for postoperative alar collapse of the external nasal valve. The dictum of "…it is not what you take, but what you leave behind that matters…" is applicable for nasal surgery in general and especially applicable to surgery of the lower lateral cartilages. While leaving an intact strip of lower lateral cartilage is advisable with tip modification, dome division (for narrowing) can be practiced especially when carried out medial to the dome [33, 95]. Should preservation of an intact strip of lower lateral cartilage be impossible and should it become necessary to disrupt the continuity of the lower lateral cartilage, then an approximation of the cartilage remnants with permanent suture (5–0 or 6–0 nylon) is recommended to avoid alar collapse of the external nasal valve. Utilizing a columellar strut (autogenous septal cartilage) has been a very successful contribution to maintaining tip projection and rotation.

Excision of the caudal (inferior) edge of the upper lateral cartilage occasionally becomes necessary when the upper lateral cartilage is excessively protuberant, either in the native (unoperated) state or after nasal shortening; if necessary, excision should always be performed conservatively since excessive resection of this region can result in loss of structural support and inspiratory collapse. Concomitant excision of the corresponding skin of the internal or external nasal valve along with the caudal (inferior) edge of the upper lateral cartilage should be avoided because skin excision markedly increases the likelihood of secondary cicatricial scarring and valvular stenosis leading to breathing dysfunction after surgery. When performing an "M" plasty to widen the internal valve angle (to compensate for infracture) in rhinoplasty [97], the skin is incised (to open the internal valve angle), NOT excised. If skin excision is required for some reason, which is highly unlikely, in cosmetic rhinoplasty, suture approximation is always advisable.

Valve problems may therefore include structural changes causing fixed encroachment into the internal nasal valve region and/or a loss of structural support leading to flaccidity and collapse during inspiration.

Numerous papers have appeared in the literature to help the surgeon decide which approaches can best be utilized to "prevent" or "treat" (repair) valve collapse [43, 48, 50, 62, 80, 98–109].

37.6.4 Hump Reduction

Dorsal nasal hump resection generally involves the nasal valve region in two frequent circumstances. First, excision of an excessive dorsal hump may involve removal of the attachment of the upper lateral cartilage to the septum, and this excision may be carried down into the valve angle (and area) in reduction rhinoplasty [26, 33, 110]. Trimming this dorsal cartilaginous region may involve just the cartilage or both the cartilage and its underlying nasal mucosa and skin. In either instance, but particularly with dehiscence of the intranasal skin at the internal nasal valve angle, awareness and care are needed for the accurate repositioning and suturing (if possible) of the remaining upper lateral cartilage in continuity with the dorsal septum, and if "spreader" grafts are contemplated, then careful introduction, placement, and fixation with precise suturing are mandatory to avoid obstructing the internal nasal valve angle and producing postoperative breathing problems [75].

The second circumstance involves patients with short nasal bones and a dorsal hump requiring resection. Occasionally, hump resection effectively removes a significant portion of the important upper lateral cartilage supports allowing the upper lateral to "flail," which can lead to valvular and associated breathing dysfunction. Avoidance of this untoward result rests on appropriate preoperative diagnosis as well as a careful approximation of the upper lateral cartilaginous back to the dorsal septum with a permanent suture (3–0 or 4–0 nylon suture with horizontal mattress placement). Approximation reestablishes support for the upper lateral cartilage, thereby minimizing potential postsurgical collapse. Most often, in my career, I performed dorsal nasal modification by either rasping the bony dorsum or lowering the cartilaginous dorsum by incremental shaving of the upper lateral cartilage (and dorsal septum when needed) using a number 15 or 15 C knife blade and when required a sharp pair of "super cut" scissors. I usually modified the combined bony and cartilaginous dorsal hump by the "push-down" or "let-down" operation supplemented by rasping of the bone and incremental shaving of the cartilage (when needed), and these procedures are covered in the literature [26, 32, 33, 110].

37.6.5 Osteotomies

Medial, intermediate, and lateral osteotomies of the nasal bone and the frontal (ascending) process of the maxilla may all impact the internal nasal valve area via alteration in the position of the upper lateral cartilage. In addition, lateral osteotomies with infracture may impact directly and result in the narrowing of the internal nasal valve area; although, I have seen this complication of rhinoplasty on a number of occasions, the literature states that breathing disturbances after infracture in rhinoplasty are less common in practice than intuitive and theoretical reasoning would have you conclude [42, 43, 73, 82, 83, 94, 99, 111, 112]. Studies have shown that the risk to the upper lateral cartilage after lateral osteotomy with the inward collapse and narrowing of the internal nasal valve angle is real and the cross-sectional diameter of the nasal airway is altered (narrowed). If this occurs, then the surgeon may consider performing an outfracture [113] as illustrated in Fig. 37.10a, b. Because lateral osteotomy with infracture, to close an open roof after hump removal, can narrow the valve angle producing symptoms, the frequency of this complication can be avoided or reduced by performing a curved lateral osteotomy as described by Webster and associates [114]. By leaving an intact triangular piece of bone at the

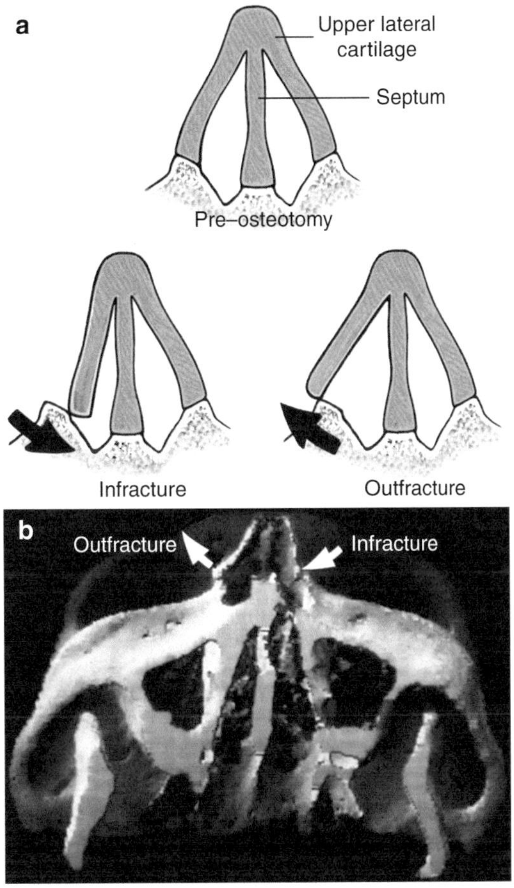

Fig. 37.10 (**a**) Illustration depicts the concepts of infracture and outfracture with the concomitant narrowing (infracture) of the internal nasal valve area and widening (outfracture) of the internal nasal valve area (By permission of Mayo Foundation for Medical Education and Research. All Rights Reserved.) (**b**) This reconstruction clearly demonstrates both an outfracture of the frontal (ascending) process of the maxilla with a widening of the internal valve area and infracture of the frontal (ascending) process of the maxilla with narrowing of the internal valve area in a symptomatic patient (By permission of Mayo Foundation for Medical Education and Research. All Rights Reserved)

piriform aperture, just superior to the level of the inferior turbinate, the surgeon accomplishes the esthetic narrowing and closing of the open roof at the nasal dorsum while minimizing the risk of compromising the integrity of the internal valve region. I have used this technique successfully.

37.6.6 Turbinates

Surgeons have long been tempted, and many have succumbed to that temptation, to remove the inferior turbinate (or at least the head of the inferior turbinate in its position as the posterior guardian of the internal nasal valve) to improve the nasal airway after infracture to close the open roof after hump resection. Remember that *the mucosa is the organ of the nose* and the removal of turbinate tissue carries with it the possibility of producing an "*empty nose*" *syndrome* [115–117]. Some authors have demonstrated that total inferior turbinectomy "carries significant morbidity and should be condemned" [118–121]. Other authors think resecting turbinate tissues carries little or no morbidity whatsoever [49, 122]. My experience with symptoms seen in patients with the "*empty nose*" *syndrome* (Table 37.3) has biased me in favor of very conservative treatment of turbinate mucosal hypertrophy, and I condemn aggressive removal of inferior turbinate tissue in nonmalignant cases [118] (after allergic, vasomotor rhinitis and other more serious medical conditions have been ruled out, some of which may require biopsy [56] with microscopic examination and other tests for accurate diagnosis of the cause of the turbinate hypertrophy).

Passali et al. [123] in a randomized clinical trial divided 382 patients into six treatment groups who had therapy for their inferior turbinate hypertrophy including:

1. Turbinectomy.
2. Laser cautery.
3. Electrocautery.
4. Cryotherapy.
5. Submucosal resection.
6. Submucosal resection with lateral displacement (outfracture).

After following patients for 6 years, these authors found that nasal patency, normalized breathing, restored mucociliary clearance, and local secretory IgA production (back to physiologic levels) with the fewest postoperative complications ($p < 0.001$) occurred with *conservative* submucosal resection combined with lateral displacement (outfracture) "as the first-choice technique for the treatment of nasal obstruction due to hypertrophy of the inferior turbinates." Turbinate surgery is indicated for obstruction after perennial allergic rhinitis and vasomotor rhinitis have been ruled out, and all other medical treatments have failed to successfully treat the inferior turbinate hypertrophy. As a general rule of approach, and because of the specter of a very distraught "empty" nose symptomatic patient, I favor outfracture of the inferior turbinates while maintaining intact mucosa using gauze packing (impregnated with steroid and antibiotic *solution*) to maintain the outfractured turbinate in the lateralized position for 1 week, effectively putting the nose to "rest" bathed in physiologic nasal secretions and away from the desiccating (drying) effects of nasal airflow.

Table 37.3 Range of symptoms seen in patients with the "*empty nose*" *syndrome*

1	Crusting
2	Bleeding
3	Anosmia or hyposmia
4	Difficulty breathing
5	Pain (localized posttraumatic neurogenic)
6	Headache
7	Nasal malodor
8	Disturbed sleep (associated with fatigue and lethargy)
9	Aprosexia nasalis (inability to concentrate)
10	Disturbed sense of Well-being
11	Emotional changes (anxiety, reactive depression)

37.7 Closing Thoughts

At this time, there are no final conclusions, merely observations and a current consensus of thinking regarding the surgical importance of the internal and external nasal valves to a patient's breathing function. With a better scientific understanding of nasal breathing and classification of nasal valve abnormalities that has evolved over the past 35 years, surgeons are better able to successfully perform cosmetic rhinoplasty while minimizing the risks to breathing function. The weight of evidence and my clinical and surgical experience essentially state that complete physiologic understanding and systematic delicate handling of the structures of *both* the *external*

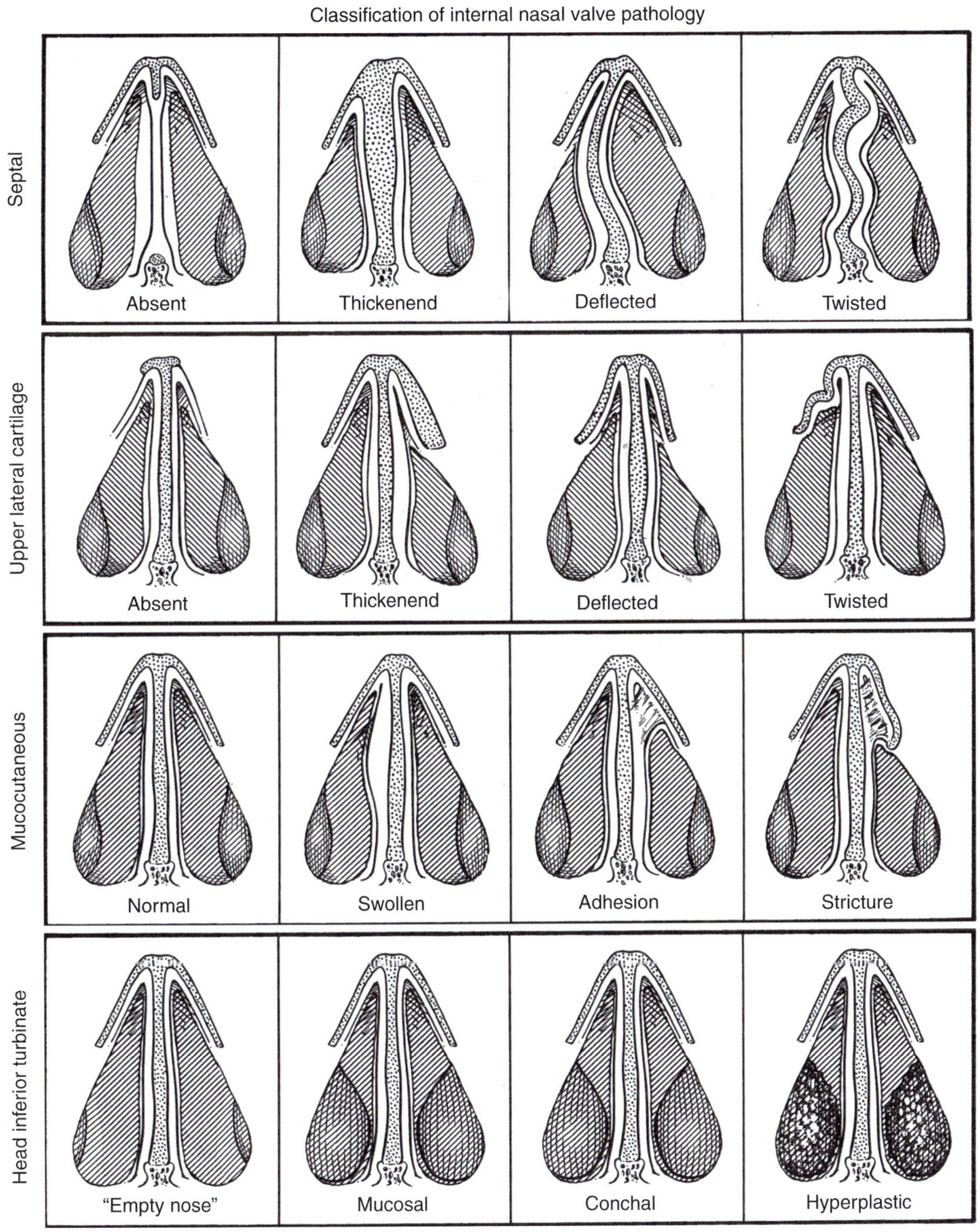

Fig. 37.11 Illustration is a summary and classification of somc of thc possible causes of internal nasal valve obstruction (By permission of Mayo Foundation for Medical Education and Research. All Rights Reserved)

and the *internal nasal valve* is crucial to avoiding postoperative breathing complications. The structures of the *internal nasal valve* include the upper lateral cartilage, the nasal septum (including the premaxillary wing region), the floor of the nose, the piriform aperture, and the head of the inferior turbinate, and their mucocutaneous coverings. Various abnormalities in the diverse component parts of the internal nasal valve are summarized in Fig. 37.11.

Table 37.4 is a *classification* of the most current thinking regarding the causes of *nasal valve* (*internal and external*) obstruction and collapse. The surgical techniques and principles suggested in this chapter have evolved over the past half-century of rhinologic surgery for nasal airway dysfunction by a variety of surgeons performing primary and secondary rhinoplasty operations.

The ultimate goal of rhinoplasty is to produce an esthetically pleasing external nose without dis-

Table 37.4 Classification of many causes of nasal valve (internal and external) obstruction and collapse

1. Internal nasal valve
- (a) *Intramural*
  - • Anatomic
    - – Mucosal
      - Inflammatory
      - Hypertrophy
    - – Submucosal
      - Scar
      - Hematoma
      - Abscess
    - – Cutaneous
      - Synechia (adhesions)
      - Stricture
    - – Cartilage
      - Septal
        - Absent (complete, incomplete)
        - Thickened
        - Deflected
        - Twisted
      - Upper lateral
        - Absent (complete, incomplete)
        - Thickened
        - Deflected
        - Twisted
        - Fixed collapse (secondary to pyramid trauma)
        - Physiologic collapse
    - – Turbinate
      - Bone (concha)
      - Mucosal
        - Physiologic nasal cycle
        - Dependent (sleep positions)
        - Vasomotor rhinitis
        - Allergic rhinitis
        - Hyperplastic
        - Tumor
        - Systemic disease
- (b) *Extramural*
  - • Fixed collapse (due to surgical, nonsurgical pyramid trauma)
  - • External pressure glasses, goggles, etc.
  - • Intranasal space-occupying lesion
    - – Foreign body
    - – Growth (poly, tumor)

2. External nasal valve
- (a) *Anatomic*
  - • Lower lateral cartilage

Table 37.4 (continued)

– Absent (complete, incomplete)
– Thickened
– Deflected
– Twisted
(b) *Pathophysiologic*
• Atrophy (aging)
• Facial palsy
• Neurologic disorders

turbing nasal breathing function. To achieve one without the other, or worse yet, at the expense of the other, represents a failed rhinoplasty. I strongly advise conservatism since most of the breathing complications treated over my past almost 50 years in otorhinolaryngology were secondary to over-resection of one or more of the varied components of the internal or external nasal valve including over-resection of functioning turbinate tissue.

After many years of experience in rhinology, and having read extensively in preparing the bibliography for this chapter, it is still humbling to realize that I did not study or read all of the literature, but I selected texts that I thought were important to construct a complete, honest, and meaningful understanding of the science and surgery of the nasal valve for the comprehensive writing of this chapter. With the exclusion of some texts from my bibliography, I have chosen to respect other serious authors and include a section entitled *Supplemental Readings*.

In nasal surgery as in life, it is not what you take, but what you leave behind that matters; therefore, I suggest that we all must strive to leave behind an enhanced (undisturbed) functional (breathing) nose with an esthetic improvement for our patients. One final suggestion is contained in this maxim "be a maximist in the office (developing rapport with the patient and the family) and a minimalist (in resecting tissue) in the operating room."

37.8 Conclusion

While the ultimate goal of cosmetic rhinoplasty is producing an esthetically pleasing external nose, it must be accomplished without disturbing nasal breathing function and it is the location, dimensions, and functions of the nasal valve (internal and external) that have been covered in this chapter which provide the surgeon with the required understanding of the critical nasal valve area in order to accomplish the goals of improved appearance without producing a postoperative breathing disturbance.

Acknowledgments This work was partially funded by an educational grant from the Gromo Foundation for Medical Education and Research, Buffalo, New York.

References

1. Gordon AS, McCaffrey TV, Kern EB, Pallanch JF. Rhinomanometry for preoperative and postoperative assessment of nasal obstruction. Otolaryngol Head Neck Surg. 1989;101:120.
2. Kern EB. Rhinomanometry. Otolaryngol Clin North Am. 1973;6:863.
3. Kern EB. Standardization of rhinomanometry. Rhinology. 1977;15:115.
4. Kern EB. Rhinomanometry. In: English GM, editor. Otolaryngology, vol. 2. Hagerstown: Harper and Row Publishers; 1979.
5. Kern EB. Examination of nasal breathing: an objective method. In: Rees TD, Baker DC, Tabbal N, editors. Rhinoplasty: problems and controversies – a discussion with the experts. St. Louis: C.V. Mosby; 1988.
6. Kern EB. Surgery of the nasal valve. In: Rees TD, Baker DC, Tabbal N, editors. Rhinoplasty: problems and controversies – a discussion with the experts. St. Louis: C.V. Mosby; 1988.
7. McCaffrey TV, Kern EB. Clinical evaluation of nasal obstruction: a study of 1000 patients. Arch Otolaryngol. 1979;105:542.
8. McCaffrey TV, Kern EB. Rhinomanometry. In: Mackay I, editor. Facial plastic surgery. International quarterly monographs, Rhinoplasty and airway, vol. 3. New York, NY: Thieme, Inc.; 1986. p. 217.
9. Mertz JS, McCaffrey TV, Kern EB. Objective evaluation of anterior septal surgical reconstruction. Otolaryngol Head Neck Surg. 1984;92:308.

10. Pallanch JF, McCaffrey TV, Kern EB. Normal nasal resistance. Otolaryngol Head Neck Surg. 1985;93(6):778.
11. Pallanch JF, McCaffrey TV, Kern EB. Evaluation of nasal breathing function. Otolaryngol Head Neck Surg. 1993;37:665.
12. Santiago-Diez DJ, McCaffrey TV, Kern EB. The nasal valve: a rhinomanometric evaluation of maximum nasal inspiratory flow and pressure curves. Ann Otol Rhinol Laryngol. 1986;95:229.
13. Bridger GP. Physiology of the nasal valve. Arch Otolaryngol. 1970;92:543.
14. Bridger GP, Proctor DF. Maximum nasal inspiratory flow and nasal resistance. Ann Otol Rhinol Laryngol. 1970;79:481.
15. de Wit G, Kapteyn TS, van Bochove WM. Some remarks on the physiology, the anatomy and the radiology of the vestibulum and the isthmus nasi. Int Rhinol. 1965;3:37.
16. Kern EB. The nose: structure and function. Postgrad Med. 1975;57:101.
17. Kern EB. Applied physiology of the nasal cavities. Rhinology. 1984;22:91.
18. Knops JL, McCaffrey TV, Kern EB. Physiology: clinical applications. Otolaryngol Clin N Am. 1993;26:517.
19. McCaffrey TV, Kern EB. Laryngeal regulation of airway resistance. I. Chemoreceptor reflexes. Ann Otol Rhinol Laryngol. 1980;89(3):209.
20. Whicker JH, Kern EB. The nasopulmonary reflex in the awake animal. Ann Otol Rhinol Laryngol. 1973;82:355.
21. Whicker JH, Kern EB. Effect of denervation of nasal mucosa on pulmonary mechanics. Ann Otol Rhinol Laryngol. 1973;82:724.
22. Olsen KD, Kern EB. Nasal influences on snoring and obstructive sleep apnea. In symposium on sleep disorders. Mayo Clin Proc. 1990;65:1095.
23. Olsen KD, Kern EB, Westbrook PR. Sleep and breathing disturbances secondary to nasal obstruction. Otolaryngol Head Neck Surg. 1981;89:804.
24. Kern EB. Nasal obstruction. In: Meyerhoff WL, Rice DH, editors. Otolaryngology – head and neck surgery. Philadelphia: W.B. Saunders Company; 1992.
25. Kern EB. Wegener's granulomatosis and lethal midline granuloma. Summary of the fireside conference. In: Rhinology, supplement 14. Proceedings of the 1991 international congress of rhinology, 1992.
26. Barelli PA, Kern EB, Loch WEE, et al. Rhinology. The collected writings of Maurice H. Cottle, M.D. Kansas City: American Rhinologic Society; 1987.
27. Cole P. Biophysics of nasal airflow: a review. Am J Rhinol. 2000;14:245.
28. Cole P. The four components of the nasal valve. Am J Rhinol. 2003;17:107.
29. Eccles R. Nasal airflow in health and disease. Acta Otolaryngol. 2000;120:580–95.
30. Haight JSJ, Cole P. The site and function of the nasal valve. Laryngoscope. 1983;93:49.
31. Hilberg O, Jackson AC, Swift DL, et al. Acoustic rhinometry: evaluation of nasal cavity geometry by acoustic reflections. J Appl Physiol. 1989;66:295.
32. Hinderer KH. Fundamentals of anatomy and surgery of the nose. Birmingham: Aesculapius Publishing Company; 1971.
33. Huizing EH, de Groot JAM. Functional nasal surgery. Stuttgart/New York: Thieme; 2003.
34. Kasperbauer JL, Kern EB. Nasal valve physiology: implications in nasal surgery. Otolaryngol Clin N Am. 1987;20(4):699.
35. O'Neill G, Tolley NS. Theoretical considerations of nasal airflow mechanics and surgical implications. Clin Otolaryngol. 1988;13:273.
36. Van Dishoeck HAE. Inspiratory nasal resistance. Acta Otolaryngol. 1942;30:431.
37. Van Dishoeck HAE. The part of the valve and the turbinates in total nasal resistance. Int Rhinol. 1965;3:19.
38. Wexler DB, Davidson TM. The nasal valve: a review of the anatomy, imaging and physiology. Am J Rhinol. 2004;18:143.
39. Gray VD. Physiologic returning of the upper lateral cartilage. Int Rhinol. 1970;8:56.
40. Miman MC, Deliktas H, Oxturan O, et al. Internal nasal valve: revisited with objective facts. Otolaryngol Head Neck Surg. 2006;134:41.
41. Arbour P, Kern EB. Paradoxical nasal obstruction. Can. J Otolaryngol. 1975;4:333.
42. Adamson PA, Smith O, Cole P. The effects of cosmetic rhinoplasty on nasal patency. Laryngoscope. 1990;100:357.
43. Araco A, Gravante G, Gentile P, et al. Iatrogenic collapse of the nasal valve in aesthetic rhinoplasty. Scand J Plast Reconstr Hand Surg. 2007;41:293.
44. Becker SS, Dobratz EJ, Stowell N, et al. Revision septoplasty: review of sources of persistent nasal obstruction. Am J Rhinol. 2008;22:440.
45. Beekhuis GJ. Nasal obstruction after rhinoplasty: etiology, and techniques for correction. Laryngoscope. 1976;86:540.
46. Berry RB. Nasal resistance before and after rhinoplasty. Br J Plast Surg. 1981;34:105.
47. Bruno JR, Papay FA, Papay FA. Dispelling the propagated myth: a quizzical look at the internal nasal valve (letters and viewpoints). Plast Reconstr Surg. 2005;116:685.
48. Constantian M, Clardy RB. The relative importance of septal and nasal valvular surgery in correcting airway obstruction in primary and secondary rhinoplasty. Plast Reconstr Surg. 1996;98:38.
49. Courtiss E, Goldwyn R. The effects of nasal surgery on air flow. Plast Reconstr Surg. 1983;7:16.
50. Fischer H, Gubisch W. Nasal valves-importance and surgical procedures. Facial Plast Surg. 2006;22:266.
51. Goode RL. Surgery of the incompetent nasal valve. Laryngoscope. 1985;95:546.
52. Hinderer KH. Surgery of the valve. Int Rhinol. 1970;8:60.
53. Jessen MD, Jacobsson S, Malm L. On rhinomanometry in rhinoplasty. Plast Reconstr Surg. 1988;81:506.

54. Kern EB. Surgery of the nasal valve. Plastic and reconstructive surgery of the face and neck. In: Sisson GA, Tardy Jr ME, editors. Proceedings of the second international symposium, vol. 2. New York: Grune and Stratton; 1977.
55. Kern EB. Surgical approaches to abnormalities of the nasal valve. Rhinology. 1978;16:165.
56. Kern EB. Nasal valve surgery. In: Krause CJ, Mangat S, Pastorek N, editors. Aesthetic facial surgery. Philadelphia: J.B. Lippincott; 1991.
57. Kern EB, Wang TD. Nasal valve surgery. In: Daniel RK, editor. Rhinoplasty. New York: Little, Brown and Company; 1993.
58. Kocer U. Effect of aesthetic rhinoplasty on respiratory functions. Aesthestic Plast Surg. 2001;25:202.
59. Courtiss EH, Goldwyn RM. The effects of nasal surgery on airflow. Plast Reconstr Surg. 1983;72(1):9–21.
60. McKee GJ, O'Neill G, Roberts C, et al. Nasal airflow after septorhinoplasty. Clin Otolaryngol. 1994;19:254.
61. Teichgraeber JF, Wainwright DJ. The treatment of nasal valve obstruction. Plast Reconstr Surg. 1994;93:1174.
62. Constantian M. The incompetent external nasal valve: pathophysiology and treatment in primary and secondary rhinoplasty. Plast Reconstr Surg. 1994;93:919.
63. Kern EB. "Alar Collapse" – an opinion in the collected letters of the international correspondence society of ophthalmologists and otolaryngologists. Feb 15, 1980; Series XXV, p. 17.
64. Rizvi SS, Gauthier MG. Lateralizing the collapsed nasal valves simplified: 10-year survey of a simple concealed suture technique. Laryngoscope. 2011;121:558.
65. Schlosser RJ, Park SS. Surgery for the dysfunctional nasal valve: cadaveric analysis and clinical outcomes. Arch Facial Plast Surg. 1999;1:105.
66. Williams HL. A reconsideration of the relation of the mechanics of nasal airflow to the function of the nose in respiration. Rhinology. 1972;10:145.
67. Cottle MH, Loring RM, Fischer GG, et al. The "maxilla-premaxilla" approach to extensive nasal septum surgery. Arch Otolaryngol. 1958;68:301.
68. Zambetti G, Moresi M, Romeo R, et al. Study and application of a mathematical model for the provisional assessment of areas and nasal resistance, obtained using acoustic rhinometry and active anterior rhinomanometry. Clin Otolaryngol. 2001;26:286.
69. Hasegawa M, Kern EB. The human nasal cycle. Mayo Clin Proc. 1977;52:28.
70. Kern EB. The noncycle nose. Rhinology. 1981;19:59.
71. Parkes MH, Kanodia R. Avulsion of the upper lateral cartilage: etiology, diagnosis, surgical anatomy and management. Laryngoscope. 1981;91:758.
72. Barrs DM, Kern EB. Acute nasal trauma: emergency room care of 250 patients. J Fam Pract. 1980;10:225.
73. Helal MZ, El-Tarabishi M, Sabry SM, et al. Effects of rhinoplasty on the internal nasal valve: a comparison between internal continuous and external perforating osteotomy. Ann Plast Surg. 2010;64:649.
74. Nunez-Fernandez D, Vokurka J, Fernandez-Munoz G. Rhinoplasty, internal valve stenosis. http://emedicine.medscape.com/article/877468-overview.
75. Sheen JH. Spreader graft: a method of reconstructing the roof of the middle nasal vault following rhinoplasty. Plast Reconstr Surg. 1983;73:230.
76. Javanbakht M, Nazari A, Javanbakht A, et al. Body dysmorphic factors and mental health problems in people seeking rhinoplastic surgery. Acta Otorhinolaryngol Ital. 2012;32:37.
77. Aksoy A, Veyseller B, Yildirim YS, et al. Role of nasal muscles in nasal valve collapse. Otolaryngol Head Neck Surg. 2010;142:365.
78. Kern EB. Use of a questionnaire for patients with nasal symptoms. Rhinology. 1972;10:133.
79. Heinberg CE, Kern EB. The Cottle sign: an aid in the physical diagnosis of nasal airflow disturbances. Rhinology. 1973;11:89.
80. Goin JG, Goin MK. Changing the body: psychological effects of plastic surgery. Baltimore/London: Williams and Wilkins; 1981.
81. Broms P. Rhinomanometry: III procedures and criteria for distinction between skeletal stenosis and mucosal swelling. Acta Otolaryngol. 1982;94:361.
82. Corey JP. Acoustic rhinometry: should we be using it? Curr Opin Otolaryngol Head Neck Surg. 2006;14:29.
83. Grymer FL. Reduction rhinoplasty and nasal patency: change in the cross-sectional area of the nose evaluated by acoustic rhinometry. Laryngoscope. 1995;105:429.
84. Roithmann R, Chapnik J, Zamel N, et al. Acoustic rhinometric assessment of the nasal valve. Am J Rhinol. 1997;11:379.
85. Kern EB. Nasal septal reconstruction versus submucous resection. In: Snow JB, editor. Controversy in otolaryngology. Philadelphia: W.B. Saunders; 1980.
86. Lipton RJ, Kern EB. Nasal septal reconstruction. In: Pillsbury HC, Goldsmith MM, et al., editors. Operative challenges in otolaryngology – head and neck surgery. Part II: nasal and sinus surgery. Chicago: Year Book Medical Publishers; 1990.
87. Breadon GB, Kern EB, Neel HBIIII. Autografts of uncrushed and crushed bone and cartilage: experimental observations and clinical implications. Arch Otolaryngol. 1979;105:75.
88. Pirsig W, Kern EB, Verse T. Reconstruction of anterior nasal septum: back-to-back autogenous ear cartilage graft. Laryngoscope. 2004;114:627.
89. Sherris DA, Kern EB. The versatile autogenous rib graft in septorhinoplasty. Am J Rhinol. 1998;12:221.
90. Slavit DH, Bansberg SF, Facer GW, et al. Reconstruction of caudal end of the septum: a case for transplantation. Arch Otolaryngol Head Neck Surg. 1995;121:1091.

91. Hellings PW, Nolst Trenite GP. Long-term patient satisfaction after revision rhinoplasty. Laryngoscope. 2007;117:985.
92. Saleh AM, Younes A, Friedman O. Cosmetics and function: quality-of-life changes after rhinoplasty surgery. Laryngoscope. 2012;122:254.
93. Kern EB. The preoperative discussion as a prelude to managing a complication. Arch Otolaryngol Head Neck Surg. 2003;129:1163.
94. Adamson JE. Constriction of the internal nasal valve in rhinoplasty. Treatment and prevention. Ann Plast Surg. 1987;18:114.
95. Parkes MH, Kanodia R, Kern EB. The universal tip: a systematic approach to aesthetic problems of the lower lateral cartilage. Plast Reconstr Surg. 1988;81:878.
96. Anderson JR. A new approach to rhinoplasty. Trans Am Acad Opthalmol Otolaryngol. 1966;70:183.
97. Schulte DL, Sherris DA, Kern EB. M-Plasty correction of nasal valve obstruction. In: Larrabee WF, Thomas RT, editors. Facial plastic surgery clinics of North America. Toronto: W. B. Saunders; 1999.
98. Bottini DJ, Gentile P, Arpino A, et al. Reconstruction of the nasal valve. J Craniofac Surg. 2007;18:516.
99. Camirand A, Doucet J, Harris J. Nose surgery: how to prevent a middle vault collapse-a review of 50 patients 3 to 21 years after surgery. Plast Reconstr Surg. 2004;114:527.
100. Hage J. Collapsed alae strengthened by conchal cartilage (the butterfly cartilage graft). Br J Plast Surg. 1965;18:92.
101. Khosh MM, Jen A, Honrado C, et al. Nasal valve reconstruction. Arch Facial Plast Surg. 2004;6:167.
102. Ozturan O. Techniques for the improvement of the internal nasal valve in functional-cosmetic nasal surgery. Acta Otolaryngol. 2000;120:312.
103. Paniello RC. Nasal valve suspension: an effective treatment for nasal valve collapse. Arch Otolaryngol Head Neck Surg. 1996;122:1342.
104. Park SS. The flaring suture to augment the repair of the dysfunctional nasal valve. Plast Reconstr Surg. 1998;101:1120.
105. Slavit DH, Lipton RJ, Kern EB, et al. Rhinolift operation in the treatment of the aging nose. Laryngoscope. 1990;103:462.
106. Spielmann PM, White PS, Hussain SSM. Surgical techniques for the treatment of nasal valve collapse: a systematic review. Laryngoscope. 2009;119:1281.
107. Stucker FJ, Lian T, Sanders K. Management of sever bilateral nasal wall collapse. Am J Rhinol. 2002;16:243.
108. Toriumi DM, Josen J, Weinberger M, et al. Use of alar batten grafts for correction of nasal valve collapse. Arch Otolaryngol Head Neck Surg. 1997;123:802.
109. Wittkopf M, Wittkopf J, Ries WR. The diagnosis and treatment of nasal valve collapse. Curr Opin Otolaryngol Head Neck Surg. 2008;16:10.
110. Kienstra MA, Sherris DA, Kern EB. The Cottle vs Joseph rhinoplasty. In: Larrabee WF, Thomas RT, editors. Facial plastic surgery clinics of North America. Philadelphia: W. B. Saunders; 1999.
111. Guenthner TA, Sathe AH, Kern EB. The effect of Le Fort I maxillary impaction on nasal airway resistance. Am J Orthod. 1984;85:308.
112. Lewin ML. Prevention and correction of cicatricial intranasal adhesions in rhinoplastic surgery. Arch Otolaryngol. 1954;60:415.
113. Pontell J, Slavit DH, Kern EB. The role of outfracture in correcting post rhinoplasty nasal obstruction. Ear Nose Throat J. 1998;77(2):106.
114. Webster RC, Davidson TM, Smith RC. Curved lateral osteotomy for airway protection in rhinoplasty. Arch Otolaryngol. 1977;103:454.
115. Chhabra N, Houser SM. The diagnosis and management of empty nose syndrome. Otolaryngol Clin N Am. 2009;42(2):311.
116. Houser SM. Surgical treatment for empty nose syndrome. Arch Otolaryngol Head Neck Surg. 2007;133:858.
117. Martin C. n.d.. http://www.emptynosesyndrome.org.
118. Moore EJ, Kern EB. Atrophic rhinitis: a review of 242 cases. Am J Rhinol. 2001;15:355.
119. Moore GF, Freeman TJ, Ogren FP, Yonkers AJ. Extended follow-up of total inferior turbinate resection for relief of chronic nasal obstruction. Laryngoscope. 1985;95:1095.
120. Rice DH, Kern EB, Marple BF, et al. The turbinates in nasal and sinus surgery: a consensus statement. Ear Nose Throat J. 2003;82(2):82.
121. White RG, Jones AS, Beckingham E. Trimming the inferior turbinates: a prospective long-term study. Clin Otolaryngol. 1990;15:347.
122. Ophir D, Schindel D, Halperin D, et al. Long-term follow-up of the effectiveness and safety of inferior turbinectomy. Plast Reconstr Surg. 1992;90:980.
123. Passali D, Passali FM, Passali GC, et al. Treatment of inferior turbinate hypertrophy: a randomized clinical trial. Ann Otol Rhinol Laryngol. 2003;112:683.

Further Reading

Aiach G. Atlas de Rhinoplastic et de la Voie d'Abord Externe. Paris: Masson; 1993.

Johnson CM, Toriumi DM. Open structure rhinoplasty. Philadelphia: Saunders; 1991.

Peck GC. Techniques in aesthetic rhinoplasty. 2nd ed. Philadelphia: Lippincott; 1990.

Rees TD. Aesthetic plastic surgery. Philadelphia: Saunders; 1980.

Sheen JH, Sheen AP. Aesthetic rhinoplasty. 2nd ed. St. Louis: Mosby; 1987.

Nasal Pulmonary Interactions

38

Jim Bartley

Abbreviations

CO_2	Carbon dioxide
F_ECO_2	Fraction of expired carbon dioxide
F_EO_2	Fraction of expired oxygen
miRNA	MicroRNA
NO	Nitric oxide
PaO_2	Arterial oxygen levels

Key Points

- Epidemiological, physiological, and clinical evidence support an integrated upper and lower respiratory airway or "unified airway" model.
- Nasal breathing significantly improves arterial oxygen concentrations, lung oxygen extraction, and carbon dioxide excretion.
- Nasal mucosal inflammation results in lower airway inflammation, and vice versa.
- Many patients with asthma, chronic obstructive airways disease, and bronchiectasis have significant upper respiratory disease.
- Nitric oxide from the nose and sinuses may have a role in sterilizing incoming air, and in improving lung ventilation-perfusion.
- Nasal resistance is inversely related to end-tidal carbon dioxide levels.

38.1 Introduction

The upper and lower respiratory tracts form a *continuum.* Physiological, epidemiological, and clinical evidence support an integrated upper and lower respiratory airway, or "unified airway", model [1–4]. Important, well-known nasal functions include the filtering, warming, and humidification of inspired air before inhalation into the lungs. Nasal mucosal inflammation results in lower airway inflammation, and vice versa [5, 6]. Inflammatory mediators and/or infectious pathogens may also be carried along the respiratory mucosa or along air currents between the upper and lower respiratory airways [1]. Neuronal responses may play a role, although the existence of nasobronchial reflexes remains controversial [7]. Nitric oxide (NO) from the nose and sinuses may have a role in the sterilization of incoming air [8], and in improving ventilation-perfusion in the lungs [9]. An inverse relationship between nasal resistance and end-tidal carbon dioxide (CO_2) levels has been described [10, 11]. NO and CO_2

J. Bartley (✉)
Department of Otolaryngology – Head and Neck Surgery, Counties Manukau District Health Board, Auckland, New Zealand

Ö. Ö. Celebi, T. M. Önerci (eds.), *Nasal Physiology and Pathophysiology of Nasal Disorders*,
https://doi.org/10.1007/978-3-031-12386-3_38

may act as aerocrine messengers [9, 12]. Olfaction is linked to the limbic system [13], which can independently control our breathing pattern and rate [14].

38.2 Physiological Interactions

The nose has an important role in filtering, warming, and humidifying inspired air before inhalation into the lungs. On expiration, as air passes through the nose, it gives up heat to the cooler nasal mucosa. This cooling causes water vapor condensation and a 33% return of both heat and moisture to the mucosal surfaces [15]. The nose also provides resistance to both inspiration and expiration that is twice that of the open mouth. This increased resistance appears to have several physiological benefits. In a study of arterial oxygen levels (PaO_2) before and after jaw wiring, which forced patients to breathe continuously through their noses, PaO_2 increased by nearly 10% [16]. After nasal surgery, nasal packing forcing patients to breathe through their mouths is associated with a reduction in arterial O_2 saturation [17]. At rest, expired end-tidal CO_2 levels increase with nasal breathing indicating that nasal breathing improves the efficiency of CO_2 excretion from the lungs [18]. During exercise, nasal breathing reduces the fraction of expired oxygen (F_EO_2), indicating that on expiration the percentage of O_2 extracted from the air by the lungs is increased, and increases the fraction of expired carbon dioxide (F_ECO_2), indicating an increase in the percentage of expired air that is CO_2 [19]. This equates to more efficient O_2 extraction and CO_2 excretion during exercise.

Explanations for these observations remain largely hypothetical. Nasal breathing increases total lung volume [16]. The corresponding increase in functional residual capacity (volume of air present in the lungs present after passive expiration) is thought to improve gas exchange leading to improved PaO_2. NO derived from the nose and sinuses might also improve ventilation-perfusion relationships in the lungs [20]. The nose also provides an inspiratory resistance forcing the diaphragm to contract against resistance. On a long-term basis, this might also have an important role in maintaining respiratory muscle strength [21, 22]. Regardless, nasal breathing appears important in aiding O_2 absorption and in facilitating CO_2 excretion in the lungs.

The breathing cycle is divided into inspiratory and expiratory phases. When the respiratory rate increases, the expiratory phase shortens. Increasing the expiratory phase of the respiratory cycle increases the body's relaxation response [23]. Nasal breathing slows the respiratory rate increasing the length of the expiratory phase [24]. Changes in nasal resistance could potentially influence breathing patterns and rate.

38.3 Respiratory Inflammation

Upper and lower airway inflammatory processes often co-exist and share common pathogenic mechanisms [3]. Based on the predominant cell type, chronic rhinosinusitis (CRS) is classified as being either eosinophilic or neutrophilic [25]. The eosinophilic group includes CRS with polyps, a subset of CRS without polyps, aspirin hypersensitivity, asthma and nasal polyps (Samter's triad), and allergic fungal rhinosinusitis. Eosinophilic CRS and allergic rhinitis are associated with asthma [3].

Lower airway inflammation has also been classified according to the cell profile of induced sputum as being either eosinophilic or non-eosinophilic [26]. In asthma patients, eosinophils are the dominant inflammatory cells in middle meatal lavage [27]. In small airway disease patients, neutrophils are the dominant inflammatory cells in middle meatal lavage [27].

Asthma and allergic rhinitis are strongly interrelated [3, 4, 28]. The severity of asthma symptoms correlates closely with rhinitis symptoms [4]. The presence of allergic rhinitis increases the risk of subsequent asthma development [29]. The prevalence of asthma is 20% in CRS patients, and asthma severity also correlates with CRS disease severity [30]. In patients having functional endoscopic sinus surgery, the asthma prevalence is 42%, rising to 50% in those with nasal polyps [31].

Neutrophilic inflammation is a feature of both chronic obstructive pulmonary disease (COPD) and bronchiectasis [26]. A neutrophilic picture may be seen in asthma [26]. In COPD patients, inflammatory cells are found both in the sputum and in lung biopsy specimens [2]. COPD patients commonly report nasal symptoms, the most common being rhinorrhea [2, 32]. Nasal symptoms increase the risk of developing COPD [33]. Increased levels of the neutrophil chemoattractant protein IL-8 are found in the upper airways of COPD patients when compared to control subjects [32]. Upper airway IL-8 concentrations correlate with those in the lower airway, and the concentration at both sites is related to bacterial colonization [32]. Many bronchiectasis patients also have nasal and sinus disease [1, 34]. Bronchiectasis severity also correlates with CRS severity [1].

38.3.1 Inflammatory Interactions

The inflammatory interactions between the upper and lower respiratory tract are numerous and complex and our understanding continues to evolve. A microRNA (miRNA) is a small non-coding RNA molecule (containing about 22 nucleotides) regulating the gene expression that is involved in many inflammatory diseases. Subsets of circulating miRNAs are expressed in patients with allergic rhinitis and asthmatic patients. MiRNA influences post-transcriptional regulation of gene expression. MiRNA's role in allergic rhinitis and asthma continues to be clarified [35].

Nasal provocation antigen challenge increases bronchial hyperresponsiveness to methacholine challenge [36]. Similarly, segmental bronchial provocation in patients with allergic rhinitis induces blood eosinophilia and mucosal inflammation in both the upper and lower airways. This inflammation is characterized by increased numbers of eosinophils, IL-5+ cells, and eotaxin-positive cells. Local allergen exposure in both the upper and lower airways results in generalized airway inflammation; this would appear to occur through vascular mechanisms. The most likely mechanism is that localized inflammatory changes in the upper and lower airways lead to a systemic response, with bone marrow involvement, which results in the release of progenitor cells that are then recruited to tissue sites [37]. Micro-aspiration and bacterial and inflammatory mediator transmission could also occur (Sect. 38.3.2).

38.3.2 Microbial and Inflammatory Mediator Transmission

Nasopharyngeal secretions may be aspirated into the lower respiratory airway [38–40]. Radionucleotide scanning in humans shows that gross aspiration of nasopharyngeal secretions into the lower respiratory airway does not occur [38]. However during sleep, in animals [39] and in healthy young males, aspiration of upper airway secretions into the lungs occurs [40].

Our understanding of the microbiome and the interrelationships between upper respiratory, oral, lung, and gut microbiome continue to evolve [41]. The healthy lung microbiome is similar to the mouth microbiome, but at lower concentrations, with lower membership and different community composition. In healthy subjects, the nasal microbiome is distinct from the oral microbiome and contributes little to the composition of the lung microbiome [42].

COPD patients with lower airway bacterial colonization have a higher total nasal bacterial load [43]. Children with bronchiectasis have a high nasopharyngeal carriage of *Streptococcus pneumoniae,* non-typable *Haemophilus influenzae* (NTHi), and *Moraxella catarrhalis*, and lower airway infection by NTHi. A high level of similarity in cultures from the nasopharynx and from bronchoalveolar lavage is seen [44]. A similar agreement is also seen in children with protracted bacterial bronchitis [44], and in patients with cystic fibrosis [45]. However, throat swabs are not representative of lower airway microbiome and metagenome in cystic fibrosis children [46]. In the critical care situation, the treatment of upper respiratory disease reduces the risk of ventilator-associated pneumonia [47]. People with a high nasal bacterial load (and associated greater nasal inflammation) appear more likely to pass bacteria into the lower respiratory tract. However traditional culture procedures allow the isolation and characterization of a very low percentage of the microorganisms of a microbiome. The standardization of culture-independent DNA-based sequencing methods continues to improve [41].

38.3.3 Influence of Upper Respiratory Interventions on the Lower Respiratory Tract

In the majority of trials, allergic rhinitis treatment with nasal steroids reduces asthma severity [4, 48]. Immunotherapy for allergic rhinitis appears to reduce the risk of developing asthma [29, 49]. Similarly, several studies have shown that the surgical treatment of CRS helps asthma patients through both an improvement in asthma symptoms and the decreased use of asthma medication [4]. Unfortunately, these studies have been uncontrolled. No randomized controlled studies have investigated the effect of CRS medical therapy on asthma [4].

38.4 Nasobronchial Reflexes

Nasobronchial reflexes have been implicated in the interactions between the upper and lower respiratory airways. Mechanical or chemical stimulation of nasal, tracheal, and laryngeal receptors could produce sneezing, coughing, or bronchoconstriction, thus preventing deeper penetration of allergens or irritants into the airway [7]. Unilateral models of nasal provocation show that secretory responses can be measured in both nostrils [7]. The mechanism appears to be neural [7]. When asthmatic patients exercise with their noses occluded, a 20% decline in forced expiratory flow occurs, compared with less than a 5% reduction among patients allowed to exercise while nose breathing [50]. Bursts of cold air on the nasal mucosa increase nasal resistance. This effect was blocked by anesthetizing the nose or by inhaling atropine, an anti-cholinergic drug, before the provocation [51, 52]. Other researchers have not confirmed these results [53]. Immersion of the head into cold water leads to immediate suppression of respiration (apnea), laryngospasm, and bronchoconstriction (the diving reflex) [54]. However, the existence of nasobronchial reflexes secondary to inflammatory exposure in the upper respiratory airway remains controversial [7].

38.5 Olfaction and the Limbic System

Evolutionary theory teaches that in primitive life forms, the olfactory brain was probably a layer of cells above the brain stem that registered a smell and then simply categorized it. New layers of the olfactory brain then developed into what was initially called the rhinencephalon (nose brain) or limbic system [55]. Our limbic system coordinates the stress "fight or flight" response. As part of this response, the respiratory rate also goes up; the limbic system is able to override normal pCO_2 homeostasis [14]. The physiological evidence suggests that fragrances such as rosemary and lavender may have influenced human memory and mood [56]. Fragrances such as lavender by influencing the limbic system, and the stress response could potentially affect breathing patterns and rate.

38.6 Aerocrine Communication

38.6.1 Nitric Oxide

NO is a gas that is produced by the nose and paranasal sinuses. NO has bacteriostatic and antiviral actions and may have a role in sterilizing incoming air [8], improving ventilation-perfusion in the lungs, and in facilitating capillary oxygen delivery [9]. NO also stimulates mucociliary clearance [57]. Humming increases nasal NO levels [58]. One case report and a randomized controlled study involving sixty people indicate that regular humming is beneficial in CRS [59, 60].

38.6.2 Carbon Dioxide

An inverse relationship between nasal resistance and end-tidal pCO_2 levels has been described [10, 11]. A reduction of end-tidal expired pCO_2 from 40 to 35 mmHg (5.3 to 4.7 kPa) corresponds to an increase in nasal resistance of 10% [10]. Breathing is controlled largely by independent voluntary and metabolic pathways. However, the limbic system is able to override metabolic respiratory control systems [14]. People practicing yoga set their pCO_2 receptors to a higher response level. End-tidal pCO_2 concentrations are nearly 4 mmHg higher in yogic breathers [61]. In contrast, people who are prone to anxiety attacks have lower arterial pCO_2 levels (5 mmHg on average) as compared with controls [62]. The mechanisms of the relationship between expired pCO_2 levels and nasal resistance are unknown. This could be either a systemic vascular effect or an aerocrine effect.

CO_2 irrigation and inhalation may also influence allergy. CO_2-enriched water irrigation improves allergic rhinitis symptom scores and reduces the concentrations of several allergic chemokines and cells [63]. Rapid relief from allergic rhinitis symptoms has been seen after single CO_2 irrigation doses in symptomatic allergic rhinitis patients [64]. People, who practice relaxed diaphragmatic breathing may be less likely to complain of nasal congestion and allergic rhinitis symptoms [12].

38.7 Conclusions

From a physiological perspective, the nose has an important role in the preparation of inspired air before inhalation into the lungs. Improvements in O_2 transfer and CO_2 excretion also occur. Inflammation, both allergic and infective, affects both the upper and lower respiratory airways. Vascular mechanisms would appear involved. Microtransmission of bacteria and inflammatory mediators between the upper and lower respiratory airways may occur. Medical and surgical treatments of upper respiratory disease appear to help asthmatic patients. NO and CO_2 may have roles in aerocrine communication. The influence of nasobronchial reflexes remains undecided. Optimal management of disease processes in both the upper and lower respiratory airways needs to consider a "unified airway" model.

References

1. Guilemany JM, Angrill J, Alobid I, Centellas S, Pujols L, Bartra J, et al. United airways again: high prevalence of rhinosinusitis and nasal polyps in bronchiectasis. Allergy. 2009;64:790–7.
2. Hurst JR. Upper airway. 3: Sinonasal involvement in chronic obstructive pulmonary disease. Thorax. 2009;65:85–90.
3. Marple BF. Allergic rhinitis and inflammatory airway disease: interactions within the unified airspace. Am J Rhinol Allergy. 2010;24:249–54.
4. Krouse J, Brown R, Fineman S, Han J, Heller A, Joe S, et al. Asthma and the unified airway. Otolaryngol Head Neck Surg. 2007;136:S75–S106.
5. Braunstahl G, Overbeek S, Fokkens W, Kleinjan A, McEuen A, Walls A, et al. Segmental bronchoprovocation in allergic rhinitis patients affects mast cell and basophil numbers in nasal and bronchial mucosa. Am J Respir Crit Care Med. 2001;164:858–65.
6. Braunstahl G, Overbeek S, KleinJan A, Prins J-B, Hoogsteden H, Fokkens W. Nasal allergen provocation induces adhesion molecule expression and tissue eosinophilia in upper and lower airways. J Allergy Clin Immunol. 2001;107:469–76.
7. Sarin S, Undem B, Sanico A, Togias A. The role of the nervous system in rhinitis. J Allergy Clin Immunol. 2006;118:999–1014.
8. Workman AD, Carey RM, Kohanski MA, Kennedy DW, Palmer JN, Adappa ND, et al. Relative susceptibility of airway organisms to antimicrobial effects of nitric oxide. Int Forum Allergy Rhinol. 2017;7:770–6.
9. Selimoglu E. Nitric oxide in health and disease from the point of view of the otorhinolaryngologist. Curr Pharm Des. 2005;11:3051–60.
10. Mertz J, McCaffrey T, Kern E. Role of the nasal airway in regulation of airway resistance during hypercapnia and exercise. Otolaryngol Head Neck Surg. 1984;92:302–7.
11. Shi Y-X, Seto-Poon M, Wheatley J. Alae nasi activation decreases nasal resistance during hyperoxic hypercapnia. J Appl Physiol. 1988;85:294–300.
12. Bartley J. Nasal congestion and hyperventilation syndrome. Am J Rhinol. 2005;19:607–11.
13. Soudry Y, Lemogne C, Malinvaud D, Consoli S, Bonfils P. Olfactory system and emotion: common substrates. Eur Ann Otorhinolaryngol Head Neck Dis. 2011;128:18–23.

14. Plum F. Breathing is controlled independently by voluntary, emotional and metabolically related pathways. Arch Neurol. 1992;49:441.
15. White DE, Bartley J, Nates RJ. Model demonstrates functional purpose of the nasal cycle. Biomed Eng Online. 2015;14:38.
16. Swift A, Campbell I, McKown T. Oronasal obstruction, lung volumes, and arterial oxygenation. Lancet. 1988;1(8577):73–5.
17. Ogretmenoglu O, Yilmaz T, Rahimi K, S A. The effect on arterial blood gases and heart rate of bilateral nasal packing. Eur Arch Otorhinolaryngol. 2002;259:63–6.
18. Tanaka Y, Morikawa T, Honda Y. An assessment of nasal functions in control of breathing. J Appl Physiol. 1988;65:1520–4.
19. Morton A, King K, Papalia S, Goodman C, Turley K, Wilmore J. Comparison of maximal oxygen consumption with oral and nasal breathing. Aust J Sci Med Sport. 1995;27:51–5.
20. Della Rocca G, Coccia C. Nitric oxide in thoracic surgery. Minerva Anestesiol. 2005;71:313–8.
21. Illi SK, Held U, Frank I, Spengler CM. Effect of respiratory muscle training on exercise performance in healthy individuals: a systematic review and meta-analysis. Sports Med. 2012;42:707–24.
22. Shei RJ, Paris HL, Wilhite DP, Chapman RF, Mickleborough TD. The role of inspiratory muscle training in the management of asthma and exercise-induced bronchoconstriction. Phys Sportsmed. 2016;44:327–34.
23. Cappo B, Holmes D. The utility of prolonged respiratory exhalation for reducing physiological and psychological arousal in non-threatening and threatening situations. J Psychosom Res. 1984;28:265–73.
24. Ayoub J, Cohendy R, Dauzat M, Targhetta R, De la Coussaye J, Bourgeois J, et al. Non-invasive quantification of diaphragm kinetics using M-mode sonography. Can J Anaesth. 1997;44:739–44.
25. Meltzer E, Hamilos D. Rhinosinusitis diagnosis and management for the clinician: a synopsis of recent consensus guidelines. Mayo Clin Proc. 2011;86:427–43.
26. Hargreave F. Quantitative sputum cell counts as a marker of airway inflammation in clinical practice. Curr Opin Allergy Clin Immunol. 2007;7:102–6.
27. Ragab A, Clement P, Vincken W. Correlation between the cytology of the nasal middle meatus and BAL in chronic rhinosinusitis. Rhinology. 2005;43:11–7.
28. Corren J. Allergic rhinitis and asthma: how important is the link? J Allergy Clin Immunol. 1997;99:S781–6.
29. Morjaria JB, Caruso M, Emma R, Russo C, Polosa R. Treatment of allergic rhinitis as a strategy for preventing asthma. Curr Allergy Asthma Rep. 2018;18:23.
30. Rosati MG, Peters AT. Relationships among allergic rhinitis, asthma, and chronic rhinosinusitis. Am J Rhinol Allergy. 2016;3:44–7.
31. Senior B, Kennedy DW, Tanabodee J. Long-term impact of functional endoscopic sinus surgery of asthma. Otolaryngol Head Neck Surg. 1999;121:66–8.
32. Hurst J, Perera W, Wilkinson T, Donaldson G, Wedzicha J. Systemic and upper and lower airway inflammation at exacerbation of chronic obstructive pulmonary disease. Am J Respir Crit Care Med. 2006;173:71–8.
33. Nihlen U, Montgomery P, Anderson M, Persson CG, Nyberg P, Löfdahl CG, et al. Specific nasal symptoms and symptom producing factors may predict increased risk of developing COPD. Clin Physiol Funct Imaging. 2008;28:240–50.
34. Bartley J, Garrett J, Camargo CA Jr, Scragg R, Vandal A, Sisk R, et al. Vitamin D3 supplementation in adults with bronchiectasis: A pilot study. Chron Respir Dis. 2018;15:384–92.
35. Liu Z, Zhang XH, Callejas-Diaz B, Mullol J. MicroRNA in united airway diseases. Int J Mol Sci. 2016;17:716.
36. Bonay M, Neukirch C, Grandsaigne M, Leçon-Malas V, Ravaud P, Dehoux M, et al. Changes in airway inflammation following nasal allergic challenge in patients with seasonal rhinitis. Allergy. 2006;61:111–8.
37. Licari A, Caimmi S, Bosa L, Marseglia A, Marseglia GL, Caimmi D. Rhinosinusitis and asthma: a very long engagement. Int J Immunopathol Pharmacol. 2014;27:499–508.
38. Bardin PG, Van Heerden BB, Joubert JR. Absence of pulmonary aspiration of sinus contents in patients with asthma and sinusitis. J Allergy Clin Immunol. 1990;86:82–3.
39. Kogahara T, Kanai K, Asano K, Suzaki H. Evidence for passing down of postnasal drip into respiratory organs. In Vivo. 2009;23:297–301.
40. Gleeson K, Eggli DF, Maxwell SL. Quantitative aspiration during sleep in normal subjects. Chest. 1997;111:1266–72.
41. Beule AG. The microbiome - the unscheduled parameter for future therapies. Laryngorhinootologie. 2018;97:S279–311.
42. Bassis CM, Erb-Downward JR, Dickson RP, Freeman CM, Schmidt TM, Young VB, et al. Analysis of the upper respiratory tract microbiotas as the source of the lung and gastric microbiotas in healthy individuals. MBio. 2015;6:e00037.
43. Hurst J, Wilkinson T, Perera W, Donaldson GC, Wedzicha JA. Relationships among bacteria, upper airway, lower airway and systemic inflammation in COPD. Chest. 2005;127:1219–26.
44. Hare KM, Grimwood K, Leach AJ, Smith-Vaughan H, Torzillo PJ, Morris PS, et al. Respiratory bacterial pathogens in the nasopharynx and lower airways of Australian indigenous children with bronchiectasis. J Pediatr. 2010;157:1001–5.

45. Godoy J, Godoy A, Ribalta G, Largo I. Bacterial pattern in chronic sinusitis and cystic fibrosis. Otolaryngol Head Neck Surg. 2011;145:673–6.
46. Kirst ME, Baker D, Li E, Abu-Hasan M, Wang GP. Upper versus lower airway microbiome and metagenome in children with cystic fibrosis and their correlation with lung inflammation. PLoS One. 2019;14:e0222323.
47. De Smet A, Kluytmans T, Cooper B, Mascini EM, Benus RF, van der Werf TS, et al. Decontamination of the digestive tract and oropharynx in ICU patients. N Engl J Med. 2009;360:20–31.
48. Stelmach R, do Patrocinio T, Nunes M, Ribeiro M, Cukier A. Effect of treating allergic rhinitis with corticosteroids in patients with mild-to moderate persistent asthma. Chest. 2005;128:3140–7.
49. Moller C, Dreborg S, Ferdousi H, Halken S, Høst A, Jacobsen L, et al. Pollen immunotherapy reduces the development of asthma in children with seasonal rhinoconjunctivitis (the PAT study). J Allergy Clin Immunol. 2002;109:251–6.
50. Shturman-Ellstein R, Zeballos R, Buckley J, Souhrada J. The beneficial effect of nasal breathing on exercise-induced bronchoconstriction. Am Rev Respir Dis. 1978;118:65–73.
51. Fontanari P, Burnet H, Zattara-Hartman M, Jammes Y. Changes in airway resistance induced by nasal inhalation of cold dry, or moist air in normal individuals. J Appl Physiol. 1996;81:1739–43.
52. Fontanari P, Zattara-Hartmann M-C, Burnet H, Jammes Y. Nasal eupnoeic inhalation of cold, dry air increases airway resistance in asthmatic patients. Eur Respir J. 1997;10:2250–4.
53. Johansson A, Bende M, Millqvist E, Bake B. Nasobronchial relationship after cold air provocation. Respir Med. 2000;94:1119–22.
54. Baraniuk JN, Merck SJ. Nasal reflexes: implications for exercise, breathing, and sex. Curr Allergy Asthma Rep. 2008;8:147–53.
55. Le Doux J. The emotional brain. Phoenix, New York: Simon & Schuster; 2003.
56. Herz R. Aromatherapy facts and fictions: a scientific analysis of olfactory effects on mood, physiology and behavior. Int J Neurosci. 2009;119:263–90.
57. Lindberg S, Cervin A, Runer T. Low levels of nasal nitric oxide (NO) correlate to impaired mucociliary function in the upper airways. Acta Otolaryngol. 1997;117:728–34.
58. Weitzberg E, Lundberg JO. Humming greatly increases nasal nitric oxide. Am J Respir Crit Care Med. 2002;166:144–5.
59. Eby GA. Strong humming for one hour daily to terminate chronic rhinosinusitis in four days: a case report and hypothesis for action by stimulation of endogenous nasal nitric oxide production. Med Hypotheses. 2006;66:851–4.
60. Abishek K, Bakshi SS, Bhavanani AB. The efficacy of yogic breathing exercise Bhramari Pranayama in relieving symptoms of chronic rhinosinusitis. Int J Yoga. 2019;12:120–3.
61. Stanescu D, Nemery B, Verityer C, Marechal C. Pattern of breathing and ventilatory response to CO_2 in subjects practicing Hatha Yoga. J Appl Physiol. 1981;51:1625–9.
62. Papp L, Martinez J, Klein D, Ross D, Liebowitz M, Fyer A, et al. Arterial blood gas changes in panic disorder and lactate-induced panic. Psychiatry Res. 1989;28:171–80.
63. Pagani D, Galliera E, Dogliotti G, Bernardi D, di Valserra M, Torretta S, Solimene U, et al. Carbon dioxide-enriched water inhalation in patients with allergic rhinitis and its relationship with nasal fluid cytokine/chemokine release. Arch Med Res. 2011;42:329–33.
64. Casale TB, Onder RF, Berkowitz RB, Korenblat PE. Nasal carbon dioxide used as needed in the symptomatic treatment of seasonal allergic rhinitis. J Allergy Clin Immunol Pract. 2018;6:183–9.

39 Physiologic and Dentofacial Effects of Mouth Breathing Compared to Nasal Breathing

Tulin Taner and Banu Saglam-Aydinatay

Core Messages

- Mouth breathing is a serious ailment which occurs in the presence of an obstruction in the nasal and nasopharyngeal regions of the upper airway.
- Chronic mouth breathing can cause problems in facial structures and oral health.
- Changes in facial structures include long anterior facial height, narrow facial width, a retrognathic mandible, open-mouthed posture, an incompetent and short upper lip, loss of tonus in perioral muscles, a pinched-looking nose and a dull appearance.
- The intraoral consequences of mouth breathing are open bite, a Class II molar occlusion with increased overjet, protruding maxillary anterior teeth, and a V-shaped maxillary arch.
- Not every patient has the same growth changes due to oral breathing. The heritable characteristics of anatomical structures also seem to play a role in determining which patients will be most affected.
- A careful evaluation of the patient by otolaryngologists, pediatricians, and orthodontists is needed before treatment decisions.
- Medical, surgical, and/or orthodontic treatment of the patients may be necessary depending on the cause of airway obstruction and severity of changes in dentofacial structures.
- Myofunctional therapy, maxillary expansion, functional appliances, dentofacial orthopedics, or orthognathic treatment are available in the treatment of these patients.

T. Taner (✉)
Department of Orthodontics, Faculty of Dentistry, Hacettepe University, Ankara, Turkey

B. Saglam-Aydinatay
Private Practice, Ankara, Turkey

39.1 Introduction

Mouth breathing is the act of inhaling and exhaling using the mouth. This respiratory phenomenon is considered normal under increased physical activity because the increased need for oxygen under such circumstances can only be supplied by breathing through the mouth and nose simultaneously. However, using one's mouth to breathe in daily life or during sleep is a serious ailment which occurs in the presence of an obstruction in nasal and nasopharyngeal regions of the upper airway. Nasal obstructions can be due to physiologic factors like allergic rhinitis and polyps or anatomical factors such as a deviated septum and a narrow nasal area. Enlarged adenoids and tonsils are the most common causes of obstruction in the nasopharyngeal area, especially at ages 5–6 years.

If chronic mouth breathing develops due to obstruction of the airways, this can cause a multitude of problems in facial structures and oral

Ö. Ö. Celebi, T. M. Önerci (eds.), *Nasal Physiology and Pathophysiology of Nasal Disorders*,
https://doi.org/10.1007/978-3-031-12386-3_39

health. Such patients have been described as having special facial characteristics generally referred to as *adenoid facies* [1–3], *long face syndrome* [4], and *respiratory obstruction syndrome* [5]. This type of face reportedly features long anterior facial height, narrow facial width, a retrognathic mandible, open-mouthed posture, an incompetent and short upper lip, loss of tonus in perioral muscles, a pinched-looking nose, and a dull appearance (Figs. 39.1 and 39.2). The intraoral consequences of mouth breathing are said to include an open bite, a Class II molar occlusion with increased overjet, protruding maxillary anterior teeth, and a V-shaped maxillary arch with a deep palatal vault (Figs. 39.3, 39.4, and 39.5). Mouth breathers may also extend their heads in order to maintain a patent airway (Figs. 39.6 and 39.7). In addition to these structural changes, mouth breathing can cause halitosis (breath malodor), increased incidence of caries, and marginal gingivitis around the maxillary anterior teeth. If diagnosed early, orthodontic and dentofacial orthopedic treatment of the patients is possible. However, in adult patients, orthognathic surgical treatment in addition to orthodontic treatment may be necessary.

Considering the potential serious side effects of mouth breathing, patients need to be carefully evaluated by an orthodontist and otolaryngologist to determine the etiology and proper treatment of nasal obstruction and the associated dentofacial changes.

Despite many attempts to establish a cause-and-effect relationship between nasal respiratory impairments and dentofacial deformities, the issue is still controversial within orthodontics and otolaryngology. The most prevalent view among clinicians is that a change in the mode of respiration, such as mouth breathing due to an inadequate nasal airway, could cause changes in craniocervical posture, maxillomandibular relationship, and position of the tongue. This in turn could affect dentofacial growth and positions of the teeth. However, clinicians from

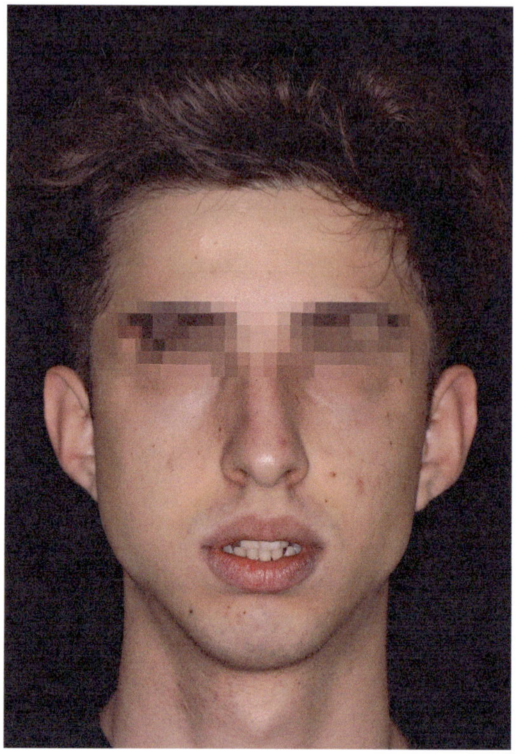

Fig. 39.1 A patient with increased anterior facial height, open-mouthed posture, and an incompetent and short upper lip. However, these facial features are not always diagnostic of the respiratory pattern since patients who breathe with their lips separated may actually be oral-nasal or nasal breathers as well as oral breathers

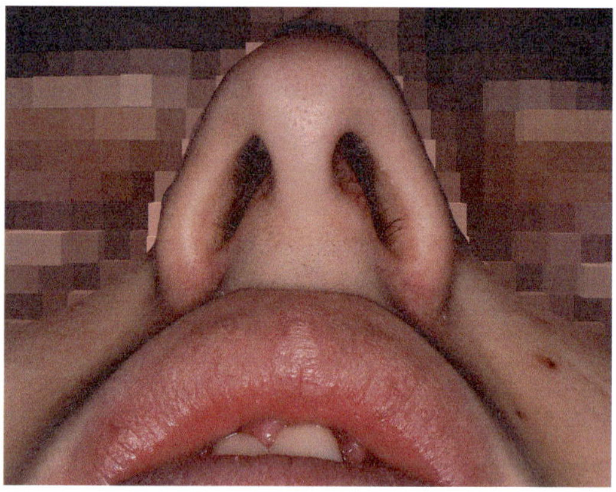

Fig. 39.2 The typical pinched nose appearance in a mouth-breathing patient

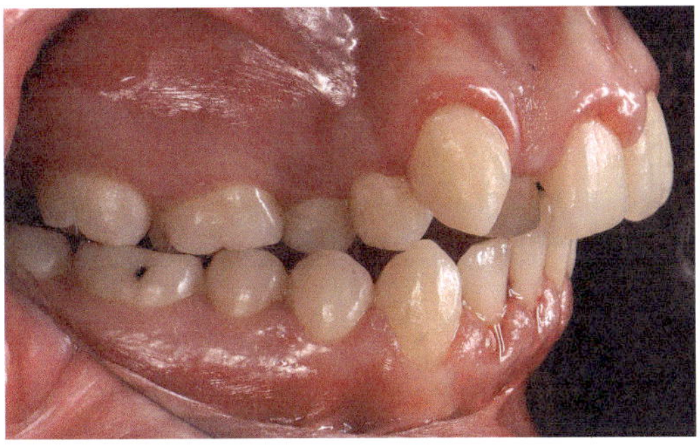

Fig. 39.3 Sagittal occlusal relationship in mouth breathing patients is said to include a Class II molar occlusion with increased overjet

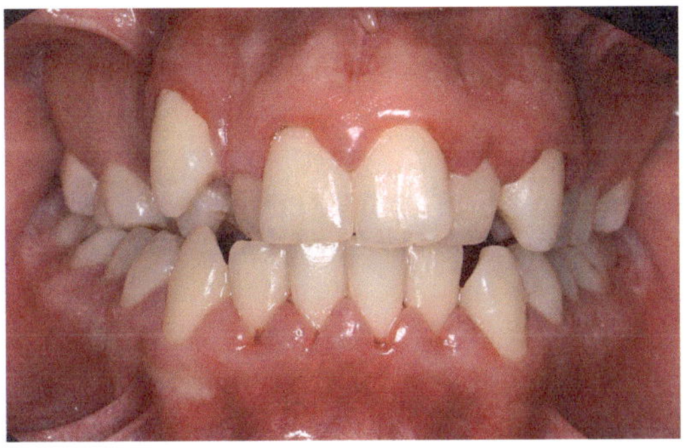

Fig. 39.4 A posterior crossbite malocclusion in a mouth-breathing patient. The posterior crossbite is suggested to be due to transverse maxillary deficiency caused by an imbalance of forces between the muscle forces acting on these structures

both sides of controversy can find ample evidence in the literature supporting their opinions, and the results of studies relating dentofacial features with the respiratory pattern are far from being conclusive.

Since there are treatment decisions that revolve around the degree of interplay between nasal obstruction and dentofacial development, the relationship must be further elucidated. If the nasal obstruction has an effect on dentofacial development, early treatment for removal of the cause of this obstruction would be necessary. On the other hand, if dentofacial growth and development are not significantly affected by the respiratory mode, then treatment of nasal obstruction in order to prevent abnormal orofacial development would not be indicated.

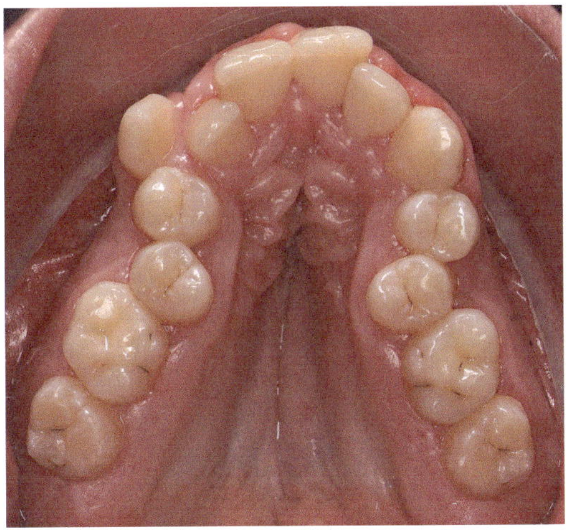

Fig. 39.5 The V-shaped maxillary arch with a deep palatal vault

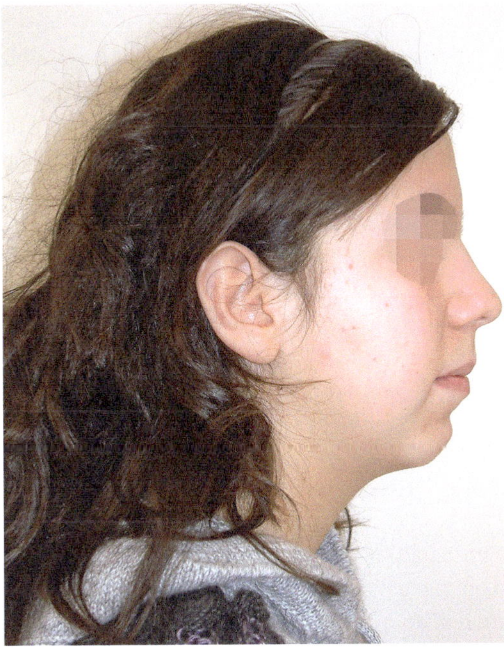

Fig. 39.6 The extended head position of a mouth-breathing child. The necessity of airway maintenance dictates the head and neck posture of these patients

In this chapter, we examine the possible interactions among respiratory mode and dentofacial growth and development by reviewing both sides of the controversy with special emphasis on questions such as:

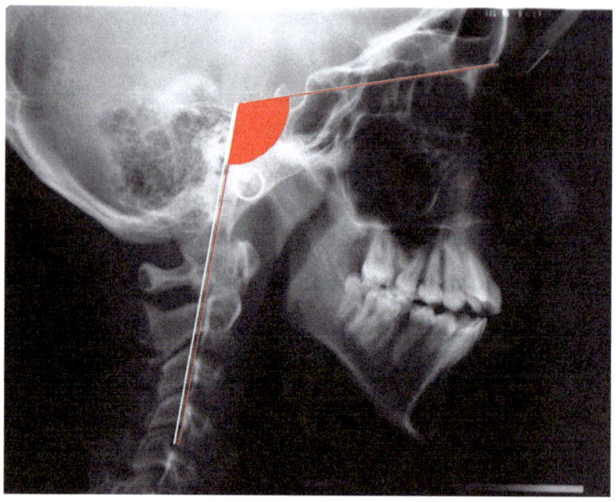

Fig. 39.7 Lateral cephalometric radiograph of the patient in Fig. 39.6. Note the dental protrusion and downward-backward rotation of the mandible in addition to the extended head position

- Does impaired nasal breathing always result in complete mouth breathing with possible negative effects on the dentofacial complex?
- What are the effects of mouth breathing on craniocervical posture, maxillomandibular relationships, facial morphology, tongue position, and occlusal features?
- Does reversing the mode of respiration from oral to nasal allow proper growth and of craniofacial and dentofacial complex?
- Do orthodontic treatment mechanics have a role in normalizing the breathing pattern?

39.2 Respiratory Pattern

The relationship between mouth breathing and dentofacial growth is still being debated after more than a century. Despite the existence of an extensive body of literature on this subject, the inconclusive results so far may be explained by different population selection criteria and the various diagnostic methods used for differentiating mouth breathers from nasal breathers. In order to determine this assumed relationship, the meaning of the term *mouth breathing* needs to be clearly established.

Human neonates are obligatory nasal breathers [6]. After birth, infant feeding requires the coordination of sucking, swallowing, and breathing so that the infant will be able to eat and breathe at the same time [7]. Thus, the maintenance of the airway by correct posturing of the mandible and the tongue is necessary for survival. James and Hastings [8] examined 53 infants, aged from 1 to 14 days, and found that most of the infants in their study with lips-apart posture achieved an oral seal by putting the tip of their tongue between the lips and their respiration was through the nose. Later, growth changes make it possible to breathe through the mouth. Although humans are primarily nasal breathers when the need for air increases, the mode of breathing changes from nasal to partially oral. In healthy adults, the switch from nasal to oronasal breathing occurs when ventilatory exchange rates above 40–45 L/min are reached [9]. This shift is transient, and when the need for air decreases back to normal, the individual goes back to nasal breathing. Human nasal passages also exhibit spontaneous changes in unilateral nasal airway resistance as a result of alternate nasal congestion and decongestion on opposing sides of the nose. This nasal cycle is said to last between 4 and 6 h, but fluctuations in nasal patency from 10 min to several days have been shown to occur [10, 11].

Mouth breathing can be defined as a shift from nasal respiration to oral respiration or mixed respiration due to obstruction or restriction on any part of the airway. Causes of such a nasal obstruction followed by a transition to mouth breathing can be adenoid or tonsil hypertrophy, chronic and allergic rhinitis, nasal traumas, congenital nasal deformities, foreign bodies, polyps, and neoplasms [12]. It has been reported that when any of these factors increase the nasal resistance and pressure, the patients sometimes break the anterior and posterior oral seals resulting in oral respiration [13]. It is also possible for mouth breathing to occur as the result of habit, with or without any impairment of the upper airway [14].

> Treatment of dentofacial deformities caused by mouth breathing should be preceded by a thorough ENT examination to determine whether the cause is *habitual* or *obstructive*. If the cause is habitual, the treatment goal is to obtain a lip seal and force the patient to breathe through the nose. However, in cases of mouth breathing due to respiratory obstruction, the treatment should focus on relieving the obstructive cause.

Various diagnostic methods have been used in research to determine the respiratory mode [8, 15–21]. Others have used the presence of adenoids [22, 23] without determining the respiratory mode. Most of these diagnostic tests have been indicated to be inconsistent and lacking in sensitivity and specificity [24, 25]. With the improvements in physiologic diagnostic methods, rhinometric measurements such as nasal resistance came to be widely used in research. However, associations between nasal resistance and respiratory mode were reported to be variable and weak [26–28]. Advances in respirometric techniques made it possible to measure the volume of nasal and oral airflow simultaneously which provided valuable information in this controversial area. Gurley and Vig [29] suggested a technique for the simultaneous measurement of nasal and oral respiration which is called the Simultaneous Nasal and Oral Respirometric Technique (SNORT). This technique measured the ratio of oral to nasal airflow and made the quantification of normal and abnormal respiratory modes possible. Other techniques were used to determine the dimensions of nasal airway impairment [30–32]. Hairfield et al. [33] reported a mean cross-sectional nasal area of 0.65 cm^2 in adults. Warren et al. [32] suggested that if the nasal size falls below 0.4 cm^2, most individuals will become oral breathers to some extent. These values would probably be smaller in children since several studies have indicated that nasal cross-sectional area increases with growth [27, 34, 35].

Perhaps the most significant finding of these studies is that one should be careful when classifying patients as oral breathers. Even in a population referred with an impaired airway, there will be oral, nasal-oral, or total nasal breathers as well as habitual oral breathers with adequate nasal airways. There are no clear cutoff points to determine the mode of breathing, and oral or nasal respiration seems to be a transient phenomenon in many individuals [36]. Since the intensity of functional changes is important in the magnitude of growth alterations, the question becomes if partial airway obstruction can lead to reflex adaptive changes that cause dentofacial deformities.

Another important factor to consider is the age at which dentofacial growth changes and breathing mode are evaluated. In order for any functional change to have a significant effect on facial growth, it must start early and be effective for a long time, especially during the peak growth period. However, cross-sectional studies that investigate subjects before or a long time after their growth spurts may not show any differences between groups because in younger subjects the growth changes would not have occurred yet, whereas in older subjects nasal airway may no longer be compromised due to growth changes in nasopharynx and lymphoid tissue.

Individuals also differ from each other in terms of adaptation and compensation processes. Several morphological factors may determine the extent of the postural response to an inadequate airway. Subjects with compromised oropharyngeal airways due to the size and shape of the soft palate or tongue may give more exaggerated responses to a decrease in nasal airflow. The dentofacial changes observed in these individuals might be more pronounced. However, if the oropharyngeal airway is clear, only a slight parting of the lips may be enough to increase airflow [37].

The extent of dentofacial changes due to any functional disturbance will be determined by multiple factors. Among these factors the age of the patient, the duration of the habit, and individual variations should be considered while making treatment decisions.

39.3 Physiologic and Dentofacial Effects

39.3.1 A Review of Early Literature

The influence of nasal respiratory function on the growth and development of craniofacial structures was first generated over a century ago by anecdotal reports describing the effects of "mouth breathing" on dental and facial morphology.

In 1872, Tomes [38] described the dentofacial changes associated with chronic nasal airway obstruction, citing the lips-apart posture as the cause for decreased pressure on the incisors and proclined anterior teeth. He also coined the term "adenoid faces" to describe the associated facial changes.

In 1925, Dr. Edward H. Angle reprinted a book by George Catlin entitled "The Breath of Life (All Life on Earth is Breath, All Else on Earth is Death)." This book by Catlin advocated the superiority of nasal breathing over mouth breathing in sleep and described the consequences of habitual mouth breathing on teeth and facial features [39]. Dr. Angle [40] was also a firm believer in these concepts and stated in his textbook:

> When there is normal nasal respiration and normal relations of the dental arches, the teeth, and the muscles, the conditions are such as to perfectly maintain the equilibrium and the mutual support necessary to the normal development of the teeth and jaws. Should nasal obstruction occur in the developing child, inducing habitual mouth-breathing, immediately the equilibrium is disturbed, the lips and muscles are placed on a different tension, and pressure upon the arches, instead of being equal, is localized, being greater than normal at some points and less at others. No matter how strenuously it may be denied, malocclusion of the teeth and abnormalities in the formation of the bones of the jaws naturally follow. The undeveloped nose and adjacent region of the face, the vacant look, the short upper lip, the open mouth, and irregular teeth of the mouth-breather are common sights familiar to all.

Later other reports reiterated the relationship between altered dentofacial form and mouth breathing. Chronic nasal allergies were suggested to cause paranasal depression, V-shaped palate, and proclination of the maxillary incisors [41, 42]. Cohen [43] attributed mouth breathing due to allergies to high-arched and narrow palates, a flat and narrow face, and some type of orthodontic deformity (although he did not classify which type). Subtelny [44] theorized that enlarged adenoids will cause a separation of the lips, downward and forward movement of the tongue away from the soft palate, depression in the position of the mandible, constriction of the maxillary arch, and a Class II division 1 type of malocclusion due to procumbent maxillary anterior teeth. Paul and Nanda [15] also found a tendency toward Class II malocclusion with an increase in overbite and overjet in mouth-breathing subjects when compared to nasal breathers.

Different theories were proposed to explain the relationship between respiratory mode and possible dentofacial changes. Morrison [45] suggested that the oral airstream in mouth-breathing individuals caused the negative pressure between the tongue and palate to be lost hindering normal downward palatal descent. James and Hastings [8] proposed that the loss of the normal pressure in the mouth due to an impairment of the airway is associated with impaired action of the tongue, lips, cheeks, and other forces acting on the jaws. They considered the impaired and misdirected action of these forces as the reason for the deformities of the jaws. A clearer understanding of these effects was possible when Moss [46] developed the "functional matrix theory." This theory was based on the principle that the skeletal unit in the functional craniofacial component provides protection and mechanical support for its specific functional matrix and grows in response to the functional demands of surrounding tissues and structures. According to Moss's theory, nasal breathing allowed proper growth and development of craniofacial and dentofacial complex [47]. Later, Solow and Tallgren [48] found an association between the posture of the head and cervical column and craniofacial alterations seen in mouth-breathing patients. These results led to an alternative explanation known as "Soft Tissue Stretching Hypothesis" [49]. This theory suggested that nasal obstruction caused extension of the head and that this postural change causes soft tissue stretching causing differential forces to act on the skeleton causing morphological changes. Hence, the existence of a relationship between respiratory mode and dentofacial growth changes was widely accepted, and early treatment of allergies and removal of adenoid tissue before the eruption of permanent teeth were suggested to prevent the alterations in dental arches [44, 50].

However, there were others who were skeptical of such wide acceptance of a possible relationship between oral respiration and growth changes. Kingsley [51] considered the deep and narrow palate a congenital morphological trait rather than the results of a muscle imbalance due to mouth breathing. Gwynne-Evans and Ballard [52] observed the relationship between jaw form, soft tissue morphology, and upper respiratory conditions for more than 15 years and concluded that mouth breathing does not result in deformities of the jaws, malocclusion, or adenoidal facies. Other authors reported that more than 50% of the patients who were characterized as mouth breathers had Class I occlusions and no specific malocclusion pattern was correlated with mouth breathing [53–55].

Since then, investigators have attempted to look more critically at the issue through experimental models and clinical research with objectively defined criteria for mouth breathing. Harvold [56] used the rhesus monkey, *Macaca mulatta*, as a model in his experiments to test the connection between neuromuscular activity and skeletal morphogenesis. He anchored a piece of plastic in the palatal vault between the molars and found that the mandible moved lower and the tongue moved forward resulting in an anterior open bite. This study was followed by another one in which the animals were induced to lower their mandibles by fitting an acrylic block in the palatal vault [57]. After 6 months there was a significant increase in face height in experimental groups. They concluded that any factor, such as mouth breathing, that lowers the postural position of the mandible could also increase the face height. They tested their hypothesis by blocking nasal inhalation with silicone nose plugs in growing monkeys [58, 59]. Radiographic, electromyographic, and dental cast measurements showed increased face height, decreased maxillary and mandibular intercanine distance, decreased maxillary dental arch length, steeper mandibular plane, larger gonial angle, altered muscle activity, and changes in the morphology of the tongue in the experimental group. The pattern of maintaining an airway differed between monkeys. Those that kept their mouths constantly open by lowering the mandible and protruding their tongues developed more severe malocclusions than those that rhythmically opened and closed their mouths with respiration.

The altered muscle activity in the rhesus monkeys due to forced oral breathing was further explored in other studies [60–62]. These experiments documented changes in neuromuscular recruitment patterns resulting in changed function and posture of the mandible, tongue, and upper lip with considerable variation among the animals and concluded that nasal obstruction can induce neuromuscular changes which extend beyond the period of obstruction and remain even after nasal breathing is established.

More recent animal studies also found that nasopharyngeal respiratory obstruction was associated with downward and backward rotation of the mandible, upward and backward growth of the condyle, divergent gonial angle, anterior open bite, spaced dental arch in the lower anterior region, and a reduction in the height of the maxilla [63, 64].

These experimental findings corroborated the previous data on the relationship between mouth breathing and an increase in facial height. However, it should be kept in mind that the results of these animal studies cannot be readily extrapolated to humans. Total nasal obstruction, as induced by researchers in these experiments, is extremely rare in humans. The oropharyngeal structures also differ between the species. Thus, the postural changes that occur in animals due to forced oral breathing may not be the same in humans.

Early clinical studies on humans generally used lateral cephalograms to evaluate dentofacial changes and the upper airway. Linder-Aronson [2] studied 162 children aged 6–12 years. Half of these children were determined to need adenoidectomies, while the other half were controls. The patients were examined with respect to nasal airflow, breathing pattern, and dentofacial morphologic variables. He reported a significant relationship between enlarged adenoids determined by cephalometric x-rays and certain craniofacial changes, including low tongue position, mouth breathing, narrow maxillary arch, crossbite, and retroclined maxillary and mandibular incisors. In 1974, he examined a group of Swedish children before and after adenoidectomies and compared them to controls with respect to changes in cephalometric measurements. He reported that the patients in adenoidectomy group had increased anterior facial height, maxillary constriction, and retroclined incisors compared to controls. After their adenoidectomies, upper airway obstruction was resolved in these patients and their growth pattern became horizontal [65].

Hannuksela [66] compared allergic and nonallergic children and reported a tendency toward clockwise rotation of the mandible and retroclined mandibular incisors. However, further evaluation of this population revealed no significant differences in the occlusion between groups [67]. Other cephalometric studies also failed to show a relationship between the airway space or adenoid size and malocclusion [23, 68].

Lateral cephalograms were readily available to orthodontists and valuable in determining dentofacial changes. They were also recommended for assessing adenoid size [69]. However, since lateral cephalograms are two-dimensional and unable to provide volumetric data, their reliability in determining airway size and adenoids have been criticized [70, 71].

With the development of respirometric techniques, more objective classification of respiratory patterns became possible. Vig et al. [72] reported higher nasal resistance in increased facial height groups compared with controls, but they found no significant differences in the nasal volume flow rate between groups. Fields et al. [14] also concluded that subjects with long faces had a significantly smaller component of nasal airflow. Warren and Spalding [73] suggested that high airway resistance due to nasal impairment may cause an exaggerated postural response if there is a low drape to the velum, the pillars form a posterior curtain, the tonsils are enlarged, or the posterior portion of the tongue is carried high.

The early clinical studies that used lateral cephalograms and/or other respirometric techniques showed that the relationship between dentofacial growth and development and respiratory mode is variable and multifactorial.

39.3.2 Current Perspective

Heredity and environmental factors are both effective in the development of dental arches and postnatal determination of craniofacial features. One of the most important environmental factors is the predominant respiratory pattern. Nasal breathing is associated with a normal posture of the tongue and lips and normal muscle activity. If there is nasal obstruction, this would likely affect the muscle forces acting on the dentofacial region. This change in muscular action may cause abnormal facial growth and development. It has been shown that during oral breathing masseter muscle activity is inhibited [74]. Increased airway resistance also stimulates mechanoreceptors in the upper airway and increases the activities of the genioglossus and mylohyoid muscles due to the forward positioning of the tongue and opening of the mouth to maintain the airway [75].

> Today, it is accepted that growth occurs under the control of both genetic and environmental factors. Genetics act on cartilages and the bones respond to the changes in these cartilaginous structures. The growth and needs of the soft tissue matrix also cause reactive changes in both the bone and the cartilage.

Long-term mouth breathing seems to affect the occlusion and facial morphology during periods of rapid growth. However, not every patient has the same growth changes due to oral breathing. The heritable characteristics of anatomical structures also seem to play a role in determining which patients will be most affected. In some patients, a slight opening of the lips may be enough to provide the necessary airway, while in others a more exaggerated postural response of the mandible, tongue, and head will be necessary. Children with narrow facial patterns also may be more susceptible to growth changes due to mouth breathing than children with broad facial features. It is also possible for patients with a vertical facial growth pattern to be more likely to be mouth breathers. The severity of the obstruction will also determine if the child is a chronic mouth breather or a partial one. If the obstruction is severe, the changed postural responses will be in place longer causing more extensive growth changes in dentofacial structures.

> If the patient has morphologic risk factors in the oropharynx such as enlarged tonsils, a large tongue, or a long soft palate, the postural response necessary to open the airway will be more pronounced and the risk for developmental disturbance will be greater.

The most common cause of oral breathing in children is enlarged pharyngeal lymphoid tissue. The enlargement of these tissues may adversely affect pharyngeal patency [76]. Normally, the size of the adenoid tissue is dependent on the associated skeletal structures. However, abnormal growth of this tissue may predispose the patient to upper airway obstruction causing oral respiration. As a matter of fact, any reason that causes nasal resistance to increase for long periods of time, such as allergies or nasal septal deformity, has the potential to cause chronic oral respiration.

> When a tooth erupts in the mouth, it is subject to chewing forces as well as forces from soft tissues such as lips, cheeks, and tongue. This creates an equilibrium of forces acting on the dentition. When this equilibrium is disturbed, there are changes in tooth positions.

During oral respiration, the mandible rotates open and the tongue is positioned lower in the mouth and no longer contacts the palate causing an eruption of the molars and the transverse maxillary deficiency (Figs. 39.8 and 39.9). Thus, the mandible rotates in a clockwise direction, losing contact with the soft palate, causing an open bite with an increase in the mandibular plane angle, a higher palatal plane, narrowing of both maxillary and mandibular arches, and mandibular retrognathism with an increased overjet [77, 78]. However, this is only a mechanistic view of the possible interactions between mandibular growth and oral breathing. Complex epigenetic events may also be responsible for the growth changes in the mandible. It has been hypothesized that children with significantly enlarged adenoids will develop obstructive sleep apnea causing abnormal nocturnal growth hormone secretion and causing somatic growth impairment [79]. Due to this

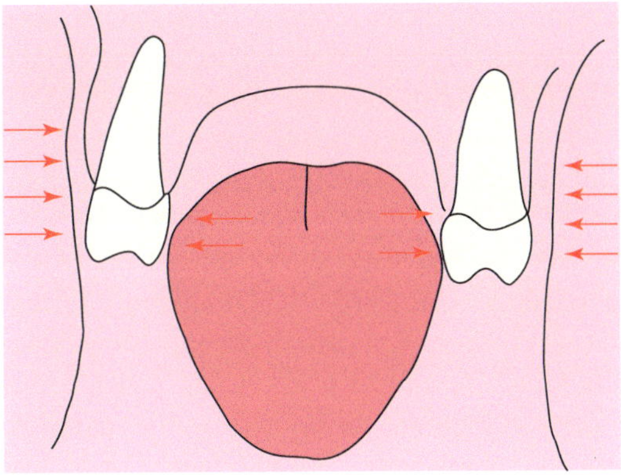

Fig. 39.8 When the tongue is in its correct position, it exerts transverse pressure on the teeth and alveolus, allowing proper growth

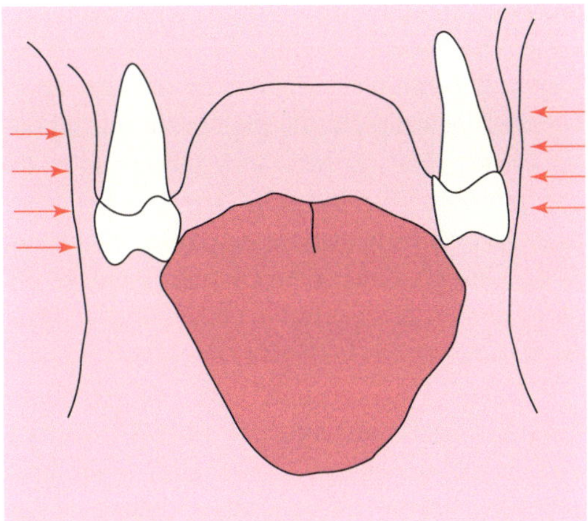

Fig. 39.9 If the tongue is in a lower position, the equilibrium of forces is disturbed and the transverse growth of the maxilla is deficient

abnormal hormonal balance, mandibular ramus growth may be less than in healthy subjects causing the observed mandibular rotation in these children. This mandibular rotation also causes backward and downward displacement of the tongue [80]. The breathing pattern seems to affect not only the position but also the pressure exerted by the tongue. The average tongue pressure in mouth-breathing children is lower than that in nasal-breathing children [81]. These altered muscle forces caused by the postural changes of the tongue lead to a constricted maxillary arch, high palate, and posterior crossbite in the transverse direction. Maxillary retrusion [2] and an increase in palatinal inclination in relation to the cranial base [2, 82] have also been reported.

The above-mentioned structural changes are similar to those reported in children with obstructive sleep apnea (OSA). Recent studies and systematic reviews have shown differences in the craniofacial structures of children with obstructive sleep apnea compared with children without OSA such as a more Class II and dolichofacial mandibular growth direction as well as a narrow maxilla, retrognathia, anterior open bite and hypoplasia of the midfacial region [83–86]. Moreover, the severity of OSA is suggested to be correlated with the severity of maxillomandibular deficiency [87]. Even though mouth breathing and OSA do not always coexist, it has been suggested that mouth breathing increases upper airway collapsibility by failing to activate upper airway dilator muscles through the negative pressure reflex. This, in turn, may contribute to the occurrence of OSA [88–93]. Thus, it is important to evaluate OSA patients to determine their mode of breathing in order to better understand the etiology of their disease and to make effective treatment choices.

Craniocervical posture and oral breathing have also been the subject of various investigations. Oral breathing is associated with head extension [94–96]. This postural change moves the hyoid bone upward establishing an adequate airway [97]. Since there is a relationship between head posture and altered muscle activity, long-term craniocervical changes may influence craniofacial growth as well as putting undue load on the neck and upper shoulders.

Mouth breathing in cleft palate patients is a clinically relevant subject as well [98]. Cleft lip and palate deformities are one of the most common congenital malformations. The airway size is reduced and nasal resistance is higher in cleft patients compared to noncleft subjects, in both children and adults [30, 99–102]. Changes in the minimal cross-sectional area of the airway, nasal volume, and nasal flow resistance are suggested to cause impairment in nasal function [102–107]. The large craniocervical angulation also indicates an extended head position in these cases [108]. The high prevalence of mouth breathing in these patients may be caused by various factors, such as septal deviation and the effects of surgical techniques. Thus, a normal breathing pattern may not be established ever because of the open communication between the nose and mouth at birth [32]. Besides, the positive OSA screening ratio of unilateral and bilateral cleft lip and palate patients (12.2%) was found to be significantly higher than the controls (4.5%) [109].

Traditionally, the changes in craniofacial and airway structures of mouth breathing patients have been evaluated using lateral cephalograms. However, the advances in technology made it possible to use Cone Beam Computed Tomography (CBCT) to evaluate the airway in 3D and make volumetric measurements; by using significantly less radiation which is necessary for conventional CT imaging. Although studies are ongoing to increase the accuracy of airway measurements in CBCT by evaluating the measurement software and threshold values, these images are currently the most simple and reliable way of determining the 3D anatomy of the airway [110–112]. In their study assessing the pharyngeal airway space in nasal and mouth breathing children using CBCT, Alves Jr. et al. [113] found that airway volume and minimum axial area were significantly greater in nasal breathers than in mouth breathers. Unfortunately, the 3D data on airway morphology in mouth-breathing patients is still scarce and further research is needed.

In addition to these structural changes in the dentofacial region, mouth breathing may cause other oral diseases. Mouth-breathing patients often have inflamed labial gingival tissues around the maxillary incisors. The gingiva becomes inflamed and hyperplastic because the mouth remains open and the salivary flow is reduced. Since saliva performs essential roles including antimicrobial action and protection of oral tissues, a reduction in salivary flow will have a negative impact on teeth and gingival tissues as well as generate odoriferous volatile compounds. Clinically, the gingiva appears swollen, red, and shiny with the classic rolled-up appearance (Figs. 39.10 and 39.11). There can be bone loss and pocket formation in the interproximal area if

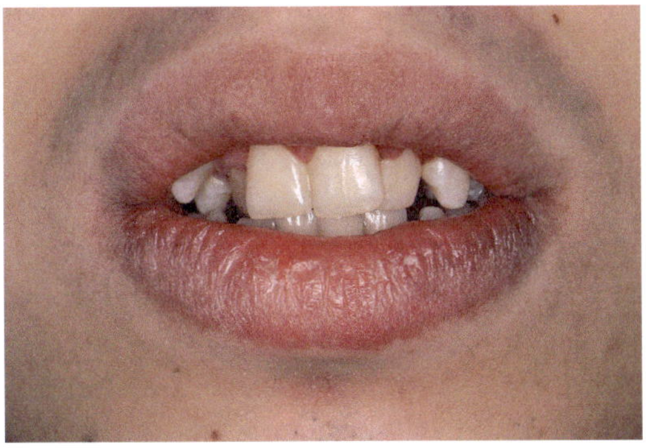

Fig. 39.10 At repose, the lips are open in mouth-breathing patients. This lack of lip seal causes the mouth to become dry. Maxillary anterior teeth and gingiva are most at risk for the negative effects of this open-mouthed posture

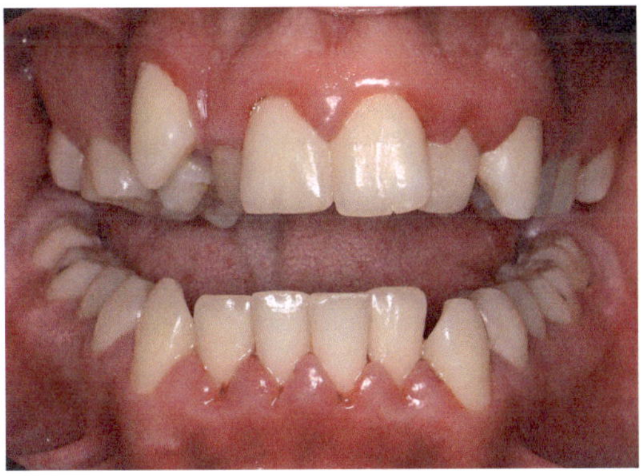

Fig. 39.11 Maxillary anterior region of the patient in Fig. 39.10. The gingiva is inflamed due to poor oral hygiene and mouth breathing. Note the classic shiny red, swollen look of the gingiva

proper oral hygiene is not maintained. There is also an increased incidence of caries and halitosis (breath malodor). Correction of mouth breathing along with necessary dental treatments will improve the health of the oral cavity.

> It is likely that the oral cavity becomes dry because the mouth remains open most of the time in chronic mouth breathers. Clinical manifestations of this negative impact will be halitosis, an increased incidence of caries, and gingivitis, especially around maxillary anterior teeth. These oral health problems will have significant negative social, economic, and psychological consequences causing a decrease in the quality of life.

It has been difficult to determine the relationship between dentofacial morphology and oral breathing because respiration is a complex act which cannot be easily classified with the current techniques. There are multiple factors which determine if the dentofacial growth of a person will be affected by the changes in the nasal airway. Thus, a careful evaluation of the patient by otolaryngologists, pediatricians, and orthodontists is needed before treatment decisions.

39.3.3 Treatment

Treatment of patients with a diagnosis of mouth breathing needs to be preceded by an evaluation by an otolaryngologist, pediatrician, and orthodontist. Treatment of upper airway obstruction may be provided by either of these specialists or by all, depending on the etiology and severity of the clinical manifestations. The case report illustrated in Figs. 39.12, 39.13, 39.14, 39.15, 39.16, 39.17, and 39.18 demonstrates the type of treatment result that can be obtained for a patient with dentofacial deformity due to mouth breathing.

In cases of habitual mouth breathing without nasal, nasopharyngeal, or oropharyngeal obstruction, the treatment is directed toward establishing a lip seal and forcing the patient to breathe nasally. This is known as *myofunctional therapy*. Myofunctional therapy can be defined as a group of targeted isotonic and isometric exercises which improve orofacial muscle strength and coordination and has been used for over a century to improve nasal breathing [114, 115], and more recently as an adjunct treatment in OSA [116, 117]. Some devices, which can help to strengthen oral and tongue muscles, have also been described [118].

If there are dentofacial changes in the patient due to mouth breathing, functional appliances or other orthopedic devices can also be used to stimulate growth. The type of appliance to be used depends on the growth pattern, which jaw is underdeveloped and in which plane the deficiency is (sagittal, vertical, transversal).

If the cause of this pattern of breathing is obstructive, the location of the obstruction and its cause must first be diagnosed and treatment must be directed accordingly. The surgical and medi-

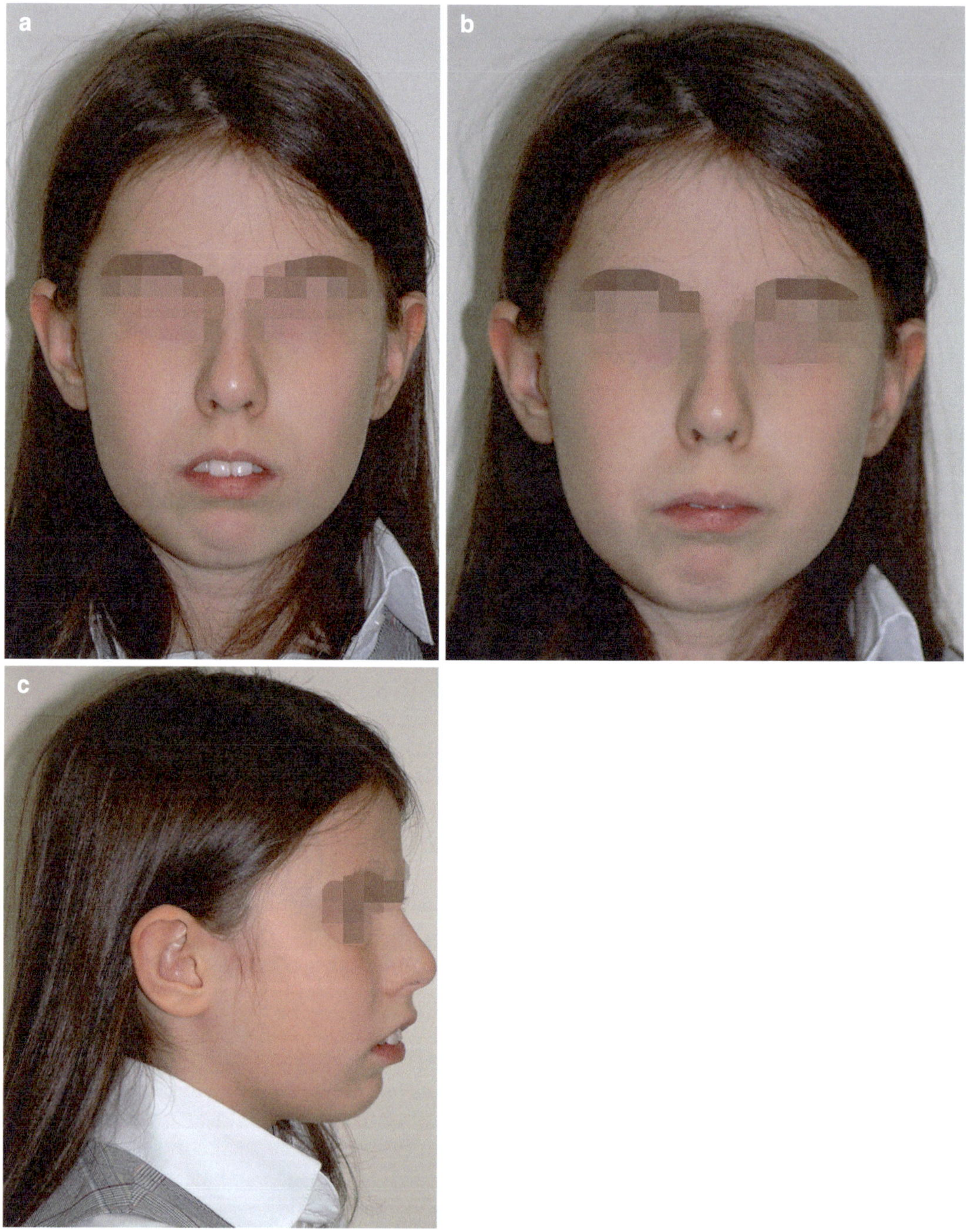

Fig. 39.12 (**a**–**c**) Pre-treatment (**a**) frontal view at rest, (**b**) frontal view with forced lip seal, and (**c**) profile view that indicates a high lip line, increased lower anterior facial height, and increased muscle activity during lip closure. The lower lip rests behind the protruded incisors during the rest position

cal treatments of such obstructions are beyond the scope of this chapter. However, the effects of treatments on dentofacial growth must be known before making any treatment decisions.

Adeneidoctomy and/or tonsillectomy are generally used to remove the obstructive pathology to normalize the breathing pattern and promote normal growth of the craniofacial complex [5].

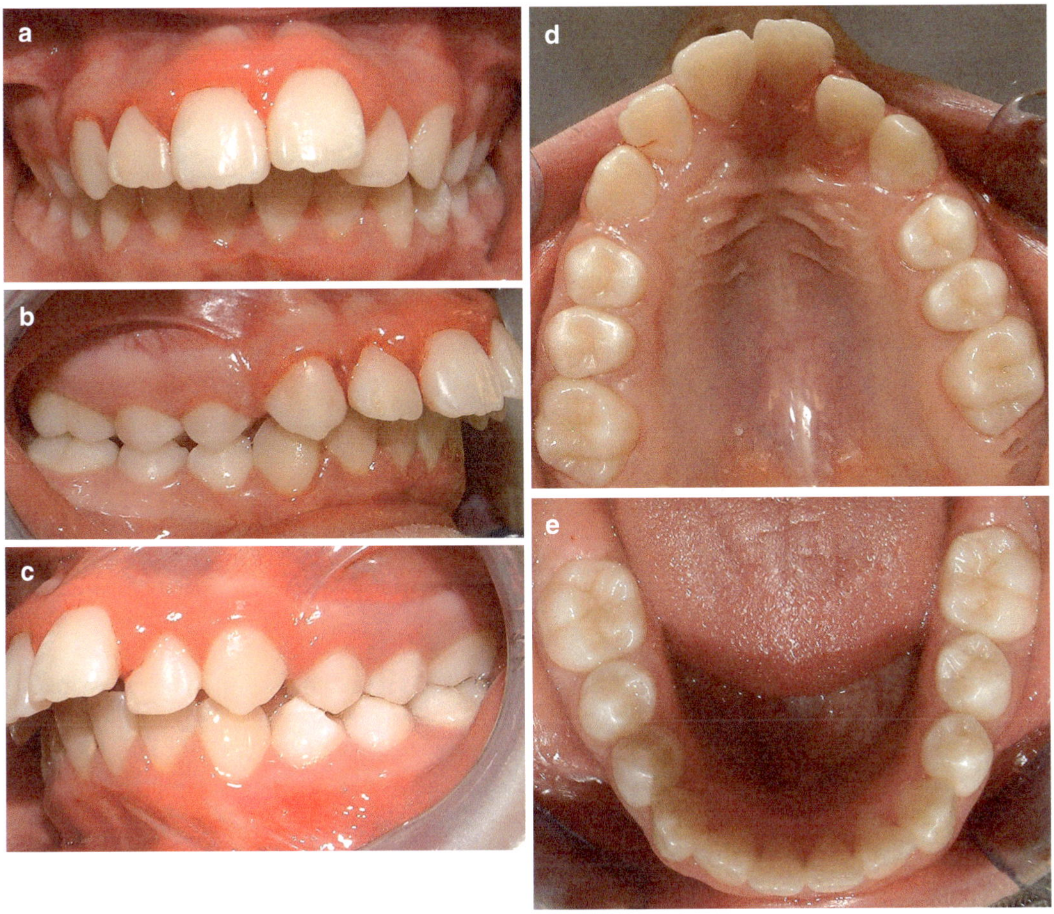

Fig. 39.13 (**a**–**e**) Pre-treatment intraoral (**a**) frontal, (**b**) left side, (**c**) right side, (**d**) upper arch, and (**e**) lower arch views of the patient demonstrating a Class II malocclusion with mild maxillary transverse deficiency, and arch length deficiency

These surgeries have been shown to increase the airway volume and minimal cross-section [119, 120]. This is followed by a change of breathing pattern from mouth breathing to nasal breathing [65, 121–123]. Most researchers report that the change in the mode of breathing is stable after surgery [121, 123]. Others have suggested that a relapse may occur during the follow-up [122].

After adenoidectomy/tonsillectomy and establishment of nasal breathing, mandibular plane angle changes toward a more horizontal growth direction though individual variation in response has been reported [124, 125]. Peltomaki [79], in a systematic review, reported that restoration of normal physiologic nasal breathing following adenotonsillectomy can induce acceleration in the secretion of growth hormone. This may in part explain the mandibular growth changes in these children.

A recent meta-analysis by Zhu et al. [126] found that after these surgeries, the malocclusion and the transverse deficiency in arch dimensions associated with airway obstruction could not be completely reversed in children. As such, other treatment modalities such as myofunctional therapy or arch expansion should be considered after the removal of the obstructive pathology. However, they cautioned that the research in this field is insufficient to rule out the influence of follow-up duration and the time of surgery.

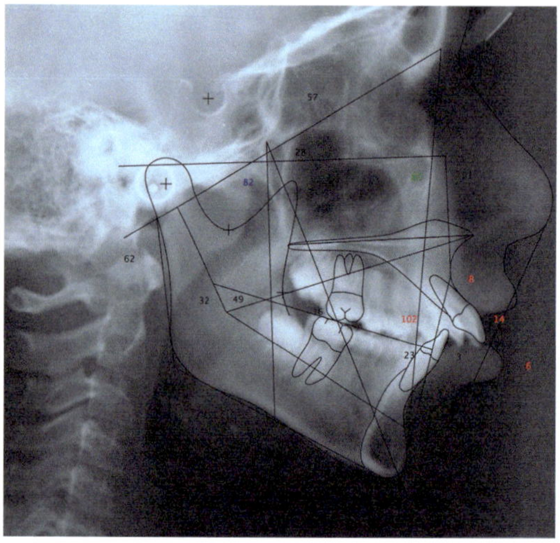

Fig. 39.14 Pre-treatment lateral cephalometric view and analysis of the patient

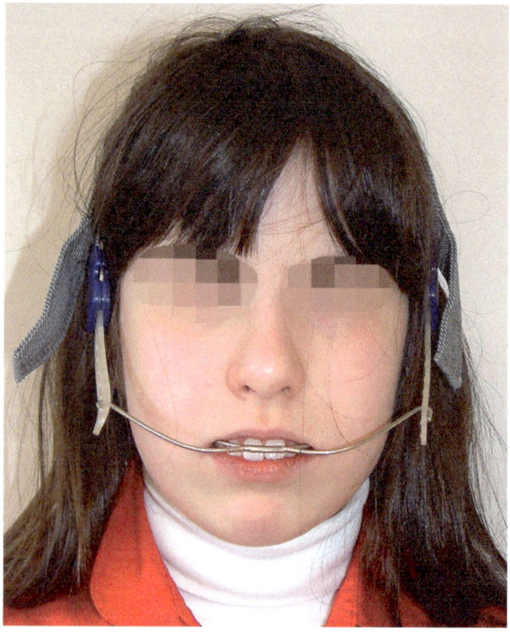

Fig. 39.15 The patient was treated with a high-pull headgear and fixed appliances. The high-pull headgear places a distal and upward force on the maxillary dentition and maxilla. This maintains the vertical position of the maxilla and inhibits eruption of the maxillary posterior teeth

A flexion in head posture always occurs with a reduction in craniocervical angle in response to the elimination of airway obstruction after adenoidectomy and has also been reported after medical treatment of children with asthma and perennial rhinitis [127–129] and after orthodontic rapid maxillary expansion (RME) [130].

Rapid maxillary expansion is an orthopedic procedure that produces sutural expansion in the maxilla, and it is widely used in orthodontics to widen the palate without the movement of teeth through alveolar bone. Transverse dimensions of the maxilla can be effectively increased by RME appliances [131]. RME appliance has a screw in the middle and is applied on the maxillary arch. Separation of midpalatal suture is accomplished by turning the screw once a day for approximately a month. A retention period of 3–6 months is recommended after the active expansion is ended (Figs. 39.19, 39.20, 39.21, and 39.22).

In cases of maxillary constriction, when the maxilla is expanded by RME treatment, an increase in the nasal floor close to the midpalatal suture occurs as well. Numerous studies have radiographically investigated nasal cavity changes after maxillary expansion on posterior-anterior cephalometric films and reported an increase in nasal cavity width [132–134]. It has even been reported that bone-borne or hybrid (tooth-and-bone borne) appliances might achieve the increased sutural opening and lower nasal airway resistance [135].

Orthopedic expansion of the maxilla by RME appliances plays an important role to improve nasal airflow. A decrease in nasal resistance and a change in breathing pattern from oral to nasal have been reported after RME treatment [132, 136, 137]. A decrease in the pharyngeal collapse was also reported in patients with obstructive sleep apnea syndrome (OSAS) following RME treatment [138]. However, in about one-third to one-half of the patients, mouth breathing could not be treated [136, 139]. Small clinical changes and wide individual variation were reported in other studies using acoustic rhinometry to test the differences in nasal volume before and after RME [140–142]. It was suggested that in the presence of nasal concha hyperplasia, nasal polyps, adenoidal hypertrophy, and septal deviation, an increase in nasal airflow may not be enough to achieve nasal breathing [136]. In general, RME is indicated in patients with maxillary constriction. Even if there is evidence that the width of the

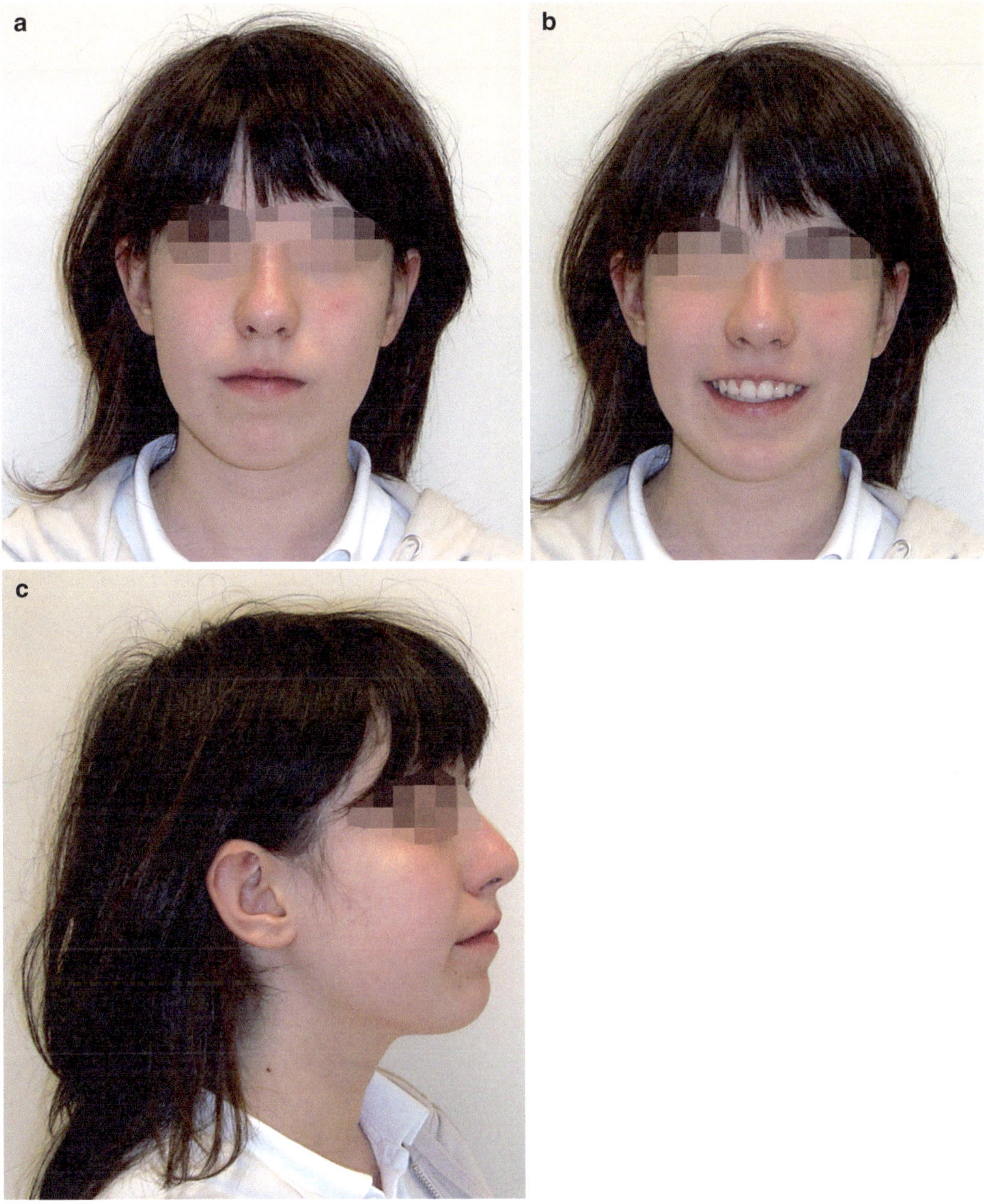

Fig. 39.16 (**a**–**c**) Post-treatment (**a**) frontal view at rest, (**b**) frontal view during a smile, and (**c**) profile view shows the improved aesthetic results obtained by orthodontic and orthopedic treatment

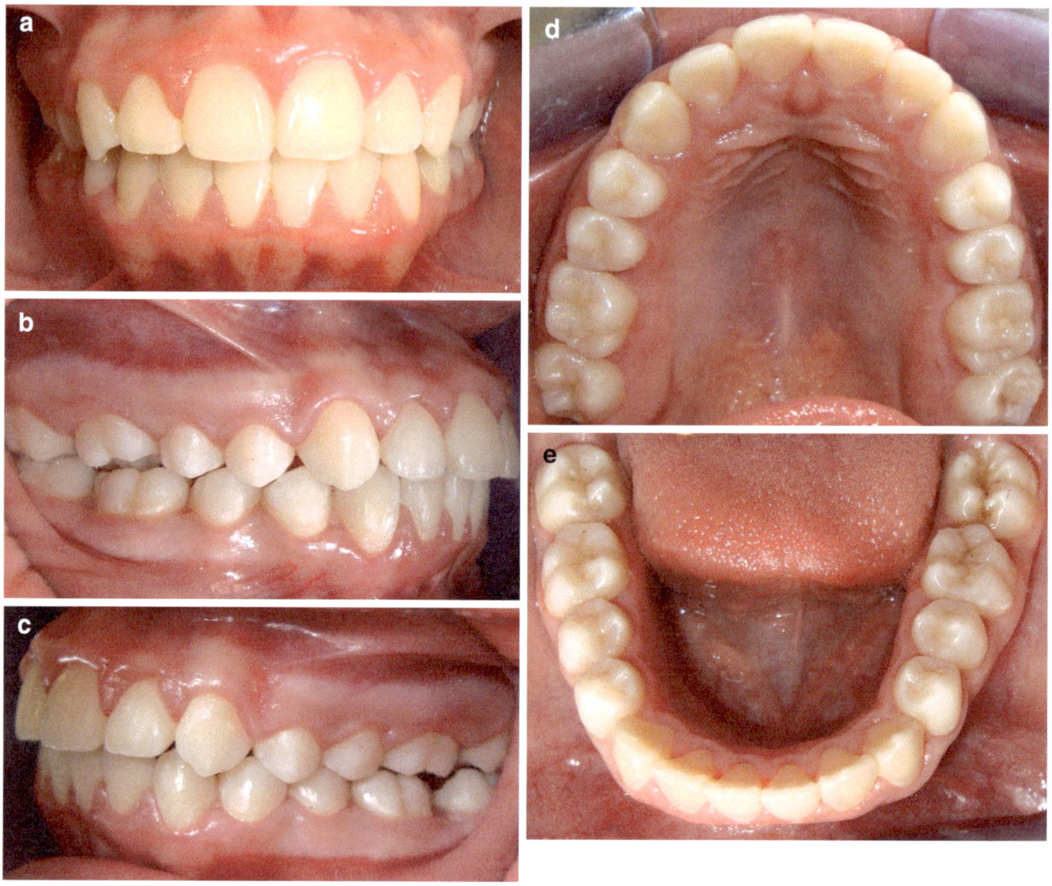

Fig. 39.17 (**a**–**e**) Post-treatment intraoral (**a**) frontal, (**b**) left side, (**c**) right side, (**d**) upper arch, and (**e**) lower arch views of the patient demonstrating improved functional results with a Class I molar and canine occlusion as well as ideal overjet, overbite and transverse relationship of the dentition

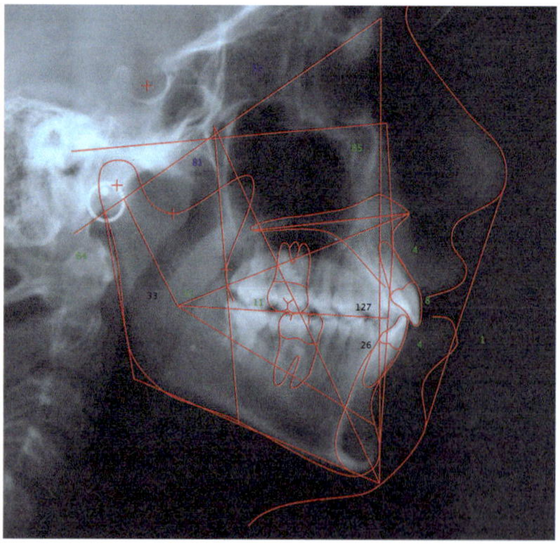

Fig. 39.18 Post-treatment lateral cephalometric view and analysis of the patient

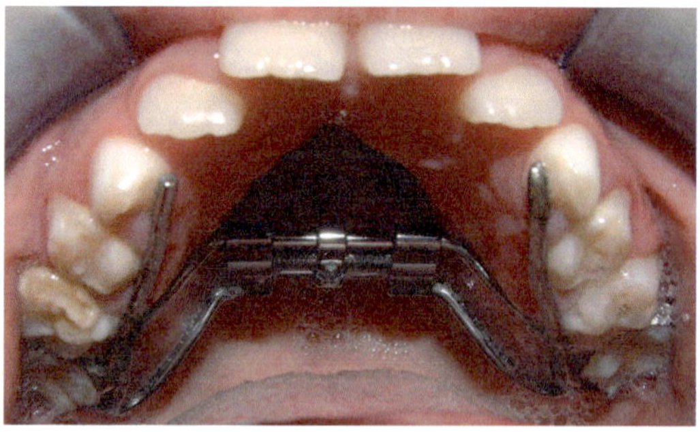

Fig. 39.19 The RME appliance in the maxillary arch is used to widen the maxilla

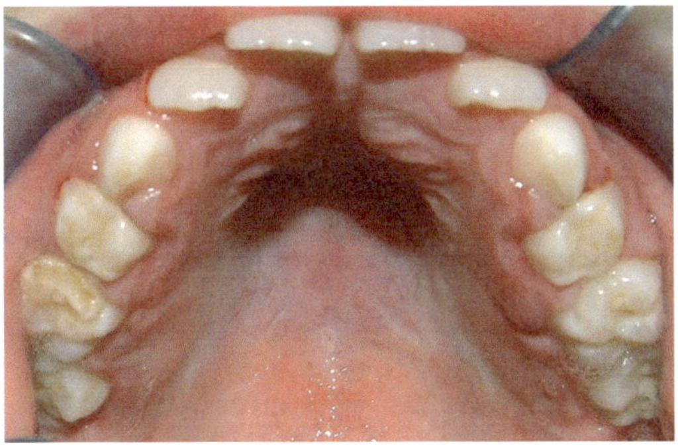

Fig. 39.20 Occlusal view of the maxillary arch after RME treatment

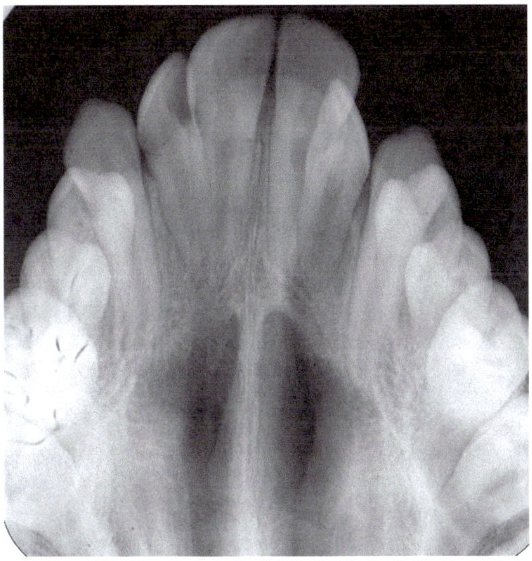

Fig. 39.21 Maxillary occlusal radiograph of a patient before RME

nasal cavity increases and improvement in nasal airflow occurs to some extent, it cannot yet be offered solely for the treatment of respiratory impairment [136, 139, 140, 143].

Even though it has been shown that midpalatal sutural expansion increases nasal cavity width and nasal airflow, this may not be enough to change the respiratory pattern. Thus, the use of RME should be limited to patients with transverse maxillary deficiency.

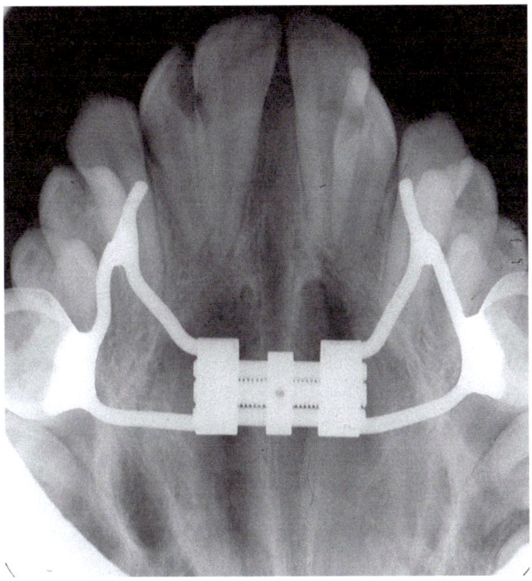

Fig. 39.22 The separation at the midpalatal suture is clearly seen on this occlusal film taken after RME is completed. A retention period of 3–6 months is needed to prevent relapse while bone fills in the space between the right and left segments of the maxilla

39.3.4 Conclusion

Mouth breathing has the potential to adversely affect the growth and development of the individual, depending on the frequency, severity, and duration of the action with the inclusion of other confounding factors such as genetic features, variability in functional demands, and age. In the worst case scenario, an adenoid facies type of facial form can occur with a long face height, open-mouthed posture, short upper lip, small nostrils, a narrow maxillary arch, and extended head posture. To determine the etiology, objective evaluation of mouth breathing is important; techniques measuring the amount of oral and nasal airflow provide more valuable information compared to nasal resistance measurements and subjective evaluation of the patient. Mouth breathing may occur due to an obstruction in the nose and upper airway, but it is also possible for mouth breathing to be habitual without accompanying airway obstruction. Open-mouthed posture is not a good indicator of mouth breathing as people can breathe through their nose by creating a posterior seal. In treatment planning, correct diagnosis and defining the etiology of the problem will eliminate unnecessary treatment approaches.

In the presence of an obstruction that impedes physiologic nasal breathing, elimination of the obstruction is necessary to prevent abnormal facial and dental growth. In cases of obstructive adenoids and tonsils or allergic rhinitis, surgical or medical treatment is frequently indicated. Favorable changes in dentofacial structures and head posture are seen in these patients, especially when treatment is performed at an earlier age. Some of these patients may further need orthodontic treatment if they have already developed malocclusions. Depending on the defined treatment goals, several different therapeutic approaches are available in the treatment of these patients which include myofunctional therapy, maxillary expansion, functional appliances, dentofacial orthopedics, or orthognathic treatment.

References

1. Emslie RD, Massler M, Zwemer JD. Mouth breathing: etiology and effects. J Am Dent Assoc. 1952;44:506–21.
2. Linder-Aronson S. Adenoids. Their effect on mode of breathing and nasal airflow and their relationship to characteristics of the facial skeleton and the dentition. A biometric, rhino-manometric and cephalometro-radiographic study on children with and without adenoids. Acta Otolaryngol Suppl. 1970;265:1–132.
3. Koski K, Lähdemäki P. Adaptation of the mandible in children with adenoids. Am J Orthod. 1975;68:660–5.
4. Schendel SA, Eisenfeld J, Bell WH, Epker BN, Mishelevich DJ. The long face syndrome: vertical maxillary excess. Am J Orthod. 1976;70:398–408.
5. Ricketts RM. Respiratory obstruction syndrome. Am J Orthod. 1968;54:495–507.
6. Moss ML. The veloepiglottic sphincter and obligate nose breathing in the neonate. J Pediatr. 1965;67:330–1.
7. Goldfield EC, Richardson MJ, Lee KG, Margetts S. Coordination of sucking, swallowing, and breathing and oxygen saturation during early infant breast-feeding and bottle-feeding. Pediatr Res. 2006;60:450–5.
8. James WW, Hastings S. Discussion on mouth-breathing and nasal obstruction. Proc R Soc Med. 1932;25:1343–55.
9. Niinimaa V, Cole P, Mintz S, Shephard RJ. Oronasal distribution of respiratory airflow. Respir Physiol. 1981;43:69–75.
10. Huang ZL, Ong KL, Goh SY, Liew HL, Yeoh KH, Wang DY. Assessment of nasal cycle by acoustic rhinometry and rhinomanometry. Otolaryngol Head Neck Surg. 2003;128:510–6.
11. Chaaban M, Corey JP. Assessing nasal air flow: options and utility. Proc Am Thorac Soc. 2011;8:70–8.
12. Schlenker WL, Jennings BD, Jeiroudi MT, Caruso JM. The effects of chronic absence of active nasal respiration on the growth of the skull: a pilot study. Am J Orthod Dentofac Orthop. 2000;117:706–13.
13. Rodenstein DO, Stanescu DC. Soft palate and oronasal breathing in humans. J Appl Physiol. 1984;57:651–7.
14. Fields HW, Warren DW, Black K, Phillips CL. Relationship between vertical dentofacial morphology and respiration in adolescents. Am J Orthod Dentofac Orthop. 1991;99:147–54.
15. Paul JL, Nanda RS. Effect of mouth breathing on dental occlusion. Angle Orthod. 1973;43:201–6.
16. Humphreys HF, Leighton BC. A survey of anteroposterior abnormalities of the jaws in children between the ages of 2 and 5 1/2 years of age. Br Dent J. 1950;88:3–15.
17. Rasmus RL, Jacobs RM. Mouth breathing and malocclusion: quantitative technique for measurement of oral and nasal air-flow velocities. Angle Orthod. 1969;39:296–302.
18. Melsen B, Attina L, Santuari M, Attina A. Relationships between swallowing pattern, mode of respiration, and development of malocclusion. Angle Orthod. 1987;57:113–20.
19. Woodside DG, Linder-Aronson S. The channelization of upper and lower anterior face heights compared to population standard in males between ages 6 to 20 years. Eur J Orthod. 1979;1:25–40.
20. Bresolin D, Shapiro PA, Shapiro GG, Chapko MK, Dassel S. Mouth breathing in allergic children: its relationship to dentofacial development. Am J Orthod. 1983;83:334–40.
21. Miller HI. The relation of long-continued respiratory allergy to occlusion. Am J Orthod. 1949;35:780–9.
22. Tarvonen PL, Koski K. Craniofacial skeleton of 7-year-old children with enlarged adenoids. Am J Orthod. 1987;91:300–4.
23. Sosa FA, Graber TM, Muller TP. Postpharyngeal lymphoid tissue in Angle class I and class II malocclusions. Am J Orthod. 1982;81:299–309.
24. Ung N, Koenig J, Shapiro PA, Shapiro G, Trask G. A quantitative assessment of respiratory patterns

and their effects on dentofacial development. Am J Orthod Dentofac Orthop. 1990;98:523–32.
25. Vig PS, Spalding PM, Lints RR. Sensitivity and specificity of diagnostic tests for impaired nasal respiration. Am J Orthod Dentofac Orthop. 1991;99:354–60.
26. Warren DW, Hairfield WM, Dalston ET. Nasal airway impairment: the oral response in cleft palate patients. Am J Orthod. 1991;99:346–53.
27. Vig PS, Zajac DJ. Age and gender effects on nasal respiratory function in normal subjects. Cleft Palate Craniofac J. 1993;30:279–84.
28. Ellingsen R, Vandevanter C, Shapiro P, Shapiro G. Temporal variation in nasal and oral breathing in children. Am J Orthod Dentofac Orthop. 1995;107:411–7.
29. Gurley WH, Vig PS. A technique for the simultaneous measurement of nasal and oral respiration. Am J Orthod. 1982;82:33–41.
30. Warren DW, Lehman MD, Hinton V. Analysis of simulated upper airway breathing. Am J Orthod. 1984;86:197–206.
31. Warren DW, Hinton VA, Hairfield WM. Measurement of nasal and oral respiration using inductive plethysmography. Am J Orthod. 1986;89:480–4.
32. Warren DW, Hairfield WM, Seaton D, Hinton VA. The relationship between size of the nasal airway and nasal-oral breathing. Am J Orthod Dentofac Orthop. 1988;93:289–93.
33. Hairfield WM, Warren DW, Hinton VA, Seaton D. Inspiratory and expiratory effects on nasal breathing. Cleft Palate J. 1987;24:183–9.
34. Laine T, Warren DW. Effects of age, gender and body size on nasal cross-sectional area in children. Eur J Orthod. 1991;13:311–6.
35. Crouse U, Laine-Alava MT, Warren DW, Wood CL. A longitudinal study of nasal airway size from age 9 to age 13. Angle Orthod. 1999;69:413–8.
36. Shanker S, Fields HW, Beck FM, Vig PS, Vig KWL. A longitudinal assessment of upper respiratory function and dentofacial morphology in 8- to 12-year-old children. Semin Orthod. 2004;10:45–53.
37. Bailey LJ, Simmons KE, Warren DW. Orthodontic problems in children. In: Bluestone CD, Stool SE, Kenna MA, editors. Pediatric otolaryngology, vol. 2. 3rd ed. Philadelphia: W.B. Saunders Company; 1996. p. 1036.
38. Tomes CS. On the developmental origin of the v-shaped contracted maxilla. Monthly Rev Dental Surg. 1872;1:2–5.
39. Goldsmith JL, Stool SE. George Catlin's concepts on mouth-breathing, as presented by Dr. Edward H. Angle. Angle Orthod. 1994;64:75–8.
40. Angle EH. Treatment of malocclusion of the teeth and fractures of the maxillae: Angle's system. 6th ed. Philadelphia: The SS White Dental Manufacturing Co; 1900. p. 32.
41. Duke WW. Deformity of the face caused by nasal allergy in childhood. Arch Otolaryngol. 1930;12:493–8.
42. Balyeat RM, Bowen R. Facial and dental deformities due to perennial nasal allergy in children. Int J Orthod Dent Child. 1934;20:445–60.
43. Cohen MB. Orthodontic problems associated with allergy. Angle Orthod. 1937;7:150–4.
44. Subtelny JD. The significance of adenoid tissue in orthodontia. Angle Orthod. 1954;24:59–69.
45. Morrison WW. The interrelationship between nasal obstruction and oral deformities. Int J Orthod. 1931;17:453–8.
46. Moss ML. The functional matrix. In: Kraus B, Riedel R, editors. Vistas in orthodontics. Philadelphia: Lea & Febiger; 1962. p. 85–98.
47. Moss ML. The primary role of functional matrices in facial growth. Am J Orthod. 1969;55:566–77.
48. Solow B, Tallgren A. Head posture and craniofacial morphology. Am J Phys Anthropol. 1976;44:417–36.
49. Solow B, Kreiberg S. Soft-tissue stretching: a possible control factor in craniofacial morphogenesis. Scand J Dent Res. 1977;85:505–7.
50. Marks MB. Allergy in relation to orofacial dental deformities in children: a review. J Allergy. 1965;36:293–302.
51. Kingsley WS. A treatise on oral deformities as a branch of mechanical surgery. New York: D. Appleton Company; 1889.
52. Gwynne-Evans E, Ballard CF. Discussion on the mouth-breather. Proc R Soc Med. 1959;51:279–85.
53. Howard CC. Inherent growth and its influence on malocclusion. J Am Dent Assoc. 1932;19:642–8.
54. Leech HL. A clinical analysis of orofacial morphology and behavior of 500 patients attending an upper respiratory research clinic. Dent Pract. 1958;9:57–68.
55. Huber RE, Reynolds JW. A dentofacial study of male students at the University of Michigan in the physical hardening program. Am J Orthod Oral Surg. 1946;32:1–21.
56. Harvold EP. The role of function in the etiology and treatment of malocclusion. Am J Orthod. 1968;54:883–98.
57. Harvold EP, Chierici G, Vargervik K. Experiments on the development of dental malocclusions. Am J Orthod. 1972;61:38–44.
58. Harvold EP, Vargervik K, Chierici G. Primate experiments on oral sensation and dental malocclusions. Am J Orthod. 1973;63:494–508.
59. Harvold EP, Tomer BS, Vargervik K, Chierici G. Primate experiments on oral respiration. Am J Orthod. 1981;79:359–72.
60. Miller AJ, Vargervik K, Chierici G. Sequential neuromuscular changes in rhesus monkeys during the initial adaptation to oral respiration. Am J Orthod. 1982;81:99–107.
61. Miller AJ, Vargervik K, Chierici G. Experimentally induced neuromuscular changes during and after nasal airway obstruction. Am J Orthod. 1984;85:385–92.
62. Vargervik K, Miller AJ, Chierici G, Harvold E, Tomer BS. Morphologic response to changes in

neuromuscular patterns experimentally induced by altered modes of respiration. Am J Orthod. 1984;85:115–24.

63. Yamada T, Tanne K, Miyamoto K, Yamauchi K. Influences of nasal respiratory obstruction on craniofacial growth in young Macaca fuscata monkeys. Am J Orthod Dentofac Orthop. 1997;111:38–43.
64. Scarano E, Ottaviani F, Di Girolamo S, Galli A, Deli R, Paludetti G. Relationship between chronic nasal obstruction and craniofacial growth: an experimental model. Int J Pediatr Otorhinolaryngol. 1998;45:125–31.
65. Linder-Aronson S. Effects of adenoidectomy on dentition and nasopharynx. Am J Orthod. 1974;65:1–15.
66. Hannuksela A. The effect of moderate and severe atrophy on the facial skeleton. Eur J Orthod. 1981;3:187–93.
67. Hannuksela A. The effect of atrophy on the dentition. Eur J Orthod. 1983;5:279–85.
68. Mergen DC, Jacobs RM. The size of nasopharynx associated with normal and class II malocclusions. Angle Orthod. 1970;40:342–6.
69. Shapiro GG, Shapiro PA. Nasal airway obstruction and facial development. Clin Rev. 1984;2:225–35.
70. Vig P, Hall D. The inadequacy of cephalometric radiographs for airway assessment. Letter to the editor. Am J Orthod. 1980;77:230–2.
71. Maw AR, Jeans WD, Fernando DCJ. Inter-observer variability in the clinical and radiological assessment of adenoid size, and the correlation with adenoid volume. Clin Otolaryngol. 1981;2:317–22.
72. Vig PS, Sarver DM, Hall DJ, Warren DW. Quantitative evaluation of nasal airflow in relation to facial morphology. Am J Orthod. 1981;79:263–72.
73. Warren DW, Spalding PM. Dentofacial morphology and breathing: a century of controversy. In: Melsen B, editor. Current controversies in orthodontics. 1st ed. Chicago: Quintessence Publishing Company; 1991. p. 45.
74. Ono T, Ishiwata Y, Kuroda T. Inhibition of masseteric electromyographic activity during oral respiration. Am J Orthod Dentofac Orthop. 1998;113:18–25.
75. Song HG, Pae EK. Changes in orofacial muscle activity in response to changes in respiratory resistance. Am J Orthod Dentofac Orthop. 2001;119:436–42.
76. Gross CW, Harrison SE. Tonsils and adenoids. Pediatr Rev. 2000;21:75–8.
77. Rubin RM. Mode of respiration and facial growth. Am J Orthod. 1980;78:504–10.
78. Harari D, Redlich M, Miri S, Hamud T, Gross M. The effect of mouth breathing versus nasal breathing on dentofacial and craniofacial development in orthodontic patients. Laryngoscope. 2010;120:2089–93.
79. Peltomaki T. The effect of mode of breathing on craniofacial growth-revisited. Eur J Orthod. 2007;29:426–9.
80. L'Estrange PR, Battagel JM, Harkness B, Spratley MH, Nolan PJ, Jorgensen GI. A method of studying adaptive changes of the oropharynx to variation in mandibular position in patients with obstructive sleep apnoea. J Oral Rehabil. 1996;23:699–711.
81. Azevedo ND, Lima JC, Furlan RMMM, Motta AR. Tongue pressure measurement in children with mouth-breathing behaviour. J Oral Rehabil. 2018;45:612–7.
82. Trotman CA, McNamara JA, Dibbets JMH, Van der Weele LT. Association of lip posture and the dimensions of the tonsils and sagittal airway with facial morphology. Angle Orthod. 1997;67:425–32.
83. Defabianis P. Impact of nasal airway obstruction on dentofacial development and sleep disturbances in children: preliminary notes. J Clin Pediatr Dent. 2003;27:95–100.
84. Zucconi M, Caprioglio A, Calori G, Ferini-Strambi L, Oldani A, Castronovo C, Smirne S. Craniofacial modifications in children with habitual snoring and obstructive sleep apnoea: a case-control study. Eur Respir J. 1999;13:411–7.
85. Flores-Mir C, Korayem M, Heo G, Witmans M, Major MP, Major PW. Craniofacial morphological characteristics in children with obstructive sleep apnea syndrome: a systematic review and meta-analysis. J Am Dent Assoc. 2013;144:269–77.
86. Katyal V, Pamula Y, Martin AJ, Daynes CN, Kennedy JD, Sampson WJ. Craniofacial and upper airway morphology in pediatric sleep-disordered breathing: systematic review and meta-analysis. Am J Orthod Dentofac Orthop. 2013;143:20–30.
87. Galeotti A, Festa P, Viarani V, Pavone M, Sitzia E, Piga S, Cutrera R, De Vincentiis GC, D'Antò V. Correlation between cephalometric variables and obstructive sleep apnoea severity in children. Eur J Paediatr Dent. 2019;20:43–7.
88. Meurice JC, Marc I, Carrier G, Sériès F. Effects of mouth opening on upper airway collapsibility in normal sleeping subjects. Am J Respir Crit Care Med. 1996;153:255–9.
89. Wilhelm CP, deShazo RD, Tamanna S, Ullah MI, Skipworth LB. The nose, upper airway, and obstructive sleep apnea. Ann Allergy Asthma Immunol. 2015;115:96–102.
90. Fitzpatrik MF, McLean H, Urton AM, Tan A, O'Donnel D, Driver HS. Effect of nasal or oral breathing route on upper airway resistance during sleep. Eur Respir J. 2003;22:827–32.
91. Koo SK, Park GH, Koh TK, Jung SH, Lee HB, Ji CL. Effect of mouth closure on upper airway obstruction in patients with obstructive sleep apnoea exhibiting mouth breathing: a drug-induced sleep endoscopy study. Eur Arch Otorhinolaryngol. 2020;277:1823–8.
92. Gleeson K, Zwillich C, Braier K, White D. Breathing route during sleep. Am Rev Respir Dis. 1986;134:115–20.
93. Lofaso F, Coste A, D'Ortho M, et al. Nasal obstruction as a risk factor for sleep apnoea syndrome. Eur Respir J. 2000;16:639–43.
94. Cuccia AM, Lotti M, Caradonna D. Oral breathing and head posture. Angle Orthod. 2008;78:77–82.

95. Chaves TC, de Andrade e Silva TS, Monteiro SA, Watanabe PC, Oliveira AS, Grossi DB. Craniocervical posture and hyoid bone position in children with mild and moderate asthma and mouth breathing. Int J Pediatr Otorhinolaryngol. 2010;74:1021–7.
96. Neiva PD, Kirkwood RN, Godinho R. Orientation and position of head posture, scapula and thoracic spine in mouth-breathing children. Int J Pediatr Otorhinolaryngol. 2009;73:227–36.
97. Gonzalez HE, Manns A. Forward head posture: its structural and functional influence on the stomatognathic system, a conceptual study. Cranio. 1996;14:71–80.
98. Hairfield WM, Warren DW, Seaton DL. Prevalence of mouthbreathing in cleft lip and palate. Cleft Palate J. 1988;25:135–8.
99. Warren DW, Duany LF, Fischer ND. Nasal airway resistance in normal and cleft lip and palate subjects. Cleft Palate J. 1969;6:134–40.
100. Drake AF, Davis JU, Warren DW. Nasal airway size in cleft and noncleft children. Laryngoscope. 1993;103:915–7.
101. Grossmann N, Brin I, Aizenbud D, Sichel JY, Gross-Isseroff R, Steiner J. Nasal airflow and olfactory function after the repair of cleft palate (with and without cleft lip). Oral Surg Oral Med Oral Pathol Oral Radiol Endod. 2005;100:539–44.
102. Kunkel M, Wahlmann U, Wagner W. Acoustic airway profiles in unilateral cleft palate patients. Cleft Palate Craniofac J. 1999;36:434–40.
103. Mani M, Morén S, Thorvardsson O, Jakobsson O, Skoog V, Holmström M. Objective assessment of the nasal airway in unilateral cleft lip and palate—a long-term study. Cleft Palate Craniofac J. 2010;47:217–24.
104. Sandham A, Solow B. Nasal respiratory resistance in cleft lip and palate. Cleft Palate J. 1987;24:278–85.
105. Hairfield WM, Warren DW. Dimensions of the cleft nasal airway in adults: a comparison with subjects without cleft. Cleft Palate J. 1989;26:9–13.
106. Duskova M, Kristen M, Hronkova K, Rakosnik P. Rhinomanometry in patients with a cleft deformity of the nose. Acta Chir Plast. 2002;44:124–8.
107. Fukushiro AP, Trindadel E. Nasal airway dimensions of adults with cleft lip and palate: differences among cleft types. Cleft Palate Craniofac J. 2005;42:396–402.
108. Oosterkamp BC, Remmelink HJ, Pruim GJ, Hoekema A, Dijkstra PU. Craniocervical, and pharyngeal morphology in bilateral cleft lip and palate and obstructive sleep apnea patients. Cleft Palate Craniofac J. 2007;44:1–7.
109. Gorucu-Coskuner H, Saglam-Aydinatay B, Aksu M, Ozgur FF, Taner T. Comparison of positive screening for obstructive sleep apnea in patients with and without cleft lip and palate. Cleft Palate Craniofac J. 2020;57:364–70.
110. Alves M Jr, Baratieri C, Mattos CT, Brunetto D, Fontes Rda C, Santos JR, Ruellas AC. Is the airway volume being correctly analyzed? Am J Orthod Dentofac Orthop. 2012;141:657–61.
111. El H, Palomo JM. Measuring the airway in 3 dimensions: a reliability and accuracy study. Am J Orthod Dentofacial Orthop. 2010;137:S50.e1–9; discussion S50–2.
112. Ghoneima A, Kula K. Accuracy and reliability of cone-beam computed tomography for airway volume analysis. Eur J Orthod. 2013;35:256–61.
113. Alves M Jr, Baratieri C, Nojima LI, Nojima MC, Ruellas AC. Three-dimensional assessment of pharyngeal airway in nasal- and mouth-breathing children. Int J Pediatr Otorhinolaryngol. 2011;75:1195–9.
114. Rogers AP. Exercises for the development of muscles of face with view to increasing their functional activity. Dental Cosmos LX. 1918;59:857–76.
115. Guimaraes KC, Drager LF, Genta PR, Marcondes BF, Lorenzi-Filho G. Effects of oropharyngeal exercises on patients with moderate obstructive sleep apnea syndrome. Am J Respir Crit Care Med. 2009;179:962–6.
116. Guilleminault C, Huang YS, Monteyrol PJ, Sato R, Quo S, Lin CH. Critical role of myofascial reeducation in pediatric sleep-disordered breathing. Sleep Med. 2013;14:518–25.
117. Camacho M, Certal V, Abdullatif J, Zaghi S, Ruoff CM, Capasso R, Kushida CA. Myofunctional therapy to treat obstructive sleep apnea: a systematic review and meta-analysis. Sleep. 2015;38:669–75.
118. Suzuki H, Watanabe A, Akihiro Y, et al. Pilot study to assess the potential of oral myofunctional therapy for improving respiration during sleep. J Prosthodont Res. 2013;57:195–9.
119. Reddy JT, Korath VA, Adamala NR, Adusumilli G, Pichai S, Varma KV. Cephalometric evaluation of oropharyngeal airway dimension changes in pre- and postadenoidectomy cases. J Contemp Dent Pract. 2012;13:764–8.
120. Pinheiro de Magalhães Bertoz A, Souki BQ, Lione R, Theresa Webber SA, Bigliazzi R, Oliveira PM, Moro A, Cozza P. Three-dimensional airway changes after adenotonsillectomy in children with obstructive apnea: do expectations meet reality? Am J Orthod Dentofacial Orthop. 2019;155:791–800.
121. Hultcrantz E, Larson M, Hellquist R, Ahlquist-Rastad J, Svanholm H, Jakobsson OP. The influence of tonsillar obstruction and tonsillectomy on facial growth and dental arch morphology. Int J Pediatr Otorhinolaryngol. 1991;22:125–34.
122. Lofstrand-Tideström B, Hultcrantz E. Development of craniofacial and dental arch morphology in relation to sleep disordered breathing from 4 to 12 years. Effects of adenotonsillar surgery. Int J Pediatr Otorhinolaryngol. 2010;74:137–43.
123. Linder-Aronson S, Woodside DG, Hellsing E, Emerson W. Normalization of incisor position after adenoidectomy. Am J Orthod Dentofac Orthop. 1993;103:412–27.

124. Linder-Aronson S, Woodside DG, Lundström A. Mandibular growth direction following adenoidectomy. Am J Orthod. 1986;89:273–84.
125. Woodside DG, Linder-Aronson S, Lundström A, McWilliam J. Mandibular and maxillary growth after changed mode of breathing. Am J Orthod Dentofac Orthop. 1991;100:1–18.
126. Zhu Y, Li J, Tang Y, Wang X, Xue X, Sun H, Nie P, Qu X, Zhu M. Dental arch dimensional changes after adenoidectomy or tonsillectomy in children with airway obstruction: A meta-analysis and systematic review under PRISMA guidelines. Medicine (Baltimore). 2016;95:e4976.
127. Solow B, Sandham A. Cranio-cervical posture: a factor in the development and function of the dentofacial structures. Eur J Orthod. 2002;24:447–56.
128. Wenzel A, Henriksen J, Melsen B. Nasal respiratory resistance and head posture: effect of intranasal corticosteroid (Budesonide) in children with asthma and perennial rhinitis. Am J Orthod. 1983;84:422–6.
129. Wenzel A, Højensgaard E, Hendriksen JM. Craniofacial morphology and head posture in children with asthma and perennial rhinitis. Eur J Orthod. 1985;7:83–9.
130. Tecco S, Festa F, Tete S, Longhi V, D'Attilio M. Changes in head posture after rapid maxillary expansion in mouth-breathing girls: a controlled study. Angle Orthod. 2005;75:171–6.
131. Haas AJ. Rapid expansion of the maxillary dental arch and nasal cavity by opening the midpalatal suture. Angle Orthod. 1961;31:73–90.
132. Hershey HG, Stewart BL, Warren DW. Changes in nasal airway resistance associated with rapid maxillary expansion. Am J Orthod. 1976;69:274–84.
133. Cross DL, McDonald JP. Effect of rapid maxillary expansion on skeletal dental, and nasal structures: a postero-anterior cephalometric study. Eur J Orthod. 2002;22:519–28.
134. Basciftci FA, Mutlu N, Karaman AI, Malkoc S, Küçükkolbasi H. Does the timing and method of rapid maxillary expansion have an effect on the changes in nasal dimensions? Angle Orthod. 2002;72:118–23.
135. Krüsi M, Eliades T, Papageorgiou SN. Are there benefits from using bone-borne maxillary expansion instead of tooth-borne maxillary expansion? A systematic review with meta-analysis. Prog Orthod. 2019;20:9.
136. Warren DW, Turvey TA, Hairfield WM. The nasal airway following maxillary expansion. Am J Orthod Dentofac Orthop. 1987;91:111–6.
137. Monini S, Malagola C, Villa MP, Tripodi C, Tarentini S, Malagnino I, et al. Rapid maxillary expansion for the treatment of nasal obstruction in children younger than 12 years. Arch Otolaryngol Head Neck Surg. 2009;135:22–7.
138. Baik UB, Suzuki M, Ikeda K, Sugawara J, Mitani H. Relationship between cephalometric characteristics and obstructive sites in obstructive sleep apnea syndrome. Angle Orthod. 2002;72:124–34.
139. Compadretti GC, Tasca I, Bonetti GA. Nasal airway measurements in children treated by rapid maxillary expansion. Am J Rhinol. 2006;20:385–93.
140. Doruk C, Sökücü O, Sezer H, Canbay E. Evaluation of nasal airway resistance during rapid maxillary expansion using acoustic rhinometry. Eur J Orthod. 2004;26:397–401.
141. Bicakci AA, Agar U, Sökücü O, Babacan H, Doruk C. Nasal airway changes due to rapid maxillary expansion timing. Angle Orthod. 2005;75:1–6.
142. Babacan H, Sokucu O, Doruk C, Ay S. Rapid maxillary expansion and surgically assisted maxillary expansion effects on nasal volume. Angle Orthod. 2006;76:66–71.
143. Gordon JM, Rosenblatt M, Witmans M, Carey JP, Heo G, Major PW, et al. Rapid palatal expansion effects on nasal airway dimensions as measured by acoustic rhinometry. A systematic review. Angle Orthod. 2009;79:1000–7.

The Nose and the Eustachian Tube

40

Özlem Önerci Celebi

Core Messages

- Normal functioning of the ear is closely related to the health status of the nose and paranasal sinuses.
- Pathologies of the nose, sinuses, and nasopharynx play an important role in the cause, treatment, and sequelae of ear disease.
- Sniffing creates negative pressure in the nose, in the nasopharynx, and in the middle ear, causing middle ear pathologies in patients with hyperpatent eustachian tubes.
- Nose blowing increases intranasal propelling viscous fluid into the paranasal sinuses and middle ear.
- Sneezing elevates intranasal pressure tenfold times less compared to nose blowing. However, sneezing while the nasal passages are blocked may lead to an increase in nasopharyngeal pressure, causing a failure of the valve which protects the entrance to the eustachian tube.
- The positive middle ear pressures with bilateral nasal obstruction are caused by tubal openings synchronized into the positive phase of nasopharyngeal pressure generation.
- An increase in the environmental pressure in the presence of nasal obstruction interferes with eustachian tube functioning; thus, the tube may remain closed and may be "locked." The continued increase causes barotrauma.

The nose is located in the middle of the face and acts as an air conditioning unit, making the air that we breathe harmless for the body. Awareness of the interrelationship between the nose, middle ear, and lower airways has increased; the respiratory tract is considered to be an integrated system, and any process that affects one part of the system affects the other parts as well.

The middle ear is connected to the other air-filled spaces of the upper respiratory tract via the eustachian tube. Politzer first suggested the abnormal function of the ET as a cause of ear pathology more than 100 years ago. The eustachian tube is closed in its resting stage. It opens periodically during swallowing, yawning, and possibly at times during normal respiration, causing frequent alterations in middle ear pressure and thus equalizing middle ear pressure to atmospheric pressure.

Normal functioning of the ear is closely related to, and depends on, the health status of the nose, paranasal sinuses, and throat. Pathologies of the nose, sinuses, and nasopharynx play a very important role in the cause, treatment, and sequelae of ear disease. Upper respiratory tract infection is the most frequent cause of otitis media [1]. In some patients with nasal septal deviation, ET function may be impaired.

Ö. Önerci Celebi (✉)
Department of Otolaryngology Head and Neck Surgery, University of Health Sciences, Istanbul Training and Research Hospital, Istanbul, Turkey

Ö. Ö. Celebi, T. M. Önerci (eds.), *Nasal Physiology and Pathophysiology of Nasal Disorders*,
https://doi.org/10.1007/978-3-031-12386-3_40

McNicholl [2] reported that the equilibration of the middle ear pressure by the Valsalva maneuver in patients with septal deviation may be difficult during diving or flying in an airplane, and this situation turns to normal after correcting the septal deviation. Anatomically, the nose and paranasal sinuses can affect the function of the eustachian tube (ET) due to its localization anterior to the nasopharynx (c).

The exact mechanism between the nose, the function of ET, and middle ear pathologies is not yet well understood. There are some explanations, all related to the nose [1]:

1. Nasal obstruction followed by sniffing results in ET dysfunction or blocking, causing high negative middle ear pressure and fluid accumulation.
2. Toynbee phenomenon (following swallowing in an obstructed nose) leads to the flow of secretions to the ET.
3. Vigorous nose blowing with increased nasopharyngeal pressure pushes the secretions up to the ET.

40.1 Sniffing

Recent investigations have demonstrated an association between sniffing, sniff-induced negative middle ear pressure, and otitis media with effusion. Sniffing creates negative pressure in the nose, in the nasopharynx, and in the middle ear. The negative nasopharyngeal pressure can close a hyperpatent eustachian tube, and abnormally negative middle ear pressures can occur. Sakakihara et al. found that most of the ETs with sniff-induced OME seemed to have excessive patency and poor active opening ability but may not be hypercompliant or "floppy" [3]. Bylander-Groth and Stenström [4] reported that habitual sniffing in combination with closing failure and poor active function of ET may be a possible mechanism for the development of otitis media. Falk and Magnusson reported that sniffing can evacuate the middle ear, causing high negative intratympanic pressure [5].

The purpose of the tube should be considered primarily as protecting the middle ear from the extensive pressure variations that physiologically take place in the nasopharynx, for example, during sniffing. A number of studies of diseased ears have shown that tubal malfunction was characterized mainly by a reduced ability to withstand negative pressure in the nasopharynx. Eustachian tube malfunction in these subjects is characterized by a reduced protective function, a condition denoted as "eustachian tube closing failure" [6]. On the other hand, Knight and Eccles found no evidence to support the hypothesis that negative middle ear pressures are associated with sniffing [7]. Hauser and Münker [8] stated that the traditional concept of opening failure of ET is no longer sufficient to explain tubal dysfunction. They suggested that sniffing can cause negative pressure in the middle ear space, and sniff-induced negative pressure is a further possible cause of tubal dysfunction and plays an important factor in the development of cholesteatoma. Dempster and Browning [9] found that a high percentage of children with otitis media with effusion are capable of inducing a negative middle ear pressure by sniffing.

Miura et al. [10] reported that the recovery of negative middle ear pressure in 5 min without swallowing was less than 10 mm H_2O in the ears with sniff-induced middle ear disease, whereas in the ears with the normal eardrum, negative middle ear pressure recovered by more than 20 mm H_2O in 5 min. They concluded that sniffing plays an important role in the development of ET dysfunction and middle ear disease.

Ohta et al. [11] found that habitual sniffing was significantly higher in cholesteatoma than in COM and in otosclerosis. The coexistence of diseases on the contralateral side was significantly higher in cases with habitual sniffing than in those without habitual sniffing. After the canal wall-up method, postoperative retraction of the eardrum was significantly related to habitual sniffing continuing after the surgery. They concluded that patulous eustachian tubes and habitual sniffing might play a role in the pathogenesis of middle ear cholesteatoma.

40.2 Nose Blowing

Nose blowing generates high nasopharyngeal and intranasal pressures. Rapid increases in middle ear pressure and the generation of a positive middle ear pressure were associated with nose blowing [7]. Fluid dynamic simulations revealed that the high intranasal pressures generated by nose blowing would propel viscous fluid into the paranasal sinuses. The CT experiments confirmed fluid deposition in the paranasal sinuses after nose blowing [12].

40.3 Sneezing

In contrast to nose blowing, sneezing (with the nostrils open) and coughing only elevated intranasal pressure slightly and tenfold less than the mean pressure produced by nose blowing. Fluid modeling indicated that sneezing and coughing would not result in the nasal fluid deposition in the paranasal sinuses. This was supported by the negative findings in the CT scan experiments [12].

However, sneezing while the nasal passages are blocked can increase nasopharyngeal pressure, causing a failure of the valve which protects the entrance to the eustachian tube. Fluids and pathogens can be forced into the tube by applying strong positive pressure to the entrance of the tube during a sneeze when the nasal passages are blocked by swelling or as we pinch our nose closed.

40.4 Valsalva

The Valsalva technique is the technique used to equalize pressure in the tympanic cavities. The technique is to close the mouth, pinch the nose, and try to breathe out through the nose. The Valsalva creates a strong positive pressure in the nasopharynx, sufficient to force air into our middle ear, thus equalizing pressure and allowing fluid to drain. The pressure has to be high enough to overcome the valve created by the redundant membrane of the nasopharyngeal orifice [13].

40.5 Nasal Obstruction

Swallowing causes an initial positive meso- and hypopharyngeal pressure followed by a negative pressure resulting from the interactive motions of the pharyngeal wall, soft palate, and tongue [14]. These pressures which are produced in the meso- and hypopharynx during swallowing are not reflected in the nasopharynx when the nose is open. However, when the nose is obstructed, these pressures which are reflected in the nasopharynx cannot be equalized through the nose because of the nasal obstruction. Finkelstein et al. [15] reported that nasopharyngeal pressure can be increased from −340 to +450 mm H_2O during swallowing in adults with complete nasal obstruction. The positive middle ear pressures with bilateral nasal obstruction are caused by tubal openings synchronized with the positive phase of nasopharyngeal pressure generation. Thus, the usual positive middle ear to nasopharyngeal pressure gradient established by the test methods is diminished or reversed in cases of bilateral nasal obstruction [16]. Alternatively, or complementarily, the airflow resulting from the high nasopharyngeal pressures may affect the clearance function of the eustachian tube by opposing the prograde, middle ear–nasopharynx mucociliary clearance mechanism and thus forcing the materials back to their origin in the tympanum [17]. Therefore, bilateral nasal obstruction plays a very important role in the function of the eustachian tube. However, this is not true for unilateral nasal obstruction. Unilateral nasal obstruction causes only low positive pressures with swallowing and consequently a minimal effect on the pressure gradient [17].

Toynbee maneuver manually equalizes ear pressure on the ascent. While the nose is closed, swallowing creates a negative pressure and will help to suck extra air pressure out of the middle ear.

40.6 Toynbee Test

Measurement of middle ear pressures as the patient swallows with the nose pinched is called the Toynbee test. This is a test of the eustachian tube.

40.7 Toynbee Phenomenon

Swallowing causes an initial positive meso- and hypopharyngeal pressure followed by negative pressure. These pressures which are produced in the meso- and hypopharynx during swallowing activity are reflected in the nasopharynx during swallowing with the nose closed/pinched. Therefore, swallowing while the nose is closed or obstructed causes an initial positive nasopharyngeal pressure followed by negative pressure. Positive nasopharyngeal pressure might insufflate infected secretions into the middle ear, especially when the middle ear has high negative pressure. With negative pressure, a pliant tube could be impeded from opening and could be further obstructed functionally. This effect of swallowing when the nose is obstructed could be related to eustachian tube dysfunction and called as Toynbee phenomenon (Fig. 40.1) [13, 18].

40.7.1 Balloon Eustachian Tuboplasty

Balloon Eustachian Tuboplasty is a new procedure in which a balloon device is introduced through the nose and the balloon is placed into the Eustachian tube and then briefly inflated.

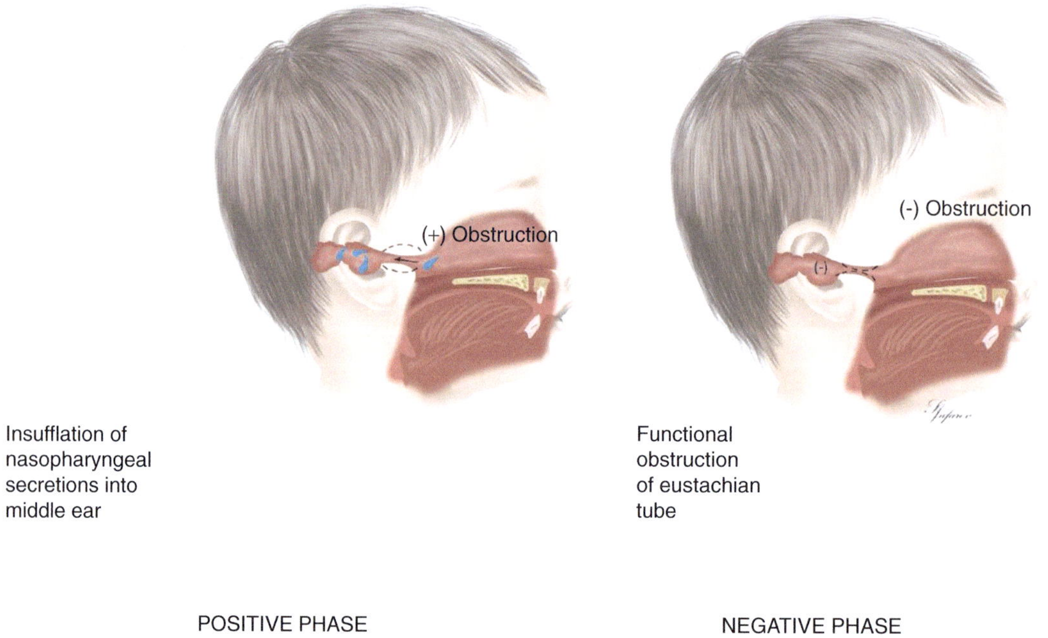

Fig. 40.1 Toynbee phenomenon. *Swallowing*: when there is nasal obstruction causing an initial positive nasopharyngeal air pressure, followed by a negative pressure phase. If the ET is not normal, positive nasopharyngeal pressure might insufflate the secretions into the middle ear, especially when the middle ear has a high negative pressure. During negative nasopharyngeal pressure, the tube could be prevented from opening and be further obstructed (With kind permission of TESAV)

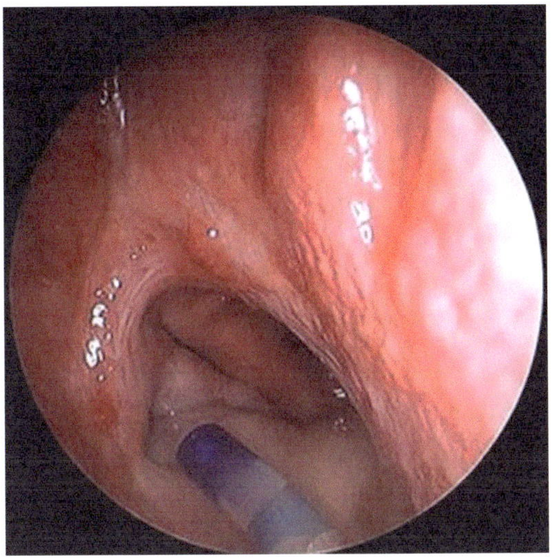

Fig. 40.2 Under direct visualization, a balloon guiding catheter is positioned in the orifice of the right Eustachian tube (With kind permission of Aytuğ Altundağ)

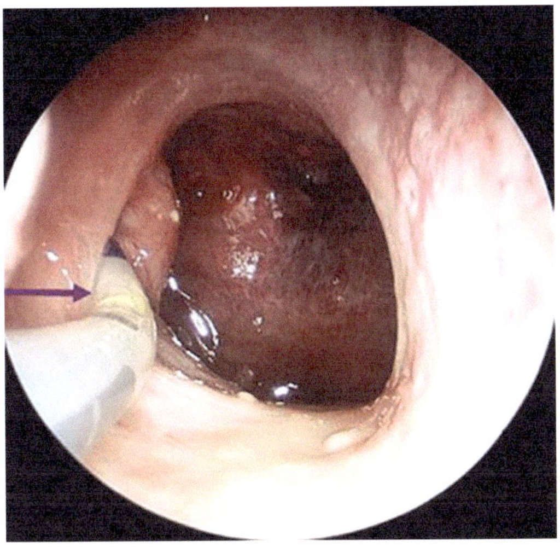

Fig. 40.3 Balloon catheter is fully inserted into the lumen of the cartilaginous Eustachian tube. Once the balloon is fully inserted into the cartilaginous Eustachian tube, the yellow mark indicates the end of the balloon and should be visible along the medial edge of the anterior cushion of the Eustachian tube (purple arrow) (With kind permission of Aytuğ Altundağ)

40.7.2 Technique

Under direct visualization, a balloon guiding catheter is positioned and gently advanced in the orifice of the Eustachian tube. The catheter is in close relationship with the medial edge of the anterior cushion of the Eustachian tube. The catheter is fully inserted into the lumen of the cartilaginous Eustachian tube. The yellow mark indicates the end of the balloon and should be visible along the medial edge of the anterior cushion of the Eustachian tube.

The assistant then inflates the balloon at approximately 1 ATM per second until the pressure reaches 12 ATM. The balloon is held in the inflated position for 2 min, then deflated and slowly retracted into the guide. The entire system is then withdrawn from the nose (Figs. 40.2, 40.3, 40.4, and 40.5).

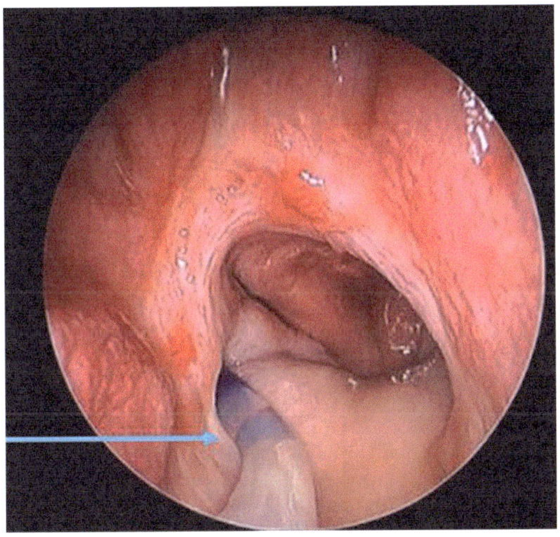

Fig. 40.4 The catheter is gently advanced to the orifice of the Eustachian tube. Note the relationship of the catheter with the medial edge of the anterior cushion of the Eustachian tube (blue arrow) (With kind permission of Altuğ Altundağ)

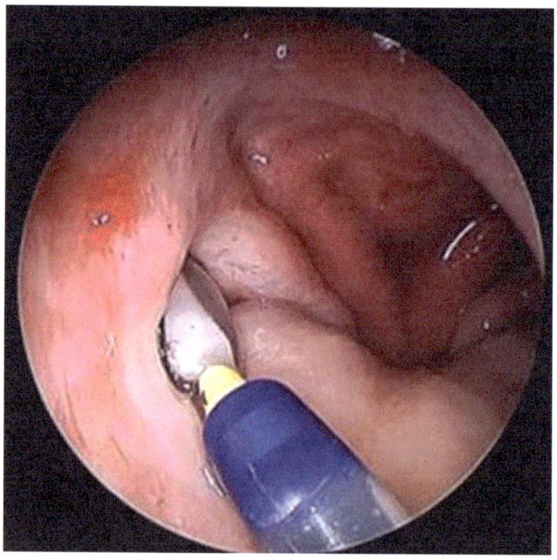

Fig. 40.5 The assistant then inflates the balloon at approximately 1 ATM per second until the pressure reaches 12 ATM. The balloon should then be held in the inflated position for 2 min then deflated and slowly retracted into the guide. The entire system is then withdrawn from the nose (With kind permission of Aytuğ Altundağ)

40.8 Otitic Barotrauma

Boyle's law states that the product of pressure times volume is constant for a given mass of confined gas as long as the temperature is constant. It is very important to understand the response of gases when subjected to pressure in order to know what happens during barotrauma because the tympanic cavity and the nasal sinuses are closed spaces full of gases. These closed spaces are aerated through the ostia in the sinuses and through the auditory tube in the middle ear. Unobstructed ventilation of the middle ear will produce no changes and no symptoms.

During an *ascent,* there is a progressive reduction in pressure. According to Boyle's law when the pressure decreases, the volume becomes bigger. Therefore, during ascent, a given mass of gas contained within an elastic structure will expand. In the middle ear, this gaseous expansion will push the tympanic membrane laterally, and the flow of air along the eustachian tube will follow [19, 20].

During a *descent*, due to an increase in pressure, there is a decrease in the volume of the middle ear gas. The eustachian tube must be opened by swallowing movements to adjust the volume. If this mechanism fails or if it is delayed, an increasing differential pressure will act on the soft nasopharyngeal end of the tube to close it. When this pressure is greater than can be generated by the tubal dilator muscles, the tube will stay closed and is said to be "locked." Thereafter, with continued descent the pathophysiological changes of barotrauma are inevitable [18–20].

To maintain equal pressure on both sides of the tympanic membrane (TM), the gas must move freely between the nasopharynx and middle ear. When acute and chronic infections in the nose, nasal allergy and vasomotor rhinitis, and malformations of the nasal skeleton or other mechanisms interfere with eustachian tube functioning during changes in environmental pressure, the pressure in the middle ear either falls below ambient pressure, causing retraction of the TM or rises above it, causing bulging. In the situation of descent without ventilation of the middle ear, the vessels will become passively engorged due to the decreased pressure in the middle ear surrounding the vessels. As the middle ear pressure decreases, the congestion and swelling of the mucosa of the middle ear space increases. Ultimately, the blood vessels become over-distended and rupture, bleeding into the eardrum and the middle ear space. A very large pressure differential may cause bleeding into the middle ear [18, 20] (see Figs. 40.6, 40.7, and 40.8).

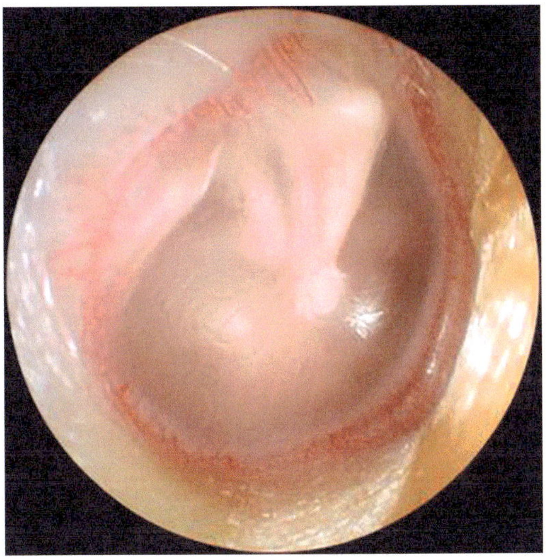

Fig. 40.6 Normal tympanic membrane, right ear

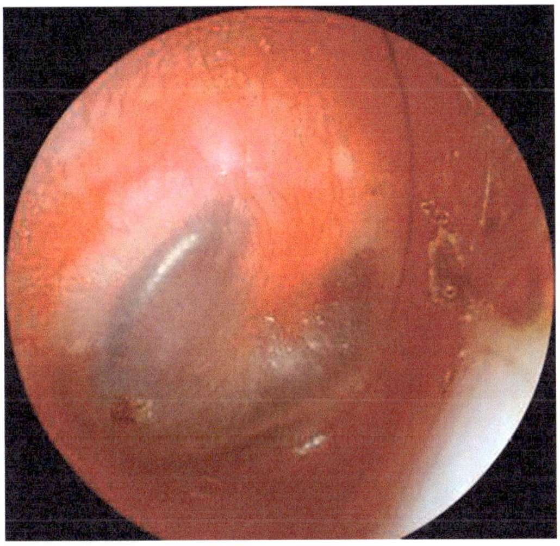

Fig. 40.7 Vascularization around the manubrium mallei due to Eustachian tube dysfunction, right ear

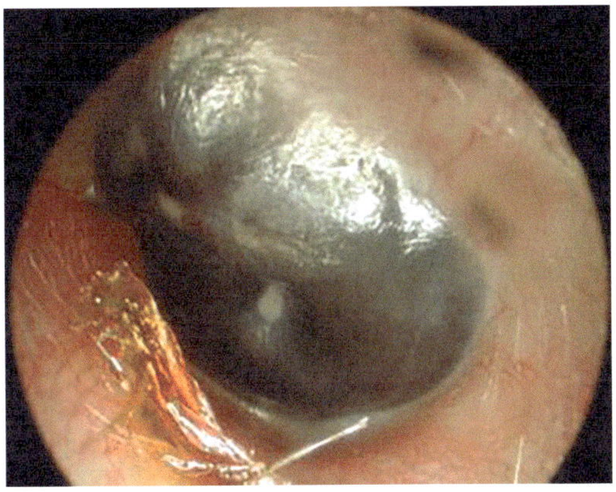

Fig. 40.8 Barotrauma during the descent of the airplane, left ear, the tympanic membrane is dark blue and bulging

40.9 Sinus Barotrauma

The paranasal sinuses also contain air, with the consequence that barotrauma of these structures may occur in flying and diving. As the water pressure changes during a dive, the sinuses normally equalize automatically by the free passage of gas into or out of their openings. Problems are inevitable, however, if these openings become obstructed.

The production of pathology as a result of a change of pressure is directly related to the degree of patency of the sinus ostium and to the nasal function. Factors creating edema of the nasal mucosa are commonly found as predisposing causes, for example, chronic sinus inflammation (sinusitis), nasal inflammation (rhinitis), folds of tissue (polyps), coryza and its infective sequelae, vasomotor rhinitis, seasonal allergy, the effects of mechanical obstruction secondary to nasal injury, and deflection of the nasal septum. Unlike the middle ear, which can be ventilated through the eustachian tube voluntarily, there is no voluntary control over the diameter of the sinus ostium so that equalization of pressure may be difficult [21].

During descent (the compression phase) when the ostium is obstructed, the gas in the sinus will be compressed, causing barotrauma of descent (according to Boyles' law). The shrinking volume is replaced by swelling of the sinus lining, tissue fluid, or bleeding—partly filling the sinus. If the descent continues mucosal vessels rupture and a subepithelial hematoma may develop.

If the sinus opening becomes obstructed during ascent, the gas contained in the sinus expands, causing the symptoms. This is called sinus barotrauma of ascent. Sinus barotrauma of descent is more common than the ascent, but they often coexist. The degree of pathological change is proportional to the magnitude of the pressure differential and the period of time over which the pressure imbalance is unrelieved.

40.10 Conclusions

To maintain equal pressure on both sides of the tympanic membrane (TM), the gas must move freely between the nasopharynx and middle ear. Sniffing, sneezing, and nose blowing may not have any significant effect on normal eustachian tubes; however, these may be the trigger factor and the cause of middle ear pathologies. Normal functioning of the ear is closely related to, and depends on, the health status of the nose, paranasal sinuses, and throat. Pathology of the nose, sinuses, and nasopharynx has a very important role in the cause, treatment, and sequelae of ear disease. The pathologies of the nose should be treated to have normal physiology of the ear. In other words, if the patient has an ear pathology, the nose and sinuses should be very carefully evaluated.

Bilateral nasal obstruction plays a very important role in the function of the eustachian tube. Nasal obstruction may cause high positive pressure in the middle ear and can also affect the clearance function of the eustachian tube and force the materials back to the tympanum. Nasal pathologies may interfere with eustachian tube function during changes in environmental pressure and may cause otitic barotrauma.

References

1. Shah A. The ear and sinusitis. Bombay Hosp J. 1999;41:681.
2. McNicoll WD. Eustachian tube dysfunction in submariners and divers. Arch Otolaryngol Head Neck Surg. 1982;108:279–83.
3. Sakakihara J, Honjo I, Fujita A, Kurata K, Takahashi H. Eustachian tube compliance in sniff-induced otitis media with effusion. A preliminary study. Acta Otolaryngol. 1993;113:187–90.
4. Bylander-Groth A, Stenström C. Eustachian tube function and otitis media in children. Ear Nose Throat J. 1998;77(9):762–4, 766, 768–9
5. Falk B, Magnuson B. Evacuation of the middle ear by sniffing: a cause of high negative pressure and development of middle ear disease. Otolaryngol Head Neck Surg. 1984;92:312–8.
6. Falk B, Magnuson B. Eustachian tube closing failure in children with persistent middle ear effusion. Int J Pediatr Otorhinolaryngol. 1984;7:97–106.
7. Knight LC, Eccles R. The relationship between nasal airway resistance and middle ear pressure in subjects with acute upper respiratory tract infection. Acta Otolaryngol. 1993;113:196–200.
8. Hauser R, Münker G. Sniff-induced negative pressure – a cause for the development of middle ear diseases? HNO. 1989;37(6):242–7.
9. Dempster JH, Browning GG. Eustachian tube function following adenoidectomy: an evaluation by sniffing. Clin Otolaryngol Allied Sci. 1989;14(5):411–4.
10. Miura M, Takahashi H, Honjo I, Hasebe S, Tanabe M. Influence of the gas exchange function through the middle ear mucosa on the development of sniff-induced middle ear diseases. Laryngoscope. 1998;108(5):683–6.
11. Ohta S, Sakagami M, Suzuki M, Mishiro Y. Eustachian tube function and habitual sniffing in middle ear cholesteatoma. Otol Neurotol. 2009;30(1):48–53.
12. Gwaltney JM Jr, Hendley JO, Phillips CD, et al. Nose blowing propels nasal fluid into the paranasal sinuses. Clin Infect Dis. 2000;30:387–91.
13. Bluestone CD, Klein JO. Otitis media in infants and children. Philadelphia: W. B. Saunders Company; 1988.
14. Gramiak R, Kelley ML Jr. Nasal pressure during swallowing. A combined cineradiographic and manometric study. Investig Radiol. 1966;1:225–36.
15. Finkelstein Y, Talmi YP, Zohar Y, Laurian N. Study of Toynbee phenomenon by combined intranasopharyngeal and tympanometric measurements. Ann Otol Rhinol Laryngol. 1988;97:119–206.

16. Buchman CA, Doyle WJ, Swarts JD. Eustachian tube function in the ferret. Acta Otolaryngol. 1993;113:75–80.
17. Buchman CA, Doyle WJ, Swarts JD, et al. Effects of nasal obstruction on Eustachian tube function and middle ear pressure. Acta Otolaryngol. 1999;119:351–5.
18. Bluestone CD. Eustachian tube. Hamilton: BC Decker Inc.; 2005.
19. Benson AJ, King PF. Pathophysiology of the ears and nasal sinuses in flying and diving. In: Gleeson M, editor. Scott-Brown's otorhinolaryngology, head and neck surgery. 7th ed. London: Hodder Arnold; 2008.
20. Edmonds C. Ear barotrauma. In: Edmonds C, et al., editors. Diving medicine for scuba divers. 4th edn. 2012. www.divingmedicine.info.
21. Edmonds C. Sinus barotrauma. In: Edmonds C, et al., editors. Diving medicine for scuba divers. 4th edn. 2012. www.divingmedicine.info.

41 Nanomedicine and the Nose

Gürer G. Budak✝, Cengiz S. Ozkan, Mihrimah Ozkan, and T. Metin Önerci

Core Message

- Nanotechnology revealed that organic structures have different physical, chemical, and biological features in macroscopic and nanometric forms.
- 'Nanomedicine' is emerged as a new scientific area in nanotechnology and make important conceptual changes in medical methods all over the world due to the different diagnostic and therapeutic alternatives.
- The use of nanomedicine in oto-rhino-laryngology is very important to us to meet the changing and increasing expectations of health.
- A bioelectronic nose can simulate the human olfactory system and detects target odor molecules with high sensitivity and selectivity.
- To develop a bioelectronic nose, the following steps should be well defined:
  - Standardization of the bioelectronic nose circuit for molecular measurement.
  - Digital codification of specific odors.
  - Selection of standard primary odor molecules that play basic roles in the electrophysiological mechanism.
  - Classification of odor chemical structures.

Gürer G. Budak died before publication of this work was completed.

G. G. Budak (Deceased)
NanoMedicine & Advanced Technologies Research Center, Ankara, Turkey

C. S. Ozkan
Department of Mechanical Engineering, Nanomaterials and Devices Laboratory, University of California Riverside, Riverside, CA, USA
e-mail: cozkan@engr.ucr.edu

M. Ozkan
Department of Electrical Engineering, Biomaterials and Nanotechnology Laboratory, University of California Riverside, Riverside, CA, USA
e-mail: mihri@ee.ucr.edu

T. M. Önerci (✉)
Faculty of Medicine, Hacettepe University, Ankara, Turkey

41.1 Introduction

Developments in nanotechnology have revealed that macroscopic and nanometric forms of organic structures possess different features in physical, chemical, and biological aspects. By proving that nano-devices that are produced in the laboratory can interact with biomolecules, both physiological processes in healthy tissues and the physiopathologic basis of diseases began to be understood in a clearer way.

'**Nanomedicine**' which appeared as a new scientific interest parallel to the above-mentioned developments in nanotechnology became one of the most studied topics in the world by the reason of the fact that it leads conceptual changes in

Ö. Ö. Celebi, T. M. Önerci (eds.), *Nasal Physiology and Pathophysiology of Nasal Disorders*,
https://doi.org/10.1007/978-3-031-12386-3_41

accepted and applicated medical methods up to now and presents different diagnosis-treatment alternatives.

Although Nanotechnology is a commonly studied field all around the world, there is still no clear consensus about what nanoscale really is. 1 nanometer is calculated as one billionth (10^{-9}). It is possible to fit 5 carbon atoms on this scale in three-dimensional forms. According to BSI (PAS 71) applications, less than 100 nm or even smaller scales are evaluated within the concept of nanotechnology. While at the beginning of 2000s, studies less than 200 nm and on smaller scale were considered as nanomedicine, today this range is accepted between 5 and100 nm.

Oto-rhino-laryngology is one of the basic disciplines of medicine which closely follows and implements medical innovations and advancements. In this regard, Oto-rhino-laryngology is one of the leading areas which heavily utilizes microscopic and endoscopic treatments. Nowadays, treatment and rehabilitation methodologies like vocal prosthesis, cochlear implant, brainstem implants, etc. are the top and best-known implementations examples of the micro nano circuits.

Nanomedicine will offer significant opportunities in meeting the treatment expectations of otolaryngology patients with the new and advanced nanomedical diagnosis and treatment technologies in coming years. Such treatment solutions with advanced technologies will offer important treatment solutions for many different types of medical problems from hearing loss to facial paralysis, from nasal plastic surgery to early diagnosis and treatment of head and neck cancers.

41.2 Clinical Nanomedicine Perspectives

Currently, the most important aim to approach patients and diseases is to diagnose, if possible, when pathologic change is only at the single-cell level and to start treatment. However, this could only be possible by increasing the efficiency of in vivo and in vitro diagnosis methods. Although Nanomedicine is a field presenting great opportunities in this regard, it also brings along disadvantages because it is a new developing discipline.

Regarding the literature, it is possible to come across a wide range of research topics from the discovery of new nano-bio materials to using these materials in clinics. While searching for physical, chemical, and biological principles for nanomaterials on one hand, on the other hand, it is attempted to be understood how to use these materials on living creatures, what could be the adverse effects caused by the use of these materials and the effects of nanomaterials on human health and environmental health. In addition, possible social and legal problems have been discussed and new ethical rules have been introduced.

A certain part of the studies is more in detail, more specific, and more focusing on developing safer diagnosis devices. Also, there are studies for performing different biological measuring methods with one integrated device. By means of very precise biosensors which are tried to be developed with the use of nano-electronic circuits, it is attempted to establish micro mobile laboratories which can be easily used by patients and, if necessary, can transmit multiple data to an external user.

The other research topic in the related literature which requires advanced technology is about the combination of above-mentioned in vitro monitoring techniques and in vivo nano-medical devices. In those studies, conducted in this regard, it basically tried to be developed nanostructures which are able to carry specific contrast substances and be directed from the outside. Thus, it will be possible to take a detailed molecular image of the target tissues.

In another conducted group study, it is researched how to combine these nanostructures with pharmacological agents. By means of nanostructures that carry therapeutic and diagnostic agents at the same time, especially in cancer cases, it would be possible to administer treatment to target tissue directly. With this approach defined as Theragnostic (therapy+diagnose), again primarily for cancer patients, it is aimed to follow up the efficiency of the treatment by taking images of the target tissue at different times.

Lastly, intense studies has also been conducted on the successful regeneration of diseased or injured tissues by means of nanografts and reproducing needed artificial organs by means of nanoscafolds in in vitro conditions and then replacing diseased or injured organs with the artificial ones.

Methods which have been developed by using nanotechnology have the potential to be effective in all medical equipment. For example, developing new materials to be used in surgical implants, nanometric systems or minimal invasive sensors which can be used in monitoring metabolic activities can be considered in this regard. Nano pumps, injectable/implantable polymer systems, liposomal drug applications, and cell/gene therapy methods can be considered with regard to developed-controlled drug delivery systems (Source: Wei et al. [1]).

41.3 Interdisciplinary Frameworks

All those efforts for understanding the development of disease at the molecular level and for treatment are very important to spread all the developments in Nanomedicine to society. Since the topic has a wide scale, different disciplines have to work together in the nanomedicine area. It can be said that for now, neither any scientific field nor areas of expertise possess the capacity of scientific and technical infrastructure to conduct such research by itself. To manage scientific research in such a field, it is a must to establish a well-organized 'team'. Within such a team, conventional disciplines such as basic-clinic medical scientists, pharmacologists, physics-chemistry-electric-electronic-biomedical-computer engineers, etc., and new fields such as genom-proteom science, pharmacokinetic modeling and microscope designing, etc. should be included.

In addition to self-disciplinary nature of Nanomedicine, the more the number of studies increase in this field the better new sub-disciplines appear. Some of these sub-disciplines are mentioned below and many studies have been conducted on each specific topic:

- Imaging: molecular, vascular, neurological, etc.
- In vitro diagnosis.
- In vivo diagnosis and biosensors.
- Advanced biomedical materials, including 'smart' and functionalized materials and surfaces.
- Regenerative medicine and tissue engineering.
- Infection control.
- Drug design and targeted drug delivery.
- Gene and cell therapy.
- Man-machine interfaces.
- Nanotoxicology.
- Nanomedicine and risk management.
- Nanomedicine and ethics.

41.4 Clinical Nanomedicine Applications

Today, research centers which conduct experimental and clinical studies are focused mainly on three fields (Source: Vision Paper for NanoMedicine [2]).

41.4.1 Regenerative Nanomedicine

Current 'traditional treatment' approaches lead to limited results in many diseases or cause the success of training to change from patient to patient. Both for improving the efficiency of the treatment and minimizing the side effects, the methods to be used should be *patient-specific characteristics*.

As a result of '*tissue engineering*' studies, it has been started to form the basis of patient-specific treatments which can be used in the regeneration and reparation of in situ tissues.

The main implementation fields of *tissue engineering*, which is an interdisciplinary field, are maintaining, improving, and repairing the functions of biological structures through the collaboration of engineering and life sciences. By means of *tissue engineering*, future therapy methods will be more focused on the treat of chronic disorders by use of self-healing mechanisms of the body than targeting symptoms or reducing the development of diseases.

It is possible to evaluate Regenerative Nanomedicine studies into two topics as **Therapy** and **Biomimesis.**

41.4.1.1 Therapy

Within the scope of regenerative nanomedicine, it has been conducted studies on protecting from such pathologies as diabetes, osteoarthritis, cardiovascular system diseases, and degenerative central neuronal system diseases and therapies for related disorders and functional loss after injuries. Regeneration of cartilage in articulation with osteoarthritis, production of mechanically stable and elastic scaffolds, vascularisation following implantation, and creating of a physiologic oscillation in diabetic pancreas islets or starting self-reparation mechanisms in the heart and nervous system are the other examples to give (Fig. 41.1).

Both in terms of mortality-morbidity rates and prevalence, there has been more intensified research on certain disease groups. This is also the case for clinical studies within the scope of nanomedicine. The most researched diseases are primarily Cardiovascular Diseases, Cancer, Musculoskeletal Disorders, Neurodegenerative Diseases, Diabetes, and Bacterial-Viral Infectious Disease.

41.4.1.2 Design an Artificial Nose (Bioelectronic Nose)

One of the areas with high implementation potential for nanosensors is the artificial nose (bioelectronic nose) studies. With the widespread use of interdisciplinary technologies, the relationships between organic structures and artificial nanomaterials have been better understood. Thus, sensitive and analytical devices have been developed that consists of a combination of biomolecules and micro-nano-mechanical systems. Studies are continuing systems to capture and process odor molecules (Fig. 41.2). Systems designed for this purpose involve the use of receptors, transmitters, receivers, and a processor. Receptors used in these systems are manufactured with nanofabrication or NEMS (nano electro-mechanical systems) technology. Recently, different bioelectronic noses have been developed that selectively detect different molecular structures such as 'heptanl', a serum biomarker of lung cancer, and 'geosmin and 2-methylisoborneol' for water contamination monitoring. Solid-phase micro-extraction gas chromatography-mass spectrometry (SPME-GC/MS) technique is typically used in the analysis of odor molecules.

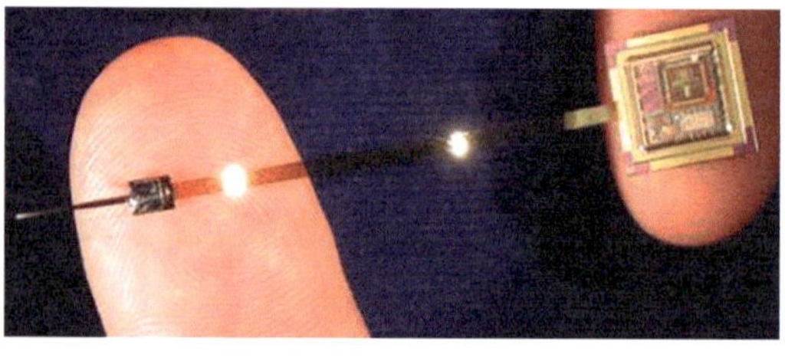

Fig. 41.1 Thin-film electrodes process the sound from 128 different channels in comparison to standard 16–22 channel implants (source: http://www.engadget.com/2006/02/09/new-type-of-cochlear-implant-to-improve-hearing/)

Currently, bioelectronic noses using human odor receptors and nanomaterials as secondary converters have been developed. Nanowire and graphene sensors can also use to produce bioelectronic noses (Figs. 41.3 and 41.4). In an artificial nose, human olfactory receptors selectively separate target ligands at a single atomic resolution, such as in the human nose, and electrical potential changes are transformed into a field by the transistor sensor. The transistor circuit is based on nanomaterials such as carbon nanotubes, graphene, and conductive polymers, which amplify very small biological signals and achieve high precision.

Multisegmented nanowires via the surface functionalization method can be used for the detection of biological or odor molecules. P-type multisegmented nanostructures are in a back-to-back Schottky diode configuration. Au-part of the multisegment can be surface functionalized to increase the sensitivity of the nanowire sensor with specific sensing molecules. Attachment of odor molecules on Au-portion of the nanowire can modulate the Fermi-level of the heterojunction device which leads to the detection of the desired molecules. Such multisegmented nanowire sensors can be fabricated into an array format

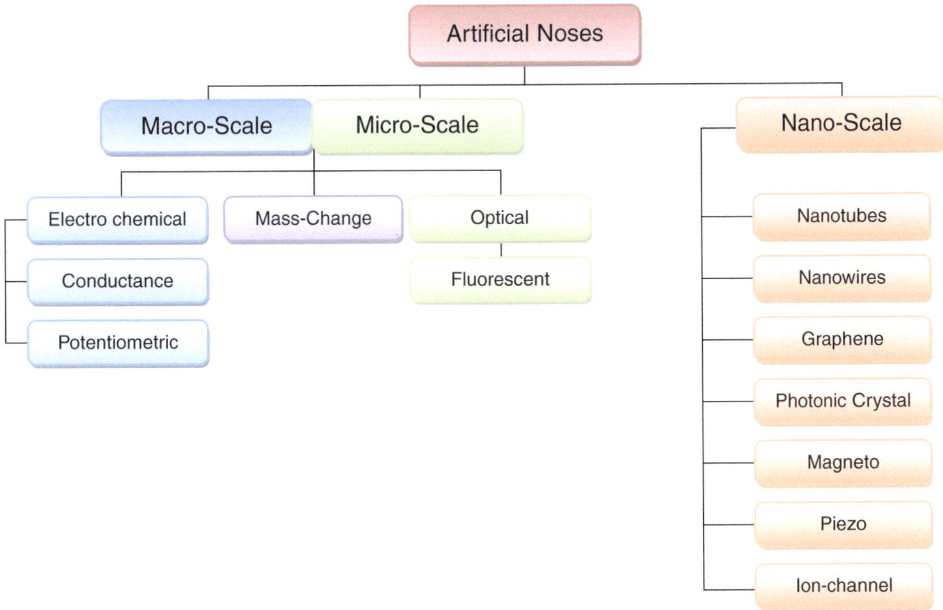

Fig. 41.2 Taxonomy of nose-like sensors (adapted from The MITRE Corporation, McLean, ©1997–2006)

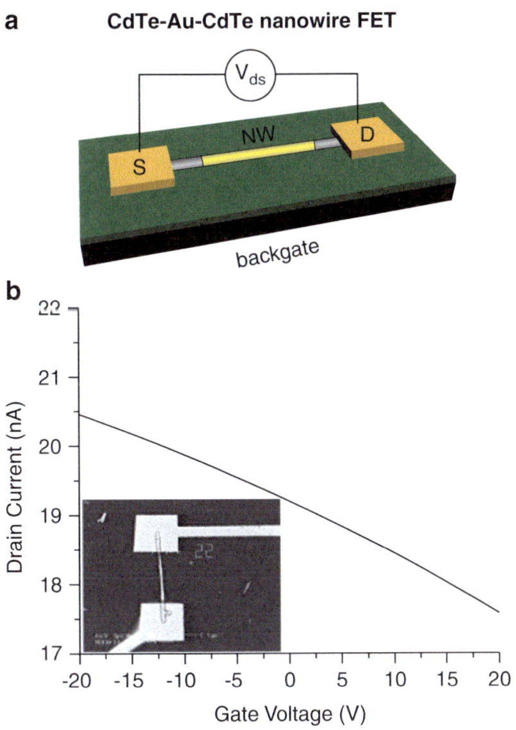

Fig. 41.3 (**a**) Schematic of CdTe-Au-CdTe nanowire field-effect transistor. (**b**) SEM image of CdTe-Au-CdTe nanowire FET. (Source: Xu Wang and Cengiz S. Ozkan, Multisegment Nanowire Sensors for the Detection of DNA Molecules NanoLetters, 2008, Vol. 8, No:2, 398–404)

for the detection of different types of odor molecules or biological agents.

Similarly, graphene field effect transistors arrays can be used for sensing as well. Using graphene sensors pH, DNA, and vapor detection have been demonstrated. The stability and reliability of such sensors became questionable due to graphene's fast response to its environment. For this purpose, extraordinary electrical property of a graphene transistor is combined with a cover of block copolymers where nanometer scale openings provide access to the graphene surface where sensing can occur (Fig. 41.4). Gas/odor molecules can penetrate through the small holes across the block copolymer layer and reach the graphene layer underneath. Interaction with these molecules and graphene surface leads to a shift in dirac point of the transistor. The electrical shift in the signal is basically the senor output.

Arrays of graphene field-effect transistors with block copolymer on the surface can be fabricated and used for the detection of different molecules. Block copolymers in appropriate volume fractions and molecular weights can provide control over the morphology and size/separation distance of cylindrical microdomains. These domains can be adjusted based on the size of the

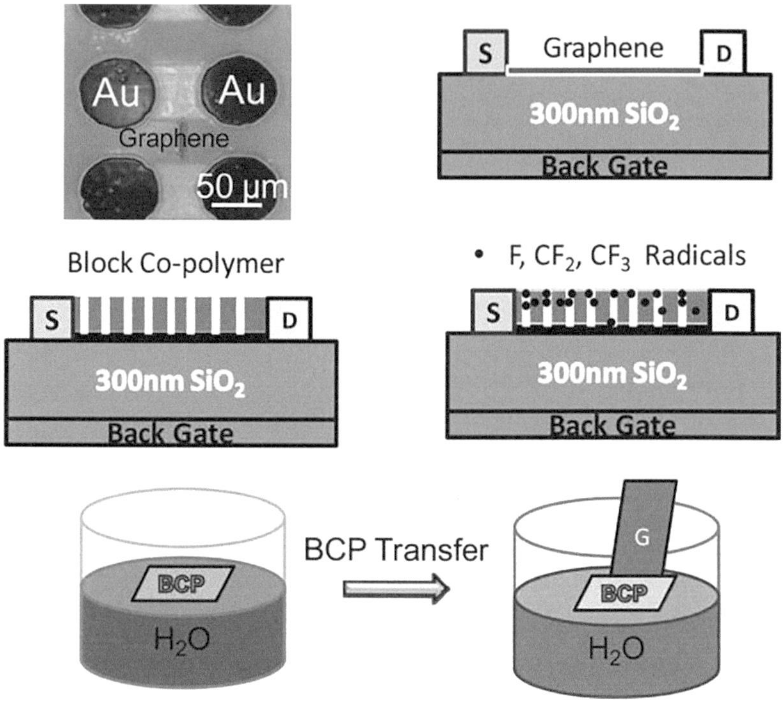

Fig. 41.4 Graphene-based electronic sensors are successfully used for chemical, biological, and gas sensing. (Source: Shirui Guo, Maziar Ghazinejad, Xiangdong Qin, Huaxing Sun, Wei Wang, Francisco Zaera, Mihrimah Ozkan, Cengiz S. Ozkan, Small, 2012)

molecules or biological agents to be detected. As sensor output, the amount of dirac point shift for different molecules is expected to be different which is important for the array-formatted sensors.

Today electro-physiological basis of the odor detection process is better understood and the upcoming period, the most important goal is to create a different type of nanosensors structure that will bring together an integrated system.

With the successful results obtained from studies based on this type of nanosensor will be able to develop an artificial nose.

41.4.2 Diagnosis and Imaging Methods Based on Nanomedicine

The most important aim in the diagnosis of diseases is to diagnose the disease when it is at the earliest stage, at the one-cell level. To reach this aim, it is required to develop new in vivo and in vitro diagnosis methods based on nanotechnology. Within the scope of in vitro applications, it has been made studies on chemo-bio nanosensors, ultra-sensitive biochips ('lab-on-a-chip' and 'cells-on-chips' devices), and it has been prepared products for routine medical applications.

To-be-produced nano-analyzer devices can be used by patients and at the same time will be able to transmit multiple data to clinicians. More important than that, by means of nano-biosensors it will be possible to increase the accuracy of already used test methods. Biosensors (*Photonic Crystal Nanobiosensors, Magneto Nano-Immunosensor, Piezoelectric Nanosensors, Resonating Beam Sensors, Ion-channel Biosensora,* etc.) harness the immensely powerful molecular recognition properties of living systems and engineer these into electronic devices to provide easy-to-use sensing devices. The most successful biosensor developed to date is the home blood glucose sensor which is now ubiquitous word wide. Biosensors can be used to measure disease markers, food safety, and environmental quality, to ensure safety and security.

Developments in microscopic scanning-imaging methods (*quantitative-PET, MRS, d-MRI, f-MRI,* etc.) and spectroscopic techniques provide ultra-high spatial resolutions and give detailed information about the complex 'functionality' of cells. Data acquired by the use of Quantum dots and fluorescent nanoparticles will lead development of more innovative and stronger in vivo diagnosis devices. Nanodevices produced as accompanying this functional molecular imaging will be more effective and much safer.

41.4.3 Targeting Delivery and Releasing

Long-term aims of controlled drug delivery systems are to develop diagnosis agents with a high level of efficiency and safety, and to perform treatment, application, and follow-up with the same nanosystem. 'Find, fight and follow', as is the concept determined; includes early diagnosis, treatment, and monitor of the results and also is stated as 'theragnostic' (diagnosis+treatment).

Drug delivery techniques suitable for the theragnostic definition are prepared in accordance with two needs. The first one is drugs targeting more effectively where the disease is located, with high patient tolerance and cost effectiveness, the other is to detect new methods for distribution of new types of pharmacologic agents which cannot be distributed effectively by conventional methods.

The main aim of pharmaceutic studies in this regard is to address any medication to specific target tissue at the right time, in a convenient amount, and with a safe-repeatable-controllable method. Today 13% of products on the pharmaceutics market are related to controlled drug distribution systems. Nano-particle formulations are still used to increase activity without increasing surface/volume proportion. In addition, nanoparticles act as drug carriers to effectively transmit therapeutics which have weak liquidity. If a therapeutic active substance is suitably encapsulated in a nano-particle, carrying this drug anywhere requested, controlled oscillation of the drug and protection from early stage activity decreases can be managed. These results will both increase the efficiency of drugs and decrease side effects dramatically. These types of nanoparticle delivery systems can be used for the treatment of cancer and many other diseases.

Controlled drug delivery is based on the principle of turning pathophysiological changes, which appear basically in diseased tissue, into an advantage for treatment. Because in tissues in which pathological process has already started, all physiological functionality disorders related to cell homeostasis are observed, accumulation of carriers which distribute drugs in a controlled manner will be easier. The anatomic barrier between normal and pathologic tissues and vascularisation differences will make it easier to reach nanocarriers to diseased tissue. Thus, nanocarriers which carry therapeutic agents will reach a much higher concentration in target tissue compared to doses applicated with normal drug treatment.

As a result of the decrease in vascular permeability and lymphatic drainage appeared especially in tissue which developed tumor and inflammatory diseases, on one hand, reach of nano-structures to target tissue will be facilitated, on the other hand, it will be more difficult to withdraw. By means of the opportunity created by this pathophysiological change, nanostructures can easily be accumulated in extravasations and the target tissue.

By means of localization tendency of nanocarrying systems especially in RES will be considered as a huge advantage in terms of both controlled and passive distribution of drugs. This natural distribution method managed by macrophages can be used for intracellular infections of the liver and spleen.

Patient-specific therapies have a critical role in nanosystems performing controlled distributions to reach the target. It is possible to find many nano-carrying systems having such an aim in the literature (Liposomes, Micellular and micro-emulsion Systems, Liquid crystal based formulations, Nanocrystals, Antibodies and conjugates, Naturally occurring proteins as delivery systems, Polymer conjugates, and bio-conjugates, Biodegradable nanoparticles/nanocapsules, Virus-like particles for gene delivery, Delivery of small nucleic acids or mimetics, Delivery of vaccines, Synthetic biomimetics, Dendimers, Carbon Nano Tubes, etc.) (Fig. 41.5).

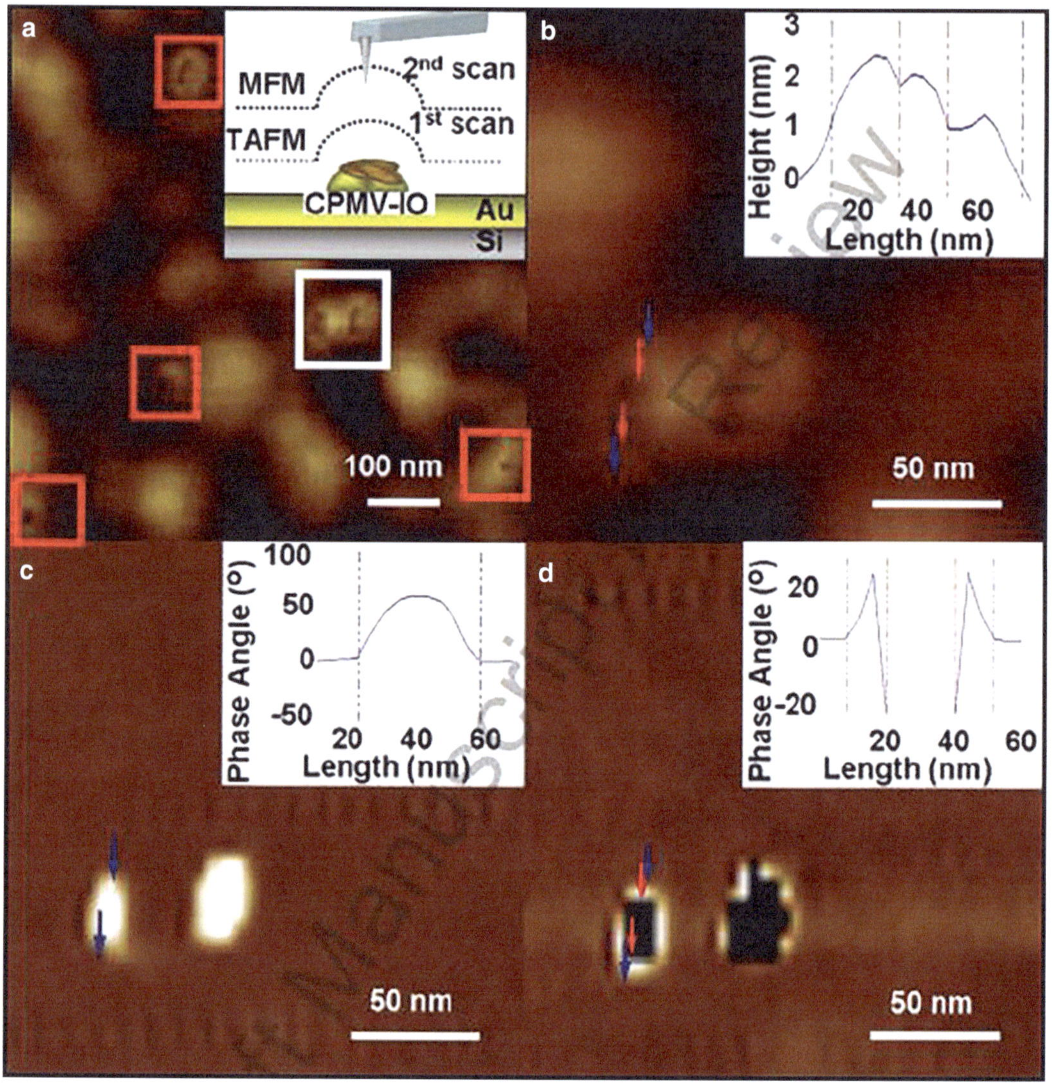

Fig. 41.5 Virus-like particles: AFM and MFM imaging of single CPMV-IO hybrids. (**a**) AFM topography showing single hybrids (white squares). AFM/MFM schematic of dynamic lift-mode operation (inset). (**b**) AFM topography, (**c**) AFM phase detection, (**d**) MFM phase detection of two adjacent CPMV-IO hybrids and their corresponding cross-sections. (Source: Alfredo A. Martinez-Morales, Nathaniel Portney, Yu Zhang, Giuseppe Destito, Gurer Budak, Ekmel Ozbay, Marianne Manchester, Cengiz Ozkan, Mihrimah Ozkan, "Synthesis and Characterization of Iron Oxide derivatized Mutant Cowpea Mosaic Virus Hybrid Nanoparticles", *Advanced Materials*, 20, 1–5 (2008) [3])

Although there have been many successful experimental studies on the topic existing today, strategies for developing new drug-carrying systems aren't completely accepted yet. Efforts on this topic have been proceeding slowly because of the uncertainties about regulation and toxic side effects. It should be accepted that drug safety has to be attached as much importance as drug efficiency considering all nano-particles.

41.5 Conclusions

As in all areas related to medicine, nanomedicine applications have many technological, legal, administrative, environmental, toxicological, social, and economic problems. However, human health-related preventive and curative health services in achieving the objectives and the high-tech solutions on the whole of society through dissemination, nanomedicine will be one of the science to shape the world soon.

With the advances in nanotechnology, it became possible to integrate olfactory receptors with nanostructured systems. Thus, bioelectronic noses with high sensitivity and specificity could be produced. The innate high sensing capacity of the olfactory system in mammals offers great information about high-performance biomimetic odor sensors. Understanding this complex system gave important clues on how to distinguish different scent molecules. Despite this, all functions of the receptors in the olfactory epithelium are not fully understood yet (Fig. 41.6).

With a better understanding of the structural and functional properties of olfactory physiology, the application area of bioelectronic noses will grow. In the near future, biosensors based on

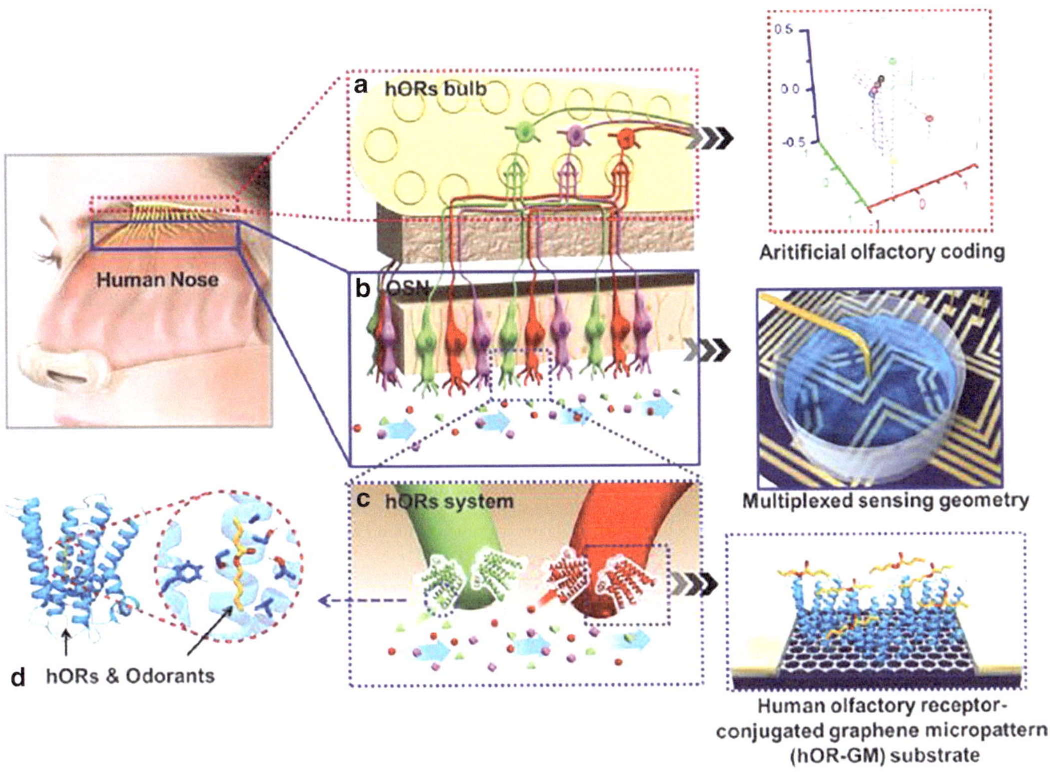

Fig. 41.6 Functional anatomy of the human olfactory system and components of bioelectronic nose. (**a**) Olfactory bulb, where the olfactory signals generated by olfactory sensory neurons are combined for the generation of combinatorial olfactory codes, matching with artificial olfactory codes generated by Multiplexed superbioelectronic nose. (**b**) Olfactory sensory neurons, where olfactory signals triggered by the specific binding of human olfactory receptors and odorants, matching with graphene micropatterns functionalized with human olfactory receptors. (**c**) Human olfactory receptors for the specific recognition of odorants. (**d**) Illumination of specific interaction between human olfactory receptors and odorant (Source-1: Kwon, O. S. et al. (2015) An ultrasensitive, selective, multiplexed superbioelectronic nose that mimics the human sense of smell. Nano Lett. 15, 6559–6567 [4]) (Source 2: Tran Thi Dung et al. (2018), Applications and Advances in Bioelectronic Noses for Odor Sensing. J. Sensors 2018, 18, 103; doi: 10.3390/s18010103 [5])

olfactory receptors are not only used in medicine; They will also be used in food analysis, environmental pollution monitoring, and border security applications.

References

1. Wei C, Lyubchenko YL, Ghandehari H, et al. New technology and clinical applications of nanomedicine: highlights of the second annual meeting of the American Academy of nanomedicine (part I). Nanomedicine. 2006;2(4):253–63.
2. Vision Paper and Basis for a Strategic Research Agenda for NanoMedicine. ETP-NM Luxembourg: Office for Official Publications of the European Communities, September 2005.
3. Martinez-Morales AA, Portney N, Zhang Y, Destito G, Budak G, Ozbay E, Manchester M, Ozkan C, Ozkan M. Synthesis and characterization of iron oxide derivatized mutant cowpea mosaic virus hybrid nanoparticles. Adv Mater. 2008;20:1–5.
4. Kwon OS, et al. An ultrasensitive, selective, multiplexed superbioelectronic nose that mimics the human sense of smell. Nano Lett. 2015;15:6559–67.
5. Dung TT, et al. Applications and advances in bioelectronic noses for odor sensing. J Sensors. 2018;18:103. https://doi.org/10.3390/s18010103.

MIX
Papier aus verantwortungsvollen Quellen
Paper from responsible sources
FSC® C105338

If you have any concerns about our products,
you can contact us on
ProductSafety@springernature.com

In case Publisher is established outside the EU,
the EU authorized representative is:
Springer Nature Customer Service Center GmbH
Europaplatz 3, 69115 Heidelberg, Germany

Printed by Libri Plureos GmbH
in Hamburg, Germany